College Algebra

with Corequisite Support

Julie Miller
Daytona State College

Donna Gerken
Miami Dade College

COLLEGE ALGEBRA WITH COREQUISITE SUPPORT

Published by McGraw-Hill Education, 2 Penn Plaza, New York, NY 10121. Copyright © 2021 by McGraw-Hill Education. All rights reserved. Printed in the United States of America. No part of this publication may be reproduced or distributed in any form or by any means, or stored in a database or retrieval system, without the prior written consent of McGraw-Hill Education, including, but not limited to, in any network or other electronic storage or transmission, or broadcast for distance learning.

Some ancillaries, including electronic and print components, may not be available to customers outside the United States.

This book is printed on acid-free paper.

1 2 3 4 5 6 7 8 9 LWI 24 23 22 21 20

ISBN 978-1-260-86727-5 (bound edition)
MHID 1-260-86727-7 (bound edition)
ISBN 978-1-260-86716-9 (loose-leaf edition)
MHID 1-260-86716-1 (loose-leaf edition)
ISBN 978-1-260-86694-0 (Annotated Instructor's Edition)
MHID 1-260-86694-7 (Annotated Instructor's Edition)

Portfolio Manager: *Ian Townsend*
Lead Product Developer: *Elaine Page*
Marketing Manager: *Noah Evans*
Content Project Managers: *Jane Mohr and Sandra Schnee*
Buyer: *Sandy Ludovissy*
Design: *David Hash*
Content Licensing Specialist: *Lorraine Buczek*
Cover Image: *©Buena Vista Images/Getty Images*
Compositor: *Aptara®, Inc.*

All credits appearing on page or at the end of the book are considered to be an extension of the copyright page.

Library of Congress Cataloging-in-Publication Data

Names: Miller, Julie, 1962- author. | Gerken, Donna, author.
Title: College algebra with corequisite support / Julie Miller (professor
   emerita, Daytona State College), Donna Gerken.
Description: New York, NY : McGraw-Hill Education, [2021] | Includes index.
Identifiers: LCCN 2019031556 (print) | LCCN 2019031557 (ebook) | ISBN
   9781260867275 (hardcover) | ISBN 9781260867169 (spiral bound) | ISBN
   9781260866940 (hardcover) | ISBN 9781260867107 (ebook) | ISBN
   9781260867213 (ebook other)
Subjects: LCSH: Algebra–Textbooks.
Classification: LCC QA152.3 .M5725 2021 (print) | LCC QA152.3 (ebook) |
   DDC 512.9–dc23
LC record available at https://lccn.loc.gov/2019031556
LC ebook record available at https://lccn.loc.gov/2019031557

The Internet addresses listed in the text were accurate at the time of publication. The inclusion of a website does not indicate an endorsement by the authors or McGraw-Hill Education, and McGraw-Hill Education does not guarantee the accuracy of the information presented at these sites.

mheducation.com/highered

About the Authors

Julie Miller is from Daytona State College where she taught developmental and upper-level mathematics courses for 20 years. Prior to her work at DSC, she worked as a software engineer for General Electric in the area of flight and radar simulation. Julie earned a bachelor of science in applied mathematics from Union College in Schenectady, New York, and a master of science in mathematics from the University of Florida. In addition to this textbook, she has authored textbooks in developmental mathematics, trigonometry, and precalculus, as well as several short works of fiction and nonfiction for young readers.

"My father is a medical researcher, and I got hooked on math and science when I was young and would visit his laboratory. I remember doing simple calculations with him and using graph paper to plot data points for his experiments. He would then tell me what the peaks and features in the graph meant in the context of his experiment. I think that applications and hands-on experience made math come alive for me, and I'd like to see math come alive for my students."

Donna Gerken is from Miami Dade College where she taught developmental courses, honors classes, and upper-level mathematics classes for decades. Throughout her career, she has been actively involved with many projects at Miami Dade, including those on computer learning, curriculum design, and the use of technology in the classroom. Donna's bachelor of science in mathematics and master of science in mathematics are both from the University of Miami.

Letter from the Authors

For many students, corequisite college algebra serves as a gateway course to the higher levels of mathematics needed for a variety of careers. We designed this book from the ground up with the corequisite student in mind and for the purpose of supporting students of all levels. Thoughtful interweaving of prerequisite and credit-level content, unique pedagogical features, and a seamless integration of print and digital content offer students the opportunity for success, regardless of where they start their journey. The clear, concise writing style and pedagogical features of our textbook are mirrored throughout the homework problems in ALEKS and in our instructional videos.

The main objectives of this corequisite college algebra textbook and our digital content are threefold:

- Provide students with a clear and logical presentation of fundamental concepts that will prepare them for continued study in mathematics.
- Help students develop logical thinking and problem-solving skills that will benefit them in all aspects of life.
- Motivate students by demonstrating the significance of mathematics in their lives through practical applications.

Julie Miller   julie.miller.math@gmail.com

donna gerken   donna.gerken.math@gmail.com

Dedications

In memory of my mother, Joanne Miller.   —**Julie Miller**

For Linda, Lindsey, and Tom.   —**Donna Gerken**

Table of Contents

CHAPTER R   Review of Prerequisites   1

Source: JPL-Caltech/MSSS/NASA

CHAPTER 1   Linear Equations and Inequalities   37

Tetra Images/SuperStock

CHAPTER 2   Polynomials   119

Parnianto parnianto/ndoeljindoel/123RF

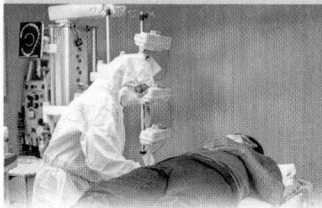

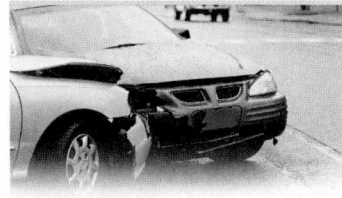

Textbook Features for Corequisite Success

Compared to a traditional College Algebra text, *College Algebra with Corequisite Support* contains a number of new pedagogical features designed for the corequisite audience based on extensive feedback from today's corequisite math instructors. These include:

Expanded Examples

To support students of all levels, previously implicit steps have been added to expand out example problems. Combined with additional annotation, these examples provide a clear and understandable template to apply on practice problems. Additional examples have also been added to ensure a more gradual increase in rigor across a chapter.

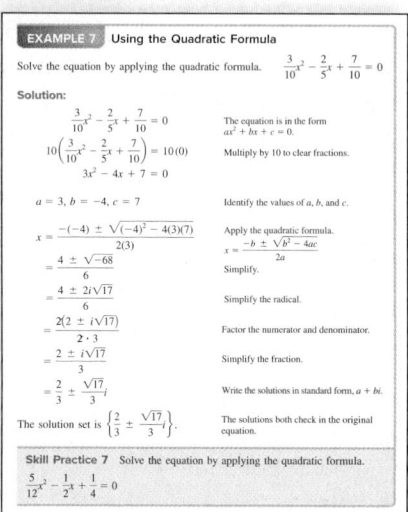

Original

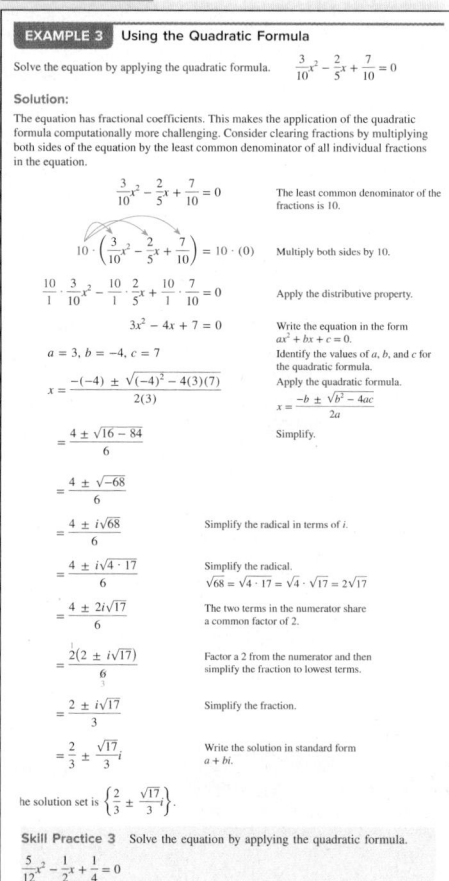

Expanded

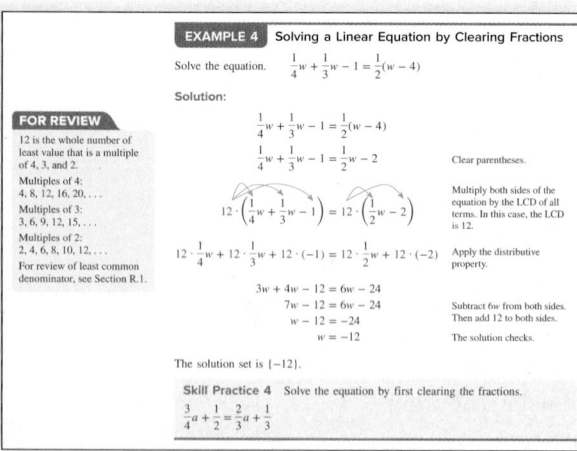

"For Review" Boxes

Throughout the text, just-in-time tips and reminders of prerequisite skills appear in the margin alongside the concepts for which they are needed. References to prior sections are given for cases where more comprehensive review is available earlier in the text.

Prerequisite Review Exercises

Within the end-of-section exercise sets, additional prerequisite review exercises have been added to ensure sufficient practice on the skills required for success within the section exercises.

Detailed Chapter Summaries

In order to solidify retention of key concepts and ensure that students can efficiently and effectively prepare for exams, care has been taken to add details and examples to every chapter summary along with references to where in the chapter the topic was covered.

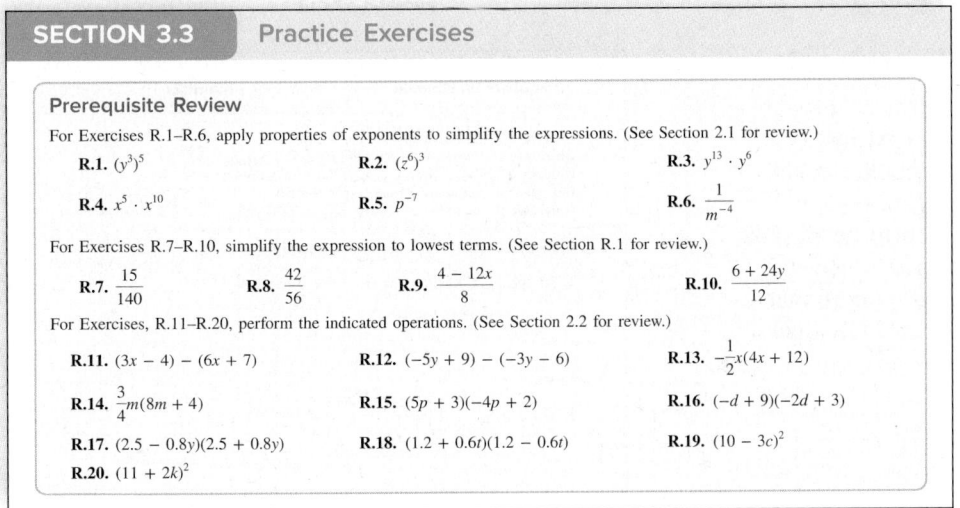

Supplementary Resources for Corequisite Success

In addition to ancillary materials such as author-created videos, in-class presentation materials, and lecture notes, additional resources have been created specifically to support corequisite courses, such as:

Corequisite Workbook

Available in print form or through downloadable files in ALEKS, the corequisite workbook contains just-in-time prerequisite review exercise sets designed to be worked prior to the section for which the worksheet is intended. Additional worksheets focused on individual prerequisite concepts are also available.

Corequisite Workbook:	Factoring Binomials	Section 2.5

Objective 1: Factor the Difference of Squares
Objective 2: Use a Difference of Squares in Grouping
Objective 3: Factor the Sum and Difference of Cubes
Objective 4: Summarize Factoring Binomials
Objective 5: Factor Binomials of the Form $x^6 - y^6$

Objective 1: Factor the Difference of Squares

1. Multiply. $(2x+5)(2x-5)$ 2. Factor. $4x^2 - 25$ 3. Factor. $4x^6 - 25$

Factor.

4. $16y^2 - 25z^2$ 5. $16y^4 - z^4$ 6. $3x^3y - 48xy$

Objective 2: Use a Difference of Squares in Grouping

Factor.

7. $x^3 + 3x^2 - 16x - 48$ 8. $x^3 - 3x^2 - 9x + 27$ 9. $x^3 - 3x^2 + 9x - 27$

Video 7: Changing Positive and Negative Improper Fractions to Mixed Numbers (1:19 min)

▶ Play the video and work along.

Changing an Improper Fraction to a Mixed Number
Step 1 Divide the numerator by the denominator to obtain the quotient and remainder.

Step 2 The mixed number is then given by: $Quotient + \dfrac{remainder}{divisor}$

Write the improper fraction as a mixed number.

$\dfrac{13}{4}$

⏸ Pause the video and try this yourself.

$-\dfrac{39}{5}$

▶ Play the video and check your answer.

Corequisite Video Study Guide

For fundamental concepts, worksheets tied to video lessons give students a guided framework for self-study, filling gaps in knowledge and preparation.

ALEKS 360—*now with Enhanced Homework*

ALEKS is the ideal corequisite companion; it provides instructors the flexibility to determine what content is critical for a student to succeed in the credit-bearing course and helps deliver individualized preparation in the corequisite support course. In addition to the dynamic learning path personalized to each student, ALEKS now offers homework assignments with textbook-specific problems. Blending these two learning environments in one platform provides the utmost flexibility to support your teaching style. No matter where students start their corequisite journey, ALEKS helps you guide them down the most efficient and effective learning path possible.

Corequisite Implementation Guide

Implementing a corequisite model is challenging! We teamed up with professors from around the country who have implemented corequisites to build an online, interactive quiz that serves as a resource to provide the ideal corequisite model based on your institution's needs. You'll also find downloadable videos, PDFs, and various best practices and lessons-learned to serve as guidelines directly from your peers. Explore for yourself today: http://bit.ly/mhcoreqguide

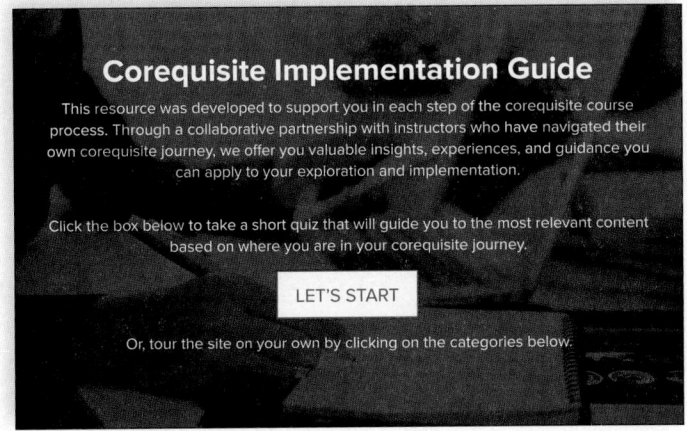

Shutterstock/Paisit Teeraphatsakool

Key Features

Clear, Precise Writing

Because a diverse group of students take this course, Julie Miller and Donna Gerken have written this text in simple and accessible language. Through their friendly and engaging writing style, students can understand the material easily.

Flexible Just-in-Time Remediation

For Review tips placed in the margin guide students back to related prerequisite skills needed for full understanding of course-level topics.

Exercise Sets

The exercises at the end of each section are graded, varied, and carefully organized to maximize student learning:

- **Prerequisite Review Exercises** begin the section-level exercises and ensure that students have the foundational skills to complete the homework sets successfully.
- **Concept Connections** prompt students to review the vocabulary and key concepts presented in the section.
- **Core Exercises** are presented next and are grouped by objective. These exercises are linked to examples in the text and direct students to similar problems whose solutions have been stepped-out in detail.
- **Mixed Exercises** do *not* refer to specific examples so that students can dip into their mathematical toolkit and decide on the best technique to use.
- **Write About It** exercises are designed to emphasize mathematical language by asking students to explain important concepts.
- **Technology Connections** require the use of a graphing utility and are found at the end of exercise sets. They can be easily skipped for those who do not encourage the use of calculators.
- **Expanding Your Skills Exercises** challenge and broaden students' understanding of the material.

Problem Recognition Exercises

Problem Recognition Exercises appear in strategic locations in each chapter of the text. These exercises provide students with an opportunity to synthesize multiple concepts and decide which problem-solving technique to apply to a given problem.

Examples

- The examples in the textbook are stepped-out in greater detail for the corequisite audience, with thorough annotations at the right explaining each step.
- Following each example is a similar **Skill Practice** exercise to engage students by practicing what they have just learned.
- For the instructor, references to an even-numbered exercise are provided next to each example. These exercises are highlighted with blue circles in the exercise sets and mirror the related examples. With increased demands on faculty time, this has been a popular feature that helps faculty write their lectures and develop their presentation of material. If an instructor presents all of the highlighted exercises, then each objective of that section of text will be covered.

Modeling and Applications

One of the most important tools to motivate our students is to make the mathematics they learn meaningful in their lives. The textbook is filled with robust applications and numerous opportunities for mathematical modeling for those instructors looking to incorporate these features into their course.

Callouts

Throughout the text, popular tools are included to highlight important ideas. These consist of:

- **Tip** boxes that offer additional insight to a concept or procedure.
- **Avoiding Mistakes** boxes that fend off common mistakes.
- **Point of Interest** boxes that offer interesting and historical mathematical facts.
- **Instructor Notes** to assist with lecture preparation.

Graphing Calculator Coverage

Material is presented throughout the book illustrating how a graphing utility can be used to view a concept in a graphical manner. The goal of the calculator material is not to replace algebraic analysis, but rather, to enhance understanding with a visual approach. Graphing calculator examples are placed in self-contained boxes and may be skipped by instructors who choose not to implement the calculator. Similarly, the graphing calculator exercises are found at the end of the exercise sets and may also be easily skipped.

End-of-Chapter Materials

The textbook has the following end-of-chapter materials for students to review before test time:

- Detailed summary with key concepts and worked-out examples.
- Chapter review exercises.
- Chapter test.

Supplement Package

Supplements for the Instructor

Author-Created Digital Media

Digital assets were created exclusively by the author team to ensure that the author voice is present and consistent throughout the supplement package.

- Donna Gerken, ensures that each algorithm in the online homework has a stepped-out solution that matches the textbook's writing style.
- Julie Miller created **video content** (lecture videos, exercise videos, graphing calculator videos, and Excel videos) to give students access to classroom-like instruction by the author.
- Julie Miller developed **math animations** to accompany *College Algebra with Corequisite Support*. The animations are diverse in scope and give students an interactive approach to conceptual learning. The animated content illustrates difficult concepts by leveraging the use of on-screen movement where static images in the text may fall short. They are organized in Connect hosted by ALEKS by chapter and section.
- Donna Gerken developed guided lecture notes. These editable handouts outline key concepts and offer recommended classroom examples, but leave blank space for students to fill in the work.

TestGen is a computerized test bank utilizing algorithm-based testing software to create customized exams quickly. This user-friendly program enables instructors to search for questions by topic, format, or difficulty level; to edit existing questions, or to add new ones; and to scramble questions and answer keys for multiple versions of a single test. Hundreds of text-specific, open-ended, and multiple-choice questions are included in the question bank.

Annotated Instructor's Edition

- Answers to exercises appear adjacent to each exercise set, in a color used only for annotations.
- Instructors will find helpful notes within the margins to consider while teaching.
- References to even-numbered exercises appear in the margin next to each example for the instructor to use as Classroom Examples.

PowerPoints present key concepts and definitions with fully editable slides that follow the textbook. An instructor may project the slides in class or post to a website in an online course.

Supplements for the Student

Student Worksheets that provide additional practice on core concepts and prerequisite skills.

Problem Recognition Exercise Worksheets for students to work on in groups or individually.

Corequisite Workbook pages contain additional exercises on prerequisite concepts that ensure students have the skills they need to tackle topics in College Algebra.

The **Corequisite Video Study Guide** provides students with video lessons on basic concepts with the opportunity to perform guided work on accompanying worksheet problems.

ALEKS 360—now with Enhanced Homework

ALEKS is the ideal corequisite companion; it provides instructors the flexibility to determine what content is critical for a student to succeed in the credit-bearing course and helps deliver individualized preparation in the corequisite support course. In addition to the dynamic learning path personalized to each student, ALEKS now offers homework assignments with textbook-specific problems. Blending these two learning environments in one platform provides the utmost flexibility to support your teaching style. No matter where students start their corequisite journey, ALEKS helps you guide them down the most efficient and effective learning path possible.

Create More Lightbulb Moments.

Every student has different needs and enters your course with varied levels of preparation. ALEKS® pinpoints what students already know, what they don't and, most importantly, what they're ready to learn next. Optimize your class engagement by aligning your course objectives to ALEKS® topics and layer on our textbook as an additional resource for students.

ALEKS® Creates a Personalized and Dynamic Learning Path

ALEKS® creates an optimized path with an ongoing cycle of learning and assessment, celebrating students' small wins along the way with positive real-time feedback. Rooted in research and analytics, ALEKS® improves student outcomes by fostering better preparation, increased motivation and knowledge retention.

*visit **bit.ly/whatmakesALEKSunique** to learn more about the science behind the most powerful adaptive learning tool in education!

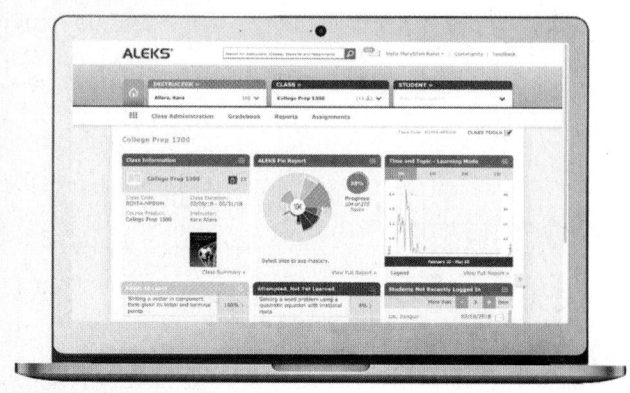

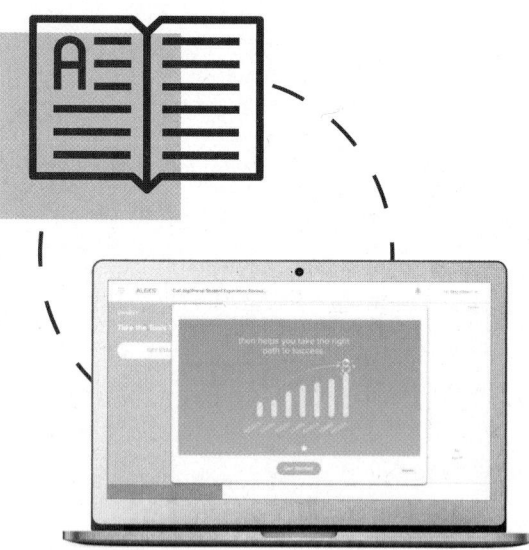

Preparation & Retention

The more prepared your students are, the more effective your instruction is. Because ALEKS® understands the prerequisite skills necessary for mastery, students are better prepared when a topic is presented to them. ALEKS® provides personalized practice and guides students to what they need to learn next to achieve mastery. ALEKS® improves knowledge and student retention through periodic knowledge checks and personalized learning paths. This cycle of learning and assessment ensures that students remember topics they have learned, are better prepared for exams, and are ready to learn new content as they continue into their next course.

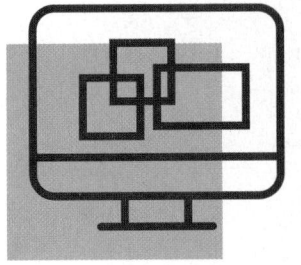

Flexible Implementation: Your Class Your Way!

ALEKS® enables you to structure your course regardless of your instruction style and format. From a traditional classroom, to various co-requisite models, to an online prep course before the start of the term, ALEKS® can supplement your instruction or play a lead role in delivering the content.

*visit **bit.ly/ALEKScasestudies** to see how your peers are delivering better outcomes across various course models!

Outcomes & Efficacy

Our commitment to improve student outcomes services a wide variety of implementation models and best practices, from lecture-based to labs and co-reqs to summer prep courses. Our case studies illustrate our commitment to help you reach your course goals and our research demonstrates our drive to support all students, regardless of their math background and preparation level.

*visit **bit.ly/outcomesandefficacy** to review empirical data from ALEKS® users around the country

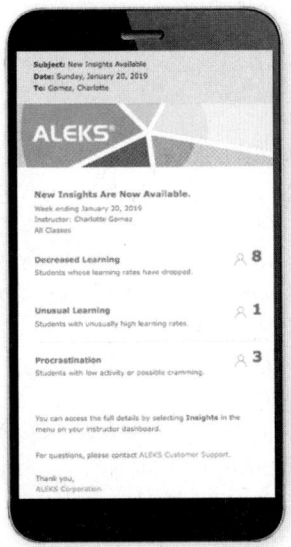

Turn Data Into Actionable Insights

ALEKS® Reports are designed to inform your instruction and create more meaningful interactions with your students when they need it the most. ALEKS® Insights alert you when students might be at risk of falling behind so that you can take immediate action. Insights summarize students exhibiting at least one of four negative behaviors that may require intervention including Failed Topics, Decreased Learning, Unusual Learning and Procrastination & Cramming.

DIGITAL EDGE50 AWARDS ‹2019› | Winner of 2019 Digital Edge 50 Award for Data Analytics!

bit.ly/ALEKS_MHE

Our Commitment to Market Development and Accuracy

Acknowledgments

Paramount to the development of *College Algebra with Corequisite Support* was the invaluable feedback provided by the instructors from around the country who reviewed the manuscript or attended a market development event over the course of the several years the text was in development.

A Special Thanks to All of the Event Attendees Who Helped Shape *College Algebra with Corequisite Support.*

Focus groups and symposia were conducted with instructors from around the country to provide feedback to editors and the authors and to ensure that the direction of the text was meeting the needs of students and instructors.

Nicole Aldridge, *Gadsen State Community College*

Brandon Bartley, *Jefferson Community and Technical College*

Troy Bryant, *Baton Rouge Community College*

Kristine Buddemeyer, *Seminole State College*

Michelle Carmel, *Broward College—North Campus*

William Davidson, *Meridian Community College*

Evelyn Edstrom, *Seminole State College*

Dena Frickey, *Delgado Community College*

Mehrnaz Ghaffarian, *Tarrant County College, Northeast Campus*

Sarah Hand, *College of Southern Maryland*

Tarcia Hubert, *Lone Star College—Montgomery*

Kanetra Jones, *Baton Rouge Community College*

Mickie Karcher, *Jefferson Community and Technical College*

Laurie Keatts, *Catawba Valley Community College*

Jennifer Kennett, *Johnson County Community College*

Thomas Kinzeler, *Tarrant County College, South Campus*

Heather Liggett, *Onondaga Community College*

Jayne Martin, *Lone Star College—Cyfair*

Sharareh Masooman, *Santa Barbara City College*

Meghan McIntyre, *Wake Technical Community College*

Trinity Mecklenburg, *Victor Valley College*

Tammy Morgan, *Gadsen State Community College*

Amy Nabors, *Lone Star College—Montgomery*

Cengiz Ozgener, *State College of Florida*

Kevin Pipkin, *Folsom Lake College*

Azar Raiszadeh, *Chattanooga State Community College*

Shanda Sampson, *College of Southern Maryland*

Christina Thompson, *Seminole State College*

Eboness Williams, *Delgado Community College*

Chris Wyniawskyj, *College of Lake County*

Manuscript Reviewers

Zalmond Abbondanza, *Palm Beach State College*

Edith Aguirre, *El Paso Community College*

Sage Bentley, *Navarro College*

Mark Billiris, *St. Petersburg College*

Renee Boggan, *East Central Community College*

Karen Bond, *Pearl River Community College*

Daniel Brock, *Arkansas State University—Beebe*

Rebecca Calahan, *Middle Tennessee State University*

Michelle Carmel, *Broward College*

Eun Cha, *College of Southern Nevada*

Fan Chen, *El Paso Community College*

Gyuheui Choi, *Atlanta Metropolitan State College*

Christina Christina, *Northern Arizona University*

Ivette Chuca, *El Paso Community College*

Beth Clickner, *Monroe Community College*

Robert Cohen, *Western State Colorado University*

KaraLynne Cook, *Bluegrass Community & Technical College*

William Davidson, *Meridian Community College*

Melanie Devine, *Washington State Community College*

Brandon Elmes, *Jefferson Community & Technical College*

Keith Erickson, *Georgia Gwinnett College*

Cassandra Firth, *Northern Oklahoma College*

Dayna Ford, *Grayson College*

Jennifer Forrester, *Wallace Community College—Dothan*

Jason Geary, *William Rainey Harper College*

Aliakbar Haghighi, *Prairie View A&M University*

Karl Havlak, *Angelo State University*

Monica Hennessy, *University of Cincinnati Blue Ash*

Justin Hill, *Temple College*

Aziza Hina, *Massasoit Community College*

Lisa Hollman, *Quinnipiac University*

Daniel Ingram, *Lindenwood University*

Steve Irons, *College of Southern Idaho*

Lynn Irons, *College of Southern Idaho*

Patrick Kelly, *Penn State Behrend*

Minsu Kim, *University of North Georgia*

Ron Koehn, *Southwestern Oklahoma State University*

Shannon Kratzmeyer, *Cedar Valley College*

Perry Lee, *Kutztown University*

Meghan McIntyre, *Wake Technical Community College*

Beth McNamee, *Washburn University*

Christopher Mizell, *Northwest Florida State College*

Eduardo Morales, *El Camino College*

Bill Morgan, *Metropolitan Community College*

Lamies Nazzal, *California State University, San Bernardino*

Sam Obeid, *Richland College*

Kaan Ozmeral, *Central Carolina Community College*

Kim Page, *Middle Tennessee State University*

Robert Plant, *South Plains College*

Michael Puente, *Richland College*

Doug Ray, *Texas State University*

Miguel San Miguel, *Texas A&M International University*

Polly Schulle, *Richland College*

Wendiann Sethi, *Seton Hall University*

Mukta Sharma, *Yuba College*

Stacey Sivley, *Wallace State Hanceville*

Bridgett Smith, *Southwest Tennessee Community College*

Emileigh Sones, *University of Southern Mississippi*

Kristi Spittler-Brown, *Arkansas Tech University*

Jeffery Swords, *Albany State University*

Ivan Temesvari, *Oakton Community College*

Brian Theroux, *National Park College*

Christina Thompson, *Seminole State College of Florida*

Justin Turner, *University of Arkansas Little Rock*

Gwen Vastine, *Lone Star College—Atascocita Center*

Diane Veneziale, *Rowan College at Burlington County*

Salvador Vera, *Northern Arizona University*

Charles Watson, *University of Central Arkansas*

Lance Williams, *Youngstown State University*

Brooks Ziegler, *Pellissippi State Community College*

Author Acknowledgments

A textbook is not just a book, but part of a larger learning package consisting of both print and digital materials to suit a variety of learning styles for the student and a variety of teaching styles for the instructor. The development of this book and its supplements was an enormous undertaking and wouldn't have been possible without the help of faculty contributors and many people at McGraw-Hill.

First and foremost, the authors want to recognize their colleague, Alina Coronel, who has worked by our side not only on this project, but on the development of all of our textbooks. Alina, thank you for the long hours spent proofreading, making PowerPoints, making videos, writing stepped-out solutions, class-testing our content, and working on accessibility. You've been an integral part of our team, and we are genuinely grateful for your day-to-day help when you'd rather have been fishing.

At McGraw-Hill, we want to thank our portfolio manager, Ian Townsend. As the pilot on this long journey, you set the standard for leadership. Each day, you lead by example to unlock our full potential and inspire our best work. You're strong but not rude; kind, but not weak; humble, but not timid. We're also indebted to the lead product developer Elaine Page, for her unwavering, day-to-day support. To Caroline Celano, Kathleen McMahon, and Mike Ryan, we are forever grateful for the amazing opportunities you and McGraw-Hill have given us.

Our heartfelt gratitude goes to the project manager Jane Mohr for steering the ship and keeping us all on task. To the designer David Hash, many thanks for a clean, student-friendly layout. The book is gorgeous. And to the content licensing specialist, Lorraine Buczek, thank you for securing permissions and enabling us to use the many beautiful photos and art in the book. And to the indexer, Bernice Eisen, thank you for preparing the index. No doubt, many students and instructors will be especially grateful. To the compositor, Ira Chawla and her team—thank you for building this book from our manuscript. You've been responsive to our crazy requests and have done a wonderful job putting the pieces together.

Special thanks to Cynthia Northrup, the director of digital content, for managing the digital content and ensuring consistency of the author voice. Along these lines, we must express our utmost gratitude to digital content contributor Alina Coronel. Thanks to the team at ALEKS for merging the goals of this book with the capabilities of the ALEKS platform, and to RonRon Lin for her management of the digital resources.

To Kelly Kohlmetz, many, many thanks for the beautiful and thorough PowerPoint presentations of our material. To our colleague and friend Kimberly Alacan, we're so grateful for your assistance and creativity with the chapter openers and the group activities.

To our talented and incredibly meticulous copy editor, Kevin Campbell—thank you for your keen eye and for ensuring consistency throughout our work. To Jennifer Blue and Celeste Hernandez, many thanks for doing multiple levels of accuracy checking. Your talents are absolutely amazing. Additional thanks to Julie Kennedy and Anand Singh for their tireless attention to detail proofreading pages.

To the math specialists Leigh Jacka, Simon Wong, Wes Black, Michelle Cook, Mitzi DeWolfe, LaMar Hester, Brittani Longoria, Justin McCord, Ashley McDonald, Michelle Payne, and Amy Wosencroft: Your artistry and creative ideas for our project are 90% inspiration, 90% innovation, and 100% perspiration. Thank you for making us shine.

Finally, to the dedicated people in the McGraw-Hill sales force, thank you so much for your continued confidence, encouragement, and support.

Most importantly, we want to give special thanks to all the students and instructors who use this textbook in their classes.

—Julie Miller and Donna Gerken

Review of Prerequisites

Source: JPL–Caltech/MSSS/NASA

R

A Trip to Mars?

In the 1950s and 1960s scientists had a dream to send men to the moon, and on July 20, 1969, that dream was realized. Astronauts Neil Armstrong and Buzz Aldrin became the first humans to step on the lunar surface. Today, the United States and its global partners are investigating the possibility of a manned trip to Mars. The distance between the two planets varies depending on the relative position of each planet in its orbit. The shortest distance between the Earth and Mars is approximately 54.6 million kilometers (33.9 million miles). This is about 7050 times as far as the distance between the Earth and the moon.

Using existing technology, a manned spacecraft would take roughly 6 months to travel to Mars and 6 months for the return trip. In addition, astronauts would have to stay between 18 and 20 months on Mars to give the two planets time to realign in their orbits to decrease the distance. Thus, the total time for the mission would be approximately $2\frac{1}{2}$ years. During this time, the astronauts would orbit the Sun while on Mars, a distance of $2 \cdot \pi \cdot$ (161 million miles) $\approx$ 1 billion miles. They would contend with temperatures at night that drop to $-73°C$ ($-100°F$), an atmosphere with only 0.13% oxygen, and giant dust storms that can last for weeks at a time and bring hurricane-force winds.

In this discussion, several numerical values were presented in the form of integers, fractions, and decimals that fall into the sets of rational and irrational numbers. In this chapter, we learn about different sets of numbers and how to manipulate numerical values with algebraic operations.

| **SECTION R.1** | Operations on Fractions |

1. Identify Parts of a Fraction

Romolo Tavani/Shutterstock

In day-to-day life, the set of numbers that are used to represent whole units are called **whole numbers**. For example, there may be 24 children at a party and together they eat a total of 8 pizzas. However, to determine the amount of pizza eaten per child, we would need to use a fraction. Fractions are important because they represent portions of a whole unit. For example,

- A child at a party may eat $\frac{1}{3}$ of a pizza.
- An architect may create a plan that is drawn to scale. On the plan we read $\frac{1}{4}$ inch represents 1 foot. Thus, an accurate representation of a large object can be displayed on a small piece of paper.
- You may have $\frac{3}{8}$ of a tank of gas and want to know if you have enough gas to make it to the next gas station.

Understanding the meaning of fractions and how to manipulate them is very important.

A Fraction and Its Parts

Fractions are numbers of the form $\frac{a}{b}$, where $\frac{a}{b} = a \div b$ and b does not equal zero.

In the fraction $\frac{a}{b}$, the **numerator** is a, and the **denominator** is b.

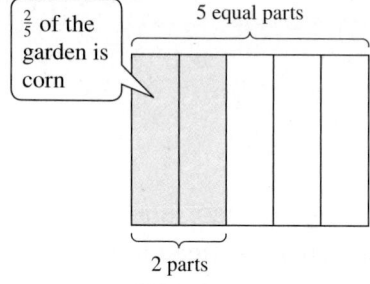

$\frac{2}{5}$ of the garden is corn

5 equal parts

2 parts

Figure R-1

The denominator of a fraction indicates how many equal parts divide the whole. The numerator indicates how many parts are being represented. For example, suppose that a farmer has tilled the soil for five rows of crops. If he plants corn in two of the five rows, $\frac{2}{5}$ then of the garden is corn (Figure R-1).

Proper Fractions, Improper Fractions, and Mixed Numbers

1. If the numerator of a fraction is less than the denominator, the fraction is a **proper fraction**. A proper fraction represents a quantity that is less than a whole unit. For example,

   Proper Fractions: $\frac{3}{5}$ 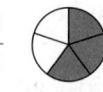 $\frac{1}{8}$

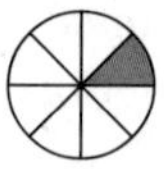

2. If the numerator of a fraction is greater than or equal to the denominator, then the fraction is an **improper fraction**. An improper fraction represents a quantity greater than or equal to a whole unit.

   Improper Fractions: $\frac{7}{5}$ $\frac{8}{8}$

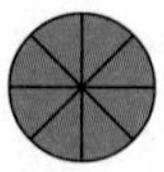

3. A **mixed number** is a proper fraction added to a whole number.

   Mixed numbers: $1\frac{1}{5}$ $2\frac{3}{8}$

2. Simplify Fractions

A fractional portion of a whole unit can be represented by infinitely many fractions. For example, $\frac{1}{2}$ is equivalent to $\frac{2}{4}, \frac{3}{6}, \frac{4}{8}$, and so on (Figure R-2).

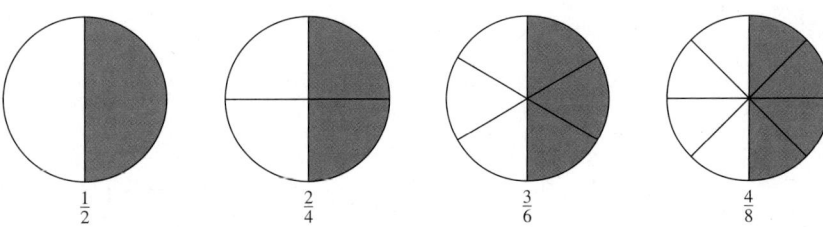

$\frac{1}{2}$ $\frac{2}{4}$ $\frac{3}{6}$ $\frac{4}{8}$

Figure R-2

The fraction $\frac{1}{2}$ is said to be in **lowest terms** because the numerator and denominator share no common factor other than 1. To simplify a fraction to lowest terms, we use the fundamental principle of fractions.

> **TIP** To denote the product of two numbers a and b, the notation $a \times b$ can be used. But to avoid confusion with the variable x, we have the following alternative forms.
>
> $a \cdot b$ ab $a(b)$
>
> $(a)b$ $(a)(b)$

Fundamental Principle of Fractions

Suppose that a number c is a common factor in the numerator and denominator of a fraction. Then, $\dfrac{a \cdot c}{b \cdot c} = \dfrac{a}{b} \cdot \dfrac{c}{c} = \dfrac{a}{b} \cdot 1 = \dfrac{a}{b}$.

One method to simplify a fraction is to factor the numerator and denominator into prime factors. This will help identify the common factors.

EXAMPLE 1   Simplifying a Fraction to Lowest Terms

Simplify. $\dfrac{45}{30}$

Solution:

$\dfrac{45}{30} = \dfrac{3 \cdot 3 \cdot 5}{2 \cdot 3 \cdot 5}$ Factor the numerator and denominator each as a product of prime factors. **Prime numbers** are whole numbers greater than 1 whose only factors are 1 and the number itself. The first several prime numbers are 2, 3, 5, 7, 11, and 13.

$= \dfrac{3}{2} \cdot \dfrac{3}{3} \cdot \dfrac{5}{5}$ Apply the fundamental principle of fractions.

$= \dfrac{3}{2} \cdot 1 \cdot 1$ The ratio $\frac{3}{3}$ simplifies to 1 and $\frac{5}{5}$ simplifies to 1.

$= \dfrac{3}{2}$ Any number multiplied by 1 is the number itself.

Skill Practice 1   Simplify. $\dfrac{20}{55}$

Answer

1. $\dfrac{4}{11}$

In Example 1, we showed numerous steps to reduce fractions to lowest terms. However, the process is often simplified. Notice that the same result can be obtained by dividing out the greatest common factor from the numerator and denominator.

The **greatest common factor** is the largest factor that is common to both numerator and denominator.

$$\frac{45}{30} = \frac{3 \cdot 15}{2 \cdot 15}$$ The greatest common factor of 45 and 30 is 15.

$$= \frac{3 \cdot \overset{1}{\cancel{15}}}{2 \cdot \cancel{15}_{1}}$$ The symbol / is often used to show that a common factor has been divided out.

$$= \frac{3}{2}$$ Notice that "dividing out" the common factor of 15 has the same effect as dividing the numerator and denominator by 15. This is often done mentally.

$$\frac{\overset{3}{\cancel{45}}}{\underset{2}{\cancel{30}}} = \frac{3}{2} \longleftarrow \text{ 45 divided by 15 equals 3.}$$
$$\longleftarrow \text{ 30 divided by 15 equals 2.}$$

3. Multiply and Divide Fractions

Suppose that one slice of pie is one-quarter of one-half of a pie. What portion of a whole pie is the slice? The answer can be found by multiplying $\frac{1}{4}$ by $\frac{1}{2}$. To multiply fractions, multiply the numerators and multiply the denominators (Figure R-3).

$$\frac{1}{4} \cdot \frac{1}{2} = \frac{1}{8}$$

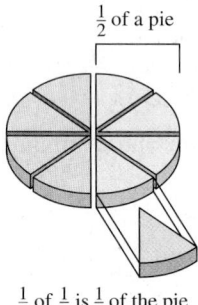

$\frac{1}{2}$ of a pie

$\frac{1}{4}$ of $\frac{1}{2}$ is $\frac{1}{8}$ of the pie

Figure R-3

Multiplying Fractions

To multiply fractions, multiply the numerators and multiply the denominators.

If b is not zero and d is not zero, then $\dfrac{a}{b} \cdot \dfrac{c}{d} = \dfrac{a \cdot c}{b \cdot d}$.

EXAMPLE 2   **Multiplying Fractions**

Multiply.   **a.** $\dfrac{7}{10} \cdot \dfrac{15}{14}$ **b.** $\dfrac{2}{13} \cdot \dfrac{13}{2}$ **c.** $5 \cdot \dfrac{1}{5}$

Solution:

TIP In Example 2(a), the same result can be obtained by dividing out common factors before multiplying.

$$\frac{\overset{1}{7}}{\underset{2}{10}} \cdot \frac{\overset{3}{\cancel{15}}}{\underset{2}{14}} = \frac{3}{4}$$

a. $\dfrac{7}{10} \cdot \dfrac{15}{14} = \dfrac{7 \cdot 15}{10 \cdot 14}$ Multiply the numerators. Multiply the denominators.

$$= \frac{\overset{1}{7} \cdot \overset{3}{\cancel{15}}}{\underset{2}{\cancel{10}} \cdot \underset{2}{14}}$$ Divide out common factors.

$$= \frac{3}{4}$$

b. $\dfrac{2}{13} \cdot \dfrac{13}{2} = \dfrac{\overset{1}{2} \cdot \overset{1}{\cancel{13}}}{\underset{1}{\cancel{13}} \cdot \underset{1}{2}} = \dfrac{1}{1} = 1$ Divide out common factors.

c. $5 \cdot \dfrac{1}{5} = \dfrac{5}{1} \cdot \dfrac{1}{5}$ Write the whole number 5 as $\frac{5}{1}$.

$$= \dfrac{\overset{1}{\cancel{5}} \cdot \overset{1}{\cancel{1}}}{\underset{1}{\cancel{1}} \cdot \underset{1}{\cancel{5}}} = \dfrac{1}{1} = 1$$ Divide out common factors.

Skill Practice 2   Multiply.   **a.** $\dfrac{8}{9} \cdot \dfrac{3}{4}$ **b.** $\dfrac{4}{5} \cdot \dfrac{5}{4}$ **c.** $10 \cdot \dfrac{1}{10}$

From Examples 2(b) and 2(c), notice that $\frac{2}{13} \cdot \frac{13}{2} = 1$ and $5 \cdot \frac{1}{5} = 1$. The numbers $\frac{2}{13}$ and $\frac{13}{2}$ are said to be reciprocals because their product is 1. Likewise, the numbers 5 and $\frac{1}{5}$ are reciprocals. The reciprocal of a number is important for dividing fractions.

The Reciprocal of a Nonzero Real Number

Two nonzero real numbers are **reciprocals** if their product is 1.

Therefore, for a and b not equal to zero, the reciprocal of $\dfrac{a}{b}$ is $\dfrac{b}{a}$ because $\dfrac{a}{b} \cdot \dfrac{b}{a} = 1$.

To understand the concept of dividing fractions, consider a pie that is half-eaten. Suppose the remaining half must be divided among three people, that is, $\frac{1}{2} \div 3$. However, dividing by 3 is equivalent to taking $\frac{1}{3}$ of the remaining $\frac{1}{2}$ of the pie (Figure R-4).

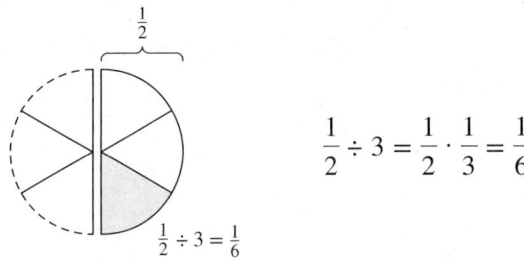

$$\frac{1}{2} \div 3 = \frac{1}{2} \cdot \frac{1}{3} = \frac{1}{6}$$

$$\frac{1}{2} \div 3 = \frac{1}{6}$$

Figure R-4

This example illustrates that dividing two numbers is equivalent to multiplying the first number by the reciprocal of the second number.

Dividing Fractions

If a, b, c, and d are real numbers where b, c, and d are nonzero, then

$$\frac{a}{b} \div \frac{c}{d} = \frac{a}{b} \cdot \frac{d}{c}$$

multiply

reciprocal

Answers

2. a. $\dfrac{2}{3}$ **b.** 1   **c.** 1

EXAMPLE 3   Dividing Fractions

Divide.   **a.** $\dfrac{8}{5} \div \dfrac{3}{10}$ **b.** $\dfrac{12}{13} \div 6$

Solution:

a. $\dfrac{8}{5} \div \dfrac{3}{10} = \dfrac{8}{5} \cdot \dfrac{10}{3}$ Multiply by the reciprocal of $\frac{3}{10}$, which is $\frac{10}{3}$.

$= \dfrac{8 \cdot \overset{2}{\cancel{10}}}{\underset{1}{\cancel{5}} \cdot 3} = \dfrac{16}{3}$ Divide out the common factors and multiply.

b. $\dfrac{12}{13} \div 6 = \dfrac{12}{13} \div \dfrac{6}{1}$ Write the whole number 6 as $\frac{6}{1}$.

$= \dfrac{12}{13} \cdot \dfrac{1}{6}$ Multiply by the reciprocal of $\frac{6}{1}$, which is $\frac{1}{6}$.

$= \dfrac{\overset{2}{\cancel{12}} \cdot 1}{13 \cdot \underset{1}{\cancel{6}}} = \dfrac{2}{13}$ Divide out the common factors and multiply.

Skill Practice 3   Divide.   **a.** $\dfrac{12}{25} \div \dfrac{8}{15}$ **b.** $\dfrac{1}{4} \div 2$

---

4. Add and Subtract Fractions

Two fractions can be added or subtracted if they have the same denominator.

> **Adding and Subtracting Fractions**
>
> If a, b, and c, are real numbers and b is nonzero, then
>
> $$\frac{a}{b} + \frac{c}{b} = \frac{a+c}{b} \quad \text{and} \quad \frac{a}{b} - \frac{c}{b} = \frac{a-c}{b}$$
>
> To add or subtract fractions with the same denominator, add or subtract the numerators and write the result over the common denominator.

EXAMPLE 4   Adding and Subtracting Fractions with the Same Denominator

Simplify.   **a.** $\dfrac{5}{18} + \dfrac{11}{18}$ **b.** $\dfrac{13}{x} - \dfrac{3}{x}$

Solution:

a. $\dfrac{5}{18} + \dfrac{11}{18} = \dfrac{5+11}{18}$ The fractions have the same denominator. Therefore, add the numerators and write the result over the common denominator.

$= \dfrac{16}{18}$ The resulting fraction is not in lowest terms.

$= \dfrac{\overset{8}{\cancel{16}}}{\underset{9}{\cancel{18}}} = \dfrac{8}{9}$ Divide out the greatest common factor of 2.

TIP The sum $\frac{5}{18} + \frac{11}{18}$ can be visualized as the sum of the blue and pink sections of the figure.

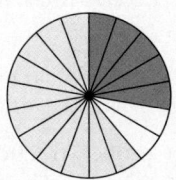

The fraction $\frac{16}{18}$ is equivalent to $\frac{8}{9}$.

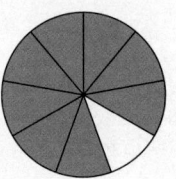

b. $\dfrac{13}{x} - \dfrac{3}{x} = \dfrac{13 - 3}{x}$

The fractions have the same denominator. Therefore, subtract the numerators and write the result over the common denominator.

$= \dfrac{10}{x}$

Skill Practice 4   Simplify.   **a.** $\dfrac{2}{3} + \dfrac{5}{3}$ **b.** $\dfrac{5}{y} - \dfrac{1}{y}$

To add or subtract fractions with *different* denominators, we must first write each fraction as an equivalent fraction with a common denominator. A common denominator may be *any* common multiple of the denominators. However, we will use the least common denominator. The **least common denominator (LCD)** of two or more fractions is the least common multiple (LCM) of the denominators. This is demonstrated in Example 5.

EXAMPLE 5   **Adding and Subtracting Fractions**

Simplify. $\dfrac{5}{12} + \dfrac{3}{4} - \dfrac{1}{2}$

Solution:

$\dfrac{5}{12} + \dfrac{3}{4} - \dfrac{1}{2}$

The fractions have different denominators. To find the least common denominator, we need to find the least common multiple of 12, 4, and 2. One method is to list several multiples of each number and then identify the smallest common value.

Multiply by $\frac{3}{3}$ to make the denominator 12.	Multiply by $\frac{6}{6}$ to make the denominator 12.

Multiples of 12:   12, 24, 36, . . .
Multiples of 4:   4, 8, 12, 16, . . .
Multiples of 2:   2, 4, 6, 8, 10, 12, 14, 16 . . .

$= \dfrac{5}{12} + \dfrac{3}{4} \cdot \dfrac{3}{3} - \dfrac{1}{2} \cdot \dfrac{6}{6}$

The least common multiple of 12, 4, and 2 is 12. Therefore, the least common denominator of the fractions is 12.

$= \dfrac{5}{12} + \dfrac{9}{12} - \dfrac{6}{12}$

The fractions now have the same denominator.

$= \dfrac{5 + 9 - 6}{12}$

Add and subtract numbers in the numerator.

$= \dfrac{8}{12}$

To simplify the numerator, add or subtract in order from left to right. $5 + 9$ is 14. Then $14 - 6$ is 8.

$= \dfrac{\overset{2}{8}}{\underset{3}{12}} = \dfrac{2}{3}$

Simplify $\frac{8}{12}$ by dividing out the greatest common factor of 4 from the numerator and denominator.

Skill Practice 5   Simplify. $\dfrac{2}{3} + \dfrac{1}{2} + \dfrac{5}{6}$

To find the least common denominator between two or more fractions, we also have the option to list the prime factors of each denominator. The LCD is the product of the unique factors, each used the greatest number of times it appears in any single denominator. For example, for the fractions $\frac{5}{12}$, $\frac{3}{4}$, and $\frac{1}{2}$, we have

$12 = 2 \cdot 2 \cdot 3$ ←   The maximum number of times that the factor 2 appears is two times.

$4 = 2 \cdot 2$ The maximum number of times the factor 3 occurs is one time.

$2 = 2$ Therefore, the LCD of $\frac{5}{12}$, $\frac{3}{4}$, and $\frac{1}{2}$ is $2 \cdot 2 \cdot 3$ which is 12.

5. Perform Operations on Mixed Numbers

Recall that a mixed number is a whole number added to a proper fraction. The number $3\frac{1}{2}$ represents the sum of three wholes plus a half, that is, $3\frac{1}{2} = 3 + \frac{1}{2}$. For this reason, any mixed number can be converted to an improper fraction by using addition.

$$3\frac{1}{2} = 3 + \frac{1}{2} = \frac{6}{2} + \frac{1}{2} = \frac{7}{2}$$

TIP   A shortcut to writing a mixed number as an improper fraction is to multiply the whole number by the denominator of the fraction. Then add this value to the numerator of the fraction, and write the result over the denominator.

$3\frac{1}{2}$ ⟶ Multiply the whole number by the denominator: $3 \cdot 2 = 6$
Add the numerator: $6 + 1 = 7$
Write the result over the denominator: $\frac{7}{2}$

     To add, subtract, multiply, or divide mixed numbers, we will first write the mixed number as an improper fraction.

EXAMPLE 6   **Subtracting Mixed Numbers**

Subtract. $4\frac{3}{5} - 2\frac{1}{4}$

Solution:

$$4\frac{3}{5} - 2\frac{1}{4} = \frac{23}{5} - \frac{9}{4}$$

Convert each mixed number to a fraction.
$4 \cdot 5$ is 20. Then $20 + 3 = 23$. Thus, $4\frac{3}{5} = \frac{23}{5}$.
$2 \cdot 4$ is 8. Then $8 + 1 = 9$. Thus, $2\frac{1}{4} = \frac{9}{4}$.

> Multiply by $\frac{4}{4}$ to make the denominator 20.

> Multiply by $\frac{5}{5}$ to make the denominator 20.

$5 = 5^1$ and $4 = 2^2$. Therefore, the least common multiple of 5 and 4 is $5 \cdot 2^2 = 20$.

$$= \frac{23}{5} \cdot \frac{4}{4} - \frac{9}{4} \cdot \frac{5}{5}$$

Write each fraction as an equivalent fraction with a denominator of 20.

$$= \frac{92}{20} - \frac{45}{20}$$

The fractions have the same denominator.

$$= \frac{92 - 45}{20}$$

Subtract the numerators and write the result over the common denominator.

$$= \frac{47}{20} \text{ or } 2\frac{7}{10}$$

To write the result as a mixed number, divide the numerator by the denominator.

$$\frac{47}{20} = \frac{40}{20} + \frac{7}{20} = 2\frac{7}{20}$$

$$\begin{array}{r} 2 \\ 20\overline{)47} \\ -40 \\ \hline 7 \end{array}$$

Answer

6. $11\frac{1}{6}$

Skill Practice 6   Add. $4\frac{2}{3} + 6\frac{1}{2}$

EXAMPLE 7	Dividing Mixed Numbers

Divide. $7\dfrac{1}{2} \div 3$

Solution:

$$7\dfrac{1}{2} \div 3 = \dfrac{15}{2} \div \dfrac{3}{1}$$

Convert the mixed number to a fraction.
Convert the whole number to a fraction.

$$= \dfrac{15}{2} \cdot \dfrac{1}{3}$$

Multiply by the reciprocal of $\frac{3}{1}$, which is $\frac{1}{3}$.

$$= \dfrac{\overset{5}{\cancel{15}}}{2} \cdot \dfrac{1}{\underset{1}{\cancel{3}}}$$

"Cancel" common factors.

$$= \dfrac{5}{2} \quad \text{or} \quad 2\dfrac{1}{2}$$

The answer may be written as an improper fraction or as a mixed number.

Answer

7. $\dfrac{35}{22}$ or $1\dfrac{13}{22}$

Skill Practice 7	Divide. $5\dfrac{5}{6} \div 3\dfrac{2}{3}$

SECTION R.1	Practice Exercises

Prerequisite Review

For Exercises R.1–R.8, identify each number as either a prime number or a composite number.

R.1. 5   **R.2.** 9   **R.3.** 4   **R.4.** 2

R.5. 39   **R.6.** 23   **R.7.** 53   **R.8.** 51

For Exercises R.9–R.16, write each number as a product of prime factors.

R.9. 36   **R.10.** 70   **R.11.** 42   **R.12.** 35

R.13. 110   **R.14.** 136   **R.15.** 135   **R.16.** 105

Concept Connections

1. Given a fraction $\frac{a}{b}$ with $b \neq 0$, the value a is the _____ and _____ is the denominator.

2. A fraction is said to be in _____ terms if the numerator and denominator share no common factor other than 1.

3. The fraction $\frac{4}{4}$ can also be written as the whole number _____, and the fraction $\frac{4}{1}$ can be written as the whole number _____.

4. Two nonzero numbers $\frac{a}{b}$ and $\frac{b}{a}$ are _____ because their product is 1.

5. The least common multiple (LCM) of two numbers is the smallest whole number that is a _____ of both numbers.

6. The _____ common denominator of two or more fractions is the LCM of their denominators.

Objective 1: Identify Parts of a Fraction

For Exercises 7–14, identify the numerator and denominator of each fraction. Then determine if the fraction is a proper fraction or an improper fraction.

7. $\dfrac{7}{8}$ **8.** $\dfrac{2}{3}$ **9.** $\dfrac{9}{5}$

10. $\dfrac{5}{2}$ **11.** $\dfrac{6}{6}$ **12.** $\dfrac{4}{4}$

13. $\dfrac{12}{1}$ **14.** $\dfrac{5}{1}$

Objective 2: Simplify Fractions

For Exercises 15–26, simplify each fraction to lowest terms. **(See Example 1)**

15. $\dfrac{3}{15}$ **16.** $\dfrac{8}{12}$ **17.** $\dfrac{16}{6}$ **18.** $\dfrac{20}{12}$

19. $\dfrac{42}{48}$ **20.** $\dfrac{35}{80}$ **21.** $\dfrac{48}{64}$ **22.** $\dfrac{32}{48}$

23. $\dfrac{110}{176}$ **24.** $\dfrac{70}{120}$ **25.** $\dfrac{200}{150}$ **26.** $\dfrac{210}{119}$

Objective 3: Multiply and Divide Fractions

For Exercises 27–46, multiply or divide as indicated. **(See Examples 2–3)**

27. $\dfrac{10}{13} \cdot \dfrac{26}{15}$ **28.** $\dfrac{15}{28} \cdot \dfrac{7}{9}$ **29.** $\dfrac{3}{7} \div \dfrac{9}{14}$ **30.** $\dfrac{7}{25} \div \dfrac{1}{5}$

31. $\dfrac{9}{10} \cdot 5$ **32.** $\dfrac{3}{7} \cdot 14$ **33.** $\dfrac{12}{5} \div 4$ **34.** $\dfrac{20}{6} \div 5$

35. $\dfrac{5}{2} \cdot \dfrac{10}{21} \cdot \dfrac{7}{5}$ **36.** $\dfrac{55}{9} \cdot \dfrac{18}{32} \cdot \dfrac{24}{11}$ **37.** $\dfrac{9}{100} \div \dfrac{13}{1000}$ **38.** $\dfrac{1000}{17} \div \dfrac{10}{3}$

39. $\dfrac{4}{5} \cdot \dfrac{5}{4}$ **40.** $\dfrac{9}{7} \cdot \dfrac{7}{9}$ **41.** $6 \cdot \dfrac{1}{6}$ **42.** $\dfrac{1}{4} \cdot 4$

43. $8 \div \dfrac{4}{5}$ **44.** $10 \div \dfrac{2}{7}$ **45.** $\dfrac{2}{9} \div \dfrac{4}{3}$ **46.** $\dfrac{8}{11} \div \dfrac{4}{7}$

47. Gus decides to save $\frac{1}{3}$ of his pay each month. If his monthly pay is $2112, how much will he save each month?

48. Stephen's take-home pay is $4200 a month. If he budgeted $\frac{1}{4}$ of his pay for rent, how much is his rent?

49. In Professor Foley's Beginning Algebra class, $\frac{5}{6}$ of the students passed the first test. If there are 42 students in the class, how many passed the test?

50. Shontell had only enough paper to print $\frac{3}{5}$ of her book report before school. If the report is 10 pages long, how many pages did she print?

51. Marty will reinforce a concrete walkway by cutting a steel rod (called rebar) that is 4 yd long. How many pieces can he cut if each piece must be $\frac{1}{2}$ yd in length?

52. There are 4 cups of oatmeal in a box. If each serving is $\frac{1}{3}$ of a cup, how many servings are contained in the box?

53. Anita buys 6 lb of mixed nuts to be divided into decorative jars that will each hold $\frac{3}{4}$ lb of nuts. How many jars will she be able to fill?

54. Beth has a $\frac{7}{8}$-in. nail that she must hammer into a board. Each strike of the hammer moves the nail $\frac{1}{16}$ in. into the board. How many strikes of the hammer must she make to drive the nail completely into the board?

Objective 4: Add and Subtract Fractions

For Exercises 55–78, add or subtract as indicated. **(See Examples 4–5)**

55. $\dfrac{5}{14} + \dfrac{1}{14}$ **56.** $\dfrac{2}{15} + \dfrac{7}{15}$ **57.** $\dfrac{17}{24} - \dfrac{5}{24}$ **58.** $\dfrac{11}{18} - \dfrac{5}{18}$

59. $\dfrac{41}{12} - \dfrac{5}{12}$

60. $\dfrac{77}{24} - \dfrac{29}{24}$

61. $\dfrac{11}{y} + \dfrac{5}{y}$

62. $\dfrac{10}{a} + \dfrac{3}{a}$

63. $\dfrac{1}{8} + \dfrac{3}{4}$

64. $\dfrac{3}{16} + \dfrac{1}{2}$

65. $\dfrac{5}{4} - \dfrac{1}{20}$

66. $\dfrac{7}{6} - \dfrac{1}{24}$

67. $\dfrac{11}{8} - \dfrac{3}{10}$

68. $\dfrac{12}{35} - \dfrac{1}{10}$

69. $\dfrac{7}{26} - \dfrac{2}{13}$

70. $\dfrac{25}{24} - \dfrac{5}{16}$

71. $\dfrac{7}{18} + \dfrac{5}{12}$

72. $\dfrac{3}{16} + \dfrac{9}{20}$

73. $\dfrac{5}{12} + \dfrac{5}{16}$

74. $\dfrac{3}{25} + \dfrac{8}{35}$

75. $\dfrac{1}{6} + \dfrac{3}{4} - \dfrac{5}{8}$

76. $\dfrac{1}{2} + \dfrac{2}{3} - \dfrac{5}{12}$

77. $\dfrac{4}{7} + \dfrac{1}{2} + \dfrac{3}{4}$

78. $\dfrac{9}{10} + \dfrac{4}{5} + \dfrac{3}{4}$

79. For his famous brownie recipe, Chef Alfonso combines $\frac{2}{3}$ cup granulated sugar with $\frac{1}{4}$ cup brown sugar. What is the total amount of sugar in his recipe?

80. Chef Alfonso eats too many of his brownies and his waistline increased by $\frac{3}{4}$ in. during one month and $\frac{3}{8}$ in. the next month. What was his total increase for the 2-month period?

81. Currently the most popular smartphone has a thickness of $\frac{9}{25}$ in. The second most popular is $\frac{1}{2}$ in. thick. How much thicker is the second most popular smartphone?

82. The diameter of a penny is $\frac{3}{4}$ in. while the dime is $\frac{7}{10}$ in. How much larger in diameter is the penny than the dime?

Objective 5: Perform Operations on Mixed Numbers

For Exercises 83–96, perform the indicated operations. **(See Examples 6–7)**

83. $3\frac{1}{5} \cdot 2\frac{7}{8}$

84. $2\frac{1}{2} \cdot 1\frac{4}{5}$

85. $1\frac{2}{9} \div 7\frac{1}{3}$

86. $2\frac{2}{5} \div 1\frac{2}{7}$

87. $1\frac{2}{9} \div 6$

88. $2\frac{2}{5} \div 2$

89. $2\frac{1}{8} + 1\frac{3}{8}$

90. $1\frac{3}{14} + 1\frac{1}{14}$

91. $3\frac{1}{2} - 1\frac{7}{8}$

92. $5\frac{1}{3} - 2\frac{3}{4}$

93. $1\frac{1}{6} + 3\frac{3}{4}$

94. $4\frac{1}{2} + 2\frac{2}{3}$

95. $1 - \dfrac{7}{8}$

96. $2 - \dfrac{3}{7}$

97. A board $26\frac{3}{8}$ in. long must be cut into three pieces of equal length. Find the length of each piece.

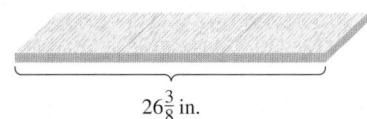

$26\frac{3}{8}$ in.

98. A futon, when set up as a sofa, measures $3\frac{5}{6}$ ft wide. When it is opened to be used as a bed, the width is increased by $1\frac{3}{4}$ ft. What is the total width of this bed?

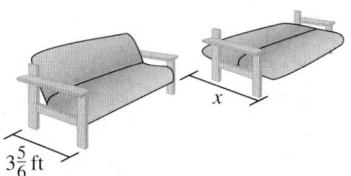

$3\frac{5}{6}$ ft x

99. A plane trip from Orlando to Detroit takes $2\frac{3}{4}$ hr. If the plane traveled for $1\frac{1}{6}$ hr, how much time remains for the flight?

100. Silvia manages a sub shop and needs to prepare smoked turkey sandwiches. She has $3\frac{3}{4}$ lb of turkey in the cooler, and each sandwich requires $\frac{3}{8}$ lb of turkey. How many sandwiches can she make?

101. José's catering company plans to prepare two different shrimp dishes for an upcoming event. One dish requires $1\frac{1}{2}$ lb of shrimp and the other requires $\frac{3}{4}$ lb of shrimp. How much shrimp should José order for the two dishes?

102. Ayako took a trip to the store $5\frac{1}{2}$ mi away. If she rode the bus for $4\frac{5}{6}$ mi and walked the rest of the way, how far did she have to walk?

103. If Tampa, Florida, averages $6\frac{1}{4}$ in. of rain during each summer month, how much total rain would be expected in June, July, August, and September?

104. Pete started working out and found that he lost approximately $\frac{3}{4}$ in. off his waistline every month. How much would he lose around his waist in 6 months?

SECTION R.2   The Set of Real Numbers

OBJECTIVES

1. Identify Subsets of the Set of Real Numbers
2. Determine the Opposite and Absolute Value of a Real Number
3. Add and Subtract Real Numbers
4. Multiply and Divide Real Numbers

1. Identify Subsets of the Set of Real Numbers

"The universe is written in the language of mathematics."
—Galileo

Mathematics is a journey that has been developed and used over the ages to improve people's understanding of complex natural phenomena. From the study of geometry by the ancient Greeks, to the construction of the pyramids by early Egyptians, to NASA's modern-day quest to send astronauts to Mars, mathematics is the foundation. Mathematics gives us a deeper, more robust understanding of the world around us and gives us the tools to solve complicated problems. Mathematics is integral to the way we reason; it is exercise for the mind; it is logic; it is art.

Point of Interest

Romanesco broccoli is a beautiful vegetable with pinecone-like protrusions that follow a spiral pattern. Furthermore, this pattern is repeated at different scales. That is, when viewed up close or from afar, the pattern looks the same. In mathematics, such a "self-similar" pattern is called a **fractal**. Mathematicians use computers that generate two- and three-dimensional fractals to model irregularly shaped objects found in nature and to produce digital art.

Jan Holm/Shutterstock

Numbers represent the building blocks of mathematics, yet we may take their presence for granted. The numbers we are familiar with today were conceived out of necessity. The natural numbers 1, 2, 3, and so on were used for counting. Zero is necessary to denote "the absence of quantity." Negative numbers are necessary to understand what happens when John has $100 in the bank but writes a check for $150. And splitting 10 cookies equally among four people wouldn't be possible without fractions. Thus, to begin our study of mathematics, we formally categorize different types of numbers.

The numbers used in day-to-day life come from the set of real numbers, denoted by $\mathbb{R}$. A **set** is a collection of items called **elements**. The braces { and } are used to enclose the elements of a set. For example,

- The set of **natural numbers**, $\mathbb{N}$ (or counting numbers) is {1, 2, 3, 4, . . .}.
- The set of **whole numbers**, $\mathbb{W}$ is {0, 1, 2, 3, . . .}.
- The set of **integers**, $\mathbb{Z}$ includes the whole numbers and their negative counterparts: {. . . −3, −2, −1, 0, 1, 2, 3, . . .}.
- Numbers that can be written as the ratio of two integers, such as $\frac{1}{2}$, $-\frac{2}{3}$, and 18 $\left(18 = \frac{18}{1}\right)$, form the set of **rational numbers**, $\mathbb{Q}$. In decimal form, a rational number is either a terminating decimal (such as 18 and 0.5) or a repeating decimal (such as $-0.\overline{6} = -0.6666\ldots$).
- The set of **irrational numbers**, which we will denote by $\mathbb{H}$, is made up of numbers that cannot be represented by the ratio of two integers. In decimal form, irrational numbers neither terminate nor repeat. Examples of irrational numbers are π and the square roots of nonperfect squares, such as $\sqrt{2}$, $\sqrt{3}$, $\sqrt{5}$, and so on.
- Together, the rational numbers and irrational numbers make up the set of **real numbers**, which is often denoted by $\mathbb{R}$.

Table R-1 summarizes these six important subsets of the real numbers.

TIP Notice that the first five letters of the word *rational* spell *ratio*. This will help you remember that a rational number is a *ratio* of integers.

Table R-1   Subsets of the Set of Real Numbers, $\mathbb{R}$

Set	Definition
Natural numbers, $\mathbb{N}$	$\{1, 2, 3, \ldots\}$
Whole numbers, $\mathbb{W}$	$\{0, 1, 2, 3, \ldots\}$
Integers, $\mathbb{Z}$	$\{\ldots, -3, -2, -1, 0, 1, 2, 3, \ldots\}$
Rational numbers, $\mathbb{Q}$	Rational numbers can be expressed as a ratio of integers where the denominator is not zero.
Irrational numbers, $\mathbb{H}$	Irrational numbers are real numbers that cannot be expressed as a ratio of integers.

Real Numbers ($\mathbb{R}$)

Rational Numbers

Integers

Whole Numbers

Irrational Numbers

Natural Numbers

Figure R-5

Notice that the whole numbers are included within the set of integers. For this reason, we say that the set of whole numbers is a **subset** of the set of integers. Likewise, the set of integers is a subset of the set of rational numbers. The rational and irrational numbers are **mutually exclusive**, meaning that their elements do not overlap. The relationships among the subsets of real numbers are shown in Figure R-5. A set that contains no elements is called the **empty set** (or **null set**) and is denoted by { } or ∅.

When referring to individual elements of a set, the symbol $\in$ means "is an element of," and the symbol $\notin$ means "is not an element of." For example,

$5 \in \{1, 3, 5, 7\}$ is read as "5 is an element of the set of elements 1, 3, 5, and 7."
$6 \notin \{1, 3, 5, 7\}$ is read as "6 is *not* an element of the set of elements 1, 3, 5, and 7."

A set can be defined in several ways. Listing the elements in a set within braces is called the **roster method**. Using the roster method, the set of the even numbers between 0 and 10 is represented by {2, 4, 6, 8}. Another method to define this set is by using **set-builder notation**. This uses a description of the elements of the set. For example,

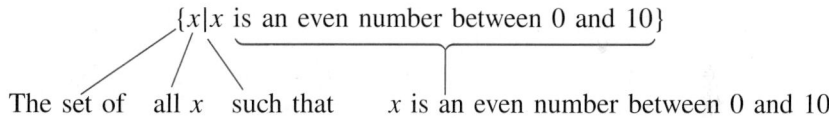

$$\{x \,|\, x \text{ is an even number between 0 and 10}\}$$

The set of   all x such that x is an even number between 0 and 10

EXAMPLE 1   Identifying Elements of a Set

Given $A = \left\{ \sqrt{3}, 0.\overline{83}, -\frac{19}{7}, 0.39, -16, 0, 11, 0.2020020002\ldots, 0.444 \right\}$, determine which elements belong to the following sets.

a. Natural numbers, $\mathbb{N}$ **b.** Whole numbers, $\mathbb{W}$ **c.** Integers, $\mathbb{Z}$

d. Rational numbers, $\mathbb{Q}$ **e.** Irrational numbers, $\mathbb{H}$ **f.** Real numbers, $\mathbb{R}$

Solution:

a) $11 \in \mathbb{N}$

b) $0, 11 \in \mathbb{W}$

c) $0, 11, -16 \in \mathbb{Z}$

d) $0, 11, -16, 0.\overline{83}, -\frac{19}{7}, 0.39, 0.444 \in \mathbb{Q}$

e) $\sqrt{3}, 0.2020020002\ldots \in \mathbb{H}$

f) $\sqrt{3}, 0.\overline{83}, -\frac{19}{7}, 0.39, -16, 0, 11, 0.2020020002\ldots, 0.444 \in \mathbb{R}$

> **Skill Practice 1** Given set B, determine which elements belong to the following sets.
>
> $$B = \left\{-\tfrac{11}{7}, \sqrt{59}, 4.3, 0, 23, -13, \pi, 4.\overline{9}\right\}$$
>
> **a.** The set of natural numbers, $\mathbb{N}$ **b.** The set of whole numbers, $\mathbb{W}$
> **c.** The set of integers, $\mathbb{Z}$ **d.** The set of rational numbers, $\mathbb{Q}$
> **e.** The set of irrational numbers, $\mathbb{H}$ **f.** The set of real numbers, $\mathbb{R}$

2. Determine the Opposite and Absolute Value of a Real Number

Every real number x has an **opposite** denoted by $-x$. For example, $-(4)$ is the opposite of 4 and simplifies to -4. Likewise, $-(-2.1)$ is the opposite of -2.1 and simplifies to 2.1. Two numbers that are opposites are the same distance from 0 on the number line, but on opposite sides of zero.

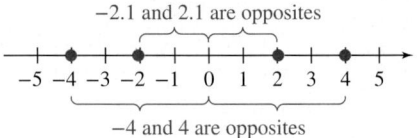

The **absolute value** of a real number x, denoted by $|x|$, is the distance between x and zero on the number line. For example:

$|-5| = 5$ because -5 is 5 units from zero on the number line.
$|5| = 5$ because 5 is 5 units from zero on the number line.

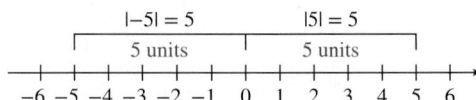

Notice that if a number is negative, its absolute value is the opposite of the number. If a number is positive, its absolute value is the number itself.

Definition of Absolute Value	
Let x be a real number. Then $\|x\| = \begin{cases} x \text{ if } x \geq 0 \\ -x \text{ if } x < 0 \end{cases}$	
Verbal Interpretation	**Numerical Example**
• If x is positive or zero, then $\|x\|$ is just x itself.	$\|4\| = 4$ $\|0\| = 0$
• If x is negative, then $\|x\|$ is the opposite of x.	$\|-4\| = -(-4) = 4$

Answers
1. a. $23 \in \mathbb{N}$ **b.** $0, 23 \in \mathbb{W}$
   c. $0, 23, -13 \in \mathbb{Z}$
   d. $0, 23, -13, -\tfrac{11}{7}, 4.3, 4.\overline{9} \in \mathbb{Q}$
   e. $\sqrt{59}, \pi \in \mathbb{H}$
   f. $-\tfrac{11}{7}, \sqrt{59}, 4.3, 0, 23, -13, \pi, 4.\overline{9} \in \mathbb{R}$

> **EXAMPLE 2**   **Finding the Opposite and Absolute Value of Real Numbers**

Determine the opposite or absolute value as indicated.

a. $-(8)$ **b.** $|-10|$ **c.** $-\left|\dfrac{3}{4}\right|$

Solution:

a. $-(8) = -8$ The expression $-(8)$ is the opposite of 8, which simplifies as -8.

b. $|-10| = -(-10)$ The expression $|-10|$ is the absolute value of -10. To evaluate
$\quad\quad\;\; = 10$ the absolute value of a negative number, take the opposite of the number. The opposite of -10 is 10.

c. $-\left|\dfrac{3}{4}\right| = -\left(\dfrac{3}{4}\right)$ This expression represents the opposite of the absolute value of $\frac{3}{4}$. First simplify the absolute value of $\frac{3}{4}$. We have $\left|\frac{3}{4}\right| = \frac{3}{4}$.

$\quad\quad\quad = -\dfrac{3}{4}$ The opposite of a positive number is a negative number.

Skill Practice 2   Determine the opposite or absolute value as indicated.

a. $-(-6)$ **b.** $\left|\dfrac{7}{13}\right|$ **c.** $-|-14|$

---

3. Add and Subtract Real Numbers

We conclude this section of the text with a refresher on adding, subtracting, multiplying, and dividing real numbers. In particular, we need to be careful of applying the correct sign to the result of each arithmetic operation.

We can visualize the addition of two numbers on a number line. Place a dot on the number line for the first addend as a starting point. Then if we add a positive number, move to the right on the number line. If we add a negative number, move to the left.

$1 + 4 = 5$
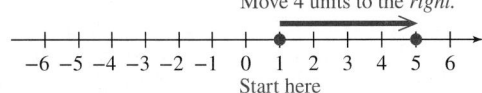

$-1 + (-4) = -5$
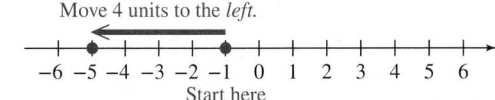

$1 + (-4) = -3$
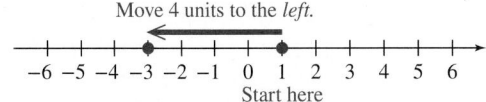

$-1 + 4 = 3$
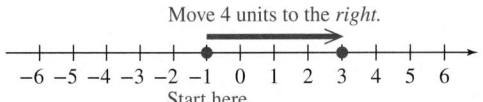

We can generalize the patterns shown in these examples with the following rules.

Addition of Real Numbers

- To add two numbers with the *same sign*, add their absolute values and apply the common sign to the sum.
- To add two numbers with *different signs*, subtract the smaller absolute value from the larger absolute value. Then apply the sign of the number having the larger absolute value.

EXAMPLE 3   **Adding Real Numbers**

Add.

a. $-2 + (-6)$ **b.** $-10.3 + 13.8$ **c.** $\dfrac{5}{6} + \left(-1\dfrac{1}{4}\right)$

Solution:

a. $-2 + (-6)$

First find the absolute value of the addends.
$|-2| = 2$ and $|-6| = 6$

$= -(2 + 6)$

Add their absolute values and apply the common sign. In this case, the common sign is negative.

Common sign is negative.

$= -8$

The sum is -8.

b. $-10.3 + 13.8$

First find the absolute value of the addends.
$|-10.3| = 10.3$ and $|13.8| = 13.8$

The absolute value of 13.8 is greater than the absolute value of -10.3. Therefore, the sum is positive.

$= +(13.8 - 10.3)$

Subtract the smaller absolute value from the larger absolute value.

Apply the sign of the number with the larger absolute value.

$= 3.5$

c. $\dfrac{5}{6} + \left(-1\dfrac{1}{4}\right)$

$= \dfrac{5}{6} + \left(-\dfrac{5}{4}\right)$

Write $-1\dfrac{1}{4}$ as a fraction.

$= \dfrac{5 \cdot 2}{6 \cdot 2} + \left(-\dfrac{5 \cdot 3}{4 \cdot 3}\right)$

The LCD is 12. Write each fraction with the LCD.

$= \dfrac{10}{12} + \left(-\dfrac{15}{12}\right)$

Find the absolute value of the addends.

$\left|\dfrac{10}{12}\right| = \dfrac{10}{12}$ and $\left|-\dfrac{15}{12}\right| = \dfrac{15}{12}$

The absolute value of $-\frac{15}{12}$ is greater than the absolute value of $\frac{10}{12}$. Therefore, the sum is negative.

$= -\left(\dfrac{15}{12} - \dfrac{10}{12}\right)$

Subtract the smaller absolute value from the larger absolute value.

Apply the sign of the number with the larger absolute value.

$= -\dfrac{5}{12}$

Skill Practice 3   Perform the indicated operations.

a. $-4 + (-1)$ **b.** $-2.6 + 1.8$ **c.** $-1 + \left(-\dfrac{3}{7}\right)$

Subtraction of real numbers is defined in terms of the addition process. To subtract two real numbers, add the opposite of the second number to the first number.

Subtraction of Real Numbers

If a and b are real numbers, then $a - b = a + (-b)$.

EXAMPLE 4   **Subtracting Real Numbers**

Subtract.   **a.** $-13 - 5$ **b.** $2.7 - (-3.8)$

Solution:

a. $-13 - 5$

$= -13 + (-5)$ Add the opposite of the second number to the first number.

$= -18$ The addends have the same sign. Add the absolute values $(13 + 5)$ and apply the common negative sign.

b. $2.7 - (-3.8)$

$= 2.7 + (3.8)$ Add the opposite of the second number to the first number.

$= 6.5$ Add.

Skill Practice 4   Subtract.   **a.** $-9 - 8$ **b.** $-1.1 - (-4.2)$

4. Multiply and Divide Real Numbers

The sign of the product of two real numbers is determined by the signs of the factors. Likewise, the sign of the quotient of two real numbers is determined by the signs of the divisor and dividend.

Multiplication and Division of Real Numbers

Assume that a and b are real numbers and b is not equal to zero.

1. If a and b have the *same* sign, then the product ab is *positive*.
   If a and b have the *same* sign, then the quotient $\frac{a}{b}$ is *positive*.
2. If a and b have the *different* signs, then the product ab is *negative*.
   If a and b have *different* signs, then the quotient $\frac{a}{b}$ is *negative*.
3. The product of any real number and zero is zero. $a \cdot 0 = 0$
4. Zero divided by any nonzero real number is zero. $\dfrac{0}{b} = 0$
5. Any nonzero real number divided by zero is undefined. $\dfrac{b}{0}$ is undefined.

Answers

3 a. -5 **b.** -0.8 **c.** $-\dfrac{10}{7}$

4. a. -17 **b.** 3.1

The relationship between multiplication and division can be used to investigate properties 4 and 5. For example,

$$\frac{0}{6} = 0 \qquad\qquad \text{Because } 6 \cdot 0 = 0 \checkmark$$

$$\frac{6}{0} \text{ is } \textit{undefined} \qquad \begin{array}{l}\text{Because there is no number that when multiplied} \\ \text{by 0 will equal 6}\end{array}$$

Note: The quotient of 0 and 0 *cannot be determined.* The statement $\frac{0}{0} = ?$ is equivalent to asking, "What number times zero will equal 0?" That is, $(0)(?) = 0$. Any real number will satisfy this requirement; however, expressions involving $\frac{0}{0}$ are usually discussed in advanced mathematics courses.

EXAMPLE 5   **Multiplying Real Numbers**

Multiply.   **a.** $2 \cdot (-5.1)$ **b.** $\left(-3\frac{1}{3}\right)\left(-\frac{3}{10}\right)$

Solution:

a. $2 \cdot (-5.1)$

$= -10.2$ The signs of the two factors are different. The product is negative.

b. $\left(-3\frac{1}{3}\right)\left(-\frac{3}{10}\right)$ Write the mixed number as a fraction.

$= \left(-\frac{10}{3}\right)\left(-\frac{3}{10}\right)$ The factors have the same sign. The product is positive.

$= \frac{30}{30}$ Multiply.

$= 1$ The values $-\frac{10}{3}$ and $-\frac{3}{10}$ are reciprocals because their product is 1.

Skill Practice 5   Multiply.   **a.** $(-5)(-2.2)$ **b.** $\left(-5\frac{1}{4}\right)\left(\frac{8}{3}\right)$

EXAMPLE 6   **Dividing Real Numbers**

Divide.   **a.** $\frac{-42}{7}$ **b.** $\frac{-96}{-144}$ **c.** $3\frac{1}{10} \div \left(-\frac{2}{5}\right)$ **d.** $\frac{-8}{-7}$

Solution:

a. $\frac{-42}{7} = -6$ The numbers have different signs. The quotient is negative.

b. $\frac{-96}{-144} = \frac{96}{144}$ The numbers have the same sign. The quotient is positive.

$= \frac{\overset{2}{\cancel{96}}}{\underset{3}{\cancel{144}}} = \frac{2}{3}$ Simplify by dividing out the greatest common factor of 48 from the numerator and denominator.

TIP Multiplication may be used to check a division problem.

$$\frac{-42}{7} = -6$$

Check: $(7)(-6) = -42 \checkmark$

Answers
5. a. 11   **b.** −14

Avoiding Mistakes

If the numerator and denominator of a fraction have opposite signs, then the quotient will be negative. Therefore, a fraction has the same value whether the negative sign is written in the numerator, in the denominator, or in front of the fraction.

$$-\frac{31}{4} = \frac{-31}{4} = \frac{31}{-4}$$

c. $3\frac{1}{10} \div \left(-\frac{2}{5}\right)$ The fractions have different signs. The quotient is negative.

$$= \frac{31}{10} \cdot \left(-\frac{5}{2}\right)$$ Write the mixed number as a fraction and multiply by the reciprocal of the second number.

$$= \frac{31}{\overset{}{\underset{2}{10}}} \cdot \left(-\frac{\overset{1}{5}}{2}\right)$$ Divide out the common factor of 5 and multiply the fractions.

$$= -\frac{31}{4}$$ The factors have different signs. The result is negative.

d. $\frac{-8}{-7} = \frac{8}{7}$ The numbers have the same sign. The quotient is positive.

Answers

6. a. -21 **b.** 7 **c.** $-\frac{1}{6}$ **d.** $\frac{1}{2}$

Skill Practice 6   Divide.   **a.** $\frac{42}{-2}$ **b.** $\frac{-28}{-4}$ **c.** $-\frac{2}{3} \div 4$ **d.** $\frac{-1}{-2}$

SECTION R.2   Practice Exercises

Prerequisite Review

For Exercises R.1–R.8, determine whether the decimal form of the given number is terminating or nonterminating.

R.1. 0.378 **R.2.** -19.56 **R.3.** $\frac{2}{3}$ **R.4.** $\frac{4}{9}$

R.5. $\frac{1}{4}$ **R.6.** $\frac{3}{8}$ **R.7.** $0.2\overline{38}$ **R.8.** $1.\overline{2}$

For Exercises R.9–R.12, round the decimals to the indicated place value.

R.9. $2.\overline{45}$; hundredths   **R.10.** 13.5796; thousandths   **R.11.** $2.\overline{45}$; thousandths   **R.12.** 13.5796; hundredths

Concept Connections

1. A(n) _____ is a collection of items called elements.

2. $\mathbb{Z} = \{\ldots, -3, -2, -1, 0, 1, 2, 3, \ldots\}$ is called the set of _____.

3. $\mathbb{R}$ is the notation used to denote the set of _____ _____.

4. A(n) _____ number is a real number that cannot be expressed as a ratio of two integers.

5. Two numbers that are the same distance from 0 but on opposite sides of 0 on the number line are called _____.

6. The absolute value of a real number, a, is denoted by _____ and is the distance between a and _____ on the number line.

7. Write the set in roster form. $\{x \mid x$ is an integer that exceeds 4 but is not more than $10\}$

8. The symbol $\in$ translates as "_____."

Objective 1: Identify Subsets of the Set of Real Numbers

For Exercises 9–14, let $A = \left\{-\frac{3}{2}, \sqrt{11}, -4, 0.\overline{6}, 0, \sqrt{7}, 1\right\}$.

9. Are all of the numbers in set A real numbers?

10. List all of the rational numbers in set A.

11. List all of the whole numbers in set A.

12. List all of the natural numbers in set A.

13. List all of the irrational numbers in set A.

14. List all of the integers in set A.

For Exercises 15–30, describe each number as (a) a terminating decimal, (b) a repeating decimal, or (c) a nonterminating, nonrepeating decimal. Then classify the number as a rational number or as an irrational number.

15. 0.29

16. 3.8

17. $\dfrac{1}{9}$

18. $\dfrac{1}{3}$

19. $\dfrac{1}{8}$

20. $\dfrac{1}{5}$

21. 2π

22. 3π

23. -0.125

24. -3.24

25. -3

26. -6

27. $0.\overline{2}$

28. $0.\overline{6}$

29. $\sqrt{6}$

30. $\sqrt{10}$

For Exercises 31–32, write an English sentence to represent the algebraic statement. Refer to Table R-1.

31. a. $3 \in \mathbb{N}$

b. $-3.1 \notin \mathbb{W}$

32. a. $\dfrac{2}{5} \in \mathbb{Q}$

b. $\pi \notin \mathbb{Q}$

For Exercises 33–34, determine whether the statement is true or false. Refer to Table R-1.

33. a. $-5 \in \mathbb{N}$

b. $-5 \in \mathbb{W}$

c. $-5 \in \mathbb{Z}$

d. $-5 \in \mathbb{Q}$

34. a. $\dfrac{1}{3} \in \mathbb{N}$

b. $\dfrac{1}{3} \in \mathbb{W}$

c. $\dfrac{1}{3} \in \mathbb{Z}$

d. $\dfrac{1}{3} \in \mathbb{Q}$

35. Refer to $A = \left\{\sqrt{5}, 0.\overline{3}, 0.33, -0.9, -12, \frac{11}{4}, 6, \frac{\pi}{6}\right\}$. Determine which elements belong to the given set. Refer to Table R-1 on page 13. (**See Example 1**)

a. $\mathbb{N}$ **b.** $\mathbb{W}$

c. $\mathbb{Z}$ **d.** $\mathbb{Q}$

e. $\mathbb{H}$ **f.** $\mathbb{R}$

36. Refer to $B = \left\{\frac{\pi}{2}, 0, -4, 0.\overline{48}, 1, -\sqrt{13}, 9.4\right\}$. Determine which elements belong to the given set. Refer to Table R-1 on page 13.

a. $\mathbb{N}$ **b.** $\mathbb{W}$

c. $\mathbb{Z}$ **d.** $\mathbb{Q}$

e. $\mathbb{H}$ **f.** $\mathbb{R}$

Objective 2: Determine the Opposite and Absolute Value of a Real Number

For Exercises 37–44, find the opposite of each number.

37. 18

38. 2

39. -6.1

40. -2.5

41. $-\dfrac{5}{8}$

42. $-\dfrac{1}{3}$

43. $\dfrac{7}{3}$

44. $\dfrac{1}{9}$

The opposite of a is denoted as $-a$. For Exercises 45–52, simplify. (**See Example 2**)

45. $-(-3)$

46. $-(-5.1)$

47. $-\left(\dfrac{7}{3}\right)$

48. $-(-7)$

49. $-(-8)$

50. $-(36)$

51. $-(72.1)$

52. $-\left(\dfrac{9}{10}\right)$

For Exercises 53–64, simplify. (**See Example 2**)

53. $|-2|$

54. $|-7|$

55. $|-1.5|$

56. $|-3.7|$

57. $-|-1.5|$

58. $-|-3.7|$

59. $\left|\dfrac{3}{2}\right|$

60. $\left|\dfrac{7}{4}\right|$

61. $-|10|$

62. $-|20|$

63. $-\left|-\dfrac{1}{2}\right|$

64. $-\left|-\dfrac{11}{3}\right|$

Objective 3: Add and Subtract Real Numbers

For Exercises 65–86, perform the indicated operations. (**See Examples 3–4**)

65. $-8 + 4$

66. $3 + (-7)$

67. $-12 + (-7)$

68. $-5 + (-11)$

69. $-17 - (-10)$

70. $-14 - (-2)$

71. $5 - (-9)$

72. $8 - (-4)$

73. $-6.3 - 15.8$

74. $-21.9 - 4.7$

75. $1.5 - 9.6$

76. $4.8 - 10$

77. $\dfrac{2}{3} + \left(-2\dfrac{1}{3}\right)$

78. $-\dfrac{4}{7} + \left(1\dfrac{4}{7}\right)$

79. $-\dfrac{5}{9} - \dfrac{14}{15}$

80. $-6 - \dfrac{2}{9}$

81. a. $14 + (-8)$ **b.** $-14 + 8$ **c.** $-14 + (-8)$ **d.** $14 - (-8)$ **e.** $-14 - 8$

82. a. $-5 - (-3)$ **b.** $-5 + (-3)$ **c.** $-5 - 3$ **d.** $-5 + 3$ **e.** $5 - (-3)$

83. a. $-25 + 25$ **b.** $25 - 25$ **c.** $25 - (-25)$ **d.** $-25 - (-25)$ **e.** $-25 + (-25)$

84. a. $\dfrac{1}{2} + \left(-\dfrac{2}{3}\right)$ **b.** $-\dfrac{1}{2} + \left(\dfrac{2}{3}\right)$ **c.** $-\dfrac{1}{2} + \left(-\dfrac{2}{3}\right)$ **d.** $\dfrac{1}{2} - \left(-\dfrac{2}{3}\right)$ **e.** $-\dfrac{1}{2} - \dfrac{2}{3}$

85. a. $3.5 - 7.1$ **b.** $3.5 - (-7.1)$ **c.** $-3.5 + 7.1$ **d.** $-3.5 - (-7.1)$ **e.** $-3.5 + (-7.1)$

86. a. $-8 - 11$ **b.** $-8 - (-11)$ **c.** $8 + (-11)$ **d.** $-8 + (-11)$ **e.** $8 - (-11)$

Objective 4: Multiply and Divide Real Numbers

For Exercises 87–106, perform the indicated operation. **(See Examples 5–6)**

87. $4(-8)$

88. $-21(3)$

89. $\dfrac{2}{9} \cdot \dfrac{12}{7}$

90. $\left(-\dfrac{5}{9}\right) \cdot \left(-1\dfrac{7}{11}\right)$

91. $\dfrac{-6}{-10}$

92. $\dfrac{-15}{-24}$

93. $-2\dfrac{1}{4} \div \dfrac{5}{8}$

94. $-\dfrac{2}{3} \div \left(-1\dfrac{5}{7}\right)$

95. $7 \div 0$

96. $\dfrac{1}{16} \div 0$

97. $0 \div (-3)$

98. $0 \div 11$

99. $(-1.2)(-3.1)$

100. $(4.6)(-2.25)$

101. $\dfrac{-5}{-11}$

102. $\dfrac{-3}{-13}$

103. $\dfrac{-81}{3}$

104. $\dfrac{20}{-5}$

105. $\dfrac{1}{5} \div (-5)$

106. $-\dfrac{2}{3} \div 3$

Mixed Exercises

107. List three numbers that are real numbers but not rational numbers.

108. List three numbers that are integers but not natural numbers.

109. List three numbers that are integers but not whole numbers.

110. List three numbers that are rational numbers but not integers.

For Exercises 111–112, answer true or false. If a statement is false, explain why.

111. If n is positive, then $|n|$ is negative.

112. If m is negative, then $|m|$ is negative.

113. For what numbers, a, is $-a$ positive?

114. For what numbers, a, is $|a| = a$?

TECHNOLOGY CONNECTIONS

The decimal forms of many rational and irrational numbers are nonterminating. For example, the number π represents the ratio of the circumference of a circle to the diameter of a circle. The decimal form of π is approximately 3.14159 ($\pi \approx 3.14159$). However, the decimal digits go on indefinitely with no repeated pattern.

```
NORMAL FLOAT AUTO REAL RADIAN CL
-(6/11)
            -.5454545455
π
             3.141592654
```

	A	B
1	**Number**	**Decimal Approximation**
2	-(6/11)	-0.545454545
3	Pi()	3.141592654

The number $-\dfrac{6}{11}$ is a rational number, and its decimal form is $-0.5454...$ which we represent as the repeating decimal $-0.\overline{54}$. It is important to realize that for nonterminating decimals, a calculator or spreadsheet will only give approximate values, not exact values.

Point of Interest

Computers have approximated π to over 22 trillion digits, but for most applications, the approximation 3.14159 is sufficient. Some mathematics enthusiasts have named March 14 (that is, the date 3.14) to be World Pi Day.

For Exercises 115–116, use a calculator to find a decimal approximation for the given value. Then round the value to the hundredths place.

115. a. $\sqrt{7}$ **b.** $\sqrt{19}$ **c.** 2π

116. a. -4π **b.** $3\sqrt{5}$ **c.** $7\sqrt{13}$

PROBLEM RECOGNITION EXERCISES

Adding, Subtracting, Multiplying, and Dividing Real Numbers

For Exercises 1–16, perform the indicated operations.

1. a. $-8 - (-4)$ **b.** $-8(-4)$ **c.** $-8 + (-4)$ **d.** $-8 \div (-4)$

2. a. $12 + (-2)$ **b.** $12 - (-2)$ **c.** $12(-2)$ **d.** $12 \div (-2)$

3. a. $-36 + 9$ **b.** $-36(9)$ **c.** $-36 \div 9$ **d.** $-36 - 9$

4. a. $27 - (-3)$ **b.** $27 + (-3)$ **c.** $27(-3)$ **d.** $27 \div (-3)$

5. a. $-5(-10)$ **b.** $-5 + (-10)$ **c.** $-5 \div (-10)$ **d.** $-5 - (-10)$

6. a. $-20 \div 4$ **b.** $-20 - 4$ **c.** $-20 + 4$ **d.** $-20(4)$

7. a. $-4(-16)$ **b.** $-4 - (-16)$ **c.** $-4 \div (-16)$ **d.** $-4 + (-16)$

8. a. $-21 \div 3$ **b.** $-21 - 3$ **c.** $-21(3)$ **d.** $-21 + 3$

9. a. $80(-5)$ **b.** $80 - (-5)$ **c.** $80 \div (-5)$ **d.** $80 + (-5)$

10. a. $-14 - (-21)$ **b.** $-14(-21)$ **c.** $-14 \div (-21)$ **d.** $-14 + (-21)$

11. a. $|-6| + |2|$ **b.** $|-6 + 2|$ **c.** $|-6| - |-2|$ **d.** $|-6 - 2|$

12. a. $-|9| - |-7|$ **b.** $|-9| - |-7|$ **c.** $-|9 - 7|$ **d.** $|-9 - 7|$

13. a. $-\dfrac{3}{4} + \dfrac{8}{3}$ **b.** $-\dfrac{3}{4} - \dfrac{8}{3}$ **c.** $-\dfrac{3}{4} \cdot \dfrac{8}{3}$ **d.** $-\dfrac{3}{4} \div \dfrac{8}{3}$

14. a. $\dfrac{7}{10} \cdot \left(-\dfrac{5}{6}\right)$ **b.** $\dfrac{7}{10} + \left(-\dfrac{5}{6}\right)$ **c.** $\dfrac{7}{10} - \left(-\dfrac{5}{6}\right)$ **d.** $\dfrac{7}{10} \div \left(-\dfrac{5}{6}\right)$

15. a. $4\dfrac{2}{5} \div 1\dfrac{7}{10}$ **b.** $4\dfrac{2}{5} - 1\dfrac{7}{10}$ **c.** $4\dfrac{2}{5} + 1\dfrac{7}{10}$ **d.** $4\dfrac{2}{5} \cdot \left(1\dfrac{7}{10}\right)$

16. a. $-2\dfrac{7}{8} - 4\dfrac{3}{4}$ **b.** $-2\dfrac{7}{8} \div 4\dfrac{3}{4}$ **c.** $-2\dfrac{7}{8} + 4\dfrac{3}{4}$ **d.** $-2\dfrac{7}{8} \cdot \left(4\dfrac{3}{4}\right)$

SECTION R.3   Simplifying Numerical Expressions

OBJECTIVES

1. Evaluate Exponential Expressions
2. Evaluate *n*th Roots
3. Apply the Order of Operations
4. Evaluate Formulas

1. Evaluate Exponential Expressions

Suppose that you invest P dollars in an account that earns annual compound interest. If r is the annual interest rate, then the formula $A = P(1 + r)^t$ gives the amount in the account t years after the investment was made. For example, if $2000 is invested and the interest rate is 4% compounded annually, then

After 1 year, the amount in the account is $A = \$2000(1 + 0.04)^1 = \$2080$.

After 2 years, the amount in the account is $A = \$2000(1 + 0.04)^2 = \$2163.20$.

After 3 years, the amount in the account is $A = \$2000(1 + 0.04)^3 = \$2249.73$.

Notice that the formula $A = P(1 + r)^t$ has four variables: A, P, r, and t. A **variable** is a letter that may represent any numerical value. Thus, to calculate the value A of the account, we would need to substitute numerical values for P, r, and t. Then we would need to know the proper order in which to perform the arithmetic operations. In this section, we discuss the proper mathematical order of operations along with a review of exponential expressions and nth roots.

Repeated multiplication can be written by using exponential notation. For example, the product $5 \cdot 5 \cdot 5$ can be written as 5^3. In this case, 5 is called the base of the expression and 3 is the exponent (or power). The exponent indicates how many times the base is used as a factor.

Definition of b^n

Let b represent any real number and n represent a positive integer. Then

$$b^n = \underbrace{b \cdot b \cdot b \cdot b \cdot \ldots b}_{n \text{ factors of } b}$$

b^n is read as "b to the nth power."
b is called the **base** and n is called the **exponent**, or **power**.
b^2 is read as "b squared," and b^3 is read as "b cubed."

EXAMPLE 1   **Evaluating Exponential Expressions**

Simplify the expression.

a. 5^3 **b.** $(-2)^4$ **c.** -2^4 **d.** $\left(-\dfrac{1}{3}\right)^3$

Solution:

a. $5^3 = 5 \cdot 5 \cdot 5$ The base is 5, and the exponent is 3.
$\quad\ = 125$

b. $(-2)^4 = (-2)(-2)(-2)(-2)$ The base is −2, and the exponent is 4. The exponent 4 applies to the entire contents of the parentheses.
$\quad\quad\ = 16$

c. $-2^4 = -(2 \cdot 2 \cdot 2 \cdot 2)$ The base is 2, and the exponent is 4. Because no parentheses enclose the negative sign, the exponent applies only to 2.
$\quad\ = -16$

TIP The quantity -2^4 can also be interpreted as $-1 \cdot 2^4$.
$$-2^4 = -1 \cdot 2^4 = -1 \cdot (2 \cdot 2 \cdot 2 \cdot 2) = -16$$

d. $\left(-\dfrac{1}{3}\right)^3 = \left(-\dfrac{1}{3}\right)\left(-\dfrac{1}{3}\right)\left(-\dfrac{1}{3}\right)$ The base is $-\frac{1}{3}$, and the exponent is 3.
$\quad\quad\quad = -\dfrac{1}{27}$

Skill Practice 1   Simplify.

a. 2^3 **b.** $(-10)^2$ **c.** -10^2 **d.** $\left(-\dfrac{3}{4}\right)^3$

TECHNOLOGY CONNECTIONS

Topic: Using the Exponent Keys

On many calculators, the $\boxed{x^2}$ key is used to square a number. The $\boxed{\wedge}$ key is used to raise a base to any power.

```
5^3
              125
(-2)^4
               16
-2^4
              -16
```

Answers
1. a. 8   **b.** 100
   c. −100   **d.** $-\dfrac{27}{64}$

2. Evaluate nth Roots

The inverse operation to squaring a number is to find its square roots. For example, finding a square root of 9 is equivalent to asking, "What number when squared equals 9?" One obvious answer is 3, because $(3)^2 = 9$. However, -3 is also a square root of 9 because $(-3)^2 = 9$. For now, we will focus on the **principal square root**, which is always taken to be nonnegative.

The symbol $\sqrt{}$, called a **radical sign**, is used to denote the principal square root of a number. Therefore, the principal square root of 9 can be written as $\sqrt{9}$. The expression $\sqrt{64}$ represents the principal square root of 64.

| **EXAMPLE 2** | **Evaluating Square Roots** |

Evaluate the expressions, if possible.

a. $\sqrt{81}$ **b.** $\sqrt{\dfrac{25}{64}}$ **c.** $\sqrt{-16}$ **d.** $-\sqrt{16}$

Solution:

a. $\sqrt{81} = 9$ because $(9)^2 = 81$

b. $\sqrt{\dfrac{25}{64}} = \dfrac{5}{8}$ because $\left(\dfrac{5}{8}\right)^2 = \dfrac{25}{64}$

c. $\sqrt{-16}$ is *not a real number* because no real number when squared will be negative.

d. $-\sqrt{16} = -4$ because $-\sqrt{16} = -\left(\sqrt{16}\right) = -4$.

Skill Practice 2 Evaluate the expressions, if possible.

a. $\sqrt{25}$ **b.** $\sqrt{\dfrac{49}{100}}$ **c.** $\sqrt{-4}$ **d.** $-\sqrt{9}$

The concept of finding the principal square root of a real number can be extended to roots of higher order. The notation $\sqrt[n]{a}$ represents the **principal nth root** of a real number a. For example,

- $\sqrt[3]{-64}$ is the cube root (third root) of -64. Its value is a number that when cubed equals -64. Thus, $\sqrt[3]{-64} = -4$ because $(-4)^3 = (-4) \cdot (-4) \cdot (-4) = -64$.

- $\sqrt[4]{16}$ is the principal fourth root of 16. Its value is a positive number that when raised to the fourth power equals 16. Thus, $\sqrt[4]{16} = 2$ because $2^4 = 2 \cdot 2 \cdot 2 \cdot 2 = 16$.

Given $\sqrt[n]{a}$, the value of n is called the **index** and the value of a is called the **radicand**. For a root with an even index and a positive radicand, such as $\sqrt[4]{16}$, we take the principal root to be positive. Thus $\sqrt[4]{16} = 2$ (not -2).

It is also important to note that some restrictions apply when taking nth roots. In Example 2c we saw that the square root of a negative number is not a real number. That is,

- $\sqrt{-16}$ is not a real number because no real number when squared is negative.

Likewise, an even-indexed root of a negative number is not a real number.

- $\sqrt[4]{-625}$ is not a real number because no real number when raised to the fourth power is negative.

TECHNOLOGY CONNECTIONS

Topic: Using the Square Root Key

The $\boxed{\sqrt{}}$ key is used to find the square root of a nonnegative real number.

```
√(81)
            9
√(25/64)▶Frac
          5/8
```

Answers

2. a. 5 **b.** $\dfrac{7}{10}$
  c. Not a real number
  d. -3

| EXAMPLE 3 | Evaluating *n*th Roots |

Simplify.   **a.** $\sqrt[3]{-8}$ **b.** $\sqrt[4]{81}$ **c.** $\sqrt[3]{\dfrac{1}{125}}$ **d.** $\sqrt[4]{-81}$

Solution:

a. $\sqrt[3]{-8} = -2$ because $(-2)(-2)(-2) = -8$.

b. $\sqrt[4]{81} = 3$ because 3 is positive and $(3)(3)(3)(3) = 81$.

c. $\sqrt[4]{-81}$ is not a real number because no real number when raised to the fourth power equals -81.

d. $\sqrt[3]{\dfrac{1}{125}} = \dfrac{1}{5}$ because $\left(\dfrac{1}{5}\right)\left(\dfrac{1}{5}\right)\left(\dfrac{1}{5}\right) = \dfrac{1}{125}$.

> **Avoiding Mistakes**
>
> Note that
> $$3^4 = 3 \cdot 3 \cdot 3 \cdot 3 = 81$$
> and
> $$(-3)^4 = (-3)(-3)(-3)(-3).$$
> However, the notation $\sqrt[4]{81}$ represents the *principal* fourth root of 81, which we define to be positive. Thus, $\sqrt[4]{81} = 3$ (not -3).

Skill Practice 3   Simplify.

a. $\sqrt[4]{625}$ **b.** $\sqrt[3]{-216}$ **c.** $\sqrt[3]{\dfrac{1}{1000}}$ **d.** $\sqrt[4]{-10,000}$

3. Apply the Order of Operations

When algebraic expressions contain numerous operations, it is important to use the proper **order of operations**. Parentheses ( ), brackets [ ], and braces { } are used for grouping numbers and algebraic expressions. It is important to recognize that operations must be done first within parentheses and other grouping symbols.

> **Order of Operations**
>
> **Step 1** First, simplify expressions within parentheses and other grouping symbols. These include absolute value bars, fraction bars, and radicals. If embedded parentheses are present, start with the innermost parentheses.
>
> **Step 2** Evaluate expressions involving exponents, radicals, and absolute values.
>
> **Step 3** Perform multiplication or division in the order in which they occur from left to right.
>
> **Step 4** Perform addition or subtraction in the order in which they occur from left to right.

| EXAMPLE 4 | Applying the Order of Operations |

Simplify. $10 - 4(9 - 6)^2 + 8$

Solution:

$$10 - 4(9 - 6)^2 + 8$$

$$= 10 - 4(3)^2 + 8 \qquad \text{Perform the subtraction within parentheses first. } 9 - 6 = 3$$

$$= 10 - 4 \cdot 9 + 8 \qquad \text{Evaluate the expression involving exponents. } (3)^2 = 9$$

$$= 10 - 36 + 8 \qquad \text{Perform multiplication before either addition or subtraction.}$$

$$= -26 + 8 \qquad \text{Subtract and add in order from left to right. Therefore, evaluate } 10 - 36 \text{ to get } -26. \text{ Then add } -26 + 8.$$

$$= -18$$

Answers

3. a. 5   **b.** −6   **c.** $\dfrac{1}{10}$

  d. Not a real number

4. 42

Skill Practice 4   Simplify. $(8 - 2)^2 - 3(-4 + 2)$

EXAMPLE 5   Applying the Order of Operations

Simplify. $10 - [2 - 4(6 - 8)]^2 + \sqrt{16 - 7}$

Solution:

$10 - [2 - 4(6 - 8)]^2 + \sqrt{16 - 7}$

$= 10 - [2 - 4(-2)]^2 + \sqrt{9}$ First simplify innermost parentheses.
Subtract $6 - 8 = -2$.
The radical is a grouping symbol. Subtract $16 - 7 = 9$.

$= 10 - [2 + 8]^2 + \sqrt{9}$ Continue simplifying inside the square brackets.
Multiply -4 times -2 to get 8.

$= 10 - [10]^2 + \sqrt{9}$

$= 10 - 100 + 3$ Simplify the exponent and square root.

$= -90 + 3$ Perform the subtraction before the addition because the subtraction occurs first from left to right.

$= -87$

Skill Practice 5   Simplify. $36 \div 2^2 \cdot 3 - [(18 - 5) \cdot 2 + 6]$

EXAMPLE 6   Applying the Order of Operations

Simplify. $\dfrac{|8 - 6 \cdot 2|}{\sqrt[3]{10^2 - 6^2}}$

Solution:

$\dfrac{|8 - 6 \cdot 2|}{\sqrt[3]{10^2 - 6^2}}$ The fraction bar groups the numerator and denominator separately.

$= \dfrac{|8 - 12|}{\sqrt[3]{100 - 36}}$ In the numerator, multiply $6 \cdot 2$ before subtracting.
In the denominator, simplify exponents before subtracting.

$= \dfrac{|-4|}{\sqrt[3]{64}}$ Simplify within grouping symbols in the numerator and denominator.

$= \dfrac{4}{4}$ The absolute value of -4 is 4. The principal cube root of 64 is 4.

$= 1$

Skill Practice 6   Simplify. $\dfrac{|3 - 5| - 14}{\sqrt[3]{3^2 - 1}}$

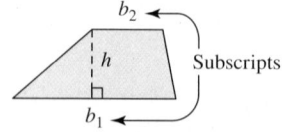

Figure R-6

4. Evaluate Formulas

In Examples 7 and 8, we apply the order of operations to evaluate formulas.

It is important to note that some formulas from geometry use Greek letters (such as π) and some use variables with subscripts. A **subscript** is a number or letter written to the right of and below a variable. For example, the area of a trapezoid is given by $A = \frac{1}{2}(b_1 + b_2)h$. The values b_1 and b_2 (read as "b sub 1" and "b sub 2") represent two different bases of the trapezoid (Figure R-6).

Answers

5. -5 **6.** -6

EXAMPLE 7   Evaluating a Formula

A homeowner in North Carolina wants to buy protective film for a trapezoid-shaped window. The film will adhere to shattered glass in the event that the glass breaks during a bad storm. Find the area of the window with the given dimensions.

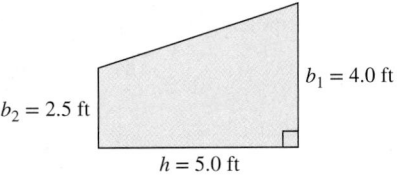

TIP Subscripts should not be confused with *superscripts*, which are written above a variable. Superscripts are used to denote exponents.

$$b_2 \neq b^2$$

Solution:

$$A = \frac{1}{2}(b_1 + b_2)h$$

$$= \frac{1}{2}(4.0 \text{ ft} + 2.5 \text{ ft})(5.0 \text{ ft}) \qquad \text{Substitute } b_1 = 4.0 \text{ ft}, b_2 = 2.5 \text{ ft}, \text{ and } h = 5.0 \text{ ft.}$$

$$= \frac{1}{2}(6.5 \text{ ft})(5.0 \text{ ft}) \qquad \text{Simplify inside parentheses.}$$

$$= 16.25 \text{ ft}^2 \qquad \text{Multiply from left to right.}$$

The area of the window is 16.25 ft^2.

Skill Practice 7   Use the formula given in Example 7 to find the area of the trapezoid.

EXAMPLE 8   Evaluating a Formula

Use the formula $A = P(1 + r)^t$ to determine the amount in an account in which $4500 was invested for 6 years compounded annually at 5% interest.

Solution:

$$A = P(1 + r)^t \qquad \text{In this formula, the principal amount invested is denoted by } P, \text{ the annual interest rate is denoted by } r, \text{ and the time of the investment in years is given by } t.$$

$$A = \$4500(1 + 0.05)^6 \qquad \text{Substitute } P = \$4500, r = 0.05, \text{ and } t = 6.$$
$$\text{Perform the addition within parentheses first.}$$

$$= \$4500(1.05)^6 \qquad \text{Next, simplify } (1.05)^6. \text{ This can be done by using a calculator and the} \bigwedge \text{key.}$$

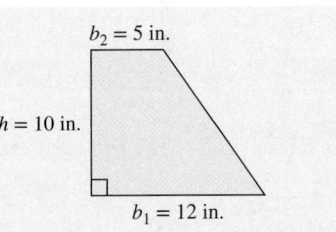

$$= \$4500(1.340095641) \qquad \text{Multiply using a calculator.}$$

$$= \$6030.43 \qquad \text{Rounding should not be done until the last step. Round to the nearest cent (hundredth of a dollar).}$$

The amount in the account is $6030.43 after 6 years.

Skill Practice 8   Use the formula $A = P(1 + r)^t$ to determine the amount in an account if $8000 in principal was invested at 4.5% interest compounded annually for 10 years.

Answers

7. The area is 85 in.^2

8. $12,423.76

SECTION R.3   Practice Exercises

Prerequisite Review

For Exercises R.1–R.6, perform the indicated operations. (See Section R.1 for review.)

R.1. $\dfrac{3}{8} - \dfrac{7}{20}$ **R.2.** $\dfrac{3}{4} + \dfrac{5}{6}$ **R.3.** $\dfrac{6}{25} \div \dfrac{2}{5}$ **R.4.** $\dfrac{8}{9} \div \dfrac{10}{3}$ **R.5.** $\dfrac{21}{5} \cdot \dfrac{10}{3}$ **R.6.** $\dfrac{15}{2} \cdot \dfrac{6}{5}$

Concept Connections

1. In the expression b^n, the value b is called the _____ and n is called the _____ or _____.

2. The symbol $\sqrt{}$ is called a _____ sign and is used to find the principal _____ root of a nonnegative real number.

3. Given $\sqrt[n]{a}$, the value of a is called the _____ and _____ is called the index. The expression $\sqrt[n]{a}$ represents the _____-root of a.

4. The set of rules that tell us the order in which to perform operations to simplify an algebraic expression is called the _____.

Objective 1: Evaluate Exponential Expressions

For Exercises 5–6, write each product using exponents.

5. $\dfrac{1}{6} \cdot \dfrac{1}{6} \cdot \dfrac{1}{6} \cdot \dfrac{1}{6}$ **6.** $10 \cdot 10 \cdot 10 \cdot 10 \cdot 10 \cdot 10$

For Exercises 7–12, write each expression in expanded form using the definition of an exponent.

7. x^3 **8.** y^4 **9.** $(2b)^3$

10. $(8c)^2$ **11.** $10y^5$ **12.** $x^2 y^3$

For Exercises 13–26, simplify each expression. **(See Example 1)**

13. 6^2 **14.** 7^3 **15.** $\left(\dfrac{1}{7}\right)^2$ **16.** $\left(\dfrac{1}{2}\right)^5$

17. $(0.2)^3$ **18.** $(0.8)^2$ **19.** 2^6 **20.** 13^2

21. -7^2 **22.** -3^4 **23.** $(-7)^2$ **24.** $(-5)^2$

25. $\left(\dfrac{5}{3}\right)^3$ **26.** $\left(\dfrac{10}{9}\right)^2$

Objective 2: Evaluate *n*th Roots

For Exercises 27–46, evaluate the expression, if possible. **(See Examples 2–3)**

27. $\sqrt{36}$ **28.** $\sqrt{144}$ **29.** $\sqrt{9}$ **30.** $\sqrt{1}$

31. $\sqrt{-4}$ **32.** $\sqrt{-36}$ **33.** $\sqrt{\dfrac{1}{4}}$ **34.** $\sqrt{\dfrac{9}{4}}$

35. $-\sqrt{49}$ **36.** $-\sqrt{100}$ **37.** $\sqrt{0.04}$ **38.** $\sqrt{0.25}$

39. $\sqrt[3]{-27}$ **40.** $\sqrt[3]{-343}$ **41.** $\sqrt[4]{10,000}$ **42.** $\sqrt[6]{1,000,000}$

43. $\sqrt[4]{-81}$ **44.** $\sqrt[6]{-1}$ **45.** $\sqrt[3]{\dfrac{27}{64}}$ **46.** $\sqrt[4]{\dfrac{16}{625}}$

Objective 3: Apply the Order of Operations

For Exercises 47–78, simplify the expressions. (**See Examples 4–6**)

47. $5 + 3^3$

48. $10 - 2^4$

49. $5 \cdot 2^3$

50. $12 \div 2^2$

51. $(2 + 3)^2$

52. $(4 - 1)^3$

53. $2^2 + 3^2$

54. $4^3 - 1^3$

55. $4 + 2 \div 2 \cdot 3 + 1$

56. $5 + 12 \div 2 \cdot 6 - 1$

57. $81 - 4 \cdot 3 + 3^2$

58. $100 - 25 \cdot 2 - 5^2$

59. $\dfrac{1}{4} \cdot \dfrac{2}{3} - \dfrac{1}{6}$

60. $\dfrac{3}{4} \cdot \dfrac{2}{3} + \dfrac{2}{3}$

61. $\left(\dfrac{11}{6} - \dfrac{3}{8} \right) \cdot \dfrac{4}{5}$

62. $\left(\dfrac{9}{8} - \dfrac{1}{3} \right) \cdot \dfrac{3}{4}$

63. $2 - 5(9 - 4\sqrt{25})^2$

64. $5^2 - (\sqrt{9} + 4 \div 2)$

65. $\left(-\dfrac{3}{5} \right)^2 - \dfrac{3}{5} \cdot \dfrac{5}{9} + \dfrac{7}{10}$

66. $\dfrac{1}{2} - \left(\dfrac{2}{3} \div \dfrac{5}{9} \right) + \dfrac{5}{6}$

67. $1.75 \div 0.25 - (1.25)^2$

68. $5.4 - (0.3)^2 \div 0.09$

69. $\dfrac{\sqrt{10^2 - 8^2}}{3^2}$

70. $\dfrac{\sqrt{16 - 7} + 3^2}{\sqrt{16} - \sqrt{4}}$

71. $-|-11 + 5| + |7 - 2|$

72. $-|-8 - 3| - (-8 - 3)$

73. $25 - 2[(7 - 3)^2 \div 4] + \sqrt{18 - 2}$

74. $\sqrt{29 - 2^2} + [8 - 3(6 - 2)] \div 4 \cdot 5$

75. $\dfrac{|(10 - 7) - 2^3|}{6 - 16 \div 8 \cdot 3}$

76. $\dfrac{|-12 - (7 - 3^2)^2|}{40 - 6^2 - 8 \div 2}$

77. $\left(\dfrac{1}{2} \right)^2 + \left(\dfrac{6 - 4}{5} \right)^2 + \left(\dfrac{5 + 2}{10} \right)^2$

78. $\left(\dfrac{2^3}{2^3 + 1} \right)^2 \div \left[\dfrac{8 - (-2)}{3^2} \right]^2$

Objective 4: Evaluate Formulas

For Exercises 79–82, find the area. Refer to the geometry formulas at the end of the text. (**See Example 7**)

79. Trapezoid

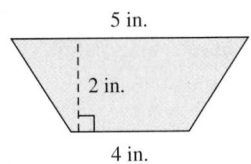

5 in.

2 in.

4 in.

80. Parallelogram

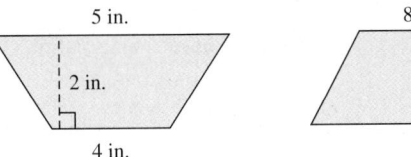

8.5 m

6 m

81. Triangle

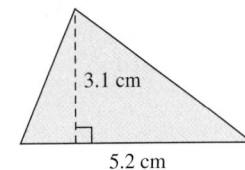

3.1 cm

5.2 cm

82. Rectangle

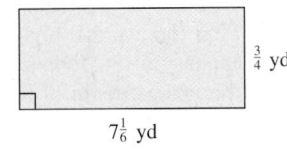

$\frac{3}{4}$ yd

$7\frac{1}{6}$ yd

For Exercises 83–88, find the volume. Refer to the geometry formulas at the end of the text. (Use the π key on your calculator, and round the final answer to one decimal place.)

83. Sphere

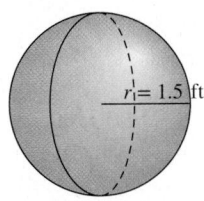

$r = 1.5$ ft

84. Sphere

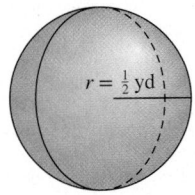

$r = \frac{1}{2}$ yd

85. Right circular cone

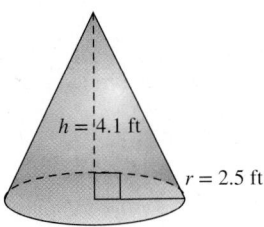

$h = 4.1$ ft

$r = 2.5$ ft

86. Right circular cone

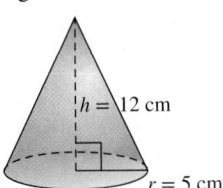

$h = 12$ cm

$r = 5$ cm

87. Right circular cylinder

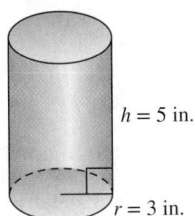

$h = 5$ in.

$r = 3$ in.

88. Right circular cylinder

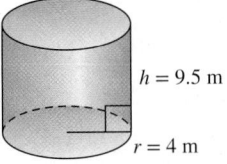

$h = 9.5$ m

$r = 4$ m

89. The formula $C = \frac{5}{9}(F - 32)$ converts temperatures in the Fahrenheit scale to the Celsius scale. Find the equivalent Celsius temperature for each Fahrenheit temperature. **(See Example 8)**

    a. 77°F   **b.** 212°F   **c.** 32°F   **d.** −40°F

90. The formula $F = \frac{9}{5}C + 32$ converts Celsius temperatures to Fahrenheit temperatures. Find the equivalent Fahrenheit temperature for each Celsius temperature.

    a. −5°C   **b.** 0°C   **c.** 37°C   **d.** −40°C

The equation $G_E = \frac{1}{22}c + \frac{1}{30}h$ represents the amount of gasoline used (in gal) for an economy car to drive c miles in the city and h miles on the highway. The equation $G_T = \frac{1}{12}c + \frac{1}{18}h$ represents the amount of gasoline used for a truck to make the same trip. Use these formulas for Exercises 91–92.

91. Determine the amount of gas used by an economy car that travels 33 mi in the city and 80 mi on the highway.

92. Determine the amount of gas used by a truck that travels 33 mi in the city and 80 mi on the highway.

Mixed Exercises

For Exercises 93–96, simplify the expressions.

93. a. 4^2 **b.** $(-4)^2$ **c.** -4^2 **d.** $\sqrt{4}$ **e.** $-\sqrt{4}$ **f.** $\sqrt{-4}$

94. a. 9^2 **b.** $(-9)^2$ **c.** -9^2 **d.** $\sqrt{9}$ **e.** $-\sqrt{9}$ **f.** $\sqrt{-9}$

95. a. $\sqrt[3]{8}$ **b.** $\sqrt[3]{-8}$ **c.** $-\sqrt[3]{8}$ **d.** $\sqrt{100}$ **e.** $\sqrt{-100}$ **f.** $-\sqrt{100}$

96. a. $\sqrt[3]{27}$ **b.** $\sqrt[3]{-27}$ **c.** $-\sqrt[3]{27}$ **d.** $\sqrt{49}$ **e.** $\sqrt{-49}$ **f.** $-\sqrt{49}$

For Exercises 97–98, find the average of the set of data values by adding the values and dividing by the number of values.

97. Find the average low temperature for a week in January in St. John's, Newfoundland. Round to the nearest tenth of a degree.

Day	Mon.	Tues.	Wed.	Thur.	Fri.	Sat.	Sun.
Low temperature	−18°C	−16°C	−20°C	−11°C	−4°C	−3°C	1°C

98. Find the average high temperature for a week in January in St. John's, Newfoundland. Round to the nearest tenth of a degree.

Day	Mon.	Tues.	Wed.	Thur.	Fri.	Sat.	Sun.
High temperature	−2°C	−6°C	−7°C	0°C	1°C	8°C	10°C

For Exercises 99–102, evaluate each expression for the given values of the variables.

99. $\dfrac{-b}{2a}$ for $a = -1$, $b = -6$

100. $\sqrt{b^2 - 4ac}$ for $a = 2$, $b = -6$, $c = 4$

101. $\sqrt{(x_2 - x_1)^2 + (y_2 - y_1)^2}$ for $x_1 = 2$, $x_2 = -1$, $y_1 = -4$, $y_2 = 1$

102. $\dfrac{y_2 - y_1}{x_2 - x_1}$ for $x_1 = -1.4$, $x_2 = 2$, $y_1 = 3.1$, $y_2 = -3.7$

Technology Connections

103. Which expression when entered into a graphing calculator will yield the correct value of $\dfrac{12}{6-2}$?

        12/6 − 2   or   12/(6 − 2)

104. Which expression when entered into a graphing calculator will yield the correct value of $\dfrac{24-6}{3}$?

        (24 − 6)/3   or   24 − 6/3

105. Verify your solution to Exercise 69 by entering the expression into a graphing calculator:

$$\left(\sqrt{(10^2 - 8^2)}\right)/3^2$$

106. Verify your solution to Exercise 70 by entering the expression into a graphing calculator:

$$\left(\sqrt{(16 - 7)} + 3^2\right)/\left(\sqrt{(16)} - \sqrt{(4)}\right)$$

For Exercises 107–110, use a calculator to approximate the expression to 2 decimal places.

107. $5000\left(1+\dfrac{0.06}{12}\right)^{(12)(5)}$

108. $8500\left(1+\dfrac{0.05}{4}\right)^{(4)(30)}$

109. $\dfrac{-3+5\sqrt{2}}{7}$

110. $\dfrac{6-3\sqrt{5}}{4}$

CHAPTER R   Detailed Summary

SECTION R.1 Operations on Fractions

Key Concepts	Examples	Page
Simplifying Fractions Divide the numerator and denominator by their greatest common factor.	**Example 1:** $\dfrac{60}{84}=\dfrac{5\cdot\overset{1}{\cancel{12}}}{7\cdot\underset{1}{\cancel{12}}}=\dfrac{5}{7}$	p. 3
Multiplication of Fractions Multiply the numerators and multiply the denominators. $\dfrac{a}{b}\cdot\dfrac{c}{d}=\dfrac{a\cdot c}{b\cdot d}$	**Example 2:** $\dfrac{25}{108}\cdot\dfrac{27}{40}=\dfrac{\overset{5}{\cancel{25}}}{\underset{4}{\cancel{108}}}\cdot\dfrac{\overset{1}{\cancel{27}}}{\underset{8}{\cancel{40}}}$ $=\dfrac{5\cdot1}{4\cdot8}=\dfrac{5}{32}$	p. 4
Division of Fractions Multiply the first fraction by the reciprocal of the second fraction. $\dfrac{a}{b}\div\dfrac{c}{d}=\dfrac{a}{b}\cdot\dfrac{d}{c}$	**Example 3:** $\dfrac{95}{49}\div\dfrac{65}{42}=\dfrac{\overset{19}{\cancel{95}}}{\underset{7}{\cancel{49}}}\cdot\dfrac{\overset{6}{\cancel{42}}}{\underset{13}{\cancel{65}}}$ $=\dfrac{19\cdot6}{7\cdot13}=\dfrac{114}{91}$ or $1\dfrac{23}{91}$	p. 5
Addition and Subtraction of Fractions Write the fractions with a common denominator. Then add or subtract the numerators and write the result over the common denominator. $\dfrac{a}{b}+\dfrac{c}{b}=\dfrac{a+c}{b}$ and $\dfrac{a}{b}-\dfrac{c}{b}=\dfrac{a-c}{b}$	**Example 4:** $\dfrac{8}{9}+\dfrac{2}{15}=\dfrac{8\cdot5}{9\cdot5}+\dfrac{2\cdot3}{15\cdot3}$ The least common denominator (LCD) of 9 and 15 is 45. $=\dfrac{40}{45}+\dfrac{6}{45}=\dfrac{46}{45}$ or $1\dfrac{1}{45}$	p. 6
To perform operations on mixed numbers, convert to improper fractions.	**Example 5:** $2\dfrac{5}{6}-1\dfrac{1}{3}=\dfrac{17}{6}-\dfrac{4}{3}$ The LCD is 6. $=\dfrac{17}{6}-\dfrac{4\cdot2}{3\cdot2}$ $=\dfrac{17}{6}-\dfrac{8}{6}$ $=\dfrac{9}{6}$ $=\dfrac{3}{2}$ or $1\dfrac{1}{2}$	p. 8

SECTION R.2 The Set of Real Numbers

Key Concepts	Examples	Page
Natural numbers: $\{1, 2, 3, \ldots\}$ **Whole numbers:** $\{0, 1, 2, 3, \ldots\}$ **Integers:** $\{\ldots -3, -2, -1, 0, 1, 2, 3, \ldots\}$ **Rational numbers:** The set of numbers that can be expressed as the ratio of two integers where the denominator is not equal to zero. In decimal form, rational numbers are terminating or repeating decimals. **Irrational numbers:** A subset of the real numbers whose elements cannot be written as a ratio of two integers. In decimal form, irrational numbers are nonterminating, nonrepeating decimals. **Real numbers:** The set of both the rational numbers and the irrational numbers.	**Example 1:** -5, 0, and 4 are integers. **Example 2:** $-\dfrac{5}{2}$, -0.5, and $0.\overline{3}$ are rational numbers. **Example 3:** $\sqrt{7}$, $-\sqrt{2}$, and π are irrational numbers.	p. 13
Absolute Value and Opposite Two numbers that are the same distance from zero but on opposite sides of zero on the number line are called **opposites**. The opposite of a is denoted $-a$. The **absolute value** of a real number, a, denoted $|a|$, is the distance between a and 0 on the number line. If $a \geq 0$, $|a| = a$ If $a < 0$, $|a| = -a$	**Example 4:** 5 and -5 are opposites. **Example 5:** $|7| = 7$ $|-7| = 7$	p. 14
Addition of Two Real Numbers **Same Signs:** Add the absolute values of the numbers and apply the common sign to the sum. **Different Signs:** Subtract the smaller absolute value from the larger absolute value. Then apply the sign of the number having the larger absolute value.	**Example 6:** $-3 + (-4) = -7$ $-1.3 + (-9.1) = -10.4$ **Example 7:** $-5 + 7 = 2$ $\dfrac{2}{3} + \left(-\dfrac{7}{3}\right) = -\dfrac{5}{3}$ or $-1\dfrac{2}{3}$	p. 16
Subtraction of Two Real Numbers Add the opposite of the second number to the first number. That is, $a - b = a + (-b)$	**Example 8:** $7 - (-5) = 7 + (5) = 12$ $-3 - 5 = -3 + (-5) = -8$ $-11 - (-2) = -11 + (2) = -9$	p. 17
Multiplication and Division of Two Real Numbers **Same Signs:** Product and quotient are positive. **Different Signs:** Product and quotient are negative.	**Example 9:** $(-5)(-2) = 10 \qquad \dfrac{-20}{-4} = 5$ **Example 10:** $(-3)(7) = -21 \qquad \dfrac{-4}{8} = -\dfrac{1}{2}$	p. 17
Multiplication and Division Involving Zero The product of any real number and 0 is 0. The quotient of 0 and any nonzero real number is 0. The quotient of any nonzero real number and 0 is undefined.	**Example 11:** $4 \cdot 0 = 0$ $0 \div 4 = 0$ $4 \div 0$ is undefined.	p. 17

SECTION R.3 Simplifying Numerical Expressions

Key Concepts	Examples	Page
$b^n = \underbrace{b \cdot b \cdot b \cdot b \cdot \ldots b}_{n \text{ factors of } b}$ b is the **base**, n is the **exponent**	**Example 1:** $5^3 = 5 \cdot 5 \cdot 5 = 125$	p. 23
$\sqrt{x}$ is the principal **square root** of x.	**Example 2:** $\sqrt{49} = 7$	p. 24
$\sqrt[n]{x}$ is the principal nth root of x.	**Example 3:** $\sqrt[3]{64} = 4$ because $4 \cdot 4 \cdot 4 = 64$.	p. 24
The Order of Operations 1. Simplify expressions within parentheses and other grouping symbols first. 2. Evaluate expressions involving exponents, radicals, and absolute values. 3. Perform multiplication or division in the order that they occur from left to right. 4. Perform addition or subtraction in the order that they occur from left to right.	**Example 4:** $10 + 5(3 - 1)^2 - \sqrt{5 - 1}$ $= 10 + 5(2)^2 - \sqrt{4}$ Work within grouping symbols. $= 10 + 5(4) - 2$ Simplify exponents and radicals. $= 10 + 20 - 2$ Perform multiplication. $= 30 - 2$ Add and subtract, left to right. $= 28$	p. 25

CHAPTER R   Review Exercises

SECTION R.1

For Exercises 1–4, identify as a proper or improper fraction.

1. $\dfrac{14}{5}$ **2.** $\dfrac{1}{6}$ **3.** $\dfrac{3}{3}$ **4.** $\dfrac{7}{1}$

For Exercises 5–6, simplify the fractions.

5. $\dfrac{70}{42}$ **6.** $\dfrac{72}{112}$

For Exercises 7–14, perform the indicated operations.

7. $\dfrac{2}{9} + \dfrac{3}{4}$ **8.** $\dfrac{7}{8} - \dfrac{1}{16}$ **9.** $\dfrac{21}{24} \cdot \dfrac{16}{49}$

10. $\dfrac{4}{9} \cdot \dfrac{15}{8}$ **11.** $\dfrac{68}{34} \div \dfrac{20}{12}$ **12.** $5\dfrac{1}{3} \div 1\dfrac{7}{9}$

13. $3\dfrac{4}{5} - 2\dfrac{1}{10}$ **14.** $2\dfrac{2}{3} + 7\dfrac{1}{2}$

15. The surface area of the Earth is approximately 510 million km^2. If water covers about $\frac{7}{10}$ of the surface, how many square kilometers of the Earth is covered by water?

16. Before a storm, a rain gauge has $\frac{3}{8}$ in. of water. After the storm, the gauge has $\frac{15}{16}$ in. How many inches of rain did the storm deliver?

SECTION R.2

17. Given the set $\left\{ 7, \frac{1}{3}, -4, 0, -\sqrt{3}, -0.\overline{2}, \pi, 1 \right\}$,

   a. List the natural numbers.

   b. List the integers.

   c. List the whole numbers.

   d. List the rational numbers.

   e. List the irrational numbers.

   f. List the real numbers.

18. Find the opposite of each value.

   a. $-\dfrac{7}{3}$ **b.** 4

The opposite of a is $-a$. For Exercises 19–20, simplify.

19. $-(-2)$ **20.** $-(3.7)$

For Exercises 21–24, simplify the expressions.

21. $\left| \dfrac{1}{2} \right|$ **22.** $|-6|$

23. $-|-4|$ **24.** $-|8|$

For Exercises 25–56, perform the indicated operations.

25. $-6 + 8$ **26.** $14 + (-10)$

27. $21 + (-6)$ **28.** $-12 + (-5)$

29. $\dfrac{2}{7} + \left(-\dfrac{1}{9}\right)$

30. $\left(-\dfrac{8}{11}\right) + \left(\dfrac{1}{2}\right)$

31. $\left(-\dfrac{1}{10}\right) + \left(-\dfrac{5}{6}\right)$

32. $\left(-\dfrac{5}{2}\right) + \left(-\dfrac{1}{5}\right)$

33. $-8.17 + 6.02$

34. $2.9 + (-7.18)$

35. $13 - 25$

36. $31 - (-2)$

37. $-8 - (-7)$

38. $-2 - 15$

39. $\left(-\dfrac{7}{9}\right) - \dfrac{5}{6}$

40. $\dfrac{1}{3} - \dfrac{9}{8}$

41. $7 - 8.2$

42. $-1.05 - 3.2$

43. $-16.1 - (-5.9)$

44. $7.09 - (-5)$

45. $10(-17)$

46. $(-7)13$

47. $(-52) \div 26$

48. $(-48) \div (-16)$

49. $\dfrac{7}{4} \div \left(-\dfrac{21}{2}\right)$

50. $\dfrac{2}{3}\left(-\dfrac{12}{11}\right)$

51. $-\dfrac{21}{5} \cdot 0$

52. $\dfrac{3}{4} \div 0$

53. $0 \div (-14)$

54. $(-0.45)(-5)$

55. $\dfrac{-21}{14}$

56. $\dfrac{-13}{-52}$

SECTION R.3

For Exercises 57–72, simplify the expressions if possible.

57. 6^3

58. 15^2

59. $\sqrt{36}$

60. $\dfrac{1}{\sqrt{100}}$

61. $\left(\dfrac{1}{4}\right)^2$

62. $\left(\dfrac{3}{2}\right)^3$

63. $\sqrt{-100}$

64. $\sqrt{-49}$

65. $-\sqrt{100}$

66. $-\sqrt{49}$

67. -5^4

68. -6^2

69. $(-5)^4$

70. $(-6)^2$

71. $\sqrt[3]{-\dfrac{8}{125}}$

72. $\sqrt[4]{\dfrac{81}{10{,}000}}$

For Exercises 73–86, perform the indicated operations.

73. $9 - 4[-2(4 - 8) - 5(3 - 1)]$

74. $\dfrac{8(-3) - 6}{-7 - (-2)}$

75. $\dfrac{2}{3} - \left(\dfrac{3}{8} + \dfrac{5}{6}\right) \div \dfrac{5}{3}$

76. $5.4 - (0.3)^2 \div 0.09$

77. $\dfrac{5 - [3 - (-4)^2]}{36 \div (-2)(3)}$

78. $|-8 + 5| - \sqrt{5^2 - 3^2}$

79. $\dfrac{2 - 4(3 - 7)}{-4 - 5(1 - 3)}$

80. $\dfrac{12(2) - 8}{4(-3) + 2(5)}$

81. $24 \div 8 \cdot 2$

82. $40 \div 5 \cdot 6$

83. $3^2 + 2(|-10 + 5| \div 5)$

84. $-91 + \sqrt{4}(\sqrt{25} - 13)^2$

85. $\dfrac{3(3 - 8)^2}{|8 - 3^2|}$

86. $\dfrac{4(5 - 2)^2}{|3 - 7 - 5|}$

87. Given $h = \tfrac{1}{2}gt^2 + v_0 t + h_0$, find h if $g = -32$, $v_0 = 64$, $h_0 = 256$, and $t = 4$.

88. Find the area of a parallelogram with base 42 in. and height 18 in.

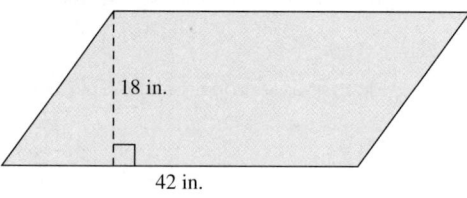

89. In statistics, the formula $x = \mu + z\sigma$ is used to find cutoff values for data that follow a bell-shaped curve. Find x if $\mu = 100$, $z = -1.96$, and $\sigma = 15$.

CHAPTER R   Test

1. Simplify. $\dfrac{135}{36}$

2. Add and subtract. $\dfrac{5}{4} - \dfrac{5}{12} + \dfrac{2}{3}$

3. Divide. $4\dfrac{1}{12} \div 1\dfrac{1}{3}$

4. Multiply. $\dfrac{7}{12} \cdot \dfrac{3}{14}$

5. Subtract. $4\dfrac{1}{4} - 1\dfrac{7}{8}$

6. Given the set $\left\{-6, \tfrac{3}{5}, 8, 0, 4\pi, 0.\overline{5}, \sqrt{7}\right\}$,

   a. List the natural numbers.

   b. List the integers.

   c. List the whole numbers.

   d. List the rational numbers.

   e. List the irrational numbers.

   f. List the real numbers.

7. a. Determine the opposite of -13.

   b. Determine the absolute value of -13.

   c. Determine the reciprocal of -13.

For Exercises 8–13, simplify the expression if possible.

8. a. $-(-9)$ **b.** $-|-9|$

9. a. $5 \cdot \dfrac{1}{5}$ **b.** $-\dfrac{2}{3} \cdot \left(-\dfrac{3}{2}\right)$

10. a. -7^2 **b.** $(-7)^2$

11. a. $\sqrt{4}$ **b.** 4^2

12. a. $\sqrt{64}$ **b.** $\sqrt[3]{64}$

13. a. $\sqrt[3]{-125}$ **b.** $\sqrt{-25}$

14. Use the formula $A = \dfrac{1}{2}(b_1 + b_2)h$ to find the area of the trapezoid.

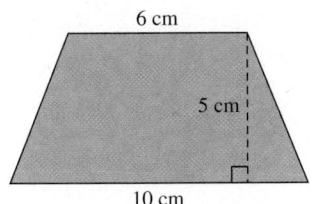

6 cm

5 cm

10 cm

For Exercises 15–32, perform the indicated operations.

15. $-\dfrac{1}{8} + \left(-\dfrac{3}{4}\right)$ **16.** $-84 \div 7$

17. $21 - (-7)$ **18.** $-15 - (-3)$

19. $-14 + (-2) - 16$ **20.** $(-16)(-2)(-1)(-3)$

21. $-22 \cdot 0$ **22.** $38 \div 0$

23. $18 + (-12)$ **24.** $-10.06 - (-14.72)$

25. $7(-4)$ **26.** $\dfrac{2}{5} \div \left(-\dfrac{7}{10}\right) \cdot \left(-\dfrac{7}{6}\right)$

27. $\dfrac{\sqrt{5^2 - 4^2}}{|-12 + 3|}$ **28.** $8 - [(2 - 4) - (8 - 9)]$

29. $(8 - 10) \cdot \dfrac{3}{2} + (-5)$ **30.** $|-8| - 4(2 - 3)^2 \div \sqrt{4}$

31. $\dfrac{-6^2 - 10^2}{-1 + 3^2}$ **32.** $\left(-\dfrac{1}{6} + \sqrt{\dfrac{4}{9}}\right)^2$

Linear Equations and Inequalities

1

Tetra Images/SuperStock

Athletes know that in order to optimize their performance they need to pace themselves and be mindful of their target heart rate. According to the American Heart Association (www.americanheart.org), an athlete should strive for a target heart rate zone of between 50% and 85% of her maximum heart rate. Maximum heart rate (in beats per minute) for an adult is given by $220 - a$, where a is the individual's age. Thus, for a 25-year-old, the maximum heart rate is $220 - 25$, which is 195 beats per minute. In turn, the target heart rate x should be found on the interval $0.50(195) < x < 0.85(195)$ or simply $98 < x < 165$ (rounded to the nearest beat per minute).

The mathematics used to find an individual's maximum heart rate and target heart rate zone include writing a linear model relating age and resting heart rate and solving a linear inequality. In this chapter, we present an introduction to algebraic modeling along with important tools to simplify algebraic expressions and solve linear equations and inequalities.

Algebraic Expressions and Models

1. Apply Properties of Real Numbers

After college, a business major is offered a sales job in which she is paid a base salary of $15,000 per year for office work and an 8% commission on sales. Her yearly salary is given by $0.08x + 15,000$ where x is the dollar amount in sales. The mathematical expression $0.08x + 15,000$ consists of two terms. The term $0.08x$ is a **variable term** because its value changes with different values of x. The term 15,000 is a **constant term** because its value does not change.

Stockasso/123RF

In general, an algebraic **term** is a numerical value or a product of factors that may include constants and variables. An algebraic **expression** is a single term or the sum of two or more terms. For example, consider the expression:

$$x^2 - 2xy + 4$$

Writing this as a sum we have:

$$x^2 + (-2xy) + 4$$

This is a three-term expression, and the individual terms are x^2, $-2xy$, and 4.

The **coefficient** of each term is the numerical factor within the term. Therefore, the term $-2xy$ has a coefficient of -2, and the coefficient of the term 4 is 4. The term x^2 is equivalent to $1x^2$, so its coefficient is 1.

For an expression such as $x^2 - 2xy + 4$, when we substitute real numbers for the variables, the value of the expression will be a real number. Therefore, to manipulate and simplify mathematical expressions, we need to review important properties of real numbers summarized in Table 1-1.

Table 1-1   Properties of Real Numbers

Let a, b, and c represent real numbers or real-valued expressions.

Property	In Symbols and Words	Examples
Commutative property of addition	$a + b = b + a$ The order in which real numbers are added does not affect the sum.	ex: $4 + (-7) = -7 + 4$ ex: $6 + w = w + 6$
Commutative property of multiplication	$a \cdot b = b \cdot a$ The order in which real numbers are multiplied does not affect the product.	ex: $5 \cdot (-4) = -4 \cdot 5$ ex: $x \cdot 12 = 12x$
Associative property of addition	$(a + b) + c = a + (b + c)$ The order in which real numbers are grouped under addition does not affect the sum.	ex: $(3 + 5) + 2 = 3 + (5 + 2)$ ex: $-9 + (2 + t) = (-9 + 2) + t$ $= -7 + t$
Associative property of multiplication	$(a \cdot b) \cdot c = a \cdot (b \cdot c)$ The order in which real numbers are grouped under multiplication does not affect the product.	ex: $(6 \cdot 7) \cdot 3 = 6 \cdot (7 \cdot 3)$ ex: $8 \cdot \left(\frac{1}{8} \cdot y\right) = \left(8 \cdot \frac{1}{8}\right) \cdot y$ $= 1y$
Identity property of addition	$a + 0 = a$ and $0 + a = a$ The number 0 is called the **identity element of addition** because any number plus 0 is the number itself.	ex: $-5 + 0 = -5$ ex: $0 + \sqrt{z} = \sqrt{z}$
Identity property of multiplication	$a \cdot 1 = a$ and $1 \cdot a = a$ The number 1 is called the **identity element of multiplication** because any number times 1 is the number itself.	ex: $\sqrt{2} \cdot 1 = \sqrt{2}$ ex: $1 \cdot (2w + 3) = 2w + 3$

(Continued)

Property	In Symbols and Words	Examples
Inverse property of addition	$a + (-a) = 0$ and $(-a) + a = 0$ For any real number a, the value $-a$ is called the **additive inverse of a** (also called the **opposite of a**). The sum of any number and its additive inverse is the identity element for addition, 0.	ex: $4\pi + (-4\pi) = 0$ ex: $-ab^2 + ab^2 = 0$
Inverse property of multiplication	$a \cdot \frac{1}{a} = 1$ and $\frac{1}{a} \cdot a = 1$ where $a \neq 0$ For any nonzero real number a, the value $\frac{1}{a}$ is called the **multiplicative inverse of a** (also called the **reciprocal of a**). The product of any nonzero number and its multiplicative inverse is the identity element of multiplication, 1.	ex: $-5 \cdot \left(-\frac{1}{5}\right) = 1$ ex: $\frac{1}{x^2} \cdot x^2 = 1$ for $x \neq 0$ *Note*: The number zero does not have a multiplicative inverse (reciprocal).
Distributive property of multiplication over addition	$a \cdot (b + c) = a \cdot b + a \cdot c$ The product of a number and a sum equals the sum of the products of the number and each term in the sum.	ex: $4 \cdot (5 + x) = 4 \cdot 5 + 4 \cdot x$ $= 20 + 4x$ ex: $2(x + \sqrt{3}) = 2x + 2\sqrt{3}$

It is very important to understand that the commutative and associative properties of real numbers apply only to addition and multiplication, not to subtraction or division. For example,

$3 + 5 = 5 + 3$,   but $3 - 5 \neq 3 - 5$ Subtraction is not commutative.

$3 \cdot 5 = 5 \cdot 3$ but $3 : 5 \neq 5 \div 3$ Division is not commutative.

$(2 + 3) + 5 = 2 + (3 + 5)$ but $(2 - 3) - 5 \neq 2 - (3 - 5)$ Subtraction is not associative.

$(2 \cdot 3) \cdot 5 = 2 \cdot (3 \cdot 5)$ but $(2 \div 3) \div 5 \neq 2 \div (3 \div 5)$ Division is not associative.

The properties of real numbers outlined in Table 1-1 give us important tools to simplify algebraic expressions. Furthermore,

- It is often customary to write an expression with the variable terms written before the constant term. Thus $4 + x$ can be written as $x + 4$ by using the commutative property of addition.
- An expression such as $-5(2y)$ can be written in a simpler form by applying the associative property of multiplication: $-5(2y) = (-5 \cdot 2)y = -10y$
- The distributive property is used to multiply a term by an algebraic expression containing more than one term. This is shown in Example 1.

EXAMPLE 1   **Applying the Distributive Property**

Apply the distributive property.

a. $4(2x + 5)$ **b.** $-(-3.4q + 5.7r)$

c. $-3(a + 2b - 5c)$ **d.** $-\dfrac{2}{3}\left(-9x + \dfrac{3}{8}y - 5\right)$

Solution:

a. $4(2x + 5)$

$= 4(2x) + 4(5)$ Apply the distributive property.

$= 8x + 20$ Simplify, using the associative property of multiplication.

TIP When applying the distributive property, a negative factor preceding the parentheses will change the signs of the terms within the parentheses.

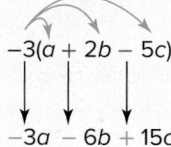

$-3a - 6b + 15c$

b. $-(-3.4q + 5.7r)$

The negative sign preceding the parentheses can be interpreted as a factor of -1.

$= -1(-3.4q + 5.7r)$

$= -1(-3.4q) + (-1)(5.7r)$ Apply the distributive property.

$= 3.4q - 5.7r$

c. $-3(a + 2b - 5c)$

$= -3(a) + (-3)(2b) + (-3)(-5c)$ Apply the distributive property.

$= -3a - 6b + 15c$ Simplify.

d. $-\dfrac{2}{3}\left(-9x + \dfrac{3}{8}y - 5\right)$

$= -\dfrac{2}{3}(-9x) + \left(-\dfrac{2}{3}\right)\left(\dfrac{3}{8}y\right) + \left(-\dfrac{2}{3}\right)(-5)$ Apply the distributive property.

$= \dfrac{18}{3}x - \dfrac{6}{24}y + \dfrac{10}{3}$ Simplify.

$= 6x - \dfrac{1}{4}y + \dfrac{10}{3}$ Simplify to lowest terms.

Skill Practice 1 Apply the distributive property.

a. $10(30y - 40)$ **b.** $-(7t - 1.6s + 9.2)$

c. $-2(4x - 3y - 6)$ **d.** $-\dfrac{1}{2}(-4a + 7)$

In Example 1, notice that parentheses are removed after the distributive property is applied. Sometimes this is referred to as *clearing parentheses*.

2. Simplify Expressions by Clearing Parentheses and Combining Like Terms

Two terms can be added or subtracted only if they are like terms. **Like terms** are terms whose variable factors are the same (the coefficients may be different). For example,

- $-6t$ and $8t$ are like terms, but the terms $-6t$ and $8v$ are not like terms.
- $2.7x^2y$ and $3.1x^2y$ are like terms, but $2.7x^2y$ and $3.1xy^2$ are not like terms.

To add or subtract like terms, we use the distributive property as shown in Example 2.

EXAMPLE 2 Using the Distributive Property to Add and Subtract Like Terms

Add and subtract as indicated.

a. $-8x + 3x$ **b.** $4.75y^2 - 9.25y^2 + y^2$

Answers

1. a. $300y - 400$
  b. $-7t + 1.6s - 9.2$
  c. $-8x + 6y + 12$
  d. $2a - \dfrac{7}{2}$

Solution:

a. $-8x + 3x$

$= (-8 + 3)x$ Apply the distributive property.

$= (-5)x$ Simplify.

$= -5x$

b. $4.75y^2 - 9.25y^2 + y^2$

$= 4.75y^2 - 9.25y^2 + 1y^2$ Notice that y^2 is interpreted as $1y^2$.

$= (4.75 - 9.25 + 1)y^2$ Apply the distributive property.

$= (-3.5)y^2$ Simplify.

$= -3.5y^2$

Skill Practice 2   Combine like terms.

    a. $-4y + 7y$ **b.** $a^2 - 6.2a^2 + 2.8a^2$

Although the distributive property is used to add and subtract like terms, it is tedious to write each step. Observe that adding or subtracting like terms is a matter of combining the coefficients and leaving the variable factors unchanged. This can be shown in one step. This shortcut will be used throughout the text. For example:

$$-8x + 3x = -5x \qquad\qquad 4.75y^2 - 9.25y^2 + 1y^2 = -3.5y^2$$

In Examples 3 and 4, we present the important skill of simplifying expressions by clearing parentheses and combining like terms.

EXAMPLE 3   **Clearing Parentheses and Combining Like Terms**

Simplify by clearing parentheses and combining like terms.

    a. $4 - 3(2x - 8) - 1$ **b.** $-(3s - 11t) - 5(2t + 8s) - 10s$

Solution:

a. $4 - 3(2x - 8) - 1$

$= 4 + (-3)[2x + (-8)] - 1$ To apply the distributive property, write subtraction as addition of the opposite.

$= 4 + (-3)(2x) + (-3)(-8) - 1$ Apply the distributive property.

$= 4 - 6x + 24 - 1$ Simplify each term.

$= -6x + 4 + 24 - 1$ Group like terms.

$= -6x + 27$ Simplify.

b. $-(3s - 11t) - 5(2t + 8s) - 10s$

$= -1[3s + (-11t)] + (-5)(2t + 8s) - 10s$ To apply the distributive property, write subtraction as addition of the opposite.

$= -1(3s) + (-1)(-11t) + (-5)(2t) +$
$\quad (-5)(8s) - 10s$ Apply the distributive property.

$= -3s + 11t - 10t - 40s - 10s$ Simplify each term.

$= -3s - 40s - 10s + 11t - 10t$ Group like terms.

$= -53s + t$ Simplify.

Answers

2. a. $3y$ **b.** $-2.4a^2$

3. a. $6x + 5$ **b.** $-28y - 11z$

Skill Practice 3   Simplify by clearing parentheses and combining like terms.

    a. $8 - 2(4 - 3x) + 5$ **b.** $-2(9y + 4z) - (6y + 3z) - 4y$

EXAMPLE 4 Clearing Parentheses and Combining Like Terms

Simplify by clearing parentheses and combining like terms.

a. $2\left[1.5x + 4.7\left(x^2 - 5.2x\right) - 3x\right]$ **b.** $-\dfrac{1}{3}(3w - 6) - \left(\dfrac{1}{4}w + 4\right)$

Solution:

a. $2\left[1.5x + 4.7\left(x^2 - 5.2x\right) - 3x\right]$

$= 2\left(1.5x + 4.7x^2 - 24.44x - 3x\right)$ Apply the distributive property to the inner parentheses.

$= 2\left(1.5x - 24.44x - 3x + 4.7x^2\right)$ Group like terms.

$= 2\left(-25.94x + 4.7x^2\right)$ Combine like terms.

$= -51.88x + 9.4x^2$ Apply the distributive property.

$= 9.4x^2 - 51.88x$

b. $-\dfrac{1}{3}(3w - 6) - \left(\dfrac{1}{4}w + 4\right)$

$= -\dfrac{3}{3}w + \dfrac{6}{3} - \dfrac{1}{4}w - 4$ Apply the distributive property.

$= -w + 2 - \dfrac{1}{4}w - 4$ Simplify fractions.

$= -\dfrac{4}{4}w - \dfrac{1}{4}w + 2 - 4$ Group like terms and find a common denominator.

$= -\dfrac{5}{4}w - 2$ Combine like terms.

> **TIP** By using the commutative property of addition, the expression $-51.88x + 9.4x^2$ can also be written as $9.4x^2 + (-51.88x)$ or simply $9.4x^2 - 51.88x$. Although the expressions are all equivalent, it is customary to write the terms in descending order of the powers of the variable.

Skill Practice 4 Simplify by clearing parentheses and combining like terms.

a. $4\left[1.4a + 2.2\left(a^2 - 6a\right)\right] - 5.1a^2$ **b.** $-\dfrac{1}{2}(4p - 1) - \dfrac{5}{2}(p - 2)$

---

3. Write Algebraic Models

Algebraic expressions are important because they can be used to give mathematical representations (or models) of real-world applications. For example, given a sales tax rate of 6.5%, the expression $0.065x$ represents the amount of sales tax owed on x dollars of merchandise. To write a mathematical relationship, it is important to know key terms that represent the four basic operations on real numbers (Table 1-2).

Table 1-2

Addition: $a + b$	Subtraction: $a - b$
• the **sum** of a and b	• the **difference** of a and b
• a plus b	• a minus b
• b added to a	• b subtracted from a
• b more than a	• a decreased by b
• a increased by b	• b less than a
• the total of a and b	
Multiplication: $a \cdot b$	Division: $a \div b, \frac{a}{b}$
• the **product** of a and b	• the **quotient** of a and b
• a times b	• a divided by b
• a multiplied by b	• b divided into a
	• the ratio of a and b
	• a over b
	• a per b

Answers

4. a. $3.7a^2 - 47.2a$ **b.** $-\dfrac{9}{2}p + \dfrac{11}{2}$

Example 5 offers practice writing algebraic expressions based on verbal statements.

EXAMPLE 5   Writing Algebraic Models

a. The maximum recommended heart rate during exercise for adults is numerically 20 less than the person's age. Write a model to represent an adult's maximum recommended heart rate M in terms of age a.

b. After eating at a restaurant, a customer leaves a 15% tip. Write a model to represent the amount of the tip t based on the cost of the meal c.

Solution:

a. $M = a - 20$ 20 less than a implies that 20 is subtracted from a. The order of the subtraction is important.

b. $t = 0.15c$ 15% of c implies multiplication. Be sure to convert the percent to its decimal form.

Skill Practice 5

a. The sale price S on a lawn mower is the difference of the original price P and the amount of the discount D. Write a model to represent the sale price.

b. The amount of simple interest I earned on a 1-year certificate of deposit is 4.65% of the amount of principal invested P. Write a model for the amount of interest.

Answers
5. a. $S = P - D$ **b.** $I = 0.0465P$

SECTION 1.1   Practice Exercises

Prerequisite Review

For Exercises R.1–R.16, perform the indicated operations.

R.1. $\dfrac{1}{8} \cdot 8$ **R.2.** $-9 \cdot \left(-\dfrac{1}{9}\right)$ **R.3.** $\left(-\dfrac{3}{7}\right)\left(-\dfrac{7}{3}\right)$ **R.4.** $\dfrac{4}{5} \cdot \dfrac{5}{4}$

R.5. $-2.6 + 2.6$ **R.6.** $8.9 + (-8.9)$ **R.7. a.** $-25 + 31$ **R.8. a.** $14 + (-12)$

   b. $31 + (-25)$ **b.** $-12 + 14$

R.9. a. $3(-6)$ **R.10. a.** $-9(-7)$ **R.11. a.** $-6 + (-4 + 2)$ **R.12. a.** $5 + [7 + (-3)]$

   b. $-6 \cdot 3$ **b.** $-7(-9)$ **b.** $[-6 + (-4)] + 2$ **b.** $(5 + 7) + (-3)$

R.13. a. $-5(4 \cdot 2)$ **R.14. a.** $-8(-3 \cdot 2)$ **R.15. a.** $6(-3 + 8)$ **R.16. a.** $-5(-2 + 10)$

   b. $(-5 \cdot 4) \cdot 2$ **b.** $[-8 \cdot (-3)] \cdot 2$ **b.** $6(-3) + 6(8)$ **b.** $-5(-2) + (-5)(10)$

Concept Connections

1. Given the expression $8x + cd - 3y + 90$, the terms $8x$, cd, and $-3y$ are variable terms, whereas 90 is a _____ term.

2. The constant factor in a term is called the _____.

3. Given the expression x, the value of the coefficient is _____, and the exponent is _____.

4. Terms that have the same variables, with corresponding variables raised to the same powers, are called _____ terms.

Objective 1: Apply Properties of Real Numbers

For Exercises 5–8:

   a. Determine the number of terms in the expression.

   b. Identify the constant term.

   c. List the coefficients of each term, separated by commas.

5. $2x^3 - 5xy + 6$

6. $a^2 - 4ab - b^2 + 8$

7. $pq - 7 + q^2 - 4q + p$

8. $7x - 1 + 3xy$

For Exercises 9–26, match each expression with the appropriate property.

9. $3 + \dfrac{1}{2} = \dfrac{1}{2} + 3$

10. $7.2(4 + 1) = 7.2(4) + 7.2(1)$

a. Commutative property of addition

11. $10 + 0 = 10$

12. $7 \cdot 1 = 7$

b. Associative property of multiplication

13. $(6 + 8) + 2 = 6 + (8 + 2)$

14. $(4 + 19) + 7 = (19 + 4) + 7$

c. Distributive property of multiplication over addition

15. $6 \cdot \dfrac{1}{6} = 1$

16. $2 + (-2) = 0$

d. Commutative property of multiplication

17. $9(4 \cdot 12) = (9 \cdot 4)12$

18. $\left(\dfrac{1}{4} + 2\right)20 = 5 + 40$

e. Associative property of addition

19. $42 \cdot 1 = 42$

20. $4 \cdot \dfrac{1}{4} = 1$

f. Identity property of addition

21. $(13 \cdot 41)6 = (41 \cdot 13)6$

22. $6(x + 3) = 6x + 18$

g. Identity property of multiplication

23. $8 + (-8) = 0$

24. $21 + 0 = 21$

h. Inverse property of addition

25. $3(y + 10) = 3(10 + y)$

26. $5(3 \cdot 7) = (5 \cdot 3)7$

i. Inverse property of multiplication

For Exercises 27–38, clear parentheses by applying the distributive property. **(See Example 1)**

27. $2(x - 3y + 8)$

28. $5(-2a + 4b - 9c)$

29. $-10(4s - 9t - 3)$

30. $-4(-8x + 6y + 3z)$

31. $-(-7w + 5z)$

32. $-(-22a - 17b)$

33. $-\dfrac{1}{5}\left(-\dfrac{5}{2}a + 10b - 8\right)$

34. $-\dfrac{3}{4}\left(6x - 4y + \dfrac{4}{9}\right)$

35. $3(2.6x - 4.1)$

36. $5(-7.2y + 2.3)$

37. $2(7c - 8) - 5(6d - f)$

38. $-2(-3q + r) - 7(5s + 2t)$

Objective 2: Simplify Expressions by Clearing Parentheses and Combining Like Terms

For Exercises 39–78, clear parentheses and combine like terms. **(See Examples 2–4)**

39. $8y - 2x + y + 5y$

40. $-9a + a - b + 5a$

41. $4p^2 - 2p + 3p - 6 + 2p^2$

42. $6q - 9 + 3q^2 - q^2 + 10$

43. $2p - 7p^2 - 5p + 6p^2$

44. $5a^2 - 2a - 7a^2 + 6a + 4$

45. $m - 4n^3 + 3 + 5n^3 - 9$

46. $x + 2y^3 - 2x - 8y^3$

47. $5ab + 2ab + 8a$

48. $-6m^2n - 3mn^2 - 2m^2n$

49. $14xy^2 - 5y^2 + 2xy^2$

50. $9uv + 3u^2 + 5uv + 4u^2$

51. $-2(c + 3) - 2c$

52. $4(z - 4) - 3z$

53. $-(10w - 1) + 9 + w$

54. $-(2y + 7) - 4 + 3y$

55. $-9 - 4(2 - z) + 1$

56. $3 + 3(4 - w) - 11$

57. $4(2s - 7) - (s - 2)$

58. $2(t - 3) - (t - 7)$

59. $-3(-5 + 2w) - 8w + 2(w - 1)$

60. $-2(-6 + 2x) - 9x + 5(4x - 1)$

61. $8x - 4(x - 2) - 2(2x + 1) - 6$

62. $6(y - 2) - 3(2y - 5) - 3$

63. $\dfrac{1}{2}(4 - 2c) + 5c$

64. $\dfrac{2}{3}(3d + 6) - 4d$

65. $3.1(2x + 2) - 4(1.2x - 1)$

66. $4.5(5 - y) + 3(1.9y + 1)$

67. $2\left[5\left(\dfrac{1}{2}a + 3\right) - (a^2 + a) + 4\right]$

68. $-3\left[3\left(b - \dfrac{2}{3}\right) - 2(b + 4) - 6b^2\right]$

69. $(2y - 5) - 2(y - y^2) - 3y$

70. $-(x + 6) + 3(x^2 + 1) + 2x$

71. $2.2\{4 - 8[6x - 1.5(x + 4) - 6] + 7.5x\}$

72. $-3.2 - \{6.1y - 4[9 - (2y + 2.5)] + 7y\}$

73. $\dfrac{1}{8}(24n - 16m) - \dfrac{2}{3}(3m - 18n - 2) + \dfrac{2}{3}$

74. $\dfrac{1}{5}(25a - 20b) - \dfrac{4}{7}(21a - 14b + 2) + \dfrac{1}{7}$

75. $12 - 4[(8 - 2v) + 5(-3w - 4v)] - w$

76. $6 - 2[(9z + 6y) - 8(y - z)] - 11$

77. $2y^2 - \left[13 - \dfrac{2}{3}(6y^2 - 9) - 10\right] + 9$

78. $6 - \left[5t^2 - \dfrac{3}{4}(12 - 8t^2) + 5\right] + 11t^2$

Objective 3: Write Algebraic Models

For Exercises 79–94, write an expression that represents the English phrase.

79. The difference of t and 5

80. Eight subtracted from x

81. Twenty-five less than x

82. x decreased by 9

83. x increased by twice t

84. Twice the sum of m and n

85. Twice the product of 3 and t

86. Seven more than y

87. The product of 8 and the quantity 3 less than y

88. The product of -2 and the quantity x more than 5

89. The quotient of 5 and the quantity t plus 2

90. The quotient of x and the difference of 5 and y

91. Twelve percent of t

92. Fifty-four percent of x

93. One-third the difference of m and 10

94. Two-fifths the product of 3 and x

For Exercises 95–102, write a mathematical statement that represents the given verbal model. **(See Example 5)**

95. Jake is 1 year younger than Charlotte.
  a. Write a statement for Jake's age J in terms of Charlotte's age C.
  b. Write a statement for Charlotte's age C in terms of Jake's age J.

96. A farmer grew 4 acres more corn than wheat.
  a. Write a statement for the number of acres of corn C in terms of the number of acres of wheat W.
  b. Write a statement for the number of acres of wheat W in terms of the number of acres of corn C.

97. Tonya's income is four times as much as Nora's income.
  a. Write a statement representing Tonya's income T in terms of Nora's income N.
  b. Write a statement representing Nora's income N in terms of Tanya's income T.

98. The width of a room is one-third the length of the room.
  a. Write a statement representing the width w of the room in terms of the length l of the room.
  b. Write a statement representing the length l of the room in terms of the width w of the room.

99. At the end of the summer, a store discounts an outdoor grill for 25% off the original price. Write a statement giving the discounted price D in terms of the original price P.

100. When Ms. Celano has excellent service at a restaurant, she leaves a tip of 20% of the cost of the meal. Write a statement giving the amount of the tip t in terms of the cost of the meal c.

101. Suppose that an object is dropped from a height h. Its velocity v at impact with the ground is given by the square root of twice the product of the acceleration due to gravity g and the height h. Write a model to represent the velocity of the object at impact.

102. The height of a sunflower plant can be determined by the time t in weeks after the seed has germinated. Write a model to represent the height h if the height is given by the product of 8 and the square root of t.

Mixed Exercises

For Exercises 103–106, simplify the expression in terms of π.

103. $\dfrac{\pi}{2} + \dfrac{\pi}{3}$

104. $\dfrac{\pi}{4} + \dfrac{\pi}{6}$

105. $\dfrac{19\pi}{6} - \dfrac{8\pi}{3}$

106. $\dfrac{19\pi}{2} - \dfrac{13\pi}{6}$

107. What is the identity element for addition? Use it in an example.

108. What is the identity element for multiplication? Use it in an example.

109. What is another name for a multiplicative inverse?

110. What is another name for an additive inverse?

111. Is the operation of subtraction commutative? If not, give an example.

112. Is the operation of division commutative? If not, give an example.

Expanding Your Skills

113. A power company charges one household $0.12 per kilowatt-hour (kWh) and $14.89 in monthly taxes.

    a. Write a formula for the monthly charge C for this household if it uses k kilowatt-hours.

    b. Compute the monthly charge if the household uses 1200 kWh.

114. A utility company charges a base rate for water usage of $19.50 per month, plus $4.58 for every additional 1000 gal of water used over a 2000-gal base.

    a. Write a formula for the monthly charge C for a household that uses n thousand gallons over the 2000-gal base.

    b. Compute the cost for a family that uses a total of 6000 gal of water for a given month.

115. The cost C (in $) to rent an apartment is $640 per month, plus a $500 nonrefundable security deposit, plus a $200 nonrefundable deposit for each dog or cat.

    a. Write a formula for the total cost to rent an apartment for m months with n cats/dogs.

    b. Determine the cost to rent the apartment for 12 months, with 2 cats and 1 dog.

116. For a certain college, the cost C (in $) for taking classes the first semester is $105 per credit-hour, $35 for each lab, plus a one-time admissions fee of $40.

    a. Write a formula for the total cost to take n credit-hours and L labs the first semester.

    b. Determine the cost for the first semester if a student takes 12 credit-hours with 2 labs.

117. A hotel charges $159 per night plus an 11% nightly room tax.

    a. Write a formula to represent the total cost C for n nights in the hotel. (*Hint:* The total cost is the cost for n nights, plus the tax on the cost for n nights.)

    b. Determine the cost to stay in the hotel for four nights.

118. A hotel charges $149 per night plus a 16% nightly room tax. In addition, there is a one-time parking fee of $40.

    a. Write a formula to represent the total cost C for n nights in the hotel.

    b. Determine the cost to stay in the hotel for two nights.

SECTION 1.2   Linear Equations in One Variable

OBJECTIVES

1. Solve Linear Equations in One Variable
2. Clear Fractions and Decimals
3. Identify Conditional Equations, Identities, and Contradictions
4. Solve Literal Equations for a Specified Variable

1. Solve Linear Equations in One Variable

An **equation** is a statement that indicates that two quantities are equal. As you work through this text, you will encounter a variety of different types of equations and their methods of solution. Furthermore, you will use equations to solve applications.

Kick Images/Photodisc/Getty Images

    For example, suppose that a student preparing for college puts aside $2000 in a mutual fund to use 1 year later for school. At the end of the year, the account is worth $2150. The student would like to know the annual rate of return x for the investment during this time period. To do so, the student can solve the equation $2000 + 2000x = 2150$. This equation is called a *linear equation in one variable*. A solution to an equation is a value of the variable that when substituted into the equation makes a true statement. In Exercise 89, we solve this equation and find that the rate of return for the investment is 7.5%.

> ### Definition of a Linear Equation in One Variable
>
> A **linear equation in one variable** is an equation that can be written in the form $ax + b = 0$, where a and b are real numbers, $a \neq 0$, and x is the variable.

Linear equation in one variable	Not a linear equation in one variable	
$5x + 35 = 0$	$5x^2 + 35 = 0$	(Exponent on x is 2 rather than 1.)
$\dfrac{x}{4} - 5 = 0$	$\dfrac{4}{x} - 5 = 0$	(Variable is in the denominator, not in the numerator.)
$3x + 4 = 7$	$3x + 4y = 7$	(Contains two variables.)
$0.7x - 0.8 = 0.1$	$0.7x - 0.8 - 0.1$	(This is an expression, not an equation.)

A **solution** to an equation is a value of the variable that makes the equation a true statement. The set of all solutions to an equation is called the **solution set** of the equation. **Equivalent equations** have the same solution set. To solve a linear equation in one variable, we form simpler, equivalent equations until we obtain an equation whose solution is obvious. The properties used to produce equivalent equations include the addition and multiplication properties of equality.

TIP Given $x + 4 = 9$, we can add the opposite of 4 to both sides to isolate x or subtract 4 from both sides.

$$x + 4 = 9$$
$$x + 4 - 4 = 9 - 4$$
$$x = 5$$

Given $4x = 24$ we can multiply both sides by the reciprocal of 4 to isolate x or divide both sides by 4.

$$4x = 24$$
$$\frac{4x}{4} = \frac{24}{4}$$
$$x = 6$$

Properties of Equality

Let a, b, and c be real-valued expressions.

Property Name	Statement of Property	Example
Addition property of equality	$a = b$ is equivalent to $a + c = b + c$.	$x + 4 = 9$ $x + 4 + (-4) = 9 + (-4)$ $x = 5$
Multiplication property of equality	$a = b$ is equivalent to $ac = bc$ $(c \neq 0)$.	$4x = 24$ $\dfrac{1}{4} \cdot 4x = \dfrac{1}{4} \cdot 24$ $x = 6$

In Examples 1 and 2, we apply the properties of equality.

EXAMPLE 1   Applying the Properties of Equality

Solve each equation.

a. $-\dfrac{3}{5}p = \dfrac{4}{15}$ **b.** $4 = \dfrac{w}{2.2}$ **c.** $-x = 6$ **d.** $12 = y + 16$

Solution:

a.
$$-\frac{3}{5}p = \frac{4}{15}$$

TIP The solution $-\frac{4}{9}$ checks in the original equation.

$$-\frac{3}{5}p = \frac{4}{15}$$

$$-\frac{3}{5}\left(-\frac{4}{9}\right) \stackrel{?}{=} \frac{4}{15}$$

$$-\frac{\overset{1}{3}}{5}\left(-\frac{4}{\underset{3}{9}}\right) \stackrel{?}{=} \frac{4}{15}$$

$$\frac{4}{15} = \frac{4}{15} \checkmark$$

$$\left(-\frac{5}{3}\right)\left(-\frac{3}{5}p\right) = \left(-\frac{5}{3}\right)\left(\frac{4}{15}\right) \qquad \text{To isolate } p, \text{ multiply both sides by the reciprocal of } -\frac{3}{5}.$$

$$p = \left(-\frac{\overset{1}{5}}{3}\right)\left(\frac{4}{\underset{3}{15}}\right) \qquad \text{Multiply fractions.}$$

$$p = -\frac{4}{9} \qquad \text{The value } -\frac{4}{9} \text{ checks in the original equation.}$$

The solution set is $\left\{-\dfrac{4}{9}\right\}$.

b. $\quad 4 = \dfrac{w}{2.2}$

$\quad 2.2(4) = 2.2\left(\dfrac{w}{2.2}\right)$ $\qquad$ To isolate w, multiply both sides by 2.2.

$\quad 8.8 = w$ $\qquad$ The value 8.8 checks in the original equation.

The solution set is $\{8.8\}$.

c. $\quad -x = 6$

$\quad -1(-x) = -1(6)$ $\qquad$ To isolate x, multiply both sides by -1.

$\quad x = -6$ $\qquad$ The value -6 checks in the original equation.

The solution set is $\{-6\}$.

d. $\quad\quad\quad 12 = y + 16$

$\quad 12 + (-16) = y + 16 + (-16)$ $\qquad$ To isolate y, add the opposite of 16 to both sides.

$\quad\quad\quad -4 = y + 0$ $\qquad$ Simplify.

$\quad\quad\quad -4 = y$

The solution set is $\{-4\}$.

> **TIP** In Example 1(d) we could also have subtracted 16 from both sides to isolate y.
>
> $12 = y + 16$
> $12 - 16 = y + 16 - 16$
> $-4 = y + 0$
> $-4 = y$

Skill Practice 1 Solve the equations.

a. $-\dfrac{6}{5}y = -\dfrac{3}{5}$ $\qquad$ **b.** $5 = \dfrac{t}{16}$ $\qquad$ **c.** $-a = -2$ $\qquad$ **d.** $-8 = t - 13$

The equations in Example 1 were easily solved in one step. However, to solve complicated linear equations, we offer the following guidelines.

Solving a Linear Equation in One Variable

Step 1 Simplify both sides of the equation.
- Use the distributive property to clear parentheses.
- Combine like terms.
- Consider clearing fractions or decimals by multiplying both sides of the equation by the least common denominator (LCD) of all terms.

Step 2 Use the addition property of equality to collect the variable terms on one side of the equation and the constant terms on the other side.

Step 3 Use the multiplication property of equality to make the coefficient of the variable term equal to 1.

Step 4 Check the potential solution in the original equation.

Step 5 Write the solution set.

EXAMPLE 2 Solving a Linear Equation

Solve the linear equation and check the answer.

$$3x + 1 = -7$$

Solution:

$$3x + 1 = -7$$

$$3x + 1 - 1 = -7 - 1 \qquad \text{Subtract 1 from both sides.}$$

$$3x = -8 \qquad \text{Combine like terms.}$$

$$\frac{3x}{3} = \frac{-8}{3} \qquad \text{To isolate } x, \text{ divide both sides of the equation by 3.}$$

$$x = -\frac{8}{3} \qquad \text{Simplify.}$$

Check: $3x + 1 \overset{?}{=} -7$

$$3\left(-\frac{8}{3}\right) + 1 \overset{?}{=} -7$$

$$\overset{1}{3}\left(-\frac{8}{3}\right) + 1 \overset{?}{=} -7$$

$$-8 + 1 \overset{?}{=} -7$$

$$-7 \overset{?}{=} -7 \checkmark \quad \text{True}$$

The solution set is $\left\{-\dfrac{8}{3}\right\}$.

Skill Practice 2   Solve the linear equation and check the answer.

$$5x - 19 = -23$$

In Example 3, we solve a linear equation in which we first simplify both sides of the equation.

EXAMPLE 3   **Solving a Linear Equation**

Solve. $-3(w - 4) + 5 = 10 - (w + 1)$

Solution:

$$-3(w - 4) + 5 = 10 - (w + 1)$$

$$-3w + 12 + 5 = 10 - w - 1 \qquad \text{Apply the distributive property.}$$

$$-3w + 17 = 9 - w \qquad \text{Combine like terms.}$$

$$-3w + w + 17 = 9 - w + w \qquad \text{Add } w \text{ to both sides of the equation.}$$

$$-2w + 17 = 9 \qquad \text{Combine like terms.}$$

$$-2w + 17 - 17 = 9 - 17 \qquad \text{Subtract 17 from both sides.}$$

$$-2w = -8 \qquad \text{Combine like terms.}$$

$$\frac{-2w}{-2} = \frac{-8}{-2} \qquad \text{Divide both sides by } -2.$$

$$w = 4$$

Check: $-3(w - 4) + 5 = 10 - (w + 1)$

$$-3[(4) - 4] + 5 \overset{?}{=} 10 - [(4) + 1]$$

$$-3(0) + 5 \overset{?}{=} 10 - (5)$$

$$5 \overset{?}{=} 5 \quad \checkmark \text{ true}$$

The solution set is $\{4\}$.

Skill Practice 3   Solve. $5(v - 4) - 2 = 2(v + 7) - 3$

Answers

2. $\left\{-\dfrac{4}{5}\right\}$

3. $\{11\}$

2. Clear Fractions and Decimals

When an equation contains fractions or decimals, it is sometimes helpful to clear the fractions and decimals. This is accomplished by multiplying both sides of the equation by the least common denominator (LCD) of all terms within the equation. This is demonstrated in Examples 4–6.

> **EXAMPLE 4**   **Solving a Linear Equation by Clearing Fractions**

Solve the equation. $\dfrac{1}{4}w + \dfrac{1}{3}w - 1 = \dfrac{1}{2}(w - 4)$

Solution:

FOR REVIEW

12 is the whole number of least value that is a multiple of 4, 3, and 2.

Multiples of 4:
4, 8, 12, 16, 20, . . .

Multiples of 3:
3, 6, 9, 12, 15, . . .

Multiples of 2:
2, 4, 6, 8, 10, 12, . . .

For review of least common denominator, see Section R.1.

$$\dfrac{1}{4}w + \dfrac{1}{3}w - 1 = \dfrac{1}{2}(w - 4)$$

$$\dfrac{1}{4}w + \dfrac{1}{3}w - 1 = \dfrac{1}{2}w - 2 \qquad \text{Clear parentheses.}$$

$$12 \cdot \left(\dfrac{1}{4}w + \dfrac{1}{3}w - 1\right) = 12 \cdot \left(\dfrac{1}{2}w - 2\right) \qquad \begin{array}{l}\text{Multiply both sides of the}\\ \text{equation by the LCD of all}\\ \text{terms. In this case, the LCD}\\ \text{is 12.}\end{array}$$

$$12 \cdot \dfrac{1}{4}w + 12 \cdot \dfrac{1}{3}w + 12 \cdot (-1) = 12 \cdot \dfrac{1}{2}w + 12 \cdot (-2) \qquad \begin{array}{l}\text{Apply the distributive}\\ \text{property.}\end{array}$$

$$3w + 4w - 12 = 6w - 24$$
$$7w - 12 = 6w - 24 \qquad \begin{array}{l}\text{Subtract } 6w \text{ from both sides.}\\ \text{Then add 12 to both sides.}\end{array}$$
$$w - 12 = -24$$
$$w = -12 \qquad \text{The solution checks.}$$

The solution set is $\{-12\}$.

> **Skill Practice 4**   Solve the equation by first clearing the fractions.
>
> $$\dfrac{3}{4}a + \dfrac{1}{2} = \dfrac{2}{3}a + \dfrac{1}{3}$$

> **EXAMPLE 5**   **Solving a Linear Equation by Clearing Fractions**

Solve. $\dfrac{x - 2}{5} - \dfrac{x - 4}{2} = 2 + \dfrac{x + 4}{10}$

Solution:

TIP   Clearing fractions is an application of the multiplication property of equality. We are multiplying both sides of the equation by the same number.

$$\dfrac{x - 2}{5} - \dfrac{x - 4}{2} = \dfrac{2}{1} + \dfrac{x + 4}{10} \qquad \begin{array}{l}\text{The LCD of all}\\ \text{terms in the}\\ \text{equation is 10.}\end{array}$$

$$10\left(\dfrac{x - 2}{5} - \dfrac{x - 4}{2}\right) = 10\left(\dfrac{2}{1} + \dfrac{x + 4}{10}\right) \qquad \begin{array}{l}\text{Multiply both}\\ \text{sides by 10.}\end{array}$$

$$\dfrac{\overset{2}{10}}{1} \cdot \left(\dfrac{x - 2}{\underset{1}{5}}\right) - \dfrac{\overset{5}{10}}{1} \cdot \left(\dfrac{x - 4}{\underset{1}{2}}\right) = \dfrac{10}{1} \cdot \left(\dfrac{2}{1}\right) + \dfrac{\overset{1}{10}}{1} \cdot \left(\dfrac{x + 4}{\underset{1}{10}}\right) \qquad \begin{array}{l}\text{Apply the}\\ \text{distributive}\\ \text{property.}\end{array}$$

Answer

4. $\{-2\}$

$$2(x-2) - 5(x-4) = 20 + 1(x+4)$$ Clear fractions.

$$2x - 4 - 5x + 20 = 20 + x + 4$$ Apply the distributive property.

$$-3x + 16 = x + 24$$ Simplify both sides of the equation.

$$-4x + 16 = 24$$ Subtract x from both sides.

$$-4x = 8$$ Subtract 16 from both sides.

$$x = -2$$ The value -2 checks in the original equation.

The solution set is $\{-2\}$.

Skill Practice 5   Solve the equation. $\dfrac{1}{8} - \dfrac{x+3}{4} = \dfrac{3x-2}{2}$

If the terms of an equation have decimal coefficients, we can clear the decimals by multiplying both sides of the equation by a convenient power of 10.

EXAMPLE 6   **Solving a Linear Equation by Clearing Decimals**

Solve the equation. $0.55x - 0.6 = 2.05x$

Solution:

Recall that any terminating decimal can be written as a fraction. Therefore, the equation $0.55x - 0.6 = 2.05x$ is equivalent to

$$\frac{55}{100}x - \frac{6}{10} = \frac{205}{100}x$$

A convenient common denominator for all terms in this equation is 100. Multiplying both sides of the equation by 100 will have the effect of "moving" the decimal point 2 places to the right.

$$100(0.55x - 0.6) = 100(2.05x)$$ Multiply both sides by 100 to clear decimals.

$$55x - 60 = 205x$$

$$-60 = 150x$$ Subtract $55x$ from both sides.

$$-\frac{60}{150} = x$$ To isolate x, divide both sides by 150.

$$x = -\frac{2}{5}$$ Simplify the fraction.

$$x = -0.4$$ The solution checks.

The solution set is $\{-0.4\}$.

Skill Practice 6   Solve the equation by first clearing decimals.

$$2.2x + 0.5 = 1.6x + 0.2$$

Answers

5. $\left\{\dfrac{3}{14}\right\}$

6. $\{-0.5\}$

3. Identify Conditional Equations, Identities, and Contradictions

The solution to a linear equation is the value of x that makes the equation a true statement. While linear equations have one unique solution, some equations have no solution, and others have infinitely many solutions.

I. Conditional Equations

An equation that is true for some values of the variable but false for other values is called a **conditional equation**. The equation $x + 4 = 6$ is a conditional equation because it is true on the *condition* that $x = 2$. For other values of x, the statement $x + 4 = 6$ is false. The solution set is $\{2\}$.

II. Contradictions

Some equations have no solution, such as $x + 1 = x + 2$. There is no value of x that when increased by 1 will equal the same value increased by 2. If we tried to solve the equation by subtracting x from both sides, we get the contradiction $1 = 2$.

$$x + 1 = x + 2$$
$$x - x + 1 = x - x + 2$$
$$1 = 2 \quad \text{(contradiction)}$$

This indicates that the equation has no solution. An equation that has no solution is called a **contradiction**. The solution set for a contradiction is the empty set. The **empty set** is the set with no elements and is denoted by $\{\ \}$ or $\varnothing$.

III. Identities

An equation that is true for all real values of the variable for which the expressions within an equation are defined is called an **identity**. For example, consider the equation $x + 4 = x + 4$. Because the left- and right-hand sides are *identical*, any real number substituted for x will result in equal quantities on both sides. If we subtract x from both sides, we get the identity $4 = 4$. In such a case, the solution set is the set of real numbers.

$$x + 4 = x + 4$$
$$x - x + 4 = x - x + 4$$
$$4 = 4 \quad \text{(identity)} \qquad \text{The solution set is the set of real numbers } \mathbb{R}.$$

EXAMPLE 7 **Identifying Conditional Equations, Contradictions, and Identities**

Identify each equation as a conditional equation, a contradiction, or an identity. Then give the solution set.

a. $3[x - (x + 1)] = -2$
b. $5(3 + c) + 2 = 2c + 3c + 17$
c. $4x - 3 = 17$

Solution:

a. $3[x - (x + 1)] = -2$

$3(x - x - 1) = -2$	Clear parentheses.
$3(-1) = -2$	Combine like terms.
$-3 = -2$	Contradiction

This equation is a contradiction.
The solution set is $\{\ \}$.

b. $5(3 + c) + 2 = 2c + 3c + 17$

$\quad\quad 15 + 5c + 2 = 5c + 17$ Clear parentheses and combine like terms.

$\quad\quad\quad 5c + 17 = 5c + 17$ Identity

$\quad\quad\quad\quad\quad\quad 0 = 0$

This equation is an identity. The solution set is $\{c \mid c \text{ is a real number}\}$.

c. $4x - 3 = 17$

$\quad\quad 4x = 20$ Add 3 to both sides.

$\quad\quad\; x = 5$ Divide by 4.

This equation is a conditional equation.

The solution set is $\{5\}$.

Skill Practice 7   Identify each equation as a conditional equation, an identity, or a contradiction. Then give the solution set.

    a. $2(-5x - 1) = 2x - 12x + 6$ **b.** $2(3x - 1) = 6(x + 1) - 8$

    c. $4x + 1 - x = 6x - 2$

---

4. Solve Literal Equations for a Specified Variable

Literal equations are equations that contain several variables. A formula is a literal equation with a specific application. For example, the perimeter P of a rectangle can be found by the formula $P = 2l + 2w$. In this equation, P is expressed in terms of the length l and the width w. However, in science and other branches of applied mathematics, formulas may be more useful in alternative forms.

For example, the formula $P = 2l + 2w$ can be manipulated to solve for either l or w:

Solve for l

$P = 2l + 2w$

$P - 2w = 2l$ Subtract $2w$.

$\dfrac{P - 2w}{2} = l$ Divide by 2.

$l = \dfrac{P - 2w}{2}$

Solve for w

$P = 2l + 2w$

$P - 2l = 2w$ Subtract $2l$.

$\dfrac{P - 2l}{2} = w$ Divide by 2.

$w = \dfrac{P - 2l}{2}$

To solve a literal equation for a specified variable, the goal is to isolate that variable.

EXAMPLE 8   **Solving an Equation for a Specified Variable**

Given $d = rt$, solve for t.

Solution:

$d = rt$

$\dfrac{d}{r} = \dfrac{rt}{r}$ The relationship between r and t is multiplication. Therefore, we want to reverse this operation to isolate t. Divide both sides by r.

$\dfrac{d}{r} = t \text{ or } t = \dfrac{d}{r}$

Skill Practice 8   Given $I = Prt$, solve for t.

Answers

7. a. Contradiction; { }

   b. Identity; $\{x \mid x \text{ is a real number}\}$

   c. Conditional equation; $\{1\}$

8. $t = \dfrac{I}{Pr}$

When solving a literal equation for a specified variable, sometimes there is more than one form for the final answer. For example, consider the equation $a = \dfrac{b + c}{d}$. Both terms in the numerator are being divided by d. Therefore, we can write this equation as

$$a = \frac{b + c}{d} \quad \text{or} \quad a = \frac{b}{d} + \frac{c}{d} \qquad \text{The equations are equivalent.}$$

EXAMPLE 9   **Solving an Equation for a Specified Variable**

Solve for the indicated variable.

a. $3x + 2y = 6$ for y **b.** $A = \dfrac{1}{2}h(B + b)$ for B

Solution:

a. $3x + 2y = 6$ for y

$\qquad 2y = -3x + 6$ Subtract $3x$ from both sides to isolate the y term on one side of the equation.

$\qquad \dfrac{2y}{2} = \dfrac{-3x + 6}{2}$ Divide both sides by 2 to isolate y.

$\qquad y = \dfrac{-3x + 6}{2}$ The variable y is isolated. We can stop here; however, there are alternative forms of the answer.

$\qquad y = \dfrac{-3x}{2} + \dfrac{6}{2}$ We have the option of dividing each term in the numerator on the right by the denominator 2.

$\qquad y = -\dfrac{3}{2}x + 3$ Simplify each term.

Thus, $\quad y = \dfrac{-3x + 6}{2}$ or $y = -\dfrac{3}{2}x + 3$

TIP   Alternatively, after clearing fractions we can solve for B by first clearing parentheses.

$2A = h(B + b)$
$2A = hB + hb$
$2A - hb = hB$
$\dfrac{2A - hb}{h} = B$

b. $A = \dfrac{1}{2}h(B + b)$ for B First note that letters in algebra are case sensitive. The letters b and B represent different variables.

$\qquad 2(A) = 2\left[\dfrac{1}{2}h(B + b)\right]$ Multiply by 2 to clear fractions.

$\qquad 2A = h(B + b)$

$\qquad \dfrac{2A}{h} = \dfrac{h(B + b)}{h}$ Divide by h.

$\qquad \dfrac{2A}{h} = B + b$

$\qquad \dfrac{2A}{h} - b = B$ or $B = \dfrac{2A}{h} - b$ Subtract b from both sides to isolate B.

Skill Practice 9   Solve for the indicated variable.

a. $4x + 3y = 12$ for y **b.** $A = \dfrac{1}{2}h(B + b)$ for b

Answers

9. a. $y = \dfrac{-4x + 12}{3}$ or

$\quad y = -\dfrac{4}{3}x + 4$

b. $b = \dfrac{2A}{h} - B$

SECTION 1.2	Practice Exercises

Prerequisite Review

For Exercises R.1–R.4, simplify. (See Section 1.1 for review.)

R.1. $x + 2 + (-2)$ **R.2.** $-3 + 3 + y$ **R.3.** $\dfrac{9}{2} \cdot \dfrac{2}{9}t$ **R.4.** $-\dfrac{5}{11} \cdot \left(-\dfrac{11}{5}m\right)$

For Exercises R.5–R.10, clear parentheses and combine like terms. (See Section 1.1 for review.)

R.5. $5(4z - 8) - (z + 3)$ **R.6.** $9(2t + 1) - (3t - 5)$ **R.7.** $100(0.45x - 0.4) - 100(0.25x + 0.6)$

R.8. $10(0.9y - 0.4) + 10(0.1y + 0.3)$ **R.9.** $12\left(\dfrac{1}{3}x + \dfrac{1}{6}\right) + 12\left(\dfrac{3}{4}x - 1\right)$ **R.10.** $18\left(\dfrac{1}{3}t - \dfrac{1}{2}\right) - 18\left(\dfrac{2}{9}t + 2\right)$

For Exercises R.11–R.12, evaluate the expression for the given value of the variable. (See Section R.3 for review.)

R.11. $2(4x + 3) - 5(2x - 7)$ for $x = -2$ **R.12.** $3 - [2 - (5x + 1)] + 4x$ for $x = -3$

Concept Connections

1. An _____ is a statement that indicates that two quantities are equal.

2. A _____ to an equation is a value of the variable that makes the equation a true statement.

3. An equation that can be written in the form $ax + b = c$ ($a \neq 0$) is called a _____ equation in one variable.

4. Which values could be used to clear fractions in the equation $\frac{1}{2}y - \frac{3}{4} = 2y - \frac{5}{6}$? Choose from: 2, 3, 6, 12, 18, 24, 30, 32, 36.

5. The set of all solutions to an equation is called the _____ _____.

6. Two equations are equivalent if they have the same _____ set.

7. A _____ equation is true for some values of the variable, but false for other values.

8. An equation that has no solution is called a _____.

9. The set containing no elements is called the _____ and is denoted by _____.

10. An equation that has all real numbers as its solution set is called an _____.

Objective 1: Solve Linear Equations in One Variable

For Exercises 11–16, label the equation as linear or nonlinear.

11. $2x + 1 = 5$ **12.** $10 = x + 6$ **13.** $x^2 + 7 = 9$

14. $3 + x^3 - x = 4$ **15.** $-3 = x$ **16.** $5.2 - 7x = 0$

17. Use substitution to determine which value is the solution to $2x - 1 = 5$.

    a. 2   **b.** 3   **c.** 0   **d.** -1

18. Use substitution to determine which value is the solution to $2y - 3 = -2$.

    a. 1   **b.** $\dfrac{1}{2}$ **c.** 0   **d.** $-\dfrac{1}{2}$

For Exercises 19–44, solve the equation. **(See Examples 1–3)**

19. $x + 7 = 19$ **20.** $-3 + y = -28$ **21.** $-x = 2$ **22.** $-t = \dfrac{3}{4}$

23. $-\dfrac{7}{8} = -\dfrac{5}{6}z$ **24.** $-\dfrac{12}{13} = \dfrac{4}{3}b$ **25.** $\dfrac{a}{5} = -8$ **26.** $\dfrac{x}{8} = \dfrac{1}{2}$

27. $-4.8 = 6.1 + y$ **28.** $p - 2.9 = 3.8$ **29.** $6q - 4 = 62$ **30.** $2w - 15 = 15$

31. $4y - 17 = 35$ **32.** $6z - 25 = 83$ **33.** $3(x - 6) = 2x - 5$ **34.** $13y + 4 = 5(y - 4)$

35. $6 - (t + 2) = 5(3t - 4)$

36. $1 - 5(p + 2) = 2(p + 13)$

37. $6(a + 3) - 10 = -2(a - 4)$

38. $8(b - 2) + 3b = -9(b - 1)$

39. $-2[5 - (2z + 1)] - 4 = 2(3 - z)$

40. $3[w - (10 - w)] = 7(w + 1)$

41. $6(-y + 4) - 3(2y - 3) = -y + 5 + 5y$

42. $13 + 4w = -5(-w - 6) + 2(w + 1)$

43. $14 - 2x + 5x = -4(-2x - 5) - 6$

44. $8 - (p + 2) + 6p + 7 = p + 13$

Objective 2: Clear Fractions and Decimals

For Exercises 45–56, solve the equations. (**See Examples 4–6**)

45. $\frac{2}{3}x - \frac{1}{6} = -\frac{5}{12}x + \frac{3}{2} - \frac{1}{6}x$

46. $-\frac{1}{2}y + 4 = -\frac{9}{10}y + \frac{2}{5}$

47. $\frac{1}{5}(p - 5) = \frac{3}{5}p + \frac{1}{10}p + 1$

48. $\frac{5}{6}(q + 2) = -\frac{7}{9}q - \frac{1}{3} + 2$

49. $\frac{3x - 7}{2} + \frac{3 - 5x}{3} = \frac{3 - 6x}{5}$

50. $\frac{2y - 4}{5} = \frac{5y + 13}{4} + \frac{y}{2}$

51. $\frac{4}{3}(2q + 6) - \frac{5q - 6}{6} - \frac{q}{3} = 0$

52. $\frac{-3a + 9}{15} - \frac{2a - 5}{5} - \frac{a + 2}{10} = 0$

53. $6.3w - 1.5 = 4.8$

54. $0.2x + 53.6 = x$

55. $0.75(m - 2) + 0.25m = 0.5$

56. $0.4(n + 10) + 0.6n = 2$

Objective 3: Identify Conditional Equations, Identities, and Contradictions

57. What is a conditional equation?

58. Explain the difference between a contradiction and an identity.

For Exercises 59–64, identify the equation as a conditional equation, a contradiction, or an identity. Then give the solution set. (**See Example 7**)

59. $4x + 1 = 2(2x + 1) - 1$

60. $3x + 6 = 3x$

61. $-11x + 4(x - 3) = -2x - 12$

62. $5(x + 2) - 7 = 3$

63. $2x - 4 + 8x = 7x - 8 + 3x$

64. $-7x + 8 + 4x = -3(x - 3) - 1$

Objective 4: Solve Literal Equations for a Specified Variable

For Exercises 65–82, solve for the indicated variable. (**See Examples 8–9**)

65. $I = Prt$ for P

66. $V = lwh$ for h

67. $A = lw$ for l

68. $E = IR$ for R

69. $P = a + b + c$ for c

70. $W = K - T$ for K

71. $\Delta s = s_2 - s_1$ for s_1

72. $\Delta t = t_f - t_i$ for t_i

73. $7x + 2y = 8$ for y

74. $3x + 5y = 15$ for y

75. $5x - 4y = 2$ for y

76. $7x - 2y = 5$ for y

77. $\frac{1}{2}x + \frac{1}{3}y = 1$ for y

78. $\frac{1}{4}x - \frac{2}{3}y = 2$ for y

79. $S = \frac{n}{2}(a + d)$ for d

80. $S = \frac{n}{2}[2a + (n - 1)d]$ for a

81. $V = \frac{1}{3}\pi r^2 h$ for h

82. $V = \frac{1}{3}Bh$ for B

Mixed Exercises

For Exercises 83–88, let x represent the unknown number. Write an equation that represents the given statement. Then find the value of the unknown number.

83. Negative six more than twice a number is -24. Find the number.

84. Ten less than three times a number is 26. Find the number.

85. The quotient of a number and 12 is $\frac{1}{3}$. Find the number.

86. Eighteen is equal to a number divided by 2. Find the number.

87. Five times the sum of a number and 3 equals 7 more than the number. Find the number.

88. Negative three times the difference of a number and 4 equals the opposite of the number. Find the number.

89. A student preparing for college puts aside $2000 in a mutual fund to use 1 year later for school. If the annual rate of return is denoted by x, then the amount earned on the account for 1 year is given by $2000x$. The total value A of the account is given by $A = 2000 + 2000x$. If at the end of the year, the account is worth $2150, find the annual rate of return.

90. A power company charges $0.12 per kilowatt-hour (kWh) and $14.89 in monthly taxes. The monthly charge C (in $) is given by $C = 0.12h + 14.89$ where h is the number of kilowatt-hours used. If a family's bill comes to $137.77, determine the number of kilowatt-hours used.

91. Brianna's SUV gets 22 mpg in the city and 30 mpg on the highway. The amount of gas she uses A (in gal) is given by $A = \frac{1}{22}c + \frac{1}{30}h$, where c is the number of city miles driven and h is the number of highway miles driven. If Brianna drove 165 mi on the highway and used 7 gal of gas, how many city miles did she drive?

92. Dexter's truck gets 32 mpg on the highway and 24 mpg in the city. The amount of gas he uses A (in gal) is given by $A = \frac{1}{24}c + \frac{1}{32}h$, where c is the number of city miles driven and h is the number of highway miles driven. If Dexter drove 60 mi in the city and used 9 gal of gas, how many highway miles did he drive?

93. A motorist drives on State Road 417 to and from work each day and pays $3.50 in tolls one-way.

   a. Write a model for the cost for tolls C (in $) for x working days.

   b. The department of transportation has a prepaid toll program that discounts tolls for high-volume use. The motorist can buy a pass for $105 per month. How many working days are required for the motorist to save money by buying the pass? **(See Example 3)**

94. A subway ride is $2.25 per ride.

   a. Write a model for the cost C (in $) for x rides on the subway.

   b. A commuter can purchase an unlimited-ride MetroCard for $89 per month. How many rides are required for a commuter to save money by buying the MetroCard?

95. Helene considers two jobs. One pays $45,000/year with an anticipated yearly raise of $2250. A second job pays $48,000/year with yearly raises averaging $2000.

   a. Write a model representing the salary S_1 (in $) for the first job in x years.

   b. Write a model representing the salary S_2 (in $) for the second job in x years.

   c. In how many years will the salary from the first job equal the salary from the second?

96. Tasha considers two sales jobs for different pharmaceutical companies. One pays a base salary of $25,000 with a 16% commission on sales. The other pays $30,000 with a 15% commission on sales.

   a. Write a model representing the salary S_1 (in $) for the first job based on x dollars in sales.

   b. Write a model representing the salary S_2 (in $) for the second job based on x dollars in sales.

   c. For how much in sales will the two jobs result in equal salaries?

Write About It

97. Explain why the value 2 is not the only solution to the equation $2x + 4 = 2(x - 3) + 10$.

98. Explain why the equation $x + 1 = x + 2$ has no solution.

Expanding Your Skills

For Exercises 99–102, find the value of a so that the equation has the given solution set.

99. $ax + 6 = 4x + 14$ $\{4\}$

100. $ax - 3 = 2x + 9$ $\{3\}$

101. $a(2x - 5) + 6 = 5x + 7$ $\{16\}$

102. $a(2x + 4) + 12x = 3(2 - x)$ $\{34\}$

PROBLEM RECOGNITION EXERCISES

Equations Versus Expressions

For Exercises 1–20, identify each exercise as an expression or an equation. Then simplify the expressions and solve the equations.

1. $4x - 2 + 6 - 8x$

2. $-3y - 3 - 4y + 8$

3. $7b - 1 = 2b + 4$

4. $10t + 2 = 2 - 7t$

5. $4(a - 8) - 7(2a + 1)$

6. $10(2x + 3) - 8(5 - x)$

7. $7(2 - w) = 5w + 8$

8. $15(3 - 2y) = 21 + 2y$

9. $2(3x - 4) - 4(5x + 1) = -8x + 7$

10. $6(2 - 3a) - 2(8a + 3) = -12a - 19$

11. $\frac{1}{2}v + \frac{3}{5} - \frac{2}{3}v - \frac{7}{10}$

12. $-\frac{7}{8}t - \frac{4}{3}u - \frac{5}{4}t + \frac{11}{6}u$

13. $20x - 8 + 7x + 28 = 27x - 9$

14. $7 + 8w - 12 = 3w - 8 + 5w$

15. $\frac{5}{6}y - \frac{7}{8} = \frac{1}{2}y + \frac{3}{4}$

16. $\frac{4}{5} + 3z = \frac{1}{2}z + 1$

17. $0.29c + 4.495 - 0.12c$

18. $0.45k - 1.67 + 0.89 - 1.456k$

19. $0.125(2p - 8) = 0.25(p - 4)$

20. $0.5u + 1.2 - 0.74u = 0.8 - 0.24u + 0.4$

SECTION 1.3   Applications of Linear Equations in One Variable

OBJECTIVES

1. Use Linear Equations in One Variable for Problem Solving
2. Solve Applications Involving Percent and Rates
3. Solve Applications Involving Simple Interest
4. Solve Applications Involving Mixtures
5. Solve Applications Involving Uniform Motion

1. Use Linear Equations in One Variable for Problem Solving

Suppose that a bat and a ball cost $1.10 in total and that the bat costs $1.00 more than the ball. How much does the ball cost? At first glance, many people answer that the ball costs $0.10 and the bat costs $1.00. Although the total cost is $0.10 + $1.00 = $1.10, the bat costs only $0.90 more than the ball. So, the second condition of the puzzle is not met.

    With a bit of trial and error, we can change our guess so that there is a greater difference in price between the bat and ball. Decreasing the price of the ball to $0.05 and increasing the price of the bat to $1.05 still gives us a total of $1.10. But with this combination the bat costs exactly $1.00 more than the ball.

    While this mind teaser is relatively straightforward, most significant applications of mathematics cannot be computed mentally. More structure is needed. We will solve this puzzle using a linear equation in Example 1.

    To solve an application using algebra, relevant information must be extracted from the wording of the problem and then translated into a mathematical equation. This is a skill that requires practice, and the key strategy is to stick with it and not get discouraged. While there is no magic formula or "recipe" to solve all word problems, we offer the following guidelines.

Problem-Solving Strategy

1. Read the problem carefully. Determine what the problem is asking for and assign variables to the unknown quantities. Make an appropriate figure or table if applicable.
2. Write a verbal model (equation in words) to describe the given scenario.
3. Write an algebraic equation that represents the verbal model.
4. Solve the equation from step 3.
5. Interpret the solution and check that the answer is reasonable.

EXAMPLE 1   **Applying a Linear Equation**

Suppose that a bat and a ball cost $1.10 in total and that the bat costs $1.00 more than the ball. How much does the ball cost?

Solution:

Step 1:   Read the problem carefully. Assign variables to the unknown quantities.

Let x represent the cost of the ball.

Then $x + 1.00$ represents the cost of the bat.

Step 2: $\left(\begin{array}{c}\text{Cost of} \\ \text{the ball}\end{array}\right) + \left(\begin{array}{c}\text{Cost of} \\ \text{the bat}\end{array}\right) = \text{Total cost}$

Step 3:   Replace the verbal model with an algebraic equation.

$\left(\begin{array}{c}\text{Cost of} \\ \text{the ball}\end{array}\right) + \left(\begin{array}{c}\text{Cost of} \\ \text{the bat}\end{array}\right) = \text{Total cost}$

$x + (x + 1.00) = 1.10$

Step 4:   Solve for x.

$$x + (x + 1.00) = 1.10$$
$$2x + 1.00 = 1.10 \qquad \text{Combine like terms.}$$
$$2x = 0.10 \qquad \text{Subtract 1.00 from both sides.}$$
$$x = 0.05 \qquad \text{Divide both sides by 2.}$$

Step 5:   Interpret the results.

$x = 0.05$ means that the ball costs $0.05.

$x + 1.00 = 0.05 + 1.00 = 1.05$ means that the bat costs $1.05.

Skill Practice 1   At a ballpark, the cost of a hot dog and drink is $1.80. The hot dog costs $0.50 more than the drink. What is the cost for each item?

---

2. Solve Applications Involving Percent and Rates

In many real-world applications, percents are used to represent rates.

- The sales tax rate for a certain county is 6%.
- An ice cream machine is discounted 20%.
- A real estate sales broker receives a $4\frac{1}{2}\%$ commission on sales.
- A savings account earns 7% simple interest.

Answer

1. A hot dog costs $1.15 and a drink costs $0.65.

The following models are used to compute sales tax, commission, and simple interest. In each case the value is found by multiplying the base by the percentage.

Sales tax = (Cost of merchandise)(tax rate)

Example: A blouse costs $44 and the tax rate is 6.5%.
Sales tax = ($44)(0.065) = $2.86

Commission = (Amount in sales)(commission rate)

Example: A Realtor makes a 5% commission on a $180,000 house.
Commission = ($180,000)(0.05) = $9000

Simple interest = (Principal)(annual interest rate)(time in years)

Example: $4000 is invested at 2.5% interest for 3 years.
Simple interest = ($4000)(0.025)(3) = $300

EXAMPLE 2　Applying a Linear Equation to Sales Tax

A refrigerator sells for $1250, but after adding sales tax the total cost is $1325. Find the tax rate.

Solution:

	Read the problem carefully.
Let r represent the sales tax rate.	Label the variable(s).
$\left(\begin{array}{c}\text{Cost of}\\\text{refrigerator}\end{array}\right) + \left(\begin{array}{c}\text{Sales}\\\text{tax}\end{array}\right) = \text{Total cost}$	Write a verbal model.
$1250 + 1250r = 1325$	Write a mathematical equation. Note that the sales tax is the cost of the merchandise times the sales tax rate: ($1250) $\cdot$ r
$1250r = 75$	Subtract 1250 from both sides.
$r = 0.06$	Divide both sides by 1250.
The sales tax rate is 6%.	Interpret the result.

FOR REVIEW

In Example 2, the value $r = 0.06$ means $\frac{6}{100}$. This translates to 6 per 100 or 6%.

> **Skill Practice 2**　A new air-conditioning unit costs $8600. If the total bill after sales tax is $9030, what is the sales tax rate?

As consumers, we often encounter situations in which merchandise has been marked up or marked down from its original cost. It is important to note that percent increase and percent decrease are based on the original cost. For example, suppose a microwave oven originally priced at $305 is marked down 20%.

The discount is determined by 20% of the original price: (0.20)($305) = $61.00. The new price is $305.00 − $61.00 = $244.00.

EXAMPLE 3　Solving a Percent Increase Application

A college bookstore uses a standard markup of 40% on all books purchased wholesale from the publisher. If the bookstore sells a calculus book for $179.20, what was the original wholesale cost?

Solution:

Let x = original wholesale cost.   Label the variables.

The selling price of the book is based on the original cost of the book plus the bookstore's markup.

$$\text{(Selling price)} = \text{(original price)} + \text{(markup)} \qquad \text{Verbal model}$$

$$\text{(Selling price)} = \text{(original price)} + \text{(original price} \cdot \text{markup rate)}$$

$$179.20 = \qquad x \qquad + (x)(0.40) \qquad \text{Mathematical equation}$$

$$179.20 = x + 0.40x$$

$$179.20 = 1.40x \qquad \text{Combine like terms.}$$

$$x = 128 \qquad \text{Divide both sides by 1.40.}$$

The original wholesale cost of the textbook was \$128.00.   Interpret the results.

> **Skill Practice 3**   An online bookstore gives a 20% discount on paperback books. Find the original price of a book that has a selling price of \$5.28 after the discount.

3. Solve Applications Involving Simple Interest

EXAMPLE 4   **Solving an Investment Growth Application**

Miguel had \$10,000 to invest in two different mutual funds. One was a relatively safe bond fund that averaged 4% return on his investment at the end of 1 year. The other fund was a riskier stock fund that averaged 7% return in 1 year. If at the end of the year Miguel's portfolio grew to \$10,625 (\$625 above his \$10,000 investment), how much money did Miguel invest in each fund?

Solution:

This type of word problem is sometimes categorized as a mixture problem. Miguel is "mixing" his money between two different investments. We have to determine how the money was divided to earn \$625.

The information in this problem can be organized in a chart. (*Note*: There are two sources of money: the amount invested and the amount earned.)

	4% Bond Fund	7% Stock Fund	Total
Amount invested (\$)	x	$(10{,}000 - x)$	10,000
Amount earned (\$)	$0.04x$	$0.07(10{,}000 - x)$	625

Because the amount of principal is unknown for both accounts, we can let x represent the amount invested in the bond fund. If Miguel spends x dollars in the bond fund, then he has $(10{,}000 - x)$ left over to spend in the stock fund. The return for each fund is found by multiplying the principal and the percent growth rate.

Answer

3. \$6.60

To establish a mathematical model, we know that the total return ($625) must equal the earnings from the bond fund plus the earnings from the stock fund:

(Earnings from bond fund) + (earnings from stock fund) = (total earnings)

$$0.04x \quad + \quad 0.07(10{,}000 - x) \quad = \quad 625$$

$0.04x + 0.07(10{,}000 - x) = 625$	Mathematical equation
$4x + 7(10{,}000 - x) = 62{,}500$	Multiply by 100 to clear decimals.
$4x + 70{,}000 - 7x = 62{,}500$	
$-3x + 70{,}000 = 62{,}500$	Combine like terms.
$-3x = -7500$	Subtract 70,000 from both sides.
$x = 2500$	Divide both sides by -3.

The amount invested in the bond fund is $2500.
The amount invested in the stock fund is $10{,}000 - x$, or $10{,}000 - \$2500 = \$7500$.

> **Skill Practice 4**   Jonathan borrowed $4000 in two loans. One loan charged 7% interest, and the other charged 1.5% interest. After 1 year, Jonathan paid $225 in interest. Find the amount borrowed in each loan.

4. Solve Applications Involving Mixtures

In Example 5, we solve an application in which two antifreeze solutions of different strengths are combined to form a third solution. To understand the role of the concentration rate within a mixture, consider this example. Suppose you have 30 gal of a 10% antifreeze mixture. This means that 10% of the mixture is pure antifreeze and the other part of the solution is a mixing agent such as water. To find the amount of pure antifreeze, multiply the concentration rate (10%) times the amount of mixture (30 gal).

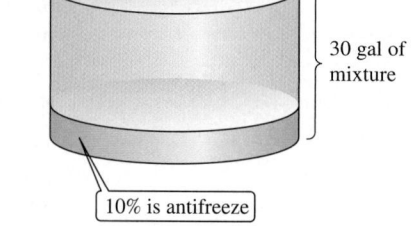

Amount of pure antifreeze = 0.10(30 gal)

= 3 gal

EXAMPLE 5   Solving a Mixture Application

How many liters (L) of a 40% antifreeze solution must be added to 4 L of a 10% antifreeze solution to produce a 35% antifreeze solution?

Solution:

The given information is illustrated in Figure 1-1.

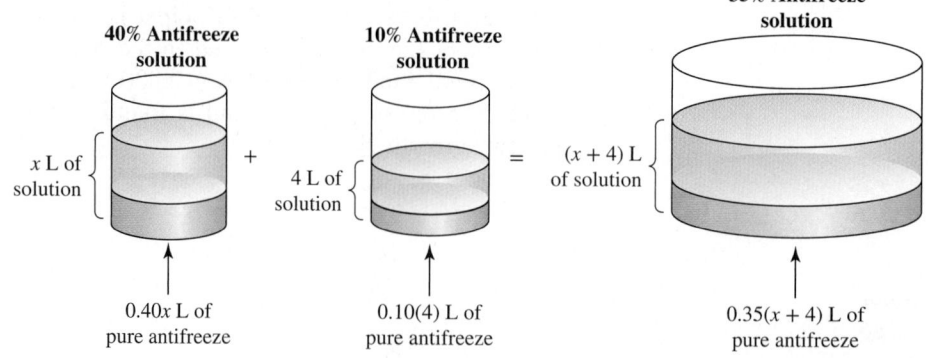

Figure 1-1

The information can also be organized in a table.

	40% Antifreeze	10% Antifreeze	Final Solution: 35% Antifreeze
Number of liters of solution	x	4	$(x + 4)$
Number of liters of pure antifreeze	0.40x	0.10(4)	0.35$(x + 4)$

Notice that an algebraic equation is obtained from the second row of the table relating the number of liters of pure antifreeze in each container.

$$\begin{pmatrix} \text{Pure antifreeze} \\ \text{from solution 1} \end{pmatrix} + \begin{pmatrix} \text{pure antifreeze} \\ \text{from solution 2} \end{pmatrix} = \begin{pmatrix} \text{pure antifreeze} \\ \text{in the final solution} \end{pmatrix}$$

$$0.40x \qquad + \qquad 0.10(4) \qquad = \qquad 0.35(x + 4)$$

$0.40x + 0.10(4) = 0.35(x + 4)$	Mathematical equation
$0.4x + 0.4 = 0.35x + 1.4$	Apply the distributive property.
$0.05x + 0.4 = 1.4$	Subtract $0.35x$ from both sides.
$0.05x = 1.0$	Subtract 0.4 from both sides.
$x = 20$	Divide both sides by 0.05.

Therefore, 20 L of a 40% antifreeze solution is needed.

Skill Practice 5   Find the number of ounces (oz) of 30% alcohol solution that must be mixed with 10 oz of a 70% solution to obtain a solution that is 40% alcohol.

5. Solve Applications Involving Uniform Motion

The fundamental relationship among the variables distance, rate, and time is given by

$$\text{Distance} = (\text{rate})(\text{time}) \quad \text{or} \quad d = rt$$

For example, a motorist traveling 65 mph (miles per hour) for 3 hr (hours) will travel a distance of

$$d = \left(\frac{65 \text{ mi}}{\text{hr}} \right)(3 \text{ hr}) = 195 \text{ mi}$$

EXAMPLE 6   Solving a Distance, Rate, Time Application

A hiker can hike 1 mph faster downhill to Moose Lake than she can hike uphill back to the campsite. If it takes her 3 hr to hike to the lake and 4.5 hr to hike back, what is her speed hiking back to the campsite?

Digital Stock/Corbis

Answer

5. 30 oz of the 30% solution is needed.

Solution:

The information given in the problem can be organized in a table.

	Distance (mi)	Rate (mph)	Time (hr)
Trip to the lake		$x + 1$	3
Return trip		x	4.5

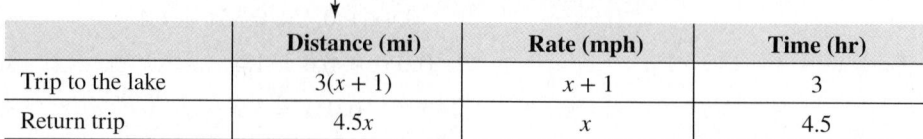

Column 2: Let the rate of the return trip be represented by x. Then the trip to the lake is 1 mph faster and can be represented by $x + 1$.

Column 3: The times hiking to and from the lake are given in the problem.

Column 1: To express the distance, we use the relationship $d = rt$. That is, multiply the quantities in the second and third columns.

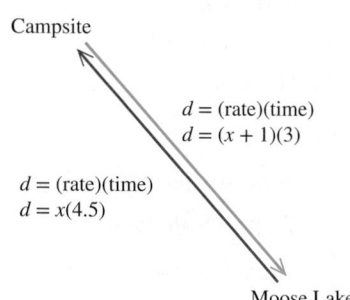

Campsite

$d = $ (rate)(time)
$d = (x + 1)(3)$

$d = $ (rate)(time)
$d = x(4.5)$

Moose Lake

	Distance (mi)	Rate (mph)	Time (hr)
Trip to the lake	$3(x + 1)$	$x + 1$	3
Return trip	$4.5x$	x	4.5

To create a mathematical model, note that the distances to and from the lake are equal. Therefore,

(Distance to lake) = (return distance)	Verbal model
$3(x + 1) = 4.5x$	Mathematical equation
$3x + 3 = 4.5x$	Apply the distributive property.
$3 = 1.5x$	Subtract $3x$ from both sides.
$2 = x$	Divide both sides by 1.5.

The hiker's speed on the return trip to the campsite is 2 mph.

> **Skill Practice 6** During a bad rainstorm, Jody drove 15 mph slower on a trip to her mother's house than she normally would when the weather is clear. If a trip to her mother's house takes 3.75 hr in clear weather and 5 hr in a bad storm, what is her normal driving speed during clear weather?

EXAMPLE 7 Solving a Distance, Rate, Time Application

Donna participated in a 41-mi biathlon that included running and bicycling. She spent 1 hr 45 min on the bike and 45 min running. If her average speed on the bicycle was 12 mph faster than her average speed running, find her average speed running and her average speed riding.

Solution:

There are two unknowns: Donna's average speed on the bike and her average speed running.

Let x represent Donna's average speed running.

Then $x + 12$ represents her speed on the bicycle.

Answer

6. Jody normally drives 60 mph.

The remaining information can be organized in a table.

	Distance (mi)	Rate (mph)	Time (hr)
Run	$0.75x$	x	0.75
Bike	$1.75(x + 12)$	$x + 12$	1.75

The expressions in this column are found by $d = rt$.

Note that consistency in the units of measurement is important. The speed is given in miles per *hour*. Therefore, we want the time to be in hours.

1 hr 45 min = 1.75 hr
45 min = 0.75 hr

$$\left(\begin{array}{c}\text{Total} \\ \text{distance}\end{array}\right) = \left(\begin{array}{c}\text{Distance} \\ \text{running}\end{array}\right) + \left(\begin{array}{c}\text{Distance} \\ \text{riding}\end{array}\right)$$

$$41 = 0.75x + 1.75(x + 12)$$
$$41 = 0.75x + 1.75x + 21$$
$$20 = 2.5x$$
$$8 = x$$

To build an equation, note that the total distance equals the sum of the distance running and the distance riding.

Solve the equation.

Donna's speed running is 8 mph.
Her speed on the bicycle is $8 + 12 = 20$ mph.

Interpret the solution in the context of the problem.

Avoiding Mistakes

Check that the answer is reasonable by verifying that the total distance traveled is 41 mi.

Distance running:
(8 mph)(0.75 hr) = 6 mi

Distance riding:
(20 mph)(1.75 hr) = 35 mi

Total: 6 mi + 35 mi = 41 mi

Answer

7. Rene drove 45 mph in the city and 65 mph on the highway.

Skill Practice 7   Rene drove from Miami to Orlando, a total distance of 240 mi. He drove for 1 hr in city traffic and for 3 hr on the highway. If his average speed on the highway was 20 mph faster than his speed in the city, determine his average speed driving in the city and his average speed driving on the highway.

SECTION 1.3   Practice Exercises

Prerequisite Review

For Exercises R.1–R.6, write each English phrase as an algebraic expression. Then simplify the result.

R.1. Five more than the sum of −7 and −2

R.2. Negative six more than the sum of 4 and −1

R.3. Negative thirteen subtracted from −1

R.4. Negative thirty-one subtracted from −19

R.5. Five times the difference of −9 and 2

R.6. Negative three times the sum of −6 and 10

R.7. **a.** What is 32% of 60?
   b. Write an expression for 32% of x.

R.8. **a.** Find 84% of 240.
   b. Write an expression for 84% of y.

For Exercises R.9–R.12, write a mathematical expression to represent the statement. (See Section 1.1 for review.)

R.9. Twice the sum of x and 10

R.10. Three times the difference of y and 6

R.11. Thirteen less than t

R.12. Negative eleven subtracted from m

Concept Connections

1. If simple interest is earned on $2500 in principal at an annual interest rate of 3.5% for 5 years, how much interest is earned?

2. How much sales tax is paid on a car that costs $32,500 if the tax rate is 6.5%?

3. How much does a real estate agent earn in commission for selling a $260,000 house if her commission rate is 4%?

4. By how much is a blouse discounted if it originally cost $65 and is discounted 25%?

5. Suppose that 15% of a 12-gal solution is bleach and the rest is water. How much is bleach and how much is water?

6. Suppose that 40% of a 15-L solution is acid and the rest is water. How much is acid and how much is water?

7. Suppose that a container has x cups of nuts and that 33% by weight of the mixture is peanuts. Write an expression for the amount of peanuts in the container.

8. Suppose that $x + 2000$ dollars is borrowed for a loan. If the annual simple interest rate is 5.5%, write an expression for the amount of interest owed after 1 year.

9. If $d = rt$, then $r = \dfrac{\square}{\square}$ and $t = \dfrac{\square}{\square}$.

10. If Jerry travels $x + 2$ mph for 6 hr, write an expression for the distance that he travels.

Objective 1: Use Linear Equations in One Variable for Problem Solving

11. The combined cost for a computer and a printer is $738. If the computer costs $240 more than the printer, find the cost of each item. **(See Example 1)**

12. The cost for a hot dog and soft drink at a ballpark is $5.00. If the soft drink costs $2.50 less than the hot dog, find the cost of each item.

13. A 50-ft piece of tubing is to be cut into three pieces for a drip irrigation system. If the longest piece is 4 ft less than twice the shortest piece, and the middle piece is 2 ft more than the shortest piece, find the length of each piece.

14. A plumber cuts a 12-ft copper pipe into three pieces. Two pieces are the same length and the third piece is 1.5 ft longer than each of the other two. Find the length of each piece.

15. Sam and Jenna work for a shipping company packing boxes. Together they prepare 510 boxes for shipment. If Jenna prepared 50 more boxes than Sam, how many boxes did each person prepare?

16. Felix and Carlos are two servers at a restaurant. During one shift, Felix made $19 less than Carlos. If their combined total in tips came to $163, how much in tips did each person make?

For Exercises 17–24, refer to the geometry formulas at the end of the text.

17. George built a rectangular pen for his rabbit such that the length is 7 ft less than twice the width. If the perimeter is 40 ft, what are the dimensions of the pen?

18. The length of a rectangular picture frame is 4 in. less than twice the width. The perimeter is 112 in. Find the length and the width.

19. Antoine wants to put edging in the form of a square around a tree in his front yard. He has enough money to buy 18 ft of edging. Find the dimensions of the square that will use all the edging.

20. A volleyball court is twice as long as it is wide. If the perimeter is 177 ft, find the dimensions of the court.

21. The measures of two angles in a triangle are equal. The third angle measures 2 times the sum of the equal angles. Find the measures of the three angles.

22. The smallest angle in a triangle is one-half the measure of the largest. The middle angle measures 25° less than the largest. Find the measures of the three angles.

23. Two angles are complementary. One angle is 5 times as large as the other angle. Find the measure of each angle.

24. Two angles are supplementary. One angle measures 12° less than 3 times the other. Find the measure of each angle.

Objective 2: Solve Applications Involving Percent and Rates

25. Belle had the choice of taking out a 4-year car loan at 8.5% simple interest or a 5-year car loan at 7.75% simple interest. If she borrows $15,000, how much interest would she pay for each loan? Which option will require less interest?

26. Robert can take out a 3-year loan at 8% simple interest or a 2-year loan at $8\frac{1}{2}$% simple interest. If he borrows $7000, how much interest will he pay for each loan? Which option will require less interest?

27. An account executive earns $600 per month plus a 3% commission on sales. The executive's goal is to earn $2400 this month. How much must she sell to achieve this goal?

28. A salesperson earns $50 a day plus 12% commission on sales over $200. If her daily earnings are $76.88, how much money in merchandise did she sell?

29. J. W. is an artist and sells his pottery each year at a local Renaissance Festival. He keeps track of his sales and the 8.05% sales tax he collects by making notations in a ledger. Every evening he checks his records by counting the total money in his cash drawer. After a day of selling pottery, the cash totaled $1293.38. How much is from the sale of merchandise and how much is sales tax? **(See Example 2)**

30. Wayne County has a sales tax rate of 7%. How much does Mike's used car cost before tax if the total cost of the car *plus tax* is $13,888.60?

31. The price of a swimsuit after a 20% markup is $43.08. What was the price before the markup? **(See Example 3)**

32. The price of a used textbook after a 35% markdown is $29.25. What was the original price?

Objective 3: Solve Applications Involving Simple Interest

33. Tony has a total of $12,500 in two accounts. One account pays 2% simple interest per year and the other pays 5% simple interest. If he earned $370 in interest in the first year, how much did he invest in each account? **(See Example 4)**

34. Lillian had $15,000 invested in two accounts, one paying 9% simple interest and one paying 10% simple interest. How much was invested in each account if the interest after 1 year is $1432?

35. Jason borrowed $18,000 in two loans. One loan charged 11% simple interest and the other charged 6% simple interest. After 1 year, Jason paid a total of $1380 in interest. Find the amount borrowed in each loan.

36. Amanda borrowed $6000 from two sources: her parents and a credit union. Her parents charged 3% simple interest and the credit union charged 8% simple interest. If after 1 year, Amanda paid $255 in interest, how much did she borrow from her parents, and how much did she borrow from the credit union?

37. Donna invested money in two accounts: one paying 4% simple interest and the other paying 3% simple interest. She invested $4000 more in the 4% account than in the 3% account. If she received $720 in interest at the end of 1 year, how much did she invest in each account?

38. Mr. Hall had some money in his bank earning 4.5% simple interest. He had $5000 more deposited in a credit union earning 6% simple interest. If his total interest for 1 year was $1140, how much did he deposit in each account?

39. Fernando invested money in a 3-year CD (certificate of deposit) that returned the equivalent of 4.4% simple interest. He invested $2000 less in an 18-month CD that had a 3% return. If the total amount of interest from these investments was $706.50, determine how much was invested in each CD.

40. Ebony bought a 5-year Treasury note that paid the equivalent of 2.8% simple interest. She invested $5000 more in a 10-year bond earning 3.6% than she did in the Treasury note. If the total amount of interest from these investments was $5300, determine the amount of principal for each investment.

Objective 4: Solve Applications Involving Mixtures

41. Ahmed mixes two plant fertilizers. How much fertilizer with 15% nitrogen should be mixed with 2 oz of fertilizer with 10% nitrogen to produce a fertilizer that is 14% nitrogen? **(See Example 5)**

42. How much 8% saline solution should Kent mix with 80 cc (cubic centimeters) of an 18% saline solution to produce a 12% saline solution?

43. Jacque has 3 L of a 50% antifreeze mixture. How much 75% mixture should be added to get a mixture that is 60% antifreeze?

44. One fruit punch has 40% fruit juice and another is 70% fruit juice. How much of the 40% punch should be mixed with 10 gal of the 70% punch to create a fruit punch that is 45% fruit juice?

45. How many liters of an 18% alcohol solution must be added to a 10% alcohol solution to get 20 L of a 15% alcohol solution?

46. How many milliliters of a 2.5% bleach solution must be mixed with a 10% bleach solution to produce 600 mL of a 5% bleach solution?

47. Ethanol fuel mixtures have "E" numbers that indicate the percentage of ethanol in the mixture by volume. For example, E10 is a mixture of 10% ethanol and 90% gasoline. How much E5 should be mixed with 5000 gal of E10 to make an E9 mixture?

48. A nurse mixes 60 cc of a 50% saline solution with a 10% saline solution to produce a 25% saline solution. How much of the 10% solution should he use?

49. The density and strength of concrete are determined by the ratio of cement and aggregate (aggregate is sand, gravel, or crushed stone). Suppose that a contractor has 480 ft³ of a dry concrete mixture that is 70% sand by volume. How much pure sand must be added to form a new mixture that is 75% sand by volume?

Mrkob/iStock/Getty Images

50. Antifreeze is a compound added to water to reduce the freezing point of a mixture. In extreme cold (less than −35°F), one car manufacturer recommends that a mixture of 65% antifreeze be used. How much 50% antifreeze solution should be drained from a 4-gal tank and replaced with pure antifreeze to produce a 65% antifreeze mixture?

Objective 5: Solve Applications Involving Uniform Motion

51. An airplane travels 60 mph faster from Atlanta to Fort Lauderdale than it travels on the return trip from Fort Lauderdale to Atlanta. If it takes 2 hr from Atlanta to Fort Lauderdale and 2.5 hr for the return trip, determine the speed of each trip. **(See Example 6)**

52. A woman can hike 1 mph faster down a trail to Archuletta Lake than she can on the return trip uphill. It takes her 3 hr to get to the lake and 6 hr to return. What is her speed hiking down to the lake?

53. Two cars are 192 mi apart and travel toward each other on the same road. They meet in 2 hr. One car travels 4 mph faster than the other. What is the average speed of each car?

54. Two cars are 190 mi apart and travel toward each other along the same road. They meet in 2 hr. One car travels 5 mph slower than the other car. What is the average speed of each car?

55. Two passengers leave the airport at Kansas City, Missouri. One flies to Los Angeles, California, in 3.4 hr and the other flies in the opposite direction to New York City in 2.4 hr. With prevailing westerly winds, the speed of the plane to New York City is 60 mph faster than the speed of the plane to Los Angeles. If the total distance traveled by both planes is 2464 mi, determine the average speed of each plane. **(See Example 7)**

56. Two planes leave from Atlanta, Georgia. One makes a 5.2-hr flight to Seattle, Washington, and the other makes a 2.5-hr flight to Boston, Massachusetts. The plane to Boston averages 44 mph slower than the plane to Seattle. If the total distance traveled by both planes is 3124 mi, determine the average speed of each plane.

57. Darren drives to school in rush hour traffic and averages 32 mph. He returns home in mid-afternoon when there is less traffic and averages 48 mph. What is the distance between his home and school if the total traveling time is 1 hr 15 min?

58. Peggy competes in a biathlon by running and bicycling around a large loop through a city. She runs the loop one time and bicycles the loop five times. She can run 8 mph and she can ride 16 mph. If the total time it takes her to complete the race is 1 hr 45 min, determine the distance of the loop.

Mixed Exercises

59. Seismographs can record two types of wave energy (P waves and S waves) that travel through the Earth after an earthquake. Traveling through granite, P waves travel approximately 5 km/sec and S waves travel approximately 3 km/sec. If a geologist working at a seismic station measures a time difference of 40 sec between an earthquake's P waves and S waves, how far from the epicenter of the earthquake is the station?

60. Suppose that a shallow earthquake occurs in which the P waves travel 8 km/sec and the S waves travel 4.8 km/sec. If a seismologist measures a time difference of 20 sec between the arrival of the P waves and the S waves, how far is the seismologist from the epicenter of the earthquake?

61. Suppose that a merchant buys a patio set from the wholesaler for $180. At what price should the merchant mark the patio set so that it may be offered at a discount of 25% but still give the merchant a 40% profit on his $180 investment?

62. Suppose that a bookstore buys a textbook from the publisher for $80. At what price should the bookstore mark the textbook so that it may be offered at a discount of 10% but still give the bookstore a 35% profit on the $80 investment?

63. Henri needs to have a toilet repaired in his house. The cost of the new plumbing fixtures is $110 and labor is $60/hr.

    a. Write a model that represents the cost of the repair C (in $) in terms of the number of hours of labor x.

    b. After how many hours of labor would the cost of the repair job equal the cost of a new toilet of $350?

64. After a hurricane, repairs to a roof will cost $2400 for materials and $80/hr in labor.

    a. Write a model that represents the cost of the repair C (in $) in terms of the number of hours of labor x.

    b. If an estimate for a new roof is $5520, after how many hours of labor would the cost to repair the roof equal the cost of a new roof?

65. A tank contains 40 L of a mixture of plant fertilizer and water in which 20% of the mixture is fertilizer. How much of the mixture should be drained and replaced by an equal amount of water to dilute the mixture to 15% fertilizer?

66. How much water must be evaporated from 200 mL of a 5% salt solution to produce a 25% salt solution?

67. The perimeter of a rectangular lot of land is 440 ft. This includes an easement of x feet of uniform width inside the lot on which no building can be done. If the buildable area is 128 ft by 60 ft, determine the width of the easement.

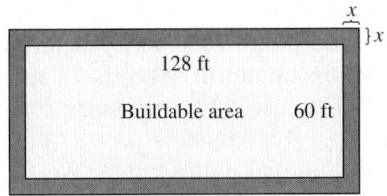

68. The Arthur Ashe Stadium tennis court is center court to the U.S. Open tennis tournament. The dimensions of the court are 78 ft by 36 ft, with a uniform border of x feet around the outside for additional play area. If the perimeter of the entire play area is 396 ft, determine the value of x.

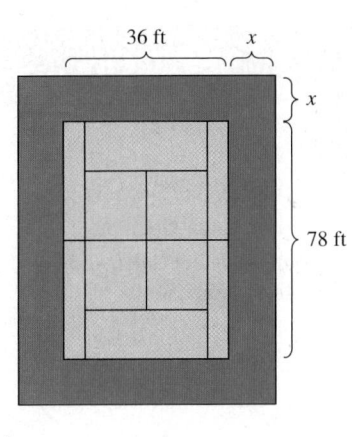

69. A contractor must tile a rectangular kitchen that is 4 ft longer than it is wide, and the perimeter of the kitchen is 48 ft.

   a. Find the dimensions of the kitchen.

   b. How many square feet of tile should be ordered if the contractor adds an additional 10% to account for waste?

   c. Determine the total cost if the tile costs $12/ft^2 and sales tax is 6%.

70. Max and Molly plan to put down all-weather carpeting on their porch. The length of the porch is 2 ft longer than twice the width, and the perimeter is 64 ft.

   a. Find the dimensions of the porch.

   b. How many square feet of carpeting should they buy if they add an additional 10% for waste?

   c. Determine the total cost if the carpeting costs $5.85/ft^2 and sales tax is 7.5%.

71. Aliyah earned an $8000 bonus from her sales job for exceeding her sales goals. After paying taxes at a 28% rate, she invested the remaining money in two stocks. One stock returned the equivalent of 11% simple interest after 1 year, and the other returned 5% at the end of 1 year. If her investments returned $453.60 (excluding commissions), how much did she invest in each stock?

72. Caitlin invested money in two mutual funds—a stock fund and a balanced fund. She invested twice as much in the stock fund as in the balanced fund. At the end of 1 year, the stock fund earned the equivalent of 17% simple interest and the balanced fund earned 3.5%. If her total gain was $1125, determine how much she invested in each fund.

Expanding Your Skills

73. The sum of the digits of a two-digit number is 14. If the digits are reversed, the new number is 18 more than the original number. Determine the original number.

74. The sum of the digits of a two-digit number is 9. If the digits are reversed, the new number is 45 less than the original number. Determine the original number.

Consider a seesaw with two children of masses m_1 and m_2 on either side. Suppose that the position of the fulcrum (pivot point) is labeled as the origin, $x = 0$. Further suppose that the position of each child relative to the origin is x_1 and x_2, respectively. The seesaw will be in equilibrium if $m_1 x_1 + m_2 x_2 = 0$. Use this equation for Exercises 75–78.

75. Find x_2 so that the system of masses is in equilibrium.

   $m_1 = 30$ kg, $x_1 = -1.2$ m and $m_2 = 20$ kg, $x_2 = $?

76. Find x_1 so that the system of masses is in equilibrium.

   $m_1 = 64$ kg, $x_1 = $? and $m_2 = 80$ kg, $x_2 = 2$ m

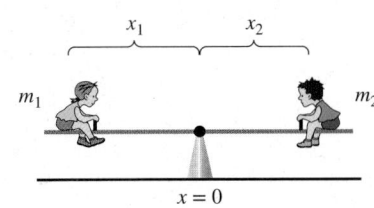

77. Find the missing mass so that the system is in equilibrium. (*Hint*: Recall that positions to the left of 0 on the number line are negative.)

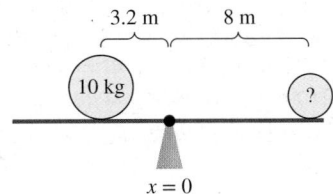

78. Find the missing mass so that the system is in equilibrium.

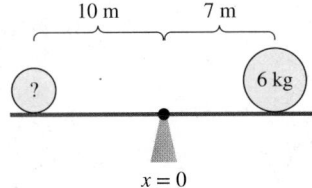

OBJECTIVES

1. Use Inequality Symbols
2. Write Interval Notation
3. Determine the Union and Intersection of Sets

1. Use Inequality Symbols

Often in our day-to-day lives we encounter situations in which we need to compare numerical values. For example, one brand of laundry detergent costs more than another. Or the wind speed of a tropical storm exceeds 74 mph, thus elevating the storm to hurricane status. In each case, it is necessary for us to compare real numbers.

All real numbers can be located on the real number line. We say that a is less than b (written symbolically as $a < b$) if a lies to the left of b on the number line. This is equivalent to saying that b is greater than a (written symbolically as $b > a$) because b lies to the right of a.

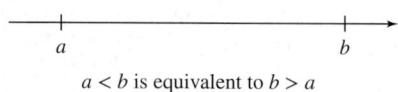

$a < b$ is equivalent to $b > a$

In Table 1-3, we summarize algebraic symbols used to compare two real numbers.

Table 1-3 **Summary of Inequality Symbols and Their Meanings**

Inequality	Verbal Interpretation	Other Implied Meanings	Numerical Examples
$a < b$	a is less than b	b exceeds a b is greater than a	$5 < 7$
$a > b$	a is greater than b	a exceeds b b is less than a	$-3 > -6$
$a \leq b$	a is less than or equal to b	a is at most b a is no more than b	$4 \leq 5$ $5 \leq 5$
$a \geq b$	a is greater than or equal to b	a is no less than b a is at least b	$9 \geq 8$ $9 \geq 9$
$a = b$	a is equal to b		$-4.3 = -4.3$
$a \neq b$	a is not equal to b		$-6 \neq -7$
$a \approx b$	a is approximately equal to b		$-12.99 \approx -13$
$a < x < b$*	x is between a and b	x is greater than a *and* less than b	$0 < 1 < 3$

*Note: Other forms of this inequality include the "$\leq$" symbol that indicates that either a or b is included in the inequality: $a \leq x \leq b$, $a < x \leq b$, and $a \leq x < b$

From Table 1-3, phrases such as *at least, at most, no more than, no less than,* and *between* can be translated into mathematical terms by using inequality signs. In Example 1, we practice translating from an English phrase to an inequality.

EXAMPLE 1 **Translating English Phrases to Inequalities**

The intensity of a hurricane is often defined according to its maximum sustained winds. Translate the italicized phrases into mathematical inequalities.

a. A tropical storm is updated to hurricane status if the sustained wind speed *w is at least 74 mph.*

b. Hurricanes are categorized according to intensity by the Saffir-Simpson Wind Scale. On a scale of 1 to 5, a category 5 hurricane has the most destructive winds. A category 5 hurricane has sustained winds *w exceeding 155 mph.*

c. A category 4 hurricane has sustained winds *w of at least 131 mph but no more than 155 mph.*

Solution:

a. $w \geq 74$ w is *at least* 74 mph. This means that 74 is the lower bound. The winds are 74 or more miles per hour.

b. $w > 155$ w exceeds (is greater than) 155 mph.

c. $131 \leq w \leq 155$ The statements *at least 131 mph* but *no more than 155 mph* represents *two* simultaneous conditions on w. The value of w is greater than or equal to 131, but simultaneously less than or equal to 155. These are wind speeds between 131 and 155, inclusive.

> **TIP** The word "inclusive" is used to describe an interval in which the endpoints are included.

Skill Practice 1   Translate the italicized phrases into mathematical inequalities.

  a. The gas mileage m for an economy car *is at least 30 mpg.*

  b. The gas mileage m for a motorcycle *is more than 45 mpg.*

  c. The gas mileage m for a truck *is at least 10 mpg but no more than 20 mpg.*

In Example 2, we directly compare two real numbers and locate their relative positions on the real number line.

EXAMPLE 2   **Ordering Real Numbers**

Fill in the blank with the appropriate inequality sign: $<$ or $>$

  a. -2 _____ -5 **b.** $\dfrac{4}{7}$ _____ $\dfrac{3}{5}$ **c.** -1.3 _____ $-1.\overline{3}$

Solution:

  a. $-2 \underline{>} -5$

  b. To compare $\frac{4}{7}$ and $\frac{3}{5}$, write the fractions as equivalent fractions with a common denominator. The LCD is 35.

$$\frac{4}{7} \cdot \frac{5}{5} = \frac{20}{35} \quad \text{and} \quad \frac{3}{5} \cdot \frac{7}{7} = \frac{21}{35}$$

Because $\dfrac{20}{35} < \dfrac{21}{35}$, then $\dfrac{4}{7} \underline{\phantom{<}<\phantom{<}} \dfrac{3}{5}$

  c. $-1.3 \underline{\geq} -1.33333\ldots$

Skill Practice 2   Fill in the blanks with the appropriate sign, $<$ or $>$.

  a. 2 _____ -12 **b.** $\dfrac{1}{4}$ _____ $\dfrac{2}{9}$ **c.** $-7.\overline{2}$ _____ -7.2

---

2. Write Interval Notation

In Chapter R, we learned that a **set** is a collection of elements. We can express a set by listing its elements within set braces or by using set-builder notation. **Set-builder notation** uses a description to define the elements of a set. For example, suppose that set L is defined as:

$$L = \{x \mid x \text{ is an integer between } -4 \text{ and } 3, \text{ inclusive}\}$$

Answers

1. a. $m \geq 30$ **b.** $m > 45$

  c. $10 \leq m \leq 20$

2. a. $>$ **b.** $>$ **c.** $<$

The word *inclusive* means that the "endpoints" −4 and 3 are *included*. Therefore, *L* can be written in roster form as {−4, −3, −2, −1, 0, 1, 2, 3}. *L* can also be visualized on a number line (Figure 1-2).

$$L = \{-4, -3, -2, -1, 0, 1, 2, 3\}$$

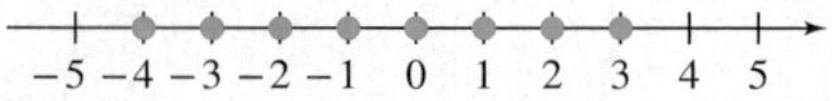

Figure 1-2

Set *L* is called a **finite set** because it has a finite number of elements. However, some sets, such as the set of whole numbers, the set of integers, and the set of real numbers, are **infinite sets**. They have infinitely many elements that cannot all be listed. Therefore, how can we visualize the elements of these sets?

For example, suppose that *x* is a real number and that set *M* is defined as

$$M = \{x \mid x \geq -2\}.$$

Clearly, we cannot list every element of set *M*, but we can visualize the elements on a number line (Figure 1-3). Notice that every point −2 or greater is shaded on the number line. To indicate that the "endpoint" −2 is included in the set, we use a square bracket [. If the endpoint were *not* included, then we would use a parenthesis (at −2.

M:

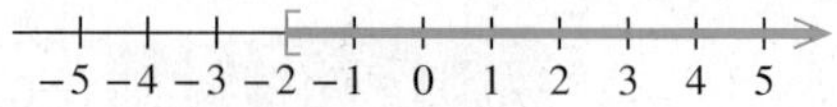

Figure 1-3

Point of Interest

The infinity symbol ∞ is called a lemniscate from the Latin *lemniscus* meaning "ribbon." English mathematician John Wallis is credited with introducing the symbol in the seventeenth century. The symbols −∞ and ∞ are not themselves real numbers, but instead refer to quantities without bound or end.

Note that set *M* defines an interval on the real number line. We can also use **interval notation** to define *M*. To understand interval notation, first consider the real number line, which extends infinitely far to the left and right. The symbol ∞ is used to represent infinity. The symbol −∞ is used to represent negative infinity.

$$-\infty \longleftarrow \quad \longrightarrow \infty$$
$$0$$

Writing Interval Notation

1. To express an interval on the number line using interval notation, first sketch the graph.
2. Write the left "endpoint" of the interval, followed by a comma, followed by the right "endpoint."
3. At the "endpoints" place either a parenthesis, (or), to denote that the endpoint *is not* included in the interval, or a square bracket, [or], to denote that the endpoint *is* included in the interval.

Note: Parentheses are always used with ∞ and −∞.

To write $M = \{x \mid x \geq -2\}$ in interval notation, we have:

Graph of *M*:

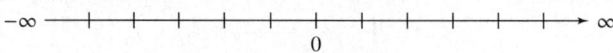

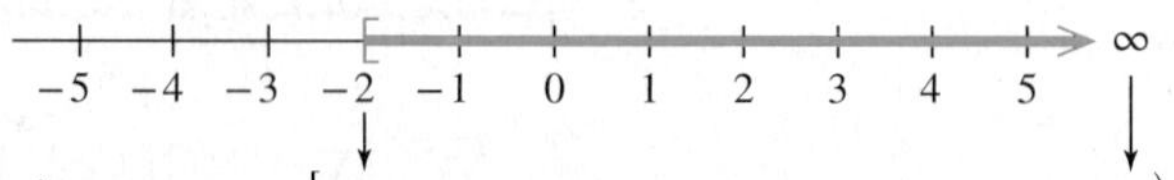

Interval notation: [−2, ∞)   [−2,   ∞)

For comparison, we show the graph and interval notation for $N = \{x \mid x < -2\}$.

Graph of *N*:

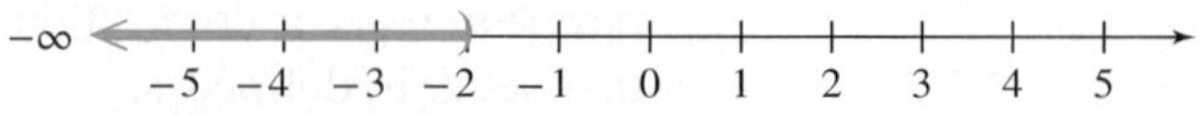

Interval notation: (−∞, −2)

In Table 1-4, we summarize the interval notation and graphs for various sets.

Table 1-4   Summary of Interval Notation and Set-Builder Notation

Let a, b, and x represent real numbers.

Set-Builder Notation	Verbal Interpretation	Graph	Interval Notation
$\{x \mid x > a\}$	the set of real numbers greater than a		(a, ∞)
$\{x \mid x \geq a\}$	the set of real numbers greater than or equal to a		$[a, \infty)$
$\{x \mid x < b\}$	the set of real numbers less than b		$(-\infty, b)$
$\{x \mid x \leq b\}$	the set of real numbers less than or equal to b		$(-\infty, b]$
$\{x \mid a < x < b\}$	the set of real numbers between a and b		(a, b)
$\{x \mid a \leq x < b\}$	the set of real numbers greater than or equal to a and less than b		$[a, b)$
$\{x \mid a < x \leq b\}$	the set of real numbers greater than a and less than or equal to b		$(a, b]$
$\{x \mid a \leq x \leq b\}$	the set of real numbers between a and b, inclusive		$[a, b]$
$\{x \mid x \text{ is a real number}\}$ $\mathbb{R}$	the set of all real numbers		$(-\infty, \infty)$

TIP On a graph, an alternative to using parentheses and brackets to represent the endpoints of an interval is to use an open or closed dot. For example, $\{x \mid a \leq x < b\}$ would be represented as follows.

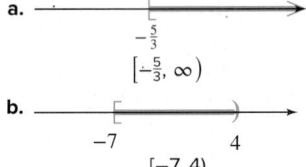

EXAMPLE 3   Expressing Sets by Using Interval Notation

Graph each set on the number line, and express the set in interval notation.

a. $\left\{z \mid z \leq -\frac{3}{2}\right\}$ **b.** $\{x \mid -4 < x \leq 2\}$

Solution:

a. Set-builder notation: $\left\{z \mid z \leq -\frac{3}{2}\right\}$

Graph:

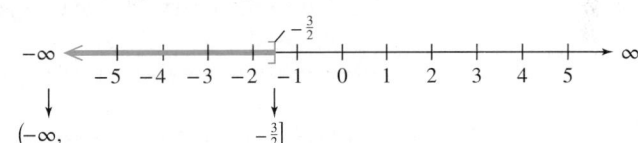

Interval notation: $\left(-\infty, -\frac{3}{2}\right]$

The graph of $\left\{z \mid z \leq -\frac{3}{2}\right\}$ extends infinitely far to the left. Interval notation is always written from left to right. Therefore, $-\infty$ is written first, followed by a comma, and then followed by the right-hand endpoint $-\frac{3}{2}$.

b. The inequality $-4 < x \leq 2$ means that x is greater than -4 and also less than or equal to 2. More concisely, we can say that x represents the real numbers *between* -4 and 2, including the endpoint, 2.

Set-builder notation: $\{x \mid -4 < x \leq 2\}$

Graph:

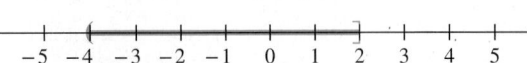

Interval notation: $(-4, 2]$

Answers

3. a.

$-\frac{5}{3}$

$\left[-\frac{5}{3}, \infty\right)$

b.

-7 4

$[-7, 4)$

Skill Practice 3   Graph the set on the number line, and express the set in interval notation.

a. $\left\{w \mid w \geq -\frac{5}{3}\right\}$ **b.** $\{y \mid -7 \leq y < 4\}$

Intervals on the real number line can be expressed in set-builder notation, interval notation, and as graphs.

EXAMPLE 4   **Expressing Sets in Interval Notation and Set-Builder Notation**

Complete the table.

Graph	Interval Notation	Set-Builder Notation
$\xleftarrow{\begin{array}{ccccccccccc} & & & & & &] & & & & \\ -5 & -4 & -3 & -2 & -1 & 0 & 1 & 2 & 3 & 4 & 5 \end{array}}$		
	$\left(\frac{7}{2}, \infty\right)$	
		$\{y \mid -4 \leq y < 2.3\}$

Solution:

Graph	Interval Notation	Set-Builder Notation	Comments
$\xleftarrow{\begin{array}{ccccccccccc} & & & & & &] & & & & \\ -5 & -4 & -3 & -2 & -1 & 0 & 1 & 2 & 3 & 4 & 5 \end{array}}$	$(-\infty, 2]$	$\{x \mid x \leq 2\}$	The bracket at 2 indicates that 2 is included in the set.
$\begin{array}{ccccccccccc} & & & & & & & & & (& \\ -5 & -4 & -3 & -2 & -1 & 0 & 1 & 2 & 3 & 4 & 5 \end{array}\rightarrow$	$\left(\frac{7}{2}, \infty\right)$	$\{x \mid x > \frac{7}{2}\}$	The parenthesis at $\frac{7}{2} = 3.5$ indicates that $\frac{7}{2}$ is *not* included in the set.
$\begin{array}{ccccccccccc} & [& & & & &) & & & & \\ -5 & -4 & -3 & -2 & -1 & 0 & 1 & 2 & 3 & 4 & 5 \end{array}\rightarrow$	$[-4, 2.3)$	$\{y \mid -4 \leq y < 2.3\}$	The set includes the real numbers between −4 and 2.3, including the endpoint −4.

Skill Practice 4

a. Write the set represented by the graph in interval notation and set-builder notation.

$\begin{array}{ccccccccccc} & & (& & & & & & & & \\ -5 & -4 & -3 & -2 & -1 & 0 & 1 & 2 & 3 & 4 & 5 \end{array}\rightarrow$

b. Given the interval, $\left(-\infty, -\frac{4}{3}\right]$, graph the set and write the set-builder notation.

c. Given the set, $\{x \mid 1.6 < x \leq 5\}$, graph the set and write the interval notation.

---

3. Determine the Union and Intersection of Sets

Two or more sets can be combined by the operations of union and intersection.

Union and Intersection of Sets

The **union** of sets A and B, denoted $A \cup B$, is the set of elements that belong to set A or to set B or to both sets A and B.

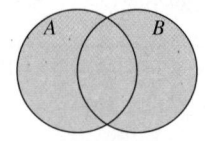

$A \cup B$
A union B
The elements in A or B or both

The **intersection** of sets A and B, denoted $A \cap B$, is the set of elements common to both set A and set B.

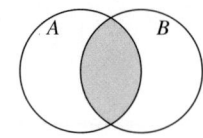

$A \cap B$
A intersection B
The elements common to A and B

| EXAMPLE 5 | **Finding the Intersection and Union of Sets** |

Given $A = \{1, 3, 5, 7, 9, 11\}$ and $B = \{1, 2, 5, 7, 12\}$, find

a. $A \cap B$ **b.** $A \cup B$

Solution:

a. $\{1, 3, 5, 7, 9, 11\} \cap \{1, 2, 5, 7, 12\} = \{1, 5, 7\}$ The intersection of sets A and B is the set of elements common to both A and B. The common elements are 1, 5, and 7.

b. $\{1, 3, 5, 7, 9, 11\} \cup \{1, 2, 5, 7, 12\}$
$ = \{1, 2, 3, 5, 7, 9, 11, 12\}$ The union of sets A and B consists of all the elements of A along with the elements of B. Notice that the elements common to both set A and set B are not listed twice.

Skill Practice 5   Given $C = \{a, e, i, o, u\}$ and $D = \{b, c, e, f, h, i, m\}$, find

a. $C \cup D$ **b.** $C \cap D$

A useful tool to visualize the relationship between sets is the Venn diagram, named after British mathematician John Venn (1834–1923). From Example 5, a circle is used to denote each set A and B. The elements in each set are placed in the diagram according to whether they fall exclusively to one set or in the overlap. Notice that the elements 1, 5, and 7 appear in the region of overlap of A and B.

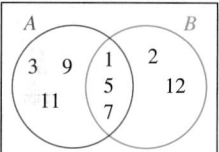

There are four basic relationships between two sets A and B.

Case 1: A and B share no common elements. In this case, A and B are said to be **disjoint** *or* **mutually exclusive.**

Case 2: The sets have some (but not all) elements in common.

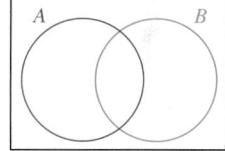

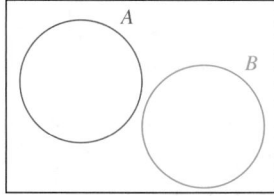

Case 3: The sets are **equal sets**. That is, the sets have exactly the same elements.

Case 4: One set is a **proper subset** of the other set. This means that one set is a subset of the other, but the sets are not equal. In the illustration, set A is contained in set B but is not equal to B.

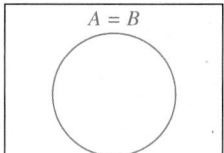

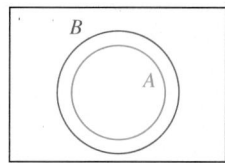

Answers
5. a. $\{a, b, c, e, f, h, i, m, o, u\}$
    b. $\{e, i\}$

EXAMPLE 6   Finding the Intersection and Union of Sets

Given $C = \{x \mid x < 4\}$ and $D = \{x \mid x \le -2\}$, find the intersection and union. Write the result in set-builder notation and interval notation.

a. $C \cap D$ **b.** $C \cup D$

Solution:

a.

Graph each individual set.

The intersection is the interval of overlap.

$C \cap D = \{x \mid x \le -2\}$

Interval notation: $(-\infty, -2]$

b.

Graph each individual set.

The union consists of all elements from *both* sets.

$C \cup D = \{x \mid x < 4\}$

Interval notation: $(-\infty, 4)$

> **TIP** From the graphs of sets C and D in Example 6, note that set D is contained in set C. Therefore, D is a proper subset of C.

Skill Practice 6   Given $A = \{x \mid x \ge 3\}$ and $B = \{x \mid x > -1\}$, find the intersection and union. Write the result in set-builder notation and interval notation.

a. $A \cap B$ **b.** $A \cup B$

EXAMPLE 7   Finding the Intersection and Union of Sets

Given $A = \{x \mid x > 0\}$ and $B = \{x \mid x \le -3\}$, find the intersection and union. Write the result in set-builder notation and interval notation if possible.

a. $A \cap B$ **b.** $A \cup B$

Solution:

a.

Graph each individual set.

The two sets do not overlap. Therefore, the intersection is the empty set and does not define an interval on the real number line.

$A \cap B = \{\ \}$

b.

Graph each individual set.

$A \cup B = \{x \mid x \le -3 \text{ or } x > 0\}$

The union consists of all elements from A or B or *both A and B*.

Interval notation: $(-\infty, -3] \cup (0, \infty)$

> **TIP** From the graphs of A and B in Example 7, we see that the sets are disjoint. The intersection is the empty set. Furthermore, the empty set cannot be represented by interval notation because there is no interval or "range" of values on the number line.

Skill Practice 7   Given $E = \{x \mid x \ge -5\}$ and $F = \{x \mid x \le -6\}$, find the intersection and union. Write the result in set-builder notation and interval notation if possible.

a. $E \cap F$ **b.** $E \cup F$

Answers

6. a. $\{x \mid x \ge 3\}$; $[3, \infty)$
   b. $\{x \mid x > -1\}$; $(-1, \infty)$
7. a. $\{\ \}$ **b.** $\{x \mid x \le -6 \text{ or } x \ge -5\}$;
   $(-\infty, -6] \cup [5, \infty)$

In Example 8, we find the intersection and union of two sets defined using interval notation.

EXAMPLE 8   Finding the Intersection and Union of Sets

Find the intersection and union of the intervals. Write the answers in interval notation.

a. $(-\infty, 4) \cap [1, \infty)$ **b.** $(-\infty, 4) \cup [1, \infty)$

Solution:

a.

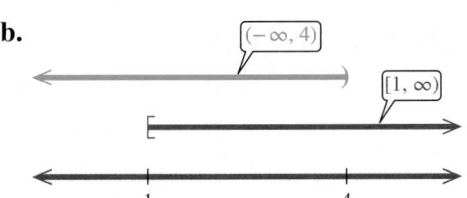

Graph each individual set.

The intersection is the interval of overlap.

The interval in purple is the intersection.

The intersection is $[1, 4)$.

> **TIP** In Example 8(b), interval $(-\infty, 4)$ extends infinitely far to the left. The interval $[1, \infty)$ extends infinitely far to the right. Furthermore, the individual intervals overlap so there are no real numbers "left out." Therefore, the union of the two intervals is all real numbers, $\mathbb{R}$.

b.

Graph each individual set.

The union consists of all elements in *both* sets, which in this case is the entire real number line.

The union is $(-\infty, \infty)$.

Skill Practice 8   Find the intersection and union of the intervals. Write the answers in interval notation.

a. $(-\infty, -3) \cap [-7, \infty)$ **b.** $(-\infty, -3) \cup [-7, \infty)$

Answers

8. a. $[-7, -3)$ **b.** $(-\infty, \infty)$

SECTION 1.4   Practice Exercises

Prerequisite Review

For Exercises R.1–R.4, let $A = \left\{-2, \frac{3}{5}, \sqrt{10}, 5\pi, 18, 1, 3.\overline{5}, 9.6\right\}$. (See Section R.2 for review.)

R.1. List the whole numbers in set A.

R.2. List the integers in set A.

R.3. List the rational numbers in set A.

R.4. List the irrational numbers in set A.

For Exercises R.5–R.14, refer to set B where $B = \{x \mid x \text{ is an integer between } -2 \text{ and } 10\}$. Recall from Section R.2 that the symbol $\in$ means "is an element of" and the symbol $\notin$ means "is not an element of." Answer true or false.

R.5. $-1 \in B$

R.6. $3 \in B$

R.7. $\dfrac{7}{3} \in B$

R.8. $-\dfrac{21}{4} \in B$

R.9. $\pi \in B$

R.10. $\sqrt{9} \in B$

R.11. $-2 \in B$

R.12. $10 \in B$

R.13. $4.1 \notin B$

R.14. $-1.1 \notin B$

Concept Connections

1. The statement $c \geq d$ is read as "_____."

2. The symbol ∞ represents _____ and $-\infty$ represents _____.

3. The set of real numbers greater than 5 can be written in set-builder notation as _____ and in _____ notation as $(5, \infty)$.

4. The interval $(-2, 5]$ (includes/excludes) the value -2 and (includes/excludes) the value 5.

5. When expressing interval notation, use a (parenthesis/bracket) with infinity.

6. The _____ of two sets A and B, denoted by _____, is the set of elements that belong to A or B or both A and B.

7. The _____ of two sets A and B, denoted by _____, is the set of elements common to both A and B.

8. Use set-builder notation to write the set of real numbers greater than or equal to a and less than b.

9. Use interval notation to write the set of real numbers, $\mathbb{R}$.

10. Which statement describes the relationship between sets A and B?
    a. A is a proper subset of B.
    b. A and B are mutually exclusive.

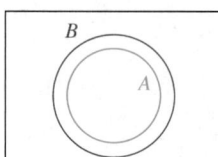

Objective 1: Use Inequality Symbols

For Exercises 11–26, write each statement as an inequality. **(See Example 1)**

11. a is at least 5.

12. b is at most -6.

13. $3c$ is no more than 9.

14. $8d$ is no less than 16.

15. The quantity $(m + 4)$ exceeds 70.

16. The quantity $(n - 7)$ is approximately equal to 4.

17. The age, a, to get in to see a certain movie is at least 18 years old.

18. Winston is a cat that was picked up at the Humane Society. His age, a, at the time was no more than 2 years.

19. The cost, c, to have dinner at Jack's Café is at most $25.

20. The number of hours, h, that Katlyn spent studying was no less than 40.

21. The wind speed, s, for an F-5 tornado is no less than 261 mph.

22. The high temperature, t, for a certain December day in Albany is at most $26°F$.

23. After a summer drought, the total rainfall, r, for June, July, and August was no more than 4.5 in.

24. Jessica works for a networking firm. Her salary, s, is at least $85,000 per year.

25. To play in a certain division of a tennis tournament, a player's age, a, must be at least 18 years but not more than 25 years.

26. The average age, a, of students at Central Community College is estimated to be between 25 years and 29 years.

For Exercises 27–34, fill in the blanks with the appropriate symbol: $<$ or $>$. **(See Example 2)**

27. -9 _____ -1

28. 0 _____ -6

29. $0.1\overline{5}$ _____ 0.15

30. $-2.\overline{5}$ _____ -2.5

31. $\dfrac{5}{3}$ _____ $\dfrac{10}{7}$

32. $-\dfrac{21}{5}$ _____ $-\dfrac{17}{4}$

33. $-\dfrac{5}{8}$ _____ $-\dfrac{1}{8}$

34. $-\dfrac{13}{15}$ _____ $-\dfrac{17}{12}$

For Exercises 35–38, determine whether the statement is true or false.

35. $3.14 < \pi$

36. $-7 < -\sqrt{7}$

37. $6.7 \geq 6.7$

38. $-2.1 \leq -2.1$

Objective 2: Write Interval Notation

For Exercises 39–44, write the interval notation and set-builder notation for each given graph. **(See Examples 3–4)**

39.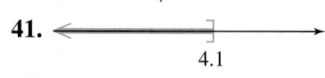
-7

40.
2

41.
4.1

42.
-2.93

43.
-6 0

44.
2 8

For Exercises 45–50, graph the given set and write the corresponding interval notation. (**See Examples 3–4**)

45. $\{x \mid x \le 6\}$

46. $\{x \mid x < -4\}$

47. $\left\{ x \mid -\dfrac{7}{6} < x \le \dfrac{1}{3} \right\}$

48. $\left\{ x \mid -\dfrac{4}{3} \le x < \dfrac{7}{4} \right\}$

49. $\{x \mid 4 < x\}$

50. $\{x \mid -3 \le x\}$

For Exercises 51–56, interval notation is given for several sets of real numbers. Graph the set and write the corresponding set-builder notation. (**See Examples 3–4**)

51. $(-3, 7]$

52. $[-4, -1)$

53. $(-\infty, 6.7]$

54. $(-\infty, -3.2)$

55. $\left[-\dfrac{3}{5}, \infty \right)$

56. $\left(\dfrac{7}{8}, \infty \right)$

For Exercises 57–66, graph the set of numbers and write the set in interval notation.

57. All real numbers less than -3.

58. All real numbers greater than 2.34.

59. All real numbers greater than $\frac{5}{2}$.

60. All real numbers less than $\frac{4}{7}$.

61. All real numbers not less than 2.

62. All real numbers no more than 5.

63. All real numbers between -4 and 4.

64. All real numbers between -7 and -1.

65. All real numbers between -3 and 0, inclusive.

66. All real numbers between -1 and 6, inclusive.

Objective 3: Determine the Union and Intersection of Sets

For Exercises 67–70, refer to sets A, B, C, X, Y, and Z and find the union or intersection of sets as indicated. (**See Example 5**)

$A = \{0, 4, 8, 12\}$, $B = \{0, 3, 6, 9, 12\}$, $C = \{-2, 4, 8\}$
$X = \{1, 2, 3, 4, 5\}$, $Y = \{1, 2, 3\}$, $Z = \{6, 7, 8\}$

67. a. $A \cup B$ **b.** $A \cap B$ **c.** $A \cup C$
   d. $A \cap C$ **e.** $B \cup C$ **f.** $B \cap C$

68. a. $X \cup Z$ **b.** $Y \cup Z$ **c.** $Y \cap Z$
   d. $X \cup Y$ **e.** $X \cap Y$ **f.** $X \cap Z$

69. a. $A \cup X$ **b.** $A \cap Z$ **c.** $C \cap Y$
   d. $B \cap Y$ **e.** $C \cup Z$ **f.** $A \cup Z$

70. a. $A \cap X$ **b.** $B \cup Z$ **c.** $C \cup Y$
   d. $B \cap Z$ **e.** $C \cap Z$ **f.** $A \cup Y$

71. Refer to sets C, D, and F and find the union or intersection of sets as indicated. Write the answers in set notation. (**See Examples 6–7**)

$C = \{x \mid x < 9\}$, $D = \{x \mid x \ge -1\}$, $F = \{x \mid x < -8\}$

   a. $C \cup D$ **b.** $C \cap D$ **c.** $C \cup F$
   d. $C \cap F$ **e.** $D \cup F$ **f.** $D \cap F$

72. Refer to sets M, N, and P and find the union or intersection of sets as indicated. Write the answers in set notation.

$M = \{y \mid y \ge -3\}$, $N = \{y \mid y \ge 5\}$, $P = \{y \mid y < 0\}$

   a. $M \cup N$ **b.** $M \cap N$ **c.** $M \cup P$
   d. $M \cap P$ **e.** $N \cup P$ **f.** $N \cap P$

For Exercises 73–84, refer to sets A, B, C, and D and find the union or intersection of sets as indicated. Write the answers in interval notation. **(See Examples 6–7)**

$$A = \{x \mid x < -4\}, \quad B = \{x \mid x > 2\}, \quad C = \{x \mid x \geq -7\}, \quad D = \{x \mid 0 \leq x < 5\}$$

73. $A \cap C$ **74.** $B \cap C$ **75.** $A \cup B$

76. $A \cup D$ **77.** $A \cap B$ **78.** $A \cap D$

79. $B \cup C$ **80.** $B \cup D$ **81.** $C \cap D$

82. $B \cap D$ **83.** $C \cup D$ **84.** $A \cup C$

For Exercises 85–94, find the union or intersection of the given intervals. Write the answers in interval notation. **(See Example 8)**

85. a. $(-\infty, 4) \cup (-2, 1]$ **b.** $(-\infty, 4) \cap (-2, 1]$

86. a. $[0, 5) \cup [-1, \infty)$ **b.** $[0, 5) \cap [-1, \infty)$

87. a. $(-\infty, 5) \cup [3, \infty)$ **b.** $(-\infty, 5) \cap [3, \infty)$

88. a. $(-\infty, -1] \cup [-4, \infty)$ **b.** $(-\infty, -1] \cap [-4, \infty)$

89. a. $(-2, 5) \cap [-1, \infty)$ **b.** $(-2, 5) \cup [-1, \infty)$

90. a. $(-\infty, 4) \cap [-1, 5)$ **b.** $(-\infty, 4) \cup [-1, 5)$

91. a. $\left(-\dfrac{5}{2}, 3\right) \cap \left(-1, \dfrac{9}{2}\right)$ **b.** $\left(-\dfrac{5}{2}, 3\right) \cup \left(-1, \dfrac{9}{2}\right)$

92. a. $(-3.4, 1.6) \cap (-2.2, 4.1)$ **b.** $(-3.4, 1.6) \cup (-2.2, 4.1)$

93. a. $(-4, 5] \cap (0, 2]$ **b.** $(-4, 5] \cup (0, 2]$

94. a. $[-1, 5) \cap (0, 3)$ **b.** $[-1, 5) \cup (0, 3)$

Mixed Exercises

For Exercises 95–102, write an expression in words that describes the set of numbers given by each interval. (Answers may vary.)

95. $(-\infty, -4)$ **96.** $[2, \infty)$

97. $(-2, 7]$ **98.** $(-3.9, 0)$

99. $[-180, 90]$ **100.** $(3.2, \infty)$

101. $(-\infty, \infty)$ **102.** $(-\infty, -1]$

The following chart defines the ranges for normal blood pressure, high normal blood pressure, and high blood pressure (*hypertension*). All values are measured in millimeters of mercury (mm Hg). (Source: American Heart Association.)

Normal	Systolic less than 130	Diastolic less than 85
High normal	Systolic 130–139, inclusive	Diastolic 85–89, inclusive
Hypertension	Systolic 140 or greater	Diastolic 90 or greater

For Exercises 103–106, write an inequality using the variable p that represents each condition.

103. Normal systolic blood pressure   **104.** Diastolic pressure in hypertension

105. High normal range for systolic pressure   **106.** Systolic pressure in hypertension

A pH scale determines whether a solution is acidic or alkaline. The pH scale runs from 0 to 14, with 0 being the most acidic and 14 being the most alkaline. A pH of 7 is neutral (distilled water has a pH of 7).

For Exercises 107–110, write the pH ranges as inequalities and label the substances as acidic or alkaline.

107. Lemon juice: 2.2 through 2.4, inclusive   **108.** Eggs: 7.6 through 8.0, inclusive

109. Carbonated soft drinks: 3.0 through 3.5, inclusive   **110.** Milk: 6.6 through 6.9, inclusive

Write About It

111. When is a parenthesis used when writing interval notation?

112. When is a bracket used when writing interval notation?

Expanding Your Skills

For Exercises 113–116, write the set as a single interval.

113. $(-\infty, 2) \cap (-3, 4] \cap [1, 3]$

114. $(-\infty, 5) \cap (-1, \infty) \cap [0, 3)$

115. $[(-\infty, -2) \cup (4, \infty)] \cap [-5, 3)$

116. $[(-\infty, 6) \cup (10, \infty)] \cap [8, 12)$

117. In a sample of 50 students taken from an honors college, it was determined that a total of 35 students were taking Anatomy, and of these 12 were taking both Anatomy and Biology. Let A represent the set of students who are taking Anatomy and let B represent the set of students who are taking Biology. Fill in the Venn diagram to represent the *number of* students who are taking Anatomy but not Biology and the number who are taking Biology but not Anatomy. (Assume that all 50 students were taking Anatomy, Biology, or both.)

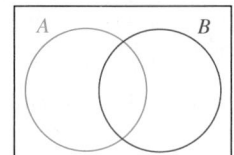

118. In a sample of 60 visitors at a ballpark, a total of 33 bought a hot dog and of these, 20 bought both a drink and a hot dog. Let D represent the set of visitors who bought a drink. Let H represent the set of visitors who bought a hot dog. Fill in the Venn diagram to represent the *number of* visitors who bought a drink only and the number who bought a hot dog only. (Assume that all 60 visitors bought a hot dog, a drink, or both.)

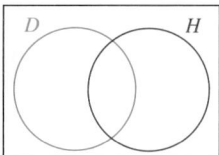

SECTION 1.5   Linear and Compound Inequalities

OBJECTIVES

1. Solve Linear Inequalities in One Variable
2. Solve Compound Linear Inequalities
3. Solve Inequalities of the Form $a < x < b$
4. Solve Applications of Inequalities

1. Solve Linear Inequalities in One Variable

Emily wants to earn an "A" in her College Algebra course and knows that the average of her tests and assignments must be at least 90. She has five test grades of 96, 84, 80, 98, and 88. She also has a score of 100 for online homework, and this carries the same weight as a test grade. She still needs to take the final exam and the final is weighted as two test grades. To determine the scores on the final exam that would result in an average of 90 or more, Emily would solve the following inequality (see Example 9):

$$\frac{96 + 84 + 80 + 98 + 88 + 100 + 2x}{8} \geq 90, \text{ where } x \text{ is Emily's score on the final exam.}$$

A linear equation in one variable is an equation that can be written as $ax + b = 0$, where a and b are real numbers and $a \neq 0$. A **linear inequality** in one variable is any relationship of the form $ax + b < 0$, $ax + b \leq 0$, $ax + b > 0$, or $ax + b \geq 0$. The solution set to a linear equation consists of a single element that can be represented by a point on the number line. The solution set to a linear inequality contains an infinite number of elements and can be expressed in set-builder notation or in interval notation.

Equation/Inequality	Solution Set	Graph
$x + 4 = 0$	$\{-4\}$	
$x + 4 \geq 0$	$\{x \mid x \geq -4\}$ or $[-4, \infty)$	
$x + 4 < 0$	$\{x \mid x < -4\}$ or $(-\infty, -4)$	

To solve a linear inequality in one variable, we use the following properties of inequality.

Properties of Inequality

Let a, b, and c represent real numbers.

1. If $x < a$, then $a > x$.
2. If $a < b$ and $b < c$, then $a < c$.
3. If $a < b$ and $c < d$, then $a + c < b + d$.
4. If $a < b$, then $a + c < b + c$ and $a - c < b - c$.
5. If c is *positive* and $a < b$, then $ac < bc$ and $\dfrac{a}{c} < \dfrac{b}{c}$.
6. If c is *negative* and $a < b$, then $ac > bc$ and $\dfrac{a}{c} > \dfrac{b}{c}$.

These statements are also true expressed with the symbols $>$, $\leq$, and $\geq$.

EXAMPLE 1   Solving a Linear Inequality

Solve the inequality. Graph the solution set and write the solution set in set-builder notation and in interval notation.

$$-6x + 4 < 34$$

Solution:

$$-6x + 4 < 34$$

$$-6x < 30 \qquad \text{Subtract 4 from both sides.}$$

$$\frac{-6x}{-6} > \frac{30}{-6} \qquad \text{Divide both sides by } -6. \text{ Reverse the inequality sign.}$$

$$x > -5$$

The solution set is $\{x \mid x > -5\}$.
Interval notation: $(-5, \infty)$

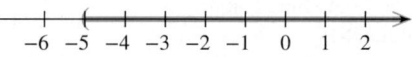

Skill Practice 1   Solve the inequality. Graph the solution set and write the solution set in set-builder notation and in interval notation.

$$-5t - 6 \geq 24$$

TIP   In Example 1, the solution set to the inequality $-6x + 4 < 34$ is $\{x \mid x > -5\}$. This means that all numbers greater than -5 make the inequality a true statement. You can check by taking an arbitrary test point from the interval $(-5, \infty)$. For example, the value $x = -4$ makes the original inequality true.

Test point

Check: $x = -4$

$$-6(-4) + 4 \overset{?}{<} 34$$

$$24 + 4 \overset{?}{<} 34 \checkmark \quad \text{true}$$

Answer

1.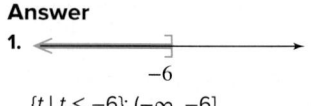

$\{t \mid t \leq -6\}; \; (-\infty, -6]$

EXAMPLE 2   **Solving a Linear Inequality**

Solve the inequality. Graph the solution set and write the solution set in set-builder notation and in interval notation.

$$-6(x - 3) \geq 2 - 2(x - 8)$$

Solution:

$$-6(x - 3) \geq 2 - 2(x - 8)$$

$-6x + 18 \geq 2 - 2x + 16$	Apply the distributive property.
$-6x + 18 \geq 18 - 2x$	Combine like terms.
$-4x + 18 \geq 18$	Add $2x$ to both sides.
$-4x \geq 0$	Subtract 18 from both sides.
$\dfrac{-4x}{-4} \leq \dfrac{0}{-4}$	Divide by -4 (reverse the inequality sign).
$x \leq 0$	

The solution set is $\{x \mid x \leq 0\}$.

Interval notation: $(-\infty, 0]$

Skill Practice 2   Solve the inequality. Graph the solution set and write the solution set in set-builder notation and in interval notation.

$$5(3x + 1) < 4(5x - 5)$$

EXAMPLE 3   **Solving a Linear Inequality Containing Fractions**

Solve the inequality. Graph the solution set and write the solution set in set-builder notation and in interval notation.

$$\frac{x + 1}{3} - \frac{2x - 4}{6} \leq -\frac{x}{2}$$

Solution:

$$\frac{x + 1}{3} - \frac{2x - 4}{6} \leq -\frac{x}{2}$$

$6\left(\dfrac{x + 1}{3} - \dfrac{2x - 4}{6}\right) \leq 6\left(-\dfrac{x}{2}\right)$	Multiply both sides by the LCD of 6 to clear fractions.
$\dfrac{6}{1} \cdot \left(\dfrac{x + 1}{3}\right) - \dfrac{6}{1} \cdot \left(\dfrac{2x - 4}{6}\right) \leq \dfrac{6}{1} \cdot \left(-\dfrac{x}{2}\right)$	Apply the distributive property.
$\dfrac{\overset{2}{6}}{1} \cdot \left(\dfrac{x + 1}{\underset{1}{3}}\right) - \dfrac{\overset{1}{6}}{1} \cdot \left(\dfrac{2x - 4}{\underset{1}{6}}\right) \leq \dfrac{\overset{3}{6}}{1} \cdot \left(-\dfrac{x}{\underset{1}{2}}\right)$	Simplify the fractions.
$2(x + 1) - (2x - 4) \leq -3x$	
$2x + 2 - 2x + 4 \leq -3x$	Apply the distributive property.
$6 \leq -3x$	
$\dfrac{6}{-3} \geq \dfrac{-3x}{-3}$	Divide both sides by -3. Since -3 is a negative number, reverse the inequality sign.
$-2 \geq x$ or $x \leq -2$	

Avoiding Mistakes

In Example 3, keep the parentheses around the quantity $(2x - 4)$ until you distribute the factor of -1.

Answer

2.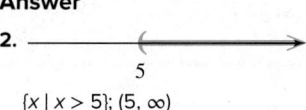
$\{x \mid x > 5\}; (5, \infty)$

The solution set is $\{x \mid x \leq -2\}$.

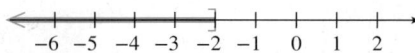

Interval notation: $(-\infty, -2]$

Skill Practice 3   Solve the inequality. Graph the solution set and write the solution set in set-builder notation and in interval notation.

$$\frac{m-4}{2} - \frac{3m+4}{10} > -\frac{3m}{5}$$

---

2. Solve Compound Linear Inequalities

In Examples 4–5, we solve **compound inequalities**. These are statements with two or more inequalities joined by the word "and" or the word "or." For example, suppose that x represents the glucose level measured from a fasting blood sugar test.

- The normal glucose range is given by $x \geq 70$ mg/dL and $x \leq 100$ mg/dL.
- An abnormal glucose level is given by $x < 70$ mg/dL or $x > 100$ mg/dL.

To find the solution sets for compound inequalities, follow these guidelines.

Solving a Compound Inequality

Step 1   To solve a compound inequality, first solve the individual inequalities.

Step 2   • If two inequalities are joined by the word "and," the solutions are the values of the variable that simultaneously satisfy each inequality. That is, we take the *intersection* of the individual solution sets.

   • If two inequalities are joined by the word "or," the solutions are the values of the variable that satisfy either inequality. Therefore, we take the *union* of the individual solution sets.

EXAMPLE 4   Solving a Compound Inequality "Or"

Solve. $x - 2 \leq 5$ or $\frac{1}{2}x > 6$

Solution:

$x - 2 \leq 5$ or $\frac{1}{2}x > 6$ First solve the individual inequalities.
Then take the *union* of the individual solution sets.

$x \leq 7$ or $x > 12$

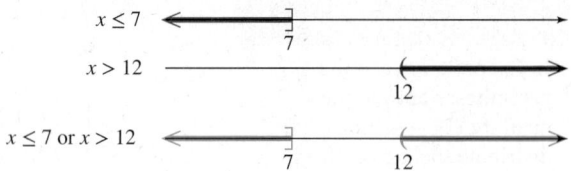

The solution set is $\{x \mid x \leq 7 \text{ or } x > 12\}$.

Interval notation: $(-\infty, 7] \cup (12, \infty)$

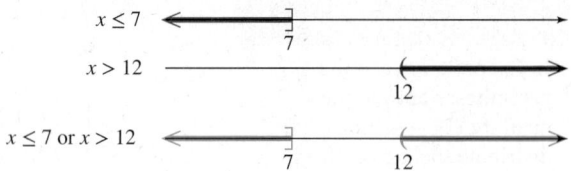
Skill Practice 4   Solve. $\frac{1}{4}y < -1$ or $3 + y \geq 5$

In Example 5, we solve a compound inequality in which the individual inequalities are joined by the word "and." In this case, we take the intersection of the individual solution sets.

EXAMPLE 5   Solving a Compound Inequality "And"

Solve. $-\dfrac{1}{4}t < 2$ and $0.52t \geq 1.3$

Solution:

$-\dfrac{1}{4}t < 2$ and $0.52t \geq 1.3$	First solve the individual inequalities. Multiply both sides of the first inequality by -4 (reverse the inequality sign). Divide the second inequality by 0.52.
$t > -8$ and $t \geq 2.5$	Take the *intersection* of the individual solution sets. The intervals overlap for values of t greater than or equal to 2.5.

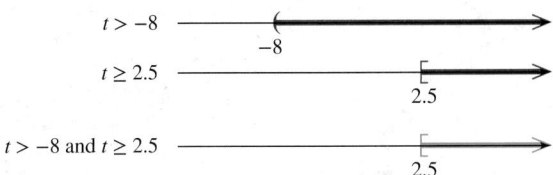

The solution set is $\{t \mid t \geq 2.5\}$. Interval notation: $[2.5, \infty)$

Skill Practice 5   Solve. $0.36w \leq 0.54$ and $-\dfrac{1}{2}w > 3$

---

3. Solve Inequalities of the Form $a < x < b$

Sometimes a compound inequality joined by the word "and" is written as a three-part inequality. For example:

$5 < -2x + 7$ and $-2x + 7 \leq 11$ In this example, two simultaneous conditions are imposed on the

$5 < \boxed{-2x + 7}$ and $\boxed{-2x + 7} \leq 11$ quantity $-2x + 7$.

$$5 < -2x + 7 \leq 11$$

To solve a three-part inequality, the goal is to isolate x in the middle region. This is demonstrated in Example 6.

EXAMPLE 6 Solving a Three-Part Compound Inequality

Solve. $5 < -2x + 7 \leq 11$

Solution:

$$5 < -2x + 7 \leq 11$$

$$5 - 7 < -2x + 7 - 7 \leq 11 - 7 \qquad \text{Subtract 7 from all three parts of the inequality.}$$

$$-2 < -2x \leq 4$$

$$\frac{-2}{-2} > \frac{-2x}{-2} \geq \frac{4}{-2} \qquad \text{Divide all three parts by } -2.$$

$1 > x \geq -2$ or equivalently $-2 \leq x < 1$.

The solution set is $\{x \mid -2 \leq x < 1\}$.

Interval notation: $[-2, 1)$

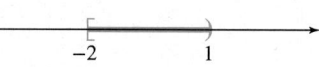

Skill Practice 6 Solve. $-16 \leq -3y - 4 < 2$

Avoiding Mistakes

A "three-part" inequality of the form $a < x < b$ is used to imply that x is greater than a and simultaneously less than b. However, note that the following statements are not valid.

$6 < x < 2$ No number is greater than 6 and simultaneously less than 2.

$-5 > x > 4$ No number is less than -5 and simultaneously greater than 4.

EXAMPLE 7 Solving a Three-Part Compound Inequality

Solve. $2 \geq \dfrac{p - 2}{-3} \geq -1$

Solution:

$$2 \geq \frac{p - 2}{-3} \geq -1 \qquad \text{Isolate the variable in the middle part.}$$

$$-3(2) \leq -3\left(\frac{p - 2}{-3}\right) \leq -3(-1) \qquad \text{Multiply all three parts by } -3. \text{ Remember to reverse the inequality signs.}$$

$$-6 \leq p - 2 \leq 3 \qquad \text{Simplify.}$$

$$-6 + 2 \leq p - 2 + 2 \leq 3 + 2 \qquad \text{Add 2 to all three parts to isolate } p.$$

$$-4 \leq p \leq 5$$

The solution set is $\{p \mid -4 \leq p \leq 5\}$, or equivalently in interval notation $[-4, 5]$.

Skill Practice 7 Solve. $8 > \dfrac{t + 4}{-2} > -5$

Answers

6. $\{y \mid -2 < y \leq 4\}$; $(-2, 4]$

7. $\{t \mid -20 < t < 6\}$; $(-20, 6)$

It is important to note that the inequality in Example 7 is equivalent to the compound inequality $2 \geq \dfrac{p-2}{-3}$ and $\dfrac{p-2}{-3} \geq -1$. The solution set can be found by solving each individual inequality and then taking the intersection of their solution sets.

4. Solve Applications of Inequalities

Compound inequalities are used in many applications, as shown in Examples 8 and 9.

EXAMPLE 8   Translating Compound Inequalities

The normal level of thyroid-stimulating hormone (TSH) for adults ranges from 0.4 to 4.8 microunits per milliliter (μU/mL), inclusive. Let x represent the amount of TSH measured in microunits per milliliter.

a. Write an inequality representing the normal range of TSH.
b. Write a compound inequality representing abnormal TSH levels.

Solution:

a. $0.4 \leq x \leq 4.8$ **b.** $x < 0.4$ or $x > 4.8$

Skill Practice 8   The length of a normal human pregnancy, w, is from 37 to 41 weeks, inclusive.

a. Write an inequality representing the normal length of a pregnancy.
b. Write a compound inequality representing an abnormal length for a pregnancy.

EXAMPLE 9   Using a Linear Inequality in an Application of Grades

Emily has test scores of 96, 84, 80, 98, and 88. Her score for online homework is 100 and is weighted as one test grade. Emily still needs to take the final exam, which counts as two test grades. What score does she need on the final exam to have an average of at least 90? (This is the minimum average to earn an "A" in the class.)

Solution:

Let x represent the grade needed on the final exam.

$$\left(\begin{array}{c} \text{Average of} \\ \text{all scores} \end{array} \right) \geq 90 \qquad \text{To earn an "A," Emily's average must be at least 90.}$$

$$\frac{96 + 84 + 80 + 98 + 88 + 100 + 2x}{8} \geq 90 \qquad \text{Take the sum of all grades. Divide by a total of eight grades.}$$

$$\frac{546 + 2x}{8} \geq 90$$

$$8\left(\frac{546 + 2x}{8} \right) \geq 8(90) \qquad \text{Multiply by 8 to clear fractions.}$$

$$546 + 2x \geq 720$$

$$2x \geq 174 \qquad \text{Subtract 546 from both sides.}$$

$$x \geq 87 \qquad \text{Divide by 2.}$$

Emily must earn a score of at least 87 to earn an "A" in the class.

Answer
8. a. $37 \leq w \leq 41$
   b. $w < 37$ or $w > 41$

Answer

9. Chicago would need more than 6.9 in. of snow in March.

Skill Practice 9   For a recent year, the monthly snowfall (in inches) for Chicago, Illinois, for November, December, January, and February was 2, 8.4, 11.2, and 7.9, respectively. How much snow would be necessary in March for Chicago to exceed its monthly average snowfall of 7.28 in. for these five months?

SECTION 1.5   Practice Exercises

Prerequisite Review

For Exercises R.1–R.8, fill in the blanks with the appropriate symbol, $<$ or $>$. (See Section 1.4 for review.)

R.1. $-7 \ \square \ -2$ **R.2.** $-16 \ \square \ -17$ **R.3.** $0 \ \square \ -4$ **R.4.** $-5 \ \square \ 0$

R.5. $-\dfrac{7}{5} \ \square \ -\dfrac{9}{7}$ **R.6.** $-\dfrac{8}{3} \ \square \ -\dfrac{9}{4}$ **R.7.** $0.1\overline{7} \ \square \ 0.17$ **R.8.** $-0.25 \ \square \ -0.\overline{25}$

For Exercises R.9–R.14, write the set in interval notation. (See Section 1.4 for review.)

R.9. $\{a \mid a < 5\}$ **R.10.** $\{x \mid x > -2\}$ **R.11.** $\{y \mid -1 \le y\}$

R.12. $\{t \mid 7 \ge t\}$ **R.13.** $\left\{c \,\middle|\, -1 \le c < \dfrac{5}{6}\right\}$ **R.14.** $\left\{c \,\middle|\, \dfrac{1}{5} < c \le 2\right\}$

For Exercises R.15–R.20, refer to sets C, D, and F and find the union or intersection as indicated. Write the answers in set-builder notation. (See Section 1.4 for review.)

$C = \{x \mid x < 5\}$, $D = \{x \mid -2 \le x < 3\}$, $F = \{x \mid x \ge 0\}$

R.15. $C \cup D$ **R.16.** $C \cap D$ **R.17.** $D \cap F$

R.18. $D \cup F$ **R.19.** $C \cup F$ **R.20.** $C \cap F$

Concept Connections

1. The multiplication and division properties of inequality indicate that if both sides of an inequality are multiplied or divided by a negative real number, the direction of the _____ sign must be reversed.

2. If a compound inequality consists of two inequalities joined by the word "and," the solution set is the _____ of the solution sets of the individual inequalities.

3. The compound inequality $a < x$ and $x < b$ can be written as the three-part inequality _____.

4. If a compound inequality consists of two inequalities joined by the word "or," the solution set is the _____ of the solution sets of the individual inequalities.

Objective 1: Solve Linear Inequalities in One Variable

For Exercises 5–6, solve the equation or inequality. **(See Example 1)**

		Set Notation	Interval Notation	Graph
5. a.	$-2x + 4 = 10$			
b.	$-2x + 4 < 10$			
c.	$-2x + 4 > 10$			
6. a.	$-4x + 2 = -6$			
b.	$-4x + 2 < -6$			
c.	$-4x + 2 > -6$			

For Exercises 7–26, solve the inequality. Graph the solution set, and write the solution set in set-builder notation and interval notation. (**See Examples 1–3**)

7. $2y + 6 \leq 4$

8. $3y + 11 > 5$

9. $-2x - 5 > 17$

10. $-8t + 1 < 17$

11. $-3 \leq -\dfrac{4}{3}w + 1$

12. $8 \geq -\dfrac{5}{2}y - 2$

13. $-1.2 + 0.6a \leq 0.4a + 0.5$

14. $-0.7 + 0.3x \leq 0.9x - 0.4$

15. $-5 > 6(c - 4) + 7$

16. $-14 < 3(m - 7) + 7$

17. $\dfrac{4 + x}{2} - \dfrac{x - 3}{5} < -\dfrac{x}{10}$

18. $\dfrac{y + 3}{4} - \dfrac{3y + 1}{6} > -\dfrac{1}{12}$

19. $\dfrac{1}{3}(x + 4) - \dfrac{5}{6}(x - 3) \geq \dfrac{1}{2}x + 1$

20. $\dfrac{1}{2}(t - 6) - \dfrac{4}{3}(t + 2) \geq -\dfrac{3}{4}t - 2$

21. $5(7 - x) + 2x < 6x - 2 - 9x$

22. $2(3x + 1) - 4x > 2(x + 8) - 5$

23. $5 - 3[2 - 4(x - 2)] \geq 6\{2 - [4 - (x - 3)]\}$

24. $8 - [6 - 10(x - 1)] \geq 2\{1 - 3[2 - (x + 4)]\}$

25. $4 - 3k > -2(k + 3) - k$

26. $2x - 9 < 6(x - 1) - 4x$

Objective 2: Solve Compound Linear Inequalities

For Exercises 27–34, solve the compound inequality. Graph the solution set, and write the solution set in interval notation. (**See Examples 4–5**)

27. a. $x < 4$ and $x \geq -2$

    b. $x < 4$ or $x \geq -2$

28. a. $y \leq -2$ and $y > -5$

    b. $y \leq -2$ or $y > -5$

29. a. $m + 1 \leq 6$ or $\dfrac{1}{3}m < -2$

    b. $m + 1 \leq 6$ and $\dfrac{1}{3}m < -2$

30. a. $n - 6 > 1$ or $\dfrac{3}{4}n \geq 6$

    b. $n - 6 > 1$ and $\dfrac{3}{4}n \geq 6$

31. a. $-\dfrac{2}{3}y > -12$ and $2.08 \geq 0.65y$

    b. $-\dfrac{2}{3}y > -12$ or $2.08 \geq 0.65y$

32. a. $-\dfrac{4}{5}m < 8$ and $0.85 \leq 0.34m$

    b. $-\dfrac{4}{5}m < 8$ or $0.85 \leq 0.34m$

33. a. $3(x - 2) + 2 \leq x - 8$ or $4(x + 1) + 2 > -2x + 4$

    b. $3(x - 2) + 2 \leq x - 8$ and $4(x + 1) + 2 > -2x + 4$

34. a. $5(t - 4) + 2 > 3(t + 1) - 3$ or $2t - 6 > 3(t - 4) - 2$

    b. $5(t - 4) + 2 > 3(t + 1) - 3$ and $2t - 6 > 3(t - 4) - 2$

Objective 3: Solve Inequalities of the Form $a < x < b$

35. Write $-2.8 < y \leq 15$ as two separate inequalities joined by "and."

36. Write $-\frac{1}{2} \leq z < 2.4$ as two separate inequalities joined by "and."

For Exercises 37–46, graph the solution set, and write the solution set in interval notation. (**See Examples 6–7**)

37. $0 \leq 2b - 5 < 9$

38. $-6 < 3k - 9 \leq 0$

39. $-1 < \dfrac{a}{6} \leq 1$

40. $-3 \leq \dfrac{1}{2}x < 0$

41. $-3 < -2x + 1 \leq 9$

42. $-6 \leq -3x + 9 < 0$

43. $1 \leq \dfrac{5x - 4}{2} < 3$

44. $-2 \leq \dfrac{4x - 1}{3} \leq 5$

45. $-2 \leq \dfrac{-2x + 1}{-3} \leq 4$

46. $-4 < \dfrac{-5x - 2}{-2} < 4$

Mixed Exercises

For Exercises 47–62, solve the inequalities and compound inequalities. Write the solution set in interval notation or indicate that the solution set is the empty set.

47. $-1.2b - 0.4 \geq -0.4b$

48. $-0.4t + 1.2 < -2$

49. $-\dfrac{3}{4}c - \dfrac{5}{4} \geq 2c$

50. $-\dfrac{2}{3}q - \dfrac{1}{3} > \dfrac{1}{2}q$

51. $\dfrac{2}{3}(2p - 1) \geq 10$ and $\dfrac{4}{5}(3p + 4) \geq 20$

52. $\dfrac{5}{2}(a + 2) < -6$ and $\dfrac{3}{4}(a - 2) < 1$

53. $-2 < -x - 12$ and $-14 < 5(x - 3) + 6x$

54. $-8 \geq -3y - 2$ and $3(y - 7) + 16 > 4y$

55. $5(x - 1) \geq -5$ or $5 - x \leq 11$

56. $-p + 7 \geq 10$ or $3(p - 1) \leq 12$

57. $\dfrac{5}{3}v \leq 5$ or $-v - 6 < 1$

58. $\dfrac{3}{8}u + 1 > 0$ or $-2u \geq -4$

59. $-4 \leq \dfrac{2 - 4x}{3} < 8$

60. $-1 < \dfrac{3 - x}{2} \leq 0$

61. $\dfrac{-x + 3}{2} > \dfrac{4 + x}{5}$ or $\dfrac{1 - x}{4} > \dfrac{2 - x}{3}$

62. $\dfrac{y - 7}{-3} < \dfrac{1}{4}$ or $\dfrac{y + 1}{-2} > -\dfrac{1}{3}$

Objective 4: Solve Applications of Inequalities

63. The normal number of white blood cells for human blood is between 4800 and 10,800 cells per cubic millimeter, inclusive. Let x represent the number of white blood cells per cubic millimeter. **(See Example 8)**

  a. Write an inequality representing the normal range of white blood cells per cubic millimeter.

  b. Write a compound inequality representing abnormal levels of white blood cells per cubic millimeter.

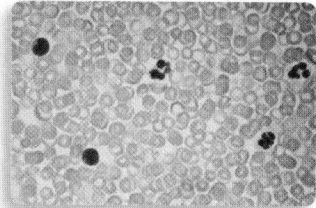

Image Source/Getty Images

64. Normal hemoglobin levels in human blood for adult males are between 13 and 16 grams per deciliter (g/dL), inclusive. Let x represent the level of hemoglobin measured in grams per deciliter.

  a. Write an inequality representing normal hemoglobin levels for adult males.

  b. Write a compound inequality representing abnormal levels of hemoglobin for adult males.

65. A polling company estimates that a certain candidate running for office will receive between 44% and 48% of the votes. Let x represent the percentage of votes for this candidate.

  a. Write a strict inequality representing the expected percentage of votes for this candidate.

  b. Write a compound inequality representing the percentage of votes that would fall outside the polling company's prediction.

66. A machine is calibrated to cut a piece of wood between 2.4 in. thick and 2.6 in. thick. Let x represent the thickness of the wood after it is cut.

  a. Write a strict inequality representing the expected range of thickness of the wood after it has been cut.

  b. Write a compound inequality representing the thickness of wood that would fall outside the normal range for this machine.

For Exercises 67–70, write a three-part inequality to represent the given statement.

67. The normal range for the hemoglobin level x for an adult female is greater than or equal to 12.0 g/dL and less than or equal to 15.2 g/dL.

68. A tennis player must play in the "open" division of a tennis tournament if the player's age a is over 18 years and under 25 years.

69. The distance d that Zina hits a 9-iron is at least 90 yd, but no more than 110 yd.

70. A small plane's average speed s is at least 220 mph but not more than 410 mph.

71. Marilee wants to earn an "A" in a class and needs an overall average of at least 92. Her test grades are 88, 92, 100, and 80. The average of her quizzes is 90 and counts as one test grade. The final exam counts as 2.5 test grades. What scores on the final exam would result in Marilee's overall average of 92 or greater? (**See Example 9**)

73. Rita earns scores of 78, 82, 90, 80, and 75 on her five chapter tests for a certain class and a grade of 85 on the class project. The overall average for the course is computed as follows: the average of the five chapter tests makes up 60% of the course grade; the project accounts for 10% of the grade; and the final exam accounts for 30%. What scores can Rita earn on the final exam to earn a "B" in the course if the cutoff for a "B" is an overall score greater than or equal to 80, but less than 90? Assume that 100 is the highest score that can be earned on the final exam and that only whole-number scores are given.

72. A 10-year-old competes in gymnastics. For several competitions she received the following "All-Around" scores: 36, 36.9, 37.1, and 37.4. Her coach recommends that gymnasts whose "All-Around" scores average at least 37 move up to the next level. What "All-Around" scores in the next competition would result in the child being eligible to move up?

74. Trent earns scores of 66, 84, and 72 on three chapter tests for a certain class. His homework grade is 60 and his grade for a class project is 85. The overall average for the course is computed as follows: the average of the three chapter tests makes up 50% of the course grade; homework accounts for 20% of the grade; the project accounts for 10%; and the final exam accounts for 20%. What scores can Trent earn on the final exam to pass the course if he needs a "C" or better? A "C" or better requires an overall score of 70 or better, and 100 is the highest score that can be earned on the final exam. Assume that only whole-number scores are given.

For Exercises 75–78, use the graph that shows the average height for boys based on age. Let a represent a boy's age (in years) and let h represent his height (in inches).

75. Determine the age range for which the average height of boys is at least 51 in.

76. Determine the age range for which the average height of boys is greater than or equal to 41 in.

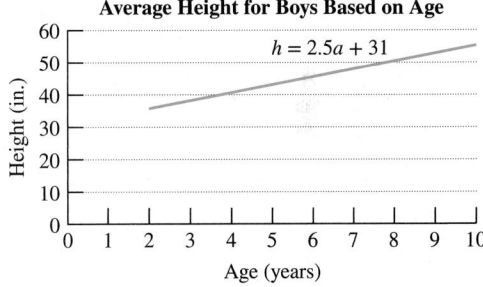

77. Determine the age range for which the average height of boys is no more than 46 in.

78. Determine the age range for which the average height of boys is at most 53.5 in.

79. Nolvia sells copy machines, and her salary is $25,000 plus a 4% commission on sales. The equation $S = 25,000 + 0.04x$ represents her salary S in dollars in terms of her total sales x in dollars.

  a. How much money in sales does Nolvia need to earn a salary that exceeds $40,000?

  b. How much money in sales does Nolvia need to earn a salary that exceeds $80,000?

  c. Why is the money in sales required to earn a salary of $80,000 more than twice the money in sales required to earn a salary of $40,000?

81. A car travels 50 mph and passes a truck traveling 40 mph. How long will it take the car to be more than 16 mi ahead?

80. The amount of money A in a savings account depends on the principal P, the interest rate r, and the time in years t that the money is invested. The equation $A = P + Prt$ shows the relationship among the variables for an account earning simple interest. If an investor deposits $5000 at $6\frac{1}{2}$% simple interest, the account will grow according to the formula $A = 5000 + 5000(0.065)t$.

  a. How many years will it take for the investment to exceed $10,000? (Round to the nearest tenth of a year.)

  b. How many years will it take for the investment to exceed $15,000? (Round to the nearest tenth of a year.)

82. A work-study job in the library pays $10.75/hr and a job in the tutoring center pays $16.25/hr. How long would it take for a tutor to make over $500 more than a student working in the library? Round to the nearest hour.

Write About It

83. How is the process to solve a linear inequality different from the process to solve a linear equation?

84. Explain why $8 < x < 2$ has no solution.

85. Explain why $-5 > y > -2$ has no solution.

86. Explain why $-3 > w > -1$ has no solution.

Expanding Your Skills

87. The revenue R for selling x fleece jackets is given by the equation $R = 49.95x$. The cost C (in dollars) to produce x jackets is $C = 2300 + 18.50x$. Find the number of jackets that the company needs to sell to produce a profit. (*Hint*: A profit occurs when revenue exceeds cost.)

88. The revenue R for selling x mountain bikes is $R = 249.95x$. The cost C (in dollars) to produce x bikes is $C = 56,000 + 140x$. Find the number of bikes that the company needs to sell to produce a profit.

89. A rectangular garden is to be constructed so that the width is 100 ft. What are the possible values for the length of the garden if at most 800 ft of fencing is to be used?

90. The lengths of the sides of a triangle are given by three consecutive integers greater than 1. What are the possible values for the shortest side if the perimeter is not to exceed 24 ft?

91. For a certain bowling league, a beginning bowler computes her handicap by taking 90% of the difference between 220 and her average score in league play. Determine the average scores that would produce a handicap of 72 or less. Also assume that a negative handicap is not possible in this league.

92. Body temperature is usually maintained between 36.5°C and 37.5°C, inclusive. Determine the corresponding range of temperatures in Fahrenheit. Use the relationship between degrees Celsius C and degrees Fahrenheit F: $C = \dfrac{5}{9}(F - 32)$.

93. Donovan has offers for two sales jobs. Job A pays a base salary of $25,000 plus a 10% commission on sales. Job B pays a base salary of $30,000 plus 8% commission on sales.

   a. How much would Donovan have to sell for the salary from Job A to exceed the salary from Job B?

   b. If Donovan routinely sells more than $500,000 in merchandise, which job would result in a higher salary?

94. Nancy wants to vacation in Austin, Texas. Hotel A charges $179 per night with a 14% nightly room tax and free parking. Hotel B charges $169 per night with an 18% nightly room tax plus a one-time $40 parking fee. After how many nights will Hotel B be less expensive?

SECTION 1.6 Absolute Value Equations

OBJECTIVES

1. Simplify Absolute Value Expressions
2. Use Absolute Value to Represent Distance
3. Solve Absolute Value Equations

1. Simplify Absolute Value Expressions

Recall that the absolute value of a real number x is the distance between x and zero on the number line. Formally,

$$|x| = \begin{cases} x & \text{if } x \geq 0 \text{ (if } x \text{ is positive or zero, then } |x| \text{ is just } x \text{ itself).} \\ -x & \text{if } x < 0 \text{ (if } x \text{ is negative, then } |x| \text{ is the opposite of } x\text{).} \end{cases}$$

For example: $|12| = 12$

$$|-12| = -(-12) = 12$$

Using the formal definition of absolute value, we can sometimes simplify an absolute value expression to an equivalent expression without absolute value bars. For example, consider the expression $|x - 6|$.

- If x is greater than or equal to 6, then $x - 6$ would be greater than or equal to zero.

  Thus, if $x \geq 6$, then $|x - 6| = x - 6$.

  > The absolute value bars can be dropped because the expression inside the absolute value bars is not negative.

- If x is less than 6, then $x - 6$ would be negative.

  Thus, if $x < 6$, then $|x - 6| = -(x - 6)$

  > The expression inside the absolute value bars is negative. Therefore, we must take the *opposite* of the expression when we drop the absolute value bars.

In Example 1, we use the definition of absolute value to simplify several expressions by removing absolute value bars.

> **TIP** When evaluating an absolute value, you must consider the sign of the entire expression inside the absolute value bars, not just the value of x. For example, consider the expression $|3 - x|$.
>
> If $x > 3$, $|3 - x| = -(3 - x)$.
> If $x \leq 3$, $|3 - x| = 3 - x$.

EXAMPLE 1   **Writing Expressions Without Absolute Value Bars**

Write the expressions without absolute value bars.

a. $|x - 2|$ for $x \geq 2$ **b.** $|x - 2|$ for $x < 2$ **c.** $\dfrac{|x + 3|}{x + 3}$ for $x < -3$

Solution:

a. $|x - 2|$ for $x \geq 2$

$= x - 2$

If $x \geq 2$, then $x - 2 \geq 0$. Therefore, the expression inside absolute value bars is greater than or equal to zero, and the absolute value bars can be dropped.

b. $|x - 2|$ for $x < 2$

$= -(x - 2)$

$= -x + 2$

If $x < 2$, then $x - 2 < 0$. Therefore, the expression inside absolute value bars is negative. Take the opposite of the expression inside absolute value bars.

c. $\dfrac{|x + 3|}{x + 3}$ for $x < -3$

$= \dfrac{-(x + 3)}{x + 3}$

$= -1$

If $x < -3$, then $x + 3 < 0$. Therefore, the expression inside absolute value bars is negative. Thus, $|x + 3| = -(x + 3)$.

The quotient of any real number and its opposite is -1.

Thus, $\dfrac{-(x + 3)}{(x + 3)} = -\dfrac{\overset{1}{\cancel{(x + 3)}}}{\cancel{(x + 3)}} = -1.$

> **TIP** To check your answers, try substituting a representative value of x into both the original expression and the simplified expression. The results should match. In Example 1(b), substitute $x = -5$.
>
> Original expression
> $|x - 2| = |(-5) - 2|$
> $= |-7| = 7$
> Simplified form
> $-x + 2 = -(-5) + 2$
> $= 5 + 2 = 7$ ✓

Skill Practice 1   Write the expressions without absolute value bars.

a. $|x + 7|$ for $x < -7$ **b.** $|x + 7|$ for $x \geq -7$ **c.** $\dfrac{x - 8}{|x - 8|}$ for $x > 8$

---

2. Use Absolute Value to Represent Distance

Suppose that two hikers in a national park leave a ranger station and walk down a straight trail. After 3 hr, one hiker has walked 6.8 mi and the other has walked 4.3 mi (Figure 1-4). To find the distance between the hikers, we can subtract the shorter distance from the longer distance.

Distance between hikers: 6.8 mi − 4.3 mi = 2.5 mi

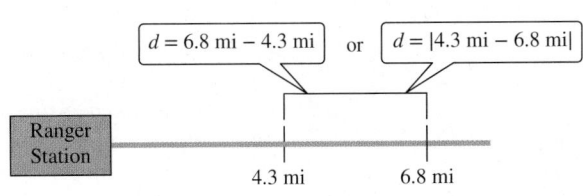

Figure 1-4

Alternatively, we could have subtracted the distances in the reverse order and taken the absolute value.

Distance between hikers: $|4.3 \text{ mi} - 6.8 \text{ mi}| = |-2.5 \text{ mi}| = 2.5 \text{ mi}$

This example illustrates that we can determine the distance between two points a and b on the real number line by taking the absolute value of their difference in either order.

Distance Between Two Points on a Number Line

The distance between two points a and b on a number line is given by

$$|a - b| \quad \text{or} \quad |b - a|$$

That is, the distance between two points on a number line is the absolute value of their difference.

EXAMPLE 2   Determining the Distance Between Two Points

Write an absolute value expression that represents the distance between 4 and -1 on the number line. Then simplify.

Solution:

$|4 - (-1)| = |5| = 5$ The distance between 4 and -1 is represented by
$|-1 - 4| = |-5| = 5$ $|4 - (-1)|$ or by $|-1 - 4|$.

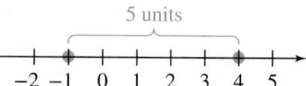

Skill Practice 2   Write an absolute value expression that represents the distance between -9 and 2 on the number line. Then simplify.

EXAMPLE 3   Determining the Distance Between Two Points

Write an absolute value expression that represents the distance between 2 and $\sqrt{2}$ on the number line. Then simplify.

Solution:

The distance is represented by $|2 - \sqrt{2}|$ or $|\sqrt{2} - 2|$. Note that $\sqrt{2} \approx 1.4$, which is less than 2. Therefore, the expression $2 - \sqrt{2}$ is positive and the expression $\sqrt{2} - 2$ is negative. Thus,

$|2 - \sqrt{2}| = 2 - \sqrt{2}$ Since $2 - \sqrt{2}$ is positive, we can drop the absolute value bars.

$|\sqrt{2} - 2| = -(\sqrt{2} - 2) = -\sqrt{2} + 2$ Since $\sqrt{2} - 2$ is negative, take the opposite of the quantity $\sqrt{2} - 2$ and drop the absolute value bars.

Notice that the expressions $2 - \sqrt{2}$ and $-\sqrt{2} + 2$ are equivalent.

Skill Practice 3   Write an absolute value expression that represents the distance between $\sqrt{7}$ and 5 on the number line. Then simplify.

Answers

2. $|-9 - 2|$ or $|2 - (-9)|$;
    The distance is 11 units.

3. $5 - \sqrt{7}$ or $-\sqrt{7} + 5$

3. Solve Absolute Value Equations

The concept of distance between two points on a number line can help us solve absolute value equations. For example, the equation $|x| = 5$ can be written as $|x - 0| = 5$, which means that the distance between x and 0 on the number line is 5 units. The values of x that satisfy this condition are 5 and -5. Likewise, the equation $|x - 4| = 2$ represents the set of real numbers x whose distance from 4 is 2 units.

Equation	Graph
Distance between x and 0 is 5. $|x - 0| = 5$	5 units 5 units -5 -4 -3 -2 -1 0 1 2 3 4 5 Solution set: $\{-5, 5\}$
Distance between x and 4 is 2. $|x - 4| = 2$	2 units 2 units -3 -2 -1 0 1 2 3 4 5 6 7 Solution set: $\{2, 6\}$

In general, given a nonnegative real number k, the generic absolute value equation $|u| = k$ can be solved directly from the definition of absolute value.

$$|u| = \begin{cases} u \text{ if } u \geq 0 \\ -u \text{ if } u < 0 \end{cases}$$

Thus, $|u| = k$ means that $u = k$ or $-u = k$. Solving for u, we have $u = k$ or $u = -k$. This and three other properties follow directly from the definition of absolute value.

Properties of Absolute Value Equations

Let k represent a positive real number.

1. $|u| = k$ is equivalent to $u = k$ or $u = -k$.
2. $|u| = 0$ is equivalent to $u = 0$.
3. $|u| = -k$ has no solution.
4. $|u| = |w|$ is equivalent to $u = w$ or $u = -w$.

To solve an absolute value equation, first isolate the absolute value. Then solve the equation by rewriting the equation in its equivalent form without absolute value bars.

EXAMPLE 4   Solving Absolute Value Equations

Solve the equations.

a. $|w| - 2 = 12$ **b.** $2|x| + 32 = 20$ **c.** $|z + 8| -3 = -3$

Solution:

a. $|w| - 2 = 12$ | Isolate the absolute value by adding 2 to both sides.
$|w| = 14$ | The equation is in the form $|u| = k$, where $u = w$ and $k = 14$.

$w = 14$ or $w = -14$ | Since $k > 0$, rewrite the equation in the form $u = k$ or $u = -k$.

The solution set is $\{14, -14\}$.

b. $2|x| + 32 = 20$ — Isolate the absolute value by first subtracting 32 from both sides.

$$2|x| = -12$$ Divide both sides by 2.

$$|x| = -6$$ The equation is in the form $|u| = k$, where $u = x$ and $k = -6$.

The solution set is { }. — Since $k < 0$, the equation has no solution. There are no real numbers x whose absolute value equals a negative number.

> **TIP** In Example 4(c), the absolute value is equated to zero. Since the equations $z + 8 = 0$ and $z + 8 = -0$ are equivalent, only one equation is needed.

c. $|z + 8| - 3 = -3$ — Isolate the absolute value by adding 3 to both sides.

$$|z + 8| = 0$$ The equation is in the form $|u| = k$, where $u = z + 8$ and $k = 0$. Since $k = 0$, rewrite the equation as $z + 8 = 0$.

$$z + 8 = 0$$

$$z = -8$$

The solution set is $\{-8\}$.

Skill Practice 4   Solve the equations.

    a. $|v| + 6 = 10$ **b.** $3|y| - 5 = -14$ **c.** $4 + |9 - x| = 4$

EXAMPLE 5   Solving Absolute Value Equations

Solve the equations.   **a.** $2|3 - 2t| = 6$ **b.** $2 = |7w - 3| + 8$

Solution:

a. $2|3 - 2t| = 6$ — Divide by 2 to isolate the absolute value.

$$|3 - 2t| = 3$$ The equation is in the form $|u| = k$, where $u = 3 - 2t$ and $k > 0$.

$$3 - 2t = 3 \quad \text{or} \quad 3 - 2t = -3$$ Rewrite the equation in the form $u = k$ or $u = -k$.

$$-2t = 0 \quad \text{or} \quad -2t = -6$$

$$t = 0 \quad \text{or} \quad t = 3$$

Check: $t = 0$

$$2|3 - 2(0)| \overset{?}{=} 6$$
$$2|3| \overset{?}{=} 6$$
$$2 \cdot 3 = 6 \checkmark$$

Check: $t = 3$

$$2|3 - 2(3)| \overset{?}{=} 6$$
$$2|3 - 6| \overset{?}{=} 6$$
$$2|-3| \overset{?}{=} 6$$
$$2 \cdot 3 = 6 \checkmark$$

The solution set is $\{3, 0\}$.

b. $2 = |7w - 3| + 8$ — Subtract 8 from both sides to isolate the absolute value.

$$-6 = |7w - 3|$$ The equation is in the form $|u| = k$, where $u = 7w - 3$ and $k < 0$. By definition, an absolute value cannot be negative. There is no solution.

The solution set is { }.

Skill Practice 5   Solve the equations.

    a. $5|2 - 4t| = 50$ **b.** $5 = |6c - 7| + 9$

In Example 6, we solve equations involving two absolute values by writing $|u| = |w|$ in the equivalent form $u = w$ or $u = -w$.

Answers

4. a. $\{-4, 4\}$ **b.** { }   **c.** $\{9\}$
5. a. $\{-2, 3\}$ **b.** { }

EXAMPLE 6   **Solving Equations with Two Absolute Values**

Solve the equations.   **a.** $|2x - 5| = |x + 1|$ **b.** $|6 - x| = |x - 6|$

Solution:

a. $|2x - 5| = |x + 1|$

$2x - 5 = x + 1$ or $2x - 5 = -(x + 1)$ The equation is in the form $|u| = |w|$. Rewrite the equation in the form $u = w$ or $u = -w$.

$x - 5 = 1$ or $2x - 5 = -x - 1$ Solve each individual equation.

$x = 6$ or $3x = 4$

$x = \dfrac{4}{3}$ Both solutions check in the original equation.

The solution set is $\left\{ 6, \dfrac{4}{3} \right\}$.

b. $|6 - x| = |x - 6|$ The equation is in the form $|u| = |w|$.

$6 - x = x - 6$ or $6 - x = -(x - 6)$ Rewrite the equation in the form $u = w$ or $u = -w$.

$-2x = -12$ or $6 - x = -x + 6$ Solve each individual equation.

$x = 6$ or $6 = 6$ (identity)

The solution set is $\mathbb{R}$.

> The solution to this equation is all real numbers (including 6).

Skill Practice 6   Solve the equations.

a. $|3x - 4| = |2x + 1|$ **b.** $|4 + x| = |-4 - x|$

Answers

6. a. $\left\{ \dfrac{3}{5}, 5 \right\}$ **b.** $\mathbb{R}$

SECTION 1.6   Practice Exercises

Prerequisite Review

For Exercises R.1–R.8, simplify. (See Section R.2 for review.)

R.1. $|-8|$ **R.2.** $\left| -\dfrac{1}{4} \right|$ **R.3.** $-\left| \dfrac{3}{4} \right|$ **R.4.** $-|18|$

R.5. $-|-2.7|$ **R.6.** $-|-36.1|$ **R.7.** $-(-6)$ **R.8.** $-(-14)$

For Exercises R.9–R.14, solve the equation. (See Section 1.2 for review.)

R.9. $3x - 4 = 7$ **R.10.** $-9x + 1 = 11$ **R.11.** $4(x - 3) = 4x - 12$

R.12. $-2x - 7 = -2(x + 3) - 1$ **R.13.** $3x - 2 = 3x - 1$ **R.14.** $-5x + 8 = -5x$

Concept Connections

1. An _____ value equation is an equation of the form $|u| = k$. If k is a positive real number, then the solution set is _____.

2. What is the first step to solve the equation $|3x - 4| + 6 = 11$?

3. The absolute value equation $|u| = |w|$ implies that $u =$ _____ or $u =$ _____.

4. The solution set to the equation $|x + 4| = -2$ is _____. The solution set to the equation $|x + 4| = 0$ is _____.

5. If $x > 6$, is the expression $6 - x$ positive or negative? If $x < 6$, is the expression $6 - x$ positive or negative?

6. The distance between a and b on the number line is given by _____ or _____.

Objective 1: Simplify Absolute Value Expressions

For Exercises 7–16, simplify by writing the expression without absolute value bars. **(See Example 1)**

7. **a.** $|t - 4|$ for $t \geq 4$
   b. $|t - 4|$ for $t < 4$

8. **a.** $|m - 11|$ for $m \geq 11$
   b. $|m - 11|$ for $m < 11$

9. **a.** $|4.6 - x|$ for $x \geq 4.6$
   b. $|4.6 - x|$ for $x < 4.6$

10. **a.** $\left|\dfrac{5}{4} - y\right|$ for $y \geq \dfrac{5}{4}$
    b. $\left|\dfrac{5}{4} - y\right|$ for $y < \dfrac{5}{4}$

11. **a.** $|x + 2|$ for $x \geq -2$
    b. $|x + 2|$ for $x < -2$

12. **a.** $|t + 6|$ for $t < -6$
    b. $|t + 6|$ for $t \geq -6$

13. **a.** $\dfrac{|z - 5|}{z - 5}$ for $z > 5$
    b. $\dfrac{|z - 5|}{z - 5}$ for $z < 5$

14. **a.** $\dfrac{7 - x}{|7 - x|}$ for $x < 7$
    b. $\dfrac{7 - x}{|7 - x|}$ for $x > 7$

15. **a.** $|\pi - 3|$
    b. $|3 - \pi|$

16. **a.** $|\sqrt{6} - 6|$
    b. $|6 - \sqrt{6}|$

Objective 2: Use Absolute Value to Represent Distance

For Exercises 17–24, write an absolute value expression to represent the distance between the two points on the number line. Then simplify without absolute value bars. **(See Examples 2–3)**

17. 1 and 6

18. 2 and 9

19. 3 and -4

20. -8 and 2

21. 6 and 2π

22. π and 4

23. $\sqrt{21}$ and 4

24. $\sqrt{7}$ and 3

Objective 3: Solve Absolute Value Equations

For Exercises 25–78, solve the equations. **(See Examples 4–6)**

25. **a.** $|p| = 6$
    b. $|p| = 0$
    c. $|p| = -6$

26. **a.** $|w| = 2$
    b. $|w| = 0$
    c. $|w| = -2$

27. **a.** $|x - 3| = 4$
    b. $|x - 3| = 0$
    c. $|x - 3| = -7$

28. **a.** $|m + 1| = 5$
    b. $|m + 1| = 0$
    c. $|m + 1| = -1$

29. $|x| + 5 = 11$

30. $|x| - 3 = 20$

31. $|y| + 8 = 5$

32. $|x| + 12 = 6$

33. $|w| - 3 = -1$

34. $|z| - 14 = -10$

35. $|y| = \sqrt{2}$

36. $|y| = \sqrt{5}$

37. $|w| - 3 = -5$

38. $|w| + 4 = -8$

39. $|3q| = 0$

40. $|4p| = 0$

41. $|3x - 4| = 8$

42. $|4x + 1| = 6$

43. $5 = |2x - 4|$

44. $10 = |3x + 7|$

45. $|3x + 8| + 1 = 21$

46. $|4x - 3| - 5 = 12$

47. $2|3x - 4| + 7 = 9$

48. $4|2t + 7| + 2 = 22$

49. $-3 = -|c - 7| + 1$

50. $-4 = -|z + 8| - 3$

51. $2 = 8 + |11y + 4|$

52. $6 = 7 + |9z - 3|$

53. $\left|4 - \dfrac{1}{2}w\right| - \dfrac{1}{3} = \dfrac{1}{2}$

54. $\left|2 - \dfrac{1}{3}p\right| - \dfrac{7}{6} = \dfrac{1}{2}$

55. $-2|3b - 7| - 9 = -9$

56. $-3|5x + 1| + 4 = 4$

57. $-2|x + 3| = 5$

58. $-3|x - 5| = 7$

59. $0 = |6x - 9|$

60. $7 = |4k - 6| + 7$

61. $|3y + 5| = |y + 1|$

62. $|2a - 3| = |a + 2|$

63. $|4 - x| = |2x + 1|$

64. $|3 - 2x| = |x + 5|$

65. $\left|\dfrac{1}{4}w\right| = |4w|$

66. $|3z| = \left|\dfrac{1}{3}z\right|$

67. $|x + 4| = |x - 7|$

68. $|k - 3| = |k + 3|$

69. $|2p - 1| = |1 - 2p|$

70. $|4d - 3| = |3 - 4d|$

71. $|3.5m - 1.2| = |8.5m + 6|$

72. $|11.2n + 9| = |7.2n - 2.1|$

73. $|4x - 3| = -|2x - 1|$

74. $-|3 - 6y| = |8 - 2y|$

75. $|8 - 7w| = |7w - 8|$

76. $|4 - 3z| = |3z - 4|$

77. $|x + 2| + |x - 4| = 0$

78. $|t + 6| + |t - 1| = 0$

Mixed Exercises

For Exercises 79–82,

   a. Write an absolute value equation to represent each statement.

   b. Solve the equation.

79. The distance between a number x and 4 on the number line is 6.

80. The distance between a number x and 3 on the number line is 8.

81. The distance between a number y and -8 on the number line is 3.

82. The distance between a number t and -15 on the number line is 4.

83. For what values of x is the statement true?

$$|2x - 1| = 1 - 2x$$

84. For what values of y is the statement true?

$$|5y - 3| = 5y - 3$$

85. For what values of t is the statement true?

$$|6 - 5t| = 6 - 5t$$

86. For what values of m is the statement true?

$$|11 - 2m| = 2m - 11$$

87. For what values of n is the statement true?

$$|3n + 5| = -3n - 5$$

88. For what values of p is the statement true?

$$|7p + 3| = -7p - 3$$

Write About It

89. Explain why the equation $-|3x + 7| = 4$ has no solution.

90. Explain how to solve the equation $|2x - 1| - 7 = 12$.

Expanding Your Skills

For Exercises 91–96, assume that n is a real number. Use the definition of absolute value to simplify each expression without absolute value bars.

91. If $n > 0$, then $n - |n| =$ _____.

92. If $n < 0$, then $n - |n| =$ _____.

93. If $n > 0$, then $n + |n| =$ _____.

94. If $n < 0$, then $n + |n| =$ _____.

95. If $n > 0$, then $-|n| =$ _____.

96. If $n < 0$, then $-|n| =$ _____.

Absolute Value Inequalities

1. Solve Absolute Value Inequalities

In this section we will learn to solve absolute value inequalities. For example, the inequality $|x| < 3$ or equivalently $|x - 0| < 3$ represents the set of real numbers x that are less than 3 units from zero on the number line. Likewise, the inequality $|x| > 3$ represents the set of real numbers x that are more than 3 units from zero on the number line.

Inequality	Graph	Solution Set
$|x| < 3$	3 units 3 units −4 −3 −2 −1 0 1 2 3 4	$\{x\| -3 < x < 3\}$
$|x| > 3$	3 units 3 units −4 −3 −2 −1 0 1 2 3 4	$\{x\|x < -3 \text{ or } x > 3\}$

We can generalize these observations with the following properties involving absolute value inequalities.

> **Properties Involving Absolute Value Inequalities**
>
> For a real number $k > 0$,
>
> **1.** $|u| < k$ is equivalent to $-k < u < k$.   (1)
> **2.** $|u| > k$ is equivalent to $u < -k$ or $u > k$.   (2)
>
> *Note*: The statements also hold true for the inequality symbols $\le$ and $\ge$, respectively.

Properties (1) and (2) follow directly from the definition: $|u| = \begin{cases} u \text{ if } u \ge 0 \\ -u \text{ if } u < 0 \end{cases}$

By definition, $|u| < k$ is equivalent to

$$u < k \quad \text{and} \quad -u < k$$
$$u < k \quad \text{and} \quad u > -k$$
$$-k < u < k \qquad (1)$$

By definition, $|u| > k$ is equivalent to

$$u > k \quad \text{or} \quad -u > k$$
$$u > k \quad \text{or} \quad u < -k$$
$$u < -k \quad \text{or} \quad u > k \qquad (2)$$

EXAMPLE 1   **Solving Absolute Value Inequalities**

Solve the inequality and write the solution set in interval notation.

a. $|m - 2| + 1 < 5$ **b.** $|m - 2| + 1 \ge 5$

Solution:

a. $|m - 2| + 1 < 5$ First isolate the absolute value. Subtract 1 from both sides.

$\qquad |m - 2| < 4$ The inequality is in the form $|u| < k$, where $u = m - 2$.

$\quad -4 < m - 2 < 4$ Write the equivalent compound inequality $-k < u < k$.

$\qquad -2 < m < 6$ Add 2 to all three parts.

The solution set is $\{m| -2 < m < 6\}$.

Interval notation: $(-2, 6)$

−2   6

b. $|m - 2| + 1 \geq 5$ First isolate the absolute value. Subtract 1 from both sides.

$\qquad |m - 2| \geq 4$ The inequality is in the form $|u| \geq k$, where $u = m - 2$.

$\qquad m - 2 \leq -4 \quad \text{or} \quad m - 2 \geq 4$ Write the equivalent compound inequality $u \leq -k$ or $u \geq k$.

$\qquad\qquad m \leq -2 \quad \text{or} \quad m \geq 6$

The solution set is $\{m | m \leq -2 \text{ or } m \geq 6\}$.

Interval notation:

$\quad (-\infty, -2] \cup [6, \infty)$

Skill Practice 1   Solve the inequality and write the solution set in interval notation.

   a. $|x + 4| - 3 \leq 10$ **b.** $|x + 4| - 3 > 10$

TIP The solutions to the inequalities given in Example 1 can be interpreted geometrically in terms of distances on a number line. Once the absolute value has been isolated, we have the inequalities $|m - 2| < 4$ and $|m - 2| \geq 4$.

$|m - 2| < 4$ The set of real numbers m that are less than 4 units from 2 on the number line.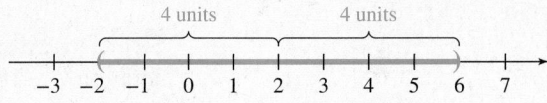

$|m - 2| \geq 4$ The set of real numbers m that are 4 or more units from 2 on the number line.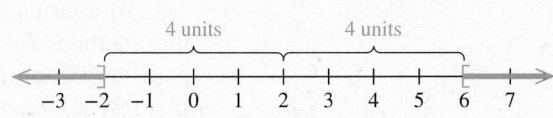

EXAMPLE 2   Solving an Absolute Value Inequality

Solve the inequality and write the solution set in interval notation.

$$2|6 - m| - 3 < 7$$

Solution:

$2|6 - m| - 3 < 7$ First isolate the absolute value. Add 3 and divide by 2.

$\qquad |6 - m| < 5$ The inequality is in the form $|u| < k$, where $u = 6 - m$.

$-5 < 6 - m < 5$ Write the equivalent compound inequality $-k < u < k$.

$-11 < -m < -1$ Subtract 6 from all three parts.

$\dfrac{-11}{-1} > \dfrac{-m}{-1} > \dfrac{-1}{-1}$ Divide by -1 and reverse the inequality signs.

$11 > m > 1$ or equivalently $1 < m < 11$.

The solution set is $\{m | 1 < m < 11\}$.
Interval notation: $(1, 11)$

Skill Practice 2   Solve the inequality and write the solution set in interval notation. $3|5 - x| + 2 \leq 14$

EXAMPLE 3   Solving an Absolute Value Inequality

Solve the inequality and write the solution set in interval notation.

$$-4 \geq -2|3x + 1|$$

Solution:

$-4 \geq -2|3x + 1|$	First isolate the absolute value.
$\dfrac{-4}{-2} \leq \dfrac{-2|3x + 1|}{-2}$	Divide both sides by -2 and reverse the inequality sign.
$2 \leq |3x + 1|$	
$|3x + 1| \geq 2$	Write the absolute value on the left. Notice that the direction of the inequality sign is also changed. The inequality is now in the form $|u| \geq k$, where $u = 3x + 1$.
$3x + 1 \leq -2$ or $3x + 1 \geq 2$	Write the equivalent form $u \leq -k$ or $u \geq k$.
$x \leq -1$ or $x \geq \dfrac{1}{3}$	Take the union of the solution sets of the individual inequalities.

The solution set is $\left\{ x \,\middle|\, x \leq -1 \ \text{ or } \ x \geq \dfrac{1}{3} \right\}$.

Interval notation: $(-\infty, -1] \cup \left[\dfrac{1}{3}, \infty \right)$

Skill Practice 3   Solve the inequality and write the solution set in interval notation. $-18 > -3|2y - 4|$

An alternative method to solve an inequality is the **test point method**. To use this method, first solve the related *equation*. Use the solutions to the equation to partition the number line into intervals. Then select test points within each interval. If a test point makes the original inequality true, then the interval from which the point was taken is a solution to the inequality. If a test point makes the inequality false, then the interval from which it was taken is *not* a solution.

$-4 \geq -2|3x + 1|$	Inequality from Example 3.
$-4 = -2|3x + 1|$	Solve the related *equation*. Change the inequality sign to an equal sign.
$2 = |3x + 1|$	Divide both sides by -2 to isolate the absolute value.
$3x + 1 = -2$ or $3x + 1 = 2$	Solve the absolute value equation by writing the equation in the form $u = -k$ or $u = k$.
$3x = -3$ or $3x = 1$	Solve for x.
$x = -1$ or $x = \dfrac{1}{3}$	

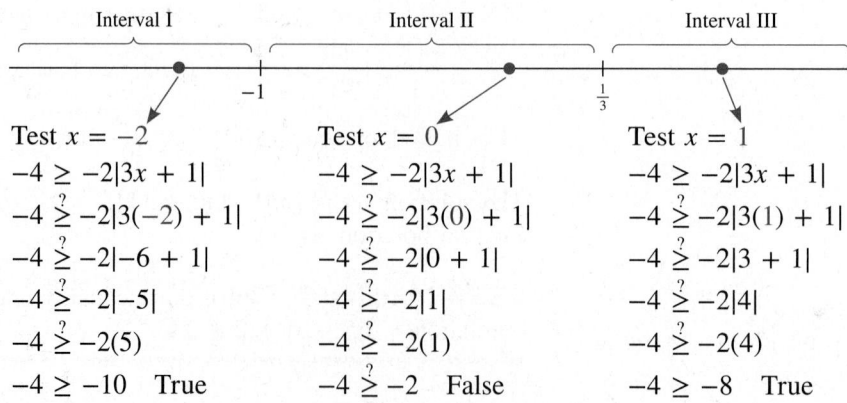

Test $x = -2$	Test $x = 0$	Test $x = 1$
$-4 \geq -2|3x + 1|$	$-4 \geq -2|3x + 1|$	$-4 \geq -2|3x + 1|$
$-4 \overset{?}{\geq} -2|3(-2) + 1|$	$-4 \overset{?}{\geq} -2|3(0) + 1|$	$-4 \overset{?}{\geq} -2|3(1) + 1|$
$-4 \overset{?}{\geq} -2|-6 + 1|$	$-4 \overset{?}{\geq} -2|0 + 1|$	$-4 \overset{?}{\geq} -2|3 + 1|$
$-4 \overset{?}{\geq} -2|-5|$	$-4 \overset{?}{\geq} -2|1|$	$-4 \overset{?}{\geq} -2|4|$
$-4 \overset{?}{\geq} -2(5)$	$-4 \overset{?}{\geq} -2(1)$	$-4 \overset{?}{\geq} -2(4)$
$-4 \geq -10$ True	$-4 \overset{?}{\geq} -2$ False	$-4 \geq -8$ True

Answer

3. $(-\infty, -1) \cup (5, \infty)$

The inequality is true for values taken from Intervals I and III.

The solution set is $\left\{ x \,\middle|\, x \le -1 \text{ or } x \ge \dfrac{1}{3} \right\}$.

Interval notation: $(-\infty, -1] \cup \left[\dfrac{1}{3}, \infty \right)$

An absolute value equation such as $|7x - 3| = -6$ has no solution because an absolute value cannot be equal to a negative number. We must also exercise caution when an absolute value is compared to a negative number or zero within an inequality. This is demonstrated in Example 4.

EXAMPLE 4   Solving Absolute Value Inequalities with Special Case Solution Sets

Solve the inequality and write the solution set in interval notation where appropriate.

a. $|x + 2| < -4$ 　　　　**b.** $|x + 2| \ge -4$

c. $|x - 5| \le 0$ 　　　　**d.** $|x - 5| > 0$

Solution:

a. $|x + 2| < -4$

The solution set is $\{ \ \}$.

> By definition an absolute value is greater than or equal to zero. Therefore, the absolute value of an expression cannot be less than zero or any negative number. This inequality has no solution.

b. $|x + 2| \ge -4$

The solution set is $\mathbb{R}$.

Interval notation: $(-\infty, \infty)$

> An absolute value of any real number is greater than or equal to zero. Therefore, it is also greater than every negative number. This inequality is true for all real numbers, x.

c. $|x - 5| \le 0$

$|x - 5| = 0$

$x - 5 = 0$

$x = 5$

The solution set is $\{5\}$.

> The absolute value of x minus 5 cannot be less than zero, but it can be *equal* to zero.

d. $|x - 5| > 0$

> $|x - 5| > 0$ for all values of x except 5. When $x = 5$, we have $|5 - 5| = 0$, and this is not greater than zero. The solution set is all real numbers excluding 5.

The solution set is $\{x | x < 5 \text{ or } x > 5\}$.

Interval notation: $(-\infty, 5) \cup (5, \infty)$

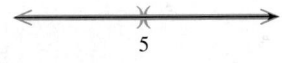

Skill Practice 4   Solve.

a. $|x - 3| < -2$ 　　　　**b.** $|x - 3| > -2$

c. $|x + 1| \le 0$ 　　　　**d.** $|x + 1| > 0$

Answers

4. a. $\{ \}$ 　**b.** $\mathbb{R}$; $(-\infty, \infty)$

　c. $\{-1\}$ 　**d.** $(-\infty, -1) \cup (-1, \infty)$

2. Solve Applications with Absolute Value Inequalities

Recall that the distance between two points a and b on a number line is given by $|a - b|$ or $|b - a|$. In Example 5, we use absolute value expressions to write mathematical models to represent English statements.

EXAMPLE 5   Expressing Distances with Absolute Value

Write an absolute value inequality to represent the following phrases.

a. All real numbers x, whose distance from zero is greater than 5 units

b. All real numbers x, whose distance from -7 is less than 3 units

Solution:

> **TIP** In Example 5(a), the inequality can also be expressed as $|0 - x| > 5$ or equivalently $|-x| > 5$.
>
> The simplifies as
>
> $$|-1 \cdot x| > 5$$
> $$|-1| \cdot |x| > 5$$
> $$|x| > 5$$
>
> Likewise, in Example 5(b), the inequality $|-7 - x| < 3$ is also correct, but we've chosen the simpler form $|x + 7| < 3$.

a. All real numbers x, whose distance from zero is greater than 5 units

$|x - 0| > 5$ or simply $|x| > 5$

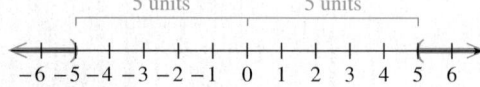

b. All real numbers x, whose distance from -7 is less than 3 units

$|x - (-7)| < 3$ or simply $|x + 7| < 3$

Skill Practice 5   Write an absolute value inequality to represent the following phrases.

a. All real numbers whose distance from zero is greater than 10 units

b. All real numbers whose distance from 4 is less than 6 units

---

Absolute value expressions can also be used to describe boundaries for measurement error.

EXAMPLE 6   Applying an Absolute Value Inequality

Suppose that a machine is calibrated to dispense 8 fl oz of orange juice into a plastic bottle, with a measurement error of no more than 0.05 fl oz. Let x represent the actual amount of orange juice poured into the bottle.

a. Write an absolute value inequality that represents an interval in which to estimate x.

b. Solve the inequality and interpret the answer.

Solution:

> **TIP** In Example 6, the inequality $|8 - x| \leq 0.05$ is also correct.

a. The measurement error is ± 0.05 fl oz. This means that the value of x can deviate from 8 fl oz by as much as ± 0.05 fl oz.

$|x - 8| \leq 0.05$ The distance between x and 8 is no more than 0.05 unit.

b. $|x - 8| \leq 0.05$

$-0.05 \leq x - 8 \leq 0.05$ The amount of orange juice in the bottle is between

$7.95 \leq x \leq 8.05$ 7.95 fl oz and 8.05 fl oz, inclusive.

Answers

5. **a.** $|x| > 10$ **b.** $|x - 4| < 6$
6. **a.** $|x - 24| \leq 0.02$
   b. $23.98 \leq x \leq 24.02$; The actual length of the board is between 23.98 in. and 24.02 in., inclusive.

Skill Practice 6   A board is to be cut to a length of 24 in. The measurement error is no more than 0.02 in. Let x represent the actual length of the board.

a. Write an absolute value inequality that represents an interval in which to estimate x.

b. Solve the inequality from part (a) and interpret the meaning.

SECTION 1.7   Practice Exercises

Prerequisite Review

For Exercises R.1–R.4, graph the solution set. (See Section 1.5 for review.)

R.1. $x \geq -7$ and $x \leq 7$ **R.2.** $x > -4$ and $x < 4$ **R.3.** $x < -7$ or $x > 7$ **R.4.** $x \leq -4$ or $x \geq 4$

For Exercises R.5–R.10, write the solution set in interval notation. (See Section 1.5 for review.)

R.5. $2x - 3 < -1$ or $2x - 3 > 1$ **R.6.** $5x + 1 < -4$ or $5x + 1 > 4$ **R.7.** $-1 \leq 2x - 3$ and $2x - 3 \leq 1$

R.8. $-4 \leq 5x + 1$ and $5x + 1 \leq 4$ **R.9.** $-\dfrac{1}{4} < 2y - 9 < \dfrac{1}{4}$ **R.10.** $-\dfrac{5}{3} < t + 3 < \dfrac{5}{3}$

Concept Connections

1. If k is a positive real number, then the inequality $|x| < k$ is equivalent to _____ $< x <$ _____.

2. If k is a positive real number, then the inequality $|x| > k$ is equivalent to $x <$ _____ or x _____ k.

3. If k is a positive real number, then the solution set to the inequality $|x| > -k$ is _____.

4. If k is a positive real number, then the solution set to the inequality $|x| < -k$ is _____.

5. The solution set to the inequality $|x + 2| < -6$ is _____, whereas the solution set to the inequality $|x + 2| > -6$ is

_____.

6. The solution set to the inequality $|x + 4| \leq 0$ (includes/excludes) -4, whereas the solution set to the inequality $|x + 4| < 0$ (includes/excludes) -4.

7. The solution set to the inequality $|x - 7| < 2$ is the set of real numbers that are less than _____ units from 7 on the number line. These are the real numbers between _____ and _____.

8. The solution set to the inequality $|x + 5| > 3$ is the set of real numbers that are more than 3 units from _____ on the number line. These are the real numbers less than _____ and greater than _____.

Objective 1: Solve Absolute Value Inequalities

For Exercises 9–20, solve the equations and inequalities. For each inequality, graph the solution set and express the solution set in interval notation. **(See Examples 1–4)**

9. a. $|x| = 5$
   b. $|x| > 5$
   c. $|x| < 5$

10. a. $|a| = 4$
   b. $|a| > 4$
   c. $|a| < 4$

11. a. $|x - 3| = 7$
   b. $|x - 3| > 7$
   c. $|x - 3| < 7$

12 a. $|w + 2| = 6$
   b. $|w + 2| > 6$
   c. $|w + 2| < 6$

13. a. $|p| = -2$
   b. $|p| > -2$
   c. $|p| < -2$

14. a. $|x| = -14$
   b. $|x| > -14$
   c. $|x| < -14$

15. a. $|y + 1| = -6$
   b. $|y + 1| > -6$
   c. $|y + 1| < -6$

16. a. $|z - 4| = -3$
   b. $|z - 4| > -3$
   c. $|z - 4| < -3$

17. a. $|p + 3| = 0$
   b. $|p + 3| > 0$
   c. $|p + 3| < 0$

18. a. $|k - 7| = 0$
   b. $|k - 7| > 0$
   c. $|k - 7| < 0$

19. a. $|a + 9| + 2 = 6$
   b. $|a + 9| + 2 < 6$
   c. $|a + 9| + 2 > 6$

20. a. $|b + 1| - 4 = 1$
   b. $|b + 1| - 4 \leq 1$
   c. $|b + 1| - 4 \geq 1$

For Exercises 21–50, solve the inequality, and write the solution set in interval notation if possible. **(See Examples 1–4)**

21. $|4x + 5| + 7 < 10$

22. $|8y + 3| - 4 \leq 11$

23. $|5 - t| - 2 \geq 11$

24. $|7 - m| + 4 > 10$

25. $5 \leq |2x - 1|$

26. $7 \leq |x - 2|$

27. $|y + 2| \geq 0$

28. $0 \leq |7n + 2|$

29. $|k - 7| < -3$

30. $|h + 2| < -9$

31. $12 \leq |9 - 4y| - 2$

32. $5 > |2m - 7| + 4$

33. $3|4 - x| - 2 < 16$

34. $2|7 - y| + 1 < 17$

35. $2|x + 3| - 4 \geq 6$

36. $5|x + 1| - 9 \geq -4$

37. $|4w - 5| + 6 \leq 2$

38. $|2x + 7| + 5 < 1$

39. $|5 - p| + 13 > 6$

40. $|12 - 7x| + 5 \geq 4$

41. $-11 \leq 5 - |2p + 4|$

42. $-18 \leq 6 - |3z + 3|$

43. $10 < |-5c - 4| + 2$

44. $15 < |-2d - 3| + 6$

45. $\left| \dfrac{y + 3}{6} \right| < 2$

46. $\left| \dfrac{m - 4}{2} \right| < 14$

47. $\left| \dfrac{x + 3}{2} \right| - 2 \geq 4$

48. $\left| \dfrac{8 - n}{3} \right| + 4 > 12$

49. $|0.02x + 0.06| - 0.1 < 0.05$

50. $|0.05x - 0.04| - 0.01 < 0.11$

Objective 2: Solve Applications with Absolute Value Inequalities

For Exercises 51–56, write an absolute value inequality equivalent to the given statement. **(See Example 5)**

51. All real numbers x whose distance from 0 is greater than 7

52. All real numbers x whose distance from 0 is at least 6

53. All real numbers x whose distance from -3 is less than 4

54. All real numbers x whose distance from -2 is at most 13.

55. All real numbers x whose distance from c is less than δ (δ is the Greek letter "delta").

56. All real numbers y whose distance from L is less than ε (ε is the Greek letter "epsilon").

For Exercises 57–60,

   a. Write an absolute value inequality to represent each statement.

   b. Solve the inequality. Write the solution set in interval notation.

57. The variation between the measured value v and 16 oz is less than 0.01 oz.

58. The variation between the measured value t and 60 min is less than 0.2 min.

59. The value of x differs from 4 by more than 1 unit.

60. The value of y differs from 10 by more than 2 units.

61. A refrigerator manufacturer recommends that the temperature t (in °F) inside a refrigerator be 36.5°F. If the thermostat has a margin of error of no more than 1.5°F,

   a. Write an absolute value inequality that represents an interval in which to estimate t. **(See Example 6)**

   b. Solve the inequality and interpret the answer.

62. A box of cereal is labeled to contain 16 oz. A consumer group takes a sample of 50 boxes and measures the contents of each box. The individual content of each box differs slightly from 16 oz, but by no more than 0.5 oz.

   a. If x represents the exact weight of the contents of a box of cereal, write an absolute value inequality that represents an interval in which to estimate x.

   b. Solve the inequality and interpret the answer.

63. The results of a political poll indicate that the leading candidate will receive 51% of the votes with a margin of error of no more than 3%. Let x represent the true percentage of votes received by this candidate.

   a. Write an absolute value inequality that represents an interval in which to estimate x.

   b. Solve the inequality and interpret the answer.

64. A police officer uses a radar detector to determine that a motorist is traveling 34 mph in a 25-mph school zone. The driver goes to court and argues that the radar detector is not accurate. The manufacturer claims that the radar detector is calibrated to be in error by no more than 3 mph.

   a. If x represents the motorist's actual speed, write an inequality that represents an interval in which to estimate x.

   b. Solve the inequality and interpret the answer. Should the motorist receive a ticket?

65. A 32-oz jug of orange juice may not contain exactly 32 oz of juice. The possibility of measurement error exists when the jug is filled in the factory. If the maximum measurement error is ±0.05 oz,

   a. Write an absolute value inequality representing the range of volumes x in which the orange juice mug may be filled.

   b. Solve the inequality and interpret the answer.

66. The length of a board is measured to be 32.5 in. The maximum measurement error is ±0.2 in.

   a. Write an absolute value inequality representing the range for the length x of the board.

   b. Solve the inequality and interpret the answer.

Mixed Exercises

For Exercises 67–74, solve the equations and inequalities. Write the solutions to the inequalities in interval notation where possible.

67. a. $|x| = -9$
  b. $|x| < -9$
  c. $|x| > -9$

68. a. $|y| = -2$
  b. $|y| < -2$
  c. $|y| > -2$

69. a. $18 = 4 - |y + 7|$
  b. $18 \leq 4 - |y + 7|$
  c. $18 \geq 4 - |y + 7|$

70. a. $15 = 2 - |p - 3|$
  b. $15 \leq 2 - |p - 3|$
  c. $15 \geq 2 - |p - 3|$

71. a. $|z| = 0$
  b. $|z| < 0$
  c. $|z| \leq 0$
  d. $|z| > 0$
  e. $|z| \geq 0$

72. a. $|2w| = 0$
  b. $|2w| < 0$
  c. $|2w| \leq 0$
  d. $|2w| > 0$
  e. $|2w| \geq 0$

73. a. $|k + 4| = 0$
  b. $|k + 4| < 0$
  c. $|k + 4| \leq 0$
  d. $|k + 4| > 0$
  e. $|k + 4| \geq 0$

74. a. $|c - 3| = 0$
  b. $|c - 3| < 0$
  c. $|c - 3| \leq 0$
  d. $|c - 3| > 0$
  e. $|c - 3| \geq 0$

For Exercises 75–78, write an absolute value inequality whose solution set is shown in the graph.

75.
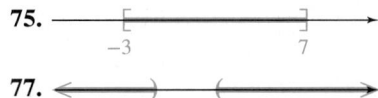
-3 7

76.
2 6

77.
4 10

78.
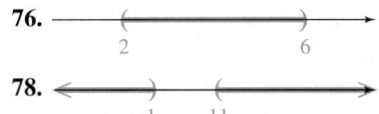
-1 11

Expanding Your Skills

For Exercises 79–84, solve the inequality and write the solution set in interval notation.

79. $|x| + x < 11$ (*Hint:* Use the definition of $|x|$ to consider two cases.)
Case 1: $x + x < 11$ if $x \geq 0$.
Case 2: $-x + x < 11$ if $x < 0$.

80. $|x| - x > 10$

81. $1 < |x| < 9$

82. $2 < |y| < 11$

83. $5 \leq |2x + 1| \leq 7$

84. $7 \leq |3x - 5| \leq 13$

85. Solve the inequality for p: $|p - \hat{p}| < z\sqrt{\dfrac{\hat{p}\hat{q}}{n}}$.

(Do not rationalize the denominator.)

86. Solve the inequality for μ: $|\mu - \bar{x}| < \dfrac{z\sigma}{\sqrt{n}}$.

(Do not rationalize the denominator.)

PROBLEM RECOGNITION EXERCISES

Identifying Equations and Inequalities

For Exercises 1–8, solve each equation or inequality. Express the solution in interval notation where appropriate.

1. a. $3x - 9 = 18$
   b. $|3x - 9| = 18$
   c. $|3x - 9| < 18$
   d. $|3x - 9| \geq 18$

2. a. $5y + 2 = -20$
   b. $|5y + 2| = -20$
   c. $|5y + 2| \leq -20$
   d. $|5y + 2| > -20$

3. a. $-2t - 14 = 0$
   b. $-2t - 14 > 0$
   c. $-2t - 14 \leq 0$

4. a. $\dfrac{x - 2}{3} = 9$
   b. $\dfrac{x - 2}{3} \geq 9$
   c. $\dfrac{x - 2}{3} < 9$

5. a. $|8t - 2| = |-2t + 3|$
   b. $8t - 2 = -2t + 3$

6. a. $-5 < x + 2$ and $x + 2 \leq 8$
   b. $-5 < x + 2 \leq 8$

7. a. $-4x - 9 < 11$ or $2 \leq x + 1$
   b. $-4x - 9 < 11$ and $2 \leq x + 1$

8. a. $4 < 2y$ or $-3(y + 2) > -2y + 1$
   b. $4 < 2y$ and $-3(y + 2) > -2y + 1$

For Exercises 9–28,

   a. Identify the type of equation or inequality. Choose from:
   - linear equation
   - absolute value equation
   - linear inequality
   - compound inequality
   - absolute value inequality

   b. Solve the equation or inequality. Express the solution set in interval notation where appropriate.

9. $-0.5y + 0.7 = 3.7$

10. $3m - 9 = 18$

11. $|2t + 8| \leq 4$

12. $|1 - 3x| < -1$

13. $-11 < 2t + 1 < 19$

14. $2z - 3 \geq 11$ or $3z + 3 < 9$

15. $\left| \dfrac{1}{2}y + 3 \right| = 5$

16. $|4x + 3| = |9 - 2x|$

17. $-\dfrac{3}{4}p \geq -9$

18. $8w + 4 \geq 5w + 1$

19. $\left| \dfrac{2x - 9}{3} \right| \geq 5$

20. $\left| \dfrac{10 - x}{5} \right| < 3$

21. $|2 - c| + 5 = 3$

22. $|10n + 2| + 7 = 7$

23. $\dfrac{w - 4}{5} - \dfrac{w + 1}{3} = 1$

24. $\dfrac{1}{3}y - \dfrac{5}{6} = \dfrac{1}{2}y + 1$

25. $2x - 7 > 9$ and $3x \leq 36$

26. $-3 + x > 2x$ and $2 \geq -\dfrac{1}{3}x$

27. $5(x - 2) + 7 = 2x + 3(x - 1)$

28. $7y - 4 = 3(y + 1) + 4y$

CHAPTER 1   Detailed Summary

SECTION 1.1 Algebraic Expressions and Models

Key Concepts	Examples	Page
Properties of real numbers: • Commutative property of addition • Commutative property of multiplication • Associative property of addition • Associative property of multiplication • Identity property of addition • Identity property of multiplication • Inverse property of addition • Inverse property of multiplication • Distributive property of multiplication over addition	**Example 1:** $a + b = b + a$ $a \cdot b = b \cdot a$ $(a + b) + c = a + (b + c)$ $(a \cdot b) \cdot c = a \cdot (b \cdot c)$ $a + 0 = a$ and $0 + a = a$ $a \cdot 1 = a$ and $1 \cdot a = a$ $a + (-a) = 0$ and $(-a) + a = 0$ $a \cdot \frac{1}{a} = 1$ and $\frac{1}{a} \cdot a = 1$, where $a \neq 0$ $a \cdot (b + c) = a \cdot b + a \cdot c$	p. 38

Simplifying algebraic expressions:
Simplify an algebraic expression by applying the distributive property to clear parentheses. Then combine like terms.
 	
A formula relating two or more variables is one type of mathematical model.
	Example 2:
	Simplify. $4(8 - y) - 2(3y + 9) + 1$
	  $= 32 - 4y - 6y - 18 + 1$
	  $= -10y + 15$
		p. 41

	Example 3: A city law states that an individual caught speeding would pay a \$70 surcharge plus \$2 for every km/hr over the posted limit. The cost C (in \$) for a ticket in this situation is given by: $C = 70 + 2x$ where x is the number of km/hr above the posted speed limit.	p. 43
	Example 4: Use the model from Example 3 to determine the cost of a speeding ticket for an individual traveling 40 km/hr over the posted speed limit. $\quad C = 70 + 2x$ $\quad C = 70 + 2(40)$ Substitute 40 km/hr for x. $\quad\quad = 70 + 80$ $\quad\quad = 150$ The driver would pay \$150.	p. 43

SECTION 1.2 Linear Equations in One Variable

Key Concepts	Examples	Page
A **linear equation in one variable** is an equation that can be written in the form $ax + b = 0$, where a and b are real numbers and $a \neq 0$.	**Example 1:** The equation $2x + 3 = 0$ is a linear equation.	p. 46
Solving a linear equation in one variable: 1. Simplify both sides of the equation. 2. Collect variable terms on one side of the equation and constants on the other side. 3. Apply the multiplication property of equality to obtain a coefficient of 1 on the variable. 4. Check the potential solution. 5. Write the solution set.	**Example 2:** Solve. $\quad -2(x - 3) + 5x = 12$ $\qquad\qquad -2x + 6 + 5x = 12$ $\qquad\qquad\qquad\quad 3x + 6 = 12$ $\qquad\qquad\qquad\qquad 3x = 6$ $\qquad\qquad\qquad\qquad\ x = 2$ The solution checks. The solution set is $\{2\}$.	p. 48
A **conditional equation** is an equation that is true for some values of the variable but false for others.	**Example 3:** $2x = 16$ is a conditional equation because it is true on the condition that $x = 8$. The solution set is $\{8\}$.	p. 52
A **contradiction** is false for all values of the variable.	$2x + 1 = 2x + 3$ or equivalently $1 = 3$ is a contradiction because this statement is false for all values of x. The solution set is $\{\ \}$.	
An **identity** is an equation that is true for all real values of the variable for which the expressions in the equation are defined.	$3x + 6 = 3(x + 2)$ or equivalently $6 = 6$ is an identity. It is true for all real numbers, x. The solution set is $\mathbb{R}$.	
To solve an equation for a specific variable when the equation contains multiple variables, use the properties of equality to isolate the variable.	**Example 4:** Solve $\ 3x - 2y = 24$ for y. $\qquad\qquad -2y = -3x + 24$ $\qquad\qquad \dfrac{-2y}{-2} = \dfrac{-3x + 24}{-2}$ $\qquad\qquad\qquad y = \dfrac{3}{2}x - 12$	p. 53

SECTION 1.3 Applications of Linear Equations in One Variable

Key Concepts	Examples	Page

Key Concepts

Equations in algebra can be used to organize information from a physical situation. The following are suggested guidelines to solve an application.

1. Read the problem carefully. Determine what the problem is asking for and assign variables to the unknown quantities. Make an appropriate figure or table if applicable.
2. Write a verbal model (equation in words) to describe the given scenario.
3. Write an algebraic equation that represents the verbal model.
4. Solve the equation from step 3.
5. Interpret the solution and check that the answer is reasonable.

Example 1 illustrates an application involving mixtures.

Example 2 illustrates an application involving uniform motion.

Examples

Example 1:

How much 20% acid solution should be mixed with 400 mL of a 5% acid solution to bring the concentration rate up to 10%?

Let x represent the amount of 20% solution.

Then $400 + x$ is the amount of 10% solution.

	20% solution	5% solution	10% solution
Mixture	x	400	$400 + x$
Pure acid	$0.20x$	$0.05(400)$	$0.10(400 + x)$

$$\begin{pmatrix} \text{Acid from} \\ \text{20\% solution} \end{pmatrix} + \begin{pmatrix} \text{Acid from} \\ \text{5\% solution} \end{pmatrix} = \begin{pmatrix} \text{Acid from} \\ \text{10\% solution} \end{pmatrix}$$

$$0.20x + 0.05(400) = 0.10(400 + x)$$
$$0.20x + 20 = 40 + 0.10x$$
$$0.10x = 20$$
$$x = 200$$

200 mL of a 20% acid solution is needed.

Example 2:

Jack and Diane participate in a bicycle race. Jack rides the first half of the race in 1.5 hr. Diane rides the second half at a rate 5 mph slower than Jack and completes her portion in 2 hr. How fast does each person ride?

	Distance	Rate	Time
Jack	$1.5x$	x	1.5
Diane	$2(x - 5)$	$x - 5$	2

$$\begin{pmatrix} \text{Distance} \\ \text{Jack rides} \end{pmatrix} = \begin{pmatrix} \text{distance} \\ \text{Diane rides} \end{pmatrix}$$

$$\begin{aligned} 1.5x &= 2(x - 5) \\ 1.5x &= 2x - 10 \\ -0.5x &= -10 \qquad \text{Subtract } 2x. \\ x &= 20 \qquad \text{Divide by } -0.5. \end{aligned}$$

Jack's speed is x. Jack rides 20 mph. Diane's speed is $x - 5$, which is 15 mph.

Page

p. 62

p. 63

SECTION 1.4 Union and Intersection of Sets

Key Concepts	Examples	Page
Set-builder and interval notation: An interval on the real number line can be represented in set-builder notation or in interval notation.	**Example 1:** Set-builder notation Graph Interval notation $\{x \mid x < 3\}$ $(-\infty, 3)$ 3 $\{x \mid -2 < x \le 5\}$ $(-2, 5]$ -2 5	p. 72
$A \cup B$ is the **union** of A and B. This is the set of elements that belong to set A or set B or to both sets A and B. $A \cap B$ is the **intersection** of A and B. This is the set of elements common to both A and B. $A \cup B$ $A \cap B$	**Example 2:** Given $A = \{x \mid x < 2\}$ and $B = \{x \mid x \le 4\}$ A: 2 B: 4 $A \cup B = \{x \mid x \le 4\}$ and $A \cap B = \{x \mid x < 2\}$	p. 74

SECTION 1.5 Linear and Compound Inequalities

Key Concepts	Examples	Page
An inequality that can be written in one of the following forms is a **linear inequality in one variable**. $ax + b < 0$, $ax + b \le 0$, $ax + b > 0$, or $ax + b \ge 0$ ***Properties of inequalities:** Let a, b, and c represent real numbers. 　1. If $x < a$, then $a > x$. 　2. If $a < b$ and $b < c$, then $a < c$. 　3. If $a < b$ and $c < d$, then $a + c < b + d$. 　4. If $a < b$, then 　　　$a + c < b + c$ and $a - c < b - c$. 　5. If c is *positive* and $a < b$, 　　　then $ac < bc$ and $\dfrac{a}{c} < \dfrac{b}{c}$. 　6. If c is *negative* and $a < b$, 　　　then $ac > bc$ and $\dfrac{a}{c} > \dfrac{b}{c}$. *These properties of inequality are also true for statements expressed with the symbols $\le$, $>$, and $\ge$.	**Example 1:** Solve.　$-4x - 7 \ge 9$ 　　　　　$-4x \ge 16$ 　　　$\dfrac{-4x}{-4} \le \dfrac{16}{-4}$ 　　　　　$x \le -4$ 　　　　　　　　-4 Solution set: $\{x \mid x \le -4\}$ Interval notation: $(-\infty, -4]$	p. 82

Solving compound inequalities:

- If two inequalities are joined by the word "and," take the *intersection* of the individual solution sets.

- The inequality $a < x < b$ is equivalent to $a < x$ and $x < b$.

- If two inequalities are joined by the word "or," take the *union* of the individual solution sets.

Example 2:

Solve. $2x < 6$ and $4 \le x + 5$

$x < 3$ and $-1 \le x$

The solution set is $[-1, 3)$.

p. 84

Example 3:

Solve. $-7 < x + 4 \le 8$

$-7 - 4 < x + 4 - 4 \le 8 - 4$

$-11 < x \le 4$

The solution set is $(-11, 4]$.

p. 85

Example 4:

Solve. $-3y > 12$ or $y + 7 \ge 9$

$y < -4$ or $y \ge 2$

The solution set is $(-\infty, -4) \cup [2, \infty)$.

p. 84

SECTION 1.6 Absolute Value Equations

Key Concepts	Examples	Page
Absolute value: For a real number x, the **absolute value of x** is $$|x| = \begin{cases} x & \text{if } x \ge 0 \\ -x & \text{if } x < 0 \end{cases}$$	**Example 1:** $\lvert 2 - \sqrt{7} \rvert = -(2 - \sqrt{7})$ because $2 - \sqrt{7} < 0$. $= \sqrt{7} - 2$	p. 92
Distance between points on a number line: The distance between two points a and b on a number line is given by $\lvert a - b \rvert$ or $\lvert b - a \rvert$.	**Example 2:** The distance between 6 and -4 on the number line is $\lvert 6 - (-4) \rvert = 10$ units or $\lvert -4 - 6 \rvert = 10$ units.	p. 94
Properties involving absolute value equations: Let k represent a positive real number. 1. $\lvert u \rvert = k$ is equivalent to $u = k$ or $u = -k$. 2. $\lvert u \rvert = 0$ is equivalent to $u = 0$. 3. $\lvert u \rvert = -k$ has no solution. 4. $\lvert u \rvert = \lvert w \rvert$ is equivalent to $u = w$ or $u = -w$.	**Example 3:** Solve. $\lvert x - 3 \rvert + 5 = 10$ $\lvert x - 3 \rvert = 5$ Isolate the absolute value. $x - 3 = 5$ or $x - 3 = -5$ $x = 8$ or $x = -2$ The solution set is $\{8, -2\}$. **Example 4:** The equation $\lvert 2x + 3 \rvert = -9$ has no solution. **Example 5:** Solve. $\lvert 2x - 10 \rvert = \lvert x + 2 \rvert$ $2x - 10 = x + 2$ or $2x - 10 = -(x + 2)$ $x = 12$ or $2x - 10 = -x - 2$ $3x = 8$ $x = \dfrac{8}{3}$ The solution set is $\left\{ 12, \dfrac{8}{3} \right\}$.	p. 95 p. 96 p. 97

SECTION 1.7 Absolute Value Inequalities

Key Concepts	Examples	Page

*Properties involving absolute value inequalities:
For a real number $k > 0$,
  1. $\lvert u \rvert < k$ is equivalent to $-k < u < k$.
 	
  2. $\lvert u \rvert > k$ is equivalent to $u < -k$ or $u > k$.
 	
*The statements also hold true for the inequality symbols $\leq$ and $\geq$, respectively.
	Example 1:
	Solve. $\lvert x + 7 \rvert - 3 \leq 4$
	  $\lvert x + 7 \rvert \leq 7$ Isolate the absolute value.
	 $-7 \leq x + 7 \leq 7$ Equivalent form
	  $-14 \leq x \leq 0$ Interval notation: $[-14, 0]$
	
	Example 2:
	Solve. $\lvert x + 1 \rvert > 5$
	$x + 1 < -5$ or $x + 1 > 5$
	  $x < -6$ or $x > 4$
	Interval notation: $(-\infty, -6) \cup (4, \infty)$
		p. 100
		
		p. 101

Absolute value inequalities with special case solution sets: $\lvert u \rvert < 0$ has no solution. $\lvert u \rvert \leq 0$ is true for $u = 0$. $\lvert u \rvert > 0$ is true for $u < 0$ or $u > 0$. $\lvert u \rvert \geq 0$ is true for all real numbers.	**Example 3:** $\lvert x - 4 \rvert < 0$ Solution set: $\{\ \}$ $\lvert x - 4 \rvert \leq 0$ Solution set: $\{4\}$ $\lvert u - 4 \rvert > 0$ Solution set: $(-\infty, 4) \cup (4, \infty)$ $\lvert u - 4 \rvert \geq 0$ Solution set: $(-\infty, \infty)$ **Example 4:** $\lvert 5x + 6 \rvert < -4$ has no solution. $\lvert 3y + 9 \rvert > -3$ is true for all real numbers, $\mathbb{R}$.	p. 103 p. 103

CHAPTER 1   Review Exercises

SECTION 1.1

For Exercises 1–9, identify the property that makes the given statement true. Choose from

  a. Commutative property of addition
  b. Commutative property of multiplication
  c. Associative property of addition
  d. Associative property of multiplication
  e. Identity property of addition
  f. Identity property of multiplication
  g. Inverse property of addition
  h. Inverse property of multiplication
  i. Distributive property of multiplication over addition

1. $5(ab) = (5a)b$ **2.** $\dfrac{1}{5} \cdot 5 = 1$

3. $p + (q + r) = (p + q) + r$

4. $p(q + r) = pq + pr$

5. $-\pi + 0 = -\pi$

6. $-\pi + \pi = 0$

7. $x + (y + z) = (y + z) + x$

8. $1 \cdot x = x$ **9.** $(ab)c = c(ab)$

For Exercises 10–13, apply the distributive property and simplify.

10. $3(x + 5y)$ **11.** $\dfrac{1}{2}(x + 8y - 5)$

12. $-(-4x + 10y - z)$ **13.** $-(13a - b - 5c)$

For Exercises 14–18, clear parentheses if necessary, and combine like terms.

14. $5 - 6q + 13q - 19$

15. $18p + 3 - 17p + 8p$

16. $7 - 3(y + 4) - 3y$

17. $\dfrac{3}{4}(8x - 4) + \dfrac{1}{2}(6x + 4)$

18. $15.2c^2 d - 11.1cd + 8.7c^2 d - 5.4cd$

19. Jesse makes $150 more per week than Ethan.
   a. Write a model for Jesse's salary J in terms of Ethan's salary E.
   b. Write a model for Ethan's salary E in terms of Jesse's salary J.

20. The width of a rectangle is 8 ft less than twice the length. Write a model for the width W in terms of the length L.

21. For a single-story home with a low-pitch roof and asphalt shingles, a roofing company charges $3.60 per square foot to tear off the old shingles and install new shingles. If the plywood underneath is damaged, the roofer charges $50 per sheet to replace it. The cost for high-impact skylights is $250 each.

    a. Write a formula for the cost C (in $) for a new roof, based on the square footage s, the number of plywood sheets replaced p, and the number of skylights n.

    b. Compute the cost to replace a 2100-ft^2 low-pitch roof that requires four sheets of new plywood and two high-impact skylights.

SECTION 1.2

For Exercises 22–26, solve the equation.

22. $-8(t - 4) + 7 = 4[t - 3(1 - t)] + 6$

23. $\dfrac{4}{5}x - \dfrac{2}{3} = \dfrac{7}{10}x - 2$

24. $\dfrac{m + 2}{3} - \dfrac{m - 4}{4} = \dfrac{m + 1}{6} - 1$

25. $x - 5 + 2(x - 4) = 3(x + 1) - 5$

26. $0.2x + 1.6 = x - 0.8(x - 2)$

For Exercises 27–29, solve for the indicated variable.

27. $4x - 3y = 6$ for y

28. $t_a = \dfrac{t_1 + t_2}{2}$ for t_2

29. $4x + 6y = c$ for x

30. Dexter's hybrid car gets 41 mpg in the city and 36 mpg on the highway. The amount of gas he uses A (in gal) is given by $A = \frac{1}{41}c + \frac{1}{36}h$, where c is the number of city miles driven and h is the number of highway miles driven. If Dexter drove 288 mi on the highway and used 11 gal of gas, how many city miles did he drive?

SECTION 1.3

31. a. Cory makes $85,200 in taxable income. If he pays an average of 28% in taxes on his income, determine the amount of tax he must pay.

    b. What is his net income (after taxes)?

32. For a recent year, approximately 7.2 million men were in college in the United States. This represents an 8% increase over the number of men in college in the year 2000. Approximately how many men were in college in 2000? (Round to the nearest tenth of a million.)

33. For a recent year, there were 17,430 deaths due to alcohol-related accidents in the United States. This was a 5% increase over the number of alcohol-related deaths in 1999. How many such deaths were there in 1999?

34. To do a rope trick, a magician needs to cut a piece of rope so that one piece is one-third the length of the other piece. If she begins with a $2\frac{2}{3}$-ft rope, what will be the lengths of the two pieces of rope?

35. Shawna invested a total of $12,000 in two mutual funds: an international fund and a real estate fund. After 1 year, the international fund earned the equivalent of 8.2% simple interest and the real estate fund returned 1.5%. If the total earnings at the end of the year were $749.50, determine the amount invested in each fund.

36. Cassandra bought a 10-year Treasury note that paid the equivalent of 3.5% simple interest. She invested $4000 more in a 15-year bond earning 4.1% than she did in the Treasury note. If the total amount of interest from these investments is $10,180, determine the amount of principal for each investment. Assume that each investment was held to maturity.

37. A chemist mixed 100 cc of a 60% acid solution with a 20% acid solution to produce a 25% acid solution. How much of the 20% solution did he use?

38. Suppose that 250 ft^3 of dry concrete mixture is 50% sand by volume. How much pure sand must be added to form a new mixture that is 70% sand by volume?

39. When Kevin commuted to work one morning, his average speed was 45 mph. He averaged only 30 mph for the return trip because of an accident on the highway. If the total time for the round trip was 50 min ($\frac{5}{6}$ hr), determine the distance between his place of work and his home.

40. Two boats leave a marina at the same time. One travels south and the other travels north. The southbound boat travels 6 mph faster than the northbound boat. After 3 hr, the distance between the boats is 66 mi. Determine the speed of each boat.

SECTION 1.4

41. Write the statement as an inequality. "$6y$ is no more than 8."

42. Complete the table.

	Graph	Interval Notation	Set-Builder Notation
a.	(graph from -3 to 7)		
b.	(graph)	$(2.1, \infty)$	
c.	(graph)		$\{x \mid 4 \geq x\}$

43. Given $A = \{10, 11, 12, 13\}$, $B = \{10, 12, 14, 16\}$, and $C = \{7, 8, 9, 10, 11\}$, find

    a. $A \cup B$ **b.** $A \cap B$ **c.** $A \cup C$

    d. $A \cap C$ **e.** $B \cup C$ **f.** $B \cap C$

44. Given $X = \{x \mid x < 7\}$, $Y = \{x \mid x \geq -2\}$, and $Z = \{x \mid x < -3\}$, write the union or intersection in set-builder notation.

    a. $X \cup Y$ **b.** $X \cap Y$ **c.** $X \cup Z$

    d. $X \cap Z$ **e.** $Y \cup Z$ **f.** $Y \cap Z$

Let $X = \{x | x \geq -10\}$, $Y = \{x | x < 1\}$, $Z = \{x | x > -1\}$, and $W = \{x | x \leq -3\}$. For Exercises 45–50, find the intersection or union of the sets X, Y, Z, and W. Write the answers in interval notation.

45. $X \cap Y$ **46.** $X \cup Y$

47. $Y \cup Z$ **48.** $Y \cap Z$

49. $Z \cup W$ **50.** $Z \cap W$

SECTION 1.5

For Exercises 51–56, solve the inequality and graph the solution set. Write the solution set in (a) set-builder notation and (b) interval notation.

51. $-6x - 2 > 6$ **52.** $-10x \leq 15$

53. $5 - 7(x + 3) > 19x$ **54.** $4 - 3x \geq 10(-x + 5)$

55. $\dfrac{5 - 4x}{8} \geq 9$ **56.** $\dfrac{3 + 2x}{4} \leq 8$

57. Dave earned the following test scores in his biology class: 82, 88, 92, and 93. How high does he have to score on the fifth test to have an average of 90 or more?

For Exercises 58–69, solve the compound inequalities. Write the solutions in interval notation.

58. $4m > -11$ and $4m - 3 \leq 13$

59. $4n - 7 < 1$ and $7 + 3n \geq -8$

60. $-3y + 1 \geq 10$ and $-2y - 5 \leq -15$

61. $\dfrac{1}{2} - \dfrac{h}{12} \leq -\dfrac{7}{12}$ and $\dfrac{1}{2} - \dfrac{h}{10} > -\dfrac{1}{5}$

62. $\dfrac{2}{3}t - 3 \leq 1$ or $\dfrac{3}{4}t - 2 > 7$

63. $2(3x + 1) < -10$ or $3(2x - 4) \geq 0$

64. $-7 < -7(2w + 3)$ or $-2 < -4(3w - 1)$

65. $5(p + 3) + 4 > p - 1$ or $4(p - 1) + 2 > p + 8$

66. $-11 \leq -4x - 1 \leq 7$

67. $0 < \dfrac{-3x + 9}{-4} < 6$

68. $2 \geq -(b - 2) - 5b \geq -6$

69. $-4 \leq \dfrac{1}{2}(x - 1) < -\dfrac{3}{2}$

70. Normal levels of total cholesterol vary according to age. For adults between 25 and 40 years old, the normal range is generally accepted to be between 140 and 225 mg/dL (milligrams per deciliter), inclusive. Let x represent cholesterol level.

   a. Write an inequality representing the normal range for total cholesterol for adults between 25 and 40 years old.

   b. Write a compound inequality representing abnormal ranges for total cholesterol for adults between 25 and 40 years old.

71. Normal levels of total cholesterol vary according to age. For adults younger than 25 years old, the normal range is generally accepted to be between 125 and 200 mg/dL, inclusive. Let x represent cholesterol level.

   a. Write an inequality representing the normal range for total cholesterol for adults younger than 25 years old.

   b. Write a compound inequality representing abnormal ranges for total cholesterol for adults younger than 25 years old.

72. Write an inequality to represent the following statement. A pilot is instructed to keep a plane at an altitude a of over 29,000 ft, but not to exceed 31,000 ft.

73. The months of June, July, August, and September are the wettest months in Miami, Florida, averaging 7.83 in. per month. If Miami gets 8.54 inches in June, 5.79 inches in July, and 8.63 inches in August, how much rain is needed in September to exceed the monthly average for these 4 months?

74. A homeowner wants to resod her 2000-ft² lawn. Sod varies in price from $0.10/ft² to $0.30/ft² depending on the type of grass. The cost of labor for her gardener to put down the sod is $400. If the homeowner has budgeted $850 for the project, determine the price range per square foot of sod that she can afford.

SECTION 1.6

For Exercises 75–77, simplify by writing the expression without absolute value bars.

75. **a.** $|w - 4|$ for $w < 4$

   b. $|w - 4|$ for $w \geq 4$

76. **a.** $|9 - t|$ for $t < 9$

   b. $|9 - t|$ for $t \geq 9$

77. **a.** $|5x + 8|$ for $x \geq -\dfrac{8}{5}$

   b. $|5x + 8|$ for $x < -\dfrac{8}{5}$

For Exercises 78–81, write an absolute value expression to represent the distance between the two points on the number line. Then simplify without absolute value bars.

78. 9 and -7 **79.** -12 and -8

80. $\sqrt{11}$ and 3   **81.** $\sqrt{19}$ and 5

For Exercises 82–95, solve the equations.

82. $|x| = 10$ **83.** $|x| = 17$

84. $|8.7 - 2x| = 6.1$ **85.** $|5.25 - 5x| = 7.45$

86. $16 = |x + 2| + 9$ **87.** $5 = |x - 2| + 4$

88. $|4x - 1| + 6 = 4$ **89.** $|3x - 1| + 7 = 3$

90. $\left|\dfrac{7x - 3}{5}\right| + 4 = 4$ **91.** $\left|\dfrac{4x + 5}{-2}\right| - 3 = -3$

92. $|3x - 5| = |2x + 1|$ **93.** $|8x + 9| = |8x - 1|$

94. $|2 + 7d| = |-7d - 2|$ **95.** $-|4y + 6| = |2y - 3|$

SECTION 1.7

For Exercises 96–111, solve the equation or inequality. Write the solution set to the inequalities in interval notation if possible.

96. a. $|w + 2| + 1 = 6$ **97. a.** $3 = |7x + 1| + 4$

    b. $|w + 2| + 1 < 6$ **b.** $3 < |7x + 1| + 4$

    c. $|w + 2| + 1 \geq 6$ **c.** $3 \geq |7x + 1| + 4$

98. a. $|y + 5| - 3 = -3$ **99. a.** $|x - 1| = |3x + 5|$

    b. $|y + 5| - 3 < -3$ **b.** $|x - 1| = |x + 5|$

    c. $|y + 5| - 3 \leq -3$ **c.** $|x - 1| = |1 - x|$

    d. $|y + 5| - 3 > -3$

    e. $|y + 5| - 3 \geq -3$

100. $2|7x - 1| + 4 > 4$ **101.** $4|5x + 1| - 3 > -3$

102. $|3x + 4| - 6 \leq -4$

103. $|5x - 3| + 3 \leq 6$

104. $\left|\dfrac{x}{2} - 6\right| < 5$ **105.** $\left|\dfrac{x}{3} + 2\right| < 2$

106. $|4 - 2x| + 8 \geq 8$ **107.** $|9 + 3x| + 1 \geq 1$

108. $-2|5.2x - 7.8| < 13$ **109.** $-|2.5x + 15| < 7$

110. $|3x - 8| < -1$ **111.** $|x + 5| < -4$

112. The Neilsen ratings estimated that the percent, p, of the television viewing audience watching a popular music show was 20% with a 3% margin of error. Solve the inequality $|p - 0.20| \leq 0.03$ and interpret the answer in the context of this problem.

113. The length, L, of a screw is supposed to be $3\frac{3}{8}$ in. Due to variation in the production equipment, there is a $\frac{1}{4}$-in. margin of error. Solve the inequality $|L - 3\frac{3}{8}| \leq \frac{1}{4}$ and interpret the answer in the context of this problem.

CHAPTER 1   Test

For Exercises 1–3, simplify the expressions.

1. $5b + 2 - 7b + 6 - 14$

2. $-3(4 - x) + 9(x - 1) - 5(2x - 4)$

3. $\dfrac{1}{2}(2x - 1) - \left(3x - \dfrac{3}{2}\right)$

For Exercises 4–12, solve the equations.

4. $\dfrac{x}{7} + 1 = 20$

5. $8 - 5(4 - 3z) = 2(4 - z) - 8z$

6. $0.12x + 0.08(60,000 - x) = 10,500$

7. $\dfrac{5 - x}{6} - \dfrac{2x - 3}{2} = \dfrac{x}{3}$

8. $\left|\dfrac{1}{2}x + 3\right| - 4 = 4$

9. $|3x + 4| = |x - 12|$

10. $-5 = -8 + |2y - 3|$

11. $|3.7x - 5| + 7 = 6.2$

12. $|8x + 11| = |8x + 5|$

13. Label each equation as a conditional equation, an identity, or a contradiction.

    a. $(5x - 9) + 19 = 5(x + 2)$

    b. $2a - 2(1 + a) = 5$

    c. $(4w - 3) + 4 = 3(5 - w)$

14. Joëlle is determined to get some exercise and walks to the store at a brisk rate of 4.5 mph. She meets her friend Yun Ling at the store, and together they walk back at a slower rate of 3 mph. Joëlle's total walking time was 1 hr.

    a. How long did it take her to walk to the store?

    b. What is the distance to the store?

15. Shawnna has money distributed between two accounts: an account that earns 5% simple interest and an account that earns 3.5% simple interest. She has $100 less invested at 3.5% than at 5%. If after 1 year her total interest is $81.50, how much did she invest at 5%?

16. How many gallons of a 20% acid solution must be mixed with 6 gal of a 30% acid solution to make a 22% solution?

For Exercises 17–18, solve the equations for the indicated variable.

17. $4x + 2y = 6$ for y **18.** $x = \mu + z\sigma$ for z

19. Max is twice as old as Jonas. If M represents Max's age and J represents Jonas's age, write an expression representing

    a. Max's age in terms of Jonas's age.

    b. Jonas's age in terms of Max's age.

20. Given $A = \{x \mid x < 2\}$, $B = \{x \mid x \geq 0\}$, and $C = (x \mid x < -1\}$, write the union or intersection in set notation.

    a. $A \cup B$ **b.** $A \cap B$

    c. $A \cup C$ **d.** $A \cap C$

    e. $B \cup C$ **f.** $B \cap C$

21. Simplify without absolute value bars. $|5 - t|$ for $t > 5$.

22. a. Write an absolute value expression that represents the distance between 3π and 9 on the number line.

    b. Simplify the expression without absolute value bars.

For Exercises 23–25, solve the inequalities. Graph the solution and write the solution set in interval notation.

23. $x + 8 > 42$ **24.** $-\dfrac{3}{2}x + 6 \geq x - 3$

25. $-1 < \dfrac{1}{2}x - 5 \leq 7$

For Exercises 26–38, solve the compound and absolute value inequalities. Write the answers in interval notation where possible.

26. $-2 \leq 3x - 1 \leq 5$ **27.** $-4 \leq \dfrac{6 - 2x}{5} < 2$

28. $-\dfrac{3}{5}x - 1 \leq 8$ or $-\dfrac{2}{3}x \geq 16$

29. $-2x - 3 > -3$ and $x + 3 \geq 0$

30. $5x + 1 \leq 6$ or $2x + 4 > -6$

31. $2x - 3 > 1$ and $x + 4 < -1$

32. $|3 - 2x| + 6 < 2$

33. $|3x - 8| \geq 9$

34. $|0.4x + 0.3| - 0.2 < 7$

35. $|7 - 3x| + 1 > -3$

36. $6 \geq |2x - 5| - 5$

37. a. $|7x + 4| + 11 = 2$ **b.** $|7x + 4| + 11 < 2$

    c. $|7x + 4| + 11 > 2$

38. a. $|x - 13| + 4 = 4$ **b.** $|x - 13| + 4 < 4$

    c. $|x - 13| + 4 \leq 4$ **d.** $|x - 13| + 4 > 4$

    e. $|x - 13| + 4 \geq 4$

39. An elevator can accommodate a maximum weight of 2000 lb. If four passengers on the elevator have an average weight of 180 lb each, how many additional passengers of the same average weight can the elevator carry before the maximum weight capacity is exceeded?

Keith Brofsky/Getty Images

40. The normal range in humans of the enzyme adenosine deaminase (ADA) is between 9 and 33 IU (international units), inclusive. Let x represent the ADA level in international units.

    a. Write an inequality representing the normal range for ADA.

    b. Write a compound inequality representing abnormal ranges for ADA.

41. The mass of a small piece of metal is measured to be 15.4 g. If the measurement error is at most ± 0.1 g,

    a. Write an absolute value inequality that represents the possible mass x of the piece of metal.

    b. Solve the inequality and interpret the result.

Polynomials

2

Parnianto parnianto/ndoeljindoel/123RF

Epidemiologists are scientists who study the spread and control of disease. Every year, epidemiologists are called into action during flu season. They help health departments all over the world to study, track, and contain the spread of the flu. During one particularly bad flu season, scientists estimated the probability of an individual getting the flu to be between 5% and 20%.

Suppose that a small manufacturing company has 5 employees. Management fears that if two or more of the employees get the flu, productivity will drop dangerously low. This in turn may prevent the company from filling its orders and staying in business. Suppose that the probability that an individual gets the flu is p. Using statistical techniques, the statement $P = -4p^5 + 15p^4 - 20p^3 + 10p^2$ gives the probability P that two or more individuals in a group of five will get the flu during flu season. The expression $-4p^5 + 15p^4 - 20p^3 + 10p^2$ is called a polynomial, and in this chapter, we will learn how to manipulate polynomials and use them in applications. For example, if $p = 0.05$ (the probability that a single individual gets the flu), then

$$P = -4(0.05)^5 + 15(0.05)^4 - 20(0.05)^3 + 10(0.05)^2 \approx 0.02$$

This means that the probability that two or more people in a group of five will get the flu is about 2%. However, if $p = 0.20$, then $P \approx 0.26$. Thus, if the individual flu rate is 20%, then there is a 26% chance that two or more people in a company of five will get the flu. In such a case, company management will probably want to hire more help during flu season.

Integer Exponents and Scientific Notation

OBJECTIVES

1. Simplify Expressions with Integer Exponents
2. Apply Scientific Notation

1. Simplify Expressions with Integer Exponents

Recall that exponents are used to represent repeated multiplication. Applications of exponents appear in many fields of study, including computer science. Computer engineers define a *bit* as a fundamental unit of information having just two possible values. These values are represented by either 0 or 1. A *byte* is usually taken as 8 bits. Computer programmers know that there are 2^n possible values for an *n*-bit variable. So 1 byte has $2^8 = 256$ possible values.

Three bytes are often used to represent color on a computer screen. The intensity of each of the colors red, green, and blue ranges from 0 to 255 (a total of 256 possible values each). So the number of colors that can be represented by this system is

$$2^8 \cdot 2^8 \cdot 2^8 = (256)(256)(256) = 16{,}777{,}216$$

There are over 16 million possible colors available using this system. For example, the color given by red 137, green 21, blue 131, is a deep pink. See Figure 2-1.

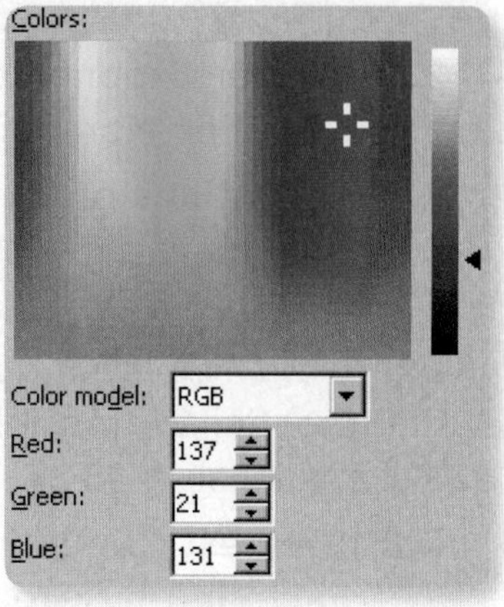

In this section, we summarize several properties of exponents that enable us to simplify expressions with exponents. For example, consider the following expressions and their expanded forms.

Figure 2-1

$$x^5 \cdot x^2 = (x \cdot x \cdot x \cdot x \cdot x)(x \cdot x) = x \cdot x \cdot x \cdot x \cdot x \cdot x \cdot x = x^7$$

$$\frac{x^5}{x^2} = \frac{x \cdot x \cdot x \cdot x \cdot x}{x \cdot x} = \frac{\cancel{x} \cdot \cancel{x} \cdot x \cdot x \cdot x}{\cancel{x} \cdot \cancel{x}} = \frac{x \cdot x \cdot x}{1} = x^3 \quad \text{(provided } x \neq 0\text{)}$$

$$(x^5)^2 = (x \cdot x \cdot x \cdot x \cdot x)^2 = (x \cdot x \cdot x \cdot x \cdot x)(x \cdot x \cdot x \cdot x \cdot x)$$
$$= x \cdot x \cdot x \cdot x \cdot x \cdot x \cdot x \cdot x \cdot x \cdot x = x^{10}$$

In each case, we could have reached the simplified form more quickly by noting the pattern that arises.

$$x^5 \cdot x^2 = x^{5+2} = x^7 \qquad \text{To multiply the same base, add the exponents.}$$

$$\frac{x^5}{x^2} = x^{5-2} = x^3 \qquad \text{To divide the same base, subtract the exponents.}$$

$$(x^5)^2 = x^{5 \cdot 2} = x^{10} \qquad \text{To raise a base to multiple powers, multiply the exponents.}$$

These examples suggest the following properties for positive integer exponents.

$$b^m \cdot b^n = b^{m+n} \qquad \frac{b^m}{b^n} = b^{m-n} \text{ for } b \neq 0 \qquad (b^m)^n = b^{m \cdot n}$$

For these properties to hold true for negative and zero exponents, we must define $b^0 = 1$ so that we have

$$b^m \cdot b^0 = b^{m+0} = b^m$$
$$\downarrow$$
$$b^m \cdot 1 = b^m \checkmark$$

Furthermore, because the product of reciprocals is 1, we define $b^{-m} = \dfrac{1}{b^m}$ so that

$$b^{-m} \cdot b^m = b^{-m+m} = b^0 = 1$$
$$\downarrow$$
$$\frac{1}{b^m} \cdot b^m = 1$$

Tables 2-1 and 2-2 summarize several important definitions and properties of exponents.

Table 2-1   Definitions Involving Exponents*

Description	Definition	Examples
Zero exponent	$b^0 = 1$ A nonzero base raised to the zero power is 1.	1. $4^0 = 1$ 2. $(3y)^0 = 1$ 3. $3y^0 = 3(1) = 3$
Negative exponent	$b^{-n} = \left(\dfrac{1}{b}\right)^n = \dfrac{1}{b^n}$ To simplify a nonzero base to a negative exponent, take the reciprocal of the base and change the exponent to positive.	4. $3^{-2} = \left(\dfrac{1}{3}\right)^2 = \dfrac{1}{3^2} = \dfrac{1}{9}$ 5. $\left(\dfrac{2}{7}\right)^{-2} = \left(\dfrac{7}{2}\right)^2 = \dfrac{49}{4}$ 6. $\left(\dfrac{1}{4}\right)^{-3} = (4)^3 = 64$

*Assume that b is a nonzero real number and that n is a positive integer.

Table 2-2   Properties Involving Exponents*

Description	Property	Example	Expanded Form
Multiplication of like bases	$b^m \cdot b^n = b^{m+n}$	$b^2 \cdot b^4 = b^{2+4}$ $= b^6$	$b^2 \cdot b^4 = (b \cdot b)(b \cdot b \cdot b \cdot b)$ $= b^6$
Division of like bases	$\dfrac{b^m}{b^n} = b^{m-n}$	$\dfrac{b^5}{b^2} = b^{5-2}$ $= b^3$	$\dfrac{b^5}{b^2} = \dfrac{b \cdot b \cdot b \cdot b \cdot b}{b \cdot b}$ $= b^3$
Power rule	$(b^m)^n = b^{m \cdot n}$	$(b^4)^2 = b^{4 \cdot 2}$ $= b^8$	$(b^4)^2 = (b \cdot b \cdot b \cdot b)(b \cdot b \cdot b \cdot b)$ $= b^8$
Power of a product	$(ab)^m = a^m b^m$	$(ab)^3 = a^3 b^3$	$(ab)^3 = (ab)(ab)(ab)$ $= (a \cdot a \cdot a)(b \cdot b \cdot b) = a^3 b^3$
Power of a quotient	$\left(\dfrac{a}{b}\right)^m = \dfrac{a^m}{b^m}$	$\left(\dfrac{a}{b}\right)^3 = \dfrac{a^3}{b^3}$	$\left(\dfrac{a}{b}\right)^3 = \left(\dfrac{a}{b}\right)\left(\dfrac{a}{b}\right)\left(\dfrac{a}{b}\right)$ $= \dfrac{a \cdot a \cdot a}{b \cdot b \cdot b} = \dfrac{a^3}{b^3}$

*Assume that a and b are real numbers ($b \neq 0$) and that m and n represent integers.

TIP The power of a product and the power of a quotient follow directly from the commutative and associative properties of real numbers.

To simplify an expression with exponents, we want to combine like bases where possible and write the result with no negative or zero exponents. In Examples 1–5, we practice simplifying expressions with exponents using the definitions and properties from Tables 2-1 and 2-2.

EXAMPLE 1   Simplifying Expressions with Zero and Negative Exponents

Simplify.

a. $(-2)^4$ **b.** -2^4 **c.** -2^{-4} **d.** $(-7x)^0$ **e.** $-7x^0$

Solution:

a. $(-2)^4 = (-2)(-2)(-2)(-2)$ The base is -2. Multiply -2 times itself 4 times.
$\qquad\quad = 16$

b. $-2^4 = -1 \cdot 2^4$

    $= -1 \cdot (2 \cdot 2 \cdot 2 \cdot 2)$

    $= -1 \cdot (16)$

    $= -16$

The exponent of 4 applies to a base of 2 (not -2). This expression is interpreted as the opposite of 2^4.

c. $-2^{-4} = -1 \cdot (2^{-4})$

    $= -1 \cdot \left(\dfrac{1}{2^4}\right)$

    $= -1 \cdot \left(\dfrac{1}{16}\right)$

    $= -\dfrac{1}{16}$

This expression is interpreted as the opposite of 2^{-4}.

Write the expression with positive exponents: $2^{-4} = \dfrac{1}{2^4}$.

d. $(-7x)^0 = 1$

This expression is of the form b^0 where $b = -7x$. By definition $b^0 = 1$. (The presence of parentheses implies that the entire expression $-7x$ is raised to the zero power.)

e. $-7x^0 = -7 \cdot x^0$

    $= -7 \cdot (1)$

    $= -7$

The exponent of zero applies only to x.

> **TIP** In Example 1(d), the parentheses enclose -7 and x so that the exponent 0 applies to both factors. In Example 1(e), there are no parentheses. Thus, the exponent 0 applies only to x.

Skill Practice 1 Simplify.

    a. $(-3)^2$ **b.** -3^2 **c.** -3^{-2} **d.** $(8y)^0$ **e.** $8y^0$

EXAMPLE 2   Applying Basic Properties of Exponents

Simplify.   **a.** $x^3 x^5 x^{-2}$ **b.** $\dfrac{y^7}{y^4}$ **c.** $(b^2)^{-5}$

Solution:

a. $x^3 x^5 x^{-2} = x^{3+5+(-2)} = x^6$

Multiply like bases. Keep the same base and add the exponents.

b. $\dfrac{y^7}{y^4} = y^{7-4} = y^3$

Divide like bases. Keep the base the same and subtract the exponent in the denominator from the exponent in the numerator.

c. $(b^2)^{-5} = b^{2 \cdot (-5)} = b^{-10}$

Apply the power rule. Keep the base the same and multiply the exponents.

    $= \dfrac{1}{b^{10}}$

Write the answer with a positive exponent.

Skill Practice 2 Simplify.

    a. $w^7 w^{-3} w$ **b.** $\dfrac{t^{11}}{t^6}$ **c.** $(p^{-4})^2$

Answers

1. a. 9   **b.** -9 **c.** $-\dfrac{1}{9}$

    d. 1   **e.** 8

2. a. w^5 **b.** t^5 **c.** $\dfrac{1}{p^8}$

EXAMPLE 3   **Simplifying an Expression with Exponents**

Simplify the expression. $\left(\dfrac{1}{5}\right)^{-3} - (2)^{-2} + 3^0$

Solution:

$$\left(\dfrac{1}{5}\right)^{-3} - (2)^{-2} + 3^0$$

$$= 5^3 - \left(\dfrac{1}{2}\right)^2 + 3^0 \qquad \text{Simplify negative exponents.}$$

$$= 125 - \dfrac{1}{4} + 1 \qquad \text{Evaluate expressions with exponents.}$$

$$= \dfrac{500}{4} - \dfrac{1}{4} + \dfrac{4}{4} \qquad \text{Write the expressions with a common denominator.}$$

$$125 = \dfrac{125}{1} \cdot \dfrac{4}{4} = \dfrac{500}{4} \quad \text{and} \quad 1 = \dfrac{1}{1} \cdot \dfrac{4}{4} = \dfrac{4}{4}$$

$$= \dfrac{503}{4} \qquad \text{Simplify.}$$

Skill Practice 3   Simplify the expression.

$$\left(\dfrac{2}{3}\right)^{-1} + 4^{-1} - \left(\dfrac{1}{4}\right)^0$$

EXAMPLE 4   **Simplifying an Expression with Exponents**

Simplify the expression. Write the answer with positive exponents only.

$$\dfrac{(2a^7b^{-4})^3}{(4a^3b^{-2})^2}$$

Solution:

$$\dfrac{(2a^7b^{-4})^3}{(4a^3b^{-2})^2}$$

$$= \dfrac{2^3 a^{21} b^{-12}}{4^2 a^6 b^{-4}} \qquad \text{Apply the power rule. Keep each base the same and multiply the exponents.}$$

$$= \dfrac{8a^{21} b^{-12}}{16a^6 b^{-4}} \qquad \text{Simplify the coefficients.}$$

$$= \dfrac{8a^{21-6} b^{-12-(-4)}}{16} \qquad \text{Divide like bases by subtracting exponents.}$$

$$= \dfrac{\overset{1}{8} a^{15} b^{-8}}{\underset{2}{16}} \qquad \text{Simplify.}$$

$$= \dfrac{a^{15}}{2b^8} \qquad \text{Simplify negative exponents.}$$

Answer

3. $\dfrac{3}{4}$

Skill Practice 4   Simplify the expression. Write the final answer with positive exponents only.

$$\frac{(3x^3y^{-4})^2}{(x^{-2}y)^{-4}}$$

EXAMPLE 5   Simplifying an Expression with Exponents

Simplify the expression. Write the answer with positive exponents only.

$$4xy^{-3}\left(\frac{8x^2}{3x^5y^2}\right)^{-2}$$

Solution:

$$4xy^{-3}\left(\frac{8x^2}{3x^5y^2}\right)^{-2} = 4xy^{-3} \cdot \left(\frac{8}{3x^3y^2}\right)^{-2}$$ Simplify within parentheses.

$$= \frac{4xy^{-3}}{1} \cdot \frac{8^{-2}}{3^{-2}x^{-6}y^{-4}}$$ Raise the expression in parentheses to the -2 power.

$$= \frac{4x}{y^3} \cdot \frac{3^2x^6y^4}{8^2}$$ Simplify negative exponents.

$$= \frac{4 \cdot 9 \cdot x \cdot x^6 \cdot y^4}{64y^3}$$ Multiply the fractions and simplify the expressions 3^2 and 8^2.

$$= \frac{\overset{1}{4} \cdot 9 \cdot x^{1+6}y^{4-3}}{\underset{16}{64}}$$ Add the exponents on x. Subtract the exponents on y.

$$= \frac{9x^7y}{16}$$ Simplify.

Skill Practice 5   Simplify the expression. Write the answer with positive exponents only.

$$-9m^5t\left(\frac{3m^{-2}t^3}{4t^6}\right)^{-3}$$

---

2. Apply Scientific Notation

In many applications of science, technology, and business we encounter very large or very small numbers. For example:

- San Francisco has some of the highest home prices in the United States. One of the most expensive homes recently sold for $45,000,000.
- The diameter of a capillary is measured as 0.000005 m.
- The mean surface temperature of the planet Saturn is $-300°F$.

Very large and very small numbers are sometimes cumbersome to write because they contain numerous zeros. Furthermore, it is difficult to determine the location of the decimal point when performing calculations with such numbers. For these reasons, scientists will often write numbers using scientific notation.

Answers

4. $\dfrac{9}{x^2y^4}$

5. $-\dfrac{64m^{11}t^{10}}{3}$

Scientific Notation

A number expressed in the form $a \times 10^n$, where $1 \leq |a| < 10$ and n is an integer is said to be in **scientific notation**.

Examples

Home Price	Capillary Size	Saturn Temp.
$45,000,000	0.000005 m	−300°F
$= \$4.5 \times 10{,}000{,}000$	$= 5.0 \times 0.000001$ m	$= -3 \times 100°$F
$= \$4.5 \times 10^7$	$= 5.0 \times 10^{-6}$ m	$= -3 \times 10^2$ °F

To write a number in scientific notation, the number of positions that the decimal point must be moved determines the power of 10. Numbers 10 or greater require a positive exponent on 10. Numbers between 0 and 1 require a negative exponent on 10.

EXAMPLE 6   **Writing Numbers in Scientific Notation**

Write the numerical values in scientific notation.

a. The size of the smallest visible object in an optical microscope is 0.0000002 m.

b. One estimate for the number of stars in the Milky Way is 230 billion.

c. The recommended daily intake of calcium is 1.2 g.

Solution:

a. 0.0000002 m $= 2.0 \times 10^{-7}$ m

7 place positions

The number 0.0000002 is between 0 and 1. Use a negative power of 10.

b. 230 billion $= 230{,}000{,}000{,}000$

First write 230 billion in standard form.

$230{,}000{,}000{,}000$ stars $= 2.3 \times 10^{11}$ stars

11 place positions

The number 230 billion is greater than 10. Use a positive power of 10.

c. 1.2 g $= 1.2 \times 10^0$ g

The decimal point is moved zero units, so the exponent on 10 is 0.

Source: NASA, ESA, and the Hubble Heritage (STScI/AURA)-ESA/Hubble Collaboration

Skill Practice 6   Write the numerical values in scientific notation.

a. Salmonella bacteria are elongated bacteria and average 0.0000035 m in length.

b. The distance from Earth to Barnard's Star is 32,000,000,000,000 mi.

c. The average weight of a newborn baby is 7.5 lb.

EXAMPLE 7   **Writing Numbers in Standard Decimal Notation**

Write the numerical values in standard decimal notation.

a. The temperature at the core of the Sun is estimated to be 1.36×10^7 °C.

b. The thickness of a dollar bill is approximately 3.9×10^{-3} in.

Answers

6. a. 3.5×10^{-6} m
b. 3.2×10^{13} mi
c. 7.5×10^0 lb

Solution:

a. $1.36 \times 10^7 \,°C = 13{,}600{,}000 \,°C$

7 place positions

10^7 is greater than 10. Move the decimal point 7 places to the right. Insert zeros to the right as needed.

b. 3.9×10^{-3} in. $= 0.0039$ in.

3 place positions

10^{-3} is between 0 and 1. Move the decimal point 3 places to the left. Insert zeros to the left as needed.

Skill Practice 7   Write the numerical values in standard decimal notation.

a. Alaska is the largest state geographically with a land area of 5.86×10^5 mi^2.

b. A doctor orders 2.0×10^{-2} g of the drug atropine given by injection.

Example 8 demonstrates the process to multiply and divide numbers written in scientific notation.

EXAMPLE 8   **Performing Calculations with Scientific Notation**

a. A light-year is the distance that light travels in 1 year. If light travels at a speed of 6.7×10^8 mph, how far will it travel in 1 year (8.76×10^3 hr)?

b. California has a land area of 1.56×10^5 mi^2. If the population of California for a recent year was 3.9×10^7, determine the population density (number of people per square mile).

Solution:

a. Distance = (Rate)(Time)

$= (6.7 \times 10^8 \text{ mph})(8.76 \times 10^3 \text{ hr})$ Regroup factors.

$= (6.7)(8.76) \times (10^8)(10^3) \text{ mi}$ Multiply and add the powers of 10.

$= 58.692 \times 10^{11} \text{ mi}$ The number 58.692 is not between 1 and 10. Rewrite this as 5.8692×10^1.

$= (5.8692 \times 10^1) \times 10^{11} \text{ mi}$

$= 5.8692 \times 10^{12} \text{ mi}$ One light-year is approximately 5.87 trillion miles.

b. $\dfrac{3.9 \times 10^7 \text{ people}}{1.56 \times 10^5 \text{ mi}^2} = \left(\dfrac{3.9}{1.56}\right) \times \left(\dfrac{10^7}{10^5}\right) \text{ people/mi}^2$ Population density is the number of people per square mile.

$= 2.5 \times 10^2 \text{ people/mi}^2$ At that time, California had a population density of 2.5×10^2 people/mi^2 or 250 people/mi^2.

Skill Practice 8

a. A satellite travels 1.72×10^4 mph. How far does it travel in 24 hr (2.4×10^1 hr)?

b. The land area of Texas is 2.6×10^5 mi^2. If the population of Texas for a recent year was 2.5×10^7, determine the population density.

Answers

7. a. 586,000 mi^2 **b.** 0.02 g

8. a. 4.128×10^5 mi

  b. Approximately
  9.6×10^1 people/mi^2 or
  96 people/mi^2

The calculations for Example 8 can also be performed on a calculator.

TECHNOLOGY CONNECTIONS

Using Scientific Notation on a Calculator

On many calculators, the EE key is used to enter the exponent for a number in scientific notation. The result on the screen shows the symbol E to indicate the exponent for the power of 10. For example, the number 5.8692E12 is read as 5.8692×10^{12}.

```
NORMAL FLOAT AUTO REAL RADIAN CL
(6.7E8)*(8.76E3)
                        5.8692E12
(3.9E7)/(1.56E5)
                              250
```

SECTION 2.1   Practice Exercises

Prerequisite Review

For Exercises R.1–R.4, write the expression using exponents. (See Section R.3 for review.)

R.1. $2 \cdot x \cdot x \cdot x \cdot y \cdot y$

R.2. $5 \cdot a \cdot b \cdot b \cdot b \cdot b \cdot c \cdot c$

R.3. $\dfrac{6 \cdot m \cdot m \cdot m \cdot m \cdot m}{n}$

R.4. $\dfrac{8 \cdot p \cdot q \cdot q}{r \cdot r \cdot r \cdot r \cdot r \cdot r \cdot r}$

For Exercises R.5–R.8, simplify the expression. (See Section R.3 for review.)

R.5. $(-4)^3$

R.6. $(-1)^5$

R.7. $\left(\dfrac{2}{3}\right)^4$

R.8. $\left(\dfrac{4}{5}\right)^3$

Concept Connections

1. For a nonzero real number b, the value of $b^0 =$ _____.

2. For a nonzero real number b, the value $b^{\boxed{}} = \dfrac{1}{b^n}$.

3. A number expressed in the form $a \times 10^n$, where $1 \le |a| < 10$ and n is an integer is said to be written in _____ notation.

4. From the properties of exponents, $b^m b^n = b^{\boxed{}}$.

5. If $b \ne 0$, then $\dfrac{b^m}{b^n} = b^{\boxed{}}$.

6. From the properties of exponents, $(b^m)^n = b^{\boxed{}}$.

Objective 1: Simplify Expressions with Integer Exponents

For Exercises 7–14, simplify the expressions. **(See Example 1)**

7. a. $(-7)^2$ **b.** -7^2 **c.** -7^{-2} **d.** $(-7)^{-2}$
    e. $(-7)^3$ **f.** -7^3

8. a. $(-3)^4$ **b.** -3^4 **c.** -3^{-4} **d.** $(-3)^{-4}$
    e. $(-3)^5$ **f.** -3^5

9. a. $\left(-\dfrac{2}{3}\right)^0$ **b.** $-\dfrac{2^0}{3}$ **c.** $-\dfrac{2}{3}p^0$ **d.** $\left(-\dfrac{2}{3}p\right)^0$

10. a. $\left(-\dfrac{3}{7}\right)^0$ **b.** $-\dfrac{3^0}{7}$ **c.** $-\dfrac{3}{7}w^0$ **d.** $\left(-\dfrac{3}{7}w\right)^0$

11. a. 8^{-2} **b.** $8x^{-2}$ **c.** $(8x)^{-2}$ **d.** -8^{-2}

12. a. 7^{-2} **b.** $7y^{-2}$ **c.** $(7y)^{-2}$ **d.** -7^{-2}

13. a. $\dfrac{1}{q^{-2}}$ **b.** q^{-2} **c.** $5p^3q^{-2}$ **d.** $5p^{-3}q^2$

14. a. $\dfrac{1}{t^{-4}}$ **b.** t^{-4} **c.** $11t^{-4}u^2$ **d.** $11t^4u^{-2}$

For Exercises 15–66, simplify. **(See Examples 2–5)**

15. $y^3 \cdot y^5$ **16.** $x^4 \cdot x^8$ **17.** $\dfrac{13^8}{13^6}$ **18.** $\dfrac{5^7}{5^3}$

19. $\left(y^2\right)^4$ **20.** $\left(z^3\right)^4$ **21.** $\left(3x^2\right)^4$ **22.** $\left(2y^5\right)^3$

23. p^{-3} **24.** q^{-5} **25.** $7^{10} \cdot 7^{-13}$ **26.** $11^{-9} \cdot 11^7$

27. $\dfrac{w^3}{w^5}$ **28.** $\dfrac{t^4}{t^8}$ **29.** $a^{-2}a^{-5}$ **30.** $b^{-1}b^{-8}$

31. $\dfrac{r}{r^{-1}}$ **32.** $\dfrac{s^{-1}}{s}$ **33.** $\dfrac{z^{-6}}{z^{-2}}$ **34.** $\dfrac{w^{-8}}{w^{-3}}$

35. $\dfrac{a^3}{b^{-2}}$ **36.** $\dfrac{c^4}{d^{-1}}$ **37.** $(6xyz^2)^0$ **38.** $(-7ab^3)^0$

39. $2^4 + 2^{-2}$ **40.** $3^2 + 3^{-1}$ **41.** $1^{-2} + 5^{-2}$ **42.** $4^{-2} + 2^{-2}$

43. $\left(\dfrac{2}{3}\right)^{-2} - \left(\dfrac{1}{2}\right)^2 + \left(\dfrac{1}{3}\right)^0$ **44.** $\left(\dfrac{1}{6}\right)^{-1} + \left(\dfrac{2}{3}\right)^0 - \left(\dfrac{1}{4}\right)^{-2}$ **45.** $\left(\dfrac{4}{5}\right)^{-1} + \left(\dfrac{3}{2}\right)^2 - \left(\dfrac{2}{7}\right)^0$ **46.** $\left(\dfrac{4}{5}\right)^0 - \left(\dfrac{2}{3}\right)^2 + \left(\dfrac{9}{5}\right)^{-1}$

47. $\dfrac{p^2q}{p^5q^{-1}}$ **48.** $\dfrac{m^{-1}n^3}{m^4n^{-2}}$ **49.** $\dfrac{-48ab^{10}}{32a^4b^3}$ **50.** $\dfrac{25x^2y^{12}}{10x^5y^7}$

51. $\left(-3x^{-4}y^5z^2\right)^{-4}$ **52.** $\left(-6a^{-2}b^3c\right)^{-2}$ **53.** $(4m^{-2}n)(-m^6n^{-3})$ **54.** $(-6pq^{-3})(2p^4q)$

55. $(p^{-2}q)^3(2pq^4)^2$ **56.** $(mn^3)^2(5m^{-2}n^2)$ **57.** $\left(\dfrac{x^2}{y}\right)^3(5x^2y)$ **58.** $\left(-\dfrac{a}{b^2}\right)^2(3a^2b^3)$

59. $\dfrac{(-8a^2b^2)^4}{(16a^3b^7)^2}$ **60.** $\dfrac{(-3x^2y^3)^2}{(-2xy^4)^3}$ **61.** $\left(\dfrac{-2x^6y^{-5}}{3x^{-2}y^4}\right)^{-3}$ **62.** $\left(\dfrac{-6a^2b^{-3}}{5a^{-1}b}\right)^{-2}$

63. $\left(\dfrac{2x^{-3}y^0}{4x^6y^{-5}}\right)^{-2}$ **64.** $\left(\dfrac{a^3b^2c^0}{a^{-1}b^{-2}c^{-3}}\right)^{-2}$ **65.** $3xy^5\left(\dfrac{2x^4y}{6x^5y^3}\right)^{-2}$ **66.** $7x^{-3}y^{-4}\left(\dfrac{3x^{-1}y^5}{9x^3y^{-2}}\right)^{-3}$

Objective 2: Apply Scientific Notation

For Exercises 67–76, write the numbers in scientific notation. **(See Example 6)**

67. a. 350,000   **b.** 0.000035   **c.** 3.5

68. a. 2710   **b.** 0.00271   **c.** 2.71

69. a. 0.86   **b.** 8.6   **c.** 86

70. a. 0.792   **b.** 7.92   **c.** 79.2

71. The speed of light is approximately 29,980,000,000 cm/sec.

72. The mean distance between the Earth and the Sun is approximately 149,000,000 km.

73. The size of an HIV particle is approximately 0.00001 cm.

74. One picosecond is 0.000000000001 sec.

75. For a test group of adult females between 18 and 20 years old, the mean blood volume was 4.2 L.

76. The longest table tennis rally ever played lasted 8.25 hr.

For Exercises 77–84, write the number in standard decimal notation. **(See Example 7)**

77. a. 2.61×10^{-6} | **b.** 2.61×10^{6} | **c.** 2.61×10^{0}

78. a. 3.52×10^{-2} | **b.** 3.52×10^{2} | **c.** 3.52×10^{0}

79. a. 6.718×10^{-1} | **b.** 6.718×10^{0} | **c.** 6.718×10^{1}

80. a. 1.87×10^{-1} | **b.** 1.87×10^{0} | **c.** 1.87×10^{1}

81. A drop of water has approximately 1.67×10^{21} molecules of H_2O.

82. A computer with a 3-terabyte hard drive can store approximately 3.0×10^{12} bytes.

83. A typical red blood cell is 7.0×10^{-6} m.

84. The blue light used to read a laser disc has a wavelength of 4.7×10^{-7} m.

For Exercises 85–90, determine which numbers are in "proper" scientific notation. If the number is not in "proper" scientific notation, correct it.

85. 35×10^{4} | **86.** 0.469×10^{-7} | **87.** 7×10^{0}

88. 8.12×10^{1} | **89.** 9×10^{1} | **90.** 6.9×10^{0}

For Exercises 91–100, perform the indicated operation. Write the answer in scientific notation. **(See Example 8)**

91. $(2 \times 10^{-3})(4 \times 10^{8})$

92. $(3 \times 10^{4})(2 \times 10^{-1})$

93. $\dfrac{8.4 \times 10^{-6}}{2.1 \times 10^{-2}}$

94. $\dfrac{6.8 \times 10^{11}}{3.4 \times 10^{3}}$

95. $(6.2 \times 10^{11})(3 \times 10^{4})$

96. $(8.1 \times 10^{6})(2 \times 10^{5})$

97. $\dfrac{3.6 \times 10^{-14}}{5 \times 10^{5}}$

98. $\dfrac{3.68 \times 10^{-8}}{4 \times 10^{2}}$

99. $\dfrac{(6.2 \times 10^{5})(4.4 \times 10^{22})}{2.2 \times 10^{17}}$

100. $\dfrac{(3.8 \times 10^{4})(4.8 \times 10^{-2})}{2.5 \times 10^{-5}}$

101. For a recent year, the United States consumed about 1.0×10^{4} gal of petroleum per second. (*Source:* U.S. Energy Information Administration, www.eia.gov)

 a. How many seconds are in a year?

 b. How many gallons of petroleum did the United States use that year?

102. Geoff's average heart rate is 65 beats/min.

 a. How many minutes are in one day?

 b. How many times will Geoff's heart beat per day?

103. Jonas has a personal music player with 80 gigabytes of memory (80 gigabytes is approximately 8×10^{10} bytes). If each song requires an average of 4 megabytes of memory (approximately 4×10^{6} bytes), how many songs can Jonas store on the device?

104. Joelle has a personal web page with 60 gigabytes of memory (approximately 6×10^{10} bytes). She stores math videos on the site for her students to watch outside of class. If each video requires an average of 5 megabytes of memory (approximately 5×10^{6} bytes), how many videos can she store on her website?

105. A typical adult human has approximately 5 L of blood in the body. If 1 μL (1 microliter) contains 5×10^{6} red blood cells, how many red blood cells does a typical adult have? (*Hint:* 1 L = 10^{6} μL.)

106. The star Proxima Centauri is the closest star (other than the Sun) to the Earth. It is approximately 4.3 light-years away. If 1 light-year is approximately 5.9×10^{12} mi, how many miles is Proxima Centauri from the Earth?

107. If the county of Queens, New York, has a population of approximately 2,200,000 and the area is 110 mi^2, how many people are there per square mile?

108. If the county of Catawba, North Carolina, has a population of approximately 150,000 and the area is 400 mi^2, how many people are there per square mile?

Mixed Exercises

For Exercises 109–112, without the assistance of a calculator, fill in the blank with the appropriate symbol <, >, or =.

109. a. 5^{15} ____ 5^{17} **b.** 5^{-15} ____ 5^{-17} **110. a.** $\left(\frac{1}{5}\right)^{15}$ ____ $\left(\frac{1}{5}\right)^{17}$ **b.** $\left(\frac{1}{5}\right)^{-15}$ ____ $\left(\frac{1}{5}\right)^{-17}$

111. a. $(-1)^{86}$ ____ $(-1)^{87}$ **b.** $(1)^{86}$ ____ $(1)^{87}$ **112. a.** $(-1)^{0}$ ____ -1^{41} **b.** $(-1)^{42}$ ____ $(-1)^{0}$

For Exercises 113–118, simplify the expressions.

113. $(3x + 5)^{14}(3x + 5)^{-2}$ **114.** $(2y - 7z)^{-4}(2y - 7z)^{13}$ **115.** $[(6v - 7)^{10}]^{9}$

116. $[(4x - 9)^{5}]^{11}$ **117.** $2^{-2} + 2^{-1} + 2^{0} + 2^{1} + 2^{2}$ **118.** $3^{-2} + 3^{-1} + 3^{0} + 3^{1} + 3^{2}$

For Exercises 119–130, simplify each expression. Assume that m and n are integers and that x and y are nonzero real numbers.

119. $x^{m}x^{4}$ **120.** $y^{n}y^{7}$ **121.** $x^{m+9}x^{m-2}$ **122.** $y^{n+9}y^{n-1}$

123. $\dfrac{x^{m}}{x^{8}}$ **124.** $\dfrac{y^{n}}{y^{3}}$ **125.** $\dfrac{x^{2m+7}}{x^{m+5}}$ **126.** $\dfrac{y^{3n+5}}{y^{2n-4}}$

127. $(x^{4m})^{3n}$ **128.** $(y^{5m})^{2n}$ **129.** $\dfrac{x^{4m-3}y^{5n+7}}{x^{m-7}y^{3n+2}}$ **130.** $\dfrac{x^{2n-4}y^{5n}}{x^{n+1}y^{3n-7}}$

Write About It

131. Explain the difference between the expressions $6x^{0}$ and $(6x)^{0}$.

132. Explain why scientific notation is used.

Expanding Your Skills

133. If $x < 0$ and m is an integer, can x^{m} be positive? If so, give an example.

134. If $x < 0$ and m is an integer, can x^{m} be negative? If so, give an example.

135. If x is a real number, can x^{-2} be negative? If so, give an example.

136. If $x > 10$ and m is an integer, can x^{m} be less than 1? If so, given an example.

For Exercises 137–138, refer to the formula $F = \dfrac{Gm_1m_2}{d^2}$. This gives the gravitational force F (in Newtons, N) between two masses m_1 and m_2 (each measured in kg) that are a distance of d meters apart. In the formula, $G = 6.6726 \times 10^{-11}$ N · m^2/kg^2.

137. Determine the gravitational force between the Earth (mass $= 5.98 \times 10^{24}$ kg) and Jupiter (mass $= 1.901 \times 10^{27}$ kg) if at one point in their orbits, the distance between them is 7.0×10^{11} m.

138. Determine the gravitational force between the Earth (mass $= 5.98 \times 10^{24}$ kg) and an 80-kg human standing at sea level. The mean radius of the Earth is approximately 6.371×10^{6} m.

139. A 20-year-old starts a savings plan for her retirement. She will put $20 per month into a mutual fund that she hopes will average 6% growth annually.

   a. If she plans to retire at age 65, for how many months will she be depositing money?

   b. By age 65, how much money will she have deposited?

   c. The value of an account built in this fashion is given by

$$A = P \cdot \left[\left(1 + \frac{r}{12}\right)^{N} - 1\right] \cdot \left(1 + \frac{12}{r}\right)$$

   where A is the final amount of money in the account, P is the amount of the monthly deposit, r is the annual interest rate as a decimal, and N is the number of months. Use a calculator to find the total amount in the woman's retirement account at age 65.

OBJECTIVES

1. Identify Key Elements of a Polynomial
2. Add and Subtract Polynomials
3. Multiply Polynomials
4. Multiply Special Case Products

1. Identify Key Elements of a Polynomial

The Environmental Protection Agency (EPA) is responsible for providing fuel economy data (gas mileage information) that is posted on the window stickers of new vehicles. Many variables contribute to fuel consumption, including the speed of the vehicle. For example, for one midsize sedan tested, the gas mileage G (in miles per gallon, mpg) can be approximated by

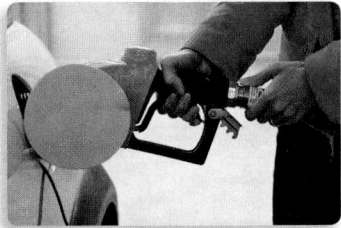

Frederic Charpentier/Alamy Stock Photo

$$G = -0.008x^2 + 0.748x + 13.5, \qquad \text{where } x \text{ is the speed of the vehicle in mph and } 15 \le x \le 75 \text{ mph.}$$

The expression on the right side of this equation is called a polynomial. A **polynomial** in the variable x is a finite sum of terms of the form ax^n. In each term, the **coefficient**, a, is a real number, and the exponent, n, is a whole number. The **degree** of ax^n is n. For example:

Term (expressed in the form ax^n)	Coefficient	Degree
$-0.008x^2$	-0.008	2
$0.748x$ rewrite as $0.748x^1$	0.748	1
13.5 rewrite as $13.5x^0$	13.5	0

If a polynomial has exactly one term, it is called a **monomial**. A two-term polynomial is called a **binomial**, and a three-term polynomial is called a **trinomial**. Usually the terms of a polynomial are written in descending order according to degree. In descending order, the highest-degree term is written first and is called the **leading term**. Its coefficient is called the **leading coefficient**. The **degree of a polynomial** is the greatest degree of all its terms. Thus, the leading term determines the degree of the polynomial.

	Expression	Descending Order	Leading Coefficient	Degree of Polynomial
Monomials	$2x^9$	$2x^9$	2	9
	-49	-49	-49	0
Binomials	$10y - 7y^2$	$-7y^2 + 10y$	-7	2
	$6 - \dfrac{2}{3}b$	$-\dfrac{2}{3}b + 6$	$-\dfrac{2}{3}$	1
Trinomials	$w + 2w^3 + 9w^6$	$9w^6 + 2w^3 + w$	9	6
	$2.5a^4 - a^8 + 1.3a^3$	$-a^8 + 2.5a^4 + 1.3a^3$	-1	8

The preceding discussion is meant as an informal introduction to polynomials and associated key vocabulary. However, as your level of mathematical sophistication increases, you should strive to understand definitions written in a more concise mathematical language. Take a minute to read the formal definition of a polynomial.

Definition of a Polynomial in x

A **polynomial in the variable x** is an expression of the form:

$$a_n x^n + a_{n-1} x^{n-1} + a_{n-2} x^{n-2} + \cdots + a_1 x + a_0$$

The coefficients a_n, a_{n-1}, a_{n-2}, ... , a_0 are real numbers, where $a_n \neq 0$, and the exponents n, $n - 1$, $n - 2$, ... , 0 are whole numbers.

The term $a_n x^n$ is called the **leading term**, the coefficient a_n is the **leading coefficient**, and the exponent n is the **degree of the polynomial**.

In the preceding definition, subscript notation a_n (read as "a sub n"), a_{n-1} (read as "a sub $n - 1$"), and so on, is used to denote the coefficients of the terms. Subscript notation is used rather than lettered variables such as a, b, c, and the like, when a large or undetermined number of terms is suggested.

Some polynomials have more than one variable:

$$-4x^4 y^7 + x^2 y^5 + 5xy^4$$

This polynomial has three terms. The degree of each term is the sum of the exponents on the variable factors.

$-4x^4 y^7$	degree 11	(sum of 4 + 7)
$x^2 y^5$	degree 7	(sum of 2 + 5)
$5xy^4$	degree 5	(sum of 1 + 4)

2. Add and Subtract Polynomials

To add or subtract two polynomials, we combine like terms. Recall that two terms are **like terms** if they each have the same variables and the corresponding variables are raised to the same powers.

EXAMPLE 1   Adding Polynomials

Add the polynomials.

a. $(3t^3 + 2t^2 - 5t) + (t^3 - 6t)$ **b.** $\left(\dfrac{2}{3}w^2 - w + \dfrac{1}{8}\right) + \left(\dfrac{4}{3}w^2 + 8w - \dfrac{1}{4}\right)$

Solution:

a. $(3t^3 + 2t^2 - 5t) + (t^3 - 6t)$

$\quad = 3t^3 + t^3 + 2t^2 + (-5t) + (-6t)$ Group like terms.

$\quad = 4t^3 + 2t^2 - 11t$ Add like terms.

b. $\left(\dfrac{2}{3}w^2 - w + \dfrac{1}{8}\right) + \left(\dfrac{4}{3}w^2 + 8w - \dfrac{1}{4}\right)$

$\quad = \dfrac{2}{3}w^2 + \dfrac{4}{3}w^2 + (-w) + 8w + \dfrac{1}{8} + \left(-\dfrac{1}{4}\right)$ Group like terms.

$\quad = \dfrac{6}{3}w^2 + 7w + \left(\dfrac{1}{8} - \dfrac{2}{8}\right)$ Add fractions with common denominators.

$\quad = 2w^2 + 7w - \dfrac{1}{8}$ Simplify.

Skill Practice 1   Add the polynomials.

a. $(2x^2 + 5x - 2) + (6x^2 - 8x - 8)$

b. $\left(-\dfrac{1}{4}m^2 - 2m + \dfrac{1}{3} \right) + \left(\dfrac{3}{4}m^2 + 7m - \dfrac{1}{12} \right)$

EXAMPLE 2   **Adding Polynomials**

Add the polynomials. $(a^2b + 7ab + 6) + (5a^2b - 2ab - 7)$

Solution:

Polynomials can be added vertically. Be sure to line up the like terms.

$$\begin{array}{r} a^2b + 7ab + 6 \\ + \; 5a^2b - 2ab - 7 \\ \hline 6a^2b + 5ab - 1 \end{array} \quad \text{Add like terms.}$$

Skill Practice 2   Add the polynomials.

$(-5a^2b - 6ab^2 + 2) + (2a^2b + ab^2 - 3)$

Subtraction of two polynomials is similar to subtracting real numbers. Add the opposite of the second polynomial to the first polynomial.

The opposite (or additive inverse) of a real number a is $-a$. Similarly, if A is a polynomial, then $-A$ is its opposite.

EXAMPLE 3   **Finding the Opposite of a Polynomial**

Find the opposite of the polynomials.

a. $5a - 2b - c$ **b.** $-5.5y^4 - 2.4y^3 + 1.1y$

Solution:

	Expression	Opposite	Simplified Form
a.	$5a - 2b - c$	$-(5a - 2b - c)$	$-5a + 2b + c$
b.	$-5.5y^4 - 2.4y^3 + 1.1y$	$-(-5.5y^4 - 2.4y^3 + 1.1y)$	$5.5y^4 + 2.4y^3 - 1.1y$

TIP   Notice that the sign of each term is changed when finding the opposite of a polynomial.

Skill Practice 3   Find the opposite of the polynomials.

a. $-7z + 6w$ **b.** $2p - 3q + r + 1$

Subtraction of Polynomials

If A and B are polynomials, then $A - B = A + (-B)$.

Answers

1. a. $8x^2 - 3x - 10$

  b. $\dfrac{1}{2}m^2 + 5m + \dfrac{1}{4}$

2. $-3a^2b - 5ab^2 - 1$

3. a. $7z - 6w$

  b. $-2p + 3q - r - 1$

EXAMPLE 4   Subtracting Polynomials

Subtract the polynomials. $(3x^2 + 2x - 5) - (4x^2 - 7x + 2)$

Solution:

$$(3x^2 + 2x - 5) - (4x^2 - 7x + 2)$$
$$= (3x^2 + 2x - 5) + (-4x^2 + 7x - 2) \qquad \text{Add the opposite of the second polynomial.}$$
$$= 3x^2 + (-4x^2) + 2x + 7x + (-5) + (-2) \qquad \text{Group like terms.}$$
$$= -x^2 + 9x - 7 \qquad \text{Combine like terms.}$$

Skill Practice 4   Subtract the polynomials.
$(6a^2 - 2a) - (-3a^2 + 2a + 3)$

EXAMPLE 5   Subtracting Polynomials

Subtract the polynomials. $(6x^2y - 2xy + 5) - (x^2y - 3)$

Solution:

$$(6x^2y - 2xy + 5) - (x^2y - 3)$$

Subtraction of polynomials can be performed vertically by vertically aligning like terms. Then add the opposite of the second polynomial. "Placeholders" (shown in red) may be used to help line up like terms.

$$
\begin{array}{r}
6x^2y - 2xy + 5 \\
-(x^2y + 0xy - 3)
\end{array}
\xrightarrow[\text{opposite.}]{\text{Add the}}
\begin{array}{r}
6x^2y - 2xy + 5 \\
+\ -x^2y - 0xy + 3 \\
\hline
5x^2y - 2xy + 8
\end{array}
$$

Skill Practice 5   Subtract the polynomials.
$(7p^2q - 6) - (2p^2q + 4pq + 4)$

3. Multiply Polynomials

The properties of real numbers and the properties of exponents can be used to simplify many algebraic expressions, including the multiplication of polynomials.

- To multiply two monomials, use the commutative and associative properties of multiplication to regroup like factors. Then apply the properties of exponents to simplify.

$$(-4x^2y^5)\left(\frac{1}{2}x^3y\right) = -4 \cdot \frac{1}{2}x^2x^3y^5y = -2x^5y^6$$

- To multiply a multi-term polynomial by a monomial, apply the distributive property.

$$-3x^3(4x^2 - 2x + 6) = -3x^3(4x^2) + (-3x^3)(-2x) + (-3x^3)(6)$$
$$= -12x^5 + 6x^4 - 18x^3$$

Answers
4. $9a^2 - 4a - 3$
5. $5p^2q - 4pq - 10$

- To multiply two polynomials each with two or more terms, we also use the distributive property. Ultimately, each term in the first polynomial must be multiplied by each term in the second. Then simplify and combine like terms.

$$(4w + 7)(2w - 3) = 4w(2w) + 4w(-3) + 7(2w) + 7(-3)$$
$$= 8w^2 - 12w + 14w - 21$$
$$= 8w^2 + 2w - 21$$

EXAMPLE 6   **Multiplying Monomials**

Multiply the monomials. $(3x^2y^7)(5x^3y)$

Solution:

$$(3x^2y^7)(5x^3y)$$
$$= (3 \cdot 5)(x^2 \cdot x^3)(y^7 \cdot y) \qquad \text{Group coefficients and like bases.}$$
$$= 15x^5y^8 \qquad \text{Add exponents and simplify.}$$

Skill Practice 6   Multiply the monomials. $(-8r^3s)(-4r^4s^4)$

The distributive property is used to multiply polynomials: $a(b + c) = ab + ac$.

EXAMPLE 7   **Multiplying a Polynomial by a Monomial**

Multiply the polynomials.

a. $5y^3(2y^2 - 7y + 6)$ **b.** $-4a^3b^7c\left(2ab^2c^4 - \dfrac{1}{2}a^5b\right)$

Solution:

a. $5y^3(2y^2 - 7y + 6)$
$$= (5y^3)(2y^2) + (5y^3)(-7y) + (5y^3)(6) \qquad \text{Apply the distributive property.}$$
$$= 10y^5 - 35y^4 + 30y^3 \qquad \text{Simplify each term.}$$

b. $-4a^3b^7c\left(2ab^2c^4 - \dfrac{1}{2}a^5b\right)$

$$= (-4a^3b^7c)(2ab^2c^4) + (-4a^3b^7c)\left(-\dfrac{1}{2}a^5b\right) \qquad \text{Apply the distributive property.}$$
$$= -8a^4b^9c^5 + 2a^8b^8c \qquad \text{Simplify each term.}$$

Skill Practice 7   Multiply the polynomials.

a. $-6b^2(2b^2 + 3b - 8)$ **b.** $8st^3\left(\dfrac{1}{2}s^2 - \dfrac{1}{4}st\right)$

Answers
6. $32r^7s^5$
7. a. $-12b^4 - 18b^3 + 48b^2$
 b. $4s^3t^3 - 2s^2t^4$

In Example 8, we multiply polynomials that each have more than one term.

EXAMPLE 8   **Multiplying Polynomials**

Multiply the polynomials. $(2x^2 + 4)(3x^2 - x + 5)$

Solution:

$(2x^2 + 4)(3x^2 - x + 5)$ Multiply each term in the first polynomial by each term in the second.

$= (2x^2)(3x^2) + (2x^2)(-x) + (2x^2)(5)$ Apply the distributive property.
$\quad + (4)(3x^2) + (4)(-x) + (4)(5)$

$= 6x^4 - 2x^3 + 10x^2 + 12x^2 - 4x + 20$ Simplify each term.

$= 6x^4 - 2x^3 + 22x^2 - 4x + 20$ Combine like terms.

TIP Multiplication of polynomials can be performed vertically by a process similar to column multiplication of real numbers.

$$
\begin{array}{r}
(2x^2 + 4)(3x^2 - x + 5) \longrightarrow 3x^2 - x + 5 \\
\times\ 2x^2 \qquad + 4 \\
\hline
12x^2 - 4x + 20 \\
6x^4 - 2x^3 + 10x^2 \\
\hline
6x^4 - 2x^3 + 22x^2 - 4x + 20
\end{array}
$$

Note: When multiplying by the column method, it is important to align like terms vertically before adding terms.

Skill Practice 8   Multiply the polynomials. $(2y - 1)(3y^2 - 2y - 1)$

---

4. Multiply Special Case Products

Two expressions of the form $a - b$ and $a + b$ are called **conjugates**. The product of two conjugates results in a **difference of squares**.

$$(a - b)(a + b) = a^2 + ab - ab - b^2$$
$$= a^2 - b^2 \quad \text{(difference of squares)}$$

TIP The product of conjugates equals the square of the first term from the binomials, minus the square of the second term from the binomials.

An expression of the form $(a + b)^2$ or $(a - b)^2$ is called a **square of a binomial**. In expanded form, the product is a **perfect square trinomial**.

$$(a + b)^2 = (a + b)(a + b) = a^2 + ab + ab + b^2$$
$$= a^2 + 2ab + b^2 \quad \text{(perfect square trinomial)}$$

TIP The square of a binomial results in the square of the first term in the binomial, plus twice the product of terms in the binomial, plus the square of the second term in the binomial.

$$(a - b)^2 = (a - b)(a - b) = a^2 - ab - ab + b^2$$
$$= a^2 - 2ab + b^2 \quad \text{(perfect square trinomial)}$$

The patterns associated with the product of conjugates and the square of a binomial are important to understand and memorize. These will be used again in many more applications of algebra, including factoring in the next section.

Answer
8. $6y^3 - 7y^2 + 1$

> **Special Case Products**
>
> Product of Conjugates: $(a + b)(a - b) = a^2 - b^2$ (difference of squares)
> Square of a Binomial: $(a + b)^2 = a^2 + 2ab + b^2$ (perfect square trinomial)
> $(a - b)^2 = a^2 - 2ab + b^2$ (perfect square trinomial)

EXAMPLE 9   **Multiplying Conjugates**

Multiply and simplify.

a. $(2x + 5)(2x - 5)$ **b.** $\left(\dfrac{1}{3}c^2 - \dfrac{1}{2}d\right)\left(\dfrac{1}{3}c^2 + \dfrac{1}{2}d\right)$

Solution:

a. $(2x + 5)(2x - 5)$ This is a product of conjugates.

$= (2x)^2 - (5)^2$ The product is a difference of squares.

$= 4x^2 - 25$ Simplify.

b. $\left(\dfrac{1}{3}c^2 - \dfrac{1}{2}d\right)\left(\dfrac{1}{3}c^2 + \dfrac{1}{2}d\right)$ This is a product of conjugates.

$= \left(\dfrac{1}{3}c^2\right)^2 - \left(\dfrac{1}{2}d\right)^2$ The product is a difference of squares.

$= \dfrac{1}{9}c^4 - \dfrac{1}{4}d^2$ Simplify.

> **Skill Practice 9**   Multiply and simplify.
>
> **a.** $(3y - 7)(3y + 7)$ **b.** $\left(\dfrac{2}{5}t - \dfrac{1}{4}w^2\right)\left(\dfrac{2}{5}t + \dfrac{1}{4}w^2\right)$

EXAMPLE 10   **Squaring Binomials**

Square the binomials.

a. $(3x - 7)^2$ **b.** $(5t^2 + 2v^2)^2$

Solution:

a. $(3x - 7)^2$ This is the square of a binomial, $(a - b)^2$, where $a = 3x$ and $b = 7$.

$= (3x)^2 - 2(3x)(7) + (7)^2$ The product is $a^2 - 2ab + b^2$.

$= 9x^2 - 42x + 49$ Simplify.

b. $(5t^2 + 2v^2)^2$ This is the square of a binomial, $(a + b)^2$, where $a = 5t^2$ and $b = 2v^2$.

$= (5t^2)^2 + 2(5t^2)(2v^2) + (2v^2)^2$ The product is $a^2 + 2ab + b^2$.

$= 25t^4 + 20t^2v^2 + 4v^4$ Simplify.

Answers

9. a. $9y^2 - 49$ **b.** $\dfrac{4}{25}t^2 - \dfrac{1}{16}w^4$

10. a. $64z^2 - 32z + 4$
    b. $9c^4 + 24c^2d^3 + 16d^6$

> **Skill Practice 10**   Square the binomials.
>
> **a.** $(8z - 2)^2$ **b.** $(3c^2 + 4d^3)^2$

In Example 11, we apply operations on polynomials to geometric formulas.

> **EXAMPLE 11**   **Applying Operations on Polynomials to Geometry**
>
> **a.** Write a polynomial that represents the area of the rectangle.
>
> **b.** Write a polynomial that represents the volume of the cube.

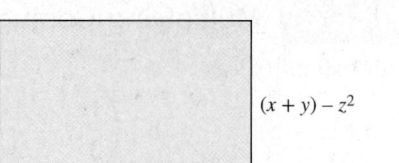

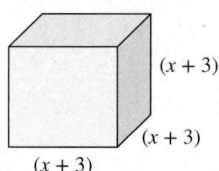

Solution:

a. $A = lw$

The area of a rectangle is length times width.

$$A = \overbrace{[(x + y) + z^2]}^{l}\overbrace{[(x + y) - z^2]}^{w}$$

Substitute $[(x + y) + z^2]$ for l, and $[(x + y) - z^2]$ for w.

$$= (x + y)^2 - (z^2)^2$$

This is a product of conjugates. The result is a difference of squares.

$$= x^2 + 2xy + y^2 - z^4$$

Square the binomial $(x + y)^2$ as $x^2 + 2xy + y^2$.

b. $V = s^3$

The volume of a cube is the product of the lengths of the sides.

$$V = (x + 3)^3$$

Substitute $(x + 3)$ for s.

$$= (x + 3)^2(x + 3)$$

This product can be written as $(x + 3)(x + 3)(x + 3)$ or as $(x + 3)^2(x + 3)$.

$$= (x^2 + 6x + 9)(x + 3)$$
$$= x^2(x) + x^2(3) + 6x(x) + 6x(3) + 9(x) + 9(3)$$
$$= x^3 + 3x^2 + 6x^2 + 18x + 9x + 27 \qquad \text{Simplify.}$$
$$= x^3 + 9x^2 + 27x + 27 \qquad \text{Combine like terms.}$$

TIP To expand $(x + 3)^2$, apply the formula $(a + b)^2 = a^2 + 2ab + b^2$.

$(x + 3)^2$
$= (x)^2 + 2(x)(3) + (3)^2$
$= x^2 + 6x + 9$

> **Skill Practice 11**
>
> **a.** Write a polynomial that represents the area of a rectangle with length $(x + 3) + y$ and width $(x + 3) - y$.
>
> **b.** Write a polynomial that represents the volume of a cube with sides of length $x + 2$.

Answers

11. a. $x^2 + 6x + 9 - y^2$
    b. $x^3 + 6x^2 + 12x + 8$

SECTION 2.2   Practice Exercises

Prerequisite Review

For Exercises R.1–R.4, apply the distributive property. (See Section 1.1 for review.)

R.1. $-2(5y^2 - 3x + 1)$

R.2. $-3(2a + 5b^2 - 6)$

R.3. $6\left(\dfrac{1}{2}m - \dfrac{2}{3}n + \dfrac{5}{6}r\right)$

R.4. $8\left(\dfrac{3}{4}r^2 - \dfrac{1}{2}r - \dfrac{7}{8}\right)$

For Exercises R.5–R.8, combine like terms. (See Section 1.1 for review.)

R.5. $-3x - 2x^2 + 4x + 7x^2$

R.6. $2q^3 - 4q^2 - 8q^3 + 4q^2$

R.7. $-1.6m^4 - 3.2m^2 - 2.2m^4 + 4.1m^2$

R.8. $10.5p - 11.6p^2 + 8.1p^2 - 1.1p$

For Exercises R.9–R.12, clear parentheses and combine like terms. (See Section 1.1 for review.)

R.9. $5(3x + 4y) - 2(-2x - y + 1)$

R.10. $-3(5c + d - 4) - 4(c + 3d)$

R.11. $-\dfrac{1}{2}(x^2 - 4x + 3) + \dfrac{1}{4}(2x^2 + 8x + 1)$

R.12. $-\dfrac{2}{3}(6y^2 + y - 9) + \dfrac{1}{6}(24y^2 + 2y + 12)$

Concept Connections

1. A _____ in the variable, x, is a single term or a sum of terms of the form ax^n, where a is a real number and n is a whole number.

2. Given the term ax^n, a is called the _____, and _____ is called the degree of the term.

3. Given the term x, the coefficient of the term is _____ and the degree is _____.

4. A monomial is a polynomial with exactly _____ term(s).

5. A _____ is a polynomial with exactly two terms.

6. A _____ is a polynomial with exactly three terms.

7. The term in a polynomial with the highest degree is called the _____ term and its coefficient is called the _____ _____.

8. The degree of a polynomial is the _____ degree of all of its terms.

9. The degree of a nonzero constant such as 7 is _____.

10. If a term of a polynomial has more than one variable, then the degree of the term is the sum of the _____ of the variables contained in the term.

11. To multiply $2(4x - 5)$, apply the _____ property.

12. The conjugate of $4x + 7$ is _____.

13. When two conjugates are multiplied, the resulting binomial is a difference of _____. This is given by the formula $(a + b)(a - b) =$ _____.

14. When a binomial is squared, the resulting trinomial is a _____ square trinomial. For example, $(a + b)^2 =$ _____.

Objective 1: Identify Key Elements of a Polynomial

For Exercises 15–16, determine if the expression is a polynomial.

15. a. $4a^2 + 7b - 3$ 　　 **b.** $\dfrac{3}{4}x^2y$ 　　 **c.** $6x + \dfrac{7}{x} + 5$ 　　 **d.** $\sqrt{p^2 + 2p - 5}$

16. a. $3x^5 - 9x^2 + \dfrac{2}{x^3}$ 　　 **b.** $\sqrt{2}ab^4$ 　　 **c.** $3|y| + 2$ 　　 **d.** $-7x^3 - 4x^2 + 2x - 5$

For Exercises 17–20, write the polynomial in descending order. Then identify the leading coefficient and degree of the polynomial.

17. $7.2x^3 - 18x^7 - 4.1$

18. $9.1y^5 + 4.6y^2 - 1.7y^8$

19. $\dfrac{1}{3}y - y^2$

20. $\dfrac{4}{5}c^2 + c^5$

For Exercises 21–22, determine the degree of the polynomial.

21. $-8p^2qr^5 + 4pq^8r^2 + 5p^3q^3r$

22. $-4.7abc^4 - 5.2a^2bc^5 + 2.6a^3c$

For Exercises 23–28, write a polynomial in one variable that is described by the following. (Answers may vary.)

23. A monomial of degree 5

24. A monomial of degree 4

25. A trinomial of degree 2

26. A trinomial of degree 3

27. A binomial of degree 4

28. A binomial of degree 2

Objective 2: Add and Subtract Polynomials

For Exercises 29–36, add the polynomials and simplify. **(See Examples 1–2)**

29. $(-4m^2 + 4m) + (5m^2 + 6m)$

30. $(3n^3 + 5n) + (2n^3 - 2n)$

31. $(3x^4 - x^3 - x^2) + (3x^3 - 7x^2 + 2x)$

32. $(6x^3 - 2x^2 - 12) + (x^2 + 3x + 9)$

33. $\left(\dfrac{1}{2}w^3 + \dfrac{2}{9}w^2 - 1.8w\right) + \left(\dfrac{3}{2}w^2 - \dfrac{1}{9}w^2 + 2.7w\right)$

34. $\left(2.9t^4 - \dfrac{7}{8}t + \dfrac{5}{3}\right) + \left(-8.1t^4 - \dfrac{1}{8}t - \dfrac{1}{3}\right)$

35.
$$\begin{array}{r} 12x^3 \qquad + 6x - 8 \\ + (-3x^3 - 5x^2 - 4x \quad) \\ \hline \end{array}$$

36.
$$\begin{array}{r} -8y^4 - 8y^3 - 6y^2 \qquad - 9 \\ + (4y^4 + 5y^3 \qquad - 10y - 3) \\ \hline \end{array}$$

For Exercises 37–42, write the opposite of the given polynomial. **(See Example 3)**

37. $-30y^3$

38. $-2x^2$

39. $4p^3 + 2p - 12$

40. $8t^2 - 4t - 3$

41. $-11ab^2 + a^2b$

42. $-23rs - 4r + 9s$

For Exercises 43–50, subtract the polynomials and simplify. **(See Examples 4–5)**

43. $(-3x^3 + 3x^2 - x + 6) - (1 - x - x^2 - x^3)$

44. $(-8x^3 + 6x + 7) - (-4 - 2x - 5x^3)$

45. $(-3xy^3 + 3x^2y - x + 6) - (-xy^3 - xy - x + 1)$

46. $(-8x^2y^2 + 6xy^2 + 7xy) - (5xy^2 - 2xy - 4)$

47.
$$\begin{array}{r} 4t^3 - 6t^2 \qquad - 18 \\ - (3t^3 + 7t^2 + 9t - 5) \\ \hline \end{array}$$

48.
$$\begin{array}{r} 5w^3 - 9w^2 + 6w + 13 \\ - (7w^3 \qquad - 10w - 8) \\ \hline \end{array}$$

49. $\left(\dfrac{1}{5}a^2 - \dfrac{1}{2}ab + \dfrac{1}{10}b^2 + 3\right) - \left(-\dfrac{3}{10}a^2 + \dfrac{2}{5}ab - \dfrac{1}{2}b^2 - 5\right)$

50. $\left(\dfrac{4}{7}a^2 - \dfrac{1}{7}ab + \dfrac{1}{14}b^2 - 7\right) - \left(\dfrac{1}{2}a^2 - \dfrac{2}{7}ab - \dfrac{9}{14}b^2 + 1\right)$

51. Subtract $(9x^2 - 5x + 1)$ from $(8x^2 + x - 15)$.

52. Subtract $(-x^3 + 5x)$ from $(10x^3 + x^2 - 10)$.

Objective 3: Multiply Polynomials

For Exercises 53–74, multiply the polynomials. **(See Examples 6–8)**

53. $(7x^4y)(-6xy^5)$

54. $(-4a^3b^7)(-2ab^3)$

55. $2m^3n^2(m^2n^3 - 3mn^2 + 4n)$

56. $3p^2q(p^3q^3 - pq^2 - 4p)$

57. $6xy^2\left(\dfrac{1}{2}x - \dfrac{2}{3}xy\right)$

58. $12ab\left(\dfrac{5}{6}a + \dfrac{1}{4}ab^2\right)$

59. $(6x - 1)(5 + 2x)$

60. $(7 + 3x)(x - 8)$

61. $(y^2 - 12)(2y^2 + 3)$

62. $(4p^2 - 1)(2p^2 + 5)$

63. $(5s + 3t)(5s - 2t)$

64. $(4a + 3b)(4a - b)$

65. $(n^2 + 10)(5n + 3)$

66. $(m^2 + 8)(3m + 7)$

67. $(x - 7)(x^2 + 7x + 49)$

68. $(x + 3)(x^2 - 3x + 9)$

69. $(2x + y)(3x^2 + 2xy + y^2)$

70. $(h - 5k)(h^2 - 2hk + 3k^2)$

71. $(4a - b)(a^3 - 4a^2b + ab^2 - b^3)$

72. $(3m + 2n)(m^3 + 2m^2n - mn^2 + 2n^3)$

73. $\left(\frac{1}{2}a - 2b + c\right)(a + 6b - c)$

74. $\left(\frac{1}{2}a^2 - 2ab + b^2\right)(2a + b)$

Objective 4: Multiply Special Case Products

For Exercises 75–92, perform the indicated operations. (**See Examples 9–10**)

75. $(a - 8)(a + 8)$

76. $(b + 2)(b - 2)$

77. $(3p + 1)(3p - 1)$

78. $(5q - 3)(5q + 3)$

79. $(3h - k)(3h + k)$

80. $(x - 7y)(x + 7y)$

81. $(t - 7)^2$

82. $(w + 9)^2$

83. $(3h - k)^2$

84. $(x - 7y)^2$

85. $(2z^2 - w^3)(2z^2 + w^3)$

86. $(a^4 - 2b^3)(a^4 + 2b^3)$

87. $(5x^2 - 3y)^2$

88. $(4p^3 - 2m)^2$

89. $\left(\frac{1}{5}c - \frac{2}{3}d^3\right)\left(\frac{1}{5}c + \frac{2}{3}d^3\right)$

90. $\left(\frac{1}{6}n - \frac{4}{5}p^4\right)\left(\frac{1}{6}n + \frac{4}{5}p^4\right)$

91. $(4t^2 + 3p^3)^2$

92. $(2a^2 + 11b^3)^2$

Mixed Exercises

93. Multiply the expressions. Explain their similarities.

   a. $(A - B)(A + B)$

   b. $[(x + y) - B][(x + y) + B]$

94. Multiply the expressions. Explain their similarities.

   a. $(A + B)(A - B)$

   b. $[A + (3h + k)][A - (3h + k)]$

For Exercises 95–100, multiply the expressions. (**See Example 11**)

95. $[(w + v) - 2][(w + v) + 2]$

96. $[(x + y) - 6][(x + y) + 6]$

97. $[2 - (x + y)][2 + (x + y)]$

98. $[a - (b + 1)][a + (b + 1)]$

99. $[(3a - 4) + b][(3a - 4) - b]$

100. $[(5p - 7) - q][(5p - 7) + q]$

101. Explain how to multiply $(x + y)^3$.

102. Explain how to multiply $(a - b)^3$.

For Exercises 103–106, multiply the expressions. (**See Example 11**)

103. $(2x + y)^3$

104. $(x - 5y)^3$

105. $(4a - b)^3$

106. $(3a + 4b)^3$

107. Explain how you would multiply the binomials.

$$(x - 2)(x + 6)(2x + 1)$$

108. Explain how you would multiply the binomials.

$$(a + b)(a - b)(2a + b)(2a - b)$$

For Exercises 109–122, simplify the expressions.

109. $2a^2(a + 5)(3a + 1)$

110. $-5y(2y - 3)(y + 3)$

111. $(x + 3)(x - 3)(x + 5)$

112. $(t + 2)(t - 3)(t + 1)$

113. $(y + 1)^2 - (2y + 3)^2$

114. $(b - 3)^2 - (3b - 1)^2$

115. $(x^n + 3)(x^n - 7)$

116. $(y^n + 4)(y^n - 5)$

117. $(z^n + w^m)^2$

118. $(w^n - y^m)^2$

119. $(a^n - 5)(a^n + 5)$

120. $(b^n + 7)(b^n - 7)$

121. $(6x + 5)(6x - 5) - (6x + 5)^2$

122. $(2y - 7)(2y + 7) - (2y - 7)^2$

For Exercises 123–130, write an expression that represents the perimeter, area, or volume as indicated, and simplify. (See Example 11)

123. Perimeter

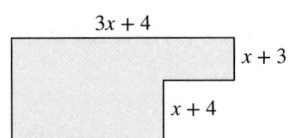

124. Perimeter

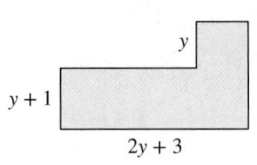

125. Area

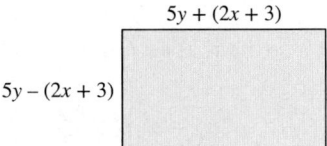

126. Area

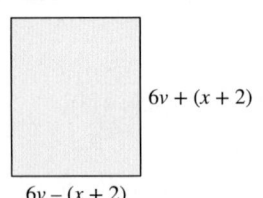

127. Volume

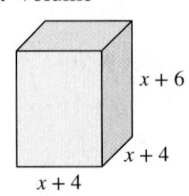

128. Volume

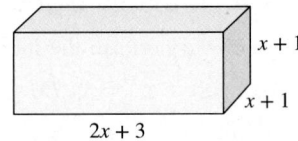

129. Area

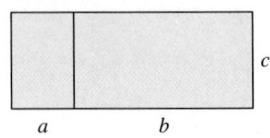

130. Area

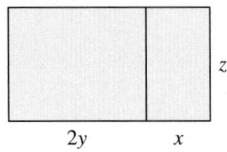

For Exercises 131–132, write an expression that represents the area of the shaded region and simplify the expression.

131.

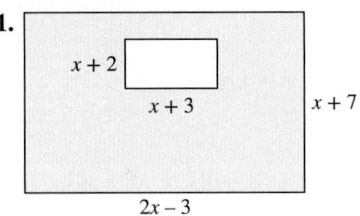

132.

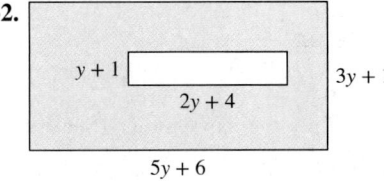

133. Suppose that x represents the smaller of two consecutive integers.

   a. Write a polynomial that represents the larger integer.

   b. Write a polynomial that represents the sum of the two integers. Then simplify.

   c. Write a polynomial that represents the product of the two integers. Then simplify.

   d. Write a polynomial that represents the sum of the squares of the two integers. Then simplify.

134. Suppose that x represents the larger of two consecutive odd integers.

   a. Write a polynomial that represents the smaller integer.

   b. Write a polynomial that represents the sum of the two integers. Then simplify.

   c. Write a polynomial that represents the product of the two integers. Then simplify.

   d. Write a polynomial that represents the difference of the squares of the two integers. Then simplify.

Expanding Your Skills

135. Multiply. $(x + 2)^4$

136. Multiply. $(x - 3)^4$

137. $(2x - 3)$ multiplied by what binomial will result in the trinomial $10x^2 - 27x + 18$? Check your answer by multiplying the binomials.

138. $(4x + 1)$ multiplied by what binomial will result in the trinomial $12x^2 - 5x - 2$? Check your answer by multiplying the binomials.

139. $(4y + 3)$ multiplied by what binomial will result in the trinomial $8y^2 + 2y - 3$? Check your answer by multiplying the binomials.

140. $(3y - 2)$ multiplied by what binomial will result in the trinomial $3y^2 - 17y + 10$? Check your answer by multiplying the binomials.

For Exercises 141–144, determine if the statement is true or false.

141. The sum of two polynomials each of degree 5 will be degree 5.

142. The sum of two polynomials each of degree 5 will be less than or equal to degree 5.

143. The product of two polynomials each of degree 4 will be degree 8.

144. The product of two polynomials each of degree 4 will be less than degree 8.

PROBLEM RECOGNITION EXERCISES

Operations on Exponential Expressions and Polynomials

For Exercises 1–32, perform the indicated operations and simplify the expression.

1. a. $(3x + 1)^2$
b. $(3x + 1)(3x - 1)$
c. $(3x + 1) - (3x - 1)$

2. a. $(9m - 5) - (9m + 5)$
b. $(9m - 5)(9m + 5)$
c. $(9m - 5)^2$

3. a. $(4x^2y)^2$
b. $(4 + x^2 + y)^2$

4. a. $(a^2b^3)^2$
b. $(a^2 - b^3)^2$

5. a. $(4c^2d^5)(-3c^3)$
b. $(4 + c^2 - d^5)(-3 + c^3)$

6. a. $x^2 \cdot x^4 \cdot x^6$
b. $[(x^2)^4]^6$

7. a. -6^2
b. $(-6)^2$
c. 6^{-2}

8. a. $10x^0$
b. $(10x)^0$
c. $-10x^0$

9. a. $(p - 5)(p + 5) - (p^2 + 5)$
b. $(p - 5)(p + 5) - (p + 5)^2$
c. $(p - 5)(p + 5) - (p^2 - 25)$

10. a. $(x + 4)(x - 4) - (x + 4)^2$
b. $(x + 4)(x - 4) - (x^2 + 4)$
c. $(x + 4)(x - 4) - (x^2 - 16)$

11. $(5t^2 - 6t + 2) - (3t^2 - 7t + 3)$

12. $-5x^2(3x^2 + x - 2)$

13. $(6z + 5)(6z - 5)$

14. $(6y^3 + 2y^2 + y - 2) + (3y^3 - 4y + 3)$

15. $(3b - 4)(2b - 1)$

16. $(5a + 2)(2a^2 + 3a + 1)$

17. $(t^3 - 4t^2 + t - 9) + (t + 12) - (2t^2 - 6t)$

18. $(k + 4)^2 + (-4k + 9)$

19. $-2t(t^2 + 6t - 3) + t(3t + 2)(3t - 2)$

20. $\left(\frac{1}{4}p^3 - \frac{1}{6}p^2 + 5\right) - \left(-\frac{2}{3}p^3 + \frac{1}{3}p^2 - \frac{1}{5}p\right)$

21. $-6w^3(1.2w - 2.6w^2 + 5.1w^3)$

22. $(6a^2 - 4b)^2$

23. $\left(\frac{1}{2}z^2 - \frac{1}{3}\right)\left(\frac{1}{2}z^2 + \frac{1}{3}\right)$

24. $(m - 3)^2 - 2(m + 8)$

25. $(2x - 5)(x + 1) - (x - 3)^2$

26. $(m^2 - 6m + 7)(2m^2 + 4m - 3)$

27. $[5 - (a + b)]^2$

28. $[a - (x - y)][a + (x - y)]$

29. $(x + y)^2 - (x - y)^2$

30. $(a - 4)^3$

31. $\left(-\frac{1}{2}x + \frac{1}{3}\right)\left(\frac{1}{4}x - \frac{1}{2}\right)$

32. $-3x^2y^3z^4\left(\frac{1}{6}x^4yzw^3\right)$

SECTION 2.3 Greatest Common Factor and Factoring by Grouping

1. Factor Out the Greatest Common Factor

In Section 2.2 we learned how to multiply polynomials. In this section, we reverse this process. The goal is to decompose a polynomial into a product of factors. This process is called **factoring**. Factoring is important in a variety of applications. In particular, factoring is often used to solve equations. For example, the two equations that follow are equivalent (that is, they have the same solution set).

$$x^2 - 19x + 88 = 0 \qquad (x - 8)(x - 11) = 0$$

Both equations ask, for what values of x will the left side equal zero? In the first equation it is difficult to determine the correct values of x. However, in the second equation with the left side factored, we can see by inspection that $x = 8$ and $x = 11$ are solutions to the equation.

There are many techniques used to factor a polynomial, but the first step is always to factor out the greatest common factor. The **greatest common factor (GCF)** of a polynomial is the expression of highest degree that divides evenly into each term of the polynomial. For example, consider the polynomial $12x^5 + 18x^4 - 24x^3$. To find the greatest common factor, we can list the factors of each term and identify the common factors.

$$\left.\begin{array}{l} 12x^5 = 2 \cdot 2 \cdot 3 \cdot x \cdot x \cdot x \cdot x \cdot x \\ 18x^4 = 3 \cdot 2 \cdot 3 \cdot x \cdot x \cdot x \cdot x \\ -24x^3 = -1 \cdot 2 \cdot 2 \cdot 2 \cdot 3 \cdot x \cdot x \cdot x \end{array}\right\} \text{GCF} = 6x^3$$

Notice that each term has common factors of 2 and 3 and three factors of x. Therefore, the GCF is $6x^3$.

x^3 is the greatest power of x common to all three terms.

$$12x^5 + 18x^4 - 24x^3 \qquad \text{The GCF is } 6x^3.$$

6 is the greatest integer that divides evenly into 12, 18, and −24.

To factor out the greatest common factor, we use the distributive property.

EXAMPLE 1   Factoring Out the Greatest Common Factor

Factor out the greatest common factor.

a. $12x^5 + 18x^4 - 24x^3$ **b.** $3y(2y - 5) + (2y - 5)$

Solution:

a. $12x^5 + 18x^4 - 24x^3$ The GCF is $6x^3$.

$\quad = 6x^3(2x^2) + 6x^3(3x) + 6x^3(-4)$ Write each term as a product of the GCF and another factor.

$\quad = 6x^3(2x^2 + 3x - 4)$ Apply the distributive property.

$\qquad\qquad$ Check: $6x^3(2x^2 + 3x - 4)$
$\qquad\qquad\qquad = 12x^5 + 18x^4 - 24x^3$ ✓

Avoiding Mistakes

In Example 1(b), there is an understood factor of 1 in the second term: $1(2y - 5)$.

Do not forget to include this in the factored form.

b. $3y(2y - 5) + (2y - 5)$

$\quad = 3y(2y - 5) + 1(2y - 5)$ The GCF is the binomial $(2y - 5)$.

$\quad = (2y - 5)(3y + 1)$ Apply the distributive property.

Skill Practice 1   Factor out the greatest common factor.

   a. $9z^6 - 27z^4 + 12z^2$ **b.** $5w(w - 3) - (w - 3)$

Sometimes it is preferable to factor out a negative factor from a polynomial. For example, consider the polynomial $-4x^2 - 8x + 12$. If we factor out -4, then the leading coefficient of the remaining polynomial will be positive.

EXAMPLE 2   **Factoring Out a Negative Factor**

Factor out -4 from the polynomial. $-4x^2 - 8x + 12$

Solution:

> **TIP** When a negative factor is factored out of a polynomial, the remaining terms in parentheses will have signs opposite to those in the original polynomial.

$$-4x^2 - 8x + 12$$

$$= (-4)(x^2) + (-4)(2x) + (-4)(-3) \quad \text{Write each term as a product of } -4 \text{ and another factor.}$$

$$= -4(x^2 + 2x - 3) \quad \text{Apply the distributive property.}$$

Skill Practice 2   Factor out -6 from the polynomial.

$-6y^4 + 24y^2 + 6$

2. Factor by Grouping

When two binomials are multiplied, the product before simplifying contains four terms. For example:

$$(3a + 2)(2b - 7) = (3a + 2)(2b) + (3a + 2)(-7)$$

$$= (3a + 2)(2b) + (3a + 2)(-7)$$

$$= 6ab + 4b - 21a - 14$$

In Example 3, we learn how to reverse this process. That is, given a four-term polynomial, we will factor it as a product of two binomials. The process is called **factoring by grouping**.

> **Factoring by Grouping**
>
> To factor a four-term polynomial by grouping:
>
> **Step 1**   Identify and factor out the GCF from all four terms.
> **Step 2**   Factor out the GCF from the first pair of terms. Factor out the GCF from the second pair of terms. (Sometimes it is necessary to factor out the *opposite* of the GCF.)
> **Step 3**   If the two terms share a common binomial factor, factor out the binomial factor.

Answers

1. a. $3z^2(3z^4 - 9z^2 + 4)$
  b. $(w - 3)(5w - 1)$
2. $-6(y^4 - 4y^2 - 1)$

EXAMPLE 3 **Factoring by Grouping**

Factor by grouping. $6ab - 21a + 4b - 14$

Solution:

$$6ab - 21a + 4b - 14$$

Step 1: Identify and factor out the GCF from all four terms. In this case the GCF is 1.

$$= 6ab - 21a \quad \vdots \quad + 4b - 14$$

Group the first pair of terms and the second pair of terms.

$$= 3a(2b - 7) + 2(2b - 7)$$

Step 2: Factor out the GCF from each pair of terms.

Note: The two terms now share a common binomial factor of $(2b - 7)$.

$$= (2b - 7)(3a + 2)$$

Step 3: Factor out the common binomial factor.

Check: $(2b - 7)(3a + 2) = 2b(3a) + 2b(2) - 7(3a) - 7(2)$

$$= 6ab + 4b - 21a - 14 \quad \checkmark$$

Skill Practice 3   Factor by grouping. $7c^2 + cd + 14c + 2d$

EXAMPLE 4 **Factoring by Grouping**

Factor by grouping. $x^3 + 3x^2 - 3x - 9$

Solution:

$$x^3 + 3x^2 - 3x - 9$$

Step 1: Identify and factor out the GCF from all four terms. In this case the GCF is 1.

$$= x^3 + 3x^2 \quad \vdots \quad - 3x - 9$$

Group the first pair of terms and the second pair of terms.

$$= x^2(x + 3) - 3(x + 3)$$

Step 2: Factor out x^2 from the first pair of terms.

Factor out -3 from the second pair of terms (this causes the signs to change in the second parentheses). The terms now contain a common binomial factor.

$$= (x + 3)(x^2 - 3)$$

Step 3: Factor out the common binomial $(x + 3)$.

Skill Practice 4   Factor by grouping. $a^3 - 4a^2 - 3a + 12$

Answers
3. $(7c + d)(c + 2)$
4. $(a^2 - 3)(a - 4)$

EXAMPLE 5   Factoring by Grouping

Factor by grouping. $24p^2q^2 - 18p^2q + 60pq^2 - 45pq$

Solution:

$24p^2q^2 - 18p^2q + 60pq^2 - 45pq$

$= 3pq(8pq - 6p + 20q - 15)$ **Step 1:** Remove the GCF $3pq$ from all four terms.

$= 3pq(8pq - 6p \ \vdots \ + 20q - 15)$ Group the first pair of terms and the second pair of terms.

$= 3pq[2p(4q - 3) + 5(4q - 3)]$ **Step 2:** Factor out the GCF from each pair of terms. The terms share the binomial factor $(4q - 3)$.

$= 3pq(4q - 3)(2p + 5)$ **Step 3:** Factor out the common binomial $(4q - 3)$.

Skill Practice 5   Factor the polynomial. $24x^2y - 12x^2 + 20xy - 10x$

Notice that in step 3 of factoring by grouping, a common binomial is factored from the two terms. These binomials must be *exactly* the same in each term. If the two binomial factors differ, try rearranging the original four terms.

EXAMPLE 6   Factoring by Grouping Where Rearranging Terms Is Necessary

Factor the polynomial. $4x + 6pa - 8a - 3px$

Solution:

$4x + 6pa - 8a - 3px$ **Step 1:** Identify and factor out the GCF from all four terms. In this case the GCF is 1.

$= 4x + 6pa \ \vdots \ - 8a - 3px$

$= 2(2x + 3pa) - 1(8a + 3px)$ **Step 2:** The binomial factors in each term are different.

> **Avoiding Mistakes**
>
> Remember that when factoring by grouping, the binomial factors must be *exactly* the same.

$= 4x - 8a \ \vdots \ - 3px + 6pa$ *Try rearranging the original four terms* in such a way that the first pair of coefficients is in the same ratio as the second pair of coefficients. Notice that the ratio 4 to -8 is the same as the ratio -3 to 6.

$= 4(x - 2a) - 3p(x - 2a)$ **Step 2:** Factor out 4 from the first pair of terms.

Factor out $-3p$ from the second pair of terms.

$= (x - 2a)(4 - 3p)$ **Step 3:** Factor out the common binomial factor.

Answers

5. $2x(6x + 5)(2y - 1)$

6. $(3r + s)(2 + y)$

Skill Practice 6   Factor the polynomial. $3ry + 2s + sy + 6r$

3. Use Factoring to Solve for a Specified Variable in an Equation

In Section 1.2 we learned how to isolate a specified variable within a literal equation. In Example 7, we will isolate the variable x from the equation $ax + by = cx + d$. However, notice the equation has multiple occurrences of x. Therefore, factoring is required to combine x terms so that we can subsequently isolate x.

EXAMPLE 7   Solving an Equation for a Specified Variable

Solve the equation for x. $ax + by = cx + d$

Solution:

$$ax + by = cx + d$$ Subtract cx from both sides to combine the x terms on one side.

$$ax - cx = d - by$$ Subtract by from both sides to combine the non-x terms on the other side.

$$x(a - c) = d - by$$ Factor out x as the GCF on the left side of the equation.

$$\frac{x(a - c)}{(a - c)} = \frac{d - by}{(a - c)}$$ Divide by $(a - c)$.

$$x = \frac{d - by}{a - c}$$

Skill Practice 7   Solve the equation for x. $3x - w = ax + z$

TIP In Example 7, the answer can be expressed in different forms. For example, if we had isolated x on the right side of the equation, the solution for x would be

$$ax + by = cx + d$$

$$by - d = cx - ax$$

$$\frac{by - d}{(c - a)} = \frac{x(c - a)}{(c - a)}$$

$$x = \frac{by - d}{c - a}$$

To show that $\dfrac{by - d}{c - a} = \dfrac{d - by}{a - c}$ multiply either expression by $\dfrac{-1}{-1}$.

$$\frac{(by - d)}{(c - a)} \cdot \frac{(-1)}{(-1)} = \frac{-by + d}{-c + a} = \frac{d - by}{a - c}$$

---

4. Factor Expressions Containing Negative Exponents

We now revisit the process to factor out the greatest common factor. In some applications, it is necessary to factor out a variable factor with a negative integer exponent. Before we demonstrate this in Example 8, take a minute to review a similar example with positive integer exponents.

$$2x^6 + 5x^5 + 7x^4$$ The GCF is x^4. This is x raised to the *smallest exponent* to which it appears in any term.

$$\boxed{6-4}\ \boxed{5-4}\ \boxed{4-4}$$

$$= x^4(2x^2 + 5x^1 + 7x^0)$$ The powers on the factors of x within the parentheses are found by subtracting 4 from the original exponents.

Answer

7. $x = \dfrac{w + z}{3 - a}$ or $x = -\dfrac{w + z}{a - 3}$

EXAMPLE 8   Factoring Out a GCF with a Negative Exponent

Factor. Write the answer with no negative exponents.

$$2x^{-6} + 5x^{-5} + 7x^{-4}$$

Solution:

$$2x^{-6} + 5x^{-5} + 7x^{-4}$$

The smallest exponent on x is -6.
Factor out x^{-6}.

$-6 - (-6)$ $-5 - (-6)$ $-4 - (-6)$

$$= x^{-6}(2x^0 + 5x^1 + 7x^2)$$

The powers on the factors of x within the parentheses are found by subtracting -6 from the original exponents.

$$= x^{-6}(2 + 5x + 7x^2)$$

$$= \frac{7x^2 + 5x + 2}{x^6}$$

Simplify the negative exponent.

$$b^{-n} = \frac{1}{b^n}$$

Skill Practice 8   Factor completely and write the answer with no negative exponents.

$$11a^{-5} - 3a^{-4} + 2a^{-3}$$

Answer

8. $\dfrac{2a^2 - 3a + 11}{a^5}$

SECTION 2.3   Practice Exercises

Prerequisite Review

For Exercises R.1–R.10, identify the greatest common factor for the given expressions.

R.1. 15 and 25

R.2. 12 and 18

R.3. 40, 36, and 12

R.4. 24, 32, and 64

R.5. x^2, x^4, and x^6

R.6. y, y^2, and y^3

R.7. $20c^2d$ and $15c^3d^2$

R.8. $8a^4b^2$ and $10a^3b$

R.9. $3(2m + n)$ and $6(2m + n)$

R.10. $5(p - 4q)$ and $25(p - 4q)$

Concept Connections

1. Factoring a polynomial means to write it as a _____ of two or more polynomials.

2. The _____ _____ _____ (GCF) of a polynomial is the greatest factor that divides each term of the polynomial evenly.

3. The first step toward factoring a polynomial is to factor out the _____ _____ _____.

4. To factor a four-term polynomial, we try the process of factoring by _____.

Objective 1: Factor Out the Greatest Common Factor

For Exercises 5–28, factor out the greatest common factor. (**See Example 1**)

5. $3x + 12$

6. $15x - 10$

7. $6z^2 + 4z$

8. $49y^3 - 35y^2$

9. $4p^6 - 4p$

10. $5q^2 - 5q$

11. $12x^4 - 36x^2$

12. $51w^4 - 34w^3$

13. $9st^2 + 27t$ **14.** $8a^2b^3 + 12a^2b$ **15.** $9a^4b^3 + 27a^3b^4 - 18a^2b^5$ **16.** $3x^5y^4 - 15x^4y^5 + 9x^2y^7$

17. $10x^2y + 15xy^2 - 5xy$ **18.** $12c^3d - 15c^2d - 3cd$ **19.** $13b^2 - 11a^2b - 12ab$ **20.** $6a^3 - 2a^2b + 5a^2$

21. $2a(3z - 2b) - 5(3z - 2b)$ **22.** $5x(3x + 4) + 2(3x + 4)$ **23.** $2x^2(2x - 3) + (2x - 3)$

24. $z(w - 9) + (w - 9)$ **25.** $y(2x + 1)^2 - 3(2x + 1)^2$ **26.** $a(b - 7)^2 + 5(b - 7)^2$

27. $3y(x - 2)^2 + 6(x - 2)^2$ **28.** $10z(z + 3)^2 - 2(z + 3)^2$

For Exercises 29–34, factor out the indicated quantity. (**See Example 2**)

29. $-x^2 - 10x + 7$: Factor out -1.   **30.** $-5y^2 + 10y + 3$: Factor out -1.

31. $-12x^3y - 6x^2y - 3xy$: Factor out $-3xy$.   **32.** $-32a^4b^2 + 24a^3b + 16a^2b$: Factor out $-8a^2b$.

33. $-2t^3 + 11t^2 - 3t$: Factor out $-t$.   **34.** $-7y^2z - 5yz - z$: Factor out $-z$.

Objective 2: Factor by Grouping

35. Factor the polynomials by grouping.

    a. $2ax - ay + 6bx - 3by$

    b. $10w^2 - 5w - 6bw + 3b$

    c. Explain why you factored out $3b$ from the second pair of terms in part (a) but factored out the quantity $-3b$ from the second pair of terms in part (b).

36. Factor the polynomials by grouping.

    a. $3xy + 2bx + 6by + 4b^2$

    b. $15ac + 10ab - 6bc - 4b^2$

    c. Explain why you factored out $2b$ from the second pair of terms in part (a) but factored out the quantity $-2b$ from the second pair of terms in part (b).

For Exercises 37–56, factor each polynomial by grouping (if possible). (**See Examples 3–6**)

37. $y^3 + 4y^2 + 3y + 12$ **38.** $ab + b + 2a + 2$

39. $6p - 42 + pq - 7q$ **40.** $2t - 8 + st - 4s$

41. $2mx + 2nx + 3my + 3ny$ **42.** $4x^2 + 6xy - 2xy - 3y^2$

43. $10ax - 15ay - 8bx + 12by$ **44.** $35a^2 - 15a + 14a - 6$

45. $x^3 - x^2 - 3x + 3$ **46.** $2rs + 4s - r - 2$

47. $6p^2q + 18pq - 30p^2 - 90p$ **48.** $5s^2t + 20st - 15s^2 - 60s$

49. $100x^3 - 300x^2 + 200x - 600$ **50.** $2x^5 - 10x^4 + 6x^3 - 30x^2$

51. $6ax - by + 2bx - 3ay$ **52.** $5pq - 12 - 4q + 15p$

53. $4a - 3b - ab + 12$ **54.** $x^2y + 6x - 3x^3 - 2y$

55. $7y^3 - 21y^2 + 5y - 10$ **56.** $5ax + 10bx - 2ac + 4bc$

57. Explain why the grouping method failed for Exercise 55.   **58.** Explain why the grouping method failed for Exercise 56.

Objective 3: Use Factoring to Solve for a Specified Variable in an Equation

For Exercises 59–68, solve for the specified variable. (**See Example 7**)

59. $6 = 4x + tx$ for x

60. $8 = 3x + kx$ for x

61. $6x + ay = bx + 5$ for x

62. $3x + 2y = cx + d$ for x

63. $A = P + Prt$ for P

64. $C = A + Ar$ for A

65. $U = Av + Acw$ for A

66. $S = rt + wt$ for t

67. $ay + bx = cy$ for y

68. $cd + 2x = ac$ for c

69. The area of a rectangle of width w is given by $A = 2w^2 + w$. Factor the right-hand side of the equation to find an expression for the length of the rectangle.

70. The area of a rectangle of length l is given by $A = l^2 - 4l$. Factor the right-hand side of the equation to find an expression for the width of the rectangle.

Objective 4: Factor Expressions Containing Negative Exponents

For Exercises 71–78, factor completely. Write the answers with positive exponents only. (**See Example 8**)

71. $2x^{-4} - 7x^{-3} + x^{-2}$

72. $5t^{-7} + 2t^{-6} - t^{-5}$

73. $5x^{-3} - 11x^{-4} + x^{-5}$

74. $9x^{-7} + 3x^{-8} + x^{-9}$

75. $3x^2(5x - 1)^{-3} - 15x^3(5x - 1)^{-4}$

76. $4x^3(2x + 5)^{-4} - 8x^4(2x + 5)^{-5}$

77. $2m(m + 3)^{-2} - 2m^2(m + 3)^{-3}$

78. $3y^2(y + 4)^{-3} - 3y^3(y + 4)^{-4}$

Mixed Exercises

79. Construct a trinomial that has a greatest common factor of $3x^2$. (Answers may vary.)

80. Construct a trinomial that has a greatest common factor of $7y^4$. (Answers may vary.)

81. Construct a binomial that has a greatest common factor of $3t - 4$. (Answers may vary.)

82. Construct a binomial that has a greatest common factor of $m + 7$. (Answers may vary.)

Write About It

83. Explain the similarity in the process to factor out the GCF in the following two expressions.
$$7x^3 + 2x^2 \quad \text{and} \quad 7(t + 5)^3 + 2(t + 5)^2$$

84. Explain the similarity in the process to factor out the GCF in the following two expressions.
$$5x^4 + 4x^3 \quad \text{and} \quad 5x^{-4} + 4x^{-3}$$

Expanding Your Skills

For Exercises 85–88, factor out the greatest common factor. Assume that n is a positive integer.

85. $x^{3n} + 2x^{2n} - 5x^n$

86. $4x^{5n} - 3x^{4n} + x^{3n}$

87. $14a^{-3n} - 6a^{-4n} + 2a^{-5n}$

88. $4t^{-4n} + 12t^{-5n} - 20t^{-6n}$

Factoring Trinomials

1. Factor Trinomials: AC-Method

In this section, we present two methods to factor trinomials. The first method is called the ac-method. The second method is called the trial-and-error method.

The product of two binomials results in a four-term expression that can sometimes be simplified to a trinomial. To factor the trinomial, we want to reverse the process.

Multiply: $(2x + 3)(x + 2) = \xrightarrow{\text{Multiply the binomials.}} 2x^2 + 4x + 3x + 6$

$= \xrightarrow{\text{Add the middle terms.}} 2x^2 + 7x + 6$

Factor: $2x^2 + 7x + 6 = \xrightarrow[\text{a sum or difference of terms.}]{\text{Rewrite the middle term as}} 2x^2 + 4x + 3x + 6$

$= \xrightarrow{\text{Factor by grouping.}} (2x + 3)(x + 2)$

To factor a trinomial $ax^2 + bx + c$ by the ac-method, we rewrite the middle term bx as a sum or difference of terms. The goal is to produce a four-term polynomial that can be factored by grouping. The process is outlined as follows.

The AC-Method to Factor $ax^2 + bx + c$ ($a \neq 0$)

Step 1 After factoring out the GCF, multiply the coefficients of the first and last terms, ac.

Step 2 Find two integers whose product is ac and whose sum is b. (If no pair of integers can be found, then the trinomial cannot be factored further and is called a **prime polynomial**.)

Step 3 Rewrite the middle term bx as the sum of two terms whose coefficients are the integers found in step 2.

Step 4 Factor by grouping.

The ac-method for factoring trinomials is illustrated in Example 1. Before we begin, however, keep these two important guidelines in mind.

- For any factoring problem you encounter, always factor out the GCF from all terms first.
- To factor a trinomial, write the trinomial in the form $ax^2 + bx + c$.

EXAMPLE 1 **Factoring a Trinomial by the AC-Method**

Factor by using the ac-method. $12x^2 - 5x - 2$

Solution:

$12x^2 - 5x - 2$ The GCF is 1.

$a = 12 \quad b = -5 \quad c = -2$ **Step 1:** The expression is written in the form $ax^2 + bx + c$. Find the product $ac = 12(-2) = -24$.

Factors of −24	**Factors of −24**
$(1)(-24)$	$(-1)(24)$
$(2)(-12)$	$(-2)(12)$
$(3)(-8)$	$(-3)(8)$
$(4)(-6)$	$(-4)(6)$

Step 2: List all the factors of −24, and find the pair whose sum equals −5.

The numbers 3 and −8 produce a product of −24 and a sum of −5.

$$12x^2 - 5x - 2$$

$$= 12x^2 + 3x - 8x - 2$$

Step 3: Write the middle term of the trinomial as two terms whose coefficients are the selected numbers 3 and −8.

$$= 12x^2 + 3x \;\vdots\; - 8x - 2$$

Step 4: Factor by grouping.

$$= 3x(4x + 1) - 2(4x + 1)$$

$$= (4x + 1)(3x - 2)$$

The check is left for the reader.

Skill Practice 1   Factor by using the ac-method. $10x^2 + x - 3.$

TIP   One frequently asked question is whether the order matters when we rewrite the middle term of the trinomial as two terms (step 3). The answer is no. From Example 1, the two middle terms in step 3 could have been reversed.

$$12x^2 - 5x - 2 = 12x^2 - 8x + 3x - 2$$
$$= 4x(3x - 2) + 1(3x - 2)$$
$$= (3x - 2)(4x + 1)$$

This example also shows that the order in which two factors are written does not matter. The expression $(3x - 2)(4x + 1)$ is equivalent to $(4x + 1)(3x - 2)$ because multiplication is a commutative operation.

EXAMPLE 2   **Factoring a Trinomial by the AC-Method**

Factor the trinomial by using the ac-method. $-20c^3 + 34c^2d - 6cd^2$

Solution:

$$-20c^3 + 34c^2d - 6cd^2$$

$$= -2c(10c^2 - 17cd + 3d^2)$$ Factor out $-2c$.

Step 1: Find the product
$a \cdot c = (10)(3) = 30$

Factors of 30	**Factors of 30**
$1 \cdot 30$	$(-1)(-30)$
$2 \cdot 15$	$(-2)(-15)$
$3 \cdot 10$	$(-3)(-10)$
$5 \cdot 6$	$(-5)(-6)$

Step 2: The numbers −2 and −15 form a product of 30 and a sum of −17.

Answer

1. $(5x + 3)(2x - 1)$

$$= -2c(10c^2 - 17cd + 3d^2)$$

$$= -2c(10c^2 - 2cd \mid -15cd + 3d^2)$$　**Step 3:** Write the middle term of the trinomial as two terms whose coefficients are -2 and -15.

$$= -2c[2c(5c - d) - 3d(5c - d)]$$　**Step 4:** Factor by grouping.

$$= -2c(5c - d)(2c - 3d)$$

Skill Practice 2　Factor by using the ac-method.

$-4wz^3 - 2w^2z^2 + 20w^3z$

TIP　In Example 2, removing the GCF from the original trinomial produced a new trinomial with smaller coefficients. This makes the factoring process simpler because the product ac is smaller.

Original trinomial	**With the GCF factored out**
$-20c^3 + 34c^2d - 6cd^2$	$-2c(10c^2 - 17cd + 3d^2)$
$ac = (-20)(-6) = 120$	$ac = (10)(3) = 30$

2. Factor Trinomials: Trial-and-Error Method

Another method that is widely used to factor trinomials of the form $ax^2 + bx + c$ is the trial-and-error method. To understand how the trial-and-error method works, first consider the multiplication of two binomials:

Product of $2 \cdot 1$　　　Product of $3 \cdot 2$

$$(2x + 3)(1x + 2) = 2x^2 \underbrace{+ 4x + 3x}_{\substack{\text{sum of products} \\ \text{of inner terms} \\ \text{and outer terms}}} + 6 = 2x^2 + 7x + 6$$

In Example 3, we will factor this trinomial by reversing this process.

EXAMPLE 3　Factoring a Trinomial by the Trial-and-Error Method

Factor by the trial-and-error method.　　$2x^2 + 7x + 6$

Solution:

To factor by the trial-and-error method, we must fill in the blanks to create the correct product.

Factors of 2

$$2x^2 + 7x + 6 = (\square x \quad \square)(\square x \quad \square)$$

Factors of 6

- The first terms in the binomials must be $2x$ and x. This creates a product of $2x^2$, which is the first term in the trinomial.

- The second terms in the binomials must form a product of 6. This means that the factors must both be positive or both be negative. Because the middle term of the trinomial is positive, we will consider only *positive* factors of 6. The options are $1 \cdot 6$, $2 \cdot 3$, $6 \cdot 1$,   and $3 \cdot 2$.
- Test each combination of factors until the correct product of binomials is found.

$$(2x + 1)(x + 6) = 2x^2 + 12x + 1x + 6$$
$$= 2x^2 + 13x + 6 \qquad \textit{Incorrect.}\quad \text{Wrong middle term.}$$

$$(2x + 2)(x + 3) = 2x^2 + 6x + 2x + 6$$
$$= 2x^2 + 8x + 6 \qquad \textit{Incorrect.}\quad \text{Wrong middle term.}$$

$$(2x + 6)(x + 1) = 2x^2 + 2x + 6x + 6$$
$$= 2x^2 + 8x + 6 \qquad \textit{Incorrect.}\quad \text{Wrong middle term.}$$

$$(2x + 3)(x + 2) = 2x^2 + 4x + 3x + 6$$
$$= 2x^2 + 7x + 6 \qquad \textit{Correct.}$$

The factored form of $2x^2 + 7x + 6$ is $(2x + 3)(x + 2)$.

Skill Practice 3   Factor by the trial-and-error method.

$5y^2 - 9y + 4$

When applying the trial-and-error method, sometimes it is not necessary to test all possible combinations of factors. For the trinomial $2x^2 + 7x + 6$, the GCF is 1. Therefore, any binomial factor that shares a common factor greater than 1 will not work and does not need to be tested. For example, the following binomials cannot work:

$$\underbrace{(2x + 2)}_{\text{Common factor of 2}}(x + 3) \qquad\qquad \underbrace{(2x + 6)}_{\text{Common factor of 2}}(x + 1)$$

Although the trial-and-error method is tedious, its principle is generally easy to remember. We reverse the process of multiplying binomials.

The Trial-and-Error Method to Factor $ax^2 + bx + c$

Step 1   Factor out the greatest common factor.

Step 2   List all pairs of positive factors of a and pairs of positive factors of c. Consider the reverse order for either list of factors.

Step 3   Construct two binomials of the form

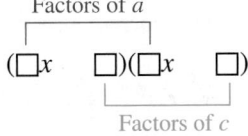

Step 4   Test each combination of factors and signs until the correct product is found.

Step 5   If no combination of factors produces the correct product, the trinomial cannot be factored further and is a **prime polynomial**.

Answer

3. $(5y - 4)(y - 1)$

> **EXAMPLE 4**   Factoring a Trinomial by the Trial-and-Error Method

Factor by the trial-and-error method. $13y - 6 + 8y^2$

Solution:

$8y^2 + 13y - 6$ Write in the form $ax^2 + bx + c$.

$(\Box y \quad \Box)(\Box y \quad \Box)$ **Step 1:** The GCF is 1.

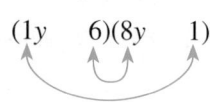

Factors of 8	**Factors of 6**
$1 \cdot 8$	$1 \cdot 6$
$2 \cdot 4$	$2 \cdot 3$
	$3 \cdot 2$
	$6 \cdot 1$	(reverse order)

Step 2: List the positive factors of 8 and positive factors of 6. Consider the reverse order in one list of factors.

$(2y \quad 1)(4y \quad 6)$
$(2y \quad 2)(4y \quad 3)$
$(2y \quad 3)(4y \quad 2)$
$(2y \quad 6)(4y \quad 1)$
$(1y \quad 1)(8y \quad 6)$
$(1y \quad 3)(8y \quad 2)$

Step 3: Construct all possible binomial factors by using different combinations of the factors of 8 and 6.

Without regard to signs, these factorizations cannot work because the terms in red in the binomial share a common factor greater than 1.

Test the remaining factorizations. Keep in mind that to produce a product of -6, the signs within the parentheses must be opposite (one positive and one negative). Also, the sum of the products of the inner terms and outer terms must be combined to form $13y$.

$(1y \quad 6)(8y \quad 1)$ *Incorrect.*   Wrong middle term.

Regardless of signs, the product of inner terms $48y$ and the product of outer terms $1y$ cannot be combined to form the middle term $13y$.

$(1y \quad 2)(8y \quad 3)$ *Correct.*   The terms $16y$ and $3y$ can be combined to form the middle term $13y$, provided the signs are applied correctly. We require $+16y$ and $-3y$.

The correct factorization of $8y^2 + 13y - 6$ is $(y + 2)(8y - 3)$.

Skill Practice 4   Factor by the trial-and-error method.

$5t - 6 + 4t^2$

In Example 4, the factors of -6 must have opposite signs to produce a negative product. Therefore, one binomial factor is a sum and one is a difference. Determining the correct signs is an important aspect of factoring trinomials. We suggest the following guidelines:

Answer
4. $(4t - 3)(t + 2)$

TIP Given the trinomial $ax^2 + bx + c$ ($a > 0$), the signs can be determined as follows:

1. If c is *positive*, then the signs in the binomials must be the same (either both positive or both negative). The correct choice is determined by the middle term. If the middle term is positive, then both signs must be positive. If the middle term is negative, then both signs must be negative.

c is positive.

Example: $20x^2 + 43x + 21$
$(4x + 3)(5x + 7)$
same signs

c is positive.

Example: $20x^2 - 43x + 21$
$(4x - 3)(5x - 7)$
same signs

2. If c is *negative*, then the signs in the binomials must be different. The middle term in the trinomial determines which factor gets the positive sign and which factor gets the negative sign.

c is negative.

Example: $x^2 + 3x - 28$
$(x + 7)(x - 4)$
different signs

c is negative.

Example: $x^2 - 3x - 28$
$(x - 7)(x + 4)$
different signs

EXAMPLE 5 **Factoring a Trinomial by the Trial-and-Error Method**

Factor by the trial-and-error method.

$$-80x^3y + 208x^2y^2 - 20xy^3$$

Solution:

$$-80x^3y + 208x^2y^2 - 20xy^3$$

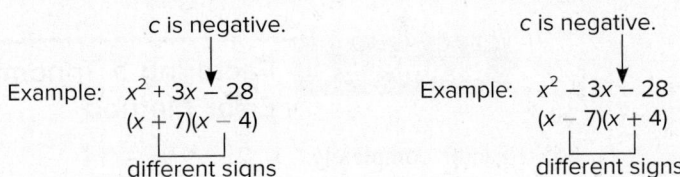

$$= -4xy(20x^2 - 52xy + 5y^2) \qquad \textbf{Step 1:}\quad \text{Factor out } -4xy.$$

$$= -4xy(\square x \ \square y)(\square x \ \square y)$$

Factors of 20	**Factors of 5**
$1 \cdot 20$	$1 \cdot 5$
$2 \cdot 10$	$5 \cdot 1$
$4 \cdot 5$	

Step 2: List the positive factors of 20 and positive factors of 5. Consider the reverse order in one list of factors.

Step 3: Construct all possible binomial factors by using different combinations of the factors of 20 and factors of 5. The signs in the parentheses must both be negative.

$$\left.\begin{array}{l} -4xy(1x - 1y)(20x - 5y) \\ -4xy(2x - 1y)(10x - 5y) \\ -4xy(4x - 1y)(5x - 5y) \end{array}\right\}$$

Incorrect. These binomials contain a common factor.

$$-4xy(1x - 5y)(20x - 1y)$$

Incorrect. Wrong middle term.
$$-4xy(x - 5y)(20x - 1y)$$
$$= -4xy(20x^2 - 101xy + 5y^2)$$

$$-4xy(4x - 5y)(5x - 1y) \qquad \textit{Incorrect.} \quad \text{Wrong middle term.}$$

$$-4xy(4x - 5y)(5x - 1y)$$
$$= -4xy(20x^2 - 29x + 5y^2)$$

$$-4xy(2x - 5y)(10x - 1y) \qquad \textbf{\textit{Correct.}} \quad -4xy(2x - 5y)(10x - 1y)$$
$$= \mathbf{-4xy(20x^2 - 52xy + 5y^2)}$$
$$= -80x^3y + 208x^2y^2 - 20xy^3$$

The correct factorization of $-80x^3y + 208x^2y^2 - 20xy^3$ is $-4xy(2x - 5y)(10x - y)$.

Skill Practice 5   Factor by the trial-and-error method.

$$-4z^3 - 22z^2 - 30z$$

EXAMPLE 6   Factoring a Trinomial by the Trial-and-Error Method

Factor completely. $2x^2 + 9x + 14$

Solution:

$2x^2 + 9x + 14$	The GCF is 1 and the trinomial is written in the form $ax^2 + bx + c$.
$(2x + 14)(x + 1)$	*Incorrect.* $(2x + 14)$ contains a common factor of 2.
$(2x + 2)(x + 7)$	*Incorrect.* $(2x + 2)$ contains a common factor of 2.
$(2x + 1)(x + 14) = 2x^2 + 29x + 14$	*Incorrect.* Wrong middle term.
$(2x + 7)(x + 2) = 2x^2 + 11x + 14$	*Incorrect.* Wrong middle term.

No combination of factors results in the correct product. Therefore, the trinomial is prime (cannot be factored).

Skill Practice 6   Factor completely. $6r^2 - 13r + 10$

If a trinomial has a leading coefficient of 1, the factoring process simplifies significantly. Consider the trinomial $x^2 + bx + c$. To produce a leading term of x^2, we can construct binomials of the form $(x + \square)(x + \square)$. The remaining terms may be satisfied by two numbers p and q whose product is c and whose sum is b:

Factors of c

$$(x + p)(x + q) = x^2 + qx + px + pq = x^2 + \underbrace{(p + q)}_{\text{Sum}\,=\,b}x + \underbrace{pq}_{\text{Product}\,=\,c}$$

This process is demonstrated in Example 7.

Answers
5. $-2z(2z + 5)(z + 3)$
6. Prime

EXAMPLE 7   Factoring a Trinomial with a Leading Coefficient of 1

Factor completely. $x^2 - 10x + 16$

Solution:

$x^2 - 10x + 16$ Factor out the GCF from all terms. In this case, the GCF is 1.

$= (x \ \square)(x \ \square)$ The trinomial is written in the form $x^2 + bx + c$. To form the product x^2, use the factors x and x.

Next, look for two numbers whose product is 16 and whose sum is -10. Because the middle term is negative, we will consider only the negative factors of 16.

Factors of 16	Sum
$-1(-16)$	$-1 + (-16) = -17$
$-2(-8)$	$-2 + (-8) = -10$
$-4(-4)$	$-4 + (-4) = -8$

The numbers are -2 and -8.

Therefore, $x^2 - 10x + 16 = (x - 2)(x - 8)$.

Skill Practice 7   Factor completely. $c^2 + 6c - 27$

3. Factor Perfect Square Trinomials

Recall that the square of a binomial always results in a **perfect square trinomial**.

$$(a + b)^2 = (a + b)(a + b) = a^2 + ab + ab + b^2 = a^2 + 2ab + b^2$$
$$(a - b)^2 = (a - b)(a - b) = a^2 - ab - ab + b^2 = a^2 - 2ab + b^2$$

For example, $(2x + 7)^2 = (2x)^2 + 2(2x)(7) + (7)^2 = 4x^2 + 28x + 49$

$$a = 2x \quad b = 7 \qquad\qquad a^2 + 2ab + b^2$$

To factor the trinomial $4x^2 + 28x + 49$, the ac-method or the trial-and-error method can be used. However, recognizing that the trinomial is a perfect square trinomial, we can use one of the following patterns to reach a quick solution.

Factored Form of a Perfect Square Trinomial

$$a^2 + 2ab + b^2 = (a + b)^2$$
$$a^2 - 2ab + b^2 = (a - b)^2$$

TIP   The following are perfect squares.

$1^2 = 1$	$(x^1)^2 = x^2$
$2^2 = 4$	$(x^2)^2 = x^4$
$3^2 = 9$	$(x^3)^2 = x^6$
$4^2 = 16$	$(x^4)^2 = x^8$
$\vdots$	$\vdots$

Any expression raised to an even power (multiple of 2) is a perfect square.

TIP   To determine if a trinomial is a perfect square trinomial, follow these steps:

1. Check if the first and third terms are both perfect squares with positive coefficients.
2. If this is the case, identify a and b, and determine if the middle term equals $2ab$ or $-2ab$.

Answer

7. $(c + 9)(c - 3)$

EXAMPLE 8   **Factoring a Perfect Square Trinomial**

Factor completely. $x^2 + 12x + 36$

Solution:

$x^2 + 12x + 36$

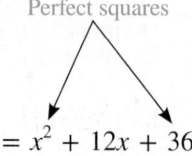

Perfect squares

$= x^2 + 12x + 36$

$(x)^2 + 2(x)(6) + (6)^2$

$= (x + 6)^2$

The GCF is 1.
- The first and third terms are positive.
- The first term is a perfect square:
$$x^2 = (x)^2$$
- The third term is a perfect square:
$$36 = (6)^2$$
- The middle term is twice the product of x and 6:
$$12x = 2(x)(6)$$

The trinomial is in the form $a^2 + 2ab + b^2$, where $a = x$ and $b = 6$.

Factor as $(a + b)^2$.

Skill Practice 8   Factor completely. $x^2 + 2x + 1$

EXAMPLE 9   **Factoring a Perfect Square Trinomial**

Factor completely. $4x^2 - 36xy + 81y^2$

Solution:

$4x^2 - 36xy + 81y^2$

Perfect squares

$= 4x^2 - 36xy + 81y^2$

$= (2x)^2 - 2(2x)(9y) + (9y)^2$

$= (2x - 9y)^2$

The GCF is 1.
- The first and third terms are positive.
- The first term is a perfect square:
$$4x^2 = (2x)^2$$
- The third term is a perfect square:
$$81y^2 = (9y)^2$$
- The middle term:
$$-36xy = -2(2x)(9y)$$

The trinomial is in the form $a^2 - 2ab + b^2$, where $a = 2x$ and $b = 9y$.

Factor as $(a - b)^2$.

Skill Practice 9   Factor completely. $9y^2 - 12yz + 4z^2$

Answers

8. $(x + 1)^2$
9. $(3y - 2z)^2$

4. Factor by Using Substitution

Sometimes it is convenient to use substitution to convert a polynomial into a simpler form before factoring.

EXAMPLE 10   Using Substitution to Factor a Polynomial

Factor by using substitution. $(2x - 7)^2 - 3(2x - 7) - 40$

Solution:

$(2x - 7)^2 - 3(2x - 7) - 40$

$= u^2 - 3u - 40$ Substitute $u = 2x - 7$. The trinomial is simpler in form.

$= (u - 8)(u + 5)$ Factor the trinomial.

$= [(2x - 7) - 8][(2x - 7) + 5]$ Reverse substitute. Replace u by $2x - 7$.

$= (2x - 7 - 8)(2x - 7 + 5)$ Simplify.

$= (2x - 15)(2x - 2)$ The second binomial has a GCF of 2.

$= (2x - 15)(2)(x - 1)$ Factor out the GCF from the second binomial.

$= 2(2x - 15)(x - 1)$

Skill Practice 10   Factor by using substitution.

$(3x + 1)^2 + 2(3x + 1) - 15$

EXAMPLE 11   Using Substitution to Factor a Polynomial

Factor by using substitution. $6y^6 - 5y^3 - 4$

Solution:

$6y^6 - 5y^3 - 4$

$= 6(y^3)^2 - 5(y^3) - 4$ Let $u = y^3$.

$= 6u^2 - 5u - 4$ Substitute u for y^3 in the trinomial.

$= (2u + 1)(3u - 4)$ Factor the trinomial.

$= (2y^3 + 1)(3y^3 - 4)$ Reverse substitute. Replace u with y^3.

Skill Practice 11   Factor by using substitution.

$2x^4 + 7x^2 + 3$

TIP The ac-method or trial-and-error method can also be used for Example 11 without using substitution.

Answers
10. $3(3x - 2)(x + 2)$
11. $(2x^2 + 1)(x^2 + 3)$

As you work through the exercises in this section, keep these guidelines in mind to factor trinomials.

Factoring Trinomials of the Form $ax^2 + bx + c$ $(a \neq 0)$

When factoring trinomials, the following guidelines should be considered:

Step 1 Factor out the greatest common factor.

Step 2 Check to see if the trinomial is a perfect square trinomial. If so, factor it as either $(a + b)^2$ or $(a - b)^2$. (With a perfect square trinomial, you do not need to use the ac-method or trial-and-error method.)

Step 3 If the trinomial is not a perfect square, use either the ac-method or the trial-and-error method to factor.

Step 4 Check the factorization by multiplication.

Note: Consider using substitution if a trinomial is in the form $au^2 + bu + c$, where u is an algebraic expression.

SECTION 2.4   Practice Exercises

Prerequisite Review

For Exercises R.1–R.6, multiply. (For review, see Section 2.2.)

R.1. $(5x - 3)(7x + 6)$ **R.2.** $(4y + 9)(2y - 5)$ **R.3.** $(t - 11)^2$

R.4. $(m + 9)^2$ **R.5.** $(3a + 5b^2)^2$ **R.6.** $(2c - 7d^3)^2$

For Exercises R.7–R.10, factor out the greatest common factor. (For review, see Section 2.3.)

R.7. $25a^3b^4 - 15a^2b^3 - 20ab^2$ **R.8.** $12m^2n^3 + 24m^3n^2 + 6m^4n$

R.9. $3a(2a + 1) - 6(2a + 1)$ **R.10.** $15x(x + 7) - 5(x + 7)$

For Exercises R.11–R.14, factor by grouping. (For review, see Section 2.3.)

R.11. $6x^2 + 9x + 14x + 21$ **R.12.** $12y^2 + 8y + 9y + 6$

R.13. $10m^3 + 40m^2 - 90m^2 - 360m$ **R.14.** $2n^3 - 16n^2 + 8n^2 - 64n$

Concept Connections

1. Given a trinomial $x^2 + bx + c$, if c is positive, then the signs in the binomial factors are either both _____ or both negative.

2. Given a trinomial $x^2 + bx + c$, if c is negative, then the signs in the binomial factors are (choose one: both positive, both negative, opposite).

3. Which is the correct factored form of $2x^2 - 5x - 12$, the product $(2x + 3)(x - 4)$ or $(x - 4)(2x + 3)$?

4. Which is the complete factorization of $6x^2 - 4x - 10$, the product $(3x - 5)(2x + 2)$ or $2(3x - 5)(x + 1)$?

5. A perfect square trinomial $a^2 + 2ab + b^2$ factors as _____.
   Likewise $a^2 - 2ab + b^2$ factors as _____.

6. Explain how to check a factoring problem.

7. A polynomial that cannot be factored is referred to as a _____ polynomial.

8. What does GCF stand for?

Objectives 1–2: Factor Trinomials

In Exercises 9–46, factor the trinomial completely by using any method. Remember to look for a common factor first. **(See Examples 1–7)**

9. $b^2 - 12b + 32$

10. $a^2 - 12a + 27$

11. $y^2 + 10y - 24$

12. $w^2 + 3w - 54$

13. $x^2 + 13x + 30$

14. $t^2 + 9t + 8$

15. $c^2 - 6c - 16$

16. $z^2 - 3z - 28$

17. $2x^2 - 7x - 15$

18. $2y^2 - 13y + 15$

19. $a + 6a^2 - 5$

20. $10b^2 - 3 - 29b$

21. $s^2 + st - 6t^2$

22. $p^2 - pq - 20q^2$

23. $3x^2 - 60x + 108$

24. $4c^2 + 12c - 72$

25. $2c^2 - 2c - 24$

26. $3x^2 + 12x - 15$

27. $2x^2 + 8xy - 10y^2$

28. $20z^2 + 26zw - 28w^2$

29. $33t^2 - 18t + 2$

30. $5p^2 - 10p + 7$

31. $3x^2 + 14xy + 15y^2$

32. $2a^2 + 15ab - 27b^2$

33. $5u^3v - 30u^2v^2 + 45uv^3$

34. $3a^3 + 30a^2b + 75ab^2$

35. $x^3 - 5x^2 - 14x$

36. $p^3 + 2p^2 - 24p$

37. $-23z - 5 + 10z^2$

38. $3 + 16y^2 + 14y$

39. $b^2 + 2b + 15$

40. $x^2 - x - 1$

41. $-2t^2 + 12t + 80$

42. $-3c^2 + 33c - 72$

43. $14a^2 + 13a - 12$

44. $12x^2 - 16x + 5$

45. $6a^2b + 22ab + 12b$

46. $6cd^2 + 9cd - 42c$

Objective 3: Factor Perfect Square Trinomials

47. **a.** Multiply the binomials $(x + 5)(x + 5)$.

   b. Factor $x^2 + 10x + 25$.

48. **a.** Multiply the binomials $(2w - 5)(2w - 5)$.

   b. Factor $4w^2 - 20w + 25$.

49. **a.** Multiply the binomials $(3x - 2y)^2$.

   b. Factor $9x^2 - 12xy + 4y^2$.

50. **a.** Multiply the binomials $(x + 7y)^2$.

   b. Factor $x^2 + 14xy + 49y^2$.

For Exercises 51–54, fill in the blank to make the trinomial a perfect square trinomial.

51. $9x^2 + (_____) + 25$

52. $16x^4 - (_____) + 1$

53. $64z^4 + (_____) + t^2$

54. $9m^4 - (_____) + 49n^2$

For Exercises 55–66, factor out the greatest common factor, if necessary. Then determine if the polynomial is a perfect square trinomial. If it is, factor it if possible. **(See Examples 8–9)**

55. $y^2 - 8y + 16$

56. $x^2 + 10x + 25$

57. $64m^2 + 80m + 25$

58. $100c^2 - 140c + 49$

59. $w^2 - 5w + 9$

60. $2a^2 + 14a + 98$

61. $9a^2 - 30ab + 25b^2$

62. $16x^4 - 48x^2y + 9y^2$

63. $16t^2 - 80tv + 20v^2$

64. $12x^2 - 12xy + 3y^2$

65. $5b^4 - 20b^2 + 20$

66. $a^4 + 12a^2 + 36$

Objective 4: Factor by Using Substitution

For Exercises 67–70, factor the polynomial in part (a). Then use substitution to help factor the polynomials in parts (b) and (c).

67. **a.** $u^2 - 10u + 25$

   b. $x^4 - 10x^2 + 25$

   c. $(a + 1)^2 - 10(a + 1) + 25$

68. **a.** $u^2 + 12u + 36$

   b. $y^4 + 12y^2 + 36$

   c. $(b - 2)^2 + 12(b - 2) + 36$

69. **a.** $u^2 + 11u - 26$

   b. $w^6 + 11w^3 - 26$

   c. $(y - 4)^2 + 11(y - 4) - 26$

70. **a.** $u^2 + 17u + 30$

   b. $z^6 + 17z^3 + 30$

   c. $(x + 3)^2 + 17(x + 3) + 30$

For Exercises 71–82, factor by using substitution. (**See Examples 10–11**)

71. $(3x - 1)^2 - (3x - 1) - 6$

72. $(2x + 5)^2 - (2x + 5) - 12$

73. $2(x - 5)^2 + 9(x - 5) + 4$

74. $4(x - 3)^2 + 7(x - 3) + 3$

75. $3(y + 4)^2 + 5(y + 4) - 2$

76. $(3t - 2)^2 - (3t - 2) - 20$

77. $3y^6 + 11y^3 + 6$

78. $3x^4 - 5x^2 - 12$

79. $4p^4 + 5p^2 + 1$

80. $t^4 + 3t^2 + 2$

81. $x^4 + 15x^2 + 36$

82. $t^6 - 16t^3 + 63$

Mixed Exercises

83. A student factored $4y^2 - 10y + 4$ as $(2y - 1)(2y - 4)$ on her factoring test. Why did her professor deduct several points, even though $(2y - 1)(2y - 4)$ does multiply out to $4y^2 - 10y + 4$?

84. A student factored $9w^2 + 36w + 36$ as $(3w + 6)^2$ on his factoring test. Why did his instructor deduct several points, even though $(3w + 6)^2$ does multiply out to $9w^2 + 36w + 36$?

For Exercises 85–104, factor completely by using an appropriate method. (Be sure to note the number of terms in the polynomial.)

85. $w^4 + 12w^2 + 36$

86. $9 - 6t^2 + t^4$

87. $81w^2 + 90w + 25$

88. $49a^2 - 28ab + 4b^2$

89. $3x(a + b) - 6(a + b)$

90. $4p(t - 8) + 2(t - 8)$

91. $12a^2bc^2 + 4ab^2c^2 - 6abc^3$

92. $18x^2z - 6xyz + 30xz^2$

93. $-20x^3 + 74x^2 - 60x$

94. $-24y^3 + 90y^2 - 75y$

95. $2y^2 - 9y - 4$

96. $3w^2 - 12w + 4$

97. $2(w^2 - 5)^2 + (w^2 - 5) - 15$

98. $5(t^2 + 3)^2 + 21(t^2 + 3) + 4$

99. $1 - 4d + 3d^2$

100. $2 - 5a + 2a^2$

101. $ax - 5a^2 + 2bx - 10ab$

102. $my + y^2 - 3xm - 3xy$

103. $8z^2 + 24zw - 224w^2$

104. $9x^2 - 18xy - 135y^2$

Expanding Your Skills

For Exercises 105–110, factor completely. Write the answers with positive exponents only.

105. $2x^{-3} - 2x^{-4} - 24x^{-5}$

106. $3x^{-2} + 15x^{-3} - 72x^{-4}$

107. $2x^{-5} - 13x^{-6} + 20x^{-7}$

108. $3x^{-4} + 19x^{-5} - 14x^{-6}$

109. $(2x + 3)^{-1} + 4(2x + 3)^{-2} + 4(2x + 3)^{-3}$

110. $(x + 6)^{-1} + 6(x + 6)^{-2} + 9(x + 6)^{-3}$

Consider the trinomial $ax^2 + bx + c$ with integer coefficients a, b, and c. The trinomial can be factored as the product of two binomials with integer coefficients if $b^2 - 4ac$ is a perfect square. For exercises 111–116, determine whether the trinomial can be factored as a product of two binomials with integer coefficients.

111. $36p^2 - 33p - 12$

112. $24w^2 - 25w + 8$

113. $8x^2 + 2x - 15$

114. $6x^2 - 7x - 20$

115. $18y^2 + 45y - 48$

116. $54z^2 - 39z - 60$

SECTION 2.5 Factoring Binomials

OBJECTIVES

1. Factor the Difference of Squares
2. Use a Difference of Squares in Grouping
3. Factor the Sum and Difference of Cubes
4. Summarize Factoring Binomials
5. Factor Binomials of the Form $x^6 - y^6$

1. Factor the Difference of Squares

Up to this point we have learned how to

- Factor out the greatest common factor from a polynomial.
- Factor a four-term polynomial by grouping.
- Recognize and factor perfect square trinomials.
- Factor trinomials by the ac-method and by the trial-and-error method.

Next, we will learn how to factor binomials that fit the pattern of a difference of squares. Recall that the product of two conjugates results in a **difference of squares**.

$$(a + b)(a - b) = a^2 - b^2$$

Therefore, to factor a difference of squares, the process is reversed. Identify a and b and construct the conjugate factors.

Factored Form of a Difference of Squares

$$a^2 - b^2 = (a + b)(a - b)$$

EXAMPLE 1   Factoring a Difference of Squares

Factor the binomial completely. $16x^2 - 9$

Solution:

$16x^2 - 9$	The GCF is 1. The binomial is a difference of squares.
$= (4x)^2 - (3)^2$	Write in the form $a^2 - b^2$, where $a = 4x$ and $b = 3$.
$= (4x + 3)(4x - 3)$	Factor as $(a + b)(a - b)$.

Skill Practice 1   Factor completely. $4z^2 - 1$

EXAMPLE 2   Factoring a Difference of Squares

Factor the binomial completely. $98c^2d - 50d^3$

Solution:

$98c^2d - 50d^3$	
$= 2d(49c^2 - 25d^2)$	The GCF is $2d$. The resulting binomial is a difference of squares.
$= 2d[(7c)^2 - (5d)^2]$	Write in the form $a^2 - b^2$, where $a = 7c$ and $b = 5d$.
$= 2d(7c + 5d)(7c - 5d)$	Factor as $(a + b)(a - b)$.

Skill Practice 2   Factor completely. $7y^3z - 63yz^3$

EXAMPLE 3   Factoring a Difference of Squares

Factor the binomial completely. $z^4 - 81$

Solution:

$z^4 - 81$	The GCF is 1. The binomial is a difference of squares.
$= (z^2)^2 - (9)^2$	Write in the form $a^2 - b^2$, where $a = z^2$ and $b = 9$.
$= (z^2 + 9)(z^2 - 9)$	Factor as $(a + b)(a - b)$. $z^2 - 9$ is also a difference of squares.
$= (z^2 + 9)(z + 3)(z - 3)$	

Skill Practice 3   Factor completely. $b^4 - 16$

Answers

1. $(2z - 1)(2z + 1)$
2. $7yz(y + 3z)(y - 3z)$
3. $(b^2 + 4)(b - 2)(b + 2)$

The difference of squares $a^2 - b^2$ factors as $(a - b)(a + b)$. However, the *sum* of squares is not factorable. To see why $a^2 + b^2$ is not factorable, consider the product of binomials:

$$(a \quad b)(a \quad b) \overset{?}{=} a^2 + b^2$$

If all possible combinations of signs are considered, none produces the correct product.

$$(a + b)(a - b) = a^2 - b^2 \qquad \text{Wrong sign}$$
$$(a + b)(a + b) = a^2 + 2ab + b^2 \qquad \text{Wrong middle term}$$
$$(a - b)(a - b) = a^2 - 2ab + b^2 \qquad \text{Wrong middle term}$$

After exhausting all possibilities, we see that if a and b share no common factors, then the sum of squares $a^2 + b^2$ is a prime polynomial.

Sum of Squares

Suppose a and b have no common factors. Then the **sum of squares $a^2 + b^2$** is *not* factorable over the real numbers.

That is, $a^2 + b^2$ is prime over the real numbers.

---

2. Use a Difference of Squares in Grouping

Sometimes a difference of squares can be used along with other factoring techniques.

EXAMPLE 4 Using a Difference of Squares in Grouping

Factor completely. $\quad y^3 - 6y^2 - 4y + 24$

Solution:

$$y^3 - 6y^2 - 4y + 24 \qquad \text{The GCF is 1.}$$
$$= y^3 - 6y^2 \mid -4y + 24 \qquad \text{The polynomial has four terms. Factor by grouping.}$$
$$= y^2(y - 6) - 4(y - 6)$$
$$= (y - 6)(y^2 - 4) \qquad y^2 - 4 \text{ is a difference of squares.}$$

$$= (y - 6)(y + 2)(y - 2)$$

Skill Practice 4 Factor completely. $\quad a^3 + 5a^2 - 9a - 45$

EXAMPLE 5	Factoring a Four-Term Polynomial by Grouping Three Terms

Factor completely. $x^2 - y^2 - 6y - 9$

Solution:

Grouping "2 by 2" will not work to factor this polynomial. However, if we factor out -1 from the last three terms, the resulting trinomial will be a perfect square trinomial.

$x^2 \mid - y^2 - 6y - 9$	Group the last three terms.
$= x^2 - 1(y^2 + 6y + 9)$	Factor out -1 from the last three terms.
$= x^2 - (y + 3)^2$	Factor the perfect square trinomial $y^2 + 6y + 9$ as $(y + 3)^2$.
	The quantity $x^2 - (y + 3)^2$ is a difference of squares, $a^2 - b^2$, where $a = x$ and $b = (y + 3)$.
$= [x - (y + 3)][x + (y + 3)]$	Factor as $a^2 - b^2 = (a + b)(a - b)$.
$= (x - y - 3)(x + y + 3)$	Apply the distributive property to clear the inner parentheses.

> **Avoiding Mistakes**
>
> When factoring the expression $x^2 - (y + 3)^2$ as a difference of squares, be sure to use parentheses around the quantity $(y + 3)$. This will help you remember to "distribute the negative" in the expression $[x - (y + 3)]$.
>
> $[x - (y + 3)] = (x - y - 3)$

Skill Practice 5   Factor completely. $x^2 + 10x + 25 - y^2$

> **TIP**   From Example 5, the expression $x^2 - (y + 3)^2$ can also be factored by using substitution. Let $u = y + 3$.
>
> | $x^2 - (y + 3)^2$ | |
> |---|---|
> | $= x^2 - u^2$ | Substitution $u = y + 3$. |
> | $= (x - u)(x + u)$ | Factor as a difference of squares. |
> | $= [x - (y + 3)][x + (y + 3)]$ | Substitute back. |
> | $= (x - y - 3)(x + y + 3)$ | Apply the distributive property. |

3. Factor the Sum and Difference of Cubes

For binomials that represent the sum or difference of cubes, factor by using the following formulas.

> **TIP**   The following are perfect cubes.
>
> | | |
> |---|---|
> | $1^3 = 1$ | $(x^1)^3 = x^3$ |
> | $2^3 = 8$ | $(x^2)^3 = x^6$ |
> | $3^3 = 27$ | $(x^3)^3 = x^9$ |
> | $4^3 = 64$ | $(x^4)^3 = x^{12}$ |
> | $\vdots$ | $\vdots$ |
>
> Any expression raised to a multiple of 3 is a perfect cube.

> **Factored Form of a Sum and Difference of Cubes**
>
> **Sum of cubes:** $a^3 + b^3 = (a + b)(a^2 - ab + b^2)$
>
> **Difference of cubes:** $a^3 - b^3 = (a - b)(a^2 + ab + b^2)$

Multiplication can be used to confirm the formulas for factoring a sum or difference of cubes.

$$(a + b)(a^2 - ab + b^2) = a^3 - a^2b + ab^2 + a^2b - ab^2 + b^3 = a^3 + b^3 \ \checkmark$$

$$(a - b)(a^2 + ab + b^2) = a^3 + a^2b + ab^2 - a^2b - ab^2 - b^3 = a^3 - b^3 \ \checkmark$$

Answer

5. $(x + 5 - y)(x + 5 + y)$

To help you remember the formulas for factoring a sum or difference of cubes, keep the following guidelines in mind.

- The factored form is the product of a binomial and a trinomial.
- The first and third terms in the trinomial are the squares of the terms within the binomial factor. Therefore, these terms are always positive.
- Without regard to sign, the middle term in the trinomial is the product of terms in the binomial factor.

$$\underbrace{\text{Square the first}}_{\text{term of the binomial.}} \qquad \underbrace{\text{Product of terms}}_{\text{in the binomial}}$$

$$x^3 + 8 = (x)^3 + (2)^3 = (x + 2)[(x)^2 - (x)(2) + (2)^2]$$

Square the last term of the binomial.

- The sign within the binomial factor is the same as the sign of the original binomial.
- The first and third terms in the trinomial are always positive.
- The sign of the middle term in the trinomial is opposite the sign within the binomial.

$$\overset{\text{Same sign}}{\qquad} \qquad \overset{\text{Positive}}{\qquad}$$

$$x^3 + 8 = (x)^3 + (2)^3 = (x + 2)[(x)^2 - (x)(2) + (2)^2]$$

Opposite signs

EXAMPLE 6 **Factoring a Difference of Cubes**

Factor. $8x^3 - 27$

Solution:

$$8x^3 - 27 \qquad\qquad 8x^3 \text{ and } 27 \text{ are perfect cubes.}$$

$$= (2x)^3 - (3)^3 \qquad\qquad \text{Write as } a^3 - b^3,$$
$$\text{where } a = 2x \text{ and } b = 3.$$

$$a^3 - b^3 = (a - b)(a^2 + ab + b^2)$$
$$(2x)^3 - (3)^3 = (2x - 3)[(2x)^2 + (2x)(3) + (3)^2] \qquad \text{Apply the difference of cubes formula.}$$

$$= (2x - 3)(4x^2 + 6x + 9) \qquad\qquad \text{Simplify.}$$

Skill Practice 6 Factor completely. $125p^3 - 8$

Answer

6. $(5p - 2)(25p^2 + 10p + 4)$

EXAMPLE 7   **Factoring a Sum of Cubes**

Factor. $125t^3 + 64z^6$

Solution:

$125t^3 + 64z^6$ $125t^3$ and $64z^6$ are perfect cubes.

$= (5t)^3 + (4z^2)^3$ Write as $a^3 + b^3$, where $a = 5t$ and $b = 4z^2$.

$a^3 + b^3 = (a + b)(a^2 - ab + b^2)$ Apply the sum of cubes formula.

$(5t)^3 + (4z^2)^3 = [(5t) + (4z^2)][(5t)^2 - (5t)(4z^2) + (4z^2)^2]$

$= (5t + 4z^2)(25t^2 - 20tz^2 + 16z^4)$ Simplify.

Skill Practice 7   Factor completely. $x^3 + 1000y^6$

4. Summarize Factoring Binomials

After factoring out the greatest common factor, the next step in any factoring problem is to recognize what type of pattern it follows. Exponents that are divisible by 2 are perfect squares, and those divisible by 3 are perfect cubes. The formulas for factoring binomials are summarized here.

Summary of Factoring Binomials

- Difference of squares: $a^2 - b^2 = (a + b)(a - b)$
- Difference of cubes: $a^3 - b^3 = (a - b)(a^2 + ab + b^2)$
- Sum of cubes: $a^3 + b^3 = (a + b)(a^2 - ab + b^2)$

EXAMPLE 8   **Review of Factoring Binomials**

Factor the binomials.

a. $m^3 - \dfrac{1}{8}$ **b.** $9k^2 + 24m^2$ **c.** $128y^6 + 54x^3$ **d.** $50y^6 - 8x^2$

Solution:

a. $m^3 - \dfrac{1}{8}$ m^3 is a perfect cube: $m^3 = (m)^3$.
                                            $\frac{1}{8}$ is a perfect cube: $\frac{1}{8} = \left(\frac{1}{2}\right)^3$.

$= (m)^3 - \left(\dfrac{1}{2}\right)^3$ This is a difference of cubes, where $a = m$ and $b = \frac{1}{2}$.

$= \left(m - \dfrac{1}{2}\right)\left(m^2 + \dfrac{1}{2}m + \dfrac{1}{4}\right)$ $a^3 - b^3 = (a - b)(a^2 + ab + b^2)$

Answer

7. $(x + 10y^2)(x^2 - 10xy^2 + 100y^4)$

b. $9k^2 + 24m^2$

$\quad = 3(3k^2 + 8m^2)$

Factor out the GCF.

The resulting binomial is not a difference of squares or a sum or difference of cubes. It cannot be factored further over the real numbers.

c. $128y^6 + 54x^3$

$\quad = 2(64y^6 + 27x^3)$

Factor out the GCF.

Both 64 and 27 are perfect cubes, and the exponents of both x and y are multiples of 3. This is a sum of cubes, where $a = 4y^2$ and $b = 3x$.

$\quad = 2[(4y^2)^3 + (3x)^3]$

$\quad = 2(4y^2 + 3x)(16y^4 - 12xy^2 + 9x^2)$

$a^3 + b^3 = (a + b)(a^2 - ab + b^2)$

d. $50y^6 - 8x^2$

$\quad = 2(25y^6 - 4x^2)$

Factor out the GCF.

Both 25 and 4 are perfect squares. The exponents of both x and y are multiples of 2. This is a difference of squares, where $a = 5y^3$ and $b = 2x$.

$\quad = 2[(5y^3)^2 - (2x)^2]$

$\quad = 2(5y^3 + 2x)(5y^3 - 2x)$

$a^2 - b^2 = (a + b)(a - b)$

Skill Practice 8 Factor the binomials.

a. $x^2 - \dfrac{1}{25}$ **b.** $16y^3 + 4y$ **c.** $24a^7 - 3a$ **d.** $18p^4 - 50t^2$

---

5. Factor Binomials of the Form $x^6 - y^6$

EXAMPLE 9 **Factoring Binomials of the Form $x^6 - y^6$**

Factor the binomial $x^6 - y^6$ as

a. A difference of cubes   **b.** A difference of squares

Solution:

Notice that the expressions x^6 and y^6 are both perfect squares and perfect cubes because the exponents are both multiples of 2 and of 3. Consequently, $x^6 - y^6$ can be interpreted initially as either a difference of cubes or a difference of squares.

a. $x^6 - y^6$

$\qquad\qquad$ Difference of cubes

$\quad = (x^2)^3 - (y^2)^3$

Write as $a^3 - b^3$, where $a = x^2$ and $b = y^2$.

$\quad = (x^2 - y^2)[(x^2)^2 + (x^2)(y^2) + (y^2)^2]$

Apply the formula $a^3 - b^3 = (a - b)(a^2 + ab + b^2)$.

$\quad = (x^2 - y^2)(x^4 + x^2y^2 + y^4)$

Factor $x^2 - y^2$ as a difference of squares.

$\quad = (x + y)(x - y)(x^4 + x^2y^2 + y^4)$

The expression $x^4 + x^2y^2 + y^4$ cannot be factored by using the skills learned thus far.

Answers

8. a. $\left(x + \dfrac{1}{5}\right)\left(x - \dfrac{1}{5}\right)$

b. $4y(4y^2 + 1)$

c. $3a(2a^2 - 1)(4a^4 + 2a^2 + 1)$

d. $2(3p^2 + 5t)(3p^2 - 5t)$

b. $x^6 - y^6$

Difference
of squares

$= (x^3)^2 - (y^3)^2$

Write as $a^2 - b^2$, where $a = x^3$ and $b = y^3$.

$= (x^3 + y^3)(x^3 - y^3)$

Apply the formula
$a^2 - b^2 = (a + b)(a - b)$

Sum of
cubes

Difference
of cubes

Factor $x^3 + y^3$ as a sum of cubes.
Factor $x^3 - y^3$ as a difference of cubes.

$= (x + y)(x^2 - xy + y^2)(x - y)(x^2 + xy + y^2)$

> **TIP** Notice that the expressions x^6 and y^6 are both perfect squares and perfect cubes because both exponents are multiples of 2 and of 3. Consequently, $x^6 - y^6$ can be factored initially as either the difference of squares or the difference of cubes. In such a case, it is recommended that you factor the expression as a difference of squares first because it factors more completely into polynomials of lower degree.
>
> $$x^6 - y^6 = (x + y)(x^2 - xy + y^2)(x - y)(x^2 + xy + y^2)$$

Answer

9. $(a - 2)(a + 2)(a^2 + 2a + 4)$
$(a^2 - 2a + 4)$

Skill Practice 9   Factor completely. $a^6 - 64$

SECTION 2.5   Practice Exercises

Prerequisite Review

For Exercises R.1–R.8, multiply. (For review, see Section 2.2.)

R.1. $(4x - 11)(4x + 11)$ **R.2.** $(7p + 2)(7p - 2)$ **R.3.** $(m^3 + 6n^2)(m^3 - 6n^2)$

R.4. $(3a^2 - 10b^4)(3a^2 + 10b^4)$ **R.5.** $(t + 5)(t^2 - 5t + 25)$ **R.6.** $(a - 4)(a^2 + 4a + 16)$

R.7. $(2x^2 - y^3)(4x^4 + 2x^2y^3 + y^6)$ **R.8.** $(3c^3 + d^4)(9c^6 - 3c^3d^4 + d^8)$

Concept Connections

1. The binomial $x^2 - 36$ is an example of a _____ of squares. A difference of squares $a^2 - b^2$ factors as _____.

2. The binomial $y^2 + 9$ is an example of a _____ of squares.

3. A sum of squares with greatest common factor 1 (is/is not) factorable over the real numbers.

4. The square of a binomial always results in a perfect _____ trinomial.

5. The binomial $x^3 + 64$ is an example of a _____ of _____.

6. The binomial $c^3 - 27$ is an example of a _____ of _____.

7. A difference of cubes $a^3 - b^3$ factors as (   )(   ).

8. A sum of cubes $a^3 + b^3$ factors as (   )(   ).

9. Identify which expressions represent perfect squares. 2, 4, 8, 16, 25, 64, x^2, x^3, x^4, x^5, x^9

10. Identify which expressions represent perfect cubes. 2, 4, 8, 16, 25, 64, x^2, x^3, x^4, x^5, x^9

Objective 1: Factor the Difference of Squares

For Exercises 11–22, factor the binomials. Identify the binomials that are prime. **(See Examples 1–3)**

11. $x^2 - 9$

12. $y^2 - 25$

13. $16 - 49w^2$

14. $81 - 64b^2$

15. $8a^2 - 162b^2$

16. $50c^2 - 72d^2$

17. $25u^2 + 1$

18. $w^2 + 4$

19. $2a^4 - 32$

20. $5y^4 - 5$

21. $49 - k^6$

22. $4 - h^6$

Objective 2: Use a Difference of Squares in Grouping

For Exercises 23–36, use the difference of squares along with factoring by grouping. **(See Examples 4–5)**

23. $x^3 - x^2 - 16x + 16$

24. $x^3 + 5x^2 - x - 5$

25. $4x^3 + 12x^2 - x - 3$

26. $5x^3 - x^2 - 45x + 9$

27. $9y^3 + 7y^2 - 36y - 28$

28. $9z^3 - 5z^2 - 36z + 20$

29. $49x^2 + 28x + 4 - y^2$

30. $100y^2 + 140y + 49 - z^2$

31. $w^2 - 9n^2 + 6n - 1$

32. $m^2 - 25c^2 + 20c - 4$

33. $p^4 - 10p^2 + 25 - t^4$

34. $m^4 - 14m^2 + 49 - z^4$

35. $9u^4 - 4v^4 + 20v^2 - 25$

36. $x^4 - 9y^4 - 42y^2 - 49$

Objective 3: Factor the Sum and Difference of Cubes

37. Explain how to identify and factor a sum of cubes.

38. Explain how to identify and factor a difference of cubes.

For Exercises 39–52, factor the sum or difference of cubes. **(See Examples 6–7)**

39. $8x^3 - 1$ (Check by multiplying.)

40. $y^3 + 64$ (Check by multiplying.)

41. $125c^3 + 27$

42. $216u^3 - v^3$

43. $x^3 - 1000$

44. $y^3 - 27$

45. $64t^6 + 1$

46. $125r^6 + 1$

47. $2000y^6 + 2x^3$

48. $3a^6 + 24b^3$

49. $16z^4 - 54z$

50. $x^5 - 64x^2$

51. $p^{12} - 125$

52. $t^9 - 8$

Objective 4: Summarize Factoring Binomials

For Exercises 53–80, factor completely. **(See Example 8)**

53. $36y^2 - \dfrac{1}{25}$

54. $16p^2 - \dfrac{1}{9}$

55. $18d^{12} - 32$

56. $3z^8 - 12$

57. $242v^2 + 32$

58. $8p^2 + 200$

59. $4x^2 - 16$

60. $9m^2 - 81n^2$

61. $25 - 49q^2$

62. $1 - 25p^2$

63. $(t + 2s)^2 - 36$

64. $(5x + 4)^2 - y^2$

65. $27 - t^3$

66. $8 + y^3$

67. $27a^3 + \dfrac{1}{8}$

68. $b^3 + \dfrac{27}{125}$

69. $2m^3 + 16$

70. $3x^3 - 375$

71. $x^4 - y^4$

72. $81u^4 - 16v^4$

73. $a^9 + b^9$

74. $27m^9 - 8n^9$

75. $\frac{1}{8}p^3 - \frac{1}{125}$

76. $1 - \frac{1}{27}d^3$

77. $4w^2 + 25$

78. $64 + a^2$

79. $\frac{1}{25}x^2 - \frac{1}{4}y^2$

80. $\frac{1}{100}a^2 - \frac{4}{49}b^2$

Objective 5: Factor Binomials of the Form $x^6 - y^6$

For Exercises 81–88, factor completely. **(See Example 9)**

81. $a^6 - b^6$ (*Hint*: First factor as a difference of squares.)

82. $64x^6 - y^6$

83. $64 - y^6$

84. $1 - p^6$

85. $h^6 + k^6$

86. $27q^6 + 125p^6$

87. $8x^6 + 125$

88. $t^6 + 1$

Mixed Exercises

89. Find a difference of squares that has $(2x + 3)$ as one of its factors.

90. Find a difference of squares that has $(4 - p)$ as one of its factors.

91. Find a difference of cubes that has $(4a^2 + 6a + 9)$ as its trinomial factor.

92. Find a sum of cubes that has $(25c^2 - 10cd + 4d^2)$ as its trinomial factor.

93. Find a sum of cubes that has $(4x^2 + y)$ as its binomial factor.

94. Find a difference of cubes that has $(3t - r^2)$ as its binomial factor.

95. Consider the shaded region.

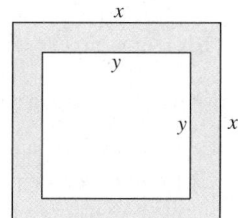

    a. Find an expression that represents the area of the shaded region.

    b. Factor the expression found in part (a).

    c. Find the area of the shaded region if $x = 6$ in. and $y = 4$ in.

96. A manufacturer needs to know the area of a metal washer. The outer radius of the washer is R and the inner radius is r.

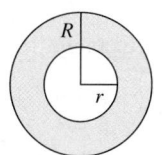

    a. Find an expression that represents the area of the washer.

    b. Factor the expression found in part (a).

    c. Find the area of the washer if $R = \frac{1}{2}$ in. and $r = \frac{1}{4}$ in. (Round to the nearest 0.01 in.^{2})

For Exercises 97–98, write the expression in factored form.

97. $\frac{4}{3}\pi R^3 - \frac{4}{3}\pi r^3$ Volume between two concentric spheres

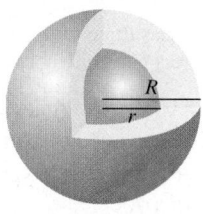

98. $2\pi rh + 2\pi r^2$ Surface area of a right circular cylinder

99. Simplify the expression $21^2 - 19^2$ by first factoring the expression. Do not use a calculator.

100. Simplify the expression $17^2 - 13^2$ by first factoring the expression. Do not use a calculator.

Expanding Your Skills

For Exercises 101–106, factor the polynomials by using the difference of squares, sum of cubes, or difference of cubes with grouping.

101. $x^2 - y^2 + x + y$

102. $64m^2 - 25n^2 + 8m + 5n$

103. $x^3 + y^3 + x + y$

104. $4pu^3 - 4pv^3 - 7yu^3 + 7yv^3$

105. $576a^5 - 9a^2 - 64a^3c^2 + c^2$

106. $32t^5 - 108t^2 - 72t^3v^2 + 243v^2$

107. Consider the following binomials and their factored forms. (The factored forms can be verified by multiplying the expressions on the right.)

$$x^2 - 1 = (x - 1)(x + 1)$$
$$x^3 - 1 = (x - 1)(x^2 + x + 1)$$
$$x^4 - 1 = (x - 1)(x^3 + x^2 + x + 1)$$

    a. Use the pattern to factor the expression $x^5 - 1$.

    b. Use the pattern to write a generic formula for $x^n - 1$ where n is a positive integer.

108. For a positive integer n, the expression $a^n - b^n$ can be factored as

$$a^n - b^n = (a - b)(a^{n-1} + a^{n-2}b + a^{n-3}b^2 + \cdots + ab^{n-2} + b^{n-1}).$$

    a. Use this formula to factor $a^5 - b^5$.

    b. Check the result to part (a) by multiplication.

PROBLEM RECOGNITION EXERCISES

A Factoring Summary

We now review the techniques of factoring presented thus far along with a general strategy for factoring polynomials.

Factoring Strategy

Step 1 Factor out the greatest common factor.

Step 2 Identify whether the polynomial has two terms, three terms, or more than three terms.

Step 3 If the polynomial has more than three terms, try factoring by grouping.

Step 4 If the polynomial has three terms, check first for a perfect square trinomial. Otherwise, factor the trinomial with the ac-method or the trial-and-error method.

Step 5 If the polynomial has two terms, determine if it fits the pattern for a difference of squares, difference of cubes, or sum of cubes. Remember, a sum of squares is not factorable over the real numbers.

Step 6 Be sure to factor the polynomial completely.

Step 7 Check by multiplying.

Four Terms (See Section 2.3 for review)

Example 1: $6x^2 + 14bx - 15ax - 35ab$ | This polynomial has four terms. Try factoring by grouping.

$= 6x^2 + 14bx \mid -15ax - 35ab$ | Group the first two terms and group the second two terms.

$= 2x(3x + 7b) - 5a(3x + 7b)$ | Factor out the GCF from the first pair of terms.
Factor out the GCF from the second pair of terms.

$= (3x + 7b)(2x - 5a)$ | Factor out the common binomial factor of $(3x + 7b)$.

Three Terms (See Section 2.4 for review)

Example 2: $9m^2 + 42mn + 49n^2$ | This polynomial has three terms. Check first to determine whether it fits the pattern of a perfect square trinomial.

$= (3m)^2 + 2(3m)(7n) + (7n)^2$

$= (3m + 7n)^2$

- The first term is a perfect square: $9m^2 = (3m)^2$
- The third term is a perfect square: $49n^2 = (7n)^2$
- The middle term is double the product of $3m$ and $7n$:
  $42mn = 2(3m)(7n)$

Therefore, $9m^2 + 42mn + 49n^2$ is a perfect square trinomial $a^2 + 2ab + b^2$ with $a = 3m$ and $b = 7n$.

Factor as: $a^2 + 2ab + b^2$

$= (a + b)^2$

Example 3: $10x^2 - 31x + 15$

$(x - 1)(10x - 15)$
$(x - 3)(10x - 5)$
$(2x - 1)(5x - 15)$
$(2x - 3)(5x - 5)$

> These cannot work because the terms in the second pair of parentheses contain a common factor.

$(x - 5)(10x - 3) = 10x^2 - 53x + 15$
$(x - 15)(10x - 1) = 10x^2 - 151x + 15$
$(2x - 15)(5x - 1) = 10x^2 - 77x + 15$

> Wrong middle terms

$(2x - 5)(5x - 3) = 10x^2 - 31x + 15$ Correct!

The polynomial has three terms, but it is not a perfect square trinomial (neither the first nor the third term is a perfect square).

Try factoring by the ac-method or by trial and error.

The pairs of factors of 10 are:

$1 \cdot 10$
$2 \cdot 5$

The pairs of factors of 15 (including the reverse orders) are:

$1 \cdot 15$
$3 \cdot 5$
$5 \cdot 3$
$15 \cdot 1$

Two Terms (See Section 2.5 for review)

Example 4: $81p^4 - 1$

$= (9p^2)^2 - (1)^2$

$= (9p^2 - 1)(9p^2 + 1)$

$= (3p - 1)(3p + 1)(9p^2 + 1)$

This is a **difference of squares**: $81p^4 = (9p^2)^2$ and $1 = (1)^2$.

Factor as $a^2 - b^2 = (a - b)(a + b)$ where $a = 9p^2$ and $b = 1$.

$9p^2 - 1$ is also a difference of squares because it can be written as $(3p)^2 - (1)^2$.

Factor $9p^2 - 1$ as $(3p - 1)(3p + 1)$.

Example 5: $125c^3 + 27d^6$

$= (5c)^3 + (3d^2)^3$

$= (5c + 3d^2)\left[(5c)^2 - (5c)(3d^2) + (3d^2)^2\right]$

$= (5c + 3d^2)(25c^2 - 15cd^2 + 9d^4)$

This is a **sum of cubes**: $125c^3 = (5c)^3$ and $27d^6 = (3d^2)^3$.

Factor as $a^3 + b^3 = (a + b)(a^2 - ab + b^2)$ where $a = 5c$ and $b = 3d^2$.

Example 6: $8t^3 - 343$

$= (2t)^3 - (7)^3$

$= (2t - 7)\left[(2t)^2 + (2t)(7) + (7)^2\right]$

$= (2t - 7)(4t^2 + 14t + 49)$

This is a **difference of cubes**: $8t^3 = (2t)^3$ and $343 = (7)^3$.

Factor as $a^3 - b^3 = (a - b)(a^2 + ab + b^2)$ where $a = 2t$ and $b = 7$.

---

1. What is meant by a prime polynomial?

2. What is the first step in factoring any polynomial?

3. When factoring a binomial, what patterns do you look for?

4. When factoring a trinomial, what pattern do you look for first?

5. What do you look for when factoring a four-term polynomial?

6. How would you use substitution to factor $3(4x^2 + 1)^2 + 20(4x^2 + 1) + 12$.

For Exercises 7–66,

    a. Factor out the GCF from each polynomial and identify the category in which the remaining polynomial best fits. Choose from

- difference of squares
- difference of cubes
- perfect square trinomial
- four terms—grouping

- sum of squares
- sum of cubes
- trinomial (ac-method or trial-and-error method)
- none of these

    b. Factor the polynomial completely.

7. $6x^2 - 21x - 45$

8. $8m^3 - 10m^2 - 3m$

9. $8a^2 - 50$

10. $ab + ay - b^2 - by$

11. $14u^2 - 11uv + 2v^2$

12. $9p^2 - 12pq + 4q^2$

13. $16x^3 - 2$

14. $9m^2 + 16n^2$

15. $27y^3 + 125$

16. $3x^2 - 16$

17. $128p^6 + 54q^3$

18. $5b^2 - 30b + 45$

19. $16a^4 - 1$

20. $81u^2 - 90uv + 25v^2$

21. $-100x^2 - 10x$

22. $4x^2 + 16$

23. $12ax - 6ay + 4bx - 2by$

24. $125y^3 - 8$

25. $5y^2 + 14y - 3$

26. $2m^4 - 128$

27. $t^2 - 100$

28. $4m^2 - 49n^2$

29. $y^3 + 27$

30. $x^3 + 1$

31. $d^2 + 3d - 28$

32. $c^2 + 5c - 24$

33. $x^2 - 12x + 36$

34. $p^2 + 16p + 64$

35. $2ax^2 - 5ax + 2bx - 5b$

36. $8x^2 - 4bx + 2ax - ab$

37. $10y^2 + 3y - 4$

38. $12z^2 + 11z + 2$

39. $10p^2 - 640$

40. $50a^2 - 72$

41. $z^4 - 64z$

42. $t^4 - 8t$

43. $b^3 - 4b^2 - 45b$

44. $y^3 - 14y^2 + 40y$

45. $9w^2 + 24wx + 16x^2$

46. $4k^2 - 20kp + 25p^2$

47. $60x^2 - 20x + 30ax - 10a$

48. $50x^2 - 200x + 10cx - 40c$

49. $w^4 - 16$

50. $k^4 - 81$

51. $t^6 - 8$

52. $p^6 + 27$

53. $8p^2 - 22p + 5$

54. $9m^2 - 3m - 20$

55. $36y^2 - 12y + 1$

56. $9a^2 + 42a + 49$

57. $2x^2 + 50$

58. $4y^2 + 64$

59. $12r^2s^2 + 7rs^2 - 10s^2$

60. $7z^2w^2 - 10zw^2 - 8w^2$

61. $x^2 + 8xy - 33y^2$

62. $s^2 - 9st - 36t^2$

63. $m^6 + n^3$

64. $a^3 - b^6$

65. $x^2 - 4x$

66. $y^2 - 9y$

To factor some polynomials, we require multiple techniques.

Factoring by Substitution (See Section 2.4 for review)

Example 7: $(x^2 - 4)^2 - 2(x^2 - 4) - 15$

Let $u = x^2 - 4$. The expression becomes:

    $u^2 - 2u - 15$

      $= (u - 5)(u + 3)$

      $= [(x^2 - 4) - 5][(x^2 - 4) + 3]$

      $= (x^2 - 9)(x^2 - 1)$

      $= (x - 3)(x + 3)(x - 1)(x + 1)$

The first and second terms share the common expression $x^2 - 4$. Therefore, the polynomial can be simplified by making the substitution $u = x^2 - 4$.

Factor the simpler expression first.

Back substitute by replacing u with $x^2 - 4$. Then simplify within the square brackets.

The expressions $x^2 - 9$ and $x^2 - 1$ are both differences of squares and can be factored further.

Using a Difference of Squares in Grouping (See Section 2.5 for review)

Example 8: $8x^3 + 20x^2 - 18x - 45$ This polynomial has four terms. Try factoring by grouping.

$8x^3 + 20x^2 \;\vert\; -18x - 45$ Group the first two terms and group the second two terms.

$= 4x^2(2x + 5) - 9(2x + 5)$ Factor out the GCF from the first pair of terms.
Factor out the GCF from the second pair of terms.

$= (2x + 5)(4x^2 - 9)$ Factor out the common binomial factor $(2x + 5)$.

$= (2x + 5)(2x - 3)(2x + 3)$ The expression $4x^2 - 9$ is a difference of squares and factors further.

Example 9: $z^2 - 4y^2 + 12y - 9$ The polynomial has four terms. However, the grouping by pairs method fails (try it to convince yourself). Another strategy is to try grouping one term with three terms.

$= z^2 \;\vert\; -4y^2 + 12y - 9$

$= z^2 - 1(\underbrace{4y^2 - 12y + 9})$ After factoring out -1 from the last three terms, we identify $4y^2 - 12y + 9$ as a perfect square trinomial.

$= z^2 - (2y - 3)^2$ Factor $a^2 - 2ab + b^2 = (a - b)^2$. Thus, $4y^2 - 12y + 9 = (2y - 3)^2$.

$= [z - (2y - 3)][z + (2y - 3)]$ The expression $z^2 - (2y - 3)^2$ is a difference of squares $a^2 - b^2$ where $a = z$ and $b = (2y - 3)$.
Factor as $a^2 - b^2 = (a - b)(a + b)$.

$= (z - 2y + 3)(z + 2y - 3)$ Simplify.

Factoring Expressions Containing Negative Exponents (See Section 2.3 for review)

Example 10: $2x^{-4} - 8x^{-5} - 90x^{-6}$ We identify the GCF as $2x^{-6}$ because each term shares a common factor of 2 and -6 is the smallest exponent to which x occurs.

$\boxed{-4 - (-6)}\;\boxed{-5 - (-6)}\;\boxed{-6 - (-6)}$

$= 2x^{-6}(x^2 - 4x^1 - 45x^0)$ Factor out $2x^{-6}$. The exponents on the x variables within the parentheses are found by subtracting -6 from the original exponents.

$= \dfrac{2(x^2 - 4x - 45)}{x^6}$ Simplify and write the expression with positive exponents.

$= \dfrac{2(x - 9)(x + 5)}{x^6}$ Factor the trinomial.

For Exercises 67–121, factor completely.

67. $x^2(x + y) - y^2(x + y)$

68. $u^2(u - v) - v^2(u - v)$

69. $(a + 3)^4 + 6(a + 3)^5$

70. $(4 - b)^4 - 2(4 - b)^3$

71. $24(3x + 5)^3 - 30(3x + 5)^2$

72. $10(2y + 3)^2 + 15(2y + 3)^3$

73. $\dfrac{1}{100}x^2 + \dfrac{1}{35}x + \dfrac{1}{49}$

74. $\dfrac{1}{25}a^2 + \dfrac{1}{15}a + \dfrac{1}{36}$

75. $(5x^2 - 1)^2 - 4(5x^2 - 1) - 5$

76. $(x^3 + 4)^2 - 10(x^3 + 4) + 24$

77. $16p^4 - q^4$

78. $s^4 t^4 - 81$

79. $y^3 + \dfrac{1}{64}$

80. $z^3 + \dfrac{1}{125}$

81. $6a^3 + a^2 b - 6ab^2 - b^3$

82. $4p^3 + 12p^2 q - pq^2 - 3q^3$

83. $\dfrac{1}{9}t^2 + \dfrac{1}{6}t + \dfrac{1}{16}$

84. $\dfrac{1}{25}y^2 + \dfrac{1}{5}y + \dfrac{1}{4}$

85. $x^2 + 12x + 36 - a^2$ **86.** $a^2 + 10a + 25 - b^2$ **87.** $p^2 + 2pq + q^2 - 81$

88. $m^2 - 2mn + n^2 - 9$ **89.** $b^2 - (x^2 + 4x + 4)$ **90.** $p^2 - (y^2 - 6y + 9)$

91. $4 - u^2 + 2uv - v^2$ **92.** $25 - a^2 - 2ab - b^2$ **93.** $6ax - by + 2bx - 3ay$

94. $5pq - 12 - 4q + 15p$ **95.** $u^6 - 64$ [*Hint:* Factor first as a difference of squares, $(u^3)^2 - (8)^2$.]

96. $1 - v^6$ **97.** $x^8 - 1$ **98.** $y^8 - 256$

99. $a^2 - b^2 + a + b$ **100.** $25c^2 - 9d^2 + 5c - 3d$ **101.** $5wx^3 + 5wy^3 - 2zx^3 - 2zy^3$

102. $(x^2 - 2)^2 - 3(x^2 - 2) - 28$ **103.** $(y^2 + 2)^2 + 5(y^2 + 2) - 24$ **104.** $(x^3 + 12)^2 - 16$

105. $(y^3 + 34)^2 - 49$ **106.** $(x + y)^3 + z^3$ **107.** $(a + 5)^3 - b^3$

108. $9m^2 + 42m(3n + 1) + 49(3n + 1)^2$ **109.** $4x^2 + 36x(7y - 1) + 81(7y - 1)^2$ **110.** $(c - 3)^2 - (2c - 5)^2$

111. $(d + 6)^2 - (4d - 3)^2$ **112.** $p^{11} - 64p^8 - p^3 + 64$ **113.** $t^7 + 27t^4 - t^3 - 27$

114. $m^6 + 26m^3 - 27$ **115.** $n^6 - 7n^3 - 8$ **116.** $16x^6z + 38x^3z - 54z$

117. $24y^7 + 21y^4 - 3y$ **118.** $x^2 - y^2 - x + y$ **119.** $a^2 - b^2 - a - b$

120. $a^2 + ac - 2c^2 - c + a$ **121.** $x^2 + 2xy - 3y^2 - y + x$

CHAPTER 2   Detailed Summary

SECTION 2.1   Integer Exponents and Scientific Notation

Key Concepts	Examples	Page
Definition of b^0 and b^{-n}: If b is a nonzero real number and n is a positive integer, then $$b^0 = 1 \quad \text{and} \quad b^{-n} = \frac{1}{b^n}$$	**Example 1:** $(-4p)^0 = 1$ $w^{-6} = \dfrac{1}{w^6}$	p. 121
Properties of exponents: Assume that a and b are real numbers $(b \neq 0)$ and that m and n are integers. 1. $b^m \cdot b^n = b^{m+n}$ 2. $\dfrac{b^m}{b^n} = b^{m-n}$ 3. $(b^m)^n = b^{m \cdot n}$ 4. $(ab)^m = a^m b^m$ 5. $\left(\dfrac{a}{b}\right)^m = \dfrac{a^m}{b^m}$	**Example 2:** 1. $x^7 \cdot x^{-2} = x^{7+(-2)} = x^5$ 2. $\dfrac{z^9}{z^5} = z^{9-5} = z^4$ 3. $(y^{-2})^4 = y^{-2 \cdot 4} = y^{-8} = \dfrac{1}{y^8}$ 4. $(2x)^4 = 2^4 \cdot x^4 = 16x^4$ 5. $\left(\dfrac{5}{x^2}\right)^3 = \dfrac{5^3}{(x^2)^3} = \dfrac{125}{x^6}$	p. 121
Scientific notation: A number expressed in the form $a \times 10^n$, where $1 \leq \lvert a \rvert < 10$ and n is an integer is said to be in **scientific notation**.	**Example 3:** $54{,}000 = 5.4 \times 10^4$ $0.00000000568 = 5.68 \times 10^{-10}$ $8.75 = 8.75 \times 10^0$ **Example 4:** $(2.5 \times 10^8)(6.0 \times 10^3) = (2.5)(6.0) \times (10^8)(10^3)$ $\qquad = 15 \times 10^{11}$ $\qquad = (1.5 \times 10^1) \times 10^{11}$ $\qquad = 1.5 \times 10^{12}$	p. 125

SECTION 2.2 Operations on Polynomials

Key Concepts	Examples	Page

A **polynomial** in x is defined by a sum of terms of the form ax^n, where a is a real number and n is a whole number.
• a is the **coefficient** of the term.
• n is the **degree of the term**.
    The **degree of a polynomial** is the greatest degree of its terms. The term of a polynomial with the greatest degree is the **leading term**. Its coefficient is the **leading coefficient**.
    A one-term polynomial is a **monomial**.
    A two-term polynomial is a **binomial**.
    A three-term polynomial is a **trinomial**.
	Example 1:
	$7y^4 - 2y^2 + 3y + 8$
	is a polynomial with leading coefficient 7 and degree 4.
	Example 2:
	$4x^3 - 6x - 11$
	The polynomial has leading term $4x^3$ and leading coefficient 4. The degree is 3.
		p. 131

To add or subtract polynomials, add or subtract like terms.	**Example 3:** $(-4x^3y + 3x^2y^2) - (7x^3y - 5x^2y^2)$ $= -4x^3y + 3x^2y^2 - 7x^3y + 5x^2y^2$ $= -11x^3y + 8x^2y^2$	p. 132
To multiply polynomials, multiply each term in the first polynomial by each term in the second polynomial. **Special Products** 1. Multiplication of **conjugates** $(a + b)(a - b) = a^2 - b^2$ The product is called a **difference of squares**.	**Example 4:** $(x - 2)(3x^2 - 4x + 11)$ $= 3x^3 - 4x^2 + 11x - 6x^2 + 8x - 22$ $= 3x^3 - 10x^2 + 19x - 22$ **Example 5:** $(3x + 5)(3x - 5)$ $= (3x)^2 - (5)^2$ $= 9x^2 - 25$	p. 134 p. 136
2. Square of a binomial $(a + b)^2 = a^2 + 2ab + b^2$ $(a - b)^2 = a^2 - 2ab + b^2$ The product is called a **perfect square trinomial**.	**Example 6:** $(4y + 3)^2$ $= (4y)^2 + (2)(4y)(3) + (3)^2$ $= 16y^2 + 24y + 9$	p. 136

SECTION 2.3 Greatest Common Factor and Factoring by Grouping

Key Concepts	Examples	Page
The **greatest common factor (GCF)** is the largest factor common to all terms of a polynomial. To factor out the GCF from a polynomial, use the distributive property.	**Example 1:** $3x^2(a + b) - 6x(a + b)$ $= 3x(a + b)x - 3x(a + b)(2)$ $= 3x(a + b)(x - 2)$	p. 144
A four-term polynomial may be **factored by grouping**. **Steps to Factor by Grouping** 1. Identify and factor out the GCF from all four terms. 2. Factor out the GCF from the first pair of terms. Factor out the GCF from the second pair of terms. (Sometimes it is necessary to factor out the *opposite* of the GCF.) 3. If the two pairs of terms share a common binomial factor, factor out the binomial factor.	**Example 2:** $60xa - 30xb - 80ya + 40yb$ $= 10[6xa - 3xb \mid - 8ya + 4yb]$ $= 10[3x(2a - b) - 4y(2a - b)]$ $= 10(2a - b)(3x - 4y)$	p. 145

Sometimes factoring is used to solve an equation for a specified variable.	**Example 3:** Given $ax + m = -bx + n$, solve for x. $ax + m = -bx + n$ $ax + bx = n - m$ Collect x terms on one side. $x(a + b) = n - m$ Factor out x. $\dfrac{x(a + b)}{(a + b)} = \dfrac{n - m}{(a + b)}$ Divide by $(a + b)$. $x = \dfrac{n - m}{a + b}$	p. 148
Factoring out negative exponents: To factor out a common variable factor raised to a negative exponent, factor out the variable raised to the smallest exponent. The exponents on the variables within parentheses are found by subtraction.	**Example 4:** $\boxed{-8 - (-8)}$ $\boxed{-7 - (-8)}$ $\boxed{-6 - (-8)}$ $7z^{-8} + 6z^{-7} - 5z^{-6} = z^{-8}(7z^0 + 6z^1 - 5z^2)$ $= z^{-8}(7 + 6z - 5z^2)$ or $\dfrac{7 + 6z - 5z^2}{z^8}$	p. 148

SECTION 2.4 Factoring Trinomials

Key Concepts	Examples	Page
AC-Method To factor trinomials of the form $ax^2 + bx + c$: 1. Factor out the GCF. Find the product ac. 2. Find two integers whose product is ac and whose sum is b. (If no pair of numbers can be found, then the trinomial is prime.) 3. Rewrite the middle term bx as the sum of two terms whose coefficients are the numbers found in step 2. 4. Factor the polynomial by grouping.	**Example 1:** $10y^2 + 35y - 20 = 5(2y^2 + 7y - 4)$ $ac = (2)(-4) = -8$ Find two integers whose product is -8 and whose sum is 7. The numbers are 8 and -1. $5[2y^2 + 8y \ \vdots \ -1y - 4]$ $= 5[2y(y + 4) - 1(y + 4)]$ $= 5(y + 4)(2y - 1)$	p. 152
Trial-and-Error Method To factor trinomials in the form $ax^2 + bx + c$: 1. Factor out the GCF. 2. List the pairs of factors of a and the pairs of factors of c. Consider the reverse order in either list. 3. Construct two binomials of the form Factors of a $(\square x \ \square)(\square x \ \square)$ Factors of c 4. Test each combination of factors and signs until the product of the outer terms and the product of inner terms add to the middle term. 5. If no combination of factors works, the polynomial is prime.	**Example 2:** $10y^2 + 35y - 20 = 5(2y^2 + 7y - 4)$ The pairs of factors of 2 are $2 \cdot 1$. The pairs of factors of -4 are $\begin{array}{cc} -1 \cdot 4 & 1 \cdot (-4) \\ -2 \cdot 2 & 2 \cdot (-2) \\ -4 \cdot 1 & 4 \cdot (-1) \end{array}$ $(2y - 2)(y + 2) = 2y^2 + 2y - 4$ No $(2y - 4)(y + 1) = 2y^2 - 2y - 4$ No $(2y + 1)(y - 4) = 2y^2 - 7y - 4$ No $(2y + 2)(y - 2) = 2y^2 - 2y - 4$ No $(2y + 4)(y - 1) = 2y^2 + 2y - 4$ No $(2y - 1)(y + 4) = 2y^2 + 7y - 4$ Yes Therefore, $10y^2 + 35y - 20$ factors as $5(2y - 1)(y + 4)$.	p. 154

The factored form of a **perfect square trinomial** is the square of a binomial: $a^2 + 2ab + b^2 = (a + b)^2$ $a^2 - 2ab + b^2 = (a - b)^2$	**Example 3:** $9w^2 - 30wz + 25z^2$ $\quad = (3w)^2 - 2(3w)(5z) + (5z)^2$ $\qquad\qquad = (3w - 5z)^2$	p. 159
Sometimes it is easier to factor a polynomial after making a substitution.	**Example 4:** $(7v^2 - 1)^2 - (7v^2 - 1) - 12$ Let $u = (7v^2 - 1)$. $\quad = u^2 - u - 12$ Substitute. $\quad = (u + 3)(u - 4)$ Factor. $\quad = (7v^2 - 1 + 3)(7v^2 - 1 - 4)$ Back substitute. $\quad = (7v^2 + 2)(7v^2 - 5)$ Simplify.	p. 161

SECTION 2.5   Factoring Binomials

Key Concepts	Examples	Page
Factoring Binomials: Summary **Difference of squares:** $a^2 - b^2 = (a + b)(a - b)$ **Sum of squares:** If a and b share no common factors, then $a^2 + b^2$ is prime.	**Example 1:** $25u^2 - 9v^4 = (5u)^2 - (3v^2)^2$ $\qquad\qquad = (5u + 3v^2)(5u - 3v^2)$ **Example 2:** $32 + 2w^2 = 2(16 + w^2)$ cannot be factored further because $16 + w^2$ is a sum of squares.	p. 165
Difference of cubes: $a^3 - b^3 = (a - b)(a^2 + ab + b^2)$ **Sum of cubes:** $a^3 + b^3 = (a + b)(a^2 - ab + b^2)$	**Example 3:** $8c^3 - d^6 = (2c - d^2)(4c^2 + 2cd^2 + d^4)$ **Example 4:** $27w^9 + 64x^3$ $\quad = (3w^3 + 4x)(9w^6 - 12w^3x + 16x^2)$	p. 167
Sometimes it is necessary to group three terms with one term.	**Example 5:** $4a^2 - 12ab + 9b^2 - c^2$ $\quad = 4a^2 - 12ab + 9b^2 \ \vdots\ - c^2$ Group 3 by 1. $\quad = (2a - 3b)^2 - c^2$ Perfect square trinomial. $\quad = (2a - 3b - c)(2a - 3b + c)$ Difference of squares.	p. 167

CHAPTER 2 Review Exercises

SECTION 2.1

For Exercises 1–7, simplify completely. Write the answers with positive exponents only.

1. a. 9^0 **b.** -9^0 **c.** $9x^0$ **d.** $(9x)^0$

2. a. $\dfrac{1}{m^{-5}}$ **b.** m^{-5} **c.** $8m^{-9}n^2$ **d.** $8m^9n^{-2}$

3. $p^{-8} \cdot p^{12} \cdot p^{-1}$ **4.** $\dfrac{m^{-4}m^{10}}{m^6}$

5. $(-12a^{-3}b^4)^2$ **6.** $\left(\dfrac{-81x^8y^5}{9x^6y^8}\right)^{-2}$

7. $\left(\dfrac{1}{2u^5v^{-2}}\right)^{-3}\left(\dfrac{4}{u^{-3}v^2}\right)^{-1}$

8. Write the numbers in scientific notation.

   a. 4920   **b.** 0.00492   **c.** 4.92

9. Write the numbers in standard decimal notation.

   a. 9.8×10^{-1} **b.** 9.8×10^0 **c.** 9.8×10^1

For Exercises 10–11, perform the indicated operations.

10. $(9.2 \times 10^4)(3.0 \times 10^5)$

11. $\dfrac{(8.6 \times 10^{-3})(4.1 \times 10^8)}{2.0 \times 10^{-6}}$

12. A healthy adult female will have approximately 5 million red blood cells per 1 μL of blood. If a woman donates 1 pint of blood to a blood bank, approximately how many red blood cells will be present? (*Hint*: 1 μL $= 10^{-6}$ L and 1 pint $\approx$ 0.47 L.)

13. The land area of California is approximately 1.56×10^5 mi^2. If there are approximately 40 million people in California, determine the number of people per acre. (*Hint*: 1 mi^2 = 640 acres.) Round to the nearest tenth of a person.

SECTION 2.2

14. Determine if the expression is a polynomial.

   a. $5a^3 - 4a^2 + 6a - 3$

   b. $\dfrac{7}{a^3} - \dfrac{4}{a} + 6$

   c. $\sqrt{6}x + 5$

15. Write the polynomial in descending order. Identify the leading coefficient and degree of the polynomial.

$$-y^5 + 7.61y^9 + 2.5y^{11}$$

16. Determine the degree of the polynomial.

$$-4ac^2d^3 + 5ac^3d^4 - a^2cd^2$$

For Exercises 17–28, add or subtract the polynomials as indicated.

17. $(x^2 - 2x - 3xy - 7) + (-3x^2 - x + 2xy + 6)$

18. $\begin{aligned}7xy - \ \ 3xz + 5yz \\ +13xy - 15xz - 8yz\end{aligned}$

19. $\begin{aligned}-4a^3 + 8a^2 - 3a \\ -(-7a^3 + 3a^2 - 9a)\end{aligned}$

20. $(3a^2 - 2a - a^3) - (5a^2 - a^3 - 8a)$

21. $\left(\dfrac{5}{8}x^4 - \dfrac{1}{4}x^2 - \dfrac{1}{2}\right) - \left(-\dfrac{3}{8}x^4 + \dfrac{3}{4}x^2 + \dfrac{1}{2}\right)$

22. $\left(\dfrac{5}{6}x^4 + \dfrac{1}{2}x^2 - \dfrac{1}{3}\right) - \left(-\dfrac{1}{6}x^4 - \dfrac{1}{4}x^2 - \dfrac{1}{3}\right)$

23. $(7x - y) - [-(2x + y) - (-3x - 6y)]$

24. $-(4x - 4y) - [(4x + 2y) - (3x + 7y)]$

25. Add $-4x + 6$ to $-7x - 5$.

26. Add $2x^2 - 4x$ to $2x^2 - 7x$.

27. Subtract $-4x + 6$ from $-7x - 5$.

28. Subtract $2x^2 - 4x$ from $2x^2 - 7x$.

For Exercises 29–46, multiply the polynomials.

29. $2x(x^2 - 7x - 4)$ **30.** $-3x(6x^2 - 5x + 4)$

31. $(x + 6)(x - 7)$ **32.** $(x - 2)(x - 9)$

33. $\left(\dfrac{1}{2}x + 1\right)\left(\dfrac{1}{2}x - 5\right)$ **34.** $\left(-\dfrac{1}{5} + 2y\right)\left(\dfrac{1}{5} + y\right)$

35. $(3x + 5)(9x^2 - 15x + 25)$

36. $(x - y)(x^2 + xy + y^2)$

37. $(2x - 5)^2$ **38.** $\left(\dfrac{1}{2}x + 4\right)^2$

39. $(3y - 11)(3y + 11)$ **40.** $(6w - 1)(6w + 1)$

41. $\left(\dfrac{2}{3}t + 4\right)\left(\dfrac{2}{3}t - 4\right)$ **42.** $\left(z + \dfrac{1}{4}\right)\left(z - \dfrac{1}{4}\right)$

43. $[(x + 2) - b][(x + 2) + b]$

44. $[c - (w + 3)][c + (w + 3)]$

45. $(2x + 1)^3$ **46.** $(y^2 - 3)^3$

47. A square garden is surrounded by a walkway of uniform width x. If the sides of the garden are given by the expression $2x + 3$, find and simplify a polynomial for the following.

   a. The area of the garden.

   b. The area of the walkway and garden.

c. The area of the walkway only.

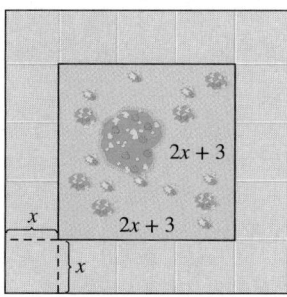

SECTION 2.3

For Exercises 48–51, factor by removing the greatest common factor.

48. $-x^3 - 4x^2 + 11x$

49. $21w^3 - 7w + 14$

50. $5x(x - 7) - 2(x - 7)$

51. $3t(t + 4) + 5(t + 4)$

For Exercises 52–55, factor by grouping (remember to take out the GCF first).

52. $m^3 - 8m^2 + m - 8$

53. $24x^3 - 36x^2 + 72x - 108$

54. $4ax^2 + 2bx^2 - 6ax - 3xb$

55. $y^3 - 6y^2 + y - 6$

For Exercises 56–67, solve for the specified variable.

56. $S = P + nrP$ for P

57. $cy - d = by + a$ for y

For Exercises 58–59, factor completely and write the answer with no negative exponents.

58. $3x^{-3} - 2x^{-4} + x^{-5}$

59. $7w^{-8} + 5w^{-7} + w^{-6}$

SECTION 2.4

For Exercises 60–74, factor the polynomials by using any method.

60. $18x^2 + 27xy + 10y^2$ **61.** $3m^2 + mt - 10t^2$

62. $60a^2 + 65a^3 - 20a^4$ **63.** $2k^2 + 7k^3 + 6k^4$

64. $49x^2 + 36 - 84x$ **65.** $80z + 32 + 50z^2$

66. $(9w + 2)^2 + 4(9w + 2) - 5$

67. $(4x + 3)^2 - 12(4x + 3) + 36$

68. $18a^4 + 39a^2 - 15$ **69.** $3w^4 - 2w^2 - 5$

70. $80m^4n^8 - 48m^5n^3 - 16m^2n$

71. $11p^2(2p + 1) - 22p(2p + 1)$

72. $15ac - 14b - 10a + 21bc$

73. $-t + 12t^2 - 6$

74. $8x^3 - 40x^2y + 50xy^2$

SECTION 2.5

For Exercises 75–81, factor the binomials.

75. $25 - y^2$ **76.** $x^3 - \dfrac{1}{27}$

77. $b^2 + 64$

78. $h^3 + 9h$ **79.** $a^3 + 64$

80. $k^4 - 16$ **81.** $9y^3 - 4y$

For Exercises 82–85, factor by grouping and by using the difference of squares.

82. $x^2 - 8xy + 16y^2 - 9$ (*Hint:* Group three terms that make up a perfect square trinomial, then factor as a difference of squares.)

83. $a^2 + 12a + 36 - b^2$

84. $t^2 + 16t + 64 - 25c^2$ **85.** $y^2 - 6y + 9 - 16x^2$

For Exercises 86–93, factor completely.

86. $256a^4 - 625$

87. $3k^4 - 81k$

88. $(c + 2)^3 + d^3$

89. $25n^2 - m^2 - 12m - 36$

90. $(x^2 - 7)^2 + 8(x^2 - 7) + 15$

91. $(2p - 5)^2 - (4p + 1)^2$

92. $m^6 + 9m^3 + 8$

93. $x^4 + 6x^2y + 9y^2 - x^2 - 3y$

CHAPTER 2   Test

For Exercises 1–4, simplify the expression, and write the answer with positive exponents only.

1. $\dfrac{20a^7}{4a^{-6}}$

2. $\dfrac{x^6x^3}{x^{-2}}$

3. $\left(\dfrac{-3x^6}{5y^7}\right)^2$

4. $\dfrac{(2^{-1}xy^{-2})^{-3}(x^{-4}y)}{(x^0y^5)^{-1}}$

5. Write the numbers in the given statement in scientific notation. "China uses 45,000,000,000 pairs of disposable chopsticks per year. This equates to approximately 1.66 million cubic meters of timber."

6. Write the number in the given statement in standard decimal notation. "The Ebola virus is approximately 8×10^{-7} m in length."

7. Perform the indicated operations.

$$\frac{(8.4 \times 10^{11})(6.0 \times 10^{-3})}{(4.2 \times 10^{-5})}$$

8. Perform the indicated operations. Write the answer in descending order.

$$(5x^2 - 7x + 3) - (x^2 + 5x - 25)$$
$$+ (4x^2 + 4x - 20)$$

For Exercises 9–12, multiply the polynomials.

9. $(2a - 5)(a^2 - 4a - 9)$

10. $\left(\frac{1}{3}x - \frac{3}{2}\right)(6x + 4)$

11. $(5x - 4y^2)(5x + 4y^2)$

12. $[(3a + b) - c][(3a + b) + c]$

13. Write and simplify an expression that describes the area of the square.

$$7x - 4$$

14. A rectangle and square are shown. Write an expression in terms of x that represents the area of the shaded region.

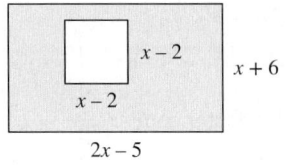

For Exercises 15–16, solve the equation for the indicated variable.

15. $C = ay - by$ for y

16. $5x - ct = dx + 7$ for x

For Exercises 17–37, factor completely.

17. $17y + 3y^2 - 28$

18. $x^2 - 5x - 4$

19. $3a^2 + 27ab + 54b^2$

20. $c^4 - 1$

21. $xy - 7x + 3y - 21$

22. $49 + p^2$

23. $-10u^2 + 30u - 20$

24. $12t^2 - 75$

25. $5y^2 - 50y + 125$

26. $21q^2 + 14q$

27. $2x^3 + x^2 - 8x - 4$

28. $y^3 - 125$

29. $x^2 + 8x + 16 - y^2$

30. $r^6 - 256r^2$

31. $(x^2 + 1)^2 + 3(x^2 + 1) + 2$

32. $12a - 6ac + 2b - bc$

33. $x^5 + 2x^4 - 81x - 162$

34. $c^2 - 4a^2 - 44a - 121$

35. $27u^3 - v^6$

36. $4w^{-6} + 2w^{-5} + 7w^{-4}$

37. $x^{-2}(2x + 1) + 3x^{-3}(2x + 1)^2$

Quadratic Equations

3

Chapter Outline

Purestock /SuperStock

A golden rectangle is a rectangle in which the ratio of its length L to its width W is equal to the ratio of the sum of its length and width to its length.

$$\frac{L}{W} = \frac{L + W}{L}$$

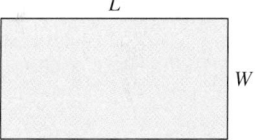

The values of L and W that meet this condition are said to be in the **golden ratio**. The golden ratio has been studied by artists and art historians for generations because the golden ratio represents an aesthetically pleasing ratio between the length and width of a figure. For example, the face of the Parthenon built in ancient Greece has the dimensions of a golden rectangle.

We can show that the length of a golden rectangle is approximately 1.62 times the width. Substituting 1 for the width, we have the proportion $\frac{L}{1} = \frac{L + 1}{L}$. Then, clearing fractions and writing the quadratic equation in standard form gives $L^2 - L - 1 = 0$. The expression on the left is not factorable, but fortunately in this chapter, we will learn two techniques to solve a quadratic equation when factoring fails. The positive solution for L in this equation is the golden ratio, $\frac{1 + \sqrt{5}}{2} \approx 1.62$.

SECTION 3.1	**Introduction to Radicals and Their Simplification**

OBJECTIVES

1. Simplify Expressions of the Form $\sqrt[n]{x^n}$
2. Use the Product and Quotient Properties to Simplify Radicals
3. Add and Subtract Radicals

1. Simplify Expressions of the Form $\sqrt[n]{x^n}$

In Chapter 3, we will apply our knowledge of factoring and nth roots to solve a new type of equation called a *quadratic equation*. A **quadratic equation** is an equation of the form $ax^2 + bx + c = 0$, where a, b, and c are real numbers and $a \neq 0$. A quadratic equation always has a term with x^2 because the coefficient a is not zero. To solve a quadratic equation, it seems reasonable that we might use the square root operation (the inverse of squaring a number). Therefore, we take time here to review operations on radicals.

Recall that the symbol $\sqrt[n]{a}$ represents the nth root of a. The expression a is called the **radicand**, and n is called the **index**. It is often advantageous to write radical expressions in their simplest form, and we begin here by simplifying radicals of the form $\sqrt[n]{x^n}$.

The expression $\sqrt[n]{x^n}$ represents the principal nth root of x^n. For an even index n, the principal nth root is defined to be nonnegative. For example, $\sqrt{25} = 5$ (not -5) and $\sqrt[4]{16} = 2$ (not -2). For a radical with an odd index, the principal nth root will be either negative or nonnegative depending on the sign of the radicand. For example, $\sqrt[3]{-64} = -4$ and $\sqrt[5]{32} = 2$. Therefore, when we simplify expressions of the form $\sqrt[n]{x^n}$, we use the following rules.

$$\sqrt[n]{x^n} = \begin{cases} x & \text{if } n \text{ is odd} \\ |x| & \text{if } n \text{ is even} \end{cases}$$

> The absolute value bars are used in the case of an even index to guarantee that the principal nth root is nonnegative.

For example:

$$\sqrt{x^2} = |x| \qquad \sqrt[3]{x^3} = x$$
$$\sqrt[4]{x^4} = |x| \qquad \sqrt[5]{x^5} = x$$
$$\sqrt[6]{x^6} = |x| \qquad \sqrt[7]{x^7} = x$$

EXAMPLE 1   **Simplifying Expressions of the Form $\sqrt[n]{x^n}$**

Simplify.

   a. $\sqrt[3]{(-12)^3}$ **b.** $\sqrt[4]{(-3)^4}$ **c.** $\sqrt{(a+b)^2}$ **d.** $\sqrt[4]{x^8}$

Solution:

   a. $\sqrt[3]{(-12)^3} = -12$ The index is 3, which is an odd number. Absolute value bars are not necessary.

   b. $\sqrt[4]{(-3)^4} = |-3| = 3$ The index is 4, which is an even number. Absolute value bars are necessary to guarantee that the answer is not negative.

   c. $\sqrt{(a+b)^2} = |a+b|$ The index is 2, which is an even number. Absolute value bars are necessary to guarantee that the answer is not negative. Since we do not know the values of a and b, and consequently the sign of the sum $a + b$, we cannot simplify further.

   d. $\sqrt[4]{x^8} = \sqrt[4]{(x^2)^4}$ The expression x^8 is a perfect square because it can be written as $(x^2)^4$.

       $= |x^2|$ The index of the radical is 4, which is even. Absolute value bars are necessary.

       $= x^2$ Because the square of a real number is never negative, we can remove the absolute value bars and write the answer as x^2.

Skill Practice 1   Simplify.

a. $\sqrt[6]{(-8)^6}$ b. $\sqrt[5]{(-1)^5}$ c. $\sqrt{(2x+1)^2}$ d. $\sqrt{x^{12}}$

In the examples and exercises that follow, we will assume that the variables within the radicals are nonnegative. In such a case, the absolute value bars are not needed to simplify expressions of the form $\sqrt[n]{x^n}$. That is, if $x \geq 0$, then $\sqrt[n]{x^n} = x$.

2. Use the Product and Quotient Properties to Simplify Radicals

To simplify radical expressions, we often use the product and quotient properties of radicals.

Product and Quotient Properties of Radicals

If $\sqrt[n]{a}$ and $\sqrt[n]{b}$ are both real numbers, then

$$\sqrt[n]{a} \cdot \sqrt[n]{b} = \sqrt[n]{ab}$$ Product property of radicals

$$\frac{\sqrt[n]{a}}{\sqrt[n]{b}} = \sqrt[n]{\frac{a}{b}}$$ Quotient property of radicals

---

One of the criteria when simplifying an nth root is to "remove" all perfect nth powers from the radicand. For example, consider the expression $\sqrt[3]{x^5}$. The radicand may be written as a product: $\sqrt[3]{x^3 \cdot x^2}$. The factor x^3 is a perfect cube and must be "removed" from the radicand. Using the product property of radicals, we can write $\sqrt[3]{x^3 \cdot x^2}$ as the cube root of a perfect cube times the cube root of the "leftover" factor.

Avoiding Mistakes

With a radicand in factored form, all exponents in the radicand should be less than the index. This will guarantee that all nth powers have been removed from the nth root.

$$\sqrt[3]{x^5} = \sqrt[3]{x^3 \cdot x^2} = \sqrt[3]{x^3} \cdot \sqrt[3]{x^2} = x \cdot \sqrt[3]{x^2} \quad \text{or simply } x\sqrt[3]{x^2}$$

Perfect cube   "leftover"

The expression $x\sqrt[3]{x^2}$ is considered simplified because the exponent in the radicand is less than the index. In general, to simplify radicals, we often use the product and quotient properties of radicals in "reverse." We decompose the nth root of a product of factors into a product of nth roots that can be easily simplified.

EXAMPLE 2   **Using the Product Property to Simplify a Radical**

Simplify. Assume that x is a positive real number. $\sqrt{x^9}$

Solution:

$$\sqrt{x^9} = \sqrt{x^8 \cdot x}$$ Write x^9 as the product of the greatest perfect square and a "leftover" factor.

$$= \sqrt{x^8} \cdot \sqrt{x}$$ Apply the product property of radicals. The first radical is the square root of a perfect square and can be simplified: $\sqrt{x^8} = \sqrt{(x^4)^2} = x^4$

$$= x^4\sqrt{x}$$

Skill Practice 2   Simplify. Assume that a is a positive real number. $\sqrt{a^{11}}$

Answers
1. a. 8   **b.** -1 **c.** $|2x+1|$ **d.** x^6
2. $a^5\sqrt{a}$

In Example 2, the expression x^9 is not a perfect square. Therefore, to simplify $\sqrt{x^9}$, it was necessary to write the expression as the product of the largest perfect square and a remaining or "leftover" factor: $\sqrt{x^9} = \sqrt{x^8 \cdot x}$. This process also applies to simplifying nth roots, as shown in Example 3.

EXAMPLE 3   Using the Product Property to Simplify Radicals

Simplify each expression. Assume all variables represent positive real numbers.

a. $\sqrt[4]{b^7}$ **b.** $\sqrt[3]{w^7 z^9}$

Solution:

The goal is to rewrite each radicand as the product of the greatest perfect square (perfect cube, perfect fourth power, and so on) and a leftover factor.

a. $\sqrt[4]{b^7} = \sqrt[4]{b^4 \cdot b^3}$ b^4 is the greatest perfect fourth power in the radicand.

$\quad = \sqrt[4]{b^4} \cdot \sqrt[4]{b^3}$ Apply the product property of radicals.

$\quad = b\sqrt[4]{b^3}$ Simplify.

b. $\sqrt[3]{w^7 z^9} = \sqrt[3]{(w^6 z^9) \cdot (w)}$ $w^6 z^9$ is the greatest perfect cube in the radicand.

$\quad = \sqrt[3]{w^6 z^9} \cdot \sqrt[3]{w}$ Apply the product property of radicals.

$\quad = w^2 z^3 \sqrt[3]{w}$ Simplify.

Skill Practice 3   Simplify the expressions. Assume all variables represent positive real numbers.

a. $\sqrt[4]{v^{25}}$ **b.** $\sqrt[3]{p^{17} q^{10}}$

Each expression in Example 3 involves a radicand that is a product of variable factors. In Example 4, we show that when a numerical factor is present, sometimes it is necessary to factor the coefficient before simplifying the radical.

EXAMPLE 4   Using the Product Property to Simplify a Radical

Simplify. $6\sqrt{56}$

Solution:

$6\sqrt{56} = 6\sqrt{2^3 \cdot 7}$ Factor the radicand into a product of prime factors.

$\quad = 6\sqrt{(2^2) \cdot (2 \cdot 7)}$ Identify the greatest perfect square in the radicand: 2^2

$\quad = 6\sqrt{2^2} \cdot \sqrt{2 \cdot 7}$ Apply the product property of radicals.

$\quad = 6 \cdot 2 \cdot \sqrt{2 \cdot 7}$ Simplify the square root of the perfect square: $\sqrt{2^2} = 2$

$\quad = 12\sqrt{14}$ Multiply.

Skill Practice 4   Simplify. $5\sqrt{24}$

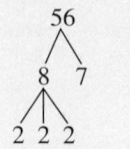
Answers
3. a. $v^6 \sqrt[4]{v}$ **b.** $p^5 q^3 \sqrt[3]{p^2 q}$
4. $10\sqrt{6}$

TECHNOLOGY CONNECTIONS

Verifying a Simplified Radical

A calculator can be used to support the solution to Example 4. The decimal approximations for $6\sqrt{56}$ and $12\sqrt{14}$ agree for the first 10 digits. This in itself does not make $6\sqrt{56} = 12\sqrt{14}$. Instead, it is the multiplication property of radicals that guarantees that the expressions are equivalent.

```
NORMAL FLOAT AUTO REAL DEGREE CL

6√(56)
                        44.89988864
12√(14)
                        44.89988864
```

In Example 4 the radical $\sqrt{56}$ can also be simplified if you can identify the largest perfect square in the radicand. In this case, 4 is the largest factor of 56 that is a perfect square. We have

$$\sqrt{56} = \sqrt{4 \cdot 14} = \sqrt{4} \cdot \sqrt{14} = 2\sqrt{14}.$$

EXAMPLE 5   Using the Product Property to Simplify a Radical

Simplify. Assume that x, y, and z represent positive real numbers.

$$\sqrt[3]{250x^3y^5z^7}$$

Solution:

$$\sqrt[3]{250x^3y^5z^7} = \sqrt[3]{5^3 \cdot 2 \cdot x^3y^5z^7} \qquad \text{Factor the radicand.}$$

$$= \sqrt[3]{(5^3x^3y^3z^6)(2y^2z)} \qquad \text{The greatest perfect cube in the radicand is the product of } 5^3, x^3, y^3, \text{ and } z^6.$$

$$= \sqrt[3]{5^3x^3y^3z^6} \cdot \sqrt[3]{2y^2z} \qquad \text{Apply the product property of radicals.}$$

$$= 5xyz^2\sqrt[3]{2y^2z} \qquad \text{Simplify the first radical as the cube root of a perfect cube.}$$

TIP In Example 5, the numerical coefficient can be written as $125 \cdot 2$ because $125 = 5^3$ is the greatest perfect cube factor of 250.

$$\sqrt[3]{250x^3y^5z^7}$$
$$= \sqrt[3]{125 \cdot 2 \cdot x^3y^5z^7}$$
$$= 5xyz^2\sqrt[3]{2y^2z}$$

Skill Practice 5   Simplify. $\sqrt[4]{32a^{10}b^{19}}$

Examples 1–5 demonstrate the process to "remove" a perfect nth power from an nth root. However, we usually abide by three additional criteria when writing radicals in simplified form. These conditions are summarized here.

Simplified Form of a Radical

Suppose that the radicand of a radical is written as a product of prime factors. Then the radical is simplified if all of the following conditions are met.

1. The radicand has no factor other than 1 that is a perfect nth power. This means that all exponents in the radicand must be less than the index.
2. No fractions may appear in the radicand.
3. No denominator of a fraction may contain a radical.
4. The exponents in the radicand may not all share a common factor with the index.

Answer
5. $2a^2b^4\sqrt[4]{2a^2b^3}$

In Example 6, we address conditions 2 and 3.

EXAMPLE 6 Using the Quotient Property to Simplify Radicals

Simplify. Assume that a is a positive real number.

a. $\sqrt{\dfrac{a^7}{a^3}}$ **b.** $\dfrac{\sqrt[3]{3}}{\sqrt[3]{81}}$

Solution:

a. $\sqrt{\dfrac{a^7}{a^3}} = \sqrt{a^4}$

The radical is not simplified because the radicand contains a fraction (fails condition 2). However, the fraction can be reduced to lowest terms.

$= a^2$

Simplify the radical.

b. $\dfrac{\sqrt[3]{3}}{\sqrt[3]{81}}$

The radical is not simplified because the denominator of the fraction contains a radical (fails condition 3).

$= \sqrt[3]{\dfrac{3}{81}}$

The numerator and denominator share a common factor of 3. Therefore, apply the quotient property of radicals to write the expression as one radical so that we can reduce the fraction.

$= \sqrt[3]{\dfrac{1}{27}}$

The expression $\frac{1}{27}$ is a perfect cube. The cube root is $\frac{1}{3}$.

$= \dfrac{1}{3}$

FOR REVIEW

Recall that to simplify a fraction, we divide common factors from the numerator and denominator.

$\dfrac{a^7}{a^3} = \dfrac{\overset{1}{\cancel{a}} \cdot \overset{1}{\cancel{a}} \cdot \overset{1}{\cancel{a}} \cdot a \cdot a \cdot a \cdot a}{\cancel{a} \cdot \cancel{a} \cdot \cancel{a}}$

$= a^4$

Alternatively, when dividing common bases, subtract the exponents (see Section 2.1).

$\dfrac{a^7}{a^3} = a^{7-3} = a^4$

Skill Practice 6   Simplify. Assume that v is a positive real number.

a. $\sqrt{\dfrac{v^{21}}{v^5}}$ **b.** $\dfrac{\sqrt[4]{2}}{\sqrt[4]{32}}$

---

3. Add and Subtract Radicals

When two radicals are added or subtracted, we can simplify the sum or difference if the radicals are like radicals. **Like radicals** have the same index and same radicand. We add and subtract like radicals as we add and subtract like terms—apply the distributive property.

$$3\sqrt{x} + 7\sqrt{x} = (3+7)\sqrt{x} = 10\sqrt{x}$$

Same index   Distributive property   Same radicand

Answers

6. a. v^8 **b.** $\dfrac{1}{2}$

To review combining like terms, see Section 1.1.

$6x - 2x = (6 - 2)x$
$= 4x$

EXAMPLE 7   **Adding and Subtracting Radicals**

Add or subtract as indicated.   **a.** $6\sqrt[3]{11} - 2\sqrt[3]{11}$ **b.** $\sqrt{3} + \sqrt{3}$

Solution:

a. $6\sqrt[3]{11} - 2\sqrt[3]{11}$ The radicals are like radicals because the index of 3 is the same, and the radicand of 11 is the same.

$= (6 - 2)\sqrt[3]{11}$ Apply the distributive property by factoring out $\sqrt[3]{11}$.

$= 4\sqrt[3]{11}$ Simplify.

b. $\sqrt{3} + \sqrt{3}$ The radicals are like radicals. They are both square roots and have the same radicand of 3.

$= 1\sqrt{3} + 1\sqrt{3}$ Note that $\sqrt{3} = 1\sqrt{3}$.

$= (1 + 1)\sqrt{3}$ Apply the distributive property by factoring out $\sqrt{3}$.

$= 2\sqrt{3}$ Simplify.

Avoiding Mistakes

The process of adding like radicals with the distributive property is similar to adding like terms. The end result is that the numerical coefficients are added and the radical factor is unchanged.

$$\sqrt{3} + \sqrt{3} = 1\sqrt{3} + 1\sqrt{3} = 2\sqrt{3}$$

Be careful: $\sqrt{3} + \sqrt{3} \neq \sqrt{6}$
In general: $\sqrt{x} + \sqrt{y} \neq \sqrt{x + y}$

Skill Practice 7   Add or subtract as indicated.

a. $5\sqrt[3]{6} - 8\sqrt[3]{6}$ **b.** $\sqrt{10} + \sqrt{10}$

Sometimes it is necessary to simplify radicals before adding or subtracting as shown in Examples 8 and 9.

EXAMPLE 8   **Adding and Subtracting Radicals**

Perform the indicated operations.

a. $3\sqrt{8} - \sqrt{2}$ **b.** $\sqrt{45} + 3\sqrt{18} - 8\sqrt{20}$

Solution:

a. $3\sqrt{8} - \sqrt{2}$ The radicands are different. Try simplifying the radicals first. Simplify. $3\sqrt{8} = 3\sqrt{2^3} = 3 \cdot 2\sqrt{2} = 6\sqrt{2}$

$= 6\sqrt{2} - 1\sqrt{2}$ The radicals are now like radicals.

$= (6 - 1)\sqrt{2}$ Apply the distributive property by factoring out $\sqrt{2}$.

$= 5\sqrt{2}$ Simplify.

Answers

7. a. $-3\sqrt[3]{6}$ **b.** $2\sqrt{10}$

TIP The radicals in Example 8 can also be simplified by recognizing the largest perfect square within each radicand.

$$\sqrt{8} = \sqrt{4} \cdot \sqrt{2} = 2\sqrt{2}$$
$$\sqrt{45} = \sqrt{9} \cdot \sqrt{5} = 3\sqrt{5}$$
$$\sqrt{18} = \sqrt{9} \cdot \sqrt{2} = 3\sqrt{2}$$
$$\sqrt{20} = \sqrt{4} \cdot \sqrt{5} = 2\sqrt{5}$$

b. $\sqrt{45} + 3\sqrt{18} - 8\sqrt{20}$ The radicands are different. Try simplifying the radicals.

$$\sqrt{45} = \sqrt{3^2 \cdot 5} = 3\sqrt{5}$$
$$3\sqrt{18} = 3\sqrt{3^2 \cdot 2} = 3 \cdot 3\sqrt{2} = 9\sqrt{2}$$
$$-8\sqrt{20} = -8\sqrt{2^2 \cdot 5} = -8 \cdot 2\sqrt{5} = -16\sqrt{5}$$

$$= 3\sqrt{5} + 9\sqrt{2} - 16\sqrt{5}$$

$$= (3 - 16)\sqrt{5} + 9\sqrt{2}$$ The radical terms with $\sqrt{5}$ are like radicals.

$$= -13\sqrt{5} + 9\sqrt{2}$$ The two terms cannot be further simplified because they are not like radicals.

Skill Practice 8 Perform the indicated operations.

a. $\sqrt{147} + 2\sqrt{3}$ **b.** $8\sqrt{75} - \sqrt{12} + 2\sqrt{8}$

EXAMPLE 9 Adding Radicals

Add the radicals. Assume that x and y represent positive real numbers.

$$x\sqrt{98x^3y} + 5\sqrt{18x^5y}$$

Solution:

$$x\sqrt{98x^3y} + 5\sqrt{18x^5y}$$ The radicals have different radicands and thus are not like radicals. However, each radical can be simplified.

$$x\sqrt{98x^3y} = x\sqrt{(7^2x^2) \cdot (2xy)} = x \cdot 7x\sqrt{2xy} = 7x^2\sqrt{2xy}$$
$$5\sqrt{18x^5y} = 5\sqrt{(3^2x^4)(2xy)} = 5 \cdot 3x^2\sqrt{2xy} = 15x^2\sqrt{2xy}$$

$$= 7x^2\sqrt{2xy} + 15x^2\sqrt{2xy}$$ The radicals are now like radicals.

$$= (7 + 15)x^2\sqrt{2xy}$$ Apply the distributive property by factoring out $x^2\sqrt{2xy}$.

$$= 22x^2\sqrt{2xy}$$ Simplify.

Skill Practice 9 Add the radicals. Assume that c and d are positive real numbers. $\sqrt{75cd^4} + 6d\sqrt{27cd^2}$

In some cases, when two radicals are added the resulting sum is written in factored form.

EXAMPLE 10 Adding Radicals

Add the radicals. Assume that x represents a positive real number.

$$3\sqrt{5x} + 2x\sqrt{5x}$$

Solution:

$$3\sqrt{5x} + 2x\sqrt{5x}$$ The radicals are like radicals.
$$= (3 + 2x)\sqrt{5x}$$ Apply the distributive property by factoring out $\sqrt{5x}$. The expression cannot be simplified further because the terms within parentheses are not like terms.

Answers

8. a. $9\sqrt{3}$ **b.** $38\sqrt{3} + 4\sqrt{2}$
9. $23d^2\sqrt{3c}$
10. $(8 + 3z)\sqrt{7z}$

Skill Practice 10 Add the radicals. Assume that z is a positive real number. $8\sqrt{7z} + 3z\sqrt{7z}$

Prerequisite Review

For Exercises R.1–R.2, determine the square roots. (See Section R.3 for review.)

R.1. a. Identify the square roots of 100.

b. Simplify $\sqrt{100}$.

R.2. a. Identify the square roots of 36.

b. Simplify $\sqrt{36}$.

For Exercises R.3–R.10, simplify the expressions. (See Section R.3 for review.)

R.3. $\sqrt{225}$

R.4. $\sqrt{121}$

R.5. $\sqrt[3]{\dfrac{1}{125}}$

R.6. $\sqrt[3]{\dfrac{8}{27}}$

R.7. $\sqrt[3]{-1}$

R.8. $\sqrt[3]{-8}$

R.9. a. $|4|$ **b.** $|-4|$

R.10. a. $|-17|$ **b.** $|17|$

For Exercises R.11–R.12, write the expression in an equivalent form without absolute value bars. (See Section 1.6 for review.)

R.11. a. $|x|$ if $x \geq 0$

b. $|x|$ if $x < 0$

R.12. a. $|y + 1|$ if $y + 1 < 0$

b. $|y + 1|$ if $y + 1 \geq 0$

For Exercises R.13–R.18, simplify the expression. (See Section 2.1 for review.)

R.13. $x^8 \cdot x$

R.14. $y^{12} \cdot y^2$

R.15. $\dfrac{y^7}{y^3}$

R.16. $\dfrac{c^2}{c^7}$

R.17. $(d^5)^2$

R.18. $(m^4)^3$

Concept Connections

1. If x represents any real number, the value of $\sqrt[n]{x^n} = x$ if n is (even/odd). The value of $\sqrt[n]{x^n} = |x|$ if n is (even/odd).

2. The product property of radicals indicates that $\sqrt[n]{a} \cdot \sqrt[n]{b} = $ _____ provided that $\sqrt[n]{a}$ and $\sqrt[n]{b}$ are real numbers.

3. The quotient property of radicals indicates that $\dfrac{\sqrt[n]{a}}{\sqrt[n]{b}} = $ ——— provided that $\sqrt[n]{a}$ and $\sqrt[n]{b}$ are real numbers.

4. Explain why the radical $\sqrt{x^3}$ is not in simplified form. Assume that $x \geq 0$.

5. To simplify the radical expression $\sqrt[3]{t^{14}}$ the radicand is rewritten as $\sqrt[3]{\_\_ \cdot t^2}$.

6. Two radical terms are called *like radicals* if they have the same _____ and the same _____.

7. The expression $\sqrt{3x} + \sqrt{3x}$ simplifies to _____.

8. The expression $\sqrt{2} + \sqrt{3}$ (can/cannot) be simplified further, whereas the expression $\sqrt{2} \cdot \sqrt{3}$ (can/cannot) be simplified further.

Objective 1: Simplify Expressions of the Form $\sqrt[n]{x^n}$

For Exercises 9–24, simplify the radical expressions. **(See Example 1)**

9. $\sqrt[3]{(-11)^3}$

10. $\sqrt[5]{(-8)^5}$

11. $\sqrt[4]{(-11)^4}$

12. $\sqrt{(-8)^2}$

13. $\sqrt{y^2}$

14. $\sqrt[4]{y^4}$

15. $\sqrt[3]{y^3}$

16. $\sqrt[5]{y^5}$

17. $\sqrt[4]{(2x - 5)^4}$

18. $\sqrt{(3z + 2)^2}$

19. $\sqrt{w^{12}}$

20. $\sqrt[4]{c^{32}}$

21. $\sqrt[4]{\dfrac{a^4}{b^8}}$

22. $\sqrt{\dfrac{p^6}{q^4}}$

23. $\sqrt[3]{c^3 d^6}$

24. $\sqrt[3]{m^9 n^{12}}$

25. a. For what values of t will the statement be true? $\sqrt{t^2} = t$

    b. For what values of t will the statement be true? $\sqrt{t^2} = |t|$

26. a. For what values of c will the statement be true? $\sqrt[4]{(c + 8)^4} = c + 8$

    b. For what values of c will the statement be true? $\sqrt[4]{(c + 8)^4} = |c + 8|$

Objective 2: Use the Product and Quotient Properties to Simplify Radicals

For Exercises 27–62, simplify the radicals. Assume that all variables represent positive real numbers. **(See Examples 2–6)**

27. a. $\sqrt{c^7}$ **b.** $\sqrt[3]{c^7}$ **c.** $\sqrt[4]{c^7}$ **d.** $\sqrt[9]{c^7}$

28. a. $\sqrt{d^{11}}$ **b.** $\sqrt[3]{d^{11}}$ **c.** $\sqrt[4]{d^{11}}$ **d.** $\sqrt[12]{d^{11}}$

29. a. $\sqrt{24}$ **b.** $\sqrt[3]{24}$

30. a. $\sqrt{54}$ **b.** $\sqrt[3]{54}$

31. $\sqrt{x^5}$ **32.** $\sqrt{p^{15}}$ **33.** $\sqrt{a^5 b^4}$ **34.** $\sqrt{c^9 d^6}$

35. $-\sqrt[4]{x^8 y^{13}}$ **36.** $-\sqrt[4]{p^{16} q^{17}}$ **37.** $\sqrt{28}$ **38.** $\sqrt{63}$

39. $5\sqrt{18}$ **40.** $2\sqrt{24}$ **41.** $\sqrt[3]{54}$ **42.** $\sqrt[3]{250}$

43. $\sqrt{25ab^3}$ **44.** $\sqrt{64m^5 n^{20}}$ **45.** $\sqrt[3]{40x^7}$ **46.** $\sqrt[3]{81y^{17}}$

47. $\sqrt[3]{-16x^6 yz^3}$ **48.** $\sqrt[3]{-192a^6 bc^2}$ **49.** $\sqrt[4]{80w^4 z^7}$ **50.** $\sqrt[4]{32p^8 qr^5}$

51. $\sqrt{\dfrac{x^3}{x}}$ **52.** $\sqrt{\dfrac{y^5}{y}}$ **53.** $\sqrt{\dfrac{p^7}{p^3}}$ **54.** $\sqrt{\dfrac{q^{11}}{q^5}}$

55. $\dfrac{\sqrt{50}}{\sqrt{2}}$ **56.** $\dfrac{\sqrt{98}}{\sqrt{2}}$ **57.** $\dfrac{\sqrt[3]{3}}{\sqrt[3]{24}}$ **58.** $\dfrac{\sqrt[3]{2}}{\sqrt[3]{250}}$

59. $\dfrac{5\sqrt[3]{16}}{6}$ **60.** $\dfrac{7\sqrt{18}}{9}$ **61.** $\dfrac{5\sqrt[3]{72}}{12}$ **62.** $\dfrac{3\sqrt[3]{250}}{10}$

Objective 3: Add and Subtract Radicals

For Exercises 63 and 64, determine if the radicals are like radicals or unlike radicals.

63. a. $\sqrt{2}$ and $\sqrt[3]{2}$ **b.** $\sqrt{2}$ and $3\sqrt{2}$ **c.** $\sqrt{2}$ and $\sqrt{5}$

64. a. $7\sqrt[3]{x}$ and $\sqrt[3]{x}$ **b.** $\sqrt[3]{x}$ and $\sqrt[4]{x}$ **c.** $2\sqrt[4]{x}$ and $x\sqrt[4]{2}$

65. Explain the similarities between the pairs of expressions.

    a. $7\sqrt{5} + 4\sqrt{5}$ and $7x + 4x$

    b. $-2\sqrt{6} - 9\sqrt{3}$ and $-2x - 9y$

66. Explain the similarities between the pairs of expressions.

    a. $-4\sqrt{3} + 5\sqrt{3}$ and $-4z + 5z$

    b. $13\sqrt{7} - 18$ and $13a - 18$

For Exercises 67–84, add or subtract the radical expressions, if possible. Assume that all variables represent positive real numbers. **(See Example 7)**

67. $3\sqrt{5} + 6\sqrt{5}$ **68.** $5\sqrt{a} + 3\sqrt{a}$ **69.** $3\sqrt[3]{tw} - 2\sqrt[3]{tw} + \sqrt[3]{tw}$

70. $6\sqrt[3]{7} - 2\sqrt[3]{7} + \sqrt[3]{7}$ **71.** $6\sqrt{10} - \sqrt{10}$ **72.** $13\sqrt{11} - \sqrt{11}$

73. $\sqrt[4]{3} + 7\sqrt[4]{3} - \sqrt[4]{14}$ **74.** $2\sqrt{11} + 3\sqrt{13} + 5\sqrt{11}$ **75.** $8\sqrt{x} + 2\sqrt{y} - 6\sqrt{x}$

76. $10\sqrt{10} - 8\sqrt{10} + \sqrt{2}$ **77.** $\sqrt[3]{ab} + a\sqrt[3]{b}$ **78.** $x\sqrt[4]{y} - y\sqrt[4]{x}$

79. $\sqrt{2t} + \sqrt[3]{2t}$ **80.** $\sqrt[4]{5c} + \sqrt[3]{5c}$ **81.** $\dfrac{5}{6}z\sqrt[3]{6} + \dfrac{7}{9}z\sqrt[3]{6}$

82. $\dfrac{3}{4}a\sqrt[4]{b} + \dfrac{1}{6}a\sqrt[4]{b}$ **83.** $0.81x\sqrt{y} - 0.11x\sqrt{y}$ **84.** $7.5\sqrt{pq} - 6.3\sqrt{pq}$

85. Explain the process for adding the two radicals. Then find the sum. $3\sqrt{2} + 7\sqrt{50}$

86. Explain the process for subtracting two radicals. Then find the difference. $\sqrt{12x} - \sqrt{75x}$

For Exercises 87–104, add or subtract the radicals. Assume that all variables represent positive real numbers. **(See Examples 8–10)**

87. $2\sqrt{12} + \sqrt{48}$

88. $5\sqrt{32} + 2\sqrt{50}$

89. $5\sqrt{18} + \sqrt{27} - 4\sqrt{50}$

90. $7\sqrt{40} - \sqrt{8} + 4\sqrt{50}$

91. $3\sqrt{2a} - \sqrt{8a} - \sqrt{72a}$

92. $\sqrt{12t} - \sqrt{27t} + 5\sqrt{3t}$

93. $7\sqrt[3]{x^4} - x\sqrt[3]{x}$

94. $6\sqrt[3]{y^{10}} - 3y^2\sqrt[3]{y^4}$

95. $5p\sqrt{20p^2} + p^2\sqrt{80}$

96. $2q\sqrt{48q^2} - \sqrt{27q^4}$

97. $\sqrt[3]{a^2b} - \sqrt[3]{8a^2b}$

98. $w\sqrt{80} - 3\sqrt{125w^2}$

99. $5x\sqrt{x} + 6\sqrt{x}$

100. $9y^2\sqrt{2} + 4\sqrt{2}$

101. $\sqrt{50x^2} - 3\sqrt{8}$

102. $\sqrt{9x^3} - \sqrt{25x}$

103. $11\sqrt[3]{54cd^3} - 2\sqrt[3]{2cd^3} + d\sqrt[3]{16c}$

104. $x\sqrt[3]{64x^5y^2} - x^2\sqrt[3]{x^2y^2} + 5\sqrt[3]{x^8y^2}$

Mixed Exercises

For Exercises 105–112, answer true or false. If an answer is false, explain why or give a counterexample. Assume that all variables represent positive real numbers.

105. $\sqrt{x} + \sqrt{y} = \sqrt{x + y}$

106. $\sqrt{x} + \sqrt{x} = 2\sqrt{x}$

107. $5\sqrt[3]{x} + 2\sqrt[3]{x} = 7\sqrt[3]{x}$

108. $6\sqrt{x} + 5\sqrt[3]{x} = 11\sqrt{x}$

109. $\sqrt{y} + \sqrt{y} = \sqrt{2y}$

110. $\sqrt{c^2 + d^2} = c + d$

111. $2w\sqrt{5} + 4w\sqrt{5} = 6w^2\sqrt{5}$

112. $7x\sqrt{3} - 2\sqrt{3} = (7x - 2)\sqrt{3}$

Expanding Your Skills

For Exercises 113–114, find the exact value of the perimeter, and then approximate the value to one decimal place.

113.

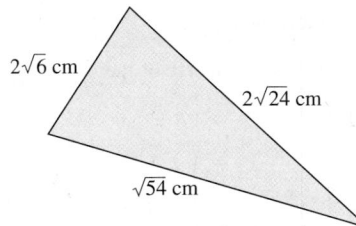

$2\sqrt{6}$ cm

$2\sqrt{24}$ cm

$\sqrt{54}$ cm

114.

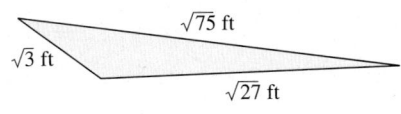

$\sqrt{75}$ ft

$\sqrt{3}$ ft

$\sqrt{27}$ ft

115. The figure has perimeter $14\sqrt{2}$ ft. Find the value of x.

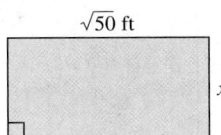

$\sqrt{50}$ ft

x

116. The figure has perimeter $12\sqrt{7}$ cm. Find the value of x.

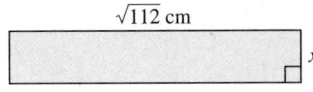

$\sqrt{112}$ cm

x

For Exercises 117–118, evaluate the expression without the use of a calculator.

117. $\sqrt{\dfrac{8.0 \times 10^{12}}{2.0 \times 10^4}}$

118. $\sqrt{\dfrac{1.44 \times 10^{16}}{9.0 \times 10^{10}}}$

For Exercises 119–120, simplify the expression.

119. $\sqrt{\sqrt[3]{6} + \sqrt[4]{16} + \sqrt{\sqrt{25} + \sqrt{16} + \sqrt{9}}}$

120. $\sqrt{\sqrt[4]{11} + \sqrt[3]{125} + \sqrt{\sqrt{81} + \sqrt[3]{1000} + \sqrt{36} + \sqrt{25}}}$

SECTION 3.2   Multiplying Radicals and Rationalizing the Denominator

1. Multiply Radicals

To multiply radical expressions, we use the product property of radicals. That is, if $\sqrt[n]{a}$ and $\sqrt[n]{b}$ are real numbers, then

$$\text{Product property of radicals: } \sqrt[n]{a} \cdot \sqrt[n]{b} = \sqrt[n]{ab}$$

Thus, we can simplify a product of radicals provided they have the same index.

EXAMPLE 1   **Multiplying Single-Term Radicals**

Multiply. Assume that x represents a positive real number.

a. $\sqrt{6} \cdot \sqrt{10}$ b. $5\sqrt{3}(2\sqrt{15})$ c. $(2\sqrt[3]{3x})(4\sqrt[3]{18x^2})$

Solution:

a. $\sqrt{6} \cdot \sqrt{10} = \sqrt{60}$ The radicals have the same index. Multiply the radicands.

$\quad = \sqrt{2^2 \cdot 3 \cdot 5}$ Factor 60. The greatest perfect square in the radicand is 2^2.

$\quad = \sqrt{2^2} \cdot \sqrt{3 \cdot 5}$ Apply the product property of radicals.

$\quad = 2\sqrt{15}$ Simplify.

> **TIP** When multiplying radicals, we have the option of factoring the individual radicands before multiplying. For example:
> $\sqrt{6} \cdot \sqrt{10}$
> $= \sqrt{2 \cdot 3} \cdot \sqrt{2 \cdot 5}$
> $= \sqrt{2^2 \cdot 3 \cdot 5}$

b. $5\sqrt{3}(2\sqrt{15})$ The radicals have the same index. Multiply the radicals.

$\quad = (5 \cdot 2)(\sqrt{3} \cdot \sqrt{15})$ Regroup factors.

$\quad = 10\sqrt{3} \cdot \sqrt{3 \cdot 5}$ Factor the radicands.

$\quad = 10\sqrt{3^2 \cdot 5}$ Apply the product property of radicals.

$\quad = 10 \cdot 3\sqrt{5}$ Simplify.

$\quad = 30\sqrt{5}$

c. $(2\sqrt[3]{3x})(4\sqrt[3]{18x^2})$

$\quad = (2\sqrt[3]{3x})(4\sqrt[3]{3^2 \cdot 2x^2})$ In this example, we factor the radicands before multiplying the radicals. This makes the perfect cubes in the radicand easier to identify.

$\quad = 2 \cdot 4 \cdot \sqrt[3]{3x} \cdot \sqrt[3]{3^2 \cdot 2x^2}$ Use the commutative and associative properties of multiplication to group the coefficients and group the radicals.

$\quad = 8\sqrt[3]{3^3 \cdot 2 \cdot x^3}$ Multiply the radicals.

$\quad = 8 \cdot 3x \cdot \sqrt[3]{2}$ Simplify.

$\quad = 24x\sqrt[3]{2}$

Skill Practice 1   Multiply. Assume that y represents a positive real number.

a. $\sqrt{15} \cdot \sqrt{21}$ b. $(7\sqrt{10})(2\sqrt{5})$ c. $(2\sqrt[3]{5y^2})(4\sqrt[3]{75y})$

When multiplying radical expressions with more than one term, use the distributive property. Every term in the first expression must be multiplied by every term in the second.

Answers
1. a. $3\sqrt{35}$ b. $70\sqrt{2}$
   c. $40y\sqrt[3]{3}$

EXAMPLE 2   **Multiplying Radical Expressions**

Multiply. Assume that x and y represent positive real numbers.

a. $3\sqrt{11}\left(\sqrt{11} + 2\right)$ **b.** $5\sqrt{x}\left(2\sqrt{x} - 3\sqrt{y}\right)$

Solution:

a. $3\sqrt{11}\left(\sqrt{11} + 2\right)$ Apply the distributive property.

$= \left(3\sqrt{11}\right)\left(\sqrt{11}\right) + \left(3\sqrt{11}\right)(2)$

$= 3\sqrt{11^2} + 6\sqrt{11}$ Multiply factors in each term.

$= 3 \cdot 11 + 6\sqrt{11}$ Note that $\sqrt{11} \cdot \sqrt{11} = \sqrt{11^2} = 11$.

$= 33 + 6\sqrt{11}$ Simplify.

b. $5\sqrt{x}\left(2\sqrt{x} - 3\sqrt{y}\right)$

$= \left(5\sqrt{x}\right)\left(2\sqrt{x}\right) + \left(5\sqrt{x}\right)\left(-3\sqrt{y}\right)$ Apply the distributive property.

$= 5 \cdot 2\sqrt{x^2} - 5 \cdot 3\sqrt{xy}$ Regroup factors and multiply.

$= 10x - 15\sqrt{xy}$ Simplify.

FOR REVIEW

Notice the similarity between multiplying polynomials and multiplying radical expressions with multiple terms.

Multiply Polynomials:

$3(x + 2) = 3 \cdot x + 3 \cdot 2$

Multiply Radicals:

$3\sqrt{11}\left(\sqrt{11} + 2\right)$
$= 3\sqrt{11} \cdot \sqrt{11} + 3\sqrt{11} \cdot 2$

Skill Practice 2   Multiply. Assume that a and b represent positive real numbers.

a. $2\sqrt{7}\left(-4\sqrt{7} + 3\right)$ **b.** $-8\sqrt{a}\left(3\sqrt{a} + 2\sqrt{b}\right)$

In Example 3, we practice multiplying two-term radical expressions.

EXAMPLE 3   **Multiplying Radical Expressions**

Multiply. Assume that x and y are positive real numbers.

a. $\left(\sqrt{5} + 3\sqrt{2}\right)\left(2\sqrt{5} - \sqrt{2}\right)$ **b.** $\left(3\sqrt{x} + 5\sqrt{y}\right)\left(2\sqrt{x} - 7\sqrt{y}\right)$

Solution:

a. $\left(\sqrt{5} + 3\sqrt{2}\right)\left(2\sqrt{5} - \sqrt{2}\right)$ Apply the distributive property.

$= \left(\sqrt{5}\right)\left(2\sqrt{5}\right) + \left(\sqrt{5}\right)\left(-\sqrt{2}\right) + \left(3\sqrt{2}\right)\left(2\sqrt{5}\right) + \left(3\sqrt{2}\right)\left(-\sqrt{2}\right)$

$= 2\sqrt{5^2} - \sqrt{10} + 6\sqrt{10} - 3\sqrt{2^2}$ Regroup like factors and multiply.

$= 2 \cdot 5 + 5\sqrt{10} - 3 \cdot 2$ Simplify radicals and combine like radicals.

$= 10 + 5\sqrt{10} - 6$

$= 4 + 5\sqrt{10}$ Combine like terms.

b. $\left(3\sqrt{x} + 5\sqrt{y}\right)\left(2\sqrt{x} - 7\sqrt{y}\right)$ Apply the distributive property.

$= \left(3\sqrt{x}\right)\left(2\sqrt{x}\right) + \left(3\sqrt{x}\right)\left(-7\sqrt{y}\right) + \left(5\sqrt{y}\right)\left(2\sqrt{x}\right) + \left(5\sqrt{y}\right)\left(-7\sqrt{y}\right)$

$= 6\sqrt{x^2} - 21\sqrt{xy} + 10\sqrt{xy} - 35\sqrt{y^2}$

$= 6x - 21\sqrt{xy} + 10\sqrt{xy} - 35y$ Since x and y are both positive, $\sqrt{x^2} = x$ and $\sqrt{y^2} = y$.

$= 6x - 11\sqrt{xy} - 35y$ The terms $-21\sqrt{xy}$ and $10\sqrt{xy}$ are like radicals and can be combined.

Answers

2. a. $-56 + 6\sqrt{7}$

   b. $-24a - 16\sqrt{ab}$

> **Skill Practice 3** Multiply. Assume that t and s are positive real numbers.
>
> **a.** $\left(2\sqrt{3} - 3\sqrt{10}\right)\left(\sqrt{3} + 2\sqrt{10}\right)$ **b.** $\left(4\sqrt{s} + 10\sqrt{t}\right)\left(3\sqrt{s} + 8\sqrt{t}\right)$

FOR REVIEW

Recall that when multiplying polynomials, we have two special cases: (1) multiplying conjugates, and (2) squaring a binomial. Let's take a minute to review the patterns presented by these cases. For an in-depth discussion, refer to Section 2.2.

1. The product of conjugates: $(a + b)(a - b) = a^2 - b^2$

2. The square of a binomial: $(a + b)^2 = a^2 + 2ab + b^2$
$$(a - b)^2 = a^2 - 2ab + b^2$$

In Example 4, we have the product of conjugates and the square of a binomial containing radical expressions.

EXAMPLE 4 Multiplying Radical Expressions

Multiply and simplify.

a. $\left(4\sqrt{5} + \sqrt{6}\right)\left(4\sqrt{5} - \sqrt{6}\right)$ **b.** $\left(3x + \sqrt{2}\right)^2$

Solution:

a. $\left(4\sqrt{5} + \sqrt{6}\right)\left(4\sqrt{5} - \sqrt{6}\right)$ — This is a product of conjugates $(a + b)(a - b)$ where $a = 4\sqrt{5}$ and $b = \sqrt{6}$.

$= \left(4\sqrt{5}\right)^2 - \left(\sqrt{6}\right)^2$ — Apply the pattern $(a + b)(a - b) = a^2 - b^2$.

$= 16\sqrt{25} - \sqrt{36}$ — Simplify. Note that $\left(4\sqrt{5}\right)^2 = \left(4\sqrt{5}\right)\left(4\sqrt{5}\right) = 16\sqrt{25}$ and $\left(\sqrt{6}\right)^2 = \left(\sqrt{6}\right)\left(\sqrt{6}\right) = \sqrt{36}$.

$= 16 \cdot 5 - 6$

$= 80 - 6$

$= 74$

b. $\left(3x + \sqrt{2}\right)^2$ — This is the square of a binomial where $a = 3x$ and $b = \sqrt{2}$.

$= \left(3x\right)^2 + 2\left(3x\right)\left(\sqrt{2}\right) + \left(\sqrt{2}\right)^2$ — Apply the pattern $(a + b)^2 = a^2 + 2ab + b^2$.

$= 9x^2 + 6x\sqrt{2} + \sqrt{4}$ — Simplify.

$= 9x^2 + 6\sqrt{2}x + 2$

> **Skill Practice 4** Multiply and simplify.
>
> **a.** $\left(3\sqrt{7} - \sqrt{5}\right)\left(3\sqrt{7} + \sqrt{5}\right)$ **b.** $\left(4y - \sqrt{3}\right)^2$

---

2. Rationalize the Denominator—One Term

Consider the expression $\dfrac{1}{\sqrt{2}}$. In decimal form, this is approximately $\dfrac{1}{1.414}$, and performing the division would be very cumbersome. However, we can show that $\dfrac{1}{\sqrt{2}}$ is equivalent to $\dfrac{\sqrt{2}}{2}$, which in decimal form is approximately $\dfrac{1.414}{2}$. This expression

Answers

3. a. $-54 + \sqrt{30}$
 b. $12s + 62\sqrt{st} + 80t$
4. a. 58 **b.** $16y^2 - 8\sqrt{3}y + 3$

TIP Rationalizing the denominator gets its name because we convert the denominator from an irrational number to a rational number.

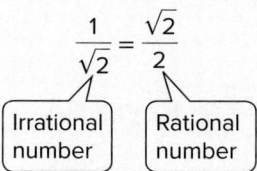

$$\frac{1}{\sqrt{2}} = \frac{\sqrt{2}}{2}$$

Irrational number   Rational number

has a divisor of 2, which makes the computation easier. By inspection, this can be estimated as approximately 0.707.

Dividing a numerical value by a radical is more difficult than dividing a radical by a whole number. For this reason, whenever possible the convention is to simplify radicals by removing the radical from the denominator of a fraction. This is called **rationalizing the denominator**. We will rationalize the denominator in two cases, one where the denominator has a single radical term and the other where the denominator has two terms with one or more square roots.

EXAMPLE 5   **Rationalizing the Denominator—One Term**

Simplify. Assume that a is a positive real number.

   a. $\dfrac{1}{\sqrt{2}}$ **b.** $\sqrt[3]{\dfrac{64}{a}}$

Solution:

a. $\dfrac{1}{\sqrt{2}} = \dfrac{1}{\sqrt{2}} \cdot \dfrac{\sqrt{2}}{\sqrt{2}}$ Multiply the original expression by $\dfrac{\sqrt{2}}{\sqrt{2}}$. We choose this convenient ratio of 1 because the product in the denominator leaves a square root of a perfect square.

   $= \dfrac{\sqrt{2}}{\sqrt{2^2}}$ The denominator is now the square root of a perfect square and simplifies without a radical.

   $= \dfrac{\sqrt{2}}{2}$ Simplify. The denominator simplifies as $\sqrt{4}$, which is 2.

b. $\sqrt[3]{\dfrac{64}{a}} = \dfrac{\sqrt[3]{64}}{\sqrt[3]{a}}$ Apply the quotient property of radicals to write the fraction as the quotient of two radicals.

   $= \dfrac{4}{\sqrt[3]{a}}$ Simplify the radical in the numerator.

   $= \dfrac{4}{\sqrt[3]{a}} \cdot \dfrac{\sqrt[3]{a^2}}{\sqrt[3]{a^2}}$ Multiply by $\dfrac{\sqrt[3]{a^2}}{\sqrt[3]{a^2}}$. We choose this convenient ratio of 1 because the product of a and a^2 makes a^3, which is a perfect cube.

   $= \dfrac{4\sqrt[3]{a^2}}{\sqrt[3]{a^3}}$ The denominator is now the cube root of a perfect cube and simplifies without a radical.

   $= \dfrac{4\sqrt[3]{a^2}}{a}$ Simplify. The denominator simplifies as $\sqrt[3]{a^3} = a$.

Avoiding Mistakes

Do not try to "cancel" $\sqrt{2}$ in the numerator with 2 in the denominator. These values are not equivalent, and therefore their ratio is not 1.

In a similar way, given $\dfrac{\sqrt{9}}{9}$, it would be incorrect to try to "cancel" the 9's. That is: $\dfrac{\sqrt{9}}{9} \neq 1$

Instead, $\dfrac{\sqrt{9}}{9} = \dfrac{3}{9} = \dfrac{1}{3}$.

Skill Practice 5   Multiply. Assume that b is a positive real number.

   a. $\dfrac{2}{\sqrt{3}}$ **b.** $\sqrt[4]{\dfrac{16}{b}}$

---

3. Rationalize the Denominator—Two Terms

In Example 6, we will demonstrate how to rationalize a two-term denominator involving square roots.

  First recall from the multiplication of polynomials that the product of two conjugates results in a difference of squares.

$$(a + b)(a - b) = a^2 - b^2$$

Answers

5. a. $\dfrac{2\sqrt{3}}{3}$ **b.** $\dfrac{2\sqrt[4]{b^3}}{b}$

If either a or b has a square root factor, the expression will simplify without a radical. That is, the expression is *rationalized*. For example:

$$\left(\sqrt{5} - \sqrt{3}\right)\left(\sqrt{5} + \sqrt{3}\right) = \left(\sqrt{5}\right)^2 - \left(\sqrt{3}\right)^2$$
$$= \sqrt{25} - \sqrt{9}$$
$$= 5 - 3$$
$$= 2$$

In Example 6, we want to simplify the expression $\dfrac{-2}{2 + \sqrt{6}}$ by eliminating the radical in the denominator. Therefore, we need to multiply the denominator by its conjugate. Of course, this means that we must also multiply the numerator by the same value.

EXAMPLE 6   Rationalizing the Denominator—Two Terms

Simplify the expression by rationalizing the denominator. $\dfrac{-2}{2 + \sqrt{6}}$

Solution:

$$\frac{-2}{2 + \sqrt{6}}$$

$$= \frac{(-2)}{\left(2 + \sqrt{6}\right)} \cdot \frac{\left(2 - \sqrt{6}\right)}{\left(2 - \sqrt{6}\right)}$$ Multiply the numerator and denominator by the conjugate of the denominator.

conjugates

$$= \frac{-2\left(2 - \sqrt{6}\right)}{(2)^2 - \left(\sqrt{6}\right)^2}$$ In the denominator, apply the formula $(a + b)(a - b) = a^2 - b^2$.

$$= \frac{-2\left(2 - \sqrt{6}\right)}{4 - 6}$$ Simplify.

$$= \frac{-2\left(2 - \sqrt{6}\right)}{-2}$$ The numerator and denominator share a common factor of -2.

$$= \frac{\overset{1}{-2}\left(2 - \sqrt{6}\right)}{-2}$$ Simplify the fraction to lowest terms.

$$= 2 - \sqrt{6}$$

> **TIP** To simplify a fraction, factor the numerator and denominator and then "cancel" common factors whose ratio is 1. For example,
>
> $$\frac{4x + 12}{8} = \frac{4(x + 3)}{8}$$
>
> $$= \frac{\overset{1}{4}(x + 3)}{8_2}$$
>
> $$= \frac{x + 3}{2}$$

Skill Practice 6   Simplify by rationalizing the denominator.

$$\frac{8}{3 + \sqrt{5}}$$

SECTION 3.2   Practice Exercises

Prerequisite Review

For Exercises R.1–R.10, multiply the polynomials. (See Section 2.2 for review.)

R.1. $\dfrac{1}{3}t(6t + 5)$

R.2. $-\dfrac{1}{4}y(9y - 1)$

R.3. $(2p - 5)(9p - 7)$

R.4. $(t - 3)(7t + 8)$

R.5. $(-2k + 5)(3k^2 - 2k + 9)$

R.6. $(-3y - 1)(5y^2 + y - 10)$

R.7. $(3d - 8)^2$

R.8. $(5x + 11y)^2$

R.9. $(5m - 7n)(5m + 7n)$

R.10. $(12a + b)(12a - b)$

For Exercises R.11–R.12, fill in the box to make a true statement.

R.11. $\sqrt[3]{x} \cdot \sqrt[3]{\square} = \sqrt[3]{x^3}$

R.12. $\sqrt[5]{x^2} \cdot \sqrt[5]{\square} = \sqrt[5]{x^5}$

For Exercises R.13–R.18, simplify the expressions to lowest terms.

R.13. $\dfrac{12y^4}{2y}$

R.14. $\dfrac{24x}{36x^3}$

R.15. $\dfrac{4x + 8}{8}$

R.16. $\dfrac{3y - 12}{6}$

R.17. $\dfrac{6 - 20\sqrt{7}}{2}$

R.18. $\dfrac{8 - 4\sqrt{2}}{4}$

Concept Connections

1. If $x \geq 0$, then $\sqrt{x} \cdot \sqrt{x} = $ _____.

2. Two binomials $(x + \sqrt{2})$ and $(x - \sqrt{2})$ are called _____ of each other, and their product is $(x)^2 - (\sqrt{2})^2$.

3. If $m \geq 0$ and $n \geq 0$, then $(\sqrt{m} + \sqrt{n})(\sqrt{m} - \sqrt{n}) = $ _____.

4. Which is the correct simplification of $(\sqrt{c} + 4)^2$? $c + 16$ or $c + 8\sqrt{c} + 16$

5. The process of removing a radical from the denominator of a fraction is called _____ the denominator.

6. The expression $\dfrac{\sqrt{3}}{3}$ (is / is not) in simplified form, whereas $\dfrac{3}{\sqrt{3}}$ (is / is not) in simplified form.

7. To rationalize the denominator of the expression $\dfrac{\sqrt{x} + 3}{\sqrt{x} + 2}$, multiply the numerator and denominator by _____.

8. To rationalize the denominator of the expression $\dfrac{1}{\sqrt[4]{y}}$, multiply the numerator and denominator by _____.

Objective 1: Multiply Radicals

For Exercises 9–34, multiply the radicals. Assume that all variables represent positive real numbers. **(See Examples 1–3)**

9. $\sqrt{5} \cdot \sqrt{3}$

10. $\sqrt[3]{7} \cdot \sqrt[3]{3}$

11. $\sqrt{2} \cdot \sqrt{10}$

12. $\sqrt{7} \cdot \sqrt{14}$

13. $(-2\sqrt{6})(4\sqrt{3})$

14. $5\sqrt{15}(2\sqrt{20})$

15. $(3\sqrt[3]{a})(-5\sqrt[3]{a^2})$

16. $(-7\sqrt[5]{x^3})(2\sqrt[5]{x^2})$

17. $\sqrt[3]{5ab^2} \cdot \sqrt[3]{25a^2}$

18. $\sqrt[3]{2x^2y^2} \cdot \sqrt[3]{4xy^2}$

19. $(-4\sqrt[4]{12m^2})(3\sqrt[4]{4m^2})$

20. $(7\sqrt[3]{6a^2})(3\sqrt[3]{9a})$

21. $\sqrt{3}(4\sqrt{3} - 6)$

22. $3\sqrt{5}(2\sqrt{5} + 4)$

23. $8\sqrt{10}(\sqrt{2} + 2\sqrt{6})$

24. $3\sqrt{14}(2\sqrt{21} - 3\sqrt{7})$

25. $-7\sqrt{a}(6\sqrt{a} - 4\sqrt{b})$

26. $-5\sqrt{c}(2\sqrt{c} + 6\sqrt{d})$

27. $(\sqrt{3} + 2\sqrt{10})(4\sqrt{3} - \sqrt{10})$

28. $(8\sqrt{7} - \sqrt{5})(\sqrt{7} + 3\sqrt{5})$

29. $(\sqrt{x} + 4)(\sqrt{x} - 9)$

30. $(\sqrt{w} - 2)(\sqrt{w} - 9)$

31. $(\sqrt{a} - 3\sqrt{b})(9\sqrt{a} - \sqrt{b})$

32. $(11\sqrt{m} + 4\sqrt{n})(\sqrt{m} + \sqrt{n})$

33. $(\sqrt{p} + 2\sqrt{q})(8 + 3\sqrt{p} - \sqrt{q})$

34. $(5\sqrt{s} - \sqrt{t})(\sqrt{s} + 5 + 6\sqrt{t})$

For Exercises 35–48, multiply the expressions. Assume that all variables represent positive real numbers. **(See Example 4)**

35. a. $(3 + x)(3 - x)$
   b. $(\sqrt{3} + x)(\sqrt{3} - x)$

36. a. $(y + 6)(y - 6)$
   b. $(y + \sqrt{6})(y - \sqrt{6})$

37. a. $(p - 7)^2$
   b. $(\sqrt{p} - \sqrt{7})^2$

38. a. $(q + 2)^2$
   b. $(\sqrt{q} + \sqrt{2})^2$

39. a. $(5c + 2d)^2$
   b. $(5\sqrt{c} + 2\sqrt{d})^2$

40. a. $(2a - 3b)^2$
   b. $(2\sqrt{a} - 3\sqrt{b})^2$

41. a. $(7x + 4y)(7x - 4y)$
   b. $(7\sqrt{3} + 4\sqrt{2})(7\sqrt{3} - 4\sqrt{2})$

42. a. $(4s + 5t)(4s - 5t)$
   b. $(4\sqrt{7} + 5\sqrt{3})(4\sqrt{7} - 5\sqrt{3})$

43. $(\sqrt{5a} - \sqrt{3b})^2$

44. $(\sqrt{3w} + \sqrt{7z})^2$

45. $(\sqrt{x + 1} + 5)(\sqrt{x + 1} - 5)$

46. $(\sqrt{y - 3} - 4)(\sqrt{y - 3} + 4)$

47. $(\sqrt[3]{x} - 2)(\sqrt[3]{x^2} + 2\sqrt[3]{x} + 4)$

48. $(\sqrt[3]{y} + 3)(\sqrt[3]{y^2} - 3\sqrt[3]{y} + 9)$

Objective 2: Rationalize the Denominator—One Term

For Exercises 49–66, rationalize the denominator and simplify. Assume that all variables represent positive real numbers. **(See Example 5)**

49. $\dfrac{1}{\sqrt{3}}$

50. $\dfrac{1}{\sqrt{7}}$

51. $\dfrac{6}{\sqrt{2y}}$

52. $\dfrac{9}{\sqrt{3t}}$

53. $\sqrt{\dfrac{a^3}{2}}$

54. $\sqrt{\dfrac{b^3}{3}}$

55. $\dfrac{6}{\sqrt{8}}$

56. $\dfrac{2}{\sqrt{48}}$

57. $\dfrac{3}{\sqrt[3]{2}}$

58. $\dfrac{2}{\sqrt[3]{7}}$

59. $\dfrac{-6}{\sqrt[4]{x}}$

60. $\dfrac{-2}{\sqrt[5]{y}}$

61. $\dfrac{7}{\sqrt[3]{4}}$

62. $\dfrac{1}{\sqrt[3]{9}}$

63. $\sqrt[3]{\dfrac{4}{w^2}}$

64. $\sqrt[3]{\dfrac{5}{z^2}}$

65. $\dfrac{2}{\sqrt[3]{4x^2}}$

66. $\dfrac{6}{\sqrt[3]{3y^2}}$

Objective 3: Rationalize the Denominator—Two Terms

67. What is the conjugate of $\sqrt{2} - \sqrt{6}$?

68. What is the conjugate of $\sqrt{11} + \sqrt{5}$?

69. What is the conjugate of $\sqrt{x} + 23$?

70. What is the conjugate of $17 - \sqrt{y}$?

For Exercises 71–80, rationalize the denominator and simplify. Assume that all variables represent positive real numbers. **(See Example 6)**

71. $\dfrac{4}{\sqrt{2} + 3}$

72. $\dfrac{6}{4 - \sqrt{3}}$

73. $\dfrac{8}{\sqrt{6} - 2}$

74. $\dfrac{-12}{\sqrt{5} - 3}$

75. $\dfrac{\sqrt{7}}{\sqrt{3} + 2}$

76. $\dfrac{\sqrt{8}}{\sqrt{3} + 1}$

77. $\dfrac{-1}{\sqrt{p} + \sqrt{q}}$

78. $\dfrac{6}{\sqrt{a} - \sqrt{b}}$

79. $\dfrac{\sqrt{w} + 2}{9 - \sqrt{w}}$

80. $\dfrac{10 - \sqrt{t}}{\sqrt{t} - 6}$

Mixed Exercises

For Exercises 81–88, identify the statement as true or false. If a statement is false, explain why. Assume that all variables represent positive real numbers.

81. $\sqrt{3} \cdot \sqrt{2} = \sqrt{6}$

82. $\sqrt{5} \cdot \sqrt[3]{2} = \sqrt{10}$

83. $(x - \sqrt{5})^2 = x - 5$

84. $3(2\sqrt{5x}) = 6\sqrt{5x}$

85. $5(3\sqrt{4x}) = 15\sqrt{20x}$

86. $\dfrac{\sqrt{5x}}{5} = \sqrt{x}$

87. $\dfrac{3\sqrt{x}}{3} = \sqrt{x}$

88. $(\sqrt{t} - 1)(\sqrt{t} + 1) = t - 1$

For Exercises 89–96, perform the indicated operations. Assume that all variable expressions represent positive real numbers.

89. $\left(-\sqrt{6x}\right)^2$

90. $\left(-\sqrt{8a}\right)^2$

91. $\left(\sqrt{3x+1}\right)^2$

92. $\left(\sqrt{x-1}\right)^2$

93. $\left(\sqrt{x+3}-4\right)^2$

94. $\left(\sqrt{x+1}+3\right)^2$

95. $\left(\sqrt{2t-3}+5\right)^2$

96. $\left(\sqrt{3w-2}-4\right)^2$

For Exercises 97–100, find the exact area. Use the geometry formulas at the end of the text.

97.

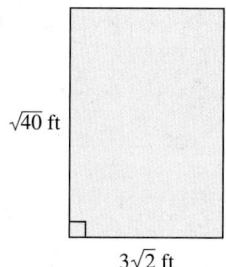

$\sqrt{40}$ ft

$3\sqrt{2}$ ft

98.

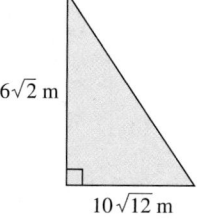

$6\sqrt{2}$ m

$10\sqrt{12}$ m

99.

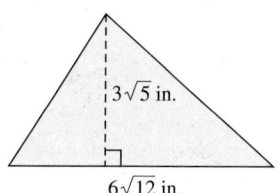

$3\sqrt{5}$ in.

$6\sqrt{12}$ in.

100.

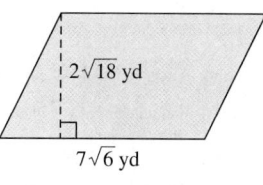

$2\sqrt{18}$ yd

$7\sqrt{6}$ yd

For Exercises 101–104, rationalize the denominator. Assume that all variable expressions represent positive real numbers.

101. a. $\dfrac{1}{\sqrt{2}}$

b. $\dfrac{1}{\sqrt[3]{2}}$

102. a. $\dfrac{1}{\sqrt[3]{x}}$

b. $\dfrac{1}{\sqrt[3]{x^2}}$

103. a. $\dfrac{1}{\sqrt{5a}}$

b. $\dfrac{1}{\sqrt{5}+a}$

104. a. $\dfrac{4}{\sqrt{2x}}$

b. $\dfrac{4}{2-\sqrt{x}}$

Expanding Your Skills

For Exercises 105–110, rationalize the *numerator* by multiplying both numerator and denominator by the conjugate of the denominator. Assume that all variable expressions represent positive real numbers.

105. $\dfrac{\sqrt{3}+6}{2}$

106. $\dfrac{\sqrt{7}-2}{5}$

107. $\dfrac{\sqrt{a}-\sqrt{b}}{\sqrt{a}+\sqrt{b}}$

108. $\dfrac{\sqrt{p}+\sqrt{q}}{\sqrt{p}-\sqrt{q}}$

109. $\dfrac{\sqrt{4+5h}-2}{h}$

110. $\dfrac{\sqrt{9+4h}-3}{h}$

111. The numbers 1, 2, 4, 5, 10, and 20 are natural numbers that are factors of 20. There are other factors of 20 within the set of rational numbers and the set of irrational numbers. For example:

a. Show that $\dfrac{14}{3}$ and $\dfrac{30}{7}$ are factors of 20 over the set of rational numbers.

b. Show that $\left(5-\sqrt{5}\right)$ and $\left(5+\sqrt{5}\right)$ are factors of 20 over the set of irrational numbers.

112. a. Show that $\dfrac{15}{2}$ and $\dfrac{4}{5}$ are factors of 6 over the set of rational numbers.

b. Show that $\left(3-\sqrt{3}\right)$ and $\left(3+\sqrt{3}\right)$ are factors of 6 over the set of irrational numbers.

PROBLEM RECOGNITION EXERCISES

Operations on Radicals

For Exercises 1–30, simplify each expression. Assume that all variables represent positive real numbers.

1. a. $\left(\sqrt{3}\right)\left(\sqrt{6}\right)$

b. $\sqrt{3}+\sqrt{6}$

c. $\dfrac{\sqrt{6}}{\sqrt{3}}$

2. a. $\dfrac{\sqrt{14}}{\sqrt{2}}$

b. $\left(\sqrt{2}\right)\left(\sqrt{14}\right)$

c. $\sqrt{2}+\sqrt{14}$

3. a. $\left(3\sqrt{z}\right)^2$

b. $\left(3+\sqrt{z}\right)^2$

c. $\left(3+\sqrt{z}\right)\left(3-\sqrt{z}\right)$

4. a. $(4 - \sqrt{x})^2$
  b. $(4 - \sqrt{x})(4 + \sqrt{x})$
  c. $(4\sqrt{x})^2$

5. a. $\dfrac{12}{\sqrt{2x}}$
  b. $\sqrt{\dfrac{12}{2x}}$
  c. $\dfrac{12}{\sqrt{2} + x}$

6. a. $\dfrac{15}{3 - \sqrt{y}}$
  b. $\dfrac{15}{\sqrt{3y}}$
  c. $\sqrt{\dfrac{15}{3y}}$

7. a. $(2\sqrt{5} + 1) + (\sqrt{5} - 2)$
  b. $(2\sqrt{5} + 1)(\sqrt{5} - 2)$
  c. $2\sqrt{5}(\sqrt{5} - 2)$

8. a. $(4\sqrt{3} - 5)(\sqrt{3} + 4)$
  b. $4\sqrt{3}(\sqrt{3} + 4)$
  c. $(4\sqrt{3} - 5) - (\sqrt{3} + 4)$

9. a. $\sqrt{16a^{15}}$
  b. $\sqrt[3]{16a^{15}}$

10. a. $\sqrt[3]{27y^9}$
  b. $\sqrt{27y^9}$

11. a. $\sqrt{24}$
  b. $\sqrt[3]{24}$

12. a. $\sqrt{54}$
  b. $\sqrt[3]{54}$

13. a. $\sqrt{200y^6}$
  b. $\sqrt[3]{200y^6}$

14. a. $\sqrt{32z^{15}}$
  b. $\sqrt[3]{32z^{15}}$

15. a. $\sqrt{80}$
  b. $\sqrt[3]{80}$
  c. $\sqrt[4]{80}$

16. a. $\sqrt{48}$
  b. $\sqrt[3]{48}$
  c. $\sqrt[4]{48}$

17. a. $\sqrt{x^5y^6}$
  b. $\sqrt[3]{x^5y^6}$
  c. $\sqrt[4]{x^5y^6}$

18. a. $\sqrt{a^{10}b^9}$
  b. $\sqrt[3]{a^{10}b^9}$
  c. $\sqrt[4]{a^{10}b^9}$

19. a. $\sqrt[3]{32s^5t^6}$
  b. $\sqrt[4]{32s^5t^6}$
  c. $\sqrt[5]{32s^5t^6}$

20. a. $\sqrt[3]{96v^7w^{20}}$
  b. $\sqrt[4]{96v^7w^{20}}$
  c. $\sqrt[5]{96v^7w^{20}}$

21. a. $\sqrt{5} + \sqrt{5}$
  b. $\sqrt{5} \cdot \sqrt{5}$

22. a. $\sqrt{10} + \sqrt{10}$
  b. $\sqrt{10} \cdot \sqrt{10}$

23. a. $2\sqrt{6} - 5\sqrt{6}$
  b. $2\sqrt{6} \cdot 5\sqrt{6}$

24. a. $3\sqrt{7} - 10\sqrt{7}$
  b. $3\sqrt{7} \cdot 10\sqrt{7}$

25. a. $\sqrt{8} + \sqrt{2}$
  b. $\sqrt{8} \cdot \sqrt{2}$

26. a. $\sqrt{12} + \sqrt{3}$
  b. $\sqrt{12} \cdot \sqrt{3}$

27. a. $5\sqrt{18} - 4\sqrt{8}$
  b. $5\sqrt{18} \cdot 4\sqrt{8}$

28. a. $\sqrt{50} - \sqrt{72}$
  b. $\sqrt{50} \cdot \sqrt{72}$

29. a. $4\sqrt[3]{24} + 6\sqrt[3]{3}$
  b. $4\sqrt[3]{24} \cdot 6\sqrt[3]{3}$

30. a. $2\sqrt[3]{54} - 5\sqrt[3]{2}$
  b. $2\sqrt[3]{54} \cdot 5\sqrt[3]{2}$

SECTION 3.3 Complex Numbers

OBJECTIVES

1. Simplify Complex Numbers in Terms of i
2. Write Complex Numbers in the Form $a + bi$
3. Simplify Powers of i
4. Perform Operations on Complex Numbers

1. Simplify Complex Numbers in Terms of i

A solution to an equation is a value of the variable that makes the right side of the equation equal to the left side of the equation. For example,

Equation:	Solution(s):	Check:	
$2x + 1 = 7$	3	$2(3) + 1 = 7$	
		$6 + 1 = 7$ ✓	
$\lvert x - 4 \rvert = 6$	10 and -2	$\lvert (10) - 4 \rvert = 6$	$\lvert (-2) - 4 \rvert = 6$
		$\lvert 6 \rvert = 6$ ✓	$\lvert -6 \rvert = 6$ ✓

Given the equation $x^2 = 1$, the solutions are 1 and -1 because $(1)^2 = 1$ and $(-1)^2 = 1$. However, what about the equation $x^2 = -1$? There is no real number that when squared equals a negative number. For this reason, mathematicians defined a new number i such that $i^2 = -1$. The number i is called an imaginary number and is used to represent $\sqrt{-1}$. Furthermore, the square root of any negative real number is an imaginary number that can be expressed in terms of i.

> **The Imaginary Number *i***
>
> - $i = \sqrt{-1}$ and $i^2 = -1$
> - If b is a positive real number, then $\sqrt{-b} = i\sqrt{b}$.

EXAMPLE 1   Writing Numbers in Terms of *i*

Write each expression in terms of *i*.

a. $\sqrt{-25}$ **b.** $\sqrt{-12}$ **c.** $\sqrt{-13}$

Solution:

a. $\sqrt{-25} = i\sqrt{25} = 5i$ For a real number $b > 0$, $\sqrt{-b} = i\sqrt{b}$.

b. $\sqrt{-12} = i\sqrt{12}$ For a real number $b > 0$, $\sqrt{-b} = i\sqrt{b}$.

$\quad\quad = i\sqrt{2^2 \cdot 3}$ Simplify $\sqrt{12}$ by first factoring the radicand.

$\quad\quad = i\sqrt{2^2} \cdot \sqrt{3}$ Apply the product property of radicals.

$\quad\quad = 2i\sqrt{3}$ or $2\sqrt{3}i$

c. $\sqrt{-13} = i\sqrt{13}$ or $\sqrt{13}i$ For a real number $b > 0$, $\sqrt{-b} = i\sqrt{b}$.

Skill Practice 1   Write each expression in terms of *i*.

a. $\sqrt{-81}$ **b.** $\sqrt{-50}$ **c.** $\sqrt{-11}$

The product and quotient properties of radicals apply only if $\sqrt[n]{a}$ and $\sqrt[n]{b}$ are both real numbers. That is,

$\sqrt[n]{a} \cdot \sqrt[n]{b} = \sqrt[n]{ab}$ provided that $\sqrt[n]{a}$ and $\sqrt[n]{b}$ are real numbers.

$\dfrac{\sqrt[n]{a}}{\sqrt[n]{b}} = \sqrt[n]{\dfrac{a}{b}}$ provided that $\sqrt[n]{a}$ and $\sqrt[n]{b}$ are real numbers and $b \neq 0$.

Therefore, in Example 2, it is important to write the radical expressions in terms of *i* first before applying the product and quotient properties of radicals.

EXAMPLE 2   Simplifying Expressions in Terms of *i*

Multiply or divide as indicated.

a. $\sqrt{-9} \cdot \sqrt{-25}$ **b.** $\sqrt{-15} \cdot \sqrt{-3}$ **c.** $\dfrac{\sqrt{-50}}{\sqrt{-2}}$

Solution:

a. $\sqrt{-9} \cdot \sqrt{-25} = i\sqrt{9} \cdot i\sqrt{25}$ Write each radical in terms of *i* first, *before* multiplying.

$\quad\quad\quad\quad\quad\quad\quad = 3i \cdot 5i$ Simplify the radicals.

$\quad\quad\quad\quad\quad\quad\quad = 15i^2$ Multiply.

$\quad\quad\quad\quad\quad\quad\quad = 15(-1)$ By definition, $i^2 = -1$.

$\quad\quad\quad\quad\quad\quad\quad = -15$

Answers

1. a. $9i$ **b.** $5i\sqrt{2}$ or $5\sqrt{2}i$
c. $i\sqrt{11}$ or $\sqrt{11}i$

b. $\sqrt{-15} \cdot \sqrt{-3} = i\sqrt{15} \cdot i\sqrt{3}$ Write each radical in terms of i first, *before* multiplying.

$$= i^2\sqrt{45}$$ Apply the multiplication property of radicals.

$$= (-1)\sqrt{3^2 \cdot 5}$$ Simplify $i^2 = -1$.

$$= -3\sqrt{5}$$

c. $\dfrac{\sqrt{-50}}{\sqrt{-2}} = \dfrac{i\sqrt{50}}{i\sqrt{2}}$ Write each radical in terms of i first, *before* dividing.

$$= \dfrac{\overset{1}{i}\sqrt{50}}{\underset{1}{i}\sqrt{2}}$$ Simplify the ratio of common factors to 1.

$$= \sqrt{\dfrac{50}{2}}$$ Apply the division property of radicals.

$$= \sqrt{25} = 5$$ Simplify.

Skill Practice 2   Multiply or divide as indicated.

a. $\sqrt{-16} \cdot \sqrt{-49}$ **b.** $\sqrt{-10} \cdot \sqrt{-2}$ **c.** $\dfrac{\sqrt{-48}}{\sqrt{-3}}$

---

2. Write Complex Numbers in the Form $a + bi$

We now define a new set of numbers that includes the real numbers and imaginary numbers. This is called the set of **complex numbers**.

Complex Numbers

Given real numbers a and b, a number written in the form $a + bi$ is called a **complex number**. The value a is called the **real part** of the complex number and the value b is called the **imaginary part**.

$$\boxed{\text{Real part: 5}} \quad \boxed{\text{Imaginary part: } -7}$$
$$5 - 7i = 5 + (-7)i$$

Notes	Examples
• If $b = 0$, then $a + bi$ equals the real number a. This tells us that all real numbers are complex numbers.	$4 + 0i$ is generally written as the real number 4.
• If $a = 0$ and $b \neq 0$, then $a + bi$ equals bi, which we say is **pure imaginary**.	The number $0 + 8i$ is a pure imaginary number and is generally written as simply $8i$.

A complex number written in the form $a + bi$ is said to be in **standard form**. That being said, we sometimes write $a - bi$ in place of $a + (-b)i$. Furthermore, a number such as $5 + \sqrt{3}i$ is sometimes written as $5 + i\sqrt{3}$ to emphasize that the factor of i is not under the radical. In Example 3, we practice writing complex numbers in standard form.

Answers

2. a. -28 **b.** $-2\sqrt{5}$ **c.** 4

EXAMPLE 3   **Writing Complex Numbers in Standard Form**

Simplify each expression and write the result in the form $a + bi$.

a. $3 - \sqrt{-100}$ **b.** $\dfrac{2 + 7i}{5}$ **c.** $\dfrac{-6 + \sqrt{-18}}{9}$

Solution:

a. $3 - \sqrt{-100} = 3 - 10i$ Simplify the expression.

$= 3 + (-10)i$ Although $3 + (-10)i$ is written in standard form, $3 - 10i$ is also acceptable.

b. $\dfrac{2 + 7i}{5} = \dfrac{2}{5} + \dfrac{7}{5}i$ Write the fraction as two separate terms.

c. $\dfrac{-6 + \sqrt{-18}}{9} = \dfrac{-6 + 3i\sqrt{2}}{9}$ Simplify the radical.
$\sqrt{-18} = i\sqrt{18} = i\sqrt{3^2 \cdot 2} = 3i\sqrt{2}$

$= \dfrac{-6}{9} + \dfrac{3i\sqrt{2}}{9}$ Write the fraction as two separate terms.

$= -\dfrac{\overset{2}{\cancel{6}}}{\underset{3}{\cancel{9}}} + \dfrac{\overset{1}{\cancel{3}}i\sqrt{2}}{\underset{3}{\cancel{9}}}$ Simplify each fraction by dividing out 3 in the numerator and denominator.

$= -\dfrac{2}{3} + \dfrac{\sqrt{2}}{3}i$ Simplify each fraction and write the result in the form $a + bi$.

Skill Practice 3   Simplify each expression and write the result in the form $a + bi$.

a. $4 + \sqrt{-49}$ **b.** $\dfrac{3 - 8i}{7}$ **c.** $\dfrac{10 + \sqrt{-75}}{20}$

FOR REVIEW

To simplify the expressions in Examples 3(b) and 3(c), use the addition property of fractions in "reverse."

$$\dfrac{a + b}{c} = \dfrac{a}{c} + \dfrac{b}{c}$$

3. Simplify Powers of i

By definition, $i^2 = -1$, but what about other powers of i? Consider the following pattern.

TIP Notice that even powers of i simplify to 1 or −1.

- If the exponent is a multiple of 4, then the expression equals 1.
- If the exponent is even but *not* a multiple of 4, then the expression equals −1.

$$
\begin{aligned}
i^1 &= i & i^1 &= i \\
i^2 &= -1 & i^2 &= -1 \\
i^3 &= i^2 \cdot i = (-1)i = -i & i^3 &= -i \\
i^4 &= i^2 \cdot i^2 = (-1)(-1) = 1 & i^4 &= 1
\end{aligned}
$$
$\left.\right\}$ Pattern: $i, -1, -i, 1$

$$
\begin{aligned}
i^5 &= i^4 \cdot i = (1)i = i & i^5 &= i \\
i^6 &= i^4 \cdot i^2 = (1)(-1) = -1 & i^6 &= -1 \\
i^7 &= i^4 \cdot i^2 \cdot i = (1)(-1)i = -i & i^7 &= -i \\
i^8 &= i^4 \cdot i^4 = (1)(1) = 1 & i^8 &= 1
\end{aligned}
$$
$\left.\right\}$ Pattern repeats: $i, -1, -i, 1, \ldots$

Notice that the fourth powers of i ($i^4, i^8, i^{12}, \ldots$) equal the real number 1. For other powers of i, we can write the expression as a product of a fourth power of i and a factor of i, i^2, or i^3, which equals i, -1, or $-i$, respectively.

Answers

3. a. $4 + 7i$ **b.** $\dfrac{3}{7} + \left(-\dfrac{8}{7}\right)i$

  c. $\dfrac{1}{2} + \dfrac{\sqrt{3}}{4}i$

| EXAMPLE 4 | Simplifying Powers of i |

Simplify.

a. i^{48} **b.** i^{23} **c.** i^{50} **d.** i^{-19}

TIP To simplify i^n, divide the exponent, n, by 4. The remainder is the exponent of the remaining factor of i once the fourth power of i has been extracted.

Example: $i^{50} = i^{48} \cdot i^2 = (1)i^2$

$$\begin{array}{r} 12 \\ 4\overline{)50} \\ 48 \\ \hline 2 \end{array}$$

So $i^{50} = (1) \cdot i^2 = -1$

Solution:

a. $i^{48} = (i^4)^{12} = (1)^{12} = 1$ The exponent 48 is a multiple of 4. Thus, i^{48} is equal to 1.

b. $i^{23} = i^{20} \cdot i^3$
$= (1) \cdot i^3 = -i$ Write i^{23} as a product of the largest fourth power of i and a remaining factor.

c. $i^{50} = i^{48} \cdot i^2$
$= (1)(-1) = -1$

d. $i^{-19} = i^{-20} \cdot i^1$
$= (i^4)^{-5} \cdot i = (1) \cdot i = i$

Skill Practice 4 Simplify.

a. i^{13} **b.** i^{103} **c.** i^{64} **d.** i^{-30}

---

4. Perform Operations on Complex Numbers

To add or subtract complex numbers, add or subtract their real parts, and add or subtract their imaginary parts. That is,

$$(a + bi) + (c + di) = (a + c) + (b + d)i$$
$$(a + bi) - (c + di) = (a - c) + (b - d)i$$

| EXAMPLE 5 | Adding and Subtracting Complex Numbers |

Add or subtract as indicated. Write the answer in the form $a + bi$.

a. $(-2 - 4i) + (5 + 2i)$ **b.** $(3 - i) - (7 - 9i)$

Solution:

a. $(-2 - 4i) + (5 + 2i)$
$= (-2 + 5) + (-4 + 2)i$ Add the real parts and add the imaginary parts.
$= 3 + (-2)i$ Simplify. The answer is written in the form $a + bi$.
or
$= 3 - 2i$ If the imaginary part of a complex number is negative, it is acceptable and customary to write the answer in terms of subtraction. Thus, $3 + (-2)i$ and $3 - 2i$ are both acceptable answers.

b. $(3 - i) - (7 - 9i)$
$= (3 - 1i) - (7 - 9i)$ The term $-i$ is equivalent to $-1i$.
$= (3 - 7) + [-1 - (-9)]i$ Subtract the real parts and subtract the imaginary parts.
$= -4 + 8i$

Skill Practice 5 Add or subtract as indicated. Write the answer in the form $a + bi$.

a. $(8 - 3i) + (5 - 7i)$ **b.** $(3 + 4i) - (-1 - 9i)$

Answers

4. a. i **b.** $-i$ **c.** 1 **d.** -1
5. a. $13 + (-10)i$ or $13 - 10i$
 b. $4 + 13i$

In Examples 6 and 7, we multiply complex numbers using a process similar to multiplying polynomials.

EXAMPLE 6   **Multiplying Complex Numbers**

Multiply. Write the results in the form $a + bi$.

a. $-\dfrac{1}{2}i(4 + 6i)$ **b.** $(-2 + 6i)(4 - 3i)$

Solution:

a. $-\dfrac{1}{2}i(4 + 6i) = \left(-\dfrac{1}{2}i\right)(4) + \left(-\dfrac{1}{2}i\right)(6i)$ Apply the distributive property.

$\qquad\qquad = -2i - 3i^2$ Multiply.

$\qquad\qquad = -2i - 3(-1)$ Recall that $i^2 = -1$.

$\qquad\qquad = 3 - 2i \quad$ or $\quad 3 + (-2)i$ Write the result in the form $a + bi$.

b. $(-2 + 6i)(4 - 3i)$ Apply the distributive property.

$\qquad = -2(4) + (-2)(-3i) + 6i(4) + 6i(-3i)$

$\qquad = -8 + 6i + 24i - 18i^2$

$\qquad = -8 + 30i - 18(-1)$ Recall that $i^2 = -1$.

$\qquad = -8 + 30i + 18$

$\qquad = 10 + 30i$ Write the result in the form $a + bi$.

Skill Practice 6   Multiply. Write the result in the form $a + bi$.

a. $-\dfrac{1}{3}i(9 - 15i)$ **b.** $(-5 + 4i)(3 - i)$

In Example 7, we make use of the special case products:

$$(a \pm b)^2 = a^2 \pm 2ab + b^2 \quad \text{and} \quad (a + b)(a - b) = a^2 - b^2$$

EXAMPLE 7   **Evaluating Special Products with Complex Numbers**

Multiply. Write the results in the form $a + bi$.

a. $(3 + 4i)^2$ **b.** $(5 + 2i)(5 - 2i)$

Solution:

a. $(3 + 4i)^2 = (3)^2 + 2(3)(4i) + (4i)^2$ Apply the property
$\qquad\qquad = 9 + 24i + 16i^2$ $(a + b)^2 = a^2 + 2ab + b^2$.

$\qquad\qquad = 9 + 24i + 16(-1)$

$\qquad\qquad = 9 + 24i - 16$

$\qquad\qquad = -7 + 24i$ Write the result in the form $a + bi$.

b. $(5 + 2i)(5 - 2i) = (5)^2 - (2i)^2$ Apply the property
$\qquad\qquad\qquad = 25 - 4i^2$ $(a + b)(a - b) = a^2 - b^2$.

$\qquad\qquad\qquad = 25 - 4(-1)$

$\qquad\qquad\qquad = 25 + 4$

$\qquad\qquad\qquad = 29 \quad$ or $\quad 29 + 0i$ Write the result in the form $a + bi$.

Skill Practice 7   Multiply. Write the result in the form $a + bi$.

a. $(4 - 7i)^2$ **b.** $(10 - 3i)(10 + 3i)$

Answers

6. a. $-5 + (-3)i$ **b.** $-11 + 17i$

7. a. $-33 + (-56)i$ **b.** 109

In Section 2.2 we noted that the expressions $(a + b)$ and $(a - b)$ are conjugates. Similarly, the expressions $(a + bi)$ and $(a - bi)$ are called **complex conjugates**. Furthermore, as illustrated in Example 7(b), the product of complex conjugates is a real number.

$$(a + bi)(a - bi) = (a)^2 - (bi)^2$$
$$= a^2 - b^2 i^2$$
$$= a^2 - b^2(-1)$$
$$= a^2 + b^2$$

Product of Complex Conjugates

If a and b are real numbers, then $(a + bi)(a - bi) = a^2 + b^2$.

Number	Standard Form	Conjugate	Product
$3 + 7i$	$3 + 7i$	$3 - 7i$	$(3 + 7i)(3 - 7i) = (3)^2 + (7)^2 = 58$
$\sqrt{-5}$	$0 + \sqrt{5}i$	$0 - \sqrt{5}i$	$\left(0 + \sqrt{5}i\right)\left(0 - \sqrt{5}i\right) = (0)^2 + \left(\sqrt{5}\right)^2 = 5$

In Example 8, we demonstrate division of complex numbers such as $\frac{8 + 2i}{3 - 5i}$. The goal is to make the denominator a real number so that the quotient can be written in standard form $a + bi$. This can be accomplished by multiplying the denominator by its complex conjugate. Of course, this means that we must also multiply the numerator by the same quantity.

EXAMPLE 8   Dividing Complex Numbers

Divide. Write the result in the form $a + bi$.

a. $\dfrac{8 + 2i}{3 - 5i}$ **b.** $\dfrac{-2}{5i}$

Solution:

a. $\dfrac{8 + 2i}{3 - 5i} = \dfrac{(8 + 2i) \cdot (3 + 5i)}{(3 - 5i) \cdot (3 + 5i)}$ Multiply numerator and denominator by the conjugate of the denominator.

$= \dfrac{24 + 40i + 6i + 10i^2}{(3)^2 + (5)^2}$ Apply the distributive property in the numerator. Multiply conjugates in the denominator.

$= \dfrac{24 + 46i + 10(-1)}{9 + 25}$ Replace i^2 by -1.

$= \dfrac{14 + 46i}{34}$

$= \dfrac{14}{34} + \dfrac{46}{34}i = \dfrac{7}{17} + \dfrac{23}{17}i$ Write the result in the form $a + bi$.

b. $\dfrac{-2}{5i} = \dfrac{-2 \cdot i}{5i \cdot i}$ In this example, it is sufficient to multiply numerator and denominator by i (rather than by the conjugate $-5i$) to produce a real number in the denominator.

$= \dfrac{-2i}{5i^2} = \dfrac{-2i}{5(-1)} = \dfrac{-2i}{-5} = \dfrac{2}{5}i$

$= 0 + \dfrac{2}{5}i$ Write the result in the form $a + bi$.

Skill Practice 8   Divide. Write the result in the form $a + bi$.

a. $\dfrac{5 + 6i}{2 - 7i}$ **b.** $\dfrac{-7}{10i}$

TECHNOLOGY CONNECTIONS

Operations on Complex Numbers

Most graphing calculators and some scientific calculators can perform operations on complex numbers. A graphing calculator may have two different modes: one for operations over the set of real numbers and one for operations over the set of complex numbers. Choose the "$a + bi$" mode on your calculator. Then evaluate the expressions.

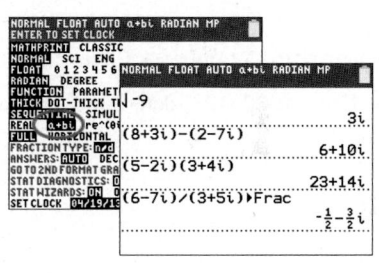

Answers

8. a. $-\dfrac{32}{53} + \dfrac{47}{53}i$ **b.** $0 + \dfrac{7}{10}i$

SECTION 3.3   Practice Exercises

Prerequisite Review

For Exercises R.1–R.6, apply properties of exponents to simplify the expressions. (See Section 2.1 for review.)

R.1. $(y^3)^5$ **R.2.** $(z^6)^3$ **R.3.** $y^{13} \cdot y^6$

R.4. $x^5 \cdot x^{10}$ **R.5.** p^{-7} **R.6.** $\dfrac{1}{m^{-4}}$

For Exercises R.7–R.10, simplify the expression to lowest terms. (See Section R.1 for review.)

R.7. $\dfrac{15}{140}$ **R.8.** $\dfrac{42}{56}$ **R.9.** $\dfrac{4 - 12x}{8}$ **R.10.** $\dfrac{6 + 24y}{12}$

For Exercises, R.11–R.20, perform the indicated operations. (See Section 2.2 for review.)

R.11. $(3x - 4) - (6x + 7)$ **R.12.** $(-5y + 9) - (-3y - 6)$ **R.13.** $-\dfrac{1}{2}x(4x + 12)$

R.14. $\dfrac{3}{4}m(8m + 4)$ **R.15.** $(5p + 3)(-4p + 2)$ **R.16.** $(-d + 9)(-2d + 3)$

R.17. $(2.5 - 0.8y)(2.5 + 0.8y)$ **R.18.** $(1.2 + 0.6t)(1.2 - 0.6t)$ **R.19.** $(10 - 3c)^2$

R.20. $(11 + 2k)^2$

Concept Connections

1. The imaginary number i is defined so that $i = \sqrt{-1}$ and $i^2 = $ _____.

2. For a positive real number b, the value $\sqrt{-b} = $ _____.

3. Given a complex number $a + bi$, the value of a is called the _____ part and the value of b is called the _____ part.

4. Given a complex number $a + bi$, the expression $a - bi$ is called the complex _____.

Objective 1: Simplify Complex Numbers in Terms of i

For Exercises 5–22, write each expression in terms of i and simplify. **(See Examples 1–2)**

5. $\sqrt{-121}$ **6.** $\sqrt{-100}$ **7.** $\sqrt{-98}$ **8.** $\sqrt{-63}$

9. $\sqrt{-19}$ **10.** $\sqrt{-23}$ **11.** $-\sqrt{-16}$ **12.** $-\sqrt{-25}$

13. $\sqrt{-4} \cdot \sqrt{-9}$ **14.** $\sqrt{-1} \cdot \sqrt{-36}$ **15.** $\sqrt{-10} \cdot \sqrt{-5}$ **16.** $\sqrt{-6} \cdot \sqrt{-15}$

17. $\sqrt{-6} \cdot \sqrt{-14}$ **18.** $\sqrt{-10} \cdot \sqrt{-15}$ **19.** $\dfrac{\sqrt{-98}}{\sqrt{-2}}$ **20.** $\dfrac{\sqrt{-45}}{\sqrt{-5}}$

21. $\dfrac{\sqrt{-63}}{\sqrt{7}}$ **22.** $\dfrac{\sqrt{-80}}{\sqrt{5}}$

Objective 2: Write Complex Numbers in the Form $a + bi$

For Exercises 23–28, determine the real and imaginary parts of the complex number.

23. $3 - 7i$ **24.** $2 - 4i$ **25.** $19i$

26. $40i$ **27.** $-\dfrac{1}{4}$ **28.** $-\dfrac{4}{7}$

For Exercises 29–40, simplify each expression and write the result in standard form, $a + bi$. **(See Example 3)**

29. $4\sqrt{-4}$ **30.** $2\sqrt{-144}$ **31.** $2 + \sqrt{-12}$ **32.** $6 - \sqrt{-24}$

33. $\dfrac{8 + 3i}{14}$ **34.** $\dfrac{4 + 5i}{6}$ **35.** $\dfrac{-4 - 6i}{-2}$ **36.** $\dfrac{9 - 15i}{-3}$

37. $\dfrac{-18 + \sqrt{-48}}{4}$ **38.** $\dfrac{-20 + \sqrt{-50}}{-10}$ **39.** $\dfrac{14 - \sqrt{-98}}{-7}$ **40.** $\dfrac{-10 + \sqrt{-125}}{5}$

Objective 3: Simplify Powers of i

For Exercises 41–44, simplify the powers of i. **(See Example 4)**

41. a. i^{20} **b.** i^{29} **c.** i^{50} **d.** i^{-41}

42. a. i^{32} **b.** i^{47} **c.** i^{66} **d.** i^{-27}

43. a. i^{37} **b.** i^{-37} **c.** i^{82} **d.** i^{-82}

44. a. i^{103} **b.** i^{-103} **c.** i^{52} **d.** i^{-52}

Objective 4: Perform Operations on Complex Numbers

For Exercises 45–62, perform the indicated operations. Write the complex numbers in standard form, $a + bi$. **(See Examples 5–6)**

45. a. $(2 - 7x) + (8 - 3x)$ **46. a.** $(6 - 10y) + (8 + 4y)$ **47. a.** $3n(5 - 8n)$

    b. $(2 - 7i) + (8 - 3i)$ **b.** $(6 - 10i) + (8 + 4i)$ **b.** $3i(5 - 8i)$

48. a. $-2m(6 + 3m)$ **49. a.** $(9 + 2x)(4 - x)$ **50. a.** $(5 + c)(2 - 3c)$

    b. $-2i(6 + 3i)$ **b.** $(9 + 2i)(4 - i)$ **b.** $(5 + i)(2 - 3i)$

51. $(15 + 21i) - (18 - 40i)$

52. $(250 + 100i) - (80 + 25i)$

53. $\left(\dfrac{1}{2} + \dfrac{2}{3}i\right) - \left(\dfrac{5}{6} + \dfrac{1}{12}i\right)$

54. $\left(\dfrac{3}{5} - \dfrac{1}{8}i\right) - \left(\dfrac{7}{10} + \dfrac{1}{6}i\right)$

55. $-\dfrac{1}{8}(16 + 24i)$

56. $-\dfrac{1}{6}(60 - 30i)$

57. $2i(5 + i)$

58. $4i(6 + 5i)$

59. $\sqrt{-3}\left(\sqrt{11} - \sqrt{-7}\right)$

60. $\sqrt{-2}\left(\sqrt{13} + \sqrt{-5}\right)$

61. $(3 - 6i)(10 + i)$

62. $(2 - 5i)(8 + 2i)$

For Exercises 63–66, for each given number, (a) identify the complex conjugate and (b) determine the product of the number and its conjugate.

63. $3 - 6i$ **64.** $4 - 5i$ **65.** $8i$ **66.** $9i$

For Exercises 67–90, perform the indicated operations. Write the complex number in standard form, $a + bi$. **(See Examples 7–8)**

67. a. $(3 + 4p)(3 - 4p)$ **68. a.** $(2 - 9y)(2 + 9y)$ **69. a.** $(6 - 5z)^2$ **70. a.** $(10 + 5x)^2$

    b. $(3 + 4i)(3 - 4i)$ **b.** $(2 - 9i)(2 + 9i)$ **b.** $(6 - 5i)^2$ **b.** $(10 + 5i)^2$

71. $(10 - 4i)(10 + 4i)$ **72.** $(3 - 9i)(3 + 9i)$ **73.** $(7i)(-7i)$

74. $(-5i)(5i)$ **75.** $\left(\sqrt{2} + \sqrt{3}i\right)\left(\sqrt{2} - \sqrt{3}i\right)$ **76.** $\left(\sqrt{5} + \sqrt{7}i\right)\left(\sqrt{5} - \sqrt{7}i\right)$

77. $(3 - 7i)^2$ **78.** $(10 - 3i)^2$ **79.** $\left(2 + \sqrt{-7}\right)^2$

80. $\left(4 - \sqrt{-5}\right)^2$ **81.** $(2 - i)^2 + (2 + i)^2$ **82.** $(3 - 2i)^2 - (3 + 2i)^2$

83. $\dfrac{6 + 2i}{3 - i}$ **84.** $\dfrac{5 + i}{4 - i}$ **85.** $\dfrac{8 - 5i}{13 + 2i}$

86. $\dfrac{10 - 3i}{11 + 4i}$ **87.** $\left(6 + \sqrt{5}i\right)^{-1}$ **88.** $\left(4 - \sqrt{3}i\right)^{-1}$

89. $\dfrac{5}{13i}$ **90.** $\dfrac{6}{7i}$

Mixed Exercises

For Exercises 91–94, evaluate $\sqrt{b^2 - 4ac}$ for the given values of a, b, and c, and simplify.

91. $a = 2$, $b = 4$, and $c = 6$ **92.** $a = 5$, $b = -5$, and $c = 10$

93. $a = 2$, $b = -6$, and $c = 5$ **94.** $a = 2$, $b = 4$, and $c = 4$

For Exercises 95–98, verify by substitution that the given values of x are solutions to the given equation.

95. $x^2 + 25 = 0$

    a. $x = 5i$

    b. $x = -5i$

96. $x^2 + 49 = 0$

    a. $x = 7i$

    b. $x = -7i$

97. $x^2 - 4x + 7 = 0$

    a. $x = 2 + i\sqrt{3}$

    b. $x = 2 - i\sqrt{3}$

98. $x^2 - 6x + 11 = 0$

    a. $x = 3 + i\sqrt{2}$

    b. $x = 3 - i\sqrt{2}$

99. Prove that $(a + bi)(c + di) = (ac - bd) + (ad + bc)i$.

100. Prove that $(a + bi)^2 = (a^2 - b^2) + (2ab)i$.

101. Give an example of a complex number that is its own conjugate.

102. Give an example of two complex numbers that are not real numbers, but whose product is a real number.

Expanding Your Skills

The variable z is often used to denote a complex number and $\bar{z}$ is used to denote its conjugate. If $z = a + bi$, simplify the expressions in Exercises 103–104.

103. $z \cdot \bar{z}$

104. $z^2 - \bar{z}^2$

For Exercises 105–110, factor the expressions over the set of complex numbers. For assistance, consider these examples.

- In Chapter 2 we saw that some expressions factor over the set of integers. For example: $x^2 - 4 = (x + 2)(x - 2)$.
- Some expressions factor over the set of irrational numbers. For example: $x^2 - 5 = (x + \sqrt{5})(x - \sqrt{5})$.
- To factor an expression such as $x^2 + 4$, we need to factor over the set of complex numbers. For example, verify that $x^2 + 4 = (x + 2i)(x - 2i)$.

105. a. $x^2 - 9$
  b. $x^2 + 9$

106. a. $x^2 - 100$
  b. $x^2 + 100$

107. a. $x^2 - 64$
  b. $x^2 + 64$

108. a. $x^2 - 25$
  b. $x^2 + 25$

109. a. $x^2 - 3$
  b. $x^2 + 3$

110. a. $x^2 - 11$
  b. $x^2 + 11$

SECTION 3.4   Solving Quadratic Equations by Factoring

OBJECTIVES

1. Solve Quadratic Equations by Factoring
2. Solve Higher-Degree Polynomial Equations by Factoring
3. Solve Applications of Quadratic Equations Involving Geometry

1. Solve Quadratic Equations by Factoring

A linear equation in one variable such as $3x + 4 = 19$ is sometimes called a first-degree polynomial equation because the highest degree of all its terms is 1. A second-degree polynomial equation is called a quadratic equation.

Definition of a Quadratic Equation

Let a, b, and c represent real numbers, where $a \neq 0$. A **quadratic equation** in the variable x is an equation of the form

$$ax^2 + bx + c = 0.$$

The following equations are quadratic because they can be written in the form $ax^2 + bx + c = 0$ where $a \neq 0$.

$$-4x^2 + 4x = 1$$
$$-4x^2 + 4x - 1 = 0$$
$$-4x^2 + 4x + (-1) = 0$$
$$a = -4, b = 4, \text{ and } c = -1$$

$$x^2 - 25 = 0$$
$$x^2 + 0x + (-25) = 0$$
$$a = 1, b = 0, \text{ and } c = -25$$

One method to solve a quadratic equation is to factor and apply the zero product property. The zero product property states that if the product of two factors is zero, then one or both of the factors is equal to zero.

Zero Product Property

If $mn = 0$, then $m = 0$ or $n = 0$.

For example, the quadratic equation $x^2 - x - 12 = 0$ can be written in factored form as $(x - 4)(x + 3) = 0$. By the zero product property, one or both factors must be zero: $x - 4 = 0$ or $x + 3 = 0$. Therefore, to solve the quadratic equation, set each factor to zero and solve for x.

$$(x - 4)(x + 3) = 0$$ Apply the zero product property.

$x - 4 = 0$ or $x + 3 = 0$ Set each factor zero.

$x = 4$ or $x = -3$ Solve each equation for x.

Quadratic equations, like linear equations, arise in many applications of mathematics, science, and business. The following steps summarize the factoring method to solve a quadratic equation.

Solving a Quadratic Equation by Factoring

Step 1 Write the equation in the form $ax^2 + bx + c = 0$.

Step 2 Factor completely.

Step 3 Apply the zero product property. That is, set each factor equal to zero and solve the resulting equations.*

*The solution(s) found in step 3 may be checked by substitution in the original equation.

EXAMPLE 1   Solving a Quadratic Equation

Solve. $2x^2 - 5x = 12$

Solution:

$$2x^2 - 5x = 12$$

$$2x^2 - 5x - 12 = 0$$ Write the equation in the form $ax^2 + bx + c = 0$.

$$(2x + 3)(x - 4) = 0$$ Factor completely.

$2x + 3 = 0$ or $x - 4 = 0$ Set each factor equal to zero.

$2x = -3$ or $x = 4$ Solve each equation.

$x = -\dfrac{3}{2}$ or $x = 4$

Check: $x = -\dfrac{3}{2}$ Check: $x = 4$

$2x^2 - 5x = 12$ $2x^2 - 5x = 12$

$2\left(-\dfrac{3}{2}\right)^2 - 5\left(-\dfrac{3}{2}\right) \overset{?}{=} 12$ $2(4)^2 - 5(4) \overset{?}{=} 12$

$2\left(\dfrac{9}{4}\right) + \dfrac{15}{2} \overset{?}{=} 12$ $2(16) - 20 \overset{?}{=} 12$

$\dfrac{9}{2} + \dfrac{15}{2} \overset{?}{=} 12$ $32 - 20 \overset{?}{=} 12$ ✓

$\dfrac{24}{2} \overset{?}{=} 12$ ✓

The solution set is $\left\{-\dfrac{3}{2}, 4\right\}$.

Avoiding Mistakes

The zero product property tells us that if $ab = 0$, then $a = 0$ or $b = 0$. This property does not hold for other numbers. For example if $ab = 12$, then it is not necessary that a or b must equal 12.

FOR REVIEW

To factor the polynomial $2x^2 - 5x - 12$, first recognize that this is a quadratic trinomial with leading coefficient not equal to 1. Consider using the trial-and-error method with two binomials of the form:

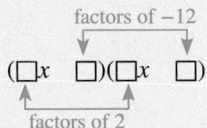

For more information on factoring trinomials, refer to Section 2.4.

Answer

1. $\{7, -5\}$

Skill Practice 1   Solve. $y^2 - 2y = 35$

EXAMPLE 2   Solving a Quadratic Equation

Solve. $6x^2 + 8x = 0$

Solution:

$$6x^2 + 8x = 0$$

$$2x(3x + 4) = 0 \qquad\qquad \text{Factor completely.}$$

$$2x = 0 \quad\text{or}\quad 3x + 4 = 0 \qquad \text{Set each factor equal to zero.}$$

$$x = 0 \qquad\qquad\qquad 3x = -4 \qquad \text{Solve each equation for } x.$$

$$x = -\frac{4}{3}$$

The solution set is $\left\{ 0, -\dfrac{4}{3} \right\}$.   The solutions check.

FOR REVIEW

Recall that the first step to factor a polynomial is to factor out the greatest common factor (GCF). For the polynomial $6x^2 + 8x$ the GCF is $2x$. The factored form is $2x(3x + 4)$.

    Factoring out the greatest common factor is presented in more detail in Section 2.3.

Skill Practice 2   Solve. $9x^2 = 21x$

EXAMPLE 3   Solving a Quadratic Equation

Solve. $9x(4x + 2) - 10x = 8x + 25$

Solution:

$$9x(4x + 2) - 10x = 8x + 25$$

$$36x^2 + 18x - 10x = 8x + 25 \qquad \text{Clear parentheses.}$$

$$36x^2 + 8x = 8x + 25 \qquad \text{Combine like terms.}$$

$$36x^2 - 25 = 0 \qquad \begin{array}{l}\text{Make one side of the equation equal to zero.}\\ \text{The equation is in the form } ax^2 + bx + c = 0.\\ (\textit{Note}: b = 0.)\end{array}$$

$$(6x - 5)(6x + 5) = 0 \qquad \text{Factor completely.}$$

$$6x - 5 = 0 \quad\text{or}\quad 6x + 5 = 0 \qquad \text{Set each factor equal to zero.}$$

$$6x = 5 \quad\text{or}\quad 6x = -5 \qquad \text{Solve each equation.}$$

$$x = \frac{5}{6} \quad\text{or}\quad x = -\frac{5}{6} \qquad \text{The check is left to the reader.}$$

The solution set is $\left\{ \dfrac{5}{6}, -\dfrac{5}{6} \right\}$.

FOR REVIEW

To factor the polynomial $36x^2 - 25$, first recognize that it is a difference of squares $a^2 - b^2$ where $a = 6x$ and $b = 5$. Apply the formula

$$a^2 - b^2 = (a - b)(a + b)$$
$$(6x)^2 - (5)^2 = (6x - 5)(6x + 5)$$

For more information on factoring binomials, refer to Section 2.5.

Skill Practice 3   Solve. $5a(2a - 3) + 4(a + 1) = 3a(3a - 2)$

Answers

2. $\left\{ 0, \dfrac{7}{3} \right\}$

3. $\{4, 1\}$

2. Solve Higher-Degree Polynomial Equations by Factoring

The zero product property can be used to solve polynomial equations of higher degree than a quadratic equation. That is, the equation may have terms of degree greater than 2. To solve a higher-degree polynomial equation, set one side equal to zero and factor the polynomial on the other side. As shown in Examples 4 and 5, a higher-degree polynomial equation may have more than two factors. We set each factor equal to zero and solve for the variable.

EXAMPLE 4   Solving a Higher-Degree Polynomial Equation

Solve. $-2(y + 7)(y - 1)(10y + 3) = 0$

Solution:

$$-2(y + 7)(y - 1)(10y + 3) = 0$$

One side of the equation is zero, and the other side is already factored.

$-2 = 0$ or $y + 7 = 0$ or $y - 1 = 0$ or $10y + 3 = 0$ Set each factor equal to zero.

No solution $y = -7$ or $y = 1$ or $y = -\dfrac{3}{10}$ Solve each equation for y.

Notice that when the constant factor is set to zero, the result is the contradiction $-2 = 0$. The constant factor does not produce a solution to the equation. Therefore, the only solutions are -7, 1, and $-\frac{3}{10}$. Each solution can be checked in the original equation.

The solution set is $\left\{-7, 1, -\dfrac{3}{10}\right\}$.

Skill Practice 4   Solve. $3(w + 2)(2w + 1)(w - 8) = 0$

FOR REVIEW

To review the technique of factoring a four-term polynomial by grouping, refer to Section 2.3.

  For the polynomial
  $z^3 + 3z^2 - 4z - 12$,

- The greatest common factor of z^2 is removed from the first pair of terms.
- The greatest common factor -4 is removed from the second pair of terms.
- Then the common binomial factor $(z + 3)$ is factored out.

Answers

4. $\left\{-2, -\dfrac{1}{2}, 8\right\}$

5. $\{-1, 3, -3\}$

EXAMPLE 5   Solving a Higher-Degree Polynomial Equation

Solve. $z^3 + 3z^2 - 4z - 12 = 0$

Solution:

$z^3 + 3z^2 - 4z - 12 = 0$ This is a higher-degree polynomial equation.

$z^3 + 3z^2 \mid - 4z - 12 = 0$ One side of the equation is zero. Now factor.

$z^2(z + 3) - 4(z + 3) = 0$ Because there are four terms, try factoring by grouping.

$(z + 3)(z^2 - 4) = 0$ $z^2 - 4$ can be factored further as a difference of squares.

$(z + 3)(z - 2)(z + 2) = 0$

$z + 3 = 0$ or $z - 2 = 0$ or $z + 2 = 0$ Set each factor equal to zero.

$z = -3$ or $z = 2$ or $z = -2$ Solve each equation.

The solution set is $\{-3, 2, -2\}$.

Skill Practice 5   Solve. $x^3 + x^2 - 9x - 9 = 0$

3. Solve Applications of Quadratic Equations Involving Geometry

Suppose a contractor is hired to tile the floor of a house. To provide the homeowner with an estimate, the contractor would first measure the area of the floor. For example, the area of a square room with sides of length s is given by $A = s^2$. So for a 10 ft by 10 ft square room, the area is $(10 \text{ ft})(10 \text{ ft}) = 100 \text{ ft}^2$. For a rectangular room with length l and width w, the area is $A = lw$. Thus, for a 20 ft by 12 ft rectangular living room, the area is $(20 \text{ ft})(12 \text{ ft}) = 240 \text{ ft}^2$.

Area is usually measured in square units such as square feet, square yards, square miles, and so on. Therefore, it seems reasonable that quadratic equations might come into play in applications involving area (recall that a quadratic equation has an x^2 term). Before we demonstrate this in Example 6, take a minute to review some common formulas from geometry involving area and volume.

TIP There are other units of area such as acres, but an acre can be converted to square feet by the conversion factor

$$1 \text{ acre} = 43,560 \text{ ft}^2$$

Area, A

Square	Rectangle	Triangle	Parallelogram
$A = s^2$	$A = lw$	$A = \dfrac{1}{2}bh$	$A = bh$

FOR REVIEW

For further review, see the list of important formulas found at the end of the text.

Volume, V

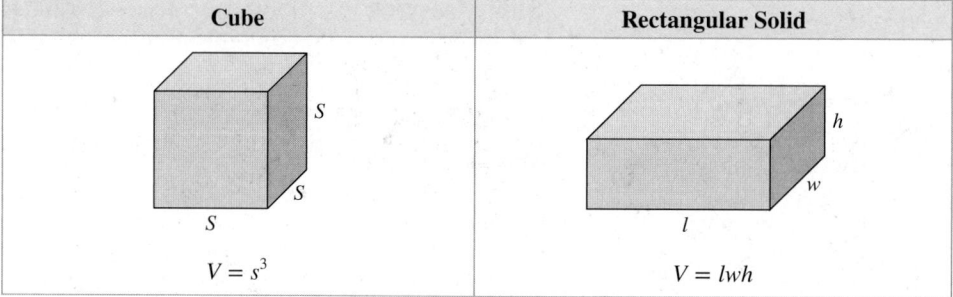

Cube	Rectangular Solid
$V = s^3$	$V = lwh$

EXAMPLE 6   **Solving an Application of a Quadratic Equation**

The length of a basketball court is 6 ft less than 2 times the width. If the total area is 4700 ft^2, find the dimensions of the court.

Solution:

If the width of the court is represented by w, then the length can be represented by $2w - 6$ (Figure 3-1).

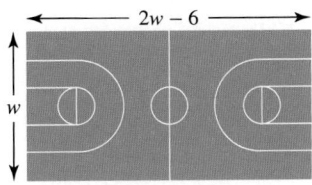

Figure 3-1

$$A = (\text{length})(\text{width}) \qquad \text{Area of a rectangle}$$
$$4700 = (2w - 6)w \qquad \text{Mathematical equation}$$
$$4700 = 2w^2 - 6w$$
$$2w^2 - 6w - 4700 = 0 \qquad \text{Set the equation equal to zero and factor.}$$
$$2(w^2 - 3w - 2350) = 0 \qquad \text{Factor out the GCF.}$$
$$2(w - 50)(w + 47) = 0 \qquad \text{Factor the trinomial.}$$

$2 \cancel{=} 0$ or $w - 50 = 0$ or $w + 47 = 0$ Set each factor equal to zero.

contradiction

$w = 50$ or $w \cancel{=} -47$ A negative width is not possible.

The width is 50 ft.

The length is $2w - 6 = 2(50) - 6 = 94$ ft.

Skill Practice 6   The width of a rectangle is 5 in. less than 3 times the length. The area is 2 in.^{2} Find the length and width.

TIP When applying the Pythagorean theorem, it does not matter which leg you label a and which you label b. Since the lengths of the legs are interchangeable, you can also write the Pythagorean theorem as $\text{leg}^2 + \text{leg}^2 = \text{hyp}^2$.

A right triangle is a triangle that contains a 90° angle. Furthermore, the sum of the squares of the two legs (the shorter sides) of a right triangle equals the square of the hypotenuse (the longest side). This important fact is known as the Pythagorean theorem. For the right triangle shown in Figure 3-2, the Pythagorean theorem is stated as

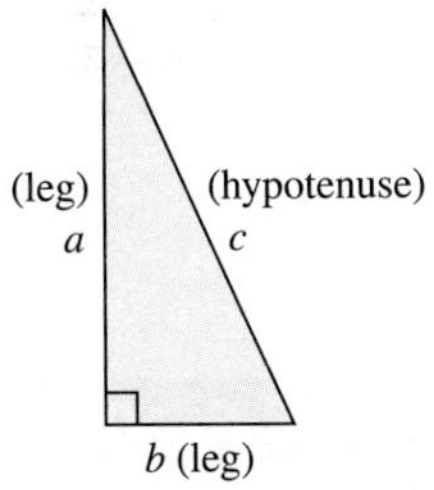

$$a^2 + b^2 = c^2$$

In this formula, a and b are the legs and c is the hypotenuse. Notice that the hypotenuse is the longest side and is opposite the right angle.

The triangle given in Figure 3-3 is a right triangle. We have

$$a^2 \;+\; b^2 \;= c^2$$
$$(5 \text{ ft})^2 + (12 \text{ ft})^2 = (13 \text{ ft})^2$$
$$25 \text{ ft}^2 + 144 \text{ ft}^2 = 169 \text{ ft}^2$$
$$169 \text{ ft}^2 = 169 \text{ ft}^2 \;\checkmark$$

Figure 3-2

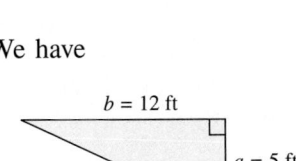

Figure 3-3

Answer

6. The width is 1 in., and the length is 2 in.

EXAMPLE 7   Applying the Pythagorean Theorem

Find the length of the missing side of the right triangle.

Solution:

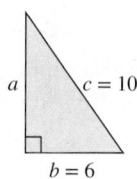

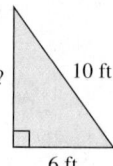

Label the triangle.

$$a^2 + b^2 = c^2$$ Apply the Pythagorean theorem.

$$a^2 + 6^2 = 10^2$$ Substitute $b = 6$ and $c = 10$.

$$a^2 + 36 = 100$$ Simplify. The equation is quadratic.

$$a^2 + 36 - 100 = 100 - 100$$ Subtract 100 from both sides.

$$a^2 - 64 = 0$$ One side is now equal to zero.

$$(a + 8)(a - 8) = 0$$ Factor.

$$a + 8 = 0 \text{ or } a - 8 = 0$$ Set each factor equal to zero.

$$a \cancel{=} -8 \text{ or } a = 8$$ Because x represents the length of a side of a triangle, reject the negative solution.

The third side is 8 ft.

Skill Practice 7   Find the length of the missing side.

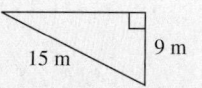

EXAMPLE 8   Applying the Pythagorean Theorem

A region of coastline off Biscayne Bay is approximately in the shape of a right angle. The corresponding triangular area has sandbars and is marked off on navigational charts as being shallow water. If one leg of the triangle is 0.5 mi shorter than the other leg, and the hypotenuse is 2.5 mi, find the lengths of the legs of the triangle (Figure 3-4).

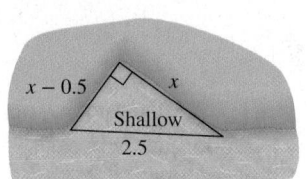

Figure 3-4

Solution:

Let x represent the longer leg.

Then $x - 0.5$ represents the shorter leg.

$$a^2 + b^2 = c^2$$ Pythagorean theorem

$$x^2 + (x - 0.5)^2 = (2.5)^2$$

$$x^2 + (x)^2 - 2(x)(0.5) + (0.5)^2 = 6.25$$

$$x^2 + x^2 - x + 0.25 = 6.25$$

$$2x^2 - x - 6 = 0$$ Write the equation in the form $ax^2 + bx + c = 0$.

$$(2x + 3)(x - 2) = 0$$ Factor.

$$2x + 3 = 0 \quad \text{or} \quad x - 2 = 0$$ Set both factors to zero.

$$x \cancel{=} -\frac{3}{2} \quad \text{or} \quad x = 2$$ Solve both equations for x.

FOR REVIEW

Recall that the square of a binomial results in a perfect square trinomial. (See Section 2.2.)

$$(a - b)^2 = a^2 - 2ab + b^2$$
$$(x - 0.5)^2 = (x)^2 - 2(x)(0.5) + (0.5)^2$$
$$= x^2 - x + 0.25$$

Answer

7. The length of the third side is 12 m.

The side of a triangle cannot be negative, so we reject the solution $x = -\frac{3}{2}$.

Therefore, one leg of the triangle is 2 mi.

The other leg is $x - 0.5 = 2 - 0.5 = 1.5$ mi.

> **Skill Practice 8**   The longer leg of a right triangle measures 7 ft more than the shorter leg. The hypotenuse is 8 ft longer than the shorter leg. Find the lengths of the sides of the triangle.

Answer

8. The sides are 5 ft, 12 ft, and 13 ft.

SECTION 3.4   Practice Exercises

Prerequisite Review

For Exercises R.1–R.4, refer to the geometry formulas given on page 218.

R.1. Find the area of a rectangle with length 6 cm and width 2 cm.

R.2. Find the area of a square with sides of 7 yd.

R.3. Find the area of a triangle with base 18 in. and height 4 in.

R.4. Find the volume of a rectangular solid with length 2.5 ft, width 8.4 ft, and height 10 ft.

R.5. If x is the smaller of two consecutive odd integers, write an expression for the next odd integer.

R.6. If x is the greater of two consecutive integers, write an expression for the smaller integer.

For Exercises R.7–R.20 factor completely.

R.7. $5x + 20$ (See Section 2.3)   **R.8.** $24x - 16$

R.9. $b^2 - 14b + 45$ (See Section 2.4)   **R.10.** $a^2 - 15a + 50$

R.11. $-39z + 9 + 12z^2$ **R.12.** $3 + 15y^2 + 14y$

R.13. $25 - 81w^2$ (See Section 2.5)   **R.14.** $81 - 100b^2$

R.15. $x^3 - x^2 - 64x + 64$ (See Section 2.3)   **R.16.** $x^3 + 10x^2 - x - 10$

R.17. $x^3 - 6x^2 - 16x$ (See Section 2.4)   **R.18.** $p^3 + 5p^2 - 66p$

R.19. $-2t^2 + 6t + 80$ **R.20.** $-3d^2 + 21d - 36$

Concept Connections

1. An equation that can be written in the form $ax^2 + bx + c = 0$ where a, b, and c are real numbers and $a \neq 0$ is called a _____ equation. It is also called a _____ -degree polynomial equation.

2. The zero product property states that if $mn = 0$, then $m = $ _____ or $n = $ _____.

3. To solve a quadratic equation or a polynomial equation, set one side of the equation equal to _____ and factor the other side.

4. The equation $2x^2 - 4x + 5 = 0$ is a quadratic equation and the equation $2x + 3 = 0$ is a _____ equation.

Objective 1: Solve Quadratic Equations by Factoring

For Exercises 5–10, identify the equation as linear, quadratic, or neither.

5. $4 - 5x = 0$ **6.** $5x^3 + 2 = 0$ **7.** $3x - 6x^2 = 0$

8. $1 - x + 2x^2 = 0$ **9.** $7x^4 + 8 = 0$ **10.** $3x + 2 = 0$

For Exercises 11–16, determine if the equation is written in the correct form to apply the zero product property directly. If the equation is not in the correct form, explain what is wrong.

11. $2x(x - 3) = 0$

12. $(u + 1)(u - 3) = 10$

13. $3p^2 - 7p + 4 = 0$

14. $t^2 - t - 12 = 0$

15. $a(a + 3)^2 = 5$

16. $\left(\dfrac{2}{3}x - 5\right)\left(x + \dfrac{1}{2}\right) = 0$

For Exercises 17–20, factor the polynomial or solve the equation as indicated.

17. a. Factor. $w^2 - 81$
    b. Solve. $w^2 - 81 = 0$

18. a. Factor. $p^2 - 25$
    b. Solve. $p^2 - 25 = 0$

19. a. Factor. $3x^2 + 14x - 5$
    b. Solve. $3x^2 + 14x - 5 = 0$

20. a. Factor. $2y^2 - y - 3$
    b. Solve. $2y^2 - y - 3 = 0$

For Exercises 21–48, solve the equation. **(See Examples 1–3)**

21. $(x + 3)(x + 5) = 0$

22. $(x + 7)(x - 4) = 0$

23. $(2w + 9)(5w - 1) = 0$

24. $(3a + 1)(4a - 5) = 0$

25. $x(x + 4)(10x - 3) = 0$

26. $t(t - 6)(3t - 11) = 0$

27. $0 = 5(y - 0.4)(y + 2.1)$

28. $0 = -4(z - 7.5)(z - 9.3)$

29. $x^2 + 6x - 27 = 0$

30. $2x^2 + x - 15 = 0$

31. $2x^2 + 5x = 3$

32. $-11x = 3x^2 - 4$

33. $10x^2 = 15x$

34. $5x^2 = 7x$

35. $6(y - 2) - 3(y + 1) = 8$

36. $4x + 3(x - 9) = 6x + 1$

37. $-9 = y(y + 6)$

38. $-62 = t(t - 16) + 2$

39. $9p^2 - 15p - 6 = 0$

40. $6y^2 + 2y = 48$

41. $(y - 3)(y + 4) = 8$

42. $(t + 10)(t + 5) = 6$

43. $(2a - 1)(a - 1) = 6$

44. $w(6w + 1) = 2$

45. $p^2 + (p + 7)^2 = 169$

46. $x^2 + (x + 2)^2 = 100$

47. $3t(t + 5) - t^2 = 2t^2 + 4t - 1$

48. $a^2 - 4a - 2 = (a + 3)(a - 5)$

Objective 2: Solve Higher-Degree Polynomial Equations by Factoring

For Exercises 49–62, solve the equation. **(See Exercises 4–5)**

49. $(x + 1)(2x - 1)(x - 3) = 0$

50. $2x(x - 4)^2(4x + 3) = 0$

51. $2x^3 - 8x^2 - 24x = 0$

52. $2p^3 + 20p^2 + 42p = 0$

53. $w^3 = 16w$

54. $12x^3 = 27x$

55. $0 = 2x^3 + 5x^2 - 18x - 45$

56. $0 = 3y^3 + y^2 - 48y - 16$

57. $-3x(2x - 1)(x + 6)^2 = 0$

58. $5y(3 - y)(4y + 1)^2 = 0$

59. $75y^3 + 100y^2 - 3y - 4 = 0$

60. $98t^3 - 49t^2 - 8t + 4 = 0$

61. $x^3 + 7x^2 = 4(x + 7)$

62. $2m^3 + 3m^2 = 9(2m + 3)$

3. Solve Applications of Quadratic Equations Involving Geometry

For Exercises 63–70,

    a. Write an equation in terms of x that represents the given relationship.

    b. Solve the equation to find the dimensions of the given shape.

63. The length of a rectangle is 3 yd more than twice the width x. The area is 629 yd^2.

64. The width of a rectangle is 2 m less than one-quarter of the length x. The area is 252 m^2.

65. The height of a triangle is 2 ft less than the base x. The area is 40 ft^2.

66. The height of a triangle is 4 yd longer than the base x. The area is 70 yd^2.

67. The width of a rectangular box is 8 in. The height is one-fifth the length x. The volume is 640 in.^{3}.

68. The height of a rectangular box is 4 ft. The length is 1 ft longer than twice the width x. The volume is 312 ft^3.

69. The length of the longer leg of a right triangle is 2 ft longer than the length of the shorter leg x. The hypotenuse is 2 ft shorter than twice the length of the shorter leg.

70. The longer leg of a right triangle is 7 cm longer than the length of the shorter leg x. The hypotenuse is 17 cm.

71. A rectangular pen has an area of 35 ft^2. If the width is 2 ft less than the length, find the dimensions of the pen. **(See Example 6)**

72. The length of a rectangular photograph is 7 in. more than the width. If the area is 78 in.^{2}, what are the dimensions of the photograph?

73. The length of a rectangular room is 5 yd more than the width. If the area is 300 yd^2, find the length and the width of the room.

74. The top of a rectangular dining room table is twice as long as it is wide. Find the dimensions of the table if the area is 18 ft^2.

75. The height of a triangle is 1 in. more than the base. If the height is increased by 2 in. while the base remains the same, the new area becomes 20 in.^{2}

 a. Find the base and height of the original triangle.

 b. Find the area of the original triangle.

76. The base of a triangle is 2 cm more than the height. If the base is increased by 4 cm while the height remains the same, the new area is 56 cm^2.

 a. Find the base and height of the original triangle.

 b. Find the area of the original triangle.

77. The area of a triangular garden is 25 ft^2. The base is twice the height. Find the base and the height of the triangle.

78. The height of a triangle is 1 in. more than twice the base. If the area is 18 in.^{2}, find the base and height of the triangle.

For Exercises 79–80, determine if the given lengths represent the lengths of the sides of a right triangle.

79. a. 20 ft, 99 ft, 101 ft

 b. 9 cm, 11 cm, 15 cm

 c. 20 yd, 21 yd, 29 yd

80. a. 8 mi, 12 mi, 15 mi

 b. 12 km, 35 km, 37 km

 c. 14 m, 15 m, 17 m

For Exercises 81–84, determine the length of the missing side of the right triangle. **(See Example 7)**

81.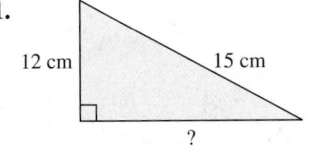

82.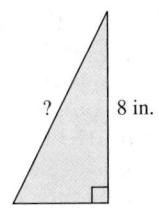

83.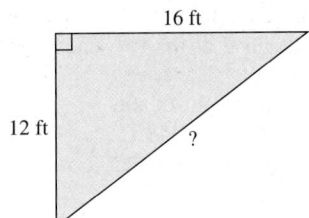

84.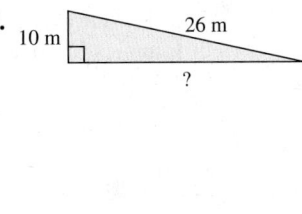

85. Roberto and Sherona began running from the same place at the same time. They ran along two different paths that formed right angles with each other. Roberto ran 4 mi and stopped, while Sherona ran 3 mi and stopped. How far apart were they when they stopped? **(See Example 7)**

86. Leine and Laura began hiking from their campground. Laura headed south while Leine headed east. Laura walked 12 mi and Leine walked 5 mi. How far apart were they when they stopped walking?

87. Justin must travel from Summersville to Clayton. He can drive 10 mi through the mountains at 40 mph, or he can drive east and then north on superhighways at 60 mph. The alternative route forms a right angle as shown in the diagram. The eastern leg is 2 mi less than the northern leg. **(See Example 8)**

 a. Find the total distance Justin would travel in going the alternative route.

 b. If Justin wants to minimize the time of the trip, which route should he take?

88. A 17-ft ladder is standing up against a wall. The distance between the bottom of the ladder and the wall is 7 ft less than the distance between the top of the ladder and the base of the wall. Find the distance between the bottom of the ladder and the wall.

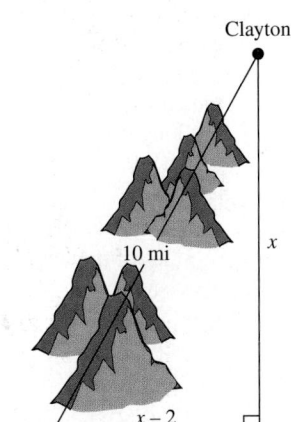

89. A right triangle has side lengths represented by three consecutive even integers. Find the lengths of the three sides, measured in meters.

90. The hypotenuse of a right triangle is 3 m more than twice the short leg. The longer leg is 2 m more than twice the shorter leg. Find the lengths of the sides.

Mixed Exercises

91. a. Write an equation representing the fact that the product of two consecutive even integers is 120.

    b. Solve the equation from part (a) to find the two integers.

93. a. Write an equation representing the fact that the sum of the squares of two consecutive integers is 113.

    b. Solve the equation from part (a) to find the two integers.

95. On moving day, Guyton needs to rent a truck. The length of the cargo space is 12 ft, and the height is 1 ft less than the width. The brochure indicates that the truck can hold 504 ft^3. What are the dimensions of the cargo space? Assume that the cargo space is in the shape of a rectangular solid.

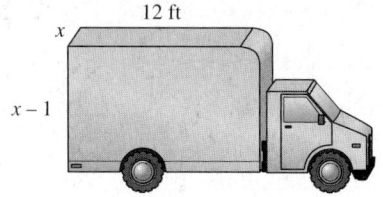

92. a. Write an equation representing the fact that the product of two consecutive odd integers is 35.

    b. Solve the equation from part (a) to find the two integers.

94. a. Write an equation representing the fact that the sum of the squares of two consecutive integers is 181.

    b. Solve the equation from part (a) to find the two integers.

96. Lorene plans to make several open-topped boxes in which to carry plants. She makes the boxes from rectangular sheets of cardboard from which she cuts out 6-in. squares from each corner. The length of the original piece of cardboard is 12 in. more than the width. If the volume of the box is 1728 in.^{3}, determine the dimensions of the original piece of cardboard.

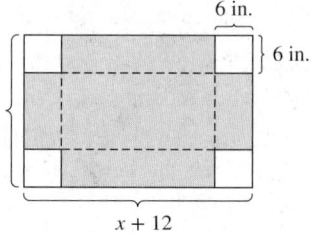

Expanding Your Skills

97. A patio is configured from a rectangle with two right triangles of equal size attached at the two ends. The length of the rectangle is 20 ft. The base of the right triangle is 3 ft less than the height of the triangle. If the total area of the patio is 348 ft^2, determine the base and height of the triangular portions.

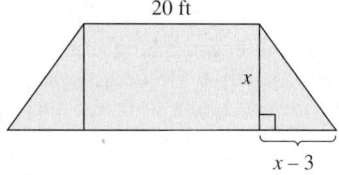

99. The sail on a sailboat is in the shape of two adjacent right triangles. In the lower triangle, the shorter leg is 2 ft less than the longer leg. The hypotenuse is 2 ft more than the longer leg.

    a. Find the lengths of the sides of the lower triangle.

    b. Find the total sail area.

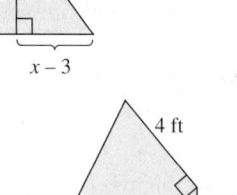

98. The front face of a house is in the shape of a rectangle with a Queen post roof truss above. The length of the rectangular region is 3 times the height of the truss. The height of the rectangle is 2 ft more than the height of the truss. If the total area of the front face of the house is 336 ft^2, determine the length and width of the rectangular region.

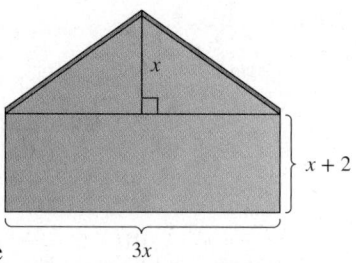

100. A portion of a roof truss is given in the figure. The triangle on the left is configured such that the longer leg is 7 ft longer than the shorter leg, and the hypotenuse is 1 ft more than twice the shorter leg.

    a. Find the lengths of the sides of the triangle on the left.

    b. Find the lengths of the sides of the triangle on the right.

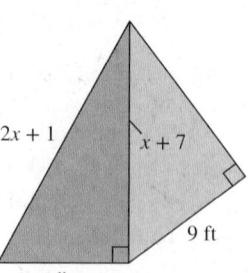

SECTION 3.5 Solving Quadratic Equations by Using the Square Root Property

OBJECTIVES

1. Solve Quadratic Equations by Using the Square Root Property
2. Complete the Square
3. Solve Quadratic Equations by Completing the Square

1. Solve Quadratic Equations by Using the Square Root Property

In Section 3.4 we learned how to solve a quadratic equation by factoring and applying the zero product property. For example, the equation $x^2 - 81 = 0$ can be written as $(x - 9)(x + 9) = 0$, and the solutions are easily identified as 9 and –9. But what happens if we cannot factor the polynomial within the equation? To address this dilemma, we bring our attention to the square root operation.

Consider a quadratic equation of the form $x^2 = k$ where k is a constant real number. This equation is of the form of a perfect square equal to a constant. We want to isolate x by "undoing" the squaring operation. Thus, the value(s) of x are the square roots of k (both positive and negative). For example,

FOR REVIEW

Given a real number x, $\sqrt{x^2} = |x|$. That is, if x represents any real number, the absolute value bars are necessary to ensure that the value of the principal square root is nonnegative.

$$x^2 - 81 = 0$$

$$x^2 = 81 \qquad \text{Write the equation as a perfect square equal to a constant.}$$

$$\sqrt{x^2} = \sqrt{81} \qquad \text{Take the principal square root of both sides.}$$

$$|x| = 9 \qquad \sqrt{x^2} \text{ represents the principal (nonnegative) square root of } x^2.$$

$$x = 9 \quad \text{or} \quad x = -9 \qquad \text{Solve the absolute value equation (see Section 1.6).}$$

Using this logic, the solutions to the equation $x^2 = k$ are $\sqrt{k}$ and $-\sqrt{k}$. This is stated formally as the **square root property**.

The Square Root Property

Given a real number k, if $x^2 = k$, then $x = \sqrt{k}$ or $x = -\sqrt{k}$.

This can be written more concisely as $x = \pm\sqrt{k}$, read as "x equals plus or minus the square root of k."

The solution set is $\{\pm\sqrt{k}.\}$

EXAMPLE 1   Solving a Quadratic Equation by Using the Square Root Property

Use the square root property to solve the equation. $4p^2 = 9$

Solution:

$$4p^2 = 9$$

$$p^2 = \frac{9}{4} \qquad \text{Isolate } p^2 \text{ by dividing both sides by 4.}$$

$$p = \pm\sqrt{\frac{9}{4}} \qquad \text{Apply the square root property.}$$

$$p = \pm\frac{3}{2} \qquad \text{Simplify the radical.}$$

The solution set is $\left\{\frac{3}{2}, -\frac{3}{2}\right\}$.

Answer

1. $\left\{\pm\frac{4}{5}\right\}$

Skill Practice 1   Solve using the square root property.

$$25a^2 = 16$$

For a quadratic equation, $ax^2 + bx + c = 0$, if $b = 0$, then the equation is easily solved by using the square root property. This is demonstrated in Example 2.

EXAMPLE 2 Solving a Quadratic Equation by Using the Square Root Property

Use the square root property to solve the equation. $3x^2 + 75 = 0$

Solution:

$$3x^2 + 75 = 0 \qquad \text{Rewrite the equation to fit the form } x^2 = k.$$
$$3x^2 = -75$$
$$x^2 = -25 \qquad \text{The equation is now in the form } x^2 = k.$$
$$x = \pm\sqrt{-25} \qquad \text{Apply the square root property.}$$
$$= \pm 5i$$

Check: $x = 5i$	Check: $x = -5i$
$3x^2 + 75 = 0$	$3x^2 + 75 = 0$
$3(5i)^2 + 75 \overset{?}{=} 0$	$3(-5i)^2 + 75 \overset{?}{=} 0$
$3(25i^2) + 75 \overset{?}{=} 0$	$3(25i^2) + 75 \overset{?}{=} 0$
$3(-25) + 75 \overset{?}{=} 0$	$3(-25) + 75 \overset{?}{=} 0$
$-75 + 75 \overset{?}{=} 0 \checkmark$	$-75 + 75 \overset{?}{=} 0 \checkmark$

The solution set is $\{\pm 5i\}$.

> **Avoiding Mistakes**
>
> A common mistake is to forget the $\pm$ symbol when solving the equation $x^2 = k$:
> $$x = \pm\sqrt{k}$$

> **FOR REVIEW**
>
> Recall that the imaginary number $i = \sqrt{-1}$ and $i^2 = -1$. For a review of imaginary numbers, see Section 3.3.

Skill Practice 2 Solve using the square root property.

$8x^2 + 72 = 0$

EXAMPLE 3 Solving a Quadratic Equation by Using the Square Root Property

Use the square root property to solve the equation. $(w + 3)^2 = 20$

Solution:

$$(w + 3)^2 = 20 \qquad \text{The equation is in the form } x^2 = k, \text{ where } x = (w + 3).$$

$$w + 3 = \pm\sqrt{20} \qquad \text{Apply the square root property.}$$
$$w + 3 = \pm\sqrt{2^2 \cdot 5} \qquad \text{Simplify the radical.}$$
$$w + 3 = \pm 2\sqrt{5}$$
$$w = -3 \pm 2\sqrt{5} \qquad \text{Isolate } w \text{ by subtracting 3 from both sides.}$$

The solution set is $\{-3 \pm 2\sqrt{5}\}$.

> **Avoiding Mistakes**
>
> Note that $-3 \pm 2\sqrt{5}$ represents two solutions:
> $-3 + 2\sqrt{5}$ and $-3 - 2\sqrt{5}$

Skill Practice 3 Solve using the square root property.

$(t - 5)^2 = 18$

Answers

2. $\{\pm 3i\}$
3. $\{5 \pm 3\sqrt{2}\}$

2. Complete the Square

In Example 3 we used the square root property to solve an equation where the square of a binomial was equal to a constant.

$$(w + 3)^2 = 20$$

$\underbrace{\qquad}$ Square of a binomial $\uparrow$ Constant

FOR REVIEW

Recall the factored forms of perfect square trinomials:

$a^2 + 2ab + b^2 = (a + b)^2$

$a^2 - 2ab + b^2 = (a - b)^2$

For review, refer to Section 2.4.

The square of a binomial is the factored form of a perfect square trinomial. For example:

Perfect Square Trinomial	Factored Form
$x^2 + 10x + 25 \longrightarrow$	$(x + 5)^2$
$t^2 - 6t + 9 \longrightarrow$	$(t - 3)^2$
$p^2 - 14p + 49 \longrightarrow$	$(p - 7)^2$

For a perfect square trinomial with a leading coefficient of 1, the constant term is the square of one-half the linear term coefficient. For example:

$$x^2 + 10x + 25$$

$$\left[\tfrac{1}{2}(10)\right]^2$$

In general an expression of the form $x^2 + bx + n$ is a perfect square trinomial if $n = \left(\tfrac{1}{2}b\right)^2$. The process to create a perfect square trinomial is called **completing the square**.

EXAMPLE 4   Completing the Square

Determine the value of n that makes the polynomial a perfect square trinomial. Then factor the expression as the square of a binomial.

a. $x^2 + 12x + n$ **b.** $x^2 - 26x + n$

c. $x^2 + 11x + n$ **d.** $x^2 - \dfrac{4}{7}x + n$

Solution:

The expressions are in the form $x^2 + bx + n$. The value of n equals the square of one-half the linear term coefficient $\left(\tfrac{1}{2}b\right)^2$.

a. $x^2 + 12x + n$

$x^2 + 12x + 36$ $n = \left[\tfrac{1}{2}(12)\right]^2 = (6)^2 = 36$

$(x + 6)^2$ Factored form

b. $x^2 - 26x + n$

$x^2 - 26x + 169$ $n = \left[\tfrac{1}{2}(-26)\right]^2 = (-13)^2 = 169$

$(x - 13)^2$ Factored form

c. $x^2 + 11x + n$

$x^2 + 11x + \dfrac{121}{4}$ $n = \left[\tfrac{1}{2}(11)\right]^2 = \left(\tfrac{11}{2}\right)^2 = \tfrac{121}{4}$

$\left(x + \dfrac{11}{2}\right)^2$ Factored form

d. $x^2 - \dfrac{4}{7}x + n$

$$x^2 - \dfrac{4}{7}x + \dfrac{4}{49}$$ $n = \left[\frac{1}{2}\left(-\frac{4}{7}\right)\right]^2 = \left(-\frac{2}{7}\right)^2 = \frac{4}{49}$

$$\left(x - \dfrac{2}{7}\right)^2$$ Factored form

Skill Practice 4   Determine the value of n that makes the polynomial a perfect square trinomial. Then factor.

a. $x^2 + 20x + n$ **b.** $y^2 - 16y + n$

c. $a^2 - 15a + n$ **d.** $w^2 + \dfrac{7}{3}w + n$

---

3. Solve Quadratic Equations by Completing the Square

The process of completing the square can be used to write a quadratic equation $ax^2 + bx + c = 0$ $(a \neq 0)$ in the form $(x - h)^2 = k$. Then the square root property can be used to solve the equation. The following steps outline the procedure.

> **Solving a Quadratic Equation $ax^2 + bx + c = 0$ by Completing the Square and Applying the Square Root Property**
>
> **Step 1** Divide both sides by a to make the leading coefficient 1.
> **Step 2** Isolate the variable terms on one side of the equation.
> **Step 3** Complete the square.
> - Add the square of one-half the linear term coefficient to both sides.
> - Factor the resulting perfect square trinomial.
> **Step 4** Apply the square root property and solve for x.

EXAMPLE 5   Solving a Quadratic Equation by Completing the Square and Applying the Square Root Property

Solve by completing the square and applying the square root property.

$$x^2 - 6x + 13 = 0$$

Solution:

$x^2 - 6x + 13 = 0$ **Step 1:** Since the leading coefficient a is equal to 1, we do not have to divide by a. We can proceed to step 2.

$x^2 - 6x = -13$ **Step 2:** Isolate the variable terms on one side.

$x^2 - 6x + 9 = -13 + 9$ **Step 3:** To complete the square, add $\left[\frac{1}{2}(-6)\right]^2 = \left[-3\right]^2 = 9$ to both sides of the equation.

$(x - 3)^2 = -4$ Factor the perfect square trinomial.

Answers

4. **a.** $n = 100; (x + 10)^2$
   b. $n = 64; (y - 8)^2$
   c. $n = \dfrac{225}{4}; \left(a - \dfrac{15}{2}\right)^2$
   d. $n = \dfrac{49}{36}; \left(w + \dfrac{7}{6}\right)^2$

$$x - 3 = \pm\sqrt{-4}$$ **Step 4:** Apply the square root property.

$$x - 3 = \pm 2i$$ Simplify the radical.

$$x = 3 \pm 2i$$ Solve for x.

The solutions are imaginary numbers and can be written as $3 + 2i$ and $3 - 2i$.

The solution set is $\{3 \pm 2i\}$.

Skill Practice 5   Solve by completing the square and applying the square root property. $z^2 - 4z + 26 = 0$

EXAMPLE 6   **Solve a Quadratic Equation by Completing the Square and Applying the Square Root Property**

Solve by completing the square and applying the square root property.

$$2m^2 + 10m = 3$$

Solution:

$$2m^2 + 10m = 3$$ The variable terms are already isolated on one side of the equation.

$$\frac{2m^2}{2} + \frac{10m}{2} = \frac{3}{2}$$ Divide by the leading coefficient of 2.

$$m^2 + 5m = \frac{3}{2}$$

$$m^2 + 5m + \frac{25}{4} = \frac{3}{2} + \frac{25}{4}$$ Add $\left[\frac{1}{2}(5)\right]^2 = \left[\frac{5}{2}\right]^2 = \frac{25}{4}$ to both sides.

$$\left(m + \frac{5}{2}\right)^2 = \frac{3}{2} \cdot \frac{2}{2} + \frac{25}{4}$$ Factor the left side. On the right side, write the terms with a common denominator.

$$\left(m + \frac{5}{2}\right)^2 = \frac{6}{4} + \frac{25}{4}$$

$$\left(m + \frac{5}{2}\right)^2 = \frac{31}{4}$$ Add the fractions on the right side.

$$m + \frac{5}{2} = \pm\sqrt{\frac{31}{4}}$$ Apply the square root property.

$$m + \frac{5}{2} = \pm\frac{\sqrt{31}}{\sqrt{4}}$$ Apply the quotient property of radicals.

$$m + \frac{5}{2} = \pm\frac{\sqrt{31}}{2}$$ Simplify the radical term.

$$m = -\frac{5}{2} \pm \frac{\sqrt{31}}{2}$$ Isolate m by subtracting $\frac{5}{2}$ from both sides.

The solution set is $\left\{-\frac{5}{2} \pm \frac{\sqrt{31}}{2}\right\}$.

TIP   The solutions to Example 6 can also be written as:

$$\frac{-5 \pm \sqrt{31}}{2}$$

Answer

5. $\{2 \pm i\sqrt{22}\}$

Skill Practice 6 Solve by completing the square and applying the square root property. $4x^2 + 12x = 5$

EXAMPLE 7 Solving a Quadratic Equation by Completing the Square and Applying the Square Root Property

Solve by completing the square and applying the square root property.

$$2x(2x - 10) = -30 + 6x$$

Solution:

$$2x(2x - 10) = -30 + 6x$$

$$4x^2 - 20x = -30 + 6x \qquad \text{Clear parentheses.}$$

$$4x^2 - 26x + 30 = 0 \qquad \text{Write the equation in the form } ax^2 + bx + c = 0.$$

$$\frac{4x^2}{4} - \frac{26x}{4} + \frac{30}{4} = \frac{0}{4} \qquad \textbf{Step 1:} \text{ Divide both sides by the leading coefficient, 4.}$$

$$x^2 - \frac{13}{2}x + \frac{15}{2} = 0$$

$$x^2 - \frac{13}{2}x = -\frac{15}{2} \qquad \textbf{Step 2:} \text{ Isolate the variable terms on one side.}$$

$$x^2 - \frac{13}{2}x + \frac{169}{16} = -\frac{15}{2} + \frac{169}{16} \qquad \textbf{Step 3:} \text{ Add } \left[\frac{1}{2}\left(-\frac{13}{2}\right)\right]^2 = \left(-\frac{13}{4}\right)^2 = \frac{169}{16} \text{ to both sides.}$$

$$\left(x - \frac{13}{4}\right)^2 = -\frac{120}{16} + \frac{169}{16} \qquad \text{Factor the perfect square trinomial. Rewrite the right side with a common denominator.}$$

$$\left(x - \frac{13}{4}\right)^2 = \frac{49}{16}$$

$$x - \frac{13}{4} = \pm\sqrt{\frac{49}{16}} \qquad \textbf{Step 4:} \text{ Apply the square root property.}$$

$$x - \frac{13}{4} = \pm\frac{7}{4} \qquad \text{Simplify the radical.}$$

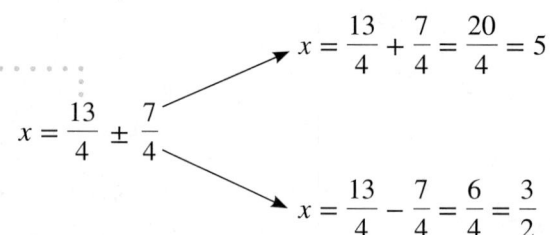

$$x = \frac{13}{4} + \frac{7}{4} = \frac{20}{4} = 5$$

$$x = \frac{13}{4} \pm \frac{7}{4}$$

$$x = \frac{13}{4} - \frac{7}{4} = \frac{6}{4} = \frac{3}{2}$$

The solution set is $\left\{5, \frac{3}{2}\right\}$.   The solutions are rational numbers.

> **TIP** In general, if the solutions to a quadratic equation are rational numbers, the equation can be solved by factoring and using the zero product rule. Consider the equation from Example 7.
>
> $$2x(2x - 10) = -30 + 6x$$
> $$4x^2 - 20x = -30 + 6x$$
> $$4x^2 - 26x + 30 = 0$$
> $$2(2x^2 - 13x + 15) = 0$$
> $$2(x - 5)(2x - 3) = 0$$
> $$x = 5 \quad \text{or} \quad x = \frac{3}{2}$$

> **Avoiding Mistakes**
>
> When the solutions are rational numbers, combine the like terms. That is, do not leave the solution with the ± sign.

Skill Practice 7 Solve by completing the square and applying the square root property. $2y(y - 1) = 3 - y$

Answers

6. $\left\{-\frac{3}{2} \pm \frac{\sqrt{14}}{2}\right\}$

7. $\left\{\frac{3}{2}, -1\right\}$

| **SECTION 3.5** | **Practice Exercises** |

Prerequisite Review

For Exercises R.1–R.4, simplify the radicals. (See Section 3.1 for review.)

R.1. $\sqrt{\dfrac{17}{4}}$ **R.2.** $\sqrt{\dfrac{23}{9}}$ **R.3.** $\sqrt{28}$ **R.4.** $\sqrt{63}$

For Exercises R.5–R.8, perform the indicated operations. (See Section R.1 for review.)

R.5. a. $\dfrac{1}{2} \cdot \dfrac{7}{3}$ **R.6. a.** $\dfrac{1}{2} \cdot \dfrac{4}{9}$ **R.7. a.** $\dfrac{8}{7} - \dfrac{4}{21}$ **R.8. a.** $\dfrac{15}{4} - \dfrac{5}{2}$

     b. $\dfrac{1}{2} + \dfrac{7}{3}$ **b.** $\dfrac{1}{2} + \dfrac{4}{9}$ **b.** $\dfrac{8}{7} \div \dfrac{4}{21}$ **b.** $\dfrac{15}{4} \div \dfrac{5}{2}$

For Exercises R.9–R.12, perform the indicated operation. (See Section 2.2 for review.)

R.9. $(a + 0.6)^2$ **R.10.** $(a - 0.8)^2$ **R.11.** $\left(h + \dfrac{1}{4}k\right)^2$ **R.12.** $\left(\dfrac{2}{3}x + 1\right)^2$

For Exercises R.13–R.16, factor completely. (See Section 2.4 for review.)

R.13. $x^2 + 20x + 100$ **R.14.** $y^2 - 14y + 49$

R.15. $\dfrac{1}{4}t^2 + \dfrac{3}{5}t + \dfrac{9}{25}$ **R.16.** $\dfrac{1}{16}y^2 + \dfrac{3}{4}y + \dfrac{9}{4}$

Concept Connections

1. The square root property states that for any real number k, if $x^2 = k$, then $x =$ _____ or $x =$ _____ .

2. To apply the square root property to the equation $t^2 + 2 = 11$, first subtract _____ from both sides. The solution set is _____ .

3. The process to create a perfect square trinomial is called _____ the square.

4. Fill in the blank to complete the square for the trinomial $x^2 + 20x +$ _____ .

5. To use completing the square to solve the equation $4x^2 + 3x + 5 = 0$, the first step is to divide by _____ so that the coefficient of the x^2 term is _____ .

6. Given the trinomial $y^2 + 8y + 16$, the coefficient of the linear term is _____ .

Objective 1: Solve Quadratic Equations by Using the Square Root Property

For Exercises 7–25, solve the equation by applying the square root property. **(See Examples 1–3)**

7. $t^2 = 144$ **8.** $x^2 = 100$ **9.** $y^2 = 4$ **10.** $a^2 = 5$

11. $k^2 - 7 = 0$ **12.** $4t^2 = 81$ **13.** $36u^2 = 121$ **14.** $3v^2 + 33 = 0$

15. $-2m^2 = 50$ **16.** $(p - 5)^2 = 9$ **17.** $(q + 3)^2 = 4$ **18.** $(3x - 2)^2 - 5 = 0$

19. $(2y + 3)^2 - 7 = 0$ **20.** $(h - 4)^2 = -8$ **21.** $(t + 5)^2 = -18$ **22.** $\left(x - \dfrac{3}{2}\right)^2 + \dfrac{7}{4} = 0$

23. $\left(m + \dfrac{4}{5}\right)^2 + \dfrac{3}{25} = 0$ **24.** $-x^2 + 4 = 13$ **25.** $-y^2 - 2 = 14$

26. Given the equation $x^2 = k$, match the following statements.

    a. If $k > 0$, then _____   **i.** there will be one real solution.

    b. If $k < 0$, then _____   **ii.** there will be two real solutions.

    c. If $k = 0$, then _____   **iii.** there will be two imaginary solutions.

Objective 2: Complete the Square

For Exercises 27–38, find the value of n so that the expression is a perfect square trinomial. Then factor the trinomial. **(See Example 4)**

27. $x^2 - 6x + n$ **28.** $x^2 + 24x + n$ **29.** $t^2 + 8t + n$ **30.** $v^2 - 18v + n$

31. $c^2 - c + n$ **32.** $x^2 + 9x + n$ **33.** $y^2 + 5y + n$ **34.** $a^2 - 7a + n$

35. $b^2 + \dfrac{2}{5}b + n$ **36.** $m^2 - \dfrac{2}{7}m + n$ **37.** $p^2 - \dfrac{2}{3}p + n$ **38.** $w^2 + \dfrac{3}{4}w + n$

Objective 3: Solve Quadratic Equations by Completing the Square

39. Summarize the steps used in solving a quadratic equation of the form $ax^2 + bx + c = 0$ by completing the square and applying the square root property.

40. What types of quadratic equations can be solved by completing the square and applying the square root property?

For Exercises 41–60, solve the quadratic equation by completing the square and applying the square root property. Write imaginary solutions in the form $a + bi$. **(See Examples 5–7)**

41. $t^2 + 8t + 15 = 0$ **42.** $m^2 + 6m + 8 = 0$ **43.** $x^2 + 6x = -16$ **44.** $x^2 - 4x = -15$

45. $p^2 + 4p + 6 = 0$ **46.** $q^2 + 2q + 2 = 0$ **47.** $-3y - 10 = -y^2$ **48.** $-24 = -2y^2 + 2y$

49. $2a^2 + 4a + 5 = 0$ **50.** $3a^2 + 6a - 7 = 0$ **51.** $9x^2 - 36x + 40 = 0$ **52.** $9y^2 - 12y + 5 = 0$

53. $25p^2 - 10p = 2$ **54.** $9n^2 - 6n = 1$ **55.** $(2w + 5)(w - 1) = 2$ **56.** $(3p - 5)(p + 1) = -3$

57. $n(n - 4) = 7$ **58.** $m(m + 10) = 2$ **59.** $2x(x + 6) = 14$ **60.** $3x(x - 2) = 24$

Mixed Exercises

For Exercises 61–70, refer to the geometry formulas at the end of the text.

61. If a square has an area of 64 in.^{2}, then what are the lengths of the sides?

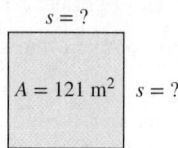

$s = ?$

$A = 64$ in.^{2} $s = ?$

62. If a square has an area of 121 m^2, then what are the lengths of the sides?

$s = ?$

$A = 121$ m^2 $s = ?$

63. A corner shelf is to be made from a triangular piece of plywood, as shown in the diagram. Find the distance x that the shelf will extend along the walls. Assume that the walls are at right angles. Round the answer to a tenth of a foot.

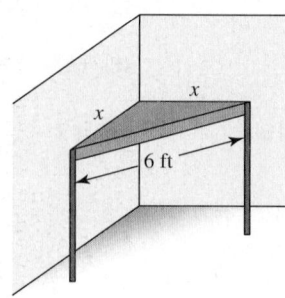

x

x

6 ft

64. The volume V (in cubic inches) of a box with a square bottom and a height of 4 in. is given by $V = 4x^2$, where x is the length (in inches) of the sides of the bottom of the box.

a. If the volume of the box is 289 in.^{3}, find the dimensions of the box.

b. Are there two possible answers to part (a)? Why or why not?

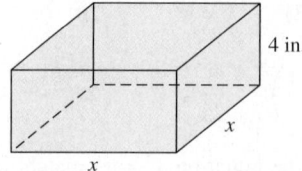

4 in.

x

x

65. Two mountain bikers take off from the same place at the same time. One travels north at 4 mph, and the other travels east at 3 mph. How far apart are they after 5 hr?

66. Professor Ortiz leaves campus on her bike, heading west at 12 ft/sec. Professor Wilson leaves campus at the same time and walks south at 5 ft/sec. How far apart are they after 40 sec?

67. A sprinkler rotates 360° to water a circular region. If the total area watered is approximately 2000 yd², determine the radius of the region (the radius is the length of the stream of water). Round the answer to the nearest yard. (*Hint*: The area A of a circle with radius r is $A = \pi r^2$).

68. An earthquake could be felt over a 46,000-mi² area. Up to how many miles from the epicenter could the earthquake be felt? Round to the nearest mile.

69. A baseball diamond is in the shape of a square with 90-ft sides. How far is it from home plate to second base? Give the exact value and give an approximation to the nearest tenth of a foot.

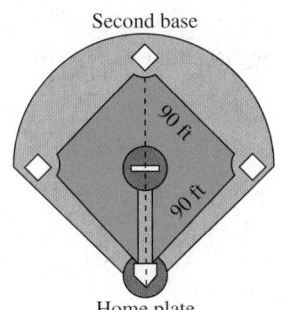

Second base

90 ft

90 ft

Home plate

70. The figure shown is a cube with 6-in. sides. Find the exact length of the diagonal through the interior of the cube d by following these steps.

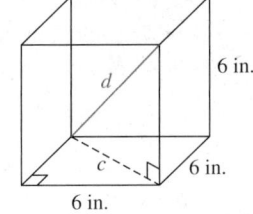

d

c

6 in.

6 in.

6 in.

a. Apply the Pythagorean theorem using the sides on the base of the cube to find the length of diagonal c.

b. Apply the Pythagorean theorem using c and the height of the cube as the legs of the right triangle through the interior of the cube.

Expanding Your Skills

71. The display area on a cell phone has a 3.5-in. diagonal.

a. If the aspect ratio of length to width is 1.5 to 1, determine the length and width of the display area. Round the values to the nearest hundredth of an inch.

b. If the phone has 326 pixels per inch, approximate the dimensions in pixels.

72. The display area on a computer has a 15-in. diagonal. If the aspect ratio of length to width is 1.6 to 1, determine the length and width of the display area. Round the values to the nearest hundredth of an inch.

73. A textbook company has discovered that the profit for selling its books is given by

$$P = -\frac{1}{8}x^2 + 5x$$

where x is the number of textbooks produced (in thousands) and P is the corresponding profit (in thousands of dollars).

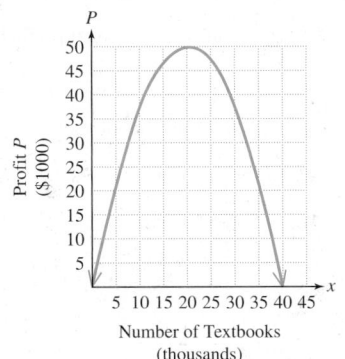

P

Profit P ($1000)

50
45
40
35
30
25
20
15
10
5

5 10 15 20 25 30 35 40 45

x

Number of Textbooks (thousands)

a. Approximate the number of books required to make a profit of $20,000. [*Hint*: Let $P = 20$. Then complete the square to solve for x.] Round to one decimal place.

b. Why are there two answers to part (a)?

74. The amount of money A in an account with an interest rate r compounded annually is given by

$$A = P(1 + r)^t$$

where P is the initial principal and t is the number of years the money is invested.

a. If a $10,000 investment grows to $11,664 after 2 years, find the interest rate.

b. If a $6000 investment grows to $7392.60 after 2 years, find the interest rate.

c. Jamal wants to invest $5000. He wants the money to grow to at least $6500 in 2 years to cover the cost of his son's first year at college. What interest rate does Jamal need for his investment to grow to $6500 in 2 years? Round to the nearest hundredth of a percent.

SECTION 3.6   Solving Quadratic Equations by Using the Quadratic Formula

OBJECTIVES

1. Solve Quadratic Equations by Using the Quadratic Formula
2. Apply the Discriminant
3. Solve Equations for a Specified Variable
4. Solve Applications Involving Quadratic Models

1. Solve Quadratic Equations by Using the Quadratic Formula

In Section 3.5 we learned how to solve a quadratic equation by completing the square and applying the square root property. This technique is important because it enables us to solve quadratic equations in which the quadratic polynomial is not factorable.

If we solve a general quadratic equation $ax^2 + bx + c = 0$ ($a \neq 0$) by completing the square, the result is a formula that gives solutions for x in terms of the coefficients a, b, and c.

$$ax^2 + bx + c = 0$$
Begin with a quadratic equation in standard form with $a > 0$.

$$\frac{ax^2}{a} + \frac{bx}{a} + \frac{c}{a} = \frac{0}{a}$$
Divide by the leading coefficient.

$$x^2 + \frac{b}{a}x + \frac{c}{a} = 0$$

$$x^2 + \frac{b}{a}x = -\frac{c}{a}$$
Isolate the terms containing x.

$$x^2 + \frac{b}{a}x + \left(\frac{1}{2} \cdot \frac{b}{a}\right)^2 = \left(\frac{1}{2} \cdot \frac{b}{a}\right)^2 - \frac{c}{a}$$
Add the square of $\frac{1}{2}$ the linear term coefficient to both sides of the equation.

$$\left(x + \frac{b}{2a}\right)^2 = \frac{b^2}{4a^2} - \frac{c}{a}$$
Factor the left side as a perfect square.

$$\left(x + \frac{b}{2a}\right)^2 = \frac{b^2 - 4ac}{4a^2}$$
Combine fractions on the right side by finding a common denominator.

$$x + \frac{b}{2a} = \pm\sqrt{\frac{b^2 - 4ac}{4a^2}}$$
Apply the square root property.

$$x + \frac{b}{2a} = \pm\frac{\sqrt{b^2 - 4ac}}{2a}$$
Simplify the denominator.

$$x = -\frac{b}{2a} \pm \frac{\sqrt{b^2 - 4ac}}{2a}$$
Subtract $\frac{b}{2a}$ from both sides.

$$= \frac{-b \pm \sqrt{b^2 - 4ac}}{2a}$$
Combine fractions.

The result is called the quadratic formula.

The Quadratic Formula

For a quadratic equation of the form $ax^2 + bx + c = 0$ ($a \neq 0$), the solutions are

$$x = \frac{-b \pm \sqrt{b^2 - 4ac}}{2a}$$

EXAMPLE 1   Using the Quadratic Formula

Solve the quadratic equation by using the quadratic formula. $2x^2 - 3x = 5$

Solution:

$$2x^2 - 3x = 5$$

$$2x^2 - 3x - 5 = 0 \qquad \text{Write the equation in the form } ax^2 + bx + c = 0.$$

$$a = 2, \qquad b = -3, \qquad c = -5 \qquad \text{Identify } a, b, \text{ and } c.$$

$$x = \frac{-b \pm \sqrt{b^2 - 4ac}}{2a} \qquad \text{Apply the quadratic formula.}$$

$$= \frac{-(-3) \pm \sqrt{(-3)^2 - 4(2)(-5)}}{2(2)} \qquad \text{Substitute } a = 2, b = -3, \text{ and } c = -5.$$

$$= \frac{3 \pm \sqrt{9 + 40}}{4} \qquad \text{Simplify.}$$

$$= \frac{3 \pm \sqrt{49}}{4}$$

$$= \frac{3 \pm 7}{4} \qquad \begin{cases} x = \dfrac{3 + 7}{4} = \dfrac{10}{4} = \dfrac{5}{2} \\[2mm] x = \dfrac{3 - 7}{4} = \dfrac{-4}{4} = -1 \end{cases}$$

The solution set is $\left\{\dfrac{5}{2}, -1\right\}$. Both solutions check in the original equation.

Skill Practice 1   Solve the equation by using the quadratic formula.

$$6x^2 - 5x = 4$$

EXAMPLE 2   Using the Quadratic Formula

Solve the equation by applying the quadratic formula. $x(x - 6) = 3$

Solution:

$$x(x - 6) = 3$$

$$x^2 - 6x - 3 = 0 \qquad \text{Write the equation in the form } ax^2 + bx + c = 0.$$

$$a = 1, b = -6, c = -3 \qquad \text{Identify the values of } a, b, \text{ and } c \text{ for the quadratic formula.}$$

$$x = \frac{-(-6) \pm \sqrt{(-6)^2 - 4(1)(-3)}}{2(1)} \qquad \text{Apply the quadratic formula.}$$

$$x = \frac{-b \pm \sqrt{b^2 - 4ac}}{2a}$$

$$= \frac{6 \pm \sqrt{36 + 12}}{2} \qquad \text{Simplify.}$$

$$= \frac{6 \pm \sqrt{48}}{2}$$

FOR REVIEW

Simplifying radicals by removing an nth power from an nth root is covered in detail in Section 3.1.

$$= \frac{6 \pm \sqrt{16 \cdot 3}}{2}$$

Simplify the radical.

$\sqrt{48} = \sqrt{16 \cdot 3} = \sqrt{16} \cdot \sqrt{3} = 4\sqrt{3}$

$$= \frac{6 \pm 4\sqrt{3}}{2}$$

The two terms in the numerator share a common factor of 2.

$$= \frac{2 \cdot (3 \pm 2\sqrt{3})}{2}$$

Factor a 2 from the numerator and then simplify the fraction to lowest terms.

$$= \frac{\overset{1}{2} \cdot (3 \pm 2\sqrt{3})}{2}$$

The ratio $\frac{2}{2} = 1$. Simplify the fraction.

$$= 3 \pm 2\sqrt{3}$$

The solution set is $\{3 \pm 2\sqrt{3}\}$.

> **Skill Practice 2** Solve the equation by applying the quadratic formula.
>
> $x(x - 8) = 3$

If a quadratic equation has fractional or decimal coefficients, we have the option of clearing fractions or decimals to create integer coefficients. This makes the application of the quadratic formula easier. This is demonstrated in Example 3.

EXAMPLE 3 Using the Quadratic Formula

Solve the equation by applying the quadratic formula. $\quad \dfrac{3}{10}x^2 - \dfrac{2}{5}x + \dfrac{7}{10} = 0$

Solution:

The equation has fractional coefficients. This makes the application of the quadratic formula computationally more challenging. Consider clearing fractions by multiplying both sides of the equation by the least common denominator of all individual fractions in the equation.

$$\frac{3}{10}x^2 - \frac{2}{5}x + \frac{7}{10} = 0$$

The least common denominator of the fractions is 10.

$$10 \cdot \left(\frac{3}{10}x^2 - \frac{2}{5}x + \frac{7}{10}\right) = 10 \cdot (0)$$

Multiply both sides by 10.

$$\frac{10}{1} \cdot \frac{3}{10}x^2 - \frac{10}{1} \cdot \frac{2}{5}x + \frac{10}{1} \cdot \frac{7}{10} = 0$$

Apply the distributive property.

$$3x^2 - 4x + 7 = 0$$

Write the equation in the form $ax^2 + bx + c = 0$.

$$a = 3, \, b = -4, \, c = 7$$

Identify the values of a, b, and c for the quadratic formula.

$$x = \frac{-(-4) \pm \sqrt{(-4)^2 - 4(3)(7)}}{2(3)}$$

Apply the quadratic formula.

$$x = \frac{-b \pm \sqrt{b^2 - 4ac}}{2a}$$

$$= \frac{4 \pm \sqrt{16 - 84}}{6}$$

Simplify.

Answer

2. $\{4 \pm \sqrt{19}\}$

$$= \frac{4 \pm \sqrt{-68}}{6}$$

$$= \frac{4 \pm i\sqrt{68}}{6} \qquad \text{Simplify the radical in terms of } i.$$

$$= \frac{4 \pm i\sqrt{4 \cdot 17}}{6} \qquad \begin{array}{l}\text{Simplify the radical.}\\ \sqrt{68} = \sqrt{4 \cdot 17} = \sqrt{4} \cdot \sqrt{17} = 2\sqrt{17}\end{array}$$

$$= \frac{4 \pm 2i\sqrt{17}}{6} \qquad \begin{array}{l}\text{The two terms in the numerator share}\\ \text{a common factor of 2.}\end{array}$$

$$= \frac{\overset{1}{2}(2 \pm i\sqrt{17})}{\underset{3}{6}} \qquad \begin{array}{l}\text{Factor a 2 from the numerator and then}\\ \text{simplify the fraction to lowest terms.}\end{array}$$

$$= \frac{2 \pm i\sqrt{17}}{3} \qquad \text{Simplify the fraction.}$$

$$= \frac{2}{3} \pm \frac{\sqrt{17}}{3}i \qquad \begin{array}{l}\text{Write the solution in standard form}\\ a + bi.\end{array}$$

The solution set is $\left\{ \dfrac{2}{3} \pm \dfrac{\sqrt{17}}{3}i \right\}$.

Skill Practice 3   Solve the equation by applying the quadratic formula.

$$\frac{5}{12}x^2 - \frac{1}{2}x + \frac{1}{4} = 0$$

Three methods have been presented to solve a quadratic equation. We offer these guidelines to choose an appropriate and efficient method to solve a given quadratic equation.

Methods to Solve a Quadratic Equation

Method/Notes	Examples
Apply the Zero Product Property • Set one side of the equation equal to zero and factor the other side. Then apply the zero product property.	Solve. $x^2 - x = 12$ $x^2 - x - 12 = 0$ $(x - 4)(x + 3) = 0$ $x = 4$ or $x = -3$
Complete the Square and Apply the Square Root Property • Good choice if the equation is in the form $x^2 = k$. • Good choice if the equation is in the form $ax^2 + bx + c = 0$, where $a = 1$ and b is an even real number.	Solve. $c^2 = -6$ $c = \pm\sqrt{-6}$ $c = \pm i\sqrt{6}$ Solve. $x^2 + 6x + 2 = 0$ $x^2 + 6x + 9 = -2 + 9$ $(x + 3)^2 = 7$ $x + 3 = \pm\sqrt{7}$ $x = -3 \pm \sqrt{7}$

Answer

3. $\left\{ \dfrac{3}{5} \pm \dfrac{\sqrt{6}}{5}i \right\}$

Apply the Quadratic Formula

- Applies in all situations.
- Consider clearing fractions or decimals if the coefficients are not integer values.

Solve. $0.2x^2 + 0.5x + 0.1 = 0$

$$10(0.2x^2 + 0.5x + 0.1) = 10(0)$$

$$2x^2 + 5x + 1 = 0$$

$$x = \frac{-(5) \pm \sqrt{(5)^2 - 4(2)(1)}}{2(2)}$$

$$x = \frac{-5 \pm \sqrt{17}}{4}$$

2. Apply the Discriminant

The solutions to a quadratic equation are given by $x = \dfrac{-b \pm \sqrt{b^2 - 4ac}}{2a}$. The radicand, $b^2 - 4ac$, is called the *discriminant*. The value of the discriminant tells us the number and type of solutions to the equation. We examine three different cases.

Using the Discriminant to Determine the Number and Type of Solutions to a Quadratic Equation

Given a quadratic equation $ax^2 + bx + c = 0$ $(a \neq 0)$, the quantity $b^2 - 4ac$ is called the **discriminant**.

Discriminant $b^2 - 4ac$	Number and Type of Solutions	Examples	Result of Quadratic Formula
$b^2 - 4ac < 0$	2 nonreal solutions	$2x^2 - 3x + 5 = 0$ $b^2 - 4ac = (-3)^2 - 4(2)(5)$ $= -31$	$x = \dfrac{3 \pm \sqrt{-31}}{4}$
$b^2 - 4ac = 0$	1 real solution	$x^2 + 6x + 9 = 0$ $b^2 - 4ac = (6)^2 - 4(1)(9)$ $= 0$	$x = \dfrac{-6 \pm \sqrt{0}}{2} = -3$
$b^2 - 4ac > 0$	2 real solutions	$2x^2 + 7x - 1 = 0$ $b^2 - 4ac = (7)^2 - 4(2)(-1)$ $= 57$	$x = \dfrac{-7 \pm \sqrt{57}}{4}$

EXAMPLE 4   Using the Discriminant

Use the discriminant to determine the number and type of solutions for each equation.

a. $5x^2 - 3x + 1 = 0$ **b.** $2x^2 = 3 - 6x$ **c.** $4x^2 + 12x = -9$

Solution:

Equation	$b^2 - 4ac$	Solution Type and Number
a. $5x^2 - 3x + 1 = 0$	$(-3)^2 - 4(5)(1)$ $= -11$	Because $-11 < 0$, there are two nonreal solutions.
b. $2x^2 = 3 - 6x$ $2x^2 + 6x - 3 = 0$	$(6)^2 - 4(2)(-3)$ $= 60$	Because $60 > 0$, there are two real solutions.

c. $4x^2 + 12x = -9$

$4x^2 + 12x + 9 = 0$ $(12)^2 - 4(4)(9)$ Because the discriminant is
                                    $= 0$ 0, there is one real solution.

Skill Practice 4   Use the discriminant to determine the number and type
of solutions for each equation.

   a. $2x^2 - 4x + 5 = 0$ **b.** $25x^2 = 10x - 1$ **c.** $x^2 + 10x = -9$

FOR REVIEW

For more practice solving a
formula for a specific
variable, see Section 1.2.
There we solve equations in
which the specified variable
is linear within the equation.

3. Solve Equations for a Specified Variable

Sometimes a formula with multiple variables may be quadratic in one or more of
the variables. For example, the equation $d = 4.9t^2$ is quadratic in the variable t. We
say this because t is raised to the second power. We can also say that the equation
$d = 4.9t^2$ is linear in the variable d because d is raised to the first power.

   In Example 5, we solve a formula for a specified variable that is raised to the
second power. To solve a formula for a specified variable, we treat the other variables
in the equations as constants.

EXAMPLE 5   **Solving an Equation for a Specified Variable**

Ignoring air resistance, the distance d (in meters) that an object falls in t sec is
given by $d = 4.9t^2$ where $t \geq 0$.

   a. Solve the equation for t.

   b. Determine the amount of time required for an object to fall 500 m. Round to
   the nearest second.

Solution:

a. $d = 4.9t^2$ In this equation we want to isolate the variable t. Therefore, we treat
                                      all other variables (in this case d) as constants.

$\dfrac{d}{4.9} = t^2$ Isolate the t term by dividing both sides by 4.9. The equation is now
                                in the form of a perfect square equal to a constant.

$t = \pm\sqrt{\dfrac{d}{4.9}}$ Apply the square root property.

$t = \sqrt{\dfrac{d}{4.9}}$ In the statement of this example, we're told that $t \geq 0$. Therefore, we
                                      dismiss the negative solution.

We have the option of rationalizing the denominator.

$$t = \sqrt{\dfrac{d}{4.9}} = \dfrac{\sqrt{d}}{\sqrt{4.9}} = \dfrac{\sqrt{d}}{\sqrt{4.9}} \cdot \dfrac{\sqrt{4.9}}{\sqrt{4.9}} = \dfrac{\sqrt{4.9d}}{\sqrt{(4.9)^2}} = \dfrac{\sqrt{4.9d}}{4.9}$$

Thus, $t = \sqrt{\dfrac{d}{4.9}}$ or $t = \dfrac{\sqrt{4.9d}}{4.9}$.

b. $t = \sqrt{\dfrac{d}{4.9}} = \sqrt{\dfrac{(500)}{4.9}} \approx 10.1$ Substitute 500 sec for d.

The object will require approximately 10.1 sec to fall 500 m.

Answers

4. a. Discriminant: -24 (2 nonreal
            solutions)
   b. Discriminant: 0 (1 real solution)
   c. Discriminant: 64 (2 real
            solutions)

5. a. $z = \sqrt{\dfrac{x}{2y}}$ or $z = \dfrac{\sqrt{2xy}}{2y}$

   b. $z = 3$

Skill Practice 5   Given $x = 2yz^2$,

   a. Solve for z where $z > 0$.

   b. Find z when $x = 54$ and $y = 3$.

EXAMPLE 6   **Solving an Equation for a Specified Variable**

Solve for r. $V = \frac{1}{3}\pi r^2 h$ $(r > 0)$

Solution:

$$V = \frac{1}{3}\pi r^2 h$$

This equation is quadratic in the variable r. The strategy in this example is to isolate r^2 and then apply the square root property.

$$3(V) = 3\left(\frac{1}{3}\pi r^2 h\right)$$

Multiply both sides by 3 to clear fractions.

$$3V = \pi r^2 h$$

$$\frac{3V}{\pi h} = \frac{\pi r^2 h}{\pi h}$$

Divide both sides by πh to isolate r^2.

$$\frac{3V}{\pi h} = r^2$$

$$r = \sqrt{\frac{3V}{\pi h}} \text{ or } r = \frac{\sqrt{3V\pi h}}{\pi h}$$

Apply the square root property. Since $r > 0$, we take the positive square root only.

Skill Practice 6   Solve for v. $E = \frac{1}{2}mv^2$ $(v > 0)$

In Example 7, we have the formula $mt^2 + nt = z$. This formula is linear in the variables m, n, and z, and quadratic in the variable t since t appears both to the first power and to the second power. To solve for t, our approach will be to write the formula as an equation written in descending order by t. The other variables are treated as constants, so the equation will be written in the form $at^2 + bt + c = 0$ and subsequently solved by applying the quadratic formula.

EXAMPLE 7   **Solving an Equation for a Specified Variable**

Solve for t. $mt^2 + nt = z$

Solution:

This equation is quadratic in the variable t. The strategy is to write the polynomial in descending order by powers of t. Then since there are two t terms with different exponents, we cannot isolate t directly. Instead we apply the quadratic formula.

$$mt^2 + nt = z$$

$$mt^2 + nt - z = 0$$

Write the polynomial in descending order by t.

$$a = m, b = n, c = -z$$

Identify the coefficients of each term.

$$t = \frac{-(n) \pm \sqrt{(n)^2 - 4(m)(-z)}}{2m}$$

Apply the quadratic formula.

$$t = \frac{-n \pm \sqrt{n^2 + 4mz}}{2m}$$

Simplify.

Answers

6. $v = \sqrt{\dfrac{2E}{m}}$ or $v = \dfrac{\sqrt{2Em}}{m}$

7. $p = \dfrac{d \pm \sqrt{d^2 + 4ck}}{2c}$

Skill Practice 7   Solve for p. $cp^2 - dp = k$

4. Solve Applications Involving Quadratic Models

Quadratic equations can be useful in solving applications involving formulas (or models) that are quadratic in one or more variables. For example, the formula $A = s^2$ gives the area of a square based on the length s of its sides. In particular, this formula is quadratic in the variable s. Consider the model given in Exercise 58 where the gas mileage m (in mpg) for a car is based on the speed of the vehicle x (in mph): $m = -0.04x^2 + 3.6x - 49$. This model is quadratic in the variable x.

In the study of physical science, a common model used to represent the vertical position of an object moving vertically under the influence of gravity is given in Table 3-1.

Table 3-1   Vertical Position of an Object

Suppose that an object has an initial vertical position of s_0 and initial velocity v_0 straight upward. The vertical position s of the object is given by

$$s = -\frac{1}{2}gt^2 + v_0t + s_0, \text{ where}$$

g	is the acceleration due to gravity. On Earth, $g = 32$ ft/sec^2 or $g = 9.8$ m/sec^2.
t	is the time of travel.
v_0	is the initial velocity.
s_0	is the initial vertical position.
s	is the vertical position of the object at time t.

For example, suppose that a child tosses a ball straight upward from a height of 1.5 ft, with an initial velocity of 48 ft/sec.

The initial height is $s_0 = 1.5$ ft.
The initial velocity is $v_0 = 48$ ft/sec.
The acceleration due to gravity is $g = 32$ ft/sec^2.
The vertical position of the ball (in feet) is given by

$$s = -\frac{1}{2}gt^2 + v_0t + s_0$$

$$s = -\frac{1}{2}(32)t^2 + (48)t + (1.5)$$

$$= -16t^2 + 48t + 1.5$$

TIP The value of g is chosen to be consistent with the units for position and velocity. In this case, the initial height is given in ft. The initial velocity is given in ft/sec. Therefore, we choose g in ft/sec^2 rather than m/sec^2.

EXAMPLE 8   Analyzing an Object Moving Vertically

A toy rocket is shot straight upward from a launch pad of 1 m above ground level with an initial velocity of 24 m/sec.

a. Write a model to express the height of the rocket s (in meters) above ground level.

b. Find the time(s) at which the rocket is at a height of 20 m. Round to 1 decimal place.

c. Find the time(s) at which the rocket is at a height of 40 m.

Solution:

a. $s = -\dfrac{1}{2}gt^2 + v_0 t + s_0$

$\qquad$ In this example,

$\qquad\qquad s_0 = 1$ m

$s = -\dfrac{1}{2}(9.8)t^2 + (24)t + (1)$

$\qquad\qquad v_0 = 24$ m/sec

$\qquad\qquad g = 9.8$ m/sec^2

$\quad = -4.9t^2 + 24t + 1$

b. $20 = -4.9t^2 + 24t + 1$ $\qquad$ Substitute 20 for s.

$4.9t^2 - 24t + 19 = 0$ $\qquad$ Set one side equal to zero.

$t = \dfrac{-(-24) \pm \sqrt{(-24)^2 - 4(4.9)(19)}}{2(4.9)}$ $\qquad$ Apply the quadratic formula.

$t = \dfrac{24 \pm \sqrt{203.6}}{9.8}$

$\qquad t = \dfrac{24 + \sqrt{203.6}}{9.8} \approx 3.9$

$\qquad t = \dfrac{24 - \sqrt{203.6}}{9.8} \approx 1.0$

The rocket will be at a height of 20 m at 1 sec and 3.9 sec after launch.

c. $40 = -4.9t^2 + 24t + 1$ $\qquad$ Substitute 40 for s.

$4.9t^2 - 24t + 39 = 0$

$t = \dfrac{-(-24) \pm \sqrt{(-24)^2 - 4(4.9)(39)}}{2(4.9)}$ $\qquad$ Apply the quadratic formula.

$t = \dfrac{24 \pm \sqrt{-188.4}}{9.8}$ $\qquad$ The solutions are not real numbers.

There is no real number t for which the height of the rocket will be 40 m. The rocket will not reach a height of 40 m.

Skill Practice 8 A fireworks mortar is launched straight upward from a pool deck 2 m off the ground at an initial velocity of 40 m/sec.

a. Write a model to express the height of the mortar s (in meters) above ground level.

b. Find the time(s) at which the mortar is at a height of 60 m. Round to 1 decimal place.

c. Find the time(s) at which the rocket is at a height of 100 m.

Answers

8. a. $s = -4.9t^2 + 40t + 2$

$\quad$ **b.** The mortar will be at a height of 60 m at 1.9 sec and 6.3 sec after launch.

$\quad$ **c.** The mortar will not reach a height of 100 m.

Prerequisite Review

For Exercises R.1–R.4, write the expressions in terms of i and simplify. (See Section 3.3 for review.)

R.1. $\sqrt{-3}$ $\qquad$ **R.2.** $\sqrt{-17}$ $\qquad$ **R.3.** $-\sqrt{-20}$ $\qquad$ **R.4.** $-\sqrt{-75}$

For Exercises R.5–R.8, simplify the expression and write the answer in the form $a + bi$. (See Section 3.3 for review.)

R.5. $\dfrac{2 + \sqrt{-16}}{8}$ $\qquad$ **R.6.** $\dfrac{6 - \sqrt{-4}}{4}$ $\qquad$ **R.7.** $\dfrac{-6 + \sqrt{-72}}{6}$ $\qquad$ **R.8.** $\dfrac{-20 + \sqrt{-500}}{10}$

For Exercises R.9–R.12, clear fractions or decimals and solve the equation. (See Section 1.2 for review.)

R.9. $\dfrac{2}{5}x - \dfrac{1}{10} = -\dfrac{2}{15}x + \dfrac{3}{2} - \dfrac{1}{6}x$

R.10. $-\dfrac{1}{3}y + 4 = -\dfrac{5}{9}y + \dfrac{2}{6}$

R.11. $0.15(m - 2) + 0.1m = 0.6$

R.12. $0.2(n + 8) + 0.8n = 7$

For Exercises R.13–R.16, solve for the indicated variable. (See Section 1.2 for review.)

R.13. $I = Prt$ for r

R.14. $K = \dfrac{1}{2}mv^2$ for v^2

R.15. $v = v_0 + at$ for a

R.16. $ax + by = c$ for y

Concept Connections

1. For the equation $ax^2 + bx + c = 0$ ($a \neq 0$), the _____ formula gives the solutions as $x =$ _____.

2. To apply the quadratic formula, a quadratic equation must be written in the form _____ where $a \neq 0$.

3. To apply the quadratic formula to solve the equation $8x^2 - 42x - 27 = 0$, the value of a is _____, the value of b is _____, and the value of c is _____.

4. To apply the quadratic formula to solve the equation $3x^2 - 7x - 4 = 0$, the value of $-b$ is _____ and the value of the radicand is _____.

5. The radicand within the quadratic formula is _____ and is called the _____.

6. If the discriminant is negative, then the solutions to a quadratic equation will be (real/imaginary) numbers.

7. If the discriminant is positive, then the solutions to a quadratic equation will be (real/imaginary) numbers.

8. If the discriminant is equal to zero, then how many unique solutions are there to the corresponding quadratic equation?

Objective 1: Solve Quadratic Equations by Using the Quadratic Formula

For Exercises 9–28, use the quadratic formula to solve the equation. (**See Examples 1–3**)

9. $x^2 + 11x - 12 = 0$

10. $5x^2 - 14x - 3 = 0$

11. $x^2 - 3x - 7 = 0$

12. $x^2 - 5x - 9 = 0$

13. $y^2 = -4y - 6$

14. $z^2 = -8z - 19$

15. $t(t - 6) = -10$

16. $m(m + 10) = -34$

17. $-7c + 3 = -5c^2$

18. $-5d + 2 = -6d^2$

19. $(6x + 5)(x - 3) = -2x(7x + 5) + x - 12$

20. $(5c + 7)(2c - 3) = -2c(c + 15) - 35$

21. $9x^2 + 49 = 0$

22. $121x^2 + 4 = 0$

23. $\dfrac{1}{2}x^2 - \dfrac{2}{7} = \dfrac{5}{14}x$

24. $\dfrac{1}{3}x^2 - \dfrac{7}{6} = \dfrac{3}{2}x$

25. $0.4y^2 = 2y - 2.5$

26. $0.09n^2 = 0.42n - 0.49$

27. $-z^2 = -2z - 35$

28. $-12x^2 + 5x = -2$

Objective 2: Apply the Discriminant

For Exercises 29–36, (a) evaluate the discriminant, and (b) determine the number and type (real or nonreal) of solutions to each equation. (**See Example 4**)

29. $3x^2 - 4x + 6 = 0$

30. $5x^2 - 2x + 4 = 0$

31. $-2w^2 + 8w = 3$

32. $-6d^2 + 9d = 2$

33. $3x(x - 4) = x - 4$

34. $2x(x - 2) = x + 3$

35. $-1.4m + 0.1 = -4.9m^2$

36. $3.6n + 0.4 = -8.1n^2$

Objective 3: Solve Equations for a Specified Variable

For Exercises 37–50, solve for the indicated variable. (**See Examples 5–7**)

37. $A = \pi r^2$ for $r > 0$

38. $V = \pi r^2 h$ for $r > 0$

39. $s = \frac{1}{2} g t^2$ for $t > 0$

40. $c = \frac{d^2 t}{2}$ for $d > 0$

41. $a^2 + b^2 = c^2$ for $a > 0$

42. $a^2 + b^2 + c^2 = d^2$ for $c > 0$

43. $L = c^2 I^2 R t$ for $I > 0$

44. $I = c N^2 r^2 s$ for $N > 0$

45. $kw^2 - cw = r$ for w

46. $dy^2 + my = p$ for y

47. $s = v_0 t + \frac{1}{2} a t^2$ for t

48. $S = 2\pi r h + \pi r^2 h$ for r

49. $L I^2 + R I + \frac{1}{C} = 0$ for I

50. $A = \pi r^2 + \pi r s$ for r

Objective 4: Solve Applications Involving Quadratic Models

For Exercises 51–54, use the model $s = -\frac{1}{2} g t^2 + v_0 t + s_0$ with $g = 32$ ft/sec² or $g = 9.8$ m/sec². (**See Example 8**)

51. NBA basketball legend Michael Jordan had a 48-in. vertical leap. Suppose that Michael jumped from ground level with an initial velocity of 16 ft/sec.

   a. Write a model to express Michael's height (in ft) above ground level t seconds after leaving the ground.

   b. Use the model from part (a) to determine how long it would take Michael to reach his maximum height of 48 in. (4 ft).

52. At the time of this printing, the highest vertical leap on record is 60 in., held by Kadour Ziani. For this record-setting jump, Kadour left the ground with an initial velocity of $8\sqrt{5}$ ft/sec.

   a. Write a model to express Kadour's height (in ft) above ground level t seconds after leaving the ground.

   b. Use the model from part (a) to determine how long it would take Kadour to reach his maximum height of 60 in. (5 ft). Round to the nearest hundredth of a second.

53. A bad punter on a football team kicks a football approximately straight upward with an initial velocity of 75 ft/sec.

   a. If the ball leaves his foot from a height of 4 ft, write an equation for the vertical height s (in ft) of the ball t seconds after being kicked.

   b. Find the time(s) at which the ball is at a height of 80 ft. Round to 1 decimal place.

54. In a classic *Seinfeld* episode, Jerry tosses a loaf of bread (a marble rye) straight upward to his friend George who is leaning out of a third-story window.

   a. If the loaf of bread leaves Jerry's hand at a height of 1 m with an initial velocity of 18 m/sec, write an equation for the vertical position of the bread s (in meters) t seconds after release.

   b. How long will it take the bread to reach George if he catches the bread on the way up at a height of 16 m? Round to the nearest tenth of a second.

55. In a round-robin tennis tournament, each player plays every other player exactly one time. The number of matches N is given by $N = \frac{1}{2} n(n - 1)$, where n is the number of players in the tournament. If 28 matches were played, how many players were in the tournament?

56. The sum of the first n natural numbers, $S = 1 + 2 + 3 + \cdots + n$, is given by $S = \frac{1}{2} n(n + 1)$. If the sum of the first n natural numbers is 171, determine the value of n.

57. The population P of a culture of *Pseudomonas aeruginosa* bacteria is given by $P = -1718 t^2 + 82{,}000 t + 10{,}000$, where t is the time in hours since the culture was started. Determine the time(s) at which the population was 600,000. Round to the nearest hour.

58. The gas mileage for a certain vehicle can be approximated by $m = -0.04 x^2 + 3.6 x - 49$, where x is the speed of the vehicle in mph. Determine the speed(s) at which the car gets 30 mpg. Round to the nearest mph.

59. The distance d (in ft) required to stop a car that was traveling at speed v (in mph) before the brakes were applied depends on the amount of friction between the tires and the road and the driver's reaction time. After an accident, a legal team hired an engineering firm to collect data for the stretch of road where the accident occurred. Based on the data, the stopping distance is given by $d = 0.05v^2 + 2.2v$.

a. Determine the distance required to stop a car going 50 mph.

b. Up to what speed (to the nearest mph) could a motorist be traveling and still have adequate stopping distance to avoid hitting a deer 330 ft away?

Gaja Snover/Alamy Stock Photo

60. Leptin is a hormone that has a central role in fat metabolism. One study published in the *New England Journal of Medicine* measured serum leptin concentrations versus the percentage of body fat for 275 individuals. The concentration of leptin c (in ng/mL) is approximated by $c = 219x^2 - 26.7x + 1.64$, where x is percentage of body fat.

a. Determine the concentration of leptin in an individual with 22% body fat ($x = 0.22$). Round to 1 decimal place.

b. If an individual has 3 ng/mL of leptin, determine the percentage of body fat. Round to the nearest whole percent.

(*Source*: "Serum Immunoreactive-Leptin Concentrations in Normal-Weight and Obese Humans," *New England Journal of Medicine*, Feb., 1996)

61. The fatality rate F (in fatalities per 100 million vehicle miles driven) for drivers x years old can be approximated by $F = 0.0036x^2 - 0.35x + 9.2$. (*Source:* U.S. Department of Transportation)

a. Approximate the fatality rate for drivers 16 years old.

b. Approximate the fatality rate for drivers 40 years old.

c. Approximate the fatality rate for drivers 80 years old.

d. For what age(s) is the fatality rate approximately 2.5?

62. The braking distance d (in feet) of a car going v mph is given by

$$d = \frac{v^2}{20} + v \qquad v \geq 0$$

a. Find the speed for a braking distance of 150 ft. Round to the nearest mile per hour.

b. Find the speed for a braking distance of 100 ft. Round to the nearest mile per hour.

Mixed Exercises

For Exercises 63–70, identify the equation as linear, quadratic, or neither. If the equation is linear or quadratic, find the solution set.

63. $2y + 4 = 0$

64. $3z - 9 = 0$

65. $2y^2 + 4y = 0$

66. $3z^2 - 9z = 0$

67. $5x(x + 6) = 5x^2 + 27x + 3$

68. $3x(x - 4) = 3x^2 - 11x + 4$

69. $2x^2(x + 7) = x^2 + 3x + 1$

70. $-x(x^2 - 5) + 4 = x^2 + 5$

For Exercises 71–86, solve the equation using any method.

71. $(3x - 4)^2 = 0$

72. $(2x + 1)^2 = 0$

73. $m^2 + 4m = -2$

74. $n^2 + 8n = -3$

75. $\dfrac{x^2 - 4x}{6} - \dfrac{5x}{3} = 1$

76. $\dfrac{m^2 + 2m}{7} - \dfrac{9m}{14} = \dfrac{3}{2}$

77. $2(x + 4) + x^2 = x(x + 2) + 8$

78. $3(y - 5) + y^2 = y(y + 3) - 15$

79. $\dfrac{3}{5}x^2 - \dfrac{1}{10}x = \dfrac{1}{2}$

80. $\dfrac{1}{12}x^2 - \dfrac{11}{24}x = -\dfrac{1}{2}$

81. $x^2 - 5x = 5x(x - 1) - 4x^2 + 1$

82. $p^2 - 4p = 4p(p - 1) - 3p^2 + 2$

83. $(2y + 7)(y + 1) = 2y^2 - 11$

84. $(3z - 8)(z + 2) = 3z^2 + 10$

85. $7d^2 + 5 = 0$

86. $11t^2 + 3 = 0$

Expanding Your Skills

For Exercises 87–88, solve for the indicated variable.

87. $x^2 - xy - 2y^2 = 0$ for x

88. $3a^2 + 2ab - b^2 = 0$ for a

For Exercises 89–98, write an equation with integer coefficients and the variable x that has the given solution set. [*Hint*: Apply the zero product property in reverse. For example, to build an equation whose solution set is $\{2, -\frac{5}{2}\}$ we have $(x - 2)(2x + 5) = 0$, or simply $2x^2 + x - 10 = 0$.]

89. $\{4, -2\}$

90. $\{7, -1\}$

91. $\left\{\dfrac{2}{3}, \dfrac{1}{4}\right\}$

92. $\left\{\dfrac{3}{5}, \dfrac{1}{7}\right\}$

93. $\left\{\sqrt{5}, -\sqrt{5}\right\}$

94. $\left\{\sqrt{2}, -\sqrt{2}\right\}$

95. $\{2i, -2i\}$

96. $\{9i, -9i\}$

97. $\{1 \pm 2i\}$

98. $\{2 \pm 9i\}$

The solutions to the equation $ax^2 + bx + c = 0$ $(a \neq 0)$ are $x_1 = \dfrac{-b + \sqrt{b^2 - 4ac}}{2a}$ and $x_2 = \dfrac{-b - \sqrt{b^2 - 4ac}}{2a}$. For Exercises 99–100, prove the given statements.

99. Prove that $x_1 + x_2 = -\dfrac{b}{a}$.

100. Prove that $x_1 x_2 = \dfrac{c}{a}$.

101. A **golden rectangle** is a rectangle in which the ratio of its length to its width is equal to the ratio of the sum of its length and width to its length: $\frac{L}{W} = \frac{L + W}{L}$ (values of L and W that meet this condition are said to be in the **golden ratio**).

    a. Suppose that a golden rectangle has a width of 1 unit. Solve the equation to find the exact value for the length. Then give a decimal approximation to 2 decimal places.

    b. To create a golden rectangle with a width of 9 ft, what should be the length? Round to 1 decimal place.

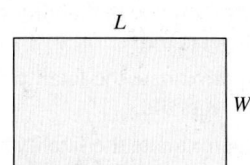

102. An artist has been commissioned to make a stained glass window in the shape of a regular octagon. The octagon must fit inside an 18-in. square space. Determine the length of each side of the octagon. Round to the nearest hundredth of an inch.

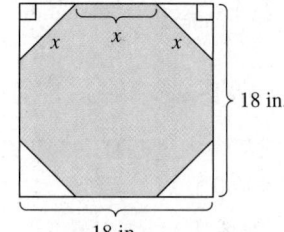

103. A farmer has 160 yd of fencing material and wants to enclose three rectangular pens. Suppose that x represents the length of each pen and y represents the width as shown in the figure.

    a. Assuming that the farmer uses all 160 yd of fencing, write an expression for y in terms of x.

    b. Write an expression in terms of x for the area of one individual pen.

    c. If the farmer wants to design the structure so that each pen encloses 250 yd^2, determine the dimensions of each pen.

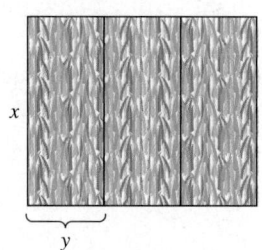

104. At noon, a ship leaves a harbor and sails south at 10 knots. Two hours later, a second ship leaves the harbor and sails east at 15 knots. When will the ships be 100 nautical miles apart? Round to the nearest minute.

CHAPTER 3   Detailed Summary

SECTION 3.1 Introduction to Radicals and Their Simplification

Key Concepts	Examples	Page
The symbol $\sqrt[n]{a}$ represents the principal nth root of a. The value a is called the **radicand** and n is the **index**. The expression $\sqrt[n]{x^n}$ represents the principal nth root of x^n. If n is even, the principal nth root is defined to be nonnegative. Thus, $$\sqrt[n]{x^n} = \begin{cases} x \text{ if } n \text{ is odd} \\ |x| \text{ if } n \text{ is even} \end{cases}$$	**Example 1:** **a.** $\sqrt[3]{(-14)^3} = -14$ **b.** $\sqrt[4]{(-14)^4} = |-14| = 14$	p. 186
Product and quotient property of radicals: Given real numbers $\sqrt[n]{a}$ and $\sqrt[n]{b}$, Product property: $\sqrt[n]{a} \cdot \sqrt[n]{b} = \sqrt[n]{ab}$ Quotient property: $\dfrac{\sqrt[n]{a}}{\sqrt[n]{b}} = \sqrt[n]{\dfrac{a}{b}}$	**Example 2:** $\sqrt{3} \cdot \sqrt{5} = \sqrt{15}$ **Example 3:** $\dfrac{\sqrt{8}}{\sqrt{2}} = \sqrt{\dfrac{8}{2}} = \sqrt{4} = 2$	p. 187
Simplified form of a radical: 1. The radicand has no factor that is a perfect nth power. 2. No fractions may appear in the radicand. 3. No denominator of a fraction may contain a radical. 4. The exponents in the radicand may not all share a common factor with the index.	**Example 4:** $\sqrt{50} = \sqrt{5^2 \cdot 2}$ $\quad = \sqrt{5^2} \cdot \sqrt{2}$ $\quad = 5\sqrt{2}$	p. 189
Adding and subtracting radicals: Like radicals have the same index and same radicand. To add or subtract like radicals, apply the distributive property.	**Example 5:** $2\sqrt[3]{5x} + 7\sqrt[3]{5x} - \sqrt[3]{5x}$ $\quad = (2 + 7 - 1)\sqrt[3]{5x}$ $\quad = 8\sqrt[3]{5x}$	p. 190

SECTION 3.2 Multiplying Radicals and Rationalizing the Denominator

Key Concepts	Examples	Page
Multiply radicals: If $\sqrt[n]{a}$ and $\sqrt[n]{b}$ are real numbers, then $$\sqrt[n]{a} \cdot \sqrt[n]{b} = \sqrt[n]{ab}.$$	**Example 1:** $\sqrt{15} \cdot \sqrt{10} = \sqrt{150} = \sqrt{5^2 \cdot 2 \cdot 3}$ $\quad = \sqrt{5^2} \cdot \sqrt{6} = 5\sqrt{6}$	p. 196
Multiply radicals with multiple terms: To multiply radical expressions with multiple terms, multiply each term in the first expression by each term in the second.	**Example 2:** $\left(3\sqrt{2} + \sqrt{y}\right)\left(4\sqrt{2} - \sqrt{y}\right)$ $= 3\sqrt{2}\left(4\sqrt{2}\right) + 3\sqrt{2}(-\sqrt{y}) + \sqrt{y}\left(4\sqrt{2}\right) + \sqrt{y}(-\sqrt{y})$ $= 12\sqrt{4} - 3\sqrt{2y} + 4\sqrt{2y} - \sqrt{y^2}$ $= 12 \cdot 2 + \sqrt{2y} - y$ $= 24 + \sqrt{2y} - y$	p. 197

Special case products:	Example 3:	p. 198
$(a + b)^2 = a^2 + 2ab + b^2$ $(a - b)^2 = a^2 - 2ab + b^2$ $(a + b)(a - b) = a^2 - b^2$	$(x + \sqrt{5})^2 = x^2 + 2x(\sqrt{5}) + (\sqrt{5})^2$ $\qquad\qquad = x^2 + 2\sqrt{5}x + 5$ **Example 4:** $(\sqrt{w} + \sqrt{7})(\sqrt{w} - \sqrt{7}) = (\sqrt{w})^2 - (\sqrt{7})^2$ $\qquad\qquad\qquad\qquad = w - 7$	

Rationalize the denominator:	Example 5:	p. 199
Eliminating a radical from the denominator of a fraction is called rationalizing the denominator.	$\dfrac{3}{\sqrt{7}} = \dfrac{3}{\sqrt{7}} \cdot \dfrac{\sqrt{7}}{\sqrt{7}} = \dfrac{3\sqrt{7}}{\sqrt{49}} = \dfrac{3\sqrt{7}}{7}$ **Example 6:** $\dfrac{11}{5 - \sqrt{3}} = \dfrac{(11)}{(5 - \sqrt{3})} \cdot \dfrac{(5 + \sqrt{3})}{(5 + \sqrt{3})}$ $\qquad = \dfrac{55 + 11\sqrt{3}}{(5)^2 - (\sqrt{3})^2} = \dfrac{55 + 11\sqrt{3}}{25 - 3}$ $\qquad = \dfrac{55 + 11\sqrt{3}}{22}$ $\qquad = \dfrac{11(5 + \sqrt{3})}{22}$ $\qquad = \dfrac{\overset{1}{11}(5 + \sqrt{3})}{\underset{2}{22}}$ $\qquad = \dfrac{5 + \sqrt{3}}{2}$	

SECTION 3.3 Complex Numbers

Key Concepts	Examples	Page
Definition of i: $i = \sqrt{-1}$ and $i^2 = -1$. For a real number $b > 0$, $\sqrt{-b} = i\sqrt{b}$.	**Example 1:** $\sqrt{-9} \cdot \sqrt{-16} = (3i) \cdot (4i) = 12i^2$ $\qquad\qquad\qquad = 12(-1) = -12$	p. 205
Complex numbers: If a and b are real numbers, then a number written in the form $a + bi$ is called a **complex number**. The value a is called the **real part**, and b is called the **imaginary part**.	**Example 2:** The number $3 + 8i$ is a complex number with real part 3 and imaginary part 8.	p. 206
Add or subtract complex numbers: To add or subtract complex numbers, combine the real parts, and combine the imaginary parts.	**Example 3:** $(4 + 7i) - (3 + 6i) + (2 + 10i)$ $\quad = (4 - 3 + 2) + (7 - 6 + 10)i$ $\quad = 3 + 11i$	p. 208
Power of i: To evaluate i raised to an integer power, we first recognize the following pattern. $i^1 = i$ $i^2 = -1$ $i^3 = i^2 \cdot i = (-1)i = -i$ $i^4 = i^2 \cdot i^2 = (-1)(-1) = 1$ Then write a power of i as a product of the largest fourth power of i and a remaining factor.	**Example 4:** $i^{63} = i^{60} \cdot i^3$ $\quad = (i^4)^{15} \cdot i^3$ $\quad = (1)^{15} \cdot (-i)$ $\quad = -i$ **Example 5:** $i^{42} = i^{40} \cdot i^2 = (1)(-1) = -1$ $i^{25} = i^{24} \cdot i = 1 \cdot i = i$	p. 207

Multiply complex numbers: Multiply complex numbers by using the distributive property.	Example 6: $$(1 - 4i)(2 + 3i) = 2 + 3i - 8i - 12i^2$$ $$= 2 - 5i - 12(-1)$$ $$= 2 - 5i + 12$$ $$= 14 - 5i$$	p. 209
Divide complex numbers: Divide complex numbers by multiplying the numerator and denominator by the conjugate of the denominator. The product of complex conjugates: $$(a + bi)(a - bi) = (a)^2 - (bi)^2$$ $$= a^2 - b^2i^2$$ $$= a^2 - b^2(-1)$$ $$= a^2 + b^2$$	Example 7: $$\frac{4}{3 + 5i} = \frac{4 \cdot (3 - 5i)}{(3 + 5i) \cdot (3 - 5i)} = \frac{12 - 20i}{(3)^2 + (5)^2}$$ $$= \frac{12 - 20i}{9 + 25} = \frac{12}{34} - \frac{20}{34}i = \frac{6}{17} - \frac{10}{17}i$$	p. 210

SECTION 3.4 Solving Quadratic Equations by Factoring

Key Concepts	Examples	Page
Quadratic equation: Let a, b, and c represent real numbers where $a \neq 0$. A quadratic equation in the variable x is an equation of the form $ax^2 + bx + c = 0$.	Example 1: $3x^2 - 2x + 4 = 0$ is a quadratic equation with $a = 3$, $b = -2$, and $c = 4$.	p. 214
Zero product property: One method to solve a quadratic equation is to set one side of the equation equal to zero and factor the other side. Then apply the zero product property. **Zero product property:** If $mn = 0$, then $m = 0$ or $n = 0$.	Example 2: $$x^2 - 3x - 28 = 0$$ $$(x - 7)(x + 4) = 0$$ $$x - 7 = 0 \quad \text{or} \quad x + 4 = 0$$ $$x = 7 \quad \text{or} \quad x = -4$$ The solution set is $\{-4, 7\}$.	p. 214
Applications of quadratic equations: Quadratic equations come up often when working with applications of area, volume, and the Pythagorean theorem.	Area of a square: $A = s^2$ Area of a rectangle: $A = lw$ Area of a triangle: $A = \frac{1}{2}bh$ Area of a parallelogram: $A = bh$ Area of a circle: $A = \pi r^2$ Volume of a rectangular solid: $V = lwh$ Pythagorean theorem: $a^2 + b^2 = c^2$ 	p. 218

SECTION 3.5 Solving Quadratic Equations by Using the Square Root Property

Key Concepts	Examples	Page
Square root property: For a real number k, if $x^2 = k$, then $x = \pm\sqrt{k}$.	**Example 1:** $\quad x^2 = 13$ $\qquad\qquad x = \pm\sqrt{13}$ The solution set is $\{\pm\sqrt{13}\}$. **Example 2:** $\quad (x-3)^2 = -4$ $\qquad\qquad x - 3 = \pm\sqrt{-4}$ $\qquad\qquad x - 3 = \pm 2i$ $\qquad\qquad\qquad x = 3 \pm 2i$ The solution set is $\{3 \pm 2i\}$.	p. 225

Complete the square:
Any quadratic equation $ax^2 + bx + c = 0$ $(a \neq 0)$ can be solved by completing the square and applying the square root property.
 	
Step 1 Divide both sides by a to make the leading coefficient 1.
Step 2 Isolate the variable terms on one side of the equation.
Step 3 Complete the square.
   • Add one-half of the linear term coefficient to both sides.
   • Factor the resulting perfect square trinomial.
Step 4 Apply the square root property and solve for x.
	Example 3: $\quad 2x^2 + 16x - 8 = 0$
	$\dfrac{2x^2}{2} + \dfrac{16x}{2} - \dfrac{8}{2} = \dfrac{0}{2}$
	$x^2 + 8x - 4 = 0$
	$x^2 + 8x = 4$
	$x^2 + 8x + 16 = 4 + 16 \quad$ Note: $\left[\frac{1}{2}(8)\right]^2 = 16$
	$(x + 4)^2 = 20$
	$x + 4 = \pm\sqrt{20}$
	$x = -4 \pm 2\sqrt{5}$
	The solution set is $\{-4 \pm 2\sqrt{5}\}$.
		p. 228

SECTION 3.6 Solving Quadratic Equations by Using the Quadratic Formula

Key Concepts	Examples	Page
Quadratic formula: Given a quadratic equation $ax^2 + bx + c = 0$ $(a \neq 0)$, the solutions are given by the quadratic formula: $$x = \frac{-b \pm \sqrt{b^2 - 4ac}}{2a}$$	**Example 1:** Given $2x^2 - 3x + 5 = 0$, $a = 2$, $b = -3$, $c = 5$. $x = \dfrac{-(-3) \pm \sqrt{(-3)^2 - 4(2)(5)}}{2(2)}$ $= \dfrac{3 \pm \sqrt{9 - 40}}{4} = \dfrac{3 \pm \sqrt{-31}}{4}$ $= \dfrac{3 \pm i\sqrt{31}}{4} = \dfrac{3}{4} \pm \dfrac{\sqrt{31}}{4}i$ The solution set is $\left\{\dfrac{3}{4} \pm \dfrac{\sqrt{31}}{4}i\right\}$.	p. 235

Discriminant:
Given a quadratic equation $ax^2 + bx + c = 0$, the quantity $b^2 - 4ac$ is called the discriminant.
   • If $b^2 - 4ac < 0$, there will be two nonreal solutions to the equation.
   • If $b^2 - 4ac = 0$, there will be one real solution to the equation.
   • If $b^2 - 4ac > 0$, there will be two real solutions to the equation.
	Example 2: $\quad x^2 + 6x + 9 = 0$
	$a = 1$, $b = 6$, $c = 9$.
	The discriminant is $b^2 - 4ac$
	$\qquad\qquad = (6)^2 - 4(1)(9)$
	$\qquad\qquad = 36 - 36$
	$\qquad\qquad = 0$
	There is one real-valued solution.
		p. 238

The quadratic formula can also be used to solve a quadratic equation for a specified variable.	**Example 3:** Solve for t. $\qquad wt^2 = zt - y$ The equation is quadratic in the variable t. $wt^2 - zt + y = 0$ $(a = w, b = -z, c = y)$ $t = \dfrac{-(-z) \pm \sqrt{(-z)^2 - 4(w)(y)}}{2(w)}$ $ = \dfrac{z \pm \sqrt{z^2 - 4wy}}{2w}$	p. 240
Quadratic equations are used to model applications with the Pythagorean theorem, volume, area, and objects moving vertically under the influence of gravity. The vertical position s of an object moving vertically under the influence of gravity is approximated by $s = -\frac{1}{2}gt^2 + v_0 t + s_0$, where • g is the acceleration due to gravity (at sea level on Earth: $g = 32$ ft/sec^2 or 9.8 m/sec^2). • t is the time after the start of the experiment. • v_0 is the initial velocity. • s_0 is the initial position (height). • s is the position of the object at time t.	**Example 4:** A diver jumps approximately straight upward from a diving board 3 m above the water with an initial velocity of 5 m/sec. **a.** Write a model to express the height s (in meters) of the diver. $\qquad s = -\dfrac{1}{2}(9.8)t^2 + 5t + 3$ $\qquad s = -4.9t^2 + 5t + 3$ **b.** Find the time required for the diver to hit the water. $\qquad 0 = -4.9t^2 + 5t + 3$ $\qquad t = \dfrac{-(5) \pm \sqrt{(5)^2 - 4(-4.9)(3)}}{2(-4.9)}$ $\qquad t = \dfrac{-5 \pm \sqrt{83.8}}{-9.8} \quad \begin{matrix} \nearrow \approx -0.42 \text{ sec (reject)} \\ \searrow \approx 1.44 \text{ sec} \end{matrix}$ Rejecting the negative solution, the diver will hit the water approximately 1.44 sec after leaving the board.	p. 241

CHAPTER 3 Review Exercises

SECTION 3.1

For Exercises 1–2, simplify the expression. Assume that x and y represent *any* real number.

1. a. $\sqrt{x^2}$ **b.** $\sqrt[3]{x^3}$

  c. $\sqrt[4]{x^4}$ **d.** $\sqrt[5]{(x+1)^5}$

2. a. $\sqrt{4y^2}$ **b.** $\sqrt[3]{27y^3}$

  c. $\sqrt[100]{y^{100}}$ **d.** $\sqrt[101]{y^{101}}$

For Exercises 3–10, simplify the radicals. Assume that the variables represent positive real numbers.

3. $\sqrt{108}$ **4.** $\sqrt[4]{x^5yz^4}$

5. $-2\sqrt[3]{250a^3b^{10}}$ **6.** $\sqrt[3]{\dfrac{-16a^4}{2ab^3}}$

7. $\sqrt[3]{54xy^{12}z^{14}}$ **8.** $\sqrt[4]{32b^3c^{15}d^8}$

9. $\sqrt{\dfrac{p^{13}}{9}}$ **10.** $\dfrac{\sqrt[3]{3xy^5}}{\sqrt[3]{81x^7y^2}}$

For Exercises 11–12, answer true or false. If a statement is false, explain why. Assume that x and y represent positive real numbers.

11. $5 + 3\sqrt{x} = 8\sqrt{x}$

12. $\sqrt{y} + \sqrt{y} = \sqrt{2y}$

For Exercises 13–18, add or subtract as indicated. Assume that all variables represent positive real numbers.

13. $4\sqrt{7} - 2\sqrt{7} + 3\sqrt{7}$

14. $2\sqrt[3]{64} + 3\sqrt[3]{54} - 16$

15. $\sqrt{50} + 7\sqrt{2} - \sqrt{8}$

16. $x\sqrt[3]{16x^2} - 4\sqrt[3]{2x^5} + 5x\sqrt[3]{54x^2}$

17. $5\sqrt{3a} + 6b\sqrt{3a}$

18. $-14y\sqrt{x} + 6\sqrt{x}$

SECTION 3.2

For Exercises 19–28, multiply the radicals and simplify the result. Assume that all variables represent positive real numbers.

19. $\sqrt{3} \cdot \sqrt{12}$

20. $\sqrt[4]{4} \cdot \sqrt[4]{8}$

21. $-2\sqrt{3}\left(\sqrt{7} - 3\sqrt{11}\right)$

22. $-3\sqrt{5}\left(2\sqrt{3} - \sqrt{5}\right)$

23. $\left(7\sqrt{2} - 2\sqrt{11}\right)\left(7\sqrt{2} + 2\sqrt{11}\right)$

24. $\left(2\sqrt{x} - 3\right)\left(2\sqrt{x} + 3\right)$

25. $\left(\sqrt{7y} - \sqrt{3x}\right)^2$

26. $\left(2\sqrt{3w} + 5\right)^2$

27. $\left(-\sqrt{z} - \sqrt{6}\right)\left(2\sqrt{z} + 7\sqrt{6}\right)$

28. $\left(3\sqrt{a} - \sqrt{5}\right)\left(\sqrt{a} + 2\sqrt{5}\right)$

For Exercises 29–34, simplify the expressions. Assume that all variables represent positive real numbers.

29. $\sqrt{\dfrac{7}{2y}}$

30. $\sqrt{\dfrac{5}{3w}}$

31. $\dfrac{4}{\sqrt[3]{9p^2}}$

32. $\dfrac{-2}{\sqrt[3]{2x}}$

33. $\dfrac{-5}{\sqrt{15} + \sqrt{10}}$

34. $\dfrac{\sqrt{x} + 3}{7 - \sqrt{x}}$

SECTION 3.3

For Exercises 35–38, rewrite the expressions in terms of i.

35. $\sqrt{-16}$

36. $-\sqrt{-5}$

37. $\sqrt{-75} \cdot \sqrt{-3}$

38. $\dfrac{-\sqrt{-24}}{\sqrt{6}}$

For Exercises 39–42, simplify the powers of i.

39. i^{38}

40. i^{101}

41. i^{19}

42. $i^{1000} + i^{1002}$

For Exercises 43–46, perform the indicated operations. Write the final answer in the form $a + bi$.

43. $(-3 + i) - (2 - 4i)$

44. $(1 + 6i)(3 - i)$

45. $(4 - 3i)(4 + 3i)$

46. $(5 - i)^2$

For Exercises 47–48, write the expressions in the form $a + bi$, and determine the real and imaginary parts.

47. $\dfrac{17 - 4i}{-4}$

48. $\dfrac{-16 - 8i}{8}$

For Exercises 49–52, divide and simplify. Write the final answer in the form $a + bi$.

49. $\dfrac{2 - i}{3 + 2i}$

50. $\dfrac{10 + 5i}{2 - i}$

51. $\dfrac{5 + 3i}{-2i}$

52. $\dfrac{4i}{4 - i}$

For Exercises 53–54, simplify the expression.

53. $\dfrac{-8 + \sqrt{-40}}{12}$

54. $\dfrac{6 - \sqrt{-144}}{3}$

SECTION 3.4

For Exercises 55–58, identify the equation as linear or quadratic.

55. $x^2 + 6x = 7$

56. $(x - 3)(x + 4) = 9$

57. $2x - 5 = 3$

58. $x + 3 = 5x^2$

59. a. Factor. $5x^2 + 6x - 8$

   b. Solve. $5x^2 + 6x - 8 = 0$

60. a. Factor. $3x^2 - 19x + 28$

   b. Solve. $3x^2 - 19x + 28 = 0$

For Exercises 61–66, solve the equation.

61. $x^2 - 2x - 15 = 0$

62. $8x^2 = 59x - 21$

63. $2t(t + 5) + 1 = 3t - 3 - t^2$

64. $3(x - 1)(x + 5)(2x - 9) = 0$

65. $x^3 + x^2 - 16x - 16 = 0$

66. $y^3 - 3y^2 - 25y + 75 = 0$

67. A moving van has the capacity to hold 1200 ft³ in volume. If the van is 10 ft high and the length is 1 ft less than twice the width, find the dimensions of the van.

68. The area of a triangular plot of land is 52 yd². The base is 5 yd longer than the height. Find the base and height.

SECTION 3.5

For Exercises 69–76, solve the equations by using the square root property.

69. $x^2 = 5$

70. $2y^2 = -8$

71. $a^2 = 81$

72. $3b^2 = -19$

73. $(x - 2)^2 = 72$

74. $(2x - 5)^2 = -9$

75. $(3y - 1)^2 = 3$

76. $3(m - 4)^2 = 15$

77. Use the square root property to find the length of the sides of a square whose area is 81 in.².

78. Use the square root property to find the exact length of the sides of a square whose area is 150 in.². Then round to the nearest tenth of an inch.

For Exercises 79–82, find the value of n so that the expression is a perfect square trinomial. Then factor the trinomial.

79. $x^2 + 16x + n$

80. $x^2 - 9x + n$

81. $y^2 + \frac{1}{2}y + n$

82. $z^2 - \frac{2}{5}z + n$

For Exercises 83–88, solve the equation by completing the square and applying the square root property.

83. $w^2 + 4w + 13 = 0$

84. $4y^2 - 8y - 20 = 0$

85. $2x^2 = 12x + 6$

86. $-t^2 + 8t - 25 = 0$

87. $3x^2 + 2x = 1$

88. $b^2 + \frac{7}{2}b = 2$

SECTION 3.6

For Exercises 89–96, solve the equations by using the quadratic formula.

89. $y^2 - 4y + 1 = 0$

90. $8y - y^2 = 10$

91. $m^2 - 5m + 25 = 0$

92. $-32 + 4x - x^2 = 0$

93. $6a(a - 1) = 10 + a$

94. $3x(x - 3) = x - 8$

95. $b^2 - \frac{4}{25} = \frac{3}{5}b$

96. $k^2 + 0.4k = 0.05$

For Exercises 97–100, (a) evaluate the discriminant and (b) determine the number and type of solutions to the equation.

97. $4x^2 - 20x + 25 = 0$

98. $-2y^2 = 5y - 1$

99. $5t(t + 1) = 4t - 11$

100. $3x^2 + 2x = -7$

101. Solve for r. $V = \pi r^2 h$ $(r > 0)$

102. Solve for s. $A = 6s^2$ $(s > 0)$

103. $(x - h)^2 + (y - k)^2 = r^2$ for y

104. $s = a_0 t^2 + v_0 t + s_0$ for t

105. A fireworks mortar is shot straight upward with an initial velocity of 200 ft/sec from a platform 2 ft off the ground.

  a. Use the formula $s = -\frac{1}{2}gt^2 + v_0 t + s_0$ to write a model for the height of the mortar s (in ft) at a time t seconds after launch. Assume that $g = 32$ ft/sec^2.

  b. How long will it take the mortar (on the way up) to clear a tree line that is 80 ft high? Round to the nearest tenth of a second.

106. The stopping distance d (in ft) for a car on a certain road is given by $d = 0.048v^2 + 2.2v$, where v is the speed of the car in mph the instant before the brakes were applied.

  a. If the car was traveling 50 mph before the brakes were applied, find the stopping distance.

  b. If the stopping distance is 390 ft, how fast was the car traveling before the brakes were applied? Round to the nearest mile per hour.

For Exercises 107–114, solve the equation by any method.

107. $3x(2x - 1) - 3 = x + 4$

108. $-2x + 3 = x(x - 1)$

109. $9y^2 = 25$

110. $16t^2 = 49$

111. $2x^2 = 15 - x$

112. $17x - 6 = -3x^2$

113. $(x - 5)^2 + 4 = 0$

114. $(x + 2)^2 + 36 = 0$

CHAPTER 3   Test

1. Simplify the expression. Assume that y represents *any* real number.

  a. $\sqrt[3]{y^3}$ **b.** $\sqrt[4]{y^4}$

For Exercises 2–9, simplify the radicals. Assume that all variables represent positive numbers.

2. $\sqrt[4]{81}$

3. $\sqrt{\frac{16}{9}}$

4. $\sqrt[3]{32}$

5. $\sqrt{a^4 b^3 c^5}$

6. $\sqrt{18x^5 y^3 z^4}$

7. $\sqrt{\frac{32w^6}{2w}}$

8. $\sqrt[3]{\frac{x^6}{125y^3}}$

9. $\frac{2\sqrt{72}}{8}$

For Exercises 10–14, perform the indicated operation. Assume that all variables represent positive real numbers.

10. $3\sqrt{5} + 4\sqrt{5} - 2\sqrt{20}$

11. $3\sqrt{x}(\sqrt{2} - \sqrt{5})$

12. $(2\sqrt{5} - 3\sqrt{x})(4\sqrt{5} + \sqrt{x})$

13. $(8\sqrt{2} - 3\sqrt{5})(8\sqrt{2} + 3\sqrt{5})$

14. $(\sqrt{7y} - 4\sqrt{z})^2$

For Exercises 15–16, rationalize the denominator. Assume that x represents a positive real number.

15. $\dfrac{-2}{\sqrt[3]{x}}$

16. $\dfrac{\sqrt{x} + 2}{3 - \sqrt{x}}$

17. Rewrite the expressions in terms of i.

    a. $\sqrt{-8}$ **b.** $2\sqrt{-16}$ **c.** $\dfrac{2 + \sqrt{-8}}{4}$

For Exercises 18–23, perform the indicated operations and simplify completely. Write the final answer in the form $a + bi$.

18. $(3 - 5i) - (2 + 6i)$

19. $(4 + i)(8 + 2i)$

20. $\sqrt{-16} \cdot \sqrt{-49}$

21. $(4 - 7i)^2$

22. $(2 - 10i)(2 + 10i)$

23. $\dfrac{3 - 2i}{3 - 4i}$

24. Find the value of n so that the expression is a perfect square trinomial. Then factor the trinomial $d^2 + 11d + n$.

For Exercises 25–26, solve the equation by completing the square and applying the square root property.

25. $2x^2 + 12x - 36 = 0$

26. $2x^2 = 3x - 7$

For Exercises 27–28, solve the equation by using the quadratic formula.

27. $3x^2 - 4x + 1 = 0$

28. $x(x + 6) = -11 - x$

For Exercises 29–31, (a) evaluate the discriminant and (b) determine the number and type of solutions to the equation.

29. $2x^2 - 4x + 7 = 0$

30. $x^2 + 25 = 10x$

31. $3x(x + 4) = 2x - 2$

For Exercises 32–33, solve for the indicated variable.

32. $-16t^2 + v_0 t + 2 = 0$ for t

33. $a^2 + b^2 + c^2 = 49$ for $c > 0$

For Exercises 34–38, solve the equation using any method.

34. $x^4 = 20x^2 - x^3$

35. $(y - 6)^2 + 10 = 1$

36. $(t - 2)(t + 5) = 4$

37. $2z^2 = 12$

38. $4x^3 + 4x^2 - 9x - 9 = 0$

39. A varsity soccer player kicks a soccer ball approximately straight upward with an initial velocity of 60 ft/sec. The ball leaves the player's foot at a height of 2 ft.

Erik Isakson/Tetra Images/
Getty Images

    a. Use the formlua $s = -\frac{1}{2}gt^2 + v_0 t + s_0$ to write a model representing the height of the ball s (in ft), t seconds after being kicked. Assume that the acceleration due to gravity is $g = 32$ ft/sec^2.

    b. Determine the times at which the ball is 52 ft in the air.

40. A garden area is configured in the shape of a rectangle with two right triangles of equal size attached at the ends. The length of the

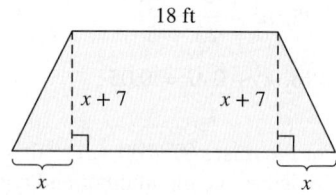

rectangle is 18 ft. The height of the right triangles is 7 ft longer than the base. If the total area of the garden is 276 ft^2, determine the base and height of the triangular portions.

More Expressions and Equations

Matt Brown/Getty Images

Suppose that Jorge plays minor league baseball and wants to improve his batting average. Thus far he has 60 hits in 240 at bats, which gives him a batting average of $\frac{60}{240} = 0.250$. How many hits x would he have to hit in a row to improve his batting average to 0.280? To answer this question, we can solve the equation $\frac{60 + x}{240 + x} = 0.280$.

The expression $\frac{60 + x}{240 + x}$ represents Jorge's new batting average. This expression is called a **rational expression** because it is the ratio of two polynomials. Likewise, the equation $\frac{60 + x}{240 + x} = 0.280$ is called a **rational equation**. We can solve the equation by first clearing fractions (see Section 1.2). We do this by multiplying both sides by the quantity $240 + x$. The resulting solution is $x = 10$. This means that Jorge would need 10 hits in a row to raise his batting average to 0.280. A daunting task!

In this chapter we present the tools to solve rational equations, equations with radicals, and equations that are quadratic in form. After completing this chapter, you will have a large repertoire of equations that you can recognize and solve in a variety of applications.

OBJECTIVES

1. Determine Restricted Values of a Rational Expression
2. Simplify Rational Expressions
3. Simplify Ratios of −1
4. Multiply and Divide Rational Expressions

1. Determine Restricted Values of a Rational Expression

Suppose that a scientist has a lab specimen that is originally at a room temperature of 35°C. After placing the specimen in a freezer, the temperature T (in °C) of the object after t hours can be approximated by the model

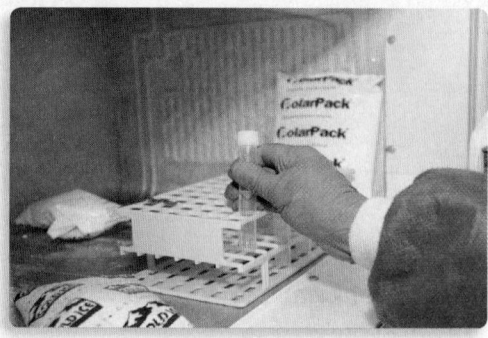

McGraw-Hill Education/Sandra Mesrine

$$T = \frac{350}{t^2 + 3t + 10}.$$

The expression on the right is called a rational expression because it is the ratio of two polynomials. That is, a **rational expression** is an expression of the form $\frac{p}{q}$, where p and q are polynomials and $q \neq 0$. To evaluate a rational expression for a given value of the variable, use substitution. For example, to find the temperature of the specimen after 2 hours, substitute $t = 2$.

$$T = \frac{350}{(2)^2 + 3(2) + 10} = \frac{350}{4 + 6 + 10} = \frac{350}{20} = 17.5°C$$ After 2 hours in the freezer, the object is 17.5°C.

It is important to note, however, that a rational expression might not be defined for all values of the variable. For example, the expression $\dfrac{x^2 - 3}{x - 2}$ is not defined for $x = 2$ because substituting 2 for x would make the denominator zero.

$$\frac{x^2 - 3}{x - 2} \xrightarrow{\text{Substitute 2 for } x} \frac{(2)^2 - 3}{(2) - 2}$$ would result in $\dfrac{1}{0}$, which is undefined.

FOR REVIEW

Recall that a **rational number** is defined as the ratio of two integers such as

$$\frac{2}{3}, -\frac{1}{5}, \text{ and } \frac{9}{1}.$$

(See Section R.2 for review.)

A **rational expression** is the ratio of two polynomials such as

$$\frac{3x - 6}{x^2 - 4}, \frac{3}{4}, \text{ and } \frac{6r^5 + 2r}{7r^3}.$$

EXAMPLE 1   Evaluating a Rational Expression

Evaluate the rational expression (if possible) for the given values of x. $\dfrac{x}{x - 3}$

a. $x = 6$ **b.** $x = -3$ **c.** $x = 0$ **d.** $x = 3$

Solution:

Substitute the given value for the variable. Use the order of operations to simplify.

a. $\dfrac{x}{x - 3}$

$\dfrac{(6)}{(6) - 3}$ Substitute $x = 6$.

$= \dfrac{6}{3}$

$= 2$

b. $\dfrac{x}{x - 3}$

$\dfrac{(-3)}{(-3) - 3}$ Substitute $x = -3$.

$= \dfrac{-3}{-6}$

$= \dfrac{1}{2}$

c. $\dfrac{x}{x-3}$

$\dfrac{(0)}{(0)-3}$ Substitute $x = 0$.

$= \dfrac{0}{-3}$

$= 0$

d. $\dfrac{x}{x-3}$

$\dfrac{(3)}{(3)-3}$ Substitute $x = 3$.

$= \dfrac{3}{0}$ Undefined.

Recall that division by zero is undefined.

Avoiding Mistakes

The numerator is 0 and the denominator is nonzero. Therefore, the fraction is equal to 0.

Avoiding Mistakes

The denominator is 0. Therefore, the fraction is undefined.

Skill Practice 1   Evaluate the expression for the given values of x. $\dfrac{x-3}{x+5}$

a. $x = 2$ **b.** $x = 0$ **c.** $x = 3$ **d.** $x = -5$

From Example 1, the value $x = 3$ is called a **restricted value of the expression** because it causes the expression to be undefined. For a rational expression, the restricted values are the values of the variable that make the denominator equal to zero.

Finding the Restricted Values of a Rational Expression

- Set the denominator equal to zero and solve the resulting equation.
- The restricted values are the solutions to the equation.

EXAMPLE 2   **Finding Restricted Values of Rational Expressions**

Identify the restricted values for each expression.

a. $\dfrac{x-7}{3x+5}$ **b.** $\dfrac{-5}{3x^2+6x}$

Solution:

a. $\dfrac{x-7}{3x+5}$

$3x + 5 = 0$ Set the denominator equal to zero.

$3x = -5$ Solve the equation. Subtract 5 from both sides.

$x = -\dfrac{5}{3}$ Divide both sides by 3.

The restricted value is $x = -\dfrac{5}{3}$.

FOR REVIEW

In Example 2(b), we apply the zero product property. This indicates that if the product of two or more factors is zero, then one or more of the individual factors must be zero. That is, if $mn = 0$, then

$$m = 0 \text{ or } n = 0.$$

See Section 3.4.

b. $\dfrac{-5}{3x^2 + 6x}$

$3x^2 + 6x = 0$	Set the denominator equal to zero.
$3x(x + 2) = 0$	The equation is quadratic. Factor on the left.
$3x = 0 \quad \text{or} \quad x + 2 = 0$	Set each factor equal to zero.
$x = 0 \quad \text{or} \quad x = -2$	Solve each equation.

The restricted values are $x = 0$ and $x = -2$.

Skill Practice 2 Identify the restricted values for each expression.

a. $\dfrac{a + 2}{2a - 8}$ **b.** $\dfrac{3x}{x^2 - 7x + 6}$

EXAMPLE 3   **Finding Restricted Values of Rational Expressions**

Identify the restricted values for each expression.

a. $\dfrac{a + 10}{a^2 - 25}$ **b.** $\dfrac{2x^3 + 5}{x^2 + 9}$

Solution:

a. $\dfrac{a + 10}{a^2 - 25}$

$a^2 - 25 = 0$	Set the denominator equal to zero. The equation is quadratic.
$(a - 5)(a + 5) = 0$	Factor.
$a - 5 = 0 \quad \text{or} \quad a + 5 = 0$	Set each factor equal to zero.
$a = 5 \quad \text{or} \quad a = -5$	

The restricted values are $a = 5$ and $a = -5$.

b. $\dfrac{2x^3 + 5}{x^2 + 9}$

$$x^2 + 9 = 0$$
$$x^2 = -9$$

The quantity x^2 cannot be negative for any real number, x. The denominator $x^2 + 9$ is the sum of two nonnegative values, and thus cannot equal zero. Therefore, there are no restricted values.

FOR REVIEW

The polynomial $a^2 - 25$ is a difference of squares. To factor a difference of squares we have:

$$a^2 - b^2 = (a - b)(a + b)$$

Thus,

$$a^2 - 25 = (a)^2 - (5)^2$$
$$= (a - 5)(a + 5)$$

See Section 2.5 for review.

Skill Practice 3 Identify the restricted values.

a. $\dfrac{w - 4}{w^2 - 9}$ **b.** $\dfrac{8}{z^4 + 1}$

2. Simplify Rational Expressions

In many cases, it is advantageous to simplify or reduce a fraction to lowest terms. The same is true for rational expressions.

The method for simplifying rational expressions mirrors the process for simplifying fractions. In each case, factor the numerator and denominator. Common factors in the numerator and denominator form a ratio of 1 and can be reduced.

Answers

2. a. $a = 4$ **b.** $x = 6$ and $x = 1$

3. a. $w = 3, w = -3$

  b. There are no restricted values.

Simplifying a fraction:
$$\frac{15}{35} \xrightarrow{\text{factor}} \frac{3 \cdot 5}{7 \cdot 5} = \frac{3}{7} \cdot \frac{5}{5} = \frac{3}{7} \cdot 1 = \frac{3}{7}$$

Simplifying a rational expression:
$$\frac{x^2 - x - 12}{x^2 - 16} \xrightarrow{\text{factor}} \frac{(x + 3)(x - 4)}{(x + 4)(x - 4)} = \frac{(x + 3)}{(x + 4)} \cdot \frac{(x - 4)}{(x - 4)}$$
$$= \frac{(x + 3)}{(x + 4)} \cdot 1$$
$$= \frac{(x + 3)}{(x + 4)}$$

Informally, to simplify a rational expression, we simplify the ratio of common factors to 1. Formally, this is accomplished by applying the fundamental principle of rational expressions.

> **TIP** In practice we often shorten the process to reduce a rational expression by dividing out common factors.
> $$\frac{p \cdot \overset{1}{\cancel{r}}}{q \cdot \cancel{r}} = \frac{p}{q}$$

Fundamental Principle of Rational Expressions

Let p, q, and r represent polynomials where $q \neq 0$ and $r \neq 0$. Then

$$\frac{pr}{qr} = \frac{p}{q} \cdot \frac{r}{r} = \frac{p}{q} \cdot 1 = \frac{p}{q}$$

EXAMPLE 4   Simplifying a Rational Expression

Given the expression $\dfrac{2x^3 + 12x^2 + 16x}{6x + 24}$

a. Factor the numerator and denominator.
b. Determine the restrictions on x.
c. Simplify the expression.

> **FOR REVIEW**
>
> Recall that the first step to factor a polynomial is to factor out the greatest common factor (see Section 2.3.) For a review on factoring trinomials, see Section 2.4.

Solution:

a. $\dfrac{2x^3 + 12x^2 + 16x}{6x + 24}$

$= \dfrac{2x(x^2 + 6x + 8)}{6(x + 4)}$ Factor the numerator and denominator.

$= \dfrac{2x(x + 4)(x + 2)}{6(x + 4)}$

> **Avoiding Mistakes**
>
> Always determine the restrictions on the variable *before* simplifying an expression.

b. The expression is not defined for all values of x for which the denominator is equal to zero.

$6(x + 4) = 0$ Solve the equation.
$x = -4$

The restriction on x is that $x \neq -4$.

> **TIP** The expression in Example 4(c) can be simplified directly from its factored form.
> $$\frac{2x(x + 4)(x + 2)}{2 \cdot 3(x + 4)}$$
> $$= \frac{2x \overset{1}{\cancel{(x + 4)}}(x + 2)}{\underset{1}{\cancel{2}} \cdot 3 \underset{1}{\cancel{(x + 4)}}}$$
> $$= \frac{x(x + 2)}{3}$$

c. $\dfrac{2x(x + 4)(x + 2)}{2 \cdot 3(x + 4)} = \dfrac{x(x + 2)}{3} \cdot \dfrac{\overset{1}{\cancel{2(x + 4)}}}{\cancel{2(x + 4)}}$ Simplify the ratio of common factors.

$= \dfrac{x(x + 2)}{3}$ (provided $x \neq -4$)

Skill Practice 4   Given $\dfrac{x^2 + 3x - 28}{2x + 14}$

    a. Factor the numerator and denominator.

    b. Determine the restrictions on x.

    c. Simplify the expression.

It is important to note that the expressions $\dfrac{2x^3 + 12x^2 + 16x}{6x + 24}$ and $\dfrac{x(x + 2)}{3}$ are equivalent except at $x = -4$. This is because the first expression would be undefined. However, at all *other* values of x, the expressions are equivalent.

    The motivation for simplifying a rational expression is to create an equivalent expression that is easier to work with. Consider the rational expression from Example 4 in its original form and its reduced form. Suppose we substitute $x = 2$ into each expression.

<table>
<tr><td></td><td>Original Expression</td><td>Simplified Expression</td></tr>
<tr><td></td><td>$\dfrac{2x^3 + 12x^2 + 16x}{6x + 24}$</td><td>$\dfrac{x(x + 2)}{3}$</td></tr>
<tr><td>Substitute $x = 2$</td><td>$\dfrac{2(2)^3 + 12(2)^2 + 16(2)}{6(2) + 24}$</td><td>$\dfrac{(2)[(2) + 2]}{3}$</td></tr>
<tr><td></td><td>$= \dfrac{2(8) + 12(4) + 32}{12 + 24}$</td><td>$= \dfrac{2(4)}{3}$</td></tr>
<tr><td></td><td>$= \dfrac{16 + 48 + 32}{36}$</td><td>$= \dfrac{8}{3}$</td></tr>
<tr><td></td><td>$= \dfrac{96}{36}$</td><td></td></tr>
<tr><td></td><td>$= \dfrac{8}{3}$</td><td></td></tr>
</table>

From this point forward, when we simplify rational expressions, we will not explicitly write the restrictions by the simplified form. Instead, the restrictions will be implied from the original expression.

EXAMPLE 5   Simplifying a Rational Expression

Simplify. $\dfrac{2x^2 y^8}{8x^4 y^3}$

Solution:

$\dfrac{2x^2 y^8}{8x^4 y^3}$ This expression has the restriction that $x \neq 0$ and $y \neq 0$.

$= \dfrac{2x^2 y^8}{2^3 x^4 y^3}$ Factor the denominator.

$= \dfrac{y^5}{2^2 x^2} \cdot \dfrac{\overset{1}{\cancel{2x^2 y^3}}}{\cancel{2x^2 y^3}}$ Simplify common factors whose ratio is 1.

$= \dfrac{y^5}{4x^2}$

Answers

4. a. $\dfrac{(x + 7)(x - 4)}{2(x + 7)}$

   b. $x \neq -7$

   c. $\dfrac{x - 4}{2}$ provided $x \neq -7$

TIP Example 5 could also have been simplified using the properties of exponents. See Section 2.1.

$$\frac{2x^2y^8}{8x^4y^3} = \frac{2x^2y^8}{2^3x^4y^3}$$

$$= 2^{1-3}x^{2-4}y^{8-3}$$

$$= 2^{-2}x^{-2}y^5 = \frac{y^5}{2^2x^2}$$

$$= \frac{y^5}{4x^2}$$

Skill Practice 5   Simplify. $\dfrac{9a^5b^3}{18a^8b}$

EXAMPLE 6   Simplifying a Rational Expression

Simplify. $\dfrac{t^3 + 8}{t^2 + 4t + 4}$

Solution:

$$\frac{t^3 + 8}{t^2 + 4t + 4} = \frac{(t+2)(t^2 - 2t + 4)}{(t+2)^2}$$

Factor the numerator and denominator. The numerator is a sum of cubes. The denominator is a perfect square trinomial. The restriction on t is $t \neq -2$.

$$= \frac{(t^2 - 2t + 4)}{(t+2)} \cdot \frac{\overset{1}{\cancel{(t+2)}}}{\cancel{(t+2)}}$$

Simplify common factors whose ratio is 1.

$$= \frac{t^2 - 2t + 4}{t + 2}$$

FOR REVIEW

To factor a sum of cubes, see Section 2.5.

$a^3 + b^3$

$\quad = (a+b)(a^2 - ab + b^2)$

To factor a perfect square trinomial, see Section 2.4.

$a^2 + 2ab + b^2$

$\quad = (a+b)^2$

Skill Practice 6   Simplify. $\dfrac{p^3 - 27}{p^2 - 6p + 9}$

Avoiding Mistakes

Because the fundamental property of rational expressions is based on the identity property of multiplication, reducing applies only to factors (remember that factors are multiplied). Therefore, terms that are added or subtracted cannot be reduced. For example:

$$\frac{3x}{3y} = \frac{\overset{1}{\cancel{3}}}{\cancel{3}} \cdot \frac{x}{y} = (1) \cdot \frac{x}{y} = \frac{x}{y}$$

↑ Reduce common factor.

However, $\dfrac{x+3}{y+3}$ cannot be simplified. ↑ Cannot reduce common terms.

3. Simplify Ratios of −1

When two factors are identical in the numerator and denominator, they form a ratio of 1 and can be simplified. Sometimes we encounter two factors that are *opposites* and form a ratio equal to −1. For example:

Simplified Form	Details/Notes
$\dfrac{-5}{5} = -1$	The ratio of a number and its opposite is −1.
$\dfrac{100}{-100} = -1$	The ratio of a number and its opposite is −1.

$$\frac{x+7}{-x-7} = -1 \qquad \frac{x+7}{-x-7} = \frac{x+7}{-1(x+7)} = \frac{\overset{1}{\cancel{x+7}}}{-1\cancel{(x+7)}} = \frac{1}{-1} = -1$$

— Factor out −1.

$$\frac{2-x}{x-2} = -1 \qquad \frac{2-x}{x-2} = \frac{-1(-2+x)}{x-2} = \frac{-1\overset{1}{\cancel{(x-2)}}}{\cancel{x-2}} = \frac{-1}{1} = -1$$

Recognizing factors that are opposites is useful when simplifying rational expressions. For example, $a - b$ and $b - a$ are opposites because the opposite of $a - b$ can be written $-(a - b) = -a + b = b - a$. Therefore, in general, $\frac{a-b}{b-a} = -1$.

Answers

5. $\dfrac{b^2}{2a^3}$

6. $\dfrac{p^2 + 3p + 9}{p - 3}$

EXAMPLE 7 Simplifying a Rational Expression

Simplify the rational expression to lowest terms. $\dfrac{x-5}{25-x^2}$

Solution:

$$\dfrac{x-5}{25-x^2}$$

$$= \dfrac{x-5}{(5-x)(5+x)}$$ Factor.

Notice that $x-5$ and $5-x$ are opposites and form a ratio of -1.

$$= \dfrac{1}{(5+x)} \cdot \dfrac{\overset{-1}{\cancel{(x-5)}}}{\cancel{(5-x)}}$$ In general, $\dfrac{a-b}{b-a} = -1$.

$$= \dfrac{1}{5+x}(-1)$$

$$= -\dfrac{1}{5+x} \quad \text{or} \quad -\dfrac{1}{x+5}$$

TIP The factor of -1 may be applied in front of the rational expression, or it may be applied to the numerator or to the denominator. Therefore, the final answer may be written in different forms.

$$\dfrac{1}{x+5} \quad \text{or} \quad \dfrac{-1}{x+5} \quad \text{or} \quad \dfrac{1}{-(x+5)}$$

Skill Practice 7 Simplify the expression. $\dfrac{20-5x}{x^2-x-12}$

4. Multiply and Divide Rational Expressions

Recall that to multiply fractions, we multiply the numerators and multiply the denominators. The same is true for multiplying rational expressions.

Multiplication Property of Rational Expressions

Let p, q, r, and s represent polynomials, such that $q \neq 0$ and $s \neq 0$. Then

$$\dfrac{p}{q} \cdot \dfrac{r}{s} = \dfrac{pr}{qs}$$

For example:

Multiply the Fractions

$$\dfrac{2}{3} \cdot \dfrac{5}{7} = \dfrac{10}{21}$$

Multiply the Rational Expressions

$$\dfrac{2x}{3y} \cdot \dfrac{5z}{7} = \dfrac{10xz}{21y}$$

Answer

7. $-\dfrac{5}{x+3}$

Sometimes it is possible to simplify a ratio of common factors to 1 *before* multiplying. To do so, we must first factor the numerators and denominators of each fraction.

$$\frac{7}{10} \cdot \frac{15}{21} \xrightarrow{\text{Factor}} \frac{7}{2 \cdot 5} \cdot \frac{3 \cdot 5}{3 \cdot 7} = \frac{\overset{1}{7} \cdot \overset{1}{3} \cdot \overset{1}{5}}{2 \cdot \underset{1}{5} \cdot \underset{1}{3} \cdot \underset{1}{7}} = \frac{1}{2}$$

The same process is also used to multiply rational expressions.

Multiplying Rational Expressions

Step 1 Factor the numerator and denominator of each expression.
Step 2 Multiply the numerators, and multiply the denominators.
Step 3 Reduce the ratios of common factors to 1 or −1 and simplify.

EXAMPLE 8   Multiplying Rational Expressions

Multiply. $\dfrac{5a - 5b}{10} \cdot \dfrac{2}{a^2 - b^2}$

Solution:

$$\frac{5a - 5b}{10} \cdot \frac{2}{a^2 - b^2}$$

$$= \frac{5(a - b)}{5 \cdot 2} \cdot \frac{2}{(a - b)(a + b)} \qquad \text{Factor numerator and denominator.}$$

$$= \frac{5(a - b) \cdot 2}{5 \cdot 2 \cdot (a - b)(a + b)} \qquad \text{Multiply.}$$

$$= \frac{\overset{1}{5}(\overset{1}{a - b}) \cdot \overset{1}{2}}{\underset{1}{5} \cdot \underset{1}{2} \cdot (\underset{1}{a - b})(a + b)} = \frac{1}{a + b} \qquad \text{Reduce common factors and simplify.}$$

Avoiding Mistakes

If all factors in the numerator simplify to 1, do not forget to write the factor of 1 in the numerator.

Skill Practice 8   Multiply. $\dfrac{3y - 6}{6y} \cdot \dfrac{y^2 + 3y + 2}{y^2 - 4}$

EXAMPLE 9   Multiplying Rational Expressions

Multiply. $\dfrac{4w - 20p}{2w^2 - 50p^2} \cdot \dfrac{2w^2 + 7wp - 15p^2}{3w + 9p}$

Solution:

$$\frac{4w - 20p}{2w^2 - 50p^2} \cdot \frac{2w^2 + 7wp - 15p^2}{3w + 9p}$$

$$= \frac{4(w - 5p)}{2(w^2 - 25p^2)} \cdot \frac{(2w - 3p)(w + 5p)}{3(w + 3p)} \qquad \text{Factor numerator and denominator.}$$

$$= \frac{2 \cdot 2(w - 5p)}{2(w - 5p)(w + 5p)} \cdot \frac{(2w - 3p)(w + 5p)}{3(w + 3p)} \qquad \text{Factor further.}$$

Answer

8. $\dfrac{y + 1}{2y}$

$$= \frac{2 \cdot 2(w - 5p)(2w - 3p)(w + 5p)}{2(w - 5p)(w + 5p) \cdot 3(w + 3p)}$$

Multiply.

$$= \frac{2 \cdot 2(w \overset{1}{\cancel{-5p}})(2w - 3p)(w \overset{1}{\cancel{+5p}})}{2(w \cancel{-5p})(w \cancel{+5p}) \cdot 3(w + 3p)}$$

Simplify common factors.

$$= \frac{2(2w - 3p)}{3(w + 3p)}$$

Notice that the expression is left in factored form to show that it has been simplified to lowest terms.

Skill Practice 9   Multiply. $\dfrac{p^2 + 8p + 16}{10p + 10} \cdot \dfrac{2p + 6}{p^2 + 7p + 12}$

Recall that to divide fractions, multiply the first fraction by the reciprocal of the second fraction.

Divide: $\dfrac{15}{14} \div \dfrac{10}{49}$ $\xrightarrow[\text{of the second fraction.}]{\text{Multiply by the reciprocal}}$ $\dfrac{15}{14} \cdot \dfrac{49}{10} = \dfrac{3 \cdot \overset{1}{\cancel{5}} \cdot \overset{1}{\cancel{7}} \cdot 7}{2 \cdot \cancel{7} \cdot 2 \cdot \cancel{5}} = \dfrac{21}{4}$

The same process is used for dividing rational expressions.

Division Property of Rational Expressions

Let p, q, r, and s represent polynomials, such that $q \neq 0$, $r \neq 0$, $s \neq 0$. Then

$$\frac{p}{q} \div \frac{r}{s} = \frac{p}{q} \cdot \frac{s}{r} = \frac{ps}{qr}$$

EXAMPLE 10   **Dividing Rational Expressions**

Divide. $\dfrac{8t^3 + 27}{9 - 4t^2} \div \dfrac{4t^2 - 6t + 9}{2t^2 - t - 3}$

Solution:

$$\frac{8t^3 + 27}{9 - 4t^2} \div \frac{4t^2 - 6t + 9}{2t^2 - t - 3}$$

$$= \frac{8t^3 + 27}{9 - 4t^2} \cdot \frac{2t^2 - t - 3}{4t^2 - 6t + 9}$$

Multiply the first fraction by the reciprocal of the second.

$$= \frac{(2t + 3)(4t^2 - 6t + 9)}{(3 - 2t)(3 + 2t)} \cdot \frac{(2t - 3)(t + 1)}{4t^2 - 6t + 9}$$

Factor numerator and denominator. Notice that $8t^3 + 27$ is a sum of cubes. Furthermore, $4t^2 - 6t + 9$ does not factor over the real numbers.

$$= \frac{(2t \overset{1}{\cancel{+ 3}})(\cancel{4t^2 - 6t + 9})(2t \overset{-1}{\cancel{- 3}})(t + 1)}{(3 \cancel{- 2t})(3 + 2t)(\cancel{4t^2 - 6t + 9})}$$

Simplify to lowest terms.

$$= (-1)\frac{(t + 1)}{1}$$

$$= -(t + 1) \quad \text{or} \quad -t - 1$$

Avoiding Mistakes

When dividing rational expressions, your first step should be to take the reciprocal of the second fraction (divisor). Do this first so that you do not forget.

Answer

9. $\dfrac{p + 4}{5(p + 1)}$

Skill Practice 10   Divide. $\dfrac{x^2 + x}{5x^3 - x^2} \div \dfrac{10x^2 + 12x + 2}{25x^2 - 1}$

TIP   In Example 10, the factors $(2t - 3)$ and $(3 - 2t)$ are opposites and form a ratio of -1. The factors $(2t + 3)$ and $(3 + 2t)$ are equal and form a ratio of 1.

$$\dfrac{2t - 3}{3 - 2t} = -1 \quad \text{whereas} \quad \dfrac{2t + 3}{3 + 2t} = 1$$

Answer

10. $\dfrac{1}{2x}$

SECTION 4.1   Practice Exercises

Prerequisite Review

For Exercises R.1–R.8, simplify. (See Sections R.1 and 2.1 for review.)

R.1. $-\dfrac{120}{18}$

R.2. $\dfrac{21}{315}$

R.3. $x^{11} \cdot x^5$

R.4. $y^4 \cdot y^{10}$

R.5. $\dfrac{m^{12}}{m^{17}}$

R.6. $\dfrac{n^3}{n^9}$

R.7. $\dfrac{-11b}{11b}$

R.8. $\dfrac{18a}{-18a}$

For Exercises R.9–R.14, perform the indicated operations. (See Section R.1 for review.)

R.9. $-\dfrac{12}{35} \cdot \dfrac{14}{3}$

R.10. $-\dfrac{11}{8} \cdot \left(-\dfrac{4}{55}\right)$

R.11. $\dfrac{12}{77} \div \dfrac{3}{22}$

R.12. $\dfrac{5}{18} \div \dfrac{35}{12}$

R.13. $-\dfrac{4}{7} \cdot \dfrac{9}{2} \div 3$

R.14. $\dfrac{2}{5} \div \dfrac{4}{15} \cdot 6$

Concept Connections

1. A _____ expression is an expression of the form $\dfrac{p}{q}$ where p and q are polynomials and $q \neq 0$.

2. Restricted values of a rational expression are values of the variable that make the _____ of the expression equal to zero.

3. For polynomials p, q, and r, where ($q \neq 0$ and $r \neq 0$), $\dfrac{pr}{qr}$ simplifies to _____.

4. The ratio $\dfrac{a - b}{a - b} =$ _____, whereas the ratio $\dfrac{a - b}{b - a} =$ _____ provided that $a \neq b$.

5. Given polynomials p, q, r, and s such that $q \neq 0$ and $s \neq 0$, $\dfrac{p}{q} \cdot \dfrac{r}{s} = \dfrac{\square}{\square}$.

6. Given polynomials p, q, r, and s such that $q \neq 0$, $r \neq 0$, and $s \neq 0$, $\dfrac{p}{q} \div \dfrac{r}{s} = \dfrac{\square}{\square}$.

Objective 1: Determine Restricted Values of a Rational Expression

For Exercises 7–14, evaluate the expression for the given value of the variable and simplify if possible. (**See Example 1**)

7. $\dfrac{5}{y - 4}$; $y = 6$

8. $\dfrac{4x}{x - 7}$; $x = 8$

9. $\dfrac{1}{x - 6}$; $x = -2$

10. $\dfrac{w - 10}{w + 6}$; $w = 0$

11. $\dfrac{t-2}{t^2-4t+8}; t=2$ **12.** $\dfrac{y-8}{2y^2+y-1}; y=8$ **13.** $\dfrac{3y-4}{y^2+3y-10}; y=2$ **14.** $\dfrac{3a-5}{a^2-5a+4}; a=1$

15. The concentration C (in ng/mL) of a drug in the bloodstream t hours after ingestion is modeled by $C=\dfrac{600t}{t^3+125}$.

    a. Determine the concentration at 1 hr, 12 hr, 24 hr, and 48 hr. Round to 1 decimal place.

    b. What appears to be the limiting concentration for large values of t?

16. An object that is originally 35°C is placed in a freezer. The temperature T (in °C) of the object can be approximated by the model $T=\dfrac{350}{t^2+3t+10}$, where t is the time in hours after the object is placed in the freezer.

    a. Determine the temperature at 2 hr, 4 hr, 12 hr, and 24 hr. Round to 1 decimal place.

    b. What appears to be the limiting temperature for large values of t?

For Exercises 17–28, identify the restricted values. **(See Examples 2–3)**

17. $\dfrac{5}{k+2}$

18. $\dfrac{-3}{h-4}$

19. $\dfrac{x+5}{(2x-5)(x+8)}$

20. $\dfrac{4y+1}{(3y+7)(y+3)}$

21. $\dfrac{m+12}{m^2+5m+6}$

22. $\dfrac{c-11}{c^2-5c-6}$

23. $\dfrac{x-4}{x^2+9}$

24. $\dfrac{x+1}{x^2+4}$

25. $\dfrac{y^2-y-12}{12}$

26. $\dfrac{z^2+10z+9}{9}$

27. $\dfrac{7}{4t^2-20t}$

28. $\dfrac{4z+1}{6z^2+18z}$

29. Construct a rational expression that is undefined for $x=2$. (Answers will vary.)

30. Construct a rational expression that is undefined for $x=5$. (Answers will vary.)

31. Construct a rational expression that is undefined for $x=-3$ and $x=7$. (Answers will vary.)

32. Construct a rational expression that is undefined for $x=-1$ and $x=4$. (Answers will vary.)

33. Evaluate the expressions for $x=-1$.

    a. $\dfrac{3x^2-2x-1}{6x^2-7x-3}$ **b.** $\dfrac{x-1}{2x-3}$

34. Evaluate the expressions for $x=4$.

    a. $\dfrac{(x+5)^2}{x^2+6x+5}$ **b.** $\dfrac{x+5}{x+1}$

35. Determine which expressions are equal to $-\dfrac{5}{x-3}$.

    a. $\dfrac{-5}{x-3}$ **b.** $\dfrac{5}{3-x}$

    c. $-\dfrac{5}{3-x}$ **d.** $\dfrac{-5}{3-x}$

36. Determine which expressions are equal to $\dfrac{-2}{a+b}$.

    a. $\dfrac{-2}{a-b}$ **b.** $-\dfrac{2}{a+b}$

    c. $\dfrac{2}{-a-b}$ **d.** $\dfrac{2}{a-b}$

Objective 2: Simplify Rational Expressions

For Exercises 37–38, simplify the expression if possible.

37. a. $\dfrac{8x}{4y}$ **b.** $\dfrac{8+x}{4+y}$

38. a. $\dfrac{a-21}{14+b}$ **b.** $\dfrac{-21a}{14b}$

For Exercises 39–42,

    a. Factor the numerator and denominator.

    b. Determine the restrictions on x.

    c. Simplify the expression. **(See Example 4)**

39. $\dfrac{x^2+6x+8}{x^2+3x-4}$ **40.** $\dfrac{x^2-6x}{2x^2-11x-6}$ **41.** $\dfrac{x^2-18x+81}{x^2-81}$ **42.** $\dfrac{x^2+14x+49}{x^2-49}$

For Exercises 43–60, simplify the rational expression. (**See Examples 5–6**)

43. $\dfrac{-3m^4n}{12m^6n^4}$

44. $\dfrac{-5x^3y^2}{20x^4y^2}$

45. $\dfrac{6a+18}{9a+27}$

46. $\dfrac{5y-15}{3y-9}$

47. $\dfrac{x-5}{x^2-25}$

48. $\dfrac{3z-6}{3z^2-12}$

49. $\dfrac{-7c}{21c^2-35c}$

50. $\dfrac{2p+3}{2p^2+7p+6}$

51. $\dfrac{2t^2+7t-4}{-2t^2-5t+3}$

52. $\dfrac{y^2+8y-9}{y^2-5y+4}$

53. $\dfrac{(p+1)(2p-1)^4}{(p+1)^2(2p-1)^2}$

54. $\dfrac{r(r-3)^5}{r^3(r-3)^2}$

55. $\dfrac{9-z^2}{2z^2+z-15}$

56. $\dfrac{2c^2+2c-12}{-8+2c+c^2}$

57. $\dfrac{2z^3+128}{16+8z+z^2}$

58. $\dfrac{p^3-1}{5-10p+5p^2}$

59. $\dfrac{10x^3-25x^2+4x-10}{-4-10x^2}$

60. $\dfrac{8x^3-12x^2+6x-9}{16x^4-9}$

Objective 3: Simplify Ratios of −1

For Exercises 61–72, simplify the rational expression. (**See Example 7**)

61. $\dfrac{r+6}{6+r}$

62. $\dfrac{a+2}{2+a}$

63. $\dfrac{b+8}{-b-8}$

64. $\dfrac{7+w}{-7-w}$

65. $\dfrac{10-x}{x-10}$

66. $\dfrac{y-14}{14-y}$

67. $\dfrac{2t-2}{1-t}$

68. $\dfrac{5p-10}{2-p}$

69. $\dfrac{c+4}{c-4}$

70. $\dfrac{b+2}{b-2}$

71. $\dfrac{y-x}{12x^2-12y^2}$

72. $\dfrac{4w^2-49z^2}{14z-4w}$

Objective 4: Multiply and Divide Rational Expressions

For Exercises 73–84, multiply the rational expression. (**See Examples 8–9**)

73. $\dfrac{8w^2}{9}\cdot\dfrac{3}{2w^4}$

74. $\dfrac{16}{z^7}\cdot\dfrac{z^4}{8}$

75. $\dfrac{5p^2q^4}{12pq^3}\cdot\dfrac{6p^2}{20q^2}$

76. $\dfrac{27r^5}{7s}\cdot\dfrac{28rs^3}{9r^3s^2}$

77. $\dfrac{3z+12}{8z^3}\cdot\dfrac{16z^3}{9z+36}$

78. $\dfrac{x^2y}{x^2-4x-5}\cdot\dfrac{2x^2-13x+15}{xy^3}$

79. $\dfrac{3y^2+18y+15}{6y+6}\cdot\dfrac{y-5}{y^2-25}$

80. $\dfrac{10w-8}{w+2}\cdot\dfrac{3w^2-w-14}{25w^2-16}$

81. $\dfrac{x-5y}{x^2+xy}\cdot\dfrac{y^2-x^2}{10y-2x}$

82. $\dfrac{3x-15}{4x^2-2x}\cdot\dfrac{10x-20x^2}{5-x}$

83. $x(x+5)^2\cdot\dfrac{2}{x^2-25}$

84. $y(y^2-4)\cdot\dfrac{y}{y+2}$

For Exercises 85–98, divide the rational expressions. (**See Example 10**)

85. $\dfrac{5x}{7}\div\dfrac{10x^2}{21}$

86. $\dfrac{2a}{7b^3}\div\dfrac{10a^5}{77}$

87. $\dfrac{6x^2y^2}{(x-2)}\div\dfrac{3xy^2}{(x-2)^2}$

88. $\dfrac{(r+3)^2}{4r^3s}\div\dfrac{r+3}{rs}$

89. $\dfrac{t^2+5t}{t+1}\div(t+5)$

90. $\dfrac{6p+7}{p+2}\div(36p^2-49)$

91. $\dfrac{a}{a-10}\div\dfrac{a^3+6a^2-40a}{a^2-100}$

92. $\dfrac{b^2-6b+9}{b^2-b-6}\div\dfrac{b^2-9}{4}$

93. $\dfrac{2x^2+5xy+2y^2}{4x^2-y^2}\div\dfrac{x^2+xy-2y^2}{2x^2+xy-y^2}$

94. $\dfrac{6s^2+st-2t^2}{6s^2-5st+t^2}\div\dfrac{3s^2+17st+10t^2}{6s^2+13st-5t^2}$

95. $\dfrac{x^4-x^3+x^2-x}{2x^3+2x^2+x+1}\div\dfrac{x^3-4x^2+x-4}{2x^3-8x^2+x-4}$

96. $\dfrac{a^3+a+a^2+1}{a^3+a^2+ab^2+b^2}\div\dfrac{a^3+a+a^2b+b}{2a^2+2ab+ab^2+b^3}$

97. $\dfrac{3y-y^2}{y^3-27}\div\dfrac{y}{y^2+3y+9}$

98. $\dfrac{8x-4x^2}{xy-2y+3x-6}\div\dfrac{3x+6}{y+3}$

Mixed Exercises

For Exercises 99–118, perform the indicated operations.

99. $\dfrac{8a^4b^3}{3c} \div \dfrac{a^7b^2}{9c}$

100. $\dfrac{3x^5}{2x^2y^7} \div \dfrac{4x^3y}{6y^6}$

101. $\dfrac{2}{25x^2} \cdot \dfrac{5x}{12} \div \dfrac{2}{15x}$

102. $\dfrac{4y}{7} \div \dfrac{y^2}{14} \cdot \dfrac{3}{y}$

103. $\dfrac{10x^2 - 13xy - 3y^2}{8x^2 - 10xy - 3y^2} \cdot \dfrac{2y + 8x}{2x^2 + 2y^2}$

104. $\dfrac{6a^2 + ab - b^2}{10a^2 + 5ab} \cdot \dfrac{2a^3 + 4a^2b}{3a^2 + 5ab - 2b^2}$

105. $(3m^2 - 12m) \div \dfrac{m^2 - 4m}{m^2 - 6m + 8}$

106. $(2x^2 + 8) \div \dfrac{x^4 - 16}{x^2 + x - 6}$

107. $\dfrac{(a + b)^2}{a - b} \cdot \dfrac{a^3 - b^3}{a^2 - b^2} \div \dfrac{a^2 + ab + b^2}{(a - b)^2}$

108. $\dfrac{m^2 - n^2}{(m - n)^2} \div \dfrac{m^2 - 2mn + n^2}{m^2 - mn + n^2} \cdot \dfrac{(m - n)^4}{m^3 + n^3}$

109. $\dfrac{x^2 - 4y^2}{x + 2y} \div (x + 2y) \cdot \dfrac{2y}{x - 2y}$

110. $\dfrac{x^2 - 6xy + 9y^2}{x^2 - 4y^2} \cdot \dfrac{x^2 - 5xy + 6y^2}{3y - x} \div \dfrac{x^2 - 9y^2}{x + 2y}$

111. $\dfrac{8x^3 - 27y^3}{4x^2 - 9y^2} \div \dfrac{8x^2 + 12xy + 18y^2}{2x + 3y}$

112. $\dfrac{25m^2 - 1}{125m^3 - 1} \div \dfrac{5m + 1}{25m^2 + 5m + 1}$

113. $\dfrac{m^3 + 2m^2 - mn^2 - 2n^2}{m^3 - m^2 - 20m} \cdot \dfrac{m^3 - 25m}{m^3 + m^2n - 4m - 4n}$

114. $\dfrac{2a^2 + ab - 8a - 4b}{2a^2 - 6a + ab - 3b} \cdot \dfrac{a^2 - 6a + 9}{a^2 - 16}$

115. $\dfrac{7}{3x + 15} \cdot (x + 5) \div \dfrac{14}{9x - 27}$

116. $\dfrac{45}{2x + 1} \cdot (8x + 4) \div \dfrac{27}{4x + 2}$

117. $\dfrac{12y + 3}{6y^2 - y - 12} \div \dfrac{4y^2 - 19y - 5}{2y^2 - y - 3}$

118. $\dfrac{2x^2 - 11x - 6}{3x - 2} \div \dfrac{2x^2 - 5x - 3}{3x^2 - 7x - 6}$

For Exercises 119–122, write an expression for the area of the figure and simplify.

119.

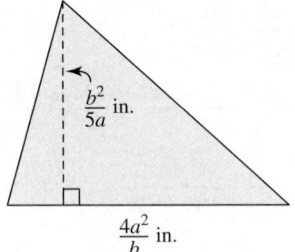

$\dfrac{b^2}{5a}$ in.

$\dfrac{4a^2}{b}$ in.

120.

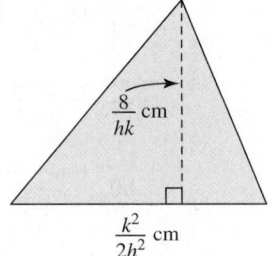

$\dfrac{8}{hk}$ cm

$\dfrac{k^2}{2h^2}$ cm

121.

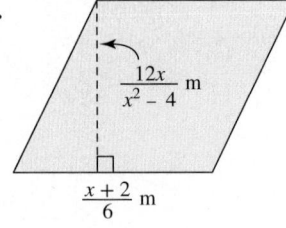

$\dfrac{12x}{x^2 - 4}$ m

$\dfrac{x + 2}{6}$ m

122.

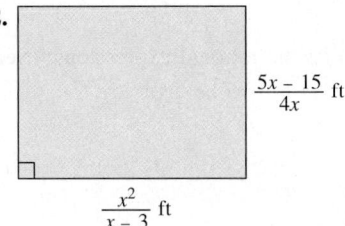

$\dfrac{5x - 15}{4x}$ ft

$\dfrac{x^2}{x - 3}$ ft

Write About It

123. Explain why the expression $\dfrac{x}{x - y}$ is not defined for $x = y$.

124. Is the statement $\dfrac{3(x - 4)}{(x + 2)(x - 4)} = \dfrac{3}{x + 2}$ true for all values of x? Explain why or why not.

Expanding Your Skills

For Exercises 125–128, simplify the expression.

125. $\dfrac{x^{2n+1} - x^{2n}}{x^{n+1} - x^n}$

126. $\dfrac{2y^{n+1} + 3y^n}{2y^{3n+1} + 3y^{3n}}$

127. $\dfrac{w^{3n+1} - w^{3n}z}{w^{n+2} - w^n z^2}$

128. $\dfrac{x^{2n+1} - x^{2n}y}{x^{n+3} - x^n y^3}$

OBJECTIVES

1. Add and Subtract Rational Expressions with Like Denominators
2. Identify the Least Common Denominator
3. Add and Subtract Rational Expressions with Unlike Denominators
4. Simplify Complex Fractions by Method I
5. Simplify Complex Fractions by Method II

1. Add and Subtract Rational Expressions with Like Denominators

To add or subtract rational expressions, the expressions must have the same denominator. As with fractions, we add or subtract rational expressions with the same denominator by combining the terms in the numerator and then writing the result over the common denominator. Then, if possible, we simplify the expression to lowest terms.

Addition and Subtraction Properties of Rational Expressions

Let p, q, and r represent polynomials where $q \neq 0$. Then

1. $\dfrac{p}{q} + \dfrac{r}{q} = \dfrac{p+r}{q}$ 2. $\dfrac{p}{q} - \dfrac{r}{q} = \dfrac{p-r}{q}$

EXAMPLE 1   Adding and Subtracting Rational Expressions with Like Denominators

Add or subtract as indicated.

a. $\dfrac{1}{8} + \dfrac{3}{8}$ **b.** $\dfrac{5x}{2x-1} + \dfrac{3}{2x-1}$ **c.** $\dfrac{x^2}{x-4} - \dfrac{x+12}{x-4}$

Solution:

a. $\dfrac{1}{8} + \dfrac{3}{8} = \dfrac{1+3}{8}$ Add the terms in the numerator.

$= \dfrac{4}{8}$

$= \dfrac{1}{2}$ Simplify the fraction.

b. $\dfrac{5x}{2x-1} + \dfrac{3}{2x-1} = \dfrac{5x+3}{2x-1}$ Add the terms in the numerator. The answer is already in lowest terms.

c. $\dfrac{x^2}{x-4} - \dfrac{x+12}{x-4}$ Combine the terms in the numerator. Use parentheses to group the terms in the numerator that follow the subtraction sign.

$= \dfrac{x^2-(x+12)}{x-4}$ This will help you remember to apply the distributive property.

$= \dfrac{x^2-x-12}{x-4}$ Apply the distributive property.

$= \dfrac{(x-4)(x+3)}{(x-4)}$ Factor the numerator and denominator.

$= \dfrac{(\overset{1}{\cancel{x-4}})(x+3)}{(\cancel{x-4})}$ Simplify the rational expression.

$= x+3$

Skill Practice 1 Add or subtract as indicated.

$$\textbf{a. } \frac{5}{12} - \frac{1}{12} \qquad \textbf{b. } \frac{4c - 3}{c - 2} + \frac{8}{c - 2} \qquad \textbf{c. } \frac{t^2}{t - 7} - \frac{5t + 14}{t - 7}$$

2. Identify the Least Common Denominator

If two rational expressions have different denominators, each expression must be rewritten with a common denominator before adding or subtracting the expressions. The **least common denominator (LCD)** of two or more rational expressions is defined as the least common multiple of the denominators.

For example, consider the fractions $\frac{1}{20}$ and $\frac{1}{8}$. By inspection, the least common denominator is 40. To understand why, find the prime factorization of both denominators.

$$20 = 2^2 \cdot 5 \qquad \text{and} \qquad 8 = 2^3$$

A common multiple of 20 and 8 must be a multiple of 5, a multiple of 2^2, and a multiple of 2^3. However, any number that is a multiple of $2^3 = 8$ is automatically a multiple of $2^2 = 4$. Therefore, it is sufficient to construct the least common denominator as the product of unique prime factors, where each factor is raised to its highest power.

$$\text{The LCD of } \frac{1}{2^2 \cdot 5} \text{ and } \frac{1}{2^3} \text{ is } 2^3 \cdot 5 = 40.$$

> **TIP** The least common multiple of 20 and 8 is 40.
>
> Multiples of 20:
> 20, 40, 60, 80, ...
>
> Multiples of 8:
> 8, 16, 24, 32, 40, ...

Finding the LCD of Two or More Rational Expressions

Step 1 Factor all denominators completely.

Step 2 The LCD is the product of unique prime factors from the denominators, where each factor is raised to the highest power to which it appears in any denominator.

EXAMPLE 2 **Finding the LCD of Rational Expressions**

Find the LCD of the rational expressions.

$$\textbf{a. } \frac{1}{12}, \frac{5}{18}, \frac{7}{30} \qquad \textbf{b. } \frac{1}{2x^3y}, \frac{5}{16xy^2z}$$

Solution:

$$\textbf{a. } \frac{1}{12}, \frac{5}{18}, \frac{7}{30}$$

$$\frac{1}{2^2 \cdot 3}, \frac{5}{2 \cdot 3^2}, \frac{7}{2 \cdot 3 \cdot 5} \qquad \text{Factor the denominators completely.}$$

$$12 = \boxed{2^2} \cdot 3 \qquad\qquad 2^2 \text{ is the greatest power of 2 that appears.}$$

$$18 = 2^1 \boxed{3^2} \qquad\qquad 3^2 \text{ is the greatest power of 3 that appears.}$$

$$30 = 2^1 \cdot 3^1 \boxed{5^1} \qquad\quad 5^1 \text{ is the greatest power of 5 that appears.}$$

$$\text{LCD} = 2^2 \cdot 3^2 \cdot 5 = 180 \qquad \text{The LCD is the product of the factors 2, 3, and 5,}$$
$$\text{where each factor is raised to its highest power.}$$

Answers

1. a. $\frac{1}{3}$ **b.** $\frac{4c + 5}{c - 2}$ **c.** $t + 2$

b. $\dfrac{1}{2x^3y}, \dfrac{5}{16xy^2z}$

$\dfrac{1}{2x^3y}, \dfrac{5}{2^4xy^2z}$ Factor the denominators completely.

$\begin{aligned} LCD &= 2^4x^3y^2z \\ &= 16x^3y^2z \end{aligned}$ The LCD is the product of the factors 2, x, y, and z, where each factor is raised to its highest power.

Skill Practice 2   Find the LCD of the rational expressions.

a. $\dfrac{7}{40}, \dfrac{1}{15}, \dfrac{5}{6}$ **b.** $\dfrac{1}{9a^3b^2}, \dfrac{5}{18a^4b}$

EXAMPLE 3   **Finding the LCD of Rational Expressions**

Find the LCD of the rational expressions.

a. $\dfrac{x^2 + 3}{x^2 + 9x + 20}, \dfrac{6}{x^2 + 8x + 16}$ **b.** $\dfrac{x + 4}{x - 3}, \dfrac{1}{3 - x}$

Solution:

a. $\dfrac{x^2 + 3}{x^2 + 9x + 20}, \dfrac{6}{x^2 + 8x + 16}$

$\dfrac{x^2 + 3}{(x + 4)(x + 5)}, \dfrac{6}{(x + 4)^2}$ Factor the denominators completely.

$LCD = (x + 5)(x + 4)^2$ The LCD is the product of the factors $(x + 5)$ and $(x + 4)$, where each factor is raised to its highest power.

b. $\dfrac{x + 4}{x - 3}, \dfrac{1}{3 - x}$ The denominators are already factored.

Notice that $x - 3$ and $3 - x$ are opposite factors. If -1 is factored from either expression, the binomial factors will be the same.

$\dfrac{x + 4}{x - 3}, \dfrac{1}{-1(-3 + x)}$ $\dfrac{x + 4}{-1(-x + 3)}, \dfrac{1}{3 - x}$

Factor out -1.   Factor out -1.

same binomial factors   same binomial factors

$\begin{aligned} LCD &= (x - 3)(-1) \\ &= -x + 3 \\ &= 3 - x \end{aligned}$ $\begin{aligned} LCD &= (-1)(3 - x) \\ &= -3 + x \\ &= x - 3 \end{aligned}$

The LCD is either $(3 - x)$ or $(x - 3)$.

Skill Practice 3   Find the LCD of the rational expressions.

a. $\dfrac{5}{x^2 + 4x + 4}, \dfrac{x + 1}{x^2 - x - 6}$ **b.** $\dfrac{6}{z - 7}, \dfrac{1}{7 - z}$

FOR REVIEW

To factor trinomials, see Section 2.4. In particular, recall that a trinomial of the form $a^2 + 2ab + b^2$ is a **perfect square trinomial** and factors as

$a^2 + 2ab + b^2 = (a + b)^2$.

Note that $x^2 + 8x + 16$ is a perfect square trinomial because the first and third terms are perfect squares:

$x^2 = (x)^2$ and $16 = (4)^2$.

The middle term is twice the product of x and 4.

Thus, $x^2 + 8x + 16$
$= (x)^2 + 2(x)(4) + (4)^2$
$= (x + 4)^2$

Answers
2. a. 120   **b.** $18a^4b^2$
3. a. $(x + 2)^2(x - 3)$
  b. $z - 7$ or $7 - z$

3. Add and Subtract Rational Expressions with Unlike Denominators

To add or subtract rational expressions with unlike denominators, we must convert each expression to an equivalent expression with the same denominator. To do this, multiply the fraction by a convenient form of 1. For example, consider adding the expressions $\frac{3}{x-2} + \frac{5}{x+1}$. The LCD is $(x-2)(x+1)$. For each expression, identify the factors from the LCD that are missing in the denominator. Then multiply the numerator and denominator of the expression by the missing factor(s):

$$\frac{(3)}{(x-2)} \cdot \frac{(x+1)}{(x+1)} + \frac{(5)}{(x+1)} \cdot \frac{(x-2)}{(x-2)}$$ The rational expressions now have the same denominator and can be added.

$$= \frac{3(x+1) + 5(x-2)}{(x-2)(x+1)}$$ Combine terms in the numerator.

$$= \frac{3x + 3 + 5x - 10}{(x-2)(x+1)}$$ Clear parentheses and simplify.

$$= \frac{8x - 7}{(x-2)(x+1)}$$

Adding and Subtracting Rational Expressions

Step 1 Factor the denominator of each rational expression.

Step 2 Identify the LCD.

Step 3 Rewrite each rational expression as an equivalent expression with the LCD as its denominator.

Step 4 Add or subtract the numerators, and write the result over the common denominator.

Step 5 Simplify, if possible.

EXAMPLE 4   Adding Rational Expressions with Unlike Denominators

Add. $\dfrac{3}{7b} + \dfrac{4}{b^2}$

Solution:

$\dfrac{3}{7b} + \dfrac{4}{b^2}$ **Step 1:** The denominators are already factored.

 Step 2: The LCD is $7b^2$.

$= \dfrac{3}{7b} \cdot \dfrac{b}{b} + \dfrac{4}{b^2} \cdot \dfrac{7}{7}$ **Step 3:** Write each expression with the LCD.

$= \dfrac{3b}{7b^2} + \dfrac{28}{7b^2}$ **Step 4:** Add the numerators and write the result over the LCD.

$= \dfrac{3b + 28}{7b^2}$ **Step 5:** Simplify.

Answer

4. $\dfrac{12y^2 + 5}{15y^3}$

Skill Practice 4   Add. $\dfrac{4}{5y} + \dfrac{1}{3y^3}$

EXAMPLE 5   **Subtracting Rational Expressions with Unlike Denominators**

Subtract. $\dfrac{3t - 2}{t^2 + 4t - 12} - \dfrac{5}{2t + 12}$

Solution:

$\dfrac{3t - 2}{t^2 + 4t - 12} - \dfrac{5}{2t + 12}$

$= \dfrac{3t - 2}{(t + 6)(t - 2)} - \dfrac{5}{2(t + 6)}$ **Step 1:** Factor the denominators.

Step 2: The LCD is $2(t + 6)(t - 2)$.

$= \dfrac{(2)}{(2)} \cdot \dfrac{(3t - 2)}{(t + 6)(t - 2)} - \dfrac{5}{2(t + 6)} \cdot \dfrac{(t - 2)}{(t - 2)}$ **Step 3:** Write each expression with the LCD.

$= \dfrac{2(3t - 2) - 5(t - 2)}{2(t + 6)(t - 2)}$ **Step 4:** Add the numerators and write the result over the LCD.

$= \dfrac{6t - 4 - 5t + 10}{2(t + 6)(t - 2)}$ **Step 5:** Simplify.

$= \dfrac{t + 6}{2(t + 6)(t - 2)}$ Combine like terms.

$= \dfrac{\overset{1}{\cancel{t + 6}}}{2\cancel{(t + 6)}(t - 2)}$ Simplify.

$= \dfrac{1}{2(t - 2)}$

Skill Practice 5   Subtract. $\dfrac{2x + 3}{x^2 + x - 2} - \dfrac{5}{3x - 3}$

EXAMPLE 6   **Adding and Subtracting Rational Expressions with Unlike Denominators**

Add and subtract as indicated. $\dfrac{2}{x} + \dfrac{x}{x + 3} - \dfrac{3x + 18}{x^2 + 3x}$

Solution:

$\dfrac{2}{x} + \dfrac{x}{x + 3} - \dfrac{3x + 18}{x^2 + 3x}$

$= \dfrac{2}{x} + \dfrac{x}{x + 3} - \dfrac{3x + 18}{x(x + 3)}$ **Step 1:** Factor the denominators.

Step 2: The LCD is $x(x + 3)$.

$= \dfrac{2}{x} \cdot \dfrac{(x + 3)}{(x + 3)} + \dfrac{x}{(x + 3)} \cdot \dfrac{x}{x} - \dfrac{3x + 18}{x(x + 3)}$ **Step 3:** Write each expression with the LCD.

Answer

5. $\dfrac{1}{3(x + 2)}$

$$= \frac{2(x+3) + x^2 - (3x+18)}{x(x+3)}$$

Step 4: Add the numerators and write the result over the LCD.

$$= \frac{2x + 6 + x^2 - 3x - 18}{x(x+3)}$$

Step 5: Simplify.

$$= \frac{x^2 - x - 12}{x(x+3)}$$

Combine like terms.

$$= \frac{(x-4)(x+3)}{x(x+3)}$$

Factor the numerator.

$$= \frac{(x-4)\overset{1}{\cancel{(x+3)}}}{x\cancel{(x+3)}}$$

Simplify.

$$= \frac{x-4}{x}$$

Skill Practice 6 Add. $\dfrac{a^2 + a + 24}{a^2 - 9} + \dfrac{5}{a+3}$

EXAMPLE 7 Adding Rational Expressions with Unlike Denominators

Add. $\dfrac{x^2}{x-y} + \dfrac{y^2}{y-x}$

Solution:

$$\frac{x^2}{x-y} + \frac{y^2}{y-x}$$

Step 1: The denominators are already factored.

Step 2: The denominators are opposites and differ by a factor of -1. The LCD can be taken as either $(x-y)$ or $(y-x)$. We will use an LCD of $(x-y)$.

$$= \frac{x^2}{(x-y)} + \frac{y^2}{(y-x)} \cdot \frac{(-1)}{(-1)}$$

Step 3: Write each expression with the LCD. Note that $(y-x)(-1) = -y + x = x - y$.

$$= \frac{x^2}{x-y} + \frac{-y^2}{x-y}$$

$$= \frac{x^2 - y^2}{x-y}$$

Step 4: Combine the numerators, and write the result over the LCD.

$$= \frac{(x+y)\overset{1}{\cancel{(x-y)}}}{\cancel{x-y}}$$

Step 5: Factor and simplify to lowest terms.

$$= x + y$$

Skill Practice 7 Add. $\dfrac{3a}{a-5} + \dfrac{15}{5-a}$

4. Simplify Complex Fractions by Method I

A **complex fraction** is an expression containing one or more fractional expressions in the numerator, denominator, or both. For example:

$$\frac{\dfrac{5x^2}{y}}{\dfrac{10x}{y^2}} \quad \text{and} \quad \frac{2 + \dfrac{1}{2} - \dfrac{1}{3}}{\dfrac{3}{4} + \dfrac{1}{6}}$$

are complex fractions.

Two methods will be presented to simplify complex fractions. The first method (Method I) follows the order of operations to simplify the numerator and denominator separately before dividing. The process is summarized as follows.

Simplifying a Complex Fraction—Method I

Step 1 Add or subtract expressions in the numerator to form a single fraction. Add or subtract expressions in the denominator to form a single fraction.

Step 2 Divide the rational expressions from Step 1.

Step 3 Simplify to lowest terms, if possible.

EXAMPLE 8   Simplifying a Complex Fraction by Method I

Simplify the expression. $\dfrac{\dfrac{1}{4x} - \dfrac{3}{2}}{3 - \dfrac{1}{2x}}$

Solution:

$$\frac{\dfrac{1}{4x} - \dfrac{3}{2}}{3 - \dfrac{1}{2x}}$$

Step 1: Combine fractions in the numerator and denominator separately.

$$= \frac{\dfrac{1}{4x} - \dfrac{2x}{2x} \cdot \dfrac{3}{2}}{\dfrac{2x}{2x} \cdot \dfrac{3}{1} - \dfrac{1}{2x}}$$

The LCD in the numerator is $4x$.
The LCD in the denominator is $2x$.

$$= \frac{\dfrac{1}{4x} - \dfrac{6x}{4x}}{\dfrac{6x}{2x} - \dfrac{1}{2x}}$$

$$= \frac{\dfrac{1 - 6x}{4x}}{\dfrac{6x - 1}{2x}}$$

Step 2: Divide the expression in the numerator of the complex fraction by the expression in the denominator.

$$= \frac{\overset{-1}{(1 - 6x)}}{\underset{2}{4x}} \cdot \frac{\overset{1}{2x}}{(6x - 1)}$$

Multiply by the reciprocal of the divisor.

$$= -\frac{1}{2}$$

Step 3: Simplify to lowest terms.

TIP The fraction bar of the complex fraction serves as the division operator between the fraction in the numerator and the fraction in the denominator.

$$\frac{\dfrac{1 - 6x}{4x}}{\dfrac{6x - 1}{2x}} \quad \xleftarrow{} \div$$

$$= \frac{1 - 6x}{4x} \div \frac{6x - 1}{2x}$$

Skill Practice 8   Simplify the expression. $\dfrac{\dfrac{1}{9m} - \dfrac{4}{3}}{4 - \dfrac{1}{3m}}$

5. Simplify Complex Fractions by Method II

We will now use a second method to simplify complex fractions—Method II. Recall that multiplying the numerator and denominator of a rational expression by the same quantity does not change the value of the expression. This is the basis for Method II.

Simplifying a Complex Fraction—Method II

Step 1 Multiply the numerator and denominator of the complex fraction by the LCD of *all* individual fractions within the expression.

Step 2 Apply the distributive property, and simplify the numerator and denominator.

Step 3 Simplify to lowest terms, if possible.

EXAMPLE 9   Simplifying a Complex Fraction by Method II

Simplify by using Method II. $\dfrac{4 - \dfrac{6}{x}}{\dfrac{2}{x} - \dfrac{3}{x^2}}$

Solution:

$\dfrac{4 - \dfrac{6}{x}}{\dfrac{2}{x} - \dfrac{3}{x^2}}$ The LCD of all individual terms is x^2.

$= \dfrac{x^2 \cdot \left(4 - \dfrac{6}{x}\right)}{x^2 \cdot \left(\dfrac{2}{x} - \dfrac{3}{x^2}\right)}$ **Step 1:** Multiply the numerator and denominator of the complex fraction by the LCD of x^2.

$= \dfrac{x^2 \cdot (4) - x^2 \cdot \left(\dfrac{6}{x}\right)}{x^2 \cdot \left(\dfrac{2}{x}\right) - x^2 \cdot \left(\dfrac{3}{x^2}\right)}$ **Step 2:** Apply the distributive property.

$= \dfrac{4x^2 - 6x}{2x - 3}$

$= \dfrac{2x(2x - 3)}{2x - 3}$ **Step 3:** Factor and simplify.

$= 2x$

Answer

8. $-\dfrac{1}{3}$

Skill Practice 9   Simplify by using Method II. $\dfrac{y - \dfrac{1}{y}}{1 - \dfrac{1}{y^2}}$

EXAMPLE 10   Simplifying a Complex Fraction by Method II

Simplify the expression by Method II. $\dfrac{\dfrac{1}{w+3} - \dfrac{1}{w-3}}{1 + \dfrac{9}{w^2 - 9}}$

Solution:

$$\dfrac{\dfrac{1}{w+3} - \dfrac{1}{w-3}}{1 + \dfrac{9}{w^2 - 9}}$$

$$= \dfrac{\dfrac{1}{w+3} - \dfrac{1}{w-3}}{1 + \dfrac{9}{(w+3)(w-3)}} \qquad \text{Factor all denominators to find the LCD.}$$

The LCD of $\dfrac{1}{1}, \dfrac{1}{w+3}, \dfrac{1}{w-3},$ and $\dfrac{9}{(w+3)(w-3)}$ is $(w+3)(w-3)$.

$$= \dfrac{(w+3)(w-3)\left(\dfrac{1}{w+3} - \dfrac{1}{w-3}\right)}{(w+3)(w-3)\left[1 + \dfrac{9}{(w+3)(w-3)}\right]} \qquad$$

Step 1: Multiply the numerator and denominator of the complex fraction by $(w+3)(w-3)$.

$$= \dfrac{\cancel{(w+3)}(w-3)\left(\dfrac{1}{\cancel{w+3}}\right) - (w+3)\cancel{(w-3)}\left(\dfrac{1}{\cancel{w-3}}\right)}{(w+3)(w-3)1 + \cancel{(w+3)}\cancel{(w-3)}\left[\dfrac{9}{\cancel{(w+3)}\cancel{(w-3)}}\right]}$$

Step 2: Distributive property.

$$= \dfrac{(w-3) - (w+3)}{(w+3)(w-3) + 9} \qquad \text{\textbf{Step 3:} Simplify.}$$

$$= \dfrac{w - 3 - w - 3}{w^2 - 9 + 9} \qquad \text{Apply the distributive property.}$$

$$= \dfrac{-6}{w^2}$$

$$= -\dfrac{6}{w^2}$$

Answer

9. y

Answer

10. $\dfrac{x-3}{x^2+1}$

Skill Practice 10   Simplify by using Method II. $\dfrac{\dfrac{2}{x+1}-\dfrac{1}{x-1}}{\dfrac{x}{x-1}-\dfrac{1}{x+1}}$

SECTION 4.2   Practice Exercises

Prerequisite Review

For Exercises R.1–R.6, perform the indicated operations. (See Section R.1 for review.)

R.1. $-\dfrac{5}{12}+\dfrac{7}{12}$

R.2. $-\dfrac{11}{18}-\dfrac{5}{18}$

R.3. $-\dfrac{11}{15}-\dfrac{5}{18}$

R.4. $-\dfrac{11}{10}+\dfrac{7}{12}$

R.5. $\dfrac{10}{9}\left(\dfrac{2}{5}-\dfrac{7}{4}\right)$

R.6. $\dfrac{6}{5}\left(\dfrac{1}{3}+\dfrac{7}{2}\right)$

For Exercises R.7–R.8, simplify the expressions. (See Section 2.1 for review.)

R.7. $\dfrac{3x^{-2}}{y^{-4}}$

R.8. $\dfrac{5a^{-1}}{b^3}$

Concept Connections

1. Given polynomials p, q, and r such that $q \neq 0$, $\dfrac{p}{q}+\dfrac{r}{q}=\dfrac{\square}{\square}$ and $\dfrac{p}{q}-\dfrac{r}{q}=\dfrac{\square}{\square}$

2. The _____ _____ _____ (LCD) of two rational expressions is defined as the least common multiple of their denominators.

Objective 1: Add and Subtract Rational Expressions with Like Denominators

For Exercises 3–10, add or subtract as indicated and simplify if possible. **(See Example 1)**

3. $\dfrac{3}{5x}+\dfrac{7}{5x}$

4. $\dfrac{1}{2x^2}-\dfrac{5}{2x^2}$

5. $\dfrac{x}{x^2-2x-3}-\dfrac{3}{x^2-2x-3}$

6. $\dfrac{x}{x^2+4x-12}+\dfrac{6}{x^2+4x-12}$

7. $\dfrac{5x-1}{(2x+9)(x-6)}-\dfrac{3x-6}{(2x+9)(x-6)}$

8. $\dfrac{4-x}{8x+1}-\dfrac{5x-6}{8x+1}$

9. $\dfrac{x+2}{x-5}+\dfrac{x-12}{x-5}$

10. $\dfrac{2x-1}{x-2}+\dfrac{x-5}{x-2}$

Objective 2: Identify the Least Common Denominator

For Exercises 11–22, find the least common denominator (LCD). **(See Examples 2–3)**

11. $\dfrac{5}{8};\dfrac{3}{20x}$

12. $\dfrac{y}{15a};\dfrac{y^2}{35}$

13. $\dfrac{-5}{6m^4};\dfrac{1}{15mn^7}$

14. $\dfrac{13}{12cd^5};\dfrac{9}{8c^3}$

15. $\dfrac{6}{(x-4)(x+2)};\dfrac{-8}{(x-4)(x-6)}$

16. $\dfrac{x}{(2x-1)(x-7)};\dfrac{2}{(2x-1)(x+1)}$

17. $\dfrac{3}{x(x-1)(x+7)^2};\dfrac{-1}{x^2(x+7)}$

18. $\dfrac{14}{(x-2)^2(x+9)}; \dfrac{41}{x(x-2)(x+9)}$

19. $\dfrac{5}{x-6}; \dfrac{x-5}{x^2-8x+12}$

20. $\dfrac{7a}{a+4}; \dfrac{a+12}{a^2-16}$

21. $\dfrac{3a}{a-4}; \dfrac{5}{4-a}$

22. $\dfrac{10}{x-6}; \dfrac{x+1}{6-x}$

Objective 3: Add and Subtract Rational Expressions with Unlike Denominators

For Exercises 23–48, add or subtract as indicated. (See Examples 4–7)

23. $\dfrac{4}{3p} - \dfrac{5}{2p^2}$

24. $\dfrac{6}{5a^2b} - \dfrac{1}{10ab}$

25. $\dfrac{s-1}{s} - \dfrac{t+1}{t}$

26. $\dfrac{x+2}{x} - \dfrac{y-2}{y}$

27. $\dfrac{4a-2}{3a+12} - \dfrac{a-2}{a+4}$

28. $\dfrac{6y+5}{5y-25} - \dfrac{y+2}{y-5}$

29. $\dfrac{10}{b(b+5)} + \dfrac{2}{b}$

30. $\dfrac{6}{w(w-2)} + \dfrac{3}{w}$

31. $\dfrac{x-2}{x-6} - \dfrac{x+2}{6-x}$

32. $\dfrac{x-10}{x-8} - \dfrac{x+10}{8-x}$

33. $\dfrac{6b}{b-4} - \dfrac{1}{b+1}$

34. $\dfrac{a}{a-3} - \dfrac{5}{a+6}$

35. $\dfrac{2}{2x+1} + \dfrac{4}{x-2}$

36. $\dfrac{3}{y+6} + \dfrac{1}{3y+1}$

37. $\dfrac{y-2}{y-4} + \dfrac{2y^2-15y+12}{y^2-16}$

38. $\dfrac{x^2+13x+18}{x^2-9} + \dfrac{x+1}{x+3}$

39. $\dfrac{x+2}{x^2-36} - \dfrac{x}{x^2+9x+18}$

40. $\dfrac{7}{x^2-x-2} + \dfrac{x}{x^2+4x+3}$

41. $\dfrac{5}{w} + \dfrac{8}{-w}$

42. $\dfrac{4}{y} + \dfrac{5}{-y}$

43. $\dfrac{n}{5-n} + \dfrac{2n-5}{n-5}$

44. $\dfrac{c}{7-c} + \dfrac{2c-7}{c-7}$

45. $\dfrac{2}{3x-15} + \dfrac{x}{25-x^2}$

46. $\dfrac{5}{9-x^2} - \dfrac{4}{x^2+4x+3}$

47. $\dfrac{m}{20+9m+m^2} - \dfrac{4}{12+7m+m^2}$

48. $\dfrac{t}{6+5t+t^2} - \dfrac{2}{2+3t+t^2}$

Objective 4: Simplify Complex Fractions by Method I

For Exercises 49–56, simplify the complex fractions by using Method I. (See Example 8)

49. $\dfrac{\dfrac{5x^2}{9y^2}}{\dfrac{3x}{y^2x}}$

50. $\dfrac{\dfrac{3w^2}{4rs}}{\dfrac{15wr}{s^2}}$

51. $\dfrac{\dfrac{x-6}{3x}}{\dfrac{3x-18}{9}}$

52. $\dfrac{\dfrac{a+4}{6}}{\dfrac{16-a^2}{3}}$

53. $\dfrac{\dfrac{2}{3} + \dfrac{1}{6}}{\dfrac{1}{2} - \dfrac{1}{4}}$

54. $\dfrac{\dfrac{7}{8} + \dfrac{3}{4}}{\dfrac{1}{3} - \dfrac{5}{6}}$

55. $\dfrac{8 - \dfrac{5}{2x}}{\dfrac{5}{8x} - 2}$

56. $\dfrac{10 - \dfrac{3}{5x}}{\dfrac{3}{10x} - 5}$

Objective 5: Simplify Complex Fractions by Method II

For Exercises 57–76, simplify the complex fractions by using Method II. (See Examples 9–10)

57. $\dfrac{\dfrac{7y}{y+3}}{\dfrac{1}{4y+12}}$

58. $\dfrac{\dfrac{6x}{x-5}}{\dfrac{1}{4x-20}}$

59. $\dfrac{1 + \dfrac{1}{3}}{\dfrac{5}{6} - 1}$

60. $\dfrac{2 + \dfrac{4}{5}}{-1 + \dfrac{3}{10}}$

61. $\dfrac{\dfrac{3q}{p} - q}{q - \dfrac{q}{p}}$

62. $\dfrac{\dfrac{b}{a} + 3b}{b + \dfrac{2b}{a}}$

63. $\dfrac{\dfrac{2}{a} + \dfrac{3}{a^2}}{\dfrac{4}{a^2} - \dfrac{9}{a}}$

64. $\dfrac{\dfrac{2}{y^2} + \dfrac{1}{y}}{\dfrac{4}{y^2} - \dfrac{1}{y}}$

65. $\dfrac{-8}{\dfrac{6w}{w - 1} - 4}$

66. $\dfrac{6}{2z - \dfrac{10}{z - 4}}$

67. $\dfrac{\dfrac{y}{y + 3}}{\dfrac{y}{y + 3} + y}$

68. $\dfrac{\dfrac{4}{w - 4}}{\dfrac{4}{w - 4} - 1}$

69. $\dfrac{1 - \dfrac{1}{x} - \dfrac{6}{x^2}}{1 - \dfrac{4}{x} + \dfrac{3}{x^2}}$

70. $\dfrac{1 + \dfrac{1}{x} - \dfrac{12}{x^2}}{\dfrac{9}{x^2} + \dfrac{3}{x} - 2}$

71. $\dfrac{2 - \dfrac{2}{t + 1}}{2 + \dfrac{2}{t}}$

72. $\dfrac{3 + \dfrac{3}{p - 1}}{3 - \dfrac{3}{p}}$

73. $\dfrac{\dfrac{2}{a} - \dfrac{3}{a + 1}}{\dfrac{2}{a + 1} - \dfrac{3}{a}}$

74. $\dfrac{\dfrac{5}{b} + \dfrac{4}{b + 1}}{\dfrac{4}{b} - \dfrac{5}{b + 1}}$

75. $\dfrac{\dfrac{1}{y + 2} + \dfrac{4}{y - 3}}{\dfrac{2}{y - 3} - \dfrac{7}{y + 2}}$

76. $\dfrac{\dfrac{1}{t - 4} + \dfrac{1}{t + 5}}{\dfrac{6}{t + 5} + \dfrac{2}{t - 4}}$

Mixed Exercises

For Exercises 77–102, simplify.

77. $w + 2 + \dfrac{1}{w - 2}$

78. $h - 3 + \dfrac{1}{h + 3}$

79. $\dfrac{t + 1}{t + 3} - \dfrac{t - 2}{t - 3} + \dfrac{6}{t^2 - 9}$

80. $\dfrac{y - 3}{y - 2} - \dfrac{y + 1}{2y - 5} + \dfrac{-4y + 7}{2y^2 - 9y + 10}$

81. $\dfrac{t^{-1} - 1}{1 - t^{-2}}$

82. $\dfrac{d^{-2} - c^{-2}}{c^{-1} - d^{-1}}$

83. $\dfrac{\dfrac{2}{x + h} - \dfrac{2}{x}}{h}$

84. $\dfrac{\dfrac{1}{2x + 2h} - \dfrac{1}{2x}}{h}$

85. $(x - 1) \cdot \left[\dfrac{3}{x^2 - 1} + \dfrac{x}{2x - 2}\right]$

86. $(3x - 2) \cdot \left[\dfrac{x}{3x^2 + x - 2} + \dfrac{2}{x + 1}\right]$

87. $-\dfrac{2x}{x^2 - y^2} - \dfrac{1}{x - y} + \dfrac{1}{y - x}$

88. $\dfrac{3w - 1}{2w^2 + w - 3} - \dfrac{2 - w}{w - 1} - \dfrac{w}{1 - w}$

89. $(2p + 1) \cdot \left[\dfrac{2p}{6p + 3} - \dfrac{1}{p + 4}\right]$

90. $(y + 8) \cdot \left[\dfrac{4}{2y + 1} - \dfrac{y}{2y^2 + 17y + 8}\right]$

91. $\dfrac{1}{x + 5} + \dfrac{3}{(x + 5)^2} - \dfrac{2}{(x + 5)^3}$

92. $\dfrac{1}{x - 2} + \dfrac{4}{(x - 2)^2} - \dfrac{3}{(x - 2)^3}$

93. $\dfrac{x^{-2}}{x + 3x^{-1}}$

94. $\dfrac{x^{-1} + x^{-2}}{5x^{-2}}$

95. $\dfrac{2a^{-1} + 3b^{-2}}{a^{-1} - b^{-1}}$

96. $\dfrac{2m^{-1} + n^{-1}}{m^{-2} - 4n^{-1}}$

97. $\dfrac{\dfrac{1}{4 + h} - \dfrac{1}{4}}{h}$

98. $\dfrac{\dfrac{1}{3 + 3h} - \dfrac{1}{3}}{h}$

99. $\dfrac{\dfrac{6}{x + h} - \dfrac{6}{x}}{h}$

100. $\dfrac{\dfrac{-3}{x + h} + \dfrac{3}{x}}{h}$

101. $\dfrac{5}{x^2 - 4} + \dfrac{2}{x^3 - 8}$

102. $\dfrac{-2}{x^2 - 9} + \dfrac{3}{x^3 - 27}$

For Exercises 103–106, write an expression that represents the perimeter of the figure and simplify.

103.

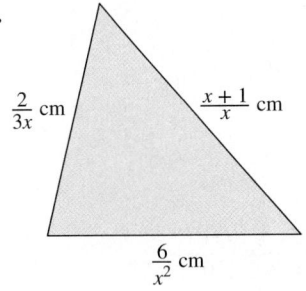

$\frac{2}{3x}$ cm $\frac{x+1}{x}$ cm

$\frac{6}{x^2}$ cm

104.

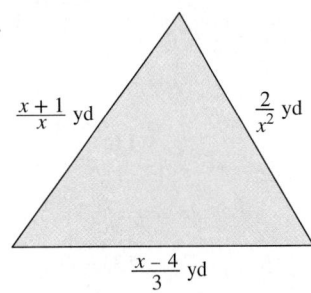

$\frac{x+1}{x}$ yd $\frac{2}{x^2}$ yd

$\frac{x-4}{3}$ yd

105.

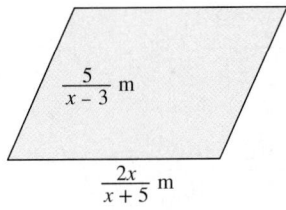

$\frac{5}{x-3}$ m

$\frac{2x}{x+5}$ m

106.

$\frac{3}{x+2}$ ft

$\frac{x}{x+1}$ ft

Expanding Your Skills

107. Simplify. $\dfrac{x}{1 - \left(1 + \dfrac{1}{x}\right)^{-1}}$

108. Simplify. $\dfrac{x}{1 - \left(1 - \dfrac{1}{x}\right)^{-1}}$

For Exercises 109–116, write the expression as a single term, factored completely. Do not rationalize the denominator.

109. $1 - x^{-2} - 2x^{-3}$

110. $1 - 8x^{-5} + 30x^{-7}$

111. $\dfrac{3}{2\sqrt{3x}} + \sqrt{3x}$

112. $\dfrac{2}{\sqrt{x}} + \sqrt{x}$

113. $\dfrac{\sqrt{x^2 + 1} + \dfrac{x^2}{\sqrt{x^2 + 1}}}{x^2 + 1}$

114. $\dfrac{\dfrac{x^2}{\sqrt{x^2 + 9}} - \sqrt{x^2 + 9}}{x^2}$

115. $2\sqrt{4x^2 + 9} + \dfrac{8x^2}{\sqrt{4x^2 + 9}}$

116. $3\sqrt{9x^2 + 1} + \dfrac{27x}{\sqrt{9x^2 + 1}}$

PROBLEM RECOGNITION EXERCISES

Operations on Rational Expressions

For Exercises 1–24, identify the operation (addition, subtraction, multiplication, or division), then simplify the expression. In each case, be sure to ask yourself if you need a common denominator.

1. $\dfrac{2}{2y - 3} - \dfrac{3}{2y} + 1$

2. $(x + 5) + \left(\dfrac{7}{x - 4}\right)$

3. $\dfrac{5x^2 - 6x + 1}{x^2 - 1} \div \dfrac{16x^2 - 9}{4x^2 + 7x + 3} - \dfrac{x}{4x - 3}$

4. $\dfrac{a^2 - 25}{3a^2 + 3ab} \cdot \dfrac{a^2 + 4a + ab + 4b}{a^2 + 9a + 20}$

5. $\dfrac{4}{y + 1} + \dfrac{y + 2}{y^2 - 1} - \dfrac{3}{y - 1}$

6. $\dfrac{8w^2}{w^3 - 16w} - \dfrac{4w}{w^2 - 4w}$

7. $\dfrac{a^2 - 16}{2x + 6} \cdot \dfrac{x + 3}{a - 4}$

8. $\dfrac{t^2 - 9}{t} \div \dfrac{t + 3}{t + 2}$

9. $\dfrac{2 + \dfrac{1}{a}}{4 - \dfrac{1}{a^2}}$

10. $\dfrac{\dfrac{6x^2 y}{5}}{\dfrac{3x}{y}}$

11. $\dfrac{6xy}{x^2 - y^2} + \dfrac{x + y}{y - x}$

12. $(x^2 - 6x + 8) \cdot \left(\dfrac{3}{x - 2}\right)$

13. $\dfrac{3}{x - 2} - \dfrac{x - 2}{6}$

14. $\dfrac{5}{x + 7} + \dfrac{x + 7}{10}$

15. $\dfrac{1}{w - 1} - \dfrac{w + 2}{3w - 3}$

16. $\dfrac{3y + 6}{y^2 - 3y - 10} \div \dfrac{27}{y - 5}$

17. $\dfrac{y + \dfrac{2}{y} - 3}{1 - \dfrac{2}{y}}$

18. $\dfrac{2}{t - 3} - \dfrac{3}{t + 2} + 5$

19. $\dfrac{4x^2 + 22x + 24}{4x + 4} \cdot \dfrac{6x + 6}{4x^2 - 9}$

20. $\dfrac{12x^3 y^5 z}{5x^4} \div \dfrac{16xy^7}{10z^2}$

21. $\dfrac{3x - 1}{4} + \dfrac{7}{6x - 2}$

22. $\dfrac{2x^{-1} + 3x^{-2}}{x^{-2} - 5x^{-1}}$

23. $(y + 2) \cdot \dfrac{2y + 1}{y^2 - 4} - \dfrac{y - 2}{y + 3}$

24. $\dfrac{a^2}{a - 10} - \dfrac{100 - 20a}{10 - a}$

SECTION 4.3   Rational Equations

1. Solve Rational Equations

Thus far we have learned to solve several types of equations, including

Equation type	Example
Linear equations	$4x - 5 = 27$
Absolute value equations	$\lvert 2x - 7 \rvert + 4 = 9$
Quadratic equations	$2x^2 - 11x + 5 = 0$
Polynomial equations	$x^3 + 3x^2 - 4x - 12 = 0$

In this section, we will study another type of equation called a rational equation.

> **Definition of a Rational Equation**
>
> An equation with one or more rational expressions is called a **rational equation**.

The following are examples of rational equations.

$$\frac{3}{5} + \frac{1}{x} = \frac{2}{3} \qquad 3 - \frac{6w}{w + 1} = \frac{6}{w + 1} \qquad \frac{36}{p^2} = \frac{2p}{p + 3} - 1$$

To solve a rational equation, we have the option of clearing the fractions. To do this, multiply both sides of the equation by the least common denominator (LCD) of all the terms of the equation. However, the terms within the rational equation may have restrictions on the variable (values of the variable that make the denominator equal to zero). Therefore, after solving a rational equation, it is important to check each potential solution.

Solving a Rational Equation

Step 1 Factor the denominators of all rational expressions. Identify any values of the variable for which any expression is undefined.

Step 2 Identify the LCD of all terms in the equation.

Step 3 Multiply both sides of the equation by the LCD.

Step 4 Solve the resulting equation.

Step 5 Check the potential solutions in the original equation. Note that any value from step 1 for which the equation is undefined cannot be a solution to the equation.

EXAMPLE 1   Solving a Rational Equation

Solve the equation. $\dfrac{3}{5} + \dfrac{1}{x} = \dfrac{2}{3}$

Solution:

$$\frac{3}{5} + \frac{1}{x} = \frac{2}{3}$$

The LCD of all terms in the equation is $15x$. Note that in this equation there is a restriction that $x \neq 0$.

$$15x\left(\frac{3}{5} + \frac{1}{x}\right) = 15x\left(\frac{2}{3}\right)$$

Multiply by $15x$ to clear fractions.

$$15x \cdot \frac{3}{5} + 15x \cdot \frac{1}{x} = 15x \cdot \frac{2}{3}$$

Apply the distributive property.

$$9x + 15 = 10x$$

Solve the resulting equation.

$$15 = x$$

Check: $x = 15$ $\dfrac{3}{5} + \dfrac{1}{x} = \dfrac{2}{3}$

$$\frac{3}{5} + \frac{1}{(15)} \overset{?}{=} \frac{2}{3}$$

$$\frac{9}{15} + \frac{1}{15} \overset{?}{=} \frac{2}{3}$$

$$\frac{10}{15} \overset{?}{=} \frac{2}{3} \checkmark$$

The solution set is $\{15\}$.

Skill Practice 1   Solve the equation. $\dfrac{3}{y} + \dfrac{4}{3} = -1$

Answer

1. $\left\{-\dfrac{9}{7}\right\}$

EXAMPLE 2   Solving a Rational Equation

Solve the equation. $3 - \dfrac{6w}{w + 1} = \dfrac{6}{w + 1}$

Solution:

$$3 - \frac{6w}{w + 1} = \frac{6}{w + 1}$$

The LCD of all terms in the equation is $w + 1$. Note that in this equation there is a restriction that $w \neq -1$.

$$(w + 1)\left(3 - \frac{6w}{w + 1}\right) = (w + 1)\left(\frac{6}{w + 1}\right)$$

Multiply by $(w + 1)$ on both sides to clear fractions.

$$(w + 1)(3) - (w + 1)\left(\frac{6w}{w + 1}\right) = (w + 1)\left(\frac{6}{w + 1}\right)$$

Apply the distributive property.

$$(w + 1)(3) - (\cancel{w + 1})\left(\frac{6w}{\cancel{w + 1}}\right) = (\cancel{w + 1})\left(\frac{6}{\cancel{w + 1}}\right)$$

$$3w + 3 - 6w = 6$$

Solve the resulting equation.

$$-3w = 3$$

$$w = -1$$

Check: $3 - \dfrac{6w}{w + 1} = \dfrac{6}{w + 1}$

$$3 - \frac{6(-1)}{(-1) + 1} \stackrel{?}{=} \frac{6}{(-1) + 1}$$

The denominator is 0 for the value of $w = -1$.

The value -1 is one of the restrictions on w found in the first step. As expected, the value $w = -1$ does not check. Since no other potential solution exists, the equation has no solution.

The solution set is the empty set, { }.

Skill Practice 2   Solve the equation. $5 - \dfrac{8}{x + 2} = \dfrac{4x}{x + 2}$

Examples 1 and 2 show that the steps to solve a rational equation mirror the process of clearing fractions. However, we must check whether the potential solutions are defined in each expression in the original equation. A potential solution that does not check is called an **extraneous solution**.

Answer

2. { } (The value -2 does not check.)

EXAMPLE 3   Solving a Rational Equation

Solve the equation. $1 + \dfrac{3}{x} = \dfrac{28}{x^2}$

Solution:

FOR REVIEW

After clearing fractions in Example 3, the resulting equation is quadratic. To solve a quadratic equation by factoring and applying the zero product property, see Section 3.4.

$$1 + \frac{3}{x} = \frac{28}{x^2}$$ The LCD of all terms in the equation is x^2. Expressions will be undefined for $x = 0$.

$$x^2\left(1 + \frac{3}{x}\right) = x^2\left(\frac{28}{x^2}\right)$$ Multiply both sides by x^2 to clear fractions.

$$x^2 \cdot 1 + x^2 \cdot \frac{3}{x} = x^2 \cdot \frac{28}{x^2}$$ Apply the distributive property.

$$x^2 + 3x = 28$$ The resulting equation is quadratic.

$$x^2 + 3x - 28 = 0$$ Set the equation equal to zero and factor.

$$(x + 7)(x - 4) = 0$$

$$x = -7 \quad \text{or} \quad x = 4$$

Check: $x = -7$

$$1 + \frac{3}{x} = \frac{28}{x^2}$$

$$1 + \frac{3}{-7} \overset{?}{=} \frac{28}{(-7)^2}$$

$$\frac{7}{7} - \frac{3}{7} \overset{?}{=} \frac{28}{49}$$

$$\frac{4}{7} \overset{?}{=} \frac{4}{7} \checkmark$$

Check: $x = 4$

$$1 + \frac{3}{x} = \frac{28}{x^2}$$

$$1 + \frac{3}{4} \overset{?}{=} \frac{28}{(4)^2}$$

$$\frac{4}{4} + \frac{3}{4} \overset{?}{=} \frac{28}{16}$$

$$\frac{7}{4} \overset{?}{=} \frac{7}{4} \checkmark$$

The solution set is $\{-7, 4\}$.

Skill Practice 3   Solve the equation. $1 + \dfrac{6}{x} = \dfrac{16}{x^2}$

EXAMPLE 4   Solving a Rational Equation

Solve. $\dfrac{36}{p^2 - 9} = \dfrac{2p}{p + 3} - 1$

Solution:

$$\frac{36}{p^2 - 9} = \frac{2p}{p + 3} - 1$$

$$\frac{36}{(p + 3)(p - 3)} = \frac{2p}{p + 3} - 1$$ The LCD is $(p + 3)(p - 3)$. Expressions will be undefined for $p = 3$ and $p = -3$.

Multiply both sides by the LCD to clear fractions.

Answer

3. $\{-8, 2\}$

$$(p + 3)(p - 3)\left[\frac{36}{(p + 3)(p - 3)}\right] = (p + 3)(p - 3)\left(\frac{2p}{p + 3} - 1\right)$$

$$(p + 3)(p - 3)\left[\frac{36}{(p + 3)(p - 3)}\right] = (p + 3)(p - 3)\left(\frac{2p}{p + 3}\right) - (p + 3)(p - 3)1$$

$$\cancel{(p + 3)(p - 3)}\left[\frac{36}{\cancel{(p + 3)(p - 3)}}\right] = \cancel{(p + 3)}(p - 3)\left(\frac{2p}{\cancel{p + 3}}\right) - (p + 3)(p - 3)1$$

$$36 = 2p(p - 3) - (p + 3)(p - 3) \qquad \text{Solve the resulting equation.}$$

$$36 = 2p^2 - 6p - (p^2 - 9) \qquad \text{The equation is quadratic.}$$

$$36 = 2p^2 - 6p - p^2 + 9$$

$$36 = p^2 - 6p + 9$$

$$0 = p^2 - 6p - 27 \qquad \text{Set the equation equal to zero and factor.}$$

$$0 = (p - 9)(p + 3)$$

$$p = 9 \quad \text{or} \quad p = -3$$

Check: $p = 9$

$$\frac{36}{p^2 - 9} = \frac{2p}{p + 3} - 1$$

$$\frac{36}{(9)^2 - 9} \overset{?}{=} \frac{2(9)}{(9) + 3} - 1$$

$$\frac{36}{72} \overset{?}{=} \frac{18}{12} - 1$$

$$\frac{1}{2} \overset{?}{=} \frac{3}{2} - 1$$

$$\frac{1}{2} \overset{?}{=} \frac{1}{2} \checkmark$$

Check: $p = -3$

$$\frac{36}{p^2 - 9} = \frac{2p}{p + 3} - 1$$

$$\frac{36}{(-3)^2 - 9} \overset{?}{=} \frac{2(-3)}{(-3) + 3} - 1$$

Denominator is zero

Here the value -3 is *not* a solution to the original equation because it is restricted in the original equation. However, 9 checks in the original equation.

The solution set is $\{9\}$.

Skill Practice 4   Solve. $\dfrac{6}{x + 2} - \dfrac{20x}{x^2 - x - 6} = \dfrac{x}{x + 2}$

2. Solve Formulas for a Specified Variable

Rational expressions also appear in formulas with multiple variables. In Example 5, we clear fractions to solve a formula for a specified variable.

EXAMPLE 5   Solving a Literal Equation Involving Rational Expressions

Avoiding Mistakes

Variables in algebra are case-sensitive. For example, M and m are different variables.

Solve for the indicated variable. $V = \dfrac{mv}{m + M}$ for m

Answer

4. $\{-9\}$ (The value -2 does not check.)

Solution:

$$V = \frac{mv}{m + M} \quad \text{for } m$$

$$V(m + M) = \left(\frac{mv}{m + M}\right)(m + M)$$

Multiply by the LCD and clear fractions.

$$V(m + M) = mv$$

$$Vm + VM = mv$$

Use the distributive property to clear parentheses.

$$Vm - mv = -VM$$

Collect all m terms on one side.

$$m(V - v) = -VM$$

Factor out m.

$$\frac{m(V - v)}{(V - v)} = \frac{-VM}{(V - v)}$$

Divide by $(V - v)$.

$$m = \frac{-VM}{V - v}$$

FOR REVIEW

In Example 5, the variable m appears multiple times in the equation. To consolidate the occurrences of m, we first collect the m terms on one side of the equation. Then we factor out m as a common factor. This in turn enables us to isolate m. See Section 2.3, Objective 3 for more practice.

TIP The factor of -1 that appears in the numerator may be written in the denominator or out in front of the expression. The following expressions are equivalent:

$$m = \frac{-VM}{V - v} \quad \text{or} \quad m = \frac{VM}{-(V - v)} = \frac{VM}{-V + v} = \frac{VM}{v - V} \quad \text{or} \quad m = -\frac{VM}{V - v}$$

Answer

5. $x = \dfrac{b - yd}{y - a} \quad \text{or} \quad x = \dfrac{yd - b}{a - y}$

Skill Practice 5   Solve the equation for x. $y = \dfrac{ax + b}{x + d}$

SECTION 4.3 — Practice Exercises

Prerequisite Review

For Exercises R.1–R.6, solve the equation. (See Section 1.2 for review.)

R.1. $\dfrac{1}{10}x - \dfrac{3}{4} = \dfrac{1}{5}x - 1$

R.2. $2 - \dfrac{2}{3}y = \dfrac{1}{6}y + \dfrac{5}{3}$

R.3. $\dfrac{m}{9} - \dfrac{m - 4}{3} = \dfrac{m + 1}{6}$

R.4. $\dfrac{t}{12} - \dfrac{2t + 1}{6} = \dfrac{t - 2}{4}$

R.5. $0.05(x + 1) - 0.04x = 10$

R.6. $0.15(y - 3) - 0.14y = 64$

For Exercises R.7–R.8, solve for the indicated variable. (See Section 2.3 for review.)

R.7. $ax - bx = t$ for x

R.8. $cm + dm = p$ for m

For Exercises R.9–R.14, factor completely.

R.9. $n^2 - 9$ (See Section 2.5.)

R.10. $t^2 - 81$

R.11. $6x^2 - 18x$ (See Section 2.3.)

R.12. $4p^2 + 20p$

R.13. $2z^2 - 11z - 40$ (See Section 2.4.)

R.14. $3m^2 + m - 44$

For Exercises R.15–R.16, evaluate the expression for the given value of the variable. (See Section 4.1 for review.)

R.15. $\dfrac{x^2 + 5}{x^2 - x - 12}$ **a.** $x = 1$ **b.** $x = -3$

R.16. $\dfrac{x - 5}{x^2 + 10x + 9}$ **a.** $x = 2$ **b.** $x = -9$

Concept Connections

1. The equation $\dfrac{5}{x+2} + \dfrac{1}{2} = \dfrac{4}{5}$ is an example of a _____ equation.

2. Why is it important to check a potential solution to a rational equation?

3. Given $\dfrac{3}{2x+1} + \dfrac{36}{2x^2 - 7x - 4} = \dfrac{4}{x-4}$, is it possible for 4 to be a solution to the equation?

4. Given $\dfrac{6}{t^2 - 7t + 12} + \dfrac{2t}{t-3} = \dfrac{3t}{t-4}$, is it possible for 3 to be a solution to the equation?

Objective 1: Solve Rational Equations

For Exercises 5–8, determine the restrictions on x.

5. $\dfrac{3}{x-5} + \dfrac{2}{x+4} = \dfrac{5}{7}$

6. $\dfrac{2}{x+1} - \dfrac{5}{x-7} = \dfrac{2}{3}$

7. $\dfrac{5}{2x-3} - \dfrac{3}{5x} = \dfrac{1}{3-x}$

8. $\dfrac{1}{2x} - \dfrac{1}{6-x} = \dfrac{2}{4x-5}$

For Exercises 9–42, solve the rational equation. **(See Examples 1–4)**

9. $\dfrac{x+2}{3} - \dfrac{x-4}{4} = \dfrac{1}{2}$

10. $\dfrac{x+6}{3} - \dfrac{x+8}{5} = 0$

11. $\dfrac{3y}{4} - 2 = \dfrac{5y}{6}$

12. $\dfrac{2w}{5} - 8 = \dfrac{4w}{2}$

13. $\dfrac{5}{4p} - \dfrac{7}{6} + 3 = 0$

14. $\dfrac{7}{15w} - \dfrac{3}{10} - 2 = 0$

15. $\dfrac{1}{2} - \dfrac{3}{2x} = \dfrac{4}{x} - \dfrac{5}{12}$

16. $\dfrac{2}{3x} + \dfrac{1}{4} = \dfrac{11}{6x} - \dfrac{1}{3}$

17. $\dfrac{3}{x-4} + 2 = \dfrac{5}{x-4}$

18. $\dfrac{5}{x+3} - 2 = \dfrac{7}{x+3}$

19. $\dfrac{1}{3} + \dfrac{2}{w-3} = 1$

20. $\dfrac{3}{5} + \dfrac{7}{p+2} = 2$

21. $\dfrac{12}{x} - \dfrac{12}{x-5} = \dfrac{2}{x}$

22. $\dfrac{25}{y} - \dfrac{25}{y-2} = \dfrac{2}{y}$

23. $\dfrac{3}{a^2} - \dfrac{4}{a} = -1$

24. $\dfrac{3}{w^2} = 2 + \dfrac{1}{w}$

25. $\dfrac{1}{4}a - 4a^{-1} = 0$

26. $\dfrac{1}{3}t - 12t^{-1} = 0$

27. $\dfrac{y}{y+3} + \dfrac{2}{y^2 + 3y} = \dfrac{6}{y}$

28. $\dfrac{-8}{t^2 - 6t} + \dfrac{t}{t-6} = \dfrac{1}{t}$

29. $\dfrac{4}{t-2} - \dfrac{8}{t^2 - 2t} = -2$

30. $\dfrac{x}{x+6} = \dfrac{72}{x^2 - 36} + 4$

31. $\dfrac{6}{5y+10} - \dfrac{1}{y-5} = \dfrac{4}{y^2 - 3y - 10}$

32. $\dfrac{-3}{x^2 - 7x + 12} - \dfrac{2}{x^2 + x - 12} = \dfrac{10}{x^2 - 16}$

33. $\dfrac{x}{x-5} + \dfrac{1}{5} = \dfrac{5}{x-5}$

34. $\dfrac{x}{x-2} + \dfrac{2}{3} = \dfrac{2}{x-2}$

35. $\dfrac{6}{x^2 - 4x + 3} - \dfrac{1}{x-3} = \dfrac{1}{4x-4}$

36. $\dfrac{1}{4x^2 - 36} - \dfrac{5}{x+3} + \dfrac{2}{x-3} = 0$

37. $\dfrac{1}{k+2} - \dfrac{4}{k-2} - \dfrac{k^2}{4-k^2} = 0$

38. $\dfrac{h}{2} - \dfrac{h}{h-4} = \dfrac{4}{4-h}$

39. $\dfrac{5}{x^2 - 7x + 12} = \dfrac{2}{x-3} + \dfrac{5}{x-4}$

40. $\dfrac{9}{x^2 + 7x + 10} = \dfrac{5}{x+2} - \dfrac{3}{x+5}$

41. $\dfrac{4}{x^2 + 7x + 12} - \dfrac{7}{x^2 + 8x + 15} = \dfrac{1}{x^2 + 9x + 20}$

42. $\dfrac{5}{x^2 - 6x + 8} - \dfrac{2}{x^2 + 3x - 10} = \dfrac{8}{x^2 + x - 20}$

Objective 2: Solve Formulas for a Specified Variable

For Exercises 43–60, solve the formula for the indicated variable. (**See Example 5**)

43. $K = \dfrac{ma}{F}$ for m

44. $K = \dfrac{ma}{F}$ for a

45. $K = \dfrac{IR}{E}$ for E

46. $K = \dfrac{IR}{E}$ for R

47. $I = \dfrac{E}{R + r}$ for R

48. $I = \dfrac{E}{R + r}$ for r

49. $h = \dfrac{2A}{B + b}$ for B

50. $\dfrac{V}{\pi h} = r^2$ for h

51. $x = \dfrac{at + b}{t}$ for t

52. $\dfrac{T + mf}{m} = g$ for m

53. $\dfrac{x - y}{xy} = z$ for x

54. $\dfrac{w - n}{wn} = P$ for w

55. $a + b = \dfrac{2A}{h}$ for h

56. $1 + rt = \dfrac{A}{P}$ for P

57. $\dfrac{1}{R} = \dfrac{1}{R_1} + \dfrac{1}{R_2}$ for R

58. $\dfrac{b + a}{ab} = \dfrac{1}{f}$ for b

59. $v = \dfrac{s_2 - s_1}{t_2 - t_1}$ for t_2

60. $a = \dfrac{v_2 - v_1}{t_2 - t_1}$ for v_1

Mixed Exercises

For Exercises 61–80, simplify the expression or solve the equation.

61. $\dfrac{2}{a^2 + 4a + 3} + \dfrac{1}{a + 3}$

62. $\dfrac{1}{c + 6} + \dfrac{4}{c^2 + 8c + 12}$

63. $\dfrac{7}{y^2 - y - 2} + \dfrac{1}{y + 1} - \dfrac{3}{y - 2} = 0$

64. $\dfrac{3}{b + 2} - \dfrac{1}{b - 1} - \dfrac{5}{b^2 + b - 2} = 0$

65. $\dfrac{x}{x - 1} - \dfrac{12}{x^2 - x}$

66. $\dfrac{3}{5t - 20} + \dfrac{4}{t - 4}$

67. $\dfrac{3}{w} - 5 = \dfrac{7}{w} - 1$

68. $\dfrac{-3}{y^2} - \dfrac{1}{y} = -2$

69. $\dfrac{4p + 1}{8p - 12} + \dfrac{p - 3}{2p - 3}$

70. $\dfrac{x + 1}{2x + 4} - \dfrac{x^2}{x + 2}$

71. $\dfrac{1}{2x^2} + \dfrac{1}{6x}$

72. $\dfrac{5}{4a} + \dfrac{1}{6a^2}$

73. $\dfrac{3}{2t} + \dfrac{2}{3t^2} = \dfrac{-1}{t}$

74. $\dfrac{-3}{b^2} + \dfrac{1}{5b} = \dfrac{1}{2b}$

75. $\dfrac{3}{c^2 + 4c + 3} - \dfrac{2}{c^2 + 6c + 9}$

76. $\dfrac{1}{y^2 - 10y + 25} - \dfrac{3}{y^2 - 7y + 10}$

77. $\dfrac{4}{w - 4} - \dfrac{36}{2w^2 - 7w - 4} = \dfrac{3}{2w + 1}$

78. $\dfrac{2}{x - 3} - \dfrac{5}{x + 2} = \dfrac{25}{x^2 - x - 6}$

79. $8t^{-1} + 2 = 3t^{-1}$

80. $6z^{-2} - 5z^{-1} = 0$

Expanding Your Skills

81. Suppose that 40 deer are introduced in a protected wilderness area. The population of the herd P can be approximated by $P = \dfrac{40 + 20x}{1 + 0.05x}$, where x is the time in years since introducing the deer. Determine the time required for the deer population to reach 200.

Raven Regan/DesignPics

82. Starting from rest, an automobile's velocity v (in ft/sec) is given by $v = \dfrac{180t}{2t + 10}$, where t is the time in seconds after the car begins forward motion. Determine the time required for the car to reach a speed of 60 ft/sec ($\approx$ 41 mph).

SECTION 4.4   Applications of Rational Equations

OBJECTIVES

1. Solve Applications of Proportions
2. Solve Applications Involving Uniform Motion
3. Solve Applications Involving Rates of Work

1. Solve Applications of Proportions

In this section, we present several applications of rational equations. We begin by defining a proportion.

Definition of Ratio and Proportion

1. The **ratio** of a to b is $\dfrac{a}{b} \, (b \neq 0)$ and can also be expressed as $a:b$ or $a \div b$.

2. An equation that equates two ratios is called a **proportion**. Therefore, if $b \neq 0$ and $d \neq 0$, then $\dfrac{a}{b} = \dfrac{c}{d}$ is a proportion.

EXAMPLE 1   Solving a Proportion

The recommended ratio of total cholesterol to HDL cholesterol is 7 to 2. If Rich's blood test revealed that he has a total cholesterol level of 210 mg/dL (milligrams per deciliter), what should his HDL level be to fit the recommended ratio?

Solution:

One method of solving this problem is to set up a proportion. Write two equivalent ratios depicting the amount of total cholesterol to HDL cholesterol. Let x represent the unknown amount of HDL cholesterol.

$$\boxed{\text{Given ratio}} \longrightarrow \frac{7}{2} = \frac{210}{x} \longleftarrow \boxed{\begin{array}{l}\text{Amount of total cholesterol} \\ \hline \text{Amount of HDL cholesterol}\end{array}}$$

$$2x\left(\frac{7}{2}\right) = 2x\left(\frac{210}{x}\right) \qquad \text{Multiply both sides by the LCD } 2x.$$

$$7x = 420 \qquad \text{Clear fractions.}$$

$$x = 60$$

Rich's HDL cholesterol level should be 60 mg/dL to fit within the recommended level.

Skill Practice 1   The ratio of cats to dogs at an animal rescue facility is 8 to 5. How many dogs are in the facility if there are 400 cats?

EXAMPLE 2   Solving a Proportion

The ratio of male to female police officers in a certain town is 11:3. If the total number of officers is 112, how many are men and how many are women?

Solution:

Let x represent the number of male police officers.

Then $112 - x$ represents the number of female police officers.

Answer

1. There are 250 dogs.

$$\boxed{\begin{array}{c}\text{Male}\\\hline\text{Female}\end{array}} \longrightarrow \frac{11}{3} = \frac{x}{112-x} \longleftarrow \boxed{\begin{array}{c}\text{Number of males}\\\hline\text{Number of females}\end{array}}$$

$$\cancel{3}(112-x)\left(\frac{11}{\cancel{3}}\right) = 3\cancel{(112-x)}\left(\frac{x}{\cancel{112-x}}\right) \qquad \text{Multiply both sides by } 3(112-x).$$

$$11(112-x) = 3x \qquad\qquad \text{The resulting equation is linear.}$$

$$1232 - 11x = 3x$$

$$1232 = 14x$$

$$\frac{1232}{14} = \frac{14x}{14}$$

$$x = 88$$

Then $112 - x = 112 - 88 = 24$

There are 88 male police officers and 24 female officers.

Skill Practice 2   Professor Wolfe has a ratio of passing students to failing students of 5 to 4. One semester he had a total of 207 students. How many students passed and how many failed?

Proportions are used in geometry with **similar triangles**. Two triangles are similar if their corresponding angles are equal. In such a case, the lengths of the corresponding sides are proportional. In Figure 4-1, triangle *ABC* is similar to triangle *XYZ*. Therefore, the following ratios are equivalent.

$$\frac{a}{x} = \frac{b}{y} = \frac{c}{z}$$

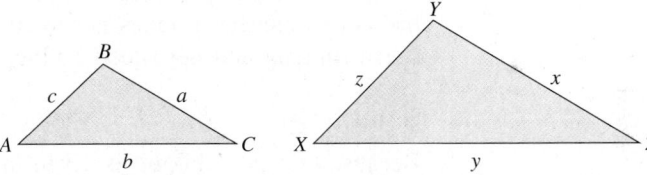

Figure 4-1

EXAMPLE 3   Using Similar Triangles in an Application

The shadow cast by a yardstick is 2 ft long. The shadow cast by a tree is 11 ft long. Find the height of the tree.

Solution:

Let *x* represent the height of the tree.

We will assume that the measurements were taken at the same time of day. Therefore, the angle of the sun is the same on both objects, and we can set up similar triangles (Figure 4-2).

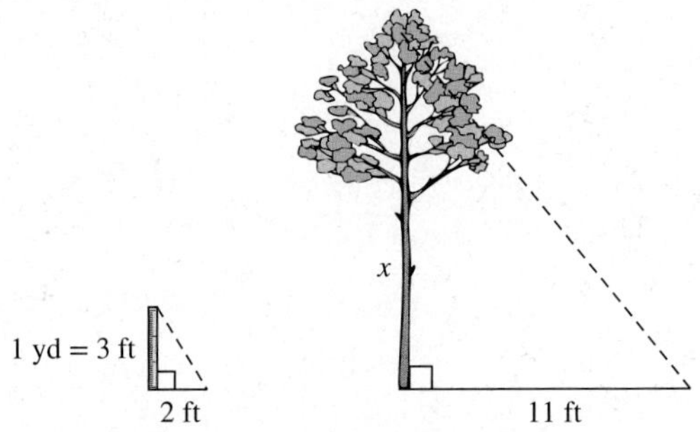

1 yd = 3 ft

2 ft

x

11 ft

Figure 4-2

Height of yardstick	→ 3 ft	2 ft ←	Length of yardstick's shadow
Height of tree	→ x ft	= 11 ft ←	Length of tree's shadow

$$\frac{3}{x} = \frac{2}{11} \qquad \text{Write an equation.}$$

$$11x \cdot \left(\frac{3}{x}\right) = 11x \cdot \left(\frac{2}{11}\right) \qquad \text{Multiply by the LCD.}$$

$$33 = 2x \qquad \text{Solve the equation.}$$

$$16.5 = x \qquad \text{Interpret the results.}$$

The tree is 16.5 ft high.

Skill Practice 3   Triangle XYZ is similar to triangle ABC. Solve for x.

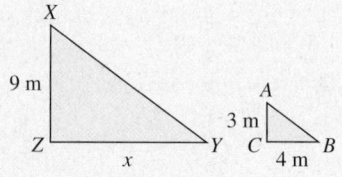

2. Solve Applications Involving Uniform Motion

EXAMPLE 4   **Solving an Application Involving Distance, Rate, and Time**

An athlete's average speed on her bike is 14 mph faster than her average speed running. She can bike 31.5 mi in the same time that it takes her to run 10.5 mi. Find her speed running and her speed biking.

Lars A. Niki

Solution:

Because the speed biking is given in terms of the speed running, let x represent the running speed.

> Let x represent the speed running.
>
> Then $x + 14$ represents the speed biking.
>
> Organize the given information in a chart.

	Distance	**Rate**	**Time**
Running	10.5	x	$\dfrac{10.5}{x}$
Biking	31.5	$x + 14$	$\dfrac{31.5}{x + 14}$

Because $d = rt$, then $t = \dfrac{d}{r}$

The time required to run 10.5 mi is the same as the time required to bike 31.5 mi, so we can equate the two expressions for time:

$$\frac{10.5}{x} = \frac{31.5}{x + 14}$$ The LCD is $x(x + 14)$.

$$\cancel{x}(x + 14)\left(\frac{10.5}{\cancel{x}}\right) = x\cancel{(x + 14)}\left(\frac{31.5}{\cancel{x + 14}}\right)$$ Multiply by $x(x + 14)$ to clear fractions.

$$10.5(x + 14) = 31.5x$$ The resulting equation is linear.

$$10.5x + 147 = 31.5x$$ Solve for x.

$$-21x = -147$$

$$x = 7$$

Then $x + 14 = 7 + 14 = 21$.

The athlete runs 7 mph and bikes 21 mph.

> **Skill Practice 4**   Devon can cross-country ski 5 km/hr faster than his sister, Shanelle. Devon skis 45 km in the same amount of time that Shanelle skis 30 km. Find their speeds.

EXAMPLE 5   Solving an Application Involving Distance, Rate, and Time

Valentina travels 70 km to Rome by train, and then takes a bus 30 km to the Coliseum. The bus travels 24 km/hr slower than the train. If the total time traveling on the bus and train is 2 hr, find the average speed of the train and the average speed of the bus.

Jennifer Barrow/jenifoto/123RF

Solution:

Because the speed of the bus is given in terms of the speed of the train, let x represent the speed of the train.

Let x represent the speed of the train.

Let $x - 24$ represent the speed of the bus.

Organize the given information in a chart.

	Distance	Rate	Time
Train	70	x	$\dfrac{70}{x}$
Bus	30	$x - 24$	$\dfrac{30}{x - 24}$

Fill in the last column with $t = \dfrac{d}{r}$.

In this problem, we are given that the total time is 2 hr. So we add the two times to equal 2.

$$\frac{70}{x} + \frac{30}{x - 24} = 2$$

The LCD is $x(x - 24)$.

$$x(x - 24)\left(\frac{70}{x}\right) + x(x - 24)\frac{30}{x - 24} = x(x - 24)(2)$$

Multiply by the LCD to clear the fractions.

$$70(x - 24) + 30x = 2x(x - 24)$$

$$70x - 1680 + 30x = 2x^2 - 48x$$

The resulting equation is quadratic.

$$0 = 2x^2 - 148x + 1680$$

Set the equation equal to 0.

$$0 = 2(x^2 - 74x + 840)$$

Factor.

$$0 = 2(x - 60)(x - 14)$$

Solve for x.

$$x - 60 = 0 \quad \text{or} \quad x - 14 = 0$$

Set each factor equal to 0.

$$x = 60 \quad \text{or} \quad x = 14$$

If $x = 14$, then the rate of the bus would be $14 - 24 = -10$. Because a negative rate of -10 km/hr is not reasonable, we reject $x = 14$ as a solution. Therefore, the solution is $x = 60$. That is, the average rate of the train is 60 km/hr and the average rate of the bus is $60 - 24$ or 36 km/hr.

Skill Practice 5   Jason drives 50 mi to a train station and then continues his trip with a 210-mi train ride. The car travels 20 mph slower than the train. If the total travel time is 4 hr, find the average speed of the car and the average speed of the train.

3. Solve Applications Involving Rates of Work

EXAMPLE 6   Solving an Application Involving "Work"

JoAn can wallpaper a bathroom in 3 hr. Bonnie can wallpaper the same bathroom in 5 hr. How long would it take them if they worked together?

Fuse/Getty Images

Solution:

Let x represent the amount of time required for both people working together to complete the job.

One method to approach this problem is to add the rates of speed at which each person works.

$$\left(\begin{array}{c}\text{JoAn's}\\\text{speed}\end{array}\right) + \left(\begin{array}{c}\text{Bonnie's}\\\text{speed}\end{array}\right) = \left(\begin{array}{c}\text{speed working}\\\text{together}\end{array}\right)$$

$$\frac{1 \text{ job}}{3 \text{ hr}} + \frac{1 \text{ job}}{5 \text{ hr}} = \frac{1 \text{ job}}{x \text{ hr}}$$

$$\frac{1}{3} + \frac{1}{5} = \frac{1}{x} \qquad\qquad \text{Set up an equation.}$$

$$15x \cdot \left(\frac{1}{3} + \frac{1}{5}\right) = 15x \cdot \left(\frac{1}{x}\right) \qquad\qquad \begin{array}{l}\text{Multiply by the LCD } 15x \text{ to clear}\\\text{fractions.}\end{array}$$

$$5x + 3x = 15 \qquad\qquad \text{Solve the resulting equation.}$$

$$8x = 15$$

$$x = \frac{15}{8} \quad \text{or} \quad x = 1\frac{7}{8}$$

JoAn and Bonnie can wallpaper the bathroom in $1\frac{7}{8}$ hr.

Skill Practice 6   Antonio can install a new roof in 4 days. Bob can install the same size roof in 6 days. How long will it take them to install a roof if they work together?

TIP   An alternative approach to solving a "work" problem is to determine the portion of the job that each person can complete in 1 hr. Let x represent the amount of time required to complete the job working together. Then

- JoAn completes $\frac{1}{3}$ of the job in 1 hr, and $\frac{1}{3}x$ jobs in x hours.
- Bonnie completes $\frac{1}{5}$ of the job in 1 hr and $\frac{1}{5}x$ jobs in x hours.

$$\left(\begin{array}{c}\text{Portion of the job}\\\text{completed by JoAn}\end{array}\right) + \left(\begin{array}{c}\text{portion of the job}\\\text{completed by Bonnie}\end{array}\right) = \left(\begin{array}{c}\text{1 whole}\\\text{job}\end{array}\right)$$

$$\frac{1}{3}x \qquad\qquad + \qquad\qquad \frac{1}{5}x \qquad\qquad = \qquad 1$$

$$15 \cdot \left(\frac{1}{3}x + \frac{1}{5}x\right) = 15 \cdot (1)$$

$$5x + 3x = 15$$

$$8x = 15$$

$$x = \frac{15}{8} \quad \text{or} \quad x = 1\frac{7}{8}$$

The time working together is $1\frac{7}{8}$ hr.

Answer

6. It will take them $\frac{12}{5}$ days or $2\frac{2}{5}$ days.

Prerequisite Review

For Exercises R.1–R.6, solve the equation.

R.1. $\frac{1}{2}x - \frac{1}{4} = \frac{5}{6}x$ (See Section 1.2)

R.2. $-\frac{4}{7}y + \frac{5}{14} = 1$

R.3. $6t^2 + 10 = -17t$ (See Section 3.4)

R.4. $5y^2 + 11y = 12$

R.5. $4x^2 = 64$ (See Section 3.5)

R.6. $4m^2 = 25$

Concept Connections

1. An equation that equates two ratios is called a _____.

2. Given similar triangles, the lengths of corresponding sides are _____.

Objective 1: Solve Applications of Proportions

For Exercises 3–10, solve the proportion.

3. $\frac{12}{18} = \frac{14}{x}$

4. $\frac{9}{75} = \frac{m}{50}$

5. $\frac{x+1}{5} = \frac{4}{15}$

6. $\frac{t-1}{7} = \frac{2}{21}$

7. $\frac{2}{y-1} = \frac{y-3}{4}$

8. $\frac{1}{x-5} = \frac{x-3}{3}$

9. $\frac{1}{49w} = \frac{w}{9}$

10. $\frac{1}{4z} = \frac{z}{25}$

11. A 3.5-oz box of candy has a total of 21.0 g of fat. How many grams of fat would a 14-oz box of candy contain? **(See Example 1)**

12. A 6-oz box of candy has 350 calories. How many calories would a 10-oz box contain?

13. Pam drives her hybrid 243 mi in city driving on 4.5 gal of gas. At this rate how many gallons of gas are required to drive 621 mi?

14. On a map, the distance from Sacramento, California, to San Francisco, California, is 8 cm. The legend gives the actual distance as 96 mi. On the same map, Fatima measured 7 cm from Sacramento to Modesto, California. What is the actual distance?

15. At a construction site, cement, sand, and gravel are mixed to make concrete. The ratio of cement to sand to gravel is 1 to 2.4 to 3.6. If a 150-lb bag of sand is used, how much cement and gravel must be used?

16. The property tax on a $180,000 house is $1296. At this rate, what is the property tax on a house that is $240,000?

17. In addition to measuring a person's individual HDL and LDL cholesterol levels, doctors also compute the ratio of total cholesterol to HDL cholesterol. Doctors recommend that the ratio of total cholesterol to HDL cholesterol be kept under 4. Suppose that the ratio of a patient's total cholesterol to HDL is 3.4 and her HDL is 60 mg/dL. Determine the patient's LDL level and total cholesterol. (Assume that total cholesterol is the sum of the LDL and HDL levels.)

18. For a recent Congress, there were 10 more Democrats than Republicans in the U.S. Senate. This resulted in a ratio of 11 Democrats to 9 Republicans. How many senators were Democrat and how many were Republican?

19. Yellowstone National Park in Wyoming has the largest population of free-roaming bison. To approximate the number of bison, 200 are captured and tagged and then left free to roam. Later, a sample of 120 bison is observed and 6 have tags. Approximate the population of bison in the park.

20. Laws have been instituted in Florida to help save the manatee. To establish the number of manatees in Florida, 150 manatees were tagged. A new sample was taken later, and among the 40 manatees in the sample, 3 were tagged. Approximate the number of manatees in Florida.

Jeff Vanuga/Fuse/Getty Images

21. The ratio of men to women enrolled in a math course to train elementary school teachers is 1 to 5. If the total enrollment in these classes is approximately 186 students per semester, how many men are enrolled? **(See Example 2)**

22. A chemist mixes water and alcohol in a 7 to 8 ratio. If she makes a 450-L solution, how much is water and how much is alcohol?

For Exercises 23–26, triangle *ABC* is similar to triangle *XYZ*. Find the lengths of the missing sides. **(See Example 3)**

23.

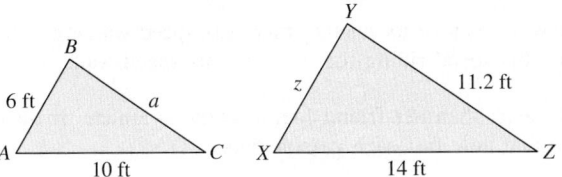

24.

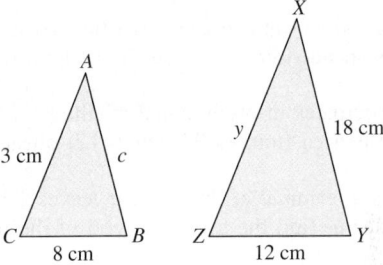

25.

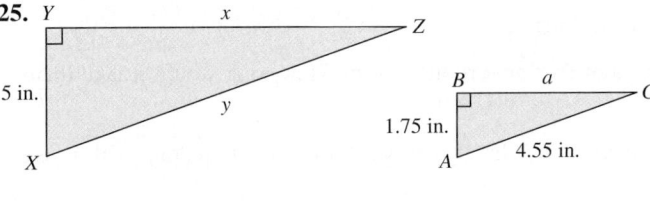

26.

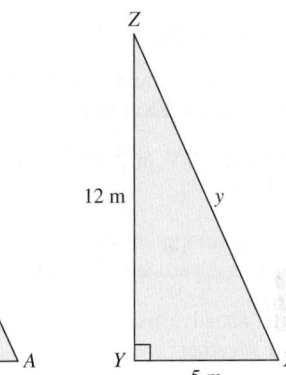

Objective 2: Solve Applications Involving Uniform Motion

For Exercises 27–28, use the fact that distance = (rate)(time).

27. A truck travels 7 mph faster than a car. Let x represent the speed of the car.

   a. Write an expression for the speed of the truck.

   b. Write an expression for the time it takes the car to travel 48 mi.

   c. Write an expression for the time it takes the truck to travel 83 mi.

28. A car travels 4 mph slower than a motorcycle. Let x represent the speed of the motorcycle.

   a. Write an expression for the speed of the car.

   b. Write an expression for the time it takes the motorcycle to travel 50 mi.

   c. Write an expression for the time it takes the car to travel 145 mi.

29. A motorist travels 80 mi while driving in a bad rainstorm. In sunny weather, the motorist drives 20 mph faster and covers 120 mi in the same amount of time. Find the speed of the motorist in the rainstorm and the speed in sunny weather. **(See Example 4)**

30. Brooke walks 2 km/hr slower than her older sister Adrianna. If Brooke can walk 12 km in the same amount of time that Adrianna can walk 18 km, find their speeds.

31. Two out-of-town firefighting crews have been called to a wildfire in the mountains. The Wescott Fire Station is 96 mi from the fire, and the Broadmoor Fire Station is 88 mi from the fire. The fire truck from the Wescott Fire Station travels 6.4 mph faster than the Broadmoor fire truck. If it takes the trucks the same amount of time to reach the fire, what is the average speed of each truck?

32. Kathy can run 3 mi to the beach in the same amount of time Dennis can ride his bike 7 mi to work. Kathy runs 8 mph slower than Dennis rides his bike. Find their speeds.

33. A bicyclist rides 30 mi against a wind and returns 30 mi with the wind. His average speed for the return trip is 5 mph faster. How fast did the cyclist ride against the wind if the total time of the trip was 5 hr? **(See Example 5)**

34. A boat travels 60 mi to an island and 60 mi back again. Changes in the wind and tide made the average speed on the return trip 3 mph slower than the speed on the way out. If the total time of the trip took 9 hr, find the speed going to the island and the speed of the return trip.

35. Celeste walked 140 ft on a moving walkway at the airport. Then she walked on the ground for 100 ft. She travels 2 ft/sec faster on the walkway than she does on the ground. If the time it takes her to travel the total distance of 240 ft is 40 sec, how fast does Celeste travel on and off the moving walkway?

36. Julio rides his bike 6 mi and gets a flat tire. Then he has to walk with the bike for another mile. His speed walking is 6 mph less than his speed riding the bike. If the total time is 1 hr, find his speed riding the bike and his speed walking.

37. Beatrice participates in professional triathlons. She runs 2 mph faster than her friend Joe, a weekend athlete. If they each run 12 mi, Beatrice finishes 30 min ($\frac{1}{2}$ hr) ahead of Joe. Determine how fast each person runs.

38. A bus leaves a terminal at 9:00. A car leaves 1 hr later and averages 10 mph faster than the bus. If the car overtakes the bus after 200 mi, find the average speed of the bus and the average speed of the car.

Objective 3: Solve Applications Involving Rates of Work

39. One painter can paint a room in 6 hr. Another painter can paint the same room in 8 hr. How long would it take them working together? **(See Example 6)**

40. Karen can wax her SUV in 2 hr. Clarann can wax the same SUV in 3 hr. If they work together, how long will it take them to wax the SUV?

41. A new housing development offers fenced-in yards that all have the same dimensions. Joel can fence a yard in 12 hr, and Michael can fence a yard in 15 hr. How long will it take if they work together?

42. Ted can change an advertisement on a billboard in 4 hr. Marie can do the same job in 5 hr. How long would it take them if they worked together?

43. A swimming pool takes 30 hr to fill using an old pump. When a new pump was installed, it took only 12 hr to fill the pool with both pumps. However, the old pump had to be repaired.

  a. Determine how long it would take for the new pump to fill the pool alone.

  b. If the new pump begins filling the empty pool at 4 P.M. on Thursday, when should the technician return to stop the pump?

44. One carpenter can complete a kitchen in 8 days. With the help of another carpenter, they can do the job together in 4 days. How long would it take the second carpenter if he worked alone?

45. Gus works twice as fast as Sid. Together they can dig a garden in 4 hr. How long would it take each person working alone?

46. It takes a child 3 times longer to vacuum a house than an adult. If it takes 1 hr for one adult and one child working together to vacuum a house, how long would it take each person working alone?

Randy Faris/Corbis /PunchStock

Mixed Exercises

47. A student 5 ft tall measures the length of the shadow of the Washington Monument to be 444 ft. At the same time, her shadow is 4 ft. Approximate the height of the Washington Monument.

48. A 6-ft man is standing 40 ft from a light post. If the man's shadow is 20 ft, determine the height of the light post.

49. Joel can run around a $\frac{1}{4}$-mi track in 66 sec, and Jason can run around the track in 60 sec. If the runners start at the same point on the track and run in opposite directions, how long will it take the runners to cover $\frac{1}{4}$ mi?

50. Marta can vacuum the house in 40 min. It takes her daughter 1 hr to vacuum the house. How long would it take them if they worked together?

51. The jet stream is a fast-flowing air current found in the atmosphere at around 36,000 ft above the surface of the Earth. During one summer day, the speed of the jet stream is 50 mph. A plane flying with the jet stream can fly 700 mi in the same amount of time that it would take to fly 500 mi against the jet stream. What is the speed of the plane in still air?

52. A fishing boat travels 9 mi downstream with the current in the same amount of time that it travels 6 mi upstream against the current. If the speed of the current is 2 mph, what is the speed at which the boat travels in still water?

53. When studying wildlife populations, biologists sometimes use a technique called "mark-recapture." For example, a researcher captured and tagged 30 deer in a wildlife management area. Several months later, the researcher observed a new sample of 80 deer and determined that 5 were tagged. What is the total number of deer in the population?

54. To estimate the number of bass in a lake, a biologist catches and tags 24 bass. Several weeks later, the biologist catches a new sample of 40 bass and finds that 4 are tagged. How many bass are in the lake?

55. One pump can fill a pool in 10 hr. Working with a second slower pump, the two pumps together can fill the pool in 6 hr. How fast can the second pump fill the pool by itself?

56. Brad and Brittney can mow their yard together with two lawn mowers in 30 min. When Brad works alone, it takes him 50 min. How long would it take Brittney to mow the lawn by herself?

SECTION 4.5   Rational Exponents

1. Simplify Expressions of the Form $a^{1/n}$ and $a^{m/n}$

In Section 2.1, we studied properties of integer exponents. We would now like to extend these properties to rational exponents. We begin by defining expressions of the form $a^{1/n}$ and $a^{m/n}$ where m and n are nonzero integers. We would like these definitions to align with the properties of integer exponents. For example, we want to define $25^{1/2}$ so that the following statement is true.

$$(25^{1/2})^2 = 25^{(1/2)\cdot 2} = 25^1 = 25$$

Define $25^{1/2}$ to be a square root of 25 so that when squared it equals 25.

Definition of $a^{1/n}$ and $a^{m/n}$

Let m and n be integers such that m/n is a rational number in lowest terms and $n > 1$. If $\sqrt[n]{a}$ is a real number, then

$$a^{1/n} = \sqrt[n]{a} \qquad \text{and} \qquad a^{m/n} = \sqrt[n]{a^m} = \left(\sqrt[n]{a}\right)^m$$

The definition of $a^{m/n}$ indicates that $a^{m/n}$ can be written as a radical whose index is the denominator of the rational exponent. The order in which the nth root and exponent m are performed within the radical does not affect the outcome. For example:

Take the 4th root first: $\quad 16^{3/4} = \left(\sqrt[4]{16}\right)^3 \quad$ or $\quad$ Cube 16 first: $\quad 16^{3/4} = \sqrt[4]{16^3}$

$$= (2)^3 \qquad\qquad = \sqrt[4]{4096}$$

$$= 8 \qquad\qquad\qquad = 8$$

EXAMPLE 1 Evaluating Expressions of the Form $a^{1/n}$

Convert the expression to radical form and simplify, if possible.

a. $(-8)^{1/3}$ **b.** $81^{1/4}$ **c.** $-100^{1/2}$ **d.** $(-100)^{1/2}$ **e.** $16^{-1/2}$

Solution:

a. $(-8)^{1/3} = \sqrt[3]{-8} = -2$

b. $81^{1/4} = \sqrt[4]{81} = 3$

c. $-100^{1/2} = -1 \cdot 100^{1/2}$ The exponent applies only to the base of 100.

$\qquad\qquad = -1\sqrt{100}$

$\qquad\qquad = -10$

d. $(-100)^{1/2}$ is not a real number because $\sqrt{-100}$ is not a real number.

e. $16^{-1/2} = \dfrac{1}{16^{1/2}}$ Write the expression with a positive exponent.

$\qquad\qquad\qquad$ Recall that $b^{-n} = \dfrac{1}{b^n}$.

$\qquad = \dfrac{1}{\sqrt{16}}$

$\qquad = \dfrac{1}{4}$

FOR REVIEW

$\sqrt[3]{-8} = -2$ because $(-2)(-2)(-2) = -8$.

$\sqrt[4]{81} = 3$ because 3 is a positive real number and $(3)(3)(3)(3) = 81$.

To review simplifying nth roots, see Section R.3.

Skill Practice 1 Convert the expression to radical form and simplify, if possible.

a. $(-64)^{1/3}$ **b.** $16^{1/4}$ **c.** $-36^{1/2}$ **d.** $(-36)^{1/2}$ **e.** $64^{-1/3}$

In Example 2, we evaluate expressions of the form $a^{m/n}$. The expression $a^{m/n}$ essentially represents two operations: the numerator m raises the base to the mth power, and the denominator takes the nth root.

EXAMPLE 2 Evaluating Expressions of the Form $a^{m/n}$

Convert each expression to radical form and simplify.

a. $8^{2/3}$ **b.** $100^{5/2}$ **c.** $\left(\dfrac{1}{25}\right)^{3/2}$ **d.** $4^{-3/2}$ **e.** $(-81)^{3/4}$

Solution:

a. $8^{2/3} = \left(\sqrt[3]{8}\right)^2$ Take the cube root of 8 and square the result.

$\qquad = (2)^2$ Simplify.

$\qquad = 4$

Answers

1. a. -4 **b.** 2 **c.** -6

d. Not a real number **e.** $\dfrac{1}{4}$

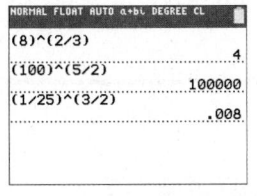
b. $100^{5/2} = \left(\sqrt{100}\right)^5$ Take the square root of 100 and raise the result to the fifth power.

$\qquad = (10)^5$ Simplify.

$\qquad = 100{,}000$

c. $\left(\dfrac{1}{25}\right)^{3/2} = \left(\sqrt{\dfrac{1}{25}}\right)^3$ Take the square root of $\dfrac{1}{25}$ and cube the result.

$\qquad = \left(\dfrac{1}{5}\right)^3$ Simplify.

$\qquad = \dfrac{1}{125}$

d. $4^{-3/2} = \left(\dfrac{1}{4}\right)^{3/2} = \dfrac{1}{4^{3/2}}$ Write the expression with positive exponents.

$\qquad = \dfrac{1}{\left(\sqrt{4}\right)^3}$ Take the square root of 4 and cube the result.

$\qquad = \dfrac{1}{2^3}$ Simplify.

$\qquad = \dfrac{1}{8}$

e. $(-81)^{3/4}$ is not a real number because $\sqrt[4]{-81}$ is not a real number.

Skill Practice 2   Convert each expression to radical form and simplify.

   a. $9^{3/2}$ **b.** $8^{5/3}$ **c.** $\left(\dfrac{1}{27}\right)^{4/3}$ **d.** $32^{-4/5}$ **e.** $(-4)^{3/2}$

---

2. Convert Between Rational Exponents and Radical Notation

EXAMPLE 3   Using Radical Notation and Rational Exponents

Convert each expression to radical notation. Assume all variables represent positive real numbers.

   a. $a^{3/5}$ **b.** $5^{1/3}x^{2/3}$ **c.** $3y^{1/4}$ **d.** $9z^{-3/4}$

Solution:

   a. $a^{3/5} = \sqrt[5]{a^3}$ or $\left(\sqrt[5]{a}\right)^3$

   b. $5^{1/3}x^{2/3} = (5x^2)^{1/3} = \sqrt[3]{5x^2}$

   c. $3y^{1/4} = 3\sqrt[4]{y}$ Note that the coefficient 3 is not raised to the $\frac{1}{4}$ power.

   d. $9z^{-3/4} = 9 \cdot \dfrac{1}{z^{3/4}} = \dfrac{9}{\sqrt[4]{z^3}}$ Note that the coefficient 9 has an implied exponent of 1, not $-\frac{3}{4}$.

Skill Practice 3   Convert each expression to radical notation. Assume all variables represent positive real numbers.

   a. $t^{4/5}$ **b.** $2^{1/4}y^{3/4}$ **c.** $10p^{1/2}$ **d.** $11q^{-2/3}$

Answers

2. a. 27   **b.** 32   **c.** $\dfrac{1}{81}$

   d. $\dfrac{1}{16}$ **e.** Not a real number

3. a. $\sqrt[5]{t^4}$ **b.** $\sqrt[4]{2y^3}$ **c.** $10\sqrt{p}$

   d. $\dfrac{11}{\sqrt[3]{q^2}}$

Using Radical Notation and Rational Exponents

Convert each expression to an equivalent expression by using rational exponents. Assume that all variables represent positive real numbers.

a. $\sqrt[4]{b^3}$ **b.** $\sqrt{7a}$ **c.** $7\sqrt{a}$

Solution:

a. $\sqrt[4]{b^3} = b^{3/4}$ **b.** $\sqrt{7a} = (7a)^{1/2}$ **c.** $7\sqrt{a} = 7a^{1/2}$

Skill Practice 4 Convert to an equivalent expression using rational exponents. Assume all variables represent positive real numbers.

a. $\sqrt[3]{x^2}$ **b.** $\sqrt{5y}$ **c.** $5\sqrt{y}$

3. Simplify Expressions with Rational Exponents

The properties and definitions for simplifying expressions with integer exponents also apply to rational exponents.

Definitions and Properties of Exponents

Let a and b be nonzero real numbers. Let m and n be rational numbers such that a^m, a^n, and b^m are real numbers.

Description	Property	Example
1. Multiplication of like bases	$a^m a^n = a^{m+n}$	$x^{1/3} x^{4/3} = x^{5/3}$
2. Division of like bases	$\dfrac{a^m}{a^n} = a^{m-n}$	$\dfrac{x^{3/5}}{x^{1/5}} = x^{2/5}$
3. Power rule	$(a^m)^n = a^{mn}$	$(2^{1/3})^{1/2} = 2^{1/6}$
4. Power of a product	$(ab)^m = a^m b^m$	$(9y)^{1/2} = 9^{1/2} y^{1/2} = 3y^{1/2}$
5. Power of a quotient	$\left(\dfrac{a}{b}\right)^m = \dfrac{a^m}{b^m}$	$\left(\dfrac{4}{25}\right)^{1/2} = \dfrac{4^{1/2}}{25^{1/2}} = \dfrac{2}{5}$

Description	Definition	Example
1. Negative exponents	$a^{-m} = \left(\dfrac{1}{a}\right)^m = \dfrac{1}{a^m}$	$(8)^{-1/3} = \left(\dfrac{1}{8}\right)^{1/3} = \dfrac{1}{2}$
2. Zero exponent	$a^0 = 1$	$5^0 = 1$

Simplifying Expressions with Rational Exponents

Use the properties of exponents to simplify the expressions. Assume all variables represent positive real numbers.

a. $y^{2/5} y^{3/5}$ **b.** $\dfrac{6a^{-1/2}}{a^{3/2}}$ **c.** $\left(\dfrac{s^{1/2} t^{1/3}}{w^{3/4}}\right)^4$

Answers
4. a. $x^{2/3}$ **b.** $(5y)^{1/2}$
c. $5y^{1/2}$

Solution:

a. $y^{2/5}y^{3/5} = y^{(2/5)+(3/5)}$ Multiply like bases by adding exponents.

$\qquad\quad = y^{5/5}$ Simplify.

$\qquad\quad = y$

b. $\dfrac{6a^{-1/2}}{a^{3/2}}$

$\quad = 6a^{(-1/2)-(3/2)}$ Divide like bases by subtracting exponents.

$\quad = 6a^{-2}$ Simplify: $-\dfrac{1}{2} - \left(\dfrac{3}{2}\right) = -\dfrac{4}{2} = -2$

$\quad = \dfrac{6}{a^2}$ Simplify the negative exponent.

c. $\left(\dfrac{s^{1/2}t^{1/3}}{w^{3/4}}\right)^4 = \dfrac{s^{(1/2)\cdot 4}t^{(1/3)\cdot 4}}{w^{(3/4)\cdot 4}}$ Apply the power rule. Multiply exponents.

$\qquad\qquad\qquad = \dfrac{s^2 t^{4/3}}{w^3}$ Simplify.

Skill Practice 5   Use the properties of exponents to simplify the expressions. Assume all variables represent positive real numbers.

a. $x^{1/2} \cdot x^{3/4}$ **b.** $\dfrac{4k^{-2/3}}{k^{1/3}}$ **c.** $\left(\dfrac{a^{1/3}b^{1/2}}{c^{5/8}}\right)^6$

---

4. Use Rational Exponents in Applications

Rational exponents may appear in applications and mathematical models in place of radical expressions.

EXAMPLE 6   **Applying Rational Exponents**

Suppose P dollars in principal is invested in an account that earns interest annually. If after t years the investment grows to A dollars, then the annual rate of return r on the investment is given by

$$r = \left(\frac{A}{P}\right)^{1/t} - 1$$

Find the annual rate of return on $5000, which grew to $6894.21 after 6 years.

Solution:

$r = \left(\dfrac{A}{P}\right)^{1/t} - 1$

$\ = \left(\dfrac{6894.21}{5000}\right)^{1/6} - 1$ where $A = \$6894.21$, $P = \$5000$, and $t = 6$

$\ \approx 0.055$ or 5.5%

The annual rate of return is 5.5%.

Answers

5. a. $x^{5/4}$ **b.** $\dfrac{4}{k}$ **c.** $\dfrac{a^2 b^3}{c^{15/4}}$

6. 3 in.

Skill Practice 6   The formula for the radius of a sphere is $r = \left(\frac{3V}{4\pi}\right)^{1/3}$ where V is the volume. Find the radius of a sphere whose volume is 113.04 in.^{3} (Use 3.14 for π.)

5. Factor Polynomials with Rational Exponents

In Section 2.3, we learned to factor out the greatest common factor (GCF) from a polynomial. For example, given $x^7 + 5x^5 + 4x^3$, the GCF is x^3 because it is the greatest power of x common to each term. The GCF is identified as x raised to its smallest power among the terms.

$$x^7 + 5x^5 + 4x^3 = x^3(x^4 + 5x^2 + 4x^0)$$

To find the exponents on the terms within parentheses, subtract 3 from the original exponents.

We use the same process to factor out a variable expression with a rational exponent. For example, given the expression $x^{7/2} + 5x^{5/2} + 4x^{3/2}$, we identify the GCF as $x^{3/2}$ because x is common to all terms and $\frac{3}{2}$ is the smallest exponent to which it occurs.

$$x^{7/2} + 5x^{5/2} + 4x^{3/2} = x^{3/2}(x^{4/2} + 5x^{2/2} + 4x^0) \quad \text{(Recall that } x^0 = 1.\text{)}$$

$$= x^{3/2}(x^2 + 5x + 4)$$

EXAMPLE 7 Factoring Out Expressions with Rational Exponents

Factor. Write the answers with no negative exponents.

a. $y^{10/3} + 6y^{7/3} - 16y^{4/3}$ **b.** $x(2x + 5)^{-1/2} + (2x + 5)^{1/2}$

Solution:

a. $y^{10/3} + 6y^{7/3} - 16y^{4/3}$

The greatest common factor is $y^{4/3}$.

$$= y^{4/3}(y^{6/3} + 6y^{3/3} - 16y^0)$$

Factor out $y^{4/3}$. Subtract $\frac{4}{3}$ from the original exponents.

$$= y^{4/3}(y^2 + 6y - 16)$$

Simplify. $16y^0 = 16(1) = 16$

$$= y^{4/3}(y + 8)(y - 2)$$

Factor the trinomial.

b. $x(2x + 5)^{-1/2} + (2x + 5)^{1/2}$

The smallest exponent on $(2x + 5)$ is $-\frac{1}{2}$.

Factor out $(2x + 5)^{-1/2}$.

$$= (2x + 5)^{-1/2}[x(2x + 5)^0 + (2x + 5)^1]$$

The powers on the factors of $(2x + 5)$ within parentheses are found by subtracting $-\frac{1}{2}$ from the original exponents.

$$= (2x + 5)^{-1/2}[x + (2x + 5)]$$

The expression $(2x + 5)^0 = 1$, for $2x + 5 \neq 0$.

$$= (2x + 5)^{-1/2}(3x + 5)$$

Simplify.

$$= \frac{3x + 5}{(2x + 5)^{1/2}}$$

Simplify the negative exponent.

Answers

7. a. $y^{3/5}(y + 5)(y + 7)$

b. $\dfrac{5x + 3}{(4x + 3)^{3/4}}$

Skill Practice 7

a. $y^{13/5} + 12y^{8/5} + 35y^{3/5}$ **b.** $x(4x + 3)^{-3/4} + (4x + 3)^{1/4}$

SECTION 4.5   Practice Exercises

Prerequisite Review

For Exercises R.1–R.6, simplify over the set of real numbers. (See Section R.3 for review.)

R.1. $\sqrt[3]{-125}$

R.2. $\sqrt[7]{-128}$

R.3. $\sqrt[4]{-625}$

R.4. $\sqrt{-144}$

R.5. $\left(\sqrt[3]{8}\right)^5$

R.6. $\left(\sqrt{16}\right)^3$

For Exercises R.7–R.16, simplify. (See Section 2.1 for review.)

R.7. $x^9 \cdot x^{11}$

R.8. $2^3 \cdot 2^5$

R.9. $\dfrac{5^{10}}{5^7}$

R.10. $\dfrac{y^{13}}{y^8}$

R.11. $\left(\dfrac{3x^2y^5}{z}\right)^3$

R.12. $\left(\dfrac{2ab^3}{c^2}\right)^4$

R.13. $7a^{-4}$

R.14. $8b^{-5}$

R.15. $\dfrac{9a^{-4}b^{-2}}{a^3}$

R.16. $\dfrac{13x^5y^{-2}}{x^{-6}}$

Concept Connections

1. The expression $a^{m/n}$ can be written in radical notation as ——————, provided that $\sqrt[n]{a}$ is a real number.

2. The expression $a^{1/n}$ can be written in radical notation as ——————, provided that $\sqrt[n]{a}$ is a real number.

3. The radical notation for $x^{-1/3}$ is —————— provided that $x \neq 0$.

4. $8^{1/3} =$ —————— and $8^{-1/3} =$ ——————.

Objective 1: Simplify Expressions of the Form $a^{1/n}$ and $a^{m/n}$

For Exercises 5–14, simplify each expression. (See Examples 1–2)

5. a. $25^{1/2}$ **b.** $(-25)^{1/2}$ **c.** $-25^{1/2}$

6. a. $36^{1/2}$ **b.** $(-36)^{1/2}$ **c.** $-36^{1/2}$

7. a. $27^{1/3}$ **b.** $(-27)^{1/3}$ **c.** $-27^{1/3}$

8. a. $125^{1/3}$ **b.** $(-125)^{1/3}$ **c.** $-125^{1/3}$

9. a. $\left(\dfrac{121}{169}\right)^{1/2}$ **b.** $\left(\dfrac{121}{169}\right)^{-1/2}$

10. a. $\left(\dfrac{49}{144}\right)^{1/2}$ **b.** $\left(\dfrac{49}{144}\right)^{-1/2}$

11. a. $16^{3/4}$ **b.** $16^{-3/4}$ **c.** $-16^{3/4}$

     d. $-16^{-3/4}$ **e.** $(-16)^{3/4}$ **f.** $(-16)^{-3/4}$

12. a. $81^{3/4}$ **b.** $81^{-3/4}$ **c.** $-81^{3/4}$

     d. $-81^{-3/4}$ **e.** $(-81)^{3/4}$ **f.** $(-81)^{-3/4}$

13. a. $64^{2/3}$ **b.** $64^{-2/3}$ **c.** $-64^{2/3}$

     d. $-64^{-2/3}$ **e.** $(-64)^{2/3}$ **f.** $(-64)^{-2/3}$

14. a. $8^{2/3}$ **b.** $8^{-2/3}$ **c.** $-8^{2/3}$

     d. $-8^{-2/3}$ **e.** $(-8)^{2/3}$ **f.** $(-8)^{-2/3}$

Objective 2: Convert Between Rational Exponents and Radial Notation

For Exercises 15–24, write the expressions using radical notation. Assume that all variables represent positive real numbers. **(See Example 3)**

15. a. $y^{4/11}$ **b.** $6y^{4/11}$ **c.** $(6y)^{4/11}$

16. a. $z^{3/10}$ **b.** $8z^{3/10}$ **c.** $(8z)^{3/10}$

17. $q^{2/3}$ **18.** $t^{3/5}$ **19.** $6y^{3/4}$ **20.** $8b^{4/9}$

21. $x^{2/3}y^{1/3}$ **22.** $c^{2/5}d^{3/5}$ **23.** $6r^{-2/5}$ **24.** $7x^{-3/4}$

For Exercises 25–32, write the expressions using rational exponents. Assume that all variables represent positive real numbers. **(See Example 4)**

25. $\sqrt[5]{a^3}$ **26.** $\sqrt[7]{z^4}$ **27.** $\sqrt{6x}$ **28.** $\sqrt{11t}$

29. $6\sqrt{x}$ **30.** $11\sqrt{t}$ **31.** $\sqrt[5]{a^5 + b^5}$ **32.** $\sqrt[3]{m^3 + n^3}$

Objective 3: Simplify Expressions with Rational Exponents

For Exercises 33–58, simplify the expressions. Assume that all variables represent positive real numbers. **(See Example 5)**

33. $x^{1/4}x^{-5/4}$ **34.** $2^{2/3}2^{-5/3}$ **35.** $\dfrac{p^{5/3}}{p^{2/3}}$ **36.** $\dfrac{q^{5/4}}{q^{1/4}}$

37. $(y^{1/5})^{10}$ **38.** $(x^{1/2})^{8}$ **39.** $6^{-1/5}6^{3/5}$ **40.** $a^{-1/3}a^{2/3}$

41. $\dfrac{4t^{-1/3}}{t^{4/3}}$ **42.** $\dfrac{5s^{-1/3}}{s^{5/3}}$ **43.** $(a^{1/3}a^{1/4})^{12}$ **44.** $(x^{2/3}x^{1/2})^{6}$

45. $(5a^2c^{-1/2}d^{1/2})^2$ **46.** $(2x^{-1/3}y^2z^{5/3})^3$ **47.** $\left(\dfrac{x^{-2/3}}{y^{-3/4}}\right)^{12}$ **48.** $\left(\dfrac{m^{-1/4}}{n^{-1/2}}\right)^{-4}$

49. $\left(\dfrac{16w^{-2}z}{2wz^{-8}}\right)^{1/3}$ **50.** $\left(\dfrac{50p^{-1}q}{2pq^{-3}}\right)^{1/2}$ **51.** $(25x^2y^4z^6)^{1/2}$ **52.** $(8a^6b^3c^9)^{2/3}$

53. $(x^2y^{-1/3})^6(x^{1/2}yz^{2/3})^2$ **54.** $(a^{-1/3}b^{1/2})^4(a^{-1/2}b^{3/5})^{10}$ **55.** $\left(\dfrac{x^{3m}y^{2m}}{z^{5m}}\right)^{1/m}$ **56.** $\left(\dfrac{a^{4n}b^{3n}}{c^n}\right)^{1/n}$

57. $\left(\dfrac{m^2}{m+n}\right)^{-1}\left(\dfrac{m^2}{m+n}\right)^{1/2}$ **58.** $\left(\dfrac{c^2}{c-d}\right)^{-2}\left(\dfrac{c^2}{c-d}\right)^{3/2}$

Objective 4: Use Rational Exponents in Applications

59. If P dollars in principal grows to A dollars after t years with annual interest, then the interest rate r is given by $r = \left(\dfrac{A}{P}\right)^{1/t} - 1$. **(See Example 6)**

     a. In one account, $10,000 grows to $16,802 after 5 years. Compute the interest rate. Round your answer to a tenth of a percent.

     b. In another account $10,000 grows to $18,000 after 7 years. Compute the interest rate. Round your answer to a tenth of a percent.

     c. Which account produced a higher average yearly return?

60. If the area A of a square is known, then the length of its sides, s, can be computed by the formula $s = A^{1/2}$.

     a. Compute the length of the sides of a square having an area of 100 in.^{2}

     b. Compute the length of the sides of a square having an area of 72 in.^{2} Round your answer to the nearest 0.1 in.

61. The radius r of a sphere of volume V is given by $r = \left(\dfrac{3V}{4\pi}\right)^{1/3}$. Find the radius of a sphere having a volume of 85 in.^{3} Round your answer to the nearest 0.1 in.

62. The radius r of a right circular cone of volume V and height h is given by $r = \left(\dfrac{3V}{\pi h}\right)^{1/2}$. Find the radius of a right circular cone having volume 1018 in.^{3} and height 12 in. Round to the nearest inch.

63. The depreciation rate for a car is given by $r = 1 - \left(\frac{S}{C}\right)^{1/n}$, where S is the value of the car after n years, and C is the initial cost. Determine the depreciation rate for a car that originally cost \$22,990 and was valued at \$11,500 after 4 years. Round to the nearest tenth of a percent.

64. For a certain oven, the baking time t (in hr) for a turkey that weighs x pounds can be approximated by the model $t = 0.84x^{3/5}$. Determine the baking time for a 15-lb turkey. Round to 1 decimal place.

Objective 5: Factor Polynomials with Rational Exponents

For Exercises 65–76, factor completely. Write the answers with positive exponents only. Assume that all variables represent positive real numbers. **(See Example 7)**

65. $2c^{7/4} + 4c^{3/4}$

66. $10y^{9/5} - 15y^{4/5}$

67. $m^{5/2} - 5m^{3/2} - 36m^{1/2}$

68. $n^{8/3} - 8n^{5/3} - 20n^{2/3}$

69. $\frac{8}{3}x^{1/3} + \frac{5}{3}x^{-2/3}$

70. $\frac{15}{2}x^{1/2} - \frac{3}{2}x^{-1/2}$

71. $5x(3x + 1)^{2/3} + (3x + 1)^{5/3}$

72. $7t(4t + 1)^{3/4} + (4t + 1)^{7/4}$

73. $x(3x + 2)^{-2/3} + (3x + 2)^{1/3}$

74. $x(5x - 8)^{-4/5} + (5x - 8)^{1/5}$

75. $10x(x - 8)^{3/5} + 3x^2(x - 8)^{-2/5}$

76. $4(x + 2)^{1/2} + 2x(x + 2)^{-1/2}$

Mixed Exercises

For Exercises 77–100, simplify the expression.

77. $64^{-3/2}$

78. $81^{-3/2}$

79. $243^{3/5}$

80. $1^{5/3}$

81. $-27^{-4/3}$

82. $-16^{-5/4}$

83. $\left(\frac{100}{9}\right)^{-3/2}$

84. $\left(\frac{49}{100}\right)^{-1/2}$

85. $(-4)^{-3/2}$

86. $(-49)^{-3/2}$

87. $(-8)^{1/3}$

88. $(-9)^{1/2}$

89. $-8^{1/3}$

90. $-9^{1/2}$

91. $\frac{1}{36^{-1/2}}$

92. $\frac{1}{16^{-1/2}}$

93. $\frac{1}{1000^{-1/3}}$

94. $\frac{1}{81^{-3/4}}$

95. $\left(\frac{1}{8}\right)^{2/3} + \left(\frac{1}{4}\right)^{1/2}$

96. $\left(\frac{1}{8}\right)^{-2/3} + \left(\frac{1}{4}\right)^{-1/2}$

97. $\left(\frac{1}{16}\right)^{-3/4} - \left(\frac{1}{49}\right)^{-1/2}$

98. $\left(\frac{1}{16}\right)^{1/4} - \left(\frac{1}{49}\right)^{1/2}$

99. $\left(\frac{1}{4}\right)^{1/2} + \left(\frac{1}{64}\right)^{-1/3}$

100. $\left(\frac{1}{36}\right)^{1/2} + \left(\frac{1}{64}\right)^{-5/6}$

Write About It

101. Explain why $\left(\sqrt[3]{8}\right)^4$ is easier to evaluate than $\sqrt[3]{8^4}$.

102. Explain the similarity in the process to factor out the GCF in the following two expressions.

$$6x^5 + 5x^2 \quad \text{and} \quad 6x^{5/3} + 5x^{2/3}$$

Expanding Your Skills

For Exercises 103–108, write each expression as a single radical for positive values of the variable. (*Hint*: Write the radicals as expressions with rational exponents and simplify. Then convert back to radical form.)

103. $\sqrt[5]{x^3 y^2} \cdot \sqrt[4]{x}$

104. $\sqrt[4]{a^2 b} \cdot \sqrt[3]{ab^2}$

105. $\sqrt[6]{m} \sqrt[3]{m^2}$

106. $\sqrt[5]{y \sqrt[4]{y^3}}$

107. $\sqrt{x \sqrt{x \sqrt{x}}}$

108. $\sqrt[3]{y \sqrt[3]{y \sqrt[3]{y}}}$

The mean surface temperature T_p (in °C) of an Earth-like planet can be approximated based on its distance from its primary star d (in km), the radius of the star r (in km), and the temperature of the star T_s (in °C) by the following formula.

$$T_p = 0.7(T_s + 273)\left(\frac{r}{d}\right)^{1/2} - 273 \qquad \text{For Exercises 109–110, use the model to find } T_p.$$

109. The star Altair is relatively close to the Earth (16.8 light-years) and has a mean surface temperature of approximately 7700°C. Although not completely spherical in shape, Altair has a mean radius of approximately 1.26×10^6 km. If a planet with an atmosphere similar to that of the Earth is 4.3×10^8 km away from Altair, will the temperature on the surface of the planet be suitable for liquid water to exist? (Recall that under pressure similar to that at sea level on Earth, water freezes at 0°C and turns to steam at 100°C.)

110. Suppose the Sun has a mean surface temperature of 5700°C and a radius of approximately 7.0×10^5 km. If the Earth is a distance of 1.49×10^8 km from the Sun, approximate the mean surface temperature for the Earth.

Technology Connections

For Exercises 111–116, use a calculator to approximate the value of the expression to four decimal places.

111. $10^{5/2}$ **112.** $20^{3/5}$ **113.** $40{,}350^{1/3}$

114. $3490^{1/2}$ **115.** $5^{-4/3}$ **116.** $7^{-3/2}$

SECTION 4.6 | **Radical Equations and Equations with Rational Exponents**

OBJECTIVES

1. Solve Radical Equations Containing One Radical
2. Solve Radical Equations Containing Two Radicals
3. Solve Applications of Radical Equations

1. Solve Radical Equations Containing One Radical

An equation with one or more radicals containing a variable (such as $\sqrt[3]{x} = 5$) is called a **radical equation**. We can eliminate the radical by raising both sides of the equation to a power equal to the index of the radical.

$$\sqrt[3]{x} = 5$$
$$\left(\sqrt[3]{x}\right)^3 = (5)^3$$
$$x = 125$$

The index is 3. Therefore, raise both sides to the third power.

By raising each side of a radical equation to a power equal to the index, a new equation is produced. However, some (or all) of the solutions to the new equation may *not* be solutions to the original equation. These are called **extraneous solutions**. For this reason, it is necessary to check all potential solutions in the original equation. For example, consider the equation $\sqrt{x} = -10$. By inspection, this equation has no solution because the principal square root of x must be non-negative. However, if we square both sides of the equation, it appears as though a solution exists:

$$\sqrt{x} = -10 \qquad \text{Solution set: } \{\ \}$$

Square both sides. $\quad \left(\sqrt{x}\right)^2 = (-10)^2 \qquad$ The value 100 does not check in the original

$$x = 100 \qquad \text{equation. Therefore, 100 is an extraneous solution.}$$

Solving a Radical Equation

Step 1   Isolate the radical. If an equation has more than one radical, choose one of the radicals to isolate.

Step 2   Raise each side of the equation to a power equal to the index of the radical.

Step 3   Solve the resulting equation. If the equation still has a radical, repeat steps 1 and 2.

*Step 4   Check the potential solutions in the original equation and write the solution set.

*In solving radical equations, extraneous solutions potentially arise when both sides of the equation are raised to an even power. Therefore, an equation with only odd-indexed roots will not have extraneous solutions. However, it is still recommended that all potential solutions be checked.

EXAMPLE 1   **Solving an Equation Containing One Radical**

Solve the equation. $\sqrt{p} + 5 = 9$

Solution:

$\sqrt{p} + 5 = 9$

$\sqrt{p} = 4$ Isolate the radical.

$\left(\sqrt{p}\right)^2 = (4)^2$ Because the index is 2, square both sides.

$p = 16$

Check: $p = 16$ Check $p = 16$ as a potential solution.

$\sqrt{p} + 5 = 9$

$\sqrt{16} + 5 \stackrel{?}{=} 9$

$4 + 5 \stackrel{?}{=} 9 \checkmark$ True, 16 is a solution to the original equation.

The solution set is $\{16\}$.

Avoiding Mistakes

When raising both sides of an equation to a power, be sure to enclose both sides of the equation in parentheses.

Skill Practice 1   Solve. $\sqrt{x} - 3 = 2$

EXAMPLE 2   **Solving an Equation Containing One Radical**

Solve the equation. $\sqrt[3]{w - 1} - 2 = 2$

Solution:

$\sqrt[3]{w - 1} - 2 = 2$

$\sqrt[3]{w - 1} = 4$ Isolate the radical.

$\left(\sqrt[3]{w - 1}\right)^3 = (4)^3$ Because the index is 3, cube both sides.

$w - 1 = 64$ Solve the resulting equation.

$w = 65$

Check: $w = 65$

$\sqrt[3]{65 - 1} - 2 \stackrel{?}{=} 2$ Check $w = 65$ as a potential solution.

$\sqrt[3]{64} - 2 \stackrel{?}{=} 2$

$4 - 2 \stackrel{?}{=} 2 \checkmark$ True, 65 is a solution to the original equation.

The solution set is $\{65\}$.

Answers

1. $\{25\}$ **2.** $\{-10\}$

Skill Practice 2   Solve. $\sqrt[3]{t + 2} + 5 = 3$

In Examples 3 and 4, we solve equations containing rational exponents. Recall that if $\sqrt[n]{a}$ is a real number, then $a^{1/n} = \sqrt[n]{a}$ and $a^{m/n} = \left(\sqrt[n]{a}\right)^m$.

EXAMPLE 3   **Solving an Equation Containing Rational Exponents**

Solve the equation. $7 = (x + 3)^{1/4} + 9$

Solution:

$$7 = (x + 3)^{1/4} + 9$$

$$7 = \sqrt[4]{x + 3} + 9 \qquad \text{Note that } (x + 3)^{1/4} = \sqrt[4]{x + 3}.$$

$$-2 = \sqrt[4]{x + 3} \qquad \text{Isolate the radical.}$$

$$(-2)^4 = \left(\sqrt[4]{x + 3}\right)^4 \qquad \text{Because the index is 4, raise both sides to the fourth power.}$$

$$16 = x + 3$$

$$x = 13 \qquad \text{Solve for } x.$$

Check: $x = 13$

$$7 = \sqrt[4]{x + 3} + 9$$

$$7 \overset{?}{=} \sqrt[4]{(13) + 3} + 9$$

$$7 \overset{?}{=} \sqrt[4]{16} + 9$$

$$7 \overset{?}{=} 2 + 9 \text{ (false)} \qquad \text{13 is } not \text{ a solution to the original equation.}$$

The equation $7 = \sqrt[4]{x + 3} + 9$ has no solution.

The solution set is the empty set, $\{\ \}$.

> **TIP** After isolating the radical in Example 3, the equation shows a fourth root equated to a negative number:
>
> $-2 = \sqrt[4]{x + 3}$
>
> By definition, a principal fourth root of any real number must be nonnegative. Therefore, there can be no real solution to this equation.

Skill Practice 3   Solve. $3 = 6 + (x - 1)^{1/4}$

EXAMPLE 4   **Solving an Equation Containing Rational Exponents**

Solve the equation. $2(x + 1)^{2/3} = 8$

Solution:

$$2(x + 1)^{2/3} = 8$$

$$2\sqrt[3]{(x + 1)^2} = 8 \qquad \text{Write the expression on the left in radical notation.}$$

$$\sqrt[3]{(x + 1)^2} = 4 \qquad \text{Divide by 2 to isolate the radical.}$$

$$\left[\sqrt[3]{(x + 1)^2}\right]^3 = (4)^3 \qquad \text{Raise both sides to a power equal to the index of the radical.}$$

$$(x + 1)^2 = 64$$

$$x + 1 = \pm\sqrt{64} \qquad \text{Apply the square root property.}$$

$$x + 1 = \pm 8$$

$$x = -1 \pm 8$$

$$x = -1 - 8 = -9 \quad \text{or} \quad x = -1 + 8 = 7$$

Check: $x = -9$

$$(-9 + 1)^{2/3} \overset{?}{=} 4$$

$$(-8)^{2/3} = 4 \checkmark$$

Check: $x = 7$

$$(7 + 1)^{2/3} \overset{?}{=} 4$$

$$(8)^{2/3} = 4 \checkmark$$

The solution set is $\{-9, 7\}$.

Answers

3. $\{\ \}$ (The value 82 does not check.)

4. $\{85\}$

Skill Practice 4   Solve. $2(x - 4)^{3/4} = 54$

In Example 5, we solve a radical equation that requires squaring a binomial. Recall that

$$(a + b)^2 = a^2 + 2ab + b^2 \quad \text{and} \quad (a - b)^2 = a^2 - 2ab + b^2$$

For example, $(x + 4)^2 = (x)^2 + 2(x)(4) + (4)^2$
$$= x^2 + 8x + 16$$

EXAMPLE 5 Solving a Radical Equation

Solve. $\sqrt{x + 10} - 4 = x$

Solution:

$$\sqrt{x + 10} - 4 = x$$

$\sqrt{x + 10} = x + 4$	Isolate the radical.
$\left(\sqrt{x + 10}\right)^2 = (x + 4)^2$	The index is 2. Therefore, raise both sides to the second power.
$x + 10 = x^2 + 8x + 16$	The resulting equation is quadratic.
$0 = x^2 + 7x + 6$	Set one side equal to zero.
$0 = (x + 6)(x + 1)$	Factor.
$x = -6 \quad \text{or} \quad x = -1$	Apply the zero product rule.

Both sides of the equation were raised to an even power. Therefore, it is necessary to check the potential solutions.

$$\underline{\text{Check: } x = -6}$$
$$\sqrt{x + 10} - 4 = x$$
$$\sqrt{(-6) + 10} - 4 \stackrel{?}{=} (-6)$$
$$\sqrt{4} - 4 \stackrel{?}{=} -6$$
$$2 - 4 \stackrel{?}{=} -6$$
$$-2 \stackrel{?}{=} -6 \quad \text{false}$$

$$\underline{\text{Check: } x = -1}$$
$$\sqrt{x + 10} - 4 = x$$
$$\sqrt{(-1) + 10} - 4 \stackrel{?}{=} (-1)$$
$$\sqrt{9} - 4 \stackrel{?}{=} -1$$
$$3 - 4 \stackrel{?}{=} -1$$
$$-1 \stackrel{?}{=} -1 \quad \checkmark \text{ true}$$

The solution set is $\{-1\}$. The value -6 does not check.

Skill Practice 5 Solve $\sqrt{t + 7} = t - 5$

---

2. Solve Radical Equations Containing Two Radicals

EXAMPLE 6 Solving an Equation Containing Two Radicals

Solve the equation. $\sqrt[3]{2x - 4} = \sqrt[3]{1 - 8x}$

Solution:

$\sqrt[3]{2x - 4} = \sqrt[3]{1 - 8x}$	
$\left(\sqrt[3]{2x - 4}\right)^3 = \left(\sqrt[3]{1 - 8x}\right)^3$	Because the index is 3, cube both sides.
$2x - 4 = 1 - 8x$	Simplify.
$10x - 4 = 1$	Solve the resulting equation.
$10x = 5$	
$x = \dfrac{1}{2}$	Solve for x.

Answer

5. $\{9\}$; The value 2 does not check.

Check: $x = \frac{1}{2}$

$$\sqrt[3]{2x - 4} = \sqrt[3]{1 - 8x}$$

$$\sqrt[3]{2\left(\frac{1}{2}\right) - 4} \stackrel{?}{=} \sqrt[3]{1 - 8\left(\frac{1}{2}\right)}$$

$$\sqrt[3]{1 - 4} \stackrel{?}{=} \sqrt[3]{1 - 4}$$

$$\sqrt[3]{-3} \stackrel{?}{=} \sqrt[3]{-3} \checkmark \text{ (True)}$$ Therefore, $\frac{1}{2}$ is a solution to the original equation.

The solution set is $\left\{\dfrac{1}{2}\right\}$.

Skill Practice 6 Solve. $\sqrt[5]{2y - 1} = \sqrt[5]{10y + 3}$

EXAMPLE 7 Solving an Equation Containing Two Radicals

Solve. $\sqrt{m - 1} - \sqrt{3m + 1} = -2$

Solution:

$$\sqrt{m - 1} - \sqrt{3m + 1} = -2$$

$$\sqrt{m - 1} = \sqrt{3m + 1} - 2 \qquad \text{Isolate one of the radicals.}$$

$$\left(\sqrt{m - 1}\right)^2 = \left(\sqrt{3m + 1} - 2\right)^2 \qquad \text{The index is 2. Therefore, raise both sides to the second power.}$$

FOR REVIEW

Apply the formula:
$(a - b)^2 = a^2 - 2ab + b^2$

$$\text{Note: } \left(\sqrt{3m + 1} - 2\right)^2$$
$$= \left(\sqrt{3m + 1}\right)^2 - 2\left(\sqrt{3m + 1}\right)(2) + (2)^2$$
$$= (3m + 1) - 4\sqrt{3m + 1} + 4$$

$$m - 1 = 3m + 1 - 4\sqrt{3m + 1} + 4$$

$$m - 1 = 3m + 5 - 4\sqrt{3m + 1}$$

$$4\sqrt{3m + 1} = 2m + 6 \qquad \text{Isolate the remaining radical.}$$

$$2\sqrt{3m + 1} = m + 3 \qquad \text{Divide both sides by 2 to simplify.}$$

$$\left(2\sqrt{3m + 1}\right)^2 = (m + 3)^2 \qquad \text{The resulting equation has another radical. Isolate the radical, and square both sides again.}$$

FOR REVIEW

Apply the formula:
$(a + b)^2 = a^2 + 2ab + b^2$

$$4(3m + 1) = m^2 + 6m + 9 \qquad (m + 3)^2 = (m)^2 + 2(m)(3) + (3)^2$$
$$12m + 4 = m^2 + 6m + 9 \qquad\qquad\quad = m^2 + 6m + 9$$

$$0 = m^2 - 6m + 5 \qquad \text{The resulting equation is quadratic.}$$

$$0 = (m - 5)(m - 1) \qquad \text{Apply the zero product property.}$$

$$m = 5 \quad \text{or} \quad m = 1 \qquad \text{Both sides of the equation were raised to an even power. Check both potential solutions.}$$

Answer

6. $\left\{-\dfrac{1}{2}\right\}$

$$\text{Check: } m = 5$$
$$\sqrt{m-1} - \sqrt{3m+1} = -2$$
$$\sqrt{(5)-1} - \sqrt{3(5)+1} \stackrel{?}{=} -2$$
$$\sqrt{4} - \sqrt{16} \stackrel{?}{=} -2$$
$$2 - 4 \stackrel{?}{=} -2 \checkmark \text{ true}$$

$$\text{Check: } m = 1$$
$$\sqrt{m-1} - \sqrt{3m+1} = -2$$
$$\sqrt{(1)-1} - \sqrt{3(1)+1} \stackrel{?}{=} -2$$
$$\sqrt{0} - \sqrt{4} \stackrel{?}{=} -2$$
$$0 - 2 \stackrel{?}{=} -2 \checkmark \text{ true}$$

Both solutions check. The solution set is $\{1, 5\}$.

Skill Practice 7 Solve. $\quad 1 + \sqrt{n+4} = \sqrt{3n+1}$

3. Solve Applications of Radical Equations

EXAMPLE 8 Using a Radical Equation in an Application

On a certain surface, the speed s (in miles per hour) of a car before the brakes were applied can be approximated from the length x (in feet) of its skid marks by $s = 3.8\sqrt{x}$.

a. Find the speed of a car before the brakes were applied if its skid marks are 361 ft long.

b. How long would you expect the skid marks to be if the car had been traveling the speed limit of 50 mph? (Round to the nearest foot.)

Solution:

a. $s = 3.8\sqrt{x}$

$\quad = 3.8\sqrt{361}$ $\qquad$ Substitute $x = 361$.

$\quad = 3.8(19)$

$\quad = 72.2$

If the skid marks are 361 ft, the car was traveling approximately 72.2 mph before the brakes were applied.

b. $\qquad s = 3.8\sqrt{x}$

$\qquad 50 = 3.8\sqrt{x}$ $\qquad$ Substitute $s = 50$ and solve for x.

$\qquad \dfrac{50}{3.8} = \sqrt{x}$ $\qquad$ Isolate the radical.

$\qquad \left(\dfrac{50}{3.8}\right)^2 = x$

$\qquad\qquad x \approx 173$

If the car had been going the speed limit (50 mph), then the length of the skid marks would have been approximately 173 ft.

Skill Practice 8 When an object is dropped from a height of 64 ft, the time t (in seconds) it takes to reach a height x (in feet) is given by

$$t = \frac{1}{4}\sqrt{64 - x}$$

a. Find the time to reach a height of 28 ft from the ground.

b. What is the height after 1 sec?

Answers

7. $\{5\}$; The value 0 does not check.

8. a. $\dfrac{3}{2}$ sec   b. 48 ft

SECTION 4.6 Practice Exercises

Prerequisite Review

For Exercises R.1–R.2, simplify the radical expression. (See Section 3.1 for review.)

R.1. $\left(\sqrt{x+4}\right)^2$ **R.2.** $\left(\sqrt[3]{2x-9}\right)^3$

For Exercises R.3–R.4, evaluate the expression for the given value of the variable. (See Section R.3 for review.)

R.3. $\sqrt{3x-9}+2$ for $x=6$ **R.4.** $\sqrt{2x-5}-4$ for $x=3$

For Exercises R.5–R.10, perform the indicated operations. (See Sections 2.2 and 3.2 for review.)

R.5. $(a+5)^2$ **R.6.** $(b+7)^2$ **R.7.** $\left(\sqrt{5a}-3\right)^2$

R.8. $\left(2+\sqrt{b}\right)^2$ **R.9.** $\left(\sqrt{r-3}+5\right)^2$ **R.10.** $\left(2-\sqrt{2t-4}\right)^2$

For Exercises R.11–R.16, solve the equations. (See Sections 1.2 and 3.4 for review.)

R.11. $7x-9=2x+6$ **R.12.** $2y-5=8y+1$ **R.13.** $m^2+44=15m$

R.14. $p^2=56-p$ **R.15.** $2x(3x-1)=28$ **R.16.** $6t(2t-1)=5-2t$

Concept Connections

1. The equation $\sqrt{x+5}+7=11$ is an example of a _____ equation.

2. The first step to solve the equation $\sqrt{x+5}+7=11$ is to _____ the radical by subtracting _____ from both sides of the equation.

3. When solving a radical equation some potential solutions may not check in the original equation. These are called _____ solutions.

4. To solve the equation $\sqrt[3]{w-1}=5$, raise both sides of the equation to the _____ power.

Objective 1: Solve Radical Equations Containing One Radical

For Exercises 5–42, solve the equations. **(See Examples 1–5)**

5. $\sqrt{x}=10$ **6.** $\sqrt{y}=7$ **7.** $\sqrt{x}+4=6$

8. $\sqrt{x}+2=8$ **9.** $\sqrt{5y+1}=4$ **10.** $\sqrt{9z-5}-2=9$

11. $6=\sqrt{2z-3}-3$ **12.** $4=\sqrt{8+3a}-1$ **13.** $\sqrt[3]{x-2}-1=2$

14. $\sqrt[3]{2x-5}-1=1$ **15.** $(15-w)^{1/3}+7=2$ **16.** $(k+18)^{1/3}+5=3$

17. $3+\sqrt{x-16}=0$ **18.** $12+\sqrt{2x+1}=0$ **19.** $2\sqrt{6a+7}-2a=0$

20. $2\sqrt{3-w}-w=0$ **21.** $(2x-5)^{1/4}=-1$ **22.** $(x+16)^{1/4}=-4$

23. $2(x+5)^{2/3}=18$ **24.** $3(x-6)^{2/3}=48$ **25.** $(3x+1)^{3/2}+2=66$

26. $(2x-1)^{3/2}-3=122$ **27.** $m^{3/4}=5$ **28.** $n^{5/6}=3$

29. $2p^{4/5}=\dfrac{1}{8}$ **30.** $5t^{2/3}=\dfrac{1}{5}$ **31.** $\sqrt{x^2+5}=x+1$

32. $\sqrt{y^2-8}=y-2$ **33.** $\sqrt{a^2+2a+1}=a+5$ **34.** $\sqrt{b^2-5b-8}=b+7$

35. $\sqrt{25w^2-2w-3}=5w-4$ **36.** $\sqrt{4p^2-2p+1}=2p-3$ **37.** $4\sqrt{p-2}-2=-p$

38. $x-3\sqrt{x-5}=5$ **39.** $\sqrt{7x+8}=x+2$ **40.** $\sqrt{9x+19}=x+3$

41. $\sqrt{m+18}+2=m$ **42.** $\sqrt{2n+29}+3=n$

Objective 2: Solve Radical Equations Containing Two Radicals

For Exercises 43–68, solve the equations. **(See Examples 6–7)**

43. $\sqrt[4]{h+4} = \sqrt[4]{2h-5}$

44. $\sqrt[4]{3b+6} = \sqrt[4]{7b-6}$

45. $\sqrt[3]{5a+3} - \sqrt[3]{a-13} = 0$

46. $\sqrt[3]{k-8} - \sqrt[3]{4k+1} = 0$

47. $\sqrt{5a-9} = \sqrt{5a} - 3$

48. $\sqrt{8+b} = 2 + \sqrt{b}$

49. $\sqrt{2h+5} - \sqrt{2h} = 1$

50. $\sqrt{3k-5} - \sqrt{3k} = -1$

51. $(t-9)^{1/2} - t^{1/2} = 3$

52. $(y-16)^{1/2} - y^{1/2} = 4$

53. $6 = \sqrt{x^2+3} - x$

54. $2 = \sqrt{y^2+5} - y$

55. $\sqrt{3t-7} = 2 - \sqrt{3t+1}$

56. $\sqrt{p-6} = \sqrt{p+2} - 4$

57. $\sqrt{z+1} + \sqrt{2z+3} = 1$

58. $\sqrt{2y+6} = \sqrt{7-2y} + 1$

59. $\sqrt{6m+7} - \sqrt{3m+3} = 1$

60. $\sqrt{5w+1} - \sqrt{3w} = 1$

61. $2 + 2\sqrt{2t+3} + 2\sqrt{3t-5} = 0$

62. $6 + 3\sqrt{3x+1} + 3\sqrt{x-1} = 0$

63. $3\sqrt{y-3} = \sqrt{4y+3}$

64. $\sqrt{5x-8} = 2\sqrt{x-1}$

65. $\sqrt{p+7} = \sqrt{2p} + 1$

66. $\sqrt{t} = \sqrt{t-12} + 2$

67. $(2v+7)^{1/3} - (v-3)^{1/3} = 0$

68. $(5u-6)^{1/5} - (3u+1)^{1/5} = 0$

Objective 3: Solve Applications of Radical Equations

69. If an object is dropped from an initial height h, its velocity at impact with the ground is given by

$$v = \sqrt{2gh}$$

where g is the acceleration due to gravity and h is the initial height. **(See Example 8)**

a. Find the initial height (in feet) of an object if its velocity at impact is 44 ft/sec. (Assume that the acceleration due to gravity is $g = 32$ ft/sec².)

b. Find the initial height (in meters) of an object if its velocity at impact is 26 m/sec. (Assume that the acceleration due to gravity is $g = 9.8$ m/sec².) Round to the nearest tenth of a meter.

70. The time T (in seconds) required for a pendulum to make one complete swing back and forth is approximated by

$$T = 2\pi\sqrt{\frac{L}{g}}$$

where g is the acceleration due to gravity and L is the length of the pendulum (in feet).

a. Find the length of a pendulum that requires 1.36 sec to make one complete swing back and forth. (Assume that the acceleration due to gravity is $g = 32$ ft/sec².) Round to the nearest tenth of a foot.

b. Find the time required for a pendulum to complete one swing back and forth if the length of the pendulum is 4 ft. (Assume that the acceleration due to gravity is $g = 32$ ft/sec².) Round to the nearest tenth of a second.

71. The airline cost for x thousand passengers to travel round trip from New York to Atlanta is given by

$$C = \sqrt{0.3x+1}$$

where C is measured in millions of dollars and $x \geq 0$. **(See Example 8)**

a. Find the airline's cost for 10,000 passengers ($x = 10$) to travel from New York to Atlanta.

b. If the airline charges $320 per passenger, find the profit made by the airline for flying 10,000 passengers from New York to Atlanta.

c. Approximate the number of passengers who traveled from New York to Atlanta if the total cost for the airline was $4 million.

72. The time t in seconds it takes an object to drop d meters is given by

$$t = \sqrt{\frac{d}{4.9}}$$

a. Approximate the height of the JP Morgan Chase Tower in Houston if it takes an object 7.89 sec to drop from the top. Round to the nearest meter.

David Forman/Image Source Limited

b. Approximate the height of the Willis Tower in Chicago if it takes an object 9.51 sec to drop from the top. Round to the nearest meter.

73. The equation $r = \sqrt[3]{\dfrac{3V}{4\pi}}$ gives the radius r of a sphere of volume V. If the radius of a sphere is 6 in., find the exact volume.

74. The distance d (in miles) that an observer can see on a clear day is approximated by $d = \dfrac{49}{40}\sqrt{h}$, where h is the height of the observer in feet. If Rita can see 24.5 mi, how far above ground is her eye level?

75. The percentage of drug released in the bloodstream t hours after being administered is affected by numerous variables, including drug solubility and filler ingredients. For a particular drug and dosage, the percentage of drug released P is given by $P = 48t^{1/5}$ ($0 \leq t \leq 35$). For example, the value $P = 50$ represents 50% of the drug released.

a. Determine the percentage of drug released after 2 hr. Round to the nearest percent.

b. How many hours is required for 75% of the drug to be released? Round to the nearest tenth of an hour.

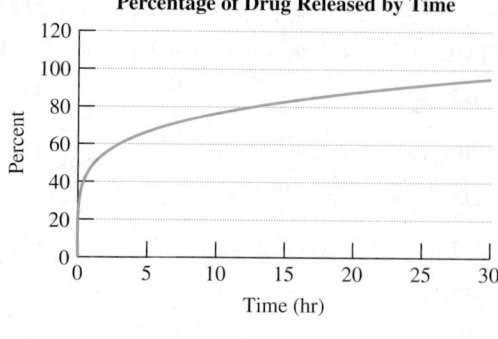

Percentage of Drug Released by Time

76. A tomato plant is purchased at a garden supply store. The initial height of the plant is 25.4 in. The height of the plant h (in inches) is approximated by $h = 16(t + 4)^{1/3}$, where t is the time in days after planting.

a. Determine the height of the plant 14 days after planting. Round to the nearest inch.

b. How long after the plant is planted will it take for the height to reach 5 ft? Round to the nearest day.

77. The yearly depreciation rate for a certain vehicle is modeled by $r = 1 - \left(\dfrac{V}{C}\right)^{1/n}$, where V is the value of the car after n years, and C is the original cost.

a. Determine the depreciation rate for a car that originally cost $18,000 and is worth $12,000 after 3 years. Round to the nearest tenth of a percent.

b. Determine the original cost of a truck that has a yearly depreciation rate of 15% and is worth $11,000 after 5 years. Round to the nearest $100.

78. The number of hours needed to cook a turkey that weighs x pounds can be approximated by

$$t = 0.90\sqrt[5]{x^3}$$

where t is the time in hours and x is the weight of the turkey in pounds.

a. Find the weight of a turkey that cooked for 4 hr. Round to the nearest pound.

b. Find the time required to cook an 18-lb turkey. Round to the nearest tenth of an hour.

Mixed Exercises

For Exercises 79–82, solve for the indicated variable.

79. $16 + \sqrt{x^2 - y^2} = z$ for x

80. $4 + \sqrt{x^2 + y^2} = z$ for y

81. $T = 2\pi\sqrt{\dfrac{L}{g}}$ for g

82. $t = \sqrt{\dfrac{2s}{g}}$ for s

For Exercises 83–86, assume that a and b are the lengths of the legs of a right triangle and c is the length of the hypotenuse. Use the Pythagorean theorem to find a, b, or c.

83. Find b when $a = 2$ and $c = y$.

84. Find b when $a = h$ and $c = 5$.

85. Find a when $b = x$ and $c = 8$.

86. Find a when $b = 14$ and $c = k$.

Expanding Your Skills

For Exercises 87–88, solve the equation.

87. $\sqrt{x + \sqrt{x + 2}} = 3$

88. $\sqrt{1 + \sqrt{x + \sqrt{x + 1}}} = 2$

89. Pam is in a canoe on a lake 400 ft from the closest point on a straight shoreline. Her house is 800 ft up the road along the shoreline. She can row 2.5 ft/sec and she can walk 5 ft/sec. If the total time it takes for her to get home is 5 min (300 sec), determine the point along the shoreline at which she landed her canoe.

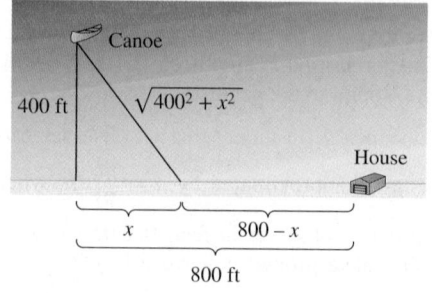

90. Martha is in a boat in the ocean 48 mi from point A, the closest point along a straight shoreline. She needs to dock the boat at a marina x miles farther up the coast, and then drive along the coast to point B, 96 mi from point A. Her boat travels 20 mph, and she drives 60 mph. If the total trip took 4 hr, determine the distance x along the shoreline.

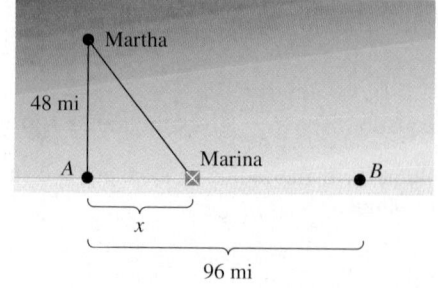

OBJECTIVE

1. Solve Equations in Quadratic Form by Using Substitution

1. Solve Equations in Quadratic Form by Using Substitution

In Sections 3.4–3.6, we learned how to solve quadratic equations by factoring, completing the square, and applying the quadratic formula. These are particularly important skills because many equations that are not quadratic can be rewritten as quadratic equations in a new variable by making an appropriate substitution. For example:

Equation in Quadratic Form		New Equation
$(2x^2 - 3)^2 + 36(2x^2 - 3) + 35 = 0$	$\xrightarrow{\text{Let } u = 2x^2 - 3}$	$u^2 + 36u + 35 = 0$
$2w^{2/3} - 3w^{1/3} - 20 = 0$	$\xrightarrow{\text{Let } u = w^{1/3}}$	$2u^2 - 3u - 20 = 0$
$x - \sqrt{x} - 12 = 0$	$\xrightarrow{\text{Let } u = \sqrt{x}}$	$u^2 - u - 12 = 0$

> **TIP** For an equation written in quadratic form, notice that the expression for u is taken to be the variable expression from the middle term.

Notice that in each equation, the terms in the equation are written in descending order of the exponents. Furthermore, the first two terms have the same base, and the exponent on the first term is exactly double the exponent on the second term. The third term is a constant. An equation that fits this pattern is said to be **quadratic in form**.

$$\underset{\text{Exponent is double}}{(2x^2 - 3)^2} + 36(2x^2 - 3)^1 + \underset{\text{Constant}}{35} = 0$$

EXAMPLE 1 Solving an Equation in Quadratic Form

Solve. $(2x^2 - 3)^2 + 36(2x^2 - 3) + 35 = 0$

Solution:

$(2x^2 - 3)^2 + 36(2x^2 - 3) + 35 = 0$	The equation is in quadratic form.
$u^2 + 36u + 35 = 0$	Let $u = 2x^2 - 3$.
$(u + 35)(u + 1) = 0$	Set one side equal to zero and factor the other side.

$u = -35$	or	$u = -1$	Apply the zero product property.
$2x^2 - 3 = -35$	or	$2x^2 - 3 = -1$	Back substitute. Replace u by $2x^2 - 3$.
$2x^2 = -32$	or	$2x^2 = 2$	Isolate the square term.
$x^2 = -16$	or	$x^2 = 1$	
$x = \pm\sqrt{-16}$	or	$x = \pm\sqrt{1}$	Apply the square root property.
$x = \pm 4i$	or	$x = \pm 1$	

The solution set is $\{\pm 4i, \pm 1\}$. The solutions all check in the original equation.

> **FOR REVIEW**
>
> The square root property indicates that the solutions to an equation of the form $x^2 = k$ are $x = \pm\sqrt{k}$. Thus, the solutions to $x^2 = -16$ are $x = \pm\sqrt{-16}$, which simplify to $x = \pm 4i$. See Section 3.5.

Skill Practice 1   Solve. $(x^2 - 6)^2 + 33(x^2 - 6) + 62 = 0$

Answer

1. $\{\pm 5i, \pm 2\}$

EXAMPLE 2   Solving an Equation in Quadratic Form

Solve. $2w^{2/3} = 3w^{1/3} + 20$

Solution:

$$2w^{2/3} = 3w^{1/3} + 20$$ Set one side equal to zero, and write the expression on the left in descending order.

$$2w^{2/3} - 3w^{1/3} - 20 = 0$$

$$2(w^{1/3})^2 - 3(w^{1/3}) - 20 = 0$$ The equation is in quadratic form.

$$2u^2 - 3u - 20 = 0$$ Let $u = w^{1/3}$.

$$(2u + 5)(u - 4) = 0$$ Factor.

$$u = -\frac{5}{2} \quad \text{or} \quad u = 4$$ Apply the zero product property.

$$w^{1/3} = -\frac{5}{2} \quad \text{or} \quad w^{1/3} = 4$$ Back substitute. Replace u by $w^{1/3}$.

$$(w^{1/3})^3 = \left(-\frac{5}{2}\right)^3 \quad \text{or} \quad (w^{1/3})^3 = (4)^3$$ Cube both sides.

$$w = -\frac{125}{8} \quad \text{or} \quad w = 64$$ Both solutions check in the original equation.

The solution set is $\left\{-\dfrac{125}{8}, 64\right\}$.

> **TIP** Consider the equation from Example 2:
>
> $2w^{2/3} - 3w^{1/3} - 20 = 0$
>
> As an alternative to using substitution, the expression on the left can be factored directly.
>
> $(2w^{1/3} + 5)(w^{1/3} - 4) = 0$
>
> Applying the zero product property results in the same solutions.

Skill Practice 2   Solve. $2t^{2/3} = 15 - 7t^{1/3}$

EXAMPLE 3   Solving an Equation in Quadratic Form

Solve. $x - \sqrt{x} - 12 = 0$

Solution:

The equation can be solved by first isolating the radical and then squaring both sides. However, this equation is also quadratic in form. By writing $\sqrt{x}$ as $x^{1/2}$, we see that the exponent on the first term is exactly double the exponent on the middle term.

$$x^1 - x^{1/2} - 12 = 0$$ The equation is in quadratic form.

$$(x^{1/2})^2 - (x^{1/2})^1 - 12 = 0$$ Let $u = x^{1/2}$.

$$u^2 - u - 12 = 0$$

$$(u - 4)(u + 3) = 0$$ Factor.

$$u = 4 \quad \text{or} \quad u = -3$$ Solve for u.

$$x^{1/2} = 4 \quad \text{or} \quad x^{1/2} = -3$$ Back substitute.

$$\sqrt{x} = 4 \quad \text{or} \quad \cancel{\sqrt{x} = -3}$$ Solve each equation for x. Recall that the principal square root of a number cannot be negative.

$$x = 16$$

> **Avoiding Mistakes**
>
> Recall that when each side of an equation is raised to an even power, we must check the potential solutions.

The solution set is $\{16\}$.   The value 16 checks in the original equation.

Answers

2. $\left\{-125, \dfrac{27}{8}\right\}$

3. $\{4\}$ (The value 1 does not check.)

Skill Practice 3   Solve. $z - \sqrt{z} - 2 = 0$

TIP The equation $x - \sqrt{x} - 12 = 0$ can also be solved as a radical equation. First isolate the radical and then square both sides.

$$x - \sqrt{x} - 12 = 0$$

$$x - 12 = \sqrt{x} \qquad \text{Isolate the radical.}$$

$$(x - 12)^2 = (\sqrt{x})^2 \qquad \text{Raise both sides to a power equal to the index of the radical.}$$

$$x^2 - 24x + 144 = x \qquad \text{Square the binomial: } (x - 12)^2 = (x)^2 - 2(x)(12) + (12)^2 = x^2 - 24x + 144$$

$$x^2 - 25x + 144 = 0 \qquad \text{Set one side equal to zero.}$$

$$(x - 16)(x - 9) = 0 \qquad \text{Factor.}$$

$$x = 16 \quad \text{or} \quad \cancel{x = 9} \qquad \text{Solve for } x.$$

The solution set is $\{16\}$.

Check $x = 16$	Check $x = 9$
$x - \sqrt{x} - 12 = 0$	$x - \sqrt{x} - 12 = 0$
$16 - \sqrt{16} - 12 \overset{?}{=} 0$	$9 - \sqrt{9} - 12 \overset{?}{=} 0$
$16 - 4 - 12 = 0$	$9 - 3 - 12 \overset{?}{=} 0$
$12 - 12 = 0 \checkmark$	$6 - 12 \overset{?}{=} 0$ (false)
	The value 9 does not check.

EXAMPLE 4   **Solving an Equation in Quadratic Form**

Solve. $\left(2 + \dfrac{3}{x}\right)^2 + 2\left(2 + \dfrac{3}{x}\right) - 35 = 0$

Solution:

$$\left(2 + \frac{3}{x}\right)^2 + 2\left(2 + \frac{3}{x}\right) - 35 = 0 \qquad \text{The equation is in quadratic form.}$$

$$u^2 + 2u - 35 = 0 \qquad \text{Let } u = 2 + \frac{3}{x}.$$

$$(u + 7)(u - 5) = 0 \qquad \text{Factor.}$$

$$u = -7 \quad \text{or} \quad u = 5 \qquad \text{Solve for } u.$$

$$2 + \frac{3}{x} = -7 \quad \text{or} \quad 2 + \frac{3}{x} = 5 \qquad \text{Back substitute.}$$

$$\frac{3}{x} = -9 \quad \text{or} \quad \frac{3}{x} = 3 \qquad \text{Isolate the } x \text{ term in each equation.}$$

$$\left(\frac{3}{\cancel{x}}\right) \cdot \cancel{x} = (-9) \cdot x \quad \text{or} \quad \left(\frac{3}{\cancel{x}}\right) \cdot \cancel{x} = (3) \cdot x \qquad \text{Clear fractions by multiplying both sides of the equation by } x.$$

$$3 = -9x \quad \text{or} \quad 3x = 3 \qquad \text{Solve for } x. \text{ Both potential solutions check in the original}$$

$$x = -3 \quad \text{or} \quad x = 1 \qquad \text{equation.}$$

The solution set is $\{-3, 1\}$.

Skill Practice 4   Solve. $\left(\dfrac{x - 4}{3}\right)^2 - 2\left(\dfrac{x - 4}{3}\right) = 3$

Answer

4. $\{1, 13\}$

Prerequisite Review

For Exercises R.1–R.6, factor completely.

R.1. $x^2 - 3x - 70$ (See Section 2.4) **R.2.** $y^2 - 11y - 26$ **R.3.** $8z^2 + 22z - 21$

R.4. $12p^2 + 8p - 15$ **R.5.** $4t^2 - 121$ (See Section 2.5) **R.6.** $9m^2 - 100$

For Exercises R.7–R.26, solve the equations.

R.7. $2x + 3 = 15$ (See Section 1.2) **R.8.** $-4x + 1 = 5$ **R.9.** $2y^2 = 288$ (See Section 3.5)

R.10. $2t^2 = 338$ **R.11.** $m^2 = -36$ (See Section 3.5) **R.12.** $n^2 = -9$

R.13. $x^2 - 8 = 0$ **R.14.** $c^2 - 75 = 0$ **R.15.** $\dfrac{5}{x} + 4 = 14$ (See Section 4.3)

R.16. $\dfrac{2}{x} - 7 = 1$ **R.17.** $\dfrac{2y + 4}{5} = 6$ **R.18.** $\dfrac{5z - 6}{3} = 1$

R.19. $\sqrt{2p + 4} - 3 = 0$ (See Section 4.6) **R.20.** $\sqrt{8q - 1} + 1 = 3$

R.21. $c^{1/3} - 4 = 0$ **R.22.** $d^{1/4} - 2 = 0$ **R.23.** $8t^2 - 6t = 5$ (See Section 3.4)

R.24. $15a^2 + 14a = 8$ **R.25.** $2p(p + 1) + 5 = 3p$ (See Section 3.6)

R.26. $3x^2 = 5(x + 3) - 9$

Concept Connections

1. The equation $m^{2/3} + 10m^{1/3} + 9 = 0$ is said to be in _____ form, because making the substitution $u = $ _____ results in a new equation that is quadratic.

2. Consider the equation $(4x^2 + 1)^2 + 4(4x^2 + 1) + 4 = 0$. If the substitution $u = $ _____ is made, then the equation becomes $u^2 + 4u + 4 = 0$.

3. To use the method of substitution to solve the equation $(3x - 1)^2 + 2(3x - 1) - 8 = 0$, let $u = $ _____.

4. To use the method of substitution to solve the equation $p^{2/3} - 2p^{1/3} - 15 = 0$, let $u = $ _____.

Objective 1: Solve Equations in Quadratic Form by Using Substitution

5. **a.** Solve the quadratic equation by factoring. $u^2 + 10u + 24 = 0$

   b. Solve the equation by using substitution. $(y^2 + 5y)^2 + 10(y^2 + 5y) + 24 = 0$

6. **a.** Solve the quadratic equation by factoring. $u^2 - 2u - 35 = 0$

   b. Solve the equation by using substitution. $(w^2 - 6w)^2 - 2(w^2 - 6w) - 35 = 0$

For Exercises 7–20, solve the equation by using substitution. **(See Examples 1–4).**

7. $(x^2 + x)^2 - 8(x^2 + x) = -12$

8. $(w^2 - 2w)^2 - 11(w^2 - 2w) = -24$

9. $(x^2 + 2)^2 + (x^2 + 2) - 42 = 0$

10. $(y^2 - 3)^2 - 9(y^2 - 3) - 52 = 0$

11. $m^{2/3} - m^{1/3} - 6 = 0$

12. $2n^{2/3} + 7n^{1/3} - 15 = 0$

13. $2t^{2/5} + 7t^{1/5} + 3 = 0$

14. $p^{2/5} + p^{1/5} - 2 = 0$

15. $y + 6\sqrt{y} = 16$

16. $p - 8\sqrt{p} = -15$

17. $2x + 3\sqrt{x} - 2 = 0$

18. $3t + 5\sqrt{t} - 2 = 0$

19. $16\left(\dfrac{x+6}{4}\right)^2 + 8\left(\dfrac{x+6}{4}\right) + 1 = 0$

20. $9\left(\dfrac{x+3}{2}\right)^2 - 6\left(\dfrac{x+3}{2}\right) + 1 = 0$

21. Rework Exercise 15 by first isolating the radical and then squaring both sides. Don't forget to check the potential solutions in the original equation.

22. Rework Exercise 16 by first isolating the radical and then squaring both sides. Don't forget to check the potential solutions in the original equation.

Mixed Exercises

For Exercises 23–52, solve the equation.

23. $t^4 + t^2 - 12 = 0$

24. $w^4 + 4w^2 - 45 = 0$

25. $x^2(9x^2 + 7) = 2$

26. $y^2(4y^2 + 17) = 15$

27. $(2x + 5)^2 - 7(2x + 5) - 30 = 0$

28. $(3x - 7)^2 - 6(3x - 7) - 16 = 0$

29. $(x^2 + 2x)^2 - 18(x^2 + 2x) = -45$

30. $(x^2 + 3x)^2 - 14(x^2 + 3x) = -40$

31. $\dfrac{y}{10} - 1 = -\dfrac{12}{5y}$

32. $1 + \dfrac{5}{x} = -\dfrac{3}{x^2}$

33. $\dfrac{2}{(n+2)^2} - \dfrac{3}{n+2} = 5$

34. $\dfrac{3}{(m-3)^2} - \dfrac{7}{m-3} = -4$

35. $\left(m - \dfrac{10}{m}\right)^2 - 6\left(m - \dfrac{10}{m}\right) - 27 = 0$

36. $\left(x + \dfrac{6}{x}\right)^2 - 12\left(x + \dfrac{6}{x}\right) + 35 = 0$

37. $\left(2 + \dfrac{3}{t}\right)^2 - \left(2 + \dfrac{3}{t}\right) = 12$

38. $\left(\dfrac{5}{y} + 3\right)^2 + 6\left(\dfrac{5}{y} + 3\right) = -8$

39. $5c^{2/5} - 11c^{1/5} + 2 = 0$

40. $3d^{2/3} - d^{1/3} - 4 = 0$

41. $y^{1/2} - y^{1/4} - 6 = 0$

42. $n^{1/2} + 6n^{1/4} - 16 = 0$

43. $9y^{-4} - 10y^{-2} + 1 = 0$

44. $100x^{-4} - 29x^{-2} + 1 = 0$

45. $4t - 25\sqrt{t} = 0$

46. $9m - 16\sqrt{m} = 0$

47. $x^2(x^2 + 5) = 7$

48. $x^2(x^2 - 2) = x^2 + 13$

49. $30k^{-2} - 23k^{-1} + 2 = 0$

50. $3q^{-2} + 16q^{-1} + 5 = 0$

51. $(x^2 - 2x)^2 - 7(x^2 - 2x) = 8$

52. $(y^2 - 4y)^2 - (y^2 - 4y) = 20$

PROBLEM RECOGNITION EXERCISES

Recognizing and Solving Equations

For Exercises 1–46,

  a. Identify the type of equation that is presented. Note that some equations will fit into more than one category. Choose from:

  - Linear equation
  - Quadratic equation
  - Higher degree polynomial equation
  - Quadratic form

  - Rational equation
  - Radical equation
  - Absolute value equation

  b. Solve the equation by using a suitable method.

1. $t^2 + 5t - 14 = 0$

2. $a^2 - 9a + 20 = 0$

3. $a^4 - 10a^2 + 9 = 0$

4. $x^4 - 3x^2 - 4 = 0$

5. $x - 3x^{1/2} - 4 = 0$

6. $x - 8\sqrt{x} - 9 = 0$

7. $8b(b + 1) + 2(3b - 4) = 4b(2b + 3)$

8. $6x(x + 1) - 3(x + 4) = 3x(2x + 5)$

9. $5a(a + 6) = 10(3a - 1)$

10. $4x(x + 3) = 6(2x - 4)$

11. $\dfrac{t}{t + 5} + \dfrac{3}{t - 4} = \dfrac{17}{t^2 + t - 20}$

12. $\dfrac{v}{v + 4} + \dfrac{12}{v^2 + 7v + 12} = \dfrac{5}{v + 3}$

13. $2u(u - 3) = 4(2 - u)$

14. $3y(y + 2) = 9(y + 1)$

15. $\sqrt{2b + 3} = b$

16. $\sqrt{5t + 6} = t$

17. $x^{2/3} + 2x^{1/3} - 15 = 0$

18. $y^{2/3} + 5y^{1/3} + 4 = 0$

19. $2|x - 4| + 1 = 11$

20. $3|2 - y| - 4 = 8$

21. $x^4 - 16 = 0$

22. $t^4 - 625 = 0$

23. $(4x + 5)^2 + 3(4x + 5) + 2 = 0$

24. $2(5x + 3)^2 - (5x + 3) - 28 = 0$

25. $4m^4 - 9m^2 + 2 = 0$

26. $x^4 - 7x^2 + 12 = 0$

27. $x^6 - 9x^3 + 8 = 0$

28. $x^6 - 26x^3 - 27 = 0$

29. $\sqrt{x^2 + 20} = 3\sqrt{x}$

30. $\sqrt{x^2 + 60} = 4\sqrt{x}$

31. $\sqrt{4t + 1} = t + 1$

32. $\sqrt{t + 10} = t + 4$

33. $6 = |7x - 3| + 9$

34. $12 = |9y + 4| + 13$

35. $2\left(\dfrac{t - 4}{3}\right)^2 - \left(\dfrac{t - 4}{3}\right) - 3 = 0$

36. $\left(\dfrac{x + 1}{5}\right)^2 - 3\left(\dfrac{x + 1}{5}\right) - 10 = 0$

37. $x^{2/3} + x^{1/3} = 20$

38. $x^{2/5} - 3x^{1/5} = -2$

39. $m^4 + 2m^2 - 8 = 0$

40. $2c^4 + c^2 - 1 = 0$

41. $a^3 + 16a - a^2 - 16 = 0$
(*Hint*: Factor by grouping first.)

42. $b^3 + 9b - b^2 - 9 = 0$

43. $x^3 + 5x - 4x^2 - 20 = 0$

44. $y^3 + 8y - 3y^2 - 24 = 0$

45. $\left(\dfrac{2}{x - 3}\right)^2 + 8\left(\dfrac{2}{x - 3}\right) + 12 = 0$

46. $\left(\dfrac{5}{x + 1}\right)^2 - 6\left(\dfrac{5}{x + 1}\right) - 16 = 0$

CHAPTER 4   Detailed Summary

SECTION 4.1   Multiplication and Division of Rational Expressions

Key Concepts	Examples	Page
A **rational expression** is a ratio of two polynomials. Values of the variable that cause the expression to be undefined are called **restricted values**. These include the values of the variable that make the denominator equal to zero.	**Example 1:** $\dfrac{x+4}{x^2+3x-40}$ is a rational expression. To find the restricted values, set the denominator equal to zero and solve the equation. $x^2+3x-40=0$ $(x+8)(x-5)=0$ $x=-8$ and $x=5$ are restricted values.	p. 257
Simplify a rational expression: To simplify a rational expression to lowest terms, factor the numerator and denominator completely. Then simplify factors whose ratio is 1 or -1. A rational expression written in lowest terms will still carry the same restrictions on the variable as the original expression.	**Example 2:** Simplify. $\dfrac{t^2-6t-16}{5t+10}$ $=\dfrac{(t-8)(t+2)}{5(t+2)}$ Note that $t\neq -2$. $=\dfrac{(t-8)\overset{1}{(t+2)}}{5(t+2)}$ $\dfrac{(t+2)}{(t+2)}$ is a ratio of 1. $=\dfrac{t-8}{5}$ provided that $t\neq -2$.	p. 259
To multiply rational expressions, factor the numerators and denominators completely. Multiply the numerators and multiply the denominators. Then simplify factors whose ratio is 1 or -1.	**Example 3:** $\dfrac{b^2-a^2}{a^2-2ab+b^2}\cdot\dfrac{a^2-3ab+2b^2}{2a+2b}$ $=\dfrac{\overset{-1}{(b-a)}(b+a)\cdot(a-2b)\overset{1}{(a-b)}}{(a-b)(a-b)\cdot 2(a+b)}$ Factor. $=-\dfrac{a-2b}{2}$ or $\dfrac{2b-a}{2}$ Simplify.	p. 263
To divide rational expressions, multiply by the reciprocal of the divisor.	**Example 4:** $\dfrac{9x+3}{x^2-4}\div\dfrac{3x+1}{4x+8}$ $=\dfrac{9x+3}{x^2-4}\cdot\dfrac{4x+8}{3x+1}$ Multiply by the reciprocal. $=\dfrac{3\overset{1}{(3x+1)}}{(x-2)(x+2)}\cdot\dfrac{4\overset{1}{(x+2)}}{3x+1}$ Factor. $=\dfrac{12}{x-2}$ Simplify.	p. 264

SECTION 4.2 Addition and Subtraction of Rational Expressions

Key Concepts	Examples	Page
To add or subtract rational expressions, the expressions must have the same denominator. The **least common denominator (LCD)** is the product of unique factors from the denominators, in which each factor is raised to its highest power.	**Example 1:** For $\dfrac{1}{3(x-1)^3(x+2)}$ and $\dfrac{-5}{6(x-1)(x+7)^2}$ $\text{LCD} = 6(x-1)^3(x+2)(x+7)^2$	p. 270
Steps to Add or Subtract Rational Expressions 1. Factor the denominator of each rational expression. 2. Identify the LCD. 3. Rewrite each rational expression as an equivalent expression with the LCD as its denominator. (This is accomplished by multiplying the numerator and denominator of each rational expression by the missing factor(s) from the LCD.) 4. Add or subtract the numerators, and write the result over the common denominator. 5. Simplify, if possible.	**Example 2:** $\dfrac{c}{c^2-c-12} - \dfrac{1}{2c-8}$ $= \dfrac{c}{(c-4)(c+3)} - \dfrac{1}{2(c-4)}$ Factor the denominators. The LCD is $2(c-4)(c+3)$ $\dfrac{2}{2} \cdot \dfrac{c}{(c-4)(c+3)} - \dfrac{1}{2(c-4)} \cdot \dfrac{(c+3)}{(c+3)}$ Write equivalent fractions with LCD. $= \dfrac{2c-(c+3)}{2(c-4)(c+3)}$ Subtract. $= \dfrac{2c-c-3}{2(c-4)(c+3)}$ Simplify. $= \dfrac{c-3}{2(c-4)(c+3)}$	p. 272
Simplifying a complex fraction: **Example 3 (Method I):** Use the order of operations to combine fractions in the numerator and denominator of the complex fraction. Then divide the resulting fractions. $\dfrac{\dfrac{1}{x}+\dfrac{1}{2}}{1+\dfrac{2}{x}} = \dfrac{\dfrac{1(2)}{x(2)}+\dfrac{1(x)}{2(x)}}{\dfrac{1(x)}{1(x)}+\dfrac{2}{x}} = \dfrac{\dfrac{2+x}{2x}}{\dfrac{x+2}{x}}$ $= \dfrac{2+x}{2x} \cdot \dfrac{x}{x+2} = \dfrac{1}{2}$	**Example 4 (Method II):** Multiply numerator and denominator of the complex fraction by the LCD of all individual fractions. $\dfrac{\dfrac{1}{x}+\dfrac{1}{2}}{1+\dfrac{2}{x}} = \dfrac{2x \cdot \left(\dfrac{1}{x}+\dfrac{1}{2}\right)}{2x \cdot \left(\dfrac{1}{1}+\dfrac{2}{x}\right)} = \dfrac{2+x}{2x+4}$ $= \dfrac{2+x}{2(x+2)} = \dfrac{1}{2}$	p. 275

SECTION 4.3 Rational Equations

Key Concepts	Examples	Page
Steps to Solve a Rational Equation 1. Factor the denominators of all rational expressions. Identify any restrictions on the variable. 2. Identify the LCD of all expressions in the equation. 3. Multiply both sides of the equation by the LCD. 4. Solve the resulting equation. 5. Check each potential solution.	**Example 1:** $\dfrac{1}{w} - \dfrac{1}{2w-1} = \dfrac{-2w}{2w-1}$ *Note:* $w = 0$ and $w = \frac{1}{2}$ are restricted. The LCD is $w(2w-1)$. $w(2w-1)\dfrac{1}{w} - w(2w-1)\cdot\dfrac{1}{2w-1} = w(2w-1)\cdot\dfrac{-2w}{2w-1}$ $(2w-1)(1) - w(1) = w(-2w)$ $2w - 1 - w = -2w^2$ (quadratic equation) $2w^2 + w - 1 = 0$ $(2w-1)(w+1) = 0$ $w = \dfrac{1}{2}$ or $w = -1$ The solution set is $\{-1\}$. (The value $\frac{1}{2}$ does not check.)	p. 283
Some rational equations have multiple variables, and we may want to solve for a specific variable.	**Example 2:** Given $P = \dfrac{vi}{i + vt}$, solve for v. $P \cdot (i + vt) = \dfrac{vi}{i + vt} \cdot (i + vt)$ $Pi + Pvt = vi$ $Pvt - vi = -Pi$ Collect v terms on one side. $v(Pt - i) = -Pi$ Factor out v. $\dfrac{v(Pt - i)}{(Pt - i)} = \dfrac{-Pi}{(Pt - i)}$ Divide by $(Pt - i)$. $v = -\dfrac{Pi}{Pt - i}$ or $v = \dfrac{Pi}{i - Pi}$	p. 286

SECTION 4.4 Applications of Rational Equations

Key Concepts	Examples	Page
Rational equations are used in a variety of applications, including solving proportions. An equation that equates two **ratios** is called a **proportion**. $\dfrac{a}{b} = \dfrac{c}{d}$ provided $b \neq 0,\ d \neq 0$	**Example 1:** A sample of 85 g of a particular ice cream contains 17 g of fat. How much fat does 324 g of the same ice cream contain? fat (g) $\longrightarrow$ $\dfrac{17}{85} = \dfrac{x}{324}$ $\longleftarrow$ fat (g) ice cream (g) $\longrightarrow$ $\longleftarrow$ ice cream (g) $(85 \cdot 324)\cdot\dfrac{17}{85} = (85 \cdot 324)\cdot\dfrac{x}{324}$ Multiply by the LCD. $5508 = 85x$ $x = 64.8$ g There would be 64.8 g of fat in 324 g of ice cream.	p. 290

Rational equations are also used in applications of rates of work.	**Example 2:** An old water pump can fill a tank in 6 hr, and a new pump can fill the tank in 4 hr. How long will it take to fill the tank if both pumps are working? Let x represent the time working together. $$\left(\begin{array}{c}\text{Speed of}\\\text{old pump}\end{array}\right) + \left(\begin{array}{c}\text{speed of}\\\text{new pump}\end{array}\right) = \left(\begin{array}{c}\text{speed working}\\\text{together}\end{array}\right)$$ $$\frac{1}{6} + \frac{1}{4} = \frac{1}{x}$$ $$12x \cdot \left(\frac{1}{6} + \frac{1}{4}\right) = 12x \cdot \left(\frac{1}{x}\right)$$ $$2x + 3x = 12$$ $$5x = 12$$ $$x = \frac{12}{5} \qquad \text{It will take } \frac{12}{5} \text{ hr.}$$	p. 294

SECTION 4.5 Rational Exponents

Key Concepts	Examples	Page
Definition of $a^{1/n}$ and $a^{m/n}$: If $n > 1$ is an integer and $\sqrt[n]{a}$ is a real number, then • $a^{1/n} = \sqrt[n]{a}$ • $a^{m/n} = \left(\sqrt[n]{a}\right)^m$ and $a^{m/n} = \sqrt[n]{a^m}$	**Example 1:** $(-64)^{1/3} = \sqrt[3]{-64} = -4$ $(-16)^{1/4}$ is undefined because $\sqrt[4]{-16}$ is not a real number. $125^{2/3} = \left(\sqrt[3]{125}\right)^2 = (5)^2 = 25$	p. 299
Properties of rational exponents: The properties of integer exponents learned in Section 2.1 can be extended to rational exponents, provided that all roots represent real numbers.	**Example 2:** **a.** $p^{1/3} \, p^{1/4} = p^{1/3+1/4} = p^{4/12+3/12} = p^{7/12}$ **b.** $\dfrac{4^{4/3}}{4^{1/3}} = 4^{4/3-1/3} = 4^{3/3} = 4$ **c.** $(y^{-1/2})^6 = y^{(-1/2)(6)} = y^{-3} = \dfrac{1}{y^3}$ **Example 3:** $\left(\dfrac{x^{1/5}x^{2/5}}{x^{4/5}}\right)^{10} = \left(\dfrac{x^{3/5}}{x^{4/5}}\right)^{10} = (x^{3/5-4/5})^{10}$ $= (x^{-1/5})^{10} = x^{(-1/5)(10)} = x^{-2} = \dfrac{1}{x^2}$	p. 302
Factoring out variables with rational exponents: To factor out a common variable factor raised to a rational exponent, factor out the variable raised to the smallest exponent. The exponents on the variables within parentheses are found by subtraction.	**Example 4:** $x^{9/2} - 7x^{7/2} + 10x^{5/2}$ $\boxed{\dfrac{9}{2} - \dfrac{5}{2}} \quad \boxed{\dfrac{7}{2} - \dfrac{5}{2}} \quad \boxed{\dfrac{5}{2} - \dfrac{5}{2}}$ $= x^{5/2}(x^{4/2} - 7x^{2/2} + 10x^0)$ $= x^{5/2}(x^2 - 7x + 10)$ $= x^{5/2}(x - 5)(x - 2)$	p. 304

SECTION 4.6 Radical Equations and Equations with Rational Exponents

Key Concepts	Examples	Page
Solving radical equations: 1. Isolate the radical. If an equation has more than one radical, choose one of the radicals to isolate. 2. Raise each side of the equation to a power equal to the index of the radical. 3. Solve the resulting equation. If the equation still has a radical, repeat steps 1 and 2. 4. Check the potential solutions in the original equation and write the solution set. **Example 1** Solve. $\qquad$ Check: $\sqrt[3]{2x+5}+7=12$ $\qquad$ $\sqrt[3]{2(60)+5}+7\stackrel{?}{=}12$ $\sqrt[3]{2x+5}=5$ $\qquad$ $\sqrt[3]{125}+7\stackrel{?}{=}12$ $\left(\sqrt[3]{2x+5}\right)^3=(5)^3$ $\qquad$ $5+7\stackrel{?}{=}12$ (true) $2x+5=125$ $2x=120$ $x=60$ The solution set is $\{60\}$.	**Example 2:** Solve. $\sqrt{b-5}-\sqrt{b+3}=2$ $\sqrt{b-5}=\sqrt{b+3}+2$ $\left(\sqrt{b-5}\right)^2=\left(\sqrt{b+3}+2\right)^2$ $b-5=b+3+4\sqrt{b+3}+4$ $b-5=b+7+4\sqrt{b+3}$ $-12=4\sqrt{b+3}$ $-3=\sqrt{b+3}$ $(-3)^2=\left(\sqrt{b+3}\right)^2$ $9=b+3$ $6=b$ Check: $\sqrt{6-5}-\sqrt{6+3}\stackrel{?}{=}2$ $\sqrt{1}-\sqrt{9}\stackrel{?}{=}2$ $1-3\stackrel{?}{=}2$ (false) No solution, $\{\ \}$	p. 309
Solving equations of the form $u^{m/n}=k$: Write the equation in radical form and solve as a radical equation.	**Example 3:** $c^{2/5}=4$ $\sqrt[5]{c^2}=4$ $\left(\sqrt[5]{c^2}\right)^5=(4)^5$ $c^2=1024$ $c=\pm\sqrt{1024}$ $c=\pm32$ Both solutions check. The solution set is $\{\pm32\}$.	p. 310

SECTION 4.7 Equations in Quadratic Form

Key Concepts	Examples	Page
Solving equations in quadratic form: Substitution can be used to solve equations that are in quadratic form.	**Example 1:** $x^{2/3}-x^{1/3}-12=0$ $\qquad$ Let $u=x^{1/3}$. $u^2-u-12=0$ $(u-4)(u+3)=0$ $\qquad$ Factor. $u-4=0$ or $u+3=0$ $u=4$ or $u=-3$ $x^{1/3}=4$ or $x^{1/3}=-3$ Back substitute. $x=64$ or $x=-27$ Cube both sides. The solution set is $\{64,-27\}$.	p. 317

CHAPTER 4 Review Exercises

SECTION 4.1

For Exercises 1–2, identify the restricted values of the variable.

1. a. $\dfrac{w-2}{w^2-4}$ b. $\dfrac{w-2}{w^2+4}$

2. a. $\dfrac{4}{5x^2+31x-28}$ b. $\dfrac{5x^2+31x-2}{4}$

For Exercises 3–10, simplify the rational expression.

3. $\dfrac{28a^3b^3}{14a^2b^3}$ 4. $\dfrac{25x^2yz^3}{125xyz}$

5. $\dfrac{x^2-4x+3}{x-3}$ 6. $\dfrac{k^2+3k-10}{k^2-5k+6}$

7. $\dfrac{x^3-27}{9-x^2}$ 8. $\dfrac{a^4-81}{3-a}$

9. $\dfrac{2t^2+3t-5}{7-6t-t^2}$ 10. $\dfrac{y^3-4y}{y^2-5y+6}$

For Exercises 11–20, multiply or divide as indicated.

11. $\dfrac{3a+9}{a^2}\cdot\dfrac{a^3}{6a+18}$ 12. $\dfrac{4-y}{5}\div\dfrac{2y-8}{15}$

13. $\dfrac{x-4y}{x^2+xy}\div\dfrac{20y-5x}{x^2-y^2}$ 14. $\dfrac{7k+28}{2k+4}\cdot\dfrac{k^2-2k-8}{k^2+2k-8}$

15. $(x^2+5x-24)\left(\dfrac{x+8}{x-3}\right)$ 16. $(9k^2-25)\cdot\left(\dfrac{k+5}{3k-5}\right)$

17. $\dfrac{2b-b^2}{b^3-8}\cdot\dfrac{b^2+2b+4}{b^2}$ 18. $\dfrac{2w}{21}\div\dfrac{3w^2}{7}\cdot\dfrac{4}{w}$

19. $\dfrac{5y^2-20}{y^3+8}\div\dfrac{7y^2-14y}{y^3+y}$

20. $\dfrac{x^2+x-20}{x^2-4x+4}\cdot\dfrac{x^2+x-6}{12+x-x^2}\div\dfrac{2x+10}{10-5x}$

SECTION 4.2

For Exercises 21–36, perform the indicated operations.

21. $\dfrac{1}{x}+\dfrac{1}{x^2}-\dfrac{1}{x^3}$ 22. $\dfrac{1}{x+2}+\dfrac{5}{x-2}$

23. $\dfrac{y}{2y-1}+\dfrac{3}{1-2y}$

24. $\dfrac{a+2}{2a+6}-\dfrac{3}{a+3}$

25. $\dfrac{4k}{k^2+2k+1}+\dfrac{3}{k^2-1}$

26. $4x+3-\dfrac{2x+1}{x+4}$

27. $\dfrac{2}{a+3}+\dfrac{2a^2-2a}{a^2-2a-15}$

28. $\dfrac{6}{x^2+4x+3}+\dfrac{7}{x^2+5x+6}$

29. $\dfrac{6a}{3a^2-7a+2}+\dfrac{2}{1-3a}+\dfrac{3a}{a-2}$

30. $4+\dfrac{2y-5}{y+2}+\dfrac{y}{3-y}$

31. $\dfrac{\dfrac{2}{x}+\dfrac{1}{xy}}{\dfrac{4}{x^2}}$ 32. $\dfrac{\dfrac{4}{y}-1}{\dfrac{1}{y}-\dfrac{4}{y^2}}$

33. $\dfrac{\dfrac{1}{a-1}+1}{\dfrac{1}{a+1}-1}$ 34. $\dfrac{\dfrac{3}{x-1}-\dfrac{1}{1-x}}{\dfrac{2}{x-1}-\dfrac{2}{x}}$

35. $\dfrac{1+xy^{-1}}{x^2y^{-2}-1}$ 36. $\dfrac{5a^{-1}+(ab)^{-1}}{3a^{-2}}$

SECTION 4.3

For Exercises 37–42, solve the equation.

37. $\dfrac{x+3}{x^2-x}-\dfrac{8}{x^2-1}=0$

38. $\dfrac{y}{y+3}-\dfrac{3}{3-y}=\dfrac{18}{y^2-9}$

39. $x-9=\dfrac{72}{x-8}$

40. $\dfrac{3x+1}{x+5}=\dfrac{x-1}{x+1}+2$

41. $5y^{-2}+1=6y^{-1}$ 42. $1+\dfrac{7}{6}m^{-1}=\dfrac{13}{6}m^{-1}$

43. Solve for x. $c=\dfrac{ax+b}{x}$

44. Solve for P. $\dfrac{A}{rt}=P+\dfrac{P}{rt}$

SECTION 4.4

For Exercises 45–48, solve the proportions.

45. $\dfrac{5}{4}=\dfrac{x}{6}$ 46. $\dfrac{x}{36}=\dfrac{6}{7}$

47. $\dfrac{x+2}{3}=\dfrac{5(x+1)}{4}$ 48. $\dfrac{x}{x+2}=\dfrac{-3}{5}$

49. In a football game, the quarterback completed 34 passes for 357 yd. At this rate how many yards would be gained for 22 passes?

50. Erik bought $108 Canadian with $100 American. At this rate, how many Canadian dollars can he buy with $235 American?

51. Tony rode 175 mi on a 2-day bicycle ride to benefit the Multiple Sclerosis Foundation. The second day he rode 5 mph slower than the first day because of a strong headwind. If Tony rode 100 mi on the first day and 75 mi on the second day in a total time of 10 hr, how fast did he ride each day?

52. Stephen drove his car 45 mi. He ran out of gas and had to walk 3 mi to a gas station. His speed driving is 15 times his speed walking. If the total time for the drive and walk was $1\frac{1}{2}$ hr, what was his speed driving?

53. Doug and Jean work as phone solicitors. They work in batches of 400 calls. Doug can finish a batch in an average of 8 hr, and Jean can finish a batch in 10 hr. How long would it take them to finish a batch if they worked together?

54. Two pipes can fill a tank in 6 hr. The larger pipe works twice as fast as the smaller pipe. How long would it take each pipe to fill the tank if they worked separately?

SECTION 4.5

For Exercises 55–58, simplify the expressions.

55. a. $10{,}000^{3/4}$ **b.** $10{,}000^{-3/4}$

      c. $-10{,}000^{3/4}$ **d.** $-10{,}000^{-3/4}$

      e. $(-10{,}000)^{3/4}$ **f.** $(-10{,}000)^{-3/4}$

56. $(-125)^{1/3}$ **57.** $16^{-1/4}$

58. $\left(\frac{1}{16}\right)^{-3/4} - \left(\frac{1}{8}\right)^{-2/3}$

For Exercises 59–60, simplify the expressions and write the answers with positive exponents only.

59. $\dfrac{p^{7/3}p^{-2/3}}{p^{2/3}}$ **60.** $(9m^{-4}n^{2/3})^{1/2}$

61. Write the expressions using radical notation.

     a. $x^{2/7}$ **b.** $9x^{2/7}$

     c. $(9x)^{2/7}$

62. Write the expressions using rational exponents.

     a. $12\sqrt{w}$ **b.** $\sqrt{12w}$

For Exercises 63–64, use a calculator to approximate the expressions to four decimal places.

63. $10^{1/3}$ **64.** $17.8^{2/3}$

For Exercises 65–68, factor completely and write the answer with no negative exponents.

65. $x^{7/2} - 19x^{5/2} + 90x^{3/2}$

66. $2x^{7/3} + x^{4/3} - 15x^{1/3}$

67. $x(2x + 5)^{-3/4} + (2x + 5)^{1/4}$

68. $2x(1 - x^2)^{1/2} + x^2 \cdot \frac{1}{2}(1 - x^2)^{-1/2}(-2x)$

SECTION 4.6

For Exercises 69–76, solve the equation.

69. $\sqrt{2y} = 7$ **70.** $\sqrt{a - 6} - 5 = 0$

71. $\sqrt[3]{2w - 3} + 5 = 2$ **72.** $\sqrt[4]{p + 12} - \sqrt[4]{5p - 16} = 0$

73. $\sqrt{t} + \sqrt{t - 5} = 5$ **74.** $\sqrt{8x + 1} = -\sqrt{x - 13}$

75. $\sqrt{2m^2 + 4} - \sqrt{9m} = 0$

76. $\sqrt{x + 2} = 1 - \sqrt{2x + 5}$

77. The velocity, v, of an ocean wave depends on the water depth d as the wave approaches land according to the equation

$$v = \sqrt{32d}$$

where v is in feet per second and d is in feet.

     a. Find the velocity at a depth of 20 ft. Round to one decimal place.

     b. Find the depth of the water at a point where a wave is traveling at 16 ft/sec.

SECTION 4.7

For Exercises 78–87, solve the equation by using substitution.

78. $x - 4\sqrt{x} - 21 = 0$

79. $n - 6\sqrt{n} + 8 = 0$

80. $y^4 - 11y^2 + 18 = 0$

81. $2m^4 - m^2 - 3 = 0$

82. $t^{2/5} + t^{1/5} - 6 = 0$

83. $p^{2/5} - 3p^{1/5} + 2 = 0$

84. $\dfrac{2t}{t + 1} + \dfrac{-3}{t - 2} = 1$

85. $\dfrac{1}{m - 2} - \dfrac{m}{m + 3} = 2$

86. $(x^2 + 5)^2 + 2(x^2 + 5) - 8 = 0$

87. $(x^2 - 3)^2 - 5(x^2 - 3) + 4 = 0$

CHAPTER 4   Test

1. Determine the restricted value of the variable for

$$\frac{x^2 - 25}{5x^2 + 24x - 5}.$$

For Exercises 2–3, simplify to lowest terms.

2. $\dfrac{12m^3n^7}{18mn^8}$

3. $\dfrac{9x^2 - 9}{3x^2 + 2x - 5}$

For Exercises 4–13, perform the indicated operations.

4. $\dfrac{2x - 5}{25 - 4x^2} \cdot (2x^2 - x - 15)$

5. $\dfrac{x^2}{x - 4} + \dfrac{8x - 16}{4 - x}$

6. $\dfrac{4x}{x + 1} + x + \dfrac{2}{x + 1}$

7. $\dfrac{3 + \dfrac{3}{k}}{4 + \dfrac{4}{k}}$

8. $\dfrac{2u^{-1} + 2v^{-1}}{4u^{-3} + 4v^{-3}}$

9. $\dfrac{ax + bx + 2a + 2b}{ax - 3a + bx - 3b} \cdot \dfrac{x - 3}{5 - x} \div \dfrac{x + 2}{ax - 5a}$

10. $\dfrac{3}{x^2 + 8x + 15} - \dfrac{1}{x^2 + 7x + 12} - \dfrac{1}{x^2 + 9x + 20}$

11. $\dfrac{3x^2}{x^3 + 14x^2 + 49x} \div \dfrac{8x - 4}{4x^2 + 26x - 14}$

12. $\dfrac{-12}{y^3 + 4y^2} + \dfrac{1}{y} - \dfrac{3}{y^2 + 4y}$

13. $\dfrac{\dfrac{x}{4} - \dfrac{9}{4x}}{\dfrac{1}{4} - \dfrac{3}{4x}}$

For Exercises 14–16, solve the equation.

14. $\dfrac{7}{z + 1} - \dfrac{z - 5}{z^2 - 1} = \dfrac{6}{z}$

15. $\dfrac{3}{y^2 - 9} + \dfrac{4}{y + 3} = 1$

16. $\dfrac{4x}{x - 4} = 3 + \dfrac{16}{x - 4}$

17. Solve for T. $\dfrac{1 + Tv}{T} = p$

18. Solve for m_1. $F = \dfrac{Gm_1m_2}{r^2}$

19. If the reciprocal of a number is added to 3 times the number, the result is $\frac{13}{2}$. Find the number.

20. Triangle ABC is similar to triangle XYZ. Find the lengths of the missing sides.

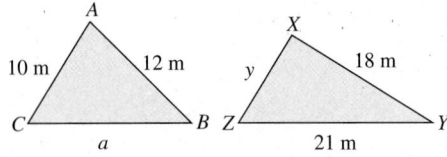

21. On a certain map, the distance between New York and Los Angeles is 8.2 in., and the actual distance is 2820 mi. What is the distance between two cities that are 5.7 in. apart on the same map? Round to the nearest mile.

22. Lance can ride 48 mi on his bike against the wind. With the wind at his back, he rides 4 mph faster and can ride 60 mi in the same amount of time. Find his speed riding against the wind and his speed riding with the wind.

23. Barbara can type a chapter in a book in 4 hr. Jack can type a chapter in a book in 10 hr. How long would it take them to type a chapter if they worked together?

For Exercises 24–25, simplify the expressions.

24. a. $1,000,000^{1/6}$ **b.** $1,000,000^{-1/2}$

c. $1,000,000^{2/3}$ **d.** $(-1,000,000)^{1/2}$

e. $(-1,000,000)^{2/3}$ **f.** $(-1,000,000)^{-1/3}$

25. $\left(\dfrac{t^{2/5} \cdot t^{7/5}}{t^3}\right)^{10}$

26. Convert $\sqrt[3]{2x}$ to exponential form.

27. Convert $5y^{1/3}$ to radical notation.

For Exercises 28–29, factor completely. Write the answers with positive exponents only.

28. $4x^{9/2} + 12x^{7/2} + 9x^{5/2}$

29. $6(x + 4)^{1/2} + 6x(x + 4)^{-1/2}$

For Exercises 30–32, solve the radical equation.

30. $\sqrt[3]{2x + 5} = -3$

31. $\sqrt{5x + 8} = \sqrt{5x - 1} + 1$

32. $\sqrt{t + 7} - \sqrt{2t - 3} = 2$

33. The equation $r = \sqrt{\dfrac{3V}{\pi h}}$ gives the radius r of a right circular cone of volume V and height h. If the radius is 9 in. for a right circular cone of volume 54π in.^{3}, determine the height of the cone.

For Exercises 34–37, solve the equation.

34. $x - \sqrt{x} - 6 = 0$

35. $y^{2/3} + 2y^{1/3} = 8$

36. $(3y - 8)^2 - 13(3y - 8) + 30 = 0$

37. $p^4 - 15p^2 = -54$

Functions and Relations

Koya979/Shutterstock

Chapter Outline

Imagine that every time you call your friend's cell phone the call is sent to the wrong device, and a different person answers the call. Suppose that one time you reach a single mother in Alabama, another time perhaps a small business owner in Georgia, and the next time a person in Kansas.

This haphazard relationship between a phone number and a target would create chaos and seriously compromise the future of your cell phone provider. Instead, when we dial 10 digits on our phone we know that the call will be routed to *only one* person's device—the person to whom the phone number is registered.

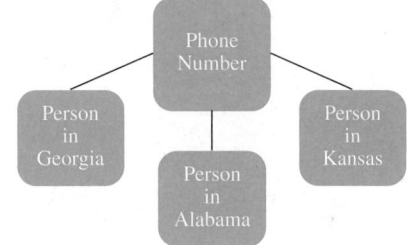

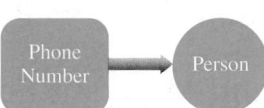

This example illustrates the importance of a **function**. That is, every item in a first set of items (in this case, a phone number being dialed) is associated with one and only one element in a second set of items (in this case, the phone number of the proper recipient). Functions are relationships that take an input value and perform an operation on that value to produce a unique and predictable output value.

In this chapter we introduce a rectangular coordinate system to visualize numerical information and to graph equations in two variables. We also give a formal definition of a function and then use functions to relate two or more variables.

The Rectangular Coordinate System and Graphing Utilities

OBJECTIVES

1. Plot Points on a Rectangular Coordinate System
2. Use the Distance and Midpoint Formulas
3. Graph Equations by Plotting Points
4. Identify *x*- and *y*-Intercepts
5. Graph Equations Using a Graphing Utility

Websites, newspapers, sporting events, and the workplace all utilize graphs and tables to present data. Therefore, it is important to learn how to create and interpret meaningful graphs. Understanding how points are located relative to a fixed origin is important for many graphing applications. For example, computer game developers use a rectangular coordinate system to define the locations of objects moving around the screen.

Gorodenkoff/Shutterstock

1. Plot Points on a Rectangular Coordinate System

Mathematician René Descartes (pronounced "day cart") (1597–1650) was the first to identify points in a plane by a pair of coordinates. He did this by intersecting two perpendicular number lines with the point of intersection called the **origin**. These lines form a **rectangular coordinate system** (also known in his honor as the **Cartesian coordinate system**) or simply a **coordinate plane**. The horizontal line is called the **x-axis** and the vertical line is called the **y-axis**. The *x*- and *y*-axes divide the plane into four **quadrants**. The quadrants are labeled counterclockwise as I, II, III, and IV (Figure 5-1).

Every point in the plane can be uniquely identified by using an ordered pair (x, y) to specify its coordinates with respect to the origin. In an ordered pair, the first coordinate is called the **x-coordinate**, and the second is called the **y-coordinate**. The origin is identified as $(0, 0)$. In Figure 5-2, six points have been graphed. The point $(-3, 5)$, for example, is placed 3 units in the negative *x* direction (to the left) and 5 units in the positive *y* direction (upward).

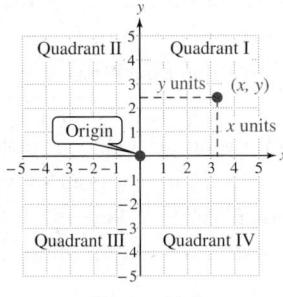

Figure 5-1

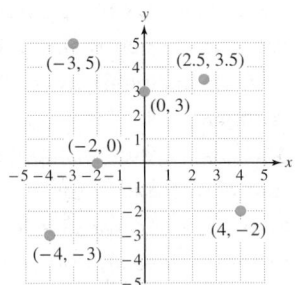

Figure 5-2

2. Use the Distance and Midpoint Formulas

Recall that the distance between two points A and B on a number line can be represented by $|A - B|$ or $|B - A|$. Now we want to find the distance between two points in a coordinate plane. For example, consider the points $(1, 5)$ and $(4, 9)$. The distance d between the points is labeled in Figure 5-3. The dashed horizontal and vertical line segments form a right triangle with hypotenuse d.

The horizontal distance between the points is $|4 - 1| = 3$.
The vertical distance between the points is $|9 - 5| = 4$.

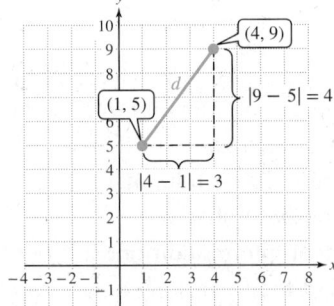

Figure 5-3

Recall that for a right triangle with legs of lengths a and b and hypotenuse of length c, the Pythagorean theorem gives a relationship among the lengths of the sides.

$$a^2 + b^2 = c^2$$

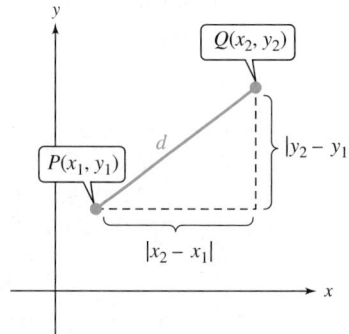

Applying the Pythagorean theorem, we have

$$d^2 = (3)^2 + (4)^2$$

Since d is a distance, reject the negative square root.

$$d = \sqrt{(3)^2 + (4)^2} = \sqrt{25} = 5$$ The distance between the points is 5 units.

We can make this process generic by labeling the points $P(x_1, y_1)$ and $Q(x_2, y_2)$. See Figure 5-4.

Figure 5-4

- The horizontal leg of the right triangle is $|x_2 - x_1|$ or equivalently $|x_1 - x_2|$.
- The vertical leg of the right triangle is $|y_2 - y_1|$ or equivalently $|y_1 - y_2|$.

Applying the Pythagorean theorem, we have

$$d^2 = (x_2 - x_1)^2 + (y_2 - y_1)^2$$
$$d = \sqrt{(x_2 - x_1)^2 + (y_2 - y_1)^2}$$

We can drop the absolute value bars because $|a|^2 = (a)^2$ for all real numbers a. Likewise $|x_2 - x_1|^2 = (x_2 - x_1)^2$ and $|y_2 - y_1|^2 = (y_2 - y_1)^2$.

Distance Formula

The distance between points (x_1, y_1) and (x_2, y_2) is given by

$$d = \sqrt{(x_2 - x_1)^2 + (y_2 - y_1)^2}$$

EXAMPLE 1   **Finding the Distance Between Two Points**

Find the distance between the points $(-5, 1)$ and $(7, -3)$. Give the exact distance and an approximation to 2 decimal places.

Solution:

$(-5, 1)$ and $(7, -3)$ Label the points. Note that the choice for
(x_1, y_1) and (x_2, y_2) (x_1, y_1) and (x_2, y_2) will not affect the outcome.

$$d = \sqrt{[7 - (-5)]^2 + (-3 - 1)^2}$$ Apply the distance formula.
$$d = \sqrt{(x_2 - x_1)^2 + (y_2 - y_1)^2}$$
$$= \sqrt{(12)^2 + (-4)^2}$$ Simplify the radical.
$$= \sqrt{160}$$
$$= \sqrt{16 \cdot 10}$$
$$= 4\sqrt{10} \approx 12.65$$ The exact distance is $4\sqrt{10}$ units.
This is approximately 12.65 units.

Skill Practice 1   Find the distance between the points $(-1, 4)$ and $(3, -6)$. Give the exact distance and an approximation to 2 decimal places.

Answer

1. $2\sqrt{29}$ units ≈ 10.77 units

Avoiding Mistakes

A statement of the form "if p, then q" is called a **conditional statement**. Its **converse** is the statement "if q, then p." The converse of a statement is not necessarily true. However, in the case of the Pythagorean theorem, the converse is a true statement.

The Pythagorean theorem tells us that if a right triangle has legs of lengths a and b and hypotenuse of length c, then $a^2 + b^2 = c^2$. The following related statement is also true: If $a^2 + b^2 = c^2$, then a triangle with sides of lengths a, b, and c is a right triangle. We use this important concept in Example 2.

EXAMPLE 2   Determining if Three Points Form the Vertices of a Right Triangle

Determine if the points $M(-2, -3)$, $P(4, 1)$, and $Q(-1, 7)$ form the vertices of a right triangle.

Solution:

Points M, P, Q, and the triangle defined by the points are shown in the figure. Our strategy is first to determine the length of each side of the triangle by applying the distance formula.

$$d = \sqrt{(x_2 - x_1)^2 + (y_2 - y_1)^2}$$

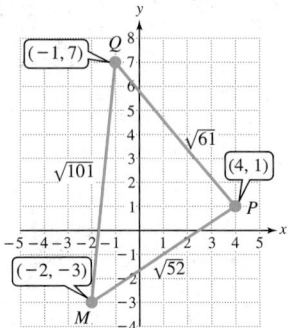

Note that for two points such as M and P we use the notation $d(M, P)$ or MP to denote the distance between the points.

To find $d(M, P)$:

Label $M(-2, -3)$ as (x_1, y_1).

Label $P(4, 1)$ as (x_2, y_2).

$$d = \sqrt{(x_2 - x_1)^2 + (y_2 - y_1)^2}$$
$$d(M, P) = \sqrt{[4 - (-2)]^2 + [1 - (-3)]^2}$$
$$= \sqrt{(6)^2 + (4)^2}$$
$$= \sqrt{36 + 16}$$
$$= \sqrt{52}$$

To find $d(P, Q)$:

Label $P(4, 1)$ as (x_1, y_1).

Label $Q(-1, 7)$ as (x_2, y_2).

$$d(P, Q) = \sqrt{(-1 - 4)^2 + (7 - 1)^2}$$
$$= \sqrt{(-5)^2 + (6)^2}$$
$$= \sqrt{25 + 36}$$
$$= \sqrt{61}$$

To find $d(M, Q)$:

Label $M(-2, -3)$ as (x_1, y_1).

Label $Q(-1, 7)$ as (x_2, y_2).

$$d(M, Q) = \sqrt{[-1 - (-2)]^2 + [7 - (-3)]^2}$$
$$= \sqrt{(1)^2 + (10)^2}$$
$$= \sqrt{1 + 100}$$
$$= \sqrt{101}$$

The line segment $\overline{MQ}$ is the longest and would potentially be the hypotenuse c. Label the shorter sides as a and b. Thus, $a = \sqrt{52}$, $b = \sqrt{61}$, and $c = \sqrt{101}$.
Check the condition that $a^2 + b^2 = c^2$.

$$\left(\sqrt{52}\right)^2 + \left(\sqrt{61}\right)^2 \stackrel{?}{=} \left(\sqrt{101}\right)^2$$

$$52 + 61 \neq 101$$

Thus, the points M, P, and Q do not form the vertices of a right triangle.

Skill Practice 2   Determine if the points $X(-6, -4)$, $Y(2, -2)$, and $Z(0, 5)$ form the vertices of a right triangle.

Figure 5-5

Now suppose that we want to find the midpoint of the line segment between the distinct points (x_1, y_1) and (x_2, y_2). The **midpoint** of a line segment is the point equidistant (the same distance) from the endpoints (Figure 5-5).

Answer

2. No

The x-coordinate of the midpoint is the average of the x-coordinates from the endpoints. Likewise, the y-coordinate of the midpoint is the average of the y-coordinates from the endpoints.

Midpoint Formula

The midpoint of the line segment with endpoints (x_1, y_1) and (x_2, y_2) is

$$M = \left(\frac{x_1 + x_2}{2}, \frac{y_1 + y_2}{2} \right)$$

average of   average of
x-coordinates y-coordinates

Avoiding Mistakes

The midpoint of a line segment is an ordered pair (with two coordinates), not a single number.

EXAMPLE 3   **Finding the Midpoint of a Line Segment**

Find the midpoint of the line segment with endpoints (4.2, –4) and (–2.8, 3).

Solution:

Label (4.2, –4) as (x_1, y_1).   Label the points.
Label (–2.8, 3) as (x_2, y_2).

$$M = \left(\frac{4.2 + (-2.8)}{2}, \frac{-4 + 3}{2} \right)$$ Apply the midpoint formula.

$$M = \left(\frac{x_1 + x_2}{2}, \frac{y_1 + y_2}{2} \right)$$

$$= \left(\frac{1.4}{2}, \frac{-1}{2} \right)$$ Simplify.

$$= (0.7, -0.5)$$

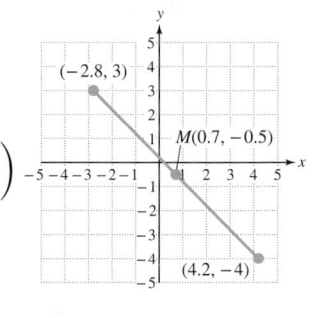

Skill Practice 3   Find the midpoint of the line segment with endpoints (–1.5, –9) and (–8.7, 4).

---

3. Graph Equations by Plotting Points

The relationship between two variables can often be expressed as a graph or expressed algebraically as an equation. For example, suppose that two variables, x and y, are related such that y is 2 more than x. An equation to represent this relationship is $y = x + 2$. A **solution to an equation** in the variables x and y is an ordered pair (x, y) that when substituted into the equation makes the equation a true statement.

For example, the following ordered pairs are solutions to the equation $y = x + 2$.

Solution	$y = x + 2$
(0, 2)	$2 = 0 + 2$ ✓
(–4, –2)	$-2 = -4 + 2$ ✓
(2, 4)	$4 = 2 + 2$ ✓

The set of all solutions to an equation is called the **solution set of the equation**. The graph of all solutions to an equation is called the **graph of the equation**. The graph of $y = x + 2$ is shown in Figure 5-6.

One of the goals of this text is to identify families of equations and the characteristics of their graphs. As we proceed through the text, we will develop tools to graph equations efficiently. For now, we present the point-plotting method to graph the solution set of an equation. In Example 4, we start by selecting several values of x and using the equation to calculate the corresponding values of y. Then we plot the points to form a general outline of the curve.

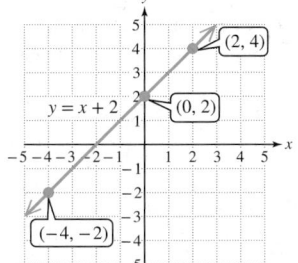

Figure 5-6

Answer

3. $\left(-5.1, -\dfrac{5}{2} \right)$ or (–5.1, –2.5)

EXAMPLE 4   Graphing an Equation by Plotting Points

Graph the equation by plotting points. $x - 2y = 0$

Solution:

$x - 2y = 0$

$-2y = -x$ Solve for y in terms of x. First subtract x from both sides.

$\dfrac{-2y}{-2} = \dfrac{-x}{-2}$ Divide both sides by -2.

$y = \dfrac{x}{2}$ Arbitrarily select negative and positive values of x and solve the equation for the corresponding y value. Notice that we have chosen values of x that are multiples of 2 so that the resulting y values are integers.

x	y
-4	
-2	
0	
2	
4	

FOR REVIEW

For more detail on solving an equation for a specified variable, see Section 1.2.

x	$y = \dfrac{x}{2}$	Ordered pair
-4	$y = \dfrac{-4}{2} = -2$	$(-4, -2)$
-2	$y = \dfrac{-2}{2} = -1$	$(-2, -1)$
0	$y = \dfrac{0}{2} = 0$	$(0, 0)$
2	$y = \dfrac{2}{2} = 1$	$(2, 1)$
4	$y = \dfrac{4}{2} = 2$	$(4, 2)$

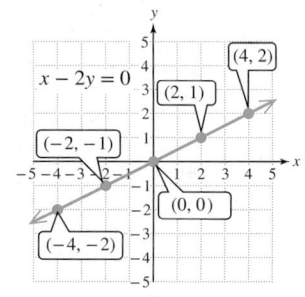

The graph of the equation $x - 2y = 0$ is a line passing through the origin.

Skill Practice 4   Graph the equation by plotting points. $x + 3y = 3$

EXAMPLE 5   Graphing an Equation by Plotting Points

Graph the equation by plotting points. $y - |x| = -1$

Solution:

$y - |x| = -1$ Solve for y in terms of x.

$y = |x| - 1$ Arbitrarily select negative and positive values for x such as -3, -2, -1, 0, 1, 2, and 3. Then use the equation to calculate the corresponding y values.

x	$y = |x| - 1$	Ordered pair
-3	$y = |-3| - 1 = 2$	$(-3, 2)$
-2	$y = |-2| - 1 = 1$	$(-2, 1)$
-1	$y = |-1| - 1 = 0$	$(-1, 0)$
0	$y = |0| - 1 = -1$	$(0, -1)$
1	$y = |1| - 1 = 0$	$(1, 0)$
2	$y = |2| - 1 = 1$	$(2, 1)$
3	$y = |3| - 1 = 2$	$(3, 2)$

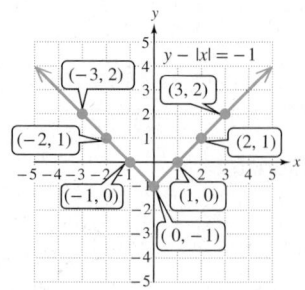

Answers

4.

$x + 3y = 3$

5.

$x^2 + y = 4$

Skill Practice 5   Graph the equation by plotting points. $x^2 + y = 4$

The graph of an equation in the variables x and y represents a relationship between a real number x and a corresponding real number y. Therefore, the values of x must be chosen so that when substituted into the equation, they produce a real number for y. Sometimes the values of x must be restricted to produce real numbers for y. This is demonstrated in Example 6.

EXAMPLE 6   Graphing an Equation by Plotting Points

Graph the equation by plotting points. $y^2 - 1 = x$

Solution:

$$y^2 - 1 = x \qquad \text{Solve for } y \text{ in terms of } x.$$
$$y^2 = x + 1$$
$$y = \pm\sqrt{x + 1} \qquad \text{Apply the square root property.}$$

Choose $x \geq -1$ so that the radicand is nonnegative.

x	$y = \pm\sqrt{x + 1}$	Ordered pairs
-1	$y = \pm\sqrt{(-1) + 1} = \pm\sqrt{0} = 0$	$(-1, 0)$
0	$y = \pm\sqrt{(0) + 1} = \pm\sqrt{1} = \pm1$	$(0, 1), (0, -1)$
1	$y = \pm\sqrt{(1) + 1} = \pm\sqrt{2}$ ≈ 1.4	$(1, \sqrt{2}), (1, -\sqrt{2})$ $\approx (1, 1.4), (1, -1.4)$
3	$y = \pm\sqrt{(3) + 1} = \pm\sqrt{4} = \pm2$	$(3, 2), (3, -2)$
8	$y = \pm\sqrt{(8) + 1} = \pm\sqrt{9} = \pm3$	$(8, 3), (8, -3)$

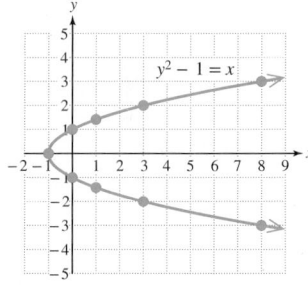

Skill Practice 6   Graph the equation by plotting points. $x + y^2 = 2$

---

4. Identify x- and y-Intercepts

When analyzing graphs, we want to examine their most important features. Two key features are the x- and y-intercepts of a graph. These are the points where a graph intersects the x- and y-axes.

Any point on the x-axis has a y-coordinate of zero. Therefore, an **x-intercept** is a point $(a, 0)$ where a graph intersects the x-axis (Figure 5-7). Any point on the y-axis has an x-coordinate of zero. Therefore, a **y-intercept** is a point $(0, b)$ where a graph intersects the y-axis (Figure 5-7).

Figure 5-7

TIP   In some applications, we may refer to an x-intercept as the *x-coordinate* of a point of intersection that a graph makes with the x-axis. For example, if an x-intercept is $(-4, 0)$, then the x-intercept may be stated simply as -4 (the y-coordinate is understood to be zero). Similarly, we may refer to a y-intercept as the *y-coordinate* of a point of intersection that a graph makes with the y-axis. For example, if a y-intercept is $(0, 2)$, then it may be stated simply as 2.

Answer

6.

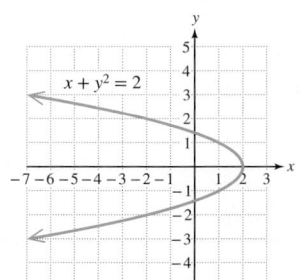

To find the x- and y-intercepts from an equation in x and y, follow these steps.

Determining x- and y-Intercepts from an Equation

Given an equation in x and y,
- Find the x-intercept(s) by substituting 0 for y in the equation and solving for x.
- Find the y-intercept(s) by substituting 0 for x in the equation and solving for y.

EXAMPLE 7 Finding *x*- and *y*-Intercepts

Given the equation $3x + 5y = 10$,

   a. Find the *x*-intercept(s).   **b.** Find the *y*-intercept(s).

Solution:

a. $3x + 5y = 10$

      $3x + 5(0) = 10$ To find the *x*-intercept(s), substitute 0 for *y* and solve for *x*.

           $3x = 10$ Simplify.

            $x = \dfrac{10}{3}$

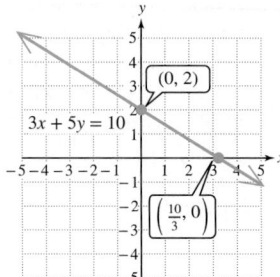

    The *x*-intercept is $\left(\dfrac{10}{3}, 0\right)$ or $\left(3\dfrac{1}{3}, 0\right)$.

b. $3x + 5y = 10$

      $3(0) + 5y = 10$ To find the *y*-intercept(s), substitute 0 for *x* and solve for *y*.

           $5y = 10$ Simplify.

            $y = 2$

    The *y*-intercept is $(0, 2)$.

Skill Practice 7   Given $2x + 3y = 4$,

   a. Find the *x*-intercept(s).   **b.** Find the *y*-intercept(s).

In Example 8, the graph of the equation has multiple intercepts.

EXAMPLE 8 Finding *x*- and *y*-Intercepts

Given the equation $y = |x| - 1$,

   a. Find the *x*-intercept(s).   **b.** Find the *y*-intercept(s).

Solution:

a. $y = |x| - 1$

      $0 = |x| - 1$ To find the *x*-intercept(s), substitute 0 for *y* and solve for *x*.

      $|x| = 1$ Isolate the absolute value.

      $x = 1$ or $x = -1$ Recall that for $k > 0$, $|x| = k$ is equivalent to $x = k$ or $x = -k$.

    The *x*-intercepts are $(1, 0)$ and $(-1, 0)$.

b. $y = |x| - 1$

      $= |0| - 1$ To find the *y*-intercept(s), substitute 0 for *x* and solve for *y*.

      $= -1$

    The *y*-intercept is $(0, -1)$.

    The intercepts $(1, 0)$, $(-1, 0)$, and $(0, -1)$ are consistent with the graph of the equation $y = |x| - 1$ found in Example 5 (Figure 5-8).

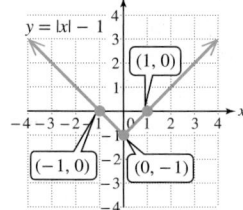

Figure 5-8

Skill Practice 8   Given the equation $y = x^2 - 4$,

   a. Find the *x*-intercept(s).   **b.** Find the *y*-intercept(s).

Answers

7. a. $(2, 0)$

  b. $\left(0, \dfrac{4}{3}\right)$

8. a. $(2, 0)$ and $(-2, 0)$

  b. $(0, -4)$

TIP Sometimes when solving for an *x*- or *y*-intercept, we encounter an equation with imaginary numbers as its only solutions. In such a case, the graph has no *x*- or *y*-intercept.

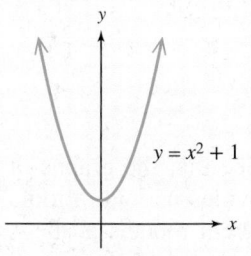

No *x*-intercept

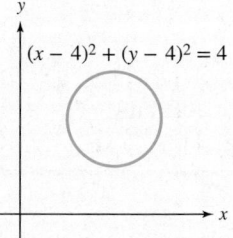

No *x*- or *y*-intercept

TIP The Greek letter Δ ("delta") written before a variable represents an increment of change in that variable. In this context, it represents the change from one value of *x* to the next.

TIP The calculator plots a large number of points and then connects the points. So instead of graphing a single smooth curve, it graphs a series of short line segments. This may give the graph a jagged look (Figure 5-10).

5. Graph Equations Using a Graphing Utility

Graphing by the point-plotting method should only be considered a beginning strategy for creating the graphs of equations in two variables. We will quickly enhance this method with other techniques that are less cumbersome and use more analysis and strategy.

One weakness of the point-plotting method is that it may be slow to execute by pencil and paper. Also, the selected points must fairly represent the shape of the graph. Otherwise the sketch will be inaccurate. Graphing utilities can help with both of these weaknesses. They can graph many points quickly, and the more points that are plotted, the greater the likelihood that we see the key features of the graph. Graphing utilities include graphing calculators, spreadsheets, specialty graphing programs, and apps on phones.

Figures 5-9 and 5-10 show a table and a graph for $y = x^2 - 3$.

TECHNOLOGY CONNECTIONS

Using the Table Feature and Graphing an Equation

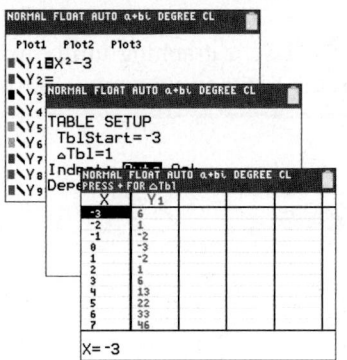

Figure 5-9

In Figure 5-9, we first enter the equation $y = x^2 - 3$ into the graphing editor. Notice that the calculator expects the equation represented with the *y* variable isolated.

To set up a table, enter the starting value for *x*, in this case, −3. Then set the increment by which to increase *x*, in this case 1. The *x*-increment is entered as ΔTbl (read "delta table"). Using the "Auto" setting means that the table of values for X and Y_1 will be automatically generated.

The table shows eleven *x*–*y* pairs but more can be accessed by using the up and down arrow keys on the keypad.

The graph in Figure 5-10 is shown between *x* and *y* values from −10 to 10. The tick marks on the axes are 1 unit apart. The viewing window with these parameters is denoted [−10, 10, 1] by [−10, 10, 1].

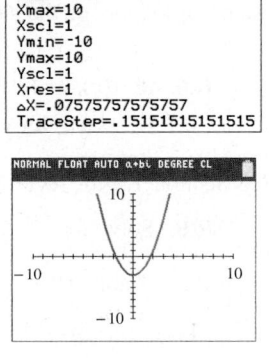

Figure 5-10

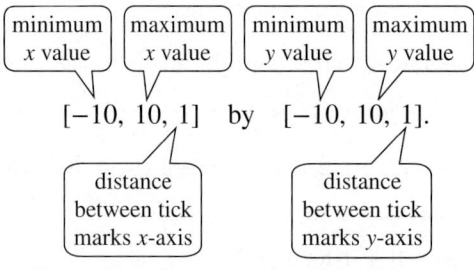

$$[-10, 10, 1] \quad \text{by} \quad [-10, 10, 1].$$

distance between tick marks *x*-axis   distance between tick marks *y*-axis

---

EXAMPLE 9   Graphing Equations Using a Graphing Utility

Use a graphing utility to graph $y = |x| - 15$ and $y = -x^2 + 12$ on the viewing window defined by [−20, 20, 2] by [−15, 15, 3].

Solution:

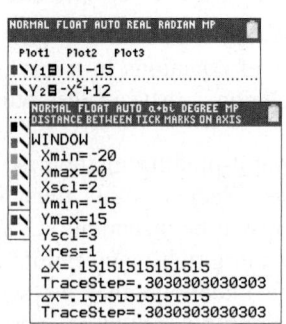

Enter the equations using the Y = editor.

Use the WINDOW editor to change the viewing window parameters. The variables Xmin, Xmax, and Xscl relate to [−20, 20, 2]. The variables Ymin, Ymax, and Yscl relate to [−15, 15, 3].

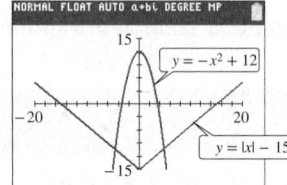

Select the GRAPH feature. Notice that the graphs of both equations appear. This provides us with a tool for visually examining two different models at the same time.

Answer

9.

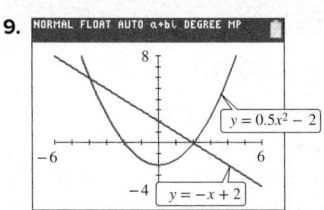

Skill Practice 9 Use a graphing utility to graph $y = -x + 2$ and $y = 0.5x^2 - 2$ on the viewing window [−6, 6, 1] by [−4, 8, 1].

SECTION 5.1   Practice Exercises

Prerequisite Review

For Exercises R.1–R.2, simplify the radicals. (See Section 3.1 for review.)

R.1. $\sqrt{48}$ 　　　　　　　　　**R.2.** $\sqrt{52}$

For Exercises R.3–R.4, determine if the expression is a real number. (See Section R.3 for review.)

R.3. a. $\sqrt{-13}$ 　　　　　　　**R.4. a.** $\sqrt{3}$
　　　b. $-\sqrt{10}$ 　　　　　　　　　　**b.** $\sqrt{-19}$

For Exercises R.5–R.6, determine if the three values represent the lengths of the sides of a right triangle. (See Section 3.4 for review.)

R.5. a. 20 ft, 21 ft, and 29 ft 　　　　　　　**R.6. a.** 10 km, 14 km, and 17 km
　　　b. 7 cm, 9 cm, and 11 cm 　　　　　　　　**b.** 28 yd, 45 yd, and 53 yd

For Exercises R.7–R.8, solve for the indicated variable. (See Section 1.2 for review.)

R.7. Solve for y.　$ax + by = c$ 　　　　　**R.8.** Solve for y.　$|x| + y - 4 = 0$

For R.9–R.12, evaluate the expressions for the given values of the variable. (See Section R.3 for review.)

R.9. $x^2 + 4x + 5$ for $x = -5$ 　　　　　　**R.10.** $-x^2 - 3x + 7$ for $x = -2$

R.11. $|x - 4| + 1$ for $x = 3$ 　　　　　　**R.12.** $|x + 3| - 4$ for $x = -6$

For Exercises R.13–R.16, solve the equations.

R.13. $|x + 1| - 3 = 7$ (See Section 1.6.) 　　　**R.14.** $|x - 6| + 2 = 4$

R.15. $y^2 - 1 = 8$ (See Section 3.5.) 　　　　**R.16.** $y^2 + 4 = 29$

For Exercises R.17–R.20, solve the inequality and write the solution set. (See Section 1.5 for review.)

R.17. $x - 7 \geq 0$ **R.18.** $2x + 1 < 0$

R.19. $3 - x > 12$ **R.20.** $10 - 4x \leq 2$

Concept Connections

1. In a rectangular coordinate system, the point where the x- and y-axes meet is called the _____.

2. The x- and y-axes divide the coordinate plane into four regions called _____.

3. The distance between two distinct points (x_1, y_1) and (x_2, y_2) is given by the formula _____.

4. The midpoint of the line segment with endpoints (x_1, y_1) and (x_2, y_2) is given by the formula _____.

5. A _____ to an equation in the variables x and y is an ordered pair (x, y) that makes the equation a true statement.

6. An x-intercept of a graph has a y-coordinate of _____.

7. A y-intercept of a graph has an x-coordinate of _____.

8. Given an equation in the variables x and y, find the y-intercept by substituting _____ for x and solving for _____.

Objective 1: Plot Points on a Rectangular Coordinate System

For Exercises 9–10, plot the points on a rectangular coordinate system.

9. $A(-3, -4)$ $B\left(\dfrac{5}{3}, \dfrac{7}{4}\right)$ $C(-1.2, 3.8)$ $D(\pi, -5)$ $E(0, 4.5)$ $F(\sqrt{5}, 0)$

10. $A(-2, -5)$ $B\left(\dfrac{9}{2}, \dfrac{7}{3}\right)$ $C(-3.6, 2.1)$ $D(5, -\pi)$ $E(3.4, 0)$ $F(0, \sqrt{3})$

For Exercises 11–12, identify in which quadrant or on which axis the point lies.

11. a. $(-2, 4)$ **b.** $(-1.3, -8.9)$ **c.** $\left(\dfrac{10}{3}, 0\right)$ **d.** $(0, -131)$

12. a. $(9, -11)$ **b.** $(0, -6.7)$ **c.** $\left(-\dfrac{15}{2}, -\dfrac{9}{7}\right)$ **d.** $(-63, 0)$

Objective 2: Use the Distance and Midpoint Formulas

For Exercises 13–18,

a. Find the exact distance between the points. **(See Example 1)**

b. Find the midpoint of the line segment whose endpoints are the given points. **(See Example 3)**

13. $(-2, 7)$ and $(-4, 11)$ **14.** $(-1, -3)$ and $(3, -7)$ **15.** $(-7, -4)$ and $(2, 5)$

16. $(3, 6)$ and $(-4, -1)$ **17.** $(2.2, -2.4)$ and $(5.2, -6.4)$ **18.** $(37.1, -24.7)$ and $(31.1, -32.7)$

For Exercises 19–22, determine if the given points form the vertices of a right triangle. **(See Example 2)**

19. $(1, 3)$, $(3, 1)$, and $(0, -2)$ **20.** $(1, 2)$, $(3, 0)$, and $(-3, -2)$

21. $(-2, 4)$, $(5, 0)$, and $(-5, 1)$ **22.** $(-6, 2)$, $(3, 1)$, and $(1, -2)$

Objective 3: Graph Equations by Plotting Points

For Exercises 23–24, determine if the given points are solutions to the equation.

23. $x^2 + y = 1$

   a. $(-2, -3)$ **b.** $(4, -17)$ **c.** $\left(\dfrac{1}{2}, \dfrac{3}{4}\right)$

24. $|x - 3| - y = 4$

   a. $(1, -2)$ **b.** $(-2, -3)$ **c.** $\left(\dfrac{1}{10}, -\dfrac{11}{10}\right)$

For Exercises 25–30, identify the set of values x for which y will be a real number.

25. $y = \dfrac{2}{x - 3}$

26. $y = \dfrac{2}{x + 7}$

27. $y = \sqrt{x - 10}$

28. $y = \sqrt{x + 11}$

29. $y = \sqrt{1.5 - x}$

30. $y = \sqrt{2.2 - x}$

For Exercises 31–48, graph the equations by plotting points. **(See Examples 4–6)**

31. $2x + y = 3$

32. $x + y = -2$

33. $x + 3y = 0$

34. $x - 4y = 0$

35. $y = x$

36. $y = x^2$

37. $y = \sqrt{x}$

38. $y = |x|$

39. $y = x^3$

40. $y = \dfrac{1}{x}$

41. $y - |x| = 2$

42. $|x| + y = 3$

43. $y^2 - x - 2 = 0$

44. $y^2 - x + 1 = 0$

45. $x = |y| + 1$

46. $x = |y| - 3$

47. $y = |x + 1|$

48. $y = |x - 2|$

Objective 4: Identify *x*- and *y*-Intercepts

For Exercises 49–54, estimate the x- and y-intercepts from the graph.

49.
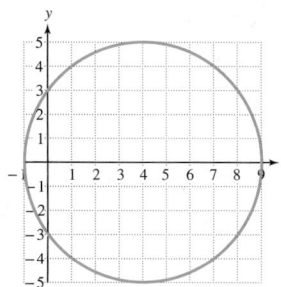

50.
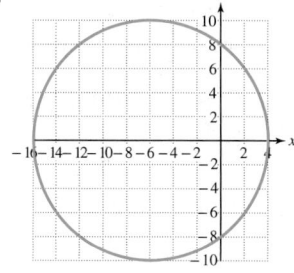

51.
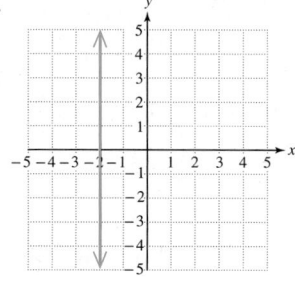

52.
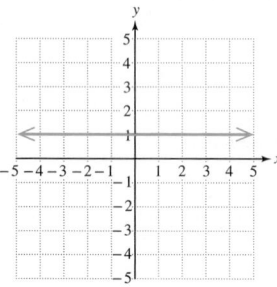

53.
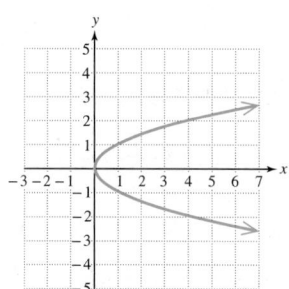

54.
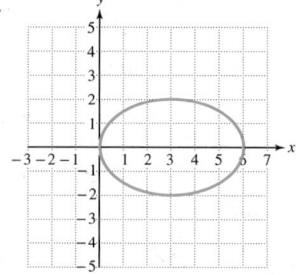

For Exercises 55–66, find the x- and y-intercepts. **(See Examples 7–8)**

55. $-2x + 4y = 12$

56. $-3x - 5y = 60$

57. $x^2 + y = 9$

58. $x^2 = -y + 16$

59. $y = |x - 5| - 2$

60. $y = |x + 4| - 3$

61. $x = y^2 - 1$

62. $x = y^2 - 4$

63. $|x| = |y|$

64. $x = |5y|$

65. $\dfrac{(x - 3)^2}{4} + \dfrac{(y - 4)^2}{9} = 1$

66. $\dfrac{(x + 6)^2}{16} + \dfrac{(y + 3)^2}{4} = 1$

Mixed Exercises

67. A map of a wilderness area is drawn with the origin placed at the parking area. Two fire observation platforms are located at points A and B. If a fire is located at point C, determine the distance to the fire from each observation platform. Assume that the units of measurement are in miles.

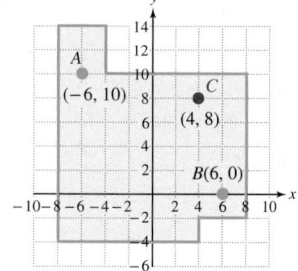

68. A map of a state park is drawn so that the origin is placed at the visitor center. The distance between grid lines is 1 mi. Suppose that two hikers are located at points A and B.

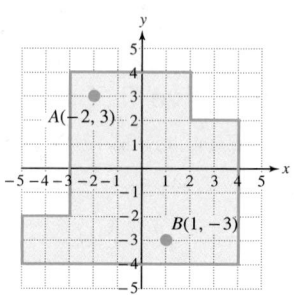

  a. Determine the distance between the hikers.

  b. If the hikers want to meet for lunch, determine the location of the midpoint between the hikers.

The position of an object in a video game is represented by an ordered pair. The coordinates of the ordered pair give the number of pixels horizontally and vertically from the origin. Use this scenario for Exercises 69–72.

69. a. Suppose that player A is located at (36, 315) and player B is located at (410, 53). How far apart are the players? Round to the nearest pixel.

   b. If the two players move directly toward each other at the same speed, where will they meet?

   c. If player A moves three times faster than player B, where will they meet? Round to the nearest pixel.

70. Suppose that a player is located at point $A(460, 420)$ and must move in a direct line to point $B(80, 210)$ and then in a direct line to point $C(120, 60)$ to pick up prizes before a 5-sec timer runs out. If the player moves at 120 pixels per second, will the player have enough time to pick up both prizes? Explain.

71. Verify that the points $A(0, 0)$, $B(x, 0)$, and $C\left(\dfrac{1}{2}x, \dfrac{\sqrt{3}}{2}x\right)$ make up the vertices of an equilateral triangle.

72. Verify that the points $A(0, 0)$, $B(x, 0)$, and $C(0, x)$ make up the vertices of an isosceles right triangle (an isosceles triangle has two sides of equal length).

For Exercises 73–74, assume that the units shown in the grid are in feet.

a. Determine the exact length and width of the rectangle shown.

b. Determine the perimeter and area.

73.

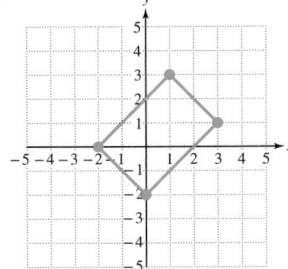

74.

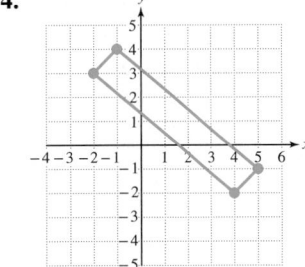

For Exercises 75–76, the endpoints of a diameter of a circle are shown. Find the center and radius of the circle.

75.

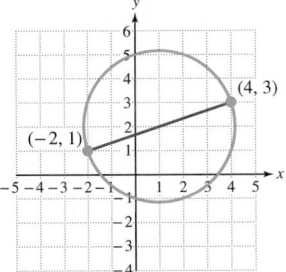

76.

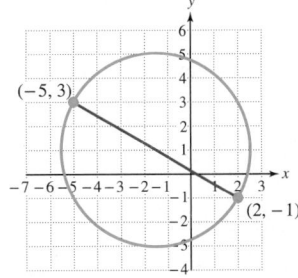

For Exercises 77–78, an isosceles triangle is shown. Find the area of the triangle. Assume that the units shown in the grid are in meters.

77.

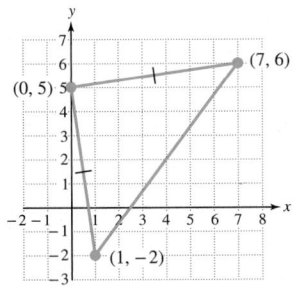

78.

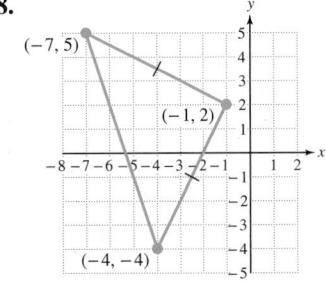

For Exercises 79–82, determine if points A, B, and C are collinear. Three points are collinear if they all fall on the same line. There are several ways that we can determine if three points, A, B, and C, are collinear. One method is to determine if the sum of the lengths of the line segments $\overline{AB}$ and $\overline{BC}$ equals the length of $\overline{AC}$.

79. (2, 2), (4, 3), and (8, 5)

80. (2, 1.5), (4, 2), and (8, 3)

81. (−2, 8), (1, 2), and (4, −3)

82. (−1, 5), (0, 3), and (5, −13)

Write About It

83. Suppose that d represents the distance between two points (x_1, y_1) and (x_2, y_2). Explain how the distance formula is developed from the Pythagorean theorem.

84. Explain how you might remember the midpoint formula to find the midpoint of the line segment between (x_1, y_1) and (x_2, y_2).

85. Explain how to find the x- and y-intercepts from an equation in the variables x and y.

86. Given an equation in the variables x and y, what does the graph of the equation represent?

Expanding Your Skills

A point in three-dimensional space can be represented in a three-dimensional coordinate system. In such a case, a z-axis is taken perpendicular to both the x- and y-axes. A point P is assigned an ordered triple $P(x, y, z)$ relative to a fixed origin where the three axes meet. For Exercises 87–90, determine the distance between the two given points in space. Use the distance formula $d = \sqrt{(x_2 - x_1)^2 + (y_2 - y_1)^2 + (z_2 - z_1)^2}$.

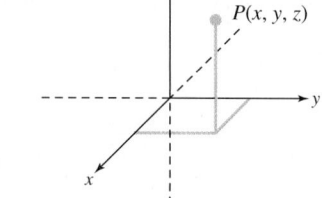

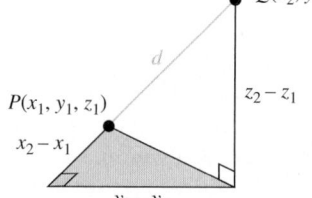

87. (5, −3, 2) and (4, 6, −1)

88. (6, −4, −1) and (2, 3, 1)

89. (3, 7, −2) and (0, −5, 1)

90. (9, −5, −3) and (2, 0, 1)

Objective 5: Graph Equations Using a Graphing Utility (Technology Connections)

91. What is meant by a viewing window on a graphing device?

92. Which of the viewing windows would show both the x- and y-intercepts of the graph of $780x - 42y = 5460$?
   a. [−20, 20, 2] by [−40, 40, 10]
   b. [−10, 10, 1] by [−10, 10, 1]
   c. [−10, 10, 1] by [−10, 150, 10]
   d. [−10, 10, 1] by [−150, 10, 10]

For Exercises 93–96, graph the equation with a graphing utility on the given viewing window. **(See Example 9)**

93. $y = 2x - 5$ on [−10, 10, 1] by [−10, 10, 1]

94. $y = -4x + 1$ on [−10, 10, 1] by [−10, 10, 1]

95. $y = 1400x^2 - 1200x$ on [−5, 5, 1] by [−1000, 2000, 500]

96. $y = -800x^2 + 600x$ on [−5, 5, 1] by [−1000, 500, 200]

For Exercises 97–98, graph the equations on the standard viewing window. **(See Example 9)**

97. a. $y = x^3$
   b. $y = |x| - 9$

98. a. $y = \sqrt{x + 4}$
   b. $y = |x - 2|$

SECTION 5.2 Circles

OBJECTIVES

1. Write an Equation of a Circle in Standard Form
2. Write the General Form of an Equation of a Circle

1. Write an Equation of a Circle in Standard Form

In addition to graphing equations by plotting points, we will learn to recognize specific categories of equations and the characteristics of their graphs. We begin by presenting the definition of a circle.

> **Definition of a Circle**
>
> A **circle** is the set of all points in a plane that are equidistant from a fixed point called the **center**. The fixed distance from any point on the circle to the center is called the **radius**.

The radius of a circle is often denoted by r, where $r > 0$. It is also important to note that the center is not actually part of the graph of a circle. It will be drawn in the text as an open dot for reference only.

Suppose that a circle is centered at the point (h, k) and has radius r (Figure 5-11). The distance formula can be used to derive an equation of the circle. Let (x, y) be an arbitrary point on the circle. Then by definition the distance between (h, k) and (x, y) must be r.

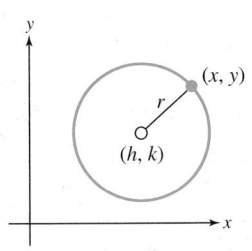

Figure 5-11

Apply the distance formula. $\sqrt{(x_2 - x_1)^2 + (y_2 - y_1)^2} = d$

$\sqrt{(x - h)^2 + (y - k)^2} = r$ Distance between (h, k) and (x, y)

$(x - h)^2 + (y - k)^2 = r^2$ Squaring both sides of the equation results in the standard form of an equation of a circle.

> **Standard Form of an Equation of a Circle**
>
> Given a circle centered at (h, k) with radius r, the **standard form** of an equation of the circle (also called **center-radius form**) is given by
>
> $$(x - h)^2 + (y - k)^2 = r^2 \quad \text{where } r > 0.$$

Examples	Standard form	Center	Radius
$(x - 4)^2 + (y + 3)^2 = 25$	$(x - 4)^2 + [y - (-3)]^2 = (5)^2$	$(4, -3)$	5
$x^2 + (y - \frac{1}{2})^2 = 12$	$(x - 0)^2 + (y - \frac{1}{2})^2 = (\sqrt{12})^2$	$(0, \frac{1}{2})$	$2\sqrt{3}$
$x^2 + y^2 = 7$	$(x - 0)^2 + (y - 0)^2 = (\sqrt{7})^2$	$(0, 0)$	$\sqrt{7}$

EXAMPLE 1   Writing an Equation of a Circle in Standard Form

a. Write the standard form of an equation of the circle with center $(-4, 6)$ and radius 2.

b. Graph the circle.

Solution:

a. $(h, k) = (-4, 6)$ and $r = 2$ Label the center (h, k) and the radius r.

$[x - (-4)]^2 + (y - 6)^2 = (2)^2$ Standard form: $(x - h)^2 + (y - k)^2 = r^2$

$(x + 4)^2 + (y - 6)^2 = 4$ Simplify.

b.

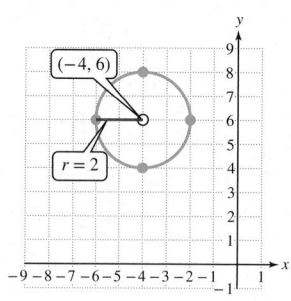

To graph the circle, first locate the center $(-4, 6)$ and draw a small open dot. Then plot points $r = 2$ units to the left, right, above, and below the center.

Draw the circle through the points.

Skill Practice 1

a. Write an equation of the circle with center $(3, -1)$ and radius 4.

b. Graph the circle.

EXAMPLE 2   Writing an Equation of a Circle in Standard Form

Write the standard form of an equation of the circle with endpoints of a diameter $(-1, 0)$ and $(3, 4)$.

Solution:

A sketch of this scenario is given in Figure 5-12. Notice that the midpoint of the diameter is the center of the circle.

$(-1, 0)$ and $(3, 4)$

(x_1, y_1) and (x_2, y_2) Label the points.

The center is $\left(\dfrac{-1 + 3}{2}, \dfrac{0 + 4}{2}\right) = (1, 2)$.

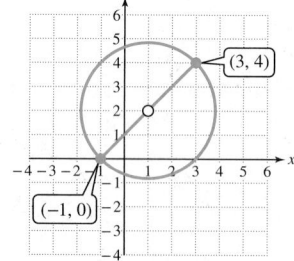

Figure 5-12

The radius of the circle is the distance between either endpoint of the diameter and the center. Using the endpoint $(-1, 0)$ as (x_1, y_1) and the center $(1, 2)$ as (x_2, y_2), apply the distance formula.

$$d = \sqrt{(x_2 - x_1)^2 + (y_2 - y_1)^2}$$

$$r = \sqrt{[1 - (-1)]^2 + (2 - 0)^2} = \sqrt{(2)^2 + (2)^2} = \sqrt{8}$$

An equation of the circle is: $(x - h)^2 + (y - k)^2 = r^2$.

$$(x - 1)^2 + (y - 2)^2 = \left(\sqrt{8}\right)^2$$

$$(x - 1)^2 + (y - 2)^2 = 8 \quad \text{(Standard form)}$$

Skill Practice 2   Write the standard form of an equation of the circle with endpoints of a diameter $(-3, 3)$ and $(-1, -1)$.

Answers

1. a. $(x - 3)^2 + (y + 1)^2 = 16$

    b.

2. $(x + 2)^2 + (y - 1)^2 = 5$

2. Write the General Form of an Equation of a Circle

In Example 2 we have the equation $(x - 1)^2 + (y - 2)^2 = 8$. If we expand the binomials and combine like terms, we can write the equation in *general form*.

$$(x - 1)^2 + (y - 2)^2 = 8 \qquad \text{Standard form (center-radius form)}$$

$$x^2 - 2x + 1 + y^2 - 4y + 4 = 8 \qquad \text{Expand the binomials}$$

$$x^2 + y^2 - 2x - 4y - 3 = 0 \qquad \text{General form}$$

Recall that the square of a binomial results in a perfect square trinomial.

$$(a + b)^2 = a^2 + 2ab + b^2$$
$$(a - b)^2 = a^2 - 2ab + b^2$$

Thus,
$$(x - 1)^2 = x^2 - 2(x)(1) + (1)^2$$
$$= x^2 - 2x + 1$$
$$(y - 2)^2 = y^2 - 2(y)(2) + (2)^2$$
$$= y^2 - 4y + 4$$

See Section 2.2 for review.

For an expression of the form
$$x^2 + bx + n$$
complete the square by adding the appropriate constant n so that the resulting expression is a perfect square trinomial. This in turn will factor as the square of a binomial.

The constant to be added is one-half the linear term coefficient, squared.

$$n = \left[\frac{1}{2}(b) \right]^2$$

For review see Section 3.5.

General Form of an Equation of a Circle

An equation of a circle written in the form $x^2 + y^2 + Ax + By + C = 0$ is called the **general form** of an equation of a circle.

By completing the square we can write an equation of a circle given in general form as an equation in standard form. The purpose of writing an equation of a circle in standard form is to identify the radius and center. This is demonstrated in Example 3.

EXAMPLE 3   Writing an Equation of a Circle in Standard Form

Write the equation of the circle in standard form. Then identify the center and radius.

$$x^2 + y^2 + 10x - 6y + 25 = 0$$

Solution:

$$x^2 + y^2 + 10x - 6y + 25 = 0$$

$(x^2 + 10x + \square) + (y^2 - 6y + \square) = -25$ Group the x terms. Group the y terms. Move the constant term to the right.

$(x^2 + 10x + 25) + (y^2 - 6y + 9) = -25 + 25 + 9$

Complete the square.
Note: $\left[\frac{1}{2}(10) \right]^2 = [5]^2 = 25$,
$\left[\frac{1}{2}(-6) \right]^2 = [-3]^2 = 9$

$(x + 5)^2 + (y - 3)^2 = 9$ Factor.

The center is $(-5, 3)$, and the radius is $\sqrt{9} = 3$. See Figure 5-13.

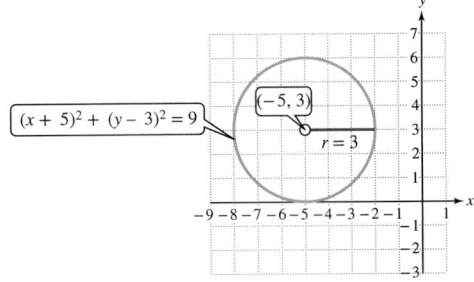

Figure 5-13

Skill Practice 3   Write the equation of the circle in standard form. Then identify the center and radius. $x^2 + y^2 - 8x + 2y - 8 = 0$

Not all equations of the form $x^2 + y^2 + Ax + By + C = 0$ represent the graph of a circle. Completing the square results in an equation of the form $(x - h)^2 + (y - k)^2 = c$, where c is a constant. In the case where $c > 0$, the graph of the equation is a circle with radius $r = \sqrt{c}$. However, if $c = 0$, or if $c < 0$, the graph will be a single point or nonexistent. These are called **degenerate cases**.

- If $c > 0$, then the graph will be a circle with radius $r = \sqrt{c}$.
- If $c = 0$, then the graph will be a single point, (h, k). The solution set is $\{(h, k)\}$.
- If $c < 0$, then the solution set is the empty set $\{ \ \}$.

Answer

3. $(x - 4)^2 + (y + 1)^2 = 25$;
  Center: $(4, -1)$; Radius: 5

EXAMPLE 4 Determining if an Equation Represents the Graph of a Circle

Write the equation in the form $(x - h)^2 + (y - k)^2 = r^2$, and identify the solution set.

$$x^2 + y^2 - 14y + 49 = 0$$

Solution:

$x^2 + y^2 - 14y + 49 = 0$

$x^2 + (y^2 - 14y \quad) = -49$ Group the y terms and complete the square. Note that the x^2 term is already a perfect square: $(x - 0)^2$.

$x^2 + (y^2 - 14y + 49) = -49 + 49$ Complete the square: $[\frac{1}{2}(-14)]^2 = [-7]^2 = 49$.

$x^2 + (y - 7)^2 = 0$ Factor.

Since $r^2 = 0$, the solution set is $\{(0, 7)\}$. The sum of two squares will equal zero only if each individual term is zero. Therefore, $x = 0$ and $y = 7$.

Skill Practice 4 Write the equation in the form $(x - h)^2 + (y - k)^2 = r^2$, and identify the solution set. $x^2 + y^2 + 2x + 5 = 0$

TECHNOLOGY CONNECTIONS

Setting a Square Viewing Window and Graphing a Circle

A graphing calculator expects an equation with the y variable isolated. Therefore, to graph an equation of a circle such as $(x + 5)^2 + (y - 3)^2 = 9$, from Example 3, we first solve for y.

$(x + 5)^2 + (y - 3)^2 = 9$

$(y - 3)^2 = 9 - (x + 5)^2$

$y - 3 = \pm\sqrt{9 - (x + 5)^2}$

$y = 3 \pm \sqrt{9 - (x + 5)^2}$

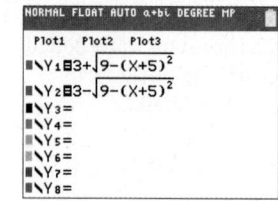

Notice that the graph looks more oval-shaped than circular. This is because the calculator has a rectangular screen. If the scaling is the same on the x- and y-axes, the graph will appear elongated horizontally. To eliminate this distortion, use a **ZSquare** option, located in the **Zoom** menu.

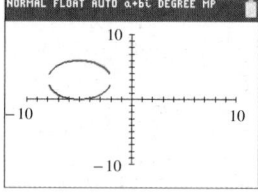

Also notice that the calculator display may not show the upper and lower semicircles connecting. The viewing window between $x = -16.1$ and $x = 16.1$ is divided by the number of pixels displayed horizontally to get the values of x used to graph the equation. These may not include x values at the leftmost and rightmost points on the circle. That is, the calculator may graph points *close* to $(-8, 3)$ and $(-2, 3)$ but not exactly at $(-8, 3)$ and $(-2, 3)$. Therefore, the upper and lower semicircles may not "hook up."

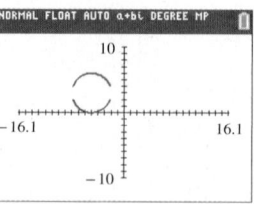

Answer

4. $(x + 1)^2 + y^2 = -4$; The solution set is $\{\ \}$.

SECTION 5.2   Practice Exercises

Prerequisite Review

For Exercises R.1–R.6, find the value of n so that the expression is a perfect square trinomial. Then factor the trinomial. (See Section 3.5 for review.)

R.1. $x^2 + 28x + n$

R.2. $v^2 - 20v + n$

R.3. $a^2 - 11a + n$

R.4. $x^2 + 3x + n$

R.5. $m^2 - \dfrac{2}{9}m + n$

R.6. $w^2 + \dfrac{3}{5}w + n$

For Exercises R.7–R.8, determine the distance between the two points. (See Section 5.1 for review.)

R.7. $(-2, 7)$ and $(4, -5)$

R.8. $(1, 10)$ and $(-2, 4)$

For Exercises R.9–R.12, multiply. (See Section 2.2 for review.)

R.9. $(x - 9)^2$

R.10. $(x + 8)^2$

R.11. $\left(x + \dfrac{2}{3}\right)^2$

R.12. $\left(x - \dfrac{5}{4}\right)^2$

For Exercises R.13–R.16, factor completely. (See Section 2.4 for review.)

R.13. $y^2 - 26y + 169$

R.14. $x^2 - 22x + 121$

R.15. $\dfrac{1}{100}x^2 + \dfrac{1}{35}x + \dfrac{1}{49}$

R.16. $\dfrac{1}{25}y^2 + \dfrac{1}{5}y + \dfrac{1}{4}$

For Exercises R.17–R.18, solve for y. (See Section 3.5 for review.)

R.17. $x^2 + y^2 = 9$

R.18. $(x - 4)^2 + y^2 = 25$

Concept Connections

1. A _____ is the set of all points in a plane equidistant from a fixed point called the _____.

2. The distance from the center of a circle to any point on the circle is called the _____ and is often denoted by r.

3. The standard form of an equation of a circle with center (h, k) and radius r is given by _____.

4. An equation of a circle written in the form $x^2 + y^2 + Ax + By + C = 0$ is called the _____ form of an equation of a circle.

Objective 1: Write an Equation of a Circle in Standard Form

5. Is the point $(2, 7)$ on the circle defined by $(x - 2)^2 + (y - 7)^2 = 4$?

6. Is the point $(3, 5)$ on the circle defined by $(x - 3)^2 + (y - 5)^2 = 36$?

7. Is the point $(-4, 7)$ on the circle defined by $(x + 1)^2 + (y - 3)^2 = 25$?

8. Is the point $(2, -7)$ on the circle defined by $(x + 6)^2 + (y + 1)^2 = 100$?

For Exercises 9–16, determine the center and radius of the circle.

9. $(x - 4)^2 + (y + 2)^2 = 81$

10. $(x + 3)^2 + (y - 1)^2 = 16$

11. $x^2 + (y - 2.5)^2 = 6.25$

12. $(x - 1.5)^2 + y^2 = 2.25$

13. $x^2 + y^2 = 20$

14. $x^2 + y^2 = 28$

15. $\left(x - \dfrac{3}{2}\right)^2 + \left(y + \dfrac{3}{4}\right)^2 = \dfrac{81}{49}$

16. $\left(x + \dfrac{1}{7}\right)^2 + \left(y - \dfrac{3}{5}\right)^2 = \dfrac{25}{9}$

For Exercises 17–32, information about a circle is given.

a. Write an equation of the circle in standard form.

b. Graph the circle. (**See Examples 1–2**)

17. Center: $(-2, 5)$; Radius: 1

18. Center: $(-3, 2)$; Radius: 4

19. Center: $(-4, 1)$; Radius: 3

20. Center: $(6, -2)$; Radius: 6

21. Center: $(-4, -3)$; Radius: $\sqrt{11}$

22. Center: $(-5, -2)$; Radius: $\sqrt{21}$

23. Center: $(0, 0)$; Radius: 2.6

24. Center: $(0, 0)$; Radius: 4.2

25. The endpoints of a diameter are $(-2, 4)$ and $(6, -2)$.

26. The endpoints of a diameter are $(7, 3)$ and $(5, -1)$.

27. The center is $(-2, -1)$ and another point on the circle is $(6, 5)$.

28. The center is $(3, 1)$ and another point on the circle is $(6, 5)$.

29. The center is $(4, 6)$ and the circle is tangent to the y-axis. (Informally, a line is tangent to a circle if it touches the circle in exactly one point.)

30. The center is $(-2, -4)$ and the circle is tangent to the x-axis.

31. The center is in Quadrant IV, the radius is 5, and the circle is tangent to both the x- and y-axes.

32. The center is in Quadrant II, the radius is 3, and the circle is tangent to both the x- and y-axes.

33. Write an equation that represents the set of points that are 5 units from $(8, -11)$.

34. Write an equation that represents the set of points that are 9 units from $(-4, 16)$.

35. Write an equation of the circle that is tangent to both axes with radius $\sqrt{7}$ and center in Quadrant I.

36. Write an equation of the circle that is tangent to both axes with radius $\sqrt{11}$ and center in Quadrant III.

Objective 2: Write the General Form of an Equation of a Circle

37. Determine the solution set for the equation $(x + 1)^2 + (y - 5)^2 = 0$.

38. Determine the solution set for the equation $(x - 3)^2 + (y + 12)^2 = 0$.

39. Determine the solution set for the equation $(x - 17)^2 + (y + 1)^2 = -9$.

40. Determine the solution set for the equation $(x + 15)^2 + (y - 3)^2 = -25$.

For Exercises 41–54, write the equation in the form $(x - h)^2 + (y - k)^2 = c$. Then if the equation represents a circle, identify the center and radius. If the equation represents a degenerate case, give the solution set. (**See Examples 3–4**)

41. $x^2 + y^2 + 6x - 2y + 6 = 0$

42. $x^2 + y^2 + 12x - 14y + 84 = 0$

43. $x^2 + y^2 - 22x + 6y + 129 = 0$

44. $x^2 + y^2 - 10x + 4y - 20 = 0$

45. $x^2 + y^2 - 20y - 4 = 0$

46. $x^2 + y^2 + 22x - 4 = 0$

47. $10x^2 + 10y^2 - 80x + 200y + 920 = 0$
(*Hint*: Divide by 10 to make the x^2 and y^2 term coefficients equal to 1.)

48. $2x^2 + 2y^2 - 32x + 12y + 90 = 0$

49. $x^2 + y^2 - 4x - 18y + 89 = 0$

50. $x^2 + y^2 - 10x - 22y + 155 = 0$

51. $4x^2 + 4y^2 - 20y + 25 = 0$

52. $4x^2 + 4y^2 - 12x + 9 = 0$

53. $x^2 + y^2 - x - \dfrac{3}{2}y - \dfrac{3}{4} = 0$

54. $x^2 + y^2 - \dfrac{2}{3}x - \dfrac{5}{3}y - \dfrac{5}{9} = 0$

Mixed Exercises

55. A cell tower is a site where antennas, transmitters, and receivers are placed to create a cellular network. Suppose that a cell tower is located at a point $A(4, 6)$ on a map and its range is 1.5 mi. Write an equation that represents the boundary of the area that can receive a signal from the tower. Assume that all distances are in miles.

56. A radar transmitter on a ship has a range of 20 nautical miles. If the ship is located at a point $(-32, 40)$ on a map, write an equation for the boundary of the area within the range of the ship's radar. Assume that all distances on the map are represented in nautical miles.

57. Suppose that three geological study areas are set up on a map at points $A(-4, 12)$, $B(11, 3)$, and $C(0, 1)$, where all units are in miles. Based on the speed of compression waves, scientists estimate the distances from the study areas to the epicenter of an earthquake to be 13 mi, 5 mi, and 10 mi, respectively. Graph three circles whose centers are located at the study areas and whose radii are the given distances to the earthquake. Then estimate the location of the earthquake.

58. Three fire observation towers are located at points $A(-6, -14)$, $B(14, 10)$, and $C(-3, 13)$ on a map where all units are in kilometers. A fire is located at distances of 17 km, 15 km, and 13 km, respectively, from the observation towers. Graph three circles whose centers are located at the observation towers and whose radii are the given distances to the fire. Then estimate the location of the fire.

Write About It

59. State the definition of a circle.

60. What are the advantages of writing an equation of a circle in standard form?

Expanding Your Skills

61. Find all values of y such that the distance between $(4, y)$ and $(-2, 6)$ is 10 units.

62. Find all values of x such that the distance between $(x, -1)$ and $(4, 2)$ is 5 units.

63. Find all points on the line $y = x$ that are 6 units from $(2, 4)$.

64. Find all points on the line $y = -x$ that are 4 units from $(-4, 6)$.

The general form of an equation of a circle is $(x - h)^2 + (y - k)^2 = r^2$. If we solve the equation for x we get equations of the form $x = h \pm \sqrt{r^2 - (y - k)^2}$. The equation $x = h + \sqrt{r^2 - (y - k)^2}$ represents the graph of the corresponding right-side semicircle, and the equation $x = h - \sqrt{r^2 - (y - k)^2}$ represents the graph of the left-side semicircle. Likewise, if we solve for y, we have $y = k \pm \sqrt{r^2 - (x - h)^2}$. These equations represent the top and bottom semicircles. For Exercises 65–68, graph the equations.

65. a. $y = \sqrt{16 - x^2}$
    b. $y = -\sqrt{16 - x^2}$
    c. $x = \sqrt{16 - y^2}$
    d. $x = -\sqrt{16 - y^2}$

66. a. $y = \sqrt{9 - x^2}$
    b. $y = -\sqrt{9 - x^2}$
    c. $x = \sqrt{9 - y^2}$
    d. $x = -\sqrt{9 - y^2}$

67. a. $x = -1 - \sqrt{9 - (y - 2)^2}$
    b. $x = -1 + \sqrt{9 - (y - 2)^2}$
    c. $y = 2 - \sqrt{9 - (x + 1)^2}$
    d. $y = 2 + \sqrt{9 - (x + 1)^2}$

68. a. $x = 3 - \sqrt{4 - (y + 2)^2}$
    b. $x = 3 + \sqrt{4 - (y + 2)^2}$
    c. $y = -2 - \sqrt{4 - (x - 3)^2}$
    d. $y = -2 + \sqrt{4 - (x - 3)^2}$

69. Find the shortest distance from the origin to a point on the circle defined by $x^2 + y^2 - 6x - 12y + 41 = 0$.

70. Find the shortest distance from the origin to a point on the circle defined by $x^2 + y^2 + 4x - 12y + 31 = 0$.

Technology Connections

For Exercises 71–74, use a graphing calculator to graph the circles on an appropriate square viewing window.

71. $x^2 + y^2 = 36$

72. $x^2 + y^2 = 49$

73. $(x - 18)^2 + (y + 20)^2 = 80$

74. $(x + 0.04)^2 + (y - 0.02)^2 = 0.01$

SECTION 5.3   Functions and Relations

1. Determine Whether a Relation Is a Function

In the physical world, many quantities that are subject to change are related to other variables. For example:

- The cost of mailing a package is related to the weight of a package.
- The minimum braking distance of a car depends on the speed of the car.
- The perimeter of a rectangle is a function of its length and width.
- The test score that a student earns is related to the number of hours of study.

Erik Isakson/Getty Images

In mathematics we can express the relationship between two values as a set of ordered pairs.

> **Definition of a Relation**
>
> A set of ordered pairs (x, y) is called a **relation** in x and y.
> - The set of x values in the ordered pairs is called the **domain** of the relation.
> - The set of y values in the ordered pairs is called the **range** of the relation.

EXAMPLE 1   Writing a Relation from Observed Data Points

Table 5-1 shows the score y that a student earned on an algebra test based on the number of hours x spent studying one week prior to the test.

a. Write the set of ordered pairs that defines the relation given in Table 5-1.

b. Write the domain.

c. Write the range.

Hours of Study, x	Test Score, y
8	92
3	58
11	98
5	72
8	86

Table 5-1

Solution:

a. Relation: $\{(8, 92), (3, 58), (11, 98), (5, 72), (8, 86)\}$

b. Domain: $\{8, 3, 11, 5\}$

c. Range: $\{92, 58, 98, 72, 86\}$

Avoiding Mistakes

Do not list the elements in a set more than once. The value 8 is listed in the domain one time only.

Skill Practice 1   For the table shown,

a. Write the set of ordered pairs that defines the relation.

b. Write the domain.

c. Write the range.

x	3	−2	5	1
y	−4	0	3	0

The data in Table 5-1 show two different test scores for 8 hr of study. That is, for $x = 8$, there are two different y values. In many applications, we prefer to work with relations that assign one and only one y value for a given value of x. Such a relation is called a function.

Answers

1. a. $\{(3, -4), (-2, 0), (5, 3), (1, 0)\}$
   b. Domain: $\{3, -2, 5, 1\}$
   c. Range: $\{-4, 0, 3\}$

> **Definition of a Function**
>
> Given a relation in x and y, we say that **y is a function of x** if for each value of x in the domain, there is exactly one value of y in the range.

Determining if a Relation Is a Function

Determine if the relation defines y as a function of x.

a. $\{(3, 1), (2, 5), (-4, 2), (-1, 0), (3, -4)\}$

b. $\{(-1, 4), (2, 3), (3, 4), (-4, 5)\}$

Solution:

a.

same x values

$\{(3, 1), (2, 5), (-4, 2), (-1, 0), (3, -4)\}$

different y values

When $x = 3$, there are two different y values: $y = 1$ and $y = -4$.

This relation is *not* a function.

b. $\{(-1, 4), (2, 3), (3, 4), (-4, 5)\}$

This relation *is* a function.

No two ordered pairs have the same x value but different y values.

TIP A function may not have the same x value paired with different y values. However, it is acceptable for a function to have two or more x values paired with the same y value, as shown in Example 2(b).

Skill Practice 2   Determine if the relation defines y as a function of x.

a. $\{(8, 4), (3, -1), (5, 4)\}$ **b.** $\{(-3, 2), (9, 5), (1, 0), (-3, 1)\}$

A relation that is not a function has at least one domain element x paired with more than one range element y. For example, the ordered pairs $(3, 1)$ and $(3, -4)$ do not make up a function. On a graph, these two points are aligned vertically. A vertical line drawn through one point also intersects the other point (Figure 5-14). This observation leads to the vertical line test.

Using the Vertical Line Test

Consider a relation defined by a set of points (x, y) graphed on a rectangular coordinate system. The graph defines y as a function of x if no vertical line intersects the graph in more than one point.

Figure 5-14

Applying the Vertical Line Test

The graphs of three relations are shown in blue. In each case, determine if the relation defines y as a function of x.

Solution:

a.

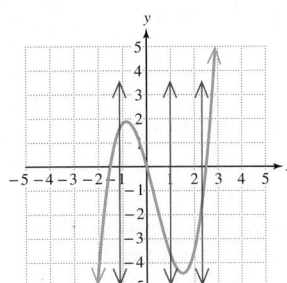

b.

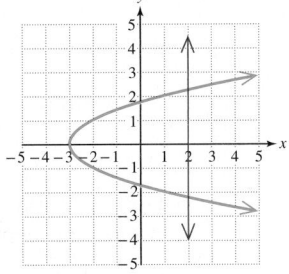

c.

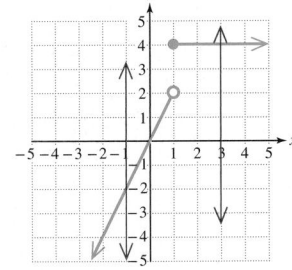

TIP In Example 3(c) there is only one y value assigned to $x = 1$. This is because the point $(1, 2)$ is *not* included in the graph of the function as denoted by the open dot.

This is a function.

No vertical line intersects the graph in more than one point.

This is not a function.

There is at least one vertical line that intersects the graph in more than one point.

This is a function.

No vertical line intersects the graph in more than one point.

Answers

2. a. Yes   **b.** No

Skill Practice 3   Determine if the given relation defines y as a function of x.

a.   **b.**   **c.**

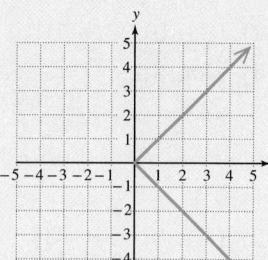   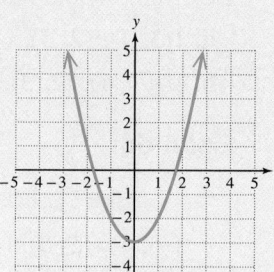   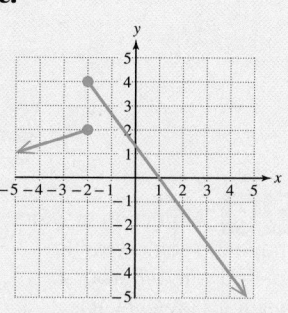

A relation can also be defined by a figure showing a "mapping" between x and y, or by an equation in x and y.

EXAMPLE 4   Determining if a Relation Is a Function

Determine if the relation defines y as a function of x.

a. x y **b.** $y^2 = x$ **c.** $(x - 2)^2 + (y + 1)^2 = 9$

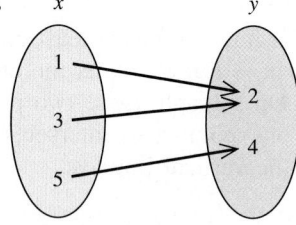

Solution:

a. This mapping defines the set of ordered pairs: $\{(1, 2), (3, 2), (5, 4)\}$.

This relation *is* a function.

No two ordered pairs have the same x value but different y values.

b. $y^2 = x$

$y = \pm \sqrt{x}$

Solve the equation for y. For any $x > 0$, there are two corresponding y values.

x	y	Ordered pairs
0	0	$(0, 0)$
1	1, −1	$(1, 1), (1, -1)$
4	2, −2	$(4, 2), (4, -2)$
9	3, −3	$(9, 3), (9, -3)$

This relation is *not* a function.

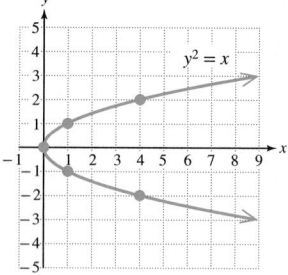

c. $(x - 2)^2 + (y + 1)^2 = 9$

This equation represents the graph of a circle with center $(2, -1)$ and radius 3.

This relation is *not* a function because it fails the vertical line test.

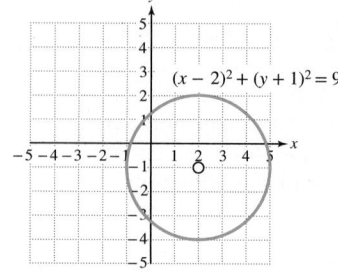

Skill Practice 4   Determine if the relation defines y as a function of x.

a.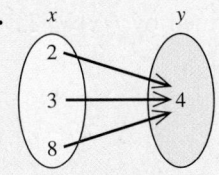
b. $|y + 1| = x$
c. $x^2 + y^2 = 25$

2. Apply Function Notation

A function may be defined by an equation with two variables. For example, the equation $y = x - 2$ defines y as a function of x. This is because for any real number x, the value of y is the unique number that is 2 less than x.

When a function is defined by an equation, we often use function notation. For example, the equation $y = x - 2$ may be written in function notation as

$$f(x) = x - 2 \text{ read as "} f \text{ of } x \text{ equals } x - 2.\text{"}$$

With function notation,

- f is the name of the function,
- x is an input variable from the domain,
- $f(x)$ is the function value (or y value) corresponding to x.

A function may be evaluated at different values of x by using substitution.

$$f(x) = x - 2$$
$$f(4) = (4) - 2 = 2 \qquad f(4) = 2 \text{ can be interpreted as } (4, 2).$$
$$f(1) = (1) - 2 = -1 \qquad f(1) = -1 \text{ can be interpreted as } (1, -1).$$

EXAMPLE 5   **Evaluating a Function**

Evaluate the function defined by $g(x) = 2x + 1$ for the given values of x.

a. $g(-2)$ **b.** $g(-1)$ **c.** $g(0)$ **d.** $g(1)$ **e.** $g(2)$

Solution:

a. $g(-2) = 2(-2) + 1$ Substitute -2 for x.
      $= -3$ $g(-2) = -3$

b. $g(-1) = 2(-1) + 1$ Substitute -1 for x.
      $= -1$ $g(-1) = -1$

c. $g(0) = 2(0) + 1$ Substitute 0 for x.
      $= 1$ $g(0) = 1$

d. $g(1) = 2(1) + 1$ Substitute 1 for x.
      $= 3$ $g(1) = 3$

e. $g(2) = 2(2) + 1$ Substitute 2 for x.
      $= 5$ $g(2) = 5$

The function values represent the ordered pairs $(-2, -3)$, $(-1, -1)$, $(0, 1)$, $(1, 3)$, and $(2, 5)$. The line through the points represents all ordered pairs defined by this function. This is the graph of the function.

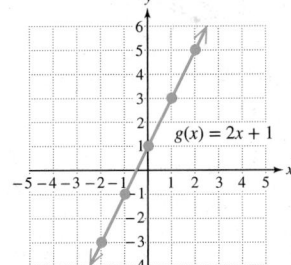

Skill Practice 5   Evaluate the function defined by $h(x) = 4x - 3$ for the given values of x.

a. $h(-3)$ **b.** $h(-1)$ **c.** $h(0)$ **d.** $h(1)$ **e.** $h(3)$

EXAMPLE 6 Evaluating a Function

Evaluate the function defined by $f(x) = 3x^2 + 2x$ for the given values of x.

a. $f(a)$ **b.** $f(x + h)$

Solution:

a. $f(a) = 3a^2 + 2a$ Substitute a for x.

b. $f(x + h) = 3(x + h)^2 + 2(x + h)$ Substitute $x + h$ for x.

$\qquad\qquad = 3(x^2 + 2xh + h^2) + 2x + 2h$ Simplify.

$\qquad\qquad\qquad\qquad\qquad\qquad\qquad\qquad$ Recall: $(a + b)^2 = a^2 + 2ab + b^2$

$\qquad\qquad = 3x^2 + 6xh + 3h^2 + 2x + 2h$

Skill Practice 6 Evaluate the function defined by $f(x) = -x^2 + 4x$ for the given values of x.

a. $f(t)$ **b.** $f(a + h)$

3. Determine x- and y-Intercepts of a Function Defined by $y = f(x)$

Recall that to find an x-intercept(s) of the graph of an equation, we substitute 0 for y in the equation and solve for x. Using function notation, $y = f(x)$, this is equivalent to finding the real solutions of the equation $f(x) = 0$. To find the y-intercept, substitute 0 for x and solve the equation for y. Using function notation, this is equivalent to finding $f(0)$.

Finding Intercepts Using Function Notation

Given a function defined by $y = f(x)$,

- The x-intercepts are the real solutions to the equation $f(x) = 0$.
- The y-intercept is given by $f(0)$.

EXAMPLE 7 Finding the x- and y-Intercepts of a Function

Find the x- and y-intercepts of the function defined by $f(x) = x^2 - 4$.

Solution:

To find the x-intercept(s), solve the equation $f(x) = 0$.

$f(x) = x^2 - 4$

$0 = x^2 - 4$

$x^2 = 4$

$x = \pm\sqrt{4}$ Apply the square root property.

$x = \pm 2$ The x-intercepts are $(2, 0)$ and $(-2, 0)$.

To find the y-intercept, evaluate $f(0)$.

$f(0) = (0)^2 - 4$

$\quad = -4$ The y-intercept is $(0, -4)$.

Skill Practice 7 Find the x- and y-intercepts of the function defined by $f(x) = |x| - 5$.

Answers

6. a. $f(t) = -t^2 + 4t$

b. $f(a + h)$

$\quad = -a^2 - 2ah - h^2 + 4a + 4h$

7. x-intercepts: $(5, 0)$ and $(-5, 0)$;

y-intercept: $(0, -5)$

4. Determine Domain and Range of a Function

Given a relation defining y as a function of x, the **domain** is the set of x values in the function, and the **range** is the set of y values in the function. In Example 8, we find the domain and range from the graph of a function.

EXAMPLE 8   **Determining Domain and Range**

Determine the domain and range for the functions shown.

a.   b.   c.

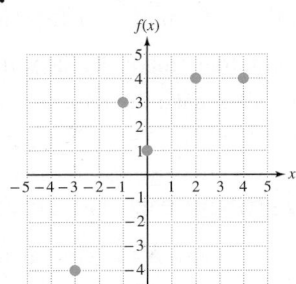   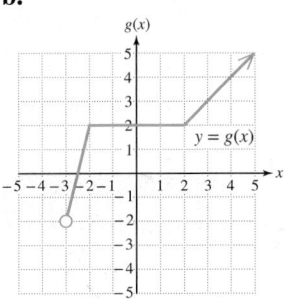   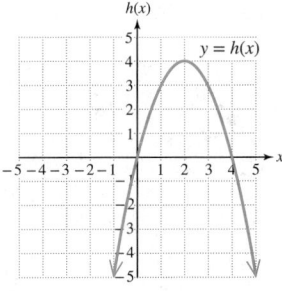

Solution:

a. The graph defines the set of ordered pairs:
   $\{(-3, -4), (-1, 3), (0, 1), (2, 4), (4, 4)\}$

   Domain: $\{-3, -1, 0, 2, 4\}$ The domain is the set of x values.

   Range: $\{-4, 1, 3, 4\}$ The range is the set of y values.

b.

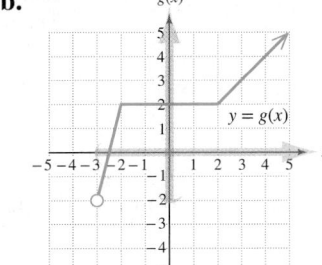

The domain is shown on the x-axis in green tint.
Domain: $\{x \mid x > -3\}$ or in interval notation: $(-3, \infty)$.

The range is shown on the y-axis in red tint.
Range: $\{y \mid y > -2\}$ or in interval notation: $(-2, \infty)$.

c.

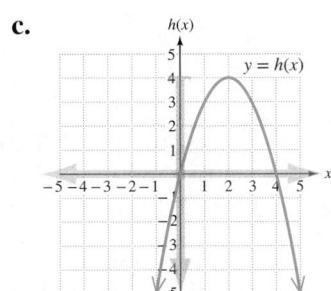

The graph extends infinitely far downward and infinitely far to the left and right. Therefore, the domain is the set of all real numbers, x.

The domain is shown on the x-axis in green tint.
Domain: $\mathbb{R}$ or in interval notation: $(-\infty, \infty)$.

The range is shown on the y-axis in red tint.
Range: $\{y \mid y \leq 4\}$ or in interval notation: $(-\infty, 4]$.

Skill Practice 8   Determine the domain and range for the functions shown.

a.   b.

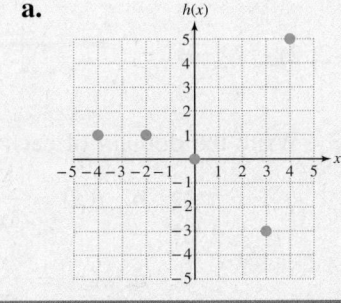   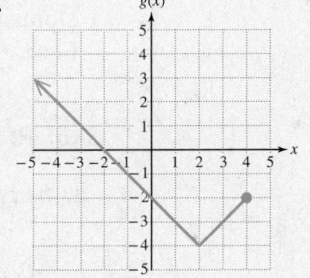

Answers

8. a. Domain: $\{-4, -2, 0, 3, 4\}$;
   Range: $\{-3, 0, 1, 5\}$
   b. Domain: $\{x \mid x \leq 4\}$ or $(-\infty, 4]$;
   Range: $\{y \mid y \geq -4\}$ or $[-4, \infty)$

In some cases, a function may have restrictions on the domain. For example, consider the function defined by

$$f(x) = x^2 + 2 \quad \text{for} \quad x \geq 0$$

The restriction on x (that is, $x \geq 0$) is explicitly stated along with the definition of the function. If no such restriction is stated, then by default, the domain is all real numbers that when substituted into the function produce real numbers in the range.

Guidelines to Find Domain of a Function

To determine the implied domain of a function defined by $y = f(x)$,

- Exclude values of x that make the denominator of a fraction zero.
- Exclude values of x that make the radicand negative within an even-indexed root.

EXAMPLE 9   Determining the Domain of a Function

Write the domain of each function in interval notation.

a. $h(x) = \dfrac{3}{x+1}$ **b.** $f(x) = \dfrac{x+3}{2x-5}$ **c.** $g(x) = \dfrac{x}{x^2+4}$ **d.** $k(x) = \dfrac{x}{x^2-4}$

Solution:

TIP Given $h(x) = \dfrac{3}{x+1}$, solve the equation $x + 1 = 0$. The domain is the set of real numbers *excluding* the solution to this equation.

Likewise, for $f(x) = \dfrac{x+3}{2x-5}$, solve the equation $2x - 5 = 0$. The domain is the set of real numbers *excluding* $\frac{5}{2}$.

a. $h(x) = \dfrac{3}{x+1}$

The variable x has the restriction that $x + 1 \neq 0$.

Therefore, $x \neq -1$.

Domain: $(-\infty, -1) \cup (-1, \infty)$

Recall that division by zero is undefined. Therefore, the domain of h is all real numbers except those that make the denominator zero. Therefore, we require $x + 1 \neq 0$.

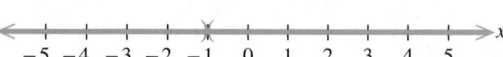

b. $f(x) = \dfrac{x+3}{2x-5}$

The variable x has the restriction that $2x - 5 \neq 0$.

Therefore, $x \neq \dfrac{5}{2}$.

Domain: $\left(-\infty, \dfrac{5}{2}\right) \cup \left(\dfrac{5}{2}, \infty\right)$

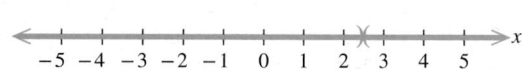

c. $g(x) = \dfrac{x}{x^2+4}$ Denominator always positive (never zero)

Domain: $(-\infty, \infty)$

The expression $x^2 \geq 0$ for all real numbers x. Therefore, $x^2 + 4 > 0$ for all real numbers x.

d. $k(x) = \dfrac{x}{x^2-4}$

Determine the value(s) of x that make the denominator zero so that we can exclude them from the domain.
$$x^2 - 4 = 0$$
$$(x-2)(x+2) = 0$$
$x = 2$ and $x = -2$ must be excluded from the domain.

Domain:
$(-\infty, -2) \cup (-2, 2) \cup (2, \infty)$

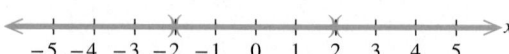

Skill Practice 9 Write the domain of each function in interval notation.

a. $h(x) = \dfrac{6}{x-7}$ **b.** $f(x) = \dfrac{x-2}{3x+1}$

c. $g(x) = \dfrac{x^2}{5}$ **d.** $k(x) = \dfrac{5}{x^2-3x-10}$

EXAMPLE 10 **Determining the Domain of a Function**

Write the domain of each function in interval notation.

a. $h(t) = \sqrt{2 - t}$ **b.** $n(t) = \dfrac{1}{\sqrt{2 - t}}$ **c.** $m(t) = |2 - t|$

Solution:

a. $h(t) = \sqrt{2 - t}$

$2 - t \geq 0$ The value $h(t)$ will be a real number provided that the radicand is nonnegative. Therefore, we require $2 - t \geq 0$.

$-t \geq -2$ Subtract 2 from both sides.

$t \leq 2$ Divide both sides by -1. Remember to reverse the direction of the inequality symbol.

Domain: $(-\infty, 2]$

b. $n(t) = \dfrac{1}{\sqrt{2 - t}}$

$2 - t > 0$

$-t > -2$

$t < 2$

There are two restrictions on the domain.

• The radicand must be nonnegative: $2 - t \geq 0$.
• The denominator must not equal zero: $\sqrt{2 - t} \neq 0$.
    The second condition is true if $2 - t \neq 0$.

The two conditions are *both* satisfied if $2 - t > 0$.

Domain: $(-\infty, 2)$

c. $m(t) = |2 - t|$

Domain: $(-\infty, \infty)$ There are no fractions or radicals that would restrict the domain. The expression $|2 - t|$ is a real number for all real numbers t.

Skill Practice 10   Write the domain of each function in interval notation.

a. $k(x) = \sqrt{x + 3}$ **b.** $r(x) = \dfrac{5}{\sqrt{x + 3}}$ **c.** $p(x) = 2x^2 + 3x$

TIP In Example 10(a) we're given $h(t) = \sqrt{2 - t}$. Notice that a value of t taken outside the domain of the function will produce an imaginary number. For example: $h(6) = \sqrt{2 - 6} = \sqrt{-4} = 2i$.

In this text, we work with *real-valued* functions only. Therefore, the domain is restricted to values that when "input" into a function, produce a real number as the "output." With this in mind, given $k(t) = \sqrt[3]{2 - t}$, the domain is all real numbers. With a cube root function, there are no restrictions on the domain because there is no danger of producing an imaginary number.

5. Interpret a Function Graphically

In Example 11, we will review the key concepts studied in this section by identifying characteristics of a function based on its graph.

FOR REVIEW

When solving an inequality, if you multiply or divide both sides by a negative number, remember to reverse the direction of the inequality symbol.

$-t \geq -2$

$\dfrac{-t}{-1} \leq \dfrac{-2}{-1}$

$t \leq 2$

If you do not multiply or divide by a negative number, then the inequality symbol remains unchanged.

$2x \geq 6$

$\dfrac{2x}{2} \geq \dfrac{6}{3}$

$x \geq 3$

Answers
10. a. $[-3, \infty)$
  b. $(-3, \infty)$
  c. $(-\infty, \infty)$

EXAMPLE 11   **Identifying Characteristics of a Function**

Use the function f pictured to answer the questions.

a. Determine $f(2)$.

b. Determine $f(-5)$.

c. Find all x for which $f(x) = 0$.

d. Find all x for which $f(x) = 3$.

e. Determine the x-intercept(s).

f. Determine the y-intercept.

g. Determine the domain of f.

h. Determine the range of f.

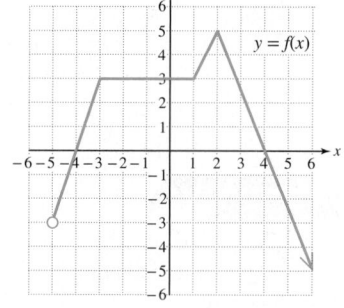

Solution:

a. $f(2) = 5$

$f(2) = 5$ because the function contains the point $(2, 5)$.

b. $f(-5)$ is not defined.

The point $(-5, -3)$ is not included in the function as indicated by the open dot.

c. $f(x) = 0$ for $x = -4$ and $x = 4$.

The points $(-4, 0)$ and $(4, 0)$ represent the points where $f(x) = 0$.

d. $f(x) = 3$ for all x on the interval $[-3, 1]$ and for $x = \frac{14}{5}$.

e. The x-intercepts are $(-4, 0)$ and $(4, 0)$.

f. The y-intercept is $(0, 3)$.

g. The domain is $(-5, \infty)$.

h. The range is $(-\infty, 5]$.

Skill Practice 11   Use the function f pictured to find:

a. $f(-2)$.

b. $f(4)$.

c. All x for which $f(x) = 3$.

d. All x for which $f(x) = 1$.

e. The x-intercept(s).

f. The y-intercept.

g. The domain of f.

h. The range of f.

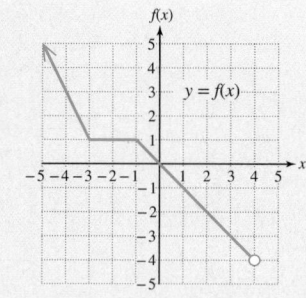

Answers

11. a. $f(-2) = 1$
   b. $f(4)$ is not defined.
   c. $x = -4$
   d. All x on the interval $[-3, -1]$
   e. $(0, 0)$
   f. $(0, 0)$
   g. $(-\infty, 4)$
   h. $(-4, \infty)$

SECTION 5.3   Practice Exercises

Prerequisite Review

For Exercises R.1–R.6, solve the equations.

R.1. $8x^2 - 40 = 0$ (See Section 3.5.)

R.2. $2x^2 - 30 = 0$

R.3. $a^2 + 5a + 4 = 0$ (See Section 3.4.)

R.4. $t^2 - 12t + 32 = 0$

R.5. $|c + 2| - 8 = 0$ (See Section 1.6.)

R.6. $|d - 3| - 9 = 0$

For Exercises R.7–R.10, solve the inequalities. Write the answer in interval notation. (See Section 1.5 for review.)

R.7. $2(x + 5) + 6 \geq 0$ **R.8.** $5(y - 4) - 2y - 1 \leq 0$

R.9. $6 - 2a > 0$ **R.10.** $15 - 3t < 0$

For Exercises R.11–R.14,

   a. Find the x-intercept(s).

   b. Find the y-intercept(s). (See Section 5.1 for review.)

R.11. $4x - 5y = 20$ **R.12.** $-2x - y = 6$

R.13. $y = |x| - 4$ **R.14.** $y = -|x| + 2$

For Exercises R.15–R.16, multiply the expression. (See Section 2.2 for review.)

R.15. $(t + h)^2$ **R.16.** $(a - 5)^2$

Concept Connections

1. A set of ordered pairs (x, y) is called a _____ in x and y. The set of x values in the relation is called the _____ of the relation. The set of _____ values is called the range of the relation.

2. Given a function defined by $y = f(x)$, the statement $f(2) = 4$ is equivalent to what ordered pair?

3. Given a function defined by $y = f(x)$, to find the _____-intercept, evaluate $f(0)$.

4. Given a function defined by $y = f(x)$, to find the x-intercept(s), substitute 0 for _____ and solve for x.

5. Given $f(x) = \dfrac{x + 1}{x + 5}$, the domain is restricted so that $x \neq$ _____.

6. Given $g(x) = \sqrt{x - 5}$, the domain is restricted so that $x \geq$ _____.

7. Consider a relation that defines the height y of a tree for a given time t after it is planted. Does this relation define y as a function of t? Explain.

8. Consider a relation that defines a time y during the course of a year when the temperature T in Fort Collins, Colorado, is 70°. Does this relation define y as a function of T? Explain.

Objective 1: Determine Whether a Relation Is a Function

For Exercises 9–12,

   a. Write a set of ordered pairs (x, y) that defines the relation.

   b. Write the domain of the relation.

   c. Write the range of the relation.

   d. Determine if the relation defines y as a function of x. **(See Examples 1–2)**

9.

Actor x	Number of Oscar Nominations y
Tom Hanks	5
Jack Nicholson	12
Sean Penn	5
Dustin Hoffman	7

10.

City x	Elevation at Airport (ft) y
Albany	285
Denver	5883
Miami	11
San Francisco	11

11.

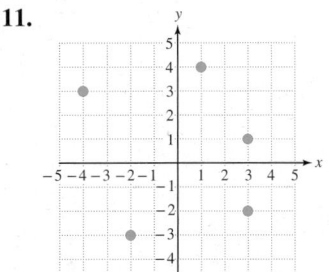

12.

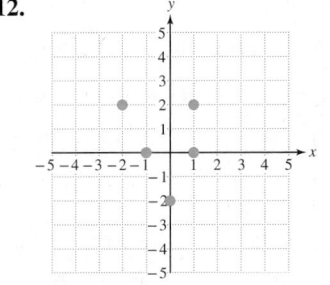

13. Answer true or false. All relations are functions.

14. Answer true or false. All functions are relations.

For Exercises 15–32, determine if the relation defines y as a function of x. (**See Examples 3–4**)

15.

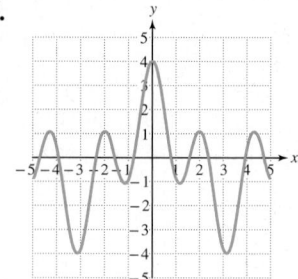

16.

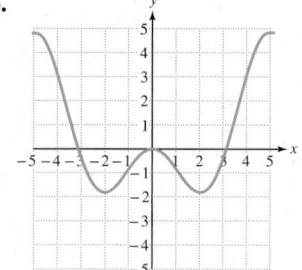

17.

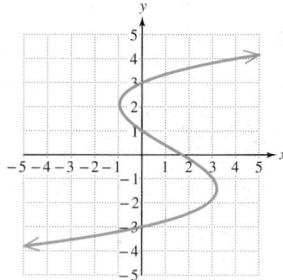

18.

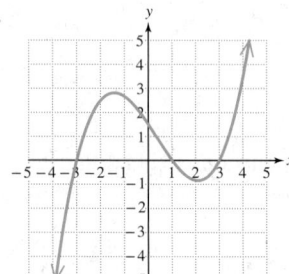

19.

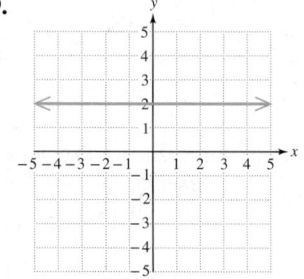

20.

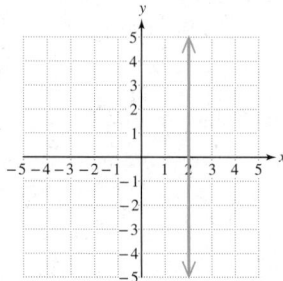

21.

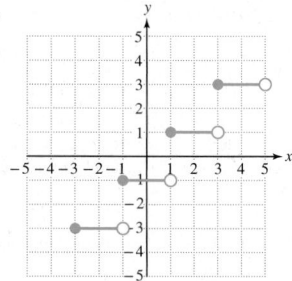

22.

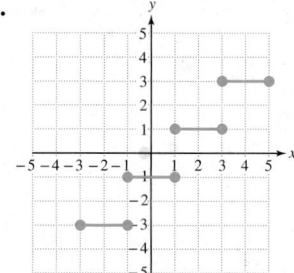

23.

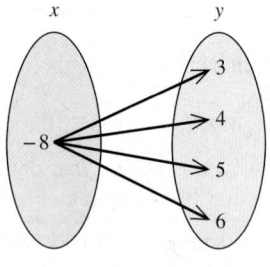

24.

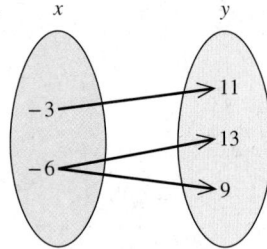

25.

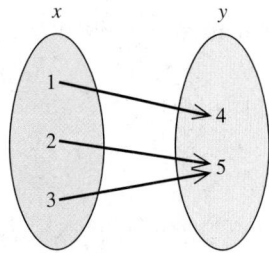

26.

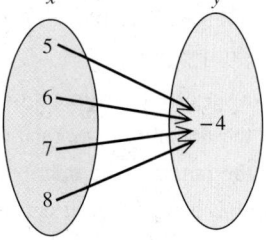

27. $(x + 1)^2 + (y + 5)^2 = 25$

28. $(x + 3)^2 + (y + 4)^2 = 1$

29. $y = x + 3$

30. $y = x - 4$

31. a. $y = x^2$

     b. $x = y^2$

32. a. $y = |x|$

     b. $x = |y|$

Objective 2: Apply Function Notation

33. The statement $f(4) = 1$ corresponds to what ordered pair?

34. The statement $g(7) = -5$ corresponds to what ordered pair?

For Exercises 35–56, evaluate the function for the given value of x. (**See Examples 5–6**)

$$f(x) = x^2 + 3x \qquad g(x) = \frac{1}{x} \qquad h(x) = 5 \qquad k(x) = \sqrt{x + 1}$$

35. a. $f(-2)$ **b.** $f(-1)$ **c.** $f(0)$ **d.** $f(1)$ **e.** $f(2)$

36. a. $g(-2)$ **b.** $g(-1)$ **c.** $g\left(-\frac{1}{2}\right)$ **d.** $g\left(\frac{1}{2}\right)$ **e.** $g(2)$

37. a. $h(-2)$ **b.** $h(-1)$ **c.** $h(0)$ **d.** $h(1)$ **e.** $h(2)$

38. a. $k(-2)$ **b.** $k(-1)$ **c.** $k(0)$ **d.** $k(1)$ **e.** $k(3)$

39. $g(3)$ **40.** $h(-7)$ **41.** $g\left(\frac{1}{3}\right)$

42. $h(7)$

43. $k(-5)$

44. $f(5)$

45. $k(8)$

46. $f(-5)$

47. $g(t)$

48. $f(a)$

49. $k(x + h)$

50. $h(x + h)$

51. $f(a + 4)$

52. $f(t - 3)$

53. $g(0)$

54. $k(-10)$

55. $f(x + h)$

56. $g(x + h)$

For Exercises 57–62, find and simplify $f(x + h)$. (**See Example 6**)

57. $f(x) = -4x^2 - 5x + 2$

58. $f(x) = -2x^2 + 6x - 3$

59. $f(x) = 7 - 3x^2$

60. $f(x) = 11 - 5x^2$

61. $f(x) = x^3 + 2x - 5$

62. $f(x) = x^3 - 4x + 2$

For Exercises 63–70, refer to the function $f = \{(2, 3), (9, 7), (3, 4), (-1, 6)\}$.

63. Determine $f(9)$.

64. Determine $f(-1)$.

65. Determine $f(3)$.

66. Determine $f(2)$.

67. For what value of x is $f(x) = 6$?

68. For what value of x is $f(x) = 7$?

69. For what value of x is $f(x) = 3$?

70. For what value of x is $f(x) = 4$?

71. Joe rides his bicycle an average of 18 mph. The distance Joe rides $d(t)$ (in mi) is given by $d(t) = 18t$, where t is the time in hours that he rides.

  a. Evaluate $d(2)$ and interpret the meaning.

  b. Determine the distance Joe travels in 40 min.

72. Frank needs to drive 250 mi from Daytona Beach to Miami. After having driven x miles, the distance remaining $r(x)$ (in mi) is given by $r(x) = 250 - x$.

  a. Evaluate $r(50)$ and interpret the meaning.

  b. Determine the distance remaining after 122 mi.

73. At a restaurant, if a party has eight or more people, the gratuity is automatically added to the bill. If x is the cost of the meal, then the total bill $C(x)$ with an 18% gratuity and a 6% sales tax is given by: $C(x) = x + 0.06x + 0.18x$. Evaluate $C(225)$ and interpret the meaning in the context of this problem.

74. A bookstore marks up the price of a book by 40% of the cost from the publisher. Therefore, the bookstore's price to the student, $P(x)$ (in \$) after a 7.5% sales tax, is given by $P(x) = 1.075(x + 0.40x)$, where x is the cost of the book from the publisher. Evaluate $P(60)$ and interpret the meaning in the context of this problem.

Objective 3: Determine *x*- and *y*-Intercepts of a Function Defined by $y = f(x)$

For Exercises 75–84, determine the *x*- and *y*-intercepts for the given function. (**See Example 7**)

75. $f(x) = 2x - 4$

76. $g(x) = 3x - 12$

77. $h(x) = |x| - 8$

78. $k(x) = -|x| + 2$

79. $p(x) = -x^2 + 12$

80. $q(x) = x^2 - 8$

81. $r(x) = |x - 8|$

82. $s(x) = |x + 3|$

83. $f(x) = \sqrt{x} - 2$

84. $g(x) = -\sqrt{x} + 3$

85. A student decides to finance a used car over a 5-year (60-month) period. After making a down payment of \$2000, the remaining cost of the car including tax and interest is \$14,820. The amount owed $y = A(t)$ (in \$) is given by $A(t) = 14,820 - 247t$, where t is the number of months after purchase and $0 \le t \le 60$. Determine the t-intercept and y-intercept and interpret their meanings in context.

86. The amount spent on video games per person in the United States has been increasing since 2006. (*Source*: www.census.gov) The function defined by $f(x) = 9.4x + 35.7$ represents the amount spent $f(x)$ (in \$) x years since 2006. Determine the y-intercept and interpret its meaning in context.

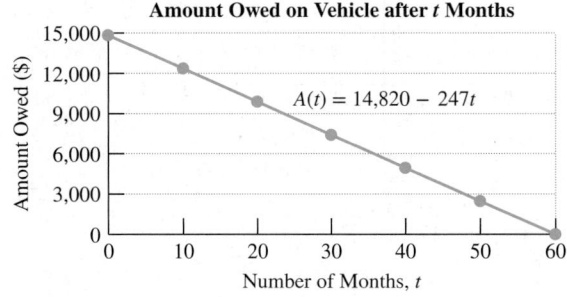

Amount Owed on Vehicle after *t* Months

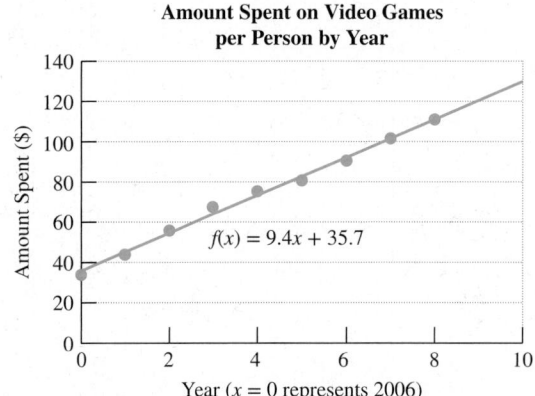

Amount Spent on Video Games per Person by Year

Objective 4: Determine Domain and Range of a Function

For Exercises 87–96, determine the domain and range of the function. (**See Example 8**)

87.

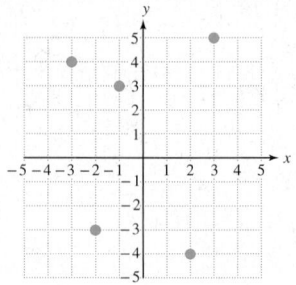

88.

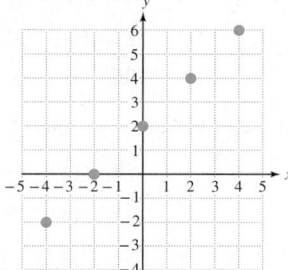

89.

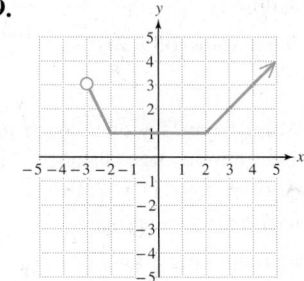

90.

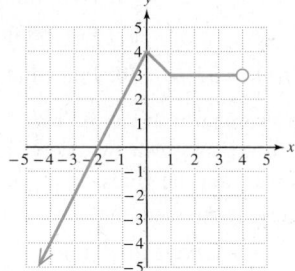

91.

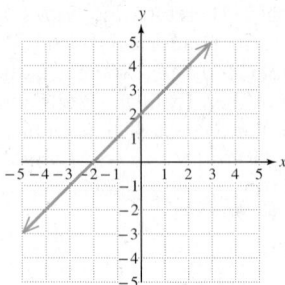

92.

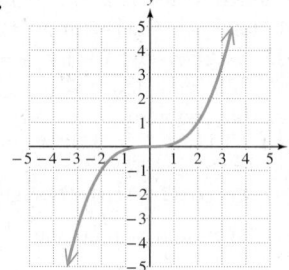

93.

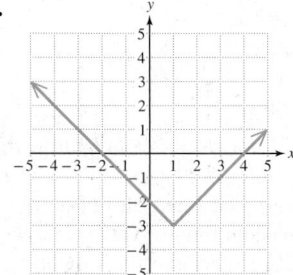

94.

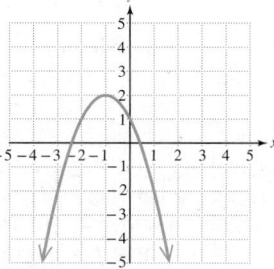

95.

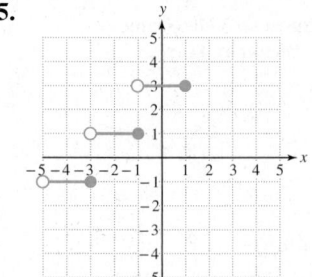

96.

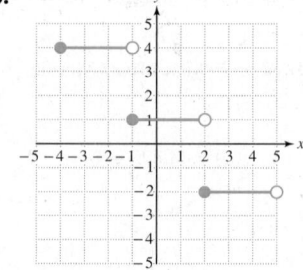

For Exercises 97–112, write the domain in interval notation. **(See Examples 9–10)**

97. a. $f(x) = \dfrac{x-3}{x-4}$

b. $g(x) = \dfrac{x-3}{x^2-4}$

c. $h(x) = \dfrac{x-3}{x^2+4}$

98. a. $k(x) = \dfrac{x+6}{x-2}$

b. $j(x) = \dfrac{x+6}{x^2+2}$

c. $p(x) = \dfrac{x+6}{x^2-2}$

99. a. $a(x) = \sqrt{x+9}$

b. $b(x) = \sqrt{9-x}$

c. $c(x) = \dfrac{1}{\sqrt{x+9}}$

100. a. $y(t) = \sqrt{16-t}$

b. $w(t) = \sqrt{t-16}$

c. $z(t) = \dfrac{1}{\sqrt{16-t}}$

101. a. $f(t) = \sqrt[3]{t-5}$

b. $g(t) = \sqrt[3]{5-t}$

c. $h(t) = \dfrac{1}{\sqrt[3]{t-5}}$

102. a. $k(x) = \sqrt[5]{3+x}$

b. $m(x) = \sqrt[5]{x-3}$

c. $n(x) = \dfrac{1}{\sqrt[5]{x-3}}$

103. a. $f(x) = x^2 - 3x - 28$

b. $g(x) = \dfrac{x+2}{x^2-3x-28}$

c. $h(x) = \dfrac{x^2-3x-28}{x+2}$

104. a. $r(x) = x^2 - 4x - 12$

b. $s(x) = \dfrac{x^2-4x-12}{x+1}$

c. $t(x) = \dfrac{x+1}{x^2-4x-12}$

105. a. $w(x) = |x+1| + 4$

b. $y(x) = \dfrac{x}{|x+1|+4}$

c. $z(x) = \dfrac{x}{|x+1|-4}$

106. a. $f(a) = 8 - |a-2|$

b. $g(a) = \dfrac{5}{8-|a-2|}$

c. $h(a) = \dfrac{5}{8+|a-2|}$

107. a. $h(x) = 7x - 5$

b. $k(x) = \sqrt{7x-5}$

c. $m(x) = \dfrac{1}{7x-5}$

108. a. $p(t) = 3t + 1$

b. $q(t) = \sqrt{3t+1}$

c. $r(t) = \dfrac{t}{3t+1}$

109. a. $f(x) = \sqrt{x+15}$

b. $g(x) = \sqrt{x+15} - 2$

c. $k(x) = \dfrac{5}{\sqrt{x+15}-2}$

110. a. $f(c) = \sqrt{c+20}$

b. $g(c) = \sqrt{c+20} - 1$

c. $h(c) = \dfrac{-4}{\sqrt{c+20}-1}$

111. a. $p(x) = 2x + 1$

b. $q(x) = 2x + 1; x \geq 0$

c. $r(x) = 2x + 1; 0 \leq x < 7$

112. a. $m(x) = 3x - 7$

b. $n(x) = 3x - 7; x < 0$

c. $p(x) = 3x - 7; -2 < x < 2$

Objective 5: Interpret a Function Graphically

For Exercises 113–116, use the graph of $y = f(x)$ to answer the following. **(See Example 11)**

a. Determine $f(-2)$.

b. Determine $f(3)$.

c. Find all x for which $f(x) = -1$.

d. Find all x for which $f(x) = -4$.

e. Determine the x-intercept(s).

f. Determine the y-intercept.

g. Determine the domain of f.

h. Determine the range of f.

113.

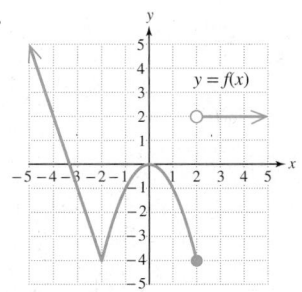

114.

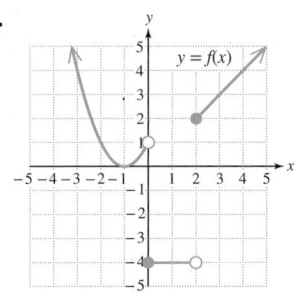

115.

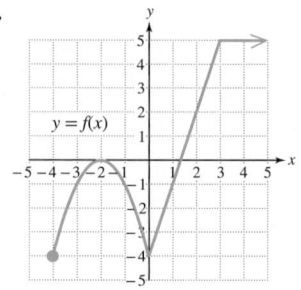

116.

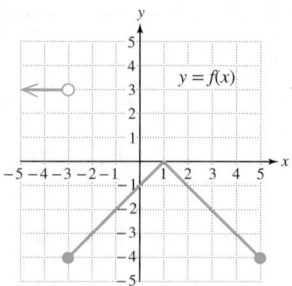

Mixed Exercises

For Exercises 117–124, write a function that represents the given statement.

117. Suppose that a phone card has 400 min. Write a relationship that represents the number of minutes remaining $r(x)$ as a function of the number of minutes already used x.

118. Suppose that a roll of wire has 200 ft. Write a relationship that represents the amount of wire remaining $w(x)$ as a function of the number of feet of wire x already used.

119. Given an equilateral triangle with sides of length x, write a relationship that represents the perimeter $P(x)$ as a function of x.

120. In an isosceles triangle, two angles are equal in measure. If the third angle is x degrees, write a relationship that represents the measure of one of the equal angles $A(x)$ as a function of x.

121. Two adjacent angles form a right angle. If the measure of one angle is x degrees, write a relationship representing the measure of the other angle $C(x)$ as a function of x.

122. Two adjacent angles form a straight angle (180°). If the measure of one angle is x degrees, write a relationship representing the measure of the other angle $S(x)$ as a function of x.

123. Write a relationship for a function whose $f(x)$ values are 2 less than three times the square of x.

124. Write a relationship for a function whose $f(x)$ values are 3 more than the principal square root of x.

Expanding Your Skills

125. Given a square with sides of length s, diagonal of length d, perimeter P, and area A,

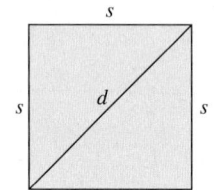

   a. Write P as a function of s.

   b. Write A as a function of s.

   c. Write A as a function of P.

   d. Write P as a function of A.

   e. Write d as a function of s.

   f. Write s as a function of d.

   g. Write P as a function of d.

   h. Write A as a function of d.

126. Given a circle with radius r, diameter d, circumference C, and area A,

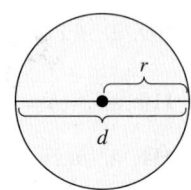

   a. Write C as a function of r.

   b. Write A as a function of r.

   c. Write r as a function of d.

   d. Write d as a function of r.

   e. Write C as a function of d.

   f. Write A as a function of d.

   g. Write A as a function of C.

   h. Write C as a function of A.

| SECTION 5.4 | Linear Equations in Two Variables and Linear Functions |

1. Graph Linear Equations in Two Variables

The median incomes for individuals for all levels of education have shown an increasing trend since 1990. However, the median income for individuals with a bachelor's degree is consistently greater than for individuals whose highest level of education is a high school degree or equivalent (Figure 5-15). (*Source*: U.S. Census Bureau, www.census.gov)

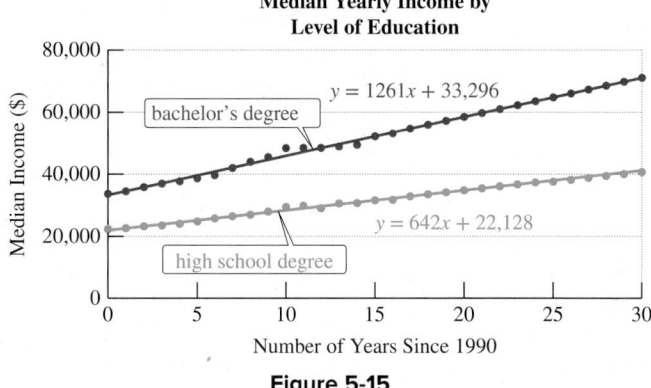

Figure 5-15

The graph in Figure 5-15 is called a scatter plot. A **scatter plot** is a visual representation of a set of points. In this case, the *x* values represent the number of years since 1990, and the *y* values represent the median income in dollars. The line that models each set of data is called a **regression line** and is found by using techniques taught in a first course in statistics. The equations that represent the two lines are called linear equations in two variables.

Linear Equation in Two Variables

Let *A*, *B*, and *C* represent real numbers such that *A* and *B* are not both zero. A **linear equation** in the variables *x* and *y* is an equation that can be written in the form:

$Ax + By = C$ This is called the **standard form** of an equation of a line.

Note: A linear equation $Ax + By = C$ has variables *x* and *y* each of first degree.

In Example 1, we demonstrate that the graph of a linear equation $Ax + By = C$ is a line. The line may be slanted, horizontal, or vertical depending on the coefficients *A*, *B*, and *C*.

| EXAMPLE 1 | Graphing Linear Equations |

Graph the line represented by each equation.

  a. $2x + 3y = 6$ **b.** $x = -3$ **c.** $2y = 4$

Solution:

  a. Solve the equation for *y*. Then substitute arbitrary values of *x* into the equation and solve for the corresponding values of *y*.

Avoiding Mistakes

The graph of a linear equation is a line. Therefore, a minimum of two points is needed to graph the line. A third point can be used to verify that the line is graphed correctly. The points must all line up.

TIP The graph of a vertical line will have no y-intercept unless the line is the y-axis itself.

TIP The graph of a horizontal line will have no x-intercept unless the line is the x-axis itself.

$$2x + 3y = 6 \qquad \text{Solve the equation for } y.$$
$$3y = -2x + 6$$
$$y = -\frac{2}{3}x + 2$$

In the table we have selected convenient values of x that are multiples of 3.

x	y
-3	4
0	2
3	0
6	-2

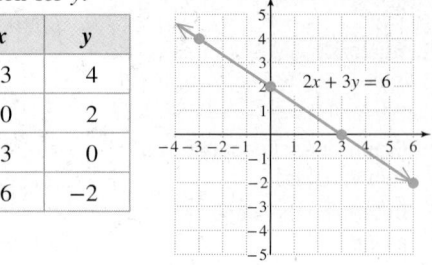

b. $x = -3$

The solutions to this equation must have an x-coordinate of -3. The y variable can be *any* real number.

x	y
-3	-2
-3	0
-3	2
-3	4

x must be -3.　y can be any real number.

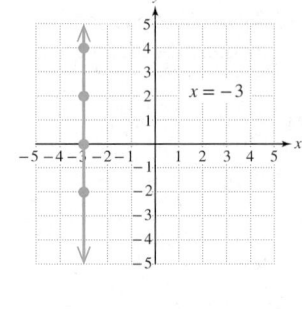

c. $2y = 4$ 　 Solve for y.
$$y = 2$$

The solutions to this equation must have a y-coordinate of 2. The x variable can be *any* real number.

x	y
-2	2
0	2
2	2
4	2

x can be any real number.　y must be 2.

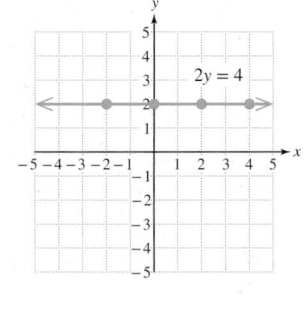

Skill Practice 1 Graph the line represented by each equation.

a. $4x + 2y = 2$ 　　**b.** $y = 1$ 　　**c.** $-3x = 12$

---

2. Determine the Slope of a Line

One of the important characteristics of a nonvertical line is that for every 1 unit of change in the horizontal variable, the vertical change is a constant m called the **slope** of the line. For example, consider the line representing the median income for individuals with a bachelor's degree, x years since the year 1990. The line in Figure 5-16

Answer

1. a.–c.

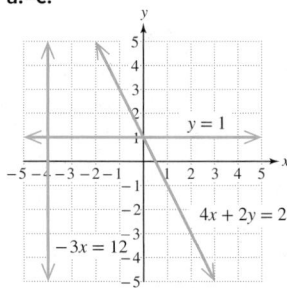

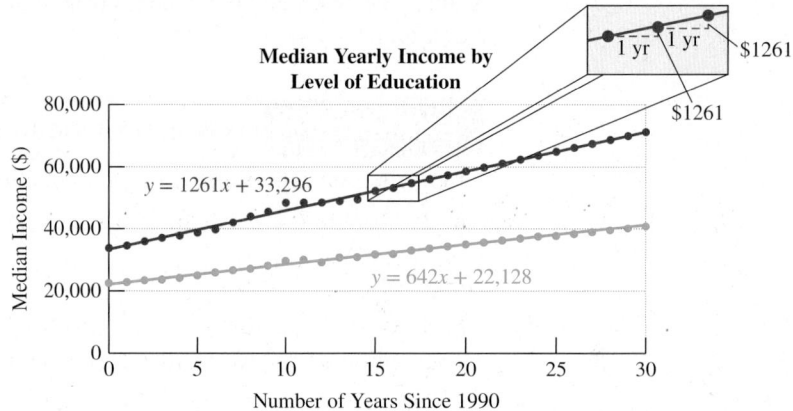

Figure 5-16

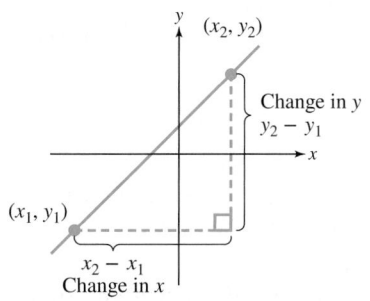

Figure 5-17

has a slope of $1261. This means that median income for individuals with a bachelor's degree increased on average by $1261 per year during this time period.

Consider any two distinct points (x_1, y_1) and (x_2, y_2) on a line (Figure 5-17). The slope m of the line through the points is the ratio between the change in the y values $(y_2 - y_1)$ and the change in the x values $(x_2 - x_1)$. In many applications in the sciences, the change in a variable is denoted by the Greek letter Δ (delta). Therefore, $(y_2 - y_1)$ can be represented by Δy and $(x_2 - x_1)$ can be represented by Δx.

Slope Formula

The **slope** of a line passing through the distinct points (x_1, y_1) and (x_2, y_2) is

change in y (rise)

$$m = \frac{\Delta y}{\Delta x} = \frac{y_2 - y_1}{x_2 - x_1} \text{ provided that } x_2 - x_1 \neq 0$$

change in x (run)

EXAMPLE 2   **Finding the Slope of a Line Through Two Points**

Find the slope of the line passing through the given points.

a. $(-3, -2)$ and $(2, 5)$ **b.** $\left(-\frac{5}{2}, 0\right)$ and $(1, -7)$

Solution:

a. $(-3, -2)$ and $(2, 5)$
   (x_1, y_1) and (x_2, y_2) Label the points.

$$m = \frac{y_2 - y_1}{x_2 - x_1} = \frac{5 - (-2)}{2 - (-3)} = \frac{7}{5}$$

A line with a positive slope "*rises*" upward from left to right.

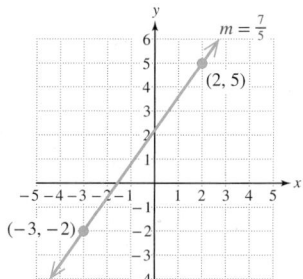

FOR REVIEW

To subtract $-\frac{5}{2}$ from 1, use the least common denominator of 2.

$1 - \left(-\frac{5}{2}\right) = 1 + \frac{5}{2}$
$\quad = \frac{1}{1} \cdot \frac{2}{2} + \frac{5}{2}$
$\quad = \frac{2}{2} + \frac{5}{2}$
$\quad = \frac{7}{2}$

To divide -7 by $\frac{7}{2}$, multiply the dividend -7 by the reciprocal of the divisor:
$-7 \cdot \frac{2}{7} = -2$
(See Section R.1.)

b. $\left(-\frac{5}{2}, 0\right)$ and $(1, -7)$
   (x_1, y_1) and (x_2, y_2) Label the points.

$$m = \frac{y_2 - y_1}{x_2 - x_1} = \frac{-7 - 0}{1 - \left(-\frac{5}{2}\right)} = \frac{-7}{\frac{7}{2}} = -7 \cdot \frac{2}{7} = -2$$

A line with a negative slope "*falls*" downward from left to right.

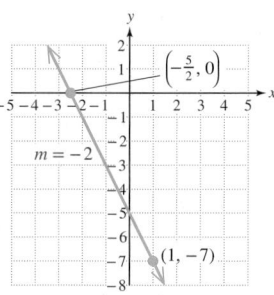

Skill Practice 2   Find the slope of the line passing through the given points.

a. $(-4, 1)$ and $(2, -2)$ **b.** $\left(\frac{3}{4}, 2\right)$ and $(-3, 17)$

Answers

2. a. $-\frac{1}{2}$ **b.** -4

EXAMPLE 3   Finding the Slope of Horizontal and Vertical Lines

Find the slope of each line.

Solution:

a.

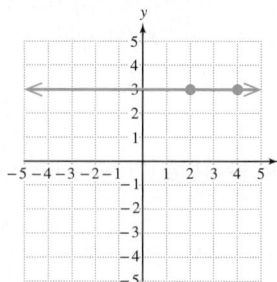

By inspection, we see that between any two points on the graph, the vertical change is zero, so the slope is zero.

To compute this numerically, select any two points on the line such as (2, 3) and (4, 3).

$$m = \frac{y_2 - y_1}{x_2 - x_1} = \frac{3 - 3}{4 - 2} = \frac{0}{2} = 0$$

b.

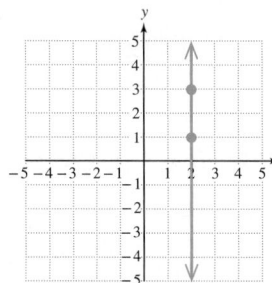

To find the slope, select any two points on the line such as (2, 1) and (2, 3).

$$m = \frac{y_2 - y_1}{x_2 - x_1} = \frac{3 - 1}{2 - 2} = \frac{2}{0} \quad \text{(undefined)}$$

By inspection, we see that between any two points on the line, the change in x is zero. This makes the slope undefined because the ratio representing the slope has a divisor of zero.

Skill Practice 3   Fill in the blank.

a. The slope of a vertical line is _____.

b. The slope of a horizontal line is _____.

From Example 1, we see that a linear equation may represent the graph of a slanted line, a horizontal line, or a vertical line. From Examples 2 and 3, we see that a line may have a positive slope, a negative slope, a zero slope, or an undefined slope.

Linear Equations and Slopes of Lines

$Ax + By = C$ ($A \neq 0, B \neq 0$) Slanted line	$y = k$ (k is a constant) Horizontal line	$x = k$ (k is a constant) Vertical line
Positive slope Negative slope	Zero slope	Undefined slope

3. Apply the Slope-Intercept Form of a Line

The slope formula can be used to develop the slope-intercept form of a line. Suppose that a line has a slope m and y-intercept $(0, b)$. Let (x, y) be any other point on the line. From the slope formula, we have:

$$\frac{y - b}{x - 0} = m \qquad \text{Slope formula}$$

$$y - b = mx \qquad \text{Multiply by } x.$$

$$y = mx + b \qquad \text{This is slope-intercept form. Slope-intercept form has the } y \text{ variable isolated.}$$

Answers
3. a. Undefined   **b.** 0

Slope-Intercept Form of a Line

Given a line with slope m and y-intercept $(0, b)$, the **slope-intercept form** of the line is given by $y = mx + b$.

The slope-intercept form of a line is particularly useful because we can identify the slope and y-intercept by inspection. For example:

Equation	Slope	y-intercept
$y = \dfrac{2}{3}x - 5$	$m = \dfrac{2}{3}$	$(0, -5)$
$y = x + 4$	$m = 1$	$(0, 4)$
$y = 2x$ (or $y = 2x + 0$)	$m = 2$	$(0, 0)$
$y = 6$ (or $y = 0x + 6$)	$m = 0$	$(0, 6)$

If the slope and y-intercept of a line are known, we can graph the line. This is demonstrated in Example 4.

EXAMPLE 4   **Using the Slope and y-Intercept to Graph a Line**

Given $3x + 4y = 4$,

  a. Write the equation in slope-intercept form.
  b. Determine the slope and y-intercept.
  c. Graph the line by using the slope and y-intercept.

Solution:

a. $3x + 4y = 4$

$4y = -3x + 4$ — To write an equation in slope-intercept form, isolate the y variable.

$\dfrac{4y}{4} = \dfrac{-3x}{4} + \dfrac{4}{4}$ — Divide both sides by 4.

$y = -\dfrac{3}{4}x + 1$ — Slope-intercept form

b. $m = -\dfrac{3}{4}$ and the y-intercept is $(0, 1)$. — The slope is the coefficient on x. The constant term gives the y-intercept.

c.

$y = -\dfrac{3}{4}x + 1$ $(0, 1)$

To graph the line, first plot the y-intercept $(0, 1)$.

Then begin at the y-intercept, and use the slope to find a second point on the line. In this case, the slope can be interpreted as the following two ratios:

$m = \dfrac{-3}{4}$ ← Move down 3 units. ← Move to the right 4 units.

$m = \dfrac{3}{-4}$ ← Move up 3 units. ← Move to the left 4 units.

Answers

4. a. $y = -\dfrac{1}{2}x + 2$

b. $m = -\dfrac{1}{2}$; y-intercept: $(0, 2)$

c.

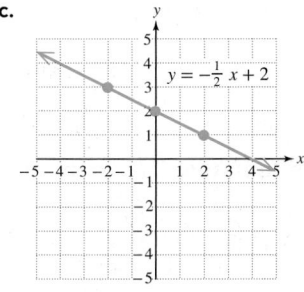

$y = -\dfrac{1}{2}x + 2$

Skill Practice 4   Given $2x + 4y = 8$,

  a. Write the equation in slope-intercept form.
  b. Determine the slope and y-intercept.
  c. Graph the line by using the slope and y-intercept.

EXAMPLE 5   **Using the Slope and *y*-Intercept to Graph a Line**

Given $5x - y = 0$,

a. Write the equation in slope-intercept form.

b. Determine the slope and *y*-intercept.

c. Graph the line by using the slope and *y*-intercept.

Solution:

a. $5x - y = 0$ To write the equation in slope-intercept form, solve for *y*.

      $-y = -5x$ Subtract $5x$ from both sides.

      $\dfrac{-y}{-1} = \dfrac{-5x}{-1}$ Divide both sides by -1.

      $y = 5x$ or $y = 5x + 0$ Slope-intercept form.

b. $m = 5$ and the *y*-intercept is $(0, 0)$.   The slope is the coefficient on *x*.

                                              The constant term gives the *y*-intercept.

c.

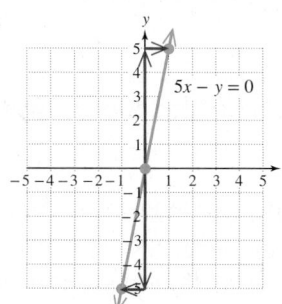

To graph the line, first plot the *y*-intercept $(0, 0)$.

Then begin at the *y*-intercept and use the slope to find a second point on the line. In this case, the slope can be interpreted as

$m = \dfrac{5}{1}$ ◄——— Move up 5 units.

             ◄——— Move to the right 1 unit.

or

$m = \dfrac{-5}{-1}$ ◄——— Move down 5 units.

              ◄——— Move to the left 1 unit.

Skill Practice 5   Given $2x + y = 0$,

a. Write the equation in slope-intercept form.

b. Determine the slope and *y*-intercept.

c. Graph the line by using the slope and *y*-intercept.

Notice that the slope-intercept form of a line $y = mx + b$ has the *y* variable isolated and defines *y* in terms of *x*. Therefore, an equation written in slope-intercept form defines *y* as a *function* of *x*. The linear equations from Examples 4 and 5 can be written using function notation as:

Example 4: $y = -\dfrac{3}{4}x + 1$ $\xrightarrow{\text{function notation}}$ $f(x) = -\dfrac{3}{4}x + 1$

Example 5: $y = 5x$ $\xrightarrow{\text{function notation}}$ $f(x) = 5x$

Furthermore, an equation of a horizontal line has the form $y = k$, where k is a constant real number. An equation of a horizontal line can also be written using function notation. Consider the equation of the horizontal line from Example 1(c).

Example 1(c): $y = 2$ $\xrightarrow{\text{function notation}}$ $f(x) = 2$

These observations lead us to the definitions of linear and constant functions.

Answers

5. a. $y = -2x$

  b. $m = -2$; *y*-intercept: $(0, 0)$

  c.

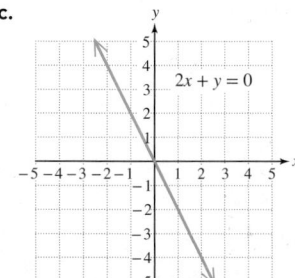

> **Definition of Linear and Constant Functions**
>
> Let *m* and *b* represent real numbers where $m \neq 0$. Then,
>
> - A function defined by $f(x) = mx + b$ is a **linear function**. The graph of a linear function is a slanted line.
> - A function defined by $f(x) = b$ is a **constant function**. The graph of a constant function is a horizontal line.

The slope-intercept form of a line can be used as a tool to define a linear function given a point on the line and the slope.

EXAMPLE 6   **Writing an Equation of a Line Given a Point and the Slope**

Write an equation of the line that passes through the point $(2, -3)$ and has slope -4. Then write the linear equation using function notation, where $y = f(x)$.

Solution:

Given $m = -4$ and $(2, -3)$.   We need to find an equation of the form $y = mx + b$.

$$y = mx + b$$
$$y = -4x + b$$ The value of m is given as -4.
$$-3 = -4(2) + b$$ Substitute $x = 2$ and $y = -3$ from the given point $(2, -3)$.

$$-3 = -8 + b$$ Solve for b.
$$b = 5$$

$$y = mx + b$$
$$y = -4x + 5$$ Substitute $m = -4$ and $b = 5$ into the equation $y = mx + b$.

$$f(x) = -4x + 5$$ Write the relation using function notation.

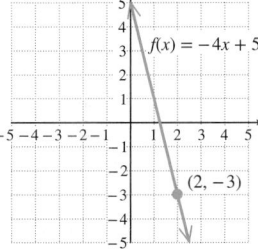

From the graph, we see that the graph of $f(x) = -4x + 5$ does indeed pass through the point $(2, -3)$ and has slope -4.

> **Skill Practice 6**   Write an equation of the line that passes through the point $(-1, -4)$ and has slope 3. Then write the equation using function notation.

EXAMPLE 7   **Finding Intercepts and Slope of a Linear Function**

Given $f(x) = -3x + 6$, find the x- and y-intercepts and slope of the line.

Solution:

The function is written in slope-intercept form, $f(x) = mx + b$.

The slope, m, is the coefficient on x, and the y-intercept is $(0, b)$.

Thus, given $f(x) = -3x + 6$,

- The slope is $m = -3$.
- The y-intercept is $(0, 6)$.

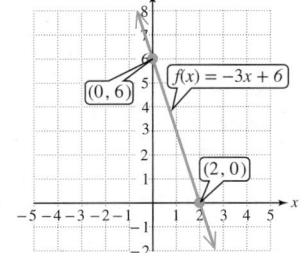

To find the x-intercept, substitute 0 for y, or in this case $f(x)$.

$$f(x) = -3x + 6$$
$$0 = -3x + 6$$
$$-6 = -3x$$
$$x = 2$$ The x-intercept is $(2, 0)$.

> **TIP** To find the y-intercept, substitute 0 for x. That is, evaluate $f(0)$.
> $$f(x) = -3x + 6$$
> $$f(0) = -3(0) + 6$$
> $$= 0 + 6$$
> $$= 6$$
> The y-intercept is $(0, 6)$.

Answers

6. $y = 3x - 1$; $f(x) = 3x - 1$

7. $m = \dfrac{1}{3}$, y-intercept: $(0, -4)$, x-intercept: $(12, 0)$

> **Skill Practice 7**   Given $g(x) = \dfrac{1}{3}x - 4$, find the x- and y-intercepts and slope of the line.

4. Solve Equations and Inequalities Graphically

In many settings, the use of technology can provide a numerical and visual interpretation of an algebraic problem. For example, consider the equation $-x - 1 = x + 5$.

$$-x - 1 = x + 5$$
$$-6 = 2x$$
$$-3 = x \qquad \text{The solution set is } \{-3\}.$$

Now suppose that we create two functions from the left and right sides of the equation. We have $Y_1 = -x - 1$ and $Y_2 = x + 5$. Figure 5-18 shows that the graphs

$$\boxed{Y_1 = -x - 1} \qquad \boxed{Y_2 = x + 5}$$
$$-x - 1 = x + 5$$

of the two lines intersect at $(-3, 2)$. The x-coordinate of the point of intersection is the solution to the equation $-x - 1 = x + 5$. That is, $Y_1 = Y_2$ when $x = -3$.

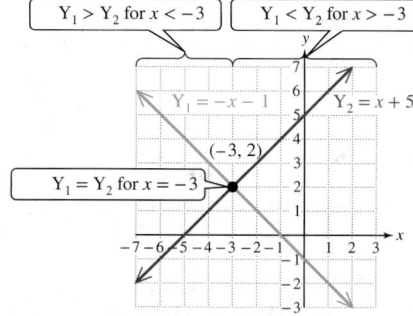

Figure 5-18

The graphs $Y_1 = -x - 1$ and $Y_2 = x + 5$ can also be used to find the solution sets to the related inequalities.

$-x - 1 < x + 5$ The solution set is the set of x values for which $Y_1 < Y_2$. This is the interval where the blue line is below the red line. The solution set is $(-3, \infty)$.

$-x - 1 > x + 5$ The solution set is the set of x values for which $Y_1 > Y_2$. This is the interval where the blue line is *above* the red line. The solution set is $(-\infty, -3)$.

EXAMPLE 8   Solving Equations and Inequalities Graphically

Solve the equations and inequalities graphically.

a. $2x - 3 = x - 1$ **b.** $2x - 3 < x - 1$ **c.** $2x - 3 > x - 1$

TIP The solution set to the inequality $2x - 3 \leq x - 1$ includes equality, so the right endpoint would be included: $(-\infty, 2]$.

The solution set to the inequality $2x - 3 \geq x - 1$ includes equality, so the left endpoint would be included: $[2, \infty)$.

Solution:

a. The left side of the equation is graphed as $Y_1 = 2x - 3$. The right side of the equation is graphed as $Y_2 = x - 1$. The point of intersection is $(2, 1)$. Therefore, $Y_1 = Y_2$ for $x = 2$.

The solution set is $\{2\}$.

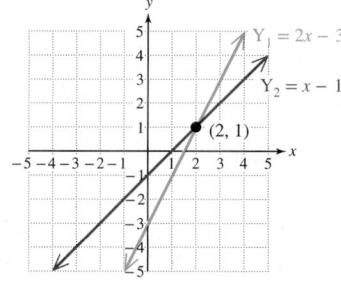

b. $Y_1 < Y_2$ to the *left* of $x = 2$. (That is, the blue line is below the red line for $x < 2$.)

In interval notation the solution set is $(-\infty, 2)$.

c. $Y_1 > Y_2$ to the *right* of $x = 2$. (That is, the blue line is above the red line for $x > 2$.)

In interval notation the solution set is $(2, \infty)$.

Skill Practice 8   Use the graph to solve the equations and inequalities.

a. $x + 1 = 2x - 2$

b. $x + 1 \leq 2x - 2$

c. $x + 1 \geq 2x - 2$

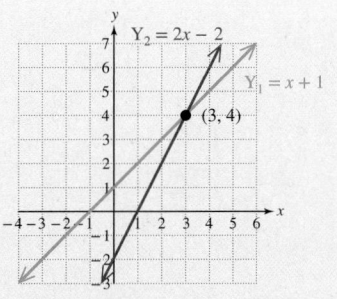

TECHNOLOGY CONNECTIONS

Verifying Solutions to an Equation

We can verify the solutions to the equations and inequalities from Example 8 on a graphing calculator.

Display the graphs of Y_1 and Y_2 and use the Intersect feature to determine the point of intersection.

$Y_1 = Y_2$ for $x = 2$

$Y_1 < Y_2$ for $x < 2$

$Y_1 > Y_2$ for $x > 2$

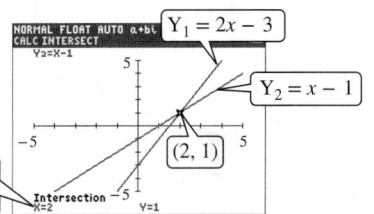

x-coordinate of the point of intersection

In Example 9 we solve the equation $6x - 2(x + 2) - 5 = 0$. Notice that one side is zero. We can check the solution graphically by determining where the related function $Y_1 = 6x - 2(x + 2) - 5$ intersects the *x*-axis.

EXAMPLE 9   Solving Equations and Inequalities Graphically

a. Solve the equation $6x - 2(x + 2) - 5 = 0$ and verify the solution graphically on a graphing utility.

b. Use the graph to find the solution set to the inequality $6x - 2(x + 2) - 5 \leq 0$.

c. Use the graph to find the solution set to the inequality $6x - 2(x + 2) - 5 \geq 0$.

Answers

8. a. $\{3\}$ **b.** $[3, \infty)$ **c.** $(-\infty, 3]$

Solution:

a. $6x - 2(x + 2) - 5 = 0$

$6x - 2x - 4 - 5 = 0$

$4x - 9 = 0$

$x = \dfrac{9}{4}$

The solution set is $\left\{ \dfrac{9}{4} \right\}$.

To verify the solution graphically enter the left side of the equation as $Y_1 = 6x - 2(x + 2) - 5$.

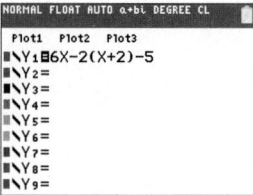

Using the Zero feature, we have $Y_1 = 0$ for $x = 2.25$. This is consistent with the solution $x = \frac{9}{4}$.

b. To solve $6x - 2(x + 2) - 5 \leq 0$ determine the values of x for which $Y_1 \leq 0$ (where the function is on or below the x-axis).

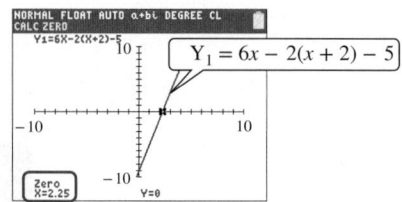

The solution set is $\left(-\infty, \dfrac{9}{4} \right]$.

c. To solve $6x - 2(x + 2) - 5 \geq 0$ determine the values of x for which $Y_1 \geq 0$ (where the function is on or above the x-axis).

The solution set is $\left[\dfrac{9}{4}, \infty \right)$.

Answers

9. a. $\left\{ \dfrac{5}{2} \right\}$

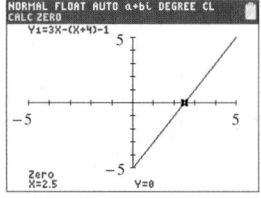

b. $\left(-\infty, \dfrac{5}{2} \right]$

c. $\left[\dfrac{5}{2}, \infty \right)$

Skill Practice 9

a. Solve the equation $3x - (x + 4) - 1 = 0$ and verify the solution graphically on a graphing utility.

b. Use the graph to find the solution set to the inequality $3x - (x + 4) - 1 \leq 0$.

c. Use the graph to find the solution set to the inequality $3x - (x + 4) - 1 \geq 0$.

SECTION 5.4   Practice Exercises

Prerequisite Review

For Exercises R.1–R.2, solve for y. (See Section 1.2 for review.)

R.1. $6x - 2y = 10$

R.2. $-7x - 8y = 1$

R.3. Given $f(x) = 5x - 4$,
 a. Evaluate $f(0)$.
 b. Determine the value(s) of x for which $f(x) = 0$. (See Section 5.3.)

R.4. Given $g(x) = -10x + 5$,
 a. Evaluate $g(0)$.
 b. Determine the value(s) of x for which $g(x) = 0$.

For Exercises R.5–R.6, determine the x- and y-intercepts. (See Section 5.3 for review.)

R.5. $f(x) = -x^2 + 9$

R.6. $g(x) = |x| - 7$

For Exercises R.7–R.10, solve the inequality. Write the solution set in interval notation. (See Section 1.5 for review.)

R.7. $-4t + 5 < 13$

R.8. $2x - 9 > 5$

R.9. $5p - 2 \geq 6p + 8$

R.10. $-8(n + 3) \leq 4 - n$

For Exercises R.11–R.12, simplify the complex fraction. (See Section 4.2 for review.)

R.11. $\dfrac{2 + \dfrac{4}{5}}{-1 + \dfrac{3}{10}}$

R.12. $\dfrac{1 + \dfrac{1}{3}}{\dfrac{5}{6} - 1}$

Concept Connections

1. A _____ equation in the variables x and y can be written in the form $Ax + By = C$, where A and B are not both zero.

2. An equation of the form $x = k$ where k is a constant represents the graph of a _____ line.

3. An equation of the form $y = k$ where k is a constant represents the graph of a _____ line.

4. Write the formula for the slope of a line between the two distinct points (x_1, y_1) and (x_2, y_2).

5. The slope of a horizontal line is _____ and the slope of a vertical line is _____.

6. A function f is a linear function if $f(x) =$ _____, where m represents the slope and $(0, b)$ represents the y-intercept.

7. Given a function defined by $f(x) = 8x - 4$, find the _____-intercept by solving the equation $f(x) = 0$.

8. Given a function defined by $y = f(x)$, find the _____-intercept by evaluating $f(0)$.

Objective 1: Graph Linear Equations in Two Variables

For Exercises 9–20, graph the equation and identify the x- and y-intercepts. **(See Example 1)**

9. $-3x + 4y = 12$

10. $-2x + y = 4$

11. $2y = -5x + 2$

12. $3y = -4x + 6$

13. $x = -6$

14. $y = 4$

15. $5y + 1 = 11$

16. $3x - 2 = 4$

17. $0.02x + 0.05y = 0.1$

18. $0.03x + 0.07y = 0.21$

19. $2x = 3y$

20. $2x = -5y$

Objective 2: Determine the Slope of a Line

21. Find the average slope of the hill.

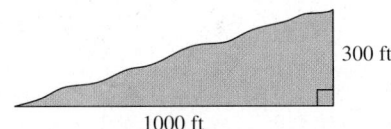

300 ft

1000 ft

22. Find the absolute value of the slope of the storm drainage pipe.

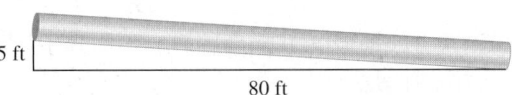

5 ft

80 ft

23. The road sign shown in the figure indicates the percent grade of a hill. This gives the slope of the road as the change in elevation per 100 horizontal feet. Given a 2.5% grade, write this as a slope in fractional form.

2.5% Grade

24. The pitch of a roof is defined as $\dfrac{\text{rafter rise}}{\text{rafter run}}$ and the fraction is typically written with a denominator of 12. Determine the pitch of the roof from point A to point C.

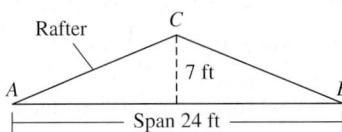

Rafter

C

7 ft

A

B

Span 24 ft

For Exercises 25–36, determine the slope of the line passing through the given points. **(See Example 2)**

25. $(4, -7)$ and $(2, -1)$

26. $(-3, -8)$ and $(4, 6)$

27. $(17, 9)$ and $(42, -6)$

28. $(-9, 4)$ and $(-1, -6)$ **29.** $(30, -52)$ and $(-22, -39)$ **30.** $(-100, -16)$ and $(84, 30)$

31. $(2.6, 4.1)$ and $(9.5, -3.7)$ **32.** $(8.5, 6.2)$ and $(-5.1, 7.9)$ **33.** $\left(\dfrac{3}{4}, 6\right)$ and $\left(\dfrac{5}{2}, 1\right)$

34. $\left(-3, \dfrac{2}{5}\right)$ and $\left(4, \dfrac{3}{10}\right)$ **35.** $(9, -7)$ and $(9, 3)$ **36.** $(-7, 9)$ and $(3, 9)$

For Exercises 37–42, determine the slope of the line. **(See Examples 2–3)**

37.
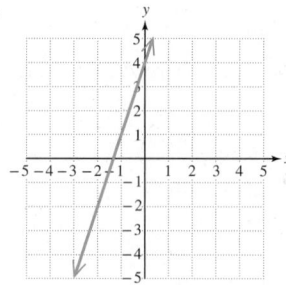

38.
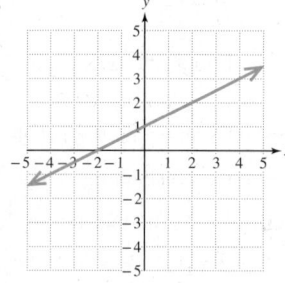

39.
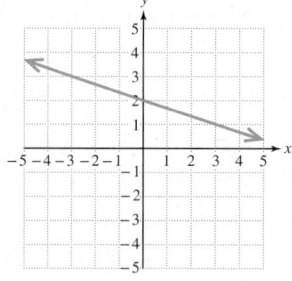

40.
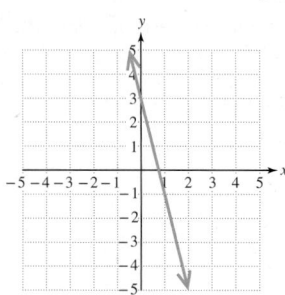

41.
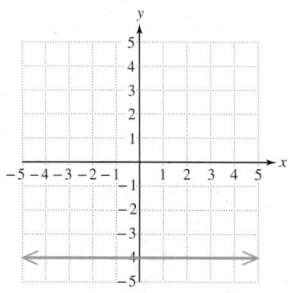

42.
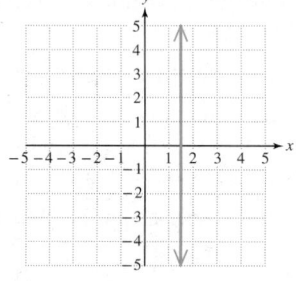

43. What is the slope of a line perpendicular to the x-axis?

44. What is the slope of a line parallel to the x-axis?

45. What is the slope of a line defined by $y = -7$?

46. What is the slope of a line defined by $x = 2$?

47. If the slope of a line is $\dfrac{4}{5}$, how much vertical change will be present for a horizontal change of 52 ft?

48. If the slope of a line is $\dfrac{5}{8}$, how much horizontal change will be present for a vertical change of 216 m?

49. Suppose that $y = P(t)$ represents the population of a city at time t. What does $\dfrac{\Delta P}{\Delta t}$ represent?

50. Suppose that $y = d(t)$ represents the distance that an object travels in time t. What does $\dfrac{\Delta d}{\Delta t}$ represent?

Objective 3: Apply the Slope-Intercept Form of a Line

For Exercises 51–62,

a. Write the equation in slope-intercept form if possible, and determine the slope and y-intercept.

b. Graph the equation using the slope and y-intercept. **(See Examples 4–5)**

51. $2x - 4y = 8$ **52.** $3x - y = 6$ **53.** $3x = 2y - 4$ **54.** $5x = 3y - 6$

55. $3x = 4y$ **56.** $-2x = 3y$ **57.** $2y - 6 = 8$ **58.** $3y + 9 = 6$

59. $0.02x + 0.06y = 0.06$ **60.** $0.03x + 0.04y = 0.12$ **61.** $\dfrac{x}{4} + \dfrac{y}{7} = 1$ **62.** $\dfrac{x}{3} + \dfrac{y}{4} = 1$

For Exercises 63–64, determine if the function is linear, constant, or neither.

63. a. $f(x) = -\dfrac{3}{4}x$ **b.** $g(x) = -\dfrac{3}{4}x - 3$ **c.** $h(x) = -\dfrac{3}{4x}$ **d.** $k(x) = -\dfrac{3}{4}$

64. a. $m(x) = 5x + 1$ **b.** $n(x) = \dfrac{5}{x} + 1$ **c.** $p(x) = 5$ **d.** $q(x) = 5x$

For Exercises 65–74,

a. Use slope-intercept form to write an equation of the line that passes through the given point and has the given slope.

b. Write the equation using function notation where $y = f(x)$. **(See Example 6)**

65. $(0, 9)$; $m = \dfrac{1}{2}$ **66.** $(0, -4)$; $m = \dfrac{1}{3}$ **67.** $(1, -6)$; $m = -3$ **68.** $(2, -8)$; $m = -5$

69. $(-5, -3)$; $m = \dfrac{2}{3}$ **70.** $(-4, -2)$; $m = \dfrac{3}{2}$ **71.** $(2, 5)$; $m = 0$ **72.** $(-1, -3)$; $m = 0$

73. $(3.6, 5.1)$; $m = 1.2$ **74.** $(1.2, 2.8)$; $m = 2.4$

For Exercises 75–78,

a. Use slope-intercept form to write an equation of the line that passes through the two given points.

b. Then write the equation using function notation where $y = f(x)$.

  75. $(4, 2)$ and $(0, -6)$ **76.** $(-8, 1)$ and $(0, -3)$ **77.** $(7, -3)$ and $(4, 1)$ **78.** $(2, -4)$ and $(-1, 3)$

For Exercises 79–84, given $y = f(x)$ find the x- and y-intercepts and slope of the line, if possible. **(See Example 7)**

  79. $f(x) = -5x - 9$ **80.** $f(x) = 2x + 7$ **81.** $f(x) = \dfrac{2}{11}x$ **82.** $f(x) = -\dfrac{1}{3}x$

  83. $f(x) = 6$ **84.** $f(x) = -12$

Objective 4: Solve Equations and Inequalities Graphically

For Exercises 85–90, refer to the graph and functions f, g, and h.

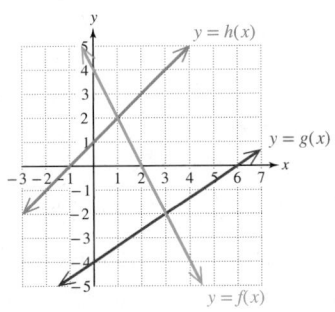

85. a. Determine the value(s) of x for which $f(x) = g(x)$.

  b. Determine the value(s) of x for which $f(x) = 0$.

  c. Find $f(0)$.

  d. Determine the value(s) of x for which $g(x) = 0$.

86. a. Determine the value(s) of x for which $f(x) = h(x)$.

  b. Determine the value(s) of x for which $h(x) = 0$.

  c. Find $h(0)$.

  d. Find $g(0)$.

87. a. Determine the interval over which $f(x) \geq 0$.

  b. Determine the interval over which $f(x) < 0$.

88. a. Determine the interval over which $g(x) > 0$.

  b. Determine the interval over which $g(x) \leq 0$.

89. a. Determine the interval over which $h(x) > f(x)$.

  b. Determine the interval over which $h(x) \leq f(x)$.

90. a. Determine the interval over which $g(x) < f(x)$.

  b. Determine the interval over which $g(x) \geq f(x)$.

For Exercises 91–98, use the graph to solve the equation and inequalities. Write the solutions to the inequalities in interval notation. **(See Examples 8–9)**

91. a. $2x + 4 = -x + 1$

  b. $2x + 4 < -x + 1$

  c. $2x + 4 \geq -x + 1$

92. a. $4x - 2 = -3x + 5$

  b. $4x - 2 < -3x + 5$

  c. $4x - 2 \geq -3x + 5$

93. a. $-3x + 1 = -x - 3$

  b. $-3x + 1 > -x - 3$

  c. $-3x + 1 \leq -x - 3$

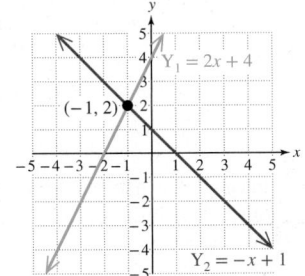   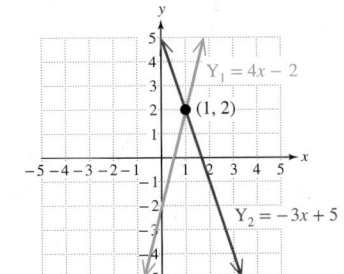   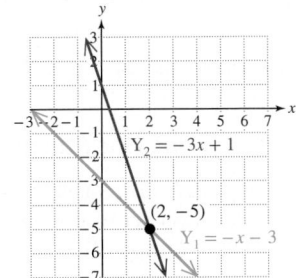

94. a. $-x - 2 = 2x - 5$

    b. $-x - 2 \leq 2x - 5$

    c. $-x - 2 > 2x - 5$

95. a. $-3(x + 2) + 1 = -x + 5$

    b. $-3(x + 2) + 1 \leq -x + 5$

    c. $-3(x + 2) + 1 \geq -x + 5$

96. a. $-4(x - 5) + 3x = -3x + 1$

    b. $-4(x - 5) + 3x \leq -3x + 1$

    c. $-4(x - 5) + 3x \geq -3x + 1$

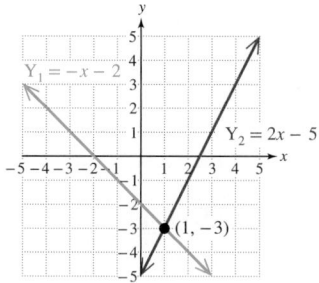
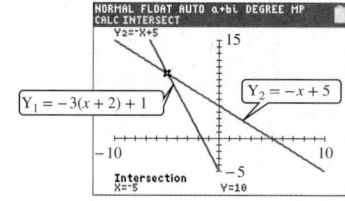
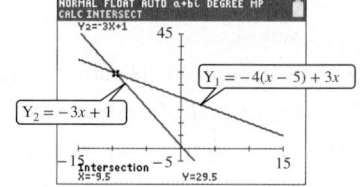

97. a. $4 - 2(x + 1) + 12 + x = 0$

    b. $4 - 2(x + 1) + 12 + x < 0$

    c. $4 - 2(x + 1) + 12 + x > 0$

98. a. $8 - 4(1 - x) - 7 - 2x = 0$

    b. $8 - 4(1 - x) - 7 - 2x < 0$

    c. $8 - 4(1 - x) - 7 - 2x > 0$

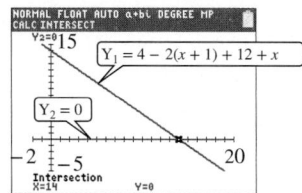
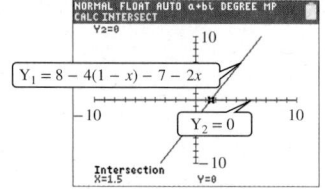

Write About It

99. Explain how you can determine from a linear equation $Ax + By = C$ (A and B not both zero) whether the line is slanted, horizontal, or vertical.

100. Explain how you can determine from a linear equation $Ax + By = C$ (A and B not both zero) whether the line passes through the origin.

Expanding Your Skills

101. Determine the area in the second quadrant enclosed by the equation $y = 2x + 4$ and the x- and y-axes.

102. Determine the area enclosed by the equations.

$$y = x + 6$$
$$y = -2x + 6$$
$$y = 0$$

103. Determine the area enclosed by the equations.

$$y = -\frac{1}{2}x - 2$$
$$y = \frac{1}{3}x - 2$$
$$y = 0$$

104. Determine the area enclosed by the equations.

$$y = \sqrt{4 - (x - 2)^2}$$
$$y = 0$$

105. Consider the standard form of a linear equation $Ax + By = C$ in the case where $B \neq 0$.

    a. Write the equation in slope-intercept form.

    b. Identify the slope in terms of the coefficients A and B.

    c. Identify the y-intercept in terms of the coefficients B and C.

106. Use the results from Exercise 105 to determine the slope and y-intercept for the graphs of the lines.

    a. $5x - 9y = 6$

    b. $0.052x - 0.013y = 0.39$

Technology Connections

For Exercises 107–110, solve the equation in part (a) and verify the solution on a graphing calculator. Then use the graph to find the solution set to the inequalities in parts (b) and (c). Write the solution sets to the inequalities in interval notation. **(See Example 9)**

107. a. $3.1 - 2.2(t + 1) = 6.3 + 1.4t$

    b. $3.1 - 2.2(t + 1) > 6.3 + 1.4t$

    c. $3.1 - 2.2(t + 1) < 6.3 + 1.4t$

108. a. $-11.2 - 4.6(c - 3) + 1.8c = 0.4(c + 2)$

    b. $-11.2 - 4.6(c - 3) + 1.8c > 0.4(c + 2)$

    c. $-11.2 - 4.6(c - 3) + 1.8c < 0.4(c + 2)$

109. a. $|2x - 3.8| - 4.6 = 7.2$

b. $|2x - 3.8| - 4.6 \geq 7.2$

c. $|2x - 3.8| - 4.6 \leq 7.2$

110. a. $|x - 1.7| + 4.95 = 11.15$

b. $|x - 1.7| + 4.95 \geq 11.15$

c. $|x - 1.7| + 4.95 \leq 11.15$

For Exercises 111–112, graph the lines in (a)–(c) on the standard viewing window. Compare the graphs. Are they exactly the same? If not, how are they different?

111. a. $y = 3x + 1$

b. $y = 2.99x + 1$

c. $y = 3.01x + 1$

112. a. $y = x + 3$

b. $y = x + 2.99$

c. $y = x + 3.01$

SECTION 5.5 Applications of Linear Equations and Modeling

OBJECTIVES

1. Apply the Point-Slope Formula
2. Determine the Slopes of Parallel and Perpendicular Lines
3. Create Linear Functions to Model Data
4. Create Models Using Linear Regression

1. Apply the Point-Slope Formula

The slope formula can be used to develop the point-slope form of an equation of a line. Suppose that a line has a slope m and passes through a known point (x_1, y_1). Let (x, y) be any other point on the line. From the slope formula, we have

$$\frac{y - y_1}{x - x_1} = m \qquad \text{Slope formula}$$

$$\left(\frac{y - y_1}{x - x_1}\right)(x - x_1) = m(x - x_1) \quad \text{Clear fractions.}$$

$$y - y_1 = m(x - x_1) \quad \text{This is called the point-slope formula for a line.}$$

The point-slope formula is useful to build an equation of a line given a point on the line and the slope of the line.

> **Point-Slope Formula**
>
> The **point-slope formula** for a line is given by $y - y_1 = m(x - x_1)$, where m is the slope of the line and (x_1, y_1) is a point on the line.

EXAMPLE 1 **Writing an Equation of a Line Given a Point on the Line and the Slope**

Use the point-slope formula to find an equation of the line passing through the point $(2, -3)$ and having slope -4. Write the answer in slope-intercept form.

Solution:

Label $(2, -3)$ as (x_1, y_1) and $m = -4$.

$$y - y_1 = m(x - x_1) \qquad \text{Apply the point-slope formula.}$$

$$y - (-3) = -4(x - 2) \qquad \text{Substitute } x_1 = 2, y_1 = -3, \text{ and } m = -4.$$

$$y + 3 = -4x + 8 \qquad \text{Simplify.}$$

$$y = -4x + 5 \quad \text{(slope-intercept form)}$$

> **Skill Practice 1** Use the point-slope formula to find an equation of the line passing through the point $(-5, 2)$ and having slope -3. Write the answer in slope-intercept form.

TIP The slope-intercept form of a line can also be used to write an equation of a line if a point on the line and the slope are known. See Example 6 in Section 5.4.

Answer

1. $y = -3x - 13$

EXAMPLE 2   **Writing an Equation of a Line Given Two Points**

Use the point-slope formula to write an equation of the line passing through the points $(4, -6)$ and $(-1, 2)$. Write the answer in slope-intercept form.

Solution:

To apply the point-slope formula, we first need to know the slope of the line.

$(4, -6)$ and $(-1, 2)$ Label the points. Either point can be

(x_1, y_1) and (x_2, y_2) labeled (x_1, y_1).

$m = \dfrac{y_2 - y_1}{x_2 - x_1} = \dfrac{2 - (-6)}{-1 - 4} = \dfrac{8}{-5} = -\dfrac{8}{5}$ Apply the slope formula.

$y - y_1 = m(x - x_1)$ Apply the point-slope formula.

> **TIP** In Example 2, the slope-intercept form of a line can also be used to find an equation of the line. Substitute $-\frac{8}{5}$ for m and $(4, -6)$ for (x, y).
>
> $$y = mx + b$$
> $$-6 = -\tfrac{8}{5}(4) + b$$
> $$-6 = -\tfrac{32}{5} + b$$
> $$-6 + \tfrac{32}{5} = b$$
> $$\tfrac{2}{5} = b$$
>
> Therefore, $y = mx + b$ is $y = -\frac{8}{5}x + \frac{2}{5}$.

$y - (-6) = -\dfrac{8}{5}(x - 4)$ Substitute $y_1 = -6$, $x_1 = 4$, and $m = -\frac{8}{5}$.

$y + 6 = -\dfrac{8}{5}x + \dfrac{32}{5}$

$y = -\dfrac{8}{5}x + \dfrac{32}{5} - 6$ Subtract 6 from both sides.

     Obtain a common denominator.

$y = -\dfrac{8}{5}x + \dfrac{32}{5} - \dfrac{30}{5}$ $6 = \dfrac{6}{1} \cdot \dfrac{5}{5} = \dfrac{30}{5}$

$y = -\dfrac{8}{5}x + \dfrac{2}{5}$

To check, we see that the graph of the line passes through $(4, -6)$ and $(-1, 2)$ as expected.

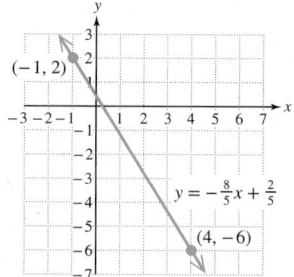

Skill Practice 2   Write an equation of the line passing through the points $(2, -5)$ and $(7, -3)$.

---

2. Determine the Slopes of Parallel and Perpendicular Lines

Lines in the same plane that do not intersect are **parallel lines**. Nonvertical parallel lines have the same slope and different y-intercepts (Figure 5-19).

     Lines that intersect at a right angle are **perpendicular lines**. If two nonvertical lines are perpendicular, then the slope of one line is the opposite of the reciprocal of the slope of the other line (Figure 5-20).

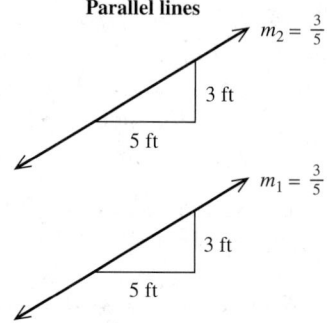

Figure 5-19

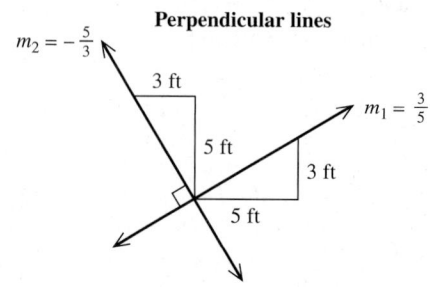

Figure 5-20

Answer

2. $y = \dfrac{2}{5}x - \dfrac{29}{5}$

Slopes of Parallel and Perpendicular Lines

- If m_1 and m_2 represent the slopes of two nonvertical parallel lines, then $m_1 = m_2$.
- If m_1 and m_2 represent the slopes of two nonvertical perpendicular lines, then $m_1 = -\dfrac{1}{m_2}$ or equivalently $m_1 m_2 = -1$.

In Examples 3 and 4, we use the point-slope formula to find an equation of a line through a specified point and parallel or perpendicular to another line.

EXAMPLE 3   Writing an Equation of a Line Parallel to Another Line

Write an equation of the line passing through the point $(-4, 1)$ and parallel to the line defined by $x + 4y = 3$. Write the answer in slope-intercept form and in standard form.

Solution:

$x + 4y = 3$ The slope of the given line can be found from its slope-intercept form. Solve for y.

$4y = -x + 3$

$y = -\dfrac{1}{4}x + \dfrac{3}{4}$ The slope of both lines is $-\frac{1}{4}$.

$y - y_1 = m(x - x_1)$ Apply the point-slope formula with $x_1 = -4$, $y_1 = 1$, and $m = -\frac{1}{4}$.

$y - 1 = -\dfrac{1}{4}[x - (-4)]$

$y - 1 = -\dfrac{1}{4}(x + 4)$

$y - 1 = -\dfrac{1}{4}x - 1$

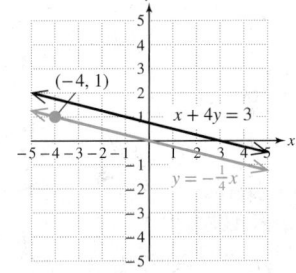

$y = -\dfrac{1}{4}x$ (slope-intercept form)

$4(y) = 4\left(-\dfrac{1}{4}x\right)$

$4y = -x$

$x + 4y = 0$ (standard form)

From the graph, we see that the line $y = -\frac{1}{4}x$ passes through the point $(-4, 1)$ and is parallel to the graph of $x + 4y = 3$.

Clearing fractions, and collecting the x and y terms on one side of the equation gives us standard form.

Skill Practice 3   Write an equation of the line passing through the point $(-3, 2)$ and parallel to the line defined by $x + 3y = 6$. Write the answer in slope-intercept form and in standard form.

EXAMPLE 4   Writing an Equation of a Line Perpendicular to Another Line

Write an equation of the line passing through the point $(2, -3)$ and perpendicular to the line defined by $y = \frac{1}{2}x - 4$. Write the answer in slope-intercept form and in standard form.

Answer

3. $y = -\dfrac{1}{3}x + 1;\ x + 3y = 3$

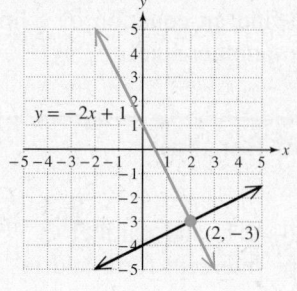

Solution:

From the slope-intercept form, $y = \frac{1}{2}x - 4$, the slope of given line is $\frac{1}{2}$.

$y - y_1 = m(x - x_1)$ The slope of a line perpendicular to the given line is -2.

$y - (-3) = -2(x - 2)$ Apply the point-slope formula with $x_1 = 2$, $y_1 = -3$, and $m = -2$.

$y + 3 = -2x + 4$ Simplify.

$y = -2x + 1$ (slope-intercept form)   Write the equation in slope-intercept form by solving for y.

$2x + y = 1$ (standard form)   Write the equation in standard form by collecting the x and y terms on one side of the equation.

> **Skill Practice 4** Write an equation of the line passing through the point $(-8, -4)$ and perpendicular to the line defined by $y = \frac{1}{6}x + 3$.

3. Create Linear Functions to Model Data

In many day-to-day applications, two variables are related linearly. By finding an equation of the line, we produce a model that relates the two variables. This is demonstrated in Example 5.

EXAMPLE 5   Using a Linear Function in an Application

A family plan for a cell phone has a monthly base price of $99 plus $12.99 for each additional family member added beyond the primary account holder.

a. Write a linear function to model the monthly cost $C(x)$ (in $) of a family plan for x additional family members added.

b. Evaluate $C(4)$ and interpret the meaning in the context of this problem.

Solution:

a. $C(x) = mx + b$ The base price $99 is the fixed cost with zero additional family members added. So the constant b is 99.

$C(x) = 12.99x + 99$ The rate of increase, $12.99 per additional family member, is the slope, m.

b. $C(4) = 12.99(4) + 99$ Substitute 4 for x.

$= 150.96$

The total monthly cost of the plan with 4 additional family members beyond the primary account holder is $150.96.

> **Skill Practice 5** A speeding ticket is $100 plus $5 for every 1 mph over the speed limit.
>
> **a.** Write a linear function to model the cost $S(x)$ (in $) of a speeding ticket for a person caught driving x mph over the speed limit.
>
> **b.** Evaluate $S(15)$ and interpret the meaning in the context of this problem.

Linear functions can sometimes be used to model the cost, revenue, and profit of producing and selling x items.

Linear Cost, Revenue, and Profit Functions

A **linear cost function** models the cost $C(x)$ to produce x items.

$$C(x) = mx + b$$

m is the variable cost per item.

b is the fixed cost.

The fixed cost does not change relative to the number of items produced. For example, the cost to rent an office is a fixed cost. The variable cost per item is the rate at which cost increases for each additional unit produced. Variable costs include labor, material, and shipping.

A **linear revenue function** models revenue $R(x)$ for selling x items.

$$R(x) = px$$

The product px represents the price per item p times the number of items sold x.

A **linear profit function** models the profit for producing and selling x items.

$$P(x) = R(x) - C(x)$$

Subtract the cost to produce x items from the revenue brought in from selling x items.

EXAMPLE 6   Writing Linear Cost, Revenue, and Profit Functions

At a summer art show a vendor sells lemonade for $2.00 per cup. The cost to rent the booth is $120. Furthermore, the vendor knows that the lemons, sugar, and cups collectively cost $0.50 for each cup of lemonade produced.

   a. Write a linear cost function to produce x cups of lemonade.

   b. Write a linear revenue function for selling x cups of lemonade.

   c. Write a linear profit function for producing and selling x cups of lemonade.

   d. How much profit will the vendor make if 50 cups of lemonade are produced and sold?

   e. How much profit will be made for producing and selling 128 cups?

   f. Determine the break-even point.

Solution:

   a. $C(x) = 0.50x + 120$

   The fixed cost is $120 because it does not change relative to the number of cups of lemonade produced. The variable cost is $0.50 per lemonade.

   b. $R(x) = 2.00x$

   The price per cup of lemonade is $2.00. Therefore, the product $2.00x$ gives the amount of revenue for x cups of lemonade sold.

   c. $P(x) = R(x) - C(x)$
   $P(x) = 2.00x - (0.50x + 120)$
   $P(x) = 1.50x - 120$

   Profit is defined as the difference of revenue and cost.

   d. $P(50) = 1.50(50) - 120$
   $= -45$

   Substitute 50 for x.
   The vendor will lose $45.

   e. $P(128) = 1.50(128) - 120$
   $= 72$

   Substitute 128 for x.
   The vendor will make $72.

f. For what value of x will $R(x) = C(x)$ or $P(x) = 0$?

The break-even point is defined as the point where revenue equals cost. Alternatively, this can be stated as the point where profit equals zero: $P(x) = 0$.

$$P(x) = 0$$
$$1.50x - 120 = 0$$
$$1.50x = 120$$
$$x = 80$$

Solve for x.

If the vendor produces and sells 80 cups of lemonade, the cost and revenue will be equal, resulting in a profit of $0. This is the break-even point.

> **Skill Practice 6** Repeat Example 6 in the case where the vendor can cut the cost to $0.40 per cup of lemonade, and sell lemonades for $1.50 per cup.

Figure 5-21 shows the graphs of the revenue and cost functions from Example 6. Notice that R and C intersect at $(80, 160)$. This means that if 80 cups of lemonade are produced and sold, the revenue and cost are both $160. That is, $R(x) = C(x)$ and the company breaks even. The graph of the profit function P is consistent with this result. The value of $P(x)$ is 0 for 80 lemonades produced and sold (Figure 5-22).

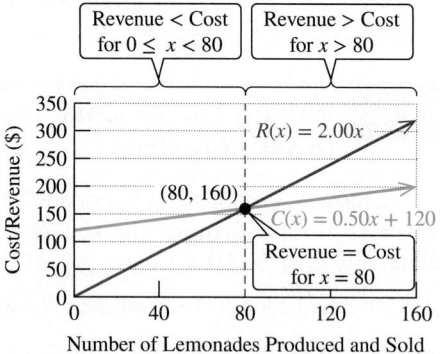

Figure 5-21

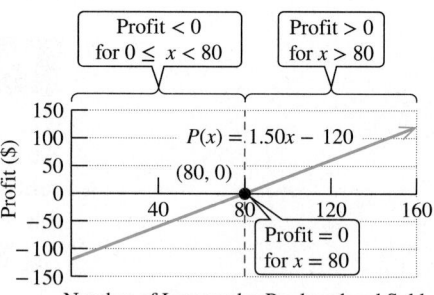

Figure 5-22

From Figures 5-21 and 5-22, we can draw the following conclusions.

- The company experiences a loss if fewer than 80 cups of lemonade are produced and sold. That is, $R(x) < C(x)$, or equivalently $P(x) < 0$.
- The company experiences a profit if more than 80 cups of lemonade are produced and sold. That is, $R(x) > C(x)$, or equivalently $P(x) > 0$.
- The company breaks even if exactly 80 cups of lemonade are produced and sold. That is, $R(x) = C(x)$, or equivalently $P(x) = 0$.

EXAMPLE 7 **Writing a Linear Model to Relate Two Variables**

The data shown in the graph represent the age and systolic blood pressure for a sample of 12 randomly selected healthy adults.

Systolic and Diastolic Blood Pressure by Age

Answers
6. a. $C(x) = 0.40x + 120$
  b. $R(x) = 1.50x$
  c. $P(x) = 1.10x - 120$
  d. $-\$65$
  e. $\$20.80$
  f. Approximately 109 cups

a. Suppose that x represents the age of an adult (in years), and y represents the systolic blood pressure (in mmHg). Use the points (21, 118) and (51, 130) to write a linear model relating y as a function of x.

b. Interpret the meaning of the slope in the context of this problem.

c. Use the model to estimate the systolic blood pressure for a 55-year-old. Round to the nearest whole unit.

TIP The equation $y = 0.4x + 109.6$ can also be expressed in function notation. For example, we can rename y as $S(x)$.

$$S(x) = 0.4x + 109.6$$

The value $S(x)$ represents the estimated systolic blood pressure for an adult of age x years.

Solution:

a. (21, 118) and (51, 130)

(x_1, y_1) and (x_2, y_2) Label the points.

$$m = \frac{130 - 118}{51 - 21} = \frac{12}{30}$$ Determine the slope of the line.

$$= \frac{2}{5}$$ In applications, we often express slope as a *unit* rate of change. In this case, we want to know how much change in blood pressure occurs per 1 year in age. To convert to a unit rate, divide 2 by 5 to get the decimal value 0.4.

$$= 0.4$$

$y - y_1 = m(x - x_1)$ Apply the point-slope formula.

$y - 118 = 0.4(x - 21)$

$y - 118 = 0.4x - 8.4$

$y = 0.4x + 109.6$ The equation $y = 0.4x + 109.6$ relates an individual's age to an estimated systolic blood pressure for that age.

b. The slope is 0.4. This means that the average increase in systolic blood pressure for adults is 0.4 mmHg per year of age.

c. $y = 0.4x + 109.6$

$y = 0.4(55) + 109.6$ Substitute 55 for x.

$y = 131.6$

Based on the sample of data, the estimated systolic blood pressure for a 55-year-old is 132 mmHg.

Skill Practice 7   Suppose that y represents the average consumer spending on television services per year (in dollars), and that x represents the number of years since 2004.

a. Use the data points (2, 308) and (6, 408) to write a linear equation relating y to x.

b. Interpret the meaning of the slope in the context of this problem.

c. Interpret the meaning of the y-intercept in the context of this problem.

d. Use the model from part (a) to estimate the average consumer spending on television services for the year 2020.

---

4. Create Models Using Linear Regression

In Example 7, we used two given data points to determine a linear model for systolic blood pressure versus age. There are two drawbacks to this method. First, the equation is not necessarily unique. If we use two different data points, we may get a different equation. Second, it is generally preferable to write a model that is based on *all* the data points, rather than just two points. One such model is called the least-squares regression line.

The procedure to find the least-squares regression line is discussed in detail in a statistics course. Here we will give the basic premise and use a graphing utility to perform the calculations. Consider a set of data points (x_1, y_1), (x_2, y_2), (x_3, y_3), ... ,

Answers

7. a. $y = 25x + 258$
   b. The slope is 25 and means that consumer spending on television services rose $25 per year during this time period.
   c. (0, 258); The average consumer spending on television services for the year 2004 was $258.
   d. $658

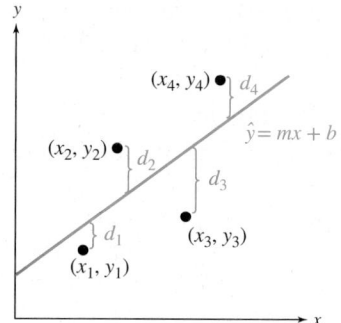

Figure 5-23

(x_n, y_n). The **least-squares regression line**, $\hat{y} = mx + b$, is the unique line that minimizes the sum of the squared vertical deviations from the observed data points to the line (Figure 5-23).

On a calculator or spreadsheet, the equation $\hat{y} = mx + b$ may be denoted as

$$y = ax + b \text{ or as } y = b_0 + b_1 x$$

In any event, the coefficient of x is the slope of the line, and the constant gives us the y-intercept. Although the exact keystrokes on different calculators and graphing utilities may vary, we will use the following guidelines to find the least-squares regression line.

Creating a Linear Regression Model

1. Graph the data in a scatter plot.
2. Inspect the data visually to determine if the data suggest a linear trend.
3. Invoke the linear regression feature on a calculator, graphing utility, or spreadsheet.
4. Check the result by graphing the line with the data points to verify that the line passes through or near the data points.

EXAMPLE 8   Finding a Least-Squares Regression Line

The data given in the table represent the age and systolic blood pressure for a sample of 12 randomly selected healthy adults.

Age (years)	17	21	27	33	35	38	43	51	58	60	64	70
Systolic blood pressure (mmHg)	110	118	121	122	118	124	125	130	132	138	134	142

a. Make a scatter plot of the data using age as the independent variable x and systolic pressure as the dependent variable y.
b. Based on the graph, does a linear model seem appropriate?
c. Determine the equation of the least-squares regression line.
d. Use the least-squares regression line to approximate the systolic blood pressure for a healthy 55-year-old. Round to the nearest whole unit.

Solution:

a. On a graphing calculator hit the STAT button and select EDIT to enter the x and y data into two lists (shown here as L1 and L2).

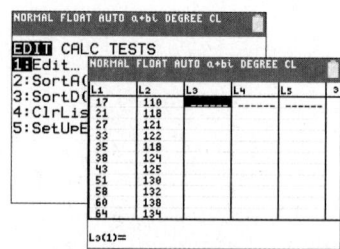

Select the STAT PLOT option and turn Plot1 to On. For the type of graph, select the scatter plot image.

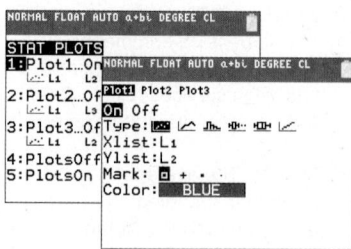

Be sure that the window is set to accommodate *x* values between 17 and 70, and *y* values between 110 and 142, inclusive. Then hit the GRAPH key. The window settings shown here are [0, 80, 10] by [0, 200, 20].

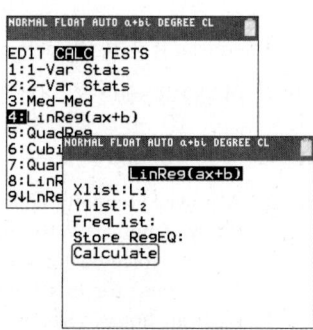

b. From the graph, the data appear to follow a linear trend.

c. Under the STAT menu, select CALC and then the LinReg(ax + b) option.

The command LinReg(ax + b) prompts the user to enter the list names (L_1 and L_2) containing the *x* and *y* data values. Then highlight Calculate and hit ENTER.

> **TIP** The linear equation found in Example 7 was based on two data points. The least-squares regression line is based on all available data points. The estimate from each model for systolic blood pressure for a 55-year-old rounds to 132 mmHg.

In the regression model $y = ax + b$, the values for the coefficients *a* and *b* are placed on the home screen.

Rounding the values of *a* and *b* gives us $y = 0.511x + 104$.

Enter the equation $Y_1 = 0.511x + 104$ into the equation editor and hit the GRAPH key. The graph of the regression line passes near or through the observed data points.

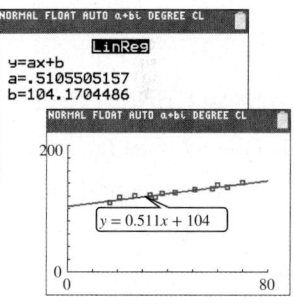

> **TIP** In Example 8(d), the value of the function at $x = 55$ can also be found by selecting the CALC menu and selecting the VALUE function. Enter 55 for *x* and press the ENTER key.

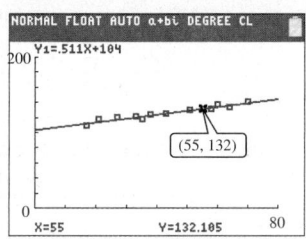

d. $y = 0.511x + 104$

$y = 0.511(55) + 104$ — To approximate the systolic blood pressure for a 55-year-old, substitute 55 for *x*.

$\quad = 132.105$

The systolic blood pressure for a healthy 55-year-old would be approximately 132 mmHg.

Skill Practice 8   The data given represent the class averages for individual students based on the number of absences from class.

Number of Absences (*x*)	3	7	1	11	2	14	2	5
Average in Class (*y*)	88	67	96	62	90	56	97	82

a. Find the equation of the least-squares regression line.

b. Use the model from part (a) to approximate the average for a student who misses 6 classes.

Answers

8. a. $y = -3.27x + 98.1$

 b. The student's average would be approximately 78.5.

SECTION 5.5   Practice Exercises

Prerequisite Review

For Exercises R.1–R.4, determine the opposite and reciprocal of each number. (See Section R.2 for review.)

R.1. -4 　　　　**R.2.** 5 　　　　**R.3.** $\dfrac{2}{3}$ 　　　　**R.4.** $-\dfrac{1}{8}$

R.5. Use slope-intercept form to write an equation of the line that passes through $(3, -7)$ with slope -5. (See Section 5.4 for review.)

R.6. Use slope-intercept form to write an equation of the line that passes through $(4, 1)$ with slope $\frac{1}{2}$.

For Exercises R.7–R.8, write the equation in slope-intercept form and determine the slope and y-intercept. (See Section 5.4 for review.)

R.7. $3x - 5y = -15$

R.8. $0.2x + 0.7y = 1.4$

For Exercises R.9–R.12, determine the slope of the line containing the given points. (See Section 5.4 for review.)

R.9. $(4, -10)$ and $(-2, 8)$

R.10. $(-6, -3)$ and $(7, -5)$

R.11. $(-4, -2)$ and $(-4, -7)$

R.12. $(3, -2)$ and $(5, -2)$

R.13. A power company charges a household $0.12 per kilowatt-hour (kWh) and $14.89 in monthly taxes. Write a formula for the monthly charge C (in dollars) for this household if it uses k kilowatt-hours. (See Section 1.1 for review.)

R.14. A utility company charges a base rate for water usage of $13.50 per month, plus $3.58 for every additional 1000 gal of water used over a 2000-gal base. Write a formula for the monthly charge C (in dollars) for a household that uses n thousand gallons over the 2000-gal base.

Concept Connections

1. Given a point (x_1, y_1) on a line with slope m, the point-slope formula is given by _____.

2. If two nonvertical lines have the same slope but different y-intercepts, then the lines are (parallel/perpendicular).

3. If m_1 and m_2 represent the slopes of two nonvertical perpendicular lines, then $m_1m_2 = $ _____.

4. Suppose that $y = C(x)$ represents the cost to produce x items, and that $y = R(x)$ represents the revenue for selling x items. The profit $P(x)$ of producing and selling x items is defined by $P(x) = $ _____.

Objective 1: Apply the Point-Slope Formula

For Exercises 5–20, use the point-slope formula to write an equation of the line having the given conditions. Write the answer in slope-intercept form (if possible). **(See Examples 1–2)**

5. Passes through $(-3, 5)$ and $m = -2$.

6. Passes through $(4, -6)$ and $m = 3$.

7. Passes through $(-1, 0)$ and $m = \frac{2}{3}$.

8. Passes through $(-4, 0)$ and $m = \frac{3}{5}$.

9. Passes through $(3.4, 2.6)$ and $m = 1.2$.

10. Passes through $(2.2, 4.1)$ and $m = 2.4$.

11. Passes through $(6, 2)$ and $(-3, 1)$.

12. Passes through $(-4, 8)$ and $(-7, -3)$.

13. Passes through $(0, 8)$ and $(5, 0)$.

14. Passes through $(0, -6)$ and $(11, 0)$.

15. Passes through $(2.3, 5.1)$ and $(1.9, 3.7)$.

16. Passes through $(1.6, 4.8)$ and $(0.8, 6)$.

17. Passes through $(3, -4)$ and $m = 0$.

18. Passes through $(-5, 1)$ and $m = 0$.

19. Passes through $\left(\frac{2}{3}, \frac{1}{5}\right)$ and the slope is undefined.

20. Passes through $\left(-\frac{4}{7}, \frac{3}{10}\right)$ and the slope is undefined.

21. Given a line defined by $x = 4$, what is the slope of the line?

22. Given a line defined by $y = -2$, what is the slope of the line?

Objective 2: Determine the Slopes of Parallel and Perpendicular Lines

For Exercises 23–28, the slope of a line is given.

a. Determine the slope of a line parallel to the given line, if possible.

b. Determine the slope of a line perpendicular to the given line, if possible.

23. $m = \frac{3}{11}$

24. $m = \frac{6}{7}$

25. $m = -6$

26. $m = -10$

27. $m = 1$

28. m is undefined

For Exercises 29–36, determine if the lines defined by the given equations are parallel, perpendicular, or neither.

29. $y = 2x - 3$

$y = -\dfrac{1}{2}x + 7$

30. $y = \dfrac{4}{3}x - 1$

$y = -\dfrac{3}{4}x + 5$

31. $8x - 5y = 3$

$2x = \dfrac{5}{4}y + 1$

32. $2x + 3y = 7$

$4x = -6y + 2$

33. $2x = 6$

$5 = y$

34. $3y = 5$

$x = 1$

35. $6x = 7y$

$\dfrac{7}{2}x - 3y = 0$

36. $5y = 2x$

$\dfrac{5}{2}x - y = 0$

For Exercises 37–44, write an equation of the line satisfying the given conditions. Write the answer in slope-intercept form (if possible) and in standard form $(Ax + By = C)$ with no fractional coefficients. **(See Examples 3–4)**

37. Passes through $(2, 5)$ and is parallel to the line defined by $2x + y = 6$.

38. Passes through $(3, -1)$ and is parallel to the line defined by $-3x + y = 4$.

39. Passes through $(6, -4)$ and is perpendicular to the line defined by $x - 5y = 1$.

40. Passes through $(5, 4)$ and is perpendicular to the line defined by $x - 2y = 7$.

41. Passes through $(6, 8)$ and is parallel to the line defined by $3x = 7y + 5$.

42. Passes through $(7, -6)$ and is parallel to the line defined by $2x = 5y - 4$.

43. Passes through $(2.2, 6.4)$ and is perpendicular to the line defined by $2x = 4 - y$.

44. Passes through $(3.6, 1.2)$ and is perpendicular to the line defined by $4x = 9 - y$.

For Exercises 45–50, write an equation of the line that satisfies the given conditions.

45. Passes through $(8, 6)$ and is parallel to the x-axis.

46. Passes through $(-11, 13)$ and is parallel to the y-axis.

47. Passes through $\left(\dfrac{5}{11}, -\dfrac{3}{4}\right)$ and is perpendicular to the y-axis.

48. Passes through $\left(-\dfrac{7}{9}, \dfrac{7}{3}\right)$ and is perpendicular to the x-axis.

49. Passes through $(-61.5, 47.6)$ and is parallel to the line defined by $x = -12$.

50. Passes through $(-0.004, 0.009)$ and is parallel to the line defined by $y = 6$.

Objective 3: Create Linear Functions to Model Data

51. A sales person makes a base salary of $400 per week plus 12% commission on sales. **(See Example 5)**

   a. Write a linear function to model the sales person's weekly salary $S(x)$ for x dollars in sales.

   b. Evaluate $S(8000)$ and interpret the meaning in the context of this problem.

52. At a parking garage in a large city, the charge for parking consists of a flat fee of $2.00 plus $1.50/hr.

   a. Write a linear function to model the cost for parking $P(t)$ for t hours.

   b. Evaluate $P(1.6)$ and interpret the meaning in the context of this problem.

53. Millage rate is the amount per $1000 that is often used to calculate property tax. For example, a home with a $60,000 taxable value in a municipality with a 19 mil tax rate would require $(0.019)(\$60,000) = \$1140$ in property taxes. In one county, homeowners pay a flat tax of $172 plus a rate of 19 mil on the taxable value of a home.

   a. Write a linear function that represents the total property tax $T(x)$ for a home with a taxable value of x dollars.

   b. Evaluate $T(80,000)$ and interpret the meaning in the context of this problem.

54. The average water level in a retention pond is 6.8 ft. During a time of drought, the water level decreases at a rate of 3 in./day.

   a. Write a linear function W that represents the water level $W(t)$ (in ft) t days after a drought begins.

   b. Evaluate $W(20)$ and interpret the meaning in the context of this problem.

For Exercises 55–56, the fixed and variable costs to produce an item are given along with the price at which an item is sold. **(See Example 6)**

   a. Write a linear cost function that represents the cost $C(x)$ to produce x items.

   b. Write a linear revenue function that represents the revenue $R(x)$ for selling x items.

   c. Write a linear profit function that represents the profit $P(x)$ for producing and selling x items.

   d. Determine the break-even point.

55. Fixed cost: $2275
    Variable cost per item: $34.50
    Price at which the item is sold: $80.00

56. Fixed cost: $5625
    Variable cost per item: $0.40
    Price at which the item is sold: $1.30

57. The profit function P is shown for producing and selling x items. Determine the values of x for which

   a. $P(x) = 0$ (the company breaks even)

   b. $P(x) < 0$ (the company experiences a loss)

   c. $P(x) > 0$ (the company makes a profit)

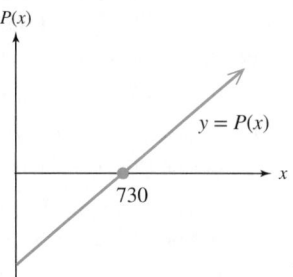

58. The cost and revenue functions C and R are shown for producing and selling x items. Determine the values of x for which

   a. $R(x) = C(x)$ (the company breaks even)

   b. $R(x) < C(x)$ (the company experiences a loss)

   c. $R(x) > C(x)$ (the company makes a profit)

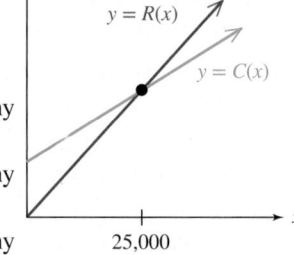

59. A small business makes cookies and sells them at the farmer's market. The fixed monthly cost for use of a Health Department–approved kitchen and rental space at the farmer's market is $790. The cost of labor, taxes, and ingredients for the cookies amounts to $0.24 per cookie, and the cookies sell for $6.00 per dozen. **(See Example 6)**

   a. Write a linear cost function representing the cost $C(x)$ to produce x dozen cookies per month.

   b. Write a linear revenue function representing the revenue $R(x)$ for selling x dozen cookies.

   c. Write a linear profit function representing the profit for producing and selling x dozen cookies in a month.

   d. Determine the number of cookies (in dozens) that must be produced and sold for a monthly profit.

   e. If 150 dozen cookies are sold in a given month, how much money will the business make or lose?

60. A lawn service company charges $60 for each lawn maintenance call. The fixed monthly cost of $680 includes telephone service and depreciation of equipment. The variable costs include labor, gasoline, and taxes and amount to $36 per lawn.

   a. Write a linear cost function representing the monthly cost $C(x)$ for x maintenance calls.

   b. Write a linear revenue function representing the monthly revenue $R(x)$ for x maintenance calls.

   c. Write a linear profit function representing the monthly profit $P(x)$ for x maintenance calls.

   d. Determine the number of lawn maintenance calls needed per month for the company to make money.

   e. If 42 maintenance calls are made for a given month, how much money will the lawn service make or lose?

61. The data in the graph show the wind speed y (in mph) for Hurricane Florence versus the barometric pressure x (in millibars, mb). Hurricane Florence struck North and South Carolina with devastating rain and flooding in 2018. (*Source*: NOAA: www.noaa.gov) **(See Example 7)**

   a. Use the points (946, 130) and (994, 70) to write a linear model for these data.

   b. Interpret the meaning of the slope in the context of this problem.

   c. Use the model from part (a) to estimate the wind speed for a hurricane if the pressure is 950 mb.

   d. The lowest barometric pressure ever recorded for an Atlantic hurricane was 882 mb for Hurricane Wilma in 2005. Would it be reasonable to use the model from part (a) to estimate the wind speed for a pressure of 800 mb?

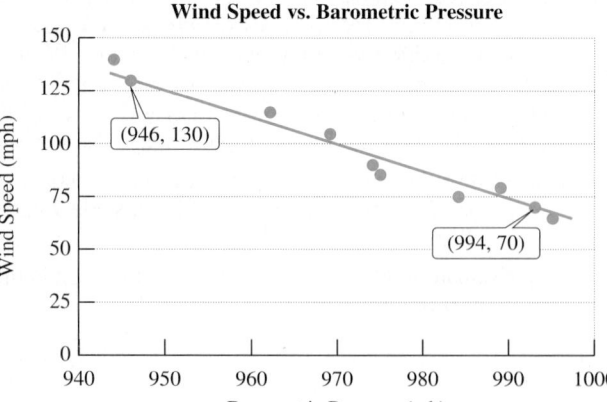

62. Caroline adopted a puppy named Dodger from an animal shelter in Chicago. She recorded Dodger's weight during the first two months. The data in the graph show Dodger's weight y (in lb), x days after adoption.

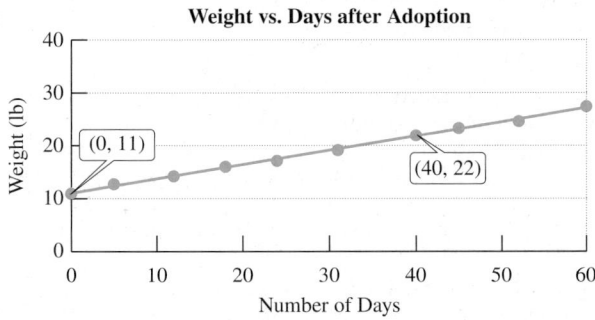

Caroline Celano

   a. Use the points (0, 11) and (40, 22) to write a linear model for these data.

   b. Interpret the meaning of the slope in context.

   c. Interpret the meaning of the y-intercept in context.

   d. If this linear trend continues during Dodger's growth period, how long will it take Dodger to reach 90% of his expected full-grown weight of 70 lb? Round to the nearest day.

   e. Is the model from part (a) reasonable long term?

63. A pediatrician records the age x (in years) and average height y (in inches) for girls between the ages of 2 and 10.

   a. Use the points (2, 35) and (6, 46) to write a linear model for these data.

   b. Interpret the meaning of the slope in context.

   c. Use the model to forecast the average height of 11-year-old girls.

   d. If the height of a girl at age 11 is 90% of her full-grown adult height, use the result of part (c) to estimate the average height of adult women. Round to the nearest tenth of an inch.

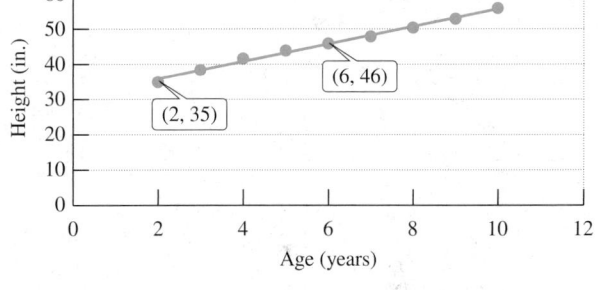

64. The graph shows the number of students enrolled in public colleges for selected years (*Source*: U.S. National Center for Education Statistics, www.nces.ed.gov). The x variable represents the number of years since 2000 and the y variable represents the number of students (in millions).

   a. Use the points (7, 13.5) and (17, 15.5) to write a linear model for these data.

   b. Interpret the meaning of the slope in the context of this problem.

   c. Interpret the meaning of the y-intercept in the context of this problem.

   d. If the linear trend continues beyond the last observed data point, use the model in part (a) to predict the number of students enrolled in public colleges for the year 2024.

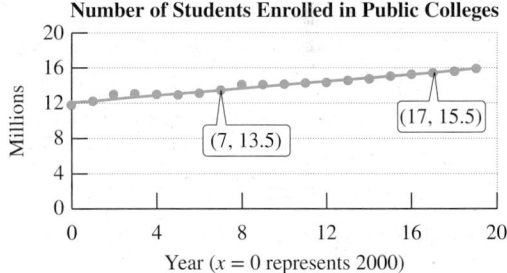

65. The table gives the number of calories and the amount of cholesterol for selected fast-food hamburgers.

   a. Graph the data in a scatter plot using the number of calories as the independent variable x and the amount of cholesterol as the dependent variable y.

   b. Use the data points (480, 60) and (720, 90) to write a linear function that defines the amount of cholesterol $c(x)$ as a linear function of the number of calories x.

   c. Interpret the meaning of the slope in the context of this problem.

   d. Use the model from part (b) to predict the amount of cholesterol for a hamburger with 650 calories.

Hamburger Calories	Cholesterol (mg)
220	35
420	50
460	50
480	60
560	70
590	105
610	65
680	80
720	90

66. The table gives the average gestation period for selected animals and their corresponding average longevity.

   a. Graph the data in a scatter plot using the number of days for gestation as the independent variable x and the longevity as the dependent variable y.

   b. Use the data points (44, 8.5) and (620, 35) to write a linear function that defines longevity $L(x)$ as a linear function of the length of the gestation period x. Round the slope to 3 decimal places and the y-intercept to 2 decimal places.

   c. Interpret the meaning of the slope in the context of this problem.

   d. Use the model from part (b) to predict the longevity for an animal with an 80-day gestation period. Round to the nearest year.

Animal	Gestation Period (days)	Longevity (years)
Rabbit	33	7.0
Squirrel	44	8.5
Fox	57	9.0
Cat	60	11.0
Dog	62	11.0
Lion	109	10.0
Pig	115	10.0
Goat	148	12.0
Horse	337	23.0
Elephant	620	35.0

Objective 4: Create Models Using Linear Regression

For Exercises 67–70, use the scatter plot to determine if a linear regression model appears to be appropriate.

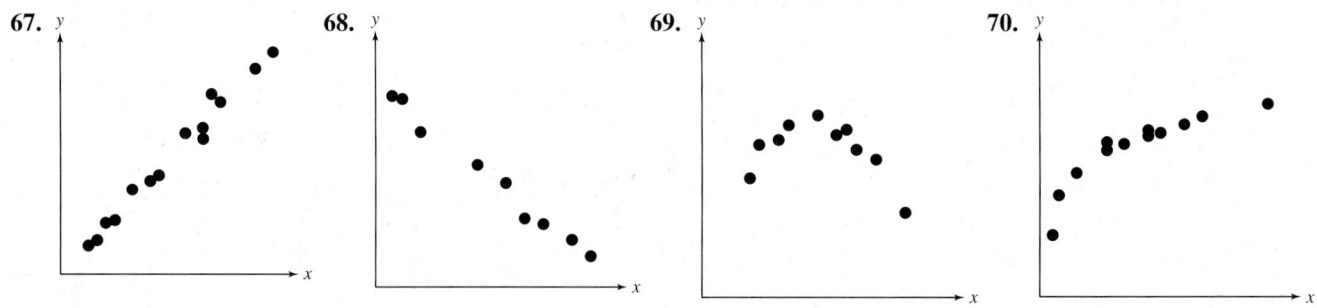

67.   **68.**   **69.**   **70.**

71. The graph in Exercise 61 shows the wind speed y (in mph) of a hurricane versus the barometric pressures x (in mb). The table shows the data from the graph. **(See Example 8)**

   a. Use the data in the table to find the least-squares regression line. Round the slope to 1 decimal place and the y-intercept to the nearest whole unit.

   b. Use a graphing utility to graph the regression line and the observed data.

   c. Use the model in part (a) to approximate the wind speed of a hurricane with a barometric pressure of 950 mb.

   d. By how much do the results of part (c) differ from the result of Exercise 61(c)?

Barometric Pressure (mb) (x)	Wind Speed (mph) (y)
989	80
993	70
995	65
984	75
975	85
974	90
969	105
962	115
946	130
944	140

72. The graph in Exercise 62 shows the weight of Dodger, a puppy recently adopted from an animal shelter. The data in the table give Dodger's weight y (in lb), x days after adoption.

   a. Use the data in the table to find the least-squares regression line. Round the slope to 2 decimal places and the y-intercept to 1 decimal place.

   b. Use a graphing utility to graph the regression line and the observed data.

   c. Use the model in part (a) to approximate the time required for Dodger to reach 90% of his full-grown weight of 70 lb. Round to the nearest day.

   d. By how much do the results of part (c) differ from the result of Exercise 62(d)?

Number of Days (x)	Weight (lb) (y)
0	11.0
5	12.8
12	14.3
18	16.1
24	17.2
31	19.2
40	22.0
45	23.4
52	24.7
60	27.5

73. The graph in Exercise 63 shows the average height of girls based on their age. The data in the table give the average height y (in inches) for girls of age x (in years).

Age (years) (x)	Height (in.) (y)
2	35.00
3	38.50
4	41.75
5	44.00
6	46.00
7	48.00
8	50.50
9	53.00
10	56.00

   a. Use the data in the table to find the least-squares regression line. Round the slope to 2 decimal places and the y-intercept to 1 decimal place.

   b. Use a graphing utility to graph the regression line and the observed data.

   c. Use the model in part (a) to approximate the average height of 11-year-old girls.

   d. If the height of a girl at age 11 is 90% of her full-grown adult height, use the result of part (c) to estimate the average height of adult women. Round to the nearest tenth of an inch.

   e. By how much do the results of part (d) differ from the result of Exercise 63(d)?

74. The graph in Exercise 64 shows the number of students y enrolled in public colleges for selected years x, where x is the number of years since 2000. The table gives a partial list of data from the graph.

Number of Years Since 2000 (x)	Enrollment (millions) (y)
0	11.8
3	12.9
6	13.2
9	14.1
12	14.4
15	15.1
18	15.7

   a. Use the data in the table to find the least-squares regression line. Round the slope and y-intercept to 1 decimal place.

   b. Use a graphing utility to graph the regression line and the observed data.

   c. Assuming that the linear trend continues, use the model from part (a) to predict the number of students enrolled in public colleges for the year 2024.

   d. By how much do the results of part (c) differ from the result of Exercise 64(d)?

75. The data in Exercise 65 give the amount of cholesterol y for a hamburger with x calories.

   a. Use these data to find the least-squares regression line. Round the slope to 3 decimal places and the y-intercept to 2 decimal places.

   b. Use a graphing utility to graph the regression line and the observed data.

   c. Use the regression line to predict the amount of cholesterol in a hamburger with 650 calories. Round to the nearest milligram.

76. The data in Exercise 66 give the average gestation period x (in days) for selected animals and their corresponding average longevity y (in years).

   a. Use these data to find the least-squares regression line. Round the slope to 3 decimal places and the y-intercept to 2 decimal places.

   b. Use a graphing utility to graph the regression line and the observed data.

   c. Use the regression line to predict the longevity for an animal with an 80-day gestation period. Round to the nearest year.

Mixed Exercises

77. Suppose that a line passes through the points $(4, -6)$ and $(2, -1)$. Where will it pass through the x-axis?

78. Suppose that a line passes through the point $(2, -5)$ and $(-4, 7)$. Where will it pass through the x-axis?

79. Write a rule for a linear function $y = f(x)$, given that $f(0) = 4$ and $f(3) = 11$.

80. Write a rule for a linear function $y = g(x)$, given that $g(0) = 7$ and $g(-2) = 4$.

81. Write a rule for a linear function $y = h(x)$, given that $h(1) = 6$ and $h(-3) = 2$.

82. Write a rule for a linear function $y = k(x)$, given that $k(-2) = 10$ and $k(5) = -18$.

Write About It

83. Explain how you can use slope to determine if two nonvertical lines are parallel or perpendicular.

84. State one application of using the point-slope formula.

85. Explain how cost and revenue are related to profit.

86. Explain how to determine the break-even point.

Expanding Your Skills

87. Find an equation of line L.

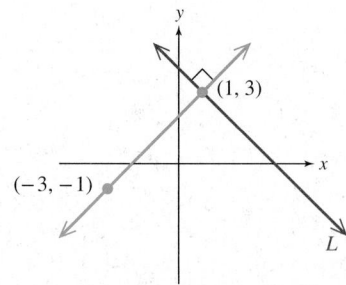

88. In geometry, it is known that the tangent line to a circle at a given point A on the circle is perpendicular to the radius drawn to point A. Suppose that line L is tangent to the given circle at the point $(4, 3)$. Write an equation representing line L.

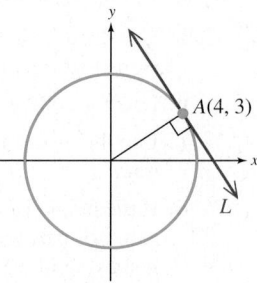

89. In calculus, we can show that the slope of the line drawn tangent to the curve $y = x^3 + 1$ at the point $(c, c^3 + 1)$ is given by $3c^2$. Find an equation of the line tangent to $y = x^3 + 1$ at the point $(-2, -7)$.

90. In calculus, we can show that the slope of the line drawn tangent to the curve $y = \frac{1}{x}$ at the point $\left(c, \frac{1}{c}\right)$ is given by $-\frac{1}{c^2}$. Find an equation of the line tangent to $y = \frac{1}{x}$ at the point $\left(2, \frac{1}{2}\right)$.

For Exercises 91–92, use the fact that a median of a triangle is a line segment drawn from a vertex of the triangle to the midpoint of the opposite side of the triangle.

91. Find an equation of the median of a triangle drawn from vertex $A(5, -2)$ to the side formed by $B(-2, 9)$ and $C(4, 7)$.

92. Find an equation of the median of a triangle drawn from vertex $A(6, -5)$ to the side formed by $B(-4, 1)$ and $C(12, 3)$.

PROBLEM RECOGNITION EXERCISES

Characteristics of Linear Equations

For Exercises 1–20, choose the equation(s) from column B whose graph satisfies the condition described in column A. Give all possible answers.

Column A		Column B
1. Line whose slope is positive.	**2.** Line whose slope is negative.	**a.** $y = -4x$
3. Line that passes through the origin.	**4.** Line that contains the point $(2, 0)$.	**b.** $2x - 4y = 4$
5. Line whose y-intercept is $(0, -3)$.	**6.** Line whose y-intercept is $(0, 0)$.	**c.** $y = -\frac{1}{3}x - 3$
7. Line whose slope is $-\frac{1}{3}$.	**8.** Line whose slope is $\frac{1}{2}$.	**d.** $3x + 5y = 10$
9. Line whose slope is 0.	**10.** Line whose slope is undefined.	**e.** $3y = -9$
11. Line that is parallel to the line with equation $x + 3y = 6$.	**12.** Line perpendicular to the line with equation $x - 4y = -4$.	**f.** $y = 5x - 1$
13. Line that is vertical.	**14.** Line that is horizontal.	**g.** $4x + 1 = 9$
15. Line whose x-intercept is $(12, 0)$.	**16.** Line whose x-intercept is $\left(\frac{1}{5}, 0\right)$.	**h.** $x + 3y = 12$
17. Line that has no x-intercept.	**18.** Line that is perpendicular to the x-axis.	
19. Line with a negative slope and positive y-intercept.	**20.** Line with a positive slope and negative y-intercept.	

PROBLEM RECOGNITION EXERCISES

Comparing Graphs of Functions

In Section 6.1, we will learn additional techniques to graph functions by recognizing characteristics of the functions. In many cases, we can also graph families of functions by relating them to one of several basic graphs. To prepare for the discussion in Section 6.1, use a graphing utility or plot points to graph the basic functions in Exercises 1–8.

1. $y = 1$ **3.** $y = x^2$ **5.** $y = \sqrt{x}$ **7.** $y = |x|$

2. $y = x$ **4.** $y = x^3$ **6.** $y = \sqrt[3]{x}$ **8.** $y = \dfrac{1}{x}$

For Exercises 9–18, graph the functions by plotting points or by using a graphing utility. Explain how the graphs are related.

9. a. $f(x) = x^2$
  b. $g(x) = x^2 + 2$
  c. $h(x) = x^2 - 4$

10. a. $f(x) = |x|$
  b. $g(x) = |x| + 2$
  c. $h(x) = |x| - 4$

11. a. $f(x) = \sqrt{x}$
  b. $g(x) = \sqrt{x - 2}$
  c. $h(x) = \sqrt{x + 4}$

12. a. $f(x) = x^2$
  b. $g(x) = (x - 2)^2$
  c. $h(x) = (x + 3)^2$

13. a. $f(x) = |x|$
  b. $g(x) = -|x|$

14. a. $f(x) = \sqrt{x}$
  b. $g(x) = -\sqrt{x}$

15. a. $f(x) = x^2$
  b. $g(x) = \dfrac{1}{2}x^2$
  c. $h(x) = 2x^2$

16. a. $f(x) = |x|$
  b. $g(x) = \dfrac{1}{3}|x|$
  c. $h(x) = 3|x|$

17. a. $f(x) = \sqrt{x}$
  b. $g(x) = \sqrt{-x}$

18. a. $f(x) = \sqrt[3]{x}$
  b. $g(x) = \sqrt[3]{-x}$

CHAPTER 5   Detailed Summary

SECTION 5.1 The Rectangular Coordinate System and Graphing Utilities

Key Concepts	Examples	Page
The **distance** between two points (x_1, y_1) and (x_2, y_2) in a rectangular coordinate system is given by: $$d = \sqrt{(x_2 - x_1)^2 + (y_2 - y_1)^2}$$ The **midpoint** between the points is given by: $$M = \left(\frac{x_1 + x_2}{2}, \frac{y_1 + y_2}{2}\right)$$	**Example 1:** Given $(-4, -5)$ and $(6, -1)$, find the distance between the points and the midpoint of the line segment between the points. $$d = \sqrt{[6 - (-4)]^2 + [-1 - (-5)]^2}$$ $$d = \sqrt{(10)^2 + (4)^2} = \sqrt{116} = 2\sqrt{29}$$ $$M = \left(\frac{-4 + 6}{2}, \frac{-5 + (-1)}{2}\right) = (1, -3)$$	p. 333 p. 335
Graphing an equation: To graph an equation in a rectangular coordinate system, plot several solutions to the equation to form a general outline of the curve. Then connect the points to form a smooth line or curve. **Determining x- and y-intercepts:** To find an x-intercept $(a, 0)$ of the graph of an equation, substitute 0 for y and solve for x. To find a y-intercept $(0, b)$ of the graph of an equation, substitute 0 for x and solve for y.	**Example 2:** Graph $x = y^2 - 4$. <table><tr><td>**x**</td><td>5</td><td>0</td><td>−3</td><td>−4</td><td>−3</td><td>0</td><td>5</td></tr><tr><td>**y**</td><td>−3</td><td>−2</td><td>−1</td><td>0</td><td>1</td><td>2</td><td>3</td></tr></table> **x-intercept:** $x = y^2 - 4$ $x = (0)^2 - 4$ $x = -4$ **y-intercepts:** $x = y^2 - 4$ $0 = y^2 - 4$ $y = 2$ or $y = -2$ 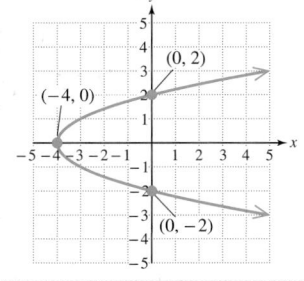	p. 336

SECTION 5.2 Circles

Key Concepts	Examples	Page
A **circle** is the set of all points in a plane that are equidistant from a fixed point called the **center**. The fixed distance between the center and any point on the circle is called the **radius** of the circle. The **standard form** of an equation of a circle with radius r and center (h, k) is $$(x - h)^2 + (y - k)^2 = r^2$$	**Example 1:** Write the standard form of an equation of the circle whose center is $\left(3, -\frac{1}{2}\right)$ and whose radius is 5. $$(x - h)^2 + (y - k)^2 = r^2$$ $$(x - 3)^2 + \left[y - \left(-\frac{1}{2}\right)\right]^2 = (5)^2$$ $$(x - 3)^2 + \left(y + \frac{1}{2}\right)^2 = 25$$	p. 345
An equation of a circle written in the form $x^2 + y^2 + Ax + By + C = 0$ is called the **general form** of an equation of a circle. **Writing an equation in standard form:** By completing the square, we can write an equation of a circle given in general form as an equation in standard form. The purpose is to identify the center and radius from the standard form.	**Example 2:** Write the equation of the circle in standard form. Then identify the center and radius. $$x^2 + y^2 - 12x + 8y + 5 = 0$$ $$x^2 - 12x + 36 + y^2 + 8y + 16 = -5 + 36 + 16$$ $$(x - 6)^2 + (y + 4)^2 = 47$$ The center is $(6, -4)$ and the radius is $\sqrt{47}$.	p. 347
Not all equations of the form $x^2 + y^2 + Ax + By + C = 0$ represent the graph of a circle. Completing the square results in an equation of the form $(x - h)^2 + (y - k)^2 = c$ where c is a constant. • If $c > 0$, then the graph will be a circle with radius $r = \sqrt{c}$. • If $c = 0$, then the graph will be a single point (h, k). The solution set is $\{(h, k)\}$. (Degenerate case) • If $c < 0$, then the solution set is the empty set $\{\ \}$. (Degenerate case)	**Example 3:** $$x^2 + y^2 - 8x + 2y + 17 = 0$$ $$x^2 - 8x + 16 + y^2 + 2y + 1 = -17 + 16 + 1$$ $$(x - 4)^2 + (y + 1)^2 = 0$$ The sum of two squares will equal zero only if each individual term is zero. Therefore, $x = 4$ and $y = -1$. The solution set is $\{(4, -1)\}$.	p. 347

SECTION 5.3 Functions and Relations

Key Concepts	Examples	Page

A set of ordered pairs (x, y) is called a **relation** in x and y. The set of x values is the **domain** of the relation, and the set of y values is the **range** of the relation.	**Example 1:**
	Given the relation $\{(8, 2), (3, 2), (9, 5)\}$,
	    The domain is $\{8, 3, 9\}$.
	    The range is $\{2, 5\}$.
		p. 352

Given a relation in x and y, we say that y **is a function of** x if for each value of x in the domain, there is exactly one value of y in the range.	**Example 2:** same x The relation $\{(4, 6), (3, 5), (4, 1)\}$ is *not* a function. different y The relation $\{(3, 5), (2, 1), (9, 0)\}$ is a function because no two ordered pairs have the same x value but different y values.	p. 352

Vertical line test:

The graph of a relation defines y as a function of x if no vertical line intersects the graph in more than one point.

Example 3:

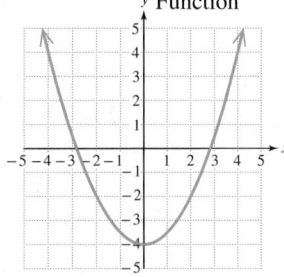 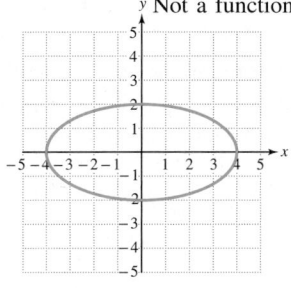

Domain: $(-\infty, \infty)$ Domain: $[-4, 4]$

Range: $[-4, \infty)$ Range: $[-2, 2]$

p. 353

Evaluating a function for different values of x:

A function may be defined by an equation with two variables. For example, $y = f(x)$ defines the variable y in terms of the variable x. Given $y = f(x)$,

- f is the name of the function,
- x is an input variable from the domain.
- $f(x)$ is the function value (or y value) corresponding to x.

A function may be evaluated at different values of x by using substitution.

Example 4:

Given $f(x) = 2x^2 + 3x$,

$f(2) = 2(2)^2 + 3(2) = 14$

$$f(x + 4) = 2(x + 4)^2 + 3(x + 4)$$
$$= 2(x^2 + 8x + 16) + 3x + 12$$
$$= 2x^2 + 16x + 32 + 3x + 12$$
$$= 2x^2 + 19x + 44$$

p. 355

Determining x- and y-intercepts:

Given a function defined by $y = f(x)$,

- The x-intercept(s) are the real solutions to $f(x) = 0$.
- The y-intercept is given by $f(0)$.

Example 5:

Given $f(x) = |x| - 2$,

- To find the x-intercept(s), substitute 0 for $f(x)$:
$$0 = |x| - 2$$
$$x = 2 \quad \text{or} \quad x = -2$$
The x-intercepts are $(2, 0)$, $(-2, 0)$.
- To find the y-intercept, evaluate $f(0)$.
$$f(0) = |0| - 2 = -2 \quad \text{The } y\text{-intercept is } (0, -2).$$

p. 356

Determining domain from $y = f(x)$:

Given $y = f(x)$, the domain of f is the set of real numbers x that when substituted into the function produce a real number. This excludes

- Values of x that make the denominator zero.
- Values of x that make a radicand negative within an even-indexed root.

Example 6:

- Given $f(x) = \dfrac{x + 5}{x - 3}$, the domain is $(-\infty, 3) \cup (3, \infty)$.
- Given $g(x) = \sqrt{x - 3}$, the domain is $[3, \infty)$.

p. 358

Function values and the domain and range of a function can be estimated from the graph of the function.

Example 7:

From the graph, find

a. the domain

b. the range

c. $f(3)$

d. the values of x for which $f(x) = 2$

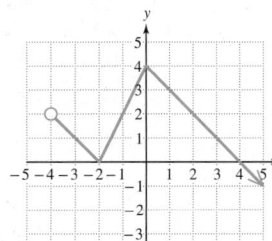

a. $(-4, \infty)$

b. $(-\infty, 4]$

c. $f(3) = 1$

d. $x = -1, x = 2$

p. 360

SECTION 5.4 Linear Equations in Two Variables and Linear Functions

Key Concepts	Examples	Page
Let A, B, and C represent real numbers where A and B are not both zero. A **linear equation** in the variables x and y is an equation that can be written as $Ax + By = C$. The graph of a linear equation is a line. • A horizontal line has an equation of the form $y = k$, where k is a constant real number. • A vertical line has an equation of the form $x = k$, where k is a constant real number.	**Example 1:**	p. 367
The slope of a line passing through the distinct points (x_1, y_1) and (x_2, y_2) is given by $$m = \frac{\Delta y}{\Delta x} = \frac{y_2 - y_1}{x_2 - x_1}$$ • The slope of a horizontal line is 0. • The slope of a vertical line is undefined.	**Example 2:** Given $(-4, 5)$ and $(6, 3)$, $\quad (x_1, y_1)$ and (x_2, y_2) Label the points. $$m = \frac{3 - 5}{6 - (-4)} = \frac{-2}{10} = -\frac{1}{5}$$	p. 369
Slope-intercept form of a line: Given a line with slope m and y-intercept $(0, b)$, the **slope-intercept form** of the line is given by $$y = mx + b$$ • A function defined by $f(x) = mx + b$ $(m \neq 0)$ is a **linear function** (graph is a slanted line). • A function defined by $f(x) = b$ is a **constant function** (graph is a horizontal line).	**Example 3:** Given $2x + 9y = 18$, write the equation in slope-intercept form. $\quad 2x + 9y = 18$ $\qquad\qquad$ Solve for y. $\qquad 9y = -2x + 18$ $\qquad y = -\frac{2}{9}x + 2$ or $f(x) = -\frac{2}{9}x + 2$ The slope is $-\frac{2}{9}$. The y-intercept is $(0, 2)$.	p. 371
Solving equations and inequalities graphically: The x-coordinates of the points of intersection between the graphs of $Y_1 = f(x)$ and $Y_2 = g(x)$ are the solutions to the equation $Y_1 = Y_2$.	**Example 4:** Solve the equation and inequalities graphically. **a.** $2x + 2 = -x + 5$ **b.** $2x + 2 < -x + 5$ **c.** $2x + 2 \geq -x + 5$ **Solutions:** **a.** $\{1\}$ **b.** $(-\infty, 1)$ **c.** $[1, \infty)$	p. 374

SECTION 5.5 Applications of Linear Equations and Modeling

Key Concepts	Examples	Page
The **point-slope formula** for a line is given by $y - y_1 = m(x - x_1)$ where m is the slope of the line and (x_1, y_1) is a point on the line.	**Example 1:** Use the point-slope formula to find an equation of the line passing through the point $(2, 1)$ and having a slope of 4. $$y - y_1 = m(x - x_1)$$ $$y - 1 = 4(x - 2)$$ $$y - 1 = 4x - 8$$ $$y = 4x - 7$$ 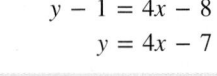	p. 381
Slopes of parallel and perpendicular lines: • If m_1 and m_2 represent the slopes of two nonvertical parallel lines, then $m_1 = m_2$. • If m_1 and m_2 represent the slopes of two nonvertical perpendicular lines, then $m_1 = -\dfrac{1}{m_2}$ or equivalently $m_1 m_2 = -1$.	**Example 2:** The slope of a line is $-\frac{1}{5}$. • The slope of a line parallel to this line is $-\frac{1}{5}$. • The slope of a line perpendicular to this line is 5.	p. 383

Applications of linear equations in two variables:

In many-day-to-day applications, two variables are related linearly.

• A linear model can be made from two data points that represent the general trend of the data.

• Alternatively, the least-squares regression line is a model that utilizes *all* observed data points. From Example 3 we have

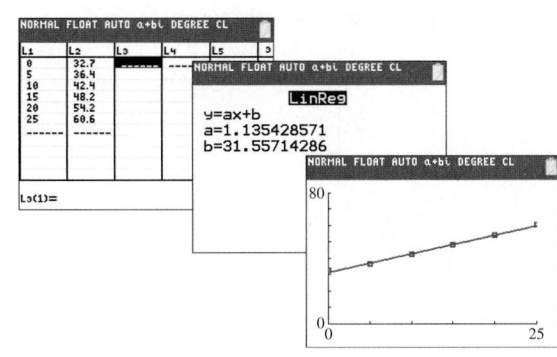

Least-squares regression line: $y = 1.14x + 31.6$

Example 3:

The yearly revenue y (in \$1000) for a small business has increased since the year the business opened. Use the points $(10, 42.4)$ and $(15, 48.2)$ to write a linear model that approximates the revenue as a function of the number of years x since the business opened.

x	0	5	10	15	20	25
y	32.7	36.4	42.4	48.2	54.2	60.6

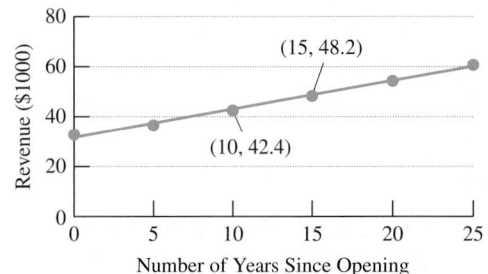

$$m = \frac{48.2 - 42.4}{15 - 10} = 1.16$$

$$y - y_1 = m(x - x_1)$$
$$y - 42.4 = 1.16(x - 10)$$
$$y = 1.16x + 30.8$$

p. 384

Linear cost, revenue, and profit functions:

A **linear cost function** models the cost $C(x)$ to produce x items.

$$C(x) = mx + b$$

where m is the variable cost per item, and b is the fixed cost.

A **linear revenue function** models the revenue $R(x)$ for selling x items.

$$R(x) = px$$

where p is the price per item.

A **profit function** models the profit for producing and selling x items.

$$P(x) = R(x) - C(x)$$

Profit is revenue minus cost. The break-even point is where profit is zero (that is, where revenue equals cost).

Example 4:   p. 385

An herbal tea company sells tea for \$3.50 per box. The tea and packaging materials cost \$0.60 per box, and the fixed monthly costs (labor, rent, and so on) are \$3800 per month.

The cost to produce x boxes of tea during a given month is

$$C(x) = 0.60x + 3800$$

The revenue for selling x boxes of tea during a given month is

$$R(x) = 3.50x$$

The profit for producing and selling x boxes during a given month is

$$P(x) = 3.50x - (0.60x + 3800)$$
$$= 2.90x - 3800$$

To make a profit,

$$P(x) > 0$$
$$2.90x - 3800 > 0$$
$$2.90x > 3800$$
$$x > 1310$$

The company will make a profit if more than 1310 boxes of tea are sold.

CHAPTER 5   Review Exercises

SECTION 5.1

For Exercises 1–2,

  a. Find the exact distance between the points.

  b. Find the midpoint of the line segment whose endpoints are the given points.

  1. $(-1, 8)$ and $(4, -2)$

  2. $(-5, -9)$ and $(0, 3)$

  3. Determine if the given ordered pair is a solution to the equation $4|x - 1| + y = 18$.

   a. $(-3, 2)$ **b.** $(5, -2)$

For Exercises 4–6, determine the x- and y-intercepts of the graph of the equation.

  4. $-3y + 4x = 6$

  5. $x = |y + 7| - 3$

  6. $\dfrac{(x + 4)^2}{9} + \dfrac{y^2}{4} = 1$

For Exercises 7–8, graph the equation by plotting points.

  7. $x = \dfrac{1}{2}y^2$ **8.** $y = |x - 3|$

9. Find the length of the diagonal shown.

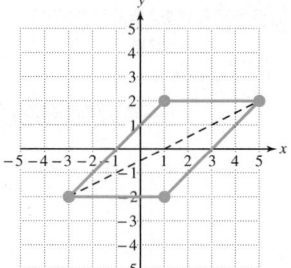

SECTION 5.2

For Exercises 10–12, determine the center and radius of the circle.

  10. $(x - 4)^2 + (y + 3)^2 = 4$

  11. $x^2 + \left(y - \dfrac{3}{2}\right)^2 = 17$

  12. $(x + 1.4)^2 + y^2 = 12$

For Exercises 13–16, information about a circle is given.

a. Write an equation of the circle in standard form.

b. Graph the circle.

13. Center: $(-3, 1)$; Radius: $\sqrt{11}$

14. Center: $(0, 0)$; Radius: 3.2

15. Endpoints of a diameter $(7, 5)$ and $(1, -3)$

16. The center is in quadrant IV, the radius is 4, and the circle is tangent to both the x- and y-axes.

For Exercises 17–18, (a) write the equation of the circle in standard form and (b) identify the center and radius.

17. $x^2 + y^2 + 10x - 2y + 17 = 0$

18. $x^2 + y^2 - 8y + 3 = 0$

For Exercises 19–20, determine the solution set to the equation.

19. $(x + 3)^2 + (y - 5)^2 = 0$

20. $x^2 + y^2 + 6x - 4y + 15 = 0$

SECTION 5.3

21. The table lists four Olympic athletes and the number of Olympic medals won by the athlete.

Athlete (x)	Number of Medals (y)
Dara Torres (swimming)	12
Carl Lewis (track and field)	10
Bonnie Blair (speed skating)	6
Michael Phelps (swimming)	16

a. Write a set of ordered pairs (x, y) that defines the relation.

b. Write the domain of the relation.

c. Write the range of the relation.

d. Determine if the relation defines y as a function of x.

For Exercises 22–25, determine if the relation defines y as a function of x.

22.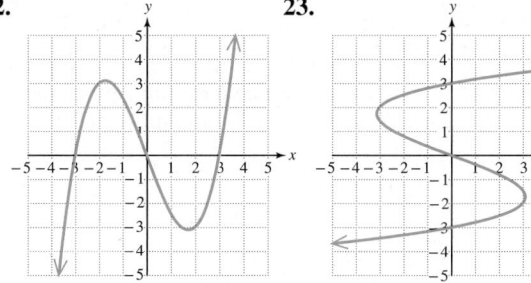

23.

24. $x^2 + (y - 3)^2 = 4$ **25.** $x^2 + y - 3 = 4$

26. Evaluate $f(x) = -2x^2 + 4x$ for the values of x given.

a. $f(0)$ b. $f(-1)$ c. $f(3)$

d. $f(t)$ e. $f(x + 4)$

27. Given $f = \{(3, -1), (1, 5), (-2, 4), (0, 4)\}$,

a. Determine $f(1)$.

b. Determine $f(0)$.

c. For what value(s) of x is $f(x) = -1$?

28. A department store marks up the price of a power drill by 32% of the price from the manufacturer. The price $P(x)$ (in \$) to a customer after a 6.5% sales tax is given by $P(x) = 1.065(x + 0.32x)$, where x is the cost of the drill from the manufacturer. Evaluate $P(189)$ and interpret the meaning in the context of this problem.

For Exercises 29–30, determine the x- and y-intercepts for the given function.

29. $p(x) = |x - 3| - 1$

30. $q(x) = -\sqrt{x} + 2$

For Exercises 31–32, determine the domain and range of the function.

31. 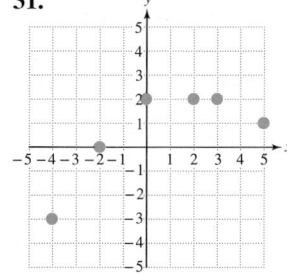   **32.**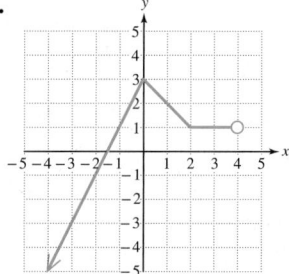

For Exercises 33–36, write the domain in interval notation.

33. $k(x) = \dfrac{4}{2x + 9}$ **34.** $g(x) = \dfrac{6}{|x| - 3}$

35. $p(x) = \sqrt{3x - 6}$ **36.** $n(x) = \dfrac{10}{\sqrt{2 - x}}$

37. Use the graph of $y = f(x)$ to

a. Determine $f(-2)$.

b. Determine $f(3)$.

c. Find all x for which $f(x) = -1$.

d. Find all x for which $f(x) = -4$.

e. Determine the x-intercept(s).

f. Determine the y-intercept.

g. Determine the domain of f.

h. Determine the range of f.

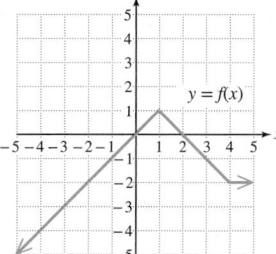

38. Write a relationship for a function whose $f(x)$ value is 4 less than two times the square of x.

SECTION 5.4

For Exercises 39–42, graph the equation and determine the *x*- and *y*-intercepts.

39. $-2x + 4y = 8$ **40.** $-4x = 5y$ **41.** $y = 2$ **42.** $3x = 5$

For Exercises 43–44, determine the slope of the line passing through the given points.

43. $(4, -2)$ and $(-12, -4)$ **44.** $\left(-3, \dfrac{2}{3}\right)$ and $\left(1, -\dfrac{4}{3}\right)$

45. What is the slope of a line parallel to the *x*-axis?

46. What is the slope of a line with equation $x = -2$?

47. What is the slope of a line perpendicular to a line with equation $y = 1$?

48. Determine if the function is linear, constant, or neither.

  a. $f(x) = -\dfrac{3}{2}x$ **b.** $g(x) = -\dfrac{3}{2x}$ **c.** $h(x) = -\dfrac{3}{2}$

For Exercises 49–50, use slope-intercept form to write an equation of the line that passes through the given point and has the given slope. Then write the equation using function notation where $y = f(x)$.

49. $(1, -5)$ and $m = -\dfrac{2}{3}$ **50.** $\left(2, \dfrac{1}{4}\right)$ and $m = 0$

For Exercises 51–52, use the graph to solve the equation and inequalities. Write the solutions to the inequalities in interval notation.

51. **a.** $2x + 1 = -x + 4$

  b. $2x + 1 < -x + 4$

  c. $2x + 1 \geq -x + 4$

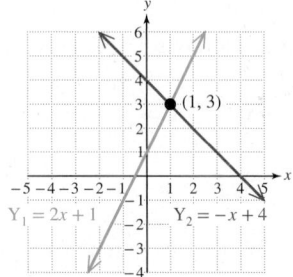

52. **a.** $3(x - 4) + 6 = 0$

  b. $3(x - 4) + 6 \leq 0$

  c. $3(x - 4) + 6 > 0$

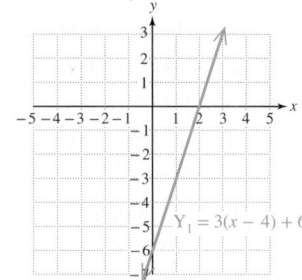

SECTION 5.5

53. If the slope of a line is $\frac{2}{3}$,

  a. Determine the slope of a line parallel to the given line.

  b. Determine the slope of a line perpendicular to the given line.

54. Given a line L_1 defined by L_1: $2x - 4y = 3$, determine if the equations given in parts (a)–(c) represent a line parallel to L_1, perpendicular to L_1, or neither parallel nor perpendicular to L_1.

  a. $12x + 6y = 6$ **b.** $3y = 1.5x - 5$

  c. $4x + 8y = 8$

For Exercises 55–61, write an equation of the line having the given conditions. Write the answer in slope-intercept form if possible.

55. Passes through $(-2, -7)$ and $m = 3$.

56. Passes through $(0, 5)$ and $m = -\dfrac{2}{5}$.

57. Passes through $(1.1, 5.3)$ and $(-0.9, 7.1)$.

58. Passes through $(5, -7)$ and the slope is undefined.

59. Passes through $(2, -6)$ and is parallel to the line defined by $2x - y = 4$.

60. Passes through $(-2, 3)$ and is perpendicular to the line defined by $5y = 2x$.

61. The line is perpendicular to the *y*-axis and the *y*-intercept is $(0, 7)$.

62. A car has a 15-gal tank for gasoline and gets 30 mpg on a highway while driving 60 mph. Suppose that the driver starts a trip with a full tank of gas and travels 450 mi on the highway at an average speed of 60 mph.

  a. Write a linear model representing the amount of gas $G(t)$ left in the tank t hours into a trip.

  b. Evaluate $G(4.5)$ and interpret the meaning in the context of this problem.

63. A dance studio has fixed monthly costs of $1500 that include rent, utilities, insurance, and advertising. The studio charges $60 for each private lesson, but has a variable cost for each lesson of $35 to pay the instructor.

  a. Write a linear cost function representing the cost to the studio $C(x)$ to hold x private lessons for a given month.

  b. Write a linear revenue function representing the revenue $R(x)$ for holding x private lessons for the month.

  c. Write a linear profit function representing the profit $P(x)$ for holding x private lessons for the month.

  d. Determine the number of private lessons that must be held for the studio to make a profit.

  e. If 82 private lessons are held during a given month, how much money will the studio make or lose?

64. The height y (in meters) of a volcano in the southeast Pacific Ocean is recorded in the table for selected years since 1960.

Number of Years Since 1960, x	Height (m) y
0	166
10	290
20	408
30	526
40	650
50	760
54	813

Graeme Knox - k-island photography/ Getty Images

a. Graph the data in a scatter plot.

b. Use the points (0, 166) and (40, 650) to write a linear function that defines the height y of the volcano, x years since 1960.

c. Interpret the meaning of the slope in the context of this problem.

d. Use the model in part (b) to predict the height of the volcano in the year 2030 assuming that the linear trend continues.

65. Refer to the data given in Exercise 64.

a. Use a graphing utility to find the least-squares regression line. Round the slope to 1 decimal place and the y-intercept to the nearest whole unit.

b. Use a graphing utility to graph the regression line and the observed data.

c. In the event that the linear trend continues, use the model from part (a) to predict the height of the volcano in the year 2030.

CHAPTER 5   Test

1. Given the points $(-1, -2)$ and $(5, -6)$,

a. Find the distance between the points.

b. Find the midpoint of the line segment whose endpoints are the given points.

2. The endpoints of a diameter of a circle are $(-2, 3)$ and $(8, -5)$.

a. Determine the center of the circle.

b. Determine the radius of the circle.

c. Write an equation of the circle in standard form.

3. Given $x = |y| - 4$,

a. Determine the x- and y-intercepts of the graph of the equation.

b. Does the equation define y as a function of x?

4. Given $x^2 + y^2 + 14x - 10y + 70 = 0$,

a. Write the equation of the circle in standard form.

b. Identify the center and radius.

For Exercises 5–6, determine if the relation defines y as a function of x.

5.

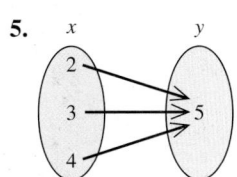

6.

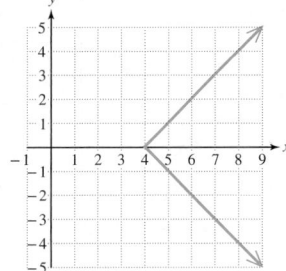

7. Given $f(x) = -2x^2 + 7x - 3$, find

a. $f(-1)$.

b. $f(x + h)$.

c. The x-intercepts of the graph of f.

d. The y-intercept of the graph of f.

8. Use the graph of $y = f(x)$ to estimate

a. $f(0)$.

b. $f(-4)$.

c. The values of x for which $f(x) = 2$.

d. The domain.

e. The range.

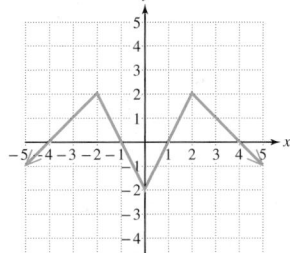

For Exercises 9–10, write the domain in interval notation.

9. $f(w) = \dfrac{2w}{3w + 7}$

10. $f(c) = \sqrt{4 - c}$

11. Given $3x = -4y + 8$,

    a. Identify the slope.

    b. Identify the y-intercept.

    c. Graph the line.

    d. What is the slope of a line perpendicular to this line?

    e. What is the slope of a line parallel to this line?

12. Write an equation of the line passing through the point $(-2, 6)$ and perpendicular to the line defined by $x + 3y = 4$.

13. Write an equation of the line passing through the points $(3, -6)$ and $(4, -2)$. Write the answer in slope-intercept form.

For Exercises 14–15, graph the equation.

14. $x^2 + \left(y + \dfrac{5}{2}\right)^2 = 9$

15. $y = |x + 2|$

16. Use the graph to solve the equation and inequalities. Write the solutions to the inequalities in interval notation.

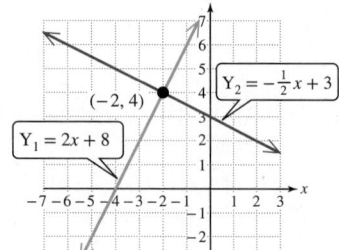

    a. $2x + 8 = -\dfrac{1}{2}x + 3$

    b. $2x + 8 < -\dfrac{1}{2}x + 3$

    c. $2x + 8 \geq -\dfrac{1}{2}x + 3$

17. The number of people y that attend a weekly bingo game at an adult recreation center is given in the table for selected weeks, x.

Week Number, x	Number of attendees, y
1	8
3	21
6	30
9	40
12	46
15	56
18	68

    a. Graph the data in a scatter plot.

    b. Use the points $(1, 8)$ and $(9, 40)$ to write a linear function that defines the number of attendees as a function of week number.

    c. Interpret the meaning of the slope in the context of this problem.

    d. Use the model in part (b) to predict the number of attendees in week 24 assuming that the linear trend continues.

18. Refer to the data given in Exercise 17.

    a. Use a graphing utility to find the least-squares regression line. Round the slope and y-intercept to 1 decimal place.

    b. Use a graphing utility to graph the regression line and the observed data.

    c. In the event that the linear trend continues, use the model from part (a) to predict the number of attendees in week 24.

Transformations and Analysis of Functions

6

Chapter Outline

lovelyday12/Shutterstock

Each year the IRS (Internal Revenue Service) publishes tax rates that tell us how much federal income tax we need to pay based on our taxable income. For example, for a recent year a single person with a taxable income of more than $38,700 but not more than $82,500 pays $4453.50 plus 22% of the amount over $38,700 in federal income tax. Thus, with a taxable income of $65,000 a single person would pay $4453.50 + 0.22($65,000 − $38,700) or $10,239.50. However, finding taxable income is not always trivial. There are numerous variables that come into play. The IRS takes into account exemptions, deductions, and tax credits among other things.

In Chapter 6, we continue our study of functions and analyze their behavior. We'll construct "new" functions from "old," and we'll use functions to create models that relate one variable to another in applications. For example, by using a tool called a **piecewise-defined function** we can model the amount of federal income tax owed based on a person's taxable income.

Schedule X–If your filing status is "Single"

If taxable income is over	But not over	The tax is	Of the amount over
$0	$9,525	$0.00 + 10%	$0
$9,525	$38,700	$952.50 + 12%	$9,525
$38,700	$82,500	$4,453.50 + 22%	$38,700
$82,500	$157,500	$14,089.50 + 24%	$82,500
$157,500	$200,000	$32,089.50 + 32%	$157,500
$200,000	$500,000	$45,689.50 + 35%	$200,000
$500,000	—	$150,689.50 + 37%	$500,000

(*Source*: Internal Revenue Service, www.irs.gov)

Transformations of Graphs

1. Recognize Basic Functions

A function defined by $f(x) = mx + b$ is a linear function, and its graph is a line in a rectangular coordinate system. In addition to linear functions, we will learn to identify other categories of functions and the shapes of their graphs (Table 6-1).

TIP The functions given in Table 6-1 were introduced in Section 5.1, Exercises 35–40, and in the Problem Recognition Exercises on page 397.

Table 6-1   Basic Functions and Their Graphs

1. Linear functions   Constant functions
   $f(x) = mx + b$ $f(x) = b$

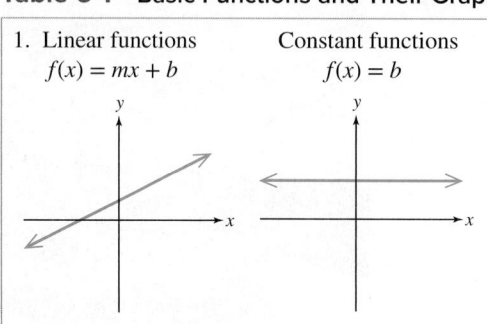

2. Identity function: $f(x) = x$

x	$f(x)$
-2	-2
-1	-1
0	0
1	1
2	2

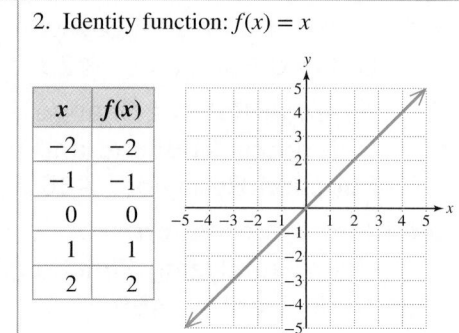

3. Quadratic function: $f(x) = x^2$
   (graph is a parabola)

x	$f(x)$
-2	4
-1	1
0	0
1	1
2	4

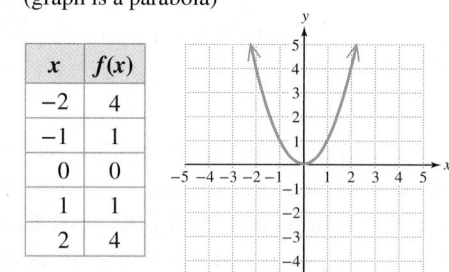

4. Cube function: $f(x) = x^3$

x	$f(x)$
-2	-8
-1	-1
0	0
1	1
2	8

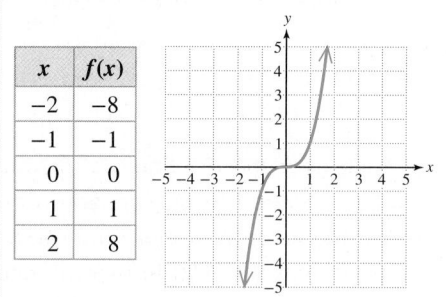

5. Square root function: $f(x) = \sqrt{x}$

x	$f(x)$
0	0
1	1
4	2
9	3
16	4

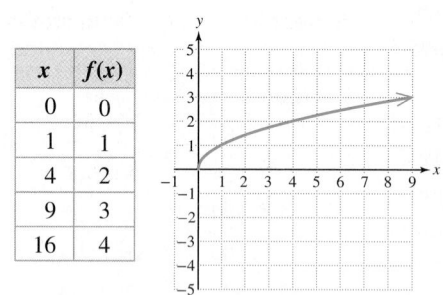

6. Cube root function: $f(x) = \sqrt[3]{x}$

x	$f(x)$
-8	-2
-1	-1
0	0
1	1
8	2

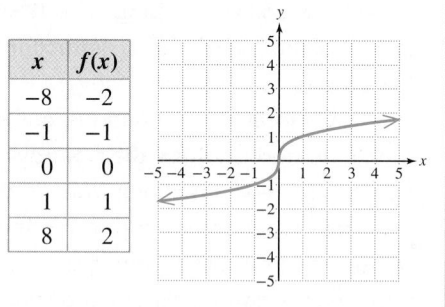

7. Absolute value function: $f(x) = |x|$

x	$f(x)$
-2	2
-1	1
0	0
1	1
2	2

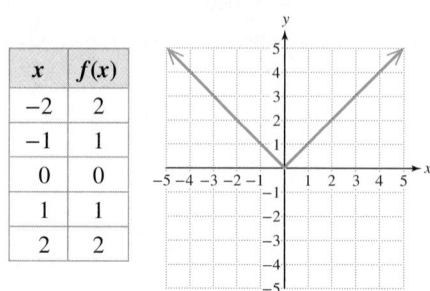

8. Reciprocal function: $f(x) = \dfrac{1}{x}$

x	$f(x)$
-2	$-\frac{1}{2}$
-1	-1
$-\frac{1}{2}$	-2
$\frac{1}{2}$	2
1	1
2	$\frac{1}{2}$

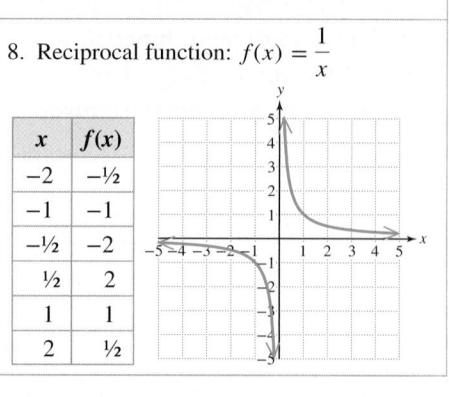

Notice that the graph of $f(x) = \frac{1}{x}$ gets close to (but never touches) the y-axis as x gets close to zero. Likewise, as x approaches ∞ and $-\infty$, the graph approaches the x-axis without touching the x-axis. The x- and y-axes are called **asymptotes** of f and will be studied in detail in Chapter 8.

2. Apply Vertical and Horizontal Translations (Shifts)

We will call the eight basic functions pictured in Table 6-1 "parent" functions. Other functions that share the characteristics of a parent function are grouped as a "family" of functions. For example, consider the functions defined by $g(x) = x^2 + 2$ and $h(x) = x^2 - 4$, pictured in Figure 6-1.

x	$f(x) = x^2$	$g(x) = x^2 + 2$	$h(x) = x^2 - 4$
-3	9	11	5
-2	4	6	0
-1	1	3	-3
0	0	2	-4
1	1	3	-3
2	4	6	0
3	9	11	5

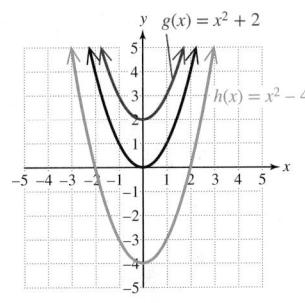

Figure 6-1

The graphs of g and h both resemble the graph of $f(x) = x^2$, but are shifted vertically upward or downward. The table of points reveals that for corresponding x values, the values of $g(x)$ are 2 more than the values of $f(x)$. Thus, the graph is shifted *upward* 2 units. Likewise, the values of $h(x)$ are 4 less than the values of $f(x)$ and the graph is shifted *downward* 4 units. Such shifts are called **translations**. These observations are consistent with the following rules.

TIP For each ordered pair (x, y) on the graph of $y = f(x)$, the corresponding point
- $(x, y + k)$ is on the graph of $y = f(x) + k$.
- $(x, y - k)$ is on the graph of $y = f(x) - k$.

Vertical Translations of Graphs

Consider a function defined by $y = f(x)$. Let k represent a positive real number.

- The graph of $y = f(x) + k$ is the graph of $y = f(x)$ shifted k units *upward*.
- The graph of $y = f(x) - k$ is the graph of $y = f(x)$ shifted k units *downward*.

EXAMPLE 1   **Translating a Graph Vertically**

Use translations to graph the given functions.

a. $g(x) = |x| - 3$ **b.** $h(x) = x^3 + 2$

Solution:

a.

The parent function for $g(x) = |x| - 3$ is $f(x) = |x|$.

The graph of g (shown in blue) is the graph of f shifted *downward* 3 units. For example the point $(0, 0)$ on the graph of $f(x) = |x|$ corresponds to $(0, -3)$ on the graph of $g(x) = |x| - 3$.

b.

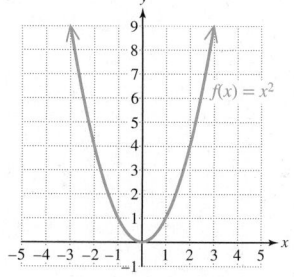

The parent function for $h(x) = x^3 + 2$ is $f(x) = x^3$.

The graph of h (shown in blue) is the graph of f shifted *upward* 2 units. For example:

The point $(0, 0)$ on the graph of $f(x) = x^3$ corresponds to $(0, 2)$ on the graph of $h(x) = x^3 + 2$.

The point $(1, 1)$ on the graph of $f(x) = x^3$ corresponds to $(1, 3)$ on the graph of $h(x) = x^3 + 2$.

Skill Practice 1 Use translations to graph the given functions.

a. $g(x) = \sqrt{x} - 2$ **b.** $h(x) = \sqrt{x} + 3$

EXAMPLE 2   Translating a Graph Horizontally

Graph the function defined by $g(x) = (x + 1)^2$.

Solution:

The graph of $f(x) = x^2$ is a parabola with x-intercept $(0, 0)$. (See Figure 6-2.) Because the constant 1 is added to the x variable, we expect the graph defined by $g(x) = (x + 1)^2$ to be the same as the graph of $f(x) = x^2$, but shifted in the x direction (horizontally). To determine the direction of the shift (left or right), we can find the x-intercept of $g(x) = (x + 1)^2$.

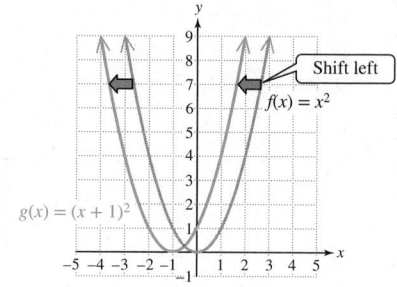

Figure 6-2

$$g(x) = (x + 1)^2$$
$$0 = (x + 1)^2 \qquad \text{To find the } x\text{-intercept, substitute 0 for } g(x).$$
$$0 = x + 1 \qquad \text{Apply the square root property.}$$
$$x = -1 \qquad \text{Isolate } x.$$

The x-intercept is $(0, -1)$, which suggests that the graph of $g(x) = (x + 1)^2$ is the same as the graph of $f(x) = x^2$ shifted to the *left* 1 unit. (See Figure 6-3.) To confirm this result, we can make a table of points.

Figure 6-3

x	$f(x) = x^2$	$g(x) = (x + 1)^2$
-4	16	9
-3	9	4
-2	4	1
-1	1	0
0	0	1
1	1	4
2	4	9
3	9	16
4	16	25

Skill Practice 2 Graph the function defined by $g(x) = |x + 2|$.

FOR REVIEW

If k is a nonnegative real number, then the square root property indicates that the solution to $x^2 = k$ is $x = \pm k$. (See Section 3.5.)

Thus, given

$$(x + 1)^2 = 0$$
$$x + 1 = \pm 0$$
$$x + 1 = 0$$
$$x = -1$$

Answers

1. a.–b.

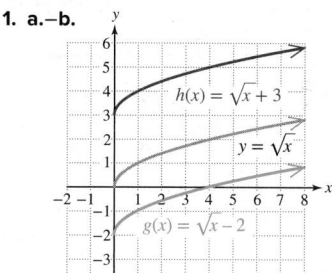

2.

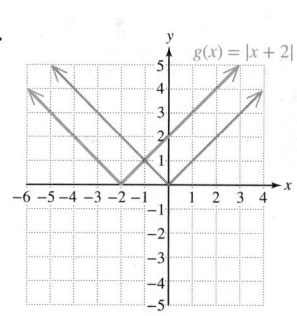

Using similar logic to that used in Example 2, we can show that the graph of $h(x) = (x - 1)^2$ is the graph of $f(x) = x^2$ shifted to the right 1 unit. These observations lead to the following rules.

Horizontal Translations of Graphs

Consider a function defined by $y = f(x)$. Let h represent a positive real number.

- The graph of $y = f(x - h)$ is the graph of $y = f(x)$ shifted h units to the *right*.
- The graph of $y = f(x + h)$ is the graph of $y = f(x)$ shifted h units to the *left*.

TIP   Consider a positive real number h. To graph $y = f(x - h)$ or $y = f(x + h)$, shift the graph of $y = f(x)$ horizontally in the opposite direction of the sign within parentheses. The graph of $y = f(x - h)$ is a shift in the positive x direction. The graph of $y = f(x + h)$ is a shift in the negative x direction.

EXAMPLE 3   Translating a Graph Horizontally and Vertically

Use translations to graph the function defined by $p(x) = \sqrt{x - 3} - 2$.

Solution:

The parent function for $p(x) = \sqrt{x - 3} - 2$ is $f(x) = \sqrt{x}$. The graph of p is the graph of f shifted to the right 3 units and downward 2 units. It can be helpful to make the translations in two stages. First, shift the graph to the right and then downward by moving several "strategic" points from function f. Second, after moving the strategic points, connect the points to sketch the new function.

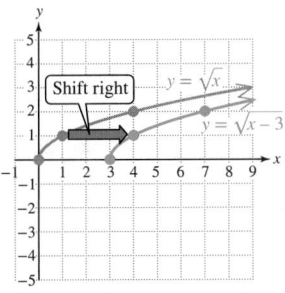

Figure 6-4

Shift the graph of f to the right 3 units:
The point $(0, 0)$ shifts to $(3, 0)$.
The point $(1, 1)$ shifts to $(4, 1)$.
The point $(4, 2)$ shifts to $(7, 2)$.
See Figure 6-4.

Shift the result downward 2 units:
The point $(3, 0)$ shifts down to $(3, -2)$.
The point $(4, 1)$ shifts down to $(4, -1)$.
The point $(7, 2)$ shifts down to $(7, 0)$.
See Figure 6-5.

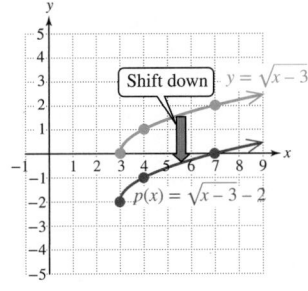

Figure 6-5

Skill Practice 3   Use translations to graph $q(x) = \sqrt{x + 2} - 5$.

Answer

3.

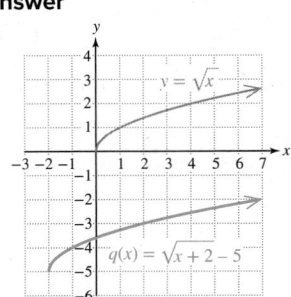

3. Apply Vertical and Horizontal Shrinking and Stretching

Horizontal and vertical translations of functions are called **rigid transformations** because the shape of the graph is not affected. We now look at **nonrigid transformations**. These operations cause a distortion of the graph (either an elongation or contraction in the horizontal or vertical direction). We begin by investigating the functions defined by $y = f(x)$ and $y = a \cdot f(x)$, where a is a positive real number.

EXAMPLE 4   Graphing a Function with a Vertical Stretch or Shrink

Graph the functions.

a. $f(x) = |x|$ **b.** $g(x) = 2|x|$ **c.** $h(x) = \dfrac{1}{2}|x|$

Solution:

x	$f(x) = |x|$	$g(x) = 2\,|x|$	$h(x) = \frac{1}{2}|x|$
-3	3	6	$\frac{3}{2}$
-2	2	4	1
-1	1	2	$\frac{1}{2}$
0	0	0	0
1	1	2	$\frac{1}{2}$
2	2	4	1
3	3	6	$\frac{3}{2}$

double

multiply by $\frac{1}{2}$

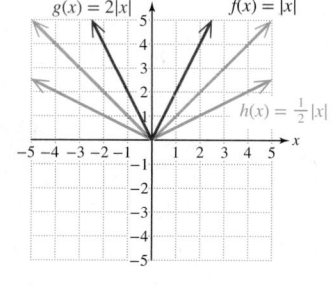

For a given value of x, the value of $g(x)$ is twice the value of $f(x)$. Therefore, the graph of g is elongated or stretched vertically by a factor of 2.

For a given value of x, the value of $h(x)$ is one-half that of $f(x)$. Therefore, the graph of h is shrunk vertically.

Skill Practice 4   Graph the functions.

a. $f(x) = x^2$ **b.** $g(x) = 3x^2$ **c.** $h(x) = \dfrac{1}{3}x^2$

Vertical Shrinking and Stretching of Graphs

Consider a function defined by $y = f(x)$. Let a represent a positive real number.

- If $a > 1$, then the graph of $y = af(x)$ is the graph of $y = f(x)$ stretched vertically by a factor of a.
- If $0 < a < 1$, then the graph of $y = af(x)$ is the graph of $y = f(x)$ shrunk vertically by a factor of a.

Note: For any point (x, y) on the graph of $y = f(x)$, the point (x, ay) is on the graph of $y = af(x)$.

Answer

4.

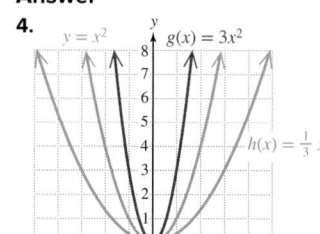

A function may also be stretched or shrunk horizontally.

Horizontal Shrinking and Stretching of Graphs

Consider a function defined by $y = f(x)$. Let a represent a positive real number.

- If $a > 1$, then the graph of $y = f(ax)$ is the graph of $y = f(x)$ shrunk horizontally by a factor of a.
- If $0 < a < 1$, then the graph of $y = f(ax)$ is the graph of $y = f(x)$ stretched horizontally by a factor of a.

Note: For any point (x, y) on the graph of $y = f(x)$, the point $\left(\frac{x}{a}, y\right)$ is on the graph of $y = f(ax)$.

A point (x, y) on the graph of $y = f(x)$ corresponds to the point $\left(\dfrac{x}{a}, y\right)$ on the graph of $y = f(ax)$. Since the x-coordinate is multiplied by the *reciprocal* of a, values of a greater than 1 actually compress (shrink) the graph horizontally toward the y-axis. Values of a between 0 and 1 *stretch* the graph horizontally away from the y-axis. This is demonstrated in Example 5.

EXAMPLE 5   **Graphing a Function with a Horizontal Shrink or Stretch**

The graph of $y = f(x)$ is shown. Graph

a. $y = f(2x)$

b. $y = f\left(\dfrac{1}{2}x\right)$

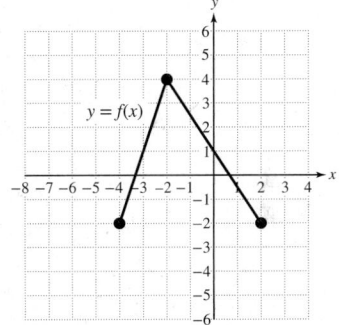

Solution:

a. $f(2x)$ is in the form $f(ax)$ with $a = 2 > 1$. The graph of $y = f(2x)$ is the graph of $y = f(x)$ horizontally compressed.

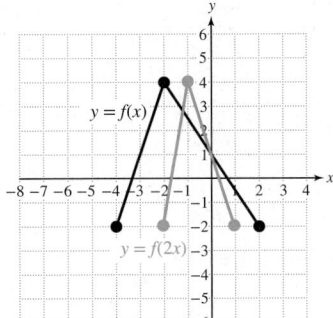

The graph of f has the following "strategic" points that define the shape of the function: $(-4, -2)$, $(-2, 4)$, and $(2, -2)$.

To graph $y = f(2x)$, divide each x value by 2.
$(-4, -2)$ becomes $\left(\frac{-4}{2}, -2\right) = (-2, -2)$.
$(-2, 4)$ becomes $\left(\frac{-2}{2}, 4\right) = (-1, 4)$.
$(2, -2)$ becomes $\left(\frac{2}{2}, -2\right) = (1, -2)$.

The graph of $y = f(2x)$ is shown in blue.

TIP Dividing the x values by $\frac{1}{2}$ is the same as multiplying the x values by 2.

b. $f\left(\frac{1}{2}x\right)$ is in the form $f(ax)$ with $a = \frac{1}{2}$. The graph of $y = f\left(\frac{1}{2}x\right)$ is the graph of $y = f(x)$ stretched horizontally.

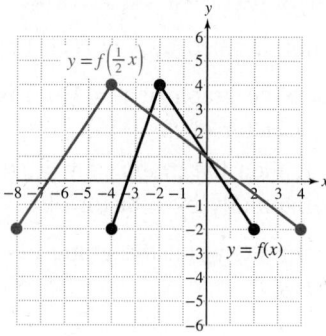

To graph $y = f\left(\frac{1}{2}x\right)$, divide each x value on the graph of $y = f(x)$ by $\frac{1}{2}$. For example:

$(-4, -2)$ becomes $\left(\dfrac{-4}{\frac{1}{2}}, -2\right) = (-8, -2)$.

$(-2, 4)$ becomes $\left(\dfrac{-2}{\frac{1}{2}}, 4\right) = (-4, 4)$.

$(2, -2)$ becomes $\left(\dfrac{2}{\frac{1}{2}}, -2\right) = (4, -2)$.

The graph of $y = f\left(\frac{1}{2}x\right)$ is shown in red.

Skill Practice 5   The graph of $y = f(x)$ is shown.

Graph.   **a.** $y = f(2x)$ **b.** $y = f\left(\dfrac{1}{2}x\right)$

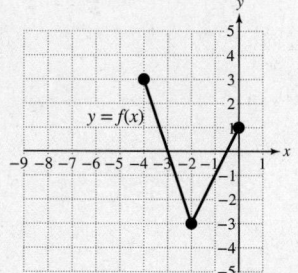

4. Apply Reflections Across the *x*- and *y*-Axes

The graphs of $f(x) = x^2$ (in black) and $g(x) = -x^2$ (in blue) are shown in Figure 6-6. Notice that a point (x, y) on the graph of f corresponds to the point $(x, -y)$ on the graph of g. Therefore, the graph of g is the graph of f reflected across the x-axis.

The graphs of $f(x) = \sqrt{x}$ (in black) and $g(x) = \sqrt{-x}$ (in blue) are shown in Figure 6-7. Notice that a point (x, y) on the graph of f corresponds to the point $(-x, y)$ on g. Therefore, the graph of g is the graph of f reflected across the y-axis.

Answers

5. a.

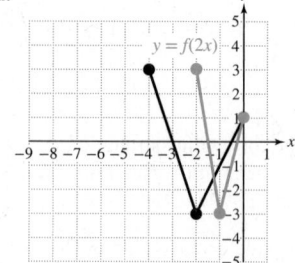

b.

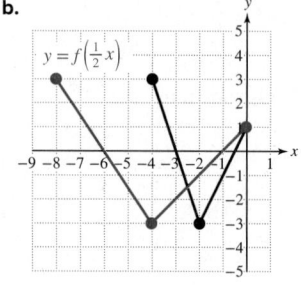

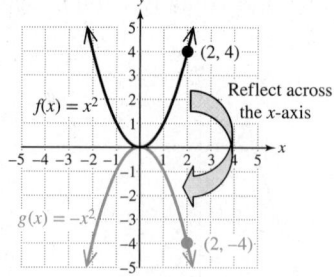

Figure 6-6

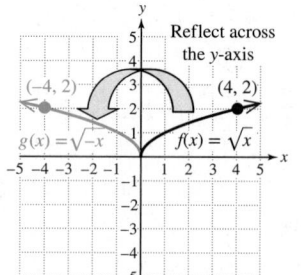

Figure 6-7

Reflections Across the x- and y-Axes

Consider a function defined by $y = f(x)$.

- The graph of $y = -f(x)$ is the graph of $y = f(x)$ reflected across the x-axis.
- The graph of $y = f(-x)$ is the graph of $y = f(x)$ reflected across the y-axis.

EXAMPLE 6   **Reflecting the Graph of a Function Across the x- and y-Axes**

The graph of $y = f(x)$ is given.

a. Graph $y = -f(x)$.

b. Graph $y = f(-x)$.

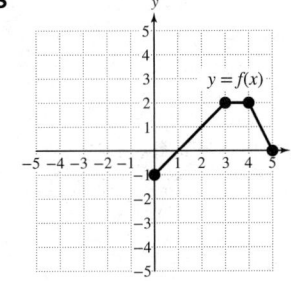

Solution:

a. Reflect $y = f(x)$ across the x-axis.   **b.** Reflect $y = f(x)$ across the y-axis.

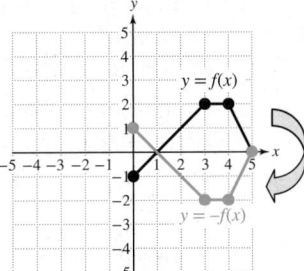   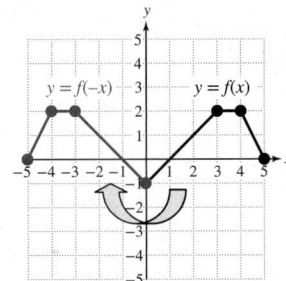

Skill Practice 6   The graph of $y = f(x)$ is given.

a. Graph $y = -f(x)$.

b. Graph $y = f(-x)$.

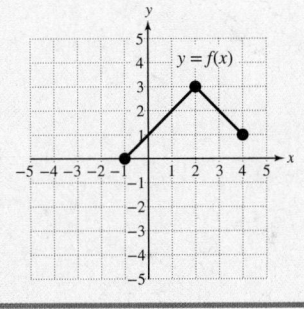

---

5. Summarize Transformations of Graphs

The operations of reflecting a graph of a function about an axis and shifting, stretching, and shrinking a graph are called **transformations**. Transformations give us tools to graph families of functions that are built from basic "parent" functions.

Answers

6. a.–b.

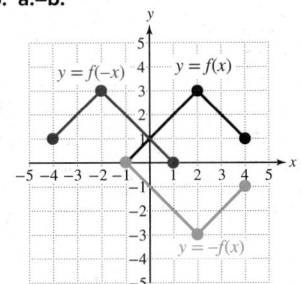

Transformations of Functions

Consider a function defined by $y = f(x)$. If h, k, and a represent positive real numbers, then the graphs of the following functions are related to $y = f(x)$ as follows.

Transformation	Effect on the Graph of f	Changes to Points on f
Vertical translation (shift) $y = f(x) + k$ $y = f(x) - k$	Shift upward k units Shift downward k units	Replace (x, y) by $(x, y + k)$. Replace (x, y) by $(x, y - k)$.
Horizontal translation (shift) $y = f(x - h)$ $y = f(x + h)$	Shift to the right h units Shift to the left h units	Replace (x, y) by $(x + h, y)$. Replace (x, y) by $(x - h, y)$.
Vertical stretch/shrink $y = a[f(x)]$	Vertical stretch (if $a > 1$) Vertical shrink (if $0 < a < 1$) Graph is stretched/shrunk vertically by a factor of a.	Replace (x, y) by (x, ay).
Horizontal stretch/shrink $y = f(a \cdot x)$	Horizontal shrink (if $a > 1$) Horizontal stretch (if $0 < a < 1$) Graph is shrunk/stretched horizontally by a factor of $\frac{1}{a}$.	Replace (x, y) by $\left(\frac{x}{a}, y\right)$.
Reflection $y = -f(x)$ $y = f(-x)$	Reflection across the x-axis Reflection across the y-axis	Replace (x, y) by $(x, -y)$. Replace (x, y) by $(-x, y)$.

When graphing a function requiring multiple transformations on the parent function, it is important to follow the correct sequence of steps.

Steps for Graphing Multiple Transformations of Functions

To graph a function requiring multiple transformations, use the following order.

1. Horizontal translation (shift)
2. Horizontal and vertical stretch and shrink
3. Reflections across the x- and y-axes
4. Vertical translation (shift)

EXAMPLE 7   Using Transformations to Graph a Function

Use transformations to graph the function defined by $n(x) = -\dfrac{1}{2}(x - 2)^2 + 3$.

Solution:

The graph of $n(x) = -\dfrac{1}{2}(x - 2)^2 + 3$ is the same as the graph of $f(x) = x^2$, with four transformations in the following order.

1. Shift the graph to the right 2 units.
2. Apply a vertical shrink (multiply the y values by $\frac{1}{2}$).
3. Reflect the graph over the x-axis.
4. Shift the graph upward 3 units.

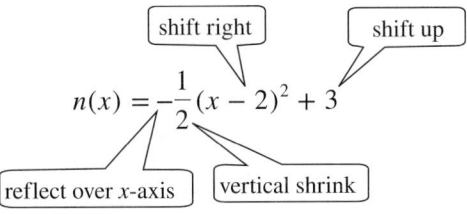

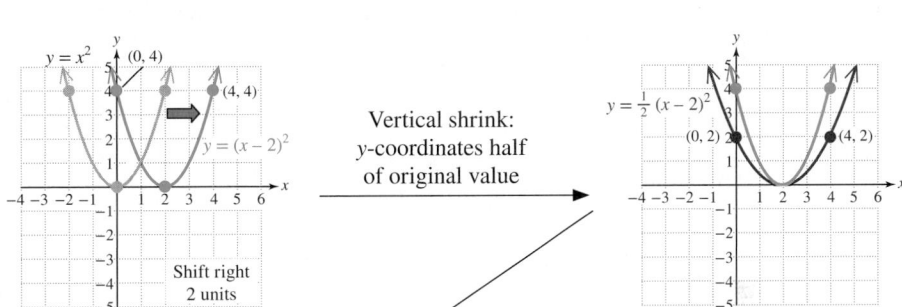

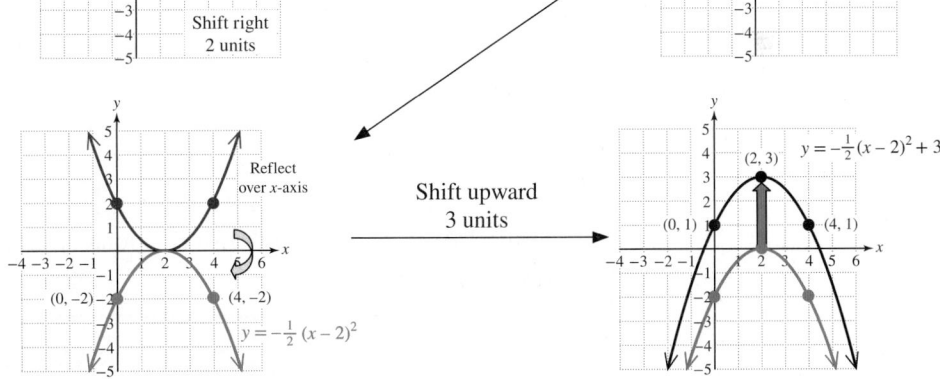

Avoiding Mistakes

As a means to check the graph of $y = n(x)$, substitute the x-coordinates of the strategic points $(0, 1)$, $(2, 3)$, and $(4, 1)$ into the function.

$n(x) = -\frac{1}{2}(x - 2)^2 + 3$
$n(0) = -\frac{1}{2}(0 - 2)^2 + 3 = 1$ ✓
$n(2) = -\frac{1}{2}(2 - 2)^2 + 3 = 3$ ✓
$n(4) = -\frac{1}{2}(4 - 2)^2 + 3 = 1$ ✓

Answer

7.

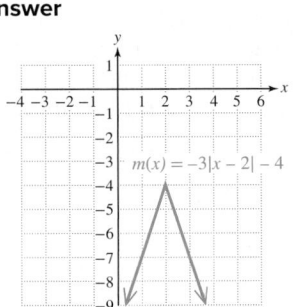

Skill Practice 7   Use transformations to graph the function defined by $m(x) = -3|x - 2| - 4$.

EXAMPLE 8   Using Transformations to Graph a Function

Use transformations to graph the function defined by $v(x) = -\sqrt{-x+2}$.

Solution:

The graph of $v(x) = -\sqrt{-x+2}$ is the same as the graph of $f(x) = \sqrt{x}$, with three transformations in the following order.

1. Shift the graph to the left 2 units.
2. Reflect the graph across the y-axis.
3. Reflect the graph across the x-axis.

   (Note that the reflections in steps 2 and 3 can be applied in either order.)

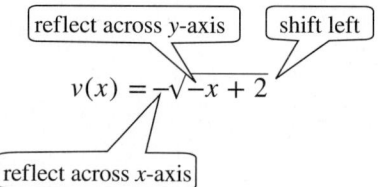

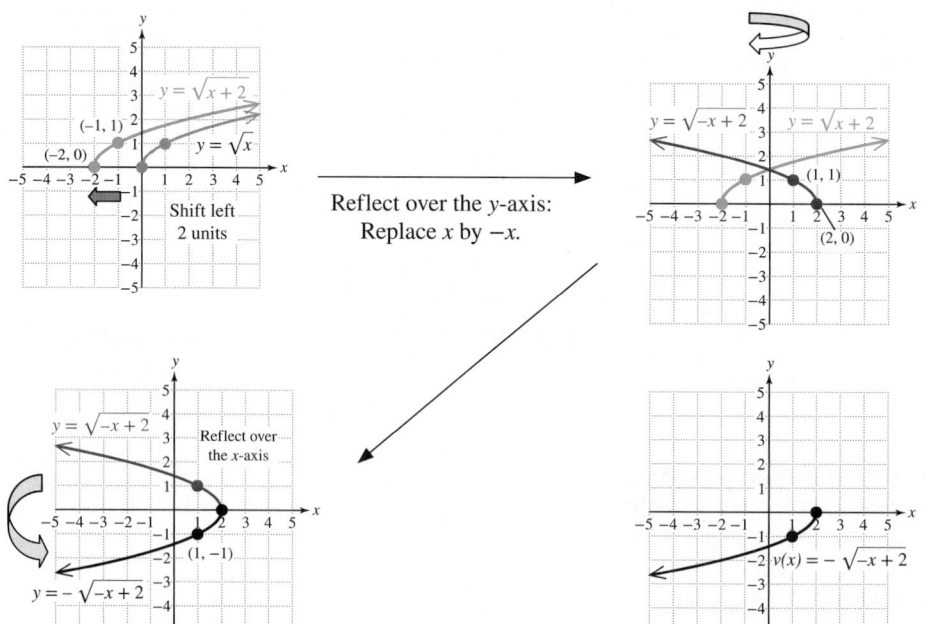

Skill Practice 8   Use transformations to graph the function defined by $r(x) = \sqrt[3]{-x+1}$.

Avoiding Mistakes

Transformations involving a horizontal shrink, stretch, or reflection often introduce confusion when coupled with a horizontal shift. To further illustrate the rationale for the order of steps taken in Example 8, begin with the parent function $y = \sqrt{x}$. Performing a horizontal shift first means that we replace x by $x + 2$. This gives us $y = \sqrt{x+2}$. Then to perform the reflection across the y-axis, we replace x by $-x$ to get $y = \sqrt{-x+2}$. Performing these two transformations in the reverse order would *not* result in the function we want. We would first have $y = \sqrt{-x}$, and then replacing x by $x + 2$ would give $y = \sqrt{-(x+2)} = \sqrt{-x-2}$ rather than $y = \sqrt{-x+2}$.

Answer
8.

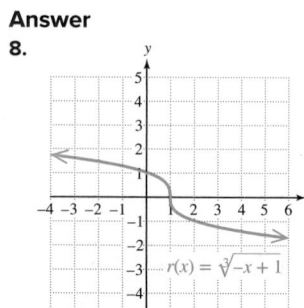

SECTION 6.1   Practice Exercises

Prerequisite Review

For Exercises R.1–R.4, graph the functions. (See Section 5.4 for review.)

R.1. $f(x) = 2x - 7$ **R.2.** $g(x) = -\dfrac{2}{3}x + 4$ **R.3.** $k(x) = -4$ **R.4.** $m(x) = \dfrac{5}{3}$

For Exercises R.5–R.6, compare the similarities and differences of the graphs of the functions defined by $y = f(x)$ and $y = g(x)$. (See Section 5.4 for review.)

R.5. $f(x) = x$ and $g(x) = x - 1$ **R.6.** $f(x) = x$ and $g(x) = 2x$

Concept Connections

1. Let c represent a positive real number. The graph of $y = f(x + c)$ is the graph of $y = f(x)$ shifted (up/down/left/right) c units.

2. Let c represent a positive real number. The graph of $y = f(x - c)$ is the graph of $y = f(x)$ shifted (up/down/left/right) c units.

3. Let c represent a positive real number. The graph of $y = f(x) - c$ is the graph of $y = f(x)$ shifted (up/down/left/right) c units.

4. The graph of $y = 3f(x)$ is the graph of $y = f(x)$ with a (choose one: vertical stretch, vertical shrink, horizontal stretch, horizontal shrink).

5. The graph of $y = f(3x)$ is the graph of $y = f(x)$ with a (choose one: vertical stretch, vertical shrink, horizontal stretch, horizontal shrink).

6. The graph of $y = f\left(\frac{1}{3}x\right)$ is the graph of $y = f(x)$ with a (choose one: vertical stretch, vertical shrink, horizontal stretch, horizontal shrink).

7. The graph of $y = \frac{1}{3}f(x)$ is the graph of $y = f(x)$ with a (choose one: vertical stretch, vertical shrink, horizontal stretch, horizontal shrink).

8. The graph of $y = -f(x)$ is the graph of $y = f(x)$ reflected across the _____ -axis.

Objective 1: Recognize Basic Functions

For Exercises 9–14, from memory match the equation with its graph.

9. $f(x) = \sqrt{x}$ 10. $f(x) = \sqrt[3]{x}$ 11. $f(x) = x^3$

12. $f(x) = x^2$ 13. $f(x) = |x|$ 14. $f(x) = \dfrac{1}{x}$

a.
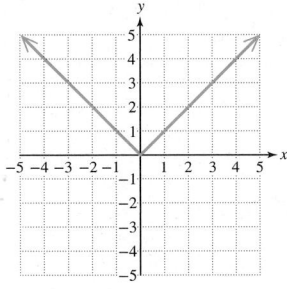

b.
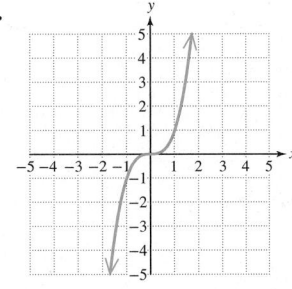

c.
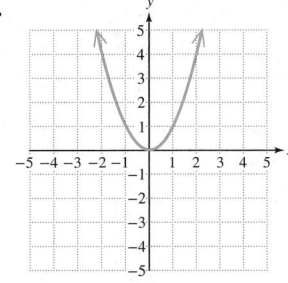

d.
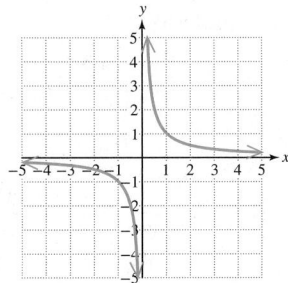

e.
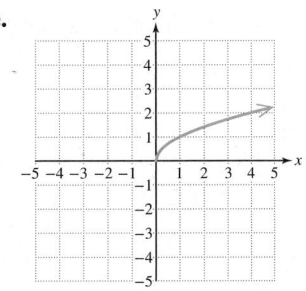

f.
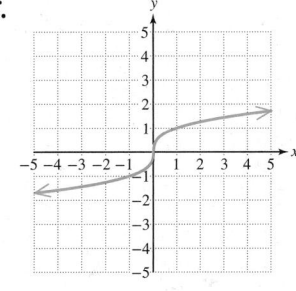

Objective 2: Apply Vertical and Horizontal Translations (Shifts)

For Exercises 15–26, use translations to graph the given functions. **(See Examples 1–3)**

15. $f(x) = |x| + 1$

16. $g(x) = \sqrt{x} + 2$

17. $k(x) = x^3 - 2$

18. $h(x) = \dfrac{1}{x} - 2$

19. $g(x) = \sqrt{x + 5}$

20. $m(x) = |x + 1|$

21. $r(x) = (x - 4)^2$

22. $t(x) = \sqrt[3]{x - 2}$

23. $a(x) = \sqrt{x + 1} - 3$

24. $b(x) = |x - 2| + 4$

25. $c(x) = \dfrac{1}{x - 3} + 1$

26. $d(x) = \dfrac{1}{x + 4} - 1$

Objective 3: Apply Vertical and Horizontal Shrinking and Stretching

For Exercises 27–32, use transformations to graph the functions. **(See Example 4)**

27. $m(x) = 4\sqrt[3]{x}$

28. $n(x) = 3|x|$

29. $r(x) = \dfrac{1}{2}x^2$

30. $t(x) = \dfrac{1}{3}|x|$

31. $p(x) = |2x|$

32. $q(x) = \sqrt{2x}$

For Exercises 33–40, use the graphs of $y = f(x)$ and $y = g(x)$ to graph the given function. **(See Example 5)**

33. $y = \dfrac{1}{3}f(x)$

34. $y = \dfrac{1}{2}g(x)$

35. $y = 3f(x)$

36. $y = 2g(x)$

37. $y = f(3x)$

38. $y = g(2x)$

39. $y = f\left(\dfrac{1}{3}x\right)$

40. $y = g\left(\dfrac{1}{2}x\right)$

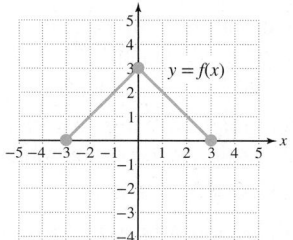
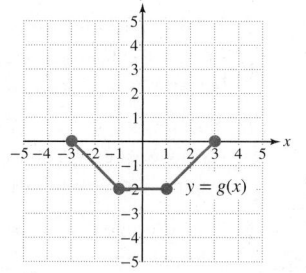

Objective 4: Apply Reflections Across the x- and y-Axes

For Exercises 41–46, graph the function by applying an appropriate reflection.

41. $f(x) = -\dfrac{1}{x}$

42. $g(x) = -\sqrt{x}$

43. $h(x) = -x^3$

44. $k(x) = -|x|$

45. $p(x) = (-x)^3$

46. $q(x) = \sqrt[3]{-x}$

For Exercises 47–50, use the graphs of $y = f(x)$ and $y = g(x)$ to graph the given function. **(See Example 6)**

47. $y = f(-x)$

48. $y = g(-x)$

49. $y = -f(x)$

50. $y = -g(x)$

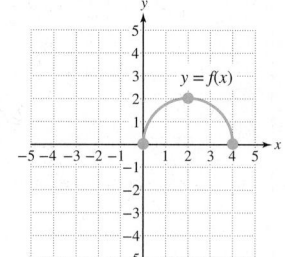
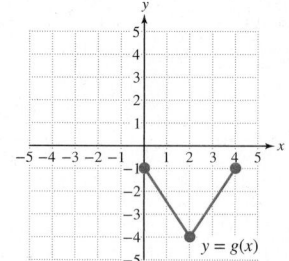

For Exercises 51–54, use the graphs of $y = f(x)$ and $y = g(x)$ to graph the given function. **(See Example 6)**

51. $y = f(-x)$

52. $y = g(-x)$

53. $y = -f(x)$

54. $y = -g(x)$

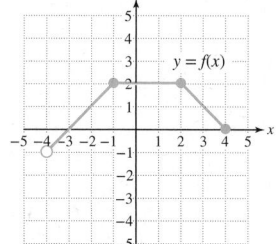
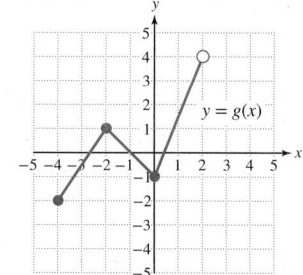

Objective 5: Summarize Transformations of Graphs

For Exercises 55–62, a function g is given. Identify the parent function from Table 6-1 on page 408. Then use the steps for graphing multiple transformations of functions on page 416 to list, in order, the transformations applied to the parent function to obtain the graph of g.

55. $g(x) = \dfrac{3}{1+x} - 2$

56. $g(x) = \dfrac{5}{x-4} + 1$

57. $g(x) = \dfrac{1}{3}(x - 2.1)^2 + 7.9$

58. $g(x) = \dfrac{1}{2}\sqrt{x + 4.3} - 8.4$

59. $g(x) = 2\sqrt{-2x + 5}$

60. $g(x) = 3\left|-\dfrac{1}{2}x - 4\right|$

61. $g(x) = -\sqrt{\dfrac{1}{3}x - 6}$

62. $g(x) = -|2x| + 8$

For Exercises 63–78, use transformations to graph the functions. **(See Examples 7–8)**

63. $v(x) = -(x + 2)^2 + 1$

64. $u(x) = -(x - 1)^2 - 2$

65. $f(x) = 2\sqrt{x + 3} - 1$

66. $g(x) = 2\sqrt{x - 1} + 3$

67. $p(x) = \dfrac{1}{2}|x - 1| - 2$

68. $q(x) = \dfrac{1}{3}|x + 2| - 1$

69. $r(x) = -\sqrt{-x} + 1$

70. $s(x) = -\sqrt{-x} - 2$

71. $f(x) = \sqrt{-x + 3}$

72. $g(x) = \sqrt{-x - 4}$

73. $n(x) = -\left|\dfrac{1}{2}x - 3\right|$

74. $m(x) = -\left|\dfrac{1}{3}x + 1\right|$

75. $f(x) = -\dfrac{1}{2}(x - 3)^2 + 8$

76. $g(x) = -\dfrac{1}{3}(x + 2)^2 + 3$

77. $p(x) = -4|x + 2| - 1$

78. $q(x) = -2|x - 1| + 4$

Mixed Exercises

For Exercises 79–86, the graph of $y = f(x)$ is given. Graph the indicated function.

79. Graph $y = -f(x - 1) + 2$.

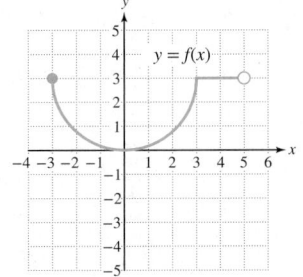

80. Graph $y = -f(x + 1) - 2$.

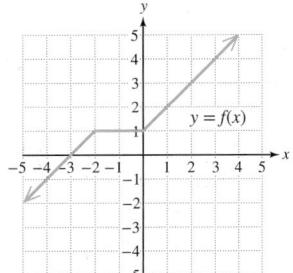

81. Graph $y = 2f(x - 2) - 3$.

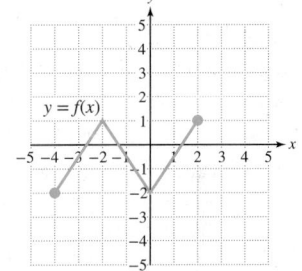

82. Graph $y = 2f(x + 2) - 4$.

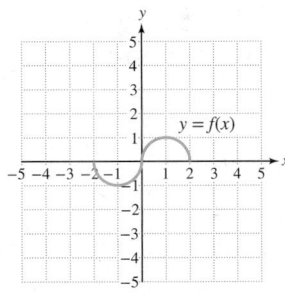

83. Graph $y = -3f(2x)$.

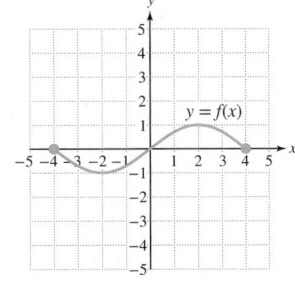

84. Graph $y = -\dfrac{1}{2}f\left(\dfrac{1}{2}x\right)$.

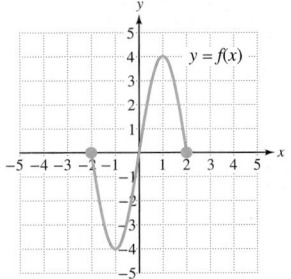

85. Graph $y = f(-x) - 2$.

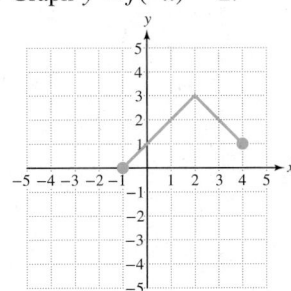

86. Graph $y = f(-x) + 3$.

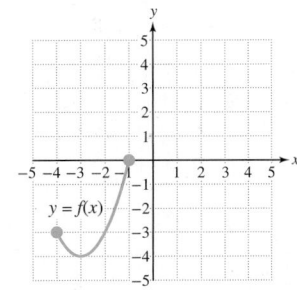

For Exercises 87–90, write a function based on the given parent function and transformations in the given order.

87. Parent function: $y = x^3$

1. Shift 4.5 units to the left.
2. Reflect across the y-axis.
3. Shift upward 2.1 units.

88. Parent function $y = \sqrt[3]{x}$

1. Shift 1 unit to the left.
2. Stretch horizontally by a factor of 4.
3. Reflect across the x-axis.

89. Parent function: $y = \dfrac{1}{x}$

1. Stretch vertically by a factor of 2.
2. Reflect across the x-axis.
3. Shift downward 3 units.

90. Parent function: $y = |x|$

1. Shift 3.7 units to the right.
2. Shrink horizontally by a factor of $\dfrac{1}{3}$.
3. Reflect across the y-axis.

Write About It

91. Explain why the graph of $g(x) = |2x|$ can be interpreted as a horizontal shrink of the graph of $f(x) = |x|$ or as a vertical stretch of the graph of $f(x) = |x|$.

92. Explain why the graph of $h(x) = \sqrt{\frac{1}{2}x}$ can be interpreted as a horizontal stretch of the graph of $f(x) = \sqrt{x}$ or as a vertical shrink of the graph of $f(x) = \sqrt{x}$.

93. Explain the difference between the graphs of $f(x) = |x - 2| - 3$ and $g(x) = |x - 3| - 2$.

94. Explain why $g(x) = \dfrac{1}{-x + 1}$ can be graphed by shifting the graph of $f(x) = \dfrac{1}{x}$ one unit to the left and reflecting across the y-axis, or by shifting the graph of f one unit to the right and reflecting across the x-axis.

Expanding Your Skills

For Exercises 95–100, use transformations on the basic functions presented in Table 6-1 to write a rule $y = f(x)$ that would produce the given graph.

95.

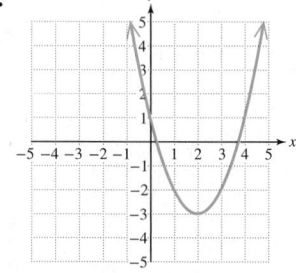

96.

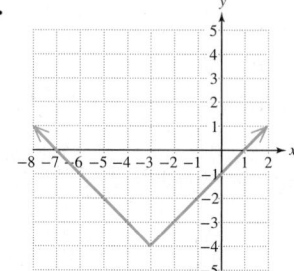

97.

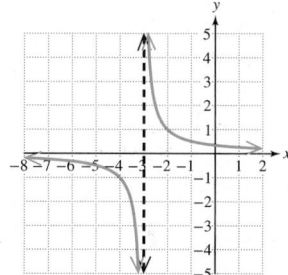

98.

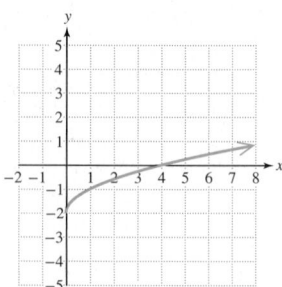

99.

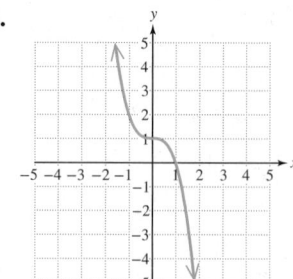

100.

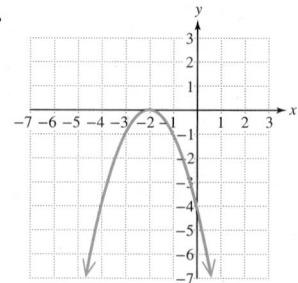

101. The graph shows the number of views y (in thousands) for a new online video, t days after it was posted. Use transformations on one of the parent functions from Table 6-1 on page 408 to model these data.

Number of Views by Day Number

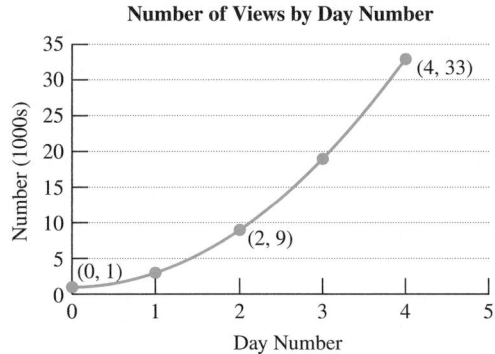

102. The graph shows the cumulative number y of flu cases among passengers on a 25-day cruise, t days after the cruise began. Use transformations on one of the parent functions from Table 6-1 on page 408 to model these data.

Cumulative Number of Flu Cases

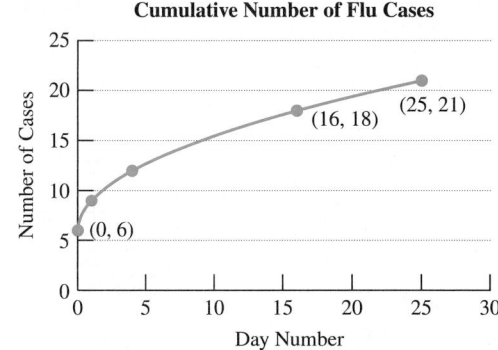

Technology Connections

103. a. Graph the functions on the viewing window $[-5, 5, 1]$ by $[-2, 8, 1]$.

$$y = x^2$$
$$y = x^4$$
$$y = x^6$$

    c. Describe the general shape of the graph of $y = x^n$ where n is an even integer greater than 1.

b. Graph the functions on the viewing window $[-4, 4, 1]$ by $[-10, 10, 1]$.

$$y = x^3$$
$$y = x^5$$
$$y = x^7$$

    d. Describe the general shape of the graph of $y = x^n$ where n is an odd integer greater than 1.

SECTION 6.2   Symmetry and Piecewise-Defined Functions

OBJECTIVES

1. **Test for Symmetry**
2. **Identify Even and Odd Functions**
3. **Graph Piecewise-Defined Functions**

1. Test for Symmetry

The photos in Figures 6-8 through 6-10 each show a type of symmetry.

MikeMcken/Getty Images
Figure 6-8

Julie Miller
Figure 6-9

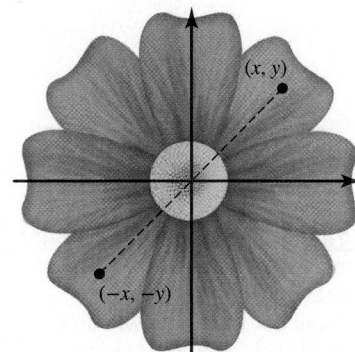

Figure 6-10

The photo of the kingfisher (Figure 6-8) shows an image of the bird reflected in the water. Suppose that we superimpose the x-axis at the waterline. Every point (x, y) on the bird has a mirror image $(x, -y)$ below the x-axis. Therefore, this image is symmetric with respect to the x-axis.

A human face is symmetric with respect to a vertical line through the center (Figure 6-9). If we place the y-axis along this line, a point (x, y) on one side has a mirror image at $(-x, y)$. This image is symmetric with respect to the y-axis.

The flower shown in Figure 6-10 is symmetric with respect to the point at its center. Suppose that we place the origin at the center of the flower. Notice that a point (x, y) on the image has a corresponding point $(-x, -y)$ on the image. This image is symmetric with respect to the origin.

Given an equation in x and y, the graph of the equation may possess symmetry with respect to the x-axis, y-axis, or origin as shown in Figures 6-11 through 6-13. In each graph, we show two points related through symmetry.

Symmetry with respect to the y-axis

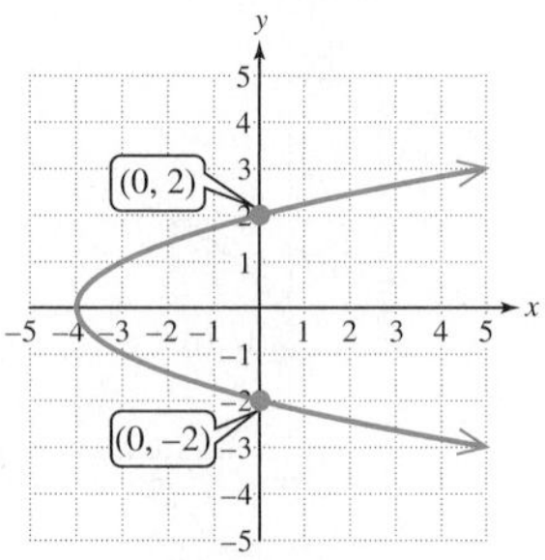

Symmetry with respect to the y-axis

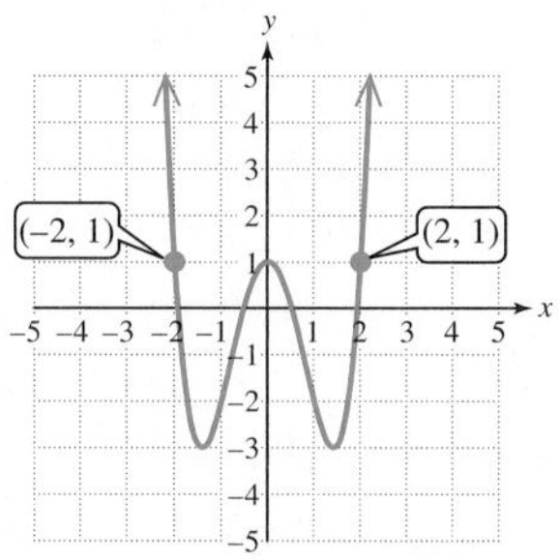

Symmetry with respect to the origin

Figure 6-11

Figure 6-12

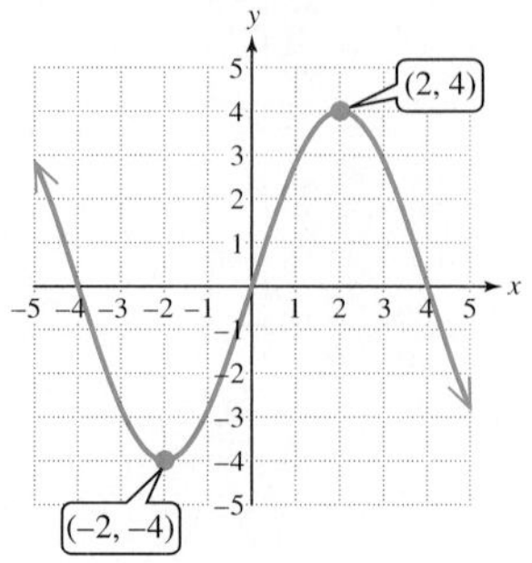

Figure 6-13

Given an equation in the variables x and y, use the following rules to determine if the graph is symmetric with respect to the x-axis, the y-axis, or the origin.

Tests for Symmetry

Consider an equation in the variables x and y.

- The graph of the equation is symmetric with respect to the y-axis if substituting $-x$ for x in the equation results in an equivalent equation.
- The graph of the equation is symmetric with respect to the x-axis if substituting $-y$ for y in the equation results in an equivalent equation.
- The graph of the equation is symmetric with respect to the origin if substituting $-x$ for x and $-y$ for y in the equation results in an equivalent equation.

EXAMPLE 1   Testing for Symmetry

Determine whether the graph is symmetric with respect to the y-axis, x-axis, or origin.

a. $y = |x|$ **b.** $x = y^2 - 4$

Solution:

TIP The graph of $y = |x|$ is one of the basic graphs presented in Section 6.1. From our familiarity with the graph we can visualize the symmetry with respect to the y-axis.

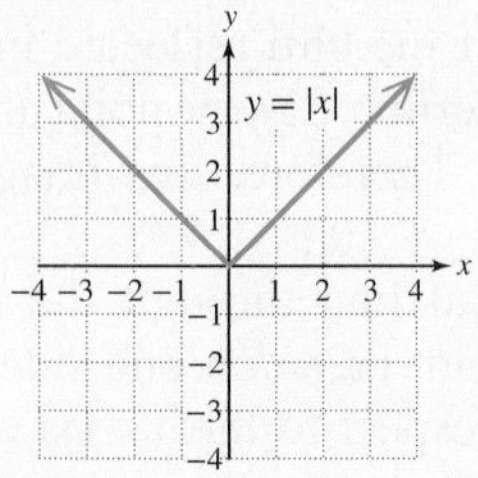

a. $y = |x|$
$y = |-x|$ Same equation: Graph is symmetric with respect to the y-axis.
$y = |x|$

Test for symmetry with respect to the y-axis.
Replace x by $-x$. Note that $|-x| = |x|$.
The resulting equation *is* equivalent to the original equation.

$y = |x|$
$-y = |x|$ not the same
$y = -|x|$

Test for symmetry with respect to the x-axis.
Replace y by $-y$. The resulting equation is *not* equivalent to the original equation.

$y = |x|$
$-y = |-x|$ not the same
$-y = |x|$
$y = -|x|$

Test for symmetry with respect to the origin.
Replace x by $-x$ and y by $-y$.
The resulting equation is *not* equivalent to the original equation.

The graph is symmetric with respect to the y-axis only.

b. $x = y^2 - 4$
   $-x = y^2 - 4$ not the same
   $x = -y^2 + 4$

Test for symmetry with respect to the y-axis.
Replace x by $-x$. The resulting equation is *not* equivalent to the original equation.

$x = y^2 - 4$ Same equation:
$x = (-y)^2 - 4$ Graph is symmetric with respect to the x-axis.
$x = y^2 - 4$

Test for symmetry with respect to the x-axis.
Replace y by $-y$. The resulting equation *is* equivalent to the original equation.

$x = y^2 - 4$
$-x = (-y)^2 - 4$ not the same
$-x = y^2 - 4$
$x = -y^2 + 4$

Test for symmetry with respect to the origin.
Replace x by $-x$ and y by $-y$.
The resulting equation is *not* equivalent to the original equation.

The graph is symmetric with respect to the x-axis only (Figure 6-14).

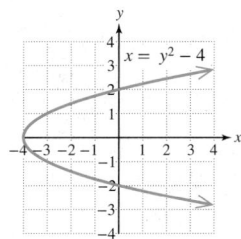

Figure 6-14

Skill Practice 1   Determine whether the graph is symmetric with respect to the y-axis, x-axis, or origin.

   a. $y = x^2$ **b.** $|y| = x + 1$

EXAMPLE 2   **Testing for Symmetry**

Determine whether the graph is symmetric with respect to the y-axis, x-axis, or origin.

$$x^2 + y^2 = 9$$

Solution:

The graph of $x^2 + y^2 = 9$ is a circle with center at the origin and radius 3. By inspection, we can see that the graph is symmetric with respect to both axes and the origin.

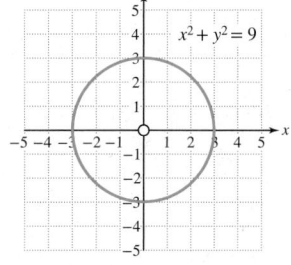

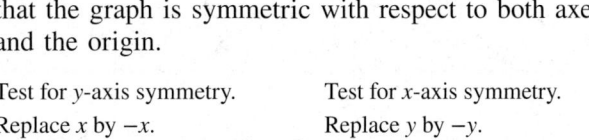

Test for y-axis symmetry.
Replace x by $-x$.

$x^2 + y^2 = 9$
$(-x)^2 + y^2 = 9$ same
$x^2 + y^2 = 9$

Test for x-axis symmetry.
Replace y by $-y$.

$x^2 + y^2 = 9$
$x^2 + (-y)^2 = 9$ same
$x^2 + y^2 = 9$

Test for origin symmetry.
Replace x by $-x$ and y by $-y$.

$x^2 + y^2 = 9$
$(-x)^2 + (-y)^2 = 9$ same
$x^2 + y^2 = 9$

The graph is symmetric with respect to the y-axis, the x-axis, and the origin.

Skill Practice 2   Determine whether the graph is symmetric with respect to the y-axis, x-axis, or origin.

$$\frac{x^2}{4} + \frac{y^2}{9} = 1$$

2. Identify Even and Odd Functions

A function may be symmetric with respect to the *y*-axis or to the origin. A function that is symmetric with respect to the *y*-axis is called an *even* function. A function that is symmetric with respect to the origin is called an *odd* function.

EXAMPLE 3	Identifying Even and Odd Functions From a Graph

By inspection, determine if the function is even, odd, or neither.

Solution:

a.

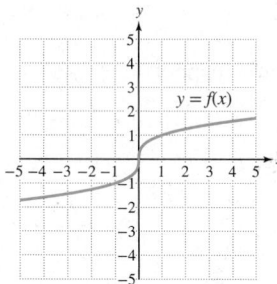

b.

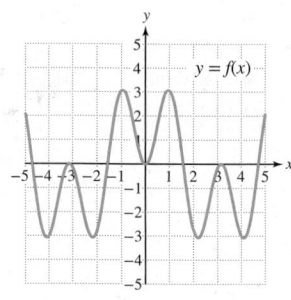

c.

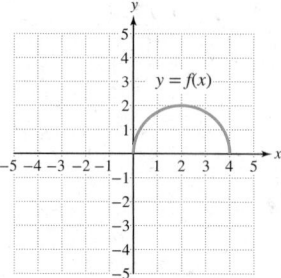

The function is symmetric with respect to the origin. Therefore, the function is an *odd* function.

The function is symmetric with respect to the *y*-axis. Therefore, the function is an *even* function.

The function is not symmetric with respect to either the *y*-axis or the origin. Therefore, the function is *neither* even nor odd.

Skill Practice 3 Determine if the function is even, odd, or neither.

a.

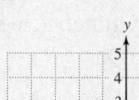

b.

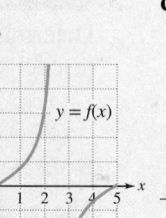

c.

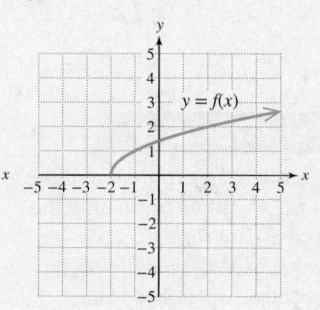

Given $y = f(x)$, we can determine if f is even, odd, or neither even nor odd by checking the following conditions.

Even and Odd Functions

- A function f is an **even function** if $f(-x) = f(x)$ for all x in the domain of f. The graph of an even function is symmetric with respect to the *y*-axis.
- A function f is an **odd function** if $f(-x) = -f(x)$ for all x in the domain of f. The graph of an odd function is symmetric with respect to the origin.

Answers

3. **a.** Even function
   b. Odd function
   c. Neither even nor odd

EXAMPLE 4   **Identifying Even and Odd Functions**

Determine if the function is even, odd, or neither.

a. $f(x) = -2x^4 + 5|x|$ **b.** $g(x) = 4x^3 - x$ **c.** $h(x) = 2x^2 + x$

Solution:

a. $f(x) = -2x^4 + 5|x|$ Determine whether the function is even.

$f(-x) = -2(-x)^4 + 5|-x|$ same   Replace x by $-x$ to determine if $f(-x) = f(x)$.

$\quad\quad = -2x^4 + 5|x|$

Since $f(-x) = f(x)$, the function   There is no need to test whether f is an odd
f is an even function.   function because a function cannot be both even
                                          and odd unless all points are on the x-axis.

> **TIP**   In Example 4(a), we suspect that f is an even function because each term is of the form x^{even} or $|x|$. In each case, replacing x by $-x$ results in an equivalent term.

b. $g(x) = 4x^3 - x$ Each term has x raised to an odd power. Therefore, replacing x by $-x$
                              will result in the *opposite* of the original term. Therefore, test
                              whether g is an odd function. That is, test whether $g(-x) = -g(x)$.

Evaluate: $g(-x)$ Evaluate: $-g(x)$

$g(-x) = 4(-x)^3 - (-x)$ $-g(x) = -(4x^3 - x)$

$\quad\quad = -4x^3 + x$ ⟵ same ⟶ $\quad\quad = -4x^3 + x$

Since $g(-x) = -g(x)$, the function g is an odd function.

> **TIP**   In Example 4(b), we suspect that g is an odd function because each term is of the form x^{odd}. In each case, replacing x by $-x$ results in the *opposite* of the original term.

c. $h(x) = 2x^2 + x$ Determine whether the function is even.

$h(-x) = 2(-x)^2 + (-x)$ not the same   Replace x by $-x$ to determine
                                          if $h(-x) = h(x)$.
$\quad\quad = 2x^2 - x$

Since $h(-x) \neq h(x)$, the function is not even.

Next, test whether h is an odd function. Test whether $h(-x) = -h(x)$.

Evaluate: $h(-x)$ Evaluate: $-h(x)$

$h(-x) = 2(-x)^2 + (-x)$ $-h(x) = -(2x^2 + x)$

$\quad\quad = 2x^2 - x$ ⟵ not the same ⟶ $\quad\quad = -2x^2 - x$

Since $h(-x) \neq -h(x)$, the function is not an odd function. Therefore, h is neither even nor odd.

> **TIP**   In Example 4(c), $h(x)$ has a mixture of terms of the form x^{odd} and x^{even}. Therefore, we might suspect that the function is neither even nor odd.

Skill Practice 4   Determine if the function is even, odd, or neither.

a. $m(x) = -x^5 + x^3$ **b.** $n(x) = x^2 - |x| + 1$ **c.** $p(x) = 2|x| + x$

3. Graph Piecewise-Defined Functions

Suppose that a car is stopped for a red light. When the light turns green, the car undergoes a constant acceleration for 20 sec until it reaches a speed of 45 mph. It travels 45 mph for 1 min (60 sec) and then decelerates for 30 sec to stop at another red light. The graph of the car's speed y (in mph) versus the time x (in sec) after leaving the first red light is shown in Figure 6-15.

Notice that the graph can be segmented into three pieces. The first 20 sec is represented by a linear function with a positive slope, $y = 2.25x$. The next 60 sec is represented by the constant function $y = 45$. And the last 30 sec is represented by a linear function with a negative slope, $y = -1.5x + 165$.

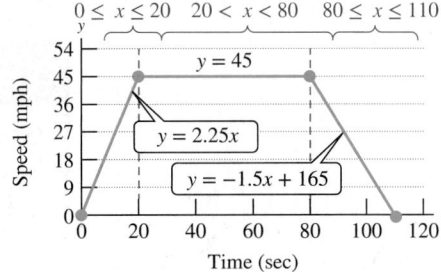

Figure 6-15

To write a rule defining this function we use a **piecewise-defined function** in which we define each "piece" on a restricted domain.

$$f(x) = \begin{cases} 2.25x & \text{for } 0 \le x \le 20 \\ 45 & \text{for } 20 < x < 80 \\ -1.5x + 165 & \text{for } 80 \le x \le 110 \end{cases}$$

EXAMPLE 5   Interpreting a Piecewise-Defined Function

Evaluate the function for the given values of x.

$$f(x) = \begin{cases} -x - 1 & \text{for } x < -1 \\ -3 & \text{for } -1 \le x < 2 \\ \sqrt{x - 2} & \text{for } x \ge 2 \end{cases}$$

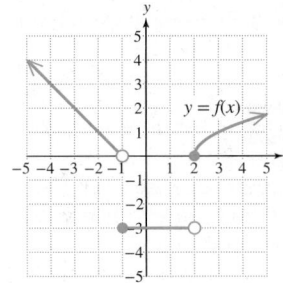

a. $f(-3)$ **b.** $f(-1)$

c. $f(2)$ **d.** $f(6)$

Solution:

a. $f(x) = -x - 1$ $x = -3$ is on the interval $x < -1$. Use the first rule
$\quad\quad f(-3) = -(-3) - 1$ in the function: $f(x) = -x - 1$.
$\quad\quad\quad\quad\;\; = 2$

b. $f(x) = -3$ $x = -1$ is on the interval $-1 \le x < 2$. Use the second rule in
$\quad\quad f(-1) = -3$ the function: $f(x) = -3$.

c. $f(x) = \sqrt{x - 2}$ $x = 2$ is on the interval $x \ge 2$. Use the third rule in the
$\quad\quad f(2) = \sqrt{2 - 2}$ function: $f(x) = \sqrt{x - 2}$.
$\quad\quad\quad\quad = 0$

d. $f(x) = \sqrt{x - 2}$ $x = 6$ is on the interval $x \ge 2$. Use the third rule in the
$\quad\quad f(6) = \sqrt{6 - 2}$ function: $f(x) = \sqrt{x - 2}$.
$\quad\quad\quad\quad = 2$

Skill Practice 5   Evaluate the function for the given values of x.

$$f(x) = \begin{cases} x + 7 & \text{for } x < -2 \\ x^2 & \text{for } -2 \le x < 1 \\ 3 & \text{for } x \ge 1 \end{cases}$$

a. $f(-3)$ **b.** $f(-2)$ **c.** $f(1)$ **d.** $f(4)$

TECHNOLOGY CONNECTIONS

Graphing a Piecewise-Defined Function

A graphing calculator can be used to graph a piecewise-defined function. The format to enter the function is as follows:

Y_1 = (first piece)/(first condition)
Y_2 = (second piece)/(second condition)
$\quad\vdots$

Each condition in parentheses is an inequality, and the calculator assigns it a value of 1 or 0 depending on whether the inequality is true or false. If an inequality is true, the function is divided by 1 on that interval and is "turned

Answers
5. a. 4   **b.** 4   **c.** 3   **d.** 3

on." If an inequality is false, then the function is divided by 0. Since division by zero is undefined, the calculator does not graph the function on that interval, and the function is effectively "turned off."

Enter the function from Example 5 as shown. Note that the inequality symbols can be found in the TEST menu.

$$f(x) = \begin{cases} -x - 1 & \text{for } x < -1 \\ -3 & \text{for } -1 \leq x < 2 \\ \sqrt{x - 2} & \text{for } x \geq 2 \end{cases}$$

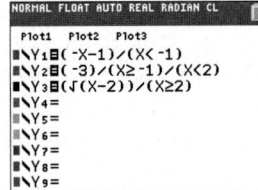 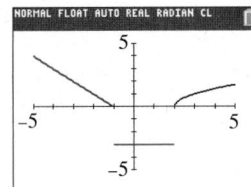

Notice that the individual "pieces" of the graph do not "hook up." For this reason, it is also a good practice to put the calculator in DOT mode in the MODE menu.

In Examples 6 and 7, we graph piecewise-defined functions. To graph a piecewise-defined function, it is helpful if you recognize the shape of each individual "piece" to be graphed. For example, before reading Example 6, recall that

* The graph of $y = mx + b$ or $f(x) = mx + b$ is a line with slope m and y-intercept $(0, b)$. If $b = 0$, then the line passes through the origin.
* The graph of $y = b$ or $f(x) = b$ is a horizontal line.

EXAMPLE 6   **Graphing a Piecewise-Defined Function**

Graph the function defined by $f(x) = \begin{cases} -3x & \text{for } x < 1 \\ -3 & \text{for } x \geq 1 \end{cases}$.

Solution:

* The first rule $f(x) = -3x$ defines a line with slope -3 and y-intercept $(0, 0)$. This line should be graphed only to the left of $x = 1$. The point $(1, -3)$ is graphed as an open dot, because the point is not part of the rule $f(x) = -3x$. See the blue portion of the graph in Figure 6-16.
* The second rule $f(x) = -3$ is a horizontal line for $x \geq 1$. The point $(1, -3)$ is a closed dot to show that it is part of the rule $f(x) = -3$. The closed dot from the red segment of the graph "overrides" the open dot from the blue segment. Taken together, the closed dot "plugs" the hole in the graph.

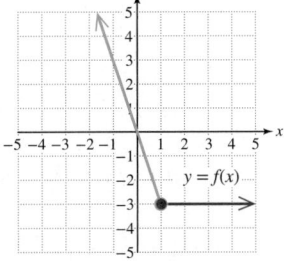

Figure 6-16

TIP The function in Example 6 has no "gaps," and therefore we say that the function is **continuous**. Informally, this means that we can draw the function without lifting our pencil from the page. The formal definition of a continuous function will be studied in calculus.

Answer

6.

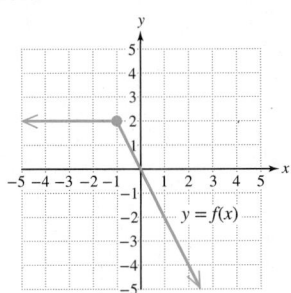

Skill Practice 6   Graph the function.

$$f(x) = \begin{cases} 2 & \text{for } x \leq -1 \\ -2x & \text{for } x > -1 \end{cases}$$

EXAMPLE 7 **Graphing a Piecewise-Defined Function**

Graph the function. $f(x) = \begin{cases} x + 3 & \text{for } x < -1 \\ x^2 & \text{for } -1 \leq x < 2 \end{cases}$

Solution:

The first rule $f(x) = x + 3$ defines a line with slope 1 and y-intercept $(0, 3)$. This line should be graphed only for $x < -1$ (that is, to the left of $x = -1$). The point $(-1, 2)$ is graphed as an open dot, because that point is not part of the function. See the red portion of the graph in Figure 6-17.

The second rule $f(x) = x^2$ is one of the basic functions learned in Section 6.1. It is a parabola with vertex at the origin. We sketch this function only for x values on the interval $-1 \leq x < 2$. The point $(-1, 1)$ is a closed dot to show that it is part of the function. The point $(2, 4)$ is displayed as an open dot to indicate that it is not part of the function.

TIP The function in Example 7 has a gap at $x = -1$, and therefore, we say that f is **discontinuous** at -1.

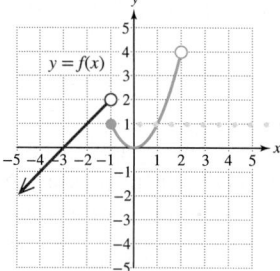

Figure 6-17

Avoiding Mistakes

Note that the function cannot have a closed dot at both $(-1, 1)$ and $(-1, 2)$ because it would not pass the vertical line test.

Skill Practice 7 Graph the function.

$f(x) = \begin{cases} |x| & \text{for } -4 \leq x < 2 \\ -x + 2 & \text{for } x \geq 2 \end{cases}$

We now look at a special category of piecewise-defined functions called **step functions**. The graph of a step function is a series of discontinuous "steps." One important step function is called the **greatest integer function** or **floor function**. It is defined by

$f(x) = [\![x]\!]$ where $[\![x]\!]$ is the greatest integer less than or equal to x.

The operation $[\![x]\!]$ may also be denoted as **int(x)** or by **floor(x)**. These alternative notations are often used in computer programming.

In Example 8, we graph the greatest integer function.

Answer

7.

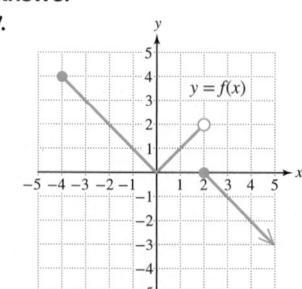

EXAMPLE 8   Graphing the Greatest Integer Function

Graph the function defined by $f(x) = [\![x]\!]$.

Solution:

x	$f(x) = [\![x]\!]$
-1.7	-2
-1	-1
-0.6	-1
0	0
0.4	0
1	1
1.8	1
2	2
2.5	2

Evaluate f for several values of x.

Greatest integer less than or equal to -1.7 is -2.

Greatest integer less than or equal to -1 is -1.

Greatest integer less than or equal to -0.6 is -1.

Greatest integer less than or equal to 0 is 0.

Greatest integer less than or equal to 0.4 is 0.

Greatest integer less than or equal to 1 is 1.

Greatest integer less than or equal to 1.8 is 1.

Greatest integer less than or equal to 2 is 2.

Greatest integer less than or equal to 2.5 is 2.

TIP On many graphing calculators, the greatest integer function is denoted by int( ) and is found under the MATH menu followed by NUM.

From the table, we see a pattern and from the pattern, we form the graph.

If $-3 \leq x < -2$, then $[\![x]\!] = -3$
If $-2 \leq x < -1$, then $[\![x]\!] = -2$
If $-1 \leq x < 0$, then $[\![x]\!] = -1$
If $0 \leq x < 1$, then $[\![x]\!] = 0$
If $1 \leq x < 2$, then $[\![x]\!] = 1$
If $2 \leq x < 3$, then $[\![x]\!] = 2$
...

Skill Practice 8   Evaluate $f(x) = [\![x]\!]$ for the given values of x.

a. $f(1.7)$ **b.** $f(5.5)$ **c.** $f(-4)$ **d.** $f(-4.2)$

In Example 9, we use a piecewise-defined function to model an application.

EXAMPLE 9   Using a Piecewise-Defined Function in an Application

A salesperson makes a monthly salary of $3000 along with a 5% commission on sales over $20,000 for the month. Write a piecewise-defined function to represent the salesperson's monthly income $I(x)$ (in $) for x dollars in sales.

Solution:

Let x represent the amount in sales.

Then $x - 20,000$ represents the amount in sales over $20,000.

There are two scenarios for the salesperson's income.

Scenario 1:   The salesperson sells $20,000 or less. In this case, the monthly income is a constant $3000. This is represented by

$$y = 3000 \quad \text{for } 0 \leq x \leq 20,000$$

Scenario 2:   The salesperson sells over $20,000. In this case, the monthly income is $3000 plus 5% of sales over $20,000. This is represented by

$$y = 3000 + 0.05(x - 20,000) \quad \text{for } x > 20,000$$

Answers
8. a. 1 **b.** 5
c. -4 **d.** -5

Therefore, a piecewise-defined function for monthly income is

$$I(x) = \begin{cases} 3000 & \text{for } 0 \le x \le 20{,}000 \\ 3000 + 0.05(x - 20{,}000) & \text{for } x > 20{,}000 \end{cases}$$

Alternatively, we can simplify to get

$$I(x) = \begin{cases} 3000 & \text{for } 0 \le x \le 20{,}000 \\ 0.05x + 2000 & \text{for } x > 20{,}000 \end{cases}$$

A graph of $y = I(x)$ is shown in Figure 6-18. Notice that for $x = \$20{,}000$, both equations within the piecewise-defined function yield a monthly salary of \$3000. Therefore, the two line segments in the graph meet at (20,000, 3000).

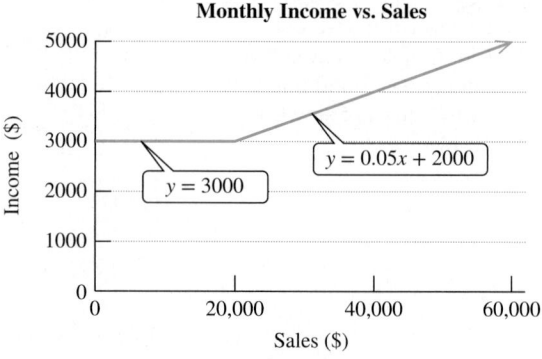

Figure 6-18

Answer

9. $C(x) =$
$\begin{cases} 7.99x & \text{for } 1 \le x \le 100 \\ 799 + 6.99(x - 100) & \text{for } x > 100 \end{cases}$

Skill Practice 9   A retail store buys T-shirts from the manufacturer. The cost is \$7.99 per shirt for 1 to 100 shirts, inclusive. Then the price is decreased to \$6.99 per shirt thereafter. Write a piecewise-defined function that expresses the cost $C(x)$ (in \$) to buy x shirts.

SECTION 6.2   Practice Exercises

Prerequisite Review

R.1. Plot the two points and explain how the points are related to the x-axis. (See Section 5.1 for review.)

(2, 4) and (2, −4)

R.2. Plot the two points and explain how the points are related to the y-axis.

(3, 1) and (−3, 1)

R.3. Plot the points and explain how the points are related to the origin.

(2, 3) and (−2, −3)

R.4. Plot the points and explain how the points are related to the origin.

(4, −3) and (−4, 3)

For Exercises R.5–R.8, evaluate the function for the given values of x. (See Section 5.3 for review.)

R.5. $f(x) = 3x^2 + x^4$

   a. $x = 1$ **b.** $x = -1$

   c. $x = 2$ **d.** $x = -2$

R.6. $g(x) = -3|x| + x^2$

   a. $x = 3$ **b.** $x = -3$

   c. $x = 4$ **d.** $x = -4$

R.7. $k(x) = 3x - 4x^3$

   a. $x = 3$ **b.** $x = -3$

   c. $x = 1$ **d.** $x = -1$

R.8. $m(x) = \dfrac{4}{x} - x^3$

   a. $x = 2$ **b.** $x = -2$

   c. $x = 4$ **d.** $x = -4$

For Exercises R.9–R.12, graph the set and express the set in interval notation. (See Section 1.5 for review.)

R.9. $\{x \mid x < 8\}$

R.10. $\left\{ x \mid x \geq -\dfrac{9}{2} \right\}$

R.11. $\{x \mid -2.4 \leq x < 5.8\}$

R.12. $\{x \mid -1 < x \leq 2.5\}$

Concept Connections

1. A graph of an equation is symmetric with respect to the _____-axis if replacing x by $-x$ results in an equivalent equation.

2. A graph of an equation is symmetric with respect to the _____-axis if replacing y by $-y$ results in an equivalent equation.

3. A graph of an equation is symmetric with respect to the _____ if replacing x by $-x$ and y by $-y$ results in an equivalent equation.

4. An even function is symmetric with respect to the _____.

5. An odd function is symmetric with respect to the _____.

6. The expression _____ represents the greatest integer, less than or equal to x.

Objective 1: Test for Symmetry

For Exercises 7–18, determine whether the graph of the equation is symmetric with respect to the x-axis, y-axis, origin, or none of these. **(See Examples 1–2)**

7. $y = x^2 + 3$

8. $y = -|x| - 4$

9. $x = -|y| - 4$

10. $x = y^2 + 3$

11. $x^2 + y^2 = 3$

12. $|x| + |y| = 4$

13. $y = |x| + 2x + 7$

14. $y = x^2 + 6x + 1$

15. $x^2 = 5 + y^2$

16. $y^4 = 2 + x^2$

17. $y = \dfrac{1}{2}x - 3$

18. $y = \dfrac{2}{5}x + 1$

Objective 2: Identify Even and Odd Functions

19. What type of symmetry does an even function have?

20. What type of symmetry does an odd function have?

For Exercises 21–26, use the graph to determine if the function is even, odd, or neither. **(See Example 3)**

21.

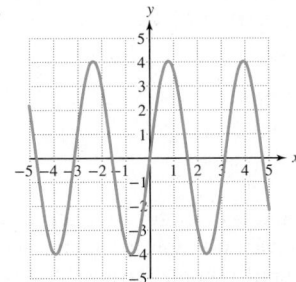

22.

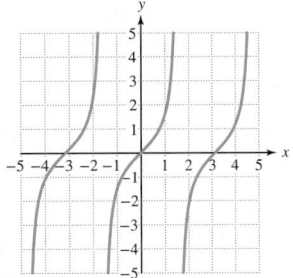

23.

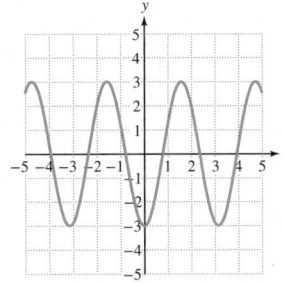

24.

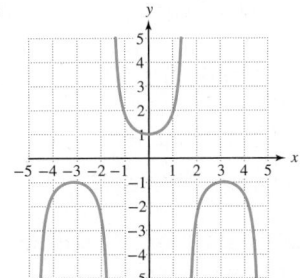

25.

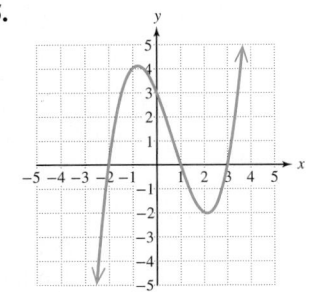

26.

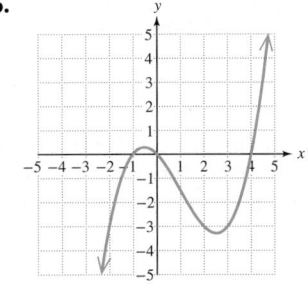

27. **a.** Given $f(x) = 4x^2 - 3|x|$, find $f(-x)$.

    b. Is $f(-x) = f(x)$?

    c. Is this function even, odd, or neither?

29. **a.** Given $h(x) = 4x^3 - 2x$, find $h(-x)$.

    b. Find $-h(x)$.

    c. Is $h(-x) = -h(x)$?

    d. Is this function even, odd, or neither?

31. **a.** Given $m(x) = 4x^2 + 2x - 3$, find $m(-x)$.

    b. Find $-m(x)$.

    c. Is $m(-x) = m(x)$?

    d. Is $m(-x) = -m(x)$?

    e. Is this function even, odd, or neither?

28. **a.** Given $g(x) = -x^8 + |3x|$, find $g(-x)$.

    b. Is $g(-x) = g(x)$?

    c. Is this function even, odd, or neither?

30. **a.** Given $k(x) = -8x^5 - 6x^3$, find $k(-x)$.

    b. Find $-k(x)$.

    c. Is $k(-x) = -k(x)$?

    d. Is this function even, odd, or neither?

32. **a.** Given $n(x) = 7|x| + 3x - 1$, find $n(-x)$.

    b. Find $-n(x)$.

    c. Is $n(-x) = n(x)$?

    d. Is $n(-x) = -n(x)$?

    e. Is this function even, odd, or neither?

For Exercises 33–46, determine if the function is even, odd, or neither. **(See Example 4)**

33. $f(x) = 3x^6 + 2x^2 + |x|$

34. $p(x) = -|x| + 12x^{10} + 5$

35. $k(x) = 13x^3 + 12x$

36. $m(x) = -4x^5 + 2x^3 + x$

37. $n(x) = \sqrt{16 - (x-3)^2}$

38. $r(x) = \sqrt{81 - (x+2)^2}$

39. $q(x) = \sqrt{16 + x^2}$

40. $z(x) = \sqrt{49 + x^2}$

41. $h(x) = 5x$

42. $g(x) = -x$

43. $f(x) = \dfrac{x^2}{3(x-4)^2}$

44. $g(x) = \dfrac{x^3}{2(x-1)^3}$

45. $v(x) = \dfrac{-x^5}{|x| + 2}$

46. $w(x) = \dfrac{-\sqrt[3]{x}}{x^2 + 1}$

Objective 3: Graph Piecewise-Defined Functions

For Exercises 47–50, evaluate the function for the given values of x. **(See Example 5)**

47. $f(x) = \begin{cases} -3x + 7 & \text{for } x < -1 \\ x^2 + 3 & \text{for } -1 \le x < 4 \\ 5 & \text{for } x \ge 4 \end{cases}$

    a. $f(3)$ **b.** $f(-2)$ **c.** $f(-1)$

    d. $f(4)$ **e.** $f(5)$

48. $g(x) = \begin{cases} -2|x| - 3 & \text{for } x \le -2 \\ 5x + 6 & \text{for } -2 < x < 3 \\ 4 & \text{for } x \ge 3 \end{cases}$

    a. $g(-3)$ **b.** $g(3)$ **c.** $g(-2)$

    d. $g(0)$ **e.** $g(4)$

49. $h(x) = \begin{cases} 2 & \text{for } -3 \le x < -2 \\ 1 & \text{for } -2 \le x < -1 \\ 0 & \text{for } -1 \le x < 0 \\ -1 & \text{for } 0 \le x < 1 \end{cases}$

    a. $h(-1.7)$ **b.** $h(-2.5)$ **c.** $h(0.05)$

    d. $h(-2)$ **e.** $h(0)$

50. $t(x) = \begin{cases} x & \text{for } 0 < x \le 1 \\ 2x & \text{for } 1 < x \le 2 \\ 3x & \text{for } 2 < x \le 3 \\ 4x & \text{for } 3 < x \le 4 \end{cases}$

    a. $t(1.99)$ **b.** $t(0.4)$ **c.** $t(3)$

    d. $t(1)$ **e.** $t(3.001)$

51. A sled accelerates down a hill and then slows down after it reaches a flat portion of ground. The speed of the sled $s(t)$ (in ft/sec) at a time t (in sec) after movement begins can be approximated by:

$$s(t) = \begin{cases} 1.5t & \text{for } 0 \le t \le 20 \\ \dfrac{30}{t - 19} & \text{for } 20 < t \le 40 \end{cases}$$

Determine the speed of the sled after 10 sec, 20 sec, 30 sec, and 40 sec. Round to 1 decimal place if necessary.

52. A car starts from rest and accelerates to a speed of 60 mph in 12 sec. It travels 60 mph for 1 min and then decelerates for 20 sec until it comes to rest. The speed of the car $s(t)$ (in mph) at a time t (in sec) after the car begins motion can be modeled by:

$$s(t) = \begin{cases} \dfrac{5}{12}t^2 & \text{for } 0 \le t \le 12 \\ 60 & \text{for } 12 < t \le 72 \\ \dfrac{3}{20}(92 - t)^2 & \text{for } 72 < t \le 92 \end{cases}$$

Determine the speed of the car 6 sec, 12 sec, 45 sec, and 80 sec after the car begins motion.

For Exercises 53–56, match the function with its graph.

53. $f(x) = x + 1$ for $x < 2$

54. $f(x) = x + 1$ for $-1 < x \leq 2$

55. $f(x) = x + 1$ for $-1 \leq x < 2$

56. $f(x) = x + 1$ for $x \geq 2$

a. 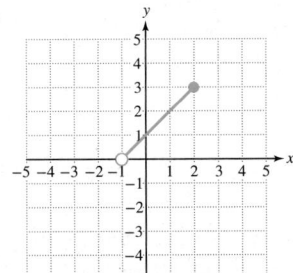   **b.** 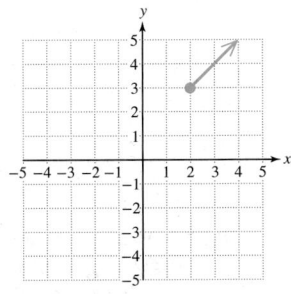   **c.** 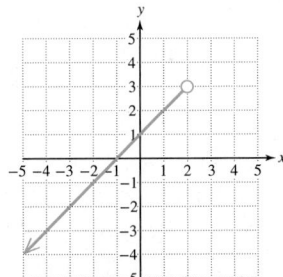   **d.**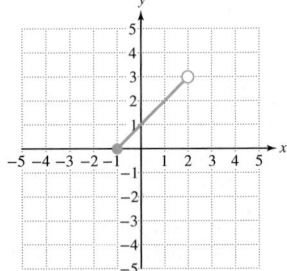

57. a. Graph $p(x) = x + 2$ for $x \leq 0$.
   (See Examples 6–7)

   b. Graph $q(x) = -x^2$ for $x > 0$.

   c. Graph $r(x) = \begin{cases} x + 2 & \text{for } x \leq 0 \\ -x^2 & \text{for } x > 0 \end{cases}$

59. a. Graph $m(x) = \dfrac{1}{2}x - 2$ for $x \leq -2$.

   b. Graph $n(x) = -x + 1$ for $x > -2$.

   c. Graph $t(x) = \begin{cases} \dfrac{1}{2}x - 2 & \text{for } x \leq -2 \\ -x + 1 & \text{for } x > -2 \end{cases}$

58. a. Graph $f(x) = |x|$ for $x < 0$.

   b. Graph $g(x) = \sqrt{x}$ for $x \geq 0$.

   c. Graph $h(x) = \begin{cases} |x| & \text{for } x < 0 \\ \sqrt{x} & \text{for } x \geq 0 \end{cases}$

60. a. Graph $a(x) = x$ for $x < 1$.

   b. Graph $b(x) = \sqrt{x - 1}$ for $x \geq 1$.

   c. Graph $c(x) = \begin{cases} x & \text{for } x < 1 \\ \sqrt{x - 1} & \text{for } x \geq 1 \end{cases}$

For Exercises 61–70, graph the function. **(See Examples 6–7)**

61. $f(x) = \begin{cases} |x| & \text{for } x < 2 \\ -x + 4 & \text{for } x \geq 2 \end{cases}$

62. $h(x) = \begin{cases} -2x & \text{for } x < 0 \\ \sqrt{x} & \text{for } x \geq 0 \end{cases}$

63. $g(x) = \begin{cases} x + 2 & \text{for } x < -1 \\ -x + 2 & \text{for } x \geq -1 \end{cases}$

64. $k(x) = \begin{cases} 3x & \text{for } x < 1 \\ -3x & \text{for } x \geq 1 \end{cases}$

65. $r(x) = \begin{cases} x^2 - 4 & \text{for } x \leq 2 \\ 2x - 4 & \text{for } x > 2 \end{cases}$

66. $s(x) = \begin{cases} -x - 1 & \text{for } x \leq -1 \\ \sqrt{x + 1} & \text{for } x > -1 \end{cases}$

67. $t(x) = \begin{cases} -3 & \text{for } -4 \leq x < -2 \\ -1 & \text{for } -2 \leq x < 0 \\ 1 & \text{for } 0 \leq x < 2 \end{cases}$

68. $z(x) = \begin{cases} -1 & \text{for } -3 < x \leq -1 \\ 1 & \text{for } -1 < x \leq 1 \\ 3 & \text{for } 1 < x \leq 3 \end{cases}$

69. $m(x) = \begin{cases} 3 & \text{for } -4 < x < -1 \\ -x & \text{for } -1 \leq x < 3 \\ \sqrt{x - 3} & \text{for } x \geq 3 \end{cases}$

70. $n(x) = \begin{cases} -4 & \text{for } -3 < x < -1 \\ x & \text{for } -1 \leq x < 2 \\ -x^2 + 4 & \text{for } x \geq 2 \end{cases}$

71. a. Graph $f(x) = \begin{cases} -x & \text{for } x < 0 \\ x & \text{for } x \geq 0 \end{cases}$ **b.** To what basic function from Section 6.1 is the graph of f equivalent?

For Exercises 72–80, evaluate the step function defined by $f(x) = [\![x]\!]$ for the given values of x. **(See Example 8)**

72. $f(-3.7)$ **73.** $f(-4.2)$ **74.** $f(-0.5)$ **75.** $f(-0.09)$ **76.** $f(0.5)$

77. $f(0.09)$ **78.** $f(6)$ **79.** $f(-9)$ **80.** $f(-5)$

For Exercises 81–84, graph the function. **(See Example 8)**

81. $f(x) = [\![x + 3]\!]$ **82.** $g(x) = [\![x - 3]\!]$ **83.** $k(x) = \text{int}\left(\dfrac{1}{2}x\right)$ **84.** $h(x) = \text{int}(2x)$

85. For a recent year, the rate for first class postage was as follows. **(See Example 9)**

Weight not Over	Price
1 oz	$0.44
2 oz	$0.61
3 oz	$0.78
3.5 oz	$0.95

Write a piecewise-defined function to model the cost $C(x)$ to mail a letter first class if the letter is x ounces.

86. The water level in a retention pond started at 5 ft (60 in.) and decreased at a rate of 2 in./day during a 14-day drought. A tropical depression moved through at the beginning of the 15th day and produced rain at an average rate of 2.5 in./day for 5 days. Write a piecewise-defined function to model the water level $L(x)$ (in inches) as a function of the number of days x since the beginning of the drought.

87. A salesperson makes a base salary of $2000 per month. Once he reaches $40,000 in total sales, he earns an additional 5% commission on the amount in sales over $40,000. Write a piecewise-defined function to model the salesperson's total monthly salary $S(x)$ (in $) as a function of the amount in sales x.

88. A cell phone plan charges $49.95 per month plus $14.02 in taxes, plus $0.40 per minute for calls beyond the 600-min monthly limit. Write a piecewise-defined function to model the monthly cost $C(x)$ (in $) as a function of the number of minutes used x for the month.

Mixed Exercises

For Exercises 89–94, produce a rule for the function whose graph is shown. (*Hint*: Consider using the basic functions learned in Section 6.1 and transformations of their graphs.)

89.

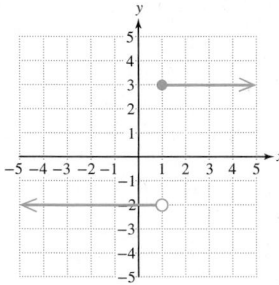

90.

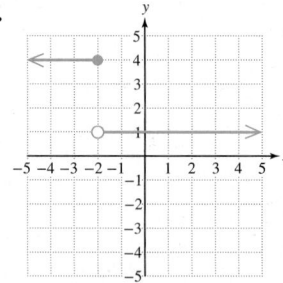

91.

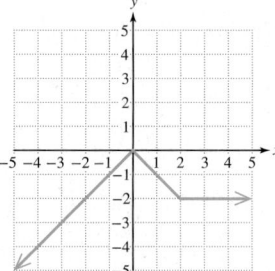

92.

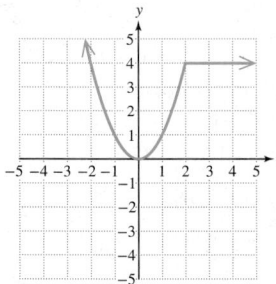

93.

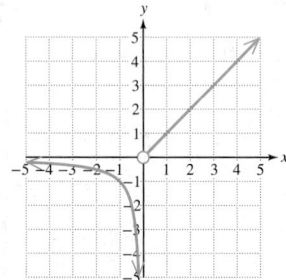

94.

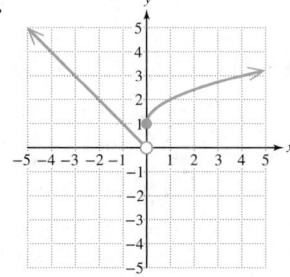

For Exercises 95–96,

a. Graph the function.

b. Write the domain in interval notation.

c. Write the range in interval notation.

d. Evaluate $f(-1)$, $f(1)$, and $f(2)$.

e. Find the value(s) of x for which $f(x) = 6$.

f. Find the value(s) of x for which $f(x) = -3$.

95. $f(x) = \begin{cases} -x^2 + 1 & \text{for } x \le 1 \\ 2x & \text{for } x > 1 \end{cases}$

96. $f(x) = \begin{cases} |x| & \text{for } x < 2 \\ -x & \text{for } x \ge 2 \end{cases}$

In computer programming, the greatest integer function is sometimes called the "floor" function. Programmers also make use of the "ceiling" function, which returns the smallest integer not less than x. For example: $\text{ceil}(3.1) = 4$. For Exercises 97–98, evaluate the floor and ceiling functions for the given value of x.

floor(x) is the greatest integer less than or equal to x.
ceil(x) is the smallest integer not less than x.

97. a. floor(2.8)   **b.** floor(−3.1)   **c.** floor(4)   **98. a.** floor(5.5)   **b.** floor(−0.1)   **c.** floor(−2)

     d. ceil(2.8)   **e.** ceil(−3.1)   **f.** ceil(4)   **d.** ceil(5.5)   **e.** ceil(−0.1)   **f.** ceil(−2)

Write About It

99. From an equation in x and y, explain how to determine whether the graph of the equation is symmetric with respect to the x-axis, y-axis, or origin.

100. From the graph of a function, how can you determine if the function is even or odd?

101. Explain why the relation defined by
$$y = \begin{cases} 2x & \text{for } x \le 1 \\ 3 & \text{for } x \ge 1 \end{cases}$$
is not a function.

102. Explain why the function is discontinuous at $x = 1$.
$$f(x) = \begin{cases} 3x & \text{for } x < 1 \\ 3 & \text{for } x > 1 \end{cases}$$

Expanding Your Skills

103. For a recent year, the federal income tax owed by a taxpayer (single—no dependents) was based on the individual's taxable income. (*Source*: Internal Revenue Service, www.irs.gov)

Schedule X–If your filing status is "Single"

If taxable income is over	But not over	The tax is	Of the amount over
$0	$9,525	$0.00 + 10%	$0
$9,525	$38,700	$952.50 + 12%	$9,525
$38,700	$82,500	$4,453.50 + 22%	$38,700

Write a piecewise-defined function that expresses an individual's federal income tax $f(x)$ (in $) as a function of the individual's taxable income x (in $).

Technology Connections

For Exercises 104–107, use a graphing utility to graph the piecewise-defined function.

104. $f(x) = \begin{cases} 2.5x + 2 & \text{for } x \le 1 \\ x^2 - x - 1 & \text{for } x > 1 \end{cases}$

105. $g(x) = \begin{cases} -3.1x - 4 & \text{for } x < -2 \\ -x^3 + 4x - 1 & \text{for } x \ge -2 \end{cases}$

106. $k(x) = \begin{cases} -2.7x - 4.1 & \text{for } x \le -1 \\ -x^3 + 2x + 5 & \text{for } -1 < x < 2 \\ 1 & \text{for } x \ge 2 \end{cases}$

107. $z(x) = \begin{cases} 2.5x + 8 & \text{for } x < -2 \\ -2x^2 + x + 4 & \text{for } -2 \le x < 2 \\ -2 & \text{for } x \ge 2 \end{cases}$

1. Compute Average Rate of Change

In many applications of real-world phenomena, the relationship between two variables is nonlinear. However, mathematicians often use linear approximations to analyze nonlinear functions on small intervals. For example, the graph in Figure 6-19 shows the blood alcohol concentration (BAC) over a 9-hour period for an individual who had been drinking.

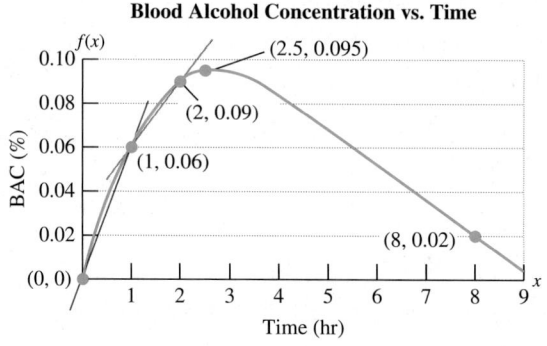

Blood Alcohol Concentration vs. Time

Figure 6-19

Point of Interest

The word *secant* is derived from the Latin *secare* meaning "to cut."

A line drawn through at least two points on a curve is called a **secant line**. In Figure 6-19, the average rate of change in BAC between two points on the graph is the slope of the secant line through the points. Notice that the slope of the secant line between $x = 0$ and $x = 1$ (shown in red) is greater than the slope of the secant line between $x = 1$ and $x = 2$ (shown in green). This means that the average increase in BAC is greater over the first hour than over the second hour.

Average Rate of Change of a Function

Suppose that the points (x_1, y_1) and (x_2, y_2) are points on the graph of a function f.
Using function notation, these are the points $(x_1, f(x_1))$ and $(x_2, f(x_2))$.
   If f is defined on the interval $[x_1, x_2]$, then the **average rate of change** of f on the interval $[x_1, x_2]$ is the slope of the secant line containing $(x_1, f(x_1))$ and $(x_2, f(x_2))$.

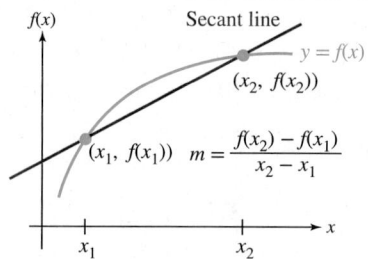

Average rate of change: $m = \dfrac{\Delta y}{\Delta x} = \dfrac{y_2 - y_1}{x_2 - x_1}$ or $m = \dfrac{f(x_2) - f(x_1)}{x_2 - x_1}$

EXAMPLE 1   **Computing Average Rate of Change**

Determine the average rate of change of blood alcohol level

a. from $x_1 = 0$ to $x_2 = 1$.

b. from $x_1 = 1$ to $x_2 = 2$.

c. Interpret the results from parts (a) and (b).

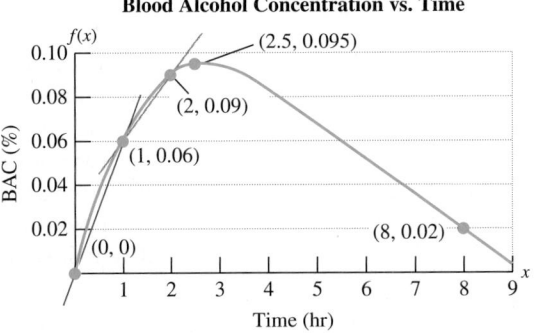

Blood Alcohol Concentration vs. Time

Solution:

a. Average rate of change $= \dfrac{f(x_2) - f(x_1)}{x_2 - x_1} = \dfrac{f(1) - f(0)}{1 - 0} = \dfrac{0.06 - 0}{1}$

$$= 0.06$$

b. Average rate of change $= \dfrac{f(x_2) - f(x_1)}{x_2 - x_1} = \dfrac{f(2) - f(1)}{2 - 1} = \dfrac{0.09 - 0.06}{1}$

$$= 0.03$$

c. The blood alcohol concentration rose by an average of 0.06% per hour during the first hour.

The blood alcohol concentration rose by an average of 0.03% per hour during the second hour.

FOR REVIEW

The slope of a line between two points (x_1, y_1) and (x_2, y_2) is given by $m = \dfrac{y_2 - y_1}{x_2 - x_1}$.

For example:
Let $(x_1, y_1) = (1, 0.06)$ and $(x_2, y_2) = (2, 0.09)$.

The slope of the line containing the points is:

$\dfrac{0.09 - 0.06}{2 - 1} = \dfrac{0.03}{1} = 0.03$

See Section 5.4 for review.

Skill Practice 1   Refer to the graph in Example 1.

a. Determine the average rate of change of blood alcohol level from $x_1 = 2.5$ to $x_2 = 8$. Round to 3 decimal places.

b. Interpret the results from part (a).

EXAMPLE 2   **Computing Average Rate of Change**

Given the function defined by $f(x) = x^2 - 1$, determine the average rate of change from $x_1 = -2$ to $x_2 = 0$.

Solution:

Average rate of change $= \dfrac{f(x_2) - f(x_1)}{x_2 - x_1}$

$$= \dfrac{f(0) - f(-2)}{0 - (-2)} = \dfrac{-1 - 3}{2} = \dfrac{-4}{2} = -2$$

The average rate of change is -2.

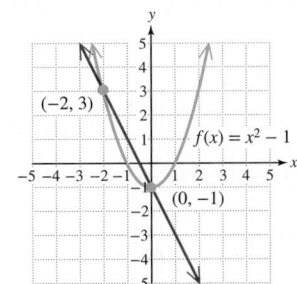

Answers

1. a. -0.014

b. The blood alcohol concentration decreased by an average of 0.014% per hour during this time interval.

2. 9

Skill Practice 2   Given the function defined by $f(x) = x^3 + 2$, determine the average rate of change from $x_1 = -3$ to $x_2 = 0$.

2. Investigate Increasing, Decreasing, and Constant Behavior of a Function

The average rate of change of a function between two points on the function gives the slope of the secant line between the two points. A positive slope indicates that the function has an overall increase between the two points and the graph trends upward. A negative slope indicates that the function has an overall decrease between the two points and the graph trends downward. We now take the opportunity to investigate increasing, decreasing, and constant (flat) behavior of a function. For example, the graph in Figure 6-20 approximates the altitude of an airplane, $f(t)$, as a function of the amount of time t (in min) after takeoff.

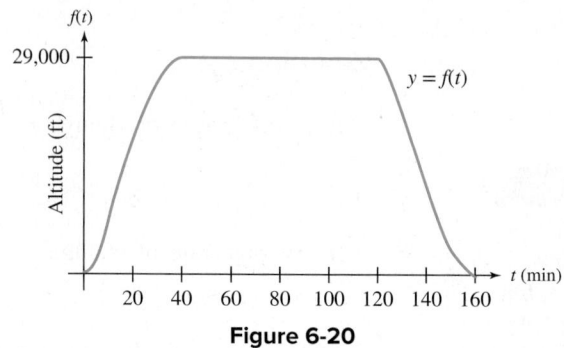

Figure 6-20

Notice that the plane's altitude increases up to the first 40 min of the flight. So we say that the function f is increasing on the interval $(0, 40)$. The plane flies at a constant altitude for the next 1 hr 20 min, so we say that f is constant on the interval $(40, 120)$. Finally, the plane's altitude decreases for the last 40 min, so we say that f is decreasing on the interval $(120, 160)$.

Informally, a function is increasing on an interval in its domain if its graph rises from left to right. A function is decreasing on an interval in its domain if the graph "falls" from left to right. A function is constant on an interval in its domain if its graph is horizontal over the interval. These ideas are stated formally using mathematical notation.

Intervals of Increasing, Decreasing, and Constant Behavior

Suppose that I is an interval contained within the domain of a function f.

- f is increasing on I if $f(x_1) < f(x_2)$ for all $x_1 < x_2$ on I.
- f is decreasing on I if $f(x_1) > f(x_2)$ for all $x_1 < x_2$ on I.
- f is constant on I if $f(x_1) = f(x_2)$ for all x_1 and x_2 on I.

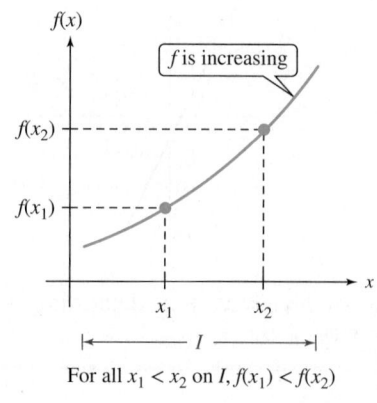

For all $x_1 < x_2$ on I, $f(x_1) < f(x_2)$

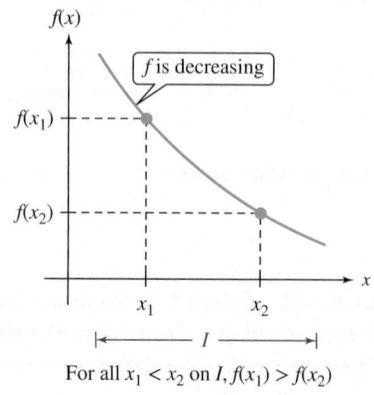

For all $x_1 < x_2$ on I, $f(x_1) > f(x_2)$

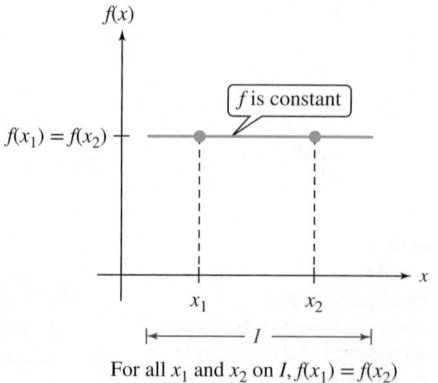

For all x_1 and x_2 on I, $f(x_1) = f(x_2)$

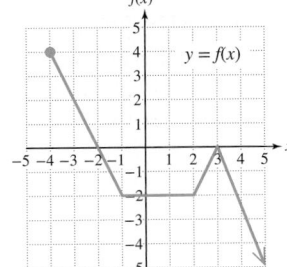

EXAMPLE 3   **Determining the Intervals Over Which a Function Is Increasing, Decreasing, and Constant**

Use interval notation to write the interval(s) over which f is

a. Increasing   **b.** Decreasing

c. Constant

FOR REVIEW

Recall that interval notation is used to express a range of values on the number line. A curved parenthesis (or) indicates that an "endpoint" is not included in the interval. A square bracket [or] indicates that an "endpoint" is included in the interval. A curved parenthesis is always used at ∞ and $-\infty$. For example, the set of real numbers strictly greater than 3 is written in interval notation as $(3, \infty)$. See Section 1.4 for review.

Solution:

a. f is increasing on the interval $(2, 3)$. (Highlighted in red tint.)

b. f is decreasing on the intervals $(-4, -1)$ and $(3, \infty)$. (Highlighted in orange tint.)

c. f is constant on the interval $(-1, 2)$. (Highlighted in green tint.)

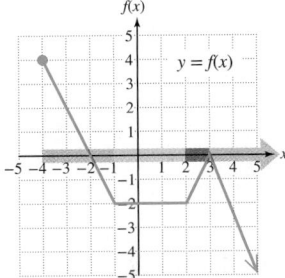

Skill Practice 3   Use interval notation to write the interval(s) over which f is

a. Increasing

b. Decreasing

c. Constant

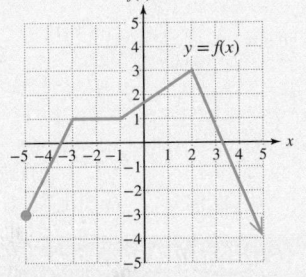

3. Determine Relative Minima and Maxima of a Function

The intervals over which a function changes from increasing to decreasing behavior or vice versa tell us where to look for relative maximum values and relative minimum values of a function. Consider the function pictured in Figure 6-21.

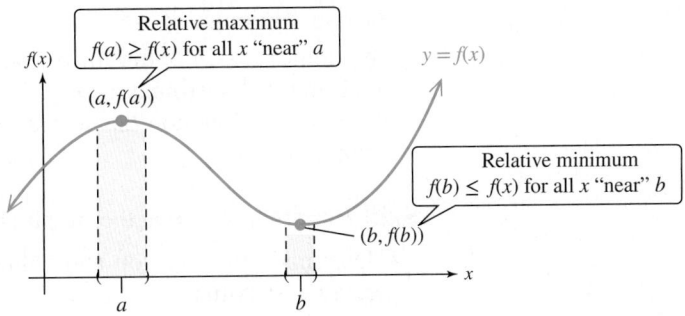

Figure 6-21

- The function has a relative maximum of $f(a)$. Informally, this means that $f(a)$ is the greatest function value relative to other points on the function nearby.
- The function has a relative minimum of $f(b)$. Informally, this means that $f(b)$ is the smallest function value relative to other points on the function nearby.

Answers

3. a. $(-5, -3)$ and $(-1, 2)$

   b. $(2, \infty)$ **c.** $(-3, -1)$

This is stated formally in the following definition.

Relative Minimum and Relative Maximum Values

- $f(a)$ is a **relative maximum** of f if there exists an open interval containing a such that $f(a) \geq f(x)$ for all x in the interval.
- $f(b)$ is a **relative minimum** of f if there exists an open interval containing b such that $f(b) \leq f(x)$ for all x in the interval.

Note: An open interval is an interval in which the endpoints are not included.

If an ordered pair $(a, f(a))$ corresponds to a relative minimum or relative maximum, we interpret the coordinates of the ordered pair as follows:

- The x-coordinate is the *location* of the relative maximum or minimum within the domain of the function.
- The y-coordinate is the *value* of the relative maximum or minimum. This tells us how "high" or "low" the graph is at that point.

EXAMPLE 4 Finding Relative Maxima and Minima

For the graph of $y = g(x)$ shown,

a. Determine the location and value of any relative maxima.

b. Determine the location and value of any relative minima.

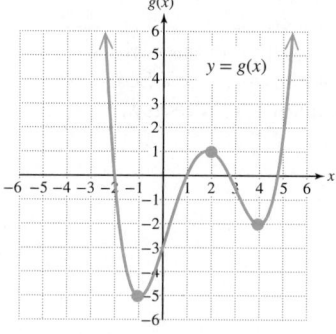

Avoiding Mistakes

Be sure to note that the value of a relative minimum or relative maximum is the y value of a function, not the x value.

Solution:

a. The point $(2, 1)$ is the highest point in a small interval surrounding $x = 2$. Therefore, at $x = 2$, the function has a relative maximum of 1.

b. The point $(-1, -5)$ is the lowest point in a small interval surrounding $x = -1$. Therefore, at $x = -1$, the function has a relative minimum of -5.

The point $(4, -2)$ is the lowest point in a small interval surrounding $x = 4$. Therefore, at $x = 4$, the function has a relative minimum of -2.

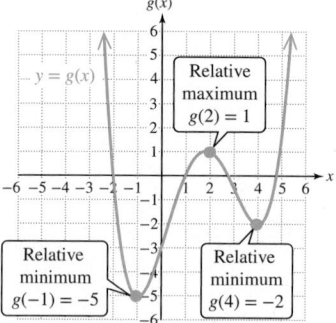

Skill Practice 4 For the graph shown,

a. Determine the location and value of any relative maxima.

b. Determine the location and value of any relative minima.

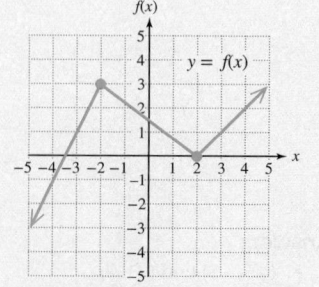

Answers

4. a. At $x = -2$, the function has a relative maximum of 3.

 b. At $x = 2$, the function has a relative minimum of 0.

TECHNOLOGY CONNECTIONS

Determining Relative Maxima and Minima

Relative maxima and relative minima are often difficult to find analytically and require techniques from calculus. However, a graphing utility can be used to approximate the location and value of relative maxima and minima. To do so, we use the Minimum and Maximum features.

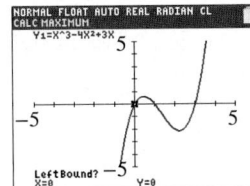

For example, enter the function defined by $Y_1 = x^3 - 4x^2 + 3x$. Then access the Maximum feature from the CALC menu.

The calculator asks for a left bound. This is a point slightly to the left of the relative maximum. Then hit ENTER.

The calculator asks for a right bound. This is a point slightly to the right of the relative maximum. Hit ENTER.

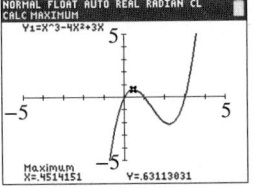

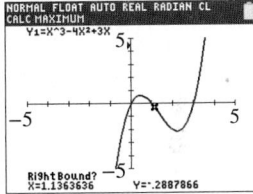

The calculator asks for a guess. This is a point close to the relative maximum. Hit ENTER and the approximate coordinates of the relative maximum point are shown (0.45, 0.63).

To find the relative minimum, repeat these steps using the Minimum feature. The coordinates of the relative minimum point are approximately (2.22, −2.11).

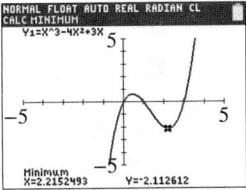

4. Evaluate a Difference Quotient

We close this section with another look at average rate of change of a function over a given interval. Suppose we select a value x from the domain of f and a second value $x + h$ in the domain. In this scenario $h \neq 0$ and h is generally taken to be very small. Suppose we label $P(x, f(x))$ and $Q(x + h, f(x + h))$ as the corresponding points on the function (Figure 6-22).

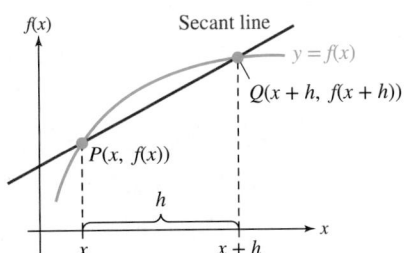

Figure 6-22

The average rate of change of f on the interval $(x, x + h)$ is the slope of the secant line through P and Q. Let $(x_1, y_1) = (x, f(x))$ and let $(x_2, y_2) = (x + h, f(x + h))$. Then the slope is

$$m = \frac{y_2 - y_1}{x_2 - x_1} = \frac{f(x + h) - f(x)}{(x + h) - x} = \frac{f(x + h) - f(x)}{\not{x} + h - \not{x}} = \frac{f(x + h) - f(x)}{h}$$

This result is called the *difference quotient* and represents the average rate of change of f between x and $x + h$.

The Difference Quotient

Suppose that $h \neq 0$ is a constant real number and that f is a smooth curve defined on the interval $[x, x + h]$. Then the **difference quotient**[*] is defined as

$$\frac{f(x + h) - f(x)}{h}$$

*The difference quotient is an extremely important concept for calculus and higher levels of mathematics. It gives a formula for the average rate of change of the function between x and $x + h$. After finding the difference quotient in a calculus class, we would "shrink" the value of h close to zero to determine the exact rate of change of the function at a given point.

In Examples 5 and 6, we practice the important skill of finding the difference quotient.

EXAMPLE 5   Finding a Difference Quotient

Given $f(x) = 3x - 5$,

  a. Find $f(x + h)$.

  b. Find the difference quotient, $\dfrac{f(x + h) - f(x)}{h}$.

TIP Finding the value of $f(x + h)$ is usually the biggest obstacle for students when evaluating a difference quotient. For this reason, we recommend finding $f(x + h)$ first as shown in Example 5(a). Then use the result to find the difference quotient as shown in Example 5(b).

Solution:

a. $f(x + h) = 3(x + h) - 5$ Substitute $(x + h)$ for x.

  $= 3x + 3h - 5$

b. $\dfrac{f(x + h) - f(x)}{h} = \dfrac{\overbrace{(3x + 3h - 5)}^{f(x+h)} - \overbrace{(3x - 5)}^{f(x)}}{h}$

  $= \dfrac{3x + 3h - 5 - 3x + 5}{h}$ Clear parentheses.

  $= \dfrac{3h}{h}$ Combine like terms.

  $= 3$ Simplify the fraction.

Skill Practice 5   Given $f(x) = 4x - 2$,

  a. Find $f(x + h)$.

  b. Find the difference quotient, $\dfrac{f(x + h) - f(x)}{h}$.

Answers
5. a. $4x + 4h - 2$ **b.** 4

EXAMPLE 6   **Finding a Difference Quotient**

Given $f(x) = -2x^2 + 4x - 1$,

a. Find $f(x + h)$.

b. Find the difference quotient, $\dfrac{f(x + h) - f(x)}{h}$.

Solution:

a. $f(x + h) = -2(x + h)^2 + 4(x + h) - 1$ Substitute $(x + h)$ for x.

$\quad\quad\quad\quad\;\; = -2(x^2 + 2xh + h^2) + 4x + 4h - 1$

$\quad\quad\quad\quad\;\; = -2x^2 - 4xh - 2h^2 + 4x + 4h - 1$

FOR REVIEW

Recall that the square of a binomial results in a perfect square trinomial.

$(x + h)^2$
$\quad = (x + h)(x + h)$
$\quad = x^2 + xh + xh + h^2$
$\quad = x^2 + 2xh + h^2$

See Section 2.2 for review.

b. $\dfrac{f(x + h) - f(x)}{h} = \dfrac{\overbrace{(-2x^2 - 4xh - 2h^2 + 4x + 4h - 1)}^{f(x+h)} - \overbrace{(-2x^2 + 4x - 1)}^{f(x)}}{h}$

$\quad = \dfrac{-2x^2 - 4xh - 2h^2 + 4x + 4h - 1 + 2x^2 - 4x + 1}{h}$ Clear parentheses.

$\quad = \dfrac{-4xh - 2h^2 + 4h}{h}$ Combine like terms.

$\quad = \dfrac{\overset{1}{h}(-4x - 2h + 4)}{\underset{}{\cancel{h}}}$ Factor numerator and denominator, and simplify the fraction.

$\quad = -4x - 2h + 4$

Skill Practice 6   Given $f(x) = -x^2 - 5x + 2$,

a. Find $f(x + h)$.

b. Find the difference quotient, $\dfrac{f(x + h) - f(x)}{h}$.

Answers

6. a. $-x^2 - 2xh - h^2 - 5x - 5h + 2$

   b. $-2x - h - 5$

SECTION 6.3   Practice Exercises

Prerequisite Review

For Exercises R.1–R.4, match the statement of the slope with the appropriate graph (a)–(d). (See Section 5.4 for review.)

a. 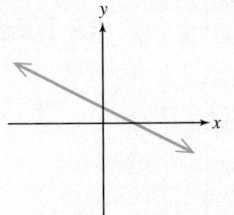   **b.** 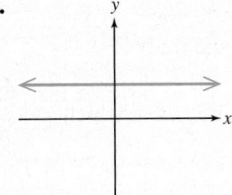   **c.** 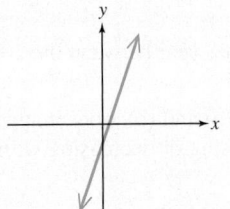   **d.**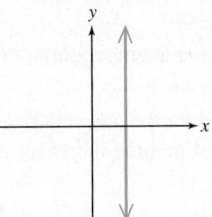

R.1. The slope is positive.   **R.2.** The slope is negative.

R.3. The slope is zero.   **R.4.** The slope is undefined.

For Exercises R.5–R.10, the graph represents the amount in sales y (in \$1000s) for a small company t years since inception.

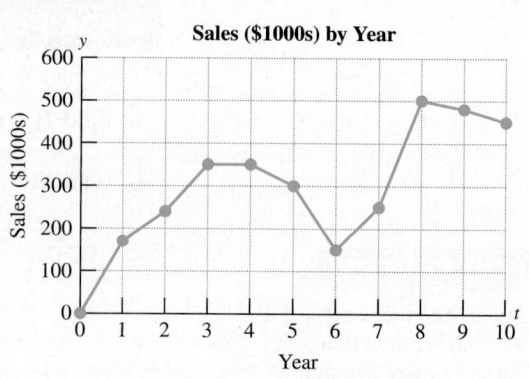

Sales (\$1000s) by Year

R.5. Did sales go up or down between years 4 and 5?

R.6. Did sales go up or down between years 1 and 2?

R.7. Between which two years did sales not change?

R.8. Between which two years did sales increase most rapidly?

R.9. Between which two years did sales decrease most rapidly?

R.10. In which year did maximum sales occur?

For Exercises R.11–R.14, find $f(x + h)$. (See Sections 5.3 and 2.2 for review.)

R.11. $f(x) = 4x - 5$

R.12. $f(x) = 6 - 3x$

R.13. $f(x) = 2x^2$

R.14. $f(x) = \dfrac{-5}{x^2}$

Concept Connections

1. If f is defined on the interval $[x_1, x_2]$, then the average rate of change of f on the interval $[x_1, x_2]$ is given by the formula _____.

2. If $f(x_1) < f(x_2)$ for all $x_1 < x_2$ on an interval I, then f is (increasing/decreasing) on I.

3. If $f(x_1) > f(x_2)$ for all $x_1 < x_2$ on an interval I, then f is (increasing/decreasing) on I.

4. Let h represent a positive real number. Given a function defined by $y = f(x)$, the difference quotient is given by _____.

Objective 1: Compute Average Rate of Change

For Exercises 5–6, find the slope of the secant line pictured in red. **(See Example 1)**

5.

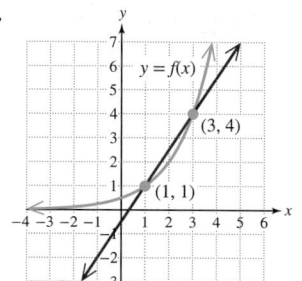

6.

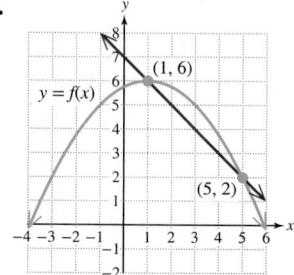

7. The function given by $y = f(x)$ shows the value of \$5000 invested at 5% interest compounded continuously, x years after the money was originally invested.

   a. Find the average amount earned per year between the 5th year and 10th year.

   b. Find the average amount earned per year between the 20th year and the 25th year.

   c. Based on the answers from parts (a) and (b), does it appear that the rate at which annual income increases is increasing or decreasing with time?

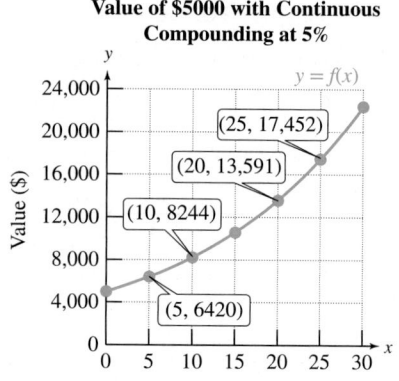

Value of \$5000 with Continuous Compounding at 5%

8. The function given by $y = f(x)$ shows the average monthly temperature (in °F) for Cedar Key. The value of x is the month number and $x = 1$ represents January.

   a. Find the average rate of change in temperature between months 3 and 5 (March and May).

   b. Find the average rate of change in temperature between months 9 and 11 (September and November).

   c. Comparing the results in parts (a) and (b), what does a positive rate of change mean in the context of this problem? What does a negative rate of change mean?

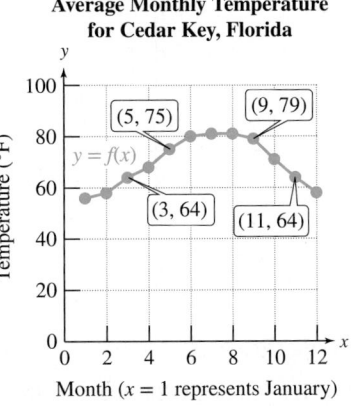

Average Monthly Temperature for Cedar Key, Florida

For Exercises 9–14, determine the average rate of change of the function on the given interval. **(See Example 2)**

9. $f(x) = x^2 - 3$

   a. on $[0, 1]$

   b. on $[1, 3]$

   c. on $[-2, 0]$

10. $g(x) = 2x^2 + 2$

   a. on $[0, 1]$

   b. on $[1, 3]$

   c. on $[-2, 0]$

11. $h(x) = x^3$

   a. on $[-1, 0]$

   b. on $[0, 1]$

   c. on $[1, 2]$

12. $k(x) = x^3 - 2$

   a. on $[-1, 0]$

   b. on $[0, 1]$

   c. on $[1, 2]$

13. $m(x) = \sqrt{x}$

   a. $[0, 1]$

   b. $[1, 4]$

   c. $[4, 9]$

14. $n(x) = \sqrt{x - 1}$

   a. $[1, 2]$

   b. $[2, 5]$

   c. $[5, 10]$

Objective 2: Investigate Increasing, Decreasing, and Constant Behavior of a Function

For Exercises 15–22, use interval notation to write the intervals over which f is (a) increasing, (b) decreasing, and (c) constant. **(See Example 3)**

15.

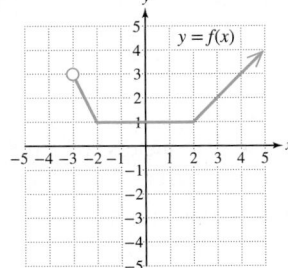

16.

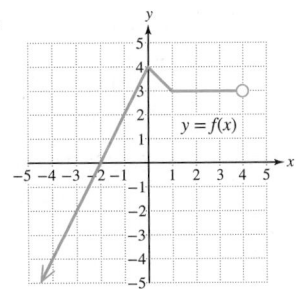

17.

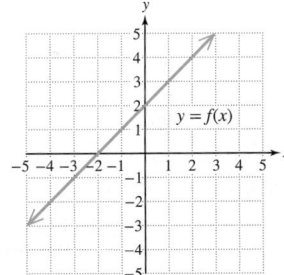

18.

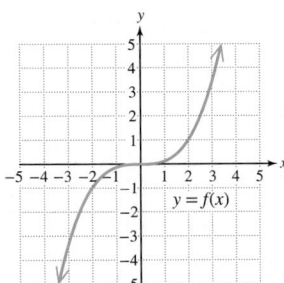

19.

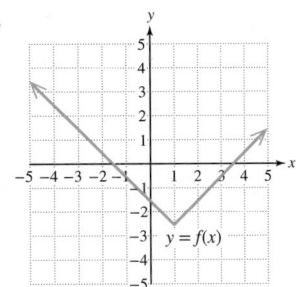

20.

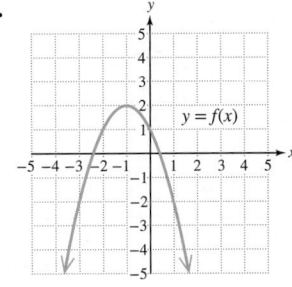

21.

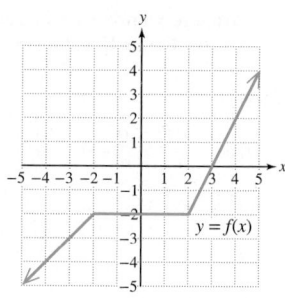

22.

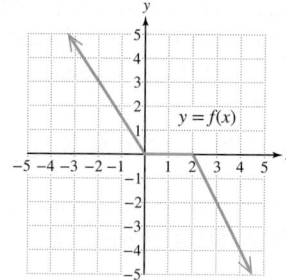

Objective 3: Determine Relative Minima and Maxima of a Function

For Exercises 23–28, identify the location and value of any relative maxima or minima of the function. (**See Example 4**)

23.

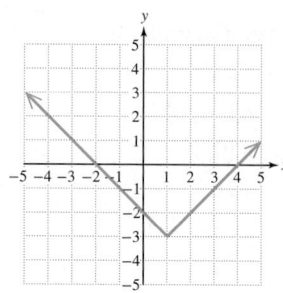

24.

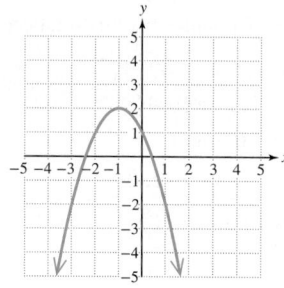

25.

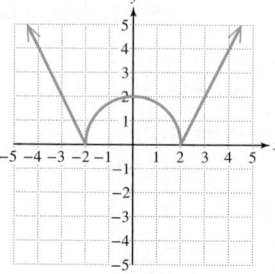

26.

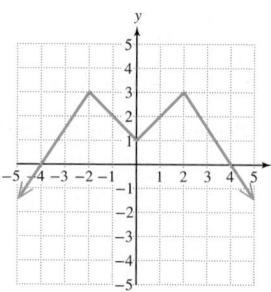

27.

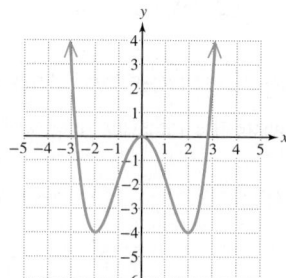

28.

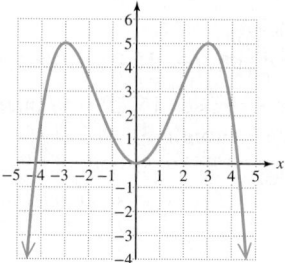

29. The graph shows the depth d (in ft) of a retention pond, t days after recording began.

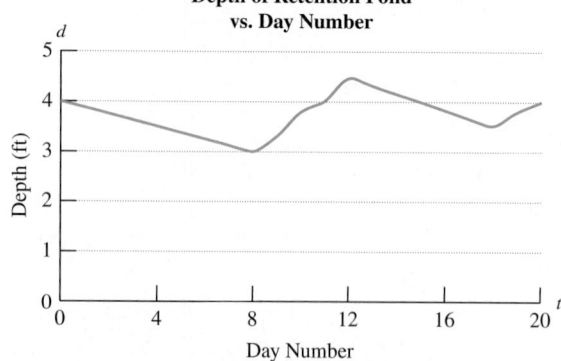

a. Over what interval(s) does the depth increase?

b. Over what interval(s) does the depth decrease?

c. Estimate the times and values of any relative maxima or minima on the interval $(0, 20)$.

d. If rain is the only water that enters the pond, explain what the intervals of increasing and decreasing behavior mean in the context of this problem.

30. The graph shows the height h (in meters) of a roller coaster t seconds after the ride starts.

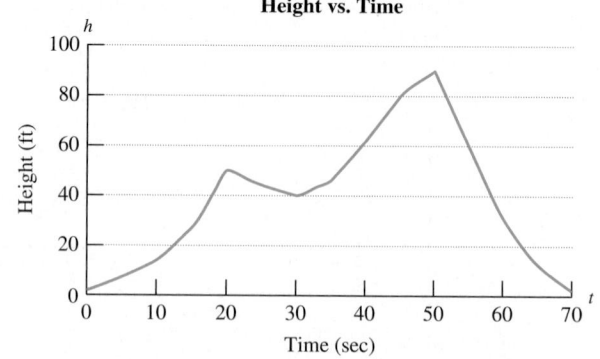

a. Over what interval(s) does the height increase?

b. Over what interval(s) does the height decrease?

c. Estimate the times and values of any relative maxima or minima on the interval $(0, 70)$.

Objective 4: Evaluate a Difference Quotient

For Exercises 31–36, a function is given. (**See Examples 5–6**)

a. Find $f(x + h)$.
b. Find $\dfrac{f(x + h) - f(x)}{h}$.

31. $f(x) = 2x - 3$ **32.** $f(x) = -x + 4$ **33.** $f(x) = 5x + 9$

34. $f(x) = 8x + 4$ **35.** $f(x) = x^2 + 4x$ **36.** $f(x) = x^2 - 3x$

For Exercises 37–44, find the difference quotient and simplify. (**See Examples 5–6**)

37. $f(x) = -2x + 5$ **38.** $f(x) = -3x + 8$ **39.** $f(x) = -5x^2 - 4x + 2$ **40.** $f(x) = -4x^2 - 2x + 6$

41. $f(x) = x^3 + 5$ **42.** $f(x) = x^3 - 2$ **43.** $f(x) = \dfrac{1}{x}$ **44.** $f(x) = \dfrac{1}{x + 2}$

45. Given $f(x) = 4\sqrt{x}$,

   a. Find the difference quotient (do not simplify).

   b. Evaluate the difference quotient for $x = 1$, and the following values of h: $h = 1$, $h = 0.1$, $h = 0.01$, and $h = 0.001$. Round to 4 decimal places.

   c. What value does the difference quotient seem to be approaching as h gets close to 0?

46. Given $f(x) = \dfrac{12}{x}$,

   a. Find the difference quotient (do not simplify).

   b. Evaluate the difference quotient for $x = 2$, and the following values of h: $h = 0.1$, $h = 0.01$, $h = 0.001$, and $h = 0.0001$. Round to 4 decimal places.

   c. What value does the difference quotient seem to be approaching as h gets close to 0?

Mixed Exercises

47. The number $N(t)$ of new cases of a flu outbreak for a given city is given by $N(t) = 5000 \cdot 2^{-0.04t^2}$, where t is the number of months since the outbreak began.

   a. Find the average rate of change in the number of new flu cases between months 0 and 2, and interpret the result. Round to the nearest whole unit.

   b. Find the average rate of change in the number of new flu cases between months 4 and 6, and between months 10 and 12.

   c. Use a graphing utility to graph the function. Use the graph and the average rates of change found in parts (a) and (b) to discuss the pattern of the number of new flu cases.

48. The speed $v(L)$ (in m/sec) of an ocean wave in deep water is approximated by $v(L) = 1.2\sqrt{L}$, where L (in meters) is the wavelength of the wave. (The wavelength is the distance between two consecutive wave crests.)

   a. Find the average rate of change in speed between waves that are between 1 m and 4 m in length.

   b. Find the average rate of change in speed between waves that are between 4 m and 9 m in length.

   c. Use a graphing utility to graph the function. Using the graph and the results from parts (a) and (b), what does the difference in the rates of change mean?

49. Suppose that the average rate of change of a continuous function between any two points to the left of $x = a$ is negative, and the average rate of change of the function between any two points to the right of $x = a$ is positive. Does the function have a relative minimum or maximum at a?

50. Suppose that the average rate of change of a continuous function between any two points to the left of $x = a$ is positive, and the average rate of change of the function between any two points to the right of $x = a$ is negative. Does the function have a relative minimum or maximum at a?

A graph is *concave up* on a given interval if it "bends" upward. A graph is *concave down* on a given interval if it "bends" downward. For Exercises 51–54, determine whether the curve is (a) concave up or concave down and (b) increasing or decreasing.

51.    **52.**    **53.**    **54.**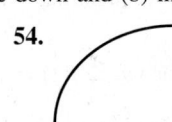

Write About It

55. Explain how the average rate of change of a function f on the interval $[x_1, x_2]$ is related to slope.

56. Provide an informal explanation of a relative maximum.

57. Explain what it means for a function to be increasing on an interval.

58. Explain what the difference quotient represents.

Expanding Your Skills

59. Given $f(x) = \sqrt{x + 3}$,

   a. Find the difference quotient.

   b. Rationalize the numerator of the expression in part (a) and simplify.

   c. Evaluate the expression in part (b) for $h = 0$.

61. A car traveling 60 mph (88 ft/sec) undergoes a constant deceleration until it comes to rest approximately 9.09 sec later. The distance $d(t)$ (in ft) that the car travels t seconds after the brakes are applied is given by $d(t) = -4.84t^2 + 88t$, where $0 \le t \le 9.09$. **(See Example 5)**

   a. Find the difference quotient $\dfrac{d(t + h) - d(t)}{h}$.

   Use the difference quotient to determine the average rate of speed on the following intervals for t.

   b. [0, 2]   (*Hint*: $t = 0$ and $h = 2$)

   c. [2, 4]   (*Hint*: $t = 2$ and $h = 2$)

   d. [4, 6]   (*Hint*: $t = 4$ and $h = 2$)

   e. [6, 8]   (*Hint*: $t = 6$ and $h = 2$)

60. Given $f(x) = \sqrt{x - 4}$,

   a. Find the difference quotient.

   b. Rationalize the numerator of the expression in part (a) and simplify.

   c. Evaluate the expression in part (b) for $h = 0$.

62. A car accelerates from 0 to 60 mph (88 ft/sec) in 8.8 sec. The distance $d(t)$ (in ft) that the car travels t seconds after motion begins is given by $d(t) = 5t^2$, where $0 \le t \le 8.8$.

   a. Find the difference quotient $\dfrac{d(t + h) - d(t)}{h}$.

   Use the difference quotient to determine the average rate of speed on the following intervals for t.

   b. [0, 2]

   c. [2, 4]

   d. [4, 6]

   e. [6, 8]

Technology Connections

For Exercises 63–66, use a graphing utility to

   a. Find the locations and values of the relative maxima and relative minima of the function on the standard viewing window. Round to 3 decimal places.

   b. Use interval notation to write the intervals over which f is increasing or decreasing.

63. $f(x) = -0.6x^2 + 2x + 3$

65. $f(x) = 0.5x^3 + 2.1x^2 - 3x - 7$

64. $f(x) = 0.4x^2 - 3x - 2.2$

66. $f(x) = -0.4x^3 - 1.1x^2 + 2x + 3$

PROBLEM RECOGNITION EXERCISES

Analyzing Functions and Their Graphs

For Exercises 1–6, graph the function defined by $y = f(x)$. Then identify,

   a. The domain in interval notation.

   b. The range in interval notation.

   c. The value $f(1)$.

   d. The values of x for which $f(x) = 1$.

   e. The x-intercept(s).

   f. The y-intercept.

   g. The intervals over which the function is increasing.

   h. The intervals over which the function is decreasing.

   i. The intervals over which the function is constant.

   j. The value and location of any relative maxima or minima.

1. $f(x) = |x - 1| - 3$

3. $f(x) = \dfrac{1}{2}x - 2$

5. $f(x) = \begin{cases} |x| & \text{if } x \le 2 \\ 2 & \text{if } x > 2 \end{cases}$

2. $f(x) = -(x + 1)^2 + 5$

4. $f(x) = -2x + 1$

6. $f(x) = \dfrac{1}{x - 1}$

Algebra of Functions and Function Composition

OBJECTIVES

1. Perform Operations on Functions
2. Compose and Decompose Functions

1. Perform Operations on Functions

In Section 5.5, we learned that a profit function can be constructed from the difference of a revenue function and a cost function according to the following rule:

$$P(x) = R(x) - C(x)$$

As this example illustrates, the difference of two functions makes up a new function. New functions can also be formed from the sum, product, and quotient of two functions.

Sum, Difference, Product, and Quotient of Functions

Given functions f and g, the functions $f + g$, $f - g$, $f \cdot g$, and $\frac{f}{g}$ are defined by

$$(f + g)(x) = f(x) + g(x)$$
$$(f - g)(x) = f(x) - g(x)$$
$$(f \cdot g)(x) = f(x) \cdot g(x)$$
$$\left(\frac{f}{g}\right)(x) = \frac{f(x)}{g(x)} \text{ provided that } g(x) \neq 0$$

The domains of the functions $f + g$, $f - g$, $f \cdot g$, and $\frac{f}{g}$ are all real numbers in the intersection of the domains of the individual functions f and g. For $\frac{f}{g}$, we further restrict the domain to exclude values of x for which $g(x) = 0$.

EXAMPLE 1   Adding Two Functions

Given $f(x) = \sqrt{25 - x^2}$ and $g(x) = 5$, find $(f + g)(x)$.

Solution:

By definition $(f + g)(x) = f(x) + g(x)$.
$$= \sqrt{25 - x^2} + 5$$

Skill Practice 1   Given $m(x) = -|x|$ and $n(x) = 4$, find $(m + n)(x)$.

In Example 1, the graph of function f is a semicircle and the graph of function g is a horizontal line (Figure 6-23). Therefore, the graph of $y = (f + g)(x)$ is the graph of f with a vertical shift (shown in blue). Notice that each y value on $f + g$ is the sum of the y values from the individual functions f and g.

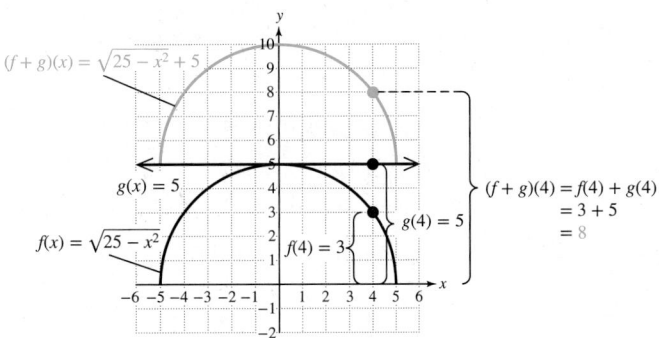

Answer

1. $(m + n)(x) = -|x| + 4$

Figure 6-23

> **EXAMPLE 2**   Evaluating Functions for a Given Value of *x*

Given $m(x) = 4x$, $n(x) = |x - 3|$, and $p(x) = \dfrac{1}{x + 1}$, determine the function values if possible.

a. $(m - n)(-2)$ **b.** $(m \cdot p)(1)$ **c.** $\left(\dfrac{p}{n}\right)(3)$

Solution:

a. $(m - n)(-2) = m(-2) - n(-2)$ Note that $(m - n)(x) = m(x) - n(x)$.

$\qquad\qquad\qquad\quad = 4(-2) - |(-2) - 3|$ Substitute -2 for x in functions m and n.
Then subtract.

$\qquad\qquad\qquad\quad = -8 - |-5|$

$\qquad\qquad\qquad\quad = -8 - 5$

$\qquad\qquad\qquad\quad = -13$

b. $(m \cdot p)(1) = m(1) \cdot p(1)$ Note that $(m \cdot p)(x) = m(x) \cdot p(x)$.

$\qquad\qquad\quad = 4(1) \cdot \dfrac{1}{(1) + 1}$ Substitute 1 for x in functions m and p.
Then multiply.

$\qquad\qquad\quad = 4 \cdot \dfrac{1}{2}$

$\qquad\qquad\quad = 2$

c. $\left(\dfrac{p}{n}\right)(3) = \dfrac{p(3)}{n(3)}$ Note that $\left(\dfrac{p}{n}\right)(x) = \dfrac{p(x)}{n(x)}$.

$\qquad\qquad = \dfrac{\dfrac{1}{(3) + 1}}{|(3) - 3|}$ Substitute 3 for x in functions p and n.
Then divide.

$\qquad\qquad = \dfrac{\frac{1}{4}}{0}$ (undefined)   The function is undefined at $x = 3$ because division by zero is undefined.

> **Skill Practice 2**   Use the functions defined in Example 2 to find
>
> **a.** $(n - m)(-6)$ **b.** $(n \cdot p)(0)$ **c.** $\left(\dfrac{p}{m}\right)(0)$

When combining two or more functions to create a new function, always be sure to determine the domain of the new function. Notice that in Example 2(c), the function $\frac{p}{n}$ is not defined for $x = -1$ or for $x = 3$.

$$\left(\frac{p}{n}\right)(x) = \frac{p(x)}{n(x)} = \frac{\dfrac{1}{x + 1}}{|x - 3|}$$

← Denominator is zero for $x = -1$.

← Denominator of the complex fraction is zero for $x = 3$.

> **EXAMPLE 3**   Combining Functions and Determining Domain

Given $g(x) = 2x$, $h(x) = x^2 - 4x$, and $k(x) = \sqrt{x - 1}$,

a. Find $(g - h)(x)$ and write the domain of $g - h$ in interval notation.

b. Find $(g \cdot k)(x)$ and write the domain of $g \cdot k$ in interval notation.

c. Find $\left(\dfrac{k}{h}\right)(x)$ and write the domain of $\dfrac{k}{h}$ in interval notation.

Answers

2. a. 33   **b.** 3   **c.** Undefined

Skill Practice 5   Given $f(x) = 3x + 4$ and $g(x) = \frac{1}{x-1}$, write a rule for each function and write the domain in interval notation.

   a. $(f \circ g)(x)$ **b.** $(g \circ f)(x)$

EXAMPLE 6   **Composing Functions and Determining Domain**

Given $m(x) = \frac{1}{x-5}$ and $p(x) = \sqrt{x-2}$, find $(m \circ p)(x)$ and write the domain in interval notation.

Solution:

$$(m \circ p)(x) = m(p(x)) = \frac{1}{p(x) - 5}$$

First note that function p has the restriction that $x \geq 2$.

$p(x) \neq 5$

The input value for function m must not be 5. Therefore, $p(x) \neq 5$. We have

$$= \frac{1}{\sqrt{x-2} - 5}$$

$$\sqrt{x-2} \neq 5$$
$$(\sqrt{x-2})^2 \neq (5)^2$$

$$(m \circ p)(x) = \frac{1}{\sqrt{x-2} - 5}$$

$$x - 2 \neq 25$$
$$x \neq 27$$

The domain is $[2, 27) \cup (27, \infty)$.

Skill Practice 6   Given $f(x) = \sqrt{x-1}$ and $g(x) = \frac{1}{x-3}$, find $(g \circ f)(x)$ and write the domain of $g \circ f$ in interval notation.

EXAMPLE 7   **Composing Functions and Determining Domain**

Given $k(x) = \frac{x}{x-2}$ and $m(x) = \frac{8}{x^2}$, find $(k \circ m)(x)$ and write the domain in interval notation.

Solution:

$$(k \circ m)(x) = k(m(x))$$

First note that $m(x) = \frac{8}{x^2}$ has the restriction that $x \neq 0$ because the denominator would be 0.

$$= \frac{m(x)}{m(x) - 2}$$

Substitute $m(x)$ for x in $k(x) = \frac{x}{x-2}$.

$$= \frac{\left(\frac{8}{x^2}\right)}{\left(\frac{8}{x^2}\right) - 2}$$

Substitute $\frac{8}{x^2}$ for x in $\frac{x}{x-2}$.

$$= \frac{x^2 \cdot \left(\frac{8}{x^2}\right)}{x^2 \cdot \left(\frac{8}{x^2} - 2\right)}$$

To simplify the complex fraction, multiply numerator and denominator by the least common denominator (LCD) of all terms in the complex fraction. The LCD is x^2.

Answers

5. a. $(f \circ g)(x) = \frac{3}{x-1} + 4$;

    Domain: $(-\infty, 1) \cup (1, \infty)$

  b. $(g \circ f)(x) = \frac{1}{3x+3}$;

    Domain: $(-\infty, -1) \cup (-1, \infty)$

6. $(g \circ f)(x) = \frac{1}{\sqrt{x-1} - 3}$;

    Domain: $[1, 10) \cup (10, \infty)$

FOR REVIEW

One method to simplify a complex fraction is to multiply numerator and denominator of the complex fraction by the LCD of all individual fractions. (See Section 4.2.)

$$= \dfrac{\dfrac{x^2}{1} \cdot \left(\dfrac{8}{x^2}\right)}{\dfrac{x^2}{1} \cdot \left(\dfrac{8}{x^2} - 2\right)}$$

Write x^2 as $\dfrac{x^2}{1}$ and apply the distributive property.

$$= \dfrac{\dfrac{\cancel{x^2}}{1} \cdot \dfrac{8}{\cancel{x^2}}}{\dfrac{\cancel{x^2}}{1} \cdot \dfrac{8}{\cancel{x^2}} - \dfrac{x^2}{1} \cdot \dfrac{2}{1}}$$

"Cancel" common factors from the numerators and denominators of the individual fractions.

$$= \dfrac{8}{8 - 2x^2}$$

Thus $(k \circ m)(x) = \dfrac{8}{8 - 2x^2}$. The domain has the added restriction that $8 - 2x^2$ must not equal zero. Thus,

$$8 - 2x^2 \neq 0$$
$$2x^2 \neq 8$$
$$x^2 \neq 4$$
$$x \neq \pm 2$$

In summary, the domain of $k \circ m$ has the restrictions that $x \neq 0$, $x \neq 2$, and $x \neq -2$.

The domain is $(-\infty, -2) \cup (-2, 0) \cup (0, 2) \cup (2, \infty)$.

Skill Practice 7   For the functions given in Example 7, find $(m \circ k)(x)$ and write the domain in interval notation.

EXAMPLE 8   Applying Function Composition

At a popular website, the cost to download individual songs is $1.49 per song. In addition, a first-time visitor to the website has a one-time coupon for $1.00 off.

a. Write a function to represent the cost $C(x)$ (in $) for a first-time visitor to purchase x songs.

b. The sales tax for online purchases depends on the location of the business and customer. If the sales tax rate on a purchase is 6%, write a function to represent the total cost $T(a)$ for a first-time visitor who buys a dollars in songs.

c. Find $(T \circ C)(x)$ and interpret the meaning in context.

d. Evaluate $(T \circ C)(10)$ and interpret the meaning in context.

Solution:

a. $C(x) = 1.49x - 1.00;\ x \geq 1$ The cost function is a linear function with $1.49 as the variable rate per song.

b. $T(a) = a + 0.06a$ The total cost is the sum of the cost of the songs plus the sales tax.
$ = 1.06a$

c. $(T \circ C)(x) = T(C(x))$
$ = 1.06(C(x))$
$ = 1.06(1.49x - 1.00)$ Substitute $1.49x - 1.00$ for $C(x)$.
$ = 1.5794x - 1.06$

Answer

7. $(m \circ k)(x) = \dfrac{8x^2 - 32x + 32}{x^2}$;

Domain: $(-\infty, 0) \cup (0, 2) \cup (2, \infty)$

$(T \circ C)(x) = 1.5794x - 1.06$ represents the total cost to buy x songs for a first-time visitor to the website.

d. $(T \circ C)(x) = 1.5794x - 1.06$

$(T \circ C)(10) = 1.5794(10) - 1.06$

$= 14.734$

The total cost for a first-time visitor to buy 10 songs is $14.73.

Skill Practice 8   An artist shops online for tubes of watercolor paint. The cost is $16 for each 14-mL tube.

a. Write a function representing the cost $C(x)$ (in $) for x tubes of paint.

b. There is a 5.5% sales tax on the cost of merchandise and a fixed cost of $4.99 for shipping. Write a function representing the total cost $T(a)$ for a dollars spent in merchandise.

c. Find $(T \circ C)(x)$ and interpret the meaning in context.

d. Evaluate $(T \circ C)(18)$ and interpret the meaning in context.

The composition of two functions creates a new function in which the output from one function becomes the input to the other. We can also reverse this process. That is, we can decompose a composite function into two or more simpler functions.

For example, consider the function h defined by $h(x) = (x - 3)^2$. To write h as a composition of two functions, we have $h(x) = (f \circ g)(x) = f(g(x))$. The function g is the "inside" function and f is the "outside" function. So one natural choice for g and f would be:

> **TIP** The decomposition of functions is not unique. For example, $h(x) = (x - 3)^2$ can also be written as $h(x) = f(g(x))$, where $g(x) = x^2 - 6x$ and $f(x) = x + 9$.

$g(x) = x - 3$ Function g subtracts 3 from the input value.

$f(x) = x^2$ Function f squares the result.

$h(x) = f(g(x)) = (g(x))^2 = (x - 3)^2$

EXAMPLE 9   **Decomposing Two Functions**

Given $h(x) = |2x^2 - 5|$, find two functions f and g such that $h(x) = (f \circ g)(x)$.

Solution:

We need to find two functions f and g such that $h(x) = (f \circ g)(x) = f(g(x))$. The function h first evaluates the expression $2x^2 - 5$, and then takes the absolute value. Therefore, it would be natural to take the absolute value of $g(x) = 2x^2 - 5$.

We have: $g(x) = 2x^2 - 5$ and $f(x) = |x|$

Check: $h(x) = (f \circ g)(x) = f(g(x)) = |g(x)|$

$= |2x^2 - 5|$

Skill Practice 9   Given $m(x) = \sqrt[3]{5x + 1}$, find two functions f and g such that $m(x) = (f \circ g)(x)$.

Answers

8. a. $C(x) = 16x$

   b. $T(a) = 1.055a + 4.99$

   c. $(T \circ C)(x) = 16.88x + 4.99$ represents the total cost to buy x tubes of paint.

   d. $(T \circ C)(18) = \$308.83$; The total cost to buy and ship 18 tubes of paint is $308.83.

9. $g(x) = 5x + 1$ and $f(x) = \sqrt[3]{x}$

In Example 10, we have the graphs of two functions, and we apply function addition, subtraction, multiplication, and composition for selected values of x.

EXAMPLE 10 **Estimating Function Values from a Graph**

The graphs of f and g are shown. Evaluate the functions at the given values of x if possible.

a. $(f + g)(1)$

b. $(fg)(0)$

c. $(g - f)(-3)$

d. $(f \circ g)(3)$

e. $(g \circ f)(4)$

f. $f(g(1))$

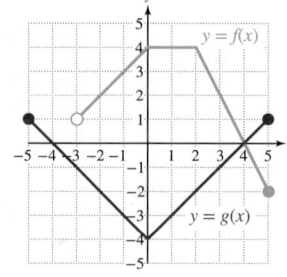

FOR REVIEW

Consider a function defined by $y = f(x)$. Given $f(3)$, the number 3 inside parentheses is an x value from the domain. The value $f(3)$ is the corresponding y value. For example, $f(3) = 2$.

See Section 5.3 for review.

Solution:

a. $(f + g)(1) = f(1) + g(1)$
$= 4 + (-3)$
$= 1$

b. $(fg)(0) = f(0) \cdot g(0)$
$= (4)(-4)$
$= -16$

c. $(g - f)(-3) = g(-3) - f(-3)$ ⟵ $f(-3)$ is undefined.

$(g - f)(-3)$ is undefined.

d. $(f \circ g)(3) = f(g(3))$
$= f(-1)$
$= 3$

e. $(g \circ f)(4) = g(f(4))$
$= g(0)$
$= -4$

f. $f(g(1)) = f(-3)$ is undefined.   The open dot at $(-3, 1)$ indicates that -3 is not in the domain of f. The value $g(1) = -3$, but $f(-3)$ is undefined. Therefore, $f(g(1))$ is undefined.

Skill Practice 10 Refer to the functions f and g pictured in Example 10. Evaluate the functions at the given values of x if possible.

a. $(f - g)(-2)$ b. $\left(\dfrac{f}{g}\right)(3)$ c. $(gf)(5)$ d. $(g \circ f)(5)$

e. $(f \circ g)(5)$ f. $f(g(0))$

Answers

10. a. 4   b. −2   c. −2
    d. −2   e. 4   f. Undefined

SECTION 6.4   Practice Exercises

Prerequisite Review

For Exercises R.1–R.8, write the domain in interval notation. (See Section 5.3 for review.)

R.1. $f(x) = \dfrac{x - 1}{x + 2}$

R.2. $g(x) = \dfrac{3x}{x - 7}$

R.3. $n(x) = \sqrt{10 - 5x}$

R.4. $r(x) = \sqrt{x + 3}$

R.5. $h(x) = \dfrac{4}{\sqrt{3 - x}}$

R.6. $k(x) = \dfrac{8}{\sqrt{x^2 + 4}}$

R.7. $p(x) = 2x^2 - 3x + 1$

R.8. $q(x) = \sqrt[3]{3x + 4}$

For Exercises 9–10, refer to the graphs of f and g and find the function values. (See Section 5.3 for review.)

R.9. a. $f(-2)$ **b.** $g(3)$ **R.10. a.** $f(5)$ **b.** $g(0)$

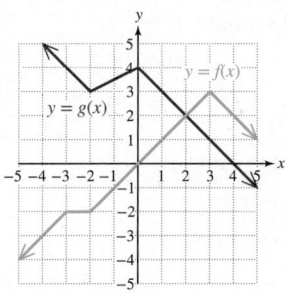

For Exercises R.11–R.12, refer to function f and find the function values. (See Section 5.3 for review.)

$$f = \{(-3, 2), (2, -3), (-1, 5), (0, -1), (1, -3)\}$$

R.11. a. $f(-1)$ **b.** $f(-3)$ **R.12. a.** $f(1)$ **b.** $f(2)$

Concept Connections

1. The function $f + g$ is defined by $(f + g)(x) = $ _____ + _____.

2. The function $\dfrac{f}{g}$ is defined by $\left(\dfrac{f}{g}\right)(x) = $ _____ provided that _____ $\neq 0$.

3. The composition of f and g, denoted by $f \circ g$, is defined by $(f \circ g)(x) = $ _____.

4. The composition of g and f, denoted by $g \circ f$, is defined by $(g \circ f)(x) = $ _____.

Objective 1: Perform Operations on Functions

For Exercises 5–8, find $(f + g)(x)$ and identify the graph of $f + g$. (See Example 1)

5. $f(x) = |x|$ and $g(x) = 3$ **6.** $f(x) = |x|$ and $g(x) = -4$

7. $f(x) = x^2$ and $g(x) = -4$ **8.** $f(x) = x^2$ and $g(x) = 3$

a.
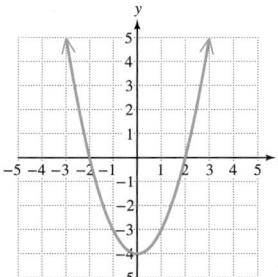

b.
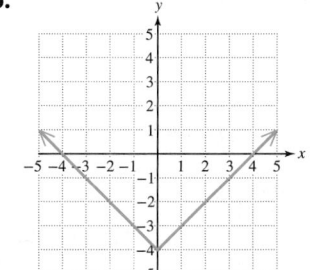

c.
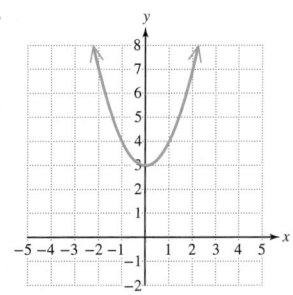

d.
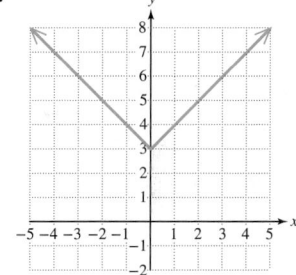

For Exercises 9–18, evaluate the functions for the given values of x. (See Example 2)

$$f(x) = -2x \qquad g(x) = |x + 4| \qquad h(x) = \frac{1}{x - 3}$$

9. $(f - g)(3)$ **10.** $(g - h)(2)$ **11.** $(f \cdot g)(-1)$ **12.** $(h \cdot g)(4)$ **13.** $(g + h)(0)$

14. $(f + h)(5)$ **15.** $\left(\dfrac{f}{g}\right)(8)$ **16.** $\left(\dfrac{h}{f}\right)(7)$ **17.** $\left(\dfrac{g}{f}\right)(0)$ **18.** $\left(\dfrac{h}{g}\right)(-4)$

For Exercises 19–26, refer to the functions r, p, and q. Find the indicated function and write the domain in interval notation. (See Example 3)

$$r(x) = -3x \qquad p(x) = x^2 + 3x \qquad q(x) = \sqrt{1 - x}$$

19. $(r - p)(x)$ **20.** $(p - r)(x)$ **21.** $(p \cdot q)(x)$ **22.** $(r \cdot q)(x)$

23. $\left(\dfrac{q}{p}\right)(x)$ **24.** $\left(\dfrac{q}{r}\right)(x)$ **25.** $\left(\dfrac{p}{q}\right)(x)$ **26.** $\left(\dfrac{r}{q}\right)(x)$

For Exercises 27–32, refer to functions s, t, and v. Find the indicated function and write the domain in interval notation. (**See Example 3**)

$$s(x) = \frac{x - 2}{x^2 - 9} \qquad t(x) = \frac{x - 3}{2 - x} \qquad v(x) = \sqrt{x + 3}$$

27. $(s \cdot t)(x)$ **28.** $\left(\dfrac{s}{t}\right)(x)$ **29.** $(s + t)(x)$ **30.** $(s - t)(x)$

31. $(s \cdot v)(x)$ **32.** $\left(\dfrac{v}{s}\right)(x)$

Objective 2: Compose and Decompose Functions

For Exercises 33–48, refer to functions f, g, and h. Evaluate the functions for the given values of x. (**See Example 4**)

$$f(x) = x^3 - 4x \qquad g(x) = \sqrt{2x} \qquad h(x) = 2x + 3$$

33. $f(g(8))$ **34.** $h(g(2))$ **35.** $h(f(1))$ **36.** $g(f(3))$

37. $(f \circ g)(18)$ **38.** $(f \circ h)(-1)$ **39.** $(g \circ f)(5)$ **40.** $(h \circ f)(-2)$

41. $(h \circ f)(-3)$ **42.** $(h \circ g)(72)$ **43.** $(g \circ f)(1)$ **44.** $(g \circ f)(-4)$

45. $(f \circ f)(3)$ **46.** $(h \circ h)(-4)$ **47.** $(f \circ h \circ g)(2)$ **48.** $(f \circ h \circ g)(8)$

49. Given $f(x) = 2x + 4$ and $g(x) = x^2$,

     a. Find $(f \circ g)(x)$.   **b.** Find $(g \circ f)(x)$.

     c. Is the operation of function composition commutative?

50. Given $k(x) = -3x + 1$ and $m(x) = \dfrac{1}{x}$,

     a. Find $(k \circ m)(x)$.   **b.** Find $(m \circ k)(x)$.

     c. Is $(k \circ m)(x) = (m \circ k)(x)$?

For Exercises 51–62, refer to functions m, n, p, q, and r. Find the indicated function and write the domain in interval notation. (**See Examples 5–6**)

$$m(x) = \sqrt{x + 8} \qquad n(x) = x - 5 \qquad p(x) = x^2 - 9x \qquad q(x) = \frac{1}{x - 10} \qquad r(x) = |2x + 3|$$

51. $(n \circ p)(x)$ **52.** $(p \circ n)(x)$ **53.** $(m \circ n)(x)$ **54.** $(n \circ m)(x)$

55. $(q \circ n)(x)$ **56.** $(q \circ p)(x)$ **57.** $(q \circ r)(x)$ **58.** $(q \circ m)(x)$

59. $(n \circ r)(x)$ **60.** $(r \circ n)(x)$ **61.** $(n \circ n)(x)$ **62.** $(p \circ p)(x)$

For Exercises 63–66, find $(f \circ g)(x)$ and write the domain in interval notation. (**See Example 7**)

63. $f(x) = \dfrac{9}{x - 1}$; $g(x) = \dfrac{9}{x^2}$

64. $f(x) = \dfrac{6}{4 - x}$; $g(x) = \dfrac{4}{x^2}$

65. $f(x) = \dfrac{2}{x - 4}$; $g(x) = \sqrt{x - 1}$

66. $f(x) = \dfrac{-5}{x - 2}$; $g(x) = \sqrt{x + 3}$

67. Given $f(x) = \dfrac{1}{x - 2}$, find $(f \circ f)(x)$ and write the domain in interval notation.

68. Given $f(x) = \dfrac{5}{x - 5}$, find $(f \circ f)(x)$ and write the domain in interval notation.

For Exercises 69–72, find the indicated functions.

$$f(x) = 2x + 1 \qquad g(x) = x^2 \qquad h(x) = \sqrt[3]{x}$$

69. $(f \circ g \circ h)(x)$ **70.** $(g \circ f \circ h)(x)$

71. $(h \circ g \circ f)(x)$ **72.** $(g \circ h \circ f)(x)$

73. A law office orders business stationery. The cost is $21.95 per box. **(See Example 8)**

   a. Write a function that represents the cost $C(x)$ (in $) for x boxes of stationery.

   b. There is a 6% sales tax on the cost of merchandise and $10.99 for shipping. Write a function that represents the total cost $T(a)$ for a dollars spent in merchandise and shipping.

   c. Find $(T \circ C)(x)$.

   d. Find $(T \circ C)(4)$ and interpret its meaning in the context of this problem.

74. The cost to buy tickets online for a dance show is $60 per ticket.

   a. Write a function that represents the cost $C(x)$ (in $) for x tickets to the show.

   b. There is a sales tax of 5.5% and a processing fee of $8.00 for tickets. Write a function that represents the total cost $T(a)$ for a dollars spent on tickets.

   c. Find $(T \circ C)(x)$.

   d. Find $(T \circ C)(6)$ and interpret its meaning in the context of this problem.

75. A bicycle wheel turns at a rate of 80 revolutions per minute (rpm).

   a. Write a function that represents the number of revolutions $r(t)$ in t minutes.

   b. For each revolution of the wheels, the bicycle travels 7.2 ft. Write a function that represents the distance traveled $d(r)$ (in ft) for r revolutions of the wheel.

   c. Find $(d \circ r)(t)$ and interpret the meaning in the context of this problem.

   d. Evaluate $(d \circ r)(30)$ and interpret the meaning in the context of this problem.

76. While on vacation in France, Sadie bought a box of almond croissants. Each croissant cost €2.40 (euros).

   a. Write a function that represents the cost $C(x)$ (in euros) for x croissants.

   b. At the time of the purchase, the exchange rate was $1 = €0.80. Write a function that represents the amount $D(C)$ (in $) for C euros spent.

   c. Find $(D \circ C)(x)$ and interpret the meaning in the context of this problem.

   d. Evaluate $(D \circ C)(12)$ and interpret the meaning in the context of this problem.

For Exercises 77–84, find two functions f and g such that $h(x) = (f \circ g)(x)$. **(See Example 9)**

77. $h(x) = (x + 7)^2$

78. $h(x) = (x - 8)^2$

79. $h(x) = \sqrt[3]{2x + 1}$

80. $h(x) = \sqrt[4]{9x - 5}$

81. $h(x) = |2x^2 - 3|$

82. $h(x) = |4 - x^2|$

83. $h(x) = \dfrac{5}{x + 4}$

84. $h(x) = \dfrac{11}{x - 3}$

For Exercises 85–88, the graphs of two functions are shown. Evaluate the function at the given values of x, if possible. **(See Example 10)**

85. a. $(f + g)(0)$

   b. $(g - f)(2)$

   c. $(g \cdot f)(-1)$

   d. $\left(\dfrac{g}{f}\right)(1)$

   e. $(f \circ g)(4)$

   f. $(g \circ f)(0)$

   g. $g(f(4))$

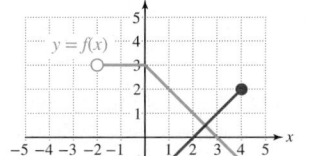

86. a. $(f + g)(0)$

   b. $(g - f)(1)$

   c. $(g \cdot f)(2)$

   d. $\left(\dfrac{f}{g}\right)(-3)$

   e. $(f \circ g)(3)$

   f. $(g \circ f)(0)$

   g. $g(f(-4))$

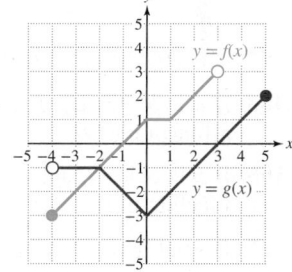

87. a. $(h + k)(-1)$

   b. $(h \cdot k)(4)$

   c. $\left(\dfrac{k}{h}\right)(-3)$

   d. $(k - h)(1)$

   e. $(k \circ h)(4)$

   f. $(h \circ k)(-2)$

   g. $h(k(3))$

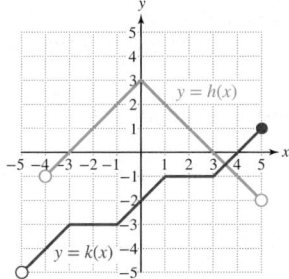

88. a. $(m + p)(1)$

   b. $(p - m)(-4)$

   c. $\left(\dfrac{m}{p}\right)(3)$

   d. $(m \cdot p)(3)$

   e. $(m \circ p)(0)$

   f. $(p \circ m)(0)$

   g. $p(m(-4))$

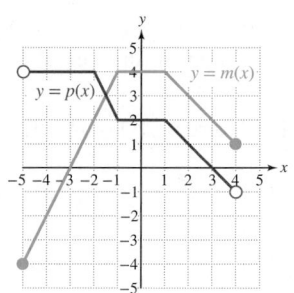

Mixed Exercises

For Exercises 89–96, refer to the functions f and g and evaluate the functions for the given values of x.

$$f = \{(2, 4), (6, -1), (4, -2), (0, 3), (-1, 6)\} \quad \text{and} \quad g = \{(4, 3), (0, 6), (5, 7), (6, 0)\}$$

89. $(f + g)(4)$ **90.** $(g \cdot f)(0)$ **91.** $(g \circ f)(2)$ **92.** $(f \circ g)(0)$

93. $(g \circ g)(6)$ **94.** $(f \circ f)(-1)$ **95.** $(f \circ g)(5)$ **96.** $(g \circ f)(0)$

97. The diameter d of a sphere is twice the radius r. The volume of the sphere as a function of its radius is given by $V(r) = \dfrac{4}{3}\pi r^3$.

     a. Write the diameter d of the sphere as a function of the radius r.

     b. Write the radius r as a function of the diameter d.

     c. Find $(V \circ r)(d)$ and interpret its meaning.

98. Consider a right circular cone with given height h. The volume of the cone as a function of its radius r is given by $V(r) = \dfrac{1}{3}\pi r^2 h$. Consider a right circular cone with fixed height $h = 6$ in.

     a. Write the diameter d of the cone as a function of the radius r.

     b. Write the radius r as a function of the diameter d.

     c. Find $(V \circ r)(d)$ and interpret its meaning. Assume that $h = 6$ in.

99. An investment earns 4.5% interest paid at the end of 1 year. If x is the amount of money initially invested, then $A(x) = 1.045x$ represents the amount of money in the account 1 year later. Find $(A \circ A)(x)$ and interpret the result.

100. The population in a certain town has been decreasing at a rate of 2% per year. If x is the population at a certain fixed time, then $P(x) = 0.98x$ represents the population 1 year later. Find $(P \circ P)(x)$ and interpret the result.

101. For the given figure,

     a. What does $A_1(x) = \pi(x + 5)^2$ represent?

     b. What does $A_2(x) = \pi x^2$ represent?

     c. Find $(A_1 - A_2)(x)$ and interpret its meaning.

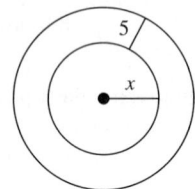

102. For the given figure,

     a. Write an expression $S_1(x)$ that represents the area of the rectangle.

     b. Write an expression $S_2(x)$ that represents the area of the semicircle.

     c. Find $(S_1 - S_2)(x)$ and interpret its meaning.

Expanding Your Skills

103. If a is b plus eight, and c is the square of a, write c as a function of b.

104. If q is r minus seven, and s is the square root of q, write s as a function of r.

105. If x is twice y, and z is four less than x, write z as a function of y.

106. If m is one-third of n, and p is two less than m, write p as a function of n.

107. Given $f(x) = \sqrt[3]{4x^2 + 1}$, define functions m, n, h, and k such that $f(x) = (m \circ n \circ h \circ k)(x)$.

108. Given $f(x) = |-2x^3 - 4|$, define functions m, n, h, and k such that $f(x) = (m \circ n \circ h \circ k)(x)$.

109. Given $g(x) = \sqrt{x - 1}$, find $(g \circ g)(x)$ and write the domain in interval notation.

110. Given $g(x) = \sqrt{x - 3}$, find $(g \circ g)(x)$ and write the domain in interval notation.

For Exercises 111–114, find $(f \circ g)(x)$ and write the domain in interval notation.

111. $f(x) = \dfrac{3}{x^2 - 16}, \; g(x) = \sqrt{2 - x}$

112. $f(x) = \dfrac{4}{x^2 - 9}, \; g(x) = \sqrt{3 - x}$

113. $f(x) = \dfrac{x}{x - 1}, \; g(x) = \dfrac{9}{x^2 - 16}$

114. $f(x) = \dfrac{x}{x + 4}, \; g(x) = \dfrac{3}{x^2 - 1}$

CHAPTER 6   Detailed Summary

SECTION 6.1 Transformations of Graphs

Key Concepts	Examples	Page

Transformations of graphs:
Consider a function defined by $y = f(x)$. Let k, h, and a represent positive real numbers. The graphs of the following functions are related to $y = f(x)$ as follows:
• Vertical translation (shift):
  $y = f(x) + k$ Shift upward
  $y = f(x) - k$ Shift downward
• Horizontal translation (shift):
  $y = f(x - h)$ Shift to the right
  $y = f(x + h)$ Shift to the left
	Example 1: **Example 2:**
	
		p. 409

• Vertical stretch/shrink
  $y = af(x)$ Vertical stretch (if $a > 1$)
  Vertical shrink (if $0 < a < 1$)
	Example 3:
	
		p. 412

• Horizontal stretch/shrink
  $y = f(ax)$ Horizontal shrink (if $a > 1$)
  Horizontal stretch (if $0 < a < 1$)
	Example 4:
	
		p. 413

• Reflection $y = -f(x)$ Reflection across the x-axis $y = f(-x)$ Reflection across the y-axis 	**Example 5:** 	p. 415

Steps for graphing multiple transformations:
1. Horizontal translation (shift)
2. Horizontal and vertical stretch and shrink
3. Reflections across the x- and y-axes
4. Vertical translation (shift)
	Example 6:
	
	To graph $f(x) = 2\sqrt[3]{-x-1} + 5$ perform the following transformations on the graph of $y = \sqrt[3]{x}$ in the following order.
	
	  1. Shift the graph to the right 1 unit.
	  2. Stretch the graph vertically by a factor of 2.
	  3. Reflect the graph across the y-axis.
	  4. Shift the graph 5 units upward.
		p. 416

SECTION 6.2 Symmetry and Piecewise-Defined Functions

Key Concepts	Examples	Page
Symmetry: Consider an equation in the variables x and y.	**Example 1:** The graph of $x^2 + y^2 = 25$ is symmetric with respect to the y-axis, the x-axis, and the origin.	p. 424
• The graph of the equation is symmetric with respect to the y-axis if substituting $-x$ for x in the equation results in an equivalent equation.	Replace x by $-x$ $x^2 + y^2 = 25$ $(-x)^2 + y^2 = 25$ Same. Thus, the graph is $x^2 + y^2 = 25$ symmetric with respect to the y-axis.	
• The graph of the equation is symmetric with respect to the x-axis if substituting $-y$ for y in the equation results in an equivalent equation.	Replace y by $-y$ $x^2 + y^2 = 25$ $x^2 + (-y)^2 = 25$ Same. Thus, the graph is $x^2 + y^2 = 25$ symmetric with respect to the x-axis.	
• The graph of the equation is symmetric with respect to the origin if substituting $-x$ for x and $-y$ for y in the equation results in an equivalent equation.	Replace x by $-x$ and y by $-y$ $x^2 + y^2 = 25$ $(-x)^2 + (-y)^2 = 25$ Same. Thus, the graph is $x^2 + y^2 = 25$ symmetric with respect to the origin.	
Even and Odd Functions: • A function f is an **even function** if $f(-x) = f(x)$ for all x in the domain of f. The graph of an even function is symmetric to the y-axis.	**Example 2:** $f(x) = x^4 + |x|$ is even. $f(x) = x^4 + |x|$ $f(-x) = (-x)^4 + |-x|$ same $= x^4 + |x|$	p. 426

- A function f is an **odd function** if $f(-x) = -f(x)$ for all x in the domain of f. The graph of an odd function is symmetric to the origin.

Example 3:　　　　　　　　　　　　　　p. 426

$f(x) = 2x^3 + x$ is odd.

$$f(x) = 2x^3 + x$$
$$f(-x) = 2(-x)^3 + (-x)$$
$$\qquad = -2x^3 - x \longleftarrow$$
$$-f(x) = -(2x^3 + x) \qquad \text{same}$$
$$\qquad = -2x^3 - x \longrightarrow$$

Piecewise-Defined Functions:

A piecewise-defined function is a function with multiple rules, each defined on a restricted domain.

Example 4:　　　　　　　　　　　　　　p. 427

$$f(x) = \begin{cases} -|x| & \text{if } x < 1 \\ 2 & \text{if } x \geq 1 \end{cases}$$

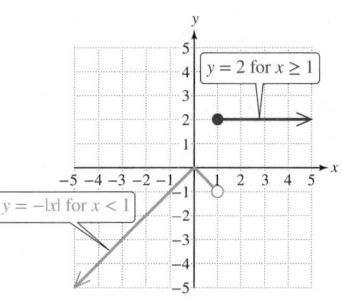

The **greatest integer function**, denoted by $f(x) = [\![x]\!]$ or $f(x) = \text{int}(x)$ or $f(x) = \text{floor}(x)$ defines $f(x)$ as the greatest integer less than or equal to x.

Example 5:　　　　　　　　　　　　　　p. 430

Given $f(x) = [\![x]\!]$,

$f(2.4) = [\![2.4]\!] = 2$　and　$f(-2.4) = [\![-2.4]\!] = -3$

SECTION 6.3 Average Rate of Change, Difference Quotient, and Function Behavior

Key Concepts	Examples	Page

Average Rate of Change:

Suppose that x_1 and x_2 are values in the domain of a function f. If f is defined on the interval $[x_1, x_2]$, then the **average rate of change** of f on the interval $[x_1, x_2]$ is given by $\dfrac{f(x_2) - f(x_1)}{x_2 - x_1}$. This is the slope of the secant line through the points $(x_1, f(x_1))$ and $(x_2, f(x_2))$.

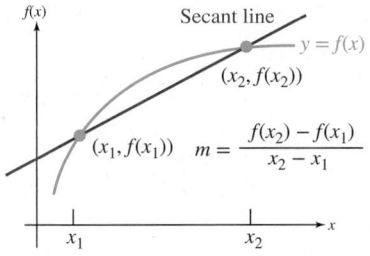

Example 1:　　　　　　　　　　　　　　p. 438

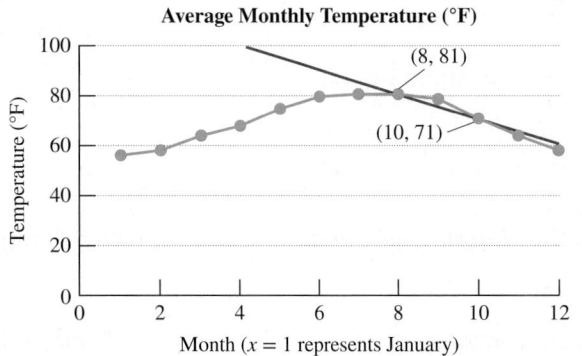

The average rate of change for the given function from $x_1 = 8$ to $x_2 = 10$ is

$$\frac{f(x_2) - f(x_1)}{x_2 - x_1}$$

$$= \frac{f(10) - f(8)}{10 - 8} = \frac{71 - 81}{2} = \frac{-10}{2} = -5$$

The average monthly temperature between August and October *decreased* by an average of 5°F per month.

Increasing, decreasing, and constant behavior:

Suppose that I is an interval contained within the domain of a function f.

- f is increasing on I if $f(x_1) < f(x_2)$ for all $x_1 < x_2$ on I.
- f is decreasing on I if $f(x_1) > f(x_2)$ for all $x_1 < x_2$ on I.
- f is constant on I if $f(x_1) = f(x_2)$ for all x_1 and x_2 on I.

Example 2:

f is increasing on $(3, \infty)$.

f is decreasing on $(-\infty, 0)$.

f is constant on $(0, 3)$.

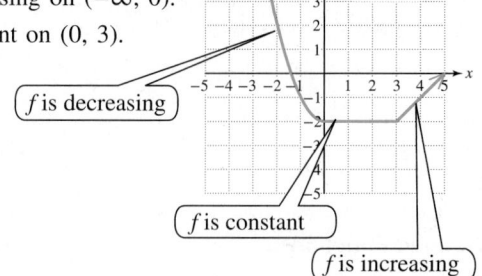

f is decreasing

f is constant

f is increasing

p. 440

Determining relative maxima and minima:

- $f(a)$ is a **relative maximum** of f if there exists an open interval containing a such that $f(a) \geq f(x)$ for all x in the interval.
- $f(b)$ is a **relative minimum** of f if there exists an open interval containing b such that $f(b) \leq f(x)$ for all x in the interval.

Example 3:

f has a relative maximum of 3 at $x = -1$.

f has a relative minimum of -1 at $x = 1$.

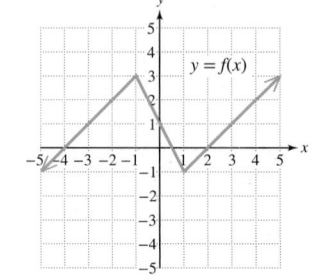

p. 442

Evaluate a difference quotient:

The **difference quotient** represents the average rate of change of a function f between two points $(x, f(x))$ and $(x + h, f(x + h))$.

$$\frac{f(x + h) - f(x)}{h} \quad \text{Difference quotient}$$

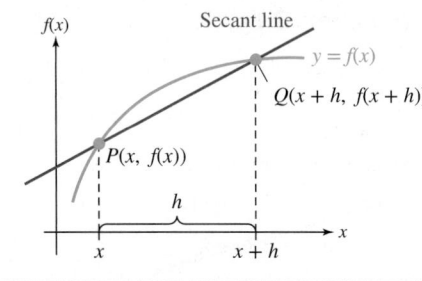

Example 4:

Given $f(x) = 4x^2 - 5x$ evaluate the difference quotient.

$$\frac{f(x + h) - f(x)}{h}$$

$$= \frac{\overbrace{[4(x + h)^2 - 5(x + h)]}^{f(x+h)} - \overbrace{(4x^2 - 5x)}^{f(x)}}{h}$$

$$= \frac{[4(x^2 + 2xh + h^2) - 5x - 5h] - (4x^2 - 5x)}{h}$$

$$= \frac{4x^2 + 8xh + 4h^2 - 5x - 5h - 4x^2 + 5x}{h}$$

$$= \frac{8xh + 4h^2 - 5h}{h}$$

$$= 8x + 4h - 5$$

p. 444

SECTION 6.4 Algebra of Functions and Function Composition

Key Concepts	Examples	Page
Operations on functions: Given functions f and g, the functions $f + g, f - g, f \cdot g,$ and $\frac{f}{g}$ are defined by $$(f + g)(x) = f(x) + g(x)$$ $$(f - g)(x) = f(x) - g(x)$$ $$(f \cdot g)(x) = f(x) \cdot g(x)$$ $$\left(\frac{f}{g}\right)(x) = \frac{f(x)}{g(x)} \text{ provided that } g(x) \neq 0$$	**Example 1:** Given $f(x) = x - 3$ and $g(x) = x^2 - 7x$ $$(f + g)(x) = f(x) + g(x) = (x - 3) + (x^2 - 7x)$$ $$= x^2 - 6x - 3$$ $$\left(\frac{g}{f}\right)(x) = \frac{g(x)}{f(x)} = \frac{x^2 - 7x}{x - 3} \text{ provided } x \neq 3.$$	p. 451

The **composition of f and g**, denoted $f \circ g$ is defined by $(f \circ g)(x) = f(g(x))$.

The domain of $f \circ g$ is the set of real numbers x in the domain of g such that $g(x)$ is in the domain of f.

Example 2:   p. 453

Given $f(x) = \dfrac{1}{x + 12}$ and $g(x) = x^2 - 7x$

$$(f \circ g)(x) = f(g(x)) = \frac{1}{(g(x)) + 12}$$

$$= \frac{1}{(x^2 - 7x) + 12} = \frac{1}{(x - 3)(x - 4)}$$

The domain of $f \circ g$ is $\{x \mid x \neq 3 \text{ and } x \neq 4\}$.

In interval notation: $(-\infty, 3) \cup (3, 4) \cup (4, \infty)$.

CHAPTER 6   Review Exercises

SECTION 6.1

1. Write a function based on the given parent function and transformations in the given order.

Parent function: $y = x^2$

  1. Shift 5 units to the left.

  2. Reflect across the y-axis.

  3. Shift downward 2 units.

For Exercises 2–11, use translations to graph the given functions.

2. $f(x) = |x| - 2$ **3.** $g(x) = \sqrt{x} + 1$

4. $h(x) = (x - 2)^2$ **5.** $k(x) = \sqrt[3]{x} + 1$

6. $r(x) = \sqrt{x - 3} + 1$ **7.** $s(x) = (x + 2)^2 - 3$

8. $t(x) = -2|x|$ **9.** $v(x) = -\dfrac{1}{2}|x|$

10. $m(x) = \sqrt{-x + 5}$ **11.** $n(x) = \sqrt{-x - 1}$

For Exercises 12–17, use the graph of $y = f(x)$ to graph the given function.

12. $y = f(2x)$

13. $y = f\left(\frac{1}{2}x\right)$

14. $y = -f(x + 1) - 3$

15. $y = -f(x - 4) - 1$

16. $y = 2f(x - 3) + 1$

17. $y = \frac{1}{2}f(x + 2) - 3$

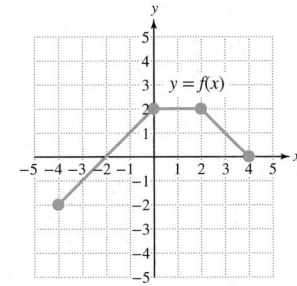

SECTION 6.2

For Exercises 18–24, determine if the graph of the equation is symmetric to the y-axis, x-axis, origin, or none of these.

18. $y = x^4 - 3$ **19.** $x = |y| + y^2$

20. $y = \dfrac{1}{3}x - 1$ **21.** $x^2 = y^2 + 1$

22. $x^2 + 2x + y^2 - 4y = 9$ **23.** $x^2 + y = 0$

24. $x + 4y^2 - 6 = 0$

For Exercises 25–30, determine if the function is even, odd, or neither.

25. $f(x) = -4x^3 + x$ **26.** $g(x) = \sqrt[3]{x}$

27. $p(x) = \sqrt{4 - x^2}$ **28.** $q(x) = -|x|$

29. $k(x) = (x - 3)^2$ **30.** $m(x) = |x + 2|$

31. Evaluate the function for the given values of x.

$$f(x) = \begin{cases} -4x + 2 & \text{for } x < -1 \\ x^2 & \text{for } -1 \leq x \leq 2 \\ 5 & \text{for } x > 2 \end{cases}$$

  a. $f(-4)$ **b.** $f(-1)$ **c.** $f(3)$ **d.** $f(2)$

For Exercises 32–34, graph the function.

32. $f(x) = \begin{cases} -4x - 3 & \text{for } x < 0 \\ x^2 & \text{for } x \geq 0 \end{cases}$

33. $g(x) = \begin{cases} |x| & \text{for } x \leq 2 \\ 2 & \text{for } x > 2 \end{cases}$

34. $h(x) = \begin{cases} -3 & \text{for } x < -2 \\ 1 & \text{for } -2 \leq x < 0 \\ \sqrt{x} & \text{for } x \geq 0 \end{cases}$

35. Evaluate $f(x) = [\![x - 1]\!]$ for the given values of x.

  a. $f(-1.5)$ **b.** $f(-2)$

  c. $f(0.1)$ **d.** $f(6.3)$

36. Write a rule for the graph of the function. Answers may vary.

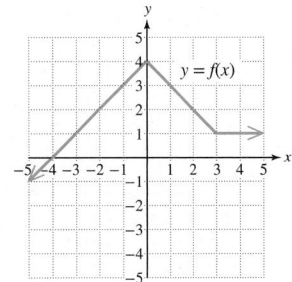

SECTION 6.3

37. Find the slope of the secant line pictured in red.

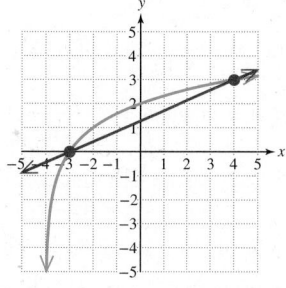

For Exercises 38–40, find the average rate of change of the function on the given interval.

38. $f(x) = -x^3 + 4$

    a. $[0, 2]$ **b.** $[2, 4]$

39. $f(x) = x^2 - x + 4$

    a. $[0, 1]$ **b.** $[1, 2]$

40. $f(x) = -3x - 5$

    a. $[1, 3]$ **b.** $[3, 5]$

41. The function given by $y = f(x)$ shows the value of $8000 invested at 6% interest compounded continuously, x years after the money was originally invested.

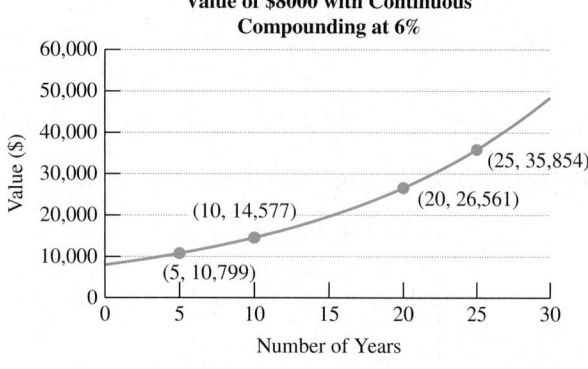

Value of $8000 with Continuous Compounding at 6%

    a. Find the average amount earned per year between the 5th year and the 10th year.

    b. Find the average amount earned per year between the 20th year and the 25th year.

    c. Based on the answers from parts (a) and (b), does it appear that the rate at which annual income increases is increasing or decreasing with time?

For Exercises 42–44, use interval notation to write the interval(s) over which f is

42. a. Increasing

    b. Decreasing

    c. Constant

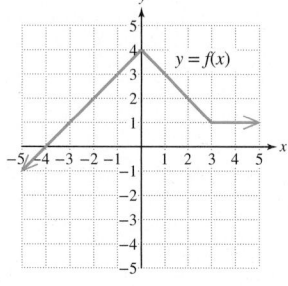

43.

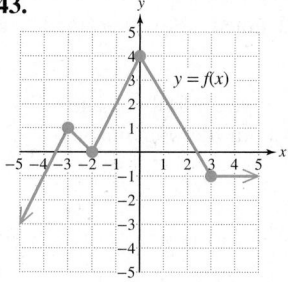

44.

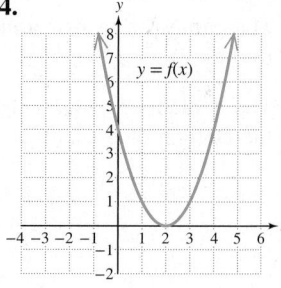

For Exercises 45–46, identify the location and value of any relative maxima or minima of the function.

45.

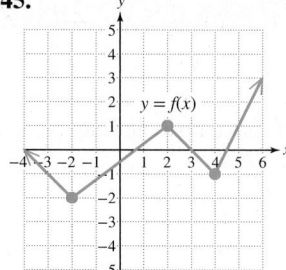

46.

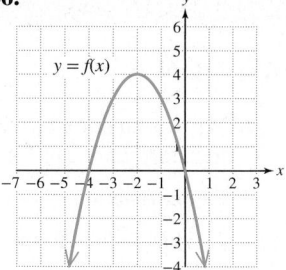

For Exercises 47–49, find the difference quotient, $\dfrac{f(x + h) - f(x)}{h}$.

47. $f(x) = -6x - 5$ **48.** $f(x) = 3x^2 - 4x + 9$

49. $f(x) = -2x^2 + 6x - 3$

SECTION 6.4

For Exercises 50–56, evaluate the function for the given values of x.

$$f(x) = -3x \qquad g(x) = |x - 2| \qquad h(x) = \frac{1}{x + 1}$$

50. $(f - h)(2)$ **51.** $(g \cdot h)(3)$

52. $\left(\dfrac{g}{h}\right)(-5)$ **53.** $(f \circ g)(5)$

54. $(g \circ f)(5)$ **55.** $(f \circ h)(2)$

56. $(h \circ g)(0)$

57. Use the graphs of f and g to find the function values for the given values of x.

    a. $(f + g)(2)$

    b. $(g \cdot f)(-4)$

    c. $\left(\dfrac{g}{f}\right)(-3)$

    d. $f[g(-4)]$

    e. $(g \circ f)(-4)$

    f. $(g \circ f)(5)$

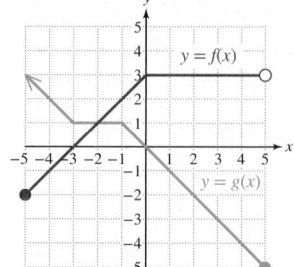

For Exercises 58–64, refer to the functions m, n, p, and q. Find the function and write the domain in interval notation.

$$m(x) = -4x \qquad n(x) = x^2 - 4x$$

$$p(x) = \sqrt{x - 2} \qquad q(x) = \frac{1}{x - 5}$$

58. $(m + p)(x)$ **59.** $(n - m)(x)$

60. $\left(\dfrac{p}{n}\right)(x)$ **61.** $\left(\dfrac{n}{p}\right)(x)$

62. $(m \cdot p)(x)$ **63.** $(q \circ n)(x)$

64. $(q \circ p)(x)$

For Exercises 65–66, find two functions, f and g, such that $h(x) = (f \circ g)(x)$.

65. $h(x) = (x - 4)^2$ **66.** $h(x) = \dfrac{12}{x + 5}$

67. A certain car traveling 60 mph gets 28 mpg.

   a. Write a function that represents the distance $d(t)$ (in miles) that the car travels in t hours.

   b. Write a function that represents the number of gallons of gasoline $n(d)$ used for d miles traveled.

   c. Find $(n \circ d)(t)$ and interpret the meaning in the context of this problem.

   d. Evaluate $(n \circ d)(7)$ and interpret the meaning in the context of this problem.

CHAPTER 6   Test

1. Use the graph of $y = f(x)$ to find

   a. $f(-4)$.

   b. $f(3)$.

   c. The values of x for which $f(x) = 1$.

   d. The domain.

   e. The range.

   f. The intervals over which f is increasing.

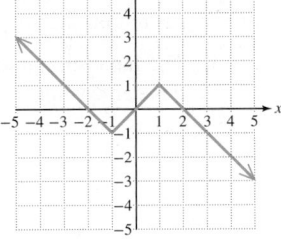

   g. The intervals over which f is decreasing.

   h. The intervals over which f is constant.

   i. The value and location of any relative maxima.

   j. The value and location of any relative minima.

   k. Whether f is an even or odd function.

   l. The x-intercept(s).

   m. The y-intercept.

For Exercises 2–5, graph the equation.

   2. $f(x) = -(x - 2)^2 + 4$

   3. $f(x) = 2|x + 3|$

   4. $g(x) = -\sqrt{x + 4} + 3$

5. $h(x) = \begin{cases} -x + 3 & \text{for } x < 1 \\ \sqrt{x - 1} & \text{for } x \geq 1 \end{cases}$

For Exercises 6–9, use the graph of $y = f(x)$ to graph the given function.

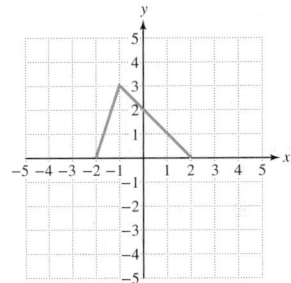

6. $y = -f(x)$

7. $y = f(-x)$

8. $y = \dfrac{1}{3}f(x)$

9. $y = f\left(\dfrac{1}{2}x\right)$

10. Given $f(x) = -x^2 + 5x - 3$,

   a. Find the average rate of change on the interval $[0, 2]$.

   b. Find the difference quotient $\dfrac{f(x + h) - f(x)}{h}$.

For Exercises 11–12, determine if the graph of the equation is symmetric to the x-axis, y-axis, origin, or none of these.

11. $x = -2y^4 + y^2$

12. $x^2 + |y| = 8$

For Exercises 13–14, determine if the function is even, odd, or neither even nor odd.

13. $f(x) = x^3 - x$

14. $g(x) = x^4 + x^3 + x$

15. Evaluate the greatest integer function for the following values of x.

    a. 4.27   **b.** −4.27

16. Determine the function values for the given function.

$$f(x) = \begin{cases} -4 & \text{for } x \le -2 \\ x^2 & \text{for } -2 < x < 1 \\ x + 3 & \text{for } x \ge 1 \end{cases}$$

    a. $f(-3)$ **b.** $f(-2)$ **c.** $f(0)$

    d. $f(1)$ **e.** $f(3)$

For Exercises 17–22, refer to the functions f, g, and h defined here.

$$f(x) = x - 4 \qquad g(x) = \frac{1}{x - 3} \qquad h(x) = \sqrt{x - 5}$$

17. Evaluate $(f - h)(6)$.

18. Evaluate $(g \cdot h)(5)$.

19. Evaluate $(h \circ f)(1)$.

20. Find $(f \cdot g)(x)$ and state the domain in interval notation.

21. Find $\left(\dfrac{g}{f}\right)(x)$ and state the domain in interval notation.

22. Find $(g \circ h)(x)$ and state the domain in interval notation.

23. Write two functions f and g such that $h(x) = (f \circ g)(x)$.

$$h(x) = \sqrt[3]{x - 7}$$

24. For f and g pictured, estimate the following.

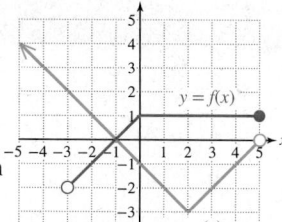

    a. $(f + g)(3)$

    b. $(f \cdot g)(0)$

    c. $g(f(3))$

    d. $(f \circ g)(2)$

    e. The interval(s) over which f is increasing.

    f. The interval(s) over which g is decreasing.

Polynomial Functions

7

Chapter Outline

Source: Laboratory for Atmospheres: NASA Geddard Space Flight Center

Meteorology and the study of weather have a strong foundation in mathematics. The factors impacting weather are not constant and change over time. For example, during the summer months, hot ocean temperatures in the Atlantic Ocean often produce breeding grounds for hurricanes off the coast of Africa or in the Caribbean. To predict the path of a hurricane, meteorologists collect data from satellites, weather stations around the world, and weather buoys in the ocean. Piecing together the data requires a variety of techniques of mathematical modeling using powerful computers. In the end, scientists combine a series of simple curves to approximate weather patterns that closely fit complicated models.

In this chapter, we study polynomial and rational functions. Both types of functions represent simple curves that can be used for modeling in a wide range of applications, including predictions for the path of a hurricane.

OBJECTIVES

1. Graph a Quadratic Function Written in Vertex Form
2. Write $f(x) = ax^2 + bx + c$ $(a \neq 0)$ in Vertex Form
3. Find the Vertex of a Parabola by Using the Vertex Formula
4. Solve Applications Involving Quadratic Functions
5. Create Quadratic Models Using Regression

1. Graph a Quadratic Function Written in Vertex Form

In Chapter 5, we defined a function of the form $f(x) = mx + b$ $(m \neq 0)$ as a linear function. The function defined by $f(x) = ax^2 + bx + c$ $(a \neq 0)$ is called a *quadratic function*. Notice that a quadratic function has a leading term of second degree. We are already familiar with the graph of $f(x) = x^2$ (Figure 7-1). The graph is a parabola opening upward with vertex at the origin. Also note that the graph is symmetric with respect to the vertical line through the vertex called the **axis of symmetry**.

We can write $f(x) = ax^2 + bx + c$ $(a \neq 0)$ in the form $f(x) = a(x - h)^2 + k$ (called *vertex form*) by completing the square. Furthermore, from Section 6.1, we know that the graph of $f(x) = a(x - h)^2 + k$ is related to the graph of $y = x^2$ by a horizontal shift determined by h and a vertical shift determined by k. Thus, the vertex of the parabola defined by $f(x) = a(x - h)^2 + k$ is (h, k). The value of a will stretch the parabola vertically if $a > 1$ and shrink the parabola vertically if $0 < a < 1$. If a is negative $(a < 0)$, the parabola is reflected across the x-axis and will open downward.

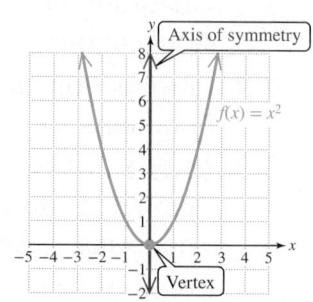

Figure 7-1

TIP A quadratic function is often used as a model for projectile motion. This is motion followed by an object influenced by an initial force and by the force of gravity.

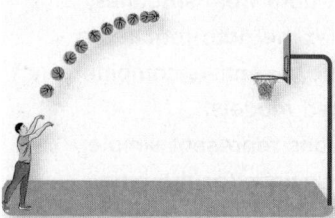

FOR REVIEW

The axis of symmetry is a vertical line. Recall that an equation of a vertical line is of the form $x = k$, where k is a real number. See Section 5.4.

Quadratic Function

A function defined by $f(x) = ax^2 + bx + c$ $(a \neq 0)$ is called a **quadratic function**. By completing the square, $f(x)$ can be expressed in **vertex form** as $f(x) = a(x - h)^2 + k$.

- The graph of f is a parabola with vertex (h, k).
- If $a > 0$, the parabola opens upward, and the vertex is the minimum point. The minimum *value* of f is k.
- If $a < 0$, the parabola opens downward, and the vertex is the maximum point. The maximum *value* of f is k.
- The axis of symmetry is $x = h$. This is the vertical line that passes through the vertex.

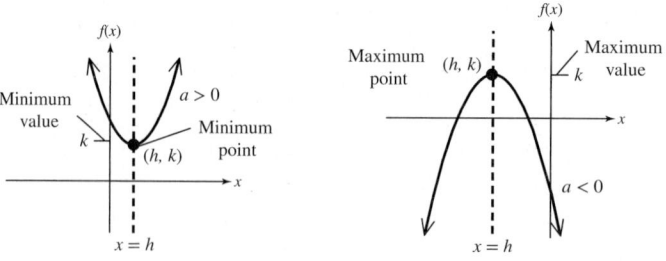

In Example 1, we analyze and graph a quadratic function by identifying the vertex, axis of symmetry, and x- and y-intercepts. From the graph, the minimum or maximum value of the function is readily apparent.

EXAMPLE 1   Analyzing and Graphing a Quadratic Function

Given $f(x) = -2(x - 1)^2 + 8$,

a. Determine whether the graph of the parabola opens upward or downward.

b. Identify the vertex.

c. Determine the x-intercept(s).

d. Determine the y-intercept.

e. Sketch the function.

f. Determine the axis of symmetry.

g. Determine the maximum or minimum value of f.

h. Write the domain and range in interval notation.

Solution:

a. $f(x) = -2(x - 1)^2 + 8$
The parabola opens downward.

> The function is written as $f(x) = a(x - h)^2 + k$, where $a = -2$, $h = 1$, and $k = 8$. Since $a < 0$, the parabola opens downward.

b. The vertex is $(1, 8)$.

> The vertex is (h, k), which is $(1, 8)$.

c.
$$f(x) = -2(x - 1)^2 + 8$$
$$0 = -2(x - 1)^2 + 8$$
$$-8 = -2(x - 1)^2$$
$$4 = (x - 1)^2$$
$$(x - 1)^2 = 4$$
$$x - 1 = \pm\sqrt{4}$$
$$x - 1 = \pm 2$$
$$x = 1 \pm 2$$
$$x = 1 + 2 \quad \text{or} \quad x = 1 - 2$$
$$x = 3 \qquad \text{or} \quad x = -1$$

> To find the x-intercept(s), find all real solutions to the equation $f(x) = 0$.
>
> Apply the square root property.
>
> Simplify.

The x-intercepts are $(3, 0)$ and $(-1, 0)$.

d. $f(0) = -2(0 - 1)^2 + 8$
$= -2(1) + 8$
$= 6$

> To find the y-intercept, evaluate $f(0)$.

The y-intercept is $(0, 6)$.

e. Plot points at the vertex $(1, 8)$, x-intercepts $(-1, 0)$ and $(3, 0)$, and y-intercept $(0, 6)$. The graph of f is shown in Figure 7-2.

f. The axis of symmetry is the vertical line through the vertex: $x = 1$.

g. The maximum value is 8.

h. The domain is $(-\infty, \infty)$.
The range is $(-\infty, 8]$.

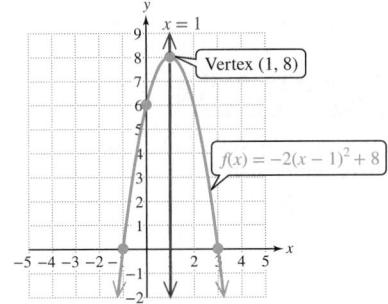

Figure 7-2

Skill Practice 1   Repeat Example 1 with $g(x) = (x + 2)^2 - 1$.

Answers

1. a. Upward   **b.** $(-2, -1)$
c. $(-3, 0)$ and $(-1, 0)$ **d.** $(0, 3)$
e.

$g(x) = (x + 2)^2 - 1$

Vertex $(-2, -1)$

f. $x = -2$
g. The minimum value is -1.
h. The domain is $(-\infty, \infty)$.
The range is $[-1, \infty)$.

2. Write $f(x) = ax^2 + bx + c$ ($a \neq 0$) in Vertex Form

In Section 5.2, we learned how to complete the square to write an equation of a circle $x^2 + y^2 + Ax + By + C = 0$ in standard form $(x - h)^2 + (y - k)^2 = r^2$. We use the same process to write a quadratic function $f(x) = ax^2 + bx + c$ ($a \neq 0$) in vertex form $f(x) = a(x - h)^2 + k$. However, we will work on the right side of the equation only. This is demonstrated in Example 2.

EXAMPLE 2 **Writing a Quadratic Function in Vertex Form**

Given $f(x) = 3x^2 + 12x + 5$,

a. Write the function in vertex form: $f(x) = a(x - h)^2 + k$.

b. Identify the vertex.

c. Identify the x-intercept(s).

d. Identify the y-intercept.

e. Sketch the function.

f. Determine the axis of symmetry.

g. Determine the minimum or maximum value of f.

h. Write the domain and range in interval notation.

Solution:

a. $f(x) = 3x^2 + 12x + 5$

      Factor out the leading coefficient of the x^2 term from the two terms containing x.

    $= 3(x^2 + 4x \qquad) + 5$

      The leading term within the parentheses now has a coefficient of 1.

    $= 3(x^2 + 4x + 4 - 4) + 5$

      Complete the square within the parentheses. Add and subtract $\left[\frac{1}{2}(4)^2\right] = [2]^2 = 4$ within the parentheses.

    $= 3(x^2 + 4x + 4) + 3(-4) + 5$

    $= 3(x + 2)^2 - 7$ (vertex form)

      Remove -4 from within the parentheses, along with the factor 3.

b. The vertex is $(-2, -7)$.

c. $f(x) = 3x^2 + 12x + 5$

      To find the x-intercept(s), find the real solutions to the equation $f(x) = 0$.

    $0 = 3x^2 + 12x + 5$

      The equation is in the form $ax^2 + bx + c = 0$, with $a = 3$, $b = 12$, and $c = 5$.

FOR REVIEW

To review the use of the quadratic formula, see Section 3.6.

    $x = \dfrac{-(12) \pm \sqrt{(12)^2 - 4(3)(5)}}{2(3)}$

      Apply the quadratic formula.

      $x = \dfrac{-b \pm \sqrt{b^2 - 4ac}}{2a}$

    $= \dfrac{-12 \pm \sqrt{144 - 60}}{6}$

      Simplify.

    $= \dfrac{-12 \pm \sqrt{84}}{6}$

      Simplify the radical. Factor the radicand and remove the largest perfect square: $\sqrt{84} = \sqrt{2^2 \cdot 7 \cdot 3} = 2\sqrt{21}$

    $= \dfrac{-12 \pm 2\sqrt{21}}{6}$

    $= \dfrac{\overset{1}{2}(-6 \pm \sqrt{21})}{\underset{3}{6}}$

      Factor the numerator and simplify the expression to lowest terms.

$$= \frac{-6 \pm \sqrt{21}}{3} \begin{cases} = \frac{-6 + \sqrt{21}}{3} \approx -0.47 \\ = \frac{-6 - \sqrt{21}}{3} \approx -3.53 \end{cases}$$

The x-intercepts are $\left(\dfrac{-6 + \sqrt{21}}{3}, 0\right)$ and $\left(\dfrac{-6 - \sqrt{21}}{3}, 0\right)$ or approximately $(-0.47, 0)$ and $(-3.53, 0)$.

d. $f(0) = 3(0)^2 + 12(0) + 5$
    $= 5$

To find the y-intercept, evaluate $f(0)$.
The y-intercept is $(0, 5)$.

e. Plot points at the vertex $(-2, -7)$, x-intercepts $(-0.47, 0)$ and $(-3.53, 0)$, and y-intercept $(0, 5)$. The graph of f is shown in Figure 7-3.

f. The axis of symmetry is $x = -2$.

g. The minimum value is -7.

h. The domain is $(-\infty, \infty)$.
The range is $[-7, \infty)$.

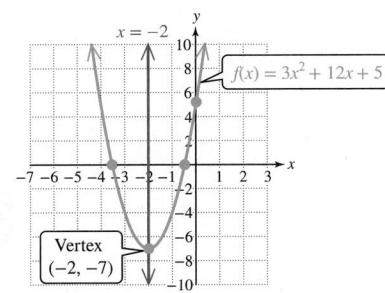

Figure 7-3

> **Skill Practice 2**   Repeat Example 2 with $f(x) = 3x^2 - 6x + 1$.

---

3. Find the Vertex of a Parabola by Using the Vertex Formula

Completing the square and writing a quadratic function in the form $f(x) = a(x - h)^2 + k$ is one method to find the vertex of a parabola. Another method is to use the vertex formula. The vertex formula can be derived by completing the square on $f(x) = ax^2 + bx + c$.

$f(x) = ax^2 + bx + c \ (a \neq 0)$

    $= a\left(x^2 + \dfrac{b}{a}x + \dfrac{b^2}{4a^2} - \dfrac{b^2}{4a^2}\right) + c$

Factor out a from the x terms, and complete the square within parentheses.
$$\left[\frac{1}{2}\left(\frac{b}{a}\right)\right]^2 = \frac{b^2}{4a^2}$$

    $= a\left(x^2 + \dfrac{b}{a}x + \dfrac{b^2}{4a^2}\right) + a\left(-\dfrac{b^2}{4a^2}\right) + c$

Remove the term $-\dfrac{b^2}{4a^2}$ from within parentheses along with the factor a.

    $= a\left(x + \dfrac{b}{2a}\right)^2 - \dfrac{b^2}{4a} + c$

Factor the trinomial.

    $= a\left(x + \dfrac{b}{2a}\right)^2 + \dfrac{4ac - b^2}{4a}$

Obtain a common denominator and add the terms outside parentheses.

    $= a\left[x - \left(\dfrac{-b}{2a}\right)\right]^2 + \dfrac{4ac - b^2}{4a}$

$f(x)$ is now written in vertex form.

$f(x) = a(x - h)^2 \quad + \quad k$

$h = \dfrac{-b}{2a}$ and $k = \dfrac{4ac - b^2}{4a}$

Answers

2. a. $f(x) = 3(x - 1)^2 - 2$ **b.** $(1, -2)$

  c. $\left(\dfrac{3 \pm \sqrt{6}}{3}, 0\right)$

  d. $(0, 1)$

  e.

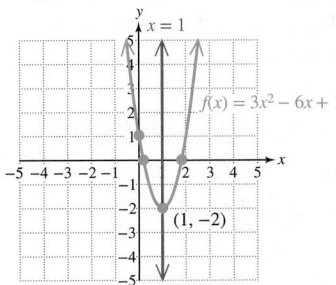

  f. $x = 1$

  g. The minimum value is -2.

  h. The domain is $(-\infty, \infty)$.
     The range is $[-2, \infty)$.

The vertex is $\left(\dfrac{-b}{2a}, \dfrac{4ac - b^2}{4a}\right)$.

The y-coordinate of the vertex is given by $\dfrac{4ac - b^2}{4a}$ and is often hard to remember. Therefore, it is usually easier to evaluate the x-coordinate first from $\dfrac{-b}{2a}$, and then evaluate $f\left(\dfrac{-b}{2a}\right)$.

Vertex Formula to Find the Vertex of a Parabola

For $f(x) = ax^2 + bx + c$ $(a \neq 0)$, the vertex of the graph of the parabola is given by $\left(\dfrac{-b}{2a},\ f\left(\dfrac{-b}{2a}\right)\right)$.

EXAMPLE 3 Using the Vertex Formula

Given $f(x) = -x^2 + 4x - 5$,

a. State whether the graph of the parabola opens upward or downward.

b. Determine the vertex of the parabola by using the vertex formula.

c. Determine the x-intercept(s).

d. Determine the y-intercept.

e. Sketch the graph.

f. Determine the axis of symmetry.

g. Determine the minimum or maximum value of f.

h. Write the domain and range in interval notation.

Solution:

a. $f(x) = -x^2 + 4x - 5$

The parabola opens downward.

> The function is written as $f(x) = ax^2 + bx + c$, where $a = -1$. Since $a < 0$, the parabola opens downward.

b. x-coordinate: $\dfrac{-b}{2a} = \dfrac{-(4)}{2(-1)} = 2$

> Apply the vertex formula with $a = -1$ and $b = 4$.

y-coordinate: $f(2) = -(2)^2 + 4(2) - 5$
$$= -4 + 8 - 5$$
$$= -1$$

The vertex is $(2, -1)$.

c. Since the vertex of the parabola is below the x-axis and the parabola opens downward, the parabola cannot cross or touch the x-axis.

Therefore, there are no x-intercepts.

> Solving the equation $f(x) = 0$ to find the x-intercepts results in imaginary solutions:
> $$0 = -x^2 + 4x - 5$$
> $$x = \dfrac{-(4) \pm \sqrt{(4)^2 - 4(-1)(-5)}}{2(-1)}$$
> $$= \dfrac{-4 \pm \sqrt{-4}}{-2}$$
>
> Negative radicand. The expression is an imaginary number.

d. To find the y-intercept, evaluate $f(0)$.

$$f(0) = -(0)^2 + 4(0) - 5 = -5$$

The y-intercept is $(0, -5)$.

e. Plot points at the vertex $(2, -1)$ and y-intercept $(0, -5)$ and sketch the parabola opening downward. Plot additional points on the graph such as $(1, -2)$, $(3, -2)$, and $(4, -5)$. The graph of f is shown in Figure 7-4.

f. The axis of symmetry is $x = 2$.

g. The maximum value of f is -1.

h. The domain is $(-\infty, \infty)$.
The range is $(-\infty, -1]$.

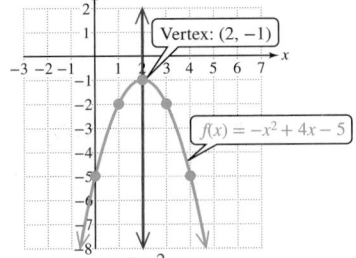

Figure 7-4

Skill Practice 3   Repeat Example 3 with $f(x) = -x^2 - 4x - 7$.

The x-intercepts of a quadratic function defined by $f(x) = ax^2 + bx + c$ are the real solutions to the equation $f(x) = 0$. The discriminant $b^2 - 4ac$ enables us to determine the number of real solutions to the equation and thus, the number of x-intercepts of the graph of the function.

Using the Discriminant to Determine the Number of x-Intercepts

Given a quadratic function defined by $f(x) = ax^2 + bx + c$ $(a \neq 0)$,

- If $b^2 - 4ac = 0$, the graph of $y = f(x)$ has one x-intercept.
- If $b^2 - 4ac > 0$, the graph of $y = f(x)$ has two x-intercepts.
- If $b^2 - 4ac < 0$, the graph of $y = f(x)$ has no x-intercept.

To illustrate the use of the discriminant to determine the number of x-intercepts of a quadratic function, consider the functions from Examples 2 and 3.

Function and related equation	Discriminant
Example 2:	$a = 3, b = 12, c = 5$
$f(x) = 3x^2 + 12x + 5$	$b^2 - 4ac = (12)^2 - 4(3)(5)$
$0 = 3x^2 + 12x + 5$	$= 84 > 0$ The function has 2 x-intercepts. See Figure 7-3.
Example 3:	$a = -1, b = 4, c = -5$
$f(x) = -x^2 + 4x - 5$	$b^2 - 4ac = (4)^2 - 4(-1)(-5)$
$0 = -x^2 + 4x - 5$	$= -4 < 0$ The function has no x-intercepts. See Figure 7-4.

4. Solve Applications Involving Quadratic Functions

Quadratic functions can be used in a variety of applications in which a variable is optimized. That is, the vertex of a parabola gives the maximum or minimum value of the dependent variable. We show three such applications in Examples 4–6.

Answers

3. a. Downward   **b.** $(-2, -3)$
 c. No x-intercepts   **d.** $(0, -7)$
 e.

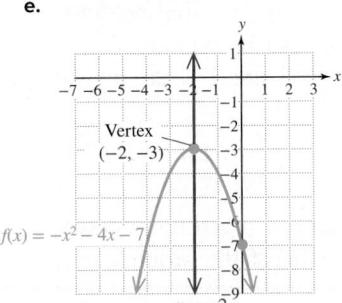

 f. $x = -2$
 g. The maximum value is -3.
 h. The domain is $(-\infty, \infty)$.
  The range is $(-\infty, -3]$.

EXAMPLE 4   Using a Quadratic Function for Projectile Motion

A stone is thrown from a 100-m cliff at an initial speed of 20 m/sec at an angle of 30° from the horizontal. The height of the stone can be modeled by $h(t) = -4.9t^2 + 10t + 100$, where $h(t)$ is the height in meters and t is the time in seconds after the stone is released.

   a. Determine the time at which the stone will be at its maximum height. Round to 2 decimal places.

   b. Determine the maximum height. Round to the nearest meter.

   c. Determine the time at which the stone will hit the ground.

Point of Interest

The movie *Apollo 13* starring Tom Hanks was filmed in part in a "Vomit Comet," an aircraft that uses a parabolic flight trajectory to produce weightlessness. As the plane climbs toward the top of the parabolic path, occupants experience a force of nearly 2 Gs (twice their body weight). Once the plane goes over the vertex of the parabola, flyers free fall inside the plane. Such motion often produces motion sickness, thus earning the aircraft its name.

Solution:

   a. The time at which the stone will be at its maximum height is the t-coordinate of the vertex.

   Given $h(t) = -4.9t^2 + 10t + 100$, the coefficients are $a = -4.9$, $b = 10$, and $c = 100$.

$$t = \frac{-b}{2a} = \frac{-10}{2(-4.9)} \approx 1.02$$

   The vertex is given by $\left(\frac{-b}{2a}, h\left(\frac{-b}{2a}\right)\right)$.

   The stone will be at its maximum height approximately 1.02 sec after release.

   b. The maximum height is the value of $h(t)$ at the vertex.

$$h(1.02) = -4.9(1.02)^2 + 10(1.02) + 100$$
$$\approx 105 \text{ The maximum height is approximately 105 m.}$$

   c. The stone will hit the ground when $h(t) = 0$.

$$h(t) = -4.9t^2 + 10t + 100$$
$$0 = -4.9t^2 + 10t + 100$$
$$t = \frac{-b \pm \sqrt{b^2 - 4ac}}{2a}$$
$$t = \frac{-10 \pm \sqrt{(10)^2 - 4(-4.9)(100)}}{2(-4.9)}$$

   $t \approx 5.65$ or $t \approx -3.61$ Reject the negative solution. The stone will hit the ground in approximately 5.65 sec.

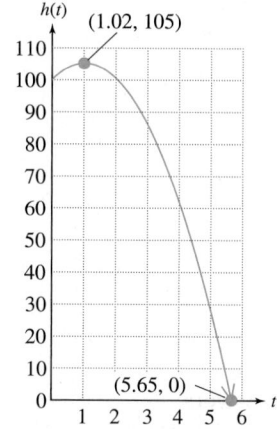

Skill Practice 4   A quarterback throws a football with an initial velocity of 72 ft/sec at an angle of 25°. The height of the ball can be modeled by $h(t) = -16t^2 + 30.4t + 5$, where $h(t)$ is the height (in ft) and t is the time in seconds after release.

   a. Determine the time at which the ball will be at its maximum height.

   b. Determine the maximum height of the ball.

   c. Determine the amount of time required for the ball to reach the receiver's hands if the receiver catches the ball at a point 3 ft off the ground.

Answers
4. a. 0.95 sec   **b.** 19.44 ft
   c. Approximately 1.96 sec

TECHNOLOGY CONNECTIONS

Compute Solutions to a Quadratic Equation

The syntax to compute the expressions from Example 4(c) is shown for a calculator in Classic mode and in Mathprint mode. In Classic mode, parentheses are required around the numerator and denominator of the fraction and around the radicand within the square root. In Mathprint mode, select the ALPHA key followed by F1 to access the fraction template.

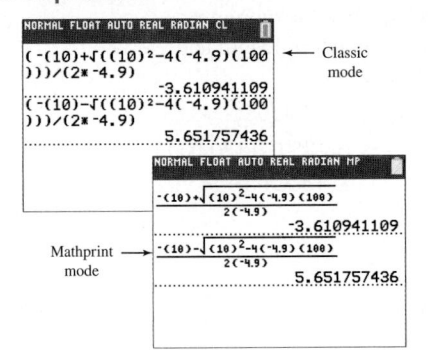

EXAMPLE 5   Applying a Quadratic Function to Geometry

A parking area is to be constructed adjacent to a road. The developer has purchased 340 ft of fencing. Determine dimensions for the parking lot that would maximize the area. Then find the maximum area.

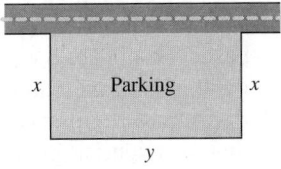

Solution:

Let x represent the width of the parking area.

Let y represent the length.

Let A represent the area.

> Read the problem carefully, draw a representative diagram, and label the unknowns.

We need to find the values of x and y that maximize the area A of the rectangular region. The area is given by $A = (\text{length})(\text{width}) = yx = xy$.

To write the area as a function of one variable only, we need an equation that relates x and y. We know that the parking area is limited by a fixed amount of fencing. That is, the sum of the lengths of the three sides to be fenced can be at most 340 ft.

$2x + y = 340$ — Solve for y.

> The equation $2x + y = 340$ is called a **constraint equation**. This equation gives an implied restriction on x and y due to the limited amount of fencing.

$y = 340 - 2x$

> Solve the constraint equation, $2x + y = 340$ for either x or y. In this case, we have solved for y.

$A = xy$

$A(x) = x(340 - 2x)$

> Substitute $340 - 2x$ for y in the equation $A = xy$.

$\quad\quad = -2x^2 + 340x$

> Function A is a quadratic function with a negative leading coefficient. The graph of the parabola opens downward, so the vertex is the maximum point on the function.

x-coordinate of vertex:

$$x = \frac{-b}{2a} = \frac{-340}{2(-2)} = 85$$

$$y = 340 - 2(85) = 170$$

The *x*-coordinate of the vertex $\frac{-b}{2a}$ is the value of *x* that will maximize the area.

The second dimension of the parking lot can be determined from the constraint equation.

The values of *x* and *y* that would maximize the area are $x = 85$ ft and $y = 170$ ft.

$$A(85) = -2(85)^2 + 340(85) = 14{,}450$$

The value of the function at $x = 85$ gives the maximum area.

The maximum area is $14{,}450 \text{ ft}^2$.

Skill Practice 5 A farmer has 200 ft of fencing and wants to build three adjacent rectangular corrals. Determine the dimensions that should be used to maximize the area, and find the area of each individual corral.

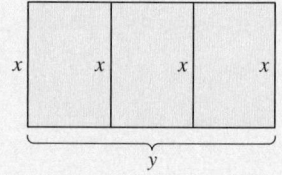

5. Create Quadratic Models Using Regression

In Section 5.5, we introduced linear regression. A regression line is a linear model based on all observed data points. In a similar fashion, we can create a quadratic function using regression. For example, suppose that a scientist growing bacteria measures the population of bacteria as a function of time. A scatter plot reveals that the data follow a curve that is approximately parabolic (Figure 7-5). In Example 6, we use a graphing calculator to find a quadratic function that models the population of the bacteria as a function of time.

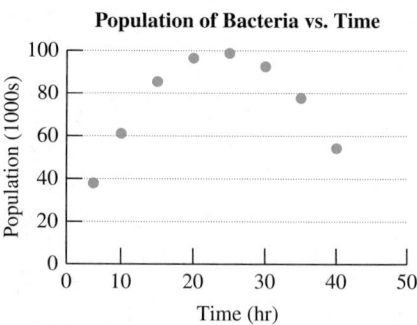

Population of Bacteria vs. Time

Figure 7-5

EXAMPLE 6 Creating a Quadratic Function Using Regression

The data in the table represent the population of bacteria $P(t)$ (in 1000s) versus the number of hours t since the culture was started.

a. Use regression to find a quadratic function to model the data. Round the coefficients to 3 decimal places.

b. Use the model to determine the time at which the population is greatest. Round to the nearest hour.

c. What is the maximum population? Round to the nearest hundred.

Time (hr) t	Population (1000s) $P(t)$
5	37.7
10	60.9
15	85.3
20	96.3
25	98.6
30	92.4
35	77.5
40	54.1

Answer

5. The dimensions should be $x = 25$ ft and $y = 50$ ft. The area of each individual corral is

$$\frac{1250}{3} = 416.\overline{6} \text{ ft}^2.$$

Solution:

a. From the graph in Figure 7-5, it appears that the data follow a parabolic curve. Therefore, a quadratic model would be reasonable.

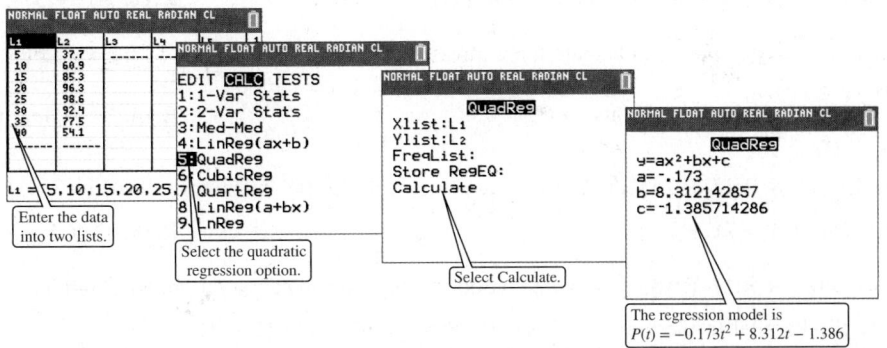

The regression model is
$P(t) = -0.173t^2 + 8.312t - 1.386$

b. From the graph, the time when the population is greatest is the t-coordinate of the vertex.

$$t = \frac{-b}{2a} = \frac{-(8.312)}{2(-0.173)} \approx 24$$

The population is greatest 24 hr after the culture is started.

c. The maximum population of the bacteria is the $P(t)$ value at the vertex.

$$P(24) = -0.173(24)^2 + 8.312(24) - 1.386$$
$$\approx 98.5 \quad \text{The maximum number of bacteria is approximately 98,500.}$$

Skill Practice 6   The funding $f(t)$ (in \$ millions) for a drug rehabilitation center is given in the table for selected years t.

t	0	3	6	9	12	15
$f(t)$	3.5	2.2	2.1	3	4.9	8

a. Use regression to find a quadratic function to model the data.

b. During what year is the funding the least? Round to the nearest year.

c. What is the minimum yearly amount of funding received? Round to the nearest million.

Answers

6. a. $f(t) = 0.060t^2 - 0.593t + 3.486$
   b. Year 5   **c.** \$2 million

Prerequisite Review

For Exercises R.1–R.4, graph the functions. (See Section 6.1 for review.)

R.1. $f(x) = |x + 4| - 1$ **R.2.** $g(x) = |x - 2| + 3$ **R.3.** $h(x) = -2|x|$ **R.4.** $k(x) = \frac{1}{3}|x|$

For Exercises R.5–R.8, determine the value of n that makes the polynomial a perfect square trinomial. Then factor the result. (See Section 3.5 for review.)

R.5. $x^2 + 18x + n$ **R.6.** $x^2 - 24x + n$ **R.7.** $x^2 - 11x + n$ **R.8.** $x^2 + 9x + n$

For Exercises R.9–R.12, solve the equations. (See Sections 3.4 and 3.6 for review.)

R.9. $5x^2 + 19x - 4 = 0$ (See Section 3.4.)

R.10. $3x^2 + 5x - 8 = 0$

R.11. $3x^2 - 2x - 4 = 0$ (See Section 3.6.)

R.12. $8x^2 + x - 3 = 0$

For Exercises R.13–R.14, evaluate the expressions for the given values of the variables.

R.13. $-\dfrac{b}{2a}$; for $a = -3$ and $b = 5$

R.14. $b^2 - 4ac$; for $a = -3$, $b = 4$, and $c = -6$

For Exercises R.15–R.16, find the x- and y-intercepts for the function defined by $y = f(x)$. (See Section 5.3 for review.)

R.15. $f(x) = 2x + 7$

R.16. $f(x) = -3x - 4$

For Exercises R.17–R.18, write an equation of the line subject to the given conditions. (See Section 5.5 for review.)

R.17. Parallel to the y-axis and passing through the point $(2, -4)$

R.18. Perpendicular to the x-axis and passing through the point $(-3, 6)$

For Exercises R.19–R.20, determine the domain and range of the function. Write the answers in interval notation. (See Section 5.3 for review.)

R.19.

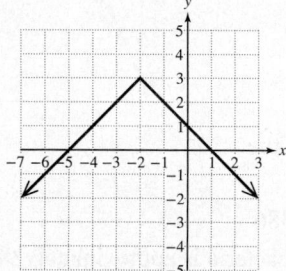

R.20.

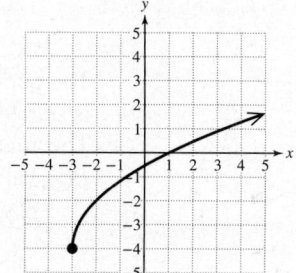

Concept Connections

1. A function defined by $f(x) = ax^2 + bx + c$ $(a \neq 0)$ is called a _____ function.

2. The vertical line drawn through the vertex of a quadratic function is called the _____ of symmetry.

3. Given $f(x) = a(x - h)^2 + k(a \neq 0)$, the vertex of the parabola is the point _____.

4. Given $f(x) = a(x - h)^2 + k$, if $a < 0$, the parabola opens (upward/downward) and the (minimum/maximum) value is _____.

5. Given $f(x) = a(x - h)^2 + k$, if $a > 0$, the parabola opens (upward/downward) and the (minimum/maximum) value is _____.

6. The graph of $f(x) = a(x - h)^2 + k$, $a \neq 0$, is a parabola and the axis of symmetry is the line given by _____.

Objective 1: Graph a Quadratic Function Written in Vertex Form

For Exercises 7–14,

a. Determine whether the graph of the parabola opens upward or downward.

b. Identify the vertex.

c. Determine the x-intercept(s).

d. Determine the y-intercept.

e. Sketch the function.

f. Determine the axis of symmetry.

g. Determine the minimum or maximum value of the function.

h. Write the domain and range in interval notation.
(See Example 1)

7. $f(x) = -(x - 4)^2 + 1$

8. $g(x) = -(x + 2)^2 + 4$

9. $h(x) = 2(x + 1)^2 - 8$

10. $k(x) = 2(x - 3)^2 - 2$

11. $m(x) = 3(x - 1)^2$

12. $n(x) = \dfrac{1}{2}(x + 2)^2$

13. $p(x) = -\dfrac{1}{5}(x + 4)^2 + 1$

14. $q(x) = -\dfrac{1}{3}(x - 1)^2 + 1$

Objective 2: Write $f(x) = ax^2 + bx + c$ $(a \neq 0)$ in Vertex Form

For Exercises 15–24,

a. Write the function in vertex form $f(x) = a(x - h)^2 + k$.

b. Identify the vertex.

c. Determine the x-intercept(s).

d. Determine the y-intercept.

e. Sketch the function.

f. Determine the axis of symmetry.

g. Determine the minimum or maximum value of the function.

h. Write the domain and range in interval notation. **(See Example 2)**

15. $f(x) = x^2 + 6x + 5$

16. $g(x) = x^2 + 8x + 7$

17. $p(x) = 3x^2 - 12x - 7$

18. $q(x) = 2x^2 - 4x - 3$

19. $c(x) = -2x^2 - 10x + 4$

20. $d(x) = -3x^2 - 9x + 8$

21. $h(x) = -2x^2 + 7x$

22. $k(x) = 3x^2 - 8x$

23. $p(x) = x^2 + 9x + 17$

24. $q(x) = x^2 + 11x + 26$

Objective 3: Find the Vertex of a Parabola by Using the Vertex Formula

For Exercises 25–32, find the vertex of the parabola by applying the vertex formula.

25. $f(x) = 3x^2 - 42x - 91$

26. $g(x) = 4x^2 - 64x + 107$

27. $k(a) = -\dfrac{1}{3}a^2 + 6a + 1$

28. $j(t) = -\dfrac{1}{4}t^2 + 10t - 5$

29. $f(c) = 4c^2 - 5$

30. $h(a) = 2a^2 + 14$

31. $P(x) = 1.2x^2 + 1.8x - 3.6$
(Write the coordinates of the vertex as decimals.)

32. $Q(x) = 7.5x^2 - 2.25x + 4.75$
(Write the coordinates of the vertex as decimals.)

For Exercises 33–42,

a. State whether the graph of the parabola opens upward or downward.

b. Identify the vertex.

c. Determine the x-intercept(s).

d. Determine the y-intercept.

e. Sketch the graph.

f. Determine the axis of symmetry.

g. Determine the minimum or maximum value of the function.

h. Write the domain and range in interval notation. **(See Example 3)**

33. $g(x) = -x^2 + 2x - 4$

34. $h(x) = -x^2 - 6x - 10$

35. $f(x) = 5x^2 - 15x + 3$

36. $k(x) = 2x^2 - 10x - 5$

37. $f(x) = 2x^2 + 3$

38. $g(x) = -x^2 - 1$

39. $f(x) = -2x^2 - 20x - 50$

40. $m(x) = 2x^2 - 8x + 8$

41. $n(x) = x^2 - x + 3$

42. $r(x) = x^2 - 5x + 7$

Objective 4: Solve Applications Involving Quadratic Functions

43. The monthly profit for a small company that makes long-sleeve T-shirts depends on the price per shirt. If the price is too high, sales will drop. If the price is too low, the revenue brought in may not cover the cost to produce the shirts. After months of data collection, the sales team determines that the monthly profit is approximated by $f(p) = -50p^2 + 1700p - 12{,}000$, where p is the price per shirt and $f(p)$ is the monthly profit based on that price. **(See Example 4)**

a. Find the price that generates the maximum profit.

b. Find the maximum profit.

c. Find the price(s) that would enable the company to break even.

44. The monthly profit for a company that makes decorative picture frames depends on the price per frame. The company determines that the profit is approximated by $f(p) = -80p^2 + 3440p - 36{,}000$, where p is the price per frame and $f(p)$ is the monthly profit based on that price.

a. Find the price that generates the maximum profit.

b. Find the maximum profit.

c. Find the price(s) that would enable the company to break even.

45. A long jumper leaves the ground at an angle of 20° above the horizontal, at a speed of 11 m/sec. The height of the jumper can be modeled by $h(x) = -0.046x^2 + 0.364x$, where h is the jumper's height in meters and x is the horizontal distance from the point of launch.

   a. At what horizontal distance from the point of launch does the maximum height occur? Round to 2 decimal places.

   b. What is the maximum height of the long jumper? Round to 2 decimal places.

   c. What is the length of the jump? Round to 1 decimal place.

46. A firefighter holds a hose 3 m off the ground and directs a stream of water toward a burning building. The water leaves the hose at an initial speed of 16 m/sec at an angle of 30°. The height of the water can be approximated by $h(x) = -0.026x^2 + 0.577x + 3$, where $h(x)$ is the height of the water in meters at a point x meters horizontally from the firefighter to the building.

   a. Determine the horizontal distance from the firefighter at which the maximum height of the water occurs. Round to 1 decimal place.

   b. What is the maximum height of the water? Round to 1 decimal place.

   c. The flow of water hits the house on the downward branch of the parabola at a height of 6 m. How far is the firefighter from the house? Round to the nearest meter.

47. The population $P(t)$ of a culture of the bacterium *Pseudomonas aeruginosa* is given by $P(t) = -1718t^2 + 82,000t + 10,000$, where t is the time in hours since the culture was started.

   a. Determine the time at which the population is at a maximum. Round to the nearest hour.

   b. Determine the maximum population. Round to the nearest thousand.

48. The gas mileage $m(x)$ (in mpg) for a certain vehicle can be approximated by $m(x) = -0.028x^2 + 2.688x - 35.012$, where x is the speed of the vehicle in mph.

   a. Determine the speed at which the car gets its maximum gas mileage.

   b. Determine the maximum gas mileage.

49. The sum of two positive numbers is 24. What two numbers will maximize the product? **(See Example 5)**

50. The sum of two positive numbers is 1. What two numbers will maximize the product?

51. The difference of two numbers is 10. What two numbers will minimize the product?

52. The difference of two numbers is 30. What two numbers will minimize the product?

53. Suppose that a family wants to fence in an area of their yard for a vegetable garden to keep out deer. One side is already fenced from the neighbor's property. **(See Example 5)**

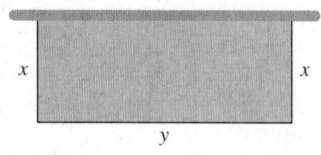

   a. If the family has enough money to buy 160 ft of fencing, what dimensions would produce the maximum area for the garden?

   b. What is the maximum area?

54. Two chicken coops are to be built adjacent to one another using 120 ft of fencing.

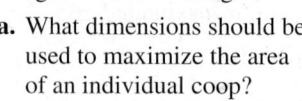

   a. What dimensions should be used to maximize the area of an individual coop?

   b. What is the maximum area of an individual coop?

55. A trough at the end of a gutter spout is meant to direct water away from a house. The homeowner makes the trough from a rectangular piece of aluminum that is 20 in. long and 12 in. wide. He makes a fold along the two long sides a distance of x inches from the edge.

   a. Write a function to represent the volume in terms of x.

   b. What value of x will maximize the volume of water that can be carried by the gutter?

   c. What is the maximum volume?

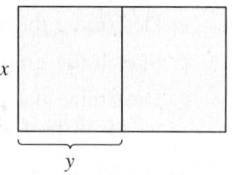

56. A rectangular frame of uniform depth for a shadow box is to be made from a 36-in. piece of wood.

   a. Write a function to represent the display area in terms of x.

   b. What dimensions should be used to maximize the display area?

   c. What is the maximum area?

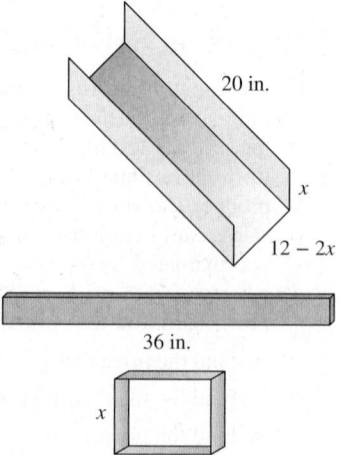

Objective 5: Create Quadratic Models Using Regression

57. *Tetanus bacillus* bacteria are cultured to produce tetanus toxin used in an inactive form for the tetanus vaccine. The amount of toxin produced per batch increases with time and then decreases as the culture becomes unstable. The variable t is the time in hours after the culture has started, and $y(t)$ is the yield of toxin in grams. **(See Example 6)**

t	8	16	24	32	40	48
$y(t)$	0.60	1.12	1.60	1.78	1.90	2.00

t	56	64	72	80	88	96
$y(t)$	1.94	1.80	1.48	1.30	0.66	0.10

a. Use regression to find a quadratic function to model the data.

b. At what time is the yield the greatest? Round to the nearest hour.

c. What is the maximum yield? Round to the nearest gram.

59. Fluid runs through a drainage pipe with a 10-cm radius and a length of 30 m (3000 cm). The velocity of the fluid gradually decreases from the center of the pipe toward the edges as a result of friction with the walls of the pipe. For the data shown, $v(x)$ is the velocity of the fluid (in cm/sec) and x represents the distance (in cm) from the center of the pipe toward the edge.

x	0	1	2	3	4
$v(x)$	195.6	195.2	194.2	193.0	191.5

x	5	6	7	8	9
$v(x)$	189.8	188.0	185.5	183.0	180.0

a. The pipe is 30 m long (3000 cm). Determine how long it will take fluid to run the length of the pipe through the center of the pipe. Round to 1 decimal place.

b. Determine how long it will take fluid at a point 9 cm from the center of the pipe to run the length of the pipe. Round to 1 decimal place.

c. Use regression to find a quadratic function to model the data.

d. Use the model from part (c) to predict the velocity of the fluid at a distance 5.5 cm from the center of the pipe. Round to 1 decimal place.

58. Gas mileage is tested for a car under different driving conditions. At lower speeds, the car is driven in stop-and-go traffic. At higher speeds, the car must overcome more wind resistance. The variable x given in the table represents the speed (in mph) for a compact car, and $m(x)$ represents the gas mileage (in mpg).

x	25	30	35	40	45
$m(x)$	22.7	25.1	27.9	30.8	31.9

x	50	55	60	65
$m(x)$	30.9	28.4	24.2	21.9

a. Use regression to find a quadratic function to model the data.

b. At what speed is the gas mileage the greatest? Round to the nearest mile per hour.

c. What is the maximum gas mileage? Round to the nearest mile per gallon.

60. The braking distance required for a car to stop depends on numerous variables such as the speed of the car, the weight of the car, reaction time of the driver, and the coefficient of friction between the tires and the road. For a certain vehicle on one stretch of highway, the braking distances $d(s)$ (in ft) are given for several different speeds s (in mph).

s	30	35	40	45	50
$d(s)$	109	134	162	191	223

s	55	60	65	70	75
$d(s)$	256	291	328	368	409

a. Use regression to find a quadratic function to model the data.

b. Use the model from part (a) to predict the stopping distance for the car if it is traveling 62 mph before the brakes are applied. Round to the nearest foot.

c. Suppose that the car is traveling 53 mph before the brakes are applied. If a deer is standing in the road at a distance of 245 ft from the point where the brakes are applied, will the car hit the deer?

Mixed Exercises

For Exercises 61–64, given a quadratic function defined by $f(x) = ax^2 + bx + c$ $(a \neq 0)$, answer true or false. If an answer is false, explain why.

61. The graph of f can have two y-intercepts.

62. The graph of f can have two x-intercepts.

63. If $a < 0$, then the vertex of the parabola is the maximum point on the graph of f.

64. The axis of symmetry of the graph of f is the line defined by $y = c$.

For Exercises 65–70, determine the number of x-intercepts of the graph of $f(x) = ax^2 + bx + c$ $(a \neq 0)$, based on the discriminant of the related equation $f(x) = 0$. (*Hint*: Recall that the discriminant is $b^2 - 4ac$.)

65. $f(x) = 4x^2 + 12x + 9$

66. $f(x) = 25x^2 - 20x + 4$

67. $f(x) = -x^2 - 5x + 8$

68. $f(x) = -3x^2 + 4x + 9$

69. $f(x) = -3x^2 + 6x - 11$

70. $f(x) = -2x^2 + 5x - 10$

For Exercises 71–78, given a quadratic function defined by $f(x) = a(x - h)^2 + k$ $(a \neq 0)$, match the graph with the function based on the conditions given.

71. $a > 0, h < 0, k > 0$

72. $a > 0, h < 0, k < 0$

73. $a < 0, h < 0, k < 0$

74. $a < 0, h < 0, k > 0$

75. $a > 0$, axis of symmetry $x = 2, k < 0$

76. $a < 0$, axis of symmetry $x = 2, k > 0$

77. $a < 0, h = 2$, maximum value equals -2

78. $a > 0, h = 2$, minimum value equals 2

a.
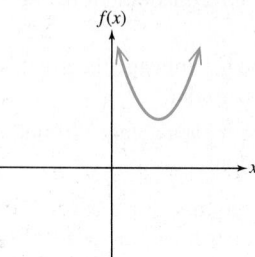

b.

c.
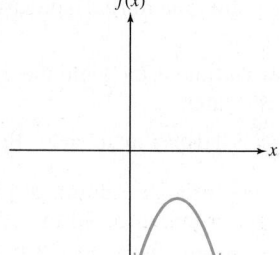

d.
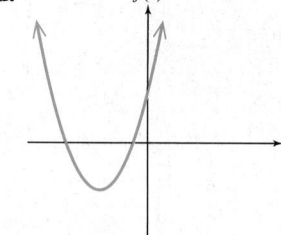

e.
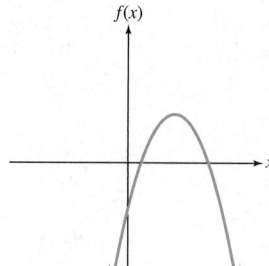

f.
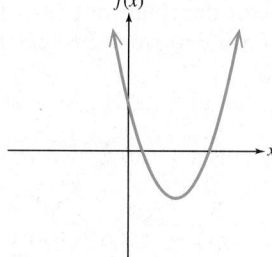

g.
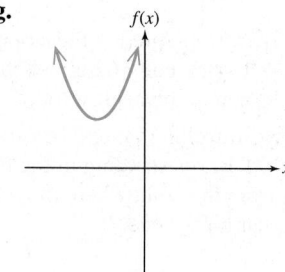

h.
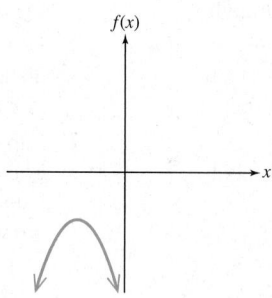

Write About It

79. Explain why a parabola opening upward has a minimum value but no maximum value. Use the graph of $f(x) = x^2$ to explain.

80. Explain why a quadratic function whose graph opens downward with vertex $(4, -3)$ has no x-intercept.

81. Explain why a quadratic function given by $f(x) = ax^2 + bx + c$ cannot have two y-intercepts.

82. Explain how to use the discriminant to determine the number of x-intercepts for the graph of $f(x) = ax^2 + bx + c$.

83. If a quadratic function given by $y = f(x)$ has x-intercepts of $(2, 0)$ and $(6, 0)$, explain why the vertex must be $(4, f(4))$.

84. Given an equation of a parabola in the form $y = a(x - h)^2 + k$, explain how to determine by inspection if the parabola has no x-intercepts.

Expanding Your Skills

For Exercises 85–88, define a quadratic function $y = f(x)$ that satisfies the given conditions.

85. Vertex $(2, -3)$ and passes through $(0, 5)$

86. Vertex $(-3, 1)$ and passes through $(0, -17)$

87. Axis of symmetry $x = 4$, maximum value 6, passes through $(1, 3)$

88. Axis of symmetry $x = -2$, minimum value 5, passes through $(2, 13)$

For Exercises 89–92, find the value of b or c that gives the function the given minimum or maximum value.

89. $f(x) = 2x^2 + 12x + c$; minimum value -9

90. $f(x) = 3x^2 + 12x + c$; minimum value -4

91. $f(x) = -x^2 + bx + 4$; maximum value 8

92. $f(x) = -x^2 + bx - 2$; maximum value 7

SECTION 7.2 — Introduction to Polynomial Functions

OBJECTIVES

1. Determine the End Behavior of a Polynomial Function
2. Identify Zeros and Multiplicities of Zeros
3. Apply the Intermediate Value Theorem
4. Sketch a Polynomial Function

1. Determine the End Behavior of a Polynomial Function

A solar oven is to be made from an open box with reflective sides. Each box is made from a 30-in. by 24-in. rectangular sheet of aluminum with squares of length x (in inches) removed from each corner. Then the flaps are folded up to form an open box.

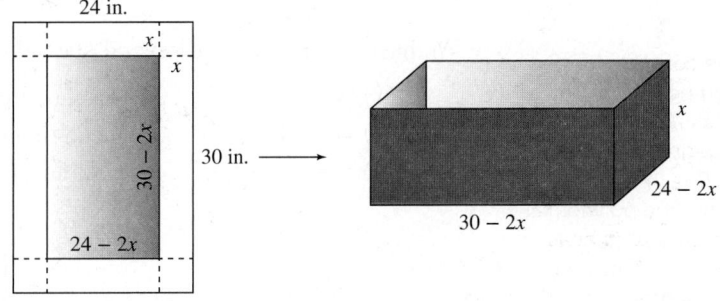

The volume $V(x)$ (in cubic inches) of the box is given by

$$V(x) = 4x^3 - 108x^2 + 720x, \text{ where } 0 < x < 12.$$

From the graph of $y = V(x)$ (Figure 7-6), the maximum volume appears to occur when squares of slightly greater than 4 inches are cut from the corners of the sheet of aluminum. See Exercise 99.

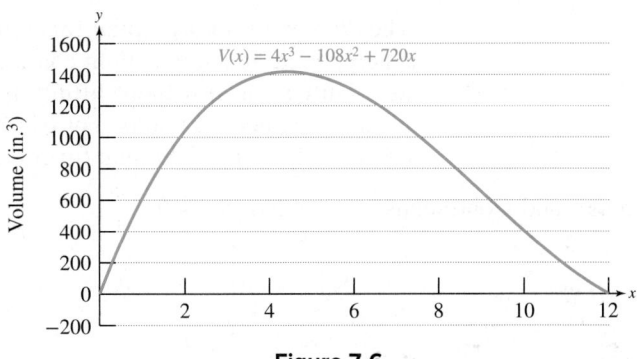

Figure 7-6

The function defined by $V(x) = 4x^3 - 108x^2 + 720x$ is an example of a polynomial function of degree 3.

Definition of a Polynomial Function

Let a_n, a_{n-1}, a_{n-2},..., a_0 represent real numbers and n, $n-1$, $n-2$,..., 0 represent whole numbers. Then a function defined by

$$f(x) = a_n x^n + a_{n-1}x^{n-1} + a_{n-2}x^{n-2} + \cdots + a_1 x + a_0$$

is called a **polynomial function**.

The term $a_n x^n$ is called the **leading term**, the coefficient a_n is the **leading coefficient**, and the exponent n is the **degree** of the polynomial function.

The coefficients of each term of a polynomial function are real numbers, and the exponents on x must be whole numbers.

Polynomial Function	**Not a Polynomial Function**
$f(x) = 4x^5 - 3x^4 + 2x^2$	$f(x) = 4\sqrt{x} - \dfrac{3}{x} + (3 + 2i)x^2$

$\sqrt{x} = x^{1/2}$	$3/x = 3x^{-1}$	$(3 + 2i)$
Exponent not a whole number	Exponent not a whole number	Coefficient not a real number

TIP The constant function defined by $f(x) = 2$ can be written as $f(x) = 2x^0$. Thus, the degree of the function is 0.

A third-degree polynomial function is referred to as a *cubic* polynomial function.

A fourth-degree polynomial function is referred to as a *quartic* polynomial function.

We have already studied several special cases of polynomial functions. For example:

$f(x) = 2$ constant function   (polynomial function, degree 0)

$g(x) = 3x + 1$ linear function   (polynomial function, degree 1)

$h(x) = 4x^2 + 7x - 1$ quadratic function   (polynomial function, degree 2)

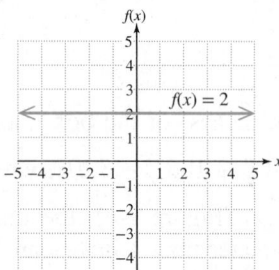 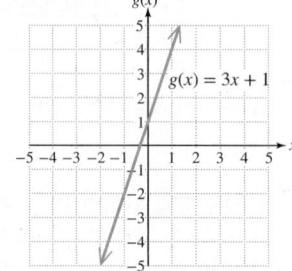 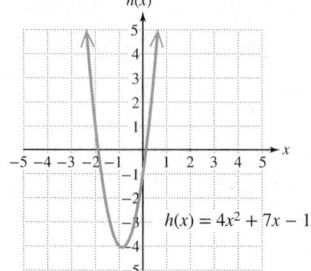

The domain of a polynomial function is all real numbers. Furthermore, the graph of a polynomial function is both continuous and smooth. Informally, a continuous function can be drawn without lifting the pencil from the paper. A smooth function has no sharp corners or points. For example, among the four curves that follow, the first curve could be a polynomial function, but the last three are not polynomial functions.

Smooth and Continuous   Not Smooth   Not Continuous   Not Continuous

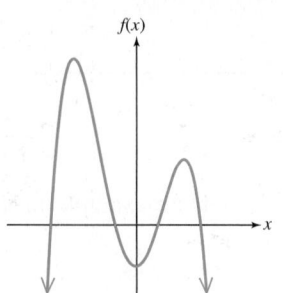 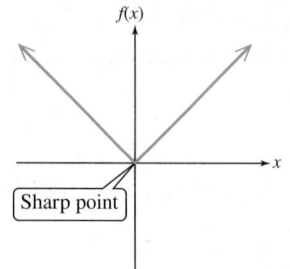 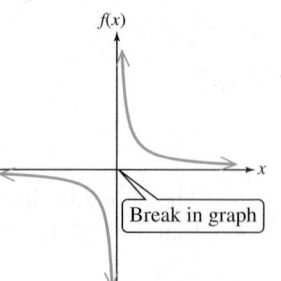 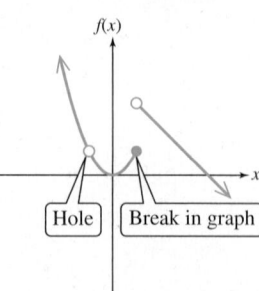

To begin our analysis of polynomial functions, we first consider the graphs of functions of the form $f(x) = ax^n$, where a is a real number and n is a positive integer. These fall into a category of functions called **power functions**. The graphs of three power functions with even degrees and positive coefficients are shown in Figure 7-7. The graphs of three power functions with odd degrees and positive coefficients are shown in Figure 7-8.

TIP For a positive integer n, the graph of the power function $y = x^n$ becomes "flatter" near the x-intercept for higher powers of n.

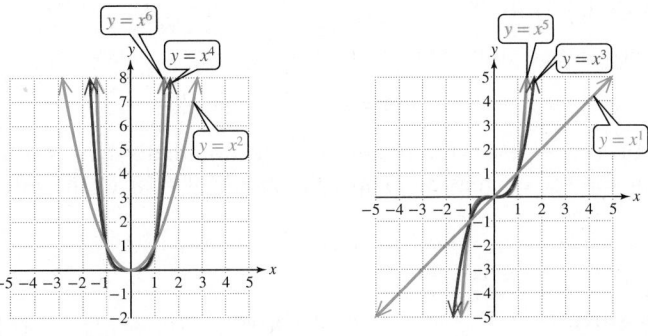

| Figure 7-7 | Figure 7-8 |

From Figure 7-7, notice that for even powers of n, the behavior of $y = x^n$ is similar to the graph of $y = x^2$ with variations on the "steepness" of the curve. Figure 7-8 shows that for odd powers, the behavior of $y = x^n$ with $n \geq 3$ is similar to the graph of $y = x^3$. For any power function $y = ax^n$, the coefficient a will impose a vertical shrink or stretch on the graph of $y = x^n$ by a factor of $|a|$. If $a < 0$, then the graph is reflected across the x-axis.

Power functions are helpful to analyze the "end behavior" of a polynomial function with multiple terms. The end behavior is the general direction that the function follows as x approaches ∞ or $-\infty$. To describe end behavior, we have the following notation.

Notation for Infinite Behavior of $y = f(x)$	
$x \to \infty$	is read as "x approaches infinity." This means that x becomes infinitely large in the positive direction.
$x \to -\infty$	is read as "x approaches negative infinity." This means that x becomes infinitely "large" in the negative direction.*
$f(x) \to \infty$	is read as "$f(x)$ approaches infinity." This means that the y value becomes infinitely large in the positive direction.
$f(x) \to -\infty$	is read as "$f(x)$ approaches negative infinity." This means that the y value becomes infinitely "large" in the negative direction.*
*"Large" in the negative direction means that $x < 0$ and $|x|$ becomes increasingly large.	

Consider the function defined by

$$f(x) = a_n x^n + a_{n-1} x^{n-1} + a_{n-2} x^{n-2} + \cdots + a_1 x + a_0$$

The leading term has the greatest exponent on x.

The leading term has the greatest exponent on x. Therefore, as $|x|$ gets large (that is, as $x \to \infty$ or as $x \to -\infty$), the leading term will be relatively larger in absolute value than all other terms. In fact, x^n will eventually be greater in absolute value than the *sum* of all other terms. Therefore, the end behavior of the function is dictated only by the leading term, and the graph of the function far to the left and far to the right will follow the general behavior of the power function $y = ax^n$.

The Leading Term Test

Consider a polynomial function given by

$$f(x) = a_n x^n + a_{n-1} x^{n-1} + a_{n-2} x^{n-2} + \cdots + a_1 x + a_0.$$

As $x \to \infty$ or as $x \to -\infty$, the graph of f will eventually have no more turns and will become forever increasing or forever decreasing. Thus, the graph of f far to the left and far to the right will follow the general behavior of $y = a_n x^n$.

n is even		*n* is odd	
a_n positive	a_n negative	a_n positive	a_n negative
As $x \to -\infty$, $f(x) \to \infty$. As $x \to \infty$, $f(x) \to \infty$. 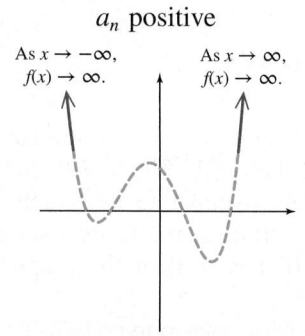 End behavior: up left/up right	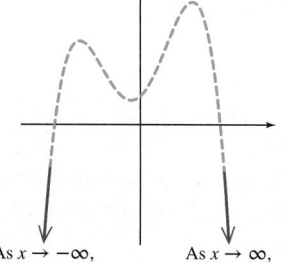 As $x \to -\infty$, $f(x) \to -\infty$. As $x \to \infty$, $f(x) \to -\infty$. End behavior: down left/down right	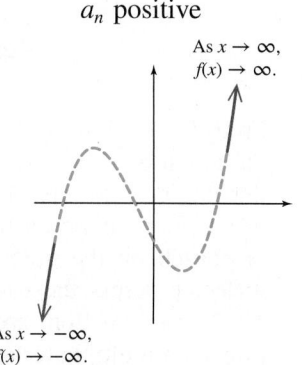 As $x \to \infty$, $f(x) \to \infty$. As $x \to -\infty$, $f(x) \to -\infty$. End behavior: down left/up right	As $x \to -\infty$, $f(x) \to \infty$. As $x \to \infty$, $f(x) \to -\infty$. 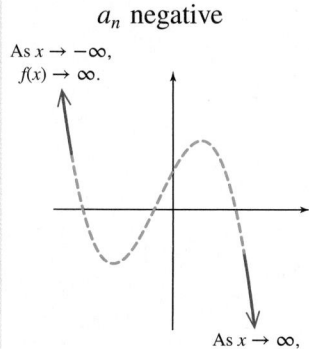 End behavior: up left/down right

---

EXAMPLE 1   Determining End Behavior

Use the leading term to determine the end behavior of the graph of the function.

a. $f(x) = -4x^5 + 6x^4 + 2x$

b. $g(x) = \dfrac{1}{4}x(2x - 3)^3(x + 4)^2$

TIP The graph of $y = f(x)$ from Example 1(a) will exhibit the same behavior as the graph of the power function $y = -4x^5$ for values of x far to the right and far to the left. This is similar to the graph of $y = x^5$ reflected across the x-axis.

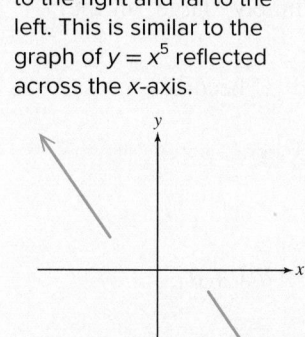

Solution:

a. $f(x) = \underbrace{-4x^5}_{} + 6x^4 + 2x$
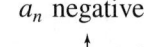
negative   odd

The leading coefficient is negative and the degree is odd. By the leading term test, the end behavior is up to the left and down to the right.

As $x \to -\infty$, $f(x) \to \infty$.
As $x \to \infty$, $f(x) \to -\infty$.

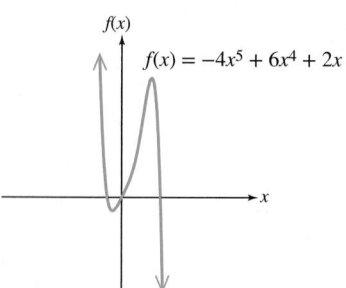
$f(x) = -4x^5 + 6x^4 + 2x$

b. $g(x) = \dfrac{1}{4}x(2x - 3)^3(x + 4)^2$

positive   even

$g(x) = \dfrac{1}{4}x(2x - 3)^3(x + 4)^2 = 2x^6 + \cdots$

To determine the leading term, multiply the leading terms from each factor. That is,

$\dfrac{1}{4}x(2x)^3(x)^2 = 2x^6.$

The leading coefficient is positive and the degree is even. By the leading term test, the end behavior is up to the left and up to the right.

As $x \to -\infty$, $f(x) \to \infty$.

As $x \to \infty$, $f(x) \to \infty$.

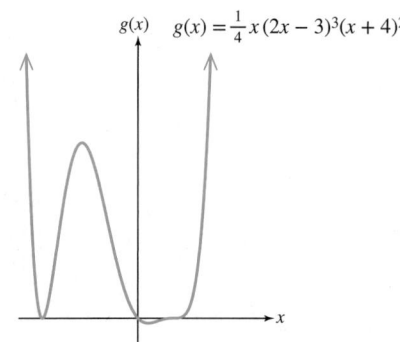
$g(x)$ $g(x) = \frac{1}{4}x(2x-3)^3(x+4)^2$

Skill Practice 1   Use the leading term to determine the end behavior of the graph of the function.

a. $f(x) = -0.3x^4 - 5x^2 - 3x + 4$ **b.** $g(x) = \frac{6}{7}(x-9)^4(x+4)^2(3x-5)$

2. Identify Zeros and Multiplicities of Zeros

Consider a polynomial function defined by $y = f(x)$. The values of x in the domain of f for which $f(x) = 0$ are called the **zeros** of the function. These are the real solutions (or **roots**) of the equation $f(x) = 0$ and correspond to the x-intercepts of the graph of $y = f(x)$.

EXAMPLE 2   **Determining the Zeros of a Polynomial Function**

Find the zeros of the function defined by $f(x) = x^3 + x^2 - 9x - 9$.

Solution:

FOR REVIEW

The equation $0 = x^3 + x^2 - 9x - 9$ is a polynomial equation. To solve the equation, factor the right side completely and then set each factor equal to zero. See Section 3.4 for review.

$$f(x) = x^3 + x^2 - 9x - 9$$
$$0 = x^3 + x^2 - 9x - 9 \qquad \text{To find the zeros of } f, \text{ set } f(x) = 0 \text{ and solve for } x.$$
$$0 = x^2(x+1) - 9(x+1) \qquad \text{Factor by grouping.}$$
$$0 = (x+1)(x^2-9)$$
$$0 = (x+1)(x-3)(x+3) \qquad \text{Factor the difference of squares.}$$
$$x + 1 = 0, \quad x - 3 = 0, \quad x + 3 = 0 \qquad \text{Set each factor equal to zero.}$$
$$x = -1, \qquad x = 3, \qquad x = -3 \qquad \text{Solve for } x.$$

The zeros of f are -1, 3, and -3.

The graph of f is shown in Figure 7-9. The zeros of the function are real numbers and correspond to the x-intercepts of the graph. By inspection, we can evaluate $f(0) = -9$, indicating that the y-intercept is $(0, -9)$.

Check:
A table of points can be used to check that $f(-1)$, $f(3)$, and $f(-3)$ all equal 0.

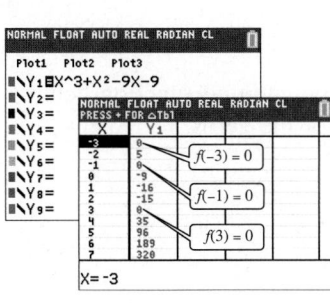

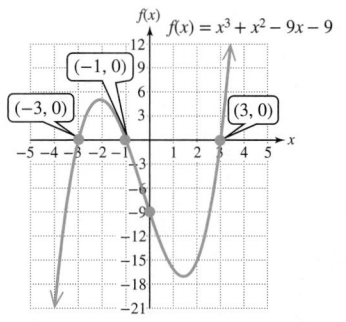
$f(x)$ $f(x) = x^3 + x^2 - 9x - 9$

Figure 7-9

Answers

1. a. Down to the left, down to the right.
As $x \to -\infty$, $f(x) \to -\infty$.
As $x \to \infty$, $f(x) \to -\infty$.

b. Down to the left, up to the right.
As $x \to -\infty$, $f(x) \to -\infty$.
As $x \to \infty$, $f(x) \to \infty$.

Skill Practice 2   Find the zeros of the function defined by
$f(x) = 4x^3 - 4x^2 - 25x + 25$.

EXAMPLE 3   Determining the Zeros of a Polynomial Function

Find the zeros of the function defined by $f(x) = -x^3 + 8x^2 - 16x$.

Solution:

$$f(x) = -x^3 + 8x^2 - 16x \qquad \text{To find the zeros of } f, \text{ set } f(x) = 0 \text{ and solve for } x.$$
$$0 = -x(x^2 - 8x + 16) \qquad \text{Factor out the GCF.}$$
$$0 = -x(x - 4)^2 \qquad \text{Factor the perfect square trinomial.}$$
$$-x = 0, \quad (x - 4)^2 = 0 \qquad \text{Set each factor equal to zero.}$$
$$x = 0, \qquad x = 4, \qquad \text{Solve for } x.$$

The zeros of f are 0 and 4.

The graph of f is shown in Figure 7-10. The zeros of
the function are real numbers and correspond to the
x-intercepts $(0, 0)$ and $(4, 0)$.

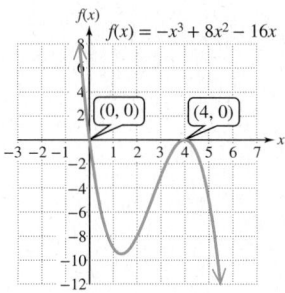

The leading term of $f(x)$ is $-x^3$. The coefficient is
negative and the exponent is odd. The graph shows
the end behavior up to the left and down to the right
as expected.

Figure 7-10

Skill Practice 3   Find the zeros of the function defined by
$f(x) = x^3 + 10x^2 + 25x$.

From Example 3, $f(x) = -x^3 + 8x^2 - 16x$ can be written as a product of linear
factors. That is:

$$f(x) = -x^3 + 8x^2 - 16x$$
$$= -x(x - 4)^2$$
$$= -x(x - 4)(x - 4)$$

Notice that the factor $(x - 4)$ appears to the second power and thus will appear twice
as a factor when written without the exponent. Therefore, we say that the corresponding zero, 4, has a multiplicity of 2. In general, we say that if a polynomial function
has a factor $(x - c)$ that appears exactly k times, then c is a **zero of multiplicity k**.
For example, consider:

$$g(x) = x^2(x - 2)^3(x + 4)^7 \qquad \text{0 is a zero of multiplicity 2.}$$
$$\text{2 is a zero of multiplicity 3.}$$
$$\text{-4 is a zero of multiplicity 7.}$$

The graph of a polynomial function behaves in the following manner based on
the multiplicity of the zeros.

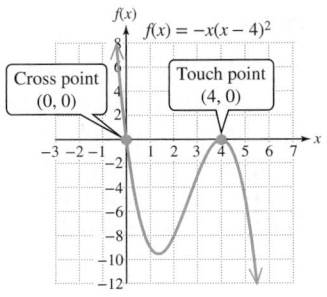

Figure 7-11

Touch Points and Cross Points

Let f be a polynomial function and let c be a real zero of f. Then the point $(c, 0)$ is an x-intercept of the graph of f. Furthermore,

- If c is a zero of odd multiplicity, then the graph *crosses* the x-axis at c. The point $(c, 0)$ is called a **cross point**.
- If c is a zero of even multiplicity, then the graph *touches* the x-axis at c and turns back around (does not cross the x-axis). The point $(c, 0)$ is called a **touch point**.

To illustrate the behavior of a polynomial function at its real zeros, consider the graph of $f(x) = -x(x - 4)^2$ from Example 3 (Figure 7-11).

- 0 has a multiplicity of 1 (odd multiplicity). The graph *crosses* the x-axis at $(0, 0)$.
- 4 has a multiplicity of 2 (even multiplicity). The graph *touches* the x-axis at $(4, 0)$ and turns back around.

EXAMPLE 4   Determining Zeros and Multiplicities

Determine the zeros and their multiplicities for the given functions.

a. $m(x) = \dfrac{1}{10}(x - 4)^2(2x + 5)^3$ **b.** $n(x) = x^4 - 2x^2$

Solution:

even   odd

a. $m(x) = \dfrac{1}{10}(x - 4)^2(2x + 5)^3$ The function is factored into linear factors. The zeros are 4 and $-\frac{5}{2}$.

The function has a zero of 4 with multiplicity 2 (even). The graph has a touch point at $(4, 0)$.

   The function has a zero of $-\frac{5}{2}$ with multiplicity 3 (odd). The graph has a cross point at $\left(-\frac{5}{2}, 0\right)$.

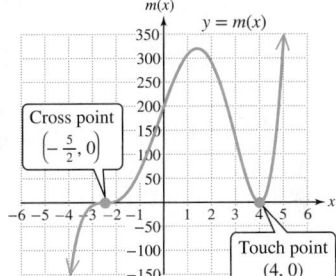

b. $n(x) = x^4 - 2x^2$
$= x^2(x^2 - 2)$
$= x^2\left(x - \sqrt{2}\right)^1\left(x + \sqrt{2}\right)^1$

The function has a zero of 0 with multiplicity 2 (even). The graph has a touch point at $(0, 0)$.

   The function has a zero of $\sqrt{2}$ with multiplicity 1 (odd). The graph has a cross point at $\left(\sqrt{2}, 0\right) \approx (1.41, 0)$.

   The function has a zero of $-\sqrt{2}$ with multiplicity 1 (odd). The graph has a cross point at $\left(-\sqrt{2}, 0\right) \approx (-1.41, 0)$.

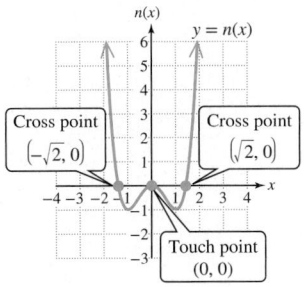

Answers

4. a. -3 (multiplicity 4) and
$\dfrac{1}{5}$ (multiplicity 5)

b. 0 (multiplicity 4),
$\sqrt{7}$ (multiplicity 1), and
$-\sqrt{7}$ (multiplicity 1)

Skill Practice 4   Determine the zeros and their multiplicities for the given functions.

a. $p(x) = -\dfrac{3}{5}(x + 3)^4(5x - 1)^5$ **b.** $q(x) = 2x^6 - 14x^4$

3. Apply the Intermediate Value Theorem

In Examples 2–4, the zeros of the functions were easily identified by first factoring the polynomial. However, in most cases, the real zeros of a polynomial are difficult or impossible to determine algebraically. For example, the function given by $f(x) = x^4 + 6x^3 - 26x + 15$ has zeros of $-1 \pm \sqrt{6}$ and $-2 \pm \sqrt{7}$. At this point, we do not have the tools to find the zeros of this function analytically. However, we can use the intermediate value theorem to help us search for zeros of a polynomial function and approximate their values.

> ### Intermediate Value Theorem
>
> Let f be a polynomial function. For $a < b$, if $f(a)$ and $f(b)$ have opposite signs, then f has at least one zero on the interval $[a, b]$.

EXAMPLE 5   Applying the Intermediate Value Theorem

Show that $f(x) = x^4 + 6x^3 - 26x + 15$ has a zero on the interval $[1, 2]$.

Solution:

$$f(x) = x^4 + 6x^3 - 26x + 15$$
$$f(1) = (1)^4 + 6(1)^3 - 26(1) + 15 = -4$$
$$f(2) = (2)^4 + 6(2)^3 - 26(2) + 15 = 27$$

$f(1)$ and $f(2)$ have opposite signs. Therefore, by the intermediate value theorem, we know that the function must have at least one zero on the interval $[1, 2]$.

The actual value of the zero on the interval $[1, 2]$ is $-1 + \sqrt{6} \approx 1.45$.

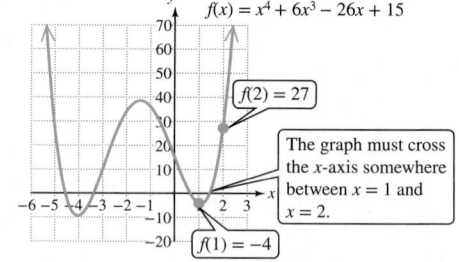

$f(x) = x^4 + 6x^3 - 26x + 15$

$f(2) = 27$

The graph must cross the x-axis somewhere between $x = 1$ and $x = 2$.

$f(1) = -4$

> **Skill Practice 5**   Show that $f(x) = x^4 + 6x^3 - 26x + 15$ has a zero on the interval $[-4, -3]$.

It is important to note the limitations of the intermediate value theorem. Consider a polynomial function defined by $y = f(x)$ on the interval $[a, b]$.

- If $f(a)$ and $f(b)$ have opposite signs, then f must have at least one zero on the interval $[a, b]$ because the graph must cross the x-axis. This includes the possibility that f may have more than one zero on $[a, b]$. In Figure 7-12, there are three zeros between a and b.

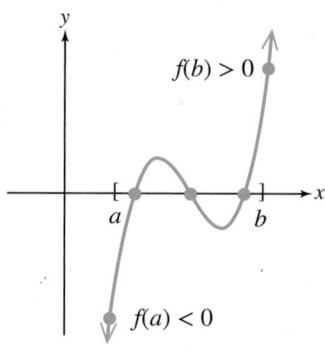

$f(b) > 0$

$f(a) < 0$

Figure 7-12

Answer

5. $f(-4) = -9$ and $f(-3) = 12$. $f(-4)$ and $f(-3)$ have opposite signs. Therefore, the intermediate value theorem guarantees the existence of at least one zero on the interval $[-4, -3]$.

- If the signs of $f(a)$ and $f(b)$ are the same, then the intermediate value theorem is inconclusive. That is, f may or may not have a zero on the interval $[a, b]$. In Figure 7-13, $f(a)$ and $f(b)$ have the same sign (both positive), and the function has no zeros on the interval $[a, b]$. In contrast, in Figure 7-14, $f(a)$ and $f(b)$ have the same sign (both positive), yet there are two zeros on the interval $[a, b]$.

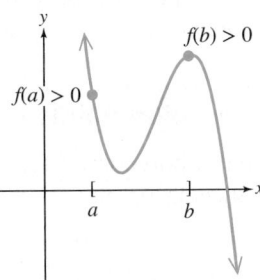

Figure 7-13

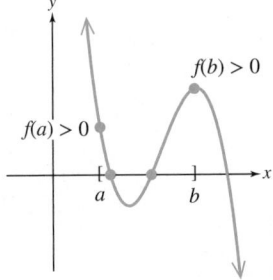

Figure 7-14

Point of Interest

The intermediate value theorem can be used repeatedly in a technique called the *bisection method* to approximate the value of a zero. The bisection method utilizes a series of smaller and smaller intervals $[a, b]$ to converge on the location of a zero. This is illustrated in the online group activity "Investigating the Bisection Method for Finding Zeros."

Graphing calculators and computers can quickly carry out the calculations involved in the bisection method to find the zeros of a polynomial. However, in the days before electronic computing, the word "computer" referred to a person who did such calculations using paper and pencil. "Human computers" were notably used in the eighteenth century to predict the path of Halley's comet and to produce astronomical tables critical to surveying and navigation. Later, during World Wars I and II, human computers developed ballistic firing tables that would describe the trajectory of a shell.

Computing tables of values was very time consuming, and the "computers" would often interpolate to find intermediate values within a table. Interpolation is a method by which intermediate values between two numbers are estimated. Often the interpolated values were based on a polynomial function.

Source: National Aeronautics and Space Administration

4. Sketch a Polynomial Function

TIP Even with advanced techniques from calculus or the use of a graphing utility, it is often difficult or impossible to find the exact location of the turning points of a polynomial function.

The graph of a polynomial function may also have "turning points." These correspond to relative maxima and minima. For example, consider $f(x) = x(x + 2)(x - 2)^2$. See Figure 7-15.

Multiplying the leading terms within the factors, the leading term of the polynomial is $(x)(x)(x)^2 = x^4$. Therefore, the end behavior of the graph is up to the left and up to the right.

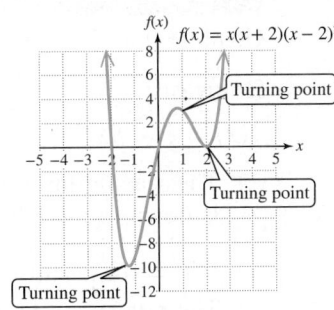

Figure 7-15

Avoiding Mistakes

A polynomial of degree n may have fewer than $n - 1$ turning points. For example, $f(x) = x^3$ is a degree 3 polynomial function (indicating that it could have a maximum of two turning points), yet the graph has no turning points.

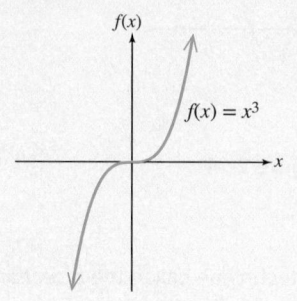

Starting from the far left, the graph of f decreases to the x-intercept of -2. Since -2 is a zero with an odd multiplicity, the graph must cross the x-axis at -2. For the same reason, the graph must cross the x-axis again at the origin. Therefore, somewhere between $x = -2$ and $x = 0$, the graph must "turn around." This point is called a "turning point."

The turning points of a polynomial function are the points where the function changes from increasing to decreasing or vice versa.

> **Number of Turning Points of a Polynomial Function**
>
> Let f represent a polynomial function of degree n. Then the graph of f has at most $n - 1$ turning points.

At this point we are ready to outline a strategy for sketching a polynomial function.

> **Graphing a Polynomial Function**
>
> To graph a polynomial function defined by $y = f(x)$,
>
> 1. Use the leading term to determine the end behavior of the graph.
> 2. Determine the y-intercept by evaluating $f(0)$.
> 3. Determine the real zeros of f and their multiplicities (these are the x-intercepts of the graph of f).
> 4. Plot the x- and y-intercepts and sketch the end behavior.
> 5. Draw a sketch starting from the left-end behavior. Connect the x- and y-intercepts in the order that they appear from left to right using these rules:
>   - The curve will cross the x-axis at an x-intercept if the corresponding zero has an odd multiplicity.
>   - The curve will touch but not cross the x-axis at an x-intercept if the corresponding zero has an even multiplicity.
> 6. If a test for symmetry is easy to apply, use symmetry to plot additional points. Recall that
>   - f is an even function (symmetric to the y-axis) if $f(-x) = f(x)$.
>   - f is an odd function (symmetric to the origin) if $f(-x) = -f(x)$.
> 7. Plot more points if a greater level of accuracy is desired. In particular, to estimate the location of turning points, find several points between two consecutive x-intercepts.

In Examples 6 and 7, we demonstrate the process of graphing a polynomial function.

EXAMPLE 6   Graphing a Polynomial Function

Graph $f(x) = x^3 - 9x$.

Solution:

$f(x) = x^3 - 9x$

1. The leading term is x^3. The end behavior is down to the left and up to the right.

   The exponent on the leading term is odd and the leading coefficient is positive.

2. $f(0) = (0)^3 - 9(0) = 0$

   The y-intercept is $(0, 0)$.

   Determine the y-intercept by evaluating $f(0)$.

3. $0 = x^3 - 9x$
$0 = x(x^2 - 9)$
$0 = x(x - 3)(x + 3)$

The zeros of the function are 0, 3, and −3, and each has a multiplicity of 1.

Find the real zeros of f by solving for the real solutions to the equation $f(x) = 0$.

The zeros are real numbers and correspond to x-intercepts on the graph. Since the multiplicity of each zero is an odd number, the graph will cross the x-axis at the zeros.

4.

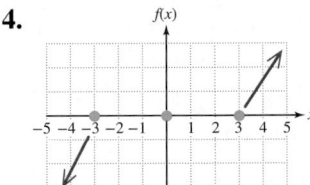

Plot the x- and y-intercepts and sketch the end behavior.

5. Moving from left to right, the curve increases from the far left and then crosses the x-axis at −3. The graph must have a turning point between $x = -3$ and $x = 0$ so that the curve can pass through the next x-intercept of (0, 0).

The graph crosses the x-axis at $x = 0$. The graph must then have another turning point between $x = 0$ and $x = 3$ so that the curve can pass through the next x-intercept of (3, 0). Finally, the graph crosses the x-axis at $x = 3$ and continues to increase to the far right.

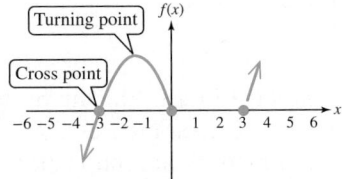

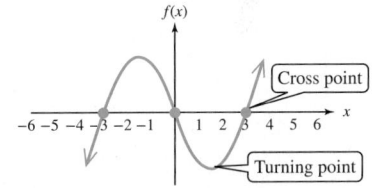

6. $f(x) = x^3 - 9x$
$f(-x) = (-x)^3 - 9(-x)$ $-f(x) = -(x^3 - 9x)$
$\qquad = -x^3 + 9x$ ⟷ $\qquad = -x^3 + 9x$
$\qquad\qquad\qquad f(-x) = -f(x)$
$\qquad\qquad\qquad\text{(same)}$

Testing for symmetry, we see that $f(-x) = -f(x)$. Therefore, f is an odd function and is symmetric with respect to the origin.

TIP Techniques of calculus can be used to find the exact coordinates of the turning points of the polynomial function in Example 6.

7. If more accuracy is desired, plot additional points. In this case, since f is symmetric to the origin, if a point (x, y) is on the graph, then so is $(-x, -y)$. The graph of f is shown in Figure 7-16.

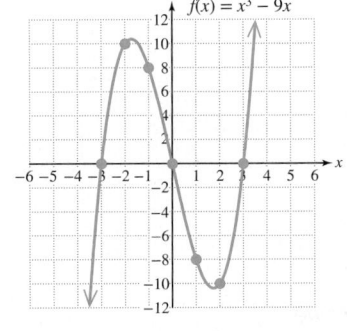

x	$f(x)$
1	−8
2	−10
4	28

Use symmetry. →

x	$f(x)$
−1	8
−2	10
−4	−28

Figure 7-16

Answer
6.

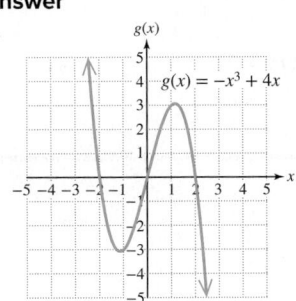

Skill Practice 6   Graph $g(x) = -x^3 + 4x$.

EXAMPLE 7 Graphing a Polynomial Function

Graph $g(x) = -0.1(x - 1)(x + 2)(x - 4)^2$.

Solution:

$g(x) = -0.1(x - 1)(x + 2)(x - 4)^2$

1. Multiplying the leading terms within the factors, we have a leading term of $-0.1(x)(x)(x)^2 = -0.1x^4$. The end behavior is down to the left and down to the right.

   The exponent on the leading term is even and the leading coefficient is negative.

2. $g(0) = -0.1(0 - 1)(0 + 2)(0 - 4)^2 = 3.2$
   The y-intercept is $(0, 3.2)$.

   Determine the y-intercept by evaluating $g(0)$.

3. $0 = -0.1(x - 1)(x + 2)(x - 4)^2$
   The zeros of the function are 1, -2, and 4.
   The multiplicity of 1 is 1.
   The multiplicity of -2 is 1.
   The multiplicity of 4 is 2.

   Find the real zeros of g by solving for the real solutions of the equation $g(x) = 0$.

   The zeros are real numbers and correspond to x-intercepts on the graph: $(1, 0)$, $(-2, 0)$, and $(4, 0)$.

4.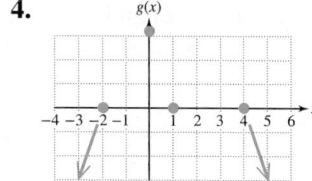

   Plot the x- and y-intercepts and sketch the end behavior.

5. Moving from left to right, the curve increases from the far left. It then crosses the x-axis at $x = -2$ and turns back around to pass through the next x-intercept at $x = 1$.

   The curve has another turning point between $x = 1$ and $x = 4$ so that it can touch the x-axis at 4. From there it turns back downward and continues to decrease to the far right.

   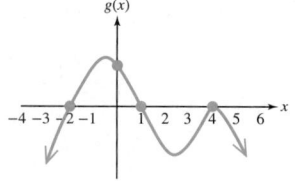

6. From our preliminary sketch in step 5, we see that the function is not symmetric with respect to either the y-axis or the origin.

7. If more accuracy is desired, plot additional points. The graph is shown in Figure 7-17.

x	$g(x)$
-3	-19.6
-1	5
2	-1.6
3	-1
5	-2.8

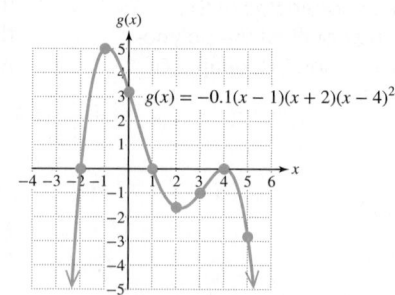

Figure 7-17

Skill Practice 7 Graph $h(x) = 0.5x(x - 1)(x + 3)^2$.

Answer

7.

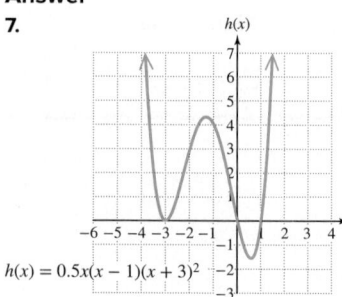

$h(x) = 0.5x(x - 1)(x + 3)^2$

TECHNOLOGY CONNECTIONS

Using a Graphing Utility to Graph a Polynomial Function

It is important to have a strong knowledge of algebra to use a graphing utility effectively. For example, consider the graph of $f(x) = 0.005(x - 2)(x + 3)(x - 5)(x + 15)$ on the standard viewing window.

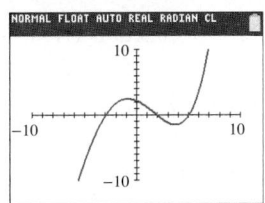

From the leading term, $0.005x^4$, we know that the end behavior should be up to the left and up to the right. Furthermore, the function has four real zeros $(2, -3, 5,$ and $-15)$, and should have four corresponding x-intercepts. Therefore, on the standard viewing window, the calculator does not show the key features of the graph.

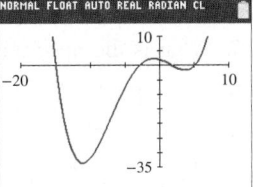

By graphing f on the window $[-20, 10, 2]$ by $[-35, 10, 5]$, we see the end behavior displayed correctly, all four x-intercepts, and the turning points (there should be at most 3).

SECTION 7.2   Practice Exercises

Prerequisite Review

For Exercises R.1–R.2, write the polynomial in descending order. Then identify the leading term and leading coefficient. (See Section 2.2 for review.)

R.1. $9x^2 - 4x^3 - 2x^5 + 7$

R.2. $5x + x^7 + 16x^2$

For Exercises R.3–R.6, solve the equation. (See Sections 3.4 and 4.7 for review.)

R.3. $3x^3 + 21x^2 - 54x = 0$ (See Section 3.4.)

R.4. $5x^3 + 6x^2 - 20x - 24 = 0$

R.5. $x^4 - 18x^2 + 32 = 0$ (See Section 4.7.)

R.6. $x^4 - 12x^2 + 27 = 0$

For Exercise R.7–R.8, find the x- and y-intercepts of the function defined by $y = f(x)$. (See Section 5.3 for review.)

R.7. $f(x) = (3x - 5)(x + 7)$

R.8. $f(x) = (x - 4)(2x + 11)$

For Exercises R.9–R.10, evaluate the expression for the given values of the variable. (See Section R.3 for review.)

R.9. $4x^3 - 5x^2 + x - 3$
    a. For $x = 1$
    b. For $x = 2$

R.10. $-2x^4 + 3x^3 - x + 1$
    a. For $x = -2$
    b. For $x = 0$

For Exercises R.11–R.16, determine whether the function is even, odd, or neither. (See Section 6.2 for review.)

R.11. $f(x) = -3x^4 + 5x^2 + 7$
R.12. $g(x) = 2x^6 - 5x^4 + 7x^2$
R.13. $h(x) = 4x^3 + 6x$

R.14. $k(x) = -9x^7 + 6x^3 + x$
R.15. $m(x) = 9x^2 + 3x - 5$
R.16. $n(x) = 4x^3 + 6x^2 + 8x + 2$

R.17. Given a function defined by $y = f(x)$, if f is an even function then the graph of f is symmetric with respect to the

_____ .

R.18. Given a function defined by $y = f(x)$, if f is an odd function then the graph of f is symmetric with respect to the

_____ .

Concept Connections

1. A function defined by $f(x) = a_n x^n + a_{n-1} x^{n-1} + a_{n-2} x^{n-2} + \cdots + a_1 x + a_0$ where $a_n, a_{n-1}, a_{n-2}, \ldots, a_1, a_0$ are real numbers and $a_n \neq 0$ is called a _____ function.

2. The function given by $f(x) = -3x^5 + \sqrt{2}x + \frac{1}{2}x$ (is/is not) a polynomial function.

3. The function given by $f(x) = -3x^5 + 2\sqrt{x} + \frac{2}{x}$ (is/is not) a polynomial function.

4. A quadratic function is a polynomial function of degree _____.

5. A linear function is a polynomial function of degree _____.

6. The values of x in the domain of a polynomial function f for which $f(x) = 0$ are called the _____ of the function.

7. What is the maximum number of turning points of the graph of $f(x) = -3x^6 - 4x^5 - 5x^4 + 2x^2 + 6$?

8. If the graph of a polynomial function has 3 turning points, what is the minimum degree of the function?

9. If c is a real zero of a polynomial function and the multiplicity is 3, does the graph of the function cross the x-axis or touch the x-axis (without crossing) at $(c, 0)$?

10. If c is a real zero of a polynomial function and the multiplicity is 6, does the graph of the function cross the x-axis or touch the x-axis (without crossing) at $(c, 0)$?

11. Suppose that f is a polynomial function and that $a < b$. If $f(a)$ and $f(b)$ have opposite signs, then what conclusion can be drawn from the intermediate value theorem?

12. What is the leading term of $f(x) = -\frac{1}{3}(x - 3)^4(3x + 5)^2$?

Objective 1: Determine the End Behavior of a Polynomial Function

For Exercises 13–20, determine the end behavior of the graph of the function. (**See Example 1**)

13. $f(x) = -3x^4 - 5x^2 + 2x - 6$

14. $g(x) = -\frac{1}{2}x^6 + 8x^4 - x^3 + 9$

15. $h(x) = 12x^5 + 8x^4 - 4x^3 - 8x + 1$

16. $k(x) = 11x^7 - 4x^2 + 9x + 3$

17. $m(x) = -4(x - 2)(2x + 1)^2(x + 6)^4$

18. $n(x) = -2(x + 4)(3x - 1)^3(x + 5)$

19. $p(x) = -2x^2(3 - x)(2x - 3)^3$

20. $q(x) = -5x^4(2 - x)^3(2x + 5)$

Objective 2: Identify Zeros and Multiplicities of Zeros

21. Given the function defined by $g(x) = -3(x - 1)^3(x + 5)^4$, the value 1 is a zero with multiplicity _____, and the value -5 is a zero with multiplicity _____.

22. Given the function defined by $h(x) = \frac{1}{2}x^5(x + 0.6)^3$, the value 0 is a zero with multiplicity _____, and the value -0.6 is a zero with multiplicity _____.

For Exercises 23–38, find the zeros of the function and state the multiplicities. (**See Examples 2–4**)

23. $f(x) = x^3 + 2x^2 - 25x - 50$

24. $g(x) = x^3 + 5x^2 - x - 5$

25. $h(x) = -6x^3 - 9x^2 + 60x$

26. $k(x) = -6x^3 + 26x^2 - 28x$

27. $m(x) = x^5 - 10x^4 + 25x^3$

28. $n(x) = x^6 + 4x^5 + 4x^4$

29. $p(x) = -3x(x + 2)^3(x + 4)$

30. $q(x) = -2x^4(x + 1)^3(x - 2)^2$

31. $t(x) = 5x(3x - 5)(2x + 9)(x - \sqrt{3})(x + \sqrt{3})$

32. $z(x) = 4x(5x - 1)(3x + 8)(x - \sqrt{5})(x + \sqrt{5})$

33. $c(x) = \left[x - (3 - \sqrt{5})\right]\left[x - (3 + \sqrt{5})\right]$

34. $d(x) = \left[x - (2 - \sqrt{11})\right]\left[x - (2 + \sqrt{11})\right]$

35. $f(x) = 4x^4 - 37x^2 + 9$

36. $k(x) = 4x^4 - 65x^2 + 16$

37. $n(x) = x^6 - 7x^4$

38. $m(x) = x^5 - 5x^3$

Objective 3: Apply the Intermediate Value Theorem

For Exercises 39–40, determine whether the intermediate value theorem guarantees that the function has a zero on the given interval. **(See Example 5)**

39. $f(x) = 2x^3 - 7x^2 - 14x + 30$

    a. $[1, 2]$ **b.** $[2, 3]$

    c. $[3, 4]$ **d.** $[4, 5]$

40. $g(x) = 2x^3 - 13x^2 + 18x + 5$

    a. $[1, 2]$ **b.** $[2, 3]$

    c. $[3, 4]$ **d.** $[4, 5]$

For Exercises 41–42, a table of values is given for $Y_1 = f(x)$. Determine whether the intermediate value theorem guarantees that the function has a zero on the given interval.

41. $Y_1 = 21x^4 + 46x^3 - 238x^2 - 506x + 77$

    a. $[-4, -3]$

    b. $[-3, -2]$

    c. $[-2, -1]$

    d. $[-1, 0]$

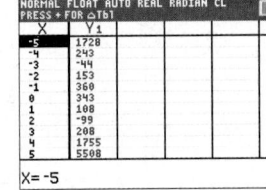

42. $Y_1 = 10x^4 + 21x^3 - 119x^2 - 147x + 343$

    a. $[-4, -3]$

    b. $[-3, -2]$

    c. $[-2, -1]$

    d. $[-1, 0]$

43. Given $f(x) = 4x^3 - 8x^2 - 25x + 50$,

    a. Determine if f has a zero on the interval $[-3, -2]$.

    b. Find a zero of f on the interval $[-3, -2]$.

44. Given $f(x) = 9x^3 - 18x^2 - 100x + 200$,

    a. Determine if f has a zero on the interval $[-4, -3]$.

    b. Find a zero of f on the interval $[-4, -3]$.

Objective 4: Sketch a Polynomial Function

For Exercises 45–52, determine if the graph can represent a polynomial function. If so, assume that the end behavior and all turning points are represented in the graph.

    a. Determine the minimum degree of the polynomial.

    b. Determine whether the leading coefficient is positive or negative based on the end behavior and whether the degree of the polynomial is odd or even.

    c. Approximate the real zeros of the function, and determine if their multiplicities are even or odd.

45.

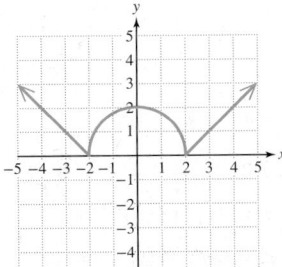

46.

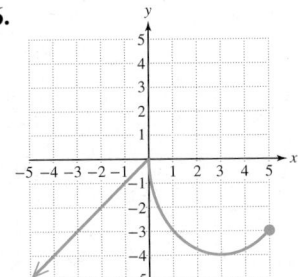

47.

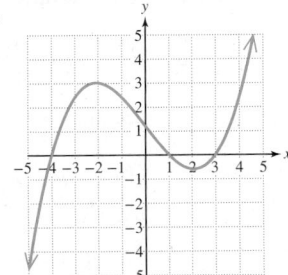

48.

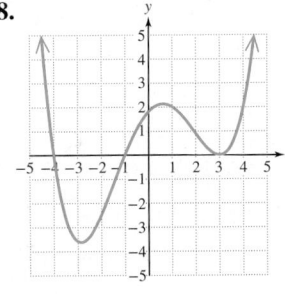

49.

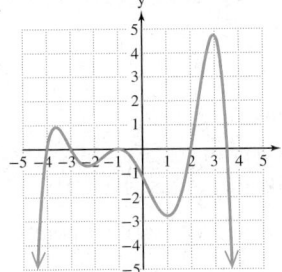

50.

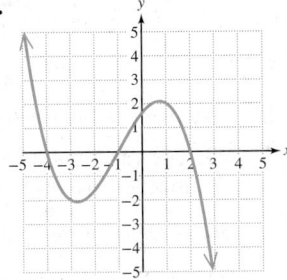

51.

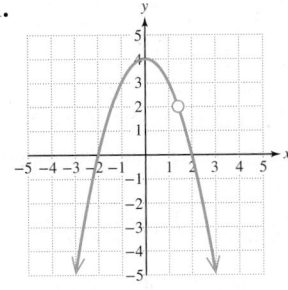

52.

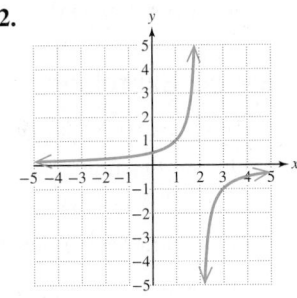

For Exercises 53–58,

   a. Identify the power function of the form $y = x^n$ that is the parent function to the given graph.

   b. In order, outline the transformations that would be required on the graph of $y = x^n$ to make the graph of the given function. See Section 6.1, page 416.

   c. Match the function with the graph of i–vi.

53. $g(x) = -\dfrac{1}{3}x^6 - 2$

54. $f(x) = -\dfrac{1}{2}(x - 3)^4$

55. $k(x) = -(x + 2)^3 + 3$

56. $p(x) = 2(x + 4)^3 - 3$

57. $m(x) = (-x - 3)^5 + 1$

58. $n(x) = (-x + 3)^4 - 1$

i.

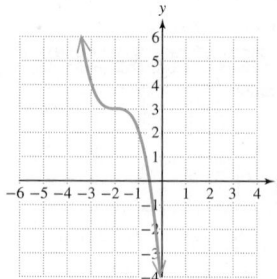

ii.

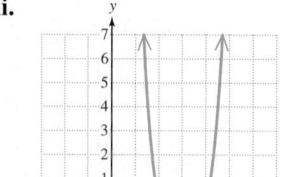

iii.

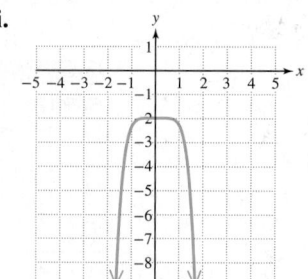

iv.

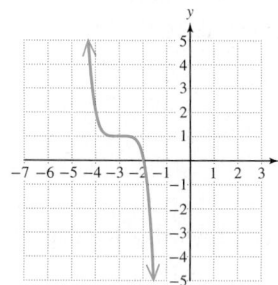

v.

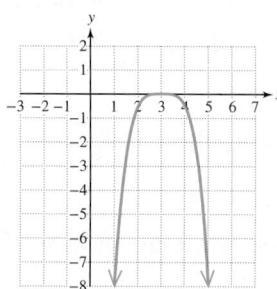

vi.

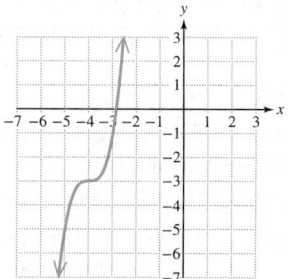

For Exercises 59–76, sketch the function. **(See Examples 6–7)**

59. $f(x) = x^3 - 5x^2$

60. $g(x) = x^5 - 2x^4$

61. $f(x) = \dfrac{1}{2}(x - 2)(x + 1)(x + 3)$

62. $h(x) = \dfrac{1}{4}(x - 1)(x - 4)(x + 2)$

63. $k(x) = x^4 + 2x^3 - 8x^2$

64. $h(x) = x^4 - x^3 - 6x^2$

65. $k(x) = 0.2(x + 2)^2(x - 4)^3$

66. $m(x) = 0.1(x - 3)^2(x + 1)^3$

67. $p(x) = 9x^5 + 9x^4 - 25x^3 - 25x^2$

68. $q(x) = 9x^5 + 18x^4 - 4x^3 - 8x^2$

69. $t(x) = -x^4 + 11x^2 - 28$

70. $v(x) = -x^4 + 15x^2 - 44$

71. $g(x) = -x^4 + 5x^2 - 4$

72. $h(x) = -x^4 + 10x^2 - 9$

73. $c(x) = 0.1x(x - 2)^4(x + 2)^3$

74. $d(x) = 0.05x(x - 2)^4(x + 3)^2$

75. $m(x) = -\dfrac{1}{10}(x + 3)(x - 3)(x + 1)^3$

76. $f(x) = -\dfrac{1}{10}(x - 1)(x + 3)(x - 4)^2$

Mixed Exercises

For Exercises 77–88, determine if the statement is true or false. If a statement is false, explain why.

77. The function defined by $f(x) = (x + 1)^5(x - 5)^2$ crosses the x-axis at 5.

78. The function defined by $g(x) = -3(x + 4)(2x - 3)^4$ touches but does not cross the x-axis at $\left(\frac{3}{2}, 0\right)$.

79. A third-degree polynomial has three turning points.

80. A third-degree polynomial has two turning points.

81. There is more than one polynomial function with zeros of 1, 2, and 6.

82. There is exactly one polynomial with integer coefficients with zeros of 2, 4, and 6.

83. The graph of a polynomial function with leading term of even degree is up to the far left and up to the far right.

84. If c is a real zero of an even polynomial function, then $-c$ is also a zero of the function.

85. The graph of $f(x) = x^3 - 27$ has three x-intercepts.

86. The graph of $f(x) = 3x^2(x - 4)^4$ has no points in Quadrants III or IV.

87. The graph of $p(x) = -5x^4(x + 1)^2$ has no points in Quadrants I or II.

88. A fourth-degree polynomial has exactly two relative minima and two relative maxima.

89. A rocket will carry a communications satellite into low Earth orbit. Suppose that the thrust during the first 200 sec of flight is provided by solid rocket boosters at different points during liftoff.

   The graph shows the acceleration in G-forces (that is, acceleration in 9.8-m/sec^2 increments) versus time after launch.

   a. Approximate the interval(s) over which the acceleration is increasing.

   b. Approximate the interval(s) over which the acceleration is decreasing.

   c. How many turning points does the graph show?

   d. Based on the number of turning points, what is the minimum degree of a polynomial function that could be used to model acceleration versus time? Would the leading coefficient be positive or negative?

   e. Approximate the time when the acceleration was the greatest.

   f. Approximate the value of the maximum acceleration.

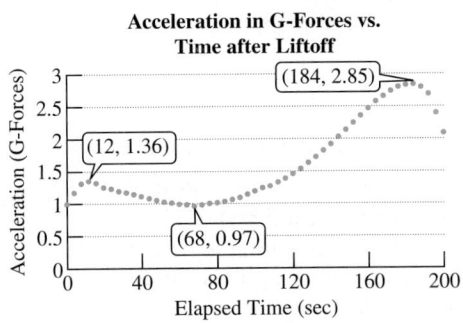

Acceleration in G-Forces vs. Time after Liftoff

90. Data from a 20-year study show the number of new AIDS cases diagnosed among 20- to 24-year-olds in the United States x years after the study began.

   a. Approximate the interval(s) over which the number of new AIDS cases among 20- to 24-year-olds increased.

   b. Approximate the interval(s) over which the number of new AIDS cases among 20- to 24-year-olds decreased.

   c. How many turning points does the graph show?

   d. Based on the number of turning points, what is the minimum degree of a polynomial function that could be used to model the data? Would the leading coefficient be positive or negative?

   e. How many years after the study began was the number of new AIDS cases among 20- to 24-year-olds the greatest?

   f. What was the maximum number of new cases diagnosed in a single year?

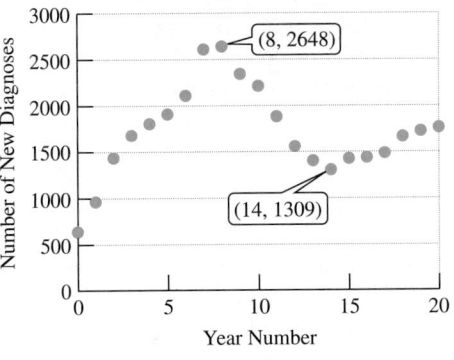

AIDS Diagnoses, 20- to 24-Year-Olds United States, 1985–2005

Write About It

91. Given a polynomial function defined by $y = f(x)$, explain how to find the x-intercepts.

92. Given a polynomial function, explain how to determine whether an x-intercept is a touch point or a cross point.

93. Write an informal explanation of what it means for a function to be continuous.

94. Write an informal explanation of the intermediate value theorem.

Expanding Your Skills

The intermediate value theorem given on page 494 is actually a special case of a broader statement of the theorem. Consider the following:

> Let f be a polynomial function. For $a < b$, if $f(a) \neq f(b)$, then f takes on every value between $f(a)$ and $f(b)$ on the interval $[a, b]$.

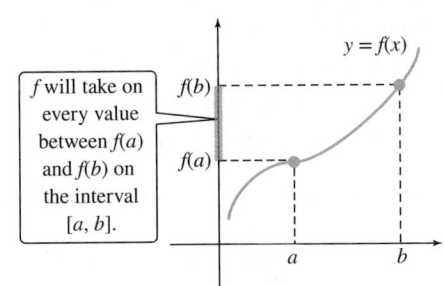

Use this broader statement of the intermediate value theorem for Exercises 95–96.

95. Given $f(x) = x^2 - 3x + 2$,

 a. Evaluate $f(3)$ and $f(4)$.

 b. Use the intermediate value theorem to show that there exists at least one value of x for which $f(x) = 4$ on the interval $[3, 4]$.

 c. Find the value(s) of x for which $f(x) = 4$ on the interval $[3, 4]$.

96. Given $f(x) = -x^2 - 4x + 3$,

 a. Evaluate $f(-4)$ and $f(-3)$.

 b. Use the intermediate value theorem to show that there exists at least one value of x for which $f(x) = 5$ on the interval $[-4, -3]$.

 c. Find the value(s) of x for which $f(x) = 5$ on the interval $[-4, -3]$.

Technology Connections

97. For a certain individual, the volume (in liters) of air in the lungs during a 4.5-sec respiratory cycle is shown in the table for 0.5-sec intervals. Graph the points and then find a third-degree polynomial function to model the volume $V(t)$ for t between 0 sec and 4.5 sec. (*Hint*: Use a CubicReg option or polynomial degree 3 option on a graphing utility.)

Time (sec)	Volume (L)
0.0	0.00
0.5	0.11
1.0	0.29
1.5	0.47
2.0	0.63
2.5	0.76
3.0	0.81
3.5	0.75
4.0	0.56
4.5	0.20

98. The torque (in ft-lb) produced by a certain automobile engine turning at x thousand revolutions per minute is shown in the table. Graph the points and then find a third-degree polynomial function to model the torque $T(x)$ for $1 \leq x \leq 5$.

Engine speed (1000 rpm)	Torque (ft-lb)
1.0	165
1.5	180
2.0	188
2.5	190
3.0	186
3.5	176
4.0	161
4.5	142
5.0	120

99. A solar oven is to be made from an open box with reflective sides. Each box is made from a 30-in. by 24-in. rectangular sheet of aluminum with squares of length x (in inches) removed from each corner. Then the flaps are folded up to form an open box.

 a. Show that the volume of the box is given by

$$V(x) = 4x^3 - 108x^2 + 720x \quad \text{for } 0 < x < 12.$$

 b. Graph the function from part (a) and use a "Maximum" feature on a graphing utility to approximate the length of the sides of the squares that should be removed to maximize the volume. Round to the nearest tenth of an inch.

 c. Approximate the maximum volume. Round to the nearest cubic inch.

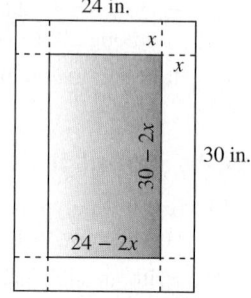

For Exercises 100–101, two viewing windows are given for the graph of $y = f(x)$. Choose the window that best shows the key features of the graph.

100. $f(x) = 2(x - 0.5)(x - 0.1)(x + 0.2)$

 a. $[-10, 10, 1]$ by $[-10, 10, 1]$

 b. $[-1, 1, 0.1]$ by $[-0.05, 0.05, 0.01]$

101. $g(x) = 0.08(x - 16)(x + 2)(x - 3)$

 a. $[-10, 10, 1]$ by $[-10, 10, 1]$

 b. $[-5, 20, 5]$ by $[-50, 30, 10]$

For Exercises 102–103, graph the function defined by $y = f(x)$ on an appropriate viewing window.

102. $k(x) = \dfrac{1}{100}(x - 20)(x + 1)(x + 8)(x - 6)$

103. $p(x) = (x - 0.4)(x + 0.5)(x + 0.1)(x - 0.8)$

SECTION 7.3 Division of Polynomials and the Remainder and Factor Theorems

1. Divide Polynomials Using Long Division

In this section, we use the notation $f(x)$, $g(x)$, and so on to represent polynomials in x. We also present two types of polynomial division: long division and synthetic division. Polynomial division can be used to factor a polynomial, solve a polynomial equation, and find the zeros of a polynomial.

When dividing polynomials, if the divisor has two or more terms we can use a long division process similar to the division of real numbers. This is demonstrated in Examples 1–3.

TIP Take a minute to review long division of whole numbers: $2273 \div 5$

$$
\begin{array}{r}
454 \longleftarrow \text{Quotient} \\
5\overline{)2273} \\
-20\downarrow \\
\overline{27} \\
-25\downarrow \\
\overline{23} \\
-20 \\
\overline{3} \longleftarrow \text{Remainder}
\end{array}
$$

Answer: $454 + \frac{3}{5}$ or $454\frac{3}{5}$

EXAMPLE 1   Dividing Polynomials Using Long Division

Use long division to divide. $(6x^3 - 5x^2 - 3) \div (3x + 2)$

Solution:

First note that the dividend can be written as $6x^3 - 5x^2 + 0x - 3$. The term $0x$ is used as a place holder for the missing power of x. The place holder is helpful to keep the powers of x lined up. We also set up long division with both the dividend and divisor written in descending order.

$$3x + 2\,\overline{)\,6x^3 - 5x^2 + 0x - 3}$$

Divide the leading term in the dividend by the leading term in the divisor.

$$
\begin{array}{r}
2x^2 \\
3x + 2\,\overline{)\,6x^3 - 5x^2 + 0x - 3} \\
-(6x^3 + 4x^2)
\end{array}
$$

$\dfrac{6x^3}{3x} = 2x^2$ This is the first term in the quotient.

Multiply the divisor by $2x^2$:
$2x^2(3x + 2) = 6x^3 + 4x^2$, and subtract the result.

Subtract.

$$
\begin{array}{r}
2x^2 \\
3x + 2\,\overline{)\,6x^3 - 5x^2 + 0x - 3} \\
-(6x^3 + 4x^2) \\
\overline{-9x^2 + 0x}
\end{array}
$$

Bring down the next term from the dividend and repeat the process.

$$
\begin{array}{r}
2x^2 - 3x \\
3x + 2\,\overline{)\,6x^3 - 5x^2 + 0x - 3} \\
-(6x^3 + 4x^2) \\
\overline{-9x^2 + 0x} \\
-(-9x^2 - 6x)
\end{array}
$$

Divide $-9x^2$ by the first term in the divisor.

$\dfrac{-9x^2}{3x} = -3x$

Multiply the divisor by $-3x$:
$-3x(3x + 2) = -9x^2 - 6x$, and subtract the result.

Subtract.

$$
\begin{array}{r}
2x^2 - 3x + 2 \\
3x + 2\,\overline{)\,6x^3 - 5x^2 + 0x - 3} \\
-(6x^3 + 4x^2) \\
\overline{-9x^2 + 0x} \\
-(-9x^2 - 6x) \\
\overline{6x - 3} \\
-(6x + 4) \\
\overline{-7}
\end{array}
$$

Bring down the next term from the dividend and repeat the process.

Divide $6x$ by the first term in the divisor.
$\frac{6x}{3x} = 2$. This is the next term in the quotient.

Multiply the divisor by 2: $2(3x + 2) = 6x + 4$, and subtract the result.

The remainder is -7.

Long division is complete when the remainder is either zero or has degree less than the degree of the divisor.

The quotient is $2x^2 - 3x + 2$.

The remainder is -7.

The divisor is $3x + 2$.

The dividend is $6x^3 - 5x^2 - 3$.

The result of a long division problem is usually written as the quotient plus the remainder divided by the divisor.

$$\boxed{\text{Dividend}} \searrow \underset{\boxed{\text{Divisor}} \nearrow}{\frac{6x^3 - 5x^2 - 3}{3x + 2}} = 2x^2 - 3x + 2 \overset{\boxed{\text{Quotient}}}{} + \frac{-7}{3x + 2} \overset{\boxed{\text{Remainder}}}{}_{\boxed{\text{Divisor}}}$$

Skill Practice 1   Use long division to divide $(4x^3 - 23x + 3) \div (2x - 5)$.

FOR REVIEW

The operations of multiplying, adding, and subtracting polynomials are covered in Section 2.2.

By clearing fractions, the result of Example 1 can be checked by multiplication.

$$\text{Dividend} = (\text{Divisor})(\text{Quotient}) \quad + \text{Remainder}$$
$$6x^3 - 5x^2 - 3 \overset{?}{=} (3x + 2)(2x^2 - 3x + 2) + (-7)$$

$$\overset{?}{=} (3x + 2)(2x^2 - 3x + 2) + (-7)$$

$$\overset{?}{=} 6x^3 - 9x^2 + 6x + 4x^2 - 6x + 4 + (-7)$$
$$\overset{?}{=} 6x^3 - 5x^2 + 4 + (-7)$$
$$\overset{?}{=} 6x^3 - 5x^2 - 3 \checkmark$$

This result illustrates the division algorithm.

Division Algorithm

Suppose that $f(x)$ and $d(x)$ are polynomials where $d(x) \neq 0$ and the degree of $d(x)$ is less than or equal to the degree of $f(x)$. Then there exist unique polynomials $q(x)$ and $r(x)$ such that

$$f(x) = d(x) \cdot q(x) + r(x)$$

where either the degree of $r(x)$ is less than $d(x)$, or $r(x)$ is the zero polynomial.

Note: The polynomial $f(x)$ is the **dividend**, $d(x)$ is the **divisor**, $q(x)$ is the **quotient**, and $r(x)$ is the **remainder**.

EXAMPLE 2   **Dividing Polynomials Using Long Division**

Use long division to divide $(-5 + x + 4x^2 + 2x^3 + 3x^4) \div (x^2 + 2)$.

Solution:

Write the dividend and divisor in descending order and insert placeholders for missing powers of x. $(3x^4 + 2x^3 + 4x^2 + x - 5) \div (x^2 + 0x + 2)$

Answer

1. $2x^2 + 5x + 1 + \dfrac{8}{2x - 5}$

$$3x^2 + 2x - 2$$
$$x^2 + 0x + 2 \overline{) 3x^4 + 2x^3 + 4x^2 + x - 5}$$
$$\underline{-(3x^4 + 0x^3 + 6x^2)}$$
$$2x^3 - 2x^2 + x$$
$$\underline{-(2x^3 + 0x^2 + 4x)}$$
$$-2x^2 - 3x - 5$$
$$\underline{-(-2x^2 + 0x - 4)}$$
$$-3x - 1$$

To begin, divide the leading term in the dividend by the leading term in the divisor.

$$\frac{3x^4}{x^2} = 3x^2$$

Multiply the divisor by $3x^2$ and subtract the result.

Bring down the next term from the dividend and repeat the process.

The process is complete when the remainder is either 0 or has degree less than the degree of the divisor.

The result is $3x^2 + 2x - 2 + \dfrac{-3x - 1}{x^2 + 2}$.

Check by using the division algorithm.

$$3x^4 + 2x^3 + 4x^2 + x - 5 \stackrel{?}{=} (x^2 + 2)(3x^2 + 2x - 2) + (-3x - 1)$$
$$\stackrel{?}{=} 3x^4 + 2x^3 - 2x^2 + 6x^2 + 4x - 4 + (-3x - 1)$$
$$\stackrel{?}{=} 3x^4 + 2x^3 + 4x^2 + x - 5 \checkmark$$

Skill Practice 2   Use long division to divide.

$(1 - 7x + 5x^2 - 3x^3 + 2x^4) \div (x^2 + 3)$

In Example 3, we discuss the implications of obtaining a remainder of zero when performing division of polynomials.

EXAMPLE 3   **Dividing Polynomials Using Long Division**

Use long division to divide. $\dfrac{2x^2 + 3x - 14}{x - 2}$

Solution:

$$2x + 7$$
$$x - 2 \overline{) 2x^2 + 3x - 14}$$
$$\underline{-(2x^2 - 4x)}$$
$$7x - 14$$
$$\underline{-(7x - 14)}$$
$$0$$

$$\frac{2x^2 + 3x - 14}{x - 2} = 2x + 7$$

To begin, divide the leading term in the dividend by the leading term in the divisor.

$$\frac{2x^2}{x} = 2x$$

Multiply the divisor by $2x$ and subtract the result.

Bring down the next term from the dividend and repeat the process.

The process is complete when the remainder is either 0 or has degree less than the degree of the divisor.

The remainder is zero. This implies that the divisor divides evenly into the dividend. Therefore, both the divisor and quotient are factors of the dividend. This is easily verified by the division algorithm.

| Dividend | Divisor | Quotient | Remainder |

$$2x^2 + 3x - 14 \stackrel{?}{=} (x - 2)(2x + 7) + 0$$
$$\stackrel{?}{=} (x - 2)(2x + 7)$$

Factored form of $2x^2 + 3x - 14$

Answer

2. $2x^2 - 3x - 1 + \dfrac{2x + 4}{x^2 + 3}$

Skill Practice 3   Use long division to divide.

$(3x^2 - 14x + 15) \div (x - 3)$

2. Divide Polynomials Using Synthetic Division

When dividing polynomials where the divisor is a binomial of the form $(x - c)$ and c is a constant, we can use synthetic division. Synthetic division enables us to find the quotient and remainder more quickly than long division. It uses an algorithm that manipulates the coefficients of the dividend, divisor, and quotient without the accompanying variable factors.

Consider the division of polynomials from Example 3 with the equivalent synthetic division shown to its right. Notice that the same coefficients are used in both cases.

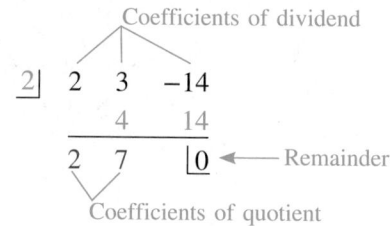

$$
\begin{array}{r}
2x + 7 \\
x - 2 \overline{\smash{)}\, 2x^2 + 3x - 14} \\
\underline{-(2x^2 - 4x)} \\
7x - 14 \\
\underline{-(7x - 14)} \\
0
\end{array}
$$

In Example 4, we demonstrate the process to divide polynomials by synthetic division.

EXAMPLE 4   Dividing Polynomials Using Synthetic Division

Use synthetic division to divide. $(-10x^2 + 2x^3 - 5) \div (x - 4)$

Solution:

As with long division, the terms of the dividend and divisor must be written in descending order with placeholders for missing powers of x.

$$(2x^3 - 10x^2 + 0x - 5) \div (x - 4)$$

To use synthetic division, the divisor must be in the form $x - c$. In this case, $c = 4$.

Avoiding Mistakes

It is important to check that the divisor is in the form $(x - c)$ before applying synthetic division. The variable x in the divisor must be of first degree, and its coefficient must be 1.

Step 1: Write the value of c in a box.

Step 2: Write the coefficients of the dividend to the right of the box.

$$
\begin{array}{c|cccc}
4 & 2 & -10 & 0 & -5 \\
& & & & \\
\hline
& 2 & & &
\end{array}
$$

Step 3: Skip a line and draw a horizontal line below the list of coefficients.

Step 4: Bring down the leading coefficient from the dividend and write it below the line.

Step 5: Multiply the value of c by the number below the line $(4 \times 2 = 8)$. Write the result in the next column above the line.

$$
\begin{array}{c|cccc}
4 & 2 & -10 & 0 & -5 \\
& & 8 & & \\
\hline
& 2 & -2 & &
\end{array}
$$

Step 6: Add the numbers in the column above the line $(-10 + 8 = -2)$, and write the result below the line.

Repeat steps 5 and 6 until all columns have been completed.

$$
\begin{array}{c|cccc}
4 & 2 & -10 & 0 & -5 \\
& & 8 & -8 & -32 \\
\hline
& 2 & -2 & -8 & \boxed{-37} \\
& \downarrow & \downarrow & \downarrow & \\
& x^2 & x & \text{constant} &
\end{array}
$$

A box is often drawn around the remainder.

Answer

3. $3x - 5$

The rightmost number below the line is the remainder. The other numbers below the line are the coefficients of the quotient in order by the degree of the term.

Since the divisor is linear (first degree), the degree of the quotient is 1 less than the degree of the dividend. In this case, the dividend is of degree 3. Therefore, the quotient will be of degree 2.

The quotient is $2x^2 - 2x - 8$ and the remainder is -37. Therefore,

$$\frac{2x^3 - 10x^2 - 5}{x - 4} = 2x^2 - 2x - 8 + \frac{-37}{x - 4}$$

Skill Practice 4   Use synthetic division to divide.

$(4x^3 - 28x - 7) \div (x - 3)$

EXAMPLE 5   **Dividing Polynomials Using Synthetic Division**

Use synthetic division to divide. $(-2x + 4x^3 + 18 + x^4) \div (x + 2)$

Solution:

TIP   Given a divisor of the form $(x - c)$, we can determine the value of c by setting the divisor equal to zero and solving for x. In Example 5, we have $x + 2 = 0$, which implies that $x = -2$. The value of c is -2.

Write the dividend and divisor in descending order and insert placeholders for missing powers of x. $(x^4 + 4x^3 + 0x^2 - 2x + 18) \div (x + 2)$

To use synthetic division, the divisor must be of the form $x - c$. In this case, we have $x + 2 = x - (-2)$. Therefore, $c = -2$.

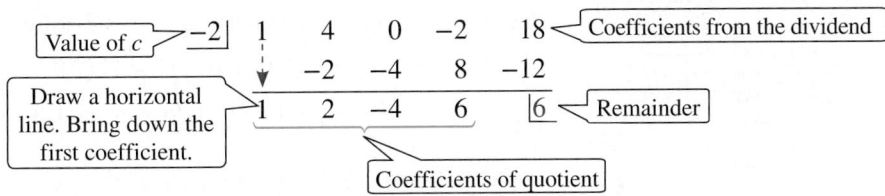

The dividend is a fourth-degree polynomial and the divisor is a first-degree polynomial. Therefore, the quotient is a third-degree polynomial. The coefficients of the quotient are found below the line: 1, 2, −4, 6. The quotient is $x^3 + 2x^2 - 4x + 6$, and the remainder is 6.

$$\frac{x^4 + 4x^3 - 2x + 18}{x + 2} = x^3 + 2x^2 - 4x + 6 + \frac{6}{x + 2}$$

Skill Practice 5   Use synthetic division to divide.

$(-3x + 7x^3 + 5 + 2x^4) \div (x + 1)$

---

3. Apply the Remainder and Factor Theorems

Consider the special case of the division algorithm where $f(x)$ is the dividend and $(x - c)$ is the divisor.

$$f(x) = (x - c) \cdot q(x) + r$$

The remainder r is constant because its degree must be one less than the degree of $x - c$.

Now evaluate $f(c)$: $f(c) = (c - c) \cdot q(c) + r$

$f(c) = 0 \cdot q(c) + r$

$f(c) = r$

This result is stated formally as the remainder theorem.

Answers

4. $4x^2 + 12x + 8 + \dfrac{17}{x - 3}$

5. $2x^3 + 5x^2 - 5x + 2 + \dfrac{3}{x + 1}$

> ### Remainder Theorem
>
> If a polynomial $f(x)$ is divided by $x - c$, then the remainder is $f(c)$.
>
> *Note*: The remainder theorem tells us that the value of $f(c)$ is the same as the remainder we get from dividing $f(x)$ by $x - c$.

EXAMPLE 6 Using the Remainder Theorem to Evaluate a Polynomial

Given $f(x) = x^4 + 6x^3 - 12x^2 - 30x + 35$, use the remainder theorem to evaluate

a. $f(2)$ **b.** $f(-7)$

Solution:

a. If $f(x)$ is divided by $x - 2$, then the remainder is $f(2)$.

$$
\begin{array}{r|rrrrr}
2 & 1 & 6 & -12 & -30 & 35 \\
  &   & 2 & 16 & 8   & -44 \\
\hline
  & 1 & 8 & 4   & -22 & \underline{-9}
\end{array}
$$

By the remainder theorem, $f(2) = -9$.

b. If $f(x)$ is divided by $x - (-7)$ or equivalently $x + 7$, then the remainder is $f(-7)$.

$$
\begin{array}{r|rrrrr}
-7 & 1 & 6 & -12 & -30 & 35 \\
   &   & -7 & 7   & 35 & -35 \\
\hline
   & 1 & -1 & -5 & 5   & \underline{0}
\end{array}
$$

By the remainder theorem, $f(-7) = 0$.

The results can be checked by direct substitution.

$$f(2) = (2)^4 + 6(2)^3 - 12(2)^2 - 30(2) + 35 = -9 \checkmark$$
$$f(-7) = (-7)^4 + 6(-7)^3 - 12(-7)^2 - 30(-7) + 35 = 0 \checkmark$$

TIP From Example 6, the values $f(2) = -9$ and $f(-7) = 0$, imply that $(2, -9)$ and $(-7, 0)$ are on the graph of $y = f(x)$.

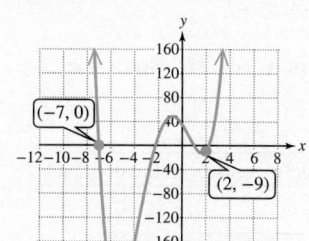

Skill Practice 6 Given $f(x) = x^4 + x^3 - 6x^2 - 5x - 15$, use the remainder theorem to evaluate

a. $f(5)$ **b.** $f(-3)$

FOR REVIEW

Recall that the imaginary number i is defined as $i = \sqrt{-1}$, and thus, $i^2 = -1$.

A **complex number** is of the form $a + bi$, where a and b are real numbers. The value a is called the **real part** and b is called the **imaginary part**. All real numbers are included in the complex numbers. For example, the real number 4 can be written as $4 + 0i$. See Section 3.3 for review.

The division algorithm and remainder theorem can be extended over the set of complex numbers. The definition of a polynomial was given in Section 7.2.

$$f(x) = a_n x^n + a_{n-1}x^{n-1} + a_{n-2}x^{n-2} + \cdots + a_1 x + a_0$$

where $a_n \neq 0$ and the coefficients $a_n, a_{n-1}, a_{n-2}, \ldots, a_0$ are real numbers. We now extend our discussion to **complex polynomials**. These are polynomials with complex coefficients, which include polynomials with real coefficients and polynomials with imaginary coefficients. For example, the following are complex polynomials.

$f(x) = (2 + 3i)x^2 + 4i$ $(2 + 3i)$ and $4i$ are imaginary numbers.

$g(x) = \sqrt{2}x^3 + 3x - 5i$ $\sqrt{2}$ and 3 are real numbers and $-5i$ is an imaginary number.

$h(x) = 2x^4 - 5x^3 - 7x + 1$ 2, −5, −7, and 1 are real numbers.

We will evaluate polynomials over the set of complex numbers rather than restricting x to the set of real numbers. A complex number $a + bi$ is a zero of a

Answers
6. a. 560   **b.** 0

polynomial $f(x)$ if $f(a + bi) = 0$. For example, given $f(x) = x - (5 + 2i)$, we see that the imaginary number $5 + 2i$ is a zero of $f(x)$.

$$f(x) = x - (5 + 2i)$$
$$f(5 - 2i) = (5 - 2i) - (5 - 2i)$$
$$= 0$$

EXAMPLE 7   **Using the Remainder Theorem to Identify Zeros of a Polynomial**

Use the remainder theorem to determine if the given number c is a zero of the polynomial.

  a. $f(x) = 2x^3 - 4x^2 - 13x - 9$; $c = 4$
  b. $f(x) = x^3 + x^2 - 3x - 3$; $c = \sqrt{3}$
  c. $f(x) = x^3 + x + 10$; $c = 1 + 2i$

Solution:

In each case, divide $f(x)$ by $x - c$ to determine the remainder. If the remainder is 0, then the value c is a zero of the polynomial.

  a. Divide $f(x) = 2x^3 - 4x^2 - 13x - 9$ by $x - 4$.

$$
\begin{array}{r|rrrr}
4 & 2 & -4 & -13 & -9 \\
  &   & 8 & 16 & 12 \\
\hline
  & 2 & 4 & 3 & \underline{|3}
\end{array}
$$

By the remainder theorem, $f(4) = 3$.
Since $f(4) \neq 0$, 4 is not a zero of $f(x)$.

  b. Divide $f(x) = x^3 + x^2 - 3x - 3$ by $x - \sqrt{3}$.

$$
\begin{array}{r|rrrr}
\sqrt{3} & 1 & 1 & -3 & -3 \\
  &   & \sqrt{3} & 3 + \sqrt{3} & 3 \\
\hline
  & 1 & 1 + \sqrt{3} & \sqrt{3} & \underline{|0}
\end{array}
$$

By the remainder theorem, $f(\sqrt{3}) = 0$.
Therefore, $\sqrt{3}$ is a zero of $f(x)$.

FOR REVIEW

To multiply radical expressions, apply the distributive property.

$$\sqrt{3}(1 + \sqrt{3})$$
$$= \sqrt{3} \cdot 1 + \sqrt{3} \cdot \sqrt{3}$$
$$= \sqrt{3} + 3$$
$$= 3 + \sqrt{3}$$

See Section 3.2 for review.

  c. Divide $f(x) = x^3 + x + 10$ by $x - (1 + 2i)$

$$
\begin{array}{r|rrrr}
1 + 2i & 1 & 0 & 1 & 10 \\
  &   & 1 + 2i & -3 + 4i & \\
\hline
  & 1 & 1 + 2i & &
\end{array}
$$

Note that $(1 + 2i)(1 + 2i)$
$= 1 + 2i + 2i + 4i^2$
$= 1 + 4i + 4(-1)$
$= 1 + 4i + (-4)$
$= -3 + 4i$

$$
\begin{array}{r|rrrr}
1 + 2i & 1 & 0 & 1 & 10 \\
  &   & 1 + 2i & -3 + 4i & -10 \\
\hline
  & 1 & 1 + 2i & -2 + 4i & \underline{|0}
\end{array}
$$

Note that $(1 + 2i)(-2 + 4i)$
$= -2 + 4i - 4i + 8i^2$
$= -2 + 8(-1)$
$= -2 - 8$
$= -10$

By the remainder theorem, $f(1 + 2i) = 0$.
Therefore, $1 + 2i$ is a zero of $f(x)$.

Skill Practice 7   Use the remainder theorem to determine if the given number, c, is a zero of the function.

  a. $f(x) = 2x^4 - 3x^2 + 5x - 11$; $c = 2$
  b. $f(x) = 2x^3 + 5x^2 - 14x - 35$; $c = \sqrt{7}$
  c. $f(x) = x^3 - 7x^2 + 16x - 10$; $c = 3 + i$

Answers

7. a. No   **b.** Yes   **c.** Yes

TECHNOLOGY CONNECTIONS

In Example 7(c), you can verify the results involving the product of imaginary numbers with a graphing calculator. First use the ⬭MODE key to place the calculator in $a + bi$ mode. Then perform the indicated operations.

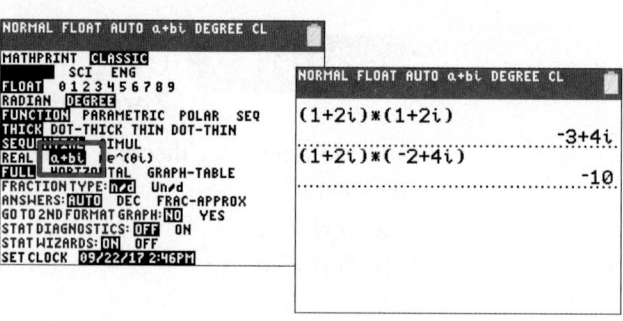

Suppose that we again apply the division algorithm to a dividend of $f(x)$ and a divisor of $x - c$, where c is a complex number.

$$f(x) = (x - c) \cdot q(x) + r$$

By the remainder theorem, $r = f(c)$.

$$= (x - c) \cdot q(x) + f(c)$$

This tells us that if $f(c)$ is a zero of $f(x)$, then $(x - c)$ is a factor of $f(x)$.

If $f(c) = 0$, then $f(x) = (x - c) \cdot q(x)$

Now suppose that $x - c$ is a factor of $f(x)$. Then for some polynomial $q(x)$,

$$f(x) = (x - c) \cdot q(x)$$
$$f(c) = (c - c) \cdot q(x)$$
$$= 0$$

This tells us that if $(x - c)$ is a factor of $f(x)$, then c is a zero of $f(x)$.

These results can be summarized in the factor theorem.

Factor Theorem

Let $f(x)$ be a polynomial.

   1. If $f(c) = 0$, then $(x - c)$ is a factor of $f(x)$.

   2. If $(x - c)$ is a factor of $f(x)$, then $f(c) = 0$.

EXAMPLE 8   Identifying Factors of a Polynomial

Use the factor theorem to determine if the given polynomials are factors of $f(x) = x^4 - x^3 - 11x^2 + 11x + 12$.

   a. $x - 3$ **b.** $x + 2$

Solution:

a. If $f(3) = 0$, then $x - 3$ is a factor of $f(x)$. Using synthetic division we have:

$$\begin{array}{r|rrrrr} 3 & 1 & -1 & -11 & 11 & 12 \\ & & 3 & 6 & -15 & -12 \\ \hline & 1 & 2 & -5 & -4 & \boxed{0} \end{array}$$

By the factor theorem, since $\boxed{f(3) = 0}$ $f(3) = 0$, $x - 3$ is a factor of $f(x)$.

b. If $f(-2) = 0$, then $x + 2$ is a factor of $f(x)$. Using synthetic division we have:

$$\begin{array}{r|rrrrr} -2 & 1 & -1 & -11 & 11 & 12 \\ & & -2 & 6 & 10 & -42 \\ \hline & 1 & -3 & -5 & 21 & \boxed{-30} \end{array}$$

By the factor theorem, since $f(-2) \neq 0$, $x + 2$ is not a factor of $f(x)$. $\boxed{f(-2) = -30}$

Skill Practice 8   Use the factor theorem to determine if the given polynomials are factors of $f(x) = 2x^4 - 13x^3 + 10x^2 - 25x + 6$.

   a. $x - 6$ **b.** $x + 3$

In Example 9, we illustrate the relationship between the zeros of a polynomial and the solutions (roots) of a polynomial equation.

EXAMPLE 9   **Factoring a Polynomial Given a Known Zero**

a. Factor $f(x) = 3x^3 + 25x^2 + 42x - 40$, given that -5 is a zero of $f(x)$.
b. Solve the equation. $3x^3 + 25x^2 + 42x - 40 = 0$

Solution:

a. The value -5 is a zero of $f(x)$, which means that $f(-5) = 0$. By the factor theorem, $x - (-5)$ or equivalently $x + 5$ is a factor of $f(x)$. Using synthetic division, we have

$$\begin{array}{r|rrrr} -5 & 3 & 25 & 42 & -40 \\ & & -15 & -50 & 40 \\ \hline & 3 & 10 & -8 & \boxed{0} \end{array}$$

divisor   quotient   remainder

This means that $3x^3 + 25x^2 + 42x - 40 = (x + 5)(3x^2 + 10x - 8) + 0$
Therefore, $f(x) = (x + 5)(3x - 2)(x + 4)$.

factors as $(3x - 2)(x + 4)$

b. $3x^3 + 25x^2 + 42x - 40 = 0$ To solve the equation, set one side equal to zero.

$(x + 5)(3x - 2)(x + 4) = 0$ Factor the left side.

$x = -5, \ x = \frac{2}{3}, \ x = -4$ Set each factor equal to zero and solve for x.

The solution set is $\left\{-5, \frac{2}{3}, -4\right\}$.

Skill Practice 9

   a. Factor $f(x) = 2x^3 + 7x^2 - 14x - 40$, given that -4 is a zero of f.
   b. Solve the equation. $2x^3 + 7x^2 - 14x - 40 = 0$

FOR REVIEW

In Example 9, once the factor $(x + 5)$ is identified, the remaining factor $(3x^2 + 10x - 8)$ is quadratic and factors as $(3x - 2)(x + 4)$. See Section 2.4 to review the techniques to factor quadratic trinomials.

Answers
8. a. Yes   **b.** No
9. a. $f(x) = (x + 4)(x + 2)(2x - 5)$
   b. $\left\{-4, -2, \frac{5}{2}\right\}$

EXAMPLE 10   Using the Factor Theorem to Build a Polynomial

Write a polynomial $f(x)$ of degree 3 that has the zeros $\frac{1}{2}$, $\sqrt{6}$, and $-\sqrt{6}$.

Solution:

By the factor theorem, if $\frac{1}{2}$, $\sqrt{6}$, and $-\sqrt{6}$ are zeros of a polynomial $f(x)$, then $\left(x - \frac{1}{2}\right)$, $(x - \sqrt{6})$, and $(x + \sqrt{6})$ are factors of $f(x)$. Therefore, $f(x) = \left(x - \frac{1}{2}\right)(x - \sqrt{6})(x + \sqrt{6})$ is a third-degree polynomial with the given zeros.

$$f(x) = \left(x - \frac{1}{2}\right)(x - \sqrt{6})(x + \sqrt{6})$$

$$= \left(x - \frac{1}{2}\right)\left[(x)^2 - (\sqrt{6})^2\right] \quad \text{Multiply conjugates. Recall that } (a-b)(a+b) = a^2 - b^2.$$

$$= \left(x - \frac{1}{2}\right)(x^2 - 6) \quad \text{Multiply the binomials.}$$

$$= x^3 - 6x - \frac{1}{2}x^2 + 3$$

Thus, $f(x) = x^3 - \frac{1}{2}x^2 - 6x + 3$

Skill Practice 10   Write a polynomial $f(x)$ of degree 3 that has the zeros $\frac{1}{3}$, $\sqrt{3}$, and $-\sqrt{3}$.

In Example 10, the polynomial $f(x)$ is not unique. If we multiply $f(x)$ by any nonzero constant a, the polynomial will still have the desired factors and zeros.

$$g(x) = a\left(x - \frac{1}{2}\right)(x - \sqrt{6})(x + \sqrt{6}) \quad \text{The zeros are still } \frac{1}{2}, \sqrt{6}, \text{ and } -\sqrt{6}.$$

If a is any nonzero multiple of 2, then the polynomial will have integer coefficients. For example:

$$g(x) = 2\left(x - \frac{1}{2}\right)(x - \sqrt{6})(x + \sqrt{6})$$

$$= 2\left(x^3 - \frac{1}{2}x^2 - 6x + 3\right)$$

$$= 2x^3 - x^2 - 12x + 6$$

The zeros of $f(x)$ and $g(x)$ are real numbers and correspond to the x-intercepts of the graphs of the related functions. The graphs of $y = f(x)$ and $y = g(x)$ are shown in Figure 7-18. Notice that the graphs have the same x-intercepts and differ only by a vertical stretch.

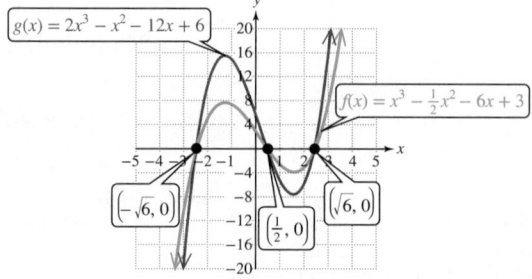

Figure 7-18

Answer

10. $f(x) = x^3 - \frac{1}{3}x^2 - 3x + 1$

Prerequisite Review

For Exercises R.1–R.2, divide without the use of a calculator and write the result as a mixed number.

R.1. $2579 \div 8$

R.2. $\dfrac{32{,}692}{43}$

For Exercises R.3–R.4, perform the indicated operations. (See Section 2.2 for review.)

R.3. $(4x^2 - 3)(x^2 + 5x + 1) + 2x$

R.4. $(2x + 7)(4x^3 - 2x + 4) + 6$

For Exercises R.5–R.6, evaluate the function for the given values of x. (See Section 5.3 for review.)

R.5. $f(x) = 3x^3 + 25x^2 + 42x - 40$

    a. $f(-5)$ **b.** $f\left(\dfrac{2}{3}\right)$ **c.** $f(1)$

R.6. $f(x) = 2x^3 - 9x^2 + 4x + 15$

    a. $f(2)$ **b.** $f(-1)$ **c.** $f\left(\dfrac{5}{2}\right)$

For Exercises R.7–R.10, solve the equation. (See Sections 3.5 and 3.6 for review.)

R.7. $x^2 + 4 = 0$ (See Section 3.5.)

R.8. $x^2 + 169 = 0$

R.9. $2x^2 - 4x + 6 = 0$ (See Section 3.6.)

R.10. $x^2 + 2x + 5 = 0$

For Exercises R.11–R.12, find the x- and y-intercepts for the function defined by $y = f(x)$. (See Section 5.3 for review.)

R.11. $f(x) = 2(x - 3)(x + 4)(2x - 1)$

R.12. $f(x) = -3(x + 1)(x - 2)(4x + 3)$

For Exercises R.13–R.20, perform the indicated operations. (See Section 3.3 for review.)

R.13. $(3 + 7i) + (2 - 4i)$

R.14. $(-3 - i) - (6 + 2i)$

R.15. $(2i)(4 - 3i)$

R.16. $(3i)(-5 + 7i)$

R.17. $(-1 + 3i)(2 + 4i)$

R.18. $(6 + 2i)(1 - 5i)$

R.19. $(3 + 4i)^2$

R.20. $(2 - 5i)^2$

For Exercises R.21–R.22, determine by substitution whether the given value is a solution to the equation.

R.21. $x^2 - 12x + 39 = 0$

    a. $6 - i\sqrt{3}$ **b.** $6 + i\sqrt{3}$

R.22. $x^2 - 8x + 17 = 0$

    a. $4 + i$ **b.** $4 - i$

Concept Connections

1. Given the division algorithm, identify the polynomials representing the dividend, divisor, quotient, and remainder.

   $$f(x) = d(x) \cdot q(x) + r(x)$$

2. Given $\dfrac{2x^3 - 5x^2 - 6x + 1}{x - 3} = 2x^2 + x - 3 + \dfrac{-8}{x - 3}$, use the division algorithm to check the result.

3. The remainder theorem indicates that if a polynomial $f(x)$ is divided by $x - c$, then the remainder is _____.

4. Given a polynomial $f(x)$, the factor theorem indicates that if $f(c) = 0$, then $x - c$ is a _____ of $f(x)$. Furthermore, if $x - c$ is a factor of $f(x)$, then $f(c) =$ _____.

5. Answer true or false. If $\sqrt{5}$ is a zero of a polynomial, then $(x - \sqrt{5})$ is a factor of the polynomial.

6. Answer true or false. If $(x + 3)$ is a factor of a polynomial, then 3 is a zero of the polynomial.

Objective 1: Divide Polynomials Using Long Division

For Exercises 7–8, **(See Example 1)**

  a. Use long division to divide.

  b. Identify the dividend, divisor, quotient, and remainder.

  c. Check the result from part (a) with the division algorithm.

7. $(6x^2 + 9x + 5) \div (2x - 5)$ **8.** $(12x^2 + 10x + 3) \div (3x + 4)$

For Exercises 9–22, use long division to divide. **(See Examples 1–3)**

9. $(3x^3 - 11x^2 - 10) \div (x - 4)$ **10.** $(2x^3 - 7x^2 - 65) \div (x - 5)$

11. $(8 + 30x - 27x^2 - 12x^3 + 4x^4) \div (x + 2)$ **12.** $(-48 - 28x + 20x^2 + 17x^3 + 3x^4) \div (x + 3)$

13. $(-20x^2 + 6x^4 - 16) \div (2x + 4)$ **14.** $(-60x^2 + 8x^4 - 108) \div (2x - 6)$

15. $(x^5 + 4x^4 + 18x^2 - 20x - 10) \div (x^2 + 5)$ **16.** $(x^5 - 2x^4 + x^3 - 8x + 18) \div (x^2 - 3)$

17. $\dfrac{6x^4 + 3x^3 - 7x^2 + 6x - 5}{2x^2 + x - 3}$ **18.** $\dfrac{12x^4 - 4x^3 + 13x^2 + 2x + 1}{3x^2 - x + 4}$

19. $\dfrac{x^3 - 27}{x - 3}$ **20.** $\dfrac{x^3 + 64}{x + 4}$

21. $(5x^3 - 2x^2 + 3) \div (2x - 1)$ **22.** $(2x^3 + x^2 + 1) \div (3x + 1)$

Objective 2: Divide Polynomials Using Synthetic Division

For Exercises 23–26, consider the division of two polynomials: $f(x) \div (x - c)$. The result of the synthetic division process is shown here. Write the polynomials representing the

  a. Dividend.   **b.** Divisor.   **c.** Quotient.   **d.** Remainder.

23.

$$\begin{array}{r|rrrrr}
3 & 2 & -5 & -5 & -4 & 29 \\
& & 6 & 3 & -6 & -30 \\
\hline
& 2 & 1 & -2 & -10 & \boxed{-1}
\end{array}$$

24.

$$\begin{array}{r|rrrrr}
2 & 1 & -5 & 2 & -1 & 20 \\
& & 2 & -6 & -8 & -18 \\
\hline
& 1 & -3 & -4 & -9 & \boxed{2}
\end{array}$$

25.

$$\begin{array}{r|rrrr}
-4 & 1 & -2 & -25 & -4 \\
& & -4 & 24 & 4 \\
\hline
& 1 & -6 & -1 & \boxed{0}
\end{array}$$

26.

$$\begin{array}{r|rrrr}
-5 & 3 & 13 & -14 & -20 \\
& & -15 & 10 & 20 \\
\hline
& 3 & -2 & -4 & \boxed{0}
\end{array}$$

For Exercises 27–38, use synthetic division to divide the polynomials. **(See Examples 4–5)**

27. $(4x^2 + 15x + 1) \div (x + 6)$ **28.** $(6x^2 + 25x - 19) \div (x + 5)$

29. $(5x^2 - 17x - 12) \div (x - 4)$ **30.** $(2x^2 + x - 21) \div (x - 3)$

31. $(4 - 8x - 3x^2 - 5x^4) \div (x + 2)$ **32.** $(-5 + 2x + 5x^3 - 2x^4) \div (x + 1)$

33. $\dfrac{4x^5 - 25x^4 - 58x^3 + 232x^2 + 198x - 63}{x - 3}$ **34.** $\dfrac{2x^5 + 13x^4 - 3x^3 - 58x^2 - 20x + 24}{x - 2}$

35. $\dfrac{x^5 + 32}{x + 2}$ **36.** $\dfrac{x^4 - 81}{x + 3}$

37. $(2x^4 - 7x^3 - 56x^2 + 37x + 84) \div \left(x - \dfrac{3}{2}\right)$ **38.** $(-5x^4 - 18x^3 + 63x^2 + 128x - 60) \div \left(x - \dfrac{2}{5}\right)$

Objective 3: Apply the Remainder and Factor Theorems

39. The value $f(-6) = 39$ for a polynomial $f(x)$. What can be concluded about the remainder or quotient of $\dfrac{f(x)}{x + 6}$?

40. Given a polynomial $f(x)$, the quotient $\dfrac{f(x)}{x - 2}$ has a remainder of 12. What is the value of $f(2)$?

41. Given $f(x) = 2x^4 - 5x^3 + x^2 - 7$,

    a. Evaluate $f(4)$.

    b. Determine the remainder when $f(x)$ is divided by $(x - 4)$.

42. Given $g(x) = -3x^5 + 2x^4 + 6x^2 - x + 4$,

    a. Evaluate $g(2)$.

    b. Determine the remainder when $g(x)$ is divided by $(x - 2)$.

For Exercises 43–46, use the remainder theorem to evaluate the polynomial for the given values of x. (**See Example 6**)

43. $f(x) = 2x^4 + x^3 - 49x^2 + 79x + 15$

    a. $f(-1)$ **b.** $f(3)$ **c.** $f(4)$ **d.** $f\left(\dfrac{5}{2}\right)$

44. $g(x) = 3x^4 - 22x^3 + 51x^2 - 42x + 8$

    a. $g(-1)$ **b.** $g(2)$ **c.** $g(1)$ **d.** $g\left(\dfrac{4}{3}\right)$

45. $h(x) = 5x^3 - 4x^2 - 15x + 12$

    a. $h(1)$ **b.** $h\left(\dfrac{4}{5}\right)$ **c.** $h(\sqrt{3})$ **d.** $h(-1)$

46. $k(x) = 2x^3 - x^2 - 14x + 7$

    a. $k(2)$ **b.** $k\left(\dfrac{1}{2}\right)$ **c.** $k(\sqrt{7})$ **d.** $k(-2)$

For Exercises 47–54, use the remainder theorem to determine if the given number c is a zero of the polynomial. (**See Example 7**)

47. $f(x) = x^4 + 3x^3 - 7x^2 + 13x - 10$

    a. $c = 2$ **b.** $c = -5$

48. $g(x) = 2x^4 + 13x^3 - 10x^2 - 19x + 14$

    a. $c = -2$ **b.** $c = -7$

49. $p(x) = 2x^3 + 3x^2 - 22x - 33$

    a. $c = -2$ **b.** $c = -\sqrt{11}$

50. $q(x) = 3x^3 + x^2 - 30x - 10$

    a. $c = -3$ **b.** $c = -\sqrt{10}$

51. $m(x) = x^3 - 2x^2 + 25x - 50$

    a. $c = 5i$ **b.** $c = -5i$

52. $n(x) = x^3 + 4x^2 + 9x + 36$

    a. $c = 3i$ **b.** $c = -3i$

53. $g(x) = x^3 - 11x^2 + 25x + 37$

    a. $c = 6 + i$ **b.** $c = 6 - i$

54. $f(x) = 2x^3 - 5x^2 + 54x - 26$

    a. $c = 1 + 5i$ **b.** $c = 1 - 5i$

For Exercises 55–60, use the factor theorem to determine if the given binomial is a factor of $f(x)$. (**See Example 8**)

55. $f(x) = x^4 + 11x^3 + 41x^2 + 61x + 30$

    a. $x + 5$ **b.** $x - 2$

56. $g(x) = x^4 - 10x^3 + 35x^2 - 50x + 24$

    a. $x - 4$ **b.** $x + 1$

57. $f(x) = x^3 + 64$

    a. $x - 4$ **b.** $x + 4$

58. $f(x) = x^4 - 81$

    a. $x - 3$ **b.** $x + 3$

59. $f(x) = 2x^3 + x^2 - 16x - 8$

    a. $x - 1$ **b.** $x - 2\sqrt{2}$

60. $f(x) = 3x^3 - x^2 - 54x + 18$

    a. $x - 2$ **c.** $x - 3\sqrt{2}$

61. Given $g(x) = x^4 - 14x^2 + 45$,

    a. Evaluate $g(\sqrt{5})$.

    b. Evaluate $g(-\sqrt{5})$.

    c. Solve $g(x) = 0$.

62. Given $h(x) = x^4 - 15x^2 + 44$,

    a. Evaluate $h(\sqrt{11})$.

    b. Evaluate $h(-\sqrt{11})$.

    c. Solve $h(x) = 0$.

63. **a.** Use synthetic division and the factor theorem to determine if $[x - (2 + 5i)]$ is a factor of $f(x) = x^2 - 4x + 29$.

    b. Use synthetic division and the factor theorem to determine if $[x - (2 - 5i)]$ is a factor of $f(x) = x^2 - 4x + 29$.

    c. Use the quadratic formula to solve the equation. $x^2 - 4x + 29 = 0$

    d. Find the zeros of the polynomial $f(x) = x^2 - 4x + 29$.

64. **a.** Use synthetic division and the factor theorem to determine if $[x - (3 + 4i)]$ is a factor of $f(x) = x^2 - 6x + 25$.

    b. Use synthetic division and the factor theorem to determine if $[x - (3 - 4i)]$ is a factor of $f(x) = x^2 - 6x + 25$.

    c. Use the quadratic formula to solve the equation. $x^2 - 6x + 25 = 0$

    d. Find the zeros of the polynomial $f(x) = x^2 - 6x + 25$.

65. **a.** Factor $f(x) = 2x^3 + x^2 - 37x - 36$, given that -1 is a zero. (**See Example 9**)

    b. Solve. $2x^3 + x^2 - 37x - 36 = 0$

66. **a.** Factor $f(x) = 3x^3 + 16x^2 - 5x - 50$, given that -2 is a zero.

    b. Solve. $3x^3 + 16x^2 - 5x - 50 = 0$

67. a. Factor $f(x) = 20x^3 + 39x^2 - 3x - 2$, given that $\frac{1}{4}$ is a zero.

   b. Solve. $20x^3 + 39x^2 - 3x - 2 = 0$

68. a. Factor $f(x) = 8x^3 - 18x^2 - 11x + 15$, given that $\frac{3}{4}$ is a zero.

   b. Solve. $8x^3 - 18x^2 - 11x + 15 = 0$

69. a. Factor $f(x) = 9x^3 - 33x^2 + 19x - 3$, given that 3 is a zero.

   b. Solve. $9x^3 - 33x^2 + 19x - 3 = 0$

70. a. Factor $f(x) = 4x^3 - 20x^2 + 33x - 18$, given that 2 is a zero.

   b. Solve. $4x^3 - 20x^2 + 33x - 18 = 0$

For Exercises 71–82, write a polynomial $f(x)$ that meets the given conditions. Answers may vary. (**See Example 10**)

71. Degree 3 polynomial with zeros 2, 3, and −4.

72. Degree 3 polynomial with zeros 1, −6, and 3.

73. Degree 4 polynomial with zeros 1, $\frac{3}{2}$ (each with multiplicity 1), and 0 (with multiplicity 2).

74. Degree 5 polynomial with zeros 2, $\frac{5}{2}$ (each with multiplicity 1), and 0 (with multiplicity 3).

75. Degree 2 polynomial with zeros $2\sqrt{11}$ and $-2\sqrt{11}$.

76. Degree 2 polynomial with zeros $5\sqrt{2}$ and $-5\sqrt{2}$.

77. Degree 3 polynomial with zeros −2, $3i$, and $-3i$.

78. Degree 3 polynomial with zeros 4, $2i$, and $-2i$.

79. Degree 3 polynomial with integer coefficients and zeros of $-\frac{2}{3}$, $\frac{1}{2}$, and 4.

80. Degree 3 polynomial with integer coefficients and zeros of $-\frac{2}{5}$, $\frac{3}{2}$, and 6.

81. Degree 2 polynomial with zeros of $7 + 8i$ and $7 - 8i$.

82. Degree 2 polynomial with zeros of $5 + 6i$ and $5 - 6i$.

Mixed Exercises

83. Given $p(x) = 2x^{452} - 4x^{92}$, is it easier to evaluate $p(1)$ by using synthetic division or by direct substitution? Find the value of $p(1)$.

84. Given $q(x) = 5x^{721} - 2x^{450}$, is it easier to evaluate $q(-1)$ by using synthetic division or by direct substitution? Find the value of $q(-1)$.

85. a. Is $(x - 1)$ a factor of $x^{100} - 1$?

   b. Is $(x + 1)$ a factor of $x^{100} - 1$?

   c. Is $(x - 1)$ a factor of $x^{99} - 1$?

   d. Is $(x + 1)$ a factor of $x^{99} - 1$?

   e. If n is a positive even integer, is $(x - 1)$ a factor of $x^n - 1$?

   f. If n is a positive odd integer, is $(x + 1)$ a factor of $x^n - 1$?

86. If a fifth-degree polynomial is divided by a second-degree polynomial, the quotient is a _____ -degree polynomial.

87. Determine if the statement is true or false: Zero is a zero of the polynomial $3x^5 - 7x^4 - 2x^3 - 14$.

88. Determine if the statement is true or false: Zero is a zero of the polynomial $-2x^4 + 5x^3 + 6x$.

89. Find m so that $x + 4$ is a factor of $4x^3 + 13x^2 - 5x + m$.

90. Find m so that $x + 5$ is a factor of $-3x^4 - 10x^3 + 20x^2 - 22x + m$.

91. Find m so that $x + 2$ is a factor of $4x^3 + 5x^2 + mx + 2$.

92. Find m so that $x - 3$ is a factor of $2x^3 - 7x^2 + mx + 6$.

93. For what value of r is the statement an identity?
$$\frac{x^2 - x - 12}{x - 4} = x + 3 + \frac{r}{x - 4} \text{ provided that } x \neq 4$$

94. For what value of r is the statement an identity?
$$\frac{x^2 - 5x - 8}{x - 2} = x - 3 + \frac{r}{x - 2} \text{ provided that } x \neq 2$$

95. A metal block is formed from a rectangular solid with a rectangular piece cut out.

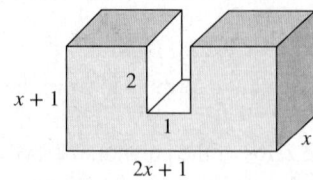

   a. Write a polynomial $V(x)$ that represents the volume of the block. All distances in the figure are in centimeters.

   b. Use synthetic division to evaluate the volume if x is 6 cm.

96. A wedge is cut from a rectangular solid.

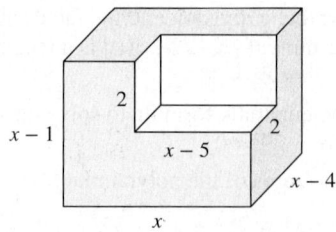

   a. Write a polynomial $V(x)$ that represents the volume of the remaining part of the solid. All distances in the figure are in feet.

   b. Use synthetic division to evaluate the volume if x is 10 ft.

Write About It

97. Under what circumstances can synthetic division be used to divide polynomials?

98. How can the division algorithm be used to check the result of polynomial division?

99. Given a polynomial $f(x)$ and a constant c, state two methods by which the value $f(c)$ can be computed.

100. Write an informal explanation of the factor theorem.

Expanding Your Skills

101. a. Factor $f(x) = x^3 - 5x^2 + x - 5$ into factors of the form $(x - c)$, given that 5 is a zero.

   b. Solve. $x^3 - 5x^2 + x - 5 = 0$

102. a. Factor $f(x) = x^3 - 3x^2 + 100x - 300$ into factors of the form $(x - c)$, given that 3 is a zero.

   b. Solve. $x^3 - 3x^2 + 100x - 300 = 0$

103. a. Factor $f(x) = x^4 + 2x^3 - 2x^2 - 6x - 3$ into factors of the form $(x - c)$, given that -1 is a zero.

   b. Solve. $x^4 + 2x^3 - 2x^2 - 6x - 3 = 0$

104. a. Factor $f(x) = x^4 + 4x^3 - x^2 - 20x - 20$ into factors of the form $(x - c)$, given that -2 is a zero.

   b. Solve. $x^4 + 4x^3 - x^2 - 20x - 20 = 0$

Technology Connections

For Exercises 105–106,

   a. Use the graph to determine a solution to the given equation.

   b. Verify your answer from part (a) using the remainder theorem.

   c. Find the remaining solutions to the equation.

105. $5x^3 + 7x^2 - 58x - 24 = 0$

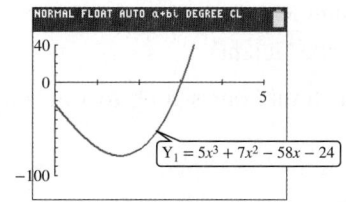

106. $2x^3 - x^2 - 41x + 70 = 0$

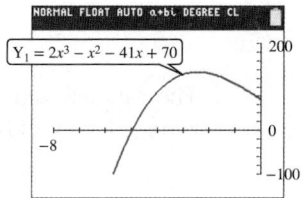

SECTION 7.4   Zeros of Polynomials

OBJECTIVES

1. Apply the Rational Zero Theorem
2. Apply the Fundamental Theorem of Algebra
3. Apply Descartes' Rule of Signs
4. Find Upper and Lower Bounds

1. Apply the Rational Zero Theorem

The **zeros of a polynomial** $f(x)$ are the solutions (roots) to the corresponding polynomial equation $f(x) = 0$. For a polynomial function defined by $y = f(x)$, the real zeros of $f(x)$ are the x-intercepts of the graph of the function. Applications of polynomials and polynomial functions arise throughout mathematics. For this reason, it is important to learn techniques to find or approximate the zeros of a polynomial.

The zeros of a polynomial may be real numbers or imaginary numbers. The real zeros can be further categorized as rational or irrational numbers. For example, consider

$$f(x) = 2x^6 - 3x^5 - 7x^4 + 102x^3 - 88x^2 - 279x + 273$$

In factored form this is:

$$f(x) = (x - 1)(2x + 7)(x - \sqrt{3})(x + \sqrt{3})[x - (2 + 3i)][x - (2 - 3i)]$$

The zeros are:

$$\underbrace{1, -\frac{7}{2},}_{\substack{\text{rational} \\ \text{zeros}}} \underbrace{\sqrt{3}, -\sqrt{3},}_{\substack{\text{irrational} \\ \text{zeros}}} \underbrace{2 + 3i, 2 - 3i}_{}$$

$$\underbrace{}_{\text{real zeros}} \quad \underbrace{}_{\text{nonreal zeros}}$$

In this section, we develop tools to search for the zeros of polynomials. First we will consider polynomials with integer coefficients. For example, the factored form of $f(x) = 6x^2 + 13x - 5$ is $f(x) = (2x + 5)(3x - 1)$, which has zeros of $-\frac{5}{2}$ and $\frac{1}{3}$. The polynomial $f(x) = 6x^2 + 13x - 5$ is in the form $f(x) = ax^2 + bx + c$, where $a = 6$, $b = 13$, and $c = -5$. Notice that the numerator of each zero is a factor of c, and the denominator of each zero is a factor of a. This observation is consistent with the following theorem to search for zeros that are rational numbers.

Rational Zero Theorem

If $f(x) = a_n x^n + a_{n-1}x^{n-1} + a_{n-2}x^{n-2} + \cdots + a_1 x + a_0$ has integer coefficients and $a_n \neq 0$, and if $\frac{p}{q}$ (written in lowest terms) is a rational zero of f, then

- p is a factor of the constant term a_0.
- q is a factor of the leading coefficient a_n.

The rational zero theorem does not guarantee the existence of rational zeros. Rather, it indicates that *if* a rational zero exists for a polynomial, then it must be of the form

$$\frac{p}{q} = \frac{\text{Factors of } a_0 \text{ (constant term)}}{\text{Factors of } a_n \text{ (leading coefficient)}}$$

The rational zero theorem is important because it limits our search to find rational zeros (if they exist) to a finite number of choices.

EXAMPLE 1 **Listing All Possible Rational Zeros**

List all possible rational zeros. $f(x) = -2x^5 + 3x^2 - 2x + 10$

Solution:

First note that the polynomial has integer coefficients.

$$f(x) = -2x^5 + 3x^2 - 2x + 10$$

The constant term is 10. $\pm 1, \pm 2, \pm 5, \pm 10$ ◁ Factors of 10

The leading coefficient is −2. $\pm 1, \pm 2$ ◁ Factors of −2

$$\frac{\text{Factors of } 10}{\text{Factors of } -2} = \frac{\pm 1, \pm 2, \pm 5, \pm 10}{\pm 1, \pm 2} = \pm\frac{1}{1}, \pm\frac{2}{1}, \pm\frac{5}{1}, \pm\frac{10}{1}, \pm\frac{1}{2}, \pm\frac{\cancel{2}}{\cancel{2}}, \pm\frac{5}{2}, \pm\frac{\cancel{10}}{\cancel{2}}$$

The values $\pm\frac{2}{2}$ and $\pm\frac{10}{2}$ are redundant. They equal ± 1 and ± 5, respectively. The possible rational zeros are ± 1, ± 2, ± 5, ± 10, $\pm\frac{1}{2}$, and $\pm\frac{5}{2}$.

Skill Practice 1 List all possible rational zeros.

$$f(x) = -4x^4 + 5x^3 - 7x^2 + 8$$

Answer

1. $\pm 1, \pm 2, \pm 4, \pm 8, \pm\frac{1}{2}$, and $\pm\frac{1}{4}$

EXAMPLE 2   **Finding the Zeros of a Polynomial**

Find the zeros. $f(x) = x^3 - 4x^2 + 3x + 2$

Solution:

We begin by first looking for rational zeros. We can apply the rational zero theorem because the polynomial has integer coefficients.

$$f(x) = 1x^3 - 4x^2 + 3x + 2$$

Possible rational zeros: $\dfrac{\text{Factors of } 2}{\text{Factors of } 1} = \dfrac{\pm 1, \pm 2}{\pm 1} = \pm 1, \pm 2$

TIP   Since $f(1) = 2$ and $f(-1) = -6$, we know from the intermediate value theorem that $f(x)$ must have a zero between -1 and 1. (See Section 7.2.)

Next, use synthetic division and the remainder theorem to determine if any of the numbers in the list is a zero of f.

Test $x = 1$:

$$\begin{array}{r|rrrr} 1 & 1 & -4 & 3 & 2 \\ & & 1 & -3 & 0 \\ \hline & 1 & -3 & 0 & \underline{2} \end{array}$$

Test $x = -1$:

$$\begin{array}{r|rrrr} -1 & 1 & -4 & 3 & 2 \\ & & -1 & 5 & -8 \\ \hline & 1 & -5 & 8 & \underline{-6} \end{array}$$

Test $x = 2$:

$$\begin{array}{r|rrrr} 2 & 1 & -4 & 3 & 2 \\ & & 2 & -4 & -2 \\ \hline & 1 & -2 & -1 & \underline{0} \end{array}$$

The remainder is not zero. Therefore, 1 is not a zero of $f(x)$.

The remainder is not zero. Therefore, -1 is not a zero of $f(x)$.

The remainder *is* zero. Therefore, 2 is a zero of $f(x)$.

By the factor theorem, since $f(2) = 0$, then $(x - 2)$ is a factor of $f(x)$. The quotient $(x^2 - 2x - 1)$ is also a factor of $f(x)$. We have

$$f(x) = x^3 - 4x^2 + 3x + 2 = (x - 2)(x^2 - 2x - 1)$$

We now have a third-degree polynomial written as the product of a first-degree polynomial and a quadratic polynomial. The quotient $x^2 - 2x - 1$ is called a **reduced polynomial** (or a **depressed polynomial**) of $f(x)$. It has degree 1 less than the degree of $f(x)$, and the remaining zeros of $f(x)$ are the zeros of the reduced polynomial.

At this point, we no longer need to test for more rational zeros. The reason is that any remaining zeros (whether they be rational, irrational, or imaginary) are the solutions to the quadratic equation $x^2 - 2x - 1 = 0$. There is no guesswork because the equation can be solved by using the quadratic formula.

TIP   The graph of $f(x) = x^3 - 4x^2 + 3x + 2$ is shown here. The zeros are $1 + \sqrt{2} \approx 2.41$, $1 - \sqrt{2} \approx -0.41$, and 2.

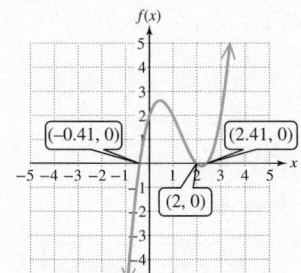

$x^2 - 2x - 1 = 0$

$\quad$ The equation is of the form $ax^2 + bx + c = 0$ with $a = 1$, $b = -2$, and $c = -1$.

$x = \dfrac{-(-2) \pm \sqrt{(-2)^2 - 4(1)(-1)}}{2(1)}$

$\quad$ Apply the quadratic formula.

$\quad x = \dfrac{-b \pm \sqrt{b^2 - 4ac}}{2a}$

$= \dfrac{2 \pm \sqrt{8}}{2}$

$\quad$ Simplify the numerator and denominator.

$= \dfrac{2 \pm 2\sqrt{2}}{2}$

$\quad$ Simplify the radical:
$\quad \sqrt{8} = \sqrt{2^3} = \sqrt{2^2 \cdot 2} = 2\sqrt{2}$

$= \dfrac{\overset{1}{2}(1 + \sqrt{2})}{2}$

$\quad$ Simplify the rational expression. Factor the numerator and denominator and cancel common factors.

$= 1 \pm \sqrt{2}$

The zeros of $f(x)$ are 2, $1 + \sqrt{2}$, and $1 - \sqrt{2}$.

Answer

2. $-1, 1 - \sqrt{3}$, and $1 + \sqrt{3}$

Skill Practice 2   Find the zeros. $f(x) = x^3 - x^2 - 4x - 2$

The polynomial in Example 2 has one rational zero. If we had continued testing for more rational zeros, we would not have found others because the remaining two zeros are irrational numbers. In Example 3, we illustrate the case where a function has multiple rational zeros.

EXAMPLE 3   **Finding the Zeros of a Polynomial**

Find the zeros and their multiplicities. $f(x) = 2x^4 + 5x^3 - 2x^2 - 11x - 6$

Solution:

Begin by searching for rational zeros.

$$f(x) = 2x^4 + 5x^3 - 2x^2 - 11x - 6$$

Possible rational zeros:

$$\frac{\text{Factors of } -6}{\text{Factors of } 2} = \frac{\pm 1, \pm 2, \pm 3, \pm 6}{\pm 1, \pm 2} = \pm 1, \pm 2, \pm 3, \pm 6, \pm\frac{1}{2}, \pm\frac{3}{2}$$

> **TIP** It does not matter in which order you test the potential rational zeros.

We can work methodically through the list of possible rational zeros to determine which if any are actual zeros of $f(x)$. After trying several possibilities, we find that -1 is a zero of $f(x)$.

$$\begin{array}{r|rrrr} -1 & 2 & 5 & -2 & -11 & -6 \\ & & -2 & -3 & 5 & 6 \\ \hline & 2 & 3 & -5 & -6 & \underline{|0} \end{array}$$

The quotient is $2x^3 + 3x^2 - 5x - 6$.

Since $f(-1) = 0$, then $x + 1$ is a factor of $f(x)$.

$$f(x) = 2x^4 + 5x^3 - 2x^2 - 11x - 6 = (x + 1)\underbrace{(2x^3 + 3x^2 - 5x - 6)}$$

Now find the zeros of the quotient.

The zeros of $f(x)$ are -1 along with the roots of the equation $2x^3 + 3x^2 - 5x - 6 = 0$. Therefore, we need to find the zeros of $g(x) = 2x^3 + 3x^2 - 5x - 6$.

> **TIP** The zeros of the polynomial in Example 3 are rational numbers (and therefore real numbers). They correspond to x-intercepts on the graph of $y = f(x)$. Also notice that the graph has a touch point at $(-1, 0)$ because -1 has an even multiplicity.

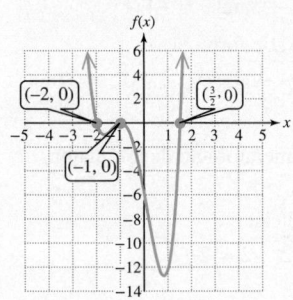

The possible rational zeros are $\pm 1, \pm 2, \pm 3, \pm 6, \pm\frac{1}{2}, \pm\frac{3}{2}$.

We will test -1 again because it may have multiplicity greater than 1.

$$\begin{array}{r|rrrr} -1 & 2 & 3 & -5 & -6 \\ & & -2 & -1 & 6 \\ \hline & 2 & 1 & -6 & \underline{|0} \end{array}$$

The quotient is $2x^2 + x - 6$.

The value -1 is a repeated zero. We have:

$$f(x) = 2x^4 + 5x^3 - 2x^2 - 11x - 6 = (x + 1)(x + 1)(2x^2 + x - 6)$$

$$= (x + 1)^2 (2x - 3)(x + 2)$$

> $2x^2 + x - 6$ factors as $(2x - 3)(x + 2)$.

The zeros of $f(x)$ are

-1 (multiplicity 2), $\frac{3}{2}$ (multiplicity 1), and -2 (multiplicity 1).

> **Skill Practice 3**   Find the zeros and their multiplicities.
> $f(x) = 2x^4 + 3x^3 - 15x^2 - 32x - 12$

Answer

3. -2 (multiplicity 2);

$-\dfrac{1}{2}$ (multiplicity 1); and

3 (multiplicity 1)

In Example 4, we have a polynomial with no rational zeros.

EXAMPLE 4   **Finding the Zeros of a Polynomial**

Find the zeros. $f(x) = x^4 - 2x^2 - 3$

Solution:

$$f(x) = 1x^4 - 2x^2 - 3$$

The possible rational zeros are $\dfrac{\text{Factors of } -3}{\text{Factors of } 1} = \dfrac{\pm 1, \pm 3}{\pm 1} = \pm 1, \pm 3.$

If we apply the rational zero theorem, we see that $f(x)$ has no *rational* zeros.

$$
\begin{array}{r|rrrrr}
1 & 1 & 0 & -2 & 0 & -3 \\
  &   & 1 & 1 & -1 & -1 \\
\hline
  & 1 & 1 & -1 & -1 & \underline{-4}
\end{array}
\qquad
\begin{array}{r|rrrrr}
-1 & 1 & 0 & -2 & 0 & -3 \\
   &   & -1 & 1 & 1 & -1 \\
\hline
   & 1 & -1 & -1 & 1 & \underline{-4}
\end{array}
$$

$$
\begin{array}{r|rrrrr}
3 & 1 & 0 & -2 & 0 & -3 \\
  &   & 3 & 9 & 21 & 63 \\
\hline
  & 1 & 3 & 7 & 21 & \underline{60}
\end{array}
\qquad
\begin{array}{r|rrrrr}
-3 & 1 & 0 & -2 & 0 & -3 \\
   &   & -3 & 9 & -21 & 63 \\
\hline
   & 1 & -3 & 7 & -21 & \underline{60}
\end{array}
$$

However, finding the zeros of $f(x)$ is equivalent to finding the roots of the equation $x^4 - 2x^2 - 3 = 0$.

$$x^4 - 2x^2 - 3 = 0$$
$$(x^2 - 3)(x^2 + 1) = 0 \qquad\qquad \text{Factor the trinomial.}$$
$$x^2 - 3 = 0 \quad \text{or} \quad x^2 + 1 = 0 \qquad \text{Set each factor equal to zero.}$$
$$x^2 = 3 \quad \text{or} \quad x^2 = -1 \qquad \text{Isolate the variable term.}$$
$$x = \pm\sqrt{3} \quad \text{or} \quad x = \pm\sqrt{-1} \qquad \text{Apply the square root property.}$$
$$x = \pm i$$

The zeros are $\sqrt{3}$, $-\sqrt{3}$, i, and $-i$.

Skill Practice 4   Find the zeros. $f(x) = x^4 - x^2 - 20$

TIP   The graph of the function defined by $f(x) = x^4 - 2x^2 - 3$ shows the real zeros of $f(x)$ as x-intercepts.

TIP   From the factor theorem,

$f(x) = (x - \sqrt{3})(x + \sqrt{3})$
$(x - i)(x + i).$

2. Apply the Fundamental Theorem of Algebra

The zeros of a polynomial may be real numbers (either rational or irrational) or non-real numbers such as $2 + 3i$ or $5i$. In any case, the zeros are all complex numbers.

To find the zeros of a polynomial, it is important to know how many zeros to expect. This is answered by the following three theorems. The first is called the fundamental theorem of algebra because it is so basic to the foundation of algebra.

Fundamental Theorem of Algebra

If $f(x)$ is a polynomial of degree $n \geq 1$ with complex coefficients, then $f(x)$ has at least one complex zero.

The fundamental theorem of algebra, first proved by German mathematician Carl Friedrich Gauss (1777–1855), guarantees that every polynomial of degree $n \geq 1$ has at least one zero.

Answer

4. $\sqrt{5}, -\sqrt{5}, 2i,$ and $-2i$

Now suppose that $f(x)$ is a polynomial of degree $n \geq 1$ with complex coefficients. The fundamental theorem of algebra guarantees the existence of at least one complex zero, call this c_1. By the factor theorem, we have

$$f(x) = (x - c_1) \cdot q_1(x) \qquad \text{where } q_1(x) \text{ is a polynomial of degree } n - 1.$$

If $q_1(x)$ is of degree 1 or more, then the fundamental theorem of algebra guarantees that $q_1(x)$ must have at least one complex zero, call this c_2. Then,

$$f(x) = (x - c_1) \cdot (x - c_2) \cdot q_2(x) \quad \text{where } q_2(x) \text{ is a polynomial of degree } n - 2.$$

We can continue with this reasoning until the quotient polynomial, $q_n(x)$, is a constant equal to the leading coefficient of $f(x)$.

TIP The set of complex numbers includes the set of real numbers. Therefore, theorems relating to polynomials with complex coefficients also apply to polynomials with real coefficients.

> ### Linear Factorization Theorem
>
> If $f(x) = a_n x^n + a_{n-1} x^{n-1} + a_{n-2} x^{n-2} + \cdots + a_1 x + a_0$, where $n \geq 1$ and $a_n \neq 0$, then
>
> $f(x) = a_n(x - c_1)(x - c_2) \ldots (x - c_n)$, where $c_1, c_2, \ldots, c_n$ are complex numbers.
>
> *Note*: The complex numbers $c_1, c_2, \ldots, c_n$ are not necessarily unique.

The linear factorization theorem tells us that a polynomial of degree $n \geq 1$ with complex coefficients has exactly n linear factors of the form $(x - c)$, where some of the factors may be repeated. The value of c in each factor is a zero of the function, so the function must also have n zeros provided that the zeros are counted according to their multiplicities.

TIP Refer to Examples 2–4. In each case, the number of zeros (including multiplicities) is the same as the degree of the polynomial.

> ### Number of Zeros of a Polynomial
>
> If $f(x)$ is a polynomial of degree $n \geq 1$ with complex coefficients, then $f(x)$ has exactly n complex zeros provided that each zero is counted by its multiplicity.

Now consider the polynomial from Example 4.

$$f(x) = x^4 - 2x^2 - 3 \quad \text{Zeros: } \sqrt{3}, -\sqrt{3}, i, -i$$

Notice that the polynomial has real coefficients. Furthermore, the zeros i and $-i$ appear as a pair. This is not a coincidence. For a polynomial with real coefficients, if $a + bi$ is a zero, then $a - bi$ is also a zero.

> ### Conjugate Zeros Theorem
>
> If $f(x)$ is a polynomial with real coefficients and if $a + bi$ $(b \neq 0)$ is a zero of $f(x)$, then its conjugate $a - bi$ is also a zero of $f(x)$.

EXAMPLE 5   Finding Zeros and Factoring a Polynomial

Given $f(x) = x^4 - 6x^3 + 28x^2 - 18x + 75$, and that $3 - 4i$ is a zero of $f(x)$,

a. Find the remaining zeros.

b. Factor $f(x)$ as a product of linear factors.

c. Solve the equation. $x^4 - 6x^3 + 28x^2 - 18x + 75 = 0$

Solution:

$f(x)$ is a fourth-degree polynomial, so we expect to find four zeros (including multiplicities). Further note that because $f(x)$ has real coefficients and because $3 - 4i$ is a zero, then the conjugate $3 + 4i$ must also be a zero. This leaves only two remaining zeros to find.

$$3 - 4i \,\big|\quad 1 \quad -6 \quad\quad 28 \quad -18 \quad\quad\quad 75$$

$$\underline{\quad\quad 3 - 4i \quad -25 \quad\quad 9 - 12i \quad -75}$$

Divide by $[x - (3 - 4i)]$. $1 \quad -3 - 4i \quad\quad 3 \quad -9 - 12i \quad\; \underline{|0}$

coefficients of the quotient

One strategy is to use synthetic division twice using the two known zeros.

Note: $(3 - 4i)(-3 - 4i)$

$\quad = -9 - 12i + 12i + 16i^2$

$\quad = -25$

Note: $(3 - 4i)(-9 - 12i)$

$\quad = -27 - 36i + 36i + 48i^2$

$\quad = -75$

Since $3 + 4i$ is a zero of $f(x)$ it must also be a zero of the quotient.

$$3 + 4i \,\big|\quad 1 \quad -3 - 4i \quad 3 \quad -9 - 12i$$

$$\underline{\quad\quad 3 + 4i \quad 0 \quad\; 9 + 12i}$$

Divide by $[x - (3 + 4i)]$. $1 \quad\quad 0 \quad\quad 3 \quad\quad \underline{|0}$

Divide the quotient by $[x - (3 + 4i)]$.

The resulting quotient is quadratic: $x^2 + 3$.

Now we have $f(x) = [x - (3 - 4i)][x - (3 + 4i)](x^2 + 3)$.
The remaining two zeros are found by solving $x^2 + 3 = 0$.

$$x^2 + 3 = 0$$
$$x^2 = -3$$
$$x = \pm i\sqrt{3}$$

a. The zeros of $f(x)$ are: $3 - 4i$, $3 + 4i$, $i\sqrt{3}$, and $-i\sqrt{3}$.

b. $f(x)$ factors as four linear factors.

$$f(x) = [x - (3 - 4i)][x - (3 + 4i)](x - i\sqrt{3})(x + i\sqrt{3})$$

c. The solution set for $x^4 - 6x^3 + 28x^2 - 18x + 75 = 0$ is $\{3 \pm 4i, \pm i\sqrt{3}\}$.

Skill Practice 5   Given $f(x) = x^4 - 2x^3 + 28x^2 - 4x + 52$, and that $1 + 5i$ is a zero of $f(x)$,

   a. Find the zeros.
   b. Factor $f(x)$ as a product of linear factors.
   c. Solve the equation. $x^4 - 2x^3 + 28x^2 - 4x + 52 = 0$

EXAMPLE 6   **Building a Polynomial with Specified Conditions**

a. Find a third-degree polynomial $f(x)$ with integer coefficients and with zeros of $\frac{2}{3}$ and $4 + 2i$.

b. Find a polynomial $g(x)$ of lowest degree with zeros of -2 (multiplicity 1) and 4 (multiplicity 3), and satisfying the condition that $g(0) = 256$.

Answers

5. a. Zeros:
   $1 + 5i, 1 - 5i, i\sqrt{2}, -i\sqrt{2}$
   b. $f(x) = [x - (1 + 5i)][x - (1 - 5i)]$
   $(x - i\sqrt{2})(x + i\sqrt{2})$
   c. $\{1 \pm 5i, \pm i\sqrt{2}\}$

Solution:

a. $f(x)$ is to be a polynomial with integer coefficients (and therefore real coefficients). If $4 + 2i$ is a zero, then $4 - 2i$ must also be a zero. By the linear factorization theorem we have:

$$f(x) = a\left(x - \frac{2}{3}\right)[x - (4 + 2i)][x - (4 - 2i)] \qquad \text{\emph{a} is nonzero number.}$$

$$= a\left(x - \frac{2}{3}\right)[x - 4 - 2i][x - 4 + 2i] \qquad \text{Simplify within square brackets.}$$

$$= a\left(x - \frac{2}{3}\right)[(x - 4) - 2i][(x - 4) + 2i] \qquad \text{Regroup terms.}$$

$$= a\left(x - \frac{2}{3}\right)[(x - 4)^2 - (2i)^2] \qquad \text{Multiply as a difference of squares.}$$

Square the binomial.
$$(x - 4)^2 = (x)^2 - 2(x)(4) + (4)^2$$
$$= x^2 - 8x + 16$$

$$= a\left(x - \frac{2}{3}\right)[x^2 - 8x + 16 - (-4)] \qquad \text{Note that } (2i)^2 = 4i^2 = 4(-1) = -4$$

$$= a\left(x - \frac{2}{3}\right)[x^2 - 8x + 20]$$

$$= 3\left(x - \frac{2}{3}\right)(x^2 - 8x + 20) \qquad \text{To give } f(x) \text{ integer coefficients,}$$
choose a to be any multiple of the denominator of $\frac{2}{3}$. We have chosen $a = 3$.

$$= (3x - 2)(x^2 - 8x + 20)$$

$$= 3x^3 - 24x^2 + 60x - 2x^2 + 16x - 40$$

$$= 3x^3 - 26x^2 + 76x - 40$$

b. $g(x) = a(x + 2)^1(x - 4)^3 \qquad$ -2 is a zero of multiplicity 1. 4 is a zero of multiplicity 3.

$$= a(x + 2)[(x - 4)(x - 4)^2] \qquad \text{Write } (x - 4)^3 \text{ as } (x - 4)(x - 4)^2.$$

$$= a(x + 2)[(x - 4)(x^2 - 8x + 16)] \qquad (x - 4)^2 \text{ expands as a perfect square trinomial.}$$
$$(x - 4)^2 = (x^2) - 2(x)(4) + (4)^2$$
$$= x^2 - 8x + 16$$

$$= a(x + 2)[(x - 4)(x^2 - 8x + 16)] \qquad \text{Multiply within square brackets.}$$

$$= a(x + 2)(x^3 - 4x^2 - 8x^2 + 32x + 16x - 64) \qquad \text{Multiply } (x + 2) \text{ by the result.}$$

$$= a(x + 2)(x^3 - 12x^2 + 48x - 64) \qquad \text{Combine like terms.}$$

$$= a(x^4 - 10x^3 + 24x^2 + 32x - 128)$$

We also have the condition that $g(0) = 256$.

$$g(0) = a[(0)^4 - 10(0)^3 + 24(0)^2 + 32(0) - 128] = 256$$

$$-128a = 256$$

$$a = -2$$

Therefore, $g(x) = -2(x^4 - 10x^3 + 24x^2 + 32x - 128)$

$$g(x) = -2x^4 + 20x^3 - 48x^2 - 64x + 256$$

FOR REVIEW

The square of a binomial results in a perfect square trinomial.

$(a + b)^2 = a^2 + 2ab + b^2$
$(a - b)^2 = a^2 - 2ab + b^2$

See Section 2.2 for review.

Skill Practice 6

a. Find a third-degree polynomial $f(x)$ with integer coefficients and with zeros of $2 + i$ and $\frac{4}{3}$.

b. Find a polynomial $g(x)$ of lowest degree with zeros of -3 (multiplicity 2) and 5 (multiplicity 2), and satisfying the condition that $g(0) = 450$.

3. Apply Descartes' Rule of Signs

Finding the zeros of a polynomial analytically can be a difficult (or impossible) task. For example, consider

$$f(x) = x^5 - 18x^4 + 128x^3 - 450x^2 + 783x - 540$$

Applying the rational zero theorem gives us the following possible rational zeros.

$$\pm 1, \pm 2, \pm 3, \pm 4, \pm 5, \pm 6, \pm 9, \pm 10, \pm 12, \pm 15, \pm 18, \pm 20,$$
$$\pm 27, \pm 30, \pm 36, \pm 45, \pm 54, \pm 60, \pm 90, \pm 108, \pm 135, \pm 180, \pm 270, \pm 540$$

However, we will soon present the upper and lower bound theorem to show that the real zeros of $f(x)$ are between -1 and 18. Furthermore, we can use a tool called Descartes' rule of signs to show that none of the zeros is negative. This eliminates all possible rational zeros except for 1, 2, 3, 4, 5, 6, 9, 10, 12, and 15.

To study Descartes' rule of signs, we need to establish what is meant by a "sign change" between consecutive terms in a polynomial. For example, the following polynomial is written in descending order and has three changes in sign between consecutive coefficients.

$$2x^6 - 3x^4 - x^3 + 5x^2 - 6x - 4 \qquad \text{(3 sign changes)}$$

| positive to negative | negative to positive | positive to negative |

TIP For a polynomial with real coefficients, the reason we reduce the number of possible real zeros in increments of 2 is because the alternative, nonreal zeros, occur in conjugate pairs.

Descartes' Rule of Signs

Let $f(x)$ be a polynomial with real coefficients and a nonzero constant term. Then,

1. The number of *positive* real zeros is either
   - the same as the number of sign changes in $f(x)$ or
   - less than the number of sign changes in $f(x)$ by a positive even integer.
2. The number of *negative* real zeros is either
   - the same as the number of sign changes in $f(-x)$ or
   - less than the number of sign changes in $f(-x)$ by a positive even integer.

Descartes' rule of signs is demonstrated in Examples 7 and 8.

EXAMPLE 7   **Applying Descartes' Rule of Signs**

Determine the number of possible positive and negative real zeros.

$$f(x) = x^5 - 6x^4 + 12x^3 - 12x^2 + 11x - 6$$

Answers
6. a. $f(x) = 3x^3 - 16x^2 + 31x - 20$
   b. $g(x) = 2x^4 - 8x^3 - 52x^2 + 120x + 450$

Solution:

$f(x)$ has real coefficients and the constant term is nonzero.

To determine the number of possible positive real zeros, determine the number of sign changes in $f(x)$.

$$f(x) = x^5 - 6x^4 + 12x^3 - 12x^2 + 11x - 6 \qquad \text{(5 sign changes)}$$

The number of possible positive real zeros is either 5, 3, or 1.

To determine the number of possible negative real zeros, determine the number of sign changes in $f(-x)$.

$$f(-x) = (-x)^5 - 6(-x)^4 + 12(-x)^3 - 12(-x)^2 + 11(-x) - 6$$
$$= -x^5 - 6x^4 - 12x^3 - 12x^2 - 11x - 6 \qquad \text{(0 sign changes)}$$

There are no sign changes in $f(-x)$. Therefore, $f(x)$ has no negative real zeros.

> **TIP** Nonreal zeros are numbers that contain the imaginary number i such as $-3i$, $3i$, $4 + 7i$, $4 - 7i$ and so on.

Number of possible positive real zeros	5	3	1
Number of possible negative real zeros	0	0	0
Number of nonreal zeros	0	2	4
Total (including multiplicities)	5	5	5

The graph of $f(x) = x^5 - 6x^4 + 12x^3 - 12x^2 + 11x - 6$ is shown in Figure 7-19. Notice that there are three positive x-intercepts and therefore, three positive real zeros. There are no negative x-intercepts, as expected. The remaining zeros of the polynomial $x^5 - 6x^4 + 12x^3 - 12x^2 + 11x - 6$ are not real numbers.

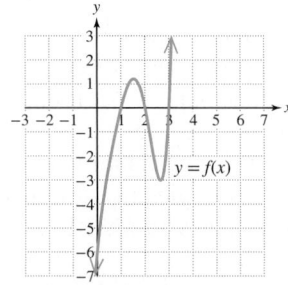

Figure 7-19

Skill Practice 7 Determine the number of possible positive and negative real zeros. $\qquad f(x) = 4x^5 + 6x^3 + 2x^2 + 6$

EXAMPLE 8 Applying Descartes' Rule of Signs

Determine the number of possible positive and negative real zeros.

$$g(x) = 2x^6 - 5x^4 - 3x^3 + 7x^2 + 2x + 5$$

Solution:

$g(x)$ has real coefficients and the constant term is nonzero.

$$g(x) = 2x^6 - 5x^4 - 3x^3 + 7x^2 + 2x + 5 \qquad \text{2 sign changes in } g(x)$$

The number of possible positive real zeros is either 2 or 0.

$$g(-x) = 2(-x)^6 - 5(-x)^4 - 3(-x)^3 + 7(-x)^2 + 2(-x) + 5$$
$$= 2x^6 - 5x^4 + 3x^3 + 7x^2 - 2x + 5 \qquad \text{4 sign changes in } g(-x)$$

The number of possible negative real zeros is either 4, 2, or 0.

Number of possible positive real zeros	2	2	2	0	0	0
Number of possible negative real zeros	4	2	0	4	2	0
Number of nonreal zeros	0	2	4	2	4	6
Total (including multiplicities)	6	6	6	6	6	6

Answers

7. Positive: 0; Negative: 1
8. Positive: 4, 2, or 0;
Negative: 2 or 0

Skill Practice 8 Determine the number of possible positive and negative real zeros. $\qquad g(x) = 8x^6 - 5x^7 + 3x^5 - x^2 - 3x + 1$

Descartes' rule of signs stipulates that the constant term of the polynomial $f(x)$ is nonzero. If the constant term is 0, we can factor out the lowest power of x and apply Descartes' rule of signs to the resulting factor.

$$f(x) = x^7 - 8x^6 + 15x^5$$

$$f(x) = x^5(x^2 - 8x + 15)$$

> Descartes' rule of signs can be applied to $x^2 - 8x + 15$ to show that there may be 2 or 0 remaining positive real zeros.

> The value 0 is a zero of $f(x)$ of multiplicity 5.

4. Find Upper and Lower Bounds

The next theorem helps us limit our search for the real zeros of a polynomial. First we define two key terms.

- A real number b is called an **upper bound** of the real zeros of a polynomial if all real zeros are less than or equal to b.
- A real number a is called a **lower bound** of the real zeros of a polynomial if all real zeros are greater than or equal to a.

Avoiding Mistakes

It is important to note that upper and lower bounds are not unique. Any number greater than b is also an upper bound for the zeros of the polynomial. Likewise any number less than a is also a lower bound.

Upper and Lower Bound Theorem for the Real Zeros of a Polynomial

Let $f(x)$ be a polynomial of degree $n \geq 1$ with real coefficients and a positive leading coefficient. Further suppose that $f(x)$ is divided by $(x - c)$.

1. If $c > 0$ and if both the remainder and the coefficients of the quotient are nonnegative, then c is an upper bound for the real zeros of $f(x)$.
2. If $c < 0$ and the coefficients of the quotient and the remainder alternate in sign (with 0 being considered either positive or negative as needed), then c is a lower bound for the real zeros of $f(x)$.

The rules for finding upper and lower bounds are stated for polynomial functions having a positive leading coefficient. However, $f(x) = 0$ and $-f(x) = 0$ are equivalent equations. Therefore, if $f(x)$ has a negative leading coefficient, we can factor out -1 from $f(x)$ and apply the rule for upper and lower bounds accordingly.

EXAMPLE 9   **Applying the Upper and Lower Bound Theorem**

Given $f(x) = 2x^5 + x^4 + 9x^2 - 32x + 20$,

a. Determine if the upper bound theorem identifies 2 as an upper bound for the real zeros of $f(x)$.

b. Determine if the lower bound theorem identifies -2 as a lower bound for the real zeros of $f(x)$.

Solution:

a. Divide $f(x)$ by $(x - 2)$.

	2	2	1	0	9	−32	20
This row nonnegative			4	10	20	58	52
		2	5	10	29	26	⎿72

First note that the leading coefficient of the polynomial is positive.

The remainder and all coefficients of the quotient are nonnegative.

2 is an upper bound for the real zeros of $f(x)$.

b. Divide $f(x)$ by $(x + 2)$.

$$
\begin{array}{r|rrrrrr}
-2 & 2 & 1 & 0 & 9 & -32 & 20 \\
 & & -4 & 6 & -12 & 6 & 52 \\
\hline
 & 2 & -3 & 6 & -3 & -26 & \boxed{72} \\
 & + & - & + & - & - & + \\
\end{array}
$$

No sign change

The signs of the quotient do not alternate. Therefore, we cannot conclude that -2 is a lower bound for the real zeros of $f(x)$.

Skill Practice 9   Given $f(x) = x^4 - 2x^3 - 13x^2 - 4x - 30$,

a. Determine if the upper bound theorem identifies 4 as an upper bound for the real zeros of $f(x)$.

b. Determine if the lower bound theorem identifies -4 as a lower bound for the real zeros of $f(x)$.

TIP   From Example 9, although we cannot conclude that -2 is a lower bound for the real zeros of $f(x)$, we can try other negative real numbers. For example, -3 is a lower bound for the real zeros of $f(x)$.

$$
\begin{array}{r|rrrrr}
-3 & 2 & 1 & 0 & 9 & -32 & 20 \\
 & & -6 & 15 & -45 & 108 & -228 \\
\hline
 & 2 & -5 & 15 & -36 & 76 & \boxed{-208} \\
\end{array}
$$

Therefore, -3 is a lower bound.

signs all alternate

In Example 10, we will use the tools presented in Sections 7.2–7.4 to find all zeros of a polynomial.

EXAMPLE 10   **Finding the Zeros of a Polynomial**

Find the zeros and their multiplicities. $f(x) = 2x^5 + x^4 + 9x^2 - 32x + 20$

Solution:

$f(x)$ is a fifth-degree polynomial and must have five zeros (including multiplicities). We begin by finding the rational zeros (if any exist). By the rational zero theorem, the possible rational zeros are

$$
\pm 1, \pm 2, \pm 4, \pm 5, \pm 10, \pm 20, \pm\frac{1}{2}, \pm\frac{5}{2}
$$

However, we also know from Example 9 that 2 is not a zero of $f(x)$, but is an upper bound for the real zeros. From the Tip following Example 9, we know that -3 is not a zero of $f(x)$, but is a lower bound for the real zeros. Therefore, we can restrict the list of possible rational zeros to those on the interval $(-3, 2)$.

$$
-\frac{5}{2}, -2, \pm\frac{1}{2}, \pm 1
$$

After testing several possible rational zeros, we find that 1 is a zero.

$$
\begin{array}{r|rrrrrr}
1 & 2 & 1 & 0 & 9 & -32 & 20 \\
 & & 2 & 3 & 3 & 12 & -20 \\
\hline
 & 2 & 3 & 3 & 12 & -20 & \boxed{0} \\
\end{array}
$$

We have $f(x) = (x - 1)(2x^4 + 3x^3 + 3x^2 + 12x - 20)$. Now look for the zeros of the reduced polynomial. We will try 1 again because it may be a repeated zero.

$$
\begin{array}{r|rrrrr}
1 & 2 & 3 & 3 & 12 & -20 \\
 & & 2 & 5 & 8 & 20 \\
\hline
 & 2 & 5 & 8 & 20 & \boxed{0} \\
\end{array}
$$

The value 1 is a repeated zero.

Answers

9. a. No   **b.** Yes

We have $f(x) = (x - 1)^2(2x^3 + 5x^2 + 8x + 20)$.

Because the polynomial $2x^3 + 5x^2 + 8x + 20$ has no sign changes, Descartes' rule of signs indicates that there are no other positive real zeros. Now the list of possible rational zeros is restricted to $-\frac{5}{2}$, -2, $-\frac{1}{2}$, and -1. We find that $-\frac{5}{2}$ is a zero of $f(x)$.

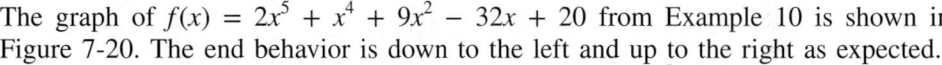

$$
\begin{array}{r|rrrr}
-\frac{5}{2} & 2 & 5 & 8 & 20 \\
 & & -5 & 0 & -20 \\
\hline
 & 2 & 0 & 8 & \underline{|0} \\
\end{array}
$$

Thus, $f(x) = (x - 1)^2\left(x + \frac{5}{2}\right)(2x^2 + 8)$.

$2x^2 + 8 = 0$ The remaining two zeros are found by solving the equation
$x^2 = -4$ $2x^2 + 8 = 0$.
$x = \pm 2i$

The zeros are: 1 (multiplicity of 2), $-\frac{5}{2}$, $2i$, and $-2i$ (each with multiplicity of 1).

> **Skill Practice 10**   Find the zeros and their multiplicities.
>
> $f(x) = x^5 + 6x^3 - 2x^2 - 27x - 18$

The graph of $f(x) = 2x^5 + x^4 + 9x^2 - 32x + 20$ from Example 10 is shown in Figure 7-20. The end behavior is down to the left and up to the right as expected.

The real zeros of $f(x)$ correspond to the x-intercepts $\left(-\frac{5}{2}, 0\right)$ and $(1, 0)$. The point $(1, 0)$ is a touch point because 1 is a zero with an even multiplicity. The graph crosses the x-axis at $-\frac{5}{2}$ because $-\frac{5}{2}$ is a zero with an odd multiplicity.

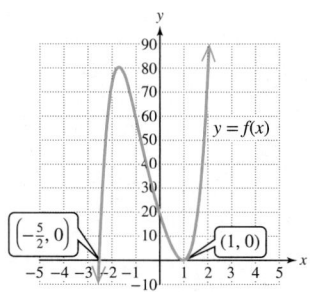

Figure 7-20

TECHNOLOGY CONNECTIONS

Applications of Graphing Utilities to Polynomials

A graphing utility can help us analyze a polynomial. For example, given $f(x) = 2x^3 - 11x^2 - 5x + 50$, the possible rational zeros are: ± 1, ± 2, ± 5, ± 10, ± 25, ± 50, $\pm\frac{1}{2}$, $\pm\frac{5}{2}$, and $\pm\frac{25}{2}$. By graphing the function f on a graphing utility, it appears that the function may cross the x-axis at -2, 2.5, and 5. So we might consider testing these values first.

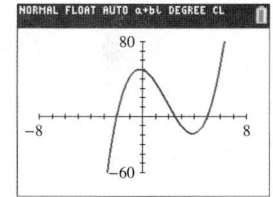

Our knowledge of algebra can also help us use a graphing device effectively. Consider $f(x) = 2x^5 + x^4 + 9x^2 - 32x + 20$ from Examples 9 and 10. We know that -3 is a lower bound for the real zeros of $f(x)$ and that 2 is an upper bound. Therefore, we can set a viewing window showing x between -3 and 2 and be guaranteed to see all the real zeros of $f(x)$.

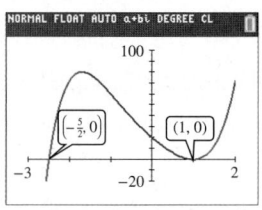

Finally, when analytical methods fail, we can use the **Zero** feature on a graphing utility to approximate the real zeros of a polynomial. Given $f(x) = x^3 - 7.14x^2 + 25.6x - 40.8$, the calculator approximates a zero of 3.1258745. The real zeros of a function can also be approximated by repeated use of the intermediate value theorem. (See the online group activity "Investigating the Bisection Method for Finding Zeros.")

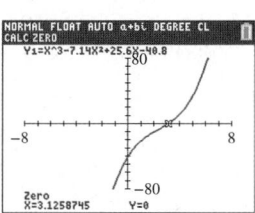

Answer

10. -1 (multiplicity 2), 2, $3i$, $-3i$ (each with multiplicity 1)

Prerequisite Review

For Exercises R.1–R.6, perform the indicated operations. Write the answers in the form $a + bi$. (See Section 3.3 for review.)

R.1. $(9i)(-9i)$ **R.2.** $(4i)(4i)$ **R.3.** $\left(6 + \sqrt{13}i\right)\left(6 - \sqrt{13}i\right)$

R.4. $\left(3 + \sqrt{11}i\right)\left(3 - \sqrt{11}i\right)$ **R.5.** $(5 - 6i)^2 + (5 + 6i)^2$ **R.6.** $(3 + 4i)^2 - (3 - 4i)^2$

For Exercises R.7–R.8, find the product. (See Section 3.3 for review.)

R.7. $(2x + 1)(x - 3i)(x + 3i)$ **R.8.** $(5x - 6)(x + i)(x - i)$

For Exercises R.9–R.12, solve the equation. (See Section 3.5 for review.)

R.9. $x^2 - 28 = 0$ **R.10.** $y^2 - 12 = 0$

R.11. $w^2 + 10 = 0$ **R.12.** $p^2 + 15 = 0$

R.13. Given $A = \left\{-\dfrac{2}{9}, 7, \sqrt{2}, 6\pi, 0, -0.\overline{9}, 6.25\right\}$, list the rational numbers in A. (See Section R.2.)

R.14. Given $B = \left\{4.72, -\dfrac{\pi}{2}, 143, -8, \sqrt{16}, 0.\overline{87}\right\}$, list the rational numbers in B.

R.15. Given $f(x) = 2x^4 - 3x^2 + 5x + 1$, find $f(-x)$.

R.16. Given $f(x) = 3x^5 + 4x^3 + 2x^2 - 4$, find $f(-x)$.

Concept Connections

1. The _____ of a polynomial $f(x)$ are the solutions (or roots) of the equation $f(x) = 0$.

2. If $f(x)$ is a polynomial of degree $n \geq 1$ with complex coefficients, then $f(x)$ has exactly _____ complex zeros, provided that each zero is counted by its multiplicity.

3. The conjugate zeros theorem states that if $f(x)$ is a polynomial with real coefficients, and if $a + bi$ is a zero of $f(x)$, then _____ is also a zero of $f(x)$.

4. A real number b is called an _____ bound of the real zeros of a polynomial $f(x)$ if all real zeros of $f(x)$ are less than or equal to b.

5. A real number a is called a lower bound of the real zeros of a polynomial $f(x)$ if all real zeros of $f(x)$ are _____ or equal to a.

6. Explain why the number 7 cannot be a rational zero of the polynomial $f(x) = 2x^3 + 5x^2 - x + 6$.

Objective 1: Apply the Rational Zero Theorem

For Exercises 7–12, list the possible rational zeros. **(See Example 1)**

7. $f(x) = x^5 - 2x^3 + 7x^2 + 4$ **8.** $g(x) = x^3 - 5x^2 + 2x - 9$ **9.** $h(x) = 4x^4 + 9x^3 + 2x - 6$

10. $k(x) = 25x^7 + 22x^4 - 3x^2 + 10$ **11.** $m(x) = -12x^6 + 4x^3 - 3x^2 + 8$ **12.** $n(x) = -16x^4 - 7x^3 + 2x + 6$

13. Which of the following is *not* a possible zero of $f(x) = 2x^3 - 5x^2 + 12$?

$$1, 7, \frac{5}{3}, \frac{3}{2}$$

14. Which of the following is *not* a possible zero of $f(x) = 4x^5 - 2x^3 + 10$?

$$3, 5, \frac{5}{2}, \frac{3}{2}$$

For Exercises 15–16, find all the rational zeros.

15. $p(x) = 2x^4 - x^3 - 5x^2 + 2x + 2$

16. $q(x) = x^4 + x^3 - 7x^2 - 5x + 10$

For Exercises 17–28, find all the zeros. **(See Examples 2–4)**

17. $c(x) = 2x^4 - 7x^3 - 17x^2 + 58x - 24$

18. $d(x) = 3x^4 - 2x^3 - 21x^2 - 4x + 12$

19. $f(x) = x^3 - 7x^2 + 6x + 20$

20. $g(x) = x^3 - 7x^2 + 14x - 6$

21. $h(x) = 5x^3 - x^2 - 35x + 7$

22. $k(x) = 7x^3 - x^2 - 21x + 3$

23. $m(x) = 3x^4 - x^3 - 36x^2 + 60x - 16$

24. $n(x) = 2x^4 + 9x^3 - 5x^2 - 57x - 45$

25. $q(x) = x^3 - 4x^2 - 2x + 20$

26. $p(x) = x^3 - 8x^2 + 29x - 52$

27. $t(x) = x^4 - x^2 - 90$

28. $v(x) = x^4 - 12x^2 - 13$

Objective 2: Apply the Fundamental Theorem of Algebra

29. Given a polynomial $f(x)$ of degree $n \geq 1$, the fundamental theorem of algebra guarantees at least _____ complex zero.

30. The number of zeros of $f(x) = 4x^3 - 5x^2 + 6x - 3$ is _____, provided that each zero is counted according to its multiplicity.

31. If $f(x)$ is a polynomial with real coefficients and zeros of 5 (multiplicity 2), −1 (multiplicity 1), $2i$, and $3 + 4i$, what is the minimum degree of $f(x)$?

32. If $g(x)$ is a polynomial with real coefficients and zeros of −4 (multiplicity 3), 6 (multiplicity 2), $1 + i$, and $2 - 7i$, what is the minimum degree of $g(x)$?

For Exercises 33–38, a polynomial $f(x)$ and one or more of its zeros is given.

    a. Find all the zeros.

    b. Factor $f(x)$ as a product of linear factors.

    c. Solve the equation $f(x) = 0$. **(See Example 5)**

33. $f(x) = x^4 - 4x^3 + 22x^2 + 28x - 203$; $2 - 5i$ is a zero

34. $f(x) = x^4 - 6x^3 + 5x^2 + 30x - 50$; $3 - i$ is a zero

35. $f(x) = 3x^3 - 28x^2 + 83x - 68$; $4 + i$ is a zero

36. $f(x) = 5x^3 - 54x^2 + 170x - 104$; $5 + i$ is a zero

37. $f(x) = 4x^5 + 37x^4 + 117x^3 + 87x^2 - 193x - 52$;

    $-3 + 2i$ and $-\dfrac{1}{4}$ are zeros

38. $f(x) = 2x^5 - 5x^4 - 4x^3 - 22x^2 + 50x + 75$;

    $-1 - 2i$ and $\dfrac{5}{2}$ are zeros

For Exercises 39–48, write a polynomial $f(x)$ that satisfies the given conditions. **(See Example 6)**

39. Degree 3 polynomial with integer coefficients with zeros $6i$ and $\frac{4}{5}$

40. Degree 3 polynomial with integer coefficients with zeros $-4i$ and $\frac{3}{2}$

41. Polynomial of lowest degree with zeros of −4 (multiplicity 1), 2 (multiplicity 3) and with $f(0) = 160$

42. Polynomial of lowest degree with zeros of 5 (multiplicity 2) and −3 (multiplicity 2) and with $f(0) = -450$

43. Polynomial of lowest degree with zeros of $-\frac{4}{3}$ (multiplicity 2) and $\frac{1}{2}$ (multiplicity 1) and with $f(0) = -16$

44. Polynomial of lowest degree with zeros of $-\frac{5}{6}$ (multiplicity 2) and $\frac{1}{3}$ (multiplicity 1) and with $f(0) = -25$

45. Polynomial of lowest degree with real coefficients and with zeros $7 - 4i$ and 0 (multiplicity 4)

46. Polynomial of lowest degree with real coefficients and with zeros $5 - 10i$ and 0 (multiplicity 3)

47. Polynomial of lowest degree with real coefficients and zeros of $5i$ and $6 - i$.

48. Polynomial of lowest degree with real coefficients and zeros of $-3i$ and $5 + 2i$.

Objective 3: Apply Descartes' Rule of Signs

For Exercises 49–56, determine the number of possible positive and negative real zeros for the given function. **(See Examples 7–8)**

49. $f(x) = x^6 - 2x^4 + 4x^3 - 2x^2 - 5x - 6$

50. $g(x) = 3x^7 + 4x^4 - 6x^3 + 5x^2 - 6x + 1$

51. $k(x) = -8x^7 + 5x^6 - 3x^4 + 2x^3 - 11x^2 + 4x - 3$

52. $h(x) = -4x^9 + 6x^8 - 5x^5 - 2x^4 + 3x^2 - x + 8$

53. $p(x) = 0.11x^4 + 0.04x^3 + 0.31x^2 + 0.27x + 1.1$

54. $q(x) = -0.6x^4 + 0.8x^3 - 0.6x^2 + 0.1x - 0.4$

55. $v(x) = \dfrac{1}{8}x^6 + \dfrac{1}{6}x^4 + \dfrac{1}{3}x^2 + \dfrac{1}{10}$

56. $t(x) = \dfrac{1}{1000}x^6 + \dfrac{1}{100}x^4 + \dfrac{1}{10}x^2 + 1$

For Exercises 57–58, use Descartes' rule of signs to determine the total number of real zeros and the number of positive and negative real zeros. (*Hint*: First factor out x to its lowest power.)

57. $f(x) = x^8 + 5x^6 + 6x^4 - x^3$

58. $f(x) = -5x^8 - 3x^6 - 4x^2 + x$

Objective 4: Find Upper and Lower Bounds

For Exercises 59–64, (**See Example 9**)

   a. Determine if the upper bound theorem identifies the given number as an upper bound for the real zeros of $f(x)$.

   b. Determine if the lower bound theorem identifies the given number as a lower bound for the real zeros of $f(x)$.

59. $f(x) = x^5 + 6x^4 + 5x^2 + x - 3$

   a. 2   **b.** −5

60. $f(x) = x^4 + 8x^3 - 4x^2 + 7x - 3$

   a. 3   **b.** −4

61. $f(x) = 8x^3 - 42x^2 + 33x + 28$

   a. 6   **b.** −1

62. $f(x) = 6x^3 - x^2 - 57x + 70$

   a. 4   **b.** −4

63. $f(x) = 2x^5 + 11x^4 - 63x^2 - 50x + 40$

   a. 3   **b.** −6

64. $f(x) = 3x^5 - 16x^4 + 5x^3 + 90x^2 - 138x + 36$

   a. 6   **b.** −3

For Exercises 65–68, determine if the statement is true or false. If a statement is false, explain why.

65. If 5 is an upper bound for the real zeros of $f(x)$, then 6 is also an upper bound.

66. If 5 is an upper bound for the real zeros of $f(x)$, then 4 is also an upper bound.

67. If −3 is a lower bound for the real zeros of $f(x)$, then −2 is also a lower bound.

68. If −3 is a lower bound for the real zeros of $f(x)$, then −4 is also a lower bound.

For Exercises 69–84, find the zeros and their multiplicities. Consider using Descartes' rule of signs and the upper and lower bound theorem to limit your search for rational zeros. (**See Example 10**)

69. $f(x) = 8x^3 - 42x^2 + 33x + 28$
   (*Hint*: See Exercise 61.)

70. $f(x) = 6x^3 - x^2 - 57x + 70$
   (*Hint*: See Exercise 62.)

71. $f(x) = 2x^5 + 11x^4 - 63x^2 - 50x + 40$
   (*Hint*: See Exercise 63.)

72. $f(x) = 3x^5 - 16x^4 + 5x^3 + 90x^2 - 138x + 36$
   (*Hint*: See Exercise 64.)

73. $f(x) = 4x^4 + 20x^3 + 13x^2 - 30x + 9$

74. $f(x) = 9x^4 + 30x^3 + 13x^2 - 20x + 4$

75. $f(x) = 2x^4 - 11x^3 + 27x^2 - 41x + 15$

76. $g(x) = 3x^4 - 20x^3 + 51x^2 - 56x + 20$

77. $h(x) = 4x^4 - 28x^3 + 73x^2 - 90x + 50$

78. $k(x) = 9x^4 - 42x^3 + 70x^2 - 34x + 5$

79. $f(x) = x^6 + 2x^5 + 11x^4 + 20x^3 + 10x^2$

80. $f(x) = x^6 + 6x^5 + 12x^4 + 18x^3 + 27x^2$

81. $f(x) = x^5 - 10x^4 + 34x^3$

82. $f(x) = x^6 - 12x^5 + 40x^4$

83. $f(x) = -x^3 + 3x^2 - 9x - 13$

84. $f(x) = -x^3 + 5x^2 - 11x + 15$

Mixed Exercises

For Exercises 85–90, determine if the statement is true or false. If a statement is false, explain why.

85. A polynomial with real coefficients of degree 4 must have at least one real zero.

86. Given $f(x) = 2ix^4 - (3 + 6i)x^3 + 5x^2 + 7$, if $a + bi$ is a zero of $f(x)$, then $a - bi$ must also be a zero.

87. The graph of a 10th-degree polynomial must cross the x-axis exactly once.

88. Suppose that $f(x)$ is a polynomial, and that a and b are real numbers where $a < b$. If $f(a) < 0$ and $f(b) < 0$, then $f(x)$ has no real zeros on the interval $[a, b]$.

89. If c is a zero of a polynomial $f(x)$, with degree $n \geq 2$ then all other zeros of $f(x)$ are zeros of $\dfrac{f(x)}{x - c}$.

90. If b is an upper bound for the real zeros of a polynomial, then $-b$ is a lower bound for the real zeros of the polynomial.

91. Given that $x - c$ divides evenly into a polynomial $f(x)$, which statements are true?

  a. $x - c$ is a factor of $f(x)$.

  b. c is a zero of $f(x)$.

  c. The remainder of $f(x) \div (x - c)$ is 0.

  d. c is a solution (root) of the equation $f(x) = 0$.

93. a. Use the intermediate value theorem to show that $f(x) = 2x^2 - 7x + 4$ has a real zero on the interval [2, 3].

  b. Find the zeros.

92. a. Use the quadratic formula to solve $x^2 - 7x + 5 = 0$.

  b. Write $x^2 - 7x + 5$ as a product of linear factors.

94. Show that $x - a$ is a factor of $x^n - a^n$ for any positive integer n and constant a.

Write About It

95. Explain why a polynomial with real coefficients of degree 3 must have at least one real zero.

96. Why is it not necessary to apply the rational zero theorem, Descartes' rule of signs, or the upper and lower bound theorem to find the zeros of a second-degree polynomial?

97. Explain why $f(x) = 5x^6 + 7x^4 + x^2 + 9$ has no real zeros.

98. Explain why the fundamental theorem of algebra does not apply to $f(x) = \sqrt{x} + 3$. That is, no complex number c exists such that $f(c) = 0$.

Expanding Your Skills

99. Let n be a positive even integer. Determine the greatest number of possible nonreal zeros of $f(x) = x^n - 1$.

101. The front face of a tent is triangular and the height of the triangle is two-thirds of the base. The length of the tent is 3 ft more than the base of the triangular face. If the tent holds a volume of 108 ft³, determine its dimensions.

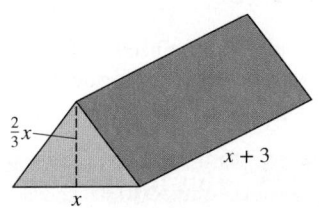

100. Let n be a positive odd integer. Determine the greatest number of possible nonreal zeros of $f(x) = x^n - 1$.

102. An underground storage tank for gasoline is in the shape of a right circular cylinder with hemispheres on each end. If the total volume of the tank is $\dfrac{104\pi}{3}$ ft³, find the radius of the tank.

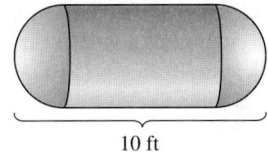

10 ft

103. A food company originally sells cereal in boxes with dimensions 10 in. by 7 in. by 2.5 in. To make more profit, the company decreases each dimension of the box by x inches but keeps the price the same. If the new volume is 81 in.³, by how much was each dimension decreased?

105. A rectangle is bounded by the x-axis and a parabola defined by $y = 4 - x^2$. What are the dimensions of the rectangle if the area is 6 cm²? Assume that all units of length are in centimeters.

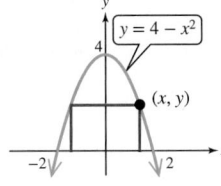

104. A truck rental company rents a 12-ft by 8-ft by 6-ft truck for $69.95 per day plus mileage. A customer prefers to rent a less expensive smaller truck whose dimensions are x ft smaller on each side. If the volume of the smaller truck is 240 ft³, determine the dimensions of the smaller truck.

106. A rectangle is bounded by the parabola defined by $y = x^2$, the x-axis, and the line $x = 5$ as shown in the figure. If the area of the rectangle is 12 in.², determine the dimensions of the rectangle.

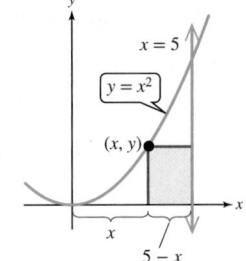

The linear factorization theorem tells us that a polynomial of degree $n \geq 1$ factors into n linear factors over the complex numbers. If we do not factor over the set of complex numbers, then a polynomial with real coefficients can be factored into linear factors and irreducible quadratic factors. An *irreducible quadratic factor* is a quadratic polynomial that does not factor further over the set of real numbers.

For example, consider the polynomial $f(x) = x^4 - 5x^3 + 5x^2 + 25x - 26$.

Factoring over the real numbers, we have two linear factors and one irreducible quadratic factor:

$$\underset{\substack{\text{2 linear} \\ \text{factors}}}{\underbrace{}} \underset{\substack{\text{irreducible} \\ \text{quadratic factor}}}{\underbrace{}}$$
$$x^4 - 5x^3 + 5x^2 + 25x - 26 = (x + 2)(x - 1)(x^2 - 6x + 13)$$

Factoring over the complex numbers, we have four linear factors as guaranteed by the linear factorization theorem.

$$x^4 - 5x^3 + 5x^2 + 25x - 26 = (x + 2)(x - 1)[x - (3 + 2i)][x - (3 - 2i)]$$

For Exercises 107–110,

    a. Factor the polynomial over the set of real numbers.   **b.** Factor the polynomial over the set of complex numbers.

107. $f(x) = x^4 + 2x^3 + x^2 + 8x - 12$

108. $f(x) = x^4 - 6x^3 + 9x^2 - 6x + 8$

109. $f(x) = x^4 + 2x^2 - 35$

110. $f(x) = x^4 + 8x^2 - 33$

111. Find all fourth roots of 1, by solving the equation $x^4 = 1$. (*Hint*: Find the zeros of the polynomial $f(x) = x^4 - 1$.)

112. Find all sixth roots of 1, by solving the equation $x^6 = 1$. [*Hint*: Find the zeros of the polynomial $f(x) = x^6 - 1$. Begin by factoring $x^6 - 1$ as $(x^3 - 1)(x^3 + 1)$.]

113. Use the rational zero theorem to show that $\sqrt{5}$ is an irrational number. (*Hint*: Show that $f(x) = x^2 - 5$ has no rational zeros.)

114. a. Given a linear equation $ax + b = 0$ $(a \neq 0)$, the solution is given by $x =$ _____.

    b. Given a quadratic equation $ax^2 + bx + c = 0$ $(a \neq 0)$, the solutions are given by $x =$ _____.

From Exercise 114, we see that linear and quadratic equations have generic formulas that can be used to find the solution sets. But what about a cubic polynomial equation? Mathematicians struggled for centuries to find such a formula. Finally, Italian mathematician Niccolo Tartaglia (1500–1557) developed a method to solve a cubic equation of the form

$$x^3 + mx = n$$

The result was later published in *Ars Magna*, by Gerolamo Cardano (1501–1576).

For Exercises 115–116, use the formula

$$x = \sqrt[3]{\sqrt{\left(\frac{n}{2}\right)^2 + \left(\frac{m}{3}\right)^3} + \frac{n}{2}} - \sqrt[3]{\sqrt{\left(\frac{n}{2}\right)^2 + \left(\frac{m}{3}\right)^3} - \frac{n}{2}}$$

to find a solution to the equation $x^3 + mx = n$.

115. $x^3 - 3x = -2$ **116.** $x^3 + 9x = 26$

Point of Interest

Early in the sixteenth century, Italian mathematicians Niccolo Tartaglia and Gerolamo Cardano solved a general cubic equation in terms of the constants appearing in the equation. Cardano's pupil, Ludovico Ferrari, then solved a general equation of fourth degree. Despite decades of work, no general solution to a fifth-degree equation was found. Finally, Norwegian mathematician Niels Abel and French mathematician Evariste Galois proved that no such solution exists.

OBJECTIVES

1. Solve Polynomial Inequalities Graphically
2. Solve Polynomial Inequalities Algebraically
3. Solve Applications Involving Polynomial Inequalities

1. Solve Polynomial Inequalities Graphically

An engineer for a food manufacturer must design an aluminum container for a hot drink mix. The container is to be a right circular cylinder 5.5 in. in height. The surface area represents the amount of aluminum used and is given by

$$S(r) = 2\pi r^2 + 11\pi r \qquad \text{where } r \text{ is the radius of the can.}$$

The engineer wants to limit the surface area so that at most 90 in.[2] of aluminum is used. To determine the restrictions on the radius, the engineer must solve the inequality $2\pi r^2 + 11\pi r \leq 90$ (see Exercise 79). This inequality is a quadratic inequality in the variable r. It is also categorized as a polynomial inequality of degree 2.

> **TIP** If $f(x)$ is a quadratic polynomial, then the inequalities $f(x) < 0$, $f(x) > 0$, $f(x) \leq 0$, and $f(x) \geq 0$ are called quadratic inequalities.

> **Definition of a Polynomial Inequality**
>
> Let $f(x)$ be a polynomial. Then an inequality of the form
>
> $f(x) < 0$, $f(x) > 0$, $f(x) \leq 0$, or $f(x) \geq 0$ is called a **polynomial inequality**.
>
> *Note:* A polynomial inequality is nonlinear if $f(x)$ is a polynomial of degree greater than 1.

Consider the polynomial inequalities $f(x) < 0$ and $f(x) > 0$. We need to determine the intervals over which $f(x)$ is negative or positive. For example, consider the graph of $f(x) = x^2 - 6x + 5$ (Figure 7-21).

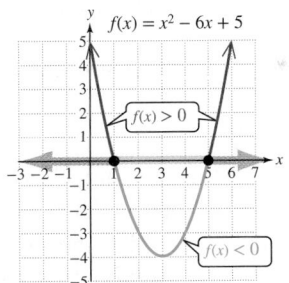

Figure 7-21

From the graph, we see that $f(x) = 0$ for $x = 1$ and $x = 5$. These are the values of x for which the graph of $y = f(x)$ intersects the x-axis. In addition, the solutions to the following inequalities can be determined from the graph.

Inequality	Solution set	Interval notation
$f(x) < 0$	The solution set is the set of x values for which the graph of $y = f(x)$ is *below* the x-axis (shown in blue.)	$(1, 5)$
$f(x) > 0$	The solution set is the set of x values for which the graph of $y = f(x)$ is *above* the x-axis (shown in red).	$(-\infty, 1) \cup (5, \infty)$

EXAMPLE 1   Using a Graph to Solve Polynomial Inequalities

Consider the graph of $y = f(x)$. Use the graph to solve the inequalities.

a. $f(x) < 0$ **b.** $f(x) \leq 0$

c. $f(x) > 0$ **d.** $f(x) \geq 0$

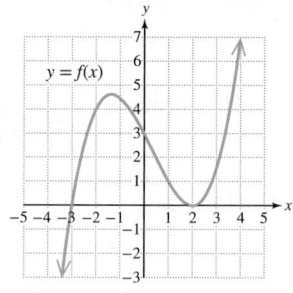

Solution:

a. The solution set to the inequality $f(x) < 0$ is the set of x values for which the graph is strictly *below* the x-axis.

The solution set is $(-\infty, -3)$.

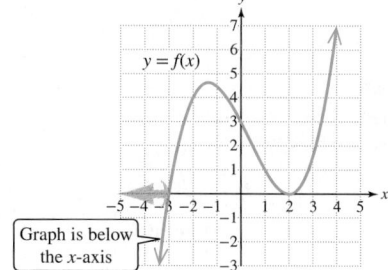

b. The solution set to the inequality $f(x) \leq 0$ is the set of x values for which the graph is *on or below* the x-axis. This includes the set of values from part (a) along with the values -3 and 2 where the graph intersects the x-axis.

The solution set is $(-\infty, -3] \cup \{2\}$.

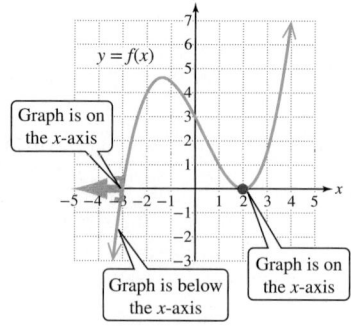

c. The solution set to the inequality $f(x) > 0$ is the set of x values for which the graph is strictly *above* the x-axis. This includes all x values greater than -3 except for 2. At $x = 2$, the graph intersects the x-axis and $f(x) = 0$.

The solution set is $(-3, 2) \cup (2, \infty)$.

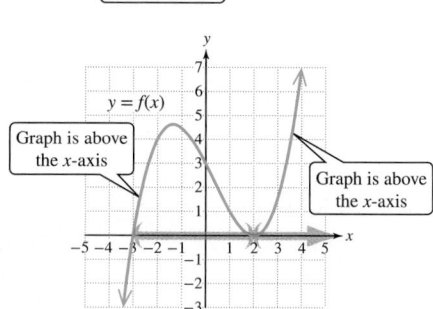

d. The solution set to the inequality $f(x) \geq 0$ is the set of x values for which the graph is *on or above* the x-axis. This includes the set of values from part (c) along with $x = -3$ and $x = 2$. Because $x = 2$ is included in the solution set, there is no interruption in the interval from -3 to infinity.

The solution set is $[-3, \infty)$.

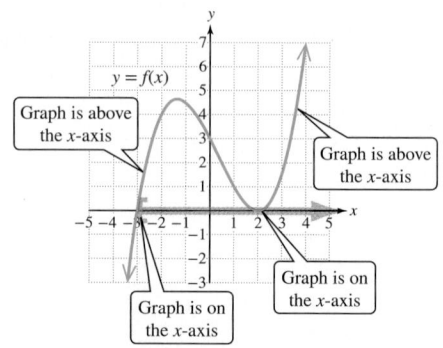

Skill Practice 1   Consider the graph of $y = f(x)$. Use the graph to solve the inequalities.

a. $f(x) < 0$ **b.** $f(x) \le 0$

c. $f(x) > 0$ **d.** $f(x) \ge 0$

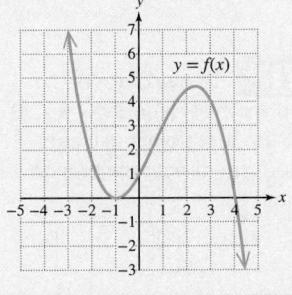

2. Solve Polynomial Inequalities Algebraically

From Example 1, notice that the x-intercepts of the graph of $y = f(x)$ define the endpoints (or "boundary" points) of the solutions to the inequalities $f(x) < 0$, $f(x) > 0$, and so on. We can find the "boundary" points analytically by solving the equation $f(x) = 0$. To solve a polynomial inequality, use the boundary points to divide the number line into distinct intervals. Then determine if the inequality is true or false on each interval. If a polynomial inequality is expressed in the form $f(x) < 0, f(x) \le 0, f(x) > 0$, or $f(x) \ge 0$ (that is with one side of the inequality set to zero), it is sufficient to test the *sign* (positive or negative) of the value of $f(x)$ on each interval. This is the basis for the procedure to solve polynomial inequalities algebraically.

Procedure to Solve a Polynomial Inequality

1. Express the inequality as $f(x) < 0, f(x) > 0, f(x) \le 0$, or $f(x) \ge 0$. That is, rearrange the terms of the inequality so that one side is set to zero.

2. Find the real solutions of the related equation $f(x) = 0$. These are the "boundary" points for the solution set to the inequality.

3. Determine the sign of $f(x)$ on the intervals defined by the boundary points.

   - If $f(x)$ is positive, then the values of x on the interval are solutions to $f(x) > 0$.

   - If $f(x)$ is negative, then the values of x on the interval are solutions to $f(x) < 0$.

4. Determine whether the boundary points are included in the solution set.

5. Write the solution set in interval notation or set-builder notation.

EXAMPLE 2   **Solving a Quadratic Inequality**

Solve the inequality. $3x(x - 1) > 10 - 2x$

Solution:

$3x(x - 1) > 10 - 2x$ **Step 1:** Write the inequality in the form $f(x) > 0$.

$3x^2 - 3x > 10 - 2x$

Answers

1. a. $(4, \infty)$ **b.** $\{-1\} \cup [4, \infty)$

   c. $(-\infty, -1) \cup (-1, 4)$

   d. $(-\infty, 4]$

To evaluate the
polynomial $f(x) = 3x^2 - x - 10$
at the test points, we can
perform direct substitution
such as:

$$f(3) = 3(3)^2 - (3) - 10$$
$$= 14$$

Or use synthetic division and
the remainder theorem.

$$
\begin{array}{r|rrr}
3 & 3 & -1 & -10 \\
  &   & 9 & 24 \\
\hline
  & 3 & 8 & \underline{14}
\end{array}
$$

$$\overbrace{3x^2 - x - 10}^{f(x)} > 0$$
$$3x^2 - x - 10 = 0$$

Step 2: Find the real solutions to the related
equation $f(x) = 0$.

$$(3x + 5)(x - 2) = 0$$

$$x = -\frac{5}{3} \quad \text{and} \quad x = 2$$

The boundary points are $-\frac{5}{3}$ and 2.

Step 3: Divide the x-axis into intervals defined by
the boundary points.

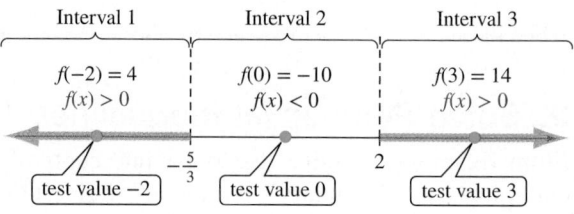

Determine the sign of
$f(x) = 3x^2 - x - 10$ on each
interval. One method is to
evaluate $f(x)$ for a test value x
on each interval.

Step 4: The solution set
does not include
the boundary
points because the
inequality is strict.

The solution set is $\left(-\infty, -\frac{5}{3}\right) \cup (2, \infty)$
or equivalently in set-builder notation
$\left\{x \mid x < -\frac{5}{3} \text{ or } x > 2\right\}$.

Step 5: Write the
solution set.

Skill Practice 2   Solve the inequality. $2x(x - 1) < 21 - x$

From Example 2, the key step is to determine the sign of $f(x)$ on the intervals
$\left(-\infty, -\frac{5}{3}\right)$, $\left(-\frac{5}{3}, 2\right)$, and $(2, \infty)$. We can avoid the arithmetic from evaluating $f(x)$ at
the test points by creating a sign chart. The inequality $3x^2 - x - 10 > 0$ is equiva-
lent to $(3x + 5)(x - 2) > 0$. We have

The sign of the
product in the bottom row of
the sign chart is determined
by the signs of the individual
factors from the rows above.

Sign of $(3x + 5)$:	$-$	$+$	$+$
Sign of $(x - 2)$:	$-$	$-$	$+$
Sign of $(3x + 5)(x - 2)$:	$+$	$-$	$+$

$-\frac{5}{3}$ 2

The sign chart organizes the signs of each factor on
the given intervals. Then the sign of the product of
factors is given in the bottom row. We see that
$f(x) = (3x + 5)(x - 2) > 0$ for $\left(-\infty, -\frac{5}{3}\right) \cup (2, \infty)$.

The result of Example 2 can also be viewed
graphically. From Section 7.1, the graph of
$f(x) = 3x^2 - x - 10$ is a parabola opening upward
(Figure 7-22).

From the factored form $f(x) = (3x + 5)(x - 2)$,
the x-intercepts $\left(-\frac{5}{3}, 0\right)$ and $(2, 0)$ mark the points of
transition between the intervals where $f(x)$ potentially
changes sign.

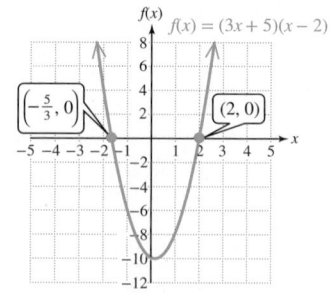

Figure 7-22

Answer

2. Interval notation: $\left(-3, \frac{7}{2}\right)$;
Set-builder notation:
$\left\{x \mid -3 < x < \frac{7}{2}\right\}$

EXAMPLE 3 **Solving a Polynomial Inequality**

Solve the inequality. $x^4 - 12x \geq 8x^2 - x^3$

Solution:

$$x^4 - 12x \geq 8x^2 - x^3$$

Step 1: Write the inequality in the form $f(x) \geq 0$.

$$\overbrace{x^4 + x^3 - 8x^2 - 12x}^{f(x)} \geq 0$$

$$x^4 + x^3 - 8x^2 - 12x = 0$$

Step 2: Find the real solutions to the related equation $f(x) = 0$.

$$x(x^3 + x^2 - 8x - 12) = 0$$

Factor the left side of the equation.

$$\underline{3|} \quad 1 \quad 1 \quad -8 \quad -12$$
$$ \quad \quad 3 \quad 12 \quad 12$$
$$ \quad 1 \quad 4 \quad 4 \quad \underline{|0}$$

The possible rational zeros of $x^3 + x^2 - 8x - 12$ are $\pm 1, \pm 2, \pm 3, \pm 4, \pm 6, \pm 12$.

After testing several potential rational zeros, we find that 3 is a zero of $f(x)$.

$$x(x - 3)(x^2 + 4x + 4) = 0$$

Now factor the quadratic polynomial.

$$x(x - 3)(x + 2)^2 = 0$$

$$x = 0, \, x = 3, \, x = -2$$

The boundary points are 0, 3, and −2.

Step 3: The inequality $x^4 + x^3 - 8x^2 - 12x \geq 0$ is equivalent to $x(x - 3)(x + 2)^2 \geq 0$. Divide the x-axis into intervals defined by the boundary points and determine the sign of $f(x)$ on each interval.

	Interval 1	Interval 2	Interval 3	Interval 4
Evaluate: $f(x) = x(x - 3)(x + 2)^2$	$f(-3) = 18$ $f(x) > 0$	$f(-1) = 4$ $f(x) > 0$	$f(1) = -18$ $f(x) < 0$	$f(4) = 144$ $f(x) > 0$
Sign of x	−	−	+	+
Sign of $(x - 3)$	−	−	−	+
Sign of $(x + 2)^2$	+	+	+	+
Sign of $x(x - 3)(x + 2)^2$	+	+	−	+

$$-2 \qquad 0 \qquad 3$$

The solution set is $(-\infty, 0] \cup [3, \infty)$. In set-builder notation this is $\{x \mid x \leq 0 \text{ or } x \geq 3\}$.

Step 4: The solution set includes the boundary points because the inequality sign includes equality. Therefore, the union of Intervals 1 and 2 becomes $(-\infty, 0]$.

Step 5: Write the solution set.

Skill Practice 3   Solve the inequality. $x^4 - 18x \geq 3x^2 - 4x^3$

The result of Example 3 can also be interpreted graphically. From Section 7.2, the graph of $f(x) = x^4 + x^3 - 8x^2 - 12x$ is up to the far left and up to the far right.

In factored form, $f(x) = x(x - 3)(x + 2)^2$. The x-intercepts are $(0, 0)$, $(3, 0)$, and $(-2, 0)$. Furthermore, the factors x and $(x - 3)$ have odd exponents. This means that the corresponding zeros have odd multiplicities, and that the graph will cross the x-axis at $(0, 0)$ and $(3, 0)$ and change sign. The factor $(x + 2)$ has an even exponent, meaning that the corresponding zero has an even multiplicity. The graph will touch the x-axis at $(-2, 0)$ but will *not* change sign. From the sketch in Figure 7-23, we see that $f(x) \geq 0$ on the intervals $(-\infty, 0]$ and $[3, \infty)$.

TIP From the graph in Figure 7-23, the solution set to the related inequality $x^4 + x^3 - 8x^2 - 12x \leq 0$ would be $\{-2\} \cup [0, 3]$. The boundary point −2 is included because $x^4 + x^3 - 8x^2 - 12x = 0$ at −2.

Answer

3. Interval notation:
$(-\infty, 0] \cup [2, \infty)$;
Set-builder notation: $\{x \mid x \leq 0 \text{ or } x \geq 2\}$

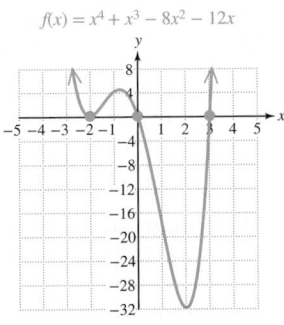

$f(x) = x^4 + x^3 - 8x^2 - 12x$

Figure 7-23

In some situations, the sign of a polynomial may be easily determined by inspection. In such a case, we can abbreviate the procedure to solve a polynomial inequality. This is demonstrated in Example 4.

EXAMPLE 4   Solving Polynomial Inequalities

Solve the inequalities.

a. $4x^2 - 12x + 9 < 0$ **b.** $4x^2 - 12x + 9 \leq 0$

c. $4x^2 - 12x + 9 > 0$ **d.** $4x^2 - 12x + 9 \geq 0$

Solution:

a. $4x^2 - 12x + 9 < 0$

   $(2x - 3)^2 < 0$

   The solution set is { }.

Factor $4x^2 - 12x + 9$ as $(2x - 3)^2$.

The square of any real number is nonnegative. Therefore, this inequality has no solution.

b. $4x^2 - 12x + 9 \leq 0$

   $(2x - 3)^2 \leq 0$

   The solution set is $\left\{\frac{3}{2}\right\}$.

The inequality in part (b) is the same as the inequality in part (a) except that equality is included.

The expression $(2x - 3)^2 = 0$ for $x = \frac{3}{2}$.

c. $4x^2 - 12x + 9 > 0$

   $(2x - 3)^2 > 0$

   The solution set is $\left(-\infty, \frac{3}{2}\right) \cup \left(\frac{3}{2}, \infty\right)$.

   In set-builder notation: $\left\{x \mid x < \frac{3}{2} \text{ or } x > \frac{3}{2}\right\}$.

The expression $(2x - 3)^2 > 0$ for all real numbers except where $(2x - 3)^2 = 0$. Therefore, the solution set is all real numbers except $\frac{3}{2}$.

d. $4x^2 - 12x + 9 \geq 0$

   $(2x - 3)^2 \geq 0$

   The solution set is $(-\infty, \infty)$.

The square of any real number is greater than or equal to zero. Therefore, the solution set is all real numbers.

Skill Practice 4   Solve the inequalities.

a. $25x^2 - 10x + 1 < 0$ **b.** $25x^2 - 10x + 1 \leq 0$

c. $25x^2 - 10x + 1 > 0$ **d.** $25x^2 - 10x + 1 \geq 0$

The graph of $f(x) = 4x^2 - 12x + 9$ or equivalently $f(x) = (2x - 3)^2$ is shown in Figure 7-24. The graph is a parabola opening upward with vertex $\left(\frac{3}{2}, 0\right)$ on the x-axis. No part of the graph is below the x-axis. That is, there are no values of x for which $f(x) < 0$. The graph of $y = f(x)$ is on the x-axis at $x = \frac{3}{2}$ and above the x-axis for all real numbers except $\frac{3}{2}$.

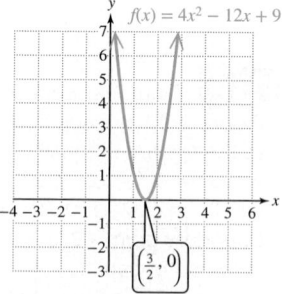

Figure 7-24

3. Solve Applications Involving Polynomial Inequalities

In Section 3.6, we studied the vertical position $s(t)$ of an object moving upward or downward under the influence of gravity. We use this model to solve the application in Example 5.

$$s(t) = -\frac{1}{2}gt^2 + v_0 t + s_0$$

where

> g is the acceleration due to gravity (32 ft/sec^2 or 9.8 m/sec^2).
>
> t is the time of travel.
>
> v_0 is the initial speed.
>
> s_0 is the initial vertical position.

EXAMPLE 5   **Solving an Application of a Polynomial Inequality**

A toy rocket is shot straight upward from a launch pad 1 ft above ground level with an initial speed of 64 ft/sec.

a. Write a model to express the vertical position $s(t)$ (in ft) of the rocket t seconds after launch.

b. Determine the times at which the rocket is above a height of 50 ft.

Solution:

TIP   Choose $g = 32$ ft/sec^2 because the height is given in feet and speed is given in feet per second.

a. $s(t) = -\frac{1}{2}gt^2 + v_0 t + s_0$ In this example,

$$s(t) = -\frac{1}{2}(32)t^2 + (64)t + (1)$$

$$s(t) = -16t^2 + 64t + 1$$

> $s_0 = 1$ ft
> $v_0 = 64$ ft/sec
> $g = 32$ ft/sec^2

b. $-16t^2 + 64t + 1 > 50$

$$\overbrace{-16t^2 + 64t - 49}^{f(t)} > 0$$ Write the inequality in the form $f(t) > 0$.

$-16t^2 + 64t - 49 = 0$ Use the quadratic formula to solve the related equation $f(t) = 0$.

$$t = \frac{-64 \pm \sqrt{(64)^2 - 4(-16)(-49)}}{2(-16)}$$ Evaluate $f(t) = -16t^2 + 64t - 49$ for test points in each interval.

$$= \frac{-64 \pm \sqrt{960}}{-32}$$

$$= \frac{-64 \pm 8\sqrt{15}}{-32}$$

$$= \frac{8 \pm \sqrt{15}}{4} \longrightarrow \begin{array}{l} \approx 2.97 \\ \approx 1.03 \end{array}$$

$f(1) = -1$ $f(2) = 15$ $f(3) = -1$
$f(t) < 0$ $f(t) > 0$ $f(t) < 0$

$\boxed{\frac{8 - \sqrt{15}}{4} \approx 1.03}$ $\boxed{\frac{8 + \sqrt{15}}{4} \approx 2.97}$

The solution set is $\left(\dfrac{8 - \sqrt{15}}{4}, \dfrac{8 + \sqrt{15}}{4} \right)$ or approximately (1.03, 2.97).

The rocket will be above 50 ft high between 1.03 sec and 2.97 sec after launch.

The graph of $s(t) = -16t^2 + 64t + 1$ is a parabola opening downward (Figure 7-25). We see that $s(t) > 50$ for t between 1.03 and 2.97 as expected.

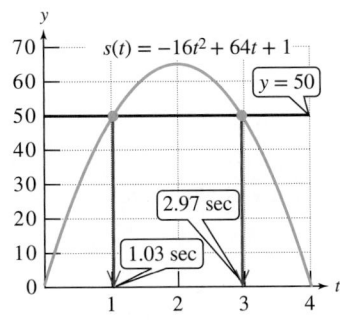

Figure 7-25

Answers

5. a. $s(t) = -16t^2 + 80t + 5$

b. $\left(\dfrac{10 - \sqrt{55}}{4}, \dfrac{10 + \sqrt{55}}{4} \right)$

Skill Practice 5   Repeat Example 5 under the assumption that the rocket is launched with an initial speed of 80 ft/sec from a height of 5 ft.

SECTION 7.5   Practice Exercises

Prerequisite Review

For Exercises R.1–R.2, perform the indicated set operations. (See Section 1.4 for review.)

a. $A \cup B$ **b.** $A \cap B$

R.1. $A = \{-4, -2, 0, 2, 4, 6\}$
$B = \{-2, -1, 0, 3, 6, 9\}$

R.2. $A = \{4, 8, 12, 16, 20, 24, 28, 32\}$
$B = \{8, 16, 24, 32\}$

For Exercises R.3–R.6, write the set in interval notation. (See Section 1.4 for review.)

R.3. $\{x \mid x < -4 \text{ or } x \geq 5\}$

R.4. $\{x \mid x \leq 1 \text{ or } x > 3\}$

R.5. $\left\{ x \mid -\dfrac{9}{5} \leq x < 2 \right\}$

R.6. $\left\{ x \mid 0 < x \leq \dfrac{4}{3} \right\}$

For Exercises R.7–R.8, solve the compound inequality and write the answer in set notation. (See Section 1.5 for review.)

R.7. $-2x - 3 \geq 7$ and $3x + 1 \geq -14$

R.8. $4x - 6 \leq 18$ and $-2x \leq -12$

R.9. Is the quantity $(3x + 5)$ positive or negative on the given interval?

     a. $\left(-\infty, -\dfrac{5}{3}\right)$ **b.** $\left(-\dfrac{5}{3}, \infty\right)$

R.10. Is the quantity $(7 - 2x)$ positive or negative on the given interval?

     a. $\left(-\infty, \dfrac{7}{2}\right)$ **b.** $\left(\dfrac{7}{2}, \infty\right)$

R.11. Is the quantity $(4 - x)^2$ positive or negative on the given interval?

     a. $(-\infty, 4)$ **b.** $(4, \infty)$

R.12. Is the quantity $(x - 3)^2$ positive or negative on the given interval?

     a. $(-\infty, 3)$ **b.** $(3, \infty)$

For Exercises R.13–R.20, solve the equations.

R.13. $2x^3 - 5x^2 - 8x + 20 = 0$ (See Section 3.4)

R.14. $3x^3 + 7x^2 - 3x - 7 = 0$

R.15. $x^3 - x^2 - 10x - 8 = 0$

R.16. $x^3 - 19x + 30 = 0$

R.17. $4x(x + 9) = -81$

R.18. $9x^2 + 49 = 42x$

R.19. $3(x^2 - x) = 7$ (See Section 3.6)

R.20. $2(x^2 + 3x) = 5$

Concept Connections

1. Let $f(x)$ be a polynomial. An inequality of the form $f(x) < 0$, $f(x) > 0$, $f(x) \geq 0$, or $f(x) \leq 0$ is called a _____ inequality. If the polynomial is of degree _____, then the inequality is also called a quadratic inequality.

2. The expression $4 - x$ is (positive/negative) for $x < 4$. The expression $4 - x$ is (positive/negative) for $x > 4$.

3. The solution set for the inequality $(x + 10)^2 \geq -4$ is _____, whereas the solution set for the inequality $(x + 10)^2 \leq -4$ is _____.

4. The solutions to an inequality $f(x) < 0$ are the values of x on the intervals where $f(x)$ is (positive/negative).

Objective 1: Solve Polynomial Inequalities Graphically

For Exercises 5–14, the graph of $y = f(x)$ is given. Solve the inequalities. **(See Example 1)**

   a. $f(x) < 0$ **b.** $f(x) \leq 0$ **c.** $f(x) > 0$ **d.** $f(x) \geq 0$

5.

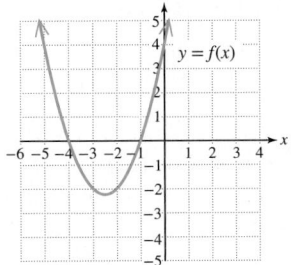

6.

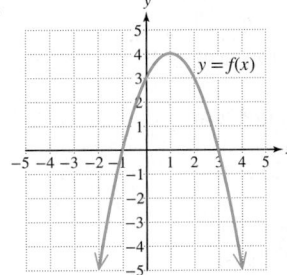

7.

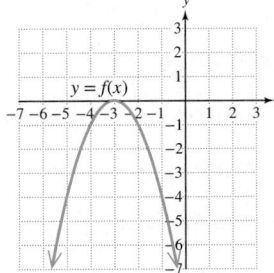

8.

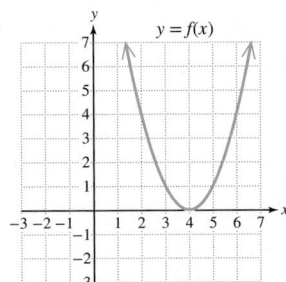

9.

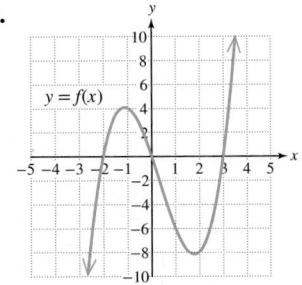

10.

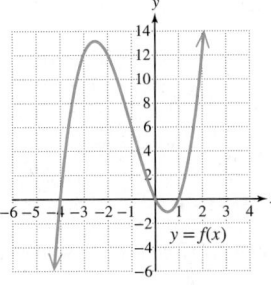

11.

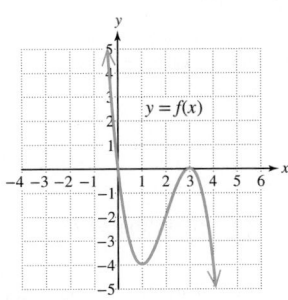

12.

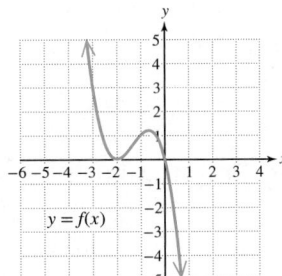

13.

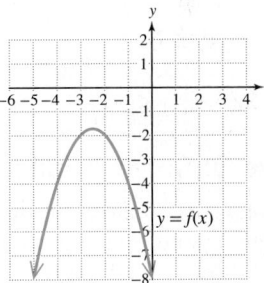

14.

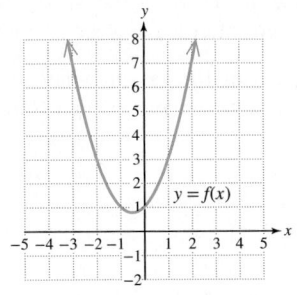

Objective 2: Solve Polynomial Inequalities Algebraically

For Exercises 15–20, solve the equations and inequalities. **(See Example 4)**

15. a. $(5x - 3)(x - 5) = 0$

    b. $(5x - 3)(x - 5) < 0$

    c. $(5x - 3)(x - 5) \leq 0$

    d. $(5x - 3)(x - 5) > 0$

    e. $(5x - 3)(x - 5) \geq 0$

16. a. $(3x + 7)(x - 2) = 0$

    b. $(3x + 7)(x - 2) < 0$

    c. $(3x + 7)(x - 2) \leq 0$

    d. $(3x + 7)(x - 2) > 0$

    e. $(3x + 7)(x - 2) \geq 0$

17. a. $-x^2 + x + 12 = 0$

    b. $-x^2 + x + 12 < 0$

    c. $-x^2 + x + 12 \leq 0$

    d. $-x^2 + x + 12 > 0$

    e. $-x^2 + x + 12 \geq 0$

18. a. $-x^2 - 10x - 9 = 0$

   b. $-x^2 - 10x - 9 < 0$

   c. $-x^2 - 10x - 9 \leq 0$

   d. $-x^2 - 10x - 9 > 0$

   e. $-x^2 - 10x - 9 \geq 0$

19. a. $a^2 + 12a + 36 = 0$

   b. $a^2 + 12a + 36 < 0$

   c. $a^2 + 12a + 36 \leq 0$

   d. $a^2 + 12a + 36 > 0$

   e. $a^2 + 12a + 36 \geq 0$

20. a. $t^2 - 14t + 49 = 0$

   b. $t^2 - 14t + 49 < 0$

   c. $t^2 - 14t + 49 \leq 0$

   d. $t^2 - 14t + 49 > 0$

   e. $t^2 - 14t + 49 \geq 0$

For Exercises 21–54, solve the inequalities. **(See Examples 2–3)**

21. $4w^2 - 9 \geq 0$

22. $16z^2 - 25 < 0$

23. $3w^2 + w < 2(w + 2)$

24. $5y^2 + 7y < 3(y + 4)$

25. $a^2 \geq 3a$

26. $d^2 \geq 6d$

27. $10 - 6x > 5x^2$

28. $6 - 4x > 3x^2$

29. $m^2 < 49$

30. $y^2 \geq 9$

31. $16p^2 \geq 2$

32. $54q^2 \leq 50$

33. $(x + 4)(x - 1)(x - 3) \geq 0$

34. $(x + 2)(x + 5)(x - 4) \geq 0$

35. $-5c(c + 2)^2(4 - c) > 0$

36. $-6u(u + 1)^2(3 - u) > 0$

37. $t^4 - 10t^2 + 9 \leq 0$

38. $w^4 - 20w^2 + 64 \leq 0$

39. $2x^3 + 5x^2 < 8x + 20$

40. $3x^3 - 3x < 4x^2 - 4$

41. $-2x^4 + 10x^3 - 6x^2 - 18x \geq 0$

42. $-4x^4 + 4x^3 + 64x^2 + 80x \geq 0$

43. $-5u^6 + 28u^5 - 15u^4 \leq 0$

44. $-3w^6 + 8w^5 - 4w^4 \leq 0$

45. $6x(2x - 5)^4(3x + 1)^5(x - 4) < 0$

46. $5x(3x - 2)^2(4x + 1)^3(x - 3)^4 < 0$

47. $(5x - 3)^2 > -2$

48. $(4x + 1)^2 > -6$

49. $-4 \geq (x - 7)^2$

50. $-1 \geq (x + 2)^2$

51. $16y^2 > 24y - 9$

52. $4w^2 > 20w - 25$

53. $(x + 3)(x + 1) \leq -1$

54. $(x + 2)(x + 4) \leq -1$

Objective 3: Solve Applications Involving Polynomial Inequalities

55. A professional fireworks team shoots an 8-in. mortar straight upward from ground level with an initial speed of 216 ft/sec. **(See Example 5)**

   a. Write a function modeling the vertical position $s(t)$ (in ft) of the shell at a time t seconds after launch.

   b. The mortar is designed to explode when the shell is at its maximum height. How long after launch will the shell explode? (*Hint*: Consider the vertex formula from Section 7.1)

   c. The spectators can see the shell rising once it clears a 200-ft tree line. For what period of time after launch is the shell visible before it explodes?

56. Suppose that a basketball player jumps straight up for a rebound.

   a. If his initial speed leaving the ground is 16 ft/sec, write a function modeling his vertical position $s(t)$ (in ft) at a time t seconds after leaving the ground.

   b. Find the times after leaving the ground when the player will be at a height of more than 3 ft in the air.

57. For a certain stretch of road, the distance d (in ft) required to stop a car that is traveling at speed v (in mph) before the brakes are applied can be approximated by $d(v) = 0.06v^2 + 2v$. Find the speeds for which the car can be stopped within 250 ft.

58. The population $P(t)$ of a bacteria culture is given by $P(t) = -1500t^2 + 60{,}000t + 10{,}000$, where t is the time in hours after the culture is started. Determine the time(s) at which the population will be greater than 460,000 organisms.

59. A rectangular quilt is to be made so that the length is 1.2 times the width. The quilt must be between 72 ft^2 and 96 ft^2 to cover the bed. Determine the restrictions on the width so that the dimensions of the quilt will meet the required area. Give exact values and the approximated values to the nearest tenth of a foot.

60. A landscaping team plans to build a rectangular garden that is between 480 yd^2 and 720 yd^2 in area. For aesthetic reasons, they also want the length to be 1.5 times the width. Determine the restrictions on the width so that the dimensions of the garden will meet the required area. Give exact values and the approximated values to the nearest tenth of a yard.

Mixed Exercises

For Exercises 61–64,

   a. Identify the x-intercepts.

   c. Sketch the function.

   e. Determine the interval(s) over which $f(x) < 0$.

   b. Identify the y-intercept.

   d. Determine the interval(s) over which $f(x) > 0$.

61. $f(x) = x^2 - 4$ **62.** $f(x) = -x^2 + 9$ **63.** $f(x) = x^3 + 3x^2 - 4$ **64.** $f(x) = x^3 - 7x^2 + 15x - 9$

For Exercises 65–66, the graph of $y = f(x)$ is given.

   a. Identify the x-intercepts.

   c. Determine the interval(s) over which $f(x) > 0$.

   b. Identify the y-intercept.

   d. Determine the interval(s) over which $f(x) < 0$.

65. $f(x) = -x^6 + 5x^4 - 6x^2$

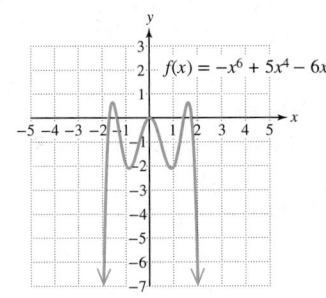

66. $f(x) = -x^5 + 4x^3 - 3x$

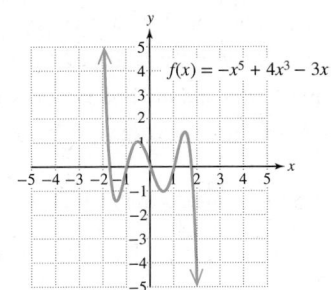

For Exercises 67–72, write the domain of the function in interval notation.

67. $f(x) = \sqrt{9 - x^2}$

69. $h(a) = \sqrt{a^2 - 5}$

71. $p(x) = \sqrt{2x^2 + 9x - 18}$

68. $g(t) = \sqrt{1 - t^2}$

70. $f(u) = \sqrt{u^2 - 7}$

72. $q(x) = \sqrt{4x^2 + 7x - 2}$

73. Let a, b, and c represent positive real numbers, where $a < b < c$, and let $f(x) = (x - a)^2(b - x)(x - c)^3$.

   a. Complete the sign chart.

Sign of $(x - a)^2$:			
Sign of $(b - x)$:			
Sign of $(x - c)^3$:			
Sign of $(x - a)^2(b - x)(x - c)^3$:			

                                                    a b c

   b. Solve $f(x) > 0$.   **c.** Solve $f(x) < 0$.

74. Let a, b, and c represent positive real numbers, where $a < b < c$, and let $f(x) = (a - x)(x - b)^4(c - x)^2$.

   a. Complete the sign chart.

Sign of $(a - x)$:			
Sign of $(x - b)^4$:			
Sign of $(c - x)^2$:			
Sign of $(a - x)(x - b)^4(c - x)^2$:			

                                                    a b c

   b. Solve $f(x) > 0$.   **c.** Solve $f(x) < 0$.

Write About It

75. Explain how the solution set to the inequality $f(x) < 0$ is related to the graph of $y = f(x)$.

76. Explain how the solution set to the inequality $f(x) \geq 0$ is related to the graph of $y = f(x)$.

Technology Connections

77. Given the inequality, $0.552x^3 + 4.13x^2 - 1.84x - 3.5 < 6.7$,

    a. Write the inequality in the form $f(x) < 0$.

    b. Graph $y = f(x)$ on a suitable viewing window.

    c. Use the Zero feature to approximate the real zeros of $f(x)$. Round to 1 decimal place.

    d. Use the graph to approximate the solution set for the inequality $f(x) < 0$.

78. Given the inequality, $0.24x^4 + 1.8x^3 + 3.3x^2 + 2.84x - 1.8 > 4.5$,

    a. Write the inequality in the form $f(x) > 0$.

    b. Graph $y = f(x)$ on a suitable viewing window.

    c. Use the Zero feature to approximate the real zeros of $f(x)$. Round to 1 decimal place.

    d. Use the graph to approximate the solution set for the inequality $f(x) > 0$.

79. An engineer for a food manufacturer designs an aluminum container for a hot drink mix. The container is to be a right circular cylinder 5.5 in. in height. The surface area represents the amount of aluminum used and is given by

    $S(r) = 2\pi r^2 + 11\pi r$, where r is the radius of the can.

    a. Graph the function $y = S(r)$ and the line $y = 90$ on the viewing window [0, 3, 1] by [0, 150, 10].

    b. Use the Intersect feature to approximate the point of intersection of $y = S(r)$ and $y = 90$. Round to 1 decimal place if necessary.

    c. Determine the restrictions on r so that the amount of aluminum used is at most 90 in.^{2}. Round to 1 decimal place.

PROBLEM RECOGNITION EXERCISES

Polynomial Equations and Inequalities

For Exercises 1–6, solve the equation or inequality.

1. a. $2(x - 3)^2 = 2$

    b. $2(x - 3)^2 \geq 2$

    c. $2(x - 3)^2 < 2$

2. a. $x^4 - 16x^2 = 0$

    b. $x^4 - 16x^2 > 0$

    c. $x^4 - 16x^2 \leq 0$

3. a. $-(x + 4) = 2(3x + 5) - 6$

    b. $-(x + 4) < 2(3x + 5) - 6$

    c. $-(x + 4) > 2(3x + 5) - 6$

4. a. $(3 - x)(2x + 5)^2(x - 7) = 0$

    b. $(3 - x)(2x + 5)^2(x - 7) > 0$

    c. $(3 - x)(2x + 5)^2(x - 7) \leq 0$

5. a. $25x^2 - 20x = -4$

    b. $25x^2 - 20x \leq -4$

    c. $25x^2 - 20x > -4$

6. a. $3x^3 - 7x^2 = 75x - 175$

    b. $3x^3 - 7x^2 \geq 75x - 175$

    c. $3x^3 - 7x^2 \leq 75x - 175$

7. a. Factor into a product of linear factors.
$$3x^4 + 8x^3 - 3x^2 - 12x + 4$$

    b. Solve.
$$3x^4 + 8x^3 - 3x^2 - 12x + 4 = 0$$

    c. Find the x-intercepts.
$$f(x) = 3x^4 + 8x^3 - 3x^2 - 12x + 4$$

8. a. Factor into a product of linear factors.
$$x^6 + 9x^5 + 28x^4 + 36x^3 + 27x^2 + 27x$$

    b. Solve.
$$x^6 + 9x^5 + 28x^4 + 36x^3 + 27x^2 + 27x = 0$$

    c. Find the x-intercepts.
$$f(x) = x^6 + 9x^5 + 28x^4 + 36x^3 + 27x^2 + 27x$$

CHAPTER 7   Detailed Summary

SECTION 7.1 Quadratic Functions and Applications

Key Concepts	Examples	Page
Quadratic function: A function defined by $f(x) = ax^2 + bx + c \ (a \neq 0)$ is called a **quadratic function**. A quadratic function can be written in **vertex form** $f(x) = a(x - h)^2 + k$ by completing the square. **Graph of a quadratic function $f(x) = a(x - h)^2 + k$:** • The graph of a quadratic function is a parabola. • The vertex is (h, k). • If $a > 0$, the parabola opens upward, and the minimum value of the function is k. • If $a < 0$, the parabola opens downward, and the maximum value of the function is k. • The axis of symmetry is the line $x = h$. • The x-intercepts are determined by the real solutions to the equation $f(x) = 0$. • The y-intercept is determined by $f(0)$.	**Example 1:** Given $f(x) = -2x^2 + 4x$, $$f(x) = -2(x^2 - 2x \qquad) + 0$$ $$= -2(x^2 - 2x + 1 - 1) + 0$$ $$= -2(x^2 - 2x + 1) - 2(-1) + 0$$ $$= -2(x - 1)^2 + 2$$ • Since $a = -2 < 0$, the parabola opens downward. • The vertex is $(1, 2)$. • The maximum value of the function is 2. • The axis of symmetry is $x = 1$. ***x*-intercepts:** Solve $\quad -2x^2 + 4x = 0$ $$-2x(x - 2) = 0$$ $$x = 0 \text{ or } x = 2$$ x-intercepts $(0, 0)$ and $(2, 0)$. ***y*-intercept:** $f(0) = -2(0)^2 + 4(0) = 0$ y-intercept $(0, 0)$. 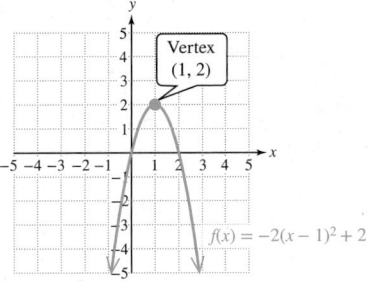	p. 472
Vertex formula: The vertex of a parabola can also be found by using the vertex formula. Given $f(x) = ax^2 + bx + c \ (a \neq 0)$, the vertex of the parabola is $$\left(\frac{-b}{2a}, f\left(\frac{-b}{2a} \right) \right)$$	**Example 2:** Given $f(x) = 2x^2 - 8x + 1$, the x-coordinate of the vertex is $\dfrac{-b}{2a} = \dfrac{-(-8)}{2(2)} = 2$. The y-coordinate is: $f(2) = 2(2)^2 - 8(2) + 1$ $$= -7$$ The vertex is $(2, -7)$.	p. 476

SECTION 7.2 Introduction to Polynomial Functions

Key Concepts	Examples	Page
Polynomial function: Let n be a whole number and a_n, a_{n-1}, a_{n-2}, ... , a_1, a_0 be real numbers, where $a_n \neq 0$. Then a function defined by $$f(x) = a_n x^n + a_{n-1}x^{n-1} + a_{n-2}x^{n-2} + \cdots + a_1 x + a_0$$ is called a **polynomial function of degree n.** The far left and far right behavior of the graph of a polynomial function is determined by the leading term of the polynomial, $a_n x^n$. n is even and $a_n > 0$ n is even and $a_n < 0$ n is odd and $a_n > 0$ n is odd and $a_n < 0$	**Example 1:** $f(x) = -3x^5 - 5x^3 + 4x^2 - 7x + 2$ is a fifth-degree polynomial function. $g(x) = 4\sqrt{x} - \dfrac{3}{x}$ is not a polynomial function. **Example 2:** Given $f(x) = -\frac{1}{2}x^2(2x - 3)^2(x + 2)$, the leading term is $-2x^5$. $n = 5$ (odd) and $a_n = -2$ (negative) The end behavior is up to the left and down to the right. That is, as $x \to -\infty, f(x) \to \infty$, and as $x \to \infty, f(x) \to -\infty$.	p. 488 p. 490
The **zeros** of a polynomial function defined by $y = f(x)$ are the values of x in the domain of f for which $f(x) = 0$. These are the real solutions (or **roots**) of the equation $f(x) = 0$.	**Example 3:** The zeros of $f(x) = -\frac{1}{2}x^2(2x - 3)^2(x + 2)$ are 0, $\frac{3}{2}$, and -2.	p. 491

If a polynomial function f has a factor $(x - c)$ that appears exactly k times, then c is a **zero of multiplicity k**.
 	
- If c is a real zero of odd multiplicity, then the graph of $y = f(x)$ *crosses* the x-axis at c.
- If c is a real zero of even multiplicity, then the graph of $y = f(x)$ *touches* the x-axis (but does not cross) at c.
 	
The graph of a polynomial function of degree n will have at most $n - 1$ turning points.
	Example 4:
	Given $f(x) = -\frac{1}{2}x^2(2x - 3)^2(x + 2)$,
	
	- 0 is a zero of multiplicity 2.
	  The graph touches the x-axis at $(0, 0)$.
	- $\frac{3}{2}$ is a zero of multiplicity 2.
	  The graph touches the x-axis at $\left(\frac{3}{2}, 0\right)$.
	- -2 is a zero of multiplicity 1.
	  The graph crosses the x-axis at $(-2, 0)$.
	
		p. 492
		
		p. 496

Intermediate value theorem: Let f be a polynomial function. For $a < b$, if $f(a)$ and $f(b)$ have opposite signs, then f has at least one zero on the interval $[a, b]$.	**Example 5:** $f(x) = x^3 - 2x^2 + 3x - 8$ has at least one zero on the interval $[2, 3]$ because $f(2)$ and $f(3)$ have opposite signs. $f(2) = (2)^3 - 2(2)^2 + 3(2) - 8 = -2$ ⟵ opposite $f(3) = (3)^3 - 2(3)^2 + 3(3) - 8 = 10$ ⟵ signs	p. 494

SECTION 7.3 Division of Polynomials and the Remainder and Factor Theorems

Key Concepts	Examples	Page
Long division can be used to divide two polynomials. In Example 1, $$\frac{6x^4 + x^3 - 7x + 1}{2x^2 - x} = 3x^2 + 2x + 1 + \frac{-6x + 1}{2x^2 - x}$$ The result can be checked by using the division algorithm: $$6x^4 + x^3 - 7x + 1$$ $$\overset{?}{=} (3x^2 + 2x + 1)(2x^2 - x) + (-6x + 1) \checkmark$$	**Example 1:** $$\begin{array}{r} 3x^2 + 2x + 1 \\ 2x^2 - x \overline{)6x^4 + x^3 + 0x^2 - 7x + 1} \\ \underline{-(6x^4 - 3x^3)} \\ 4x^3 + 0x^2 \\ \underline{-(4x^3 - 2x^2)} \\ 2x^2 - 7x \\ \underline{-(2x^2 - x)} \\ -6x + 1 \end{array}$$	p. 505
Synthetic division can be used to divide polynomials if the divisor is of the form $x - c$. In Example 2, the quotient is $4x^2 - 10x + 20$, and the remainder is -35.	**Example 2:** Divide. $(4x^3 - 2x^2 + 5) \div (x + 2)$ $$\begin{array}{r|rrrr} -2 & 4 & -2 & 0 & 5 \\ & & -8 & 20 & -40 \\ \hline & 4 & -10 & 20 & \underline{-35} \end{array}$$	p. 508
Remainder theorem: If a polynomial $f(x)$ is divided by $x - c$, then the remainder is $f(c)$.	**Example 3:** Consider $f(x) = 4x^3 - 2x^2 + 5$. From the result of Example 2, we see that $f(-2) = -35$ because when $(4x^3 - 2x^2 + 5)$ is divided by $(x + 2)$, the remainder is -35.	p. 510
Factor theorem: Let $f(x)$ be a polynomial. 1. If $f(c) = 0$, then $(x - c)$ is a factor of $f(x)$. 2. If $(x - c)$ is a factor of $f(x)$, then $f(c) = 0$.	**Example 4:** Determine if $(x - 3)$ is a factor of $f(x) = x^3 - 7x - 6$. $$\begin{array}{r|rrrr} 3 & 1 & 0 & -7 & -6 \\ & & 3 & 9 & 6 \\ \hline & 1 & 3 & 2 & \underline{0} \end{array}$$ The remainder is 0, which implies that $(x - 3)$ is a factor of $f(x)$. Furthermore, because $f(3) = 0$ we know that 3 is a zero of $f(x)$.	p. 512

SECTION 7.4 Zeros of Polynomials

Key Concepts	Examples	Page
Rational zero theorem: If $f(x) = a_n x^n + a_{n-1}x^{n-1} + \cdots + a_1 x + a_0$ has integer coefficients and $a_n \neq 0$, and if $\frac{p}{q}$ (written in lowest terms) is a rational zero of f, then • p is a factor of the constant term, a_0. • q is a factor of the leading coefficient a_n. **Fundamental theorem of algebra:** If $f(x)$ is a polynomial of degree $n \geq 1$ with complex coefficients, then $f(x)$ has at least one complex zero.	**Example 1:** Given $f(x) = 2x^3 + x^2 - 4x - 3$, the possible rational zeros are $\pm 1, \pm 3, \pm\frac{1}{2}, \pm\frac{3}{2}$. After testing several potential rational zeros, we find that -1 is a zero of $f(x)$. $$\begin{array}{r|rrrr} -1 & 2 & 1 & -4 & -3 \\ & & -2 & 1 & 3 \\ \hline & 2 & -1 & -3 & \underline{0} \end{array}$$	p. 520 p. 523

Linear factorization theorem:

If $f(x) = a_n x^n + a_{n-1} x^{n-1} + \cdots + a_1 x + a_0$, where $n \geq 1$ and $a_n \neq 0$, then

$f(x) = a_n(x - c_1)(x - c_2) \ldots (x - c_n)$,

where $c_1, c_2, \ldots, c_n$ are complex numbers.

Example 2:

From Example 1, $f(x) = 2x^3 + x^2 - 4x - 3$ factors as:

$$f(x) = (x + 1)(2x^2 - x - 3)$$
$$= (x + 1)(x + 1)(2x - 3)$$

The zeros are -1 (multiplicity 2) and $\frac{3}{2}$ (multiplicity 1).

p. 524

Number of complex zeros:

If $f(x)$ is a polynomial of degree $n \geq 1$ with complex coefficients, then $f(x)$ has exactly n complex zeros provided that each zero is counted by its multiplicity.

Example 3:

From Example 1, $f(x) = 2x^3 + x^2 - 4x - 3$ is a degree 3 polynomial, and has 3 zeros (including multiplicities).

p. 524

Conjugate zeros theorem:

If $f(x)$ is a polynomial with real coefficients and if $a + bi$ is a zero of $f(x)$, then its conjugate $a - bi$ is also a zero of $f(x)$.

Example 4:

Write a third-degree polynomial $f(x)$ with integer coefficients and with zeros of $3i$ and 4.

From the conjugate zeros theorem, if $3i$ is a zero of $f(x)$, then $-3i$ must also be a zero.

$$f(x) = (x - 3i)(x + 3i)(x - 4)$$
$$= (x^2 + 9)(x - 4)$$
$$= x^3 - 4x^2 + 9x - 36$$

p. 524

Descartes' rule of signs:

Let $f(x)$ be a polynomial with real coefficients and a nonzero constant term. Then,

1. The number of *positive* real zeros is either the same as the number of sign changes in $f(x)$ or less than the number of sign changes in $f(x)$ by a positive even integer.
2. The number of *negative* real zeros is either the same as the number of sign changes in $f(-x)$ or less than the number of sign changes in $f(-x)$ by a positive even integer.

Example 5:

Determine the number of possible positive and negative real zeros of $y = f(x)$.

$$f(x) = 3x^5 - 2x^3 - 5x^2 - 8x + 1 \quad \text{2 sign changes}$$

There are 2 or 0 possible positive real zeros.

$$f(-x) = 3(-x)^5 - 2(-x)^3 - 5(-x)^2 - 8(-x) + 1$$
$$= -3x^5 + 2x^3 - 5x^2 + 8x + 1 \quad \text{3 sign changes}$$

There are 3 or 1 possible negative real zeros.

p. 527

Upper and lower bounds:

Let $f(x)$ be a polynomial of degree $n \geq 1$ with real coefficients and a positive leading coefficient. Further suppose that $f(x)$ is divided by $(x - c)$.

1. If $c > 0$ and if both the remainder and the coefficients of the quotient are nonnegative, then c is an upper bound for the real zeros of $f(x)$.
2. If $c < 0$ and the coefficients of the quotient and the remainder alternate in sign (with 0 being considered either positive or negative as needed), then c is a lower bound for the real zeros of $f(x)$.

Example 6:

Given $f(x) = 2x^3 - 7x^2 - 3x + 18$,

a. Determine if 4 is an upper bound for the real zeros of $f(x)$.

$$\begin{array}{r|rrrr} 4 & 2 & -7 & -3 & 18 \\ & & 8 & 4 & 4 \\ \hline & 2 & 1 & 1 & 22 \end{array}$$ This row all positive

The value 4 is an upper bound.

b. Determine if -1 is a lower bound for the real zeros of $f(x)$.

$$\begin{array}{r|rrrr} -1 & 2 & -7 & -3 & 18 \\ & & -2 & 9 & -6 \\ \hline & 2 & -9 & 6 & 12 \end{array}$$

No sign change

The lower bound theorem does not identify -1 as a lower bound of the zeros of $f(x)$.

p. 529

SECTION 7.5 Polynomial Inequalities and Applications

Key Concepts	Examples	Page

Key Concepts

Polynomial inequality:

Let $f(x)$ be a polynomial. Then an inequality of the form

$$f(x) < 0, \quad f(x) > 0, \quad f(x) \le 0, \quad \text{or} \quad f(x) \ge 0$$

is called a **polynomial inequality**.

Solving polynomial inequalities:

1. Express the inequality as $f(x) < 0, f(x) > 0, f(x) \le 0$, or $f(x) \ge 0$.
2. Find the real solutions of the related equation $f(x) = 0$. These are the "boundary" points for the solution set to the inequality.
3. Determine the sign of $f(x)$ on the intervals defined by the boundary points.
   - If $f(x)$ is positive, then the values of x on the interval are solutions to $f(x) > 0$.
   - If $f(x)$ is negative, then the values of x on the interval are solutions to $f(x) < 0$.
4. Determine whether the boundary points are included in the solution set.
5. Write the solution set.

Examples

Example 1:

Solve: $-3x(2 - x)(x + 4)^2 < 0$

	+	+	−	−
Sign of $-3x$	+	+	−	−
Sign of $(2 - x)$	+	+	+	−
Sign of $(x + 4)^2$	+	+	+	+
Sign of $-3x(2 - x)(x + 4)^2$	+	+	−	+

$$-4 \qquad 0 \qquad 2$$

The solution set is $(0, 2)$.

Page

p. 539

CHAPTER 7   Review Exercises

SECTION 7.1

1. Given $f(x) = -(x + 5)^2 + 2$, identify the vertex of the graph of the parabola.

For Exercises 2–3,

a. Write the equation in vertex form: $f(x) = a(x - h)^2 + k$.

b. Determine whether the parabola opens upward or downward.

c. Identify the vertex.

d. Identify the x-intercepts.

e. Identify the y-intercept.

f. Sketch the function.

g. Determine the axis of symmetry.

h. Determine the minimum or maximum value of the function.

i. State the domain and range.

2. $f(x) = x^2 - 8x + 15$

3. $f(x) = -2x^2 + 4x + 6$

4. a. Use the vertex formula to determine the vertex of $f(x) = 2x^2 + 12x + 19$.

b. Based on the location of the vertex and the orientation of the parabola, how many x-intercepts will the graph of $f(x) = 2x^2 + 12x + 19$ have?

5. Suppose that a farmer encloses a corral for cattle adjacent to a river. No fencing is used by the river.

a. If he has 180 yd of fencing, what dimensions should he use to maximize the area?

b. What is the maximum area?

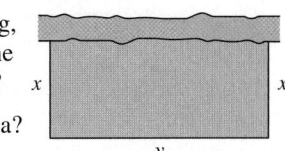

6. Suppose that p is the probability that a randomly selected person is left-handed. The value $(1 - p)$ is the probability that the person is not left-handed. In a sample of 100 people, the function $V(p) = 100p(1 - p)$ represents the variance of the number of left-handed people in a group of 100.

a. What value of p maximizes the variance?

b. What is the maximum variance?

7. The annual expenditure for cell phones and cellular service varies in part by the age of an individual. The average annual expenditure $E(a)$ (in \$) for individuals of age a (in years) is given in the table. (*Source*: U.S. Bureau of Labor Statistics, www.bls.gov)

a	20	30	40	50	60	70
$E(a)$	502	658	649	627	476	213

   a. Use regression to find a quadratic function to model the data.

   b. At what age is the yearly expenditure for cell phones and cellular service the greatest? Round to the nearest year.

   c. What is the maximum yearly expenditure? Round to the nearest dollar.

SECTION 7.2

For Exercises 8–11,

   a. Determine the end behavior of the graph of the function.

   b. Find all the zeros of the function and state their multiplicities.

   c. Determine the x-intercepts.

   d. Determine the y-intercept.

   e. Is the function even, odd, or neither?

   f. Graph the function.

   8. $f(x) = -4x^3 + 16x^2 + 25x - 100$

   9. $f(x) = x^4 - 10x^2 + 9$

  10. $f(x) = x^4 + 3x^3 - 3x^2 - 11x - 6$

  11. $f(x) = x^5 - 8x^4 + 13x^3$

12. Determine whether the intermediate value theorem guarantees that the function has a zero on the given interval.

$$f(x) = 2x^3 - 5x^2 - 6x + 2$$

   a. $[-2, -1]$ **b.** $[-1, 0]$ **c.** $[0, 1]$ **d.** $[1, 2]$

For Exercises 13–16, determine if the statement is true or false. If a statement is false, explain why.

13. A fourth-degree polynomial has exactly three turning points.

14. A fourth-degree polynomial has at most three turning points.

15. There is exactly one polynomial with zeros of 2, 3, and 4.

16. If c is a real zero of an odd polynomial function, then $-c$ is also a zero.

SECTION 7.3

For Exercises 17–18,

   a. Divide the polynomials.

   b. Identify the dividend, divisor, quotient, and remainder.

  17. $(-2x^4 + x^3 + 4x - 1) \div (x^2 + x - 3)$

  18. $\dfrac{3x^4 - 2x^3 - 15x^2 + 22x - 8}{3x - 2}$

For Exercises 19–20, use synthetic division to divide the polynomials.

19. $(2x^5 + x^2 - 5x + 1) \div (x + 2)$

20. $\dfrac{x^4 + 3x^3 - x^2 + 7x + 2}{x - 3}$

For Exercises 21–22, use the remainder theorem to evaluate the polynomial for the given values of x.

21. $f(x) = 3x^4 + 2x^2 - 4x + 1; f(-2)$

22. $f(x) = x^4 + 2x^3 - 4x^2 - 10x - 5; f(\sqrt{5})$

For Exercises 23–24, use the remainder theorem to determine if the given number c is a zero of the polynomial.

23. $f(x) = 3x^4 + 13x^3 + 2x^2 + 52x - 40$

   a. $c = 2$ **b.** $c = \frac{2}{3}$

24. $f(x) = x^4 + 6x^3 + 9x^2 + 24x + 20$

   a. $c = -5$ **b.** $c = 2i$

For Exercises 25–26, use the factor theorem to determine if the given binomial is a factor of the polynomial.

25. $f(x) = x^3 + 4x^2 + 9x + 36$

   a. $x + 4$ **b.** $x - 3i$

26. $f(x) = x^2 - 4x - 46$

   a. $x + 2$ **b.** $x - (2 - 5\sqrt{2})$

27. Factor $f(x) = 15x^3 - 67x^2 + 26x + 8$, given that $\frac{2}{3}$ is a zero of $f(x)$.

28. Write a third-degree polynomial $f(x)$ with zeros -1, $3\sqrt{2}$, and $-3\sqrt{2}$.

29. Write a third-degree polynomial $f(x)$ with integer coefficients and zeros of $\frac{1}{4}$, $-\frac{1}{2}$, and 3.

SECTION 7.4

30. Given $f(x) = 2x^5 - 7x^4 + 9x^3 - 18x^2 + 4x + 40$,

   a. How many zeros does $f(x)$ have (including multiplicities)?

   b. List the possible rational zeros of $f(x)$.

   c. Find all rational zeros of $f(x)$.

   d. Find all the zeros of $f(x)$.

31. Given $f(x) = x^4 + 4x^3 + 2x^2 - 8x - 8$,

   a. How many zeros does $f(x)$ have (including multiplicities)?

   b. List the possible rational zeros of $f(x)$.

   c. Find all rational zeros of $f(x)$.

   d. Find all the zeros of $f(x)$.

32. If $f(x)$ is a polynomial with real coefficients and zeros of 4 (multiplicity 3), -2 (multiplicity 1), and $2 + 7i$ (multiplicity 1), what is the minimum degree of $f(x)$?

33. Given $f(x) = x^4 - 22x^3 + 119x^2 + 66x - 366$ and that $11 - i$ is a zero of $f(x)$,

   a. Find all the zeros of $f(x)$.

   b. Factor $f(x)$ as a product of linear factors.

   c. Solve the equation $f(x) = 0$.

34. Write a polynomial $f(x)$ of lowest degree with real coefficients and with zeros $2 - 3i$ (multiplicity 1) and 0 (multiplicity 2).

35. Write a third-degree polynomial $f(x)$ with integer coefficients and with zeros of $-2i$ and $\frac{5}{3}$.

For Exercises 36–37, determine the number of possible positive and negative real zeros for the given function.

36. $g(x) = -3x^7 + 4x^6 - 2x^2 + 5x - 4$

37. $n(x) = x^6 + \frac{1}{3}x^4 + \frac{2}{7}x^3 + 4x^2 + 3$

38. Given $f(x) = x^4 - 3x^3 + 2x - 3$,

   a. Determine if 2 is an upper bound for the real zeros of $f(x)$.

   b. Determine if -2 is a lower bound for the real zeros of $f(x)$.

39. Given $f(x) = x^3 - 4x^2 + 2x + 1$,

   a. Determine if 5 is an upper bound for the real zeros of $f(x)$.

   b. Determine if -2 is a lower bound for the real zeros of $f(x)$.

SECTION 7.5

40. The graph of $y = f(x)$ is given. Solve the inequalities.

   a. $f(x) < 0$

   b. $f(x) \le 0$

   c. $f(x) > 0$

   d. $f(x) \ge 0$

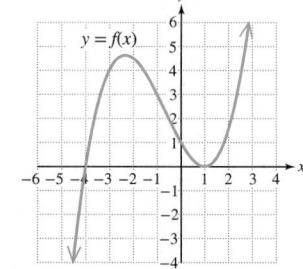

41. The graph of $y = f(x)$ is given. Solve the inequalities.

   a. $f(x) < 0$

   b. $f(x) \le 0$

   c. $f(x) > 0$

   d. $f(x) \ge 0$

42. Solve the equation and inequalities.

   a. $x^2 + 7x + 10 = 0$ **b.** $x^2 + 7x + 10 < 0$

   c. $x^2 + 7x + 10 \le 0$ **d.** $x^2 + 7x + 10 > 0$

   e. $x^2 + 7x + 10 \ge 0$

For Exercises 43–48, solve the inequalities.

43. $t(t - 3) \ge 18$

44. $w^3 + w^2 - 9w - 9 > 0$

45. $x^2 - 2x + 4 \le 3$

46. $-6x^4(3x - 4)^2(x + 2)^3 \le 0$

47. $z^3 - 3z^2 > 10z - 24$

48. $(4x - 5)^4 > 0$

49. A child throws a ball straight upward to his friend who is sitting in a tree 18 ft above ground level.

   a. If the ball leaves the child's hand at a height of 2 ft with an initial speed of 40 ft/sec, write a function representing the vertical position of the ball $s(t)$ (in ft) in terms of the time t after the ball leaves the child's hand. (*Hint*: Use the model $s(t) = -\frac{1}{2}gt^2 + v_0 t + s_0$ with $g = 32$ ft/sec^2. See page 543.)

   b. Determine the time interval for which the ball will be more than 18 ft high.

CHAPTER 7 Test

1. Given $f(x) = 2x^2 - 12x + 16$,

   a. Write the equation in vertex form: $f(x) = a(x - h)^2 + k$.

   b. Determine whether the parabola opens upward or downward.

   c. Identify the vertex.

   d. Identify the x-intercepts.

   e. Identify the y-intercept.

   f. Sketch the function.

   g. Determine the axis of symmetry.

   h. Determine the minimum or maximum value of the function.

   i. State the domain and range.

2. Given $f(x) = 2x^4 - 5x^3 - 17x^2 + 41x - 21$,

   a. Determine the end behavior of the graph of the function.

   b. List all possible rational zeros.

   c. Find all the zeros of the function and state their multiplicities.

d. Determine the x-intercepts.

e. Determine the y-intercept.

f. Is the function even, odd, or neither?

g. Graph the function.

3. Given $f(x) = -0.25x^3(x - 2)^2(x + 1)^4$,

a. Identify the leading term.

b. Determine the end behavior of the graph of the function.

c. Find all the zeros of the function and state their multiplicities.

4. Given $f(x) = x^4 + 5x^2 - 36$,

a. How many zeros does $f(x)$ have (including multiplicities)?

b. Find the zeros of $f(x)$.

c. Identify the x-intercepts of the graph of f.

d. Is the function even, odd, or neither?

5. Determine whether the intermediate value theorem guarantees that the function has a zero on the given interval.

$$f(x) = x^3 - 5x^2 + 2x + 5$$

a. $[-2, -1]$ **b.** $[-1, 0]$ **c.** $[0, 1]$ **d.** $[1, 2]$

6. a. Divide the polynomials. $\dfrac{2x^4 - 4x^3 + x - 5}{x^2 - 3x + 1}$

b. Identify the dividend, divisor, quotient, and remainder.

7. Given $f(x) = 5x^4 + 47x^3 + 80x^2 - 51x - 9$,

a. Is $\dfrac{3}{5}$ a zero of $f(x)$?

b. Is -1 a zero of $f(x)$?

c. Is $(x + 1)$ a factor of $f(x)$?

d. Is $(x + 3)$ a factor of $f(x)$?

e. Use the remainder theorem to evaluate $f(-2)$.

8. Given $f(x) = x^4 - 8x^3 + 21x^2 - 32x + 68$ and that $2i$ is a zero of $f(x)$,

a. Find all zeros of $f(x)$.

b. Factor $f(x)$ as a product of linear factors.

c. Solve the equation $f(x) = 0$.

9. Given $f(x) = 3x^4 + 7x^3 - 12x^2 - 14x + 12$,

a. How many zeros does $f(x)$ have (including multiplicities)?

b. List the possible rational zeros.

c. Determine if the upper bound theorem identifies 2 as an upper bound for the real zeros of $f(x)$.

d. Determine if the lower bound theorem identifies -4 as a lower bound for the real zeros of $f(x)$.

e. Revise the list of possible rational zeros based on the answers to parts (c) and (d).

f. Find the rational zeros.

g. Find all the zeros.

h. Graph the function.

10. Write a third-degree polynomial $f(x)$ with integer coefficients and zeros of $\frac{1}{5}$, $-\frac{2}{3}$, and 4.

11. Determine the number of possible positive and negative real zeros for $f(x) = -6x^7 - 4x^5 + 2x^4 - 3x^2 + 1$.

For Exercises 12–15, solve the inequality.

12. $c^2 < c + 20$

13. $y^3 > 13y - 12$

14. $-2x(x - 4)^2(x + 1)^3 \le 0$

15. $9x^2 + 42x + 49 > 0$

16. An agricultural school wants to determine the number of corn plants per acre that will produce the maximum yield. The model $y(n) = -0.103n^2 + 8.32n + 15.1$ represents the yield $y(n)$ (in bushels per acre) based on n thousand plants per acre.

a. Evaluate $y(20)$, $y(30)$, and $y(60)$ and interpret their meaning in the context of this problem.

b. Determine the number of plants per acre that will maximize yield. Round to the nearest hundred plants.

c. What is the maximum yield? Round to the nearest bushel per acre.

17. Suppose that a rocket is shot straight upward from ground level with an initial speed of 98 m/sec.

a. Write a model that represents the height of the rocket $s(t)$ (in meters) t seconds after launch. (*Hint*: Use the model $s(t) = -\dfrac{1}{2}gt^2 + v_0t + s_0$ with $g = 9.8$ m/sec^2. See page 543.)

b. When will the rocket reach its maximum height?

c. What is the maximum height?

d. Determine the time interval for which the rocket will be more than 200 m high. Round to the nearest tenth of a second.

18. The number of yearly visits to physicians' offices varies in part by the age of the patient. For the data shown in the table, a represents the age of patients (in years) and $n(a)$ represents the corresponding number of visits to physicians' offices per year. (*Source*: Centers for Disease Control and Prevention, www.cdc.gov)

a	8	20	35	55	65	85
$n(a)$	2.7	2.0	2.5	3.7	6.7	7.6

a. Use regression to find a quadratic function to model the data.

b. At what age is the number of yearly visits to physicians' offices the least? Round to the nearest year of age.

c. What is the minimum number of yearly visits? Round to 1 decimal place.

Rational Functions

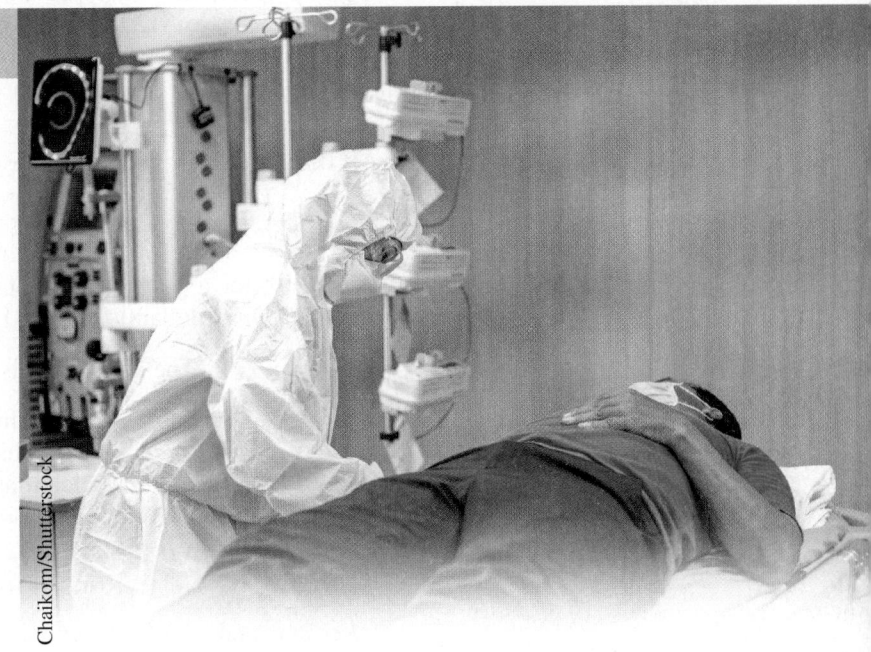

Chaikom/Shutterstock

In 2014, a world health crisis unfolded in West Africa with a widespread outbreak of the deadly Ebola virus. The 2014 outbreak infected over 28,000 people and left over 11,000 dead. The high mortality rate and the worry of future outbreaks in a world of increasing population density and international travel have prompted scientists to search for a vaccine.

Scientists measure how contagious a disease is with "the basic reproduction number" denoted by R_0. This value measures the average number of secondary cases expected from a given primary case—that is, the average number of *other* people that an infected person might sicken. The greater the value of R_0, the more contagious the virus or bacterium. For reference, Ebola has an R_0 value of between 1.5 and 2.0, whereas influenza has an R_0 of between 2 and 3. Measles, the most contagious of the childhood diseases, is transmitted through airborne particles and has an R_0 of between 16 and 18.

The model $P(R_0) = 1 - \dfrac{1}{R_0}$ measures the proportion of the population that would need to be vaccinated to contain an outbreak of a disease. This is a rational function with horizontal asymptote $y = 1$, which tells us that for diseases with high R_0 values, the vaccination rate would need to approach 100%.

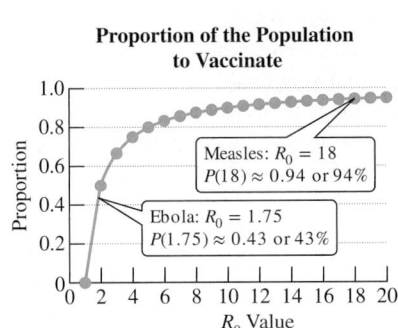

Proportion of the Population to Vaccinate

Measles: $R_0 = 18$
$P(18) \approx 0.94$ or 94%

Ebola: $R_0 = 1.75$
$P(1.75) \approx 0.43$ or 43%

SECTION 8.1   Introduction to Rational Functions

1. Apply Notation Describing Infinite Behavior of a Function

A group of friends go for pizza after a college football game. The total bill, including tax, comes to $120, and the group decides to split the bill evenly.

- If there are 12 people in the group, then each person would pay $\dfrac{\$120}{12} = \$10$.

- If there are 15 people in the group, then each person would pay $\dfrac{\$120}{15} = \$8$.

- If there are 20 people in the group, then each person would pay $\dfrac{\$120}{20} = \$6$.

In general, for a group of x people, the cost per person $C(x)$ (in dollars) for the $120 bill is given by $C(x) = \dfrac{120}{x}$. The value $C(x)$ is defined by the ratio of two polynomials. For this reason, C is called a *rational function*. Rational functions are used in a variety of applications, including in business to model average cost.

Moxie Productions/Blend Images LLC

Definition of a Rational Function

Let $p(x)$ and $q(x)$ be polynomials where $q(x) \neq 0$. A function f defined by

$$f(x) = \frac{p(x)}{q(x)} \text{ is called a } \textbf{rational function}.$$

Note: The domain of a rational function is all real numbers excluding the real zeros of $q(x)$.

Function	Factored Form	Domain
$f(x) = \dfrac{1}{x}$	$f(x) = \dfrac{1}{x}$	$\{x \mid x \neq 0\}$ $(-\infty, 0) \cup (0, \infty)$
$g(x) = \dfrac{5x^2}{2x^2 + 5x - 12}$	$g(x) = \dfrac{5x^2}{(2x - 3)(x + 4)}$	$\left\{x \mid x \neq \frac{3}{2}, x \neq -4\right\}$ $(-\infty, -4) \cup \left(-4, \frac{3}{2}\right) \cup \left(\frac{3}{2}, \infty\right)$
$k(x) = \dfrac{x + 3}{x^2 + 4}$	$k(x) = \dfrac{x + 3}{x^2 + 4}$	$\mathbb{R}$ $(-\infty, \infty)$

FOR REVIEW

Recall that a rational expression is undefined where the denominator equals zero. See Section 4.1.

In this section, we will analyze rational functions. To do so, we want to determine the behavior of the function as x approaches ∞ or $-\infty$ and as x approaches values for which the function is undefined. We will use the following notation (Table 8-1).

Table 8-1

Notation	Meaning
$x \to c^+$	x approaches c from the right (but will not equal c).
$x \to c^-$	x approaches c from the left (but will not equal c).
$x \to \infty$	x approaches infinity (x increases without bound).
$x \to -\infty$	x approaches negative infinity (x decreases without bound).

For example, consider the reciprocal function $f(x) = \frac{1}{x}$ first introduced in Section 6.1 (Figure 8-1).

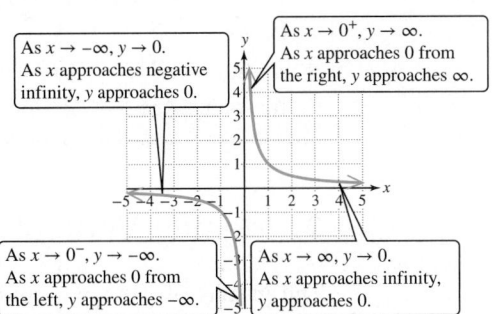

x	$f(x) = \frac{1}{x}$
-1	-1
-10	-0.1
-100	-0.01
-1000	-0.001

x	$f(x) = \frac{1}{x}$
-1	-1
-0.1	-10
-0.01	-100
-0.001	-1000

x	$f(x) = \frac{1}{x}$
1	1
0.1	10
0.01	100
0.001	1000

x	$f(x) = \frac{1}{x}$
1	1
10	0.1
100	0.01
1000	0.001

Figure 8-1

In Example 1, we study the graph of another basic rational function, $f(x) = \frac{1}{x^2}$. From the definition of the function, we make the following observations.

- The domain of $f(x) = \frac{1}{x^2}$ is all real numbers excluding zero.

- f is an even function, and the graph is symmetric to the y-axis.

- The values of $f(x)$ are positive over the domain of f.

EXAMPLE 1   **Investigating the Behavior of a Rational Function**

The graph of $f(x) = \frac{1}{x^2}$ is given.

Complete the statements.

  a. As $x \to -\infty$, $f(x) \to$ _____.

  b. As $x \to 0^-$, $f(x) \to$ _____.

  c. As $x \to 0^+$, $f(x) \to$ _____.

  d. As $x \to \infty$, $f(x) \to$ _____.

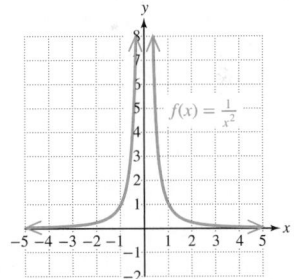

Solution:

  a. As $x \to -\infty$, $f(x) \to 0$.

  b. As $x \to 0^-$, $f(x) \to \infty$.

  c. As $x \to 0^+$, $f(x) \to \infty$.

  d. As $x \to \infty$, $f(x) \to 0$.

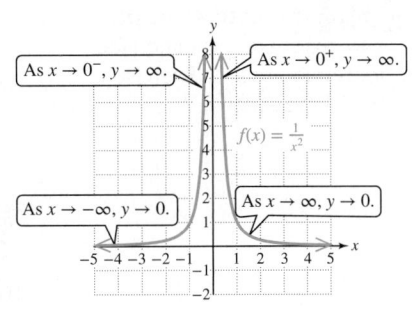

Skill Practice 1 The graph of $f(x) = \dfrac{1}{x - 2}$ is given.

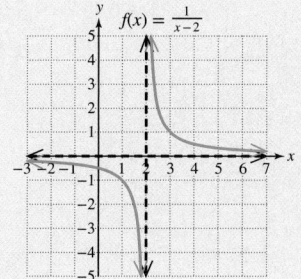

Complete the statements.

  a. As $x \to -\infty$, $f(x) \to$ _____.
  b. As $x \to 2^-$, $f(x) \to$ _____.
  c. As $x \to 2^+$, $f(x) \to$ _____.
  d. As $x \to \infty$, $f(x) \to$ _____.

2. Identify Vertical Asymptotes

The graphs of $f(x) = \dfrac{1}{x}$ and $f(x) = \dfrac{1}{x^2}$ both approach the y-axis, but do not touch the y-axis. The y-axis is called a vertical asymptote of the graphs of the functions.

Definition of a Vertical Asymptote

The line $x = c$ is a **vertical asymptote** of the graph of a function f if $f(x)$ approaches infinity or negative infinity as x approaches c from either side.

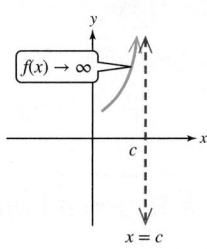 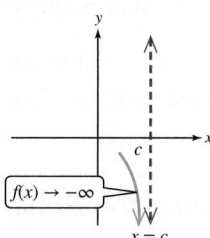 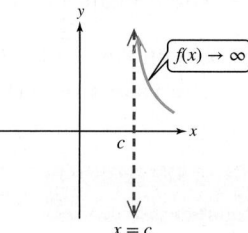 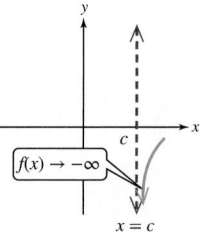

A function may have no vertical asymptotes, one vertical asymptote, or many vertical asymptotes. To locate the vertical asymptotes of a function, determine the real numbers x where the denominator is zero, but the numerator is nonzero.

Identifying Vertical Asymptotes of a Rational Function

TIP The case where $p(x)$ and $q(x)$ share a common factor is addressed in Exercises 61–64 in Section 8.2.

Consider a rational function f defined by $f(x) = \dfrac{p(x)}{q(x)}$, where $p(x)$ and $q(x)$ have no common factors other than 1. If c is a real zero of $q(x)$, then $x = c$ is a vertical asymptote of the graph of f.

EXAMPLE 2   Identifying Vertical Asymptotes

Identify the vertical asymptotes.

  a. $f(x) = \dfrac{2}{x - 3}$ **b.** $g(x) = \dfrac{x - 4}{3x^2 + 5x - 2}$ **c.** $k(x) = \dfrac{4x^2}{x^2 + 4}$

Answers
1. a. 0 **b.** $-\infty$ **c.** ∞ **d.** 0

FOR REVIEW

Recall that a vertical line is represented by an equation of the form $x = c$, where c is a constant real number. Thus, the vertical asymptote in Example 2(a) is represented by the equation $x = 3$. See Section 5.4 for review.

Solution:

a. $f(x) = \dfrac{2}{x - 3}$

f has a vertical asymptote of $x = 3$.

The expression $\frac{2}{x-3}$ is written in lowest terms. The denominator is zero for $x = 3$.

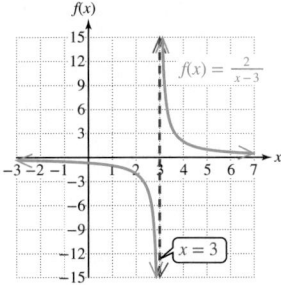

TIP The zeros of the denominator occur where $3x - 1 = 0$ and $x + 2 = 0$.

b. $g(x) = \dfrac{x - 4}{3x^2 + 5x - 2}$

$= \dfrac{x - 4}{(3x - 1)(x + 2)}$

The vertical asymptotes are

$x = \dfrac{1}{3}$ and $x = -2$.

Factor the numerator and denominator.

The numerator and denominator share no common factors other than 1. The zeros of the denominator are $\frac{1}{3}$ and -2.

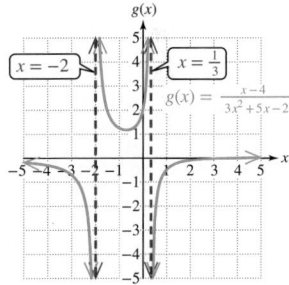

c. $k(x) = \dfrac{4x^2}{x^2 + 4}$

For any real number x, the expression $x^2 + 4$ is positive. Therefore, the denominator of $k(x) = \dfrac{4x^2}{x^2 + 4}$ has no real zeros, and the graph of k has no vertical asymptotes.

The numerator and denominator are already factored over the real numbers and the rational expression is in lowest terms.

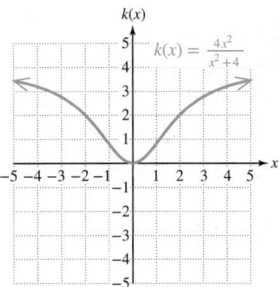

Skill Practice 2   Identify the vertical asymptotes.

a. $f(x) = \dfrac{3}{x + 1}$ **b.** $h(x) = \dfrac{x + 7}{2x^2 - x - 10}$ **c.** $m(x) = \dfrac{5x}{x^4 + 1}$

Answers

2. a. $x = -1$

b. $x = \dfrac{5}{2}$ and $x = -2$

c. No vertical asymptotes

Avoiding Mistakes

The procedure to find the vertical asymptotes of a rational function is given under the condition that the numerator and denominator share no common factors. This important observation can be illustrated by the graph of

$$f(x) = \frac{2x^2 + 5x + 3}{x + 1} \quad \text{(Figure 8-2)}.$$

The numerator and denominator share a common factor of $(x + 1)$.

$$f(x) = \frac{2x^2 + 5x + 3}{x + 1} = \frac{(2x + 3)(x + 1)}{x + 1}$$

The value $x = -1$ is not in the domain of f, but the graph of f has a "hole" at $x = -1$ rather than a vertical asymptote.

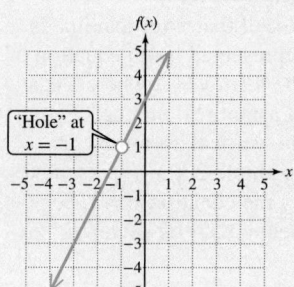

Figure 8-2

3. Identify Horizontal Asymptotes

Refer to the graph of $f(x) = \frac{1}{x}$ (Figure 8-3). Toward the far left and far right of the graph, $f(x)$ approaches the line $y = 0$ (the x-axis). The x-axis is called a horizontal asymptote of the graph of f.

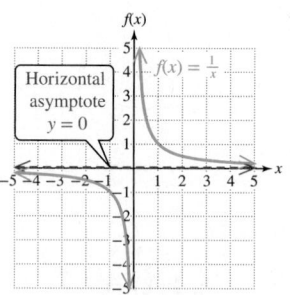

Figure 8-3

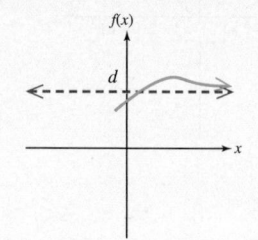

TIP While the graph of a function may not cross a vertical asymptote, it may cross a horizontal asymptote.

Definition of a Horizontal Asymptote

The line $y = d$ is a **horizontal asymptote** of the graph of a function f if $f(x)$ approaches d as x approaches infinity or negative infinity.

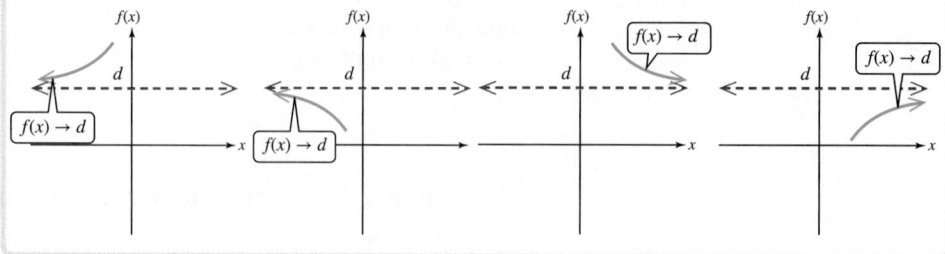

Recall that the leading term determines the far left and far right behavior of the graph of a polynomial function. Since a rational function is the ratio of two polynomials, it seems reasonable that the leading terms of the numerator and denominator determine the end behavior of a rational function.

TIP A rational function may have many vertical asymptotes, but at most one horizontal asymptote.

Identifying Horizontal Asymptotes of a Rational Function

Let f be a rational function defined by

$$f(x) = \frac{a_n x^n + a_{n-1}x^{n-1} + a_{n-2}x^{n-2} + \cdots + a_1 x + a_0}{b_m x^m + b_{m-1}x^{m-1} + b_{m-2}x^{m-2} + \cdots + b_1 x + b_0}$$

The definition of $f(x)$ indicates that n is the degree of the numerator and m is the degree of the denominator.

1. If $n > m$, then f has no horizontal asymptote.
2. If $n < m$, then the line $y = 0$ (the x-axis) is the horizontal asymptote of f.
3. If $n = m$, then the line $y = \dfrac{a_n}{b_m}$ is the horizontal asymptote of f.

To illustrate these three cases, consider the following examples with numerical values given to m and n.

1. If the degree of the numerator is greater than the degree of the denominator ($n > m$), then the numerator will "dominate" the quotient. For example:

   $$f(x) = \frac{x^4 + \cdots}{x^2 + \cdots}$$ will behave like $y = x^2$ as $|x|$ becomes large. Therefore, f has no horizontal asymptote.

   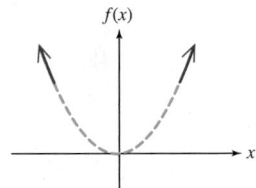

2. If the degree of the numerator is less than the degree of the denominator ($n < m$), then the denominator will "dominate" the quotient. For example:

   $$f(x) = \frac{x^2 + \cdots}{x^4 + \cdots}$$ will behave like $y = \dfrac{1}{x^2}$. The ratio $\dfrac{1}{x^2}$ tends toward 0 as $|x|$ becomes large. Therefore, f has a horizontal asymptote of $y = 0$.

   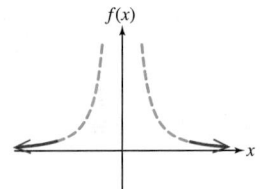

3. If the degree of the numerator is equal to the degree of the denominator ($n = m$), then the magnitude of the numerator and denominator somewhat "offset" each other. As a result, the function tends toward a constant value equal to the ratio of the leading coefficients. For example:

   $$f(x) = \frac{4x^2 + \cdots}{3x^2 + \cdots}$$ will behave like $y = \frac{4}{3}$ as $|x|$ becomes large. Therefore, f has a horizontal asymptote of $y = \frac{4}{3}$.

   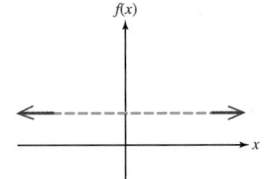

EXAMPLE 3   **Identifying Horizontal Asymptotes**

Find the horizontal asymptotes (if any) for the given functions.

a. $f(x) = \dfrac{8x^2 + 1}{x^4 + 1}$ **b.** $g(x) = \dfrac{2x^3 - 6x}{x^2 + 4}$ **c.** $h(x) = \dfrac{8x^2 + 9x - 5}{2x^2 + 1}$

Solution:

a. $f(x) = \dfrac{8x^2 + 1}{x^4 + 1}$

The degree of the numerator is 2 ($n = 2$).
The degree of the denominator is 4 ($m = 4$).

Since $n < m$, then the line $y = 0$ (the x-axis) is a horizontal asymptote of f.

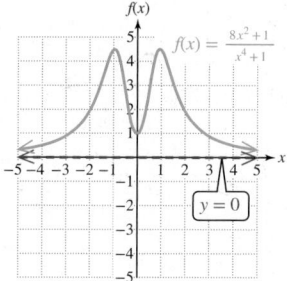

b. $g(x) = \dfrac{2x^3 - 6x}{x^2 + 4}$

The degree of the numerator is 3 ($n = 3$).
The degree of the denominator is 2 ($m = 2$).

Since $n > m$, then the function has no horizontal asymptotes.

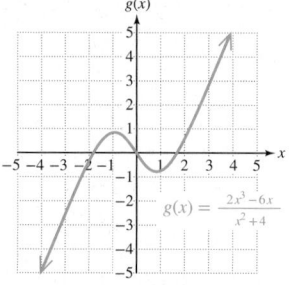

c. $h(x) = \dfrac{8x^2 + 9x - 5}{2x^2 + 1}$

The degree of the numerator is 2 ($n = 2$).
The degree of the denominator is 2 ($m = 2$).

Since $n = m$, then the line $y = \frac{8}{2}$ or equivalently $y = 4$ is a horizontal asymptote of the graph of f.

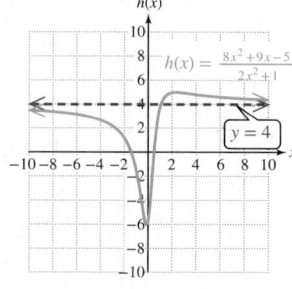

Skill Practice 3   Find the horizontal asymptotes (if any) for the given functions.

a. $f(x) = \dfrac{7x^2 + 2x}{4x^2 - 3}$ **b.** $m(x) = \dfrac{4x^3 + 2}{2x - 1}$ **c.** $n(x) = \dfrac{5}{4x^2 + 9}$

The graph of a rational function may not cross a vertical asymptote. However, as demonstrated in Example 3(c), the graph may cross a horizontal asymptote. For the purpose of graphing a rational function, it is helpful to determine where a graph crosses a horizontal asymptote.

Suppose that the line $y = d$ is a horizontal asymptote of a rational function $y = f(x)$. The solutions to the equation $f(x) = d$ are the values of x where the graph of f crosses its horizontal asymptote. If the equation has no real solution, then the graph does not cross its horizontal asymptote.

Answers

3. a. $y = \dfrac{7}{4}$

b. No horizontal asymptotes

c. $y = 0$

EXAMPLE 4 **Determining Where a Graph Crosses a Horizontal Asymptote**

Given $h(x) = \dfrac{8x^2 + 9x - 5}{2x^2 + 1}$, determine the point where the graph of h crosses its horizontal asymptote.

Solution:

$$h(x) = \frac{8x^2 + 9x - 5}{2x^2 + 1}$$

From Example 3(c), the horizontal asymptote is $y = 4$.

$$\frac{8x^2 + 9x - 5}{2x^2 + 1} = 4$$

Set $h(x)$ equal to 4.

$$\frac{8x^2 + 9x - 5}{2x^2 + 1} \cdot (2x^2 + 1) = 4 \cdot (2x^2 + 1)$$

Clear fractions by multiplying by the LCD.

$$8x^2 + 9x - 5 = 8x^2 + 4$$
$$9x - 5 = 4$$
$$9x = 9$$
$$x = 1$$

FOR REVIEW

To review the procedure of clearing fractions and solving a rational equation, see Section 4.3.

The function crosses its horizontal asymptote at $(1, 4)$.

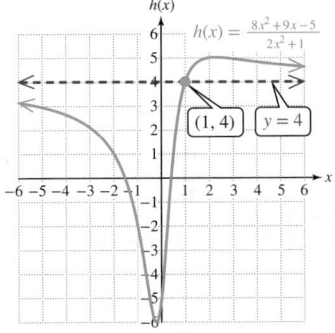

Skill Practice 4   Given $g(x) = \dfrac{3x^2 + 4x - 3}{x^2 + 3}$, determine the horizontal asymptote and the point where the graph crosses the horizontal asymptote.

---

4. Identify Slant and Nonlinear Asymptotes

In Example 3(b) we determined that the function defined by $f(x) = \dfrac{2x^3 - 6x}{x^2 + 4}$ has no horizontal asymptotes because the degree of the numerator is greater than the degree of the denominator. However, as x approaches infinity and negative infinity, the graph approaches the slanted line $y = 2x$ (Figure 8-4). The line $y = 2x$ is called a *slant asymptote* of the function. A rational function has a slant asymptote if the degree of the numerator is exactly one greater than the degree of the denominator. In this case, the degree of the numerator is 3 and the degree of the denominator is 2.

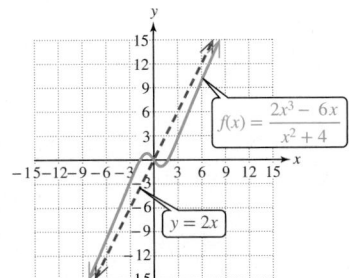

Figure 8-4

Now suppose the degree of the numerator is *more than one* greater than the degree of the denominator. In this case, the far left and far right behavior of the function will follow a nonlinear pattern. It will resemble the end behavior of the power function given by the ratio of the leading terms. For example, given $g(x) = \dfrac{x^4 + x^2 + 1}{x^2}$, the degree of the numerator (4) is two more than the degree of the denominator (2). Notice that the end behavior of g resembles the end behavior of $y = \dfrac{x^4}{x^2} = x^2$ (see Figure 8-5).

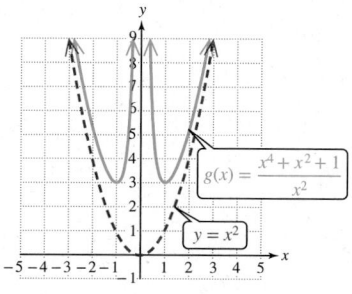

Figure 8-5

Slant and Nonlinear Asymptotes

Let f be a rational function defined by

$$f(x) = \frac{a_n x^n + a_{n-1} x^{n-1} + a_{n-2} x^{n-2} + \cdots + a_1 x + a_0}{b_m x^m + b_{m-1} x^{m-1} + b_{m-2} x^{m-2} + \cdots + b_1 x + b_0}$$

Note that n is the degree of the numerator and m is the degree of the denominator.

1. If $n = m + 1$ (that is, the degree of the numerator is exactly 1 greater than the degree of the denominator), the function has a **slant asymptote**.

2. If $n > m + 1$ (that is, the degree of the numerator is more than 1 greater than the degree of the denominator), the end behavior of the function resembles the end behavior of the power function $y = \dfrac{a_n}{b_m}$.

To find an equation of a slant or nonlinear asymptote, divide the numerator of $f(x)$ by the denominator and determine the quotient $Q(x)$. The equation $y = Q(x)$ is an equation of the asymptote.

TIP A rational function cannot have both a horizontal asymptote and a slant or nonlinear asymptote.

EXAMPLE 5   Identifying the Asymptotes of a Rational Function

Determine the asymptotes. $f(x) = \dfrac{2x^2 - 5x - 3}{x - 2}$

Solution:

$f(x) = \dfrac{2x^2 - 5x - 3}{x - 2}$

f has a vertical asymptote of $x = 2$.

f has no horizontal asymptote.

The expression $\dfrac{2x^2 - 5x - 3}{x - 2} = \dfrac{(2x + 1)(x - 3)}{x - 2}$ is in lowest terms, and the denominator is zero at $x = 2$.

The degree of the numerator is exactly one greater than the degree of the denominator. Therefore, f has no horizontal asymptote, but it does have a slant asymptote.

TIP In Example 5, the divisor is of the form $x - c$. Therefore, we can also use synthetic division.

$$\begin{array}{r|rrr} 2 & 2 & -5 & -3 \\ & & 4 & -2 \\ \hline & 2 & -1 & \boxed{-5} \end{array}$$

To find the slant asymptote, divide $(2x^2 - 5x - 3)$ by $(x - 2)$.

$$\begin{array}{r} 2x - 1 \\ x - 2 \overline{)\, 2x^2 - 5x - 3 } \\ \underline{-(2x^2 - 4x)} \\ -x - 3 \\ \underline{-(-x + 2)} \\ -5 \end{array}$$

The quotient is $2x - 1$.
The slant asymptote is given by $y = 2x - 1$.

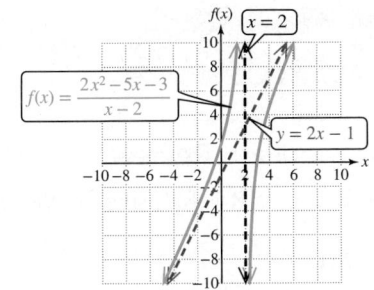

Skill Practice 5   Determine the asymptotes. $g(x) = \dfrac{2x^2 - 9}{x + 1}$

EXAMPLE 6   **Finding an Equation of a Nonlinear Asymptote**

Find an equation of the nonlinear asymptote. $f(x) = \dfrac{x^5 - 3x^3 + 1}{x^3 + x}$

Solution:

The degree of the numerator (5) is exactly two more than the degree of the denominator (3). Therefore, we expect the far left and far right behavior of f to be similar to the end behavior of $y = \dfrac{x^5}{x^3} = x^2$. That is, f has a nonlinear asymptote that is quadratic. To find an equation of the asymptote, divide the numerator by the denominator.

$$\begin{array}{r} x^2 -4 \\ x^3 + 0x^2 + x \overline{)\, x^5 + 0x^4 - 3x^3 + 0x^2 + 0x + 1} \\ \underline{-\left(x^5 + 0x^4 + x^3\right)} \\ -4x^3 + 0x^2 + 0x \\ \underline{-\left(-4x^3 + 0x^2 - 4x\right)} \\ 4x + 1 \end{array}$$

The quotient is $x^2 - 4$.
The nonlinear asymptote is $y = x^2 - 4$ (shown in red in Figure 8-6).

$f(x) = \dfrac{x^3 - 3x^3 + 1}{x^3 + x}$ $y = x^2 - 4$

Figure 8-6

Skill Practice 6   Find an equation of the nonlinear asymptote.
$$f(x) = \dfrac{-2x^4 + 8x^2 - 6}{x^2 - x}$$

Answers

5. Vertical asymptote: $x = -1$;
No horizontal asymptote; Slant asymptote: $y = 2x - 2$

6. $y = -2x^2 - 2x + 6$

SECTION 8.1   Practice Exercises

Prerequisite Review

For Exercises R.1–R.4, simplify the expression and state the restrictions on the variable. (See Section 4.1 for review.)

R.1. $\dfrac{q^2 - 36}{q^2 - 4q - 12}$

R.2. $\dfrac{2p^2 - 13p + 20}{4p^2 - 25}$

R.3. $\dfrac{16 - v^2}{5v^2 - 20v}$

R.4. $\dfrac{9 - u^2}{2u^2 - 6u}$

For Exercises R.5–R.8, divide the polynomials. Use either synthetic division or long division as appropriate. (See Section 7.3 for review.)

R.5. $(3x^4 - 2x^3 + x - 4) \div (x^2 - 3x)$

R.6. $(2x^5 + x^3 - 2x^2 - 6) \div (x^2 - 1)$

R.7. $(6x^3 - 3x^2 + x - 2) \div (x + 3)$

R.8. $(5x^3 + x^2 - 6x + 1) \div (x + 2)$

For Exercises R.9–R.14, simplify. (See Section 2.1 for review.)

R.9. $\dfrac{-3x^4}{6x^3}$

R.10. $\dfrac{4x^5}{2x^3}$

R.11. $\dfrac{2x^3}{7x^3}$

R.12. $\dfrac{-x^6}{5x^6}$

R.13. $\dfrac{8x^2}{4x^5}$

R.14. $\dfrac{12x}{2x^2}$

For Exercises R.17–R.18, find the real solutions to the equation. (See Section 4.3 for review.)

R.15. $\dfrac{2x - 3}{x^2 - x - 12} = 0$

R.16. $\dfrac{x + 6}{x^2 + 10x + 9} = 0$

R.17. $\dfrac{3x^2 - 5}{3x^2 + 2x + 1} = 1$

R.18. $\dfrac{5x^2 + 1}{10x^2 - 4x - 2} = \dfrac{1}{2}$

Concept Connections

1. The domain of a rational function defined by $f(x) = \dfrac{p(x)}{q(x)}$ is all real numbers excluding the zeros of _____.

2. The notation $x \to \infty$ is read as _____.

3. The notation $x \to 5^-$ is read as _____.

4. The line $x = c$ is a _____ asymptote of the graph of a function f if $f(x)$ approaches infinity or negative infinity as x approaches _____ from either the left or right.

5. To locate the vertical asymptotes of a function, determine the real numbers x where the denominator is zero, but the numerator is _____.

6. Consider a rational function in which the degree of the numerator is n and the degree of the denominator is m. If n _____ m, then the x-axis is the horizontal asymptote. If n _____ m, then the function has no horizontal asymptote.

Objective 1: Apply Notation Describing Infinite Behavior of a Function

For Exercises 7–12, write the domain of the function in interval notation.

7. $f(x) = \dfrac{x^2 - 25}{x - 5}$

8. $g(x) = \dfrac{x^2 - 9}{x - 3}$

9. $r(x) = \dfrac{2x - 3}{4x^2 + 3x - 1}$

10. $p(x) = \dfrac{3x - 5}{2x^2 + 5x - 7}$

11. $h(x) = \dfrac{18x}{x^2 + 100}$

12. $k(x) = \dfrac{14}{x^2 + 49}$

For Exercises 13–16, refer to the graph of the function and complete the statement. (**See Example 1**)

13. a. As $x \to -\infty, f(x) \to$ _____.
    b. As $x \to 4^-, f(x) \to$ _____.
    c. As $x \to 4^+, f(x) \to$ _____.
    d. As $x \to \infty, f(x) \to$ _____.
    e. The graph is increasing over the interval(s) _____.
    f. The graph is decreasing over the interval(s) _____.
    g. The domain is _____.
    h. The range is _____.
    i. The vertical asymptote is the line _____.
    j. The horizontal asymptote is the line _____.

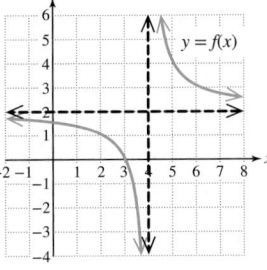

14. a. As $x \to -\infty, f(x) \to$ _____.
    b. As $x \to -3^-, f(x) \to$ _____.
    c. As $x \to -3^+, f(x) \to$ _____.
    d. As $x \to \infty, f(x) \to$ _____.
    e. The graph is increasing over the interval(s) _____.
    f. The graph is decreasing over the interval(s) _____.
    g. The domain is _____.
    h. The range is _____.
    i. The vertical asymptote is the line _____.
    j. The horizontal asymptote is the line _____.

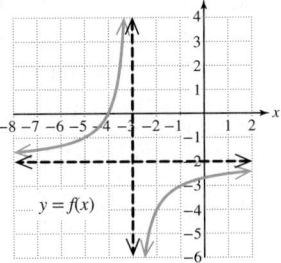

15. a. As $x \to -\infty, f(x) \to$ _____.
    b. As $x \to -3^-, f(x) \to$ _____.
    c. As $x \to -3^+, f(x) \to$ _____.
    d. As $x \to \infty, f(x) \to$ _____.
    e. The graph is increasing over the interval(s) _____.
    f. The graph is decreasing over the interval(s) _____.
    g. The domain is _____.
    h. The range is _____.
    i. The vertical asymptote is the line _____.
    j. The horizontal asymptote is the line _____.

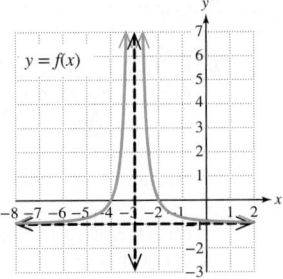

16. a. As $x \to -\infty, f(x) \to$ _____.
    b. As $x \to 1^-, f(x) \to$ _____.
    c. As $x \to 1^+, f(x) \to$ _____.
    d. As $x \to \infty, f(x) \to$ _____.
    e. The graph is increasing over the interval(s) _____.
    f. The graph is decreasing over the interval(s) _____.
    g. The domain is _____.
    h. The range is _____.
    i. The vertical asymptote is the line _____.
    j. The horizontal asymptote is the line _____.

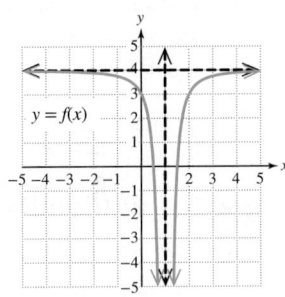

Objective 2: Identify Vertical Asymptotes

For Exercises 17–24, determine the vertical asymptotes of the graph of the function. (**See Example 2**)

17. $f(x) = \dfrac{8}{x - 4}$

18. $g(x) = \dfrac{2}{x + 7}$

19. $h(x) = \dfrac{x - 3}{2x^2 - 9x - 5}$

20. $k(x) = \dfrac{x + 2}{3x^2 + 8x - 3}$

21. $m(x) = \dfrac{x}{x^2 + 5}$

22. $n(x) = \dfrac{6}{x^4 + 1}$

23. $f(t) = \dfrac{t^2 + 2}{2t^2 + 4t - 3}$

24. $k(a) = \dfrac{5 + a^4}{3a^2 + 4a - 1}$

Objective 3: Identify Horizontal Asymptotes

25. The graph of $f(x) = \dfrac{-x^2 + 8}{2x^2 - 3}$ will behave like which function for large values of $|x|$?

    a. $y = -\dfrac{1}{2}$
    b. $y = -\dfrac{x}{2}$
    c. $y = -\dfrac{8}{3}$
    d. $y = -\dfrac{1}{2}x - \dfrac{8}{3}$

26. The graph of $f(x) = \dfrac{2x^3 + 7}{5x^3}$ will behave like which function for large values of $|x|$?

    a. $y = \dfrac{2}{5x}$
    b. $y = \dfrac{2x}{5}$
    c. $y = \dfrac{2}{5}$
    d. $y = \dfrac{2}{5}x^3$

27. The graph of $f(x) = \dfrac{-3x^4 - 2x + 5}{x^5 + x^2 - 2}$ will behave like which function for large values of $|x|$?

    a. $y = -3$ **b.** $y = -3x$

    c. $y = -\dfrac{5}{2}$ **d.** $y = 0$

28. The graph of $f(x) = \dfrac{x^2 + 7x - 3}{6x^4 + 2}$ will behave like which function for large values of $|x|$?

    a. $y = \dfrac{x^2}{6}$ **b.** $y = 0$

    c. $y = -\dfrac{3}{2}$ **d.** $y = \dfrac{1}{6}x - \dfrac{3}{2}$

For Exercises 29–36,

    a. Identify the horizontal asymptotes (if any). **(See Example 3)**

    b. If the graph of the function has a horizontal asymptote, determine the point (if any) where the graph crosses the horizontal asymptote. **(See Example 4)**

29. $p(x) = \dfrac{5}{x^2 + 2x + 1}$ **30.** $q(x) = \dfrac{8}{x^2 + 4x + 4}$ **31.** $h(x) = \dfrac{3x^2 + 8x - 5}{x^2 + 3}$ **32.** $r(x) = \dfrac{-4x^2 + 5x - 1}{x^2 + 2}$

33. $m(x) = \dfrac{x^4 + 2x + 1}{5x + 2}$ **34.** $n(x) = \dfrac{x^3 - x^2 + 1}{2x - 3}$ **35.** $t(x) = \dfrac{2x + 4}{x^2 + 7x - 4}$ **36.** $s(x) = \dfrac{x + 3}{2x^2 - 3x - 5}$

37. Consider the expression $\dfrac{x^2 + 3x + 1}{2x^2 + 5}$.

    a. Divide the numerator and denominator by the greatest power of x that appears in the denominator. That is, divide numerator and denominator by x^2.

    b. As $|x| \to \infty$, what value will $\dfrac{3}{x}$, $\dfrac{1}{x^2}$, and $\dfrac{5}{x^2}$ approach?

    (*Hint*: Substitute large values of x such as 100, 1000, 10,000, and so on to help you understand the behavior of each expression.)

    c. Use the results from parts (a) and (b) to identify the horizontal asymptote for the graph of

    $$f(x) = \dfrac{x^2 + 3x + 1}{2x^2 + 5}.$$

38. Consider the expression $\dfrac{3x^3 - 2x^2 + 7x}{5x^3 + 1}$.

    a. Divide the numerator and denominator by the greatest power of x that appears in the denominator.

    b. As $|x| \to \infty$, what value will $-\dfrac{2}{x}$, $\dfrac{7}{x^2}$, and $\dfrac{1}{x^3}$ approach?

    c. Use the results from parts (a) and (b) to identify the horizontal asymptote for the graph of

    $$f(x) = \dfrac{3x^3 - 2x^2 + 7x}{5x^3 + 1}.$$

Objective 4: Identify Slant and Nonlinear Asymptotes

For Exercises 39–46, write an equation of the slant or nonlinear asymptote. **(See Examples 5–6)**

39. $f(x) = \dfrac{x^2 - 3x + 2}{x - 4}$ **40.** $g(x) = \dfrac{x^2 + 5x + 2}{x + 3}$ **41.** $k(x) = \dfrac{2x^4 - x^2 + 1}{x^2 + 4}$

42. $m(x) = \dfrac{3x^4 + 2x^2 - 4}{x^2 + 2}$ **43.** $n(x) = \dfrac{12x^3 - 13x^2 + 6x - 1}{4x^2 - 3x + 1}$ **44.** $p(x) = \dfrac{10x^3 - 7x^2 - 2x + 8}{2x^2 - 3x + 2}$

45. $f(x) = \dfrac{-x^3 + 2x^2 - 4x + 7}{x - 2}$ **46.** $r(x) = \dfrac{-x^3 - 6x^2 - 2x + 2}{x + 5}$

For Exercises 47–56, identify the asymptotes. **(See Example 5)**

47. $f(x) = \dfrac{2x^2 + 3}{x}$ **48.** $g(x) = \dfrac{3x^2 + 2}{x}$ **49.** $h(x) = \dfrac{-3x^2 + 4x - 5}{x + 6}$ **50.** $k(x) = \dfrac{-2x^2 - 3x + 7}{x + 3}$

51. $p(x) = \dfrac{x^3 + 5x^2 - 4x + 1}{x^2 - 5}$ **52.** $q(x) = \dfrac{x^3 + 3x^2 - 2x - 4}{x^2 - 7}$

53. $r(x) = \dfrac{2x + 1}{x^3 + x^2 - 4x - 4}$ **54.** $t(x) = \dfrac{3x - 4}{x^3 + 2x^2 - 9x - 18}$

55. $f(x) = \dfrac{4x^3 - 2x^2 + 7x - 3}{2x^2 + 4x + 3}$ **56.** $a(x) = \dfrac{9x^3 - 5x + 4}{3x^2 + 2x + 1}$

Mixed Exercises

57. The distance between two cities is 120 mi.

   a. How long will the trip take if the average speed is 30 mph?

   b. How long will the trip take if the average speed is 40 mph?

   c. How long with the trip take if the average speed is 60 mph?

   d. Write a function representing the duration of the trip $D(x)$ (in hours) if the average speed is x miles per hour.

   e. Will the duration of the trip ever equal zero?

58. A dance instructor charges $60 per hour and the charge is to be split evenly among the students in the class.

   a. How much would each student pay if there were 6 students in the class?

   b. How much would each student pay if there were 10 students in the class?

   c. How much would each student pay if there were 12 students in the class?

   d. Write a function representing the amount each student would pay $A(x)$ (in dollars) for x students in the class.

   e. Will the amount that each student pays ever reach zero?

59. A power company burns coal to generate electricity. The cost $C(x)$ (in $1000) to remove $x\%$ of the air pollutants is given by

$$C(x) = \frac{600x}{100 - x}$$

   a. Compute the cost to remove 25% of the air pollutants. (*Hint*: $x = 25$.)

   b. Determine the cost to remove 50%, 75%, and 90% of the air pollutants.

   c. If the power company budgets $1.4 million for pollution control, what percentage of the air pollutants can be removed?

60. The cost $C(x)$ (in $1000) for a city to remove $x\%$ of the waste from a polluted river is given by

$$C(x) = \frac{80x}{100 - x}$$

   a. Determine the cost to remove 20%, 40%, and 90% of the waste. Round to the nearest thousand dollars.

   b. If the city has $320,000 budgeted for river cleanup, what percentage of the waste can be removed?

61. A parallel circuit is one with several paths through which electricity can travel. The total resistance in a parallel circuit is always less than the resistance in any single branch. The total resistance R for the circuit shown can be computed from the formula $\dfrac{1}{R} = \dfrac{1}{R_1} + \dfrac{1}{R_2}$, where R_1 and R_2 are the resistances in the individual branches. Suppose that a resistor with a fixed resistance of 6 Ω (ohms) is placed in parallel with a variable resistor of resistance x.

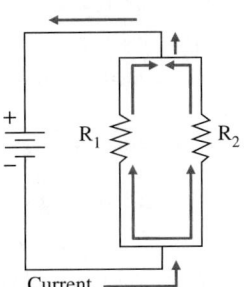

   a. Write R as a function of x.

   b. Complete the table.

x	6	12	18	30
$R(x)$				

   c. What value does $R(x)$ approach as $x \to \infty$? Discuss the significance of this result.

62. The total resistance R of three resistors in parallel is given by

$$R = \frac{R_1 R_2 R_3}{R_1 R_2 + R_1 R_3 + R_2 R_3}$$

Suppose that an 8-Ω and a 12-Ω resistor are placed in parallel with a variable resistor of resistance x.

   a. Write R as a function of x.

   b. What value does $R(x)$ approach as $x \to \infty$? Write the value in decimal form.

Write About It

63. Explain why $x = -2$ is not a vertical asymptote of the graph of $f(x) = \dfrac{x^2 + 7x + 10}{x + 2}$.

64. Write an informal definition of a horizontal asymptote of a rational function.

Expanding Your Skills

65. A chemist has a tank with 20 L of water into which 240 g of salt has been added. She adds water at a constant rate of 2 L per minute. At the same time, salt is added at a rate of 40 g per minute.

  a. Write a function that expresses the amount of water $w(t)$ (in liters) after t minutes.

  b. Write a function that expresses the amount of salt $s(t)$ (in grams) after t minutes.

  c. Write a function that gives the concentration of salt in the water $C(t)$ (in grams per liter) after t minutes.

  d. What was the concentration of salt in the water initially?

  e. What is the concentration after 10 min, 20 min, 1 hr? Round to 1 decimal place.

  f. What is the horizontal asymptote of function C, and what does it mean in the context of this problem?

66. In a certain city, a government official estimated the land area used for housing at 60 mi^2. At that time, the number of people in the city was 69,000. The amount of land used for housing increases at 3 mi^2 per year and the number of people increases by 1400 per year.

  a. Write a function that expresses the amount of land area used for housing $a(t)$ (in square miles) after t years.

  b. Write a function that expresses the number of people $n(t)$ after t years.

  c. Write a function that gives the population density $D(t)$ (in people per square mile) after t years.

  d. What was the population density initially?

  e. What is the population density after 5 years, 10 years, and 20 years? Round to the nearest whole unit.

  f. What is the horizontal asymptote of function D, and what does it mean in the context of this problem?

Technology Connections

For Exercises 67–74, graph the functions and the horizontal or slant asymptotes from the given exercises. Use the recommended window settings.

67. Exercise 47; $f(x) = \dfrac{2x^2 + 3}{x}$

  $[-10, 10, 1]$ by $[-10, 10, 1]$

68. Exercise 48; $g(x) = \dfrac{3x^2 + 2}{x}$

  $[-10, 10, 1]$ by $[-10, 10, 1]$

69. Exercise 49; $h(x) = \dfrac{-3x^2 + 4x - 5}{x + 6}$

  $[-100, 100, 10]$ by $[-80, 120, 10]$

70. Exercise 50; $k(x) = \dfrac{-2x^2 - 3x + 7}{x + 3}$

  $[-15, 15, 3]$ by $[-6, 24, 3]$

71. Exercise 53; $r(x) = \dfrac{2x + 1}{x^3 + x^2 - 4x - 4}$

  $[-10, 10, 1]$ by $[-10, 10, 1]$

72. Exercise 54; $t(x) = \dfrac{3x - 4}{x^3 + 2x^2 - 9x - 18}$

  $[-15, 15, 3]$ by $[-15, 15, 3]$

73. Exercise 41; $k(x) = \dfrac{2x^4 - x^2 + 1}{x^2 + 4}$

  $[-10, 10, 1]$ by $[-10, 10, 1]$

74. Exercise 42; $m(x) = \dfrac{3x^4 + 2x^2 - 4}{x^2 + 2}$

  $[-10, 10, 1]$ by $[-10, 10, 1]$

Graphs of Rational Functions

1. Graph Rational Functions

In Section 8.1, we learned how to identify the vertical, horizontal, slant, and non-linear asymptotes of a rational function. We now turn our attention to graphing rational functions. To begin, note that some rational functions can be graphed using transformations of the parent functions $y = \dfrac{1}{x}$ and $y = \dfrac{1}{x^2}$ shown in Figures 8-7 and 8-8.

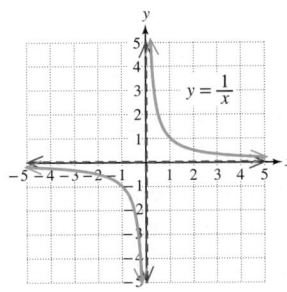

Figure 8-7

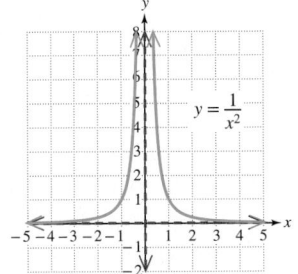

Figure 8-8

EXAMPLE 1   **Using Transformations to Graph a Rational Function**

Use transformations to graph $f(x) = \dfrac{1}{(x + 2)^2} + 3$.

Solution:

$$f(x) = \frac{1}{(x + 2)^2} + 3$$

The graph of f is the graph of $y = \dfrac{1}{x^2}$ with a shift to the left 2 units and a shift upward 3 units.

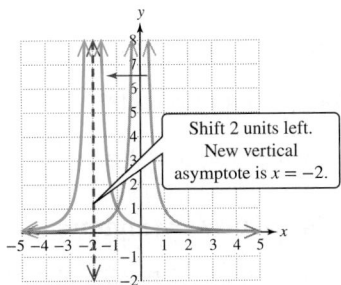

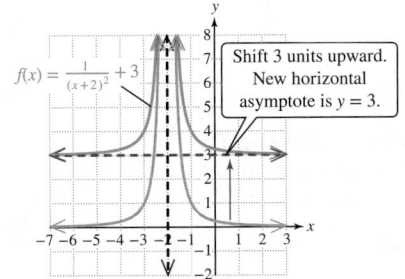

Skill Practice 1   Use transformations to graph $g(x) = \dfrac{1}{x - 3} - 2$.

Answer

1.

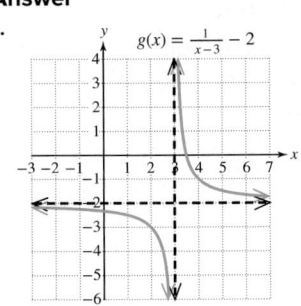

To graph a rational function that is not a simple transformation of $y = \dfrac{1}{x}$ or $y = \dfrac{1}{x^2}$, more steps must be employed. Our strategy is to find all asymptotes and key points (intercepts and points where the function crosses a horizontal asymptote). Then determine the behavior of the function on the intervals defined by these key points and the vertical asymptotes.

Graphing a Rational Function

Consider a rational function f defined by $f(x) = \dfrac{p(x)}{q(x)}$, where $p(x)$ and $q(x)$ are polynomials with no common factors.

1. Determine the y-intercept by evaluating $f(0)$.
2. Determine the x-intercept(s) by finding the real solutions of $f(x) = 0$. The value $f(x)$ equals zero when the numerator $p(x) = 0$.
3. Identify any vertical asymptotes and graph them as dashed lines.
4. Determine whether the function has a horizontal asymptote or a slant asymptote (or neither), and graph the asymptote as a dashed line.
5. Determine where the function crosses the horizontal or slant asymptote (if applicable).
6. If a test for symmetry is easy to apply, use symmetry to plot additional points. Recall:
   - f is an even function (symmetric to the y-axis) if $f(-x) = f(x)$.
   - f is an odd function (symmetric to the origin) if $f(-x) = -f(x)$.
7. Plot at least one point on the intervals defined by the x-intercepts, vertical asymptotes, and points where the function crosses a horizontal or slant asymptote.
8. Sketch the function based on the information found in steps 1–7.

EXAMPLE 2 **Graphing a Rational Function**

Graph $f(x) = \dfrac{x + 3}{x - 2}$.

Solution:

1. Determine the y-intercept.

$$f(0) = \frac{(0) + 3}{(0) - 2} = -\frac{3}{2}$$

The y-intercept is $\left(0, -\frac{3}{2}\right)$.

2. Determine the x-intercept(s).

$$\frac{x + 3}{x - 2} = 0 \text{ when } x + 3 = 0 \text{ or } x = -3.$$

The x-intercept is $(-3, 0)$.

> **TIP** In Example 2, we find the x-intercept(s) by setting $f(x)$ equal to zero. Thus, we solve the equation $\dfrac{x + 3}{x - 2} = 0$.
> A fraction will equal zero if its numerator equals zero. The solution set to the equation $x + 3 = 0$ is $\{-3\}$, and the x-intercept is $(-3, 0)$.

3. Identify the vertical asymptotes.

The denominator $x - 2$ is zero at $x = 2$, and the numerator $x + 3$ is nonzero for $x = 2$.

The graph has one vertical asymptote, $x = 2$.

The vertical asymptotes occur at the values of x for which the denominator is zero and the numerator is nonzero.

4. Determine whether f has a horizontal or slant asymptote.

The degree of the numerator is equal to the degree of the denominator. Therefore, the graph has a horizontal asymptote given by the ratio of leading coefficients of the numerator and denominator.

$$y = \frac{1}{1} = 1 \text{ is the horizontal asymptote.}$$

5. Determine where f crosses its horizontal asymptote (if at all).

Solve the equation $f(x) = 1$.

$$\frac{x+3}{x-2} = 1 \Rightarrow \cancel{(x-2)} \cdot \frac{x+3}{\cancel{x-2}} = (x-2) \cdot 1$$

$$x + 3 = x - 2$$

$$3 = -2$$

(Contradiction) The graph of f does not cross its horizontal asymptote.

6. Test for symmetry.

$f(-x) = \dfrac{-x+3}{-x-2}$ does not equal $f(x)$ or $-f(x)$. The function is neither even nor odd and is not symmetric with respect to the y-axis or origin.

7. Determine the behavior of f on each interval.

Determine the sign of the function on the intervals (shown in red) defined by the x-intercept at $x = -3$ and the vertical asymptote $x = 2$.

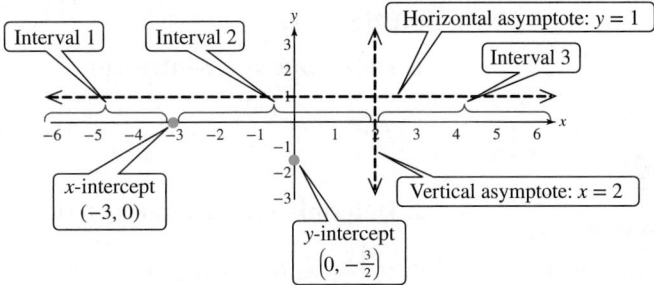

TIP $f(x) = \dfrac{x+3}{x-2}$

$f(-4) = \dfrac{1}{6}$

(positive on Interval 1)

$f(0) = -\dfrac{3}{2}$

(negative on Interval 2)

$f(3) = 6$

(positive on Interval 3)

Interval	Test Point	Comments
$(-\infty, -3)$	$\left(-4, \dfrac{1}{6}\right)$	• Since $f(x)$ is positive on this interval, $f(x)$ must approach the horizontal asymptote $y = 1$ from below as $x \to -\infty$.
$(-3, 2)$	$\left(0, -\dfrac{3}{2}\right)$	• Since $f(x)$ is negative on this interval, the graph crosses the x-axis at the intercept $(-3, 0)$ and continues downward (through the y-intercept). As x approaches the vertical asymptote $x = 2$ from the left, $f(x) \to -\infty$.
$(2, \infty)$	$(3, 6)$	• Since $f(x)$ is positive on this interval, as x approaches the vertical asymptote $x = 2$ from the right, $f(x) \to \infty$. • Since $f(x)$ is positive on this interval, $f(x)$ must approach the horizontal asymptote from above as $x \to \infty$.

8. Sketch the function (Figure 8-9).

Plot the x- and y-intercepts $(-3, 0)$ and $\left(0, -\frac{3}{2}\right)$, and the additional points $\left(-4, \frac{1}{6}\right)$ and $(3, 6)$.

Graph the horizontal asymptote ($y = 1$) and vertical asymptote $x = 2$ as dashed lines.

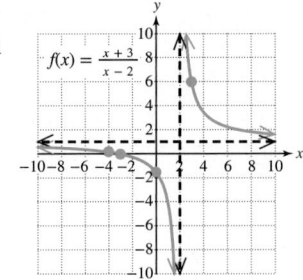

Figure 8-9

Answer

2.

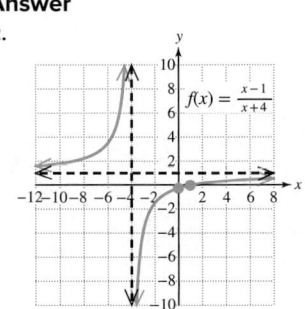

Skill Practice 2   Graph $f(x) = \dfrac{x-1}{x+4}$.

TIP In Example 2, we can rewrite $f(x) = \dfrac{x+3}{x-2}$ as $f(x) = 1 + \dfrac{5}{x-2}$ by dividing $(x+3) \div (x-2)$.

$$
\begin{array}{r}
1 \\
x-2 \overline{)\, x+3} \\
-(x-2) \\
\hline
5
\end{array}
$$

Thus, $\dfrac{x+3}{x-2} = 1 + \dfrac{5}{x-2}$

When written in the form $f(x) = 1 + \dfrac{5}{x-2}$, we can graph f by shifting the graph of $y = \dfrac{1}{x}$ two units to the right, stretching by a factor of 5, and shifting the graph upward 1 unit.

EXAMPLE 3   Graphing a Rational Function

Graph $f(x) = \dfrac{4x}{x^2 - 4}$.

Solution:

1. **Determine the y-intercept.**

   $f(0) = \dfrac{4(0)}{(0)^2 - 4} = 0$ The y-intercept is $(0, 0)$.

2. **Determine the x-intercept(s).**

   $\dfrac{4x}{x^2 - 4} = 0$ for $x = 0$.   The x-intercept is $(0, 0)$.

TIP The expression $\dfrac{4x}{x^2-4}$ equals zero when the numerator equals zero. Thus, to solve the equation $\dfrac{4x}{x^2-4} = 0$, we have:

$$(x^2-4)\left(\dfrac{4x}{x^2-4}\right) = (x^2-4)\cdot 0$$
$$4x = 0$$
$$x = 0$$

3. **Identify the vertical asymptotes.**

   The zeros of $x^2 - 4$ are 2 and -2. Vertical asymptotes: $x = 2$ and $x = -2$.

4. **Determine whether f has a horizontal or slant asymptote.**

   The degree of the numerator is less than the degree of the denominator. The horizontal asymptote is $y = 0$.

5. **Determine where f crosses its horizontal asymptote.**

   Set $f(x) = 0$. We have $\dfrac{4x}{x^2 - 4} = 0$ for $x = 0$.

   Therefore, f crosses its horizontal asymptote at $(0, 0)$.

6. **Test for symmetry.**

   f is an odd function because $f(-x) = \dfrac{4(-x)}{(-x)^2 - 4} = -\dfrac{4x}{x^2 - 4} = -f(x)$.

7. **Determine the behavior of f on each interval.**

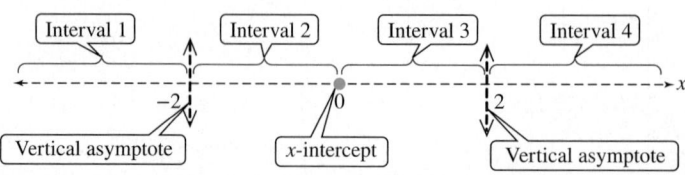

TIP The graph of *f* is symmetric to the origin because $f(-x) = -f(x)$. Therefore, if $\left(-3, -\frac{12}{5}\right)$ and $\left(-1, \frac{4}{3}\right)$ are points on the graph of *f*, then $\left(3, \frac{12}{5}\right)$ and $\left(1, -\frac{4}{3}\right)$ are also points on the graph.

Interval	Test Point	Comments
$(-\infty, -2)$	$\left(-3, -\dfrac{12}{5}\right)$	• Since $f(x)$ is negative on this interval, $f(x)$ must approach the horizontal asymptote $y = 0$ from below as $x \to -\infty$. • Since $f(x)$ is negative on this interval, as x approaches the vertical asymptote $x = -2$ from the left, $f(x) \to -\infty$.
$(-2, 0)$	$\left(-1, \dfrac{4}{3}\right)$	• Since $f(x)$ is positive on this interval, as x approaches the vertical asymptote $x = -2$ from the right, $f(x) \to \infty$.
$(0, 2)$	$\left(1, -\dfrac{4}{3}\right)$	• Since $f(x)$ is negative on this interval, as x approaches the vertical asymptote $x = 2$ from the left, $f(x) \to -\infty$.
$(2, \infty)$	$\left(3, \dfrac{12}{5}\right)$	• Since $f(x)$ is positive on this interval, as x approaches the vertical asymptote $x = 2$ from the right, $f(x) \to \infty$. • Since $f(x)$ is positive on this interval, $f(x)$ must approach the horizontal asymptote from above as $x \to \infty$.

8. Sketch the function.

Plot the *x*- and *y*-intercept $(0, 0)$.
Graph the asymptotes as dashed lines.
Plot the points.

$$\left(-3, -\frac{12}{5}\right), \left(-1, \frac{4}{3}\right), \left(1, -\frac{4}{3}\right), \text{ and } \left(3, \frac{12}{5}\right)$$

Sketch the curve.

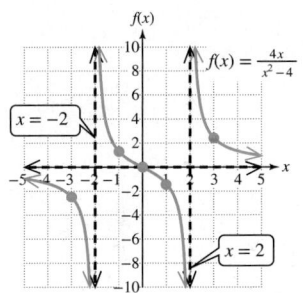

Skill Practice 3 Graph $g(x) = \dfrac{-5x}{x^2 - 9}$.

EXAMPLE 4 **Graphing a Rational Function**

Graph $g(x) = \dfrac{2x^2 - 3x - 5}{x^2 + 1}$.

Solution:

1. Determine the *y*-intercept.

$$g(0) = \frac{2(0)^2 - 3(0) - 5}{(0)^2 + 1} = -5 \qquad \text{The *y*-intercept is } (0, -5).$$

2. Determine the *x*-intercept(s).

$$\frac{2x^2 - 3x - 5}{x^2 + 1} = 0$$

$$2x^2 - 3x - 5 = 0$$

$$(2x - 5)(x + 1) = 0$$

$$x = \frac{5}{2}, \ x = -1 \qquad \text{The *x*-intercepts are } \left(\tfrac{5}{2}, 0\right) \text{ and } (-1, 0).$$

3. Identify the vertical asymptotes.

$x^2 + 1$ is nonzero for all real numbers. The graph of *g* has no vertical asymptotes.

Answer

3.

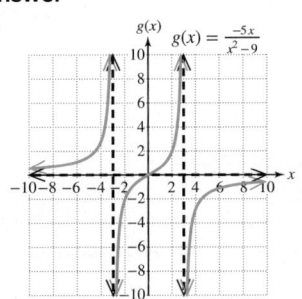

4. Determine whether g has a horizontal or slant asymptote.

The degree of the numerator is equal to the degree of the denominator.

The horizontal asymptote is $y = \frac{2}{1}$ or simply $y = 2$.

5. Determine where g crosses its horizontal asymptote.

Find the real solutions to the equation $\dfrac{2x^2 - 3x - 5}{x^2 + 1} = 2$.

$2x^2 - 3x - 5 = 2(x^2 + 1)$

$2x^2 - 3x - 5 = 2x^2 + 2$

$-3x - 5 = 2$

$x = -\dfrac{7}{3}$ g crosses its horizontal asymptote at $\left(-\dfrac{7}{3}, 2\right)$.

6. Test for symmetry.

g is neither even nor odd because $g(-x) \neq g(x)$ and $g(-x) \neq -g(x)$.

7. Determine the behavior of g on each interval.

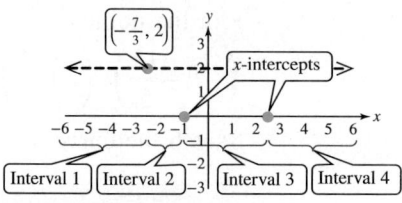

Interval	Test Point	Comments
$\left(-\infty, -\dfrac{7}{3}\right)$	$\left(-3, \dfrac{11}{5}\right)$	• Since $g(-3) = \frac{11}{5}$ is above the horizontal asymptote $y = 2$, $g(x)$ must approach the horizontal asymptote from above as $x \to -\infty$.
$\left(-\dfrac{7}{3}, -1\right)$	$\left(-2, \dfrac{9}{5}\right)$	• Plot the point $\left(-2, \frac{9}{5}\right)$ between the horizontal asymptote and the x-intercept of $(-1, 0)$.
$\left(-1, \dfrac{5}{2}\right)$	$(0, -5)$	• The point $(0, -5)$ is the y-intercept.
$\left(\dfrac{5}{2}, \infty\right)$	$\left(3, \dfrac{2}{5}\right)$	• Since $g(3) = \frac{2}{5}$ is below the horizontal asymptote $y = 2$, $g(x)$ must approach the horizontal asymptote from below as $x \to \infty$.

8. Sketch the function.

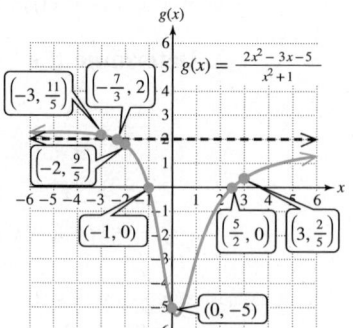

Answer

4.

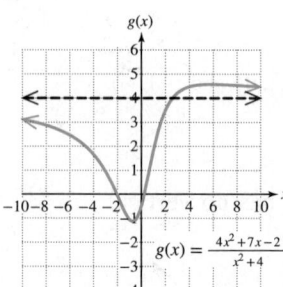

Skill Practice 4   Graph $g(x) = \dfrac{4x^2 + 7x - 2}{x^2 + 4}$.

In Example 5, we graph a rational function with a slant asymptote.

EXAMPLE 5   Graphing a Rational Function

Graph. $h(x) = \dfrac{2x^2 + 9x + 4}{x + 3}$

Solution:

1. Determine the y-intercept.

$$h(0) = \frac{2(0)^2 + 9(0) + 4}{(0) + 3} = \frac{4}{3}$$ The y-intercept is $\left(0, \frac{4}{3}\right)$.

2. Determine the x-intercept(s).

$$\frac{2x^2 + 9x + 4}{x + 3} = 0$$

$$2x^2 + 9x + 4 = 0$$

$$(2x + 1)(x + 4) = 0$$

$$x = -\frac{1}{2}, \ x = -4$$ The x-intercepts are $\left(-\frac{1}{2}, 0\right)$ and $(-4, 0)$.

3. Identify the vertical asymptotes.

$\dfrac{2x^2 + 9x + 4}{x + 3}$ is in lowest terms, and $x + 3$ is zero for $x = -3$.

4. Determine whether h has a horizontal or slant asymptote.

The degree of the numerator is one greater than the degree of the denominator. To find the slant asymptote, divide $(2x^2 + 9x + 4)$ by $(x + 3)$.

$$
\begin{array}{r|rrr}
-3 & 2 & 9 & 4 \\
   &   & -6 & -9 \\
\hline
   & 2 & 3 & \underline{|-5} \\
\end{array}
$$

The quotient is $2x + 3$.
The slant asymptote is $y = 2x + 3$.

5. Determine where h will cross the slant asymptote.

Set $h(x) = 2x + 3$. We have $\dfrac{2x^2 + 9x + 4}{x + 3} = 2x + 3$.

$$2x^2 + 9x + 4 = (2x + 3)(x + 3)$$

> The equation has no solution. Therefore, the graph does not cross the slant asymptote.

$$2x^2 + 9x + 4 = 2x^2 + 9x + 9$$

$$4 = 9 \qquad \text{(No solution)}$$

6. Test for symmetry.

h is neither even nor odd because $h(-x) \neq h(x)$ and $h(-x) \neq -h(x)$.

7. **Plot test points.** Pick values of x on the intervals defined by the x-intercepts and vertical asymptote.

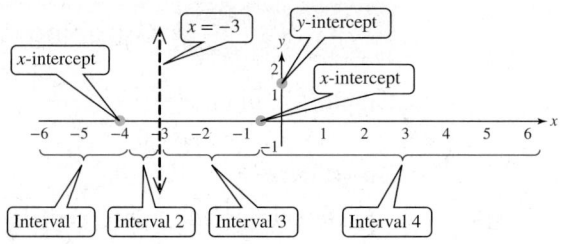

Select test points from each interval.

The graph of $y = h(x)$ passes through $(-5, -4.5)$, $(-3.5, 6)$, $(-2, -6)$, and $(2, 6)$.

8. **Sketch the graph.**

Plot the x-intercepts: $(-4, 0)$ and $\left(-\frac{1}{2}, 0\right)$.

Plot the y-intercept: $\left(0, \frac{4}{3}\right)$.

Graph the asymptotes as dashed lines.

Plot the points:

$(-5, -4.5)$, $(-3.5, 6)$, $(-2, -6)$, and $(2, 6)$.

Sketch the graph.

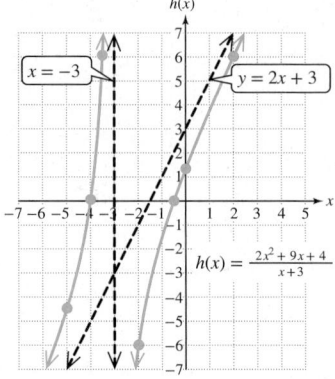

Skill Practice 5   Graph $k(x) = \dfrac{2x^2 - 7x + 3}{x - 2}$.

FOR REVIEW

Recall that variable costs include items such as materials and labor. Fixed costs include overhead costs such as rent and utilities.

2. Use Rational Functions in Applications

In Section 5.5, we presented a linear model for the cost for a business to manufacture x items. The model is $C(x) = mx + b$, where m is the variable cost to produce an individual item, and b is the fixed cost.

The average cost $\overline{C}(x)$ per item manufactured is the sum of all costs (variable and fixed) divided by the total number of items produced x. This is given by

$$\overline{C}(x) = \frac{C(x)}{x}$$

The average cost per item will decrease as more items are produced because the fixed cost will be distributed over a greater number of items. This is demonstrated in Example 6.

Answer

5.

$$k(x) = \frac{2x^2 - 7x + 3}{x - 2}$$

EXAMPLE 6   Investigating Average Cost

A cleaning service cleans homes. For each house call, the cost to the company is approximately \$40 for cleaning supplies, gasoline, and labor. The business also has fixed monthly costs of \$300 from phone service, advertising, and depreciation on the vehicles.

a. Write a cost function to represent the cost $C(x)$ (in dollars) for x house calls per month.

b. Write the average cost function that represents the average cost $\overline{C}(x)$ (in \$) for x house calls per month.

c. Evaluate $\overline{C}(5)$, $\overline{C}(20)$, $\overline{C}(30)$, and $\overline{C}(100)$.

d. The cleaning service can realistically make a maximum of 160 calls per month. However, if the number of calls were unlimited, what value would the average cost approach? What does this mean in the context of the problem?

Solution:

a. $C(x) = 40x + 300$ The variable cost is $40 per call ($m = 40$), and the fixed cost is $300 ($b = 300$). $C(x) = mx + b$.

b. $\overline{C}(x) = \dfrac{40x + 300}{x}$ The average cost per item is the total cost divided by the total number of items produced.

c. $\overline{C}(5) = 100$ The average cost per house call is $100 if 5 calls are made.

$\overline{C}(20) = 55$ The average cost per house call is $55 if 20 calls are made.

$\overline{C}(30) = 50$ The average cost per house call is $50 if 30 calls are made.

$\overline{C}(100) = 43$ The average cost per house call is $43 if 100 calls are made.

d. As x approaches infinity, $\overline{C}(x)$ will approach its horizontal asymptote $y = 40$. This is the cost per house call in the absence of other fixed costs.

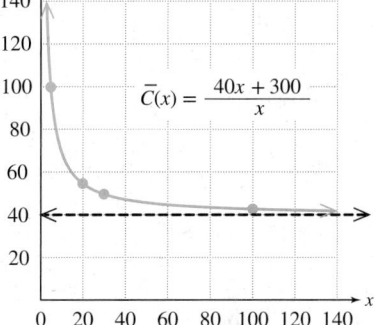

$$\overline{C}(x) = \frac{40x + 300}{x}$$

Answers

6. a. $C(x) = 50x + 200$

b. $\overline{C}(x) = \dfrac{50x + 200}{x}$

c. $\overline{C}(5) = 90$; $\overline{C}(20) = 60$;
$\overline{C}(30) = 56.67$;
$\overline{C}(100) = 52$

d. The average cost $\overline{C}(x)$ will approach $50 per house call.

Skill Practice 6   Repeat Example 6 under the assumption that the company cuts its fixed costs to $200 per month and pays its employees more, leading to a variable cost per house call of $50.

SECTION 8.2   Practice Exercises

Prerequisite Review

For Exercises R.1–R.2, determine the x- and y-intercepts for the function defined by $y = f(x)$. (See Section 5.3 for review.)

R.1. $f(x) = \dfrac{2x^2 - 5x - 7}{x - 1}$

R.2. $f(x) = \dfrac{3x^2 + x - 4}{x + 2}$

For Exercises R.3–R.4, divide the polynomials. (See Section 7.3 for review.)

R.3. $\dfrac{2x^2 - 5x - 7}{x - 1}$

R.4. $\dfrac{3x^2 + x - 4}{x + 2}$

For Exercises R.5–R.10,

a. Determine if the function is even, odd, or neither. (See Section 6.2 for review.)

b. State whether the function is symmetric with respect to the y-axis, origin, or neither.

R.5. $f(x) = \dfrac{3x^5 - 5x^3 + 8x}{4x^2 + 6}$

R.6. $g(x) = \dfrac{6x^4 + 2x^2 - 7}{x^5}$

R.7. $h(x) = \dfrac{2x^2 - 5x - 7}{x - 1}$

R.8. $k(x) = \dfrac{3x^2 + x - 4}{x + 2}$

R.9. $f(x) = \dfrac{2x^5 - 6x^3 + 8x}{2x^3 - 4x}$

R.10. $g(x) = \dfrac{6x^3 + 4x}{x^5}$

R.11. If function f is symmetric with respect to the y-axis, and the point $(-2, 5)$ is on the graph of f, name another point on the graph of f.

R.12. If function g is symmetric with respect to the origin, and the point $(8, -3)$ is on the graph of g, name another point on the graph of g.

For Exercises R.13–R.16, find the real solutions to the equations. (See Section 4.3 for review.)

R.13. a. $\dfrac{2x^2 - 8}{x^2 - x + 1} = 0$

 b. $\dfrac{2x^2 - 8}{x^2 - x + 1} = 2$

R.14. a. $\dfrac{3x^2 - 3}{x^2 + 5x - 3} = 0$

 b. $\dfrac{3x^2 - 3}{x^2 + 5x - 3} = 3$

R.15. $\dfrac{2x - 5}{x^2 + 2x + 1} = 0$

R.16. $\dfrac{3x + 11}{4x^2 + 28x + 49} = 0$

For Exercises R.17–R.18, determine the real values of x for which $f(x)$ is undefined. (See Section 4.1 for review.)

R.17. $f(x) = \dfrac{2x^5 - 6x^3 + 8x}{2x^3 - 4x}$

R.18. $f(x) = \dfrac{3x - 6}{x^3 - 5x}$

R.19. $f(x) = \dfrac{3x + 11}{4x^2 + 28x + 49}$

R.20. $f(x) = \dfrac{2x - 5}{x^2 + 2x + 1}$

Concept Connections

1. Can a rational function cross a vertical asymptote?

2. Can a rational function cross a horizontal asymptote?

3. Describe the relationship between the graph of $f(x) = \dfrac{1}{x}$ and the graph of $g(x) = \dfrac{1}{x + 4} - 5$.

4. Describe the relationship between the graph of $h(x) = \dfrac{1}{x^2}$ and the graph of $k(x) = -\dfrac{1}{(x - 3)^2}$.

5. The expression $\dfrac{x - 4}{x + 2}$ will equal zero when the (numerator/denominator) equals zero.

6. The expression $\dfrac{x - 4}{x + 2}$ is undefined when the (numerator/denominator) equals zero.

Objective 1: Graph Rational Functions

For Exercises 7–14, graph the functions by using transformations of the graphs of $y = \dfrac{1}{x}$ and $y = \dfrac{1}{x^2}$. **(See Example 1)**

7. $f(x) = \dfrac{1}{x - 3}$

8. $g(x) = \dfrac{1}{x + 4}$

9. $h(x) = \dfrac{1}{x^2} + 2$

10. $k(x) = \dfrac{1}{x^2} - 3$

11. $m(x) = \dfrac{1}{(x + 4)^2} - 3$

12. $n(x) = \dfrac{1}{(x - 1)^2} + 2$

13. $p(x) = -\dfrac{1}{x}$

14. $q(x) = -\dfrac{1}{x^2}$

For Exercises 15–20, for the graph of $y = f(x)$,

 a. Identify the x-intercepts.

 c. Identify the horizontal asymptote or slant asymptote if applicable.

 b. Identify any vertical asymptotes.

 d. Identify the y-intercept.

15. $f(x) = \dfrac{(x+3)(2x-7)}{(x+2)(4x+1)}$

16. $f(x) = \dfrac{(3x-4)(x-6)}{(2x-3)(x+5)}$

17. $f(x) = \dfrac{4x-9}{x^2-9}$

18. $f(x) = \dfrac{5x-8}{x^2-4}$

19. $f(x) = \dfrac{(5x-1)(x+3)}{x+2}$

20. $f(x) = \dfrac{(4x+3)(x+2)}{x+3}$

For Exercises 21–24, sketch a rational function subject to the given conditions. Answers may vary.

21. Horizontal asymptote: $y = 2$

Vertical asymptote: $x = 3$

y-intercept: $\left(0, \frac{8}{3}\right)$

x-intercept: $(4, 0)$

22. Horizontal asymptote: $y = 0$

Vertical asymptote: $x = -1$

y-intercept: $(0, 1)$

No x-intercepts

Range: $(0, \infty)$

23. Horizontal asymptote: $y = 0$

Vertical asymptotes: $x = -2$ and $x = 2$

y-intercept: $(0, 1)$

No x-intercepts

Symmetric to the y-axis

Passes through the point $\left(3, -\frac{4}{5}\right)$

24. Horizontal asymptote: $y = 3$

Vertical asymptotes: $x = -1$ and $x = 1$

y-intercept: $(0, 0)$

x-intercept: $(0, 0)$

Symmetric to the y-axis

Passes through the point $(2, 4)$

For Exercises 25–48, graph the function. **(See Examples 2–5)**

25. $n(x) = \dfrac{-3}{2x+7}$

26. $m(x) = \dfrac{-4}{2x-5}$

27. $f(x) = \dfrac{x-4}{x-2}$

28. $g(x) = \dfrac{x-3}{x-1}$

29. $h(x) = \dfrac{2x-4}{x+3}$

30. $k(x) = \dfrac{3x-9}{x+2}$

31. $p(x) = \dfrac{6}{x^2-9}$

32. $q(x) = \dfrac{4}{x^2-16}$

33. $r(x) = \dfrac{5x}{x^2-x-6}$

34. $t(x) = \dfrac{4x}{x^2-2x-3}$

35. $k(x) = \dfrac{5x-3}{2x-7}$

36. $h(x) = \dfrac{4x+3}{3x-5}$

37. $g(x) = \dfrac{3x^2-5x-2}{x^2+1}$

38. $c(x) = \dfrac{2x^2-5x-3}{x^2+1}$

39. $n(x) = \dfrac{x^2+2x+1}{x}$

40. $m(x) = \dfrac{x^2-4x+4}{x}$

41. $f(x) = \dfrac{x^2+7x+10}{x+3}$

42. $d(x) = \dfrac{x^2-x-12}{x-2}$

43. $w(x) = \dfrac{-4x^2}{x^2+4}$

44. $u(x) = \dfrac{-3x^2}{x^2+1}$

45. $f(x) = \dfrac{x^3+x^2-4x-4}{x^2+3x}$

46. $g(x) = \dfrac{x^3+3x^2-x-3}{x^2-2x}$

47. $v(x) = \dfrac{2x^4}{x^2+9}$

48. $g(x) = \dfrac{4x^4}{x^2+8}$

Objective 2: Use Rational Functions in Applications

49. A sports trainer has monthly costs of $69.95 for phone service and $39.99 for his website and advertising. In addition he pays a $20 fee to the gym for each session in which he trains a client. **(See Example 6)**

Denkou Images/Getty Images

a. Write a cost function to represent the cost $C(x)$ for x training sessions for a given month.

b. Write a function representing the average cost $\overline{C}(x)$ for x sessions.

c. Evaluate $\overline{C}(5)$, $\overline{C}(30)$, and $\overline{C}(120)$.

d. The trainer can realistically have 120 sessions per month. However, if the number of sessions were unlimited, what value would the average cost approach? What does this mean in the context of the problem?

50. An on-demand printing company has monthly overhead costs of $1200 in rent, $420 in electricity, $100 for phone service, and $200 for advertising and marketing. The printing cost is $40 per thousand pages for paper and ink.

a. Write a cost function to represent the cost $C(x)$ for printing x thousand pages for a given month.

b. Write a function representing the average cost $\overline{C}(x)$ for printing x thousand pages for a given month.

c. Evaluate $\overline{C}(20)$, $\overline{C}(50)$, $\overline{C}(100)$, and $\overline{C}(200)$.

d. Interpret the meaning of $\overline{C}(200)$.

e. For a given month, if the printing company could print an unlimited number of pages, what value would the average cost per thousand pages approach? What does this mean in the context of the problem?

When medications are taken orally, it often takes time for the drug to be absorbed into the bloodstream. Then after more time elapses, the drug is gradually eliminated from the body, often through respiration and urine. The graph of the concentration of the drug in the bloodstream as a function of time shows a maximum value shortly after ingestion and then a gradual dissipation. This is illustrated in Exercises 51–54.

51. The concentration of a pain medication in the bloodstream is given by $C(t) = \dfrac{4t}{0.01t^2 + 3}$ where $C(t)$ is the concentration in micrograms (mcg) per milliliter (mL) and t is the time in minutes after ingestion.

  a. What is the concentration after 10 min, 1 hr, and 2 hr? Round to one decimal place.

  b. Use a graphing utility to graph the function on the window [0, 300, 30] by [0, 12, 1].

  c. Use the *Maximum* feature on the graphing utility to approximate the maximum concentration in the bloodstream and the time at which this occurs. Round to one decimal place.

  d. The drug can be taken every 4 hr. What is the concentration in the bloodstream at that time?

52. The concentration of an antibiotic in the bloodstream is given by $C(t) = \dfrac{75t}{4t^2 + 25}$ where $C(t)$ is the concentration in micrograms per milliliter and t is the time in hours after ingestion.

  a. What is the concentration after 1 hr, 2 hr, and 4 hr? Round to two decimal places.

  b. Use a graphing utility to graph the function on the window [0, 12, 1] by [0, 5, 1].

  c. Use the *Maximum* feature on the graphing utility to approximate the maximum concentration in the bloodstream and the time at which this occurs.

  d. Doctors generally do not want the concentration of an antibiotic in the bloodstream to drop significantly until the full course of the medication has been taken. The drug can be taken every 6 hr. What is the concentration in the bloodstream at that time?

53. The concentration $C(t)$ (in milligrams per liter, mg/L) of a drug in the bloodstream t hours after the drug is administered is modeled by

$$C(t) = \frac{10t}{2t^2 + 1}$$

  a. Use a graphing utility to graph the function.

  b. What are the domain restrictions on the function?

  c. Use the graph to approximate the maximum concentration. Round to the nearest mg/L.

  d. What is the limiting concentration?

54. A certain diet pill is designed to delay the administration of the active ingredient for several hours. The concentration $C(t)$ (in mg/L) of the active ingredient in the bloodstream t hours after taking the pill is modeled by

$$C(t) = \frac{3t}{2t^2 - 20t + 51}$$

  a. Use a graphing utility to graph the function.

  b. What are the domain restrictions on the function?

  c. Use the graph to approximate the maximum concentration. Round to the nearest mg/L.

  d. What is the limiting concentration?

The Doppler effect is a change in the observed frequency of a wave (such as a sound wave or light wave) when the source of the wave and observer are in motion relative to each other. The Doppler effect explains why an observer hears a change in pitch of an ambulance siren as the ambulance passes by the observer. The frequency $F(v)$ of a sound relative to an observer is given by $F(v) = f_a \left(\dfrac{s_0}{s_0 - v} \right)$, where f_a is the actual frequency of the sound at the source, s_0 is the speed of sound in air (772.4 mph), and v is the speed at which the source of sound is moving toward the observer. Use this relationship for Exercises 55–56.

55. Suppose that an ambulance moves toward an observer.

  a. Write F as a function of v if the actual frequency of sound emitted by the ambulance is 560 Hz.

  b. Use a graphing utility to graph the function from part (a) on the window [0, 1000, 100] by [0, 5000, 1000].

  c. As the speed of the ambulance increases, what is the effect of the frequency of sound?

56. Suppose the frequency of sound emitted by a police car siren is 600 Hz.

  a. Write F as a function of v if the police car is moving toward an observer.

  b. Suppose that the frequency of the siren as heard by an observer is 664 Hz. Determine the velocity of the police car. Round to the nearest tenth of a mph.

  c. Although a police car cannot travel close to the speed of sound, interpret the meaning of the vertical asymptote.

Mixed Exercises

57. a. Write an equation for a rational function f whose graph is the same as the graph of $y = \dfrac{1}{x^2}$ shifted up 3 units and to the left 1 unit.

  b. Write the domain and range of the function in interval notation.

58. a. Write an equation for a rational function f whose graph is the same as the graph of $y = \dfrac{1}{x}$ shifted to the right 4 units and down 3 units.

  b. Write the domain and range of the function in interval notation.

For Exercises 59–60, given $y = f(x)$,

    a. Divide the numerator by the denominator to write $f(x)$ in the form $f(x) = \text{quotient} + \dfrac{\text{remainder}}{\text{divisor}}$.

    b. Use transformations of $y = \dfrac{1}{x}$ to graph the function.

59. $f(x) = \dfrac{2x + 7}{x + 3}$ **60.** $f(x) = \dfrac{5x + 11}{x + 2}$

Graphs with "Holes"

The rational functions studied in this section all have the characteristic that the numerator and denominator do not share a common variable factor. We now investigate rational functions for which this is not the case. For Exercises 61–64,

    a. Write the domain of f in interval notation.

    b. Simplify the rational expression defining the function.

    c. Identify any vertical asymptotes.

    d. Identify any other values of x (other than those corresponding to vertical asymptotes) for which the function is discontinuous.

    e. Identify the graph of the function.

61. $f(x) = \dfrac{x^2 + x - 6}{x - 2}$ **62.** $f(x) = \dfrac{-x^2 + 2x + 3}{x + 1}$ **63.** $f(x) = \dfrac{2x + 10}{x^2 + 9x + 20}$ **64.** $f(x) = \dfrac{2x - 2}{x^2 + 2x - 3}$

i.

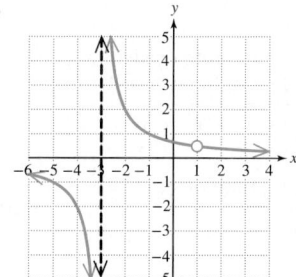

ii.

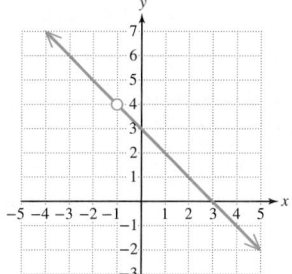

iii.

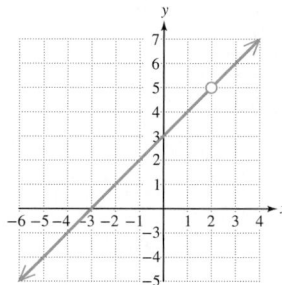

iv.

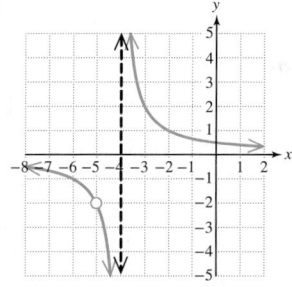

Expanding Your Skills

For Exercises 65–68, write an equation of a function that meets the given conditions. Answers may vary.

65. x-intercepts: $(-3, 0)$ and $(-1, 0)$
    vertical asymptote: $x = 2$
    horizontal asymptote: $y = 1$
    y-intercept: $\left(0, \tfrac{3}{4}\right)$

66. x-intercepts: $(4, 0)$ and $(2, 0)$
    vertical asymptote: $x = 1$
    horizontal asymptote: $y = 1$
    y-intercept: $(0, 8)$

67. x-intercept: $\left(\tfrac{3}{2}, 0\right)$
    vertical asymptotes: $x = -2$ and $x = 5$
    horizontal asymptote: $y = 0$
    y-intercept: $(0, 3)$

68. x-intercept: $\left(\tfrac{4}{3}, 0\right)$
    vertical asymptotes: $x = -3$ and $x = -4$
    horizontal asymptote: $y = 0$
    y-intercept: $(0, -1)$

PROBLEM RECOGNITION EXERCISES

Polynomial and Rational Functions

For Exercises 1–8, refer to $p(x) = x^3 + 3x^2 - 6x - 8$ and $q(x) = x^3 - 2x^2 - 5x + 6$.

1. Find the zeros of $p(x)$.

2. Find the zeros of $q(x)$.

3. Find the x-intercept(s) of the graph of $y = q(x)$.

4. Find the x-intercept(s) of the graph of $y = p(x)$.

5. Find the x-intercepts of the graph of
$$f(x) = \frac{p(x)}{q(x)} = \frac{x^3 + 3x^2 - 6x - 8}{x^3 - 2x^2 - 5x + 6}.$$

6. Find the vertical asymptotes of the graph of
$$f(x) = \frac{p(x)}{q(x)} = \frac{x^3 + 3x^2 - 6x - 8}{x^3 - 2x^2 - 5x + 6}.$$

7. Find the horizontal asymptote or slant asymptote of the graph of $f(x) = \frac{p(x)}{q(x)} = \frac{x^3 + 3x^2 - 6x - 8}{x^3 - 2x^2 - 5x + 6}.$

8. Determine where the graph of $f(x) = \frac{x^3 + 3x^2 - 6x - 8}{x^3 - 2x^2 - 5x + 6}$ crosses its horizontal or slant asymptote.

For Exercises 9–16, refer to $c(x) = x^3 - 4x^2 - 2x + 8$ and $d(x) = x^3 + 3x^2 - 4$.

9. Find the zeros of $c(x)$.

10. Find the zeros of $d(x)$.

11. Find the x-intercept(s) of the graph of $y = d(x)$.

12. Find the x-intercept(s) of the graph of $y = c(x)$.

13. Find the x-intercepts of the graph of
$$g(x) = \frac{c(x)}{d(x)} = \frac{x^3 - 4x^2 - 2x + 8}{x^3 + 3x^2 - 4}.$$

14. Find the vertical asymptotes of the graph of
$$g(x) = \frac{c(x)}{d(x)} = \frac{x^3 - 4x^2 - 2x + 8}{x^3 + 3x^2 - 4}.$$

15. Find the horizontal asymptote or slant asymptote of the graph of $g(x) = \frac{c(x)}{d(x)} = \frac{x^3 - 4x^2 - 2x + 8}{x^3 + 3x^2 - 4}.$

16. Determine where the graph of $g(x) = \frac{x^3 - 4x^2 - 2x + 8}{x^3 + 3x^2 - 4}$ crosses its horizontal or slant asymptote.

For Exercises 17–18, use the results from Exercises 5–8 and 13–16 to match the function with its graph.

17. $f(x) = \dfrac{x^3 + 3x^2 - 6x - 8}{x^3 - 2x^2 - 5x + 6}$

18. $g(x) = \dfrac{x^3 - 4x^2 - 2x + 8}{x^3 + 3x^2 - 4}$

a.

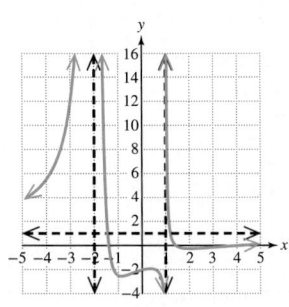

b.

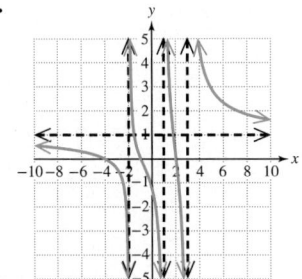

19. Divide $(2x^3 - 4x^2 - 10x + 12) \div (x^2 - 11)$ by using an appropriate method.

　　a. Identify the quotient $q(x)$.　　　　**b.** Identify the remainder $r(x)$.

20. Identify the slant asymptote of $f(x) = \dfrac{2x^3 - 4x^2 - 10x + 12}{x^2 - 11}$.

21. Identify the point where the graph of $f(x) = \dfrac{2x^3 - 4x^2 - 10x + 12}{x^2 - 11}$ crosses its slant asymptote.

22. Refer to Exercise 19. Solve the equation $r(x) = 0$. How does the solution to the equation $r(x) = 0$ relate to the point where the graph of f crosses its slant asymptote?

SECTION 8.3   Rational Inequalities

OBJECTIVES

1. Solve Rational Inequalities Graphically
2. Solve Rational Inequalities Algebraically

1. Solve Rational Inequalities Graphically

In Section 7.5, we learned how to solve polynomial inequalities such as $x^2 - x - 12 > 0$. We now turn our attention to solving rational inequalities such as $\dfrac{x^2 - x - 12}{x - 2} > 0$.

Definition of a Rational Inequality

Let $f(x)$ be a rational expression. Then an inequality of the form $f(x) < 0$, $f(x) > 0$, $f(x) \le 0$, or $f(x) \ge 0$ is called a **rational inequality**.

Avoiding Mistakes

The graph of a rational function will not always change sign to the left and right of a vertical asymptote or x-intercept. However, since the possibility exists, we must test each interval defined by these values of x.

We will first solve rational inequalities graphically by graphing the related function defined by $y = f(x)$. For example, $f(x) = \dfrac{x^2 - x - 12}{x - 2}$ or equivalently $f(x) = \dfrac{(x + 3)(x - 4)}{x - 2}$ changes sign from negative to positive (red to blue) to the left and right of the x-intercepts $(-3, 0)$ and $(4, 0)$. See Figure 8-10. The function also changes sign from positive to negative (blue to red) to the left and right of the vertical asymptote $x = 2$. Thus, to solve a rational inequality, we identify the real solutions to the equation $f(x) = 0$ and the values of x for which $f(x)$ is undefined. These are used as "boundary points" to divide the real number line into intervals. Then we determine whether the inequality is true or false on each interval.

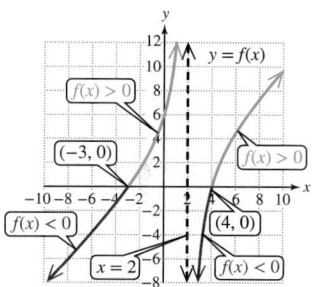

Figure 8-10

EXAMPLE 1   Using a Graph to Solve Rational Inequalities

Consider the graph of $y = f(x)$. Use the graph to solve the inequalities.

   a. $f(x) < 0$ **b.** $f(x) \le 0$

   c. $f(x) > 0$ **d.** $f(x) \ge 0$

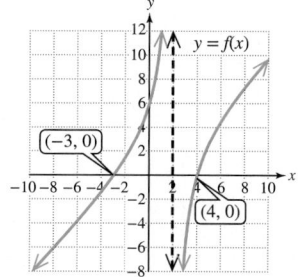

Solution:

The boundary points for the solutions to these inequalities are the x values where the function intersects the x-axis (x-intercepts) and where the function is undefined (at the vertical asymptotes). In this case, the boundary points are $x = -3$, $x = 2$, and $x = 4$.

a. The solution set to the inequality $f(x) < 0$ is the set of x values for which the graph is strictly *below* the x-axis.

The solution set is $(-\infty, -3) \cup (2, 4)$.

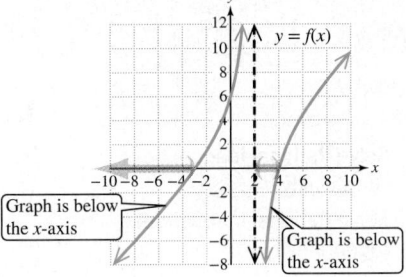

b. The solution set to the inequality $f(x) \leq 0$ is the set of x values for which the graph is *on* or *below* the x-axis. This includes the set of values from part (a) along with the values -3 and 4 where the graph intersects the x-axis. It is important to exclude the value $x = 2$ where f is undefined.

The solution set is $(-\infty, -3] \cup (2, 4]$.

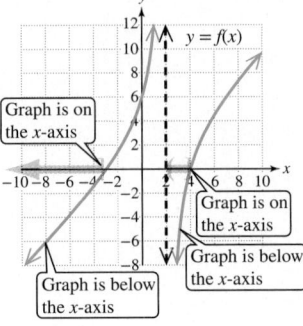

c. The solution set to the inequality $f(x) > 0$ is the set of x values for which the graph is strictly *above* the x-axis.

The solution set is $(-3, 2) \cup (4, \infty)$.

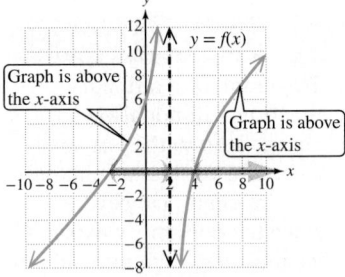

d. The solution set to the inequality $f(x) \geq 0$ is the set of x values for which the graph is *on* or *above* the x-axis. This includes the set of values from part (c) along with $x = -3$ and $x = 4$. It is important to exclude the value $x = 2$ where f is undefined.

The solution set is $[-3, 2) \cup [4, \infty)$.

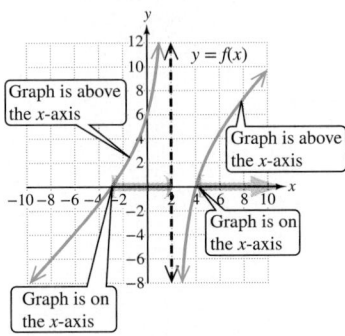

Skill Practice 1   Consider the graph of $y = f(x)$. Use the graph to solve the inequalities.

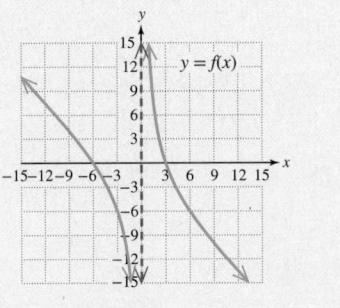

a. $f(x) < 0$ **b.** $f(x) \leq 0$

c. $f(x) > 0$ **d.** $f(x) \geq 0$

Answers

1. a. $(-6, 0) \cup (3, \infty)$
   b. $[-6, 0) \cup [3, \infty)$
   c. $(-\infty, -6) \cup (0, 3)$
   d. $(-\infty, -6] \cup (0, 3]$

2. Solve Rational Inequalities Algebraically

We now turn our attention to solving rational inequalities algebraically. From Section 7.5, we have the following general procedure to solve nonlinear inequalities.

> ### Procedure to Solve a Nonlinear Inequality
>
> 1. Express the inequality as $f(x) < 0$, $f(x) > 0$, $f(x) \leq 0$, or $f(x) \geq 0$. That is, rearrange the terms of the inequality so that one side is set to zero.
> 2. Find the real solutions of the related equation $f(x) = 0$ and any values of x that make $f(x)$ undefined. These are the "boundary" points for the solution set to the inequality.
> 3. Determine the sign of $f(x)$ on the intervals defined by the boundary points.
>   - If $f(x)$ is positive, then the values of x on the interval are solutions to $f(x) > 0$.
>   - If $f(x)$ is negative, then the values of x on the interval are solutions to $f(x) < 0$.
> 4. Determine whether the boundary points are included in the solution set.
> 5. Write the solution set in interval notation or set-builder notation.

EXAMPLE 2   Solving a Rational Inequality

Solve the inequality. $\dfrac{4x - 5}{x - 2} \leq 3$

Solution:

$$\frac{4x - 5}{x - 2} \leq 3$$

$$\overbrace{\frac{4x - 5}{x - 2}}^{f(x)} - 3 \leq 0$$

Step 1: First write the inequality in the form $f(x) \leq 0$. That is, set one side to 0.

$$\frac{4x - 5}{x - 2} - 3 \cdot \frac{x - 2}{x - 2} \leq 0$$

Write each term with a common denominator.

$$\frac{4x - 5 - 3(x - 2)}{x - 2} \leq 0$$

Simplify.

$$\frac{4x - 5 - 3x + 6}{x - 2} \leq 0$$

$$\frac{x + 1}{x - 2} \leq 0$$

The expression $\dfrac{x + 1}{x - 2}$ is undefined for $x = 2$.

Therefore, the value $x = 2$ is *not* part of the solution set.

However, 2 *is* a boundary point for the solution set.

TIP A rational expression is equal to zero where the numerator is equal to zero.
  A rational expression is undefined where the denominator is equal to zero.

The fraction equals 0 when the numerator is 0.

$$\frac{x + 1}{x - 2} = 0$$

Step 2: Solve for the real solutions to the equation $f(x) = 0$. The solution is -1, and this is another boundary point.

The boundary points are $x = 2$ and $x = -1$.

Step 3: Divide the x-axis into intervals defined by the boundary points and determine the sign of $f(x)$ on each interval.

	Interval 1	Interval 2	Interval 3
Evaluate: $f(x) = \frac{x+1}{x-2}$	$f(-2) = \frac{1}{4}$ $f(x) > 0$	$f(1) = -2$ $f(x) < 0$	$f(3) = 4$ $f(x) > 0$
Sign of $(x+1)$	$-$	$+$	$+$
Sign of $(x-2)$	$-$	$-$	$+$
Sign of $\frac{(x+1)}{(x-2)}$	$+$	$-$	$+$

$$-1 \qquad\qquad 2$$

Step 4: Substituting the boundary point $x = -1$ into the inequality $\dfrac{x+1}{x-2} \le 0$ makes a true statement, $0 \le 0$. Thus, $x = -1$ is part of the solution set. The boundary point 2 is excluded because $\dfrac{x+1}{x-2}$ is undefined for $x = 2$.

Step 5: Write the solution set.
The solution set is $[-1, 2)$.
In set-builder notation this is $\{x \mid -1 \le x < 2\}$.

Skill Practice 2   Solve the inequality. $\dfrac{5-x}{x-1} \ge -2$

EXAMPLE 3   Solving a Rational Inequality

Solve the inequality. $\dfrac{1}{x-2} \ge \dfrac{1}{x+3}$

Solution:

FOR REVIEW

In Example 3, the least common denominator (LCD) of the fractions $\dfrac{1}{x-2}$ and $\dfrac{1}{x+3}$ is $(x-2)(x+3)$.
To subtract the fractions, first write each fraction as an equivalent fraction with the LCD as the denominator. Multiply the numerator and denominator of $\dfrac{1}{x-2}$ by the missing the factor $x+3$. Multiply the numerator and denominator of $\dfrac{1}{x+3}$ by the missing factor $x-2$. See Section 4.2 for review.

$$\frac{1}{x-2} \ge \frac{1}{x+3}$$

Step 1: First write the inequality in the form $f(x) \ge 0$.

$$\frac{1}{x-2} - \frac{1}{x+3} \ge 0$$

Set one side to zero.

$$\frac{1}{x-2} \cdot \frac{x+3}{x+3} - \frac{1}{x+3} \cdot \frac{x-2}{x-2} \ge 0$$

Write each term with a common denominator.

$$\frac{x+3}{(x-2)(x+3)} - \frac{x-2}{(x-2)(x+3)} \ge 0$$

Simplify and write the numerator and denominator in factored form.

$$\frac{x+3-x+2}{(x-2)(x+3)} \ge 0$$

$$\frac{5}{(x-2)(x+3)} \ge 0$$

The expression $\frac{5}{(x-2)(x+3)}$ is undefined for $x = 2$ and $x = -3$. The values 2 and -3 are not included in the solution set, but they are boundary points for the solution set.

Answer

2. $(-\infty, -3] \cup (1, \infty)$

$$\frac{5}{(x-2)(x+3)} = 0$$

Step 2: Solve the related equation to determine any additional boundary points.

$$\frac{5}{(x-2)(x+3)} \cdot (x-2)(x+3) = 0 \cdot (x-2)(x+3)$$

$$5 = 0$$
(contradiction)

Clearing fractions results in the contradiction $5 = 0$. There are no solutions to the related equation.

Step 3: Divide the x-axis into intervals defined by the boundary points and determine the sign of $f(x) = \frac{5}{(x-2)(x+3)}$ on each interval.

	Interval 1	Interval 2	Interval 3
Evaluate: $f(x) = \frac{5}{(x-2)(x+3)}$	$f(-4) = \frac{5}{6}$ $f(x) > 0$	$f(0) = -\frac{5}{6}$ $f(x) < 0$	$f(3) = \frac{5}{6}$ $f(x) > 0$
Sign of 5	+	+	+
Sign of $(x-2)$	−	−	+
Sign of $(x+3)$	−	+	+
Sign of $f(x) = \frac{5}{(x-2)(x+3)}$	+	−	+

$$\xleftarrow{\qquad} \underset{-3}{\quad} \underset{2}{\qquad} \xrightarrow{\qquad}$$

Now interpret the results to solve $\dfrac{5}{(x-2)(x+3)} \geq 0$.

Step 4: Neither boundary point is included because $\frac{5}{(x-2)(x+3)}$ is undefined for $x = 2$ and $x = -3$.

Step 5: Write the solution set. The expression $\frac{5}{(x-2)(x+3)}$ is greater than zero for $x < -3$ and $x > 2$.

The solution set is $(-\infty, -3) \cup (2, \infty)$.

In set-builder notation, this is $\{x \mid x < -3 \text{ or } x > 2\}$.

> **Skill Practice 3**   Solve the inequality. $\dfrac{1}{x+4} \geq \dfrac{1}{x+2}$

In Example 4 we encounter an inequality in which the signs of the numerator and denominator of the rational expression can be determined by inspection.

EXAMPLE 4   **Solving Rational Inequalities**

Solve the inequalities.

a. $\dfrac{x^2}{x^2+4} \geq 0$ **b.** $\dfrac{x^2}{x^2+4} > 0$

c. $\dfrac{x^2}{x^2+4} \leq 0$ **d.** $\dfrac{x^2}{x^2+4} < 0$

Solution:

The solution to the related equation $\dfrac{x^2}{x^2+4}=0$ is $x=0$.

The denominator is nonzero for all real numbers.

Therefore, the only boundary point is $x=0$.

Sign of x^2	+	+
Sign of x^2+4	+	+
Sign of $\dfrac{x^2}{x^2+4}$	+	+

$$0$$

Therefore, $\dfrac{x^2}{x^2+4}=0$ at $x=0$ and is positive for all other real numbers.

a. $\dfrac{x^2}{x^2+4}\geq 0$ Solution set: $(-\infty,\infty)$

b. $\dfrac{x^2}{x^2+4}> 0$ Solution set: $(-\infty,0)\cup(0,\infty)$

c. $\dfrac{x^2}{x^2+4}\leq 0$ Solution set: $\{0\}$

d. $\dfrac{x^2}{x^2+4}< 0$ Solution set: $\{\ \}$

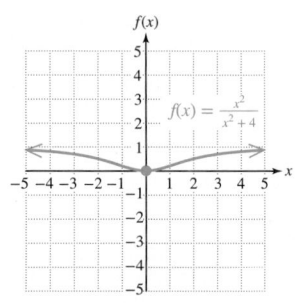

$f(x)=\dfrac{x^2}{x^2+4}$

Answers

4. a. $(-\infty,\infty)$
  b. $(-\infty,0)\cup(0,\infty)$
  c. $\{0\}$
  d. $\{\ \}$

Skill Practice 4   Solve the inequalities.

a. $\dfrac{x^2}{x^4+1}\geq 0$ **b.** $\dfrac{x^2}{x^4+1}> 0$ **c.** $\dfrac{x^2}{x^4+1}\leq 0$ **d.** $\dfrac{x^2}{x^4+1}< 0$

SECTION 8.3   Practice Exercises

Prerequisite Review

For Exercises R.1–R.8, solve the inequality. Write the solution set in interval notation. (See Sections 1.5 and 7.5 for review.)

R.1. $1-3(z-5)\geq 3+3(2z+7)$ (See Section 1.5.)

R.2. $-4(3x+2)<1-(x-4)-9x$

R.3. $-2w+3\leq -9$ or $-2w>2$

R.4. $-5t-4\leq 1$ or $-3t\geq 9$

R.5. a. $x^2+4x<21$ (See Section 7.5.)
  b. $x^2+4x\leq 21$
  c. $x^2+4x>21$
  d. $x^2+4x\geq 21$

R.6. a. $x^2+2x<24$
  b. $x^2+2x\leq 24$
  c. $x^2+2x>24$
  d. $x^2+2x\geq 24$

R.7. a. $x(x+8)<-16$
  b. $x(x+8)\leq -16$
  c. $x(x+8)>-16$
  d. $x(x+8)\geq -16$

R.8. a. $x(x-10)+25<0$
  b. $x(x-10)+25\leq 0$
  c. $x(x-10)+25>0$
  d. $x(x-10)+25\geq 0$

For Exercises R.9–R.12, solve the equation. (See Section 4.3 for review.)

R.9. $\dfrac{3x}{2x+5}=1$ **R.10.** $\dfrac{2x}{3x-8}=2$ **R.11.** $\dfrac{6}{w-2}=\dfrac{2}{w+1}$ **R.12.** $\dfrac{-4}{y+3}=\dfrac{2}{y+6}$

For Exercises R.13–R.14, determine the values of x that make the expressions undefined. (See Section 4.1 for review.)

R.13. a. $\dfrac{5x+3}{x-36}$ **b.** $\dfrac{5x+3}{x^2-36}$ **c.** $\dfrac{5x+3}{x^2+36}$

R.14. a. $\dfrac{4}{9-x}$ **b.** $\dfrac{4}{9-x^2}$ **c.** $\dfrac{4}{9+x^2}$

Concept Connections

1. Given $\dfrac{1-x}{x^2+4}$, for real values of x, the denominator is always (positive/negative). For $x<1$, the numerator is (positive/negative). For $x>1$, the numerator is (positive/negative).

2. Given $\dfrac{2x^2+6}{4-2x}$, for real values of x, the numerator is always (positive/negative). For $x<2$, the denominator is (positive/negative). For $x>2$, the denominator is (positive/negative).

3. Given $\dfrac{t+3}{5-t}$, for $t<-3$, the numerator is (positive/negative). For $t>-3$, the numerator is (positive/negative). For $t<5$, the denominator is (positive/negative). For $t>5$, the denominator is (positive/negative).

4. Given $\dfrac{6-y}{y+4}$, for $y<6$, the numerator is (positive/negative). For $y>6$, the numerator is (positive/negative). For $y<-4$, the denominator is (positive/negative). For $y>-4$, the denominator is (positive/negative).

Objective 1: Solve Rational Inequalities Graphically

For Exercises 5–12, the graph of $y=f(x)$ is given. Solve the inequalities. **(See Example 1)**

 a. $f(x)<0$ **b.** $f(x)\le 0$ **c.** $f(x)>0$ **d.** $f(x)\ge 0$

5.

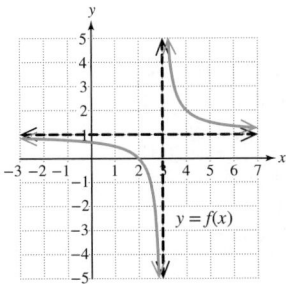

6.

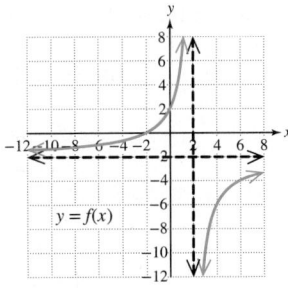

7.

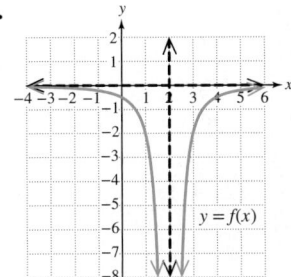

8.

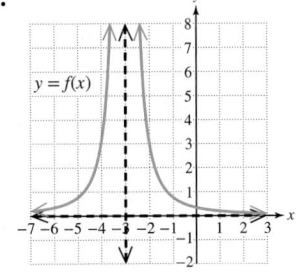

9.

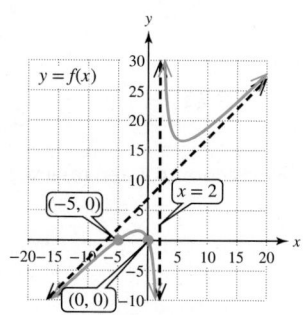

10.

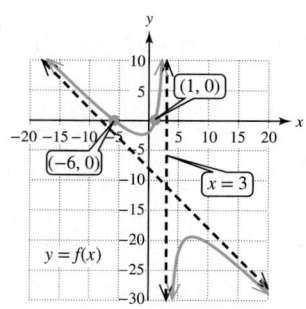

11.

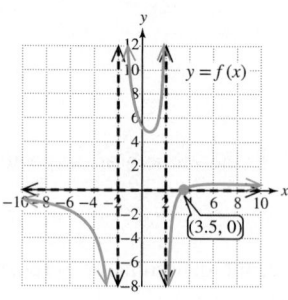

12.

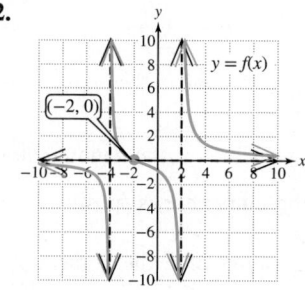

Objective 2: Solve Rational Inequalities Algebraically

For Exercises 13–16, solve the inequalities. **(See Example 4)**

13. a. $\dfrac{x+2}{x-3} \le 0$

b. $\dfrac{x+2}{x-3} < 0$

c. $\dfrac{x+2}{x-3} \ge 0$

d. $\dfrac{x+2}{x-3} > 0$

14. a. $\dfrac{x+4}{x-1} \le 0$

b. $\dfrac{x+4}{x-1} < 0$

c. $\dfrac{x+4}{x-1} \ge 0$

d. $\dfrac{x+4}{x-1} > 0$

15. a. $\dfrac{x^4}{x^2+9} \le 0$

b. $\dfrac{x^4}{x^2+9} < 0$

c. $\dfrac{x^4}{x^2+9} \ge 0$

d. $\dfrac{x^4}{x^2+9} > 0$

16. a. $\dfrac{-x^2}{x^4+16} \le 0$

b. $\dfrac{-x^2}{x^4+16} < 0$

c. $\dfrac{-x^2}{x^4+16} \ge 0$

d. $\dfrac{-x^2}{x^4+16} > 0$

For Exercises 17–38, solve the inequalities. **(See Examples 2–3)**

17. $\dfrac{5-x}{x+1} \ge 0$

18. $\dfrac{2-x}{x+6} \ge 0$

19. $\dfrac{4-2x}{x^2} \le 0$

20. $\dfrac{9-3x}{x^2} \le 0$

21. $\dfrac{w^2-w-2}{w+3} \ge 0$

22. $\dfrac{p^2-2p-8}{p-1} \ge 0$

23. $\dfrac{5}{2t-7} > 1$

24. $\dfrac{4}{3c-8} > 1$

25. $\dfrac{2x}{x-2} \le 2$

26. $\dfrac{3x}{3x-7} \le 1$

27. $\dfrac{4-x}{x+5} \ge 2$

28. $\dfrac{3-x}{x+2} \ge 4$

29. $\dfrac{a-2}{a^2+4} \le 0$

30. $\dfrac{d-3}{d^2+1} \le 0$

31. $\dfrac{10}{x+2} \ge \dfrac{2}{x+2}$

32. $\dfrac{4}{x-3} \ge \dfrac{1}{x-3}$

33. $\dfrac{4}{y+3} > -\dfrac{2}{y}$

34. $\dfrac{2}{z-1} > -\dfrac{4}{z}$

35. $\dfrac{3}{4-x} \le \dfrac{6}{1-x}$

36. $\dfrac{5}{2-x} \le \dfrac{3}{3-x}$

37. $\dfrac{(2-x)(2x+1)^2}{(x-4)^4} \le 0$

38. $\dfrac{(3-x)(4x-1)^4}{(x+2)^2} \le 0$

Mixed Exercises

39. Suppose that an object that is originally at room temperature of 32°C is placed in a freezer. The temperature $T(x)$ (in °C) of the object can be approximated by the model $T(x) = \dfrac{320}{x^2 + 3x + 10}$, where x is the time in hours after the object is placed in the freezer.

a. What is the horizontal asymptote of the graph of this function, and what does it represent in the context of this problem?

b. A chemist needs a compound cooled to less than 5°C. Determine the amount of time required for the compound to cool so that its temperature is less than 5°C.

40. The average round trip speed S (in mph) of a vehicle traveling a distance of d miles in each direction is given by

$$S = \dfrac{2d}{\dfrac{d}{r_1} + \dfrac{d}{r_2}}$$
where r_1 and r_2 are the rates of speed for the initial trip and the return trip, respectively.

a. Suppose that a motorist travels 200 mi from her home to an athletic event and averages 50 mph for the trip to the event. Determine the speeds necessary if the motorist wants the average speed for the round trip to be at least 60 mph.

b. Would the motorist be traveling within the speed limit of 70 mph?

41. A scientist measures the nicotine in the bloodstream after a volunteer begins smoking a cigarette. The blood was tested at 4-min intervals for 2 hr. The scientist fitted the data with the model $N(t) = \dfrac{600t}{t^2 + 450}$, where $N(t)$ is measured in nanograms per milliliter (ng/mL) and t is the time in minutes. If the smoker goes into withdrawal when the nicotine level falls below 10 ng/mL, when would the smoker need another cigarette? Round to the nearest minute.

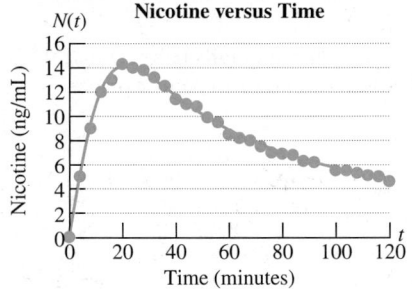

42. A small company runs a television commercial at the beginning of the month to test its effectiveness. The director of marketing monitors sales each day for a period of one month. From the graph, he notes that there is an initial "spike" in sales followed by a gradual decrease. From the data, he fits the model $S(t) = \dfrac{500t}{t^2 + 10}$, where $S(t)$ represents daily sales (in $1000s) and t represents the number of days after the commercial airs. For future commercials, the marketing manager wants to run a follow-up commercial before the sales drop below $40,000 per day. When should the follow-up commercial be run? Round to the nearest day.

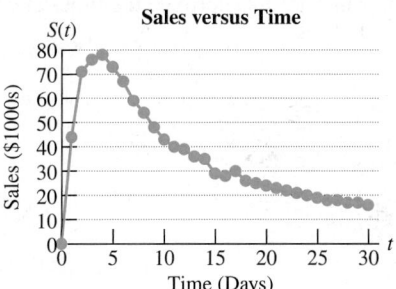

For Exercises 43–46, write the domain of each function in interval notation.

43. $r(x) = \dfrac{1}{\sqrt{2x^2 + 9x - 18}}$

44. $s(x) = \dfrac{1}{\sqrt{4x^2 + 7x - 2}}$

45. $h(x) = \sqrt{\dfrac{3x}{x + 2}}$

46. $k(x) = \sqrt{\dfrac{2x}{x + 1}}$

47. Let a, b, and c represent positive real numbers, where $a < b < c$, and let $g(x) = \dfrac{(a - x)(x - b)^2}{(c - x)^5}$.

a. Complete the sign chart.

Sign of $(a - x)$:			
Sign of $(x - b)^2$:			
Sign of $(c - x)^5$:			
Sign of $\dfrac{(a - x)(x - b)^2}{(c - x)^5}$:			
	a	b	c

b. Solve $g(x) > 0$.   **c.** Solve $g(x) < 0$.

48. Let a, b, and c represent positive real numbers, where $a < b < c$, and let $f(x) = \dfrac{(x - a)(b - x)}{(c - x)^2}$.

a. Complete the sign chart.

Sign of $(x - a)$:			
Sign of $(b - x)$:			
Sign of $(c - x)^2$:			
Sign of $\dfrac{(x - a)(b - x)}{(c - x)^2}$:			
	a	b	c

b. Solve $f(x) > 0$.   **c.** Solve $f(x) < 0$.

Write About It

49. Explain why $\dfrac{x^2 + 2}{x^2 + 1} < 0$ has no solution.

50. Given $\dfrac{x - 3}{x - 1} \leq 0$, explain why the solution set includes 3, but does not include 1.

Expanding Your Skills

The procedure to solve a polynomial or rational inequality may be applied to all inequalities of the form $f(x) > 0, f(x) < 0,$ $f(x) \geq 0,$ and $f(x) \leq 0.$ That is, find the real solutions to the related equation and determine restricted values of x. Then determine the sign of $f(x)$ on each interval defined by the boundary points. Use this process to solve the inequalities in Exercises 51–62.

51. $\sqrt{2x - 6} - 2 < 0$

52. $\sqrt{3x - 5} - 4 < 0$

53. $\sqrt{4 - x} - 6 \geq 0$

54. $\sqrt{5 - x} - 7 \geq 0$

55. $\dfrac{1}{\sqrt{x - 2} - 4} \leq 0$

56. $\dfrac{1}{\sqrt{x - 3} - 5} \leq 0$

57. $-3 < x^2 - 6x + 5 \leq 5$

58. $8 \leq x^2 + 4x + 3 < 15$

59. $|x^2 - 4| < 5$

60. $|x^2 + 1| < 17$

61. $|x^2 - 18| > 2$

62. $|x^2 - 6| > 3$

Technology Connections

63. The concentration $C(t)$ (in ng/mL) of a drug in the bloodstream t hours after ingestion is modeled by

$$C(t) = \frac{500t}{t^3 + 100}$$

   a. Graph the function $y = C(t)$ and the line $y = 4$ on the window $[0, 32, 4]$ by $[0, 15, 3]$.

   b. Use the **Intersect** feature to approximate the point(s) of intersection of $y = C(t)$ and $y = 4$. Round to 1 decimal place if necessary.

   c. To avoid toxicity, a physician may give a second dose of the medicine once the concentration falls below 4 ng/mL for increasing values of t. Determine the times at which it is safe to give a second dose. Round to 1 decimal place.

SECTION 8.4	Variation

1. Write Models Involving Direct, Inverse, and Joint Variation

The familiar relationship $d = rt$ tells us that distance traveled equals the rate of speed times the time of travel. For a car traveling 60 mph, we have $d = 60t$. From Table 8-2, notice that as the time of travel increases, the distance increases proportionally. We say that d is directly proportional to t, or that d varies directly as t. This is shown graphically in Figure 8-11.

Table 8-2

t (hr)	d (mi) $d = 60t$
1	60
2	120
3	180
4	240
5	300
6	360

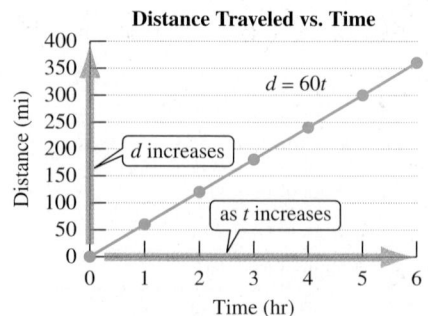

Distance Traveled vs. Time

$d = 60t$

d increases

as t increases

Figure 8-11

Now suppose that a motorist travels a fixed distance of 240 mi. We have

$$d = rt$$

$$240 = rt \longrightarrow t = \frac{240}{r}$$

The time of travel t varies *inversely* as the rate of speed. As the rate r increases, the time of travel will decrease proportionally. Likewise, for slower rates, the time of travel is greater. See Table 8-3 and Figure 8-12.

Table 8-3

r (mph)	t (hr)
10	24
20	12
30	8
40	6
50	4.8
60	4

$t = \dfrac{240}{r}$

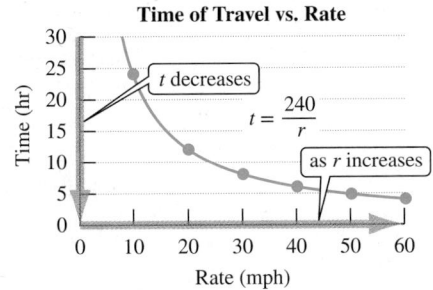

Figure 8-12

Direct and Inverse Variation

Let k be a nonzero constant real number. The statements on the left are equivalent to the equation on the right.

1. y varies **directly** as x.
 y is **directly** proportional to x. $\Big\} y = kx$

2. y varies **inversely** as x.
 y is **inversely** proportional to x. $\Big\} y = \dfrac{k}{x}$

Note: The value of k is called the **constant of variation**.

The first step in using a variation model is to write an English statement as an equivalent mathematical equation.

EXAMPLE 1   Writing a Variation Model

Write a variation model using k as the constant of variation.

Solution:

a. The amount of medicine A prescribed by a physician varies directly as the weight of the patient w.

$A = kw$

Since the variables are directly related, set up the *product* of k and w.

b. The frequency f in a vibrating string is inversely proportional to the length L of the string.

$f = \dfrac{k}{L}$

Since the variables are inversely related, set up the *quotient* of k and L.

c. The variable y varies directly as the square of x and inversely as the square root of z.

$y = \dfrac{kx^2}{\sqrt{z}}$

Since the square of the variable x is directly related to y, set up the *product* of k and x^2. And since the square root of z is inversely related to y, set up the *quotient* of k and $\sqrt{z}$.

TIP Notice that in each variation model, the constant of variation, k, is always in the numerator.

Skill Practice 1   Write a variation model using k as the constant of variation.

a. The distance d that a spring stretches varies directly as the force F applied to the spring.

b. The force F required to keep a car from skidding on a curved road varies inversely as the radius r of the curve.

c. The variable a varies directly as b and inversely as the cube root of c.

Sometimes a variable varies directly as the product of two or more other variables. In such a case we have joint variation.

Joint Variation

Let k be a nonzero constant real number. The statements on the left are equivalent to the equation on the right.

$$\left.\begin{array}{l} y \text{ varies } \textbf{jointly} \text{ as } w \text{ and } x. \\ y \text{ is } \textbf{jointly} \text{ proportional to } w \text{ and } x. \end{array}\right\} y = kwx$$

EXAMPLE 2   **Writing a Joint Variation Model**

Write a variation model using k as the constant of variation.

Solution:

a. y varies jointly as t and the cube root of u.

$y = kt\sqrt[3]{u}$

The variable t and the quantity $\sqrt[3]{u}$ are jointly related to y. Set up the product of k, t, and $\sqrt[3]{u}$.

b. The gravitational force of attraction between two planets varies jointly as the product of their masses and inversely as the square of the distance between them.

$F = \dfrac{km_1 m_2}{d^2}$

Let m_1 and m_2 represent the masses of the planets, let d represent the distance between the planets, and let F represent the gravitational force between the planets.

Skill Practice 2   Write a variation model using k as the constant of variation.

a. The kinetic energy E of an object varies jointly as the object's mass m and the square of its velocity v.

b. z varies jointly as x and y and inversely as the square root of w.

2. Solve Applications Involving Variation

Consider the variation models $y = kx$ and $y = \dfrac{k}{x}$. In either case, if values for x and y are known, we can solve for k. Once k is known, we can write a variation model and use it to find y if x is known, or to find x if y is known. This concept is the basis for solving many applications involving variation.

Answers

1. **a.** $d = kF$ **b.** $F = \dfrac{k}{r}$

   c. $a = \dfrac{kb}{\sqrt[3]{c}}$

2. **a.** $E = kmv^2$ **b.** $z = \dfrac{kxy}{\sqrt{w}}$

Procedure to Solve an Application Involving Variation

Step 1 Write a general variation model that relates the variables given in the problem. Let k represent the constant of variation.

Step 2 Solve for k by substituting known values of the variables into the model from step 1.

Step 3 Substitute the value of k into the original variation model from step 1.

Step 4 Use the variation model from step 3 to solve the application.

EXAMPLE 3   Solving an Application Involving Direct Variation

The amount of an allergy medicine that a physician prescribes for a child varies directly as the weight of the child. Clinical research suggests that 13.5 mg of the drug should be given for a 30-lb child.

  a. How much should be prescribed for a 50-lb child?

  b. How much should be prescribed for a 60-lb child?

  c. A nurse wants to double check the dosage on a doctor's order of 18 mg. For a child of what weight is this dosage appropriate?

Solution:

Let A represent the amount of medicine.   Label the variables.

Let w represent the weight of the child.

$A = kw$ **Step 1:** Write a general variation model.

$13.5 = k(30)$ **Step 2:** Substitute known values of A and w into the variation model.

$\dfrac{13.5}{30} = k$ Solve for k by dividing both sides by 30.

$k = 0.45$

$A = 0.45w$ **Step 3:** Substitute the value of k into the original variation model.

a. $A = 0.45(50)$ **Step 4:** Solve the application by substituting 50 for w.

$A = 22.5$ A 50-lb child would require 22.5 mg of the drug.

b. $A = 0.45(60)$ Substitute 60 for w.

$A = 27$ A 60-lb child would require 27 mg of the drug.

c. $A = 0.45w$

$18 = 0.45w$ Substitute 18 mg for the amount of medicine.

$40 = w$ Solve for the weight w. A 40-lb child would receive 18 mg.

TIP Notice from Example 3 that more medicine is given in proportion to a patient's weight.

A 60-lb patient weighs twice as much as a 30-lb patient and the amount of medicine given is also twice as much. This is consistent with direct variation.

Skill Practice 3   The amount of the medicine ampicillin that a physician prescribes for a child varies directly as the weight of the child. A physician prescribes 420 mg for a 35-lb child.

  a. How much should be prescribed for a 30-lb child?

  b. How much should be prescribed for a 40-lb child?

Answers

3. a. 360 mg   **b.** 480 mg

EXAMPLE 4   Solving an Application Involving Inverse Variation

The loudness of sound measured in decibels (dB) varies inversely as the square of the distance between the listener and the source of the sound. If the loudness of sound is 17.92 dB at a distance of 10 ft from a stereo speaker, what is the decibel level 20 ft from the speaker?

Solution:

Let L represent the loudness of sound in decibels and d represent the distance in feet. The inverse relationship between decibel level and the square of the distance is modeled by:

$$L = \frac{k}{d^2}$$

$$17.92 = \frac{k}{(10)^2} \qquad \text{Substitute } L = 17.92 \text{ dB and } d = 10 \text{ ft.}$$

$$17.92 = \frac{k}{100}$$

$$(17.92)100 = \frac{k}{\cancel{100}} \cdot \cancel{100} \qquad \text{Solve for } k \text{ (clear fractions).}$$

$$k = 1792$$

$$L = \frac{1792}{d^2} \qquad \text{Substitute } k = 1792 \text{ into the original model } L = \frac{k}{d^2}.$$

With the value of k known, we can find L for any value of d.

$$L = \frac{1792}{(20)^2} \qquad \text{Find the loudness when } d = 20 \text{ ft.}$$

$$= 4.48 \text{ dB}$$

Notice that the loudness of sound is 17.92 dB at a distance 10 ft from the speaker. When the distance from the speaker is increased to 20 ft, the decibel level decreases to 4.48 dB. This is consistent with an inverse relationship. For $k > 0$, as one variable is increased, the other is decreased. It also seems reasonable that the farther one moves away from the source of a sound, the softer the sound becomes.

> **Skill Practice 4**   The yield on a bond varies inversely as the price. The yield on a particular bond is 4% when the price is $100. Find the yield when the price is $80.

EXAMPLE 5   Solving an Application Involving Joint Variation

In the early morning hours of September 20, 2017, Hurricane Maria made landfall on Puerto Rico with heavy rainfall and sustained winds over 150 mph. The storm caused catastrophic damage, heavy flooding, and mudslides, leaving many people on the island without homes, power, and clean water for months.

The kinetic energy of an object is the energy that the object possesses by virtue of being in motion. Furthermore, kinetic energy varies jointly as the weight of the object at sea level and as the square of its

Source: NOAA's National Weather Service (NWS) Collection

Answer

4. 5%

velocity. During a hurricane, a 0.5-lb stone traveling 60 mph has 81 joules (J) of kinetic energy. Suppose the wind speed doubles to 120 mph. Find the kinetic energy.

Solution:

Let E represent the kinetic energy, let w represent the weight, and let v represent the velocity of the stone. The variation model is

$$E = kwv^2$$
$$81 = k(0.5)(60)^2 \qquad \text{Substitute } E = 81 \text{ J, } w = 0.5 \text{ lb, and } v = 60 \text{ mph.}$$
$$81 = k(0.5)(3600) \qquad \text{Simplify the exponent.}$$
$$81 = k(1800)$$
$$\frac{81}{1800} = \frac{k(\cancel{1800})}{\cancel{1800}} \qquad \text{Divide by 1800.}$$
$$0.045 = k \qquad \text{Solve for } k.$$

With the value of k known, the model $E = kwv^2$ can be written as $E = 0.045wv^2$. We now find the kinetic energy of a 0.5-lb stone traveling 120 mph.

$$E = 0.045(0.5)(120)^2$$
$$= 324$$

The kinetic energy of a 0.5-lb stone traveling 120 mph is 324 J.

Skill Practice 5   The amount of simple interest earned in an account varies jointly as the interest rate and time of the investment. An account earns $200 in 2 years at 4% interest. How much interest would be earned in 3 years at a rate of 5%?

Answer

5. $375

Prerequisite Review

For Exercises R.1–R.6, solve for k. (See Sections 1.2 and 4.3 for review.)

R.1. $18 = \dfrac{k}{5.5}$

R.2. $6.3 = \dfrac{k}{9}$

R.3. $\dfrac{16}{12} = \dfrac{24}{k}$

R.4. $\dfrac{31}{4} = \dfrac{k}{28}$

R.5. $44 = k(0.02)(11)^2$

R.6. $12 = k(0.3)(6)^2$

For Exercises R.7–R.10, solve for the indicated variable. (See Section 1.2 for review.)

R.7. $A = Ptr$ for r

R.8. $V = \dfrac{1}{3}\pi r^2 h$ for h

R.9. $K = \dfrac{IR}{E}$ for E

R.10. $B = Qrt^2$ for r

Concept Connections

1. If k is a nonzero constant real number, then the statement $y = kx$ implies that y varies _____ as x.

2. If k is a nonzero constant real number, then the statement $y = \frac{k}{x}$ implies that y varies _____ as x.

3. The value of k in the variation models $y = kx$ and $y = \frac{k}{x}$ is called the _____ of _____.

4. If y varies directly as two or more other variables such as x and w, then $y = kxw$, and we say that y varies _____ as x and w.

5. **a.** Given $y = 2x$, evaluate y for the given values of x: $x = 1$, $x = 2$, $x = 3$, $x = 4$, and $x = 5$.

   b. How does y change when x is doubled?

   c. How does y change when x is tripled?

   d. Complete the statement. Given $y = 2x$, when x increases, y (increases/decreases) proportionally.

   e. Complete the statement. Given $y = 2x$, when x decreases, y (increases/decreases) proportionally.

6. **a.** Given $y = \frac{24}{x}$, evaluate y for the given values of x: $x = 1$, $x = 2$, $x = 3$, $x = 4$, and $x = 6$.

   b. How does y change when x is doubled?

   c. How does y change when x is tripled?

   d. Complete the statement. Given $y = \frac{24}{x}$, when x increases, y (increases/decreases) proportionally.

   e. Complete the statement. Given $y = \frac{24}{x}$, when x decreases, y (increases/decreases) proportionally.

7. The time required to drive from Atlanta, Georgia, to Nashville, Tennessee, varies _____ as the average speed at which a vehicle travels.

8. The amount of a person's paycheck varies _____ as the number of hours worked.

9. The volume of a right circular cone varies _____ as the square of the radius of the cylinder and as the height of the cylinder.

10. A student's grade on a test varies _____ as the number of hours the student spends studying for the test.

Objective 1: Write Models Involving Direct, Inverse, and Joint Variation

For Exercises 11–20, write a variation model using k as the constant of variation. **(See Examples 1–2)**

11. The circumference C of a circle varies directly as its radius r.

12. Simple interest I on a loan or investment varies directly as the amount A of the loan.

13. The average cost per minute $\overline{C}$ for a flat rate cell phone plan is inversely proportional to the number of minutes used n.

14. The time of travel t is inversely proportional to the rate of travel r.

15. The volume V of a right circular cylinder varies jointly as the height h of the cylinder and as the square of the radius r of the cylinder.

16. The volume V of a rectangular solid varies jointly as the length l and width w of the solid.

17. The variable E is directly proportional to s and inversely proportional to the square root of n.

18. The variable n is directly proportional to the square of σ and inversely proportional to the square of E.

19. The variable c varies jointly as m and n and inversely as the cube of t.

20. The variable d varies jointly as u and v and inversely as the cube root of T.

Objective 2: Solve Applications Involving Variation

For Exercises 21–26, find the constant of variation k.

21. y varies directly as x. When x is 8, y is 20.

22. m varies directly as x. When x is 10, m is 42.

23. p is inversely proportional to q. When q is 18, p is 54.

24. T is inversely proportional to x. When x is 50, T is 200.

25. y varies jointly as w and v. When w is 40 and v is 0.2, y is 40.

26. N varies jointly as t and p. When t is 2 and p is 2.5, N is 15.

27. The value of y equals 4 when $x = 10$. Find y when $x = 5$ if

   a. y varies directly as x.

   b. y varies inversely as x.

28. The value of y equals 24 when x is $\frac{1}{2}$. Find y when $x = 3$ if

   a. y varies directly as x.

   b. y varies inversely as x.

For Exercises 29–48, use a variation model to solve for the unknown value.

29. The amount of a pain reliever that a physician prescribes for a child varies directly as the weight of the child. A physician prescribes 180 mg of the medicine for a 40-lb child. **(See Example 3)**

   a. How much medicine would be prescribed for a 50-lb child?

   b. How much would be prescribed for a 60-lb child?

   c. How much would be prescribed for a 70-lb child?

   d. If 135 mg of medicine is prescribed, what is the weight of the child?

30. The number of people that a ham can serve varies directly as the weight of the ham. An 8-lb ham feeds 20 people.

Ingram Publishing

   a. How many people will a 10-lb ham serve?

   b. How many people will a 15-lb ham serve?

   c. How many people will an 18-lb ham serve?

   d. If a ham feeds 30 people, what is the weight of the ham?

31. A rental car company charges a fixed amount to rent a car per day. Therefore, the cost per mile to rent a car for a given day is inversely proportional to the number of miles driven. If 100 mi is driven, the average daily cost is $0.80 per mile.

   a. Find the cost per mile if 200 mi is driven.

   b. Find the cost per mile if 300 mi is driven.

   c. Find the cost per mile if 400 mi is driven.

   d. If the cost per mile is $0.16, how many miles were driven?

32. A chef self-publishes a cookbook and finds that the number of books she can sell per month varies inversely as the price of the book. The chef can sell 1500 books per month when the price is set at $8 per book.

   a. How many books would she expect to sell per month if the price were $12?

   b. How many books would she expect to sell per month if the price were $15?

   c. How many books would she expect to sell per month if the price were $6?

   d. If the chef sells 1200 books, what price was set?

33. The distance that a bicycle travels in 1 min varies directly as the number of revolutions per minute (rpm) that the wheels are turning. A bicycle with a 14-in. radius travels approximately 440 ft in 1 min if the wheels turn at 60 rpm. How far will the bicycle travel in 1 min if the wheels turn at 87 rpm?

34. The amount of pollution entering the atmosphere varies directly as the number of people living in an area. If 100,000 people create 71,000 tons of pollutants, how many tons enter the atmosphere in a city with 750,000 people?

35. The stopping distance of a car is directly proportional to the square of the speed of the car.

   a. If a car traveling 50 mph has a stopping distance of 170 ft, find the stopping distance of a car that is traveling 70 mph.

   b. If it takes 244.8 ft for a car to stop, how fast was it traveling before the brakes were applied?

36. The area of a picture projected on a wall varies directly as the square of the distance from the projector to the wall.

   a. If a 15-ft distance produces a 36 ft^2 picture, what is the area of the picture when the projection unit is moved to a distance of 25 ft from the wall?

   b. If the projected image is 144 ft^2, how far is the projector from the wall?

37. The time required to complete a job varies inversely as the number of people working on the job. It takes 8 people 12 days to do a job. **(See Example 4)**

   a. How many days will it take if 15 people work on the job?

   b. If the contractor wants to complete the job in 8 days, how many people should work on the job?

38. The yield on a bond varies inversely as the price. The yield on a particular bond is 5% when the price is $120.

   a. Find the yield when the price is $100.

   b. What price is necessary for a yield of 7.5%?

39. The current in a wire varies directly as the voltage and inversely as the resistance. If the current is 9 amperes (A) when the voltage is 90 volts (V) and the resistance is 10 ohms (Ω), find the current when the voltage is 160 V and the resistance is 5 Ω.

40. The resistance of a wire varies directly as its length and inversely as the square of its diameter. A 50-ft wire with a 0.2-in. diameter has a resistance of 0.0125 Ω. Find the resistance of a 40-ft wire with a diameter of 0.1 in.

41. The amount of simple interest owed on a loan varies jointly as the amount of principal borrowed and the amount of time the money is borrowed. If $4000 in principal results in $480 in interest in 2 years, determine how much interest will be owed on $6000 in 4 years.

42. The amount of simple interest earned in an account varies jointly as the amount of principal invested and the amount of time the money is invested. If $5000 in principal earns $750 in 6 years, determine how much interest will be earned on $8000 in 4 years.

43. The body mass index (BMI) of an individual varies directly as the weight of the individual and inversely as the square of the height of the individual. The body mass index for a 150-lb person who is 70 in. tall is 21.52. Determine the BMI for an individual who is 68 in. tall and 180 lb. Round to 2 decimal places. **(See Example 5)**

44. The strength of a wooden beam varies jointly as the width of the beam and the square of the thickness of the beam, and inversely as the length of the beam. A beam that is 48 in. long, 6 in. wide, and 2 in. thick can support a load of 417 lb. Find the maximum load that can be safely supported by a board that is 12 in. wide, 72 in. long, and 4 in. thick.

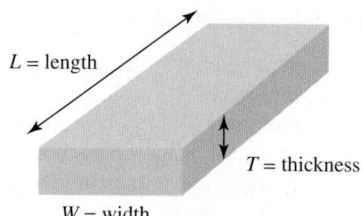

$L = $ length

$T = $ thickness

$W = $ width

45. The speed of a racing canoe in still water varies directly as the square root of the length of the canoe. A 16-ft canoe can travel 6.2 mph in still water. Find the speed of a 25-ft canoe.

46. The period of a pendulum is the length of time required to complete one swing back and forth. The period varies directly as the square root of the length of the pendulum. If it takes 1.8 sec for a 0.81-m pendulum to complete one period, what is the period of a 1-m pendulum?

47. The cost to carpet a rectangular room varies jointly as the length of the room and the width of the room. A 10-yd by 15-yd room costs $3870 to carpet. What is the cost to carpet a room that is 18 yd by 24 yd?

48. The cost to tile a rectangular kitchen varies jointly as the length of the kitchen and the width of the kitchen. A 10-ft by 12-ft kitchen costs $1104 to tile. How much will it cost to tile a kitchen that is 20 ft by 14 ft?

Mixed Exercises

For Exercises 49–52, use the given data to find a variation model relating y to x.

49.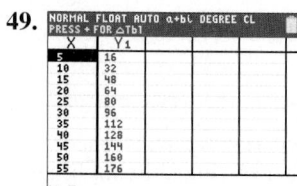

NORMAL FLOAT AUTO a+bi DEGREE CL				
PRESS + FOR △Tbl				
X	Y1			
5	16			
10	32			
15	48			
20	64			
25	80			
30	96			
35	112			
40	128			
45	144			
50	160			
55	176			
X=5				

50.

NORMAL FLOAT AUTO a+bi DEGREE CL				
PRESS + FOR △Tbl				
X	Y1			
5	24			
15	72			
25	120			
35	168			
45	216			
55	264			
65	312			
75	360			
85	408			
95	456			
105	504			
X=5				

51.

	A	B	C	D	E
1	**x**	2	4	12	48
2	**y**	6	3	1	0.25

52.

	A	B	C	D	E
1	**x**	4	8	32	100
2	**y**	2	1	0.25	0.08

53. Which formula(s) can represent a variation model?

   a. $y = kxyz$ **b.** $y = kx + yz$

   c. $y = \dfrac{kx}{yz}$ **d.** $y = kx - yz$

54. Which formula(s) can represent a variation model?

   a. $y = k\sqrt{x} - z^2$ **b.** $y = \dfrac{k\sqrt{x}}{z^2}$

   c. $y = k\sqrt{x}z^2$ **d.** $y = k + \sqrt{x}z^2$

Write About It

For Exercises 55–56, write a statement in words that describes the variation model given. Use k as the constant of variation.

55. $P = \dfrac{kv^2}{t}$

56. $E = \dfrac{kc^2}{\sqrt{b}}$

Expanding Your Skills

57. The light from a lightbulb radiates outward in all directions.

   a. Consider the interior of an imaginary sphere on which the light shines. The surface area of the sphere is directly proportional to the square of the radius. If the surface area of a sphere with a 10-m radius is 400π m^2, determine the exact surface area of a sphere with a 20-m radius.

   b. Explain how the surface area changed when the radius of the sphere increased from 10 m to 20 m.

   c. Based on your answer from part (b) how would you expect the intensity of light to change from a point 10 m from the lightbulb to a point 20 m from the lightbulb?

   d. The intensity of light from a light source varies inversely as the square of the distance from the source. If the intensity of a lightbulb is 200 lumen/m^2 (lux) at a distance of 10 m, determine the intensity at 20 m.

58. The time required for a planet to complete one orbit around the Sun is called the *period* of the planet. Kepler's third law states that the square of the period T is directly proportional to the cube of the average distance d of the planet to the Sun. For the Earth, assume that $d = 9.3 \times 10^7$ mi and $T = 365$ days.

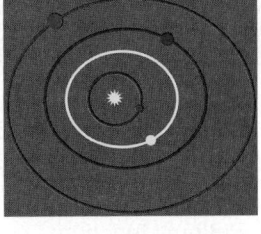

    a. Find the period of Mars, given that the distance between Mars and the Sun is 1.5 times the distance from the Earth to the Sun. Round to the nearest day.

    b. Find the average distance of Venus to the Sun, given that Venus revolves around the Sun in 223 days. Round to the nearest million miles.

59. The intensity of radiation varies inversely as the square of the distance from the source to the receiver. If the distance is increased to 10 times its original value, what is the effect on the intensity to the receiver?

60. Suppose that y varies inversely as the cube of x. If the value of x is decreased to $\frac{1}{4}$ of its original value, what is the effect on y?

61. Suppose that y varies directly as x^2 and inversely as w^4. If both x and w are doubled, what is the effect on y?

62. Suppose that y varies directly as x^5 and inversely as w^2. If both x and w are doubled, what is the effect on y?

63. Suppose that y varies jointly as x and w^3. If x is replaced by $\frac{1}{3}x$ and w is replaced by $3w$, what is the effect on y?

64. Suppose that y varies jointly as x^4 and w. If x is replaced by $\frac{1}{4}x$ and w is replaced by $4w$, what is the effect on y?

PROBLEM RECOGNITION EXERCISES

Recognizing Equations and Inequalities

At this point, you have learned how to solve a variety of equations and inequalities. Being able to distinguish the type of problem being posed is the first step in successfully solving it.

For Exercises 1–20,

    a. Identify the problem type. Choose from

- linear equation
- quadratic equation
- rational equation
- absolute value equation
- radical equation
- equation quadratic in form

- polynomial equation
- linear inequality
- polynomial inequality
- rational inequality
- absolute value inequality
- compound inequality

    b. Solve the equation or inequality. Write the solution to each inequality in interval notation if possible.

1. $(z^2 - 4)^2 - (z^2 - 4) - 12 = 0$

2. $3 + |4t - 1| < 6$

3. $2y(y - 4) \le 5 + y$

4. $\sqrt[3]{11x - 3} + 4 = 6$

5. $-5 = -|w - 4|$

6. $\dfrac{5}{x - 2} + \dfrac{3}{x + 2} = 1$

7. $m^3 + 5m^2 - 4m - 20 \ge 0$

8. $-x - 4 > -5$ and $2x - 3 \le 23$

9. $5 - 2[3 - (x - 4)] \le 3x + 14$

10. $|2x - 6| = |x + 3|$

11. $\dfrac{3}{x - 2} \le 1$

12. $9 < |x + 4|$

13. $\sqrt{t + 8} - 6 = t$

14. $(4x - 3)^2 = -10$

15. $-4 - x > 2$ or $8 < 2x$

16. $\dfrac{1}{3}x - 2 = \dfrac{3}{4} + \dfrac{5}{6}x$

17. $x^2 - 10x \le -25$

18. $\dfrac{10}{x^2 + 1} < 0$

19. $x - 13\sqrt{x} + 36 = 0$

20. $x^4 - 13x^2 + 36 = 0$

CHAPTER 8   Detailed Summary

SECTION 8.1   Introduction to Rational Functions

Key Concepts	Examples	Page
Rational function: Let $p(x)$ and $q(x)$ be polynomials where $q(x) \neq 0$. A function f defined by $f(x) = \dfrac{p(x)}{q(x)}$ is called a **rational function**. **Vertical asymptote:** The line $x = c$ is a **vertical asymptote** of the graph of $y = f(x)$ if $f(x)$ approaches infinity or negative infinity as x approaches c from either side. To locate the vertical asymptotes of a function, determine the real numbers x where the denominator is zero, but the numerator is nonzero.	**Example 1:** To find the vertical asymptotes of the graph of $f(x) = \dfrac{x-3}{x^2+x-2}$ verify that the rational expression $\dfrac{x-3}{x^2+x-2}$ is in lowest terms. Then find the real zeros of the denominator. $f(x) = \dfrac{x-3}{(x+2)(x-1)}$ The expression is in lowest terms and the real zeros of the denominator are -2 and 1. Therefore, the vertical asymptotes are $x = -2$ and $x = 1$.	p. 560
Horizontal asymptote: The line $y = d$ is a **horizontal asymptote** of the graph of $y = f(x)$ if $f(x)$ approaches d as x approaches infinity or negative infinity. Let f be a rational function defined by $f(x) = \dfrac{a_n x^n + a_{n-1}x^{n-1} + a_{n-2}x^{n-2} + \cdots + a_1 x + a_0}{b_m x^m + b_{m-1}x^{m-1} + b_{m-2}x^{m-2} + \cdots + b_1 x + b_0}$ 1. If $n > m$, then f has no horizontal asymptote. 2. If $n < m$, then the line $y = 0$ (the x-axis) is the horizontal asymptote of f. 3. If $n = m$, then the line $y = \dfrac{a_n}{b_m}$ is the horizontal asymptote of f.	**Example 2:** Given $f(x) = \dfrac{x-3}{x^2+x-2}$, the degree of the numerator (1) is less than the degree of the denominator (2). Thus, the line $y = 0$ is a horizontal asymptote of the graph of f. 	p. 563

Slant and nonlinear asymptotes:

Let f be a rational function defined by

$$f(x) = \frac{a_n x^n + a_{n-1} x^{n-1} + a_{n-2} x^{n-2} + \cdots + a_1 x + a_0}{b_m x^m + b_{m-1} x^{m-1} + b_{m-2} x^{m-2} + \cdots + b_1 x + b_0}$$

1. If $n = m + 1$ (that is, the degree of the numerator is exactly 1 greater than the degree of the denominator), the function has a **slant asymptote**.

2. If $n > m + 1$ (that is, the degree of the numerator is more than 1 greater than the degree of the denominator), the end behavior of the function resembles the end behavior of the power function $y = \frac{a_n}{b_m}$.

To find an equation of a slant or nonlinear asymptote, divide the numerator of $f(x)$ by the denominator and determine the quotient $Q(x)$. The equation $y = Q(x)$ is an equation of the asymptote.

Example 3:   p. 566

Given $f(x) = \dfrac{2x^2 - 4}{x + 1}$, the degree of the numerator is exactly 1 greater than the degree of the denominator. Therefore, the graph of the function has a slant asymptote.

To find an equation of the slant asymptote, divide: $(2x^2 - 4) \div (x + 1)$.

```
-1 | 2    0   -4        The quotient
       -2    2          is y = 2x - 2
   ─────────────
     2   -2  | -2
```

The slant asymptote is $y = 2x - 2$.

The graph of f also has a vertical asymptote of $x = -1$.

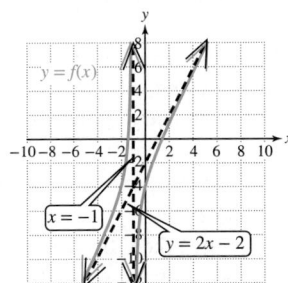

SECTION 8.2 Graphs of Rational Functions

Key Concepts	Examples	Page

Graph rational functions:

Consider a rational function f defined by $f(x) = \dfrac{p(x)}{q(x)}$, where $p(x)$ and $q(x)$ are polynomials with no common factors.

1. Determine the y-intercept by evaluating $f(0)$.
2. Determine the x-intercept(s) by finding the real solutions of $f(x) = 0$. The value $f(x)$ equals zero when the numerator $p(x) = 0$.
3. Identify any vertical asymptotes and graph them as dashed lines.
4. Determine whether the function has a horizontal asymptote or a slant asymptote (or neither), and graph the asymptote as a dashed line.
5. Determine where the function crosses the horizontal or slant asymptote (if applicable).
6. If a test for symmetry is easy to apply, use symmetry to plot additional points. Recall:
   - f is an even function (symmetric to the y-axis) if $f(-x) = f(x)$.
   - f is an odd function (symmetric to the origin) if $f(-x) = -f(x)$.
7. Plot at least one point on the intervals defined by the x-intercepts, vertical asymptotes, and points where the function crosses a horizontal or slant asymptote.
8. Sketch the function based on the information found in steps 1–7.

Example 1:   p. 574

Given $f(x) = \dfrac{2x - 3}{x + 2}$,

- $f(0) = \dfrac{2(0) - 3}{(0) + 2} = -\dfrac{3}{2}$ y-intercept: $\left(0, -\dfrac{3}{2}\right)$

- Solving the equation $\dfrac{2x - 3}{x + 2} = 0$ gives $x = \dfrac{3}{2}$. The x-intercept is $\left(\dfrac{3}{2}, 0\right)$.

- The vertical asymptote is $x = -2$.

- The horizontal asymptote is $y = \dfrac{2}{1}$ or $y = 2$.

To determine if the graph of f crosses the horizontal asymptote, solve $f(x) = 2$.

$$\frac{2x - 3}{x + 2} = 2$$
$$2x - 3 = 2(x + 2)$$
$$2x - 3 = 2x + 4$$
$$-3 = 4 \quad \text{No solution}$$

The graph of f does not cross its horizontal asymptote.

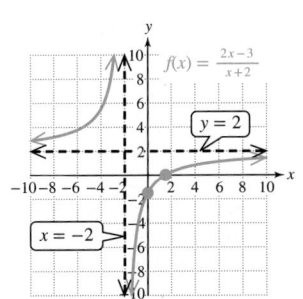

SECTION 8.3 Rational Inequalities

Key Concepts	Examples	Page
Rational inequality: Let $f(x)$ be a rational expression. Then an inequality of the form $$f(x) < 0, \quad f(x) > 0, \quad f(x) \le 0, \quad \text{or} \quad f(x) \ge 0$$ is called a **rational inequality**.	**Example 1:** $$\frac{5}{(1-x)(x+4)} \ge 0$$ The inequality is in the form $f(x) \ge 0$, where $$f(x) = \frac{5}{(1-x)(x+4)}.$$	p. 587

Solving nonlinear inequalities:

1. Express the inequality as $f(x) < 0, f(x) > 0,$ $f(x) \le 0,$ or $f(x) \ge 0$.
2. Find the real solutions of the related equation $f(x) = 0$ and the values of x where $f(x)$ is undefined. These are the "boundary" points for the solution set to the inequality.
3. Determine the sign of $f(x)$ on the intervals defined by the boundary points.
   - If $f(x)$ is positive, then the values of x on the interval are solutions to $f(x) > 0$.
   - If $f(x)$ is negative, then the values of x on the interval are solutions to $f(x) < 0$.
4. Determine whether the boundary points are included in the solution set.
5. Write the solution set.

- There are no real solutions to the equation $f(x) = 0$.
- $f(x)$ is undefined for $x = 1$ and $x = -4$.

p. 589

Test the signs of each factor and the sign of the quotient on each interval defined by the boundary points.

Sign of 5:	+	+	+
Sign of $(1-x)$:	+	+	−
Sign of $(x+4)$:	−	+	+
Sign of $\dfrac{5}{(1-x)(x+4)}$:	−	+	−

$$-4 \qquad 1$$

$f(x)$ is positive between -4 and 1. However, the values -4 and 1 themselves must not be part of the solution set because these are the values for which $f(x)$ is undefined.

The solution set is $(-4, 1)$.

SECTION 8.4 Variation

Key Concepts	Examples	Page
Direct, inverse, and joint variation: Let k be a nonzero constant real number. Then the statements on the left are equivalent to the equations on the right.	**Example 1:** The force F between two electrical charges q_1 and q_2 is jointly proportional to q_1 and q_2 and inversely proportional to the square of the distance d between them.	p. 597

1. y varies **directly** as x.
   y is **directly** proportional to x. $\Big\} y = kx$
2. y varies **inversely** as x.
   y is **inversely** proportional to x. $\Big\} y = \dfrac{k}{x}$
3. y varies **jointly** as w and x.
   y is **jointly** proportional to w and x. $\Big\} y = kwx$

The value of k is called the **constant of variation**.

A force of 0.0899 newtons (N) is measured between two charges of 4×10^{-7} coulombs (C) and 1×10^{-4} C at a distance of 2 m.

a. A general variation model is given by $F = \dfrac{kq_1q_2}{d^2}$.

p. 598

b. Solve for the constant of variation k by substituting known values of F, q_1, q_2, and d into the model.

$$0.0899 = \frac{k(4 \times 10^{-7})(1 \times 10^{-4})}{(2)^2}$$

$$k = \frac{0.0899(2)^2}{(4 \times 10^{-7})(1 \times 10^{-4})} = 8.99 \times 10^9$$

c. Find the force if the charges are 1 m apart.

$$F = \frac{(8.99 \times 10^9)(4 \times 10^{-7})(1 \times 10^{-4})}{(1)^2}$$

$$= 0.3596 \text{ N}$$

In Example 1, when the distance between the two charges is halved, the force between the charges is 2^2 or 4 times as great. Note that $4(0.0899 \text{ N}) = 0.3596 \text{ N}$.

CHAPTER 8 Review Exercises

SECTION 8.1

1. Refer to the graph of $y = f(x)$ and complete the statements.

   a. As $x \to -\infty$,
      $f(x) \to$ _____.

   b. As $x \to -2^-$,
      $f(x) \to$ _____.

   c. As $x \to -2^+$,
      $f(x) \to$ _____.

   d. As $x \to \infty$,
      $f(x) \to$ _____.

   e. The graph is increasing over the interval(s) _____.

   f. The graph is decreasing over the interval(s) _____.

   g. The domain is _____.

   h. The range is _____.

   i. The vertical asymptote is the line _____.

   j. The horizontal asymptote is the line _____.

2. Refer to the graph of $y = f(x)$ and complete the statements.

   a. As $x \to -\infty$,
      $f(x) \to$ _____.

   b. As $x \to -2^-$,
      $f(x) \to$ _____.

   c. As $x \to -2^+$,
      $f(x) \to$ _____.

   d. As $x \to 1^-$,
      $f(x) \to$ _____.

   e. As $x \to 1^+$,
      $f(x) \to$ _____.

   f. As $x \to \infty$, $f(x) \to$ _____.

   g. The vertical asymptotes are the lines _____.

   h. The horizontal asymptote is the line _____.

For Exercises 3–4, determine the vertical asymptotes of the graph of the function.

3. $f(x) = \dfrac{x + 4}{2x^2 + x - 15}$

4. $g(x) = \dfrac{5}{x^2 + 3}$

For Exercises 5–8,

   a. Determine the horizontal asymptotes (if any).

   b. If the graph of the function has a horizontal asymptote, determine the point where the graph crosses the horizontal asymptote.

5. $r(x) = \dfrac{3}{x^2 + 2x + 1}$

6. $k(x) = \dfrac{x^3 + 4}{x + 1}$

7. $q(x) = \dfrac{-2x^2 - 3x + 4}{x^2 + 1}$

8. $m(x) = \dfrac{3x^2 + x - 2}{x^2 - 2}$

For Exercises 9–12, identify all asymptotes (vertical, horizontal, slant, and nonlinear).

9. $m(x) = \dfrac{2x^3 - x^2 - 6x + 7}{x^2 - 3}$

10. $n(x) = \dfrac{-4x^2 + 5}{3x^2 - 14x - 5}$

11. $p(x) = \dfrac{-x^4 + x^2 - 1}{x^2}$

12. $q(x) = \dfrac{x^4 - 3x^2 + 2}{x^2 + 1}$

SECTION 8.2

For Exercises 13–16, graph the function.

13. $f(x) = \dfrac{1}{x - 4} + 2$

14. $k(x) = \dfrac{x^2}{x^2 - x - 12}$

15. $m(x) = \dfrac{x^2 + 6x + 9}{x}$

16. $q(x) = \dfrac{12}{x^2 + 6}$

17. After taking a certain class, the percentage of material retained $P(t)$ decreases with the number of months t after taking the class. $P(t)$ can be approximated by

$$P(t) = \dfrac{t + 90}{0.16t + 1}$$

   a. Determine the percentage retained after 1 month, 4 months, and 6 months. Round to the nearest percent.

   b. As t becomes infinitely large, what percentage of material will be retained?

18. The concentration $C(t)$ (in ng/mL) of an antibiotic in the bloodstream t hours after the drug is taken orally can be modeled by $C(t) = \dfrac{16t}{t^2 + 1}$.

   a. Determine the concentration 1 hr, 4 hr, and 12 hr after taking the drug. Round to one decimal place.

   b. Use a graphing utility to graph the function on the window [0, 12, 1] by [0, 10, 1].

   c. Use the *Maximum* feature of the graphing utility to determine the maximum concentration of the drug and the time at which the maximum concentration occurs.

SECTION 8.3

For Exercises 19–20, the graph of $y = f(x)$ is given. Solve the inequalities.

   a. $f(x) < 0$ **b.** $f(x) \leq 0$ **c.** $f(x) > 0$ **d.** $f(x) \geq 0$

19.

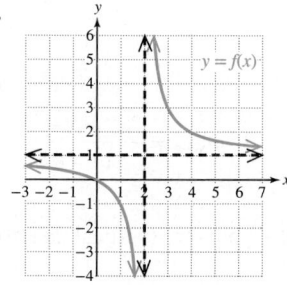

20.

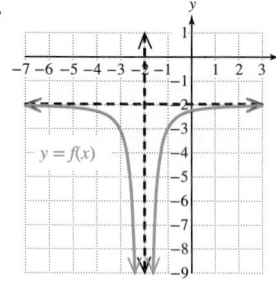

For Exercises 21–24, solve the equation and related inequalities.

21. a. $\dfrac{10}{x-5} = 5$

  b. $\dfrac{10}{x-5} < 5$

  c. $\dfrac{10}{x-5} > 5$

22. a. $\dfrac{8}{a+1} = 4$

  b. $\dfrac{8}{a+1} > 4$

  c. $\dfrac{8}{a+1} < 4$

23. a. $\dfrac{z+2}{z-6} = -3$

  b. $\dfrac{z+2}{z-6} \leq -3$

  c. $\dfrac{z+2}{z-6} \geq -3$

24. a. $\dfrac{w-8}{w+6} = 2$

  b. $\dfrac{w-8}{w+6} \leq 2$

  c. $\dfrac{w-8}{w+6} \geq 2$

25. $\dfrac{6-2x}{x^2} \geq 0$

26. $\dfrac{8}{3x-4} \leq 1$

27. $\dfrac{3}{x-2} < -\dfrac{2}{x}$

28. $\dfrac{(1-x)(3x+5)^2}{(x-3)^4} < 0$

29. A sports trainer has monthly costs of $80 for phone service and $40 for his website and advertising. In addition he pays a $15 fee to the gym for each session in which he works with a client.

  a. Write a function representing the average cost $\overline{C}(x)$ (in $) for x training sessions.

  b. Find the number of sessions the trainer needs if he wants the average cost to drop below $16 per session.

30. The concentration $C(t)$ (in ng/mL) of an antibiotic in the bloodstream t hours after the drug is taken orally can be modeled by $C(t) = \dfrac{16t}{t^2 + 1}$. The concentration in the bloodstream spikes after approximately 1 hr and then begins to decrease. During the decrease phase, the doctor wants to give a second dose so that the concentration in the bloodstream is kept above 2.5 ng/mL. When should the second dose be given? Round to the nearest hour.

SECTION 8.4

For Exercises 31–33, write a variation model using k as the constant of variation.

31. The mass m of an animal varies directly as the weight w of the animal's heart.

32. The value of x varies inversely to the square of p.

33. The variable y is jointly proportional to x and the square root of z, and inversely proportional to the cube of t.

For Exercises 34–39, determine the constant of variation k.

34. The variable Q varies jointly as p and the square root of t. The value of Q is 132 when p is 11 and t is 9.

35. The variable d is directly proportional to c and inversely proportional to the square of x. The value of d is 1.8 when c is 3 and x is 2.

36. The weight of a ball varies directly as the cube of its radius. A weighted exercise ball of radius 3 in. weighs 3.24 lb. How much would a ball weigh if its radius were 5 in.?

37. In karate, the force F required to break a board varies inversely as the length L of the board. If it takes 6.25 lb of force to break a board 1.6 ft long, determine how much force is required to break a 2-ft board.

38. The power in an electric circuit varies jointly as the current and the square of the resistance. If the power is 144 watts (W) when the current is 4 A and the resistance is 6 Ω, find the power when the current is 3 A and the resistance is 10 Ω.

39. Coulomb's law states that the force F of attraction between two oppositely charged particles varies jointly as the magnitude of their electrical charges q_1 and q_2 and inversely as the square of the distance d between the particles. Find the effect on F of doubling q_1 and q_2 and halving the distance between them.

CHAPTER 8   Test

For Exercises 1–4, determine the asymptotes (vertical, horizontal, slant, and nonlinear).

1. $r(x) = \dfrac{2x^2 - 3x + 5}{x - 7}$

2. $p(x) = \dfrac{-3x + 1}{4x^2 - 1}$

3. $n(x) = \dfrac{5x^2 - 2x + 1}{3x^2 + 4}$

4. $m(x) = \dfrac{x^4 - 4x^2 + 1}{x^2 + 2}$

For Exercises 5–7, graph the function.

5. $m(x) = -\dfrac{1}{x^2} + 3$

6. $h(x) = \dfrac{-4}{x^2 - 4}$

7. $k(x) = \dfrac{x^2 - 2x + 1}{x}$

8. The graph of $y = f(x)$ is given. Solve the inequalities.

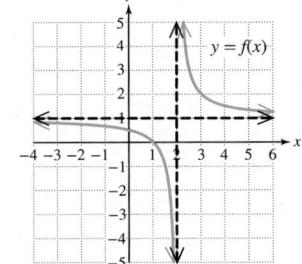

  a. $f(x) < 0$

  b. $f(x) \le 0$

  c. $f(x) > 0$

  d. $f(x) \ge 0$

9. Solve the inequalities.

  a. $\dfrac{x + 1}{x - 5} \le 0$

  b. $\dfrac{x + 1}{x - 5} < 0$

  c. $\dfrac{x + 1}{x - 5} \ge 0$

  d. $\dfrac{x + 1}{x - 5} > 0$

For Exercises 10–12, solve the inequalities.

10. $\dfrac{x + 3}{2 - x} \le 0$

11. $\dfrac{-4}{x^2 - 9} \ge 0$

12. $\dfrac{4}{x - 1} < -\dfrac{3}{x}$

13. Write a variation model using k as the constant of variation: Energy E varies directly as the square of the velocity v of the wind.

14. Solve for the constant of variation k: The variable w varies jointly as y and the square root of x, and inversely as z. The value of w is 7.2 when x is 4, y is 6, and z is 7.

15. The surface area of a cube varies directly as the square of the length of an edge. The surface area is 24 ft^2 when the length of an edge is 2 ft. Find the surface area of a cube with an edge that is 7 ft.

16. The weight of a body varies inversely as the square of its distance from the center of the Earth. The radius of the Earth is approximately 4000 mi. How much would a 180-lb man weigh 20 mi above the surface of the Earth? Round to the nearest pound.

17. The pressure of wind on a wall varies jointly as the area of the wall and the square of the velocity of the wind. If the velocity of the wind is tripled, what is the effect on the pressure on the wall?

18. The population $P(t)$ of rabbits in a wildlife area t years after being introduced to the area is given by

$$P(t) = \frac{2000t}{t + 1}$$

  a. Determine the number of rabbits after 1 year, 5 years, and 10 years. Round to the nearest whole unit.

  b. What will the rabbit population approach as t approaches infinity?

Exponential and Logarithmic Functions

9

Chapter Outline

Justin Lewis/Digital Vision/Getty Images

Visible light from the Sun is vitally important for the health of an ocean, lake, or any body of water. In particular, light penetrating through a body of water provides the energy to fuel vast amounts of microscopic plants called phytoplankton that are an essential source of food and oxygen for an aquatic ecosystem. Phytoplankton converts energy from the Sun to usable energy for plant growth. Thus, the amount of light directly affects plant productivity at the base of the food chain and ultimately animal life farther up the food chain.

With increasing depth, the percentage of visible light from the surface of a body of water drops exponentially. This means that light intensity drops quickly at first and then drops more slowly with increasing depth. (More specifically, light intensity decreases at a rate proportional to the intensity at a particular depth.) To study this phenomenon, scientists use exponential functions and their inverses, logarithmic functions. These two important categories of functions have many applications, including the study of the decay of radioactive substances, short-term population growth, and the growth of investments subject to compound interest.

SECTION 9.1   Inverse Functions

OBJECTIVES

1. Identify One-to-One Functions
2. Determine Whether Two Functions Are Inverses
3. Find the Inverse of a Function

1. Identify One-to-One Functions

Throughout our study of algebra, we have made use of the fact that the operations of addition and subtraction are inverse operations. For example, adding 5 to a number and then subtracting 5 from the result gives us the original number. Likewise, multiplication and division are inverse operations. We now look at the concept of an inverse function.

Changing currency is an important consideration when traveling abroad. For example, traveling between the United States and several countries in Europe would involve changing American dollars to Euros and then changing back for the return trip. Fortunately, we can use a function to change from one currency to the other, and then use the function's *inverse* to change back again.

Suppose that $1 (American dollar) can be exchanged for 0.8 € (Euro). Then,

$f(x) = 0.8x$ gives the number of Euros $f(x)$ that can be bought from x dollars.

$g(x) = \dfrac{x}{0.8}$ gives the number of dollars $g(x)$ that can be bought from x Euros.

Tables 9-1 and 9-2 show the values of $f(x)$ and $g(x)$ for several values of x.

Table 9-1

x (Dollars)	$f(x) = 0.8x$ (Euros)
100	80
150	120
200	160
250	200

Table 9-2

x (Euros)	$g(x) = \dfrac{x}{0.8}$ (Dollars)
80	100
120	150
160	200
200	250

FOR REVIEW

For a set of ordered pairs, (x, y), the domain is the set of x values and the range is the set of y values. From Table 9-1, the domain of f is $\{100, 150, 200, 250\}$ and the range of f is $\{80, 120, 160, 200\}$. From Table 9-2, the domain of g is $\{80, 120, 160, 200\}$ and the range of g is $\{100, 150, 200, 250\}$. See Section 5.3 for review.

In this example, functions f and g are inverses of each other, and we observe several interesting characteristics about inverse functions.

- By listing the ordered pairs from Tables 9-1 and 9-2, notice that the x and y values are reversed.

f: $\{(100, 80), (150, 120), (200, 160), (250, 200)\}$

g: $\{(80, 100), (120, 150), (160, 200), (200, 250)\}$

- For a function and its inverse, the values of x and y are interchanged. This tells us that the domain of a function is the same as the range of its inverse and vice versa.
- From the graphs of f and g (Figure 9-1), we see that the corresponding points on f and g are symmetric with respect to the line $y = x$.
- When we compose functions f and g in both directions, the result is the input value x. In a sense, what function f does to x, function g "undoes" and vice versa.

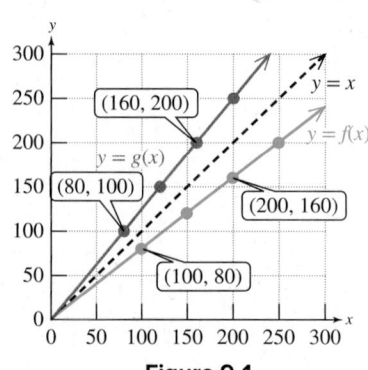

Figure 9-1

$$(f \circ g)(x) = f[g(x)] = 0.8\left(\frac{x}{0.8}\right) = x$$

$$(g \circ f)(x) = g[f(x)] = \frac{(0.8x)}{0.8} = x$$

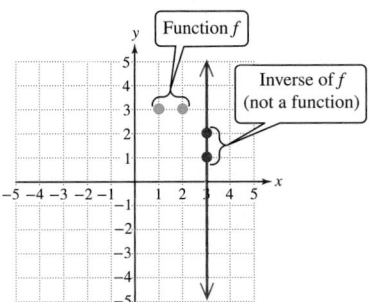

Figure 9-2

The inverse of any relation is found by interchanging the values of x and y in the relation. However, the inverse of a function may itself not be a function. For example, consider $f = \{(1, 3), (2, 3)\}$ shown in blue in Figure 9-2. The inverse is the set of ordered pairs $\{(3, 1), (3, 2)\}$ shown in red. Notice that the relation defining the inverse of f is not a function because it fails the vertical line test.

The function $f = \{(1, 3), (2, 3)\}$ has two points that are aligned horizontally. When the x and y values are reversed to form the inverse, the resulting points will be aligned vertically and will fail the vertical line test. Thus, a function will have an inverse function only if no points in the original function are aligned horizontally. That is, no two distinct points on the function may have the same y value. In such a case, the function is said to be one-to-one.

Definition of a One-to-One Function

A function f is a **one-to-one function**, if for a and b in the domain of f,

if $a \neq b$, then $f(a) \neq f(b)$, or equivalently, if $f(a) = f(b)$, then $a = b$.

FOR REVIEW

The vertical line test tells us that a relation defined by a set of points (x, y) is a function if no vertical line crosses the graph in more than one point. See Section 5.3 for review.

The definition of a one-to-one function tells us that each y value in the range is associated with only one x value in the domain. This implies that the graph of a one-to-one function will have no two points aligned horizontally.

Horizontal Line Test for a One-to-One Function

A function defined by $y = f(x)$ is a one-to-one function if no horizontal line intersects the graph in more than one point.

EXAMPLE 1   Determining Whether a Function Is One-to-One

Determine whether the function is one-to-one.

a. $f = \{(1, 4), (2, 3), (-2, 4)\}$ **b.** $g = \{(-3, 4), (1, -1), (2, 0)\}$

Solution:

same y value

a. $f = \{(1, 4), (2, 3), (-2, 4)\}$

different x value

f is not a one-to-one function.

The ordered pairs $(1, 4)$ and $(-2, 4)$ have the same y value but different x values. That is, $f(1) = f(-2)$, but $1 \neq -2$.

All points have different y-values.

b. $g = \{(-3, 4), (1, -1), (2, 0)\}$

g is a one-to-one function.

Each unique ordered pair has a different y value, so the function is one-to-one.

Skill Practice 1   Determine whether the function is one-to-one.

a. $h = \{(4, -5), (6, 1), (2, 4), (0, -3)\}$ **b.** $k = \{(1, 0), (3, 0), (4, -5)\}$

Answers

1. a. Yes   **b.** No

The graph of the function *f* from Example 1(a) is shown in Figure 9-3. Notice that a horizontal line passes through the function in more than one point. Thus, *f* is not a one-to-one function.

The graph of function *g* from Example 2(a) is shown in Figure 9-4. No points on *g* are aligned horizontally. Thus, no horizontal line can pass through the function in more than one point. Thus, *g* is a one-to-one function.

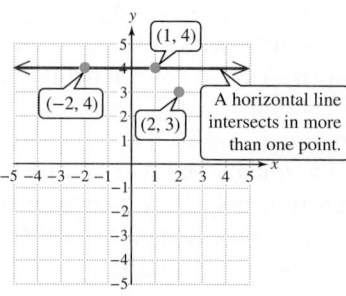

Figure 9-3

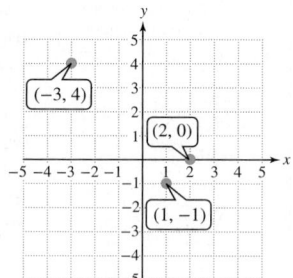

Figure 9-4

EXAMPLE 2   **Using the Horizontal Line Test**

Use the horizontal line test to determine if the graph in blue defines *y* as a one-to-one function of *x*.

Solution:

a.

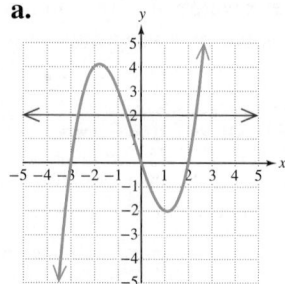

The graph does not define *y* as a one-to-one function of *x* because a horizontal line intersects the graph in more than one point.

b.

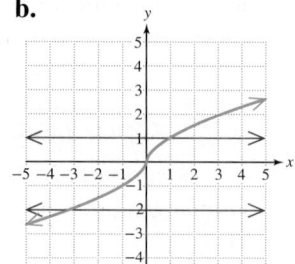

The graph does define *y* as a one-to-one function of *x* because no horizontal line intersects the graph in more than one point.

c.

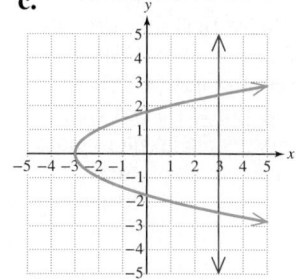

The relation does not define *y* as a function of *x*, because it fails the vertical line test. If the relation is not a function, it is not a one-to-one function.

Skill Practice 2   Use the horizontal line test to determine if the graph defines *y* as a one-to-one function of *x*.

a.

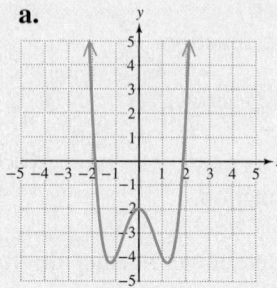

b.

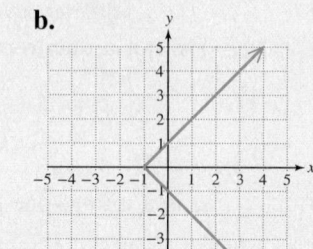

c.

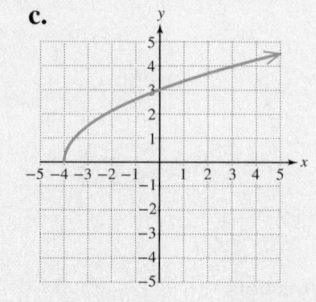

Answers

2. a. No   **b.** No   **c.** Yes

In Example 3, we use algebraic methods to determine whether a function is one-to-one.

EXAMPLE 3   Determining Whether a Function Is One-to-One

Use the definition of a one-to-one function to determine whether the function is one-to-one.

a. $f(x) = 2x - 3$ **b.** $f(x) = x^2 + 1$

Solution:

a. We must show that if $f(a) = f(b)$, then $a = b$.

Assume that $f(a) = f(b)$. That is,

$$2a - 3 = 2b - 3$$
$$2a - 3 + 3 = 2b - 3 + 3$$
$$\frac{2a}{2} = \frac{2b}{2}$$
$$a = b$$

The logic of this algebraic proof begins with the assumption that $f(a) = f(b)$, that is, that two y values are equal. For a one-to-one function, this can happen only if the x values (in this case a and b) are the same.

Otherwise, if $a \neq b$, we would have the same y value with two different x values and f would not be one-to-one.

$f(a) = f(b)$ implies that $a = b$.
Therefore, f is one-to-one.

b. $f(x) = x^2 + 1$

Assume that $f(a) = f(b)$.
$$a^2 + 1 = b^2 + 1$$
$$a^2 = b^2$$
$$a = \pm b$$

For nonzero values of b, $f(a) = f(b)$ does not necessarily imply that $a = b$. Therefore, f is not one-to-one.

From the graph of $f(x) = x^2 + 1$, we see that f is not one-to-one (Figure 9-5). We can also show this algebraically by finding two ordered pairs with the same y value but different x values. From the graph, we have arbitrarily selected $(-2, 5)$ and $(2, 5)$.

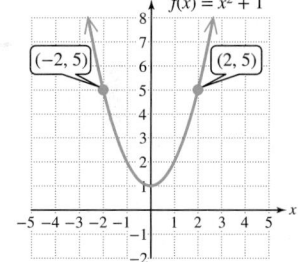

Figure 9-5

If $a = 2$ and $b = -2$, we have:

$$f(a) = f(2) = (2)^2 + 1 = 5 \longleftarrow$$
$$f(b) = f(-2) = (-2)^2 + 1 = 5 \longleftarrow$$

Same y value but different x values

We have that $f(a) = f(b)$, but $a \neq b$. Therefore, f fails to be a one-to-one function.

Skill Practice 3   Determine whether the function is one-to-one.

a. $f(x) = -4x + 1$ **b.** $f(x) = |x| - 3$

---

2. Determine Whether Two Functions Are Inverses

The inverse of a function is defined formally in terms of the composition of functions. First take a moment to review the composition of functions from Section 6.4. The composition of f and g, denoted by $f \circ g$, is defined by $(f \circ g)(x) = f[g(x)]$. Likewise, the composition of g and f is denoted by $g \circ f$ and is defined by $(g \circ f)(x) = g[f(x)]$.

Answers
3. a. Yes   **b.** No

For example, given $f(x) = \sqrt{x-2}$ and $g(x) = |x+1|$, we have

$$(f \circ g)(x) = f[g(x)] = f(|x+1|) = \sqrt{|x+1|-2}$$
$$(g \circ f)(x) = g[f(x)] = g(\sqrt{x-2}) = |\sqrt{x-2}+1|$$

We now have enough background to define an inverse function.

Definition of an Inverse Function

Let f be a one-to-one function. Then g is the **inverse of f** if the following conditions are both true.

1. $(f \circ g)(x) = x$ for all x in the domain of g.
2. $(g \circ f)(x) = x$ for all x in the domain of f.

Avoiding Mistakes

Do not confuse inverse notation f^{-1} with exponential notation. The notation f^{-1} does not mean $\frac{1}{f}$.

We should also note that if g is the inverse of f, then f is the inverse of g. Furthermore, given a function f, we often denote its inverse as f^{-1}. So given a function f and its inverse f^{-1}, the definition implies that

$$(f \circ f^{-1})(x) = x \text{ and } (f^{-1} \circ f)(x) = x$$

EXAMPLE 4   Determining Whether Two Functions Are Inverses

Determine whether the functions are inverses.

a. $f(x) = 100 + 12x$ and $g(x) = \dfrac{x-100}{12}$

b. $h(x) = \sqrt[3]{x-1}$ and $k(x) = -1 + x^3$

Solution:

a.
$$(f \circ g)(x) = f(g(x))$$
$$= f\left(\frac{x-100}{12}\right)$$
$$= 100 + 12\left(\frac{x-100}{12}\right)$$
$$= 100 + (x - 100)$$
$$= x \checkmark$$

$$(g \circ f)(x) = g(f(x))$$
$$= g(100 + 12x)$$
$$= \frac{(100 + 12x) - 100}{12}$$
$$= \frac{12x}{12}$$
$$= x \checkmark$$

$(f \circ g)(x) = (g \circ f)(x) = x$. Thus, f and g are inverses.

b.
$$(h \circ k)(x) = h(k(x))$$
$$= \sqrt[3]{(-1 + x^3) - 1}$$
$$= \sqrt[3]{x^3 - 2} \neq x$$

If either $(h \circ k)(x) \neq x$ or $(k \circ h)(x) \neq x$, then h and k are *not* inverses.

$(h \circ k)(x) \neq x$. Thus, h and k are not inverses.

Skill Practice 4   Determine whether the functions are inverses.

a. $f(x) = \dfrac{x+6}{2}$ and $g(x) = 2(x-6)$ **b.** $m(x) = \dfrac{5}{x-2}$ and $n(x) = \dfrac{2x+5}{x}$

Answers

4. a. No   **b.** Yes

3. Find the Inverse of a Function

For a one-to-one function defined by $y = f(x)$, the inverse is a function $y = f^{-1}(x)$ that performs the inverse operations in the reverse order. The function given by $f(x) = 100 + 12x$ multiplies x by 12 first, and then adds 100 to the result. Therefore, the inverse function must *subtract* 100 from x first and then *divide* by 12.

$$f^{-1}(x) = \frac{x - 100}{12}$$

To facilitate the process of finding an equation of the inverse of a one-to-one function, we offer the following steps.

Procedure to Find an Equation of an Inverse of a Function

For a one-to-one function defined by $y = f(x)$, the equation of the inverse can be found as follows.

Step 1 Replace $f(x)$ by y.
Step 2 Interchange x and y.
Step 3 Solve for y.
Step 4 Replace y by $f^{-1}(x)$.

EXAMPLE 5   Finding an Equation of an Inverse Function

Write an equation for the inverse function for $f(x) = 3x - 1$.

Solution:

Function f is a linear function, and its graph is a nonvertical line. Therefore, f is a one-to-one function.

$$f(x) = 3x - 1$$
$$y = 3x - 1 \qquad \text{**Step 1:** Replace } f(x) \text{ by } y.$$
$$x = 3y - 1 \qquad \text{**Step 2:** Interchange } x \text{ and } y.$$
$$x + 1 = 3y \qquad \text{**Step 3:** Solve for } y. \text{ Add 1 to both sides and divide by 3.}$$
$$\frac{x + 1}{3} = y$$
$$f^{-1}(x) = \frac{x + 1}{3} \qquad \text{**Step 4:** Replace } y \text{ by } f^{-1}(x).$$

To check the result, verify that $(f \circ f^{-1})(x) = x$ and $(f^{-1} \circ f)(x) = x$.

$$(f \circ f^{-1})(x) = 3\left(\frac{x + 1}{3}\right) - 1 = x \checkmark \quad \text{and} \quad (f^{-1} \circ f)(x) = \frac{(3x - 1) + 1}{3} = x \checkmark$$

Skill Practice 5   Write an equation for the inverse function for $f(x) = 4x + 3$.

Answer

5. $f^{-1}(x) = \dfrac{x - 3}{4}$

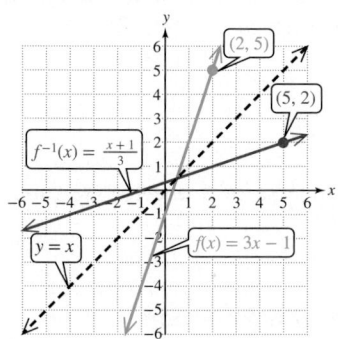

Figure 9-6

TIP We can sometimes find an equation of an inverse function by mentally reversing the operations given in the original function. In Example 5, we have $f(x) = 3x - 1$. Function f multiplies x by 3 and then subtracts 1. Therefore, f^{-1} must add 1 to x and then divide by 3.

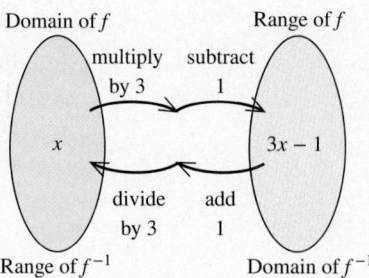

The key step in determining the equation of the inverse of a function is to interchange x and y. By so doing, a point (a, b) on f corresponds to a point (b, a) on f^{-1}. This is why the graphs of f and f^{-1} are symmetric with respect to the line $y = x$. From Example 5, notice that the point $(2, 5)$ on the graph of f corresponds to the point $(5, 2)$ on the graph of f^{-1} (Figure 9-6).

EXAMPLE 6   Finding an Equation of an Inverse Function

Write an equation for the inverse function for the one-to-one function defined by $f(x) = \dfrac{3 - x}{x + 3}$.

Solution:

TIP In Example 6, we can show that f is a one-to-one function by graphing the function (see Section 8.2). Or we can show that $f(a) = f(b)$ implies that $a = b$ by solving the equation $\dfrac{3 - a}{a + 3} = \dfrac{3 - b}{b + 3}$ for a or b to show that $a = b$.

$$f(x) = \frac{3 - x}{x + 3}$$

$$y = \frac{3 - x}{x + 3} \qquad \textbf{Step 1:} \text{ Replace } f(x) \text{ by } y.$$

$$x = \frac{3 - y}{y + 3} \qquad \textbf{Step 2:} \text{ Interchange } x \text{ and } y.$$

$$x \cdot (y + 3) = \left(\frac{3 - y}{y + 3}\right) \cdot (y + 3) \qquad \textbf{Step 3:} \text{ Solve for } y.$$
Clear fractions (multiply both sides by $y + 3$).

$$x(y + 3) = 3 - y \qquad \text{Apply the distributive property.}$$
$$xy + 3x = 3 - y$$
$$xy + y = 3 - 3x \qquad \text{Collect the } y \text{ terms on one side.}$$
$$y(x + 1) = 3 - 3x \qquad \text{Factor out } y \text{ as the greatest common factor.}$$

$$y = \frac{3 - 3x}{x + 1} \qquad \text{Divide both sides by } x + 1.$$

$$f^{-1}(x) = \frac{3 - 3x}{x + 1} \qquad \textbf{Step 4:} \text{ Replace } y \text{ by } f^{-1}(x).$$

Skill Practice 6   Write an equation for the inverse function for the one-to-one function defined by $f(x) = \dfrac{x - 2}{x + 2}$.

Answer

6. $f^{-1}(x) = -\dfrac{2x + 2}{x - 1}$

For a function that is not one-to-one, sometimes we restrict its domain to create a new function that is one-to-one. This is demonstrated in Example 7.

EXAMPLE 7   Finding an Equation of an Inverse Function

Given $m(x) = x^2 + 4$ for $x \geq 0$, write an equation of the inverse.

Solution:

The graph of $y = x^2 + 4$ is a parabola with vertex $(0, 4)$. See Figure 9-7. The function is not one-to-one. However, with the restriction on the domain that $x \geq 0$, the graph consists of only the right branch of the parabola (Figure 9-8). This *is* a one-to-one function.

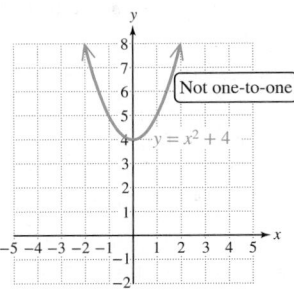

Figure 9-7

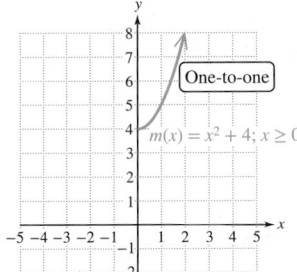

Figure 9-8

To find the inverse, we have

$$m(x) = x^2 + 4; \quad x \geq 0$$
$$y = x^2 + 4; \quad x \geq 0 \qquad \textbf{Step 1: } \text{Replace } m(x) \text{ by } y.$$
$$x = y^2 + 4 \quad y \geq 0 \qquad \textbf{Step 2: } \text{Interchange } x \text{ and } y. \text{ Notice that the restriction } x \geq 0 \text{ becomes } y \geq 0.$$

$$x - 4 = y^2 \qquad \textbf{Step 3: } \text{Solve for } y \text{ by subtracting 4 from both sides.}$$
$$y = \pm\sqrt{x - 4} \qquad \qquad \text{Apply the square root property.}$$
$$y = +\sqrt{x - 4} \qquad \qquad \text{Choose the positive square root of } (x - 4) \text{ because of the restriction } y \geq 0.$$

$$m^{-1}(x) = \sqrt{x - 4} \qquad \textbf{Step 4: } \text{Replace } y \text{ by } m^{-1}(x).$$

FOR REVIEW

Recall that the square root property indicates that for an equation of the form $x^2 = k$, where k is a constant, the solutions are $x = \pm\sqrt{k}$. See Section 3.5 for review.

The graphs of m and m^{-1} are symmetric with respect to the line $y = x$ as expected (Figure 9-9).

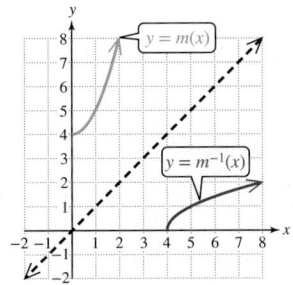

Figure 9-9

Skill Practice 7   Given $n(x) = x^2 + 1$ for $x \leq 0$, write an equation of the inverse.

Answer

7. $n^{-1}(x) = -\sqrt{x - 1}$

EXAMPLE 8 **Finding an Equation of an Inverse Function**

Given $f(x) = \sqrt{x-1}$, find an equation of the inverse.

Solution:

The function f is a one-to-one function and the graph is the same as the graph of $y = \sqrt{x}$ with a shift 1 unit to the right. The domain of f is $\{x \mid x \geq 1\}$ and the range is $\{y \mid y \geq 0\}$. When defining the inverse, we will have the conditions that $x \geq 0$ and $y \geq 1$.

TIP When finding the inverse of a function, the key step of interchanging x and y has the effect of interchanging the domain and range between the function and its inverse.

$f(x) = \sqrt{x-1}$ Note that $x \geq 1$ and $y \geq 0$.

$y = \sqrt{x-1}$

$x = \sqrt{y-1}$ Interchange x and y.
  Note that $y \geq 1$ and $x \geq 0$.

$x^2 = y - 1$ Square both sides.

$y = x^2 + 1$

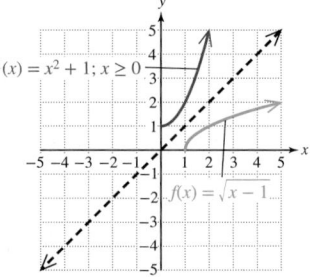

$f^{-1}(x) = x^2 + 1, \quad x \geq 0$ The restriction $x \geq 0$ on f^{-1} is necessary because f has the restriction that $y \geq 0$. Furthermore, $y = x^2 + 1$ is not a one-to-one function without a restricted domain.

Answer

8. $g^{-1}(x) = x^2 - 2; x \geq 0$

Skill Practice 8 Given $g(x) = \sqrt{x+2}$, find an equation of the inverse.

SECTION 9.1   Practice Exercises

Prerequisite Review

For Exercises R.1–R.4, find the domain. Write the answer in interval notation. (See Section 5.3 for review.)

R.1. $f(x) = \dfrac{x+4}{x+1}$

R.2. $g(x) = \dfrac{2x-5}{3-x}$

R.3. $m(x) = \dfrac{1}{\sqrt{2-8x}}$

R.4. $n(x) = \dfrac{1}{\sqrt{4x+2}}$

R.5. Given $f(x) = 5x - 4$, find the function values. (See Section 5.3 for review.)

  a. $f(-2)$ **b.** $f(t)$ **c.** $f\left(\dfrac{x+4}{5}\right)$

R.6. Given $g(x) = \sqrt[3]{x-4}$, find the function values.

  a. $g(-4)$ **b.** $g(a-2)$ **c.** $g(x^3+4)$

For Exercises R.7–R.12, refer to functions f, g, n, and p to evaluate the function. (See Section 6.4 for review.)

$f(x) = 2x - 7, \qquad g(x) = \dfrac{x+7}{2}, \qquad n(x) = x + 1, \qquad p(x) = x^2 - 3x$

  R.7. $(n \circ p)(x)$ **R.8.** $(p \circ n)(x)$ **R.9.** $(f \circ g)(x)$

  R.10. $(g \circ f)(x)$ **R.11.** $(p \circ p)(x)$ **R.12.** $(f \circ f)(x)$

For Exercises R.13–R.16, solve for y. (See Sections 1.2 and 4.3 for review.)

R.13. $x = 7y + 2$ (See Section 1.2.)

R.14. $x = \sqrt[3]{y - 2}$

R.15. $x = \dfrac{5 - y}{5 + y}$ (See Section 4.3.)

R.16. $x = \dfrac{7 - y}{y + 3}$

For Exercises R.17–R.18, determine whether the graph represents y as a function of x. (See Section 5.3 for review.)

R.17.

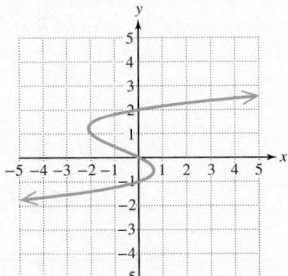

R.18.

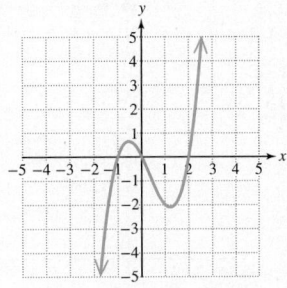

For Exercises R.19–R.20, graph the function on the given interval. (See Section 6.2 for review.)

R.19. $f(x) = -x^2 + 5$ for $x \le 0$

R.20. $g(x) = x^2 - 3$ for $x \ge 0$

Concept Connections

1. Given the function $f = \{(1, 2), (2, 3), (3, 4)\}$ write the set of ordered pairs representing f^{-1}.

2. The graphs of a function and its inverse are symmetric with respect to the line _____.

3. If no horizontal line intersects the graph of a function f in more than one point, then f is a _____ - _____ - _____ function.

4. Given a one-to-one function f, if $f(a) = f(b)$, then a _____ b.

5. Let f be a one-to-one function and let g be the inverse of f. Then $(f \circ g)(x) =$ _____ and $(g \circ f)(x) =$ _____.

6. If (a, b) is a point on the graph of a one-to-one function f, then the corresponding ordered pair _____ is a point on the graph of f^{-1}.

Objective 1: Identify One-to-One Functions

For Exercises 7–12, a relation in x and y is given. Determine if the relation defines y as a one-to-one function of x.
(See Example 1)

7. $\{(6, -5), (4, 2), (3, 1), (8, 4)\}$

8. $\{(-14, 1), (-2, 3), (7, 4), (-9, -2)\}$

9.

x	y
0.6	1.8
1	-1.1
0.5	1.8
2.4	0.7

10.

x	y
12.5	3.21
5.75	-4.5
2.34	7.25
-12.7	3.21

11.

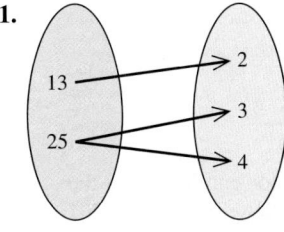

12.

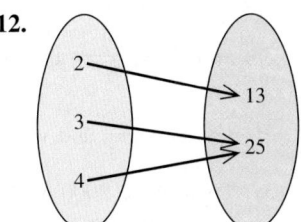

For Exercises 13–22, determine if the relation defines y as a one-to-one function of x. **(See Example 2)**

13.
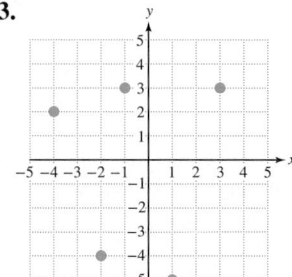

14.
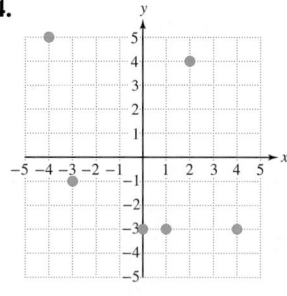

15.
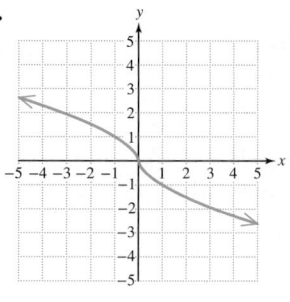

16.
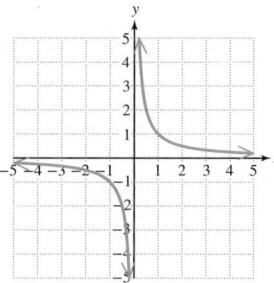

17.
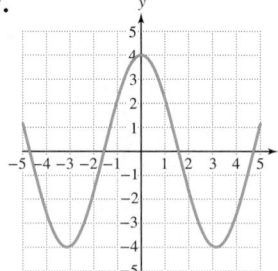

18.
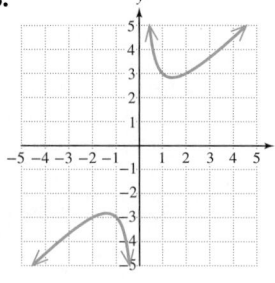

19.
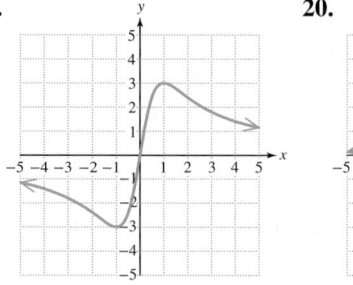

20.
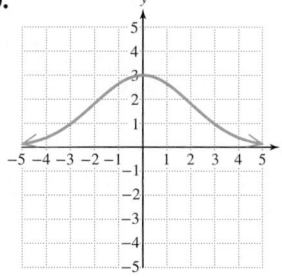

21.
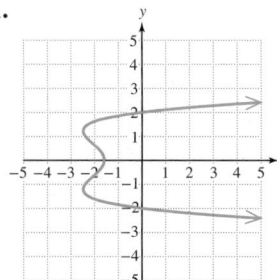

22.
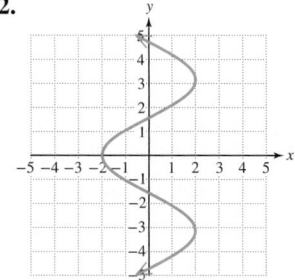

For Exercises 23–30, use the definition of a one-to-one function to determine if the function is one-to-one. **(See Example 3)**

23. $f(x) = 4x - 7$

24. $h(x) = -3x + 2$

25. $g(x) = x^3 + 8$

26. $k(x) = x^3 - 27$

27. $m(x) = x^2 - 4$

28. $n(x) = x^2 + 1$

29. $p(x) = |x + 1|$

30. $q(x) = |x - 3|$

Objective 2: Determine Whether Two Functions Are Inverses

For Exercises 31–36, determine whether the two functions are inverses. **(See Example 4)**

31. $f(x) = 5x + 4$ and $g(x) = \dfrac{x - 4}{5}$

32. $h(x) = 7x - 3$ and $k(x) = \dfrac{x + 3}{7}$

33. $m(x) = \dfrac{-2 + x}{6}$ and $n(x) = 6x - 2$

34. $p(x) = \dfrac{-3 + x}{4}$ and $q(x) = 4x - 3$

35. $t(x) = \dfrac{4}{x - 1}$ and $v(x) = \dfrac{x + 4}{x}$

36. $w(x) = \dfrac{6}{x + 2}$ and $z(x) = \dfrac{6 - 2x}{x}$

37. There were 2000 applicants for enrollment to the freshman class at a small college in the year 2010. The number of applications has risen linearly by roughly 150 per year. The number of applications $f(x)$ is given by $f(x) = 2000 + 150x$, where x is the number of years since 2010.

   a. Determine if the function $g(x) = \dfrac{x - 2000}{150}$ is the inverse of f.

   b. Interpret the meaning of function g in the context of this problem.

38. The monthly sales for January for a whole foods market was $60,000 and has increased linearly by $2500 per month. The amount in sales $f(x)$ (in $) is given by $f(x) = 60{,}000 + 2500x$, where x is the number of months since January.

   a. Determine if the function $g(x) = \dfrac{x - 60{,}000}{2500}$ is the inverse of f.

   b. Interpret the meaning of function g in the context of this problem.

Objective 3: Find the Inverse of a Function

39. a. Show that $f(x) = 2x - 3$ defines a one-to-one function.

   b. Write an equation for $f^{-1}(x)$.

   c. Graph $y = f(x)$ and $y = f^{-1}(x)$ on the same coordinate system.

40. a. Show that $f(x) = 4x + 4$ defines a one-to-one function.

   b. Write an equation for $f^{-1}(x)$.

   c. Graph $y = f(x)$ and $y = f^{-1}(x)$ on the same coordinate system.

For Exercises 41–52, a one-to-one function is given. Write an equation for the inverse function. **(See Examples 5–6)**

41. $f(x) = \dfrac{4 - x}{9}$

42. $g(x) = \dfrac{8 - x}{3}$

43. $h(x) = \sqrt[3]{x - 5}$

44. $k(x) = \sqrt[3]{x + 8}$

45. $m(x) = 4x^3 + 2$

46. $n(x) = 2x^3 - 5$

47. $c(x) = \dfrac{5}{x + 2}$

48. $s(x) = \dfrac{2}{x - 3}$

49. $t(x) = \dfrac{x - 4}{x + 2}$

50. $v(x) = \dfrac{x - 5}{x + 1}$

51. $f(x) = \dfrac{(x - a)^3}{b} - c$

52. $g(x) = b(x + a)^3 + c$

53. a. Graph $f(x) = x^2 - 3$; $x \le 0$. **(See Example 7)**

   b. From the graph of f, is f a one-to-one function?

   c. Write the domain of f in interval notation.

   d. Write the range of f in interval notation.

   e. Write an equation for $f^{-1}(x)$.

   f. Graph $y = f(x)$ and $y = f^{-1}(x)$ on the same coordinate system.

   g. Write the domain of f^{-1} in interval notation.

   h. Write the range of f^{-1} in interval notation.

54. a. Graph $f(x) = x^2 + 1$; $x \le 0$.

   b. From the graph of f, is f a one-to-one function?

   c. Write the domain of f in interval notation.

   d. Write the range of f in interval notation.

   e. Write an equation for $f^{-1}(x)$.

   f. Graph $y = f(x)$ and $y = f^{-1}(x)$ on the same coordinate system.

   g. Write the domain of f^{-1} in interval notation.

   h. Write the range of f^{-1} in interval notation.

55. a. Graph $f(x) = \sqrt{x + 1}$. **(See Example 8)**

   b. From the graph of f, is f a one-to-one function?

   c. Write the domain of f in interval notation.

   d. Write the range of f in interval notation.

   e. Write an equation for $f^{-1}(x)$.

   f. Explain why the restriction $x \ge 0$ is placed on f^{-1}.

   g. Graph $y = f(x)$ and $y = f^{-1}(x)$ on the same coordinate system.

   h. Write the domain of f^{-1} in interval notation.

   i. Write the range of f^{-1} in interval notation.

56. a. Graph $f(x) = \sqrt{x - 2}$.

   b. From the graph of f, is f a one-to-one function?

   c. Write the domain of f in interval notation.

   d. Write the range of f in interval notation.

   e. Write an equation for $f^{-1}(x)$.

   f. Explain why the restriction $x \ge 0$ is placed on f^{-1}.

   g. Graph $y = f(x)$ and $y = f^{-1}(x)$ on the same coordinate system.

   h. Write the domain of f^{-1} in interval notation.

   i. Write the range of f^{-1} in interval notation.

57. Given that the domain of a one-to-one function f is $[0, \infty)$ and the range of f is $[0, 4)$, state the domain and range of f^{-1}.

58. Given that the domain of a one-to-one function f is $[-3, 5)$ and the range of f is $(-2, \infty)$, state the domain and range of f^{-1}.

59. Given $f(x) = |x| + 3$; $x \le 0$, write an equation for f^{-1}. (*Hint*: Sketch $f(x)$ and note the domain and range.)

60. Given $f(x) = |x| - 3$; $x \ge 0$, write an equation for f^{-1}. (*Hint*: Sketch $f(x)$ and note the domain and range.)

For Exercises 61–66, fill in the blanks and determine an equation for $f^{-1}(x)$ mentally.

61. If function f adds 6 to x, then f^{-1} _____ 6 from x. Function f is defined by $f(x) = x + 6$, and function f^{-1} is defined by $f^{-1}(x) =$ _____.

62. If function f multiplies x by 2, then f^{-1} _____ x by 2. Function f is defined by $f(x) = 2x$, and function f^{-1} is defined by $f^{-1}(x) =$ _____.

63. Suppose that function f multiplies x by 7 and subtracts 4. Write an equation for $f^{-1}(x)$.

64. Suppose that function f divides x by 3 and adds 11. Write an equation for $f^{-1}(x)$.

65. Suppose that function f cubes x and adds 20. Write an equation for $f^{-1}(x)$.

66. Suppose that function f takes the cube root of x and subtracts 10. Write an equation for $f^{-1}(x)$.

For Exercises 67–70, find the inverse mentally.

67. $f(x) = 8x + 1$ **68.** $p(x) = 2x - 10$ **69.** $q(x) = \sqrt[5]{x - 4} + 1$ **70.** $m(x) = \sqrt[3]{4x} + 3$

Mixed Exercises

For Exercises 71–74, the graph of a function is given. Graph the inverse function.

71. 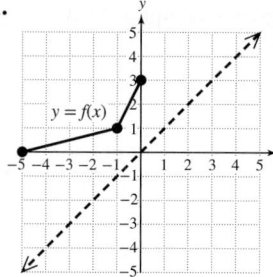   **72.** 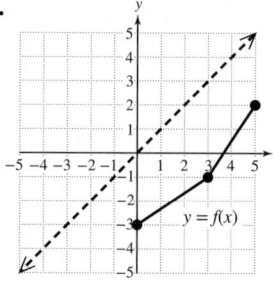   **73.** 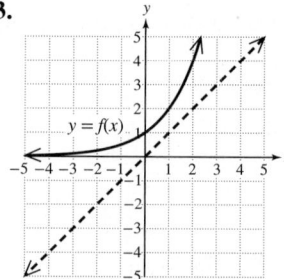   **74.**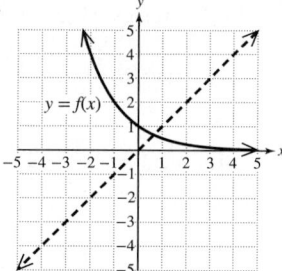

For Exercises 75–76, the table defines $Y_1 = f(x)$ as a one-to-one function of x. Find the values of f^{-1} for the selected values of x.

75. **a.** $f^{-1}(32)$
      b. $f^{-1}(-2.5)$
      c. $f^{-1}(26)$

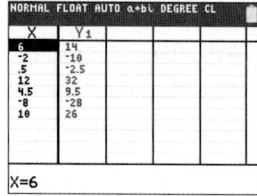

76. **a.** $f^{-1}(5)$
      b. $f^{-1}(9.45)$
      c. $f^{-1}(8)$

For Exercises 77–80, determine if the statement is true or false. If a statement is false, explain why.

77. All linear functions with a nonzero slope have an inverse function.

78. The domain of any one-to-one function is the same as the domain of its inverse function.

79. The range of a one-to-one function is the same as the range of its inverse function.

80. No quadratic function defined by $f(x) = ax^2 + bx + c$ $(a \neq 0)$ is one-to-one.

81. Based on data from Hurricane Katrina, the function defined by $w(x) = -1.17x + 1220$ gives the wind speed $w(x)$ (in mph) based on the barometric pressure x (in millibars, mb).

   a. Approximate the wind speed for a hurricane with a barometric pressure of 1000 mb.

   b. Write a function representing the inverse of w and interpret its meaning in context.

   c. Approximate the barometric pressure for a hurricane with wind speed 100 mph. Round to the nearest mb.

82. The function defined by $F(x) = \dfrac{9}{5}x + 32$ gives the temperature $F(x)$ (in degrees Fahrenheit) based on the temperature x (in Celsius).

   a. Determine the temperature in Fahrenheit if the temperature in Celsius is 25°C.

   b. Write a function representing the inverse of F and interpret its meaning in context.

   c. Determine the temperature in Celsius if the temperature in Fahrenheit is 5°F.

83. Suppose that during normal respiration, the volume of air inhaled per breath (called "tidal volume") by a mammal of any size is 6.33 mL per kilogram of body mass.

   a. Write a function representing the tidal volume $T(x)$ (in mL) of a mammal of mass x (in kg).

   b. Write an equation for $T^{-1}(x)$.

   c. What does the inverse function represent in the context of this problem?

   d. Find $T^{-1}(170)$ and interpret its meaning in context. Round to the nearest whole unit.

Malcolm Schuyl/Alamy Stock Photo

84. At a cruising altitude of 35,000 ft, a certain airplane travels 555 mph.

   a. Write a function representing the distance $d(x)$ (in mi) for x hours at cruising altitude.

   b. Write an equation for $d^{-1}(x)$.

   c. What does the inverse function represent in the context of this problem?

   d. Evaluate $d^{-1}(2553)$ and interpret its meaning in context.

85. The millage rate is the amount of property tax per $1000 of the taxable value of a home. For a certain county the millage rate is 24 mil ($24 in tax per $1000 of taxable value of the home). A city within the county also imposes a flat fee of $108 per home.

  a. Write a function representing the total amount of property tax $T(x)$ (in $) for a home with a taxable value of x thousand dollars.

  b. Write an equation for $T^{-1}(x)$.

  c. What does the inverse function represent in the context of this problem?

  d. Evaluate $T^{-1}(2988)$ and interpret its meaning in context.

86. Beginning on January 1, park rangers in Everglades National Park began recording the water level for one particularly dry area of the park. The water level was initially 2.5 ft and decreased by approximately 0.015 ft/day.

  a. Write a function representing the water level $L(x)$ (in ft), x days after January 1.

  b. Write an equation for $L^{-1}(x)$.

  c. What does the inverse function represent in the context of this problem?

  d. Evaluate $L^{-1}(1.9)$ and interpret its meaning in context.

Write About It

87. Explain the relationship between the domain and range of a one-to-one function f and its inverse f^{-1}.

88. Write an informal definition of a one-to-one function.

89. Explain why if a horizontal line intersects the graph of a function in more than one point, then the function is not one-to-one.

90. Explain why the domain of $f(x) = x^2 + k$ must be restricted to find an inverse function.

Expanding Your Skills

91. Consider a function defined as follows: Given x, the value $f(x)$ is the exponent above the base of 2 that produces x. For example, $f(16) = 4$ because $2^4 = 16$. Evaluate

  a. $f(8)$ **b.** $f(32)$

  c. $f(2)$ **d.** $f\left(\frac{1}{8}\right)$

93. Show that every increasing function is one-to-one.

92. Consider a function defined as follows: Given x, the value $f(x)$ is the exponent above the base of 3 that produces x. For example, $f(9) = 2$ because $3^2 = 9$. Evaluate

  a. $f(27)$ **b.** $f(81)$

  c. $f(3)$ **d.** $f\left(\frac{1}{9}\right)$

94. A function is said to be periodic if there exists some nonzero real number p, called the period, such that $f(x + p) = f(x)$ for all real numbers x in the domain of f. Explain why no periodic function is one-to-one.

SECTION 9.2   Exponential Functions

OBJECTIVES

1. Graph Exponential Functions
2. Evaluate the Exponential Function Base e
3. Use Exponential Functions to Compute Compound Interest
4. Use Exponential Functions in Applications

1. Graph Exponential Functions

The concept of a function was first introduced in Section 5.3. Since then we have learned to recognize several categories of functions. In this section and the next, we will define two new types of functions called exponential functions and logarithmic functions.

To introduce exponential functions, consider two salary plans for a new job. Plan A pays $1 million for 1 month's work. Plan B starts with 2¢ on the first day, and every day thereafter the salary is doubled. At first glance, the million-dollar plan appears to be more favorable. However, Table 9-3 shows otherwise. The daily payments for 30 days are listed for Plan B.

Table 9-3

Day	Payment	Day	Payment	Day	Payment
1	2¢	11	$20.48	21	$20,971.52
2	4¢	12	$40.96	22	$41,943.04
3	8¢	13	$81.92	23	$83,886.08
4	16¢	14	$163.84	24	$167,772.16
5	32¢	15	$327.68	25	$335,544.32
6	64¢	16	$655.36	26	$671,088.64
7	$1.28	17	$1310.72	27	$1,342,177.28
8	$2.56	18	$2621.44	28	$2,684,354.56
9	$5.12	19	$5242.88	29	$5,368,709.12
10	$10.24	20	$10,485.76	30	$10,737,418.24

TIP   Consider the pattern involved for the payment for day $x = 1, 2, 3, 4, 5, \dots$

$$2^1 \text{ ¢} = 2\text{¢}$$
$$2^2 \text{ ¢} = 4\text{¢}$$
$$2^3 \text{ ¢} = 8\text{¢}$$
$$2^4 \text{ ¢} = 16\text{¢}$$
$$2^5 \text{ ¢} = 32\text{¢}$$

The salary for the 30th day for Plan B is over $10 million. Taking the sum of the payments, we see that the total salary for the 30-day period is $21,474,836.46.

The daily salary $S(x)$ (in ¢) for Plan B can be represented by the function $S(x) = 2^x$, where x is the number of days on the job. An interesting characteristic of this function is that for every positive 1-unit change in x, the function value doubles. The function $S(x) = 2^x$ is called an exponential function.

Definition of an Exponential Function

Let b be a constant real number such that $b > 0$ and $b \neq 1$. Then for any real number x, a function of the form $f(x) = b^x$ is called an **exponential function of base b**.

An exponential function is recognized as a function with a constant base (positive and not equal to 1) with a variable exponent, x.

Exponential Functions	**Not Exponential Functions**	
$f(x) = 3^x$	$m(x) = x^2$	base is not constant
$g(x) = \left(\dfrac{1}{3}\right)^x$	$n(x) = \left(-\dfrac{1}{3}\right)^x$	base is negative
$h(x) = \left(\sqrt{2}\right)^x$	$p(x) = 1^x$ base is 1	

FOR REVIEW

Recall that a nonzero base raised to a negative exponent can be simplified by taking the reciprocal of the base and changing the exponent to positive. For example,

$$4^{-1} = \left(\frac{1}{4}\right)^1 = \frac{1}{4}$$

$$\left(\frac{1}{2}\right)^{-3} = 2^3 = 8$$

See Section 2.1 for review.

Avoiding Mistakes

- The base of an exponential function must not be negative to avoid situations where the function values are not real numbers. For example, $f(x) = (-4)^x$ is not defined for $x = \frac{1}{2}$ because $\sqrt{-4}$ is not a real number.
- The base of an exponential function must not equal 1 because $f(x) = 1^x = 1$ for all real numbers x. This is a constant function, not an exponential function.

At this point in the text, we have evaluated exponential expressions with integer exponents and with rational exponents. For example,

$$4^2 = 16 \qquad\qquad 4^{1/2} = \sqrt{4} = 2$$

$$4^{-1} = \frac{1}{4} \qquad\qquad 4^{10/23} = \sqrt[23]{4^{10}} \approx 1.827112184$$

However, how do we evaluate an exponential expression with an *irrational* exponent such as 4^π? In such a case, the exponent is a nonterminating, nonrepeating decimal. We define an exponential expression raised to an irrational exponent as a sequence of approximations using rational exponents. For example:

$$4^{3.14} \approx 77.7084726$$
$$4^{3.141} \approx 77.81627412$$
$$4^{3.1415} \approx 77.87023095$$
$$\cdots$$
$$4^\pi \approx 77.88023365$$

With this definition of a base raised to an irrational exponent, we can define an exponential function over the entire set of real numbers. In Example 1, we graph two exponential functions by plotting points.

EXAMPLE 1   **Graphing Exponential Functions**

Graph the functions.

a. $f(x) = 2^x$ **b.** $g(x) = \left(\dfrac{1}{2}\right)^x$

Solution:

Table 9-4 shows several values of $f(x)$ and $g(x)$ for both positive and negative values of x.

Table 9-4

x	$f(x) = 2^x$	$g(x) = \left(\frac{1}{2}\right)^x$
-3	$\frac{1}{8}$	8
-2	$\frac{1}{4}$	4
-1	$\frac{1}{2}$	2
0	1	1
1	2	$\frac{1}{2}$
2	4	$\frac{1}{4}$
3	8	$\frac{1}{8}$

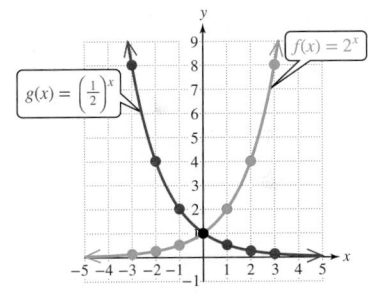

Figure 9-10

TIP   The values of $f(x)$ become closer and closer to 0 as $x \to -\infty$. This means that the x-axis is a horizontal asymptote.

Likewise, the values of $g(x)$ become closer to 0 as $x \to \infty$. The x-axis is a horizontal asymptote.

Notice that $g(x) = \left(\frac{1}{2}\right)^x$ is equivalent to $g(x) = 2^{-x}$. Therefore, the graph of $g(x) = \left(\frac{1}{2}\right)^x = 2^{-x}$ is the same as the graph of $f(x) = 2^x$ with a reflection across the y-axis (Figure 9-10).

Skill Practice 1   Graph the functions.

a. $f(x) = 5^x$ **b.** $g(x) = \left(\dfrac{1}{5}\right)^x$

Answer

1.

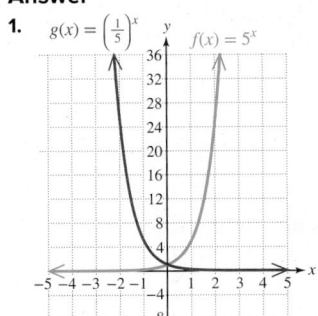

The graphs in Figure 9-10 illustrate several important features of exponential functions.

Graphs of $f(x) = b^x$

The graph of an exponential function defined by $f(x) = b^x$ ($b > 0$ and $b \neq 1$) has the following properties.

1. If $b > 1$, f is an *increasing* exponential function, sometimes called an **exponential growth function**.

If $0 < b < 1$, f is a *decreasing* exponential function, sometimes called an **exponential decay function**.

2. The domain is the set of all real numbers, $(-\infty, \infty)$.

3. The range is $(0, \infty)$.

4. The line $y = 0$ (x-axis) is a horizontal asymptote.

5. The function passes through the point $(0, 1)$ because $f(0) = b^0 = 1$.

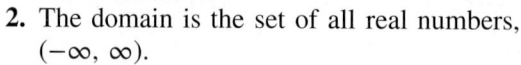

These properties indicate that the graph of an exponential function is an increasing function if the base is greater than 1. Furthermore, the base affects the rate of increase. Consider the graphs of $f(x) = 2^x$ and $k(x) = 5^x$ (Figure 9-11). For every positive 1-unit change in x, $f(x) = 2^x$ is 2 times as great and $k(x) = 5^x$ is 5 times as great (Table 9-5).

Table 9-5

x	$f(x) = 2^x$	$k(x) = 5^x$
-3	$\frac{1}{8}$	$\frac{1}{125}$
-2	$\frac{1}{4}$	$\frac{1}{25}$
-1	$\frac{1}{2}$	$\frac{1}{5}$
0	1	1
1	2	5
2	4	25
3	8	125

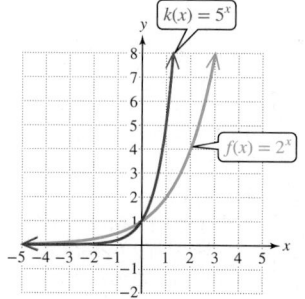

Figure 9-11

In Example 2, we use the transformations of functions learned in Section 6.1 to graph an exponential function.

If $h > 0$, shift to the right.
If $h < 0$, shift to the left.

$$f(x) = ab^{x-h} + k$$

If $a < 0$, reflect across the x-axis.
Shrink vertically if $0 < |a| < 1$.
Stretch vertically if $|a| > 1$.

If $k > 0$, shift upward.
If $k < 0$, shift downward.

EXAMPLE 2   Graphing an Exponential Function

Graph. $f(x) = 3^{x-2} + 4$

Solution:

The graph of f is the graph of the parent function $y = 3^x$ shifted 2 units to the right and 4 units upward.

The parent function $y = 3^x$ is an increasing exponential function. We can plot a few points on the graph of $y = 3^x$ and use these points and the horizontal asymptote to form the outline of the transformed graph.

x	$y = 3^x$
-2	$\frac{1}{9}$
-1	$\frac{1}{3}$
0	1
1	3
2	9

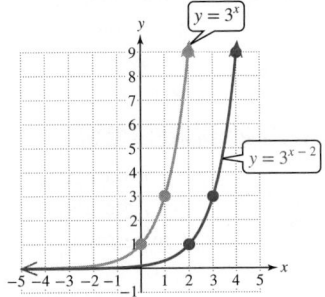

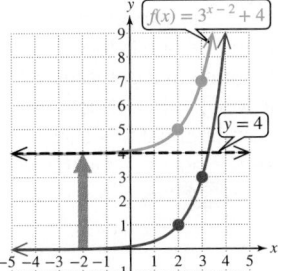

Shift 2 units to the right. For example, the point $(0, 1)$ on $y = 3^x$ corresponds to $(2, 1)$ on $y = 3^{x-2}$.

Shift the graph of $y = 3^{x-2}$ up 4 units. Notice that with the vertical shift, the new horizontal asymptote is $y = 4$.

Skill Practice 2   Graph. $g(x) = 2^{x+2} - 1$

2. Evaluate the Exponential Function Base e

We now introduce an important exponential function whose base is an irrational number called e. Consider the expression $\left(1 + \dfrac{1}{x}\right)^x$. The value of the expression for increasingly large values of x approaches a constant (Table 9-6).

As $x \to \infty$, the expression $\left(1 + \dfrac{1}{x}\right)^x$ approaches a constant value that we call e. From Table 9-6, this value is approximately 2.718281828.

$$e \approx 2.718281828$$

The value of e is an irrational number (a non-terminating, nonrepeating decimal) and like the number π, it is a universal constant. The function defined by $f(x) = e^x$ is called the exponential function base e or the **natural exponential function**.

Table 9-6

x	$\left(1 + \dfrac{1}{x}\right)^x$
100	2.70481382942
1000	2.71692393224
10,000	2.71814592683
100,000	2.71826823717
1,000,000	2.71828046932
1,000,000,000	2.71828182710

Answer

2.

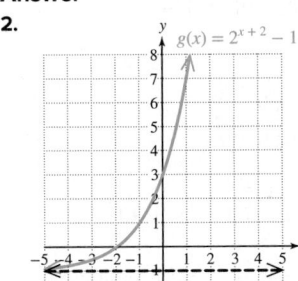

EXAMPLE 3 Graphing $f(x) = e^x$

Graph the function. $f(x) = e^x$

Solution:

Because the base e is greater than 1 ($e \approx 2.718281828$), the graph is an increasing exponential function. We can use a calculator to evaluate $f(x) = e^x$ at several values of x. On many calculators, the exponential function, base e, is invoked by selecting **2ND** **LN** or by accessing e^x on the keyboard.

```
NORMAL FLOAT AUTO a+bi DEGREE CL
e^(-3)
                    .0497870684
e^(-2)
                    .1353352832
e^(-1)
                    .3678794412
```
```
NORMAL FLOAT AUTO a+bi DEGREE CL
e^(1)
                    2.718281828
e^(2)
                    7.389056099
e^(3)
                    20.08553692
```

x	$f(x) = e^x$
-3	0.050
-2	0.135
-1	0.368
0	1.000
1	2.718
2	7.389
3	20.086

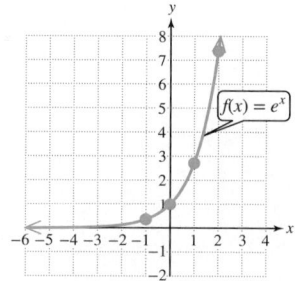

Figure 9-12

The graph of $f(x) = e^x$ is shown in Figure 9-12.

Skill Practice 3 Explain how the graph of $f(x) = -e^{x-1}$ is related to the graph of $y = e^x$.

3. Use Exponential Functions to Compute Compound Interest

Recall that simple interest is interest computed on the principal amount invested (or borrowed). Compound interest is interest computed on both the original principal and the interest already accrued.

Suppose that interest is compounded annually (one time per year) on an investment of P dollars at an annual interest rate r for t years. Then the amount A (in $) in the account after 1, 2, and 3 years is computed as follows.

After 1 year: $\begin{pmatrix} \text{Total} \\ \text{amount} \end{pmatrix} = \begin{pmatrix} \text{Initial} \\ \text{principal} \end{pmatrix} + (\text{Interest})$

> The interest is given by $I = Prt$, where $t = 1$ year. So $I = Pr$.

$A = P + Pr$

$= P(1 + r)$ Factor out P.

After 2 years: $\begin{pmatrix} \text{Total} \\ \text{amount} \end{pmatrix} = \begin{pmatrix} \text{Year 1} \\ \text{balance} \end{pmatrix} + \begin{pmatrix} \text{Interest on} \\ \text{Year 1 balance} \end{pmatrix}$

$A = P(1 + r) + [P(1 + r)]r$

$= P(1 + r)(1 + r)$ Factor out $P(1 + r)$.

$= P(1 + r)^2$

After 3 years: $\begin{pmatrix} \text{Total} \\ \text{amount} \end{pmatrix} = \begin{pmatrix} \text{Year 2} \\ \text{balance} \end{pmatrix} + \begin{pmatrix} \text{Interest on} \\ \text{Year 2 balance} \end{pmatrix}$

$A = P(1 + r)^2 + [P(1 + r)^2]r$

$= P(1 + r)^2(1 + r)$ Factor out $P(1 + r)^2$.

$= P(1 + r)^3$

$\cdots$

After t years: $A = P(1 + r)^t$

> Amount in an account with interest compounded annually.

Answer

3. The graph of $f(x) = -e^{x-1}$ is the graph of $y = e^x$ with a shift to the right 1 unit and a reflection across the x-axis.

Compound interest is often computed more frequently during the course of 1 year. Let n represent the number of compounding periods per year. For example:

$n = 1$ for interest compounded annually

$n = 4$ for interest compounded quarterly

$n = 12$ for interest compounded monthly

$n = 365$ for interest compounded daily

Each compounding period represents a fraction of a year, and the interest rate is scaled accordingly for each compounding period as $\frac{1}{n} \cdot r$ or $\frac{r}{n}$. The number of compounding periods over the course of the investment is nt. Therefore, to determine the amount in an account where interest is compounded n times per year, we have

replace t by nt

$$A = P(1 + r)^t \qquad\qquad A = P\left(1 + \frac{r}{n}\right)^{nt}$$

Amount in an account with interest compounded n times per year.

replace r by $\frac{r}{n}$

Now suppose it were possible to compute interest continuously, that is, for $n \to \infty$. If we use the substitution $x = \frac{n}{r}$ (which implies that $n = xr$), the formula for compound interest becomes

$$A = P\left(1 + \frac{r}{n}\right)^{nt} \xrightarrow{\text{Substitute } x = \frac{n}{r}} P\left(1 + \frac{1}{x}\right)^{xrt} = P\left[\left(1 + \frac{1}{x}\right)^x\right]^{rt}$$

For a fixed interest rate r, as n approaches infinity, x also approaches infinity. Since the expression $\left(1 + \frac{1}{x}\right)^x$ approaches e as $x \to \infty$, we have

$$A = Pe^{rt}$$

Amount in an account with interest compounded continuously.

Summary of Formulas Relating to Simple and Compound Interest

Suppose that P dollars in principal is invested (or borrowed) at an annual interest rate r for t years. Then

- $I = Prt$ — Amount of simple interest I (in \$).

- $A = P\left(1 + \frac{r}{n}\right)^{nt}$ — The future value A (in \$) of the account after t years with n compounding periods per year.

- $A = Pe^{rt}$ — The future value A (in \$) of the account after t years under continuous compounding.

In Example 4, we compare the value of an investment after 10 years under several different compounding options.

EXAMPLE 4   **Computing the Balance on an Account**

Suppose that \$5000 is invested and pays 6.5% per years under the following compounding options.

a. Compounded annually   **b.** Compounded quarterly

c. Compounded monthly   **d.** Compounded daily

e. Compounded continuously

Determine the total amount in the account after 10 years with each option.

Solution:

Using $A = P\left(1 + \dfrac{r}{n}\right)^{nt}$ and $A = Pe^{rt}$, we have

Compounding Option	n Value	Formula	Result
Annually	$n = 1$	$A = 5000\left(1 + \dfrac{0.065}{1}\right)^{(1 \cdot 10)}$	\$9385.69
Quarterly	$n = 4$	$A = 5000\left(1 + \dfrac{0.065}{4}\right)^{(4 \cdot 10)}$	\$9527.79
Monthly	$n = 12$	$A = 5000\left(1 + \dfrac{0.065}{12}\right)^{(12 \cdot 10)}$	\$9560.92
Daily	$n = 365$	$A = 5000\left(1 + \dfrac{0.065}{365}\right)^{(365 \cdot 10)}$	\$9577.15
Continuously	Not applicable	$A = 5000e^{(0.065 \cdot 10)}$	\$9577.70

FOR REVIEW

We first introduced the concept of simple interest $I = Prt$ in Section 1.3. For example, the simple interest I earned on \$5000 in principal P at 6.5% annual interest r for 10 years t is given by

$I = (\$5000)(0.065)(10)$
$= \$3250$

This results in a total of \$5000 + \$3250 = \$8250. Compare this amount to the results in Example 4.

Notice that there is a \$192.01 difference in the account balance between annual compounding and continuous compounding. The table also supports our finding that

$$A = P\left(1 + \frac{r}{n}\right)^{nt} \quad \text{converges to} \quad A = Pe^{rt} \quad \text{as } n \to \infty.$$

Skill Practice 4 Suppose that \$8000 is invested and pays 4.5% per year under the following compounding options.

a. Compounded annually   **b.** Compounded quarterly
c. Compounded monthly   **d.** Compounded daily
e. Compounded continuously

Determine the total amount in the account after 5 years with each option.

4. Use Exponential Functions in Applications

Increasing and decreasing exponential functions can be used in a variety of real-world applications. For example:

- Population growth can often be modeled by an exponential function.
- The growth of an investment under compound interest increases exponentially.
- The mass of a radioactive substance decreases exponentially with time.
- The temperature of a cup of coffee decreases exponentially as it approaches room temperature.

A substance that undergoes radioactive decay is said to be radioactive. The **half-life** of a radioactive substance is the amount of time it takes for one-half of the original amount of the substance to change into something else. That is, after each half-life, the amount of the original substance decreases by one-half.

Answers
4. a. \$9969.46   **b.** \$10,006.00
    c. \$10,014.37   **d.** \$10,018.44
    e. \$10,018.58

EXAMPLE 5   Using an Exponential Function in an Application

The half-life of radium 226 is 1620 years. In a sample originally having 1 g of radium 226, the amount $A(t)$ (in grams) of radium 226 present after t years is given by $A(t) = \left(\frac{1}{2}\right)^{t/1620}$ where t is the time in years after the start of the experiment. How much radium will be present after

a. 1620 years?   **b.** 3240 years?   **c.** 4860 years?

Solution:

a. $A(t) = \left(\frac{1}{2}\right)^{t/1620}$

$A(1620) = \left(\frac{1}{2}\right)^{1620/1620}$

$= \left(\frac{1}{2}\right)^{1}$

$= 0.5$

b. $A(t) = \left(\frac{1}{2}\right)^{t/1620}$

$A(3240) = \left(\frac{1}{2}\right)^{3240/1620}$

$= \left(\frac{1}{2}\right)^{2}$

$= 0.25$

c. $A(t) = \left(\frac{1}{2}\right)^{t/1620}$

$A(4860) = \left(\frac{1}{2}\right)^{4860/1620}$

$= \left(\frac{1}{2}\right)^{3}$

$= 0.125$

The half-life of radium is 1620 years. Therefore, we can interpret these results as follows:

After 1620 years (1 half-life), 0.5 g remains $\left(\frac{1}{2}\right.$ of the original amount remains$\left.\right)$.

After 3240 years (2 half-lives), 0.25 g remains $\left(\frac{1}{4}\right.$ of the original amount remains$\left.\right)$.

After 4860 years (3 half-lives), 0.125 g remains $\left(\frac{1}{8}\right.$ of the original amount remains$\left.\right)$.

Skill Practice 5   Cesium-137 is a radioactive metal with a short half-life of 30 years. In a sample originally having 2 g of cesium-137, the amount $A(t)$ (in grams) of cesium-137 present after t years is given by $A(t) = 2\left(\frac{1}{2}\right)^{t/30}$. How much cesium-137 will be present after

a. 30 years?   **b.** 60 years?   **c.** 90 years?

TECHNOLOGY CONNECTIONS

Graphing an Exponential Function

A graphing utility can be used to graph and analyze exponential functions. The table shows several values of $A(x) = \left(\frac{1}{2}\right)^{x/1620}$ for selected values of x.

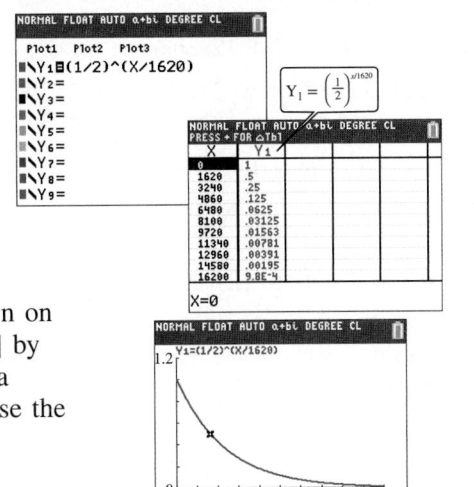

The graph of $A(x) = \left(\frac{1}{2}\right)^{x/1620}$ is shown on the viewing window [0, 10,000, 1000] by [0, 1.2, 0.2]. Notice that the graph is a decreasing exponential function because the base is between 0 and 1.

Answers
5. a. 1 g   **b.** 0.5 g   **c.** 0.25 g

SECTION 9.2   Practice Exercises

Prerequisite Review

For Exercises R.1–R.4, simplify the expressions. (See Section 2.1 for review.)

R.1. a. 5^2 **b.** 5^0 **c.** 5^{-2} **R.2. a.** 6^2 **b.** 6^0 **c.** 6^{-2}

R.3. a. $\left(\dfrac{1}{5}\right)^2$ **b.** $\left(\dfrac{1}{5}\right)^0$ **c.** $\left(\dfrac{1}{5}\right)^{-2}$ **R.4. a.** $\left(\dfrac{1}{6}\right)^2$ **b.** $\left(\dfrac{1}{6}\right)^0$ **c.** $\left(\dfrac{1}{6}\right)^{-2}$

For Exercises R.5–R.6, refer to the graph of the function and complete the statements. (See Section 8.1 for review.)

R.5.

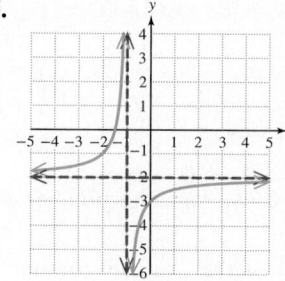

a. As $x \to -\infty, f(x) \to$ _____

b. As $x \to -1^+, f(x) \to$ _____

c. As $x \to \infty, f(x) \to$ _____

d. As $x \to -1^-, f(x) \to$ _____

e. The graph is decreasing over the interval(s)
_____.

f. The graph is increasing over the interval(s)
_____.

g. The domain is _____.

h. The range is _____.

R.6.

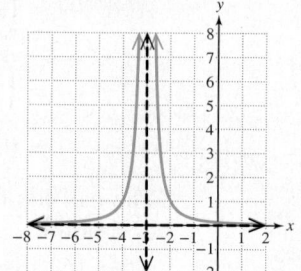

a. As $x \to -\infty, f(x) \to$ _____

b. As $x \to -3^+, f(x) \to$ _____

c. As $x \to \infty, f(x) \to$ _____

d. As $x \to -3^-, f(x) \to$ _____

e. The graph is decreasing over the interval(s)
_____.

f. The graph is increasing over the interval(s)
_____.

g. The domain is _____.

h. The range is _____.

For Exercises R.7–R.8, describe the graph of g as it relates to the graph of $y = f(x)$. (See Section 6.1 for review.)

R.7. $g(x) = 2f(x - 1) + 4$ **R.8.** $g(x) = -f(x + 1) - 3$

For Exercises R.9–R.10, graph the function. (See Section 6.1 for review.)

R.9. $q(x) = -(x - 3)^2 + 4$ **R.10.** $p(x) = 2|x + 2| - 3$

For Exercises R.11–R.12, compute the simple interest for the given scenario. (See Section 1.3 for review.)

R.11. Jorge borrows $12,000 from his dad for a used car and pays his father back at 4% simple interest for $5\dfrac{1}{2}$ years. How much will Jorge pay in interest?

R.12. Alissa invests $10,000 in a 9-month certificate of deposit that earns 1.45% annual interest. How much interest will she earn?

Concept Connections

1. The function defined by $y = x^3$ (is/is not) an exponential function, whereas the function defined by $y = 3^x$ (is/is not) an exponential function.

2. The graph of $f(x) = \left(\dfrac{5}{3}\right)^x$ is (increasing/decreasing) over its domain.

3. The graph of $f(x) = \left(\dfrac{3}{5}\right)^x$ is (increasing/decreasing) over its domain.

4. The domain of an exponential function $f(x) = b^x$ is _____.

5. The range of an exponential function $f(x) = b^x$ is _____.

6. All exponential functions $f(x) = b^x$ pass through the point _____.

7. The horizontal asymptote of an exponential function $f(x) = b^x$ is the line _____.

8. As $x \to \infty$, the value of $\left(1 + \dfrac{1}{x}\right)^x$ approaches _____.

Objective 1: Graph Exponential Functions

For Exercises 9–12, evaluate the functions at the given values of x. Round to 4 decimal places if necessary.

9. $f(x) = 5^x$
 a. $f(-1)$
 b. $f(4.8)$
 c. $f(\sqrt{2})$
 d. $f(\pi)$

10. $g(x) = 7^x$
 a. $g(-2)$
 b. $g(5.9)$
 c. $g(\sqrt{11})$
 d. $g(e)$

11. $h(x) = \left(\dfrac{1}{4}\right)^x$
 a. $h(-3)$
 b. $h(1.4)$
 c. $h(\sqrt{3})$
 d. $h(0.5e)$

12. $k(x) = \left(\dfrac{1}{6}\right)^x$
 a. $k(-3)$
 b. $k(1.4)$
 c. $k(\sqrt{0.5})$
 d. $k(0.5\pi)$

13. Which functions are exponential functions?
 a. $f(x) = 4.2^x$ **b.** $g(x) = x^{4.2}$ **c.** $h(x) = 4.2x$
 d. $k(x) = (\sqrt{4.2})^x$ **e.** $m(x) = (-4.2)^x$

14. Which functions are exponential functions?
 a. $v(x) = (-\pi)^x$ **b.** $t(x) = \pi^x$ **c.** $w(x) = \pi x$
 d. $n(x) = (\sqrt{\pi})^x$ **e.** $p(x) = x^\pi$

For Exercises 15–22, graph the functions and write the domain and range in interval notation. **(See Example 1)**

15. $f(x) = 3^x$

16. $g(x) = 4^x$

17. $h(x) = \left(\dfrac{1}{3}\right)^x$

18. $k(x) = \left(\dfrac{1}{4}\right)^x$

19. $m(x) = \left(\dfrac{3}{2}\right)^x$

20. $n(x) = \left(\dfrac{5}{4}\right)^x$

21. $b(x) = \left(\dfrac{2}{3}\right)^x$

22. $c(x) = \left(\dfrac{4}{5}\right)^x$

For Exercises 23–32,
 a. Use transformations of the graphs of $y = 3^x$ (see Exercise 15) and $y = 4^x$ (see Exercise 16) to graph the given function. **(See Example 2)**
 b. Write the domain and range in interval notation.
 c. Write an equation of the asymptote.

23. $f(x) = 3^x + 2$

24. $g(x) = 4^x - 3$

25. $m(x) = 3^{x+2}$

26. $n(x) = 4^{x-3}$

27. $p(x) = 3^{x-4} - 1$

28. $q(x) = 4^{x+1} + 2$

29. $k(x) = -3^x$

30. $h(x) = -4^x$

31. $t(x) = 3^{-x}$

32. $v(x) = 4^{-x}$

For Exercises 33–36,
 a. Use transformations of the graphs of $y = \left(\frac{1}{3}\right)^x$ (see Exercise 17) and $y = \left(\frac{1}{4}\right)^x$ (see Exercise 18) to graph the given function. **(See Example 2)**
 b. Write the domain and range in interval notation.
 c. Write an equation of the asymptote.

33. $f(x) = \left(\dfrac{1}{3}\right)^{x+1} - 3$

34. $g(x) = \left(\dfrac{1}{4}\right)^{x-2} + 1$

35. $k(x) = -\left(\dfrac{1}{3}\right)^x + 2$

36. $h(x) = -\left(\dfrac{1}{4}\right)^x - 2$

Objective 2: Evaluate the Exponential Function Base e

For Exercises 37–38, evaluate the functions for the given values of x. Round to 4 decimal places.

37. $f(x) = e^x$

    a. $f(4)$ **b.** $f(-3.2)$

    c. $f(\sqrt{13})$ **d.** $f(\pi)$

38. $f(x) = e^x$

    a. $f(-3)$ **b.** $f(6.8)$

    c. $f(\sqrt{7})$ **d.** $f(e)$

For Exercises 39–44,

    a. Use transformations of the graph of $y = e^x$ to graph the given function. (**See Example 3**)

    b. Write the domain and range in interval notation.

    c. Write an equation of the asymptote.

39. $f(x) = e^{x-4}$ **40.** $g(x) = e^{x-2}$ **41.** $h(x) = e^x + 2$

42. $k(x) = e^x - 1$ **43.** $m(x) = -e^x - 3$ **44.** $n(x) = -e^x + 4$

Objective 3: Use Exponential Functions to Compute Compound Interest

For Exercises 45–46, complete the table to determine the effect of the number of compounding periods when computing interest. (**See Example 4**)

45. Suppose that $10,000 is invested at 4% interest for 5 years under the following compounding options. Complete the table.

	Compounding Option	*n* Value	Result
a.	Annually		
b.	Quarterly		
c.	Monthly		
d.	Daily		
e.	Continuously		

46. Suppose that $8000 is invested at 3.5% interest for 20 years under the following compounding options. Complete the table.

	Compounding Option	*n* Value	Result
a.	Annually		
b.	Quarterly		
c.	Monthly		
d.	Daily		
e.	Continuously		

For Exercises 47–48, suppose that P dollars in principal is invested for t years at the given interest rates with continuous compounding. Determine the amount that the investment is worth at the end of the given time period.

47. $P = \$20,000$, $t = 10$ years

    a. 3% interest

    b. 4% interest

    c. 5.5% interest

48. $P = \$6000$, $t = 12$ years

    a. 1% interest

    b. 2% interest

    c. 4.5% interest

49. Bethany needs to borrow $10,000. She can borrow the money at 5.5% simple interest for 4 years or she can borrow at 5% with interest compounded continuously for 4 years.

    a. How much total interest would Bethany pay at 5.5% simple interest?

    b. How much total interest would Bethany pay at 5% interest compounded continuously?

    c. Which option results in less total interest?

50. Al needs to borrow $15,000 to buy a car. He can borrow the money at 6.7% simple interest for 5 years or he can borrow at 6.4% interest compounded continuously for 5 years.

    a. How much total interest would Al pay at 6.7% simple interest?

    b. How much total interest would Al pay at 6.4% interest compounded continuously?

    c. Which option results in less total interest?

51. Jerome wants to invest $25,000 as part of his retirement plan. He can invest the money at 5.2% simple interest for 30 years, or he can invest at 3.8% interest compounded continuously for 30 years. Which option results in more total interest?

52. Heather wants to invest $35,000 of her retirement. She can invest at 4.8% simple interest for 20 years, or she can choose an option with 3.6% interest compounded continuously for 20 years. Which option results in more total interest?

Objective 4: Use Exponential Functions in Applications

53. Strontium-90 (^{90}Sr) is a by-product of nuclear fission with a half-life of approximately 28.9 years. After the Chernobyl nuclear reactor accident in 1986, large areas surrounding the site were contaminated with ^{90}Sr. If 10 μg (micrograms) of ^{90}Sr is present in a sample, the function $A(t) = 10\left(\dfrac{1}{2}\right)^{t/28.9}$ gives the amount $A(t)$ (in μg) present after t years. Evaluate the function for the given values of t and interpret the meaning in context. Round to 3 decimal places if necessary. **(See Example 5)**

   a. $A(28.9)$ **b.** $A(57.8)$ **c.** $A(100)$

54. In 2006, the murder of Alexander Litvinenko, a Russian dissident, was thought to be by poisoning from the rare and highly radioactive element polonium-210 (^{210}Po). The half-life of ^{210}Po is 138.4 years. If 0.1 mg of ^{210}Po is present in a sample, then $A(t) = 0.1\left(\dfrac{1}{2}\right)^{t/138.4}$ gives the amount $A(t)$ (in mg) present after t years. Evaluate the function for the given values of t and interpret the meaning in context. Round to 3 decimal places if necessary.

   a. $A(138.4)$ **b.** $A(276.8)$ **c.** $A(500)$

55. According to the CIA's *World Fact Book*, in 2010, the population of the United States was approximately 310 million with a 0.97% annual growth rate. (*Source*: www.cia.gov) At this rate, the population $P(t)$ (in millions) can be approximated by $P(t) = 310(1.0097)^t$, where t is the time in years since 2010.

   a. Is the graph of P an increasing or decreasing exponential function?

   b. Evaluate $P(0)$ and interpret its meaning in the context of this problem.

   c. Evaluate $P(10)$ and interpret its meaning in the context of this problem. Round the population value to the nearest million.

   d. Evaluate $P(20)$ and $P(30)$.

   e. Evaluate $P(200)$ and use this result to determine if it is reasonable to expect this model to continue indefinitely.

56. The population of Canada in 2010 was approximately 34 million with an annual growth rate of 0.804%. At this rate, the population $P(t)$ (in millions) can be approximated by $P(t) = 34(1.00804)^t$, where t is the time in years since 2010. (*Source*: www.cia.gov)

   a. Is the graph of P an increasing or decreasing exponential function?

   b. Evaluate $P(0)$ and interpret its meaning in the context of this problem.

   c. Evaluate $P(5)$ and interpret its meaning in the context of this problem. Round the population value to the nearest million.

   d. Evaluate $P(15)$ and $P(25)$.

57. The atmospheric pressure on an object decreases as altitude increases. If a is the height (in km) above sea level, then the pressure $P(a)$ (in mmHg) is approximated by $P(a) = 760e^{-0.13a}$.

   a. Find the atmospheric pressure at sea level.

   b. Determine the atmospheric pressure at 8.848 km (the altitude of Mt. Everest). Round to the nearest whole unit.

58. The function defined by $A(t) = 100e^{0.0318t}$ approximates the equivalent amount of money needed t years after the year 2010 to equal $100 of buying power in the year 2010. The value 0.0318 is related to the average rate of inflation.

   a. Evaluate $A(15)$ and interpret its meaning in the context of this problem.

   b. Verify that by the year 2032, more than $200 will be needed to have the same buying power as $100 in 2010.

Tyler Stableford/Brand X Pictures/SuperStock.

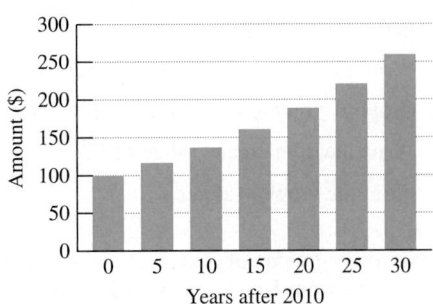

Newton's law of cooling indicates that the temperature of a warm object, such as a cake coming out of the oven, will decrease exponentially with time and will approach the temperature of the surrounding air. The temperature $T(t)$ is modeled by $T(t) = T_a + (T_0 - T_a)e^{-kt}$. In this model, T_a represents the temperature of the surrounding air, T_0 represents the initial temperature of the object, and t is the time after the object starts cooling. The value of k is a constant of proportion relating the temperature of the object to its rate of temperature change. Use this model for Exercises 59–60.

59. A cake comes out of the oven at 350°F and is placed on a cooling rack in a 78°F kitchen. After checking the temperature several minutes later, the value of k is measured as 0.046.

   a. Write a function that models the temperature $T(t)$ (in °F) of the cake t minutes after being removed from the oven.

   b. What is the temperature of the cake 10 min after coming out of the oven? Round to the nearest degree.

   c. It is recommended that the cake should not be frosted until it has cooled to under 100°F. If Jessica waits 1 hr to frost the cake, will the cake be cool enough to frost?

61. A farmer depreciates a $120,000 tractor. He estimates that the resale value $V(t)$ (in $1000) of the tractor t years after purchase is 80% of its value from the previous year. Therefore, the resale value can be approximated by $V(t) = 120(0.8)^t$.

   a. Find the resale value 5 years after purchase. Round to the nearest $1000.

   b. The farmer estimates that the cost to run the tractor is $18/hr in labor, $36/hr in fuel, and $22/hr in overhead costs (for maintenance and repair). Estimate the farmer's cost to run the tractor for the first year if he runs the tractor for a total of 800 hr. Include hourly costs and depreciation.

60. Water in a water heater is originally 122°F. The water heater is shut off and the water cools to the temperature of the surrounding air, which is 60°F. The water cools slowly because of the insulation inside the heater, and the value of k is measured as 0.00351.

   a. Write a function that models the temperature $T(t)$ (in °F) of the water t hours after the water heater is shut off.

   b. What is the temperature of the water 12 hr after the heater is shut off? Round to the nearest degree.

   c. Dominic does not like to shower with water less than 115°F. If Dominic waits 24 hr, will the water still be warm enough for a shower?

62. A veterinarian depreciates a $10,000 X-ray machine. He estimates that the resale value $V(t)$ (in $) after t years is 90% of its value from the previous year. Therefore, the resale value can be approximated by $V(t) = 10,000(0.9)^t$.

   a. Find the resale value after 4 years.

   b. If the veterinarian wants to sell his practice 8 years after the X-ray machine was purchased, how much is the machine worth? Round to the nearest $100.

Mixed Exercises

For Exercises 63–64, solve the equations in parts (a)–(c) by inspection. Then estimate the solutions to parts (d) and (e) between two consecutive integers.

63. **a.** $2^x = 4$
   b. $2^x = 8$
   c. $2^x = 16$
   d. $2^x = 7$
   e. $2^x = 10$

64. **a.** $3^x = 3$
   b. $3^x = 9$
   c. $3^x = 27$
   d. $3^x = 7$
   e. $3^x = 10$

65. **a.** Graph $f(x) = 2^x$. **(See Example 1)**
   b. Is f a one-to-one function?
   c. Write the domain and range of f in interval notation.
   d. Graph f^{-1} on the same coordinate system as f.
   e. Write the domain and range of f^{-1} in interval notation.
   f. From the graph evaluate $f^{-1}(1), f^{-1}(2)$, and $f^{-1}(4)$.

66. **a.** Graph $g(x) = 3^x$ (see Exercise 15).
   b. Is g a one-to-one function?
   c. Write the domain and range of g in interval notation.
   d. Graph g^{-1} on the same coordinate system as g.
   e. Write the domain and range of g^{-1} in interval notation.
   f. From the graph evaluate $g^{-1}(1), g^{-1}(3)$, and $g^{-1}\left(\frac{1}{3}\right)$.

67. Refer to the graphs of $f(x) = 2^x$ and the inverse function, $y = f^{-1}(x)$ from Exercise 65. Fill in the blanks.

   a. As $x \to \infty$, $f(x) \to$ _____.
   b. As $x \to -\infty$, $f(x) \to$ _____.
   c. As $x \to \infty$, $f^{-1}(x) \to$ _____.
   d. As $x \to 0^+$, $f^{-1}(x) \to$ _____.

68. Refer to the graphs of $g(x) = 3^x$ and the inverse function, $y = g^{-1}(x)$ from Exercise 66. Fill in the blanks.

   a. As $x \to \infty$, $g(x) \to$ _____.
   b. As $x \to -\infty$, $g(x) \to$ _____.
   c. As $x \to \infty$, $g^{-1}(x) \to$ _____.
   d. As $x \to 0^+$, $g^{-1}(x) \to$ _____.

Write About It

69. Explain why the equation $2^x = -2$ has no solution.

70. Explain why the $f(x) = x^2$ is not an exponential function.

Expanding Your Skills

For Exercises 71–72, find the real solutions to the equation.

71. $3x^2 e^{-x} - 6xe^{-x} = 0$

72. $x^2 e^x - e^x = 0$

73. Use the properties of exponents to simplify.

    a. $e^x e^h$ **b.** $(e^x)^2$ **c.** $\dfrac{e^x}{e^h}$

    d. $e^x \cdot e^{-x}$ **e.** e^{-2x}

74. Factor.

    a. $e^{x+h} - e^x$

    b. $e^{4x} - e^{2x}$

75. Multiply. $(e^x + e^{-x})^2$

76. Multiply. $(e^x - e^{-x})^2$

77. Show that $\left(\dfrac{e^x + e^{-x}}{2}\right)^2 - \left(\dfrac{e^x - e^{-x}}{2}\right)^2 = 1$.

78. Show that $2\left(\dfrac{e^x - e^{-x}}{2}\right)\left(\dfrac{e^x + e^{-x}}{2}\right) = \dfrac{e^{2x} - e^{-2x}}{2}$.

For Exercises 79–80, find the difference quotient $\dfrac{f(x + h) - f(x)}{h}$. Write the answers in factored form.

79. $f(x) = e^x$

80. $f(x) = 2^x$

Technology Connections

81. Graph the following functions on the window $[-3, 3, 1]$ by $[-1, 8, 1]$ and comment on the behavior of the graphs near $x = 0$.

    $Y_1 = e^x$

    $Y_2 = 1 + x + \dfrac{x^2}{2}$

    $Y_3 = 1 + x + \dfrac{x^2}{2} + \dfrac{x^3}{6}$

SECTION 9.3   Logarithmic Functions

OBJECTIVES

1. Convert Between Logarithmic and Exponential Forms
2. Evaluate Logarithmic Expressions
3. Apply Basic Properties of Logarithms
4. Graph Logarithmic Functions
5. Use Logarithmic Functions in Applications

1. Convert Between Logarithmic and Exponential Forms

Consider the following equations in which the variable is located in the exponent of an expression. In some cases, the solution can be found by inspection.

Equation	Solution
$5^x = 5$	$x = 1$
$5^x = 20$	$x = ?$
$5^x = 25$	$x = 2$
$5^x = 60$	$x = ?$
$5^x = 125$	$x = 3$

The equation $5^x = 20$ cannot be solved by inspection. However, we suspect that x is between 1 and 2 because $5^1 = 5$ and $5^2 = 25$. To solve for x explicitly, we must isolate x by performing the inverse operation of 5^x. Fortunately, all exponential functions $y = b^x$ ($b > 0$, $b \neq 1$) are one-to-one and therefore have inverse functions. The inverse of an exponential function, base b, is the *logarithmic* function base b which we define here.

> ### Definition of a Logarithmic Function
>
> If x and b are positive real numbers such that $b \neq 1$, then $y = \log_b x$ is called the **logarithmic function base b**, where
>
> $$y = \log_b x \text{ is equivalent to } b^y = x$$
>
> ---
>
> *Notes*:
>
> - Given $y = \log_b x$, the value y is the exponent to which b must be raised to obtain x.
> - The value of y is called the **logarithm**, b is called the **base**, and x is called the **argument**.
> - The equations $y = \log_b x$ and $b^y = x$ both define the same relationship between x and y. The expression $y = \log_b x$ is called the **logarithmic form**, and $b^y = x$ is called the **exponential form**.

The logarithmic function base b is defined as the inverse of the exponential function base b.

exponential function	$f(x) = b^x$	First replace $f(x)$ by y.
	$y = b^x$	Next, interchange x and y.
inverse of exponential function	$x = b^y$	This equation provides an implicit relationship between x and y. To solve for y explicitly (that is, to isolate y), we
logarithmic function	$y = \log_b x$	must use logarithmic notation.

To be able to solve equations involving logarithms, it is often advantageous to write a logarithmic expression in its exponential form.

EXAMPLE 1 Writing Logarithmic Form and Exponential Form

Write each equation in exponential form.

a. $\log_2 16 = 4$ **b.** $\log_{10}\left(\dfrac{1}{100}\right) = -2$ **c.** $\log_7 1 = 0$

Solution:

Logarithmic form $y = \log_b x$ **Exponential form** $b^y = x$

a. $\log_2 16 = 4$ $\Leftrightarrow$ $2^4 = 16$

> The logarithm is the exponent to which the base is raised to obtain x.

b. $\log_{10}\left(\dfrac{1}{100}\right) = -2$ $\Leftrightarrow$ $10^{-2} = \dfrac{1}{100}$

c. $\log_7 1 = 0$ $\Leftrightarrow$ $7^0 = 1$

Skill Practice 1 Write each equation in exponential form.

a. $\log_3 9 = 2$ **b.** $\log_{10}\left(\dfrac{1}{1000}\right) = -3$ **c.** $\log_6 1 = 0$

Answers

1. a. $3^2 = 9$ **b.** $10^{-3} = \dfrac{1}{1000}$

     c. $6^0 = 1$

In Example 2 we reverse this process and write an exponential equation in its logarithmic form.

EXAMPLE 2   Writing Exponential Form and Logarithmic Form

Write each equation in logarithmic form.

a. $3^4 = 81$ **b.** $10^6 = 1,000,000$ **c.** $\left(\dfrac{1}{5}\right)^{-1} = 5$

Solution:

Exponential form $b^y = x$ **Logarithmic form** $\log_b x = y$

logarithm   logarithm (power)

a. $3^4 = 81$ $\Leftrightarrow$ $\log_3 81 = 4$

base   argument   base   argument

b. $10^6 = 1,000,000$ $\Leftrightarrow$ $\log_{10} 1,000,000 = 6$

c. $\left(\dfrac{1}{5}\right)^{-1} = 5$ $\Leftrightarrow$ $\log_{1/5} 5 = -1$

Skill Practice 2   Write each equation in logarithmic form.

a. $2^5 = 32$ **b.** $10^4 = 10,000$ **c.** $\left(\dfrac{1}{8}\right)^{-2} = 64$

---

2. Evaluate Logarithmic Expressions

To evaluate a logarithmic expression, we can write the expression in exponential form. Then we make use of the equivalence property of exponential expressions. This states that if two exponential expressions of the same base are equal, then their exponents must be equal.

Equivalence Property of Exponential Expressions

If b, x, and y are real numbers, with $b > 0$ and $b \neq 1$, then

$$b^x = b^y \text{ implies that } x = y.$$

In Example 3, we evaluate several logarithmic expressions.

EXAMPLE 3   Evaluating Logarithmic Expressions

Evaluate each expression.

a. $\log_4 16$ **b.** $\log_2 8$ **c.** $\log_{1/2} 8$

Answers

2. a. $\log_2 32 = 5$
   b. $\log_{10} 10,000 = 4$
   c. $\log_{1/8} 64 = -2$

Solution:

Let y represent the value of the logarithm.

a. $\log_4 16$ is the exponent to which 4 must be raised to equal 16. That is, $4^{\square} = 16$.

$$\log_4 16 = y$$

$\quad\quad 4^y = 16$ or equivalently $4^y = 4^2$ Write the equivalent exponential form.

$\quad\quad y = 2$

Therefore, $\log_4 16 = 2$.   Check: $4^2 = 16$ ✓

b. $\log_2 8$ is the exponent to which 2 must be raised to obtain 8. That is, $2^{\square} = 8$.

$$\log_2 8 = y$$

$\quad\quad 2^y = 8$ or equivalently $2^y = 2^3$ Write the equivalent exponential form.

$\quad\quad y = 3$

Therefore, $\log_2 8 = 3$.   Check: $2^3 = 8$ ✓

c. $\log_{1/2} 8 = y$

$\quad\quad \left(\frac{1}{2}\right)^y = 8$ or equivalently $\left(\frac{1}{2}\right)^y = \left(\frac{1}{2}\right)^{-3}$ Write the equivalent exponential form.

$\quad\quad y = -3$

Therefore, $\log_{1/2} 8 = -3$.   Check: $\left(\frac{1}{2}\right)^{-3} = 8$ ✓

TIP Once you become comfortable with the concept of a logarithm, you can take fewer steps to evaluate a logarithm.

To evaluate the expression $\log_4 16$ we ask $4^{\square} = 16$. The exponent is 2, so $\log_4 16 = 2$.

Likewise, to evaluate $\log_2 8$ we ask $2^{\square} = 8$. So $\log_2 8 = 3$.

> **Skill Practice 3** Evaluate each expression.
>
> **a.** $\log_5 125$ **b.** $\log_3 81$ **c.** $\log_4 \left(\dfrac{1}{64}\right)$

The statement $y = \log_b x$ represents a family of logarithmic functions where the base is any positive real number except 1. Two specific logarithmic functions that come up often in applications are the logarithmic functions base 10 and base e.

Definition of Common and Natural Logarithmic Functions

- The logarithmic function base 10 is called the **common logarithmic function**. The common logarithmic function is denoted by $y = \log x$. Notice that the base 10 is not explicitly written; that is, $y = \log_{10} x$ is written simply as $y = \log x$.
- The logarithmic function base e is called the **natural logarithmic function**. The natural logarithmic function is denoted by $y = \ln x$; that is, $y = \log_e x$ is written as $y = \ln x$.

TIP To help you remember the notation $y = \ln x$, think of "ln" as "log natural."

EXAMPLE 4   Evaluating Common and Natural Logarithms

Evaluate.

a. $\log 100{,}000$ **b.** $\log 0.001$ **c.** $\ln e^4$ **d.** $\ln \left(\dfrac{1}{e}\right)$

Answers

3. a. 3   **b.** 4   **c.** −3

Solution:

Let y represent the value of the logarithm.

a. $\log 100{,}000 = y$

$\quad 10^y = 100{,}000$ or equivalently $10^y = 10^5$ Write the exponential form.

$\qquad y = 5$

Thus, $\log 100{,}000 = 5$ because $10^5 = 100{,}000$.

b. $\log 0.001 = y$

$\quad 10^y = 0.001$ or equivalently $10^y = 10^{-3}$ Write the exponential form.

$\qquad y = -3$

Thus, $\log 0.001 = -3$ because $10^{-3} = 0.001$.

c. $\ln e^4 = y$

$\quad e^y = e^4$ Write the equivalent exponential form.

$\quad y = 4$

Therefore, $\ln e^4 = 4$.

d. $\ln\left(\dfrac{1}{e}\right) = y$

$\quad e^y = \left(\dfrac{1}{e}\right)$ or equivalently $e^y = e^{-1}$ Write the equivalent exponential form.

$\quad y = -1$

Therefore, $\ln\left(\dfrac{1}{e}\right) = -1$.

> **TIP** In Example 4, to evaluate:
>
> **a.** log 100,000 we ask $10^{\square} = 100{,}000$.
>
> **b.** log 0.001 we ask $10^{\square} = 0.001$.
>
> **c.** ln e^4 we ask $e^{\square} = e^4$.
>
> **d.** $\ln\left(\dfrac{1}{e}\right)$ we ask $e^{\square} = e^{-1}$.

Skill Practice 4   Evaluate.

    a. $\log 10{,}000{,}000$ **b.** $\log 0.1$ **c.** $\ln e^5$ **d.** $\ln e$

Most scientific calculators have a key for the common logarithmic function **LOG** and a key for the natural logarithmic function **LN**. We demonstrate their use in Example 5.

EXAMPLE 5   **Approximating Common and Natural Logarithms**

Approximate the logarithms.

    a. $\log 5809$ **b.** $\log(4.6 \times 10^7)$ **c.** $\log 0.003$

    d. $\ln 472$ **e.** $\ln 0.05$ **f.** $\ln\sqrt{87}$

Solution:

For parts (a)–(c), use the **LOG** key.   For parts (d)–(f), use the **LN** key.

> **TIP** There are an infinite number of logarithmic functions. The calculator has keys for the base 10 and base e logarithmic functions only.
> To approximate a logarithm of different base, we need to use the change-of-base formula presented in Section 9.4.

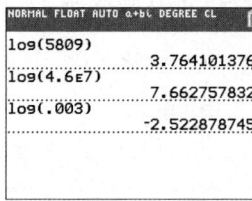

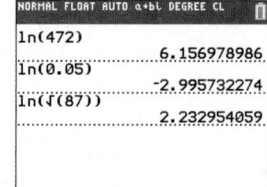

Answers

4. a. 7   **b.** −1   **c.** 5   **d.** 1

When using a calculator, there is always potential for user-input error. Therefore, it is good practice to estimate values when possible to confirm the reasonableness of an answer from a calculator. For example,

For part (a), $10^3 < 5809 < 10^4$. Therefore, $3 < \log 5809 < 4$.
For part (b), $10^7 < 4.6 \times 10^7 < 10^8$. Therefore, $7 < \log(4.6 \times 10^7) < 8$.
For part (c), $10^{-3} < 0.003 < 10^{-2}$. Therefore, $-3 < \log 0.003 < -2$.

Skill Practice 5 Approximate the logarithms. Round to 4 decimal places.
   a. $\log 229$ **b.** $\log(3.76 \times 10^{12})$ **c.** $\log 0.0216$
   d. $\ln 87$ **e.** $\ln 0.0032$ **f.** $\ln \pi$

3. Apply Basic Properties of Logarithms

From the definition of a logarithmic function, we have the following basic properties.

Basic Properties of Logarithms

Property	Example
1. $\log_b 1 = 0$ because $b^0 = 1$	$\log_5 1 = 0$ because $5^0 = 1$
2. $\log_b b = 1$ because $b^1 = b$	$\log_3 3 = 1$ because $3^1 = 3$
3. $\log_b b^x = x$ because $b^x = b^x$	$\log_2 2^x = x$ because $2^x = 2^x$
4. $b^{\log_b x} = x$ because $\log_b x = \log_b x$	$7^{\log_7 x} = x$ because $\log_7 x = \log_7 x$

Properties 3 and 4 follow from the fact that a logarithmic function is the inverse of an exponential function of the same base. Given $f(x) = b^x$ and $f^{-1}(x) = \log_b x$,

$$(f \circ f^{-1})(x) = b^{(\log_b x)} = x \qquad \text{(Property 4)}$$
$$(f^{-1} \circ f)(x) = \log_b(b^x) = x \qquad \text{(Property 3)}$$

EXAMPLE 6 Applying the Properties of Logarithms

Simplify.
   a. $\log_3 3^{10}$ **b.** $\ln e^2$ **c.** $\log_{11} 11$ **d.** $\log 10$
   e. $\log_{\sqrt{7}} 1$ **f.** $\ln 1$ **g.** $5^{\log_5(c^2+4)}$ **h.** $10^{\log(a^2+b^2)}$

Solution:

a. $\log_3 3^{10} = 10$ Property 3   **b.** $\ln e^2 = \log_e e^2 = 2$ Property 3
c. $\log_{11} 11 = 1$ Property 2   **d.** $\log 10 = \log_{10} 10 = 1$ Property 2
e. $\log_{\sqrt{7}} 1 = 0$ Property 1   **f.** $\ln 1 = 0$ Property 1
g. $5^{\log_5(c^2+4)} = c^2 + 4$ Property 4   **h.** $10^{\log(a^2+b^2)} = a^2 + b^2$ Property 4

Skill Practice 6 Simplify.
   a. $\log_{13} 13$ **b.** $\ln e$ **c.** $a^{\log_a 3}$ **d.** $e^{\ln 6}$
   e. $\log_\pi 1$ **f.** $\log 1$ **g.** $\log_9 9^{\sqrt{2}}$ **h.** $\log 10^e$

Answers
5. a. 2.3598   **b.** 12.5752
   c. −1.6655   **d.** 4.4659
   e. −5.7446   **f.** 1.1447
6. a. 1   **b.** 1   **c.** 3   **d.** 6
   e. 0   **f.** 0   **g.** $\sqrt{2}$ **h.** e

4. Graph Logarithmic Functions

Since a logarithmic function $y = \log_b x$ is the inverse of the corresponding exponential function $y = b^x$, their graphs must be symmetric with respect to the line $y = x$. See Figures 9-13 and 9-14.

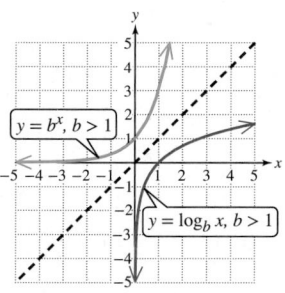

Figure 9-13

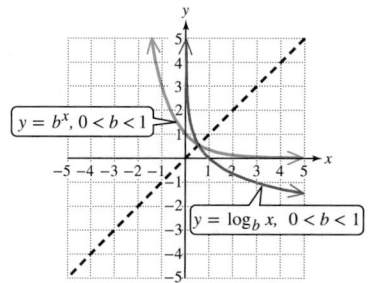

Figure 9-14

From Figures 9-13 and 9-14, the range of $y = b^x$ is the set of positive real numbers. As expected, the domain of its inverse function $y = \log_b x$ is the set of positive real numbers.

EXAMPLE 7   **Graphing Logarithmic Functions**

Graph the functions.

a. $y = \log_2 x$ **b.** $y = \log_{1/4} x$

Solution:

To find points on a logarithmic function, we can interchange the x- and y-coordinates of the ordered pairs on the corresponding exponential function.

a. To graph $y = \log_2 x$, interchange the x- and y-coordinates of the ordered pairs from its inverse function $y = 2^x$. The graph of $y = \log_2 x$ is shown in Figure 9-15.

Exponential Function

x	$y = 2^x$
-3	$\frac{1}{8}$
-2	$\frac{1}{4}$
-1	$\frac{1}{2}$
0	1
1	2
2	4
3	8

Logarithmic Function

x	$y = \log_2 x$
$\frac{1}{8}$	-3
$\frac{1}{4}$	-2
$\frac{1}{2}$	-1
1	0
2	1
4	2
8	3

Switch x and y.

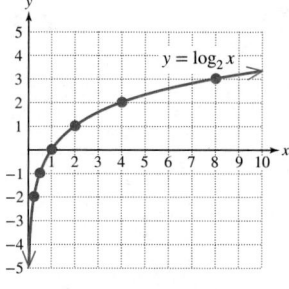

Figure 9-15

b. To graph $y = \log_{1/4} x$, interchange the x- and y-coordinates of the ordered pairs from its inverse function $y = \left(\frac{1}{4}\right)^x$. See Figure 9-16.

Exponential Function   **Logarithmic Function**

x	$y = \left(\dfrac{1}{4}\right)^x$
-3	64
-2	16
-1	4
0	1
1	$\frac{1}{4}$
2	$\frac{1}{16}$
3	$\frac{1}{64}$

x	$y = \log_{1/4} x$
64	-3
16	-2
4	-1
1	0
$\frac{1}{4}$	1
$\frac{1}{16}$	2
$\frac{1}{64}$	3

Switch x and y.

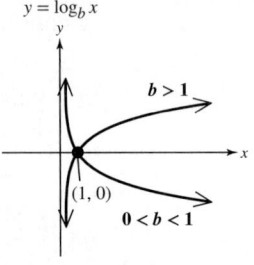

Figure 9-16

Skill Practice 7   Graph the functions.

a. $y = \log_4 x$ **b.** $y = \log_{1/2} x$

Based on the graphs in Example 7 and our knowledge of exponential functions, we offer the following summary of the characteristics of logarithmic and exponential functions.

Graphs of Exponential and Logarithmic Functions

Exponential Functions

$y = b^x$

Domain: $(-\infty, \infty)$
Range: $(0, \infty)$
Horizontal asymptote: $y = 0$
Passes through $(0, 1)$
If $b > 1$, the function is increasing.
If $0 < b < 1$, the function is decreasing.

Logarithmic Functions

$y = \log_b x$

Domain: $(0, \infty)$
Range: $(-\infty, \infty)$
Vertical asymptote: $x = 0$
Passes through $(1, 0)$
If $b > 1$, the function is increasing.
If $0 < b < 1$, the function is decreasing.

Answers

7. a.

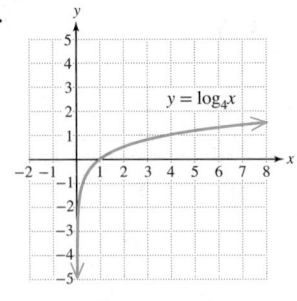

b.

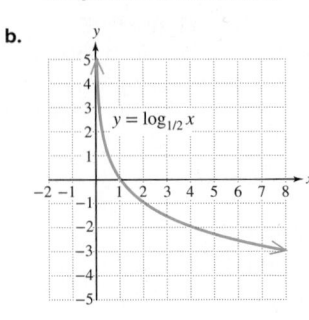

The roles of x and y are reversed between a function and its inverse. Therefore, it is not surprising that the domain and range are reversed between exponential and logarithmic functions. Furthermore, an exponential function passes through $(0, 1)$, whereas a logarithmic function passes through $(1, 0)$. An exponential function has a horizontal asymptote of $y = 0$, whereas a logarithmic function has a vertical asymptote of $x = 0$.

From Section 6.1, we know that the order in which we perform transformations on functions is:

1. Horizontal translation (shift)
2. Horizontal and vertical stretch and shrink
3. Reflections across the x- and y-axes
4. Vertical translation (shift)

In Example 8 we use the transformations of functions learned in Section 6.1 to graph a logarithmic function.

> If $h > 0$, shift to the right.   If $k > 0$, shift upward.
> If $h < 0$, shift to the left.   If $k < 0$, shift downward.

$$f(x) = a \log_b (x - h) + k$$

> If $a < 0$, reflect across the x-axis.
> Shrink vertically if $0 < |a| < 1$.
> Stretch vertically if $|a| > 1$.

EXAMPLE 8   **Using Transformations to Graph a Logarithmic Function**

Graph the function. Identify the vertical asymptote and write the domain in interval notation.

$$f(x) = \log_2(x + 3) - 2$$

Solution:

The graph of the "parent" function $y = \log_2 x$ was presented in Example 7. The graph of $f(x) = \log_2(x + 3) - 2$ is the graph of $y = \log_2 x$ shifted to the left 3 units and down 2 units.

We can plot a few points on the graph of $y = \log_2 x$ and use these points and the vertical asymptote to form an outline of the transformed graph.

x	$y = \log_2 x$
$\frac{1}{4}$	-2
$\frac{1}{2}$	-1
1	0
2	1
4	2

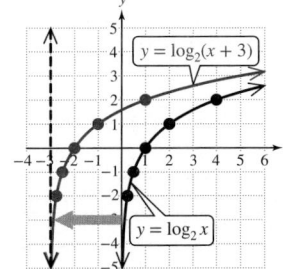 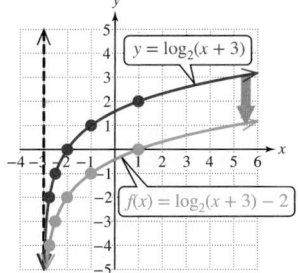

The graph of $f(x) = \log_2(x + 3) - 2$ is shown in blue.

The vertical asymptote is $x = -3$. The domain is $(-3, \infty)$.

Skill Practice 8 Graph the function. Identify the vertical asymptote and write the domain in interval notation. $g(x) = \log_3(x - 4) + 1$

Answer

8. Vertical asymptote: $x = 4$
   Domain: $(4, \infty)$

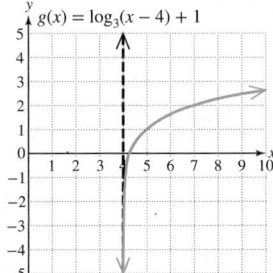

The domain of $f(x) = \log_b x$ is restricted to $x > 0$. In Example 8, this graph was shifted to the left 3 units, restricting the domain of $f(x) = \log_2(x + 3) - 2$ to $x > -3$. The domain of a logarithmic function is the set of real numbers that make the argument positive.

> **EXAMPLE 9** Identifying the Domain of a Logarithmic Function

Write the domain in interval notation.

a. $f(x) = \log_2(2x + 4)$ **b.** $g(x) = \ln(5 - x)$ **c.** $h(x) = \log(x^2 - 9)$

Solution:

a. $f(x) = \log_2(2x + 4)$

$$2x + 4 > 0 \qquad \text{Set the argument greater than zero.}$$
$$2x > -4 \qquad \text{Solve for } x.$$
$$x > -2$$

The domain is $(-2, \infty)$.

The graph of f is shown in Figure 9-17.

The vertical asymptote is $x = -2$.

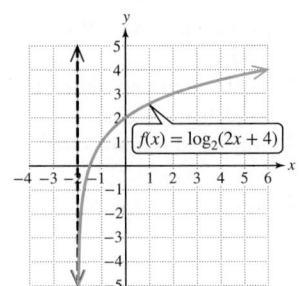

Figure 9-17

b. $g(x) = \ln(5 - x)$

$$5 - x > 0 \qquad \text{Set the argument greater than zero.}$$
$$-x > -5 \qquad \text{Subtract 5 and divide by } -1$$
$$x < 5 \qquad \text{(reverse the inequality sign).}$$

The domain is $(-\infty, 5)$.

The graph of g is shown in Figure 9-18.

The vertical asymptote is $x = 5$.

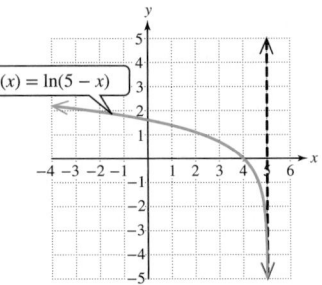

Figure 9-18

c. $h(x) = \log(x^2 - 9)$

$$x^2 - 9 > 0 \qquad\qquad \text{Set the argument greater than zero. The result is a polynomial inequality (Section 7.5).}$$

$$(x - 3)(x + 3) = 0 \qquad \text{Solve the related equation by setting one side equal to zero and factoring the other side.}$$

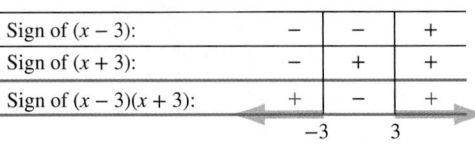

The boundary points for the solution set are the solutions to the equation: $x = 3$ and $x = -3$.

FOR REVIEW

The procedure to solve polynomial inequalities was presented in Section 7.5. To solve the inequality $x^2 - 9 > 0$, or equivalently $(x - 3)(x + 3) > 0$, we solve the related equation. The solutions $x = 3$ and $x = -3$ are the boundary points that divide the number line into intervals. Then we test the sign of each factor and analyze the product of factors on each interval. The intervals that make the original inequality true are part of the solution set.

The domain is $(-\infty, -3) \cup (3, \infty)$.

The graph of h is shown in Figure 9-19.

The vertical asymptotes are $x = -3$ and $x = 3$.

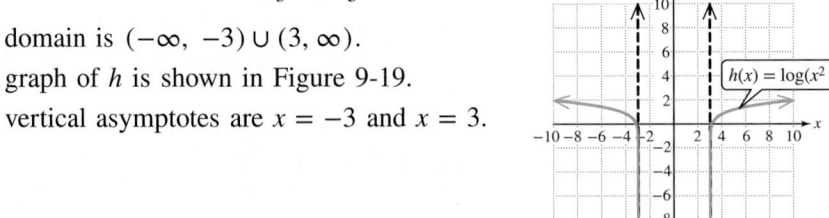

Figure 9-19

Skill Practice 9 Write the domain in interval notation.

    a. $\log_4(1 - 3x)$ **b.** $\log(2 + x)$ **c.** $m(x) = \ln(64 - x^2)$

Answers

9. a. $\left(-\infty, \dfrac{1}{3}\right)$ **b.** $(-2, \infty)$

    c. $(-8, 8)$

5. Use Logarithmic Functions in Applications

When physical quantities vary over a large range, it is often convenient to take a logarithm of the quantity to have a more manageable set of numbers. For example, suppose a set of data values consists of 10, 100, 1000, and 10,000. The corresponding common logarithms are 1, 2, 3, and 4. The latter list of numbers is easier to manipulate and to visualize on a graph. For this reason, logarithmic scales are used in applications such as

- measuring pH (representing hydrogen ion concentration from 10^{-14} to 1 mole per liter).
- measuring wave energy from an earthquake (often ranging from 10^6 J to 10^{17} J).
- measuring loudness of sound on the decibel scale (representing sound intensity from 10^{-12} to 10^2 Watts per square meter).

Point of Interest

In 1935, American geologist Charles Richter developed the local magnitude (M_L) scale, or Richter scale, for measuring the intensity of moderate-sized earthquakes ($3 < M_L < 7$) in southern California. Today, seismologists no longer follow Richter's methodology because it does not give reliable results for earthquakes of higher magnitude. The magnitudes of modern earthquakes are based on a variety of data types from numerous seismic stations. However, both the Richter scale and modern magnitude scales use a base 10 logarithmic

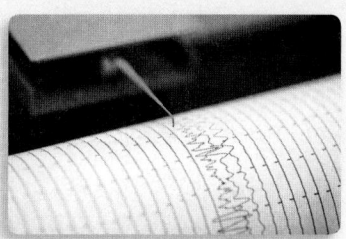

kickers/E+/Getty Images

scale to compare amplitudes of waves on a seismogram. This means that an increase of 1 unit in magnitude represents a 10-fold increase in the amplitude of the waves on a seismogram.

EXAMPLE 10   Using a Logarithmic Function in an Application

The intensity I of an earthquake is measured by a seismograph—a device that measures amplitudes of shock waves. I_0 is a minimum reference intensity of a "zero-level" earthquake against which the intensities of other earthquakes may be compared. The magnitude M of an earthquake of intensity I is given by

$$M = \log\left(\frac{I}{I_0}\right).$$

a. Determine the magnitude of the earthquake that devastated Haiti on January 12, 2010, if the intensity was approximately $10^{7.0}$ times I_0.

b. Determine the magnitude of the earthquake that occurred near Washington, D.C., on August 23, 2011, if the intensity was approximately $10^{5.8}$ times I_0.

c. How many times more intense was the earthquake that hit Haiti than the earthquake that hit Washington, D.C.? Round to the nearest whole unit.

TIP The ratio $\dfrac{I_0}{I_0}$ equals 1. Therefore, the two expressions in Example 10 simplify as the log base 10 of an exponential expression of base 10.

$$M = \log\left(\frac{10^{7.0} \cdot I_0}{I_0}\right)$$
$$= \log(10^{7.0})$$
$$= 7.0 \quad \text{(by Property 3)}$$

Solution:

a. $M = \log\left(\dfrac{I}{I_0}\right)$

$M = \log\left(\dfrac{10^{7.0} \cdot I_0}{I_0}\right)$

$= \log 10^{7.0}$

$= 7.0$

b. $M = \log\left(\dfrac{I}{I_0}\right)$

$M = \log\left(\dfrac{10^{5.8} \cdot I_0}{I_0}\right)$

$= \log 10^{5.8}$

$= 5.8$

c. Using the intensities given in parts (a) and (b) we have

$$\frac{10^{7.0} I_0}{10^{5.8} I_0} = 10^{7.0-5.8} = 10^{1.2} \approx 16$$

The earthquake in Haiti was approximately 16 times more intense.

Skill Practice 10

a. Determine the magnitude of an earthquake that is $10^{5.2}$ times I_0.

b. Determine the magnitude of an earthquake that is $10^{4.2}$ times I_0.

c. How many times more intense is a 5.2-magnitude earthquake than a 4.2-magnitude earthquake?

Answers

10. a. 5.2   **b.** 4.2

c. 10 times more intense

SECTION 9.3   Practice Exercises

Prerequisite Review

For Exercises R.1–R.4, simplify the expression. (See Section 4.5 for review.)

R.1. $(-8)^{2/3}$ **R.2.** $64^{2/3}$ **R.3.** $9^{-3/2}$ **R.4.** $\dfrac{1}{25^{-1/2}}$

For Exercises R.5–R.8, fill in the box to make a true statement. (See Section 2.1 for review.)

R.5. $\left(\dfrac{2}{3}\right)^{\square} = \dfrac{8}{27}$ **R.6.** $\left(\dfrac{4}{5}\right)^{\square} = \dfrac{256}{625}$ **R.7.** $(2)^{\square} = \dfrac{1}{64}$ **R.8.** $\left(\dfrac{1}{3}\right)^{\square} = 81$

For Exercises R.9–R.12, solve the inequality. Write the solution set in interval notation. (See Sections 1.5 and 7.5 for review.)

R.9. $-3x - 4 > 8$ (See Section 1.5.)   **R.10.** $4 - \dfrac{1}{2}x \leq 0$

R.11. $y^2 + 20y \geq -75$ (See Section 7.5.)   **R.12.** $x^2 - 8x + 16 > 0$

For Exercises R.13–R.16, write the domain in interval notation. (See Section 5.3 for review.)

R.13. $g(x) = \sqrt{6 - x}$ **R.14.** $h(x) = \sqrt{2x - 10}$

R.15. $k(x) = \dfrac{x}{2x + 7}$ **R.16.** $m(x) = \dfrac{2x - 3}{3 - 6x}$

For Exercises R.17–R.18, a one-to-one function is given. Write an equation for the inverse function. (See Section 9.1 for review.)

R.17. $m(x) = (x - 4)^3$ **R.18.** $n(x) = \dfrac{x - 2}{9}$

For Exercises R.19–R.20, find $(f \circ g)(x)$ and $(g \circ f)(x)$. (See Section 6.4 for review.)

R.19. $f(x) = \dfrac{2x - 1}{6}$ and $g(x) = \dfrac{6x + 1}{2}$ **R.20.** $f(x) = \dfrac{x + 8}{8 - x}$ and $g(x) = \dfrac{8x - 8}{1 + x}$

Concept Connections

1. Given positive real numbers x and b such that $b \neq 1$, $y = \log_b x$ is the _____ function base b and is equivalent to $b^y = x$.

2. Given $y = \log_b x$, the value y is called the _____, b is called the _____, and x is called the _____.

3. The logarithmic function base 10 is called the _____ logarithmic function, and the logarithmic function base e is called the _____ logarithmic function.

4. Given $y = \log x$, the base is understood to be _____. Given $y = \ln x$, the base is understood to be _____.

5. $\log_b 1 =$ _____ because $b^\square = 1$.

6. $\log_b b =$ _____ because $b^\square = b$.

7. $f(x) = \log_b x$ and $g(x) = b^x$ are inverse functions. Therefore, $\log_b b^x =$ _____ and $b^{\log_b x} =$ _____.

8. The graph of $y = \log_b x$ passes through the point $(1, 0)$ and the line _____ is a (horizontal/vertical) asymptote.

For the exercises in this set, assume that all variable expressions represent positive real numbers.

Objective 1: Convert Between Logarithmic and Exponential Forms

For Exercises 9–16, write the equation in exponential form. **(See Example 1)**

9. $\log_8 64 = 2$

10. $\log_9 81 = 2$

11. $\log\left(\dfrac{1}{10,000}\right) = -4$

12. $\log\left(\dfrac{1}{1,000,000}\right) = -6$

13. $\ln 1 = 0$

14. $\log_8 1 = 0$

15. $\log_a b = c$

16. $\log_x M = N$

For Exercises 17–24, write the equation in logarithmic form. **(See Example 2)**

17. $5^3 = 125$

18. $2^5 = 32$

19. $\left(\dfrac{1}{5}\right)^{-3} = 125$

20. $\left(\dfrac{1}{2}\right)^{-5} = 32$

21. $10^9 = 1,000,000,000$

22. $e^1 = e$

23. $a^7 = b$

24. $M^3 = N$

Objective 2: Evaluate Logarithmic Expressions

For Exercises 25–50, simplify the expression without using a calculator. **(See Examples 3–4)**

25. $\log_3 9$

26. $\log_2 16$

27. $\log_5 5$

28. $\log_6 6$

29. $\log 100,000,000$

30. $\log 10,000,000$

31. $\log_2\left(\dfrac{1}{16}\right)$

32. $\log_3\left(\dfrac{1}{9}\right)$

33. $\log\left(\dfrac{1}{10}\right)$

34. $\log\left(\dfrac{1}{10,000}\right)$

35. $\ln e^6$

36. $\ln e^{10}$

37. $\ln\left(\dfrac{1}{e^3}\right)$

38. $\ln\left(\dfrac{1}{e^8}\right)$

39. $\log_{1/7} 49$

40. $\log_{1/4} 16$

41. $\log_{1/2}\left(\dfrac{1}{32}\right)$

42. $\log_{1/6}\left(\dfrac{1}{36}\right)$

43. $\log 0.00001$

44. $\log 0.0001$

45. $\log_{3/2}\dfrac{4}{9}$

46. $\log_{3/2}\dfrac{9}{4}$

47. $\log_3 \sqrt[5]{3}$

48. $\log_2 \sqrt[3]{2}$

49. $\log_5 \sqrt{\dfrac{1}{5}}$

50. $\log \sqrt{\dfrac{1}{1000}}$

For Exercises 51–52, estimate the value of each logarithm between two consecutive integers. Then use a calculator to approximate the value to 4 decimal places. For example, $\log 8970$ is between 3 and 4 because $10^3 < 8970 < 10^4$. **(See Example 5)**

51. **a.** $\log 46,832$
   b. $\log 1,247,310$
   c. $\log 0.24$
   d. $\log 0.0000032$
   e. $\log(5.6 \times 10^5)$
   f. $\log(5.1 \times 10^{-3})$

52. **a.** $\log 293,416$
   b. $\log 897$
   c. $\log 0.038$
   d. $\log 0.00061$
   e. $\log(9.1 \times 10^8)$
   f. $\log(8.2 \times 10^{-2})$

For Exercises 53–54, approximate $f(x) = \ln x$ for the given values of x. Round to 4 decimal places. **(See Example 5)**

53. a. $f(94)$

    b. $f(0.182)$

    c. $f\left(\sqrt{155}\right)$

    d. $f(4\pi)$

    e. $f(3.9 \times 10^9)$

    f. $f(7.1 \times 10^{-4})$

54. a. $f(1860)$

    b. $f(0.0694)$

    c. $f\left(\sqrt{87}\right)$

    d. $f(2\pi)$

    e. $f(1.3 \times 10^{12})$

    f. $f(8.5 \times 10^{-17})$

Objective 3: Apply Basic Properties of Logarithms

For Exercises 55–64, simplify the expression without using a calculator. **(See Example 6)**

55. $\log_4 4^{11}$

56. $\log_6 6^7$

57. $\log_c c$

58. $\log_d d$

59. $5^{\log_5(x+y)}$

60. $4^{\log_4(a-c)}$

61. $\ln e^{a+b}$

62. $\ln e^{x^2+1}$

63. $\log_{\sqrt{5}} 1$

64. $\log_\pi 1$

Objective 4: Graph Logarithmic Functions

For Exercises 65–70, graph the function. **(See Example 7)**

65. $y = \log_3 x$

66. $y = \log_5 x$

67. $y = \log_{1/3} x$

68. $y = \log_{1/5} x$

69. $y = \ln x$

70. $y = \log x$

For Exercises 71–78, **(See Example 8)**

a. Use transformations of the graphs of $y = \log_2 x$ (see Example 7) and $y = \log_3 x$ (see Exercise 65) to graph the given functions.

b. Write the domain and range in interval notation.

c. Write an equation of the asymptote.

71. $y = \log_3(x + 2)$

72. $y = \log_2(x + 3)$

73. $y = 2 + \log_3 x$

74. $y = 3 + \log_2 x$

75. $y = \log_3(x - 1) - 3$

76. $y = \log_2(x - 2) - 1$

77. $y = -\log_3 x$

78. $y = -\log_2 x$

For Exercises 79–92, write the domain in interval notation. **(See Example 9)**

79. $f(x) = \log(8 - x)$

80. $g(x) = \log(3 - x)$

81. $h(x) = \log_2(6x + 7)$

82. $k(x) = \log_3(5x + 6)$

83. $m(x) = \ln(x^2 + 14)$

84. $n(x) = \ln(x^2 + 11)$

85. $f(x) = \log_4(x^2 - 16)$

86. $g(x) = \log_7(x^2 - 49)$

87. $m(x) = 3 + \ln \dfrac{1}{\sqrt{11 - x}}$

88. $n(x) = 4 - \log \dfrac{1}{\sqrt{x + 5}}$

89. $p(x) = \log(x^2 - x - 12)$

90. $q(x) = \log(x^2 + 10x + 9)$

91. $r(x) = \log_3(4 - x)^2$

92. $s(x) = \log_5(3 - x)^2$

Objective 5: Use Logarithmic Functions in Applications

93. In 1989, the Loma Prieta earthquake damaged the city of San Francisco with an intensity of approximately $10^{6.9} I_0$. Film footage of the 1989 earthquake was captured on a number of video cameras including a broadcast of Game 3 of the World Series played at Candlestick Park. **(See Example 10)**

    a. Determine the magnitude of the Loma Prieta earthquake.

    b. Smaller earthquakes occur daily in the San Francisco area and most are not detectable without a seismograph. Determine the magnitude of an earthquake with an intensity of $10^{3.2} I_0$.

    c. How many times more intense was the Loma Prieta earthquake than an earthquake with a magnitude of 3.2? Round to the nearest whole unit.

94. The intensities of earthquakes are measured with seismographs all over the world at different distances from the epicenter. Suppose that the intensity of a medium earthquake is originally reported as $10^{5.4}$ times I_0. Later this value is revised as $10^{5.8}$ times I_0.

    a. Determine the magnitude of the earthquake using the original estimate for intensity.

    b. Determine the magnitude using the revised estimate for intensity.

    c. How many times more intense was the earthquake than originally thought? Round to 1 decimal place.

Sounds are produced when vibrating objects create pressure waves in some medium such as air. When these variations in pressure reach the human eardrum, it causes the eardrum to vibrate in a similar manner and the ear detects sound. The intensity of sound is measured as power per unit area. The threshold for hearing (minimum sound detectable by a young, healthy ear) is defined to be $I_0 = 10^{-12}$ W/m^2 (watts per square meter). The sound level L, or "loudness" of sound, is measured in decibels (dB) as $L = 10 \log\left(\dfrac{I}{I_0}\right)$, where I is the intensity of the given sound. Use this formula for Exercises 95–96.

95. a. Find the sound level of a jet plane taking off if its intensity is 10^{15} times the intensity of I_0.

  b. Find the sound level of the noise from city traffic if its intensity is 10^9 times I_0.

  c. How many times more intense is the sound of a jet plane taking off than noise from city traffic?

96. a. Find the sound level of a motorcycle if its intensity is 10^{10} times I_0.

  b. Find the sound level of a vacuum cleaner if its intensity is 10^7 times I_0.

  c. How many times more intense is the sound of a motorcycle than a vacuum cleaner?

Don Hammond/DesignPics

Scientists use the pH scale to represent the level of acidity or alkalinity of a liquid. This is based on the molar concentration of hydrogen ions, [H$^+$]. Since the values of [H$^+$] vary over a large range, 1×10^0 mole per liter to 1×10^{-14} mole per liter (mol/L), a logarithmic scale is used to compute pH. The formula

$$pH = -\log[H^+]$$

represents the pH of a liquid as a function of its concentration of hydrogen ions, [H$^+$].

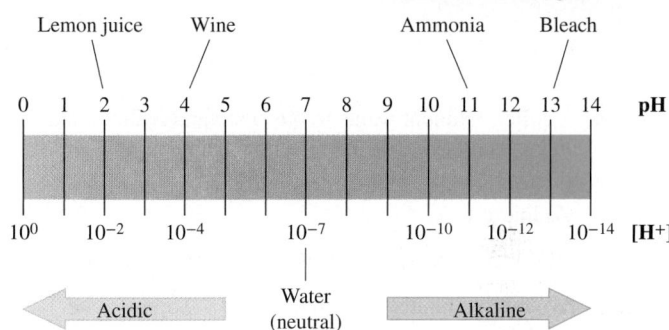

The pH scale ranges from 0 to 14. Pure water is taken as neutral having a pH of 7. A pH less than 7 is acidic. A pH greater than 7 is alkaline (or basic). For Exercises 97–98, use the formula for pH. Round pH values to 1 decimal place.

97. Vinegar and lemon juice are both acids. Their [H$^+$] values are 5.0×10^{-3} mol/L and 1×10^{-2} mol/L, respectively.

  a. Find the pH for vinegar.

  b. Find the pH for lemon juice.

  c. Which substance is more acidic?

98. Bleach and milk of magnesia are both bases. Their [H$^+$] values are 2.0×10^{-13} mol/L and 4.1×10^{-10} mol/L, respectively.

  a. Find the pH for bleach.

  b. Find the pH for milk of magnesia.

  c. Which substance is more basic?

Mixed Exercises

For Exercises 99–102,

a. Write the equation in exponential form.

b. Solve the equation from part (a).

c. Verify that the solution checks in the original equation.

99. $\log_3(x + 1) = 4$ **100.** $\log_2(x - 5) = 4$ **101.** $\log_4(7x - 6) = 3$ **102.** $\log_5(9x - 11) = 2$

For Exercises 103–106, evaluate the expressions.

103. $\log_3(\log_4 64)$ **104.** $\log_2\left[\log_{1/2}\left(\dfrac{1}{4}\right)\right]$ **105.** $\log_{16}(\log_{81} 3)$ **106.** $\log_4(\log_{16} 4)$

107. a. Evaluate $\log_2 2 + \log_2 4$

  b. Evaluate $\log_2(2 \cdot 4)$

  c. How do the values of the expressions in parts (a) and (b) compare?

108. a. Evaluate $\log_3 3 + \log_3 27$

  b. Evaluate $\log_3(3 \cdot 27)$

  c. How do the values of the expressions in parts (a) and (b) compare?

109. a. Evaluate $\log_4 64 - \log_4 4$

  b. Evaluate $\log_4\left(\dfrac{64}{4}\right)$

  c. How do the values of the expressions in parts (a) and (b) compare?

110. a. Evaluate $\log 100{,}000 - \log 100$

  b. Evaluate $\log\left(\dfrac{100{,}000}{100}\right)$

  c. How do the values of the expressions in parts (a) and (b) compare?

111. a. Evaluate $\log_2 2^5$

b. Evaluate $5 \cdot \log_2 2$

c. How do the values of the expressions in parts (a) and (b) compare?

112. a. Evaluate $\log_7 7^6$

b. Evaluate $6 \cdot \log_7 7$

c. How do the values of the expressions in parts (a) and (b) compare?

113. The time t (in years) required for an investment to double with interest compounded continuously depends on the interest rate r according to the function $t(r) = \dfrac{\ln 2}{r}$.

a. If an interest rate of 3.5% is secured, determine the length of time needed for an initial investment to double. Round to 1 decimal place.

b. Evaluate $t(0.04)$, $t(0.06)$, and $t(0.08)$.

114. The number n of monthly payments of P dollars each required to pay off a loan of A dollars in its entirety at interest rate r is given by

$$n = -\frac{\log\left(1 - \dfrac{Ar}{12P}\right)}{\log\left(1 + \dfrac{r}{12}\right)}$$

a. A college student wants to buy a car and realizes that he can only afford payments of $200 per month. If he borrows $3000 and pays it off at 6% interest, how many months will it take him to retire the loan? Round to the nearest month.

b. Determine the number of monthly payments of $611.09 that would be required to pay off a home loan of $128,000 at 4% interest.

For Exercises 115–116, use a calculator to approximate the given logarithms to 4 decimal places.

115. a. Avogadro's number is 6.022×10^{23}. Approximate $\log(6.022 \times 10^{23})$.

b. Planck's constant is 6.626×10^{-34} J · sec. Approximate $\log(6.626 \times 10^{-34})$.

c. Compare the value of the common logarithm to the power of 10 used in scientific notation.

116. a. The speed of light is 2.9979×10^8 m/sec. Approximate $\log(2.9979 \times 10^8)$.

b. An elementary charge is 1.602×10^{-19} C. Approximate $\log(1.602 \times 10^{-19})$.

c. Compare the value of the common logarithm to the power of 10 used in scientific notation.

Expanding Your Skills

For Exercises 117–122, write the domain in interval notation.

117. $t(x) = \log_4\left(\dfrac{x-1}{x-3}\right)$

118. $r(x) = \log_5\left(\dfrac{x+2}{x-4}\right)$

119. $s(x) = \ln\left(\sqrt{x+5} - 1\right)$

120. $v(x) = \ln\left(\sqrt{x-8} - 1\right)$

121. $c(x) = \log\left(\dfrac{1}{\sqrt{x-6}}\right)$

122. $d(x) = \log\left(\dfrac{1}{\sqrt{x+8}}\right)$

Technology Connections

123. a. Graph $f(x) = \ln x$ and

$$g(x) = (x-1) - \frac{(x-1)^2}{2} + \frac{(x-1)^3}{3} - \frac{(x-1)^4}{4}$$

on the viewing window $[-2, 4, 1]$ by $[-5, 2, 1]$. How do the graphs compare on the interval $(0, 2)$?

b. Use function g to approximate $\ln 1.5$. Round to 4 decimal places.

124. Compare the graphs of $Y_1 = \dfrac{e^x - e^{-x}}{2}$, $Y_2 = \ln\left(x + \sqrt{x^2 + 1}\right)$, and $Y_3 = x$ on the viewing window $[-16.1, 16.1, 1]$ by $[-10, 10, 1]$. Based on the graphs, how do you suspect that the functions are related?

125. Compare the graphs of the functions.

$Y_1 = \ln(2x)$ and $Y_2 = \ln 2 + \ln x$

126. Compare the graphs of the functions.

$Y_1 = \ln\left(\dfrac{x}{2}\right)$ and $Y_2 = \ln x - \ln 2$

PROBLEM RECOGNITION EXERCISES

Analyzing Functions

For Exercises 1–14,

    a. Write the domain.   **b.** Write the range.   **c.** Find the *x*-intercept(s).   **d.** Find the *y*-intercept.

    e. Determine the asymptotes if applicable.   **f.** Determine the intervals over which the function is increasing.

    g. Determine the intervals over which the function is decreasing.   **h.** Match the function with its graph.

1. $f(x) = 3$

2. $g(x) = 2x - 3$

3. $d(x) = (x - 3)^2 - 4$

4. $h(x) = \sqrt[3]{x - 2}$

5. $k(x) = \dfrac{2}{x - 1}$

6. $z(x) = \dfrac{3x}{x + 2}$

7. $p(x) = \left(\dfrac{4}{3}\right)^x$

8. $q(x) = -x^2 - 6x - 9$

9. $m(x) = |x - 4| - 1$

10. $n(x) = -|x| + 3$

11. $r(x) = \sqrt{3 - x}$

12. $s(x) = \sqrt{x - 3}$

13. $t(x) = e^x + 2$

14. $v(x) = \ln(x + 2)$

A.
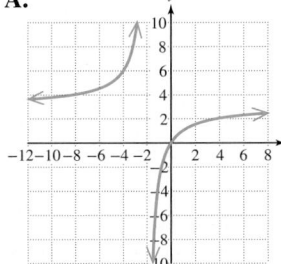

B.
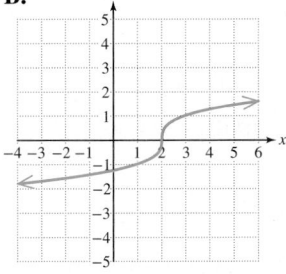

C.
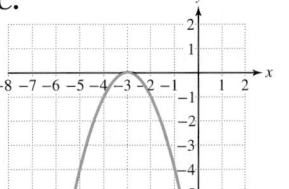

D.
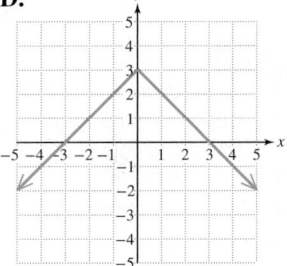

E.
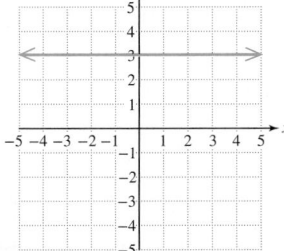

F.
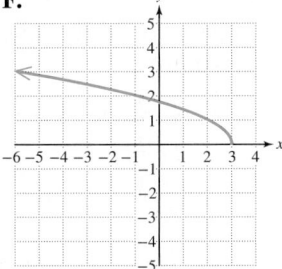

G.
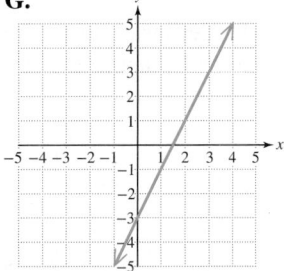

H.
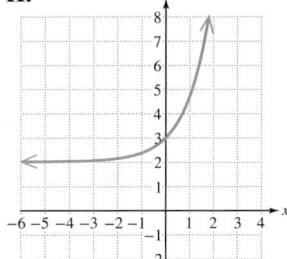

I.
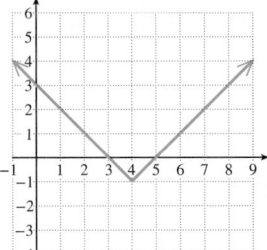

J.
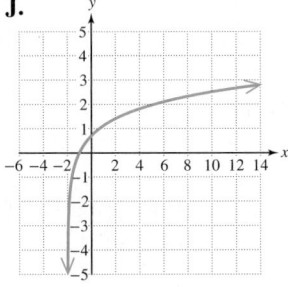

K.
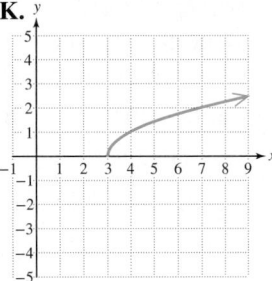

L.
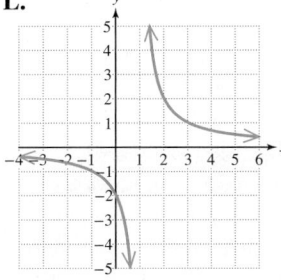

M.
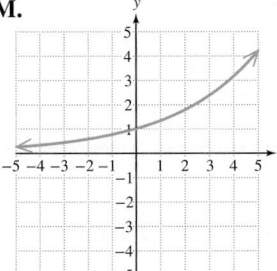

N.
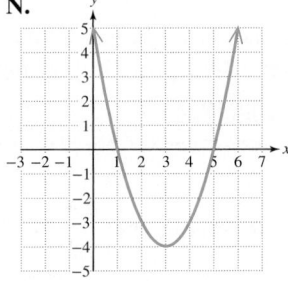

OBJECTIVES

1. Apply the Product, Quotient, and Power Properties of Logarithms
2. Write a Logarithmic Expression in Expanded Form
3. Write a Logarithmic Expression as a Single Logarithm
4. Apply the Change-of-Base Formula

1. Apply the Product, Quotient, and Power Properties of Logarithms

By definition, $y = \log_b x$ is equivalent to $b^y = x$. Because a logarithm is an exponent, the properties of exponents can be applied to logarithms. The first is called the product property of logarithms.

> **Product Property of Logarithms**
>
> Let b, x, and y be positive real numbers where $b \neq 1$. Then
> $$\log_b(xy) = \log_b x + \log_b y.$$
> The logarithm of a product equals the sum of the logarithms of the factors.

TIP When two factors of the same base are multiplied, the base is unchanged and we add the exponents. This is the underlying principle for the product property of logarithms.

Proof:

Let $M = \log_b x$, which implies $b^M = x$.
Let $N = \log_b y$, which implies $b^N = y$.
Then $xy = b^M b^N = b^{M+N}$.

Writing the expression $xy = b^{M+N}$ in logarithmic form, we have,

$$\log_b(xy) = M + N$$
$$\log_b(xy) = \log_b x + \log_b y \checkmark$$

To demonstrate the product property of logarithms, simplify the following expressions by using the order of operations.

$$\log_3(3 \cdot 9) \overset{?}{=} \log_3 3 + \log_3 9$$
$$\log_3 27 \overset{?}{=} 1 + 2$$
$$3 \overset{?}{=} 3 \checkmark \text{ True}$$

EXAMPLE 1 Applying the Product Property of Logarithms

Write the logarithm as a sum and simplify if possible. Assume that x and y represent positive real numbers.

   a. $\log_2(8x)$ **b.** $\ln(5xy)$

FOR REVIEW

$\log_2 8$ is the exponent to which 2 must be raised to make 8. That is, the value of the logarithm is the exponent in the expression $2^\square = 8$. The value is 3.

Solution:

a. $\log_2(8x) = \log_2 8 + \log_2 x$ Product property of logarithms
             $= 3 + \log_2 x$ Simplify. $\log_2 8 = \log_2 2^3 = 3$

b. $\ln(5xy) = \ln 5 + \ln x + \ln y$

> **Skill Practice 1** Write the logarithm as a sum and simplify if possible. Assume that a, c, and d represent positive real numbers.
>
>   **a.** $\log_4(16a)$ **b.** $\log(12cd)$

Answers

1. a. $2 + \log_4 a$
   b. $\log 12 + \log c + \log d$

The quotient rule of exponents tells us that $\dfrac{b^M}{b^N} = b^{M-N}$ for $b \neq 0$. This property can be applied to logarithms.

> ## Quotient Property of Logarithms
>
> Let b, x, and y be positive real numbers where $b \neq 1$. Then
>
> $$\log_b\left(\frac{x}{y}\right) = \log_b x - \log_b y.$$
>
> The logarithm of a quotient equals the difference of the logarithm of the numerator and the logarithm of the denominator.

The proof of the quotient property for logarithms is similar to the proof of the product property (see Exercise 107). To demonstrate the quotient property for logarithms, simplify the following expressions by using the order of operations.

$$\log\left(\frac{1{,}000{,}000}{100}\right) \overset{?}{=} \log 1{,}000{,}000 - \log 100$$

$$\log 10{,}000 \overset{?}{=} 6 - 2$$

$$4 \overset{?}{=} 4 \; \checkmark \; \text{True}$$

EXAMPLE 2   Applying the Quotient Property of Logarithms

Write the logarithm as the difference of logarithms and simplify if possible. Assume that the variables represent positive real numbers.

a. $\log_3\left(\dfrac{c}{d}\right)$ **b.** $\log\left(\dfrac{x}{1000}\right)$

Solution:

a. $\log_3\left(\dfrac{c}{d}\right) = \log_3 c - \log_3 d$ Quotient property of logarithms.

b. $\log\left(\dfrac{x}{1000}\right) = \log x - \log 1000$ Quotient property of logarithms.

$\phantom{\textbf{b.} \log\left(\dfrac{x}{1000}\right)} = \log x - 3$ Simplify. $\log 1000 = \log 10^3 = 3$

> **Skill Practice 2**   Write the logarithm as the difference of logarithms and simplify if possible. Assume that t represents a positive real number.
>
> **a.** $\log_6\left(\dfrac{8}{t}\right)$ **b.** $\ln\left(\dfrac{e}{12}\right)$

The last property we present here is the power property of logarithms. The power property of exponents tells us that $\left(b^M\right)^N = b^{MN}$. The same principle can be applied to logarithms.

> ## Power Property of Logarithms
>
> Let b and x be positive real numbers where $b \neq 1$. Let p be any real number. Then
>
> $$\log_b x^p = p \log_b x.$$

The power property of logarithms is proved in Exercise 108.

EXAMPLE 3   Applying the Power Property of Logarithms

Apply the power property of logarithms.   **a.** $\ln \sqrt[5]{x^2}$ **b.** $\log x^2$

Solution:

a. $\ln \sqrt[5]{x^2} = \ln x^{2/5}$ Write $\sqrt[5]{x^2}$ using rational exponents.

$\qquad = \dfrac{2}{5} \ln x$ provided that $x > 0$ Apply the power rule.

b. $\log x^2 = 2 \log x$ provided that $x > 0$ Apply the power rule.

In both parts (a) and (b), the condition that $x > 0$ is mandatory. The properties of logarithms hold true only for values of the variable for which the logarithms are defined. That is, the arguments must be positive.

From the graphs of $y = \log x^2$ and $y = 2 \log x$, we see that the domains are different. Therefore, the statement $\log x^2 = 2 \log x$ is true only for $x > 0$.

$y = \log x^2$ Domain: $(-\infty, 0) \cup (0, \infty)$ $y = 2 \log x$ Domain: $(0, \infty)$

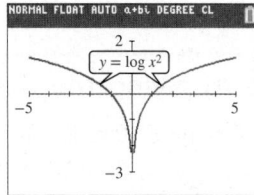   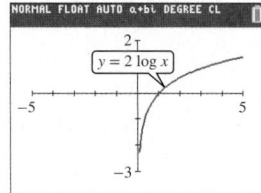

Skill Practice 3   Apply the power property of logarithms.

a. $\log_5 \sqrt[5]{x^4}$ **b.** $\ln x^4$

At this point, we have learned seven properties of logarithms. The properties hold true for values of the variable for which the logarithms are defined. **Therefore, in the examples and exercises, we will assume that the variable expressions within the logarithms represent positive real numbers.**

Properties of Logarithms

Let b, x, and y be positive real numbers where $b \neq 1$, and let p be a real number. Then the following properties of logarithms are true.

1. $\log_b 1 = 0$ **5.** $\log_b(xy) = \log_b x + \log_b y$ **Product property**

2. $\log_b b = 1$ **6.** $\log_b\left(\dfrac{x}{y}\right) = \log_b x - \log_b y$ **Quotient property**

3. $\log_b b^p = p$ **7.** $\log_b x^p = p \log_b x$ **Power property**

4. $b^{\log_b x} = x$

---

2. Write a Logarithmic Expression in Expanded Form

Properties 5, 6, and 7 can be used in either direction. For example,

$$\log\left(\frac{ab}{c}\right) = \log a + \log b - \log c \qquad \text{or} \qquad \log a + \log b - \log c = \log\left(\frac{ab}{c}\right).$$

In some applications of algebra and calculus, the "condensed" form of the logarithm is preferred. In other applications, the "expanded" form is preferred. In Examples 4–6, we practice manipulating logarithmic expressions in both forms.

EXAMPLE 4   **Writing a Logarithmic Expression in Expanded Form**

Write the expression as the sum or difference of logarithms.

a. $\log_2\left(\dfrac{z^3}{xy^5}\right)$ **b.** $\log\sqrt[3]{\dfrac{(x+y)^2}{10}}$

Solution:

a. $\log_2\left(\dfrac{z^3}{xy^5}\right) = \log_2 z^3 - \log_2(xy^5)$ Apply the quotient property.

$\qquad = \log_2 z^3 - (\log_2 x + \log_2 y^5)$ Apply the product property.

$\qquad = \log_2 z^3 - \log_2 x - \log_2 y^5$ Apply the distributive property.

$\qquad = 3\log_2 z - \log_2 x - 5\log_2 y$ Apply the power property.

b. $\log\sqrt[3]{\dfrac{(x+y)^2}{10}} = \log\left[\dfrac{(x+y)^2}{10}\right]^{1/3}$ Write the radical expression with rational exponents.

$\qquad = \dfrac{1}{3}\log\left[\dfrac{(x+y)^2}{10}\right]$ Apply the power property.

$\qquad = \dfrac{1}{3}[\log(x+y)^2 - \log 10]$ Apply the quotient property.

$\qquad = \dfrac{1}{3}[2\log(x+y) - 1]$ Apply the power property and simplify: $\log 10 = 1$.

$\qquad = \dfrac{2}{3}\log(x+y) - \dfrac{1}{3}$ Apply the distributive property.

Avoiding Mistakes

In Example 4(b) do not try to simplify $\log(x+y)$. The argument contains a sum, not a product.

$\underbrace{\log(x+y)}_{\text{sum}}$ cannot be simplified.

Compare to the logarithm of a product which can be simplified.

$\underbrace{\log(xy)}_{\text{product}} = \log x + \log y$

Skill Practice 4   Write the expression as the sum or difference of logarithms.

a. $\ln\left(\dfrac{a^4 b}{c^9}\right)$ **b.** $\log_5\sqrt[3]{\dfrac{25}{(a^2+b)^2}}$

3. Write a Logarithmic Expression as a Single Logarithm

In Examples 5 and 6, we demonstrate how to write a sum or difference of logarithms as a single logarithm. We apply Properties 5, 6, and 7 of logarithms in reverse.

EXAMPLE 5   **Writing the Sum or Difference of Logarithms as a Single Logarithm**

Write the expression as a single logarithm and simplify the result if possible.

$$\log_2 560 - \log_2 7 - \log_2 5$$

FOR REVIEW

To simplify $\dfrac{560}{7\cdot 5}$, factor the numerator and denominator and cancel common factors.

$\dfrac{2^4\cdot 7\cdot 5}{7\cdot 5} = 2^4 = 16$

Solution:

$\log_2 560 - \log_2 7 - \log_2 5$

$= \log_2 560 - (\log_2 7 + \log_2 5)$ Factor out -1 from the last two terms.

$= \log_2 560 - \log_2(7\cdot 5)$ Apply the product property.

$= \log_2\left(\dfrac{560}{7\cdot 5}\right)$ Apply the quotient property.

$= \log_2 16$ Simplify within the argument.

$= 4$ Simplify. $\log_2 16 = \log_2 2^4 = 4$

Answers

4. a. $4\ln a + \ln b - 9\ln c$

$\quad$ **b.** $\dfrac{2}{3} - \dfrac{2}{3}\log_5(a^2+b)$

Skill Practice 5 Write the expression as a single logarithm and simplify the result if possible. $\log_3 54 + \log_3 10 - \log_3 20$

EXAMPLE 6 Writing the Sum or Difference of Logarithms as a Single Logarithm

Write the expression as a single logarithm and simplify the result if possible.

a. $3 \log a - \dfrac{1}{2} \log b - \dfrac{1}{2} \log c$ 　　　**b.** $\dfrac{1}{2} \ln x + \ln(x^2 - 1) - \ln(x + 1)$

Solution:

a. $3 \log a - \dfrac{1}{2} \log b - \dfrac{1}{2} \log c$

$= 3 \log a - \dfrac{1}{2}(\log b + \log c)$ 　　　Factor out $-\frac{1}{2}$ from the last two terms.

$= 3 \log a - \dfrac{1}{2} \log(bc)$ 　　　Apply the product property.

$= \log a^3 - \log(bc)^{1/2}$

$= \log a^3 - \log \sqrt{bc}$ 　　　Apply the power property.

$= \log\left(\dfrac{a^3}{\sqrt{bc}}\right)$ 　　　Apply the quotient property.

b. $\dfrac{1}{2} \ln x + \ln(x^2 - 1) - \ln(x + 1)$

$= \ln x^{1/2} + \ln(x^2 - 1) - \ln(x + 1)$ 　　　Apply the power property.

$= \ln[x^{1/2}(x^2 - 1)] - \ln(x + 1)$ 　　　Apply the product property.

$= \ln\left[\dfrac{\sqrt{x}(x^2 - 1)}{x + 1}\right]$ 　　　Apply the quotient property.

$= \ln\left[\dfrac{\sqrt{x}(x + 1)(x - 1)}{x + 1}\right]$ 　　　Factor the numerator of the argument.

$= \ln\left[\sqrt{x}(x - 1)\right]$ 　　　Simplify the argument.

> **Avoiding Mistakes**
>
> In all examples and exercises in which we manipulate logarithmic expressions, it is important to note that the equivalences are true only for the values of the variables that make the expressions defined. In Example 6(b) we have the restriction that $x > 1$.

Skill Practice 6 Write the expression as a single logarithm and simplify the result if possible.

a. $3 \log x - \dfrac{1}{3} \log y - \dfrac{2}{3} \log z$ 　　　**b.** $\dfrac{1}{3} \ln t + \ln(t^2 - 9) - \ln(t - 3)$

EXAMPLE 7 Applying Properties of Logarithms

Given that $\log_b 2 \approx 0.356$ and $\log_b 3 \approx 0.565$, approximate the value of $\log_b 36$.

Solution:

$\log_b 36 = \log_b(2 \cdot 3)^2$ 　　　Write the argument as a product of the factors 2 and 3.

$= 2 \log_b(2 \cdot 3)$ 　　　Apply the power property of logarithms.

$= 2(\log_b 2 + \log_b 3)$ 　　　Apply the product property of logarithms.

$\approx 2(0.356 + 0.565)$ 　　　Simplify.

≈ 1.842

Answers

5. $\log_3 27 = 3$

6. a. $\log\left(\dfrac{x^3}{\sqrt[3]{yz^2}}\right)$

b. $\ln[\sqrt[3]{t}(t + 3)]$

4. Apply the Change-of-Base Formula

A calculator can be used to approximate the value of a logarithm base 10 or base e by using the **LOG** key or the **LN** key, respectively. However, to use a calculator to evaluate a logarithmic expression with a different base, we must use the change-of-base formula.

Change-of-Base Formula

Let a and b be positive real numbers such that $a \neq 1$ and $b \neq 1$. Then for any positive real number x,

$$\log_b x = \frac{\log_a x}{\log_a b}$$

Note: The change-of-base formula converts a logarithm of one base to a ratio of logarithms of a different base. For the purpose of using a calculator, we often apply the change-of-base formula with base 10 or base e.

$$\log_b x = \frac{\log x}{\log b}$$

Original base is b. Ratio of base 10 logarithms

$$\log_b x = \frac{\ln x}{\ln b}$$

Original base is b. Ratio of base e logarithms

To derive the change-of-base formula, assume that a and b are positive real numbers with $a \neq 1$ and $b \neq 1$. Begin by letting $y = \log_b x$. If $y = \log_b x$, then

$b^y = x$	Write the original logarithm in exponential form.
$\log_a b^y = \log_a x$	Take the logarithm base a on both sides.
$y \cdot \log_a b = \log_a x$	Apply the power property of logarithms.
$y = \dfrac{\log_a x}{\log_a b}$	Solve for y.
$\log_b x = \dfrac{\log_a x}{\log_a b}$	Replace y by $\log_b x$. This is the change-of-base formula.

EXAMPLE 8   Applying the Change-of-Base Formula

a. Estimate $\log_4 153$ between two consecutive integers.
b. Use the change-of-base formula to approximate $\log_4 153$ by using base 10. Round to 4 decimal places.
c. Use the change-of-base formula to approximate $\log_4 153$ by using base e.
d. Check the result by using the related exponential form.

Solution:

a. $64 < 153 < 256$
$4^3 < 153 < 4^4$
$3 < \log_4 153 < 4$ $\log_4 153$ is between 3 and 4.

Answer
7. 1.633

TIP Although the numerators and denominators in parts (b) and (c) are different, their ratios are the same.

b. $\log_4 153 = \dfrac{\log 153}{\log 4} \approx \dfrac{2.184691431}{0.6020599913} \approx 3.6287$

c. $\log_4 153 = \dfrac{\ln 153}{\ln 4} \approx \dfrac{5.030437921}{1.386294361} \approx 3.6287$

d. Check: $4^{3.6287} \approx 153$ ✓

Skill Practice 8

a. Estimate $\log_6 23$ between two consecutive integers.

b. Use the change-of-base formula to evaluate $\log_6 23$ by using base 10. Round to 4 decimal places.

c. Use the change-of-base formula to evaluate $\log_6 23$ by using base e. Round to 4 decimal places.

d. Check the result by using the related exponential form.

TECHNOLOGY CONNECTIONS

Using the Change-of-Base Formula to Graph a Logarithmic Function

The change-of-base formula can be used to graph logarithmic functions using a graphing utility. For example, to graph $Y_1 = \log_2 x$, enter the function as

$$Y = \log(x)/\log(2) \quad \text{or} \quad Y = \ln(x)/\ln(2)$$

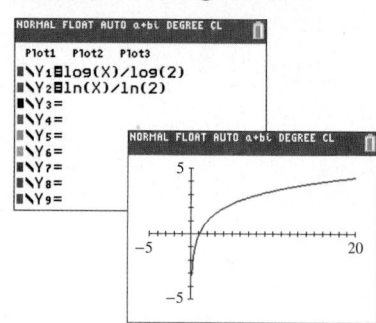

Point of Interest

The slide rule, first built in England in the early 17th century, is a mechanical computing device that uses logarithmic scales to perform operations involving multiplication, division, roots, logarithms, exponentials, and trigonometry. Amazingly, slide rules were used into the space age by engineers in the 1960s to help send astronauts to the moon. It was only with the invention of the pocket calculator that slide rules were replaced by modern computing devices.

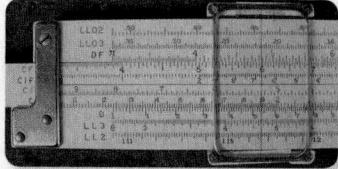

Julie Miller

Answers

8. a. Between 1 and 2
   b. 1.7500
   c. 1.7500
   d. $6^{1.7500} \approx 23$

SECTION 9.4   Practice Exercises

Prerequisite Review

For Exercises R.1–R.4, use the properties of exponents to simplify the expression. (See Section 2.1 for review.)

R.1. $x^{-3} \cdot x^5 \cdot x^7$

R.2. $\dfrac{y^{-2}y^{10}}{y^3}$

R.3. $\left(4w^{-3}z^4\right)^2$

R.4. $\left(\dfrac{7k^4}{n}\right)^{-3}$

For Exercises R.5–R.12, evaluate the logarithms. (See Section 9.3 for review.)

R.5. $\log_5 625$ **R.6.** $\log_2 128$ **R.7.** $\log_3 \dfrac{1}{27}$ **R.8.** $\log_4 \dfrac{1}{16}$

R.9. $\ln \sqrt{e}$ **R.10.** $\ln e^3$ **R.11.** $\log 10{,}000$ **R.12.** $\log 10{,}000{,}000$

For Exercises R.13–R.14, simplify the expression. (See Section 2.1 for review.)

R.13. $\dfrac{120x^5y^8}{15x^7y}$ **R.14.** $\dfrac{126ab^3}{21ab^6}$

Concept Connections

1. The product property of logarithms states that $\log_b(xy) = $ _____ for positive real numbers b, x, and y, where $b \neq 1$.

2. The _____ property of logarithms states that $\log_b\left(\dfrac{x}{y}\right) = $ _____ for positive real numbers b, x, and y, where $b \neq 1$.

3. The power property of logarithms states that for any real number p, $\log_b x^p = $ _____ for positive real numbers b, x, and y, where $b \neq 1$.

4. The change-of-base formula states that $\log_b x$ can be written as a ratio of logarithms with base a as

$$\log_b x = \frac{\Box}{\Box}$$

5. The change-of-base formula is often used to convert a logarithm to a ratio of logarithms with base _____ or base _____ so that a calculator can be used to approximate the logarithm.

6. To use a graphing utility to graph the function defined by $y = \log_5 x$, use the change-of-base formula to write the function as $y = $ _____ or $y = $ _____ .

For the exercises in this set, assume that all variable expressions represent positive real numbers.

Objective 1: Apply the Product, Quotient, and Power Properties of Logarithms

For Exercises 7–12, use the product property of logarithms to write the logarithm as a sum of logarithms. Then simplify if possible. **(See Example 1)**

7. $\log_5(125z)$ **8.** $\log_7(49k)$ **9.** $\log(8cd)$

10. $\log(24vw)$ **11.** $\log_2[(x+y) \cdot z]$ **12.** $\log_3[(a+b) \cdot c]$

For Exercises 13–18, use the quotient property of logarithms to write the logarithm as a difference of logarithms. Then simplify if possible. **(See Example 2)**

13. $\log_{12}\left(\dfrac{p}{q}\right)$ **14.** $\log_9\left(\dfrac{m}{n}\right)$ **15.** $\ln\left(\dfrac{e}{5}\right)$

16. $\ln\left(\dfrac{x}{e}\right)$ **17.** $\log\left(\dfrac{m^2+n}{100}\right)$ **18.** $\log\left(\dfrac{1000}{c^2+1}\right)$

For Exercises 19–24, apply the power property of logarithms. **(See Example 3)**

19. $\log(2x-3)^4$ **20.** $\log(8t-3)^2$ **21.** $\log_6 \sqrt[7]{x^3}$

22. $\log_8 \sqrt[4]{x^3}$ **23.** $\ln 2^{kt}$ **24.** $\ln(0.5)^{rt}$

Objective 2: Write a Logarithmic Expression in Expanded Form

For Exercises 25–44, write the logarithm as a sum or difference of logarithms. Simplify each term as much as possible. **(See Example 4)**

25. $\log_4(7yz)$ **26.** $\log_2(5ab)$ **27.** $\log_7\left(\dfrac{1}{7}mn^2\right)$

28. $\log_4\left(\dfrac{1}{16}t^3v\right)$ **29.** $\log_2\left(\dfrac{x^{10}}{yz}\right)$ **30.** $\log_5\left(\dfrac{p^5}{mn}\right)$

31. $\log_6\left(\dfrac{p^5}{qt^3}\right)$

32. $\log_8\left(\dfrac{a^4}{b^9c}\right)$

33. $\log\left(\dfrac{10}{\sqrt{a^2+b^2}}\right)$

34. $\log\left(\dfrac{\sqrt{d^2+1}}{10,000}\right)$

35. $\ln\left(\dfrac{\sqrt[3]{xy}}{wz^2}\right)$

36. $\ln\left(\dfrac{\sqrt[4]{pq}}{t^3m}\right)$

37. $\ln\sqrt[4]{\dfrac{a^2+4}{e^3}}$

38. $\ln\sqrt[5]{\dfrac{e^2}{c^2+5}}$

39. $\log\left[\dfrac{2x(x^2+3)^8}{\sqrt{4-3x}}\right]$

40. $\log\left[\dfrac{5y(4x+1)^7}{\sqrt[3]{2-7x}}\right]$

41. $\log_5\sqrt[3]{x\sqrt{5}}$

42. $\log_2\sqrt[4]{y\sqrt{2}}$

43. $\log_2\left[\dfrac{4a^2\sqrt{3-b}}{c(b+4)^2}\right]$

44. $\log_3\left[\dfrac{27x^3\sqrt{y^2-1}}{y(x-1)^2}\right]$

Objective 3: Write a Logarithmic Expression as a Single Logarithm

For Exercises 45–68, write the logarithmic expression as a single logarithm with coefficient 1, and simplify as much as possible. **(See Examples 5–6)**

45. $\ln y + \ln 4$

46. $\log 5 + \log p$

47. $\log_{15} 3 + \log_{15} 5$

48. $\log_{12} 8 + \log_{12} 18$

49. $\log_7 98 - \log_7 2$

50. $\log_6 144 - \log_6 4$

51. $\log 150 - \log 3 - \log 5$

52. $\log_3 693 - \log_3 33 - \log_3 7$

53. $2\log_2 x + \log_2 t$

54. $5\log_4 y + \log_4 w$

55. $4\log_8 m - 3\log_8 n - 2\log_8 p$

56. $8\log_3 x - 2\log_3 z - 7\log_3 y$

57. $3[\ln x - \ln(x+3) - \ln(x-3)]$

58. $2[\log(p-4) - \log(p-1) - \log(p+4)]$

59. $\dfrac{1}{2}\ln(x+1) - \dfrac{1}{2}\ln(x-1)$

60. $\dfrac{1}{3}\ln(x^2+1) - \dfrac{1}{3}\ln(x+1)$

61. $6\log x - \dfrac{1}{3}\log y - \dfrac{2}{3}\log z$

62. $15\log c - \dfrac{1}{4}\log d - \dfrac{3}{4}\log k$

63. $\dfrac{1}{3}\log_4 p + \log_4(q^2-16) - \log_4(q-4)$

64. $\dfrac{1}{4}\log_2 w + \log_2(w^2-100) - \log_2(w+10)$

65. $\dfrac{1}{2}[6\ln(x+2) + \ln x - \ln x^2]$

66. $\dfrac{1}{3}[12\ln(x-5) + \ln x - \ln x^3]$

67. $\log(8y^2 - 7y) + \log y^{-1}$

68. $\log(9t^3 - 5t) + \log t^{-1}$

For Exercises 69–78, use $\log_b 2 \approx 0.356$, $\log_b 3 \approx 0.565$, and $\log_b 5 \approx 0.827$ to approximate the value of the given logarithms. **(See Example 7)**

69. $\log_b 15$

70. $\log_b 10$

71. $\log_b 81$

72. $\log_b 125$

73. $\log_b 50$

74. $\log_b 12$

75. $\log_b\left(\dfrac{15}{2}\right)$

76. $\log_b\left(\dfrac{6}{5}\right)$

77. $\log_b 100$

78. $\log_b 225$

Objective 4: Apply the Change-of-Base Formula

For Exercises 79–84, **(See Example 8)**

a. Estimate the value of the logarithm between two consecutive integers. For example, $\log_2 7$ is between 2 and 3 because $2^2 < 7 < 2^3$.

b. Use the change-of-base formula and a calculator to approximate the logarithm to 4 decimal places.

c. Check the result by using the related exponential form.

79. $\log_2 15$

80. $\log_3 15$

81. $\log_5 3$

82. $\log_8 5$

83. $\log_2 0.3$

84. $\log_2 0.2$

For Exercises 85–88, use the change-of-base formula and a calculator to approximate the given logarithms. Round to 4 decimal places. Then check the answer by using the related exponential form. (**See Example 8**)

85. $\log_2(4.68 \times 10^7)$ **86.** $\log_2(2.54 \times 10^{10})$ **87.** $\log_4(5.68 \times 10^{-6})$ **88.** $\log_4(9.84 \times 10^{-5})$

Mixed Exercises

For Exercises 89–98, determine if the statement is true or false. For each false statement, provide a counterexample. For example, $\log(x + y) \ne \log x + \log y$ because $\log(2 + 8) \ne \log 2 + \log 8$ (the left side is 1 and the right side is approximately 1.204).

89. $\log e = \dfrac{1}{\ln 10}$

90. $\ln 10 = \dfrac{1}{\log e}$

91. $\log_5\left(\dfrac{1}{x}\right) = \dfrac{1}{\log_5 x}$

92. $\log_6\left(\dfrac{1}{t}\right) = \dfrac{1}{\log_6 t}$

93. $\log_4\left(\dfrac{1}{p}\right) = -\log_4 p$

94. $\log_8\left(\dfrac{1}{w}\right) = -\log_8 w$

95. $\log(xy) = (\log x)(\log y)$

96. $\log\left(\dfrac{x}{y}\right) = \dfrac{\log x}{\log y}$

97. $\log_2(7y) + \log_2 1 = \log_2(7y)$

98. $\log_4(3d) + \log_4 1 = \log_4(3d)$

Write About It

99. Explain why the product property of logarithms does not apply to the following statement.

$$\log_5(-5) + \log_5(-25)$$
$$= \log_5[(-5)(-25)]$$
$$= \log_5 125 = 3$$

100. Explain how to use the change-of-base formula and explain why it is important.

Expanding Your Skills

101. a. Write the difference quotient for $f(x) = \ln x$.

    b. Show that the difference quotient from part (a) can be written as $\ln\left(\dfrac{x + h}{x}\right)^{1/h}$.

102. Show that

$$-\ln\left(x - \sqrt{x^2 - 1}\right) = \ln\left(x + \sqrt{x^2 - 1}\right)$$

103. Show that

$$\log\left(\dfrac{-b + \sqrt{b^2 - 4ac}}{2a}\right) + \log\left(\dfrac{-b - \sqrt{b^2 - 4ac}}{2a}\right)$$
$$= \log c - \log a$$

104. Show that

$$\ln\left(\dfrac{c + \sqrt{c^2 - x^2}}{c - \sqrt{c^2 - x^2}}\right) = 2\ln\left(c + \sqrt{c^2 - x^2}\right) - 2\ln x$$

105. Use the change-of-base formula to write $(\log_2 5)(\log_5 9)$ as a single logarithm.

106. Use the change-of-base formula to write $(\log_3 11)(\log_{11} 4)$ as a single logarithm.

107. Prove the quotient property of logarithms:

$$\log_b\left(\dfrac{x}{y}\right) = \log_b x - \log_b y.$$

(*Hint*: Modify the proof of the product property given on page 658.)

108. Prove the power property of logarithms:
$\log_b x^p = p \log_b x.$

Technology Connections

For Exercises 109–112, graph the function.

109. $f(x) = \log_5(x + 4)$ **110.** $g(x) = \log_7(x - 3)$

111. $k(x) = -3 + \log_{1/2} x$ **112.** $h(x) = 4 + \log_{1/3} x$

113. a. Graph $Y_1 = \log|x|$ and $Y_2 = \dfrac{1}{2}\log x^2$. How are the graphs related?

    b. Show algebraically that $\dfrac{1}{2}\log x^2 = \log|x|$.

114. Graph $Y_1 = \ln(0.1x)$, $Y_2 = \ln(0.5x)$, $Y_3 = \ln x$, and $Y_4 = \ln(2x)$. How are the graphs related? Support your answer algebraically.

SECTION 9.5 | Exponential Equations and Applications

OBJECTIVES

1. Solve Exponential Equations
2. Use Exponential Equations in Applications

1. Solve Exponential Equations

A couple invests $8000 in a bond fund. The expected yield is 4.5% and the earnings are reinvested monthly. The growth of the investment is modeled by

$$A = 8000\left(1 + \frac{0.045}{12}\right)^{12t}$$ where A is the amount in the account after t years.

If the couple wants to know how long it will take for the investment to double, they would solve the equation:

$$16{,}000 = 8000\left(1 + \frac{0.045}{12}\right)^{12t}$$ (See Example 6.)

Comstock Getty Images

This equation is called an **exponential equation** because the equation contains a variable in the exponent. To solve an exponential equation, first note that all exponential functions are one-to-one. Therefore, $b^x = b^y$ implies that $x = y$. This is called the equivalence property of exponential expressions.

> **TIP** The equivalence property tells us that if two exponential expressions with the same base are equal, then their exponents must be equal.

Equivalence Property of Exponential Expressions

If b, x, and y are real numbers with $b > 0$ and $b \neq 1$, then

$$b^x = b^y \quad \text{implies that } x = y.$$

EXAMPLE 1 | Solving Exponential Equations Using the Equivalence Property

Solve.   **a.** $3^{2x-6} = 81$ **b.** $25^{4-t} = \left(\frac{1}{5}\right)^{3t+1}$

Solution:

a.
$$3^{2x-6} = 81$$
$$3^{2x-6} = 3^4 \qquad \text{Write 81 as an exponential expression with a base of 3.}$$
$$2x - 6 = 4 \qquad \text{Equate the exponents.}$$
$$x = 5$$

Check: $3^{2x-6} = 81$
$3^{2(5)-6} \overset{?}{=} 81$
$3^4 \overset{?}{=} 81$ ✓

The solution set is {5}.

> **Avoiding Mistakes**
>
> When writing the expression $(5^2)^{4-t}$ as $5^{2(4-t)}$, it is important to use parentheses around the quantity $(4 - t)$. The exponent of 2 must be multiplied by the entire quantity $(4 - t)$. Likewise, parentheses are used around $(3t + 1)$ in the expression $5^{-1(3t+1)}$.

b.
$$25^{4-t} = \left(\frac{1}{5}\right)^{3t+1}$$
$$\left(5^2\right)^{4-t} = \left(5^{-1}\right)^{3t+1} \qquad \text{Express both 25 and } \tfrac{1}{5} \text{ as integer powers of 5.}$$
$$5^{2(4-t)} = 5^{-1(3t+1)} \qquad \text{Apply the power property of exponents: } (b^m)^n = b^{m \cdot n}.$$
$$5^{8-2t} = 5^{-3t-1} \qquad \text{Apply the distributive property within the exponents.}$$
$$8 - 2t = -3t - 1 \qquad \text{Equate the exponents.}$$
$$t = -9 \qquad \text{The solution checks in the original equation.}$$

The solution set is {−9}.

Skill Practice 1   Solve.   **a.** $4^{2x-3} = 64$ **b.** $27^{2w+5} = \left(\frac{1}{3}\right)^{2-5w}$

Answers

1. a. {3}   **b.** {−17}

In Example 1, we were able to write the left and right sides of the equation with a common base. However, most exponential equations cannot be written in this form by inspection. For example:

$$7^x = 60$$
$$7^x = 7^?$$

60 is not a recognizable power of 7.

To solve such an equation, we can take a logarithm of the same base on each side of the equation, and then apply the power property of logarithms. This is demonstrated in Examples 2–4.

Steps to Solve Exponential Equations by Using Logarithms

1. Isolate the exponential expression on one side of the equation.
2. Take a logarithm of the same base on both sides of the equation.
3. Use the power property of logarithms to "bring down" the exponent.
4. Solve the resulting equation.

EXAMPLE 2   Solving an Exponential Equation Using Logarithms

Solve. $7^x = 60$

Solution:

$7^x = 60$	The exponential expression 7^x is isolated.
$\log 7^x = \log 60$	Take a logarithm of the same base on both sides of the equation. In this case, we have chosen base 10.
$x \log 7 = \log 60$	Apply the power property of logarithms.
$x = \dfrac{\log 60}{\log 7} \approx 2.1041$	Divide both sides by $\log 7$.

This equation is now linear.

It is important to note that the exact solution to this equation is $\dfrac{\log 60}{\log 7}$ or equivalently by the change-of-base formula, $\log_7 60$. The value 2.1041 is merely an approximation.

The solution set is $\left\{ \dfrac{\log 60}{\log 7} \right\}$ or $\{\log_7 60\}$.

Skill Practice 2   Solve. $5^x = 83$

To solve the equation from Example 2, we can take a logarithm of any base. For example:

$$7^x = 60$$
$$\log_7 7^x = \log_7 60$$
$$x = \log_7 60 \quad \text{(solution)}$$

Take the logarithm base 7 on both sides.

$$7^x = 60$$
$$\ln 7^x = \ln 60$$
$$x \ln 7 = \ln 60$$
$$x = \frac{\ln 60}{\ln 7} \quad \text{(solution)}$$

Take the natural logarithm on both sides.

The values $\log_7 60$, $\dfrac{\log 60}{\log 7}$, and $\dfrac{\ln 60}{\ln 7}$ are all equivalent. However, common logarithms and natural logarithms are often used to express the solution to an exponential equation so that the solution can be approximated on a calculator.

FOR REVIEW

Recall that the power property of logarithms indicates that $\log_b x^p = p \log_b x$. See Section 9.4 for review.

Avoiding Mistakes

While 2.1041 is only an approximation, it is useful to check the result.

$$7^{2.1041} \approx 60$$

Answer

2. $\left\{ \dfrac{\log 83}{\log 5} \right\}$ or $\{\log_5 83\}$

EXAMPLE 3 Solving Exponential Equations Using Logarithms

Solve. **a.** $10^{5+2x} + 820 = 49{,}600$ **b.** $2000 = 18{,}000e^{-0.4t}$

Solution:

a. $10^{5+2x} + 820 = 49{,}600$

$\quad\quad 10^{5+2x} = 48{,}780$

Isolate the exponential expression on the left by subtracting 820 on both sides.

$\quad \log 10^{5+2x} = \log 48{,}780$

Since the exponential expression on the left has a base of 10, take the log base 10 on both sides.

$\quad\quad 5 + 2x = \log 48{,}780$

On the left, $\log 10^{5+2x} = 5 + 2x$.

$\quad\quad\quad 2x = \log 48{,}780 - 5$

Solve the linear equation by subtracting 5 and dividing by 2.

$\quad\quad\quad x = \dfrac{\log 48{,}780 - 5}{2} \approx -0.1559$

The solution checks in the original equation.

The solution set is $\left\{ \dfrac{\log 48{,}780 - 5}{2} \right\}$.

> **TIP** The equation $5 + 2x = \log 48{,}780$ looks cumbersome, but note that the term $\log 48{,}780$ is simply a constant. We solve the equation $5 + 2x = \log 48{,}780$ in the same manner that we would solve the simpler-looking equation $5 + 2x = 4.7$.
> We subtract 5 from both sides and then divide by 2.

b. $\quad 2000 = 18{,}000e^{-0.4t}$

$\quad \dfrac{2000}{18{,}000} = e^{-0.4t}$

Isolate the exponential expression on the right by dividing both sides by 18,000.

$\quad\quad \dfrac{1}{9} = e^{-0.4t}$

Since the exponential expression on the right has a base of e, take the log base e on both sides.

$\quad \ln\left(\dfrac{1}{9}\right) = \ln e^{-0.4t}$

On the right, $\ln e^{-0.4t} = -0.4t$.

$\quad \ln\left(\dfrac{1}{9}\right) = -0.4t$

Solve the linear equation by dividing by -0.4.

$\quad \dfrac{\ln\left(\dfrac{1}{9}\right)}{-0.4} = t$

linear equation

The exact solution to the equation can be written in a variety of forms by applying the properties of logarithms:

$\dfrac{\ln\left(\dfrac{1}{9}\right)}{-0.4} = \dfrac{\ln 1 - \ln 9}{-0.4} = \dfrac{0 - \ln 9}{-0.4} = \dfrac{\ln 9}{0.4} \approx 5.4931$

Alternatively, $\dfrac{\ln 9}{0.4} = \dfrac{\ln 9}{\frac{2}{5}} = \dfrac{5\ln 9}{2} \approx 5.4931$

The solution set is $\left\{ \dfrac{\ln 9}{0.4} \right\}$ or $\left\{ \dfrac{5 \ln 9}{2} \right\}$.

> **TIP** The equation $\ln\left(\dfrac{1}{9}\right) = -0.4t$ is a linear equation. Compare this equation to the similar equation $-2 = -0.4t$. In each case we isolate t by dividing both sides by -0.4.

Skill Practice 3 Solve.

a. $400 + 10^{4x-1} = 63{,}000$ **b.** $100 = 700e^{-0.2k}$

Answers

3. a. $\left\{ \dfrac{\log 62{,}600 + 1}{4} \right\}$

b. $\left\{ \dfrac{\ln 7}{0.2} \right\}$ or $\{5 \ln 7\}$

In Example 4, we have an equation with two exponential expressions involving different bases.

EXAMPLE 4   Solving an Exponential Equation

Solve. $4^{2x-7} = 5^{3x+1}$

Solution:

$$4^{2x-7} = 5^{3x+1}$$

$$\ln 4^{2x-7} = \ln 5^{3x+1} \qquad \text{Take a logarithm of the same base on both sides.}$$

$$(2x - 7)\ln 4 = (3x + 1)\ln 5 \qquad \text{Apply the power property of logarithms.}$$

$$(2x - 7)\ln 4 = (3x + 1)\ln 5 \qquad \begin{array}{l}\text{The resulting equation is a first-degree equation} \\ \text{in } x. \text{ The equation is linear.}\end{array}$$

$$2x \ln 4 - 7 \ln 4 = 3x \ln 5 + \ln 5 \qquad \text{Apply the distributive property.}$$

$$2x \ln 4 - 3x \ln 5 = \ln 5 + 7 \ln 4 \qquad \begin{array}{l}\text{Collect } x \text{ terms on one side of} \\ \text{the equation.}\end{array}$$

$$x(2 \ln 4 - 3 \ln 5) = \ln 5 + 7 \ln 4 \qquad \text{Factor out } x \text{ on the left.}$$

$$x = \frac{\ln 5 + 7 \ln 4}{2 \ln 4 - 3 \ln 5} \approx -5.5034 \qquad \text{Divide by } (2 \ln 4 - 3 \ln 5).$$

The solution set is $\left\{ \dfrac{\ln 5 + 7 \ln 4}{2 \ln 4 - 3 \ln 5} \right\}$.   The solution checks in the original equation.

Skill Practice 4   Solve. $3^{5x-6} = 2^{4x+1}$

TIP   In the second step of Example 4, we took the natural logarithm (logarithm of base e) on both sides of the equation. We could have used a logarithm of any base. However, we usually use base e or base 10 because the result can be easily approximated on a calculator. For example, we could have solved the equation in Example 4 in terms of the logarithm base 10.

$$4^{2x-7} = 5^{3x+1}$$
$$\log 4^{2x-7} = \log 5^{3x+1}$$
$$(2x - 7)\log 4 = (3x + 1)\log 5$$

Using the same series of steps as in Example 4, we have

$$x = \frac{\log 5 + 7 \log 4}{2 \log 4 - 3 \log 5} \approx -5.5034.$$

In Example 5, we will solve an exponential equation that is quadratic in form. Recall from Section 4.7 that an equation is quadratic in form if after making a substitution of the form $u = f(x)$, the equation can be written as $au^2 + bu + c = 0$. For example, given $2(x^2 - 1)^2 + 11(x^2 - 1) - 21 = 0$, we can write this as a quadratic equation in u by making the substitution $u = x^2 - 1$.

$$2(x^2 - 1)^2 + 11(x^2 - 1) - 21 = 0 \xrightarrow{\text{Let } u = x^2 - 1} 2u^2 + 11u - 21 = 0$$

We also want to recall that an expression of the form $(b^m)^n$ equals $b^{m \cdot n}$. That is, for a base raised to multiple powers, we multiply the exponents. Therefore, the expression e^{2x} can also be written as $(e^x)^2$. This shows us that the expression e^{2x} is a perfect square.

Answer

4. $\left\{ \dfrac{\ln 2 + 6 \ln 3}{5 \ln 3 - 4 \ln 2} \right\}$

EXAMPLE 5   Solving an Exponential Equation
in Quadratic Form

Solve. $e^{2x} + 5e^x - 36 = 0$

Solution:

$$e^{2x} + 5e^x - 36 = 0$$

$$(e^x)^2 + 5(e^x) - 36 = 0 \qquad \text{Note that } e^{2x} = (e^x)^2.$$

$$u^2 + 5u - 36 = 0 \qquad \text{The equation is in quadratic form. Let } u = e^x.$$

$$(u - 4)(u + 9) = 0 \qquad \text{Factor.}$$

$$u = 4 \quad \text{or} \quad u = -9$$

$$e^x = 4 \quad \text{or} \quad e^x = -9 \qquad \text{Back substitute. The second equation } e^x = -9 \text{ has no}$$

$$\ln e^x = \ln 4 \qquad\qquad\qquad \text{solution.}$$

$$x = \ln 4 \approx 1.3863 \quad \boxed{\text{No solution to this equation because } \ln(-9) \text{ is undefined.}}$$

The solution set is $\{\ln 4\}$.   The solution checks in the original equation.

Skill Practice 5   Solve. $e^{2x} - 5e^x - 14 = 0$

TIP   In Example 5, we reduced the original equation into two simpler equations $e^x = 4$ and $e^x = -9$. By inspection, we know that the second equation has no solution. The values of e^x are positive for all real numbers x. This is similar to the equation $\sqrt{x} = -9$. There is no solution to this equation either because the principal square root of any nonnegative real number x is always greater than or equal to zero.

2. Use Exponential Equations in Applications

In Example 6, we solve applications involving exponential equations.

EXAMPLE 6   Using an Exponential Equation
in a Finance Application

A couple invests $8000 in a bond fund. The expected yield is 4.5% and the earnings are reinvested monthly.

TIP   Recall that monthly compounding indicates that interest is computed $n = 12$ times per year.

a. Use $A = P\left(1 + \dfrac{r}{n}\right)^{nt}$ to write a model representing the amount A (in $) in the account after t years. The value r is the interest rate and n is the number of times interest is compounded per year.

b. Determine how long it will take the initial investment to double. Round to 1 decimal place.

Solution:

a. $A = P\left(1 + \dfrac{r}{n}\right)^{nt}$

$A = 8000\left(1 + \dfrac{0.045}{12}\right)^{12t} \qquad \text{Substitute } P = 8000, r = 0.045, \text{ and } n = 12.$

Answer

5. $\{\ln 7\}$

b. $16{,}000 = 8000\left(1 + \dfrac{0.045}{12}\right)^{12t}$

The couple wants to double their money from $8000 to $16,000. Substitute $A = 16{,}000$ and solve for t.

$2 = \left(1 + \dfrac{0.045}{12}\right)^{12t}$

Isolate the exponential expression by dividing both sides by 8000.

$\ln 2 = \ln\left(1 + \dfrac{0.045}{12}\right)^{12t}$

Take a logarithm of the same base on both sides. We have chosen to use the natural logarithm.

TIP The equation can be written as

$\ln 2 = 12 \ln\left(1 + \dfrac{0.045}{12}\right)t.$

In this form, it might be easier to visualize the coefficient on the t term.

$\cdots\cdots\cdots\cdots \ln 2 = 12t \ln\left(1 + \dfrac{0.045}{12}\right)$

Apply the power property of logarithms. The equation is now linear in the variable t.

$\dfrac{\ln 2}{12 \ln\left(1 + \dfrac{0.045}{12}\right)} = t$

Divide both sides by $12 \ln\left(1 + \frac{0.045}{12}\right)$.

$t \approx 15.4$

It will take approximately 15.4 years for the investment to double.

Skill Practice 6   Determine how long it will take $8000 compounded monthly at 6% to double. Round to 1 decimal place.

Answer

6. 11.6 years

SECTION 9.5 | Practice Exercises

Prerequisite Review

For Exercises R.1–R.4, simplify the expression. (See Section 2.1 for review.)

R.1. $\left(5^2\right)^x$ **R.2.** $\left(6^3\right)^x$ **R.3.** $\left(3^5\right)^{x-4}$ **R.4.** $\left(4^2\right)^{3x+1}$

For Exercises R.5–R.8, give two consecutive integers between which the solution to the equation is found. For example, the solution to $2^x = 14$ is between 3 and 4. (See Section 2.1 for review.)

R.5. $5^x = 120$ **R.6.** $6^x = 10$ **R.7.** $2^t = 80$ **R.8.** $3^y = 70$

For Exercises R.9–R.10, write the expression with rational exponents. (See Section 4.5 for review.)

R.9. $\sqrt[3]{7}$ **R.10.** $\dfrac{1}{\sqrt{2}}$

For Exercises R.11–R.14, simplify the logarithmic expression. (See Section 9.3 for review.)

R.11. $\log 10^{3x}$ **R.12.** $\log \dfrac{1}{10}$ **R.13.** $\ln e^5$ **R.14.** $\ln \sqrt[3]{e}$

For Exercises R.15–R.16, apply the power property of logarithms. (See Section 9.4 for review.)

R.15. $\log 3^{x-4}$ **R.16.** $\ln 5^{2x+7}$

For Exercises R.17–R.18, write an equation for the inverse function. (See Section 9.1 for review.)

R.17. $g(x) = \sqrt[3]{2x - 1}$ **R.18.** $h(x) = (x + 4)^3$

For Exercises R.19–R.20, solve the equation. (See Section 4.7 for review.)

R.19. $\left(x^2 - 2x\right)^2 - 7\left(x^2 - 2x\right) - 8 = 0$ **R.20.** $\left(x^2 + 10x\right)^2 - 2\left(x^2 + 10x\right) - 99 = 0$

Concept Connections

1. An equation such as $4^x = 9$ is called an _____ equation because the equation contains a variable in the exponent.

2. The equivalence property of exponential expressions states that if $b^x = b^y$, then _____ = _____.

3. What are the steps required to solve the equation $2x + 3 = \ln 417$?

4. What are the steps required to solve the equation $4 = e^{3t}$?

Objective 1: Solve Exponential Equations

For Exercises 5–16, solve the equation. (**See Example 1**)

5. $3^x = 81$

6. $2^x = 32$

7. $\sqrt[3]{5} = 5^t$

8. $\sqrt{3} = 3^w$

9. $2^{-3y+1} = 16$

10. $5^{2z+2} = 625$

11. $11^{3c+1} = \left(\dfrac{1}{11}\right)^{c-5}$

12. $7^{2x-3} = \left(\dfrac{1}{49}\right)^{x+1}$

13. $8^{2x-5} = 32^{x-6}$

14. $27^{x-4} = 9^{2x+1}$

15. $100^{3t-5} = 1000^{3-t}$

16. $100{,}000^{2w+1} = 10{,}000^{4-w}$

For Exercises 17–34, solve the equation. Write the solution set with the exact values given in terms of common or natural logarithms. Also give approximate solutions to 4 decimal places. (**See Examples 2–5**)

17. $6^t = 87$

18. $2^z = 70$

19. $1024 = 19^x + 4$

20. $801 = 23^y + 6$

21. $10^{3+4x} - 8100 = 120{,}000$

22. $10^{5+8x} + 4200 = 84{,}000$

23. $21{,}000 = 63{,}000e^{-0.2t}$

24. $80 = 320e^{-0.5t}$

25. $4e^{2n-5} + 3 = 11$

26. $5e^{4m-3} - 7 = 13$

27. $3^{6x+5} = 5^{2x}$

28. $7^{4x-1} = 3^{5x}$

29. $2^{1-6x} = 7^{3x+4}$

30. $11^{1-8x} = 9^{2x+3}$

31. $e^{2x} - 9e^x - 22 = 0$

32. $e^{2x} - 6e^x - 16 = 0$

33. $e^{2x} = -9e^x$

34. $e^{2x} = -7e^x$

Objective 2: Use Exponential Equations in Applications

For Exercises 35–44, use the model $A = Pe^{rt}$ or $A = P\left(1 + \dfrac{r}{n}\right)^{nt}$, where A is the future value of P dollars invested at interest rate r compounded continuously or n times per year for t years. (**See Example 6**)

35. If $10,000 is invested in an account earning 5.5% interest compounded continuously, determine how long it will take the money to triple. Round to the nearest year.

36. If a couple has $80,000 in a retirement account, how long will it take the money to grow to $1,000,000 if it grows by 6% compounded continuously? Round to the nearest year.

37. A $2500 bond grows to $3729.56 in 10 years under continuous compounding. Find the interest rate. Round to the nearest whole percent.

38. $5000 grows to $5438.10 in 2 years under continuous compounding. Find the interest rate. Round to the nearest tenth of a percent.

39. An $8000 investment grows to $9289.50 at 3% interest compounded quarterly. For how long was the money invested? Round to the nearest year.

40. $20,000 is invested at 3.5% interest compounded monthly. How long will it take for the investment to double? Round to the nearest tenth of a year.

41. A $25,000 inheritance is invested for 15 years compounded quarterly and grows to $52,680. Find the interest rate. Round to the nearest percent.

42. A $10,000 investment grows to $11,273 in 4 years compounded monthly. Find the interest rate. Round to the nearest percent.

43. If $4000 is put aside in a money market account with interest compounded continuously at 2.2%, find the time required for the account to *earn* $1000. Round to the nearest month.

44. Victor puts aside $10,000 in an account with interest compounded continuously at 2.7%. How long will it take for him to *earn* $2000? Round to the nearest month.

45. Physicians often treat thyroid cancer with a radioactive form of iodine called iodine-131 (^{131}I). The radiological half-life of ^{131}I is approximately 8 days, but the biological half-life for most individuals is 4.2 days. The biological half-life is shorter because in addition to ^{131}I being lost to decay, the iodine is also excreted from the body in urine, sweat, and saliva.

For a patient treated with 100 mCi (millicuries) of ^{131}I, the radioactivity level R (in mCi) after t days is given by $R = 100(2)^{-t/4.2}$.

  a. State law mandates that the patient stay in an isolated hospital room for 2 days after treatment with ^{131}I. Determine the radioactivity level at the end of 2 days. Round to the nearest whole unit.

  b. After the patient is released from the hospital, the patient is directed to avoid direct human contact until the radioactivity level drops below 30 mCi. For how many days *after* leaving the hospital will the patient need to stay in isolation? Round to the nearest tenth of a day.

46. Caffeine occurs naturally in a variety of food products such as coffee, tea, and chocolate. The kidneys filter the blood and remove caffeine and other drugs through urine. The biological half-life of caffeine is approximately 6 hr. If one cup of coffee has 80 mg of caffeine, then the amount of caffeine C (in mg) remaining after t hours is given by $C = 80(2)^{-t/6}$.

  a. How long will it take for the amount of caffeine to drop below 60 mg? Round to 1 decimal place.

  b. Laura has trouble sleeping if she has more than 30 mg of caffeine in her bloodstream. How many hours after drinking a cup of coffee would Laura have to wait so that the coffee would not disrupt her sleep? Round to 1 decimal place.

Sunlight is absorbed in water, and as a result the light intensity in oceans, lakes, and ponds decreases exponentially with depth. The percentage of visible light, P (in decimal form), at a depth of x meters is given by $P = e^{-kx}$, where k is a constant related to the clarity and other physical properties of the water. The graph shows models for the open ocean, Lake Tahoe, and Lake Erie for data taken under similar conditions. Use these models for Exercises 47–50.

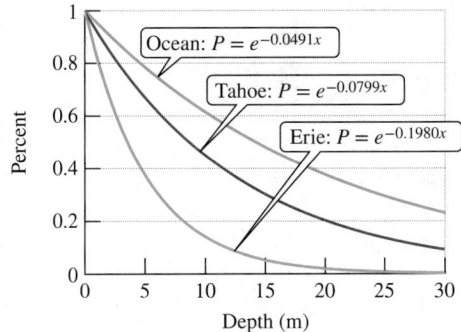

Percent of Surface Light vs. Depth

Ocean: $P = e^{-0.0491x}$

Tahoe: $P = e^{-0.0799x}$

Erie: $P = e^{-0.1980x}$

47. Determine the depth at which the light intensity is half the value from the surface for each body of water given. Round to the nearest tenth of a meter.

48. Determine the depth at which the light intensity is 20% of the value from the surface for each body of water given. Round to the nearest tenth of a meter.

49. The *euphotic* depth is the depth at which light intensity falls to 1% of the value at the surface. This depth is of interest to scientists because no appreciable photosynthesis takes place. Find the euphotic depth for the open ocean. Round to the nearest tenth of a meter.

50. Refer to Exercise 49, and find the euphotic depth for Lake Tahoe and for Lake Erie. Round to the nearest tenth of a meter.

51. Forge welding is a process in which two pieces of steel are joined together by heating the pieces of steel and hammering them together. A welder takes a piece of steel from a forge at 1600°F and places it on an anvil where the outdoor temperature is 50°F. The temperature of the steel T (in °F) can be modeled by $T = 50 + 1550e^{-0.05t}$, where t is the time in minutes after the steel is removed from the forge. How long will it take for the steel to reach a temperature of 100°F so that it can be handled without heat protection? Round to the nearest minute.

52. A pie comes out of the oven at 325°F and is placed to cool in a 70°F kitchen. The temperature of the pie T (in °F) after t minutes is given by $T = 70 + 255e^{-0.017t}$. The pie is cool enough to cut when the temperature reaches 110°F. How long will this take? Round to the nearest minute.

Mixed Exercises

For Exercises 53–80, solve the equation. Write the solution set with exact solutions.

53. $5^x = 625$

54. $3^x = 81$

55. $36^x = 6$

56. $343^x = 7$

57. $8^a = 21$

58. $6^y = 39$

59. $10^t = 0.0138$

60. $10^p = 16.8125$

61. $e^{0.07h} - 6 = 9$

62. $e^{0.03k} + 7 = 11$

63. $10^{0.03h} - 5 = 20$

64. $10^{0.04k} + 17 = 30$

65. $81^{3x-4} = \dfrac{1}{243}$

66. $4^{2x-7} = \dfrac{1}{128}$

67. $e^{2x+1} = \dfrac{1}{e^3}$

68. $10^{3x-4} = \dfrac{1}{100}$

69. $3^{x+1} = 5^x$

70. $2^{x-1} = 7^x$

71. $2^{x+2} = 6^x$

72. $5^{x-2} = 3^x$

73. $e^{2x} - 8e^x - 20 = 0$

74. $e^{2x} - 3e^x - 28 = 0$

75. $10^{2x} - 7 \cdot 10^x - 18 = 0$

76. $10^{2x} - 2 \cdot 10^x - 15 = 0$

77. $2e^{2x} - 7e^x + 3 = 0$

78. $3e^{2x} - 19e^x + 6 = 0$

79. $2 \cdot 10^{2x} - 11 \cdot 10^x + 12 = 0$

80. $9 \cdot 10^{2x} - 17 \cdot 10^x + 8 = 0$

For Exercises 81–84, find an equation for the inverse function.

81. $f(x) = 2^x - 7$

82. $f(x) = 5^x + 6$

83. $f(x) = 10^{x-3} + 1$

84. $f(x) = 10^{x+2} - 4$

Expanding Your Skills

85. $e^{2x} - 8e^x + 6 = 0$

86. $e^{2x} - 6e^x + 4 = 0$

87. $e^x - 3 + 2e^{-x} = 0$

88. $e^x - 2 - 15e^{-x} = 0$

89. $\dfrac{10^x - 13 \cdot 10^{-x}}{3} = 4$

90. $\dfrac{e^x - 9e^{-x}}{2} = 4$

91. $\dfrac{e^x + e^{-x}}{2} = 4$

92. $\dfrac{e^x - e^{-x}}{2} = 3$

93. $2xe^{2x} + 2e^{2x}x^2 = 0$

94. $e^{x/2} + \dfrac{1}{2}xe^{x/2} = 0$

95. $3x^2e^{-x} - e^{-x}x^3 = 0$

96. $8xe^{-2x} - 8x^2e^{-2x} = 0$

Technology Connections

For Exercises 97–98, an equation is given in the form $Y_1(x) = Y_2(x)$. Graph Y_1 and Y_2 on a graphing utility on the window [10, 10, 1] by [10, 10, 1]. Then approximate the point(s) of intersection to approximate the solution(s) to the equation. Round to 4 decimal places.

97. $4x - e^x + 6 = 0$

98. $x^3 - e^{2x} + 4 = 0$

SECTION 9.6   Logarithmic Equations and Applications

1. Solve Logarithmic Equations

An equation containing a variable within a logarithmic expression is called a **logarithmic equation**. For example:

$$\log_2(3x - 4) = \log_2(x + 2) \quad \text{and} \quad \ln(x + 4) = 7 \quad \text{are logarithmic equations.}$$

Given an equation in which two logarithms of the same base are equated, we can apply the equivalence property of logarithms. Since all logarithmic functions are one-to-one, $\log_b x = \log_b y$ implies that $x = y$.

TIP The equivalence property tells us that if two logarithmic expressions with the same base are equal, then their arguments must be equal.

Equivalence Property of Logarithmic Expressions

If b, x, and y are positive real numbers with $b \neq 1$, then

$$\log_b x = \log_b y \quad \text{implies that } x = y.$$

EXAMPLE 1   Solving a Logarithmic Equation Using the Equivalence Property

Solve. $\log_2(3x - 4) = \log_2(x + 2)$

Solution:

$\log_2(3x - 4) = \log_2(x + 2)$	Two logarithms of the same base are equated.
$3x - 4 = x + 2$	Equate the arguments.
$2x = 6$	Solve for x.
$x = 3$	Because the domain of a logarithmic function is restricted, it is mandatory that we check all potential solutions to a logarithmic equation.

Check: $\log_2(3x - 4) = \log_2(x + 2)$

$\log_2[3(3) - 4] \overset{?}{=} \log_2[(3) + 2]$

The solution set is $\{3\}$.

$\log_2 5 \overset{?}{=} \log_2 5 \checkmark$

Skill Practice 1   Solve. $\log_2(7x - 4) = \log_2(2x + 1)$

TECHNOLOGY CONNECTIONS

Using a Calculator to View the Potential Solutions to a Logarithmic Equation

The solution to the equation in Example 1 is the x-coordinate of the point of intersection of $Y_1 = \log_2(3x - 4)$ and $Y_2 = \log_2(x + 2)$. The domain of $Y_1 = \log_2(3x - 4)$ is $\left\{x \mid x > \frac{4}{3}\right\}$ and the domain of $Y_2 = \log_2(x + 2)$ is $\{x \mid x > -2\}$. The solution to the equation $Y_1 = Y_2$ may not lie outside the domain of either function. This is why it is mandatory to check all potential solutions to a logarithmic equation.

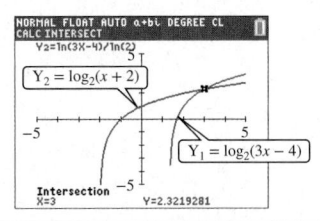

In Example 2, we illustrate that sometimes it is necessary to combine logarithms on each side of an equation before applying the equivalence property of logarithms. Furthermore, in Example 2 we encounter a situation in which one or more solutions do not check.

EXAMPLE 2   Solving a Logarithmic Equation

Solve. $\ln(x - 4) = \ln(x + 6) - \ln x$

Answer

1. $\{1\}$

FOR REVIEW

The equation

$x - 4 = \dfrac{x + 6}{x}$ is a

rational equation. To clear fractions, multiply both sides by the least common denominator x (see Section 4.3).

$$x \cdot (x - 4) = x \cdot \left(\dfrac{x + 6}{x}\right)$$
$$x^2 - 4x = x + 6$$

To solve the resulting quadratic equation, set one side of the equation to zero and factor the other side. See Section 3.4.

Solution:

$$\ln(x - 4) = \ln(x + 6) - \ln x$$

$$\ln(x - 4) = \ln\left(\dfrac{x + 6}{x}\right) \qquad \text{Combine the two logarithmic terms on the right.}$$

$$x - 4 = \dfrac{x + 6}{x} \qquad \text{Apply the equivalence property of logarithms.}$$

$$x^2 - 4x = x + 6 \qquad \text{Clear fractions by multiplying both sides by } x.$$

$$x^2 - 5x - 6 = 0 \qquad \text{The resulting equation is quadratic.}$$

$$(x - 6)(x + 1) = 0$$

$$x = 6 \quad \text{or} \quad x = -1 \qquad \text{The potential solutions are 6 and } -1.$$

Check:

$$\ln(x - 4) = \ln(x + 6) - \ln x \qquad\qquad \ln(x - 4) = \ln(x + 6) - \ln x$$
$$\ln(6 - 4) \overset{?}{=} \ln(6 + 6) - \ln 6 \qquad\qquad \ln(-1 - 4) \overset{?}{=} \ln(-1 + 6) - \ln(-1)$$
$$\ln 2 \overset{?}{=} \ln 12 - \ln 6 \qquad\qquad\qquad \ln(-5) \overset{?}{=} \ln 5 - \ln(-1)$$
$$\ln 2 \overset{?}{=} \ln\left(\dfrac{12}{6}\right) \checkmark \qquad\qquad\qquad \boxed{\text{undefined}} \quad \boxed{\text{undefined}}$$

The only solution that checks is 6.
The solution set is $\{6\}$.

Skill Practice 2 Solve. $\quad \ln x + \ln(x - 8) = \ln(x - 20)$

Many logarithmic equations, such as $4 \log_3 (2t - 7) = 8$ and $\log_2 x = 3 - \log_2 (x - 2)$, involve logarithmic terms and constant terms. In such a case, we can apply the properties of logarithms to write the equation in the form $\log_b x = k$, where k is a constant. At this point, we can solve for x by writing the equation in its equivalent exponential form $x = b^k$.

Solving Logarithmic Equations by Using Exponential Form

Step 1 Given a logarithmic equation, isolate the logarithms on one side of the equation.

Step 2 Use the properties of logarithms to write the equation in the form $\log_b x = k$, where k is a constant.

Step 3 Write the equation in exponential form.

Step 4 Solve the equation from step 3.

Step 5 Check the potential solution(s) in the original equation.

EXAMPLE 3 **Solving Logarithmic Equations**

Solve. $\quad$ **a.** $\log_5 x = -3$ $\qquad$ **b.** $\ln x = 2.4$ $\qquad$ **c.** $\log(2c - 3) = 0$

Solution:

a. $\log_5 x = -3 \qquad$ The logarithm is isolated on one side of the equation. The equation is in the form $\log_b x = k$.

$$x = 5^{-3} \qquad \text{Write the equation in exponential form.}$$

$$= \dfrac{1}{5^3} \qquad \text{Simplify the negative exponent.}$$

$$= \dfrac{1}{125}$$

Check: $\log_5 x = -3$

$$\log_5\left(\dfrac{1}{125}\right) \overset{?}{=} -3 \checkmark$$

The solution set is $\left\{\dfrac{1}{125}\right\}$.

Answer

2. $\{\ \}$; The values 4 and 5 do not check.

b. $\ln x = 2.4$ The logarithm is isolated on one side of the equation.
The equation is in the form $\log_e x = k$.

$x = e^{2.4} \approx 11.0232$ Write the equation in exponential form.
The exact solution is $e^{2.4}$ which is approximately 11.0232.

Check: $\ln x = 2.4$

$\ln\left(e^{2.4}\right) \overset{?}{=} 2.4$

The solution set is $\left\{e^{2.4}\right\}$. $2.4 \overset{?}{=} 2.4$ ✓

c. $\log(2c - 3) = 0$ The logarithm is isolated on one side of the equation.
The equation is in the form $\log_{10} x = k$, where $x = 2c - 3$.

$2c - 3 = 10^0$ Write the equation in exponential form.

$2c - 3 = 1$ Recall that for a nonzero base, $b^0 = 1$. Thus, $10^0 = 1$.

$2c = 4$ Check: $\log(2c - 3) = 0$

$c = 2$ $\log[2(2) - 3] \overset{?}{=} 0$

$\log(4 - 3) \overset{?}{=} 0$

The solution set is $\{2\}$. $\log 1 \overset{?}{=} 0$ ✓

Skill Practice 3   Solve.

a. $\log_4 x = -1$ **b.** $\ln x = 3.2$ **c.** $\log(4x - 4) = 2$

EXAMPLE 4   **Solving a Logarithmic Equation**

Solve. $4 \log_3(2t - 7) = 8$

Solution:

$4 \log_3(2t - 7) = 8$

$\log_3(2t - 7) = 2$ Isolate the logarithm by dividing both sides by 4.
The equation is in the form $\log_b x = k$, where $x = 2t - 7$.

$2t - 7 = 3^2$ Write the equation in exponential form.

$2t - 7 = 9$ Check: $4 \log_3(2t - 7) = 8$

$t = 8$ $4 \log_3[2(8) - 7] \overset{?}{=} 8$

$4 \log_3 9 \overset{?}{=} 8$

$4 \cdot 2 \overset{?}{=} 8$ ✓

The solution set is $\{8\}$.

FOR REVIEW

Recall that $\log_b x = k$ is written in exponential form as $x = b^k$. Thus, $\log_3(2t - 7) = 2$ is written as $2t - 7 = 3^2$. See Section 9.3.

Skill Practice 4   Solve. $8 \log_4(w + 6) = 24$

EXAMPLE 5   **Solving a Logarithmic Equation**

Solve. $\log(w + 47) = 2.6$

Solution:

$\log(w + 47) = 2.6$ The equation is in the form $\log_b x = k$
where $x = w + 47$ and $b = 10$.

$w + 47 = 10^{2.6}$ Write the equation in exponential form.

$w = 10^{2.6} - 47 \approx 351.1072$ Solve the resulting linear equation.

Check: $\log(w + 47) = 2.6$

$\log[(10^{2.6} - 47) + 47] \overset{?}{=} 2.6$

The solution set is $\{10^{2.6} - 47\}$. $\log 10^{2.6} \overset{?}{=} 2.6$ ✓

Answers

3. a. $\left\{\dfrac{1}{4}\right\}$

   b. $\{e^{3.2}\}$; $x \approx 24.5325$ **c.** $\{26\}$

4. $\{58\}$

5. $\{10^{1.4} + 18\}$

Skill Practice 5   Solve. $\log(t - 18) = 1.4$

Example 6 contains multiple logarithmic terms and a constant term. We apply the strategy of collecting the logarithmic terms on one side and the constant term on the other side. Then after combining the logarithmic terms, we write the equation in exponential form.

EXAMPLE 6 Solving a Logarithmic Equation

Solve. $\log_2 x = 3 - \log_2(x - 2)$

Solution:

$$\log_2 x = 3 - \log_2(x - 2)$$

$$\log_2 x + \log_2(x - 2) = 3 \qquad \text{Isolate the logarithms on one side of the equation.}$$

$$\log_2[x(x - 2)] = 3 \qquad \text{Use the product property of logarithms to write a single logarithm.}$$

$$x(x - 2) = 2^3 \qquad \text{Write the equation in exponential form.}$$

$$x^2 - 2x = 8 \qquad \text{The resulting equation is quadratic.}$$

$$x^2 - 2x - 8 = 0 \qquad \text{Set one side equal to zero.}$$

$$(x - 4)(x + 2) = 0$$

$x = 4 \quad x = -2$ ___Check:___

$$\log_2 x = 3 - \log_2(x - 2) \qquad\qquad \log_2 x = 3 - \log_2(x - 2)$$

$$\log_2 4 \stackrel{?}{=} 3 - \log_2(4 - 2) \qquad\qquad \log_2(-2) \stackrel{?}{=} 3 - \log_2(-2 - 2)$$

$$\log_2 4 \stackrel{?}{=} 3 - \log_2 2 \qquad\qquad \log_2(\underbrace{-2}) \stackrel{?}{=} 3 - \log_2(\underbrace{-4})$$

$$2 \stackrel{?}{=} 3 - 1 \checkmark \qquad\qquad\qquad\quad \text{undefined} \qquad \text{undefined}$$

The only solution that checks is $x = 4$.
The solution set is $\{4\}$.

Skill Practice 6 Solve. $2 - \log_7 x = \log_7(x - 48)$

---

2. Use Logarithmic Equations in Applications

In Example 7, we use a logarithmic equation in an application.

EXAMPLE 7 Using a Logarithmic Equation in a Medical Application

Suppose that the sound at a rock concert measures 124 dB (decibels).

a. Use the formula $L = 10 \log\left(\frac{I}{I_0}\right)$ to find the intensity of sound I (in W/m^2). The variable L represents the loudness of sound (in dB) and $I_0 = 10^{-12}$ W/m^2.

b. If the threshold at which sounds become painful is 1 W/m^2, will the music at this concert be physically painful? (Ignore the quality of the music.)

Solution:

a.
$$L = 10 \log\left(\frac{I}{I_0}\right)$$

$$124 = 10 \log\left(\frac{I}{10^{-12}}\right) \qquad \text{Substitute 124 for } L \text{ and } 10^{-12} \text{ for } I_0.$$

$$12.4 = \log\left(\frac{I}{10^{-12}}\right) \qquad \text{Divide both sides by 10. The logarithm is now isolated.}$$

$$10^{12.4} = \frac{I}{10^{-12}} \qquad \text{Write the equation in exponential form.}$$

$$10^{12.4} \cdot 10^{-12} = I \qquad \text{Multiply both sides by } 10^{-12}.$$

$$I = 10^{0.4} \approx 2.5 \text{ W/m}^2 \qquad \text{Simplify.}$$

Answer
6. $\{49\}$; The value -1 does not check.

b. The intensity of sound at the rock concert is approximately 2.5 W/m^2. This is above the threshold for pain.

Skill Practice 7

a. Find the intensity of sound from a leaf blower if the decibel level is 115 dB.

b. Is the intensity of sound from a leaf blower above the threshold for pain?

TECHNOLOGY CONNECTIONS

Using a Calculator to Approximate the Solutions to Exponential and Logarithmic Equations

There are many situations in which analytical methods fail to give a solution to a logarithmic or exponential equation. To find solutions graphically,

Enter the left side of the equation as Y_1.

Enter the right side of the equation as Y_2.

Then determine the point(s) of intersection of the graphs.

<u>Example:</u> $4 \ln x - 3x = -8$

$Y_1 = 4 \ln x - 3x$

$Y_2 = -8$ Solutions: $x \approx 0.1516$ and $x \approx 4.7419$

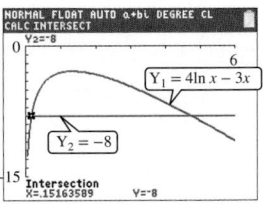 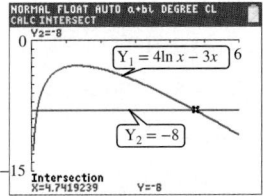

Answers

7. a. $10^{-0.5}$ W/m$^2 \approx 0.3$ W/m^2

  b. No

SECTION 9.6   Practice Exercises

Prerequisite Review

For Exercises R.1–R.2, write the domain in interval notation. (See Section 9.3 for review.)

R.1. $f(x) = \ln(3 - x)$

R.2. $g(x) = \log(x + 4)$

For Exercises R.3–R.8, write the expression as a single logarithm. (See Section 9.4 for review.)

R.3. $\log_b (x - 1) + \log_b (x + 2)$

R.4. $\log_b x + \log_b (2x + 3)$

R.5. $\log_b x - \log_b (1 - x)$

R.6. $\log_b (x + 2) - \log_b (3x - 5)$

R.7. $2 \log x + 3 \log(x - 4) - 4 \log(x + 3)$

R.8. $\dfrac{1}{2} \ln y - 3 \ln(y + 1) + 2 \ln(y - 5)$

For Exercises R.9–R.12, write the equation in exponential form. (See Section 9.3 for review.)

R.9. $\log_2(x^2 - 2x) = 3$

R.10. $\log_6(x - 4) = 2$

R.11. $\ln(x^2) = 4$

R.12. $\ln(6x) = 3$

Concept Connections

1. The equivalence property of logarithmic expressions states that if $\log_b x = \log_b y$, then _____ = _____.

2. An equation containing a variable within a logarithmic expression is called a _____ equation.

3. Explain how to solve the equation $\log(5x - 3) = \log(4x + 2)$.

4. Explain how to solve the equation $\log_2(x - 1) = 4$.

Objective 1: Solve Logarithmic Equations

For Exercises 5–6, determine if the given value of x is a solution to the logarithmic equation.

5. $\log_2(x - 31) = 5 - \log_2 x$

  a. $x = 16$

  b. $x = 32$

  c. $x = -1$

6. $\log_4 x = 3 - \log_4(x - 63)$

  a. $x = 64$

  b. $x = -1$

  c. $x = 32$

For Exercises 7–58, solve the equation. Write the solution set with the exact solutions. Also give approximate solutions to 4 decimal places if necessary. **(See Examples 1–6)**

7. $\log_2(2x - 6) = \log_2 x$

8. $\log_5(15 - 4x) = \log_5 x$

9. $\log(2x - 5) = \log(3x)$

10. $\ln(3x + 2) = \ln(2x)$

11. $\log_4(3w + 11) = \log_4(3 - w)$

12. $\log_7(12 - t) = \log_7(t + 6)$

13. $\log(x^2 + 7x) = \log 18$

14. $\log(p^2 + 6p) = \log 7$

15. $\log_3 x = 2$

16. $\log_4 x = 9$

17. $\log p = 42$

18. $\log q = \dfrac{1}{2}$

19. $\ln x = 0.08$

20. $\ln x = 9$

21. $\log(3x - 5) = 2$

22. $\log(5x - 4) = 0$

23. $6 \log_5(4p - 3) - 2 = 16$

24. $5 \log_6(7w + 1) + 3 = 13$

25. $2 \log_8(3y - 5) + 20 = 24$

26. $5 \log_3(7 - 5z) + 2 = 17$

27. $\log(p + 17) = 4.1$

28. $\log(q - 6) = 3.5$

29. $2 \ln(4 - 3t) + 1 = 7$

30. $4 \ln(6 - 5t) + 2 = 22$

31. $\log_2 w - 3 = -\log_2(w + 2)$

32. $\log_3 y + \log_3(y + 6) = 3$

33. $\log_6(7x - 2) = 1 + \log_6(x + 5)$

34. $\log_4(5x - 13) = 1 + \log_4(x - 2)$

35. $\log_5 z = 3 - \log_5(z - 20)$

36. $\log_2 x = 4 - \log_2(x - 6)$

37. $\ln x + \ln(x - 4) = \ln(3x - 10)$

38. $\ln x + \ln(x - 3) = \ln(5x - 7)$

39. $\log x + \log(x - 7) = \log(x - 15)$

40. $\log x + \log(x - 10) = \log(x - 18)$

41. $\log_8(6 - m) + \log_8(-m - 1) = 1$

42. $\log_3(n - 5) + \log_3(n + 3) = 2$

43. $\log_x 25 = 2$ $(x > 0)$

44. $\log_x 100 = 2$ $(x > 0)$

45. $\log_b 10,000 = 4$ $(b > 0)$

46. $\log_b e^3 = 3$ $(b > 0)$

47. $\log_y 5 = \dfrac{1}{2}$ $(y > 0)$

48. $\log_b 8 = \dfrac{1}{2}$ $(b > 0)$

49. $\log_3 8 - \log_3(x + 5) = 2$

50. $\log_2(x + 3) - \log_2(x + 2) = 1$

51. $\log_2(h - 1) + \log_2(h + 1) = 3$

52. $\log_3 k + \log_3(2k + 3) = 2$

53. $\log(x + 2) = \log(3x - 6)$

54. $\log x = \log(1 - x)$

55. $\ln(x + 5) - \ln x = \ln(4x)$

56. $\log(6y - 7) + \log y = \log 5$

57. $\log(4m) = \log 2 + \log(m - 3)$

58. $\log(-h) + \log 3 = \log(2h - 15)$

Objective 2: Use Logarithmic Equations in Applications

For Exercises 59–60, the formula $L = 10 \log \left(\frac{I}{I_0}\right)$ gives the loudness of sound L (in dB) based on the intensity of sound I (in W/m^2). The value $I_0 = 10^{-12}$ W/m^2 is the minimal threshold for hearing for midfrequency sounds. Hearing impairment is often measured according to the minimal sound level (in dB) detected by an individual for sounds at various frequencies. For one frequency, the table depicts the level of hearing impairment.

Category	Loudness (dB)
Mild	$26 \le L \le 40$
Moderate	$41 \le L \le 55$
Moderately severe	$56 \le L \le 70$
Severe	$71 \le L \le 90$
Profound	$L > 90$

59. a. If the minimum intensity heard by an individual is 3.4×10^{-8} W/m^2, determine if the individual has a hearing impairment.

   b. If the minimum loudness of sound detected by an individual is 30 dB, determine the corresponding intensity of sound. **(See Example 7)**

60. Determine the range that represents the intensity of sound that can be heard by an individual with severe hearing impairment.

For Exercises 61–62, use the formula pH $= -\log[\text{H}^+]$. The variable pH represents the level of acidity or alkalinity of a liquid on the pH scale, and H$^+$ is the concentration of hydronium ions in the solution. Determine the value of H$^+$ (in mol/L) for the following liquids, given their pH values.

61. a. Seawater pH $= 8.5$

   b. Acid rain pH $= 2.3$

62. a. Milk pH $= 6.2$

   b. Sodium bicarbonate pH $= 8.4$

63. A new teaching method to teach vocabulary to sixth-graders involves having students work in groups on an assignment to learn new words. After the lesson was completed, the students were tested at 1-month intervals. The average score for the class $S(t)$ can be modeled by

$$S(t) = 94 - 18 \ln(t + 1)$$

where t is the time in months after completing the assignment. If the average score is 65, how many months had passed since the students completed the assignment? Round to the nearest month.

64. A company spends x hundred dollars on an advertising campaign. The amount of money in sales $S(x)$ (in $1000) for the 4-month period after the advertising campaign can be modeled by

$$S(x) = 5 + 7 \ln(x + 1)$$

If the sales total $19,100, how much was spent on advertising? Round to the nearest dollar.

65. Radiated seismic energy from an earthquake is estimated by $\log E = 4.4 + 1.5M$, where E is the energy in Joules (J) and M is surface wave magnitude.

   a. How many times more energy does an 8.2-magnitude earthquake have than a 5.5-magnitude earthquake? Round to the nearest thousand.

   b. How many times more energy does a 7-magnitude earthquake have than a 6-magnitude earthquake? Round to the nearest whole number.

66. On August 31, 1854, an epidemic of cholera was discovered in London, England, resulting from a contaminated community water pump. By the end of September, more than 600 citizens who drank water from the pump had died. The cumulative number of deaths $D(t)$ at a time t days after August 31 is given by $D(t) = 91 + 160 \ln(t + 1)$.

   a. Determine the cumulative number of deaths by September 15. Round to the nearest whole unit.

   b. Approximately how many days after August 31 did the cumulative number of deaths reach 600?

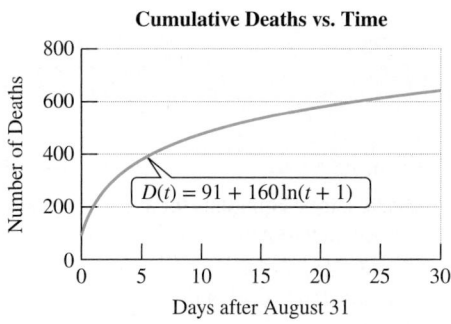

Cumulative Deaths vs. Time
$D(t) = 91 + 160 \ln(t + 1)$

Mixed Exercises

For Exercises 67–70, find an equation for the inverse function.

67. $f(x) = \ln(x + 5)$

68. $f(x) = \ln(x - 7)$

69. $f(x) = \log(x + 7) - 9$

70. $f(x) = \log(x - 11) + 8$

For Exercises 71–88, solve the equation. Write the solution set with exact solutions. Also give approximate solutions to 4 decimal places if necessary.

71. $5^{|x|} - 3 = 122$

72. $11^{|x|} + 9 = 130$

73. $\log x - 2 \log 3 = 2$

74. $\log y - 3 \log 5 = 3$

75. $6^{x^2-2} = 36$

76. $8^{y^2-7} = 64$

77. $\log_9 |x + 4| = \log_9 6$

78. $\log_8 |3 - x| = \log_8 5$

79. $x^2 e^x = 9e^x$

80. $x^2 6^x = 6^x$

81. $\log_3(\log_3 x) = 0$

82. $\log_5(\log_5 x) = 1$

83. $3|\ln x| - 12 = 0$

84. $7|\ln x| - 14 = 0$

85. $\log_3 x - \log_3(2x + 6) = \frac{1}{2}\log_3 4$

86. $\log_5 x - \log_5(x + 1) = \frac{1}{3}\log_5 8$

87. $2e^x(e^x - 3) = 3e^x - 4$

88. $3e^x(e^x - 6) = 4e^x - 7$

Expanding Your Skills

For Exercises 89–100, solve the equation. Write the solution set with exact solutions. Also give approximate solutions to 4 decimal places if necessary.

89. a. $\log x^2 = \log x$

  b. $[\log x]^2 = \log x$

90. a. $\log x^2 - 4 \log x + 3 = 0$

  b. $[\log x]^2 - 4 \log x + 3 = 0$

91. a. $\ln \sqrt{x} = \ln x$

  b. $\sqrt{\ln x} = \ln x$

92. a. $\ln x - 4 \ln \sqrt{x} + 5 = 0$

  b. $\ln x - 5\sqrt{\ln x} + 4 = 0$

93. $(\ln x)^2 - \ln x^5 = -4$

94. $(\ln x)^2 + \ln x^3 = -2$

95. $(\log x)^2 = \log x^2$

96. $(\log x)^2 = \log x^3$

97. $\log w + 4\sqrt{\log w} - 12 = 0$

98. $\ln x + 3\sqrt{\ln x} - 10 = 0$

99. $\log_5 \sqrt{6c + 5} + \log_5 \sqrt{c} = 1$

100. $\log_3 \sqrt{x - 8} + \log_3 \sqrt{x} = 1$

Technology Connections

For Exercises 101–102, an equation is given in the form $Y_1(x) = Y_2(x)$. Graph Y_1 and Y_2 on a graphing utility on the window $[10, 10, 1]$ by $[10, 10, 1]$. Then approximate the point(s) of intersection to approximate the solution(s) to the equation. Round to 4 decimal places.

101. $x^2 + 5 \log x = 6$

102. $x^2 - 0.05 \ln x = 4$

SECTION 9.7 Modeling with Exponential and Logarithmic Functions

OBJECTIVES

1. Solve Literal Equations for a Specified Variable
2. Create Models for Exponential Growth and Decay
3. Apply Logistic Growth Models
4. Create Exponential and Logarithmic Models Using Regression

1. Solve Literal Equations for a Specified Variable

A short-term model to predict the U.S. population P is $P = 310e^{0.00965t}$, where t is the number of years since 2010. If we solve this equation for t, we have

$$t = \frac{\ln\left(\dfrac{P}{310}\right)}{0.00965} \quad \text{or equivalently} \quad t = \frac{\ln P - \ln 310}{0.00965}.$$

This is a model that predicts the time required for the U.S. population to reach a value P. Manipulating an equation for a specified variable was first introduced in Section 1.2. In Example 1, we revisit this skill using exponential and logarithmic equations.

> **EXAMPLE 1**   Solving an Equation for a Specified Variable

a. Given $P = 100e^{kx} - 100$, solve for x. (Used in geology)

b. Given $L = 8.8 + 5.1 \log D$, solve for D. (Used in astronomy)

Solution:

a.
$$P = 100e^{kx} - 100$$

$$P + 100 = 100e^{kx} \qquad \text{Add 100 to both sides to isolate the } x \text{ term.}$$

$$\frac{P + 100}{100} = e^{kx} \qquad \text{Divide by 100.}$$

$$\ln\left(\frac{P + 100}{100}\right) = \ln e^{kx} \qquad \text{Take the natural logarithm of both sides.}$$

$$\ln\left(\frac{P + 100}{100}\right) = kx \qquad \text{Simplify: } \ln e^{kx} = kx$$

$$x = \frac{\ln\left(\dfrac{P + 100}{100}\right)}{k} \qquad \text{Divide by } k.$$

$$x = \frac{\ln\left(\dfrac{P + 100}{100}\right)}{k} \text{ or equivalently } x = \frac{\ln(P + 100) - \ln 100}{k}$$

b.
$$L = 8.8 + 5.1 \log D$$

$$L - 8.8 = 5.1 \log D \qquad \text{Subtract 8.8 from both sides.}$$

$$\frac{L - 8.8}{5.1} = \log D \qquad \text{Divide both sides by 5.1.}$$

$$D = 10^{(L-8.8)/5.1} \qquad \text{Write the equation in exponential form.}$$

Skill Practice 1

a. Given $T = 78 + 272e^{-kt}$, solve for k.

b. Given $S = 90 - 20 \ln(t + 1)$, solve for t.

---

2. Create Models for Exponential Growth and Decay

In Section 9.2, we defined an exponential function as $y = b^x$, where $b > 0$ and $b \neq 1$. Throughout the chapter, we have used transformations of basic exponential functions to solve a variety of applications. The following variation of the general exponential form is used to solve applications involving exponential growth and decay.

Answers

1. a. $k = -\dfrac{\ln\left(\dfrac{T - 78}{272}\right)}{t}$ or

$\quad k = \dfrac{\ln 272 - \ln(T - 78)}{t}$

b. $t = e^{(90-S)/20} - 1$

Exponential Growth and Decay Models

Let y be a variable changing exponentially with respect to t, and let y_0 represent the initial value of y when $t = 0$. Then for a constant k:

If $k > 0$, then $y = y_0e^{kt}$ is a model for exponential growth.	If $k < 0$, then $y = y_0e^{kt}$ is a model for exponential decay.

Example:

$y = 2000e^{0.06t}$ represents the value of a \$2000 investment after t years with interest compounded continuously.

(*Note*: $k = 0.06 > 0$)

Example:

$y = 100e^{-0.165t}$ represents the radioactivity level t hours after a patient is treated for thyroid cancer with 100 mCi of radioactive iodine.

(*Note*: $k = -0.165 < 0$)

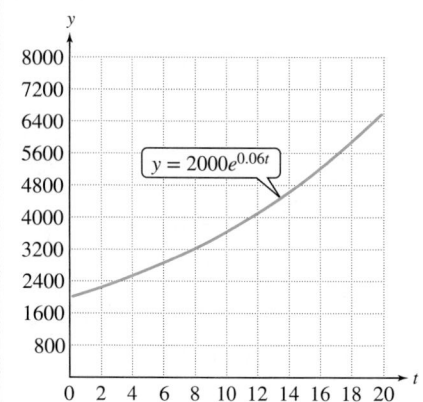

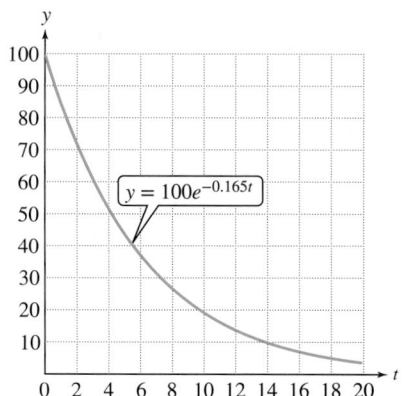

The model $y = y_0e^{kt}$ is often presented with different letters or symbols in place of y, y_0, k, and t to convey their meaning in the context of the application. For example, to compute the value of an investment under continuous compounding, we have

$$A = Pe^{rt}$$

P (for principal) is used in place of y_0.

r (for the annual interest rate) is used in place of k.

A (for the future value of the investment) is used in place of y.

We can also use function notation when expressing a model for exponential growth or decay. For example, consider the model for population growth.

$$P(t) = P_0e^{kt}$$

P_0 (for initial population) is used in place of y_0.

$P(t)$ represents the population as a function of time and is used in place of y.

EXAMPLE 2   Creating a Model for Growth of an Investment

Suppose that \$15,000 is invested and at the end of 3 years, the value of the account is \$19,356.92. Use the model $A = Pe^{rt}$ to determine the average rate of return r under continuous compounding.

Solution:

$A = Pe^{rt}$	Begin with an appropriate model.
$A = 15{,}000e^{rt}$	P represents the initial value of the account (initial principal). Substitute 15,000 for P.
$19{,}356.92 = 15{,}000e^{r(3)}$	We have a known data point where $A = 19{,}356.92$ when $t = 3$. Substituting these values into the formula enables us to solve for r.
$\dfrac{19{,}356.92}{15{,}000} = e^{3r}$	Divide both sides by 15,000.
$\ln\left(\dfrac{19{,}356.92}{15{,}000}\right) = \ln\left(e^{3r}\right)$	Take the natural logarithm of both sides.
$\ln\left(\dfrac{19{,}356.92}{15{,}000}\right) = 3r$	Simplify: $\ln e^{3r} = 3r$
$r = \dfrac{\ln\left(\dfrac{19{,}356.92}{15{,}000}\right)}{3}$	Divide by 3 to isolate r.
$r \approx 0.085$	

The average rate of return is approximately 8.5%.

> **Skill Practice 2**   Suppose that $10,000 is invested and at the end of 5 years, the value of the account is $13,771.28. Use the model $A = Pe^{rt}$ to determine the average rate of return r under continuous compounding.

EXAMPLE 3   **Creating a Model for Population Growth**

On January 1, 2010, the population of California was approximately 37.3 million. On January 1, 2019, the population was 40.0 million. Let $t = 0$ represent the year 2010.

a. Write a function defined by $P(t) = P_0e^{kt}$ to represent the population of California $P(t)$ (in millions), t years after 2010.

b. Use the function from part (a) to predict the population in 2025. Round to 1 decimal place.

c. Use the function from part (a) to determine the year for which the population of California will be twice that of the year 2010.

Solution:

> **TIP** The value of k in the model $P(t) = P_0e^{kt}$ is called a parameter and is related to the growth rate of the population being studied. The value of k will be different for different populations.

a. $P(t) = P_0e^{kt}$	Begin with an appropriate model.
$P(t) = 37.3e^{kt}$	The initial population is $P_0 = 37.3$ million.
$40.0 = 37.3e^{k(9)}$	We have a known data point $P(9) = 40.0$. Substituting these values into the function enables us to solve for k.
$\dfrac{40.0}{37.3} = e^{9k}$	Divide both sides by 37.3.
$\ln\left(\dfrac{40.0}{37.3}\right) = \ln(e^{9k})$	Take the natural logarithm of both sides.
$\ln\left(\dfrac{40.0}{37.3}\right) = 9k$	
$k = \dfrac{\ln\left(\dfrac{40.0}{37.3}\right)}{9} \approx 0.00777$	Divide both sides by 9 to isolate k.
$P(t) = 37.3e^{0.00777t}$	This model gives the population as a function of time.

Answer

2. 6.4%

b. $P(t) = 37.3e^{0.00777t}$

$P(15) = 37.3e^{0.00777(15)}$ The year 2025 is 15 years after 2010. Substitute 15 for t.

≈ 41.9

The population in California in 2025 will be approximately 41.9 million if this trend continues.

c. $P(t) = 37.3e^{0.00777t}$ Two times California's population of 2010 is 2(37.3 million) which is 74.6 million.

$74.6 = 37.3e^{0.00777t}$ Substitute 74.6 for $P(t)$.

$\dfrac{74.6}{37.3} = e^{0.00777t}$ Divide both sides by 37.3.

$2 = e^{0.00777t}$

$\ln 2 = 0.00777t$ Take the natural logarithm of both sides.

$t = \dfrac{\ln 2}{0.00777} \approx 89.2$ Divide both sides by 0.00777 to isolate t.

The population of California will reach 74.6 million in the year 2099 if this trend continues.

Skill Practice 3   On January 1, 2010, the population of Texas was 25.2 million. On January 1, 2019, the population was 29.1 million. Let $t = 0$ represent the year 2010.

a. Write a function defined by $P(t) = P_0e^{kt}$ to represent the population $P(t)$ of Texas t years after 2010.

b. Use the function in part (a) to predict the population in 2029. Round to 1 decimal place.

c. Use the function in part (a) to determine the year for which the population of Texas will reach 40 million if this trend continues.

An exponential model can be presented with a base other than base e. For example, suppose that a culture of bacteria begins with 5000 organisms and the population doubles every 4 hr. Then the population $P(t)$ can be modeled by

$P(t) = 5000(2)^{t/4}$, where t is the time in hours after the culture was started.

Notice that this function is defined using base 2. It is important to realize that any exponential function of one base can be rewritten in terms of an exponential function of another base. In particular we are interested in expressing the function with base e.

Writing an Exponential Expression Using Base e

Let t and b be real numbers, where $b > 0$ and $b \neq 1$. Then,

$$b^t \text{ is equivalent to } e^{(\ln b)t}.$$

To show that $e^{(\ln b)t} = b^t$, use the power property of exponents; that is,

$$e^{(\ln b)t} = (e^{\ln b})^t = b^t$$

EXAMPLE 4   **Writing an Exponential Function with Base e**

a. The population $P(t)$ of a culture of bacteria is given by $P(t) = 5000(2)^{t/4}$, where t is the time in hours after the culture was started. Write the rule for this function using base e.

b. Find the population after 12 hr using both forms of the function from part (a).

Solution:

a. $P(t) = 5000(2)^{t/4}$

Note that $2^{t/4} = (2^t)^{1/4}$

$$= \left[e^{(\ln 2)t}\right]^{1/4} \qquad \text{Apply the property that } e^{(\ln b)t} = b^t.$$

$$= e^{[(\ln 2)/4]t} \qquad \text{Apply the power rule of exponents.}$$

Therefore, $P(t) = 5000(2)^{t/4}$

$$= 5000e^{[(\ln 2)/4]t}$$

$$\approx 5000e^{0.17329t}$$

b. $P(t) = 5000(2)^{t/4}$ $P(t) \approx 5000e^{0.17329t}$

$P(12) = 5000(2)^{(12)/4}$ $P(12) \approx 5000e^{0.17329(12)}$

$= 40,000$ $\approx 40,000$

Skill Practice 4

a. Given $P(t) = 10,000(2)^{-0.4t}$, write the rule for this function using base e.

b. Find the function value for $t = 10$ for both forms of the function from part (a).

In Example 5, we apply an exponential decay function to determine the age of a bone through radiocarbon dating. Animals ingest carbon through respiration and through the food they eat. Most of the carbon is carbon-12 (^{12}C), an abundant and stable form of carbon. However, a small percentage of carbon is the radioactive isotope, carbon-14 (^{14}C). The ratio of carbon-12 to carbon-14 is constant for all living things. When an organism dies, it no longer takes in carbon from the environment. Therefore, as the carbon-14 decays, the ratio of carbon-12 to carbon-14 changes. Scientists know that the half-life of ^{14}C is 5730 years and from this, they can build a model to represent the amount of ^{14}C remaining t years after death. This is illustrated in Example 5.

EXAMPLE 5   **Creating a Model for Exponential Decay**

a. Carbon-14 has a half-life of 5730 years. Write a model of the form $Q(t) = Q_0 e^{-kt}$ to represent the amount $Q(t)$ of carbon-14 remaining after t years if no additional carbon is ingested.

b. An archeologist uncovers human remains at an ancient Roman burial site and finds that 76.6% of the carbon-14 still remains in the bone. How old is the bone? Round to the nearest hundred years.

Answers

4. a. $P(t) = 10,000e^{-0.27726t}$

   b. 625

TIP Given the half-life of a radioactive substance, we can also write an exponential model using base $\frac{1}{2}$. The format is

$$Q(t) = Q_0 \left(\frac{1}{2}\right)^{t/h}$$

where h is the half-life of the substance.

In Example 5, we have

$$Q(t) = Q_0 \left(\frac{1}{2}\right)^{t/5730}$$

Solution:

a. $Q(t) = Q_0 e^{-kt}$ Begin with a general exponential decay model.

$0.5Q_0 = Q_0 e^{-k(5730)}$ Substitute the known data value. One-half of the original quantity Q_0 is present after 5730 years.

$\dfrac{0.5Q_0}{Q_0} = \dfrac{Q_0 e^{-5730k}}{Q_0}$ Divide by Q_0 on both sides.

$0.5 = e^{-5730k}$

$\ln 0.5 = -5730k$ Take the natural logarithm of both sides.

$k = \dfrac{\ln 0.5}{-5730}$ Divide by -5730.

≈ 0.000121

$Q(t) = Q_0 e^{-0.000121t}$

b. $0.766Q_0 = Q_0 e^{-0.000121t}$ The quantity $Q(t)$ of carbon-14 in the bone is 76.6% of Q_0.

$\dfrac{0.766Q_0}{Q_0} = \dfrac{Q_0 e^{-0.000121t}}{Q_0}$ Divide by Q_0 on both sides.

$0.766 = e^{-0.000121t}$

$\ln 0.766 = -0.000121t$ Take the natural logarithm of both sides.

$t = \dfrac{\ln 0.766}{-0.000121} \approx 2200$ Divide by -0.000121 to isolate t.

The bone is approximately 2200 years old.

Skill Practice 5   Use the function $Q(t) = Q_0 e^{-0.000121t}$ to determine the age of a piece of wood that has 42% of its carbon-14 remaining. Round to the nearest 10 years.

3. Apply Logistic Growth Models

In Examples 3 and 4, we used a model of the form $P(t) = P_0 e^{kt}$ to predict population as an exponential function of time. However, unbounded population growth is not possible due to limited resources. A growth model that addresses this problem is called logistic growth. In particular, a logistic growth model imposes a limiting value on the dependent variable.

Logistic Growth Model

A logistic growth model is a function written in the form

$$y = \frac{c}{1 + ae^{-bt}}$$

where a, b, and c are positive constants.

The general logistic growth equation can be written with a complex fraction.

$$y = \frac{c}{1 + \dfrac{a}{e^{bt}}}$$ This term approaches 0 as t approaches ∞.

In this form, we can see that for large values of t, the term $\dfrac{a}{e^{bt}}$ approaches 0, and the function value y approaches $\frac{c}{1}$.

Answer

5. 7170 years

TIP The rate of increase of a logistic curve changes from increasing to decreasing to the left and right of a point called the *point of inflection*.

The line $y = c$ is a horizontal asymptote of the graph, and c represents the limiting value of the function (Figure 9-20).

Notice that the graph of a logistic curve is increasing over its entire domain. However, the *rate* of increase begins to decrease as the function levels off and approaches the horizontal asymptote $y = c$.

In Example 3 we created a function to approximate the population of California assuming unlimited growth. In Example 6, we use a logistic growth model.

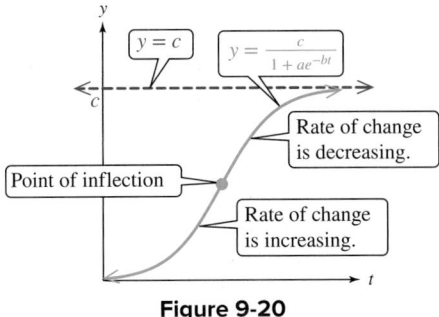

Figure 9-20

EXAMPLE 6   Using Logistic Growth to Model Population

The population of California $P(t)$ (in millions) can be approximated by the logistic growth function

$$P(t) = \frac{93.3}{1 + 1.5e^{-0.013t}} \quad \text{where } t \text{ is the number of years since the year 2010.}$$

a. Determine the population in the year 2010.

b. Use this function to determine the time required for the population of California to double from its value of 37.3 million in 2010. Compare this with the result from Example 3(c).

c. What is the limiting value of the population of California under this model?

Solution:

a. $P(t) = \dfrac{93.3}{1 + 1.5e^{-0.013t}}$

$P(0) = \dfrac{93.3}{1 + 1.5e^{-0.013(0)}}$ To find the initial population of California with this model substitute 0 for t.

$= \dfrac{93.3}{1 + 1.5e^{0}}$ Recall that $e^0 = 1$.

$= \dfrac{93.3}{1 + 1.5(1)} \approx 37.3$

The population was approximately 37.3 million in the year 2010.

b. $P(t) = \dfrac{93.3}{1 + 1.5e^{-0.013t}}$ Two times California's population of 2010 is 2(37.3 million) which is 74.6 million.

$74.6 = \dfrac{93.3}{1 + 1.5e^{-0.013t}}$ Substitute 74.6 for $P(t)$.

$74.6(1 + 1.5e^{-0.013t}) = 93.3$ Multiply both sides by $(1 + 1.5e^{-0.013t})$ to clear fractions.

$\dfrac{74.6(1 + 1.5e^{-0.013t})}{74.6} = \dfrac{93.3}{74.6}$ Divide by 74.6 on both sides.

$1 + 1.5e^{-0.013t} \approx 1.25$

$1.5e^{-0.013t} \approx 0.25$ Subtract 1 from both sides.

$$e^{-0.013t} \approx \frac{0.25}{1.5}$$ Divide by 1.5 on both sides.

$$\ln(e^{-0.013t}) \approx \ln\left(\frac{0.25}{1.5}\right)$$ Take the natural logarithm of both sides.

$$-0.013t \approx \ln\left(\frac{0.25}{1.5}\right)$$

$$t \approx \frac{\ln\left(\dfrac{0.25}{1.5}\right)}{-0.013} \approx 137.8$$ Divide by –0.013 on both sides.

The population will double in approximately 138 years. This is 49 years later than the predicted value from Example 3(c).

The graphs of

$$P(t) = \frac{93.3}{1 + 1.5e^{-0.013t}} \text{ and } P(t) = 37.3e^{0.00777t}$$

are shown in Figure 9-21. Notice that the two models agree closely for short-term population growth (out to ≈ 2040). However, in the long term, the unbounded exponential model breaks down. The logistic growth model approaches a limiting population which is reasonable due to the limited resources to sustain a large human population.

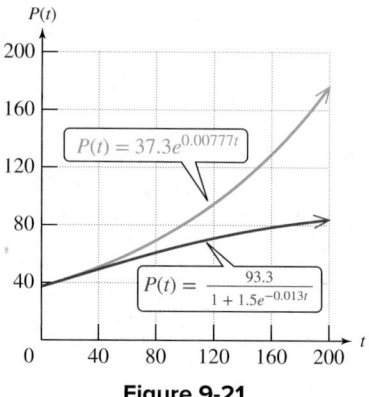

Figure 9-21

c. $P(t) = \dfrac{93.3}{1 + 1.5e^{-0.013t}} = \dfrac{93.3}{1 + \dfrac{1.5}{e^{0.013t}}}$ As $t \to \infty$, the term $\dfrac{1.5}{e^{0.013t}} \to 0$.

As t becomes large, the denominator of $\frac{1.5}{e^{0.013t}}$ also becomes large. This causes the quotient to approach zero. Therefore, as t approaches infinity, $P(t)$ approaches 93.3. Under this model, the limiting value for the population of California is 93.3 million.

Skill Practice 6   The score on a test of dexterity is given by
$$P(t) = \frac{100}{1 + 19e^{-0.354x}}, \text{ where } x \text{ is the number of times the test is taken.}$$

a. Use the function to determine the minimum number of times required for the score to exceed 90.

b. What is the limiting value of the scores?

---

4. Create Exponential and Logarithmic Models Using Regression

Linear regression was first introduced in Section 5.5 to model data that followed a linear pattern when graphed in a rectangular coordinate system. Likewise, in Section 7.1, we used regression to find quadratic models of data whose graphs appeared parabolic. In Examples 7 and 8, we use a graphing utility and regression to find an exponential model or logarithmic model based on observed data.

Answers
6. a. 15   **b.** 100

> **EXAMPLE 7**   **Creating an Exponential Model from Observed Data**

The amount of sunlight y [in langleys (Ly)—a unit used to measure solar energy in calories/cm^2] is measured for six different depths x (in meters) in Lake Lyndon B. Johnson in Texas.

x (m)	1	3	5	7	9	11
y (Ly)	300	161	89	50	27	15

a. Graph the data.

b. From visual inspection of the graph, which model would best represent the data? Choose from $y = mx + b$ (linear), $y = ab^x$ (exponential), or $y = a + b \ln x$ (logarithmic).

c. Use a graphing utility to find a regression equation that fits the data.

Solution:

a. Enter the data in two lists.

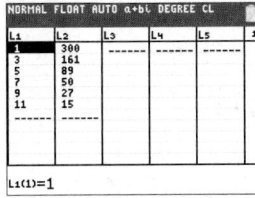   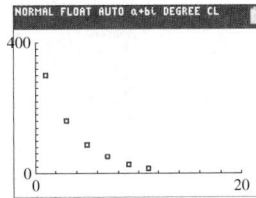

b. Note that for large depths, the amount of sunlight approaches 0. Therefore, the curve is asymptotic to the x-axis. This is consistent with a decreasing exponential model. The exponential model $y = ab^x$ appears to fit.

c. Under the STAT menu, choose CALC, ExpReg, and then Calculate.

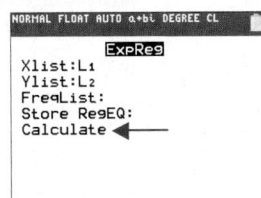   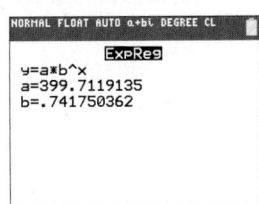   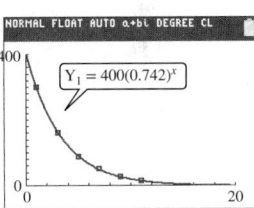

The equation $y = ab^x$ is $y = 400(0.742)^x$.

Skill Practice 7   For the given data,

x	1	3	5	7	9	11
y	2.9	5.6	11.1	22.4	43.0	85.0

a. Graph the data points.

b. Use a graphing utility to find a model of the form $y = ab^x$ to fit the data.

Answers

7. a–b.

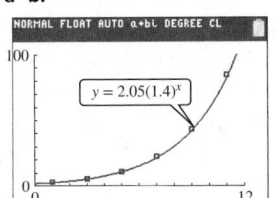

EXAMPLE 8   **Creating a Logarithmic Model from Observed Data**

The diameter x (in mm) of a sugar maple tree, along with the corresponding age y (in years) of the tree is given for six different trees.

x (mm)	1	50	100	200	300	400
y (years)	4	60	72	82	89	94

a. Graph the data.

b. From visual inspection of the graph, which model would best represent the data? Choose from $y = mx + b$ (linear), $y = ab^x$ (exponential), or $y = a + b \ln x$ (logarithmic).

c. Use a graphing utility to find a regression equation that fits the data.

Solution:

a. Enter the data into two lists.

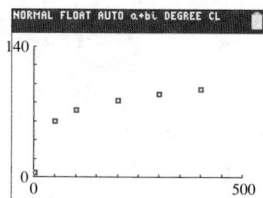

b. By inspection of the graph, the logarithmic model $y = a + b \ln x$ appears to fit.

c. Under the STAT menu, choose CALC, and then LnReg.

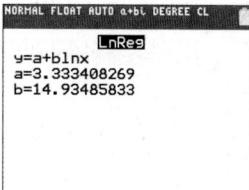
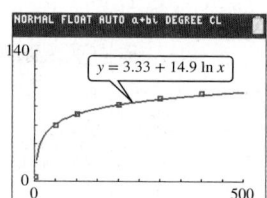

Skill Practice 8   For the given data,

x	1	5	9	13	17	21
y	11.9	19.3	21.9	23.5	24.7	25.7

a. Graph the data points.

b. Use a graphing utility to find a model of the form $y = a + b \ln x$ to fit the data.

Answers

8. a–b.

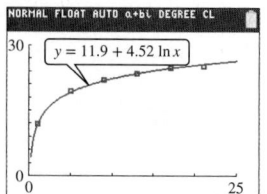

SECTION 9.7   Practice Exercises

Prerequisite Review

For Exercises R.1–R.2, use the change-of-base formula to approximate the value of the logarithm to 4 decimal places. (See Section 9.4 for review.)

R.1. $\log_3 100$

R.2. $\log_5 26{,}115$

For Exercises R.3–R.6, solve for the indicated variable. Assume all variables represent positive real numbers. (See Sections 1.2 and 3.6 for review.)

R.3. $at + b = 5c$ for t (See Section 1.2.)

R.4. $S = \dfrac{a + p}{4}$ for a

R.5. $t = cn^2 p^2 s$ for $p > 0$ (See Section 3.6.)

R.6. $V = \dfrac{1}{3}q^2 h$ for $q > 0$

For Exercises R.7–R.8, solve the equation. Write the solution set with the exact values given in terms of common or natural logarithms. Also approximate the solutions to 4 decimal places. (See Section 9.5 for review.)

R.7. $500 = 250e^{0.12t}$

R.8. $1124 = 100e^{0.0034t}$

For Exercises R.9–R.10, solve the equation. (See Section 4.3 for review.)

R.9. $12 = \dfrac{24x}{1 + 4x}$

R.10. $-10 = \dfrac{12t}{1 + 2t}$

For Exercises R.11–R.12,

   a. Determine whether the graph P is increasing or decreasing. (See Section 9.2 for review.)

   b. Evaluate P for $t = 0$ and $t = 20$. Round to 4 decimal places if necessary.

R.11. $P(t) = 10e^{-0.0126t}$

R.12. $P(t) = 45e^{0.0053t}$

Concept Connections

1. If $k > 0$, the equation $y = y_0 e^{kt}$ is a model for exponential (growth/decay), whereas if $k < 0$, the equation is a model for exponential (growth/decay).

2. A function defined by $y = ab^x$ can be written in terms of an exponential function base e as _____.

3. A function defined by $y = \dfrac{c}{1 + ae^{-bt}}$ is called a _____ growth model and imposes a limiting value on y.

4. Given a logistic growth function $y = \dfrac{c}{1 + ae^{-bt}}$, the limiting value of y is _____.

Objective 1: Solve Literal Equations for a Specified Variable

For Exercises 5–14, solve for the indicated variable. **(See Example 1)**

5. $Q = Q_0 e^{-kt}$ for k (used in chemistry)

6. $N = N_0 e^{-0.025t}$ for t (used in chemistry)

7. $M = 8.8 + 5.1 \log D$ for D (used in astronomy)

8. $\log E - 12.2 = 1.44M$ for E (used in geology)

9. $\text{pH} = -\log[\text{H}^+]$ for $[\text{H}^+]$ (used in chemistry)

10. $L = 10 \log \left(\dfrac{I}{I_0} \right)$ for I (used in medicine)

11. $A = P(1 + r)^t$ for t (used in finance)

12. $A = Pe^{rt}$ for r (used in finance)

13. $\ln \left(\dfrac{k}{A} \right) = \dfrac{-E}{RT}$ for k (used in chemistry)

14. $-\dfrac{1}{k}\ln \left(\dfrac{P}{14.7} \right) = A$ for P (used in meteorology)

Objective 2: Create Models for Exponential Growth and Decay

15. Suppose that $12,000 is invested in a bond fund and the account grows to $14,309.26 in 4 years. **(See Example 2)**

    a. Use the model $A = Pe^{rt}$ to determine the average rate of return under continuous compounding. Round to the nearest tenth of a percent.

    b. How long will it take the investment to reach $20,000 if the rate of return continues? Round to the nearest tenth of a year.

17. Suppose that P dollars in principal is invested in an account earning 3.2% interest compounded continuously. At the end of 3 years, the amount in the account has earned $806.07 in interest.

    a. Find the original principal. Round to the nearest dollar. (*Hint:* Use the model $A = Pe^{rt}$ and substitute $P + 806.07$ for A.)

    b. Using the original principal from part (a) and the model $A = Pe^{rt}$, determine the time required for the investment to reach $10,000. Round to the nearest year.

19. An initial census taken in a small Asian country found the population to be 22.9 million. Ten years later the population had increased to 23.7 million. During the same time period, a census was taken in a South Pacific nation. The initial census found a population of 19.0 million. Ten years later, the population had increased to 22.6 million. **(See Example 3)**

Country	Population (in millions) when $t = 0$	Population (in millions) when $t = 10$	$P(t) = P_0 e^{kt}$
Asian			
South Pacific			

    a. Write a function of the form $P(t) = P_0 e^{kt}$ to model the population $P(t)$ (in millions) t years after the initial census for the Asian country.

    b. Write a function of the form $P(t) = P_0 e^{kt}$ to model the population $P(t)$ (in millions) t years after the initial census for the South Pacific country.

    c. Use the models from parts (a) and (b) to predict the population 20 years after the initial census was taken for each country. Round to the nearest hundred thousand.

    d. The South Pacific country had fewer people than the Asian country in the initial census, yet from the result of part (c), the South Pacific country will have more people 20 years later. Why?

    e. Assuming that the population growth trend continues in each country, use the models from parts (a) and (b) to predict the number of years for the population to reach 30 million in each country. Round to the nearest year.

16. Suppose that $50,000 from a retirement account is invested in a large cap stock fund. After 20 years, the value is $194,809.67.

    a. Use the model $A = Pe^{rt}$ to determine the average rate of return under continuous compounding. Round to the nearest tenth of a percent.

    b. How long will it take the investment to reach one-quarter million dollars? Round to the nearest tenth of a year.

18. Suppose that P dollars in principal is invested in an account earning 2.1% interest compounded continuously. At the end of 2 years, the amount in the account has earned $193.03 in interest.

    a. Find the original principal. Round to the nearest dollar. (*Hint:* Use the model $A = Pe^{rt}$ and substitute $P + 193.03$ for A.)

    b. Using the original principal from part (a) and the model $A = Pe^{rt}$, determine the time required for the investment to reach $6000. Round to the nearest tenth of a year.

20. An initial census taken in a Scandinavian country found the population to be 7.3 million. Ten years later, the population had increased to 7.8 million. During the same time period, a census was taken in a Middle Eastern nation. The initial census found a population of 6.7 million. Ten years later, the population had increased to 7.7 million.

Country	Population (in millions) when $t = 0$	Population (in millions) when $t = 10$	$P(t) = P_0 e^{kt}$
Scandinavian			
Middle Eastern			

    a. Write a function of the form $P(t) = P_0 e^{kt}$ to model the population $P(t)$ (in millions) t years after the initial census for the Scandinavian country.

    b. Write a function of the form $P(t) = P_0 e^{kt}$ to model the population $P(t)$ (in millions) t years after the initial census for the Middle Eastern country.

    c. Use the models from parts (a) and (b) to predict the population 20 years after the initial census was taken for each country. Round to the nearest hundred thousand.

    d. The Middle Eastern country had fewer people than the Scandinavian country in the initial census, yet from the result of part (c), the Middle Eastern country will have more people 20 years later. Why?

    e. Assuming that the population growth trend continues in each country, use the models from parts (a) and (b) to predict the number of years for the population to reach 10 million in each country. Round to the nearest year.

21. The table gives two functions of the form
$P(t) = ab^t$ that represent the populations of a Central
American country and a Northern European country,
t years after a baseline census is taken.
(See Example 4)

a. For each function given in the table, write an
equivalent function using base e. That is, write a
function of the form $P(t) = P_0e^{kt}$. Also, determine the
population of each country in the baseline census.

Country	$P(t) = ab^t$	$P(t) = P_0e^{kt}$	Baseline Population
Central American	$P(t) = 4.3(1.0135)^t$		
Northern European	$P(t) = 4.6(1.0062)^t$		

b. The populations of the two given countries are very
close for the initial census, but their growth rates are
different. Use the model to approximate the number
of years for the population of each country to reach
5 million. Round to the nearest year.

c. The Central American nation had fewer people than
the Northern European country according to the initial
census. Why will the Central American country reach
a population of 5 million sooner than the Northern
European country?

22. The table gives two functions of the form $P(t) = ab^t$
that represent the populations of a Caribbean nation
and a small European country, t years after a baseline
census is taken.

a. For each function given in the table, write an
equivalent function using base e. That is, write a
function of the form $P(t) = P_0e^{kt}$. Also, determine
the population of each country in the baseline
census.

Country	$P(t) = ab^t$	$P(t) = P_0e^{kt}$	Baseline Population
Caribbean	$P(t) = 8.5(1.0158)^t$		
European	$P(t) = 9.0(1.0048)^t$		

b. The populations of the two given countries are very
close for the initial census, but their growth rates
are different. Use the model to approximate the
number of years for the population of each country
to reach 10 million. Round to the nearest year.

c. The Caribbean nation had fewer people in the initial
census than the European country. Why did the
Caribbean nation reach a population of 10 million
sooner than the European country?

For Exercises 23–24, refer to the model $Q(t) = Q_0e^{-0.000121t}$ used in Example 5 for radiocarbon dating.

23. A sample from a mummified bull was taken from a pyramid in Dashur, Egypt. The sample shows that 78% of the
carbon-14 still remains. How old is the sample? Round to the nearest year. **(See Example 5)**

24. At the "Marmes Man" archeological site in southeastern Washington State, scientists
uncovered the oldest human remains yet to be found in Washington State. A sample
from a human bone taken from the site showed that 29.4% of the carbon-14 still
remained. How old is the sample? Round to the nearest year.

25. The isotope of plutonium ^{238}Pu is used to make thermoelectric power sources for
spacecraft. Suppose that a space probe was launched in 2012 with 2.0 kg of ^{238}Pu.

a. If the half-life of ^{238}Pu is 87.7 years, write a function of the form $Q(t) = Q_0e^{-kt}$ to
model the quantity $Q(t)$ of ^{238}Pu left after t years.

b. If 1.6 kg of ^{238}Pu is required to power the spacecraft's data transmitter, for how long
after launch would scientists be able to receive data? Round to the nearest year.

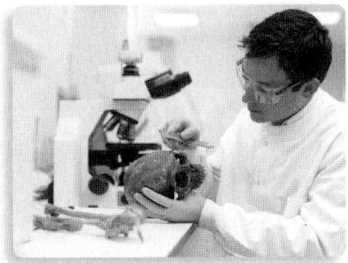

Adam Gault/agefotostock

26. Technetium-99 (^{99m}Tc) is a radionuclide used widely in nuclear medicine. ^{99m}Tc is combined with another substance that is
readily absorbed by a targeted body organ. Then, special cameras sensitive to the gamma rays emitted by the technetium
are used to record pictures of the organ. Suppose that a technician prepares a sample of ^{99m}Tc-pyrophosphate to image the
heart of a patient suspected of having had a mild heart attack.

a. At noon, the patient is given 10 mCi (millicuries) of ^{99m}Tc. If the half-life of ^{99m}Tc is 6 hr, write a function of the form
$Q(t) = Q_0e^{-kt}$ to model the radioactivity level $Q(t)$ after t hours.

b. At what time will the level of radioactivity reach 3 mCi? Round to the nearest tenth of an hour.

27. Fluorodeoxyglucose is a derivative of glucose that contains the radionuclide fluorine-18 (^{18}F). A patient is given a sample
of this material containing 300 MBq of ^{18}F (a megabecquerel is a unit of radioactivity). The patient then undergoes a PET
scan (positron emission tomography) to detect areas of metabolic activity indicative of cancer. After 174 min, one-third of
the original dose remains in the body.

a. Write a function of the form $Q(t) = Q_0e^{-kt}$ to model the radioactivity level $Q(t)$ of fluorine-18 at a time t minutes after the
initial dose.

b. What is the half-life of ^{18}F? Round to the nearest minute.

28. Painful bone metastases are common in advanced prostate cancer. Physicians often order treatment with strontium-89 (^{89}Sr), a radionuclide with a strong affinity for bone tissue. A patient is given a sample containing 4 mCi of ^{89}Sr.

    a. If 20% of the ^{89}Sr remains in the body after 90 days, write a function of the form $Q(t) = Q_0 e^{-kt}$ to model the amount $Q(t)$ of radioactivity in the body t days after the initial dose.

    b. What is the biological half-life of ^{89}Sr under this treatment? Round to the nearest tenth of a day.

29. Two million *E. coli* bacteria are present in a laboratory culture. An antibacterial agent is introduced and the population of bacteria $P(t)$ decreases by half every 6 hr. The population can be represented by $P(t) = 2{,}000{,}000\left(\frac{1}{2}\right)^{t/6}$.

    a. Convert this to an exponential function using base e.

    b. Verify that the original function and the result from part (a) yield the same result for $P(0)$, $P(6)$, $P(12)$, and $P(60)$. (*Note:* There may be round-off error.)

30. The half-life of radium-226 is 1620 years. Given a sample of 1 g of radium-226, the quantity left $Q(t)$ (in g) after t years is given by $Q(t) = \left(\frac{1}{2}\right)^{t/1620}$.

    a. Convert this to an exponential function using base e.

    b. Verify that the original function and the result from part (a) yield the same result for $Q(0)$, $Q(1620)$, and $Q(3240)$. (*Note:* There may be round-off error.)

Objective 3: Apply Logistic Growth Models

31. The population of the United States $P(t)$ (in millions) since January 1, 1900, can be approximated by

$$P(t) = \frac{725}{1 + 8.295e^{-0.0165t}}$$

where t is the number of years since January 1, 1900. **(See Example 6)**

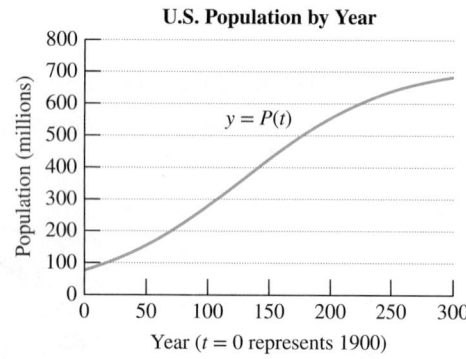

U.S. Population by Year

Population (millions) vs *Year (t = 0 represents 1900)*

$y = P(t)$

    a. Evaluate $P(0)$ and interpret its meaning in the context of this problem.

    b. Use the function to approximate the U.S. population on January 1, 2020. Round to the nearest million.

    c. Use the function to predict the U.S. population on January 1, 2050.

    d. From the model, during which year would the U.S. population reach 500 million?

    e. What value will the term $\dfrac{8.295}{e^{0.0165t}}$ approach as $t \to \infty$?

    f. Determine the limiting value of $P(t)$.

32. The population of Canada $P(t)$ (in millions) since January 1, 1900, can be approximated by

$$P(t) = \frac{55.1}{1 + 9.6e^{-0.02515t}}$$

where t is the number of years since January 1, 1900.

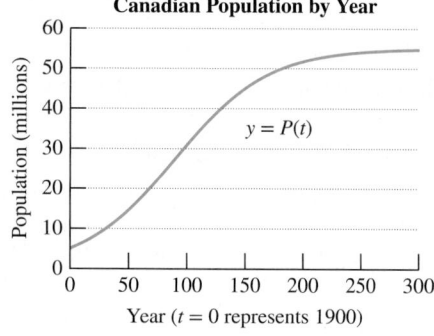

Canadian Population by Year

Population (millions) vs *Year (t = 0 represents 1900)*

$y = P(t)$

    a. Evaluate $P(0)$ and interpret its meaning in the context of this problem.

    b. Use the function to approximate the Canadian population on January 1, 2020. Round to the nearest tenth of a million.

    c. Use the function to predict the Canadian population on January 1, 2040.

    d. From the model, during which year would the Canadian population reach 45 million?

    e. What value will the term $\dfrac{9.6}{e^{0.02515t}}$ approach as $t \to \infty$?

    f. Determine the limiting value of $P(t)$.

33. The number of computers $N(t)$ (in millions) infected by a computer virus can be approximated by

$$N(t) = \frac{2.4}{1 + 15e^{-0.72t}}$$

where t is the time in months after the virus was first detected.

  a. Determine the number of computers initially infected when the virus was first detected.

  b. How many computers were infected after 6 months? Round to the nearest hundred thousand.

  c. Determine the amount of time required after initial detection for the virus to affect 1 million computers. Round to the nearest tenth of a month.

  d. What is the limiting value of the number of computers infected according to this model?

34. After a new product is launched, the cumulative sales $S(t)$ (in $1000) t weeks after launch is given by

$$S(t) = \frac{72}{1 + 9e^{-0.36t}}$$

  a. Determine the cumulative amount in sales 3 weeks after launch. Round to the nearest thousand.

  b. Determine the amount of time required for the cumulative sales to reach $70,000.

  c. What is the limiting value in sales?

Objective 4: Create Exponential and Logarithmic Models Using Regression

For Exercises 35–38, a graph of data is given. From visual inspection, which model would best fit the data? Choose from

$y = mx + b$ (linear) $y = ab^x$ (exponential)

$y = a + b \ln x$ (logarithmic) $y = \dfrac{c}{1 + ae^{-bx}}$ (logistic)

35.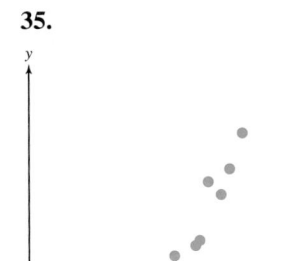

36.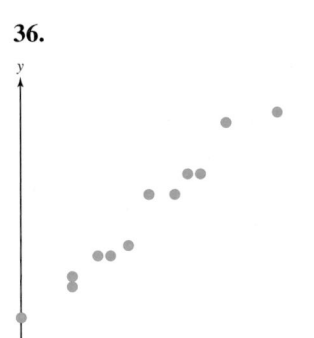

37.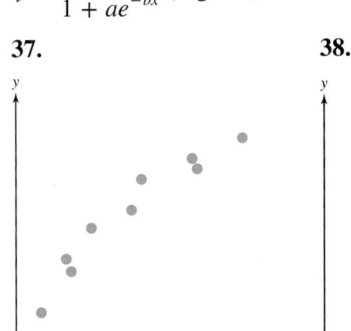

38.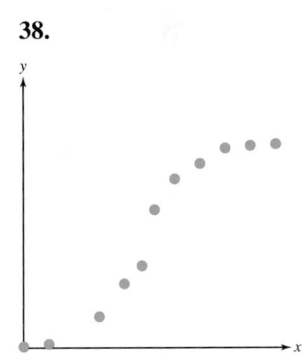

For Exercises 39–46, a table of data is given.

a. Graph the points and from visual inspection, select the model that would best fit the data. Choose from

$y = mx + b$ (linear) $y = ab^x$ (exponential)

$y = a + b \ln x$ (logarithmic) $y = \dfrac{c}{1 + ae^{-bx}}$ (logistic)

b. Use a graphing utility to find a function that fits the data. (*Hint*: For a logistic model, go to STAT, CALC, Logistic.)

39.

x	y
0	2.3
4	3.6
8	5.7
12	9.1
16	14
20	22

40.

x	y
0	52
1	67
2	87
3	114
4	147
5	195

41.

x	y
3	2.7
7	12.2
13	25.7
15	30
17	34
21	44.4

42.

x	y
0	640
20	530
40	430
50	360
80	210
100	90

43.

x	y
10	43.3
20	50
30	53
40	56.8
50	58.8
60	60.8

44.

x	y
5	29
10	40
15	45.6
20	50
25	53.3
30	56

45.

x	y
2	0.326
4	2.57
6	10.8
8	16.8
10	17.9
5	6
7	14.8

46.

x	y
0	0.05
2	0.45
4	2.94
5	5.8
6	8.8
7	10.6
8	11.5
10	11.9

47. During an outbreak of Ebola in western Africa, the cumulative number of cases y was reported t months after April 1. **(See Example 7)**

Month Number (t)	Cumulative Cases (y)
0	18
1	105
2	230
3	438
4	752
5	1437
6	2502

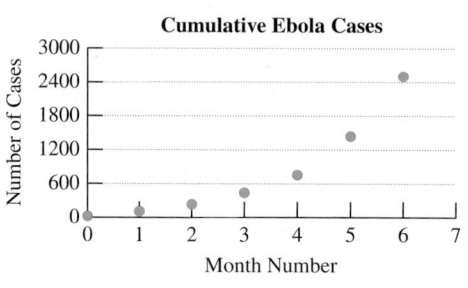

a. Use a graphing utility to find a model of the form $y = ab^t$. Round a to 1 decimal place and b to 3 decimal places.

b. Write the function from part (a) as an exponential function of the form $y = ae^{bt}$.

c. Use either model to predict the number of Ebola cases 8 months after April 1 if this trend continues. Round to the nearest thousand.

d. Would it seem reasonable for this trend to continue indefinitely?

e. Use a graphing utility to find a logistic model $y = \dfrac{c}{1 + ae^{-bt}}$. Round a and c to the nearest whole number and b to 2 decimal places.

f. Use the logistic model from part (e) to predict the number of Ebola cases 8 months after April 1. Round to the nearest thousand.

48. The monthly costs for a small company to do business has been increasing over time due in part to inflation. The table gives the monthly cost y (in \$) for the month of January for selected years. The variable t represents the number of years since 2016.

Year ($t = 0$ is 2016)	Monthly Costs (\$) y
0	12,000
1	12,400
2	12,800
3	13,300

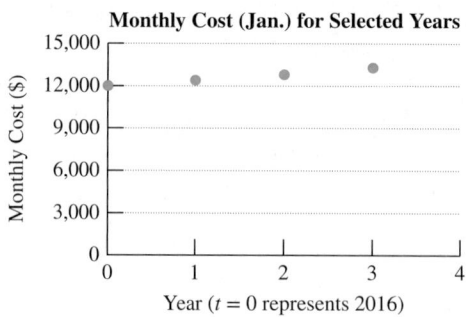

a. Use a graphing utility to find a model of the form $y = ab^t$. Round a to the nearest whole unit and b to 3 decimal places.

b. Write the function from part (a) as an exponential function with base e.

c. Use either model to predict the monthly cost for January in the year 2023 if this trend continues. Round to the nearest hundred dollars.

49. The age of a tree t (in years) and its corresponding height $H(t)$ are given in the table. **(See Example 8)**

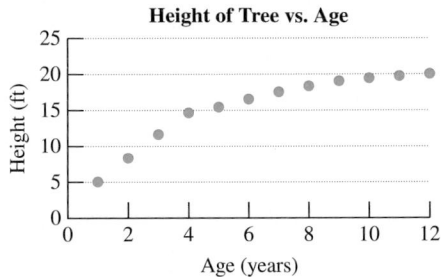

Height of Tree vs. Age

Age of Tree (years) t	Height (ft) $H(t)$
1	5
2	8.3
3	11.6
4	14.6
5	15.4
6	16.5
7	17.5
8	18.3
9	19
10	19.4
11	19.7
12	20

a. Write a model of the form $H(t) = a + b \ln t$. Round a and b to 2 decimal places.

b. Use the model to predict the age of a tree if it is 25 ft high. Round to the nearest year.

c. Is it reasonable to assume that this logarithmic trend will continue indefinitely? Why or why not?

50. The sales of a book tend to increase over the short term as word-of-mouth makes the book "catch on." The number of books sold $N(t)$ for a new novel t weeks after release at a certain bookstore is given in the table for the first 6 weeks.

Weeks t	Number Sold $N(t)$
1	20
2	27
3	31
4	35
5	38
6	39

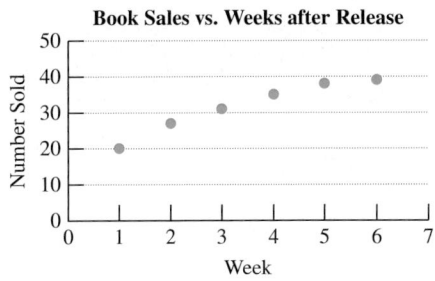

Book Sales vs. Weeks after Release

a. Find a model of the form $N(t) = a + b \ln t$. Round a and b to 1 decimal place.

b. Use the model to predict the sales in week 7. Round to the nearest whole unit.

c. Is it reasonable to assume that this logarithmic trend will continue? Why or why not?

Mixed Exercises

51. A van is purchased new for $29,200.

  a. Write a linear function of the form $y = mt + b$ to represent the value y of the vehicle t years after purchase. Assume that the vehicle is depreciated by $2920 per year.

  b. Suppose that the vehicle is depreciated so that it holds only 80% of its value from the previous year. Write an exponential function of the form $y = V_0 b^t$, where V_0 is the initial value and t is the number of years after purchase.

  c. To the nearest dollar, determine the value of the vehicle after 5 years and after 10 years using the linear model.

  d. To the nearest dollar, determine the value of the vehicle after 5 years and after 10 years using the exponential model.

52. A delivery truck is purchased new for $54,000.

  a. Write a linear function of the form $y = mt + b$ to represent the value y of the vehicle t years after purchase. Assume that the vehicle is depreciated by $6750 per year.

  b. Suppose that the vehicle is depreciated so that it holds 70% of its value from the previous year. Write an exponential function of the form $y = V_0 b^t$, where V_0 is the initial value and t is the number of years after purchase.

  c. To the nearest dollar, determine the value of the vehicle after 4 years and after 8 years using the linear model.

  d. To the nearest dollar, determine the value of the vehicle after 4 years and after 8 years using the exponential model.

Write About It

53. Why is it important to graph a set of data before trying to find an equation or function to model the data.

54. How does the average rate of change differ for a linear function versus an increasing exponential function?

55. Explain the difference between an exponential growth model and a logistic growth model.

56. Explain how to convert an exponential expression b^t to an exponential expression base e.

Expanding Your Skills

57. The monthly payment P (in \$) to pay off a loan of amount A (in \$) at an interest rate r in t years is given by

$$P = \frac{\dfrac{Ar}{12}}{1 - \left(1 + \dfrac{r}{12}\right)^{-12t}}.$$

a. Solve for t (note that there are numerous equivalent algebraic forms for the result).

b. Interpret the meaning of the resulting relationship.

58. Suppose that a population follows a logistic growth pattern, with a limiting population N. If the initial population is denoted by P_0, and t is the amount of time elapsed, then the population P can be represented by

$$P = \frac{P_0 N}{P_0 + (N - P_0)e^{-kt}}.$$

where k is a constant related to the growth rate.

a. Solve for t (note that there are numerous equivalent algebraic forms for the result).

b. Interpret the meaning of the resulting relationship.

CHAPTER 9 Detailed Summary

SECTION 9.1 Inverse Functions

Key Concepts	Examples	Page
A function f is **one-to-one** if for a and b in the domain of f, if $a \neq b$, then $f(a) \neq f(b)$ or equivalently if $f(a) = f(b)$, then $a = b$.	**Example 1:** $f = \{(1, 3), (2, 4), (6, -2)\}$ is one-to-one because no two x values have the same y value. same y $g = \{(1, 3), (2, 4), (6, 3)\}$ is not one-to-one. different x	p. 615
Horizontal line test: A function defined by $y = f(x)$ is one-to-one if no horizontal line intersects the graph in more than one point.	**Example 2:** A horizontal line intersects more than once. f is not one-to-one. No horizontal line intersects more than once. f is one-to-one.	p. 615

Function g **is the inverse of** f if
$$(f \circ g)(x) = x \text{ for all } x \text{ in the domain of } g$$
and
$$(g \circ f)(x) = x \text{ for all } x \text{ in the domain of } f.$$
The inverse of f is denoted by $f^{-1}(x)$.

Procedure to find $f^{-1}(x)$:
1. Replace $f(x)$ by y.
2. Interchange x and y.
3. Solve for y.
4. Replace y by $f^{-1}(x)$.

The graph of a function and its inverse are symmetric with respect to the line $y = x$.

Example 3:   p. 618

Given $f(x) = 3x - 2$, find $f^{-1}(x)$.

$y = 3x - 2$	1. Replace $f(x)$ by y.
$x = 3y - 2$	2. Interchange x and y.
$\dfrac{x+2}{3} = y$	3. Solve for y.
$f^{-1}(x) = \dfrac{x+2}{3}$	4. Replace y by $f^{-1}(x)$.

Check: p. 619

$$(f \circ f^{-1})(x) = 3\left(\frac{x+2}{3}\right) - 2 = x + 2 - 2 = x \checkmark$$

$$(f^{-1} \circ f)(x) = \frac{(3x-2) + 2}{3} = \frac{3x}{3} = x \checkmark$$

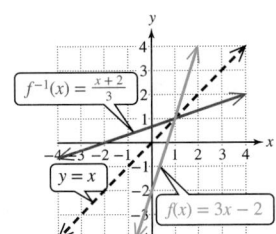

SECTION 9.2 Exponential Functions

Key Concepts	Examples	Page
Let b be a real number with $b > 0$ and $b \neq 1$. Then for any real number x, a function of the form $$f(x) = b^x \text{ is an } \textbf{exponential function of base } b.$$ For the graph of an exponential function $f(x) = b^x$, • If $b > 1$, f is an increasing function. • If $0 < b < 1$, f is a decreasing function. • The domain is $(-\infty, \infty)$. • The range is $(0, \infty)$. • The line $y = 0$ is a horizontal asymptote. • The function passes through $(0, 1)$.	**Example 1:** $f(x) = 2^x$ is an exponential function. $g(x) = x^2$ is not an exponential function. **Example 2:** 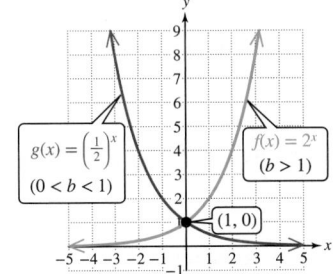	p. 628 p. 630
The irrational number e is the limiting value of the expression $\left(1 + \frac{1}{x}\right)^x$ as x approaches ∞. $$e \approx 2.71828$$ Transformations can be used to graph an exponential function.	**Example 3:** The graph of $f(x) = e^x - 3$ is the graph of $y = e^x$ shifted down 3 units. 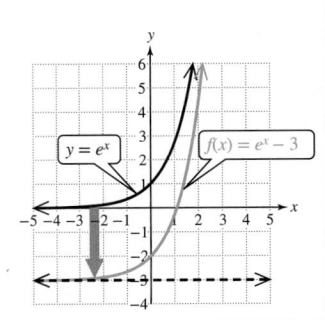	p. 631

If P dollars in principal is invested or borrowed at an annual interest rate r for t years, then $I = Prt$ Amount of simple interest I $A = P\left(1 + \dfrac{r}{n}\right)^{nt}$ Future value A with interest compounded n times per year $A = Pe^{rt}$ Future value A with interest compounded continuously	**Example 4:** Suppose that \$8000 is invested at 5.4% interest. Compute the amount in the account at the end of 3 years under the following compounding options. **a.** Quarterly ($n = 4$) $A = P\left(1 + \dfrac{r}{n}\right)^{nt} = \$8000\left(1 + \dfrac{0.054}{4}\right)^{(4 \cdot 3)}$ $\approx \$9396.69$ **b.** Continuously $A = Pe^{rt} = \$8000e^{(0.054 \cdot 3)} \approx \$9406.88$	p. 633

SECTION 9.3 Logarithmic Functions

Key Concepts	Examples	Page
If x and b are positive real numbers such that $b \neq 1$, then $y = \log_b x$ is called the **logarithmic function** base b, where $y = \log_b x$ is equivalent to $b^y = x$. logarithmic form exponential form The functions $f(x) = \log_b x$ and $g(x) = b^x$ are inverses.	**Example 1:** $y = \log_2 x$ is a logarithmic function and is equivalent to $2^y = x$. The value of y is called the **logarithm**. The value 2 is the **base**, and x is the **argument**.	p. 642
$y = \log_{10} x$ is written as $y = \log x$ and is called the **common logarithmic function**. $y = \log_e x$ is written as $y = \ln x$ and is called the **natural logarithmic function**.	**Example 2:** Evaluate the logarithms. **a.** $\log_2 16 = 4$ because $2^4 = 16$ **b.** $\log 100 = 2$ because $10^2 = 100$ **c.** $\ln\left(\dfrac{1}{e}\right) = -1$ because $e^{-1} = \dfrac{1}{e}$	p. 644
Basic properties of logarithms: 1. $\log_b 1 = 0$ 2. $\log_b b = 1$ 3. $\log_b b^x = x$ 4. $b^{\log_b x} = x$	**Example 3:** **a.** $\log_5 1 = 0$ **b.** $\log_7 7 = 1$ **c.** $\ln e^5 = 5$ **d.** $10^{\log(x+5)} = x + 5$	p. 646
The graph of a logarithmic function: Given $f(x) = \log_b x$, • If $b > 1$, f is an increasing function. • If $0 < b < 1$, f is a decreasing function. • The domain is $(0, \infty)$. • The range is $(-\infty, \infty)$. • The line $x = 0$ is a vertical asymptote. • The function passes through $(1, 0)$.	**Example 4:** 	p. 648
The argument to a logarithmic function must be a positive real number	**Example 5:** Given $f(x) = \ln(3 - x)$, write the domain of f in interval notation. $3 - x > 0$ Set the argument greater than 0. $-x > -3$ Solve for x. $x < 3$ The domain is: $(-\infty, 3)$.	p. 650

SECTION 9.4 Properties of Logarithms

Key Concepts	Examples	Page
Additional properties of logarithms: Let b, x, and y be positive real numbers with $b \neq 1$. Let p be any real number. Then, $\log_b(xy) = \log_b x + \log_b y$ (Product property) $\log_b\left(\dfrac{x}{y}\right) = \log_b x - \log_b y$ (Quotient property) $\log_b x^p = p \log_b x$ (Power property)	**Example 1:** **a.** $\log_3(9y) = \log_3 9 + \log_3 y = 2 + \log_3 y$ **b.** $\log\left(\dfrac{100}{t}\right) = \log 100 - \log t = 2 - \log t$ **c.** $\ln \sqrt[3]{w} = \ln w^{1/3} = \dfrac{1}{3}\ln w$	p. 660
A single logarithmic expression can be written as the sum or difference of logarithms.	**Example 2:** $\log\left(\dfrac{5x^7}{\sqrt{y}}\right) = \log(5x^7) - \log\sqrt{y}$ Quotient property $\qquad = \log 5 + \log x^7 - \log y^{1/2}$ Product property $\qquad = \log 5 + 7\log x - \dfrac{1}{2}\log y$ Power property	p. 661
A sum or difference of logarithms can be written as a single logarithm.	**Example 3:** $\ln c + \ln(a^2 - b^2) - \ln(a + b)$ $\quad = \ln[c(a^2 - b^2)] - \ln(a + b)$ Product property $\quad = \ln\left[\dfrac{c(a^2 - b^2)}{a + b}\right]$ Quotient property $\quad = \ln\left[\dfrac{c(a + b)(a - b)}{a + b}\right]$ Factor. $\quad = \ln[c(a - b)]$ Simplify.	p. 662
Change-of-base formula: For positive real numbers a and b, where $a \neq 1$ and $b \neq 1$, $\log_b x = \dfrac{\log_a x}{\log_a b}$ Change-of-base formula *Note*: The change-of-base formula is often used with base 10 or base e so that a calculator can be used to approximate the value of the logarithm.	**Example 4:** Use a calculator to approximate $\log_2 806$. $\log_2 806 = \dfrac{\log 806}{\log 2} \approx 9.6546$ $\log_2 806 = \dfrac{\ln 806}{\ln 2} \approx 9.6546$ Check: $2^{9.6546} \approx 806$ ✓	p. 663

SECTION 9.5 Exponential Equations and Applications

Key Concepts	Examples	Page
Equivalence property of exponential expressions: Let b, x, and y are real numbers with $b > 0$ and $b \neq 1$, then, $b^x = b^y$ implies that $x = y$.	**Example 1:** Solve. $27^{-2x+1} = \left(\dfrac{1}{9}\right)^x$ $(3^3)^{-2x+1} = (3^{-2})^x$ $3^{-6x+3} = 3^{-2x}$ implies that $-6x + 3 = -2x$ $\qquad\qquad\qquad\qquad -4x = -3$ $\qquad\qquad\qquad\qquad\quad x = \dfrac{3}{4}$ The solution set is $\left\{\dfrac{3}{4}\right\}$.	p. 668

Steps to solve an exponential equation by using logarithms:	**Example 2:**	p. 669
1. Isolate the exponential expression on one side of the equation.	$5^x - 2 = 87$	
2. Take a logarithm of the same base on both sides of the equation.	$5^x = 89$	
3. Use the power property of logarithms to "bring down" the exponent.	$\log 5^x = \log 89$	
4. Solve the resulting equation.	$x \log 5 = \log 89$	
	$x = \dfrac{\log 89}{\log 5} \approx 2.7889$	
	The solution set is $\left\{ \dfrac{\log 89}{\log 5} \right\}$.	

Example 3:	**Example 4:**	p. 672
$2^{x-3} = 5^{x+1}$	Some exponential equations are quadratic in form.	
$\ln 2^{x-3} = \ln 5^{x+1}$	$e^{2x} - 3e^x - 4 = 0 \qquad$ Let $u = e^x$.	
$(x - 3)\ln 2 = (x + 1)\ln 5$	$u^2 - 3u - 4 = 0$	
$x \ln 2 - 3 \ln 2 = x \ln 5 + \ln 5$	$(u - 4)(u + 1) = 0$	
$x \ln 2 - x \ln 5 = \ln 5 + 3 \ln 2$	$u = 4 \qquad$ or $\qquad u = -1$	
$x(\ln 2 - \ln 5) = \ln 5 + 3 \ln 2$	$e^x = 4 \qquad$ or $\qquad e^x = -1$ (no solution)	
$x = \dfrac{\ln 5 + 3 \ln 2}{\ln 2 - \ln 5}$	$\ln e^x = \ln 4$	
	$x = \ln 4$	
The solution set is $\left\{ \dfrac{\ln 5 + 3 \ln 2}{\ln 2 - \ln 5} \right\}$.	The solution set is $\{\ln 4\}$.	

SECTION 9.6 Logarithmic Equations and Applications

Key Concepts	Examples	Page
Equivalence property of logarithmic expressions: If b, x, and y are positive real numbers and $b \neq 1$, then, $$\log_b x = \log_b y \text{ implies that } x = y.$$	**Example 1:** Solve. $\qquad \log(5x - 1) = \log(4x + 6)$ $\qquad\qquad\qquad 5x - 1 = 4x + 6$ $\qquad\qquad\qquad\qquad\qquad x = 7$ <u>Check:</u> $\quad \log[5(7) - 1] \overset{?}{=} \log[4(7) + 6]$ $\qquad\qquad\qquad \log 34 = \log 34 \checkmark$ The solution set is $\{7\}$.	p. 677
Guidelines to solve a logarithmic equation: 1. Isolate the logarithms on one side of the equation. 2. Use the properties of logarithms to write the equation in the form $\log_b x = k$, where k is a constant. 3. Write the equation in exponential form. 4. Solve the equation from step 3. 5. Check the potential solution(s) in the original equation. In Example 2, the solution 6 checks. However, 2 does not check because 2 is undefined in the original equation in the expressions $\log_3(x - 5)$ and $\log_3(x - 3)$.	**Example 2:** Solve. $\qquad\qquad \log_3(x - 5) = 1 - \log_3(x - 3)$ $\quad \log_3(x - 5) + \log_3(x - 3) = 1$ $\qquad\quad \log_3[(x - 5)(x - 3)] = 1$ $\qquad\qquad \log_3(x^2 - 8x + 15) = 1$ $\qquad\qquad\qquad x^2 - 8x + 15 = 3^1$ $\qquad\qquad\qquad x^2 - 8x + 12 = 0$ $\qquad\qquad\qquad (x - 6)(x - 2) = 0$ $\qquad\quad x = 6 \quad$ or $\quad x = 2 \quad$ (2 does not check.) The solution set is $\{6\}$.	p. 678

	Example 3:	p. 679
	$5 \ln(x + 1) - 40 = 0$	
	$5 \ln(x + 1) = 40$	
	$\ln(x + 1) = 8$	
	$x + 1 = e^8$	
	$x = e^8 - 1$	
	Check: $\quad 5 \ln(x + 1) - 40 = 0$	
	$5 \ln\left[\left(e^8 - 1\right) + 1\right] - 40 \overset{?}{=} 0$	
	$5 \ln\left(e^8\right) - 40 \overset{?}{=} 0$	
	$5 \cdot 8 - 40 \overset{?}{=} 0 \checkmark$	
	The solution set is $\left\{e^8 - 1\right\}$.	

SECTION 9.7 Modeling with Exponential and Logarithmic Functions

Key Concepts	Examples	Page
Exponential growth and decay models: Let y be a variable changing exponentially with respect to time t and let y_0 represent the initial value of y when $t = 0$. Then for a constant k, the model $y = y_0 e^{kt}$ represents • exponential growth if the constant $k > 0$. • exponential decay if the constant $k < 0$.	**Example 1:** The population of Virginia was 7.1 million in the year 2000 and grew to 8.6 million by 2019. Write a model of the form $P(t) = P_0 e^{kt}$, where P_0 represents the initial population and t represents the number of years since 2000. $P(t) = P_0 e^{kt}$ $P(t) = 7.1 e^{kt}$ $\qquad$ Substitute 7.1 for P_0. $8.6 = 7.1 e^{k(19)}$ $\qquad$ $P(19) = 8.6$. $\dfrac{8.6}{7.1} = e^{19k}$ $\ln\left(\dfrac{8.6}{7.1}\right) = 19k$ $k = \dfrac{\ln\left(\dfrac{8.6}{7.1}\right)}{19} \approx 0.0101$ $\qquad$ Solve for k. $P(t) = 7.1 e^{0.0101t}$	p. 686
Changing bases in an exponential model: An exponential expression can be rewritten as an expression of a different base. In particular, to convert to base e, we have $\qquad b^t$ is equivalent to $e^{(\ln b)\, t}$	**Example 2:** $y = 400(0.5)^{0.03t} = 400\left[(0.5)^t\right]^{0.03}$ $\qquad = 400\left[e^{(\ln 0.5)t}\right]^{0.03}$ $\qquad = 400\left[e^{(\ln 0.5)(0.03)}\right]^t$ $\qquad \approx 400 e^{-0.0208t}$	p. 686
A **logistic growth function** is a function of the form: $\qquad y = \dfrac{c}{1 + ae^{-bt}}$ where a, b, and c are positive constants. A logistic growth function imposes a limiting value on the dependent variable.	**Example 3:** $y = \dfrac{42.6}{1 + 9e^{-0.35t}}$ defines a logistic function. 	p. 690

A graphing utility and regression techniques can be used to model data that exhibit an exponential or logarithmic trend.

Exponential model: $y = ab^x$

Logarithmic model: $y = a + b \ln x$

Example 4:

x	1	2	3	4	5
y	2	7.5	10.8	13.1	14.9

p. 693

```
NORMAL FLOAT AUTO REAL RADIAN CL
L1    L2    L3    L4    L5    1
1     2
2     7.5   NORMAL FLOAT AUTO REAL RADIAN CL
3     10.8
4     13.1
5     14.9  EDIT CALC TESTS
            7↑QuartReg
            8:LinReg(a+bx)
            9:LnReg
L1 ={1,2,3, 0: NORMAL FLOAT AUTO REAL RADIAN CL
            A:
            B:        LnReg
            C:  y=a+blnx
            D:  a=1.980089365
            E:  b=8.02080823
               NORMAL FLOAT AUTO REAL RADIAN CL
               20
                       y = 1.98 + 8.02 ln x

                0
               -4                          5
```

CHAPTER 9   Review Exercises

SECTION 9.1

For Exercises 1–2, determine if the relation defines y as a one-to-one function of x.

1.

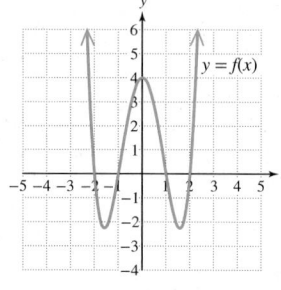

$y = f(x)$

2.

x	y
5	7
-3	1
-4	-2
6	0

For Exercises 3–4, use the definition of a one-to-one function to determine if the function is one-to-one. Recall that f is one-to-one if $a \neq b$ implies that $f(a) \neq f(b)$, or equivalently, if $f(a) = f(b)$, then $a = b$.

3. $f(x) = x^3 - 1$ **4.** $f(x) = x^2 - 1$

For Exercises 5–6, determine if the functions are inverses.

5. $f(x) = 4x - 3$ and $g(x) = \dfrac{x + 3}{4}$

6. $m(x) = \sqrt[3]{x + 1}$ and $n(x) = (x - 1)^3$

For Exercises 7–8, a one-to-one function is given. Write an equation for the inverse function.

7. $f(x) = 2x^3 - 5$ **8.** $f(x) = \dfrac{2}{x + 7}$

9. a. Graph $f(x) = x^2 - 9$, $x \leq 0$.

  b. Is f a one-to-one function?

  c. Write the domain of f in interval notation.

  d. Write the range of f in interval notation.

  e. Find an equation for f^{-1}.

  f. Graph $y = f(x)$ and $y = f^{-1}(x)$ on the same coordinate system.

  g. Write the domain of f^{-1} in interval notation.

  h. Write the range of f^{-1} in interval notation.

10. a. Graph $g(x) = \sqrt{x + 4}$.

  b. Is g a one-to-one function?

  c. Write the domain of g in interval notation.

  d. Write the range of g in interval notation.

  e. Find an equation for g^{-1}.

  f. Graph $y = g(x)$ and $y = g^{-1}(x)$ on the same coordinate system.

  g. Write the domain of g^{-1} in interval notation.

  h. Write the range of g^{-1} in interval notation.

11. The function $f(x) = 5280x$ provides the conversion from x miles to $f(x)$ feet.

  a. Write an equation for f^{-1}.

  b. What does the inverse function represent in the context of this problem?

  c. Determine the number of miles represented by 22,176 ft.

SECTION 9.2

12. Which of the functions is an exponential function?

a. $f(x) = x^4$ **b.** $h(x) = 4^{-x}$ **c.** $g(x) = \left(\dfrac{4}{3}\right)^x$

d. $k(x) = \dfrac{4x}{3}$ **e.** $n(x) = \dfrac{4}{3x}$ **f.** $r(x) = \left(-\dfrac{4}{3}\right)^x$

For Exercises 13–16,

a. Graph the function.
b. Write the domain in interval notation.
c. Write the range in interval notation.
d. Write an equation of the asymptote.

13. $f(x) = \left(\dfrac{5}{2}\right)^x$ **14.** $g(x) = \left(\dfrac{5}{2}\right)^{-x}$

15. $k(x) = -3^x + 1$ **16.** $h(x) = 2^{x-3} - 4$

17. Is the graph of $y = e^x$ an increasing or decreasing exponential function?

For Exercises 18–19, use the formulas on page 633.

18. Suppose that $24,000 is invested at the given interest rates and compounding options. Determine the amount that the investment is worth at the end of t years.

   a. 5% interest compounded monthly for 10 years

   b. 4.5% interest compounded continuously for 30 years

19. Jorge needs to borrow $12,000 to buy a car. He can borrow the money at 7.2% simple interest for 4 years or he can borrow at 6.5% interest compounded continuously for 4 years.

   a. How much total interest would Jorge pay at 7.2% simple interest?

   b. How much total interest would Jorge pay at 6.5% interest compounded continuously?

   c. Which option results in less total interest?

20. A patient is treated with 128 mCi (millicuries) of iodine-131 (^{131}I). The radioactivity level $R(t)$ (in mCi) after t days is given by $R(t) = 128(2)^{-t/4.2}$. (In this model, the value 4.2 is related to the biological half-life of radioactive iodine in the body.)

   a. Determine the radioactivity level of ^{131}I in the body after 6 days. Round to the nearest whole unit.

   b. Evaluate $R(4.2)$ and interpret its meaning in the context of this problem.

   c. After how many half-lives will the radioactivity level be 16 mCi?

SECTION 9.3

For Exercises 21–22, write the equation in exponential form.

21. $\log_b(x^2 + y^2) = 4$ **22.** $\ln x = (c + d)$

For Exercises 23–24, write the equation in logarithmic form.

23. $10^6 = 1{,}000{,}000$ **24.** $8^{-1/3} = \dfrac{1}{2}$

For Exercises 25–32, simplify the logarithmic expression without using a calculator.

25. $\log_3 81$ **26.** $\log 100{,}000$

27. $\log_2\left(\dfrac{1}{64}\right)$ **28.** $\log_{1/4}(16)$

29. $\log_{11} 1$ **30.** $\log_5 5$

31. $4^{\log_4 7}$ **32.** $\ln e^{11}$

For Exercises 33–37, write the domain of the function in interval notation.

33. $f(x) = \log(x - 4)$ **34.** $g(x) = \ln(3 - 2x)$

35. $h(x) = \log_2(x^2 + 4)$ **36.** $k(x) = \log_2(x^2 - 4)$

37. $m(x) = \log_2(x - 4)^2$

For Exercises 38–39,

a. Graph the function.
b. Write the domain in interval notation.
c. Write the range in interval notation.
d. Write an equation of the asymptote.

38. $f(x) = \log_2(x - 3)$ **39.** $g(x) = 2 + \ln x$

For Exercises 40–41, use the formula $pH = -\log[H^+]$ to compute the pH of a liquid as a function of its concentration of hydronium ions, $[H^+]$ in mol/L. If the pH is less than 7, then the substance is acidic. If the pH is greater than 7, then the substance is alkaline (or basic).

a. Find the pH. Round to 1 decimal place.

b. Determine whether the substance is acidic or alkaline.

40. Baking soda: $[H^+] = 5.0 \times 10^{-9}$ mol/L

41. Tomatoes: $[H^+] = 3.16 \times 10^{-5}$ mol/L

SECTION 9.4

For Exercises 42–48, fill in the blanks to state the basic properties of logarithms. Assume that x, y, and b are positive real numbers with $b \neq 1$.

42. $\log_b 1 = $ _____   **43.** $\log_b b = $ _____

44. $\log_b b^p = $ _____   **45.** $b^{\log_b x} = $ _____

46. $\log_b(xy) = $ _____   **47.** $\log_b\left(\dfrac{x}{y}\right) = $ _____

48. $\log_b x^p = $ _____

For Exercises 49–52, write the logarithm as a sum or difference of logarithms. Simplify each term as much as possible.

49. $\log\left(\dfrac{100}{\sqrt{c^2 + 10}}\right)$

50. $\log_2\left(\dfrac{1}{8}a^2 b\right)$

51. $\ln\left(\dfrac{\sqrt[3]{ab^2}}{cd^5}\right)$

52. $\log\left(\dfrac{x^2(2x + 1)^5}{\sqrt{1 - x}}\right)$

For Exercises 53–55, write the logarithmic expression as a single logarithm with coefficient 1, and simplify as much as possible.

53. $4\log_5 y - 3\log_5 x + \dfrac{1}{2}\log_5 z$

54. $\log 250 + \log 2 - \log 5$

55. $\dfrac{1}{4}\ln(x^2 - 9) - \dfrac{1}{4}\ln(x - 3)$

For Exercises 56–58, use $\log_b 2 \approx 0.289$, $\log_b 3 \approx 0.458$, and $\log_b 5 \approx 0.671$ to approximate the value of the given logarithms.

56. $\log_b 8$ **57.** $\log_b 45$ **58.** $\log_b\left(\dfrac{1}{9}\right)$

For Exercises 59–60, use the change-of-base formula and a calculator to approximate the given logarithms. Round to 4 decimal places. Then check the answer by using the related exponential form.

59. $\log_7 596$ **60.** $\log_4 0.982$

SECTION 9.5

For Exercises 61–74, solve the equation. Write the solution set with exact values and give approximate solutions to 4 decimal places.

61. $2^{x+1} = 16$ **62.** $4^{2y-7} = 64$

63. $3^{x-5} = \dfrac{1}{27}$ **64.** $1000^{2x+1} = \left(\dfrac{1}{100}\right)^{x-4}$

65. $7^x = 51$ **66.** $5^x = 1000$

67. $516 = 11^w - 21$ **68.** $7^x + 8 = 96$

69. $3^{2x+1} = 4^{3x}$ **70.** $2^{c+3} = 7^{2c+5}$

71. $400e^{-2t} = 2.989$ **72.** $2 \cdot 10^{1.2t} = 58$

73. $e^{2x} - 3e^x - 40 = 0$ **74.** $e^{2x} = -10e^x$

SECTION 9.6

For Exercises 75–90, solve the equation. Write the solution set with exact values and give approximate solutions to 4 decimal places.

75. $\log_5(4p + 7) = \log_5(2 - p)$

76. $\ln(3x - 5) = \ln(2x + 1)$

77. $\log_2(m^2 + 10m) = \log_2 11$

78. $\log(w^2 + 5w) = \log 14$

79. $2\log_6(4 - 8y) + 6 = 10$

80. $5 = -4\log_3(2 - 5x) + 1$

81. $3\ln(n - 8) = 6.3$

82. $5\log(m + 4) = 25$

83. $-4 + \log_2 x = -\log_2(x + 6)$

84. $\log_6(3x + 2) = \log_6(x + 4) + 1$

85. $\ln x + \ln(x + 2) = \ln(x + 6)$

86. $\log x + \log(x - 6) = \log(3x - 20)$

87. $\log_5(\log_2 x) = 1$

88. $\log_2(\log_3 x) = 2$

89. $(\log x)^2 - \log x^2 = 35$

90. $(\log x)^2 - 13\log x + 36 = 0$

For Exercises 91–92, find the inverse of the function.

91. $f(x) = 4^x$ **92.** $g(x) = \log(x - 5) - 1$

93. The percentage of visible light P (in decimal form) at a depth of x meters for Long Island Sound can be approximated by $P = e^{-0.5x}$.

   a. Determine the depth at which the light intensity is half the value from the surface. Round to the nearest hundredth of a meter. Based on your answer, would you say that Long Island Sound is murky or clear water?

   b. Determine the euphotic depth for Long Island Sound. That is, find the depth at which the light intensity falls below 1%. Round to the nearest tenth of a meter.

SECTION 9.7

For Exercises 94–95, solve for the indicated variable.

94. $\log B - 1.7 = 2.3M$ for B

95. $T = T_f + T_0 e^{-kt}$ for t

96. Suppose that $18,000 is invested in a bond fund and the account grows to $23,344.74 in 5 years.

   a. Use the model $A = Pe^{rt}$ to determine the average rate of return under continuous compounding. Round to the nearest tenth of a percent.

   b. How long will it take the investment to reach $30,000 if the rate of return continues? Round to the nearest tenth of a year.

97. The population $P(t)$ (in millions) of a European country is represented by $P(t) = 85.5e^{-0.00208t}$, where t is the number of years since a baseline census was taken.

   a. What was the population the year of the census?

   b. Based on this model, is the population of the country increasing or decreasing?

   c. If this trend continues, determine the number of years from the date of the census that will be needed for the population to reach 80 million. Round to the nearest year.

98. A census was taken in a South American country, and the population was determined to be 16.9 million with a yearly growth rate of 0.836%. The model $P(t) = 16.9(1.00836)^t$ gives the population $P(t)$ (in millions), t years after the census.

  a. Write a model of the form $P(t) = P_0e^{kt}$ to represent the population t years after the census.

  b. Based on this model, is the population of the country increasing or decreasing?

  c. If this trend continues, determine the number of years that will be required for the population to reach 20 million. Round to the nearest year.

99. A sample from human remains found near Stonehenge in England shows that 71.2% of the carbon-14 still remains. Use the model $Q(t) = Q_0e^{-0.000121t}$ to determine the age of the sample. In this model, $Q(t)$ represents the amount of carbon-14 remaining t years after death, and Q_0 represents the initial amount of carbon-14 at the time of death. Round to the nearest 100 years.

100. A lake is stocked with bass by the U.S. Park Service. The population of bass is given by $P(t) = \dfrac{3000}{1 + 2e^{-0.37t}}$, where t is the time in years after the lake was stocked.

  a. Evaluate $P(0)$ and interpret its meaning in the context of this problem.

  b. Use the function to predict the bass population 2 years after being stocked. Round to the nearest whole unit.

  c. Use the function to predict the bass population 4 years after being stocked.

  d. Determine the number of years required for the bass population to reach 2800. Round to the nearest year.

  e. What value will the term $\dfrac{2}{e^{0.37t}}$ approach as $t \to \infty$?

  f. Determine the limiting value of $P(t)$.

101. For the given data,

  a. Use a graphing utility to find an exponential function $Y_1 = ab^x$ that fits the data.

  b. Graph the data and the function from part (a) on the same coordinate system.

x	y
0	2.4
1	3.5
2	5.5
3	8.1
4	12.0
5	18.4

CHAPTER 9   Test

1. Given $f(x) = 4x^3 - 1$,

  a. Write an equation for $f^{-1}(x)$.

  b. Verify that $(f \circ f^{-1})(x) = (f^{-1} \circ f)(x) = x$.

2. The graph of f is given.

  a. Is f a one-to-one function?

  b. If f is a one-to-one function, graph f^{-1} on the same coordinate system as f.

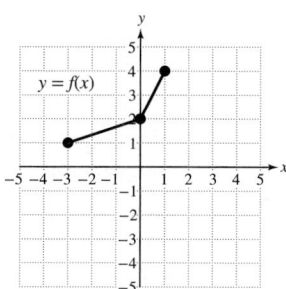

3. Given $f(x) = \dfrac{x + 3}{x - 4}$, write an equation for the inverse function.

For Exercises 4–7,

a. Write the domain and range of f in interval notation.

b. Write an equation of the inverse function.

c. Write the domain and range of f^{-1} in interval notation.

  4. $f(x) = -x^2 + 1, x \le 0$ **5.** $f(x) = \log x$

  6. $f(x) = 3^x + 1$ **7.** $f(x) = \sqrt{x + 5}$

For Exercises 8–11,

a. Graph the function.

b. Write the domain in interval notation.

c. Write the range in interval notation.

d. Write an equation of the asymptote.

  8. $f(x) = \left(\dfrac{1}{3}\right)^x + 2$ **9.** $g(x) = 2^{x-4}$

  10. $h(x) = -\ln x$ **11.** $k(x) = \log_2(x + 1) - 3$

12. Write the statement in exponential form. $\ln(x + y) = a$

For Exercises 13–18, evaluate the logarithmic expression without using a calculator.

  13. $\log_9 \dfrac{1}{81}$ **14.** $\log_6 216$ **15.** $\ln e^8$

  16. $\log 10^{-4}$ **17.** $10^{\log(a^2 + b^2)}$ **18.** $\log_{1/2} 1$

For Exercises 19–20, write the domain of the function in interval notation.

  19. $f(x) = \log(7 - 2x)$ **20.** $g(x) = \log_4(x^2 - 25)$

For Exercises 21–22, write the logarithm as a sum or difference of logarithms. Simplify each term as much as possible.

21. $\ln\left(\dfrac{x^5 y^2}{w\sqrt[3]{z}}\right)$

22. $\log\left(\dfrac{\sqrt{a^2 + b^2}}{10^4}\right)$

For Exercises 23–24, write the logarithmic expression as a single logarithm with coefficient 1, and simplify as much as possible.

23. $6\log_2 a - 4\log_2 b + \dfrac{2}{3}\log_2 c$

24. $\dfrac{1}{2}\ln(x^2 - x - 12) - \dfrac{1}{2}\ln(x - 4)$

For Exercises 25–26, use $\log_b 2 \approx 0.289$, $\log_b 3 \approx 0.458$, and $\log_b 5 \approx 0.671$ to approximate the value of the given logarithms.

25. $\log_b 72$

26. $\log_b\left(\dfrac{1}{125}\right)$

For Exercises 27–38, solve the equation. Write the solution set with exact values and give approximate solutions to 4 decimal places.

27. $10^{2x-5} = 10,000,000$

28. $2^{5y+1} = 4^{y-3}$

29. $5^{x+3} + 3 = 56$

30. $2^{c+7} = 3^{2c+3}$

31. $7e^{4x} - 2 = 12$

32. $5 \cdot 10^{x/2} + 9 = 24$

33. $e^{2x} + 7e^x - 8 = 0$

34. $\log_5(3 - x) = \log_5(x + 1)$

35. $5\ln(x + 2) + 1 = 16$

36. $\log x + \log(x - 1) = \log 12$

37. $-3 + \log_4 x = -\log_4(x + 30)$

38. $\log 3 + \log(x + 3) = \log(4x + 5)$

For Exercises 39–40, solve for the indicated variable.

39. $S = 92 - k\ln(t + 1)$ for t

40. $A = P\left(1 + \dfrac{r}{n}\right)^{nt}$ for t

41. Suppose that $10,000 is invested and the account grows to $13,566.25 in 5 years.

   a. Use the model $A = Pe^{rt}$ to determine the average rate of return under continuous compounding. Round to the nearest tenth of a percent.

   b. Using the interest rate from part (a), how long will it take the investment to reach $50,000? Round to the nearest tenth of a year.

42. The number of bacteria in a culture begins with approximately 10,000 organisms at the start of an experiment. If the bacteria double every 5 hr, the model $P(t) = 10,000(2)^{t/5}$ represents the population $P(t)$ after t hours.

   a. Write a function of the form $P(t) = P_0 e^{kt}$ to model the population.

   b. Determine the amount of time required for the population to grow to 5 million. Round to the nearest hour.

43. The population $P(t)$ of a herd of deer on an island can be modeled by $P(t) = \dfrac{1200}{1 + 2e^{-0.12t}}$, where t represents the number of years since the park service has been tracking the herd.

   a. Evaluate $P(0)$ and interpret its meaning in the context of this problem.

   b. Use the function to predict the deer population after 4 years. Round to the nearest whole unit.

   c. Use the function to predict the deer population after 8 years.

   d. Determine the number of years required for the deer population to reach 900. Round to the nearest year.

   e. What value will the term $\dfrac{2}{e^{0.12t}}$ approach as $t \to \infty$?

   f. Determine the limiting value of $P(t)$.

44. The number N of visitors to a new website is given in the table t weeks after the website was launched.

t	0	1	2	3	4
N	24	50	121	270	640

   a. Use a graphing utility to find an equation of the form $N = ab^t$ to model the data. Round a to 1 decimal place and b to 3 decimal places.

   b. Use a graphing utility to graph the data and the model from part (a).

   c. Use the model to predict the number of visitors to the website 10 weeks after launch. Round to the nearest thousand.

Systems of Equations and Inequalities

10

Chapter Outline

Michael Hitoshi/Digital Vision/Getty Images

The economy influences how people spend their money, but it also impacts how people save. When comparing options for investments such as stocks, bonds, and mutual funds, an individual can easily become overwhelmed by the possible scenarios. Collecting data on the rates of return on different investments is helpful in making an informed decision. In some scenarios, we turn to systems of linear equations to analyze such data.

In this chapter, we will solve systems involving both linear and nonlinear equations. In addition, we will use systems of linear inequalities in manufacturing applications. In such applications, the goal is to determine constraints on the production process (such as limits on the amount of material, labor, and equipment) and then maximize profit or minimize cost subject to these constraints.

SECTION 10.1 Graphs of Systems of Linear Equations in Two Variables

1. Identify Solutions to Systems of Linear Equations

Recall from Section 5.4 that a **linear equation in two variables** is an equation of the form $Ax + By = C$, where A, B, and C are real numbers and A and B are not both zero. A linear equation in two variables has an infinite number of solutions and the graph is a line in a rectangular coordinate system. Two or more linear equations form a **system of linear equations**. A **solution** to a system of equations in two variables is an ordered pair that is a solution to each individual equation in the system. The graph of a system of equations in two variables is a pair of lines. The point or points of intersection of the lines make up the **solution set** to the system.

Systems of equations are used in many applications, including the analysis of supply and demand of a product. For example, airlines offer tickets at varying prices. An airline supplies tickets to consumers but will supply only a small number of tickets at low fares. On the other hand, the airline is happy to supply a larger number of tickets at higher fares. This is illustrated in the blue supply curve in Figure 10-1. From the consumer standpoint, few people want to purchase tickets at high prices. But for low fares, the demand for tickets is high. This is illustrated in the red demand curve in Figure 10-1.

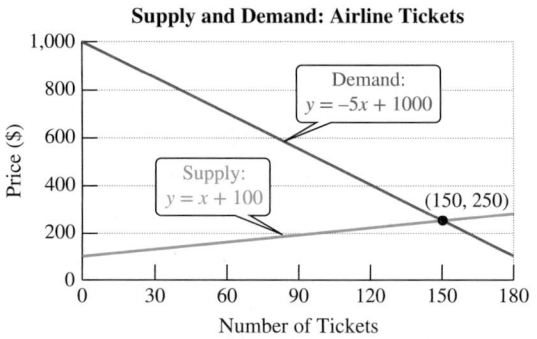

Figure 10-1

In an open market, the price of an item is dependent on the number of items supplied by the producer and demanded by the consumer. Competition between buyers and sellers steers the price toward an **equilibrium price**. This is the point where the supply curve and the demand curve intersect.

Suppose that the following linear equations represent the supply and demand curves for the airline scenario. In each equation, x represents the number of tickets supplied or demanded, and y is the corresponding price (in dollars).

$$\left.\begin{array}{ll} \text{Supply:} & y = x + 100 \\ \text{Demand:} & y = -5x + 1000 \end{array}\right\} \quad \text{System of linear equations}$$

From the graph in Figure 10-1, the point of intersection is (150, 250). To confirm that it is a solution to each individual equation, we can use direct substitution.

Supply: $y = x + 100$ $\xrightarrow{\text{Test } (150, 250)}$ $250 \stackrel{?}{=} 150 + 100$ ✓

Demand: $y = -5x + 1000$ $\xrightarrow{\text{Test } (150, 250)}$ $250 \stackrel{?}{=} -5(150) + 1000$ ✓

The point (150, 250) checks in both equations and thus is a solution to the system of equations. It is the equilibrium point balancing the number of tickets supplied and demanded (150) at a mutually agreeable price of $250.

EXAMPLE 1   **Determining Solutions to a System of Linear Equations**

Determine whether the ordered pairs are solutions to the system.

$$x + y = -6$$
$$3x - y = -2$$

a. $(-2, -4)$ **b.** $(0, -6)$

Solution:

a. Substitute the ordered pair $(-2, -4)$ into each equation:

$$x + y = -6 \longrightarrow (-2) + (-4) \stackrel{?}{=} -6 \checkmark \text{ True}$$
$$3x - y = -2 \longrightarrow 3(-2) - (-4) \stackrel{?}{=} -2 \checkmark \text{ True}$$

Because the ordered pair $(-2, -4)$ is a solution to each equation, it is a solution to the *system* of equations.

b. Substitute the ordered pair $(0, -6)$ into each equation:

$$x + y = -6 \longrightarrow (0) + (-6) \stackrel{?}{=} -6 \checkmark \text{ True}$$
$$3x - y = -2 \longrightarrow 3(0) - (-6) \stackrel{?}{=} -2 \quad \text{ False}$$

Because the ordered pair $(0, -6)$ is not a solution to the second equation, it is *not* a solution to the system of equations.

Skill Practice 1   Determine whether the ordered pairs are solutions to the system.

$$3x + 2y = -8$$
$$y = 2x - 18$$

a. $(-2, -1)$ **b.** $(4, -10)$

FOR REVIEW

Recall that a linear equation in two variables written in the form $y = mx + b$ is said to be in slope-intercept form. The value of m is the slope and the point $(0, b)$ is the y-intercept. See Section 5.4 for review.

We can graph the lines from Example 1 to support our findings. Writing each equation in slope-intercept form, we have:

First equation	Second equation
$x + y = -6$	$3x - y = -2$
$y = -x - 6$	$-y = -3x - 2$
	$y = 3x + 2$

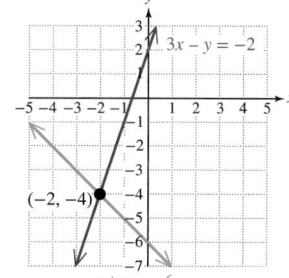

Figure 10-2

The graph of the first equation has a y-intercept of $(0, -6)$ and slope of -1 (Figure 10-2, blue line).

The graph of the second equation has a y-intercept of $(0, 2)$ and a slope of 3 (Figure 10-2, red line).

The point $(-2, -4)$ is the only point of intersection. Therefore, the solution set for the system of equations is $\{(-2, -4)\}$.

2. Solve Systems of Linear Equations by Graphing

The graph of the system of equations from Example 1 shows that there is only one point of intersection of the two lines, and thus only one solution to the system. However, there are three different possibilities regarding the number of solutions to a system of linear equations in two variables.

Answers
1. a. No   **b.** Yes

Solutions to Systems of Linear Equations in Two Variables

One Unique Solution	**No Solution**	**Infinitely Many Solutions**
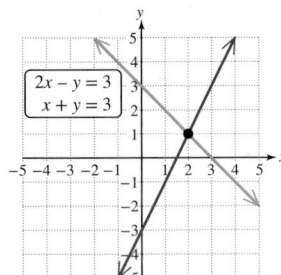	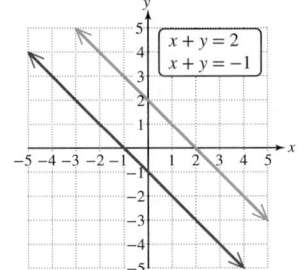	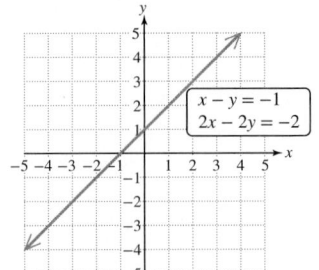
If a system of linear equations represents intersecting lines, then it has exactly one solution.	If a system of linear equations represents parallel lines, then the lines do not intersect, and the system has no solution. In such a case, we say that the system is **inconsistent**.	If a system of linear equations represents the same line, then all points on the common line satisfy each equation. Therefore, the system has infinitely many solutions. In such a case, we say that the equations are **dependent**.

Equations in a system of two variables are said to be **independent** if they represent different lines and **consistent** if they have one or more solutions.

EXAMPLE 2   Solving a System of Linear Equations by Graphing

Solve the system by graphing both linear equations and finding the point(s) of intersection.

$$y = \frac{1}{2}x - 2$$
$$4x + 2y = 6$$

Solution:

To graph each equation, write the equation in slope-intercept form $y = mx + b$.

First equation

$$y = \frac{1}{2}x - 2 \quad \text{Slope: } \tfrac{1}{2}$$

Second equation

$$4x + 2y = 6$$
$$2y = -4x + 6$$
$$\frac{2y}{2} = \frac{-4x}{2} + \frac{6}{2}$$
$$y = -2x + 3 \quad \text{Slope: } -2$$

To graph a line such as

$y = \dfrac{1}{2}x - 2$ from the slope

and y-intercept, plot the y-intercept $(0, -2)$ first. Then

use the slope $\dfrac{1}{2}$ as a ratio of

change in y to change in x. Starting from the y-intercept, move up 1 unit and to the right 2 units to plot a second point. See Section 5.4 for review.

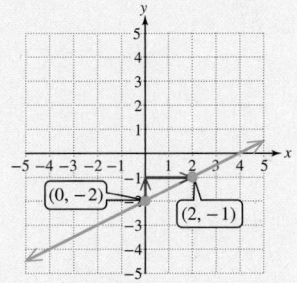

From their slope-intercept forms, we see that the lines have different slopes, indicating that the lines must intersect at exactly one point. We can graph the lines using the slope and y-intercept to find the point of intersection (Figure 10-3).

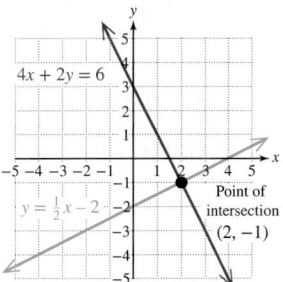

Figure 10-3

The point $(2, -1)$ appears to be the point of intersection. This can be confirmed by substituting $x = 2$ and $y = -1$ into each equation.

$$y = \frac{1}{2}x - 2$$

$$-1 \overset{?}{=} \frac{1}{2}(2) - 2$$

$$-1 \overset{?}{=} 1 - 2$$

$$-1 \overset{?}{=} -1 \; \checkmark \; \text{True}$$

$$4x + 2y = 6$$

$$4(2) + 2(-1) \overset{?}{=} 6$$

$$8 - 2 \overset{?}{=} 6$$

$$6 \overset{?}{=} 6 \; \checkmark \; \text{True}$$

The solution set is $\{(2, -1)\}$.

Skill Practice 2   Solve by using the graphing method.

$y = -3x - 5$
$x - 2y = -4$

TIP   In Example 2, the lines could also have been graphed by using the x- and y-intercepts or by using a table of points. However, the advantage of writing the equations in slope-intercept form is that we can compare the slope *and* y-intercept of each line.

1. If the slopes differ, the lines are different and nonparallel and must cross in exactly one point.
2. If the slopes are the same and the y-intercepts are different, the lines are parallel and do not intersect.
3. If the slopes are the same and the y-intercepts are the same, the two equations represent the same line.

EXAMPLE 3   **Solving a System of Linear Functions by Graphing**

A function defined by $f(x) = k$ is a constant function and its graph is a horizontal line with y-intercept $(0, k)$. In Example 3, notice that $f(x) = 3$ is a horizontal line passing through $(0, 3)$. See Section 5.4 for review.

Answer

2. $\{(-2, 1)\}$

Solve the system. $f(x) = 3$
$g(x) = 2x + 1$

Solution:

This first function can be written as $y = 3$. This is an equation of a horizontal line. Writing the second equation as $y = 2x + 1$, we have a slope of 2 and a y-intercept of $(0, 1)$.

The graphs of the functions are shown in Figure 10-4. The point of intersection is $(1, 3)$. Therefore, the solution set is $\{(1, 3)\}$.

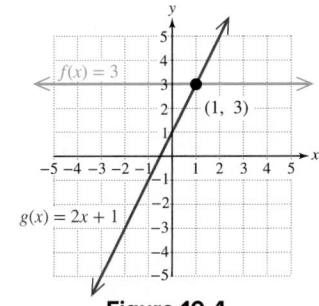

Figure 10-4

Skill Practice 3 Solve the system by graphing.

$f(x) = 1$
$g(x) = -3x + 4$

EXAMPLE 4 **Solving a System of Linear Equations by Graphing**

Solve the system by graphing. $\qquad -x + 3y = -6$
$\qquad\qquad\qquad\qquad\qquad\qquad 6y = 2x + 6$

Solution:

To graph the lines, write each equation in slope-intercept form.

First equation	Second equation
$-x + 3y = -6$	$6y = 2x + 6$
$3y = x - 6$	
$\dfrac{3y}{3} = \dfrac{x}{3} - \dfrac{6}{3}$	$\dfrac{6y}{6} = \dfrac{2x}{6} + \dfrac{6}{6}$
$y = \dfrac{1}{3}x - 2$	$y = \dfrac{1}{3}x + 1$

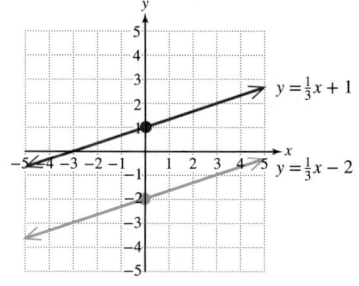

Because the lines have the same slope but different
y-intercepts, they are parallel (Figure 10-5). Two
parallel lines do not intersect, which implies
that the system has no solution. The system is
inconsistent.

The solution set is the empty set, { }.

Figure 10-5

Skill Practice 4 Solve the system by graphing.

$2y = 2x$
$-x + y = -3$

EXAMPLE 5 **Solving a System of Linear Equations by Graphing**

Solve the system by graphing. $\qquad x + 4y = 8$
$$y = -\frac{1}{4}x + 2$$

Solution:

Write the first equation in slope-intercept form. The second equation is already
in slope-intercept form.

First equation	Second equation
$x + 4y = 8$	$y = -\dfrac{1}{4}x + 2$
$4y = -x + 8$	
$\dfrac{4y}{4} = \dfrac{-x}{4} + \dfrac{8}{4}$	
$y = -\dfrac{1}{4}x + 2$	

Answers

3. {(1, 1)}
4. The solution set is { }. The
system is inconsistent.

Notice that the slope-intercept forms of the two lines are identical. Therefore, the equations represent the same line. The equations are dependent, and the solution to the system of equations is the set of all points on the line.

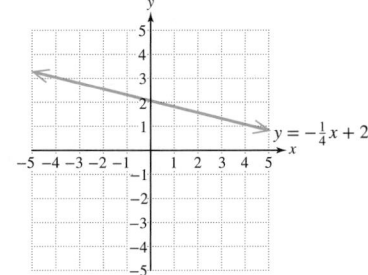

$$y = -\frac{1}{4}x + 2$$

The solution set consists of an infinite number of ordered pairs (x, y) that fall on the common line $y = -\frac{1}{4}x + 2$. Therefore, the solutions are the ordered pairs of the form $\left(x, -\frac{1}{4}x + 2\right)$.

The solution set is $\left\{\left(x, -\frac{1}{4}x + 2\right) \mid x \text{ is any real number}\right\}$.

Skill Practice 5   Solve the system by graphing.

$$y = \frac{1}{2}x + 1$$
$$x - 2y = -2$$

For the system given in Example 5, the solution set $\left\{\left(x, -\frac{1}{4}x + 2\right) \mid x \text{ is any real number}\right\}$ is called the **general solution**. By varying the value of x, we can produce any number of specific solutions to the system. For example,

	$\left(x, -\frac{1}{4}x + 2\right)$	**Solution**
If $x = 0$,	$\left(0, -\frac{1}{4}(0) + 2\right)$	$(0, 2)$
If $x = 4$,	$\left(4, -\frac{1}{4}(4) + 2\right)$	$(4, 1)$
If $x = 8$	$\left(8, -\frac{1}{4}(8) + 2\right)$	$(8, 0)$

Notice that the solutions fall on the common line of intersection as expected (Figure 10-6).

We should also note that the general solution to a system of dependent equations can be written in a variety of forms. The solutions to Example 5 fall on the line $y = -\frac{1}{4}x + 2$. By solving the equation for x, we have $x = 8 - 4y$. Thus, the solution set can also be written as $\{(8 - 4y, y) \mid y \text{ is any real number}\}$.

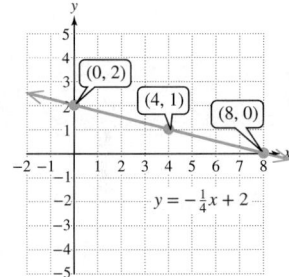

Figure 10-6

TECHNOLOGY CONNECTIONS

Using the *Intersect* Feature

The solution to a system of equations can be found by using either a *Trace* feature or an *Intersect* feature on a graphing calculator to find the point of intersection between two curves.

For example, consider the system

$$-2x + y = 6 \xrightarrow{\text{Isolate } y} y = 2x + 6$$
$$5x + y = -1 \longrightarrow y = -5x - 1$$

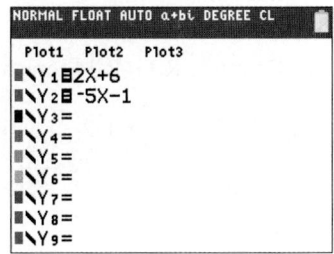

Graph the equations together on the same viewing window.

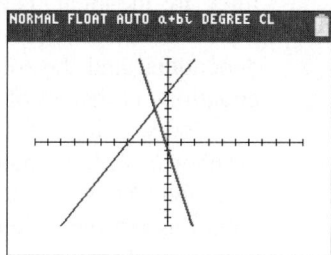

By inspection of the graph, it appears that the solution is $(-1, 4)$. The *Trace* option on the calculator may come close to $(-1, 4)$ but may not show the exact solution (Figure 10-7). However, an *Intersect* feature on a graphing calculator may provide the exact solution (Figure 10-8). See your user's manual for further details.

Using *Trace*   Using *Intersect*

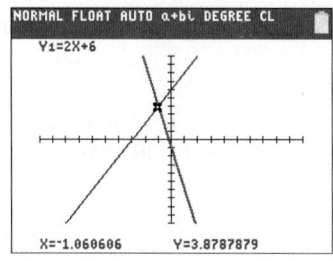   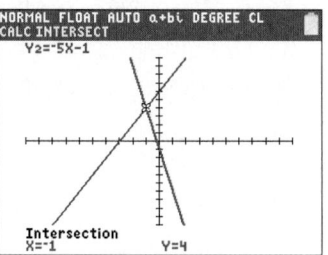

Figure 10-7   **Figure 10-8**

SECTION 10.1   Practice Exercises

Prerequisite Review

For Exercises R.1–R.2, identify the *x*- and *y*-intercepts of the given line. (See Section 5.1 for review.)

R.1. $2x + 5y = 15$ **R.2.** $-4x + 3y = 8$

For Exercises R.3–R.4, write the equation in slope-intercept form. Then identify the slope and *y*-intercept and graph the line. (See Section 5.4 for review.)

R.3. $-7x - 2y = 14$ **R.4.** $-3x + 2y = 6$

For Exercises R.5–R.10, by inspection of the equation, state whether the graph of the line is horizontal, vertical, or slanted. (See Section 5.4 for review.)

R.5. $2x = 7$ **R.6.** $-6 = 2y$ **R.7.** $-5y = -10$

R.8. $\frac{1}{2}x = 4$ **R.9.** $3x = 4y$ **R.10.** $2x + 5y = 12$

For Exercises R.11–R.14, graph the function defined by $y = f(x)$. (See Section 5.4 for review.)

R.11. $f(x) = \frac{2}{3}x + 2$ **R.12.** $f(x) = -\frac{1}{2}x - 3$ **R.13.** $f(x) = -2$ **R.14.** $f(x) = 4$

For Exercises R.15–R.22, determine whether the lines are parallel, perpendicular, or neither. (See Section 5.5 for review.)

R.15. $f(x) = 2x - 4$ and $g(x) = -\dfrac{1}{2}x - 3$

R.16. $h(x) = -\dfrac{3}{2}x + 2$ and $k(x) = \dfrac{2}{3}x - 4$

R.17. $3x - 4y = 1$ and $4x - 3y = 6$

R.18. $2x + y = 7$ and $-2x + y = 3$

R.19. $2x = 5$ and $y = 6$

R.20. $\dfrac{1}{2}y + 4 = 0$ and $3 = x$

R.21. $3x - y = 6$ and $2y = 6x + 4$

R.22. $4x = 2y + 8$ and $2x - y = 6$

Concept Connections

1. A _____ of linear equations consists of two or more linear equations.

2. A _____ to a system of linear equations is an ordered pair that is a solution to both individual equations in the system.

3. Graphically, a solution to a system of linear equations in two variables is a point where the lines _____.

4. A system of equations that has one or more solutions is said to be _____.

5. The solution set to an inconsistent system of equations is _____.

6. Two equations in a system of linear equations in two variables are said to be _____ if they represent the same line.

7. Two equations in a system of linear equations in two variables are said to be _____ if they represent different lines.

8. From the graph shown, determine the solution to the system.

$$x + y = 4$$
$$y = 2x + 1$$

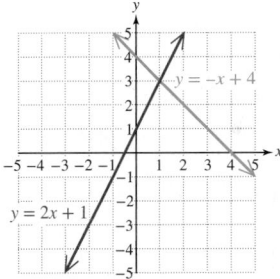

Objective 1: Identify Solutions to Systems of Linear Equations

For Exercises 9–14, determine which points are solutions to the given system. **(See Example 1)**

9. $y = 8x - 5$
$y = 4x + 3$
$(-1, 13), (-1, 1), (2, 11)$

10. $y = -\dfrac{1}{2}x - 5$
$y = \dfrac{3}{4}x - 10$
$(4, -7), (0, -10), \left(3, -\dfrac{9}{2}\right)$

11. $2x - 7y = -30$
$y = 3x + 7$
$(0, -30), \left(\dfrac{3}{2}, 5\right), (-1, 4)$

12. $x + 2y = 4$
$y = -\dfrac{1}{2}x + 2$
$(-2, 3), (4, 0), \left(3, \dfrac{1}{2}\right)$

13. $x - \quad y = 6$
$4x + 3y = -4$
$(4, -2), (6, 0), (2, 4)$

14. $x - 3y = 3$
$2x - 9y = 1$
$(0, 1), (4, -1), (9, 2)$

Objective 2: Solve Systems of Linear Equations by Graphing

For Exercises 15–20, the graph of a system of linear equations is given.

   a. Identify whether the system is consistent or inconsistent.

   b. Identify the equations as dependent or independent.

   c. Identify the number of solutions to the system.

15. $y = x + 3$

    $3x + y = -1$

16. $5x - 3y = 6$

    $3y = 2x + 3$

17. $2x = y + 4$

    $-4x + 2y = 2$

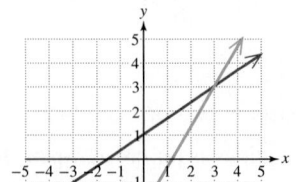

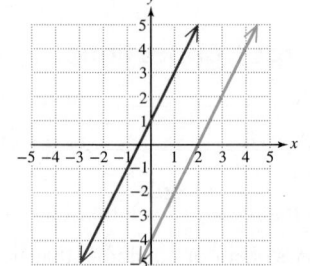

18. $y = -2x - 3$

    $-4x - 2y = 0$

19. $y = \dfrac{1}{3}x + 2$

    $-x + 3y = 6$

20. $y = -\dfrac{2}{3}x - 1$

    $-4x - 6y = 6$

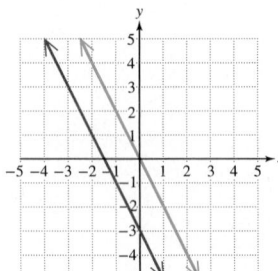

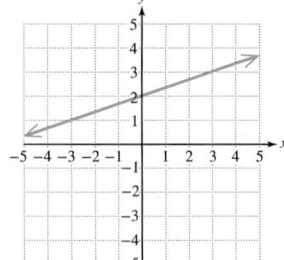

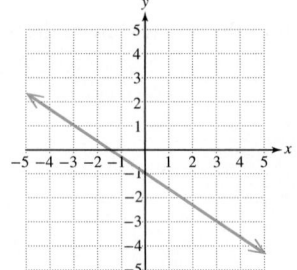

For Exercises 21–26, a system of equations is given in which each equation is written in slope-intercept form. Determine the number of solutions. If the system does not have one unique solution, state whether the system is inconsistent or whether the equations are dependent.

21. $y = \dfrac{2}{5}x - 7$

    $y = \dfrac{1}{4}x + 7$

22. $y = \dfrac{1}{2}x + 3$

    $y = 2x + \dfrac{1}{3}$

23. $y = 6x - \dfrac{2}{3}$

    $y = 6x + 4$

24. $y = -4x - 3$

    $y = -4x + 3$

25. $y = 8x - \dfrac{1}{2}$

    $y = 8x - \dfrac{1}{2}$

26. $y = -x - 5$

    $y = -x - 5$

For Exercises 27–44, solve the system by graphing. For systems that do not have one unique solution, also state the number of solutions and whether the system is inconsistent or the equations are dependent. **(See Examples 2–5)**

27. $2x + y = -3$
$-x + y = 3$

28. $4x - 3y = 12$
$3x + 4y = -16$

29. $f(x) = -2x + 3$
$g(x) = 5x - 4$

30. $h(x) = 2x + 5$
$g(x) = -x + 2$

31. $k(x) = -4$
$f(x) = -\dfrac{2}{3}x - 2$

32. $f(x) = -2$
$g(x) = \dfrac{5}{2}x - 2$

33. $x = 4 + y$
$3y = -3x$

34. $3y = 4x$
$x - y = -1$

35. $y = -2x + 3$
$-2x = y + 1$

36. $y = \dfrac{1}{3}x - 2$
$x = 3y - 9$

37. $y = \dfrac{2}{3}x - 1$
$2x = 3y + 3$

38. $4x = 16 - 8y$
$y = -\dfrac{1}{2}x + 2$

39. $2x = 4$
$\dfrac{1}{2}y = -1$

40. $y + 7 = 6$
$-5 = 2x$

41. $-x + 3y = 6$
$6y = 2x + 12$

42. $3x = 2y - 4$
$-4y = -6x - 8$

43. $2x - y = 4$
$4x + 2 = 2y$

44. $x = 4y + 4$
$-2x + 8y = -16$

Mixed Exercises

45. Write a system of linear equations with solution set $\{(-3, 5)\}$.

46. Write a system of linear equations with solution set $\{(4, -3)\}$.

47. Find C and D so that the solution set to the system is $\{(4, 1)\}$.

$$Cx + 5y = 13$$
$$-2x + Dy = -5$$

48. Find A and B so that the solution set to the system is $\{(-5, 2)\}$.

$$3x + Ay = -3$$
$$Bx - y = -12$$

Technology Connections

For Exercises 49–52, use a graphing utility to approximate the solution to the system of equations. Round the x and y values to 3 decimal places.

49. $y = -3.729x + 6.958$
$y = 2.615x - 8.713$

50. $y = -0.041x + 0.068$
$y = 0.019x - 0.053$

51. $-0.25x + 0.04y = -0.42$
$6.775x + 2.5y = -38.1$

52. $0.36x - 0.075y = -0.813$
$0.066x + 0.008y = 0.194$

<table>
<tr><td>**SECTION 10.2**</td><td>Systems of Linear Equations in Two Variables and Applications</td></tr>
</table>

SECTION 10.2 Systems of Linear Equations in Two Variables and Applications

OBJECTIVES

1. Solve Systems of Linear Equations in Two Variables Algebraically
2. Use Systems of Linear Equations in Applications

1. Solve Systems of Linear Equations in Two Variables Algebraically

Graphing a system of equations is one method to find the solution(s) to the system. However, sometimes it is difficult to determine the solution(s) using this method because of limitations in the accuracy of the graph. Instead we often use algebraic methods to solve a system of equations. The first method we present is called the **substitution method**.

Solving a System of Equations by Using the Substitution Method

Step 1 Isolate one of the variables from one equation.

Step 2 Substitute the expression found in step 1 into the *other* equation.

Step 3 Solve the resulting equation.

Step 4 Substitute the value found in step 3 back into the equation in step 1 to find the value of the remaining variable.

Step 5 Check the ordered pair in each equation and write the solution as an ordered pair in set notation.

EXAMPLE 1   **Solving a System of Equations by the Substitution Method**

Solve the system by using the substitution method. $-5x - 4y = 2$
$4x + y = 5$

Solution:

$-5x - 4y = 2$
$4x + y = 5 \longrightarrow y = \underline{-4x + 5}$ **Step 1:** Isolate one of the variables from one of the equations. A variable with coefficient 1 or -1 is easily isolated.

$-5x - 4(-4x + 5) = 2$ **Step 2:** Substitute the expression from step 1 into the other equation.

$-5x + 16x - 20 = 2$ **Step 3:** Solve for the remaining variable.
$11x = 22$
$x = 2$

$y = -4x + 5$ **Step 4:** Substitute the known value of x into
$y = -4(2) + 5$ the equation where y is isolated.
$y = -3$ From step 1, this is $y = -4x + 5$.

Step 5: Check the ordered pair $(2, -3)$ in each original equation.

Check $(2, -3)$. $-5x - 4y = 2$ $4x + y = 5$
$-5(2) - 4(-3) \stackrel{?}{=} 2$ $4(2) + (-3) \stackrel{?}{=} 5$
$-10 + 12 \stackrel{?}{=} 2$ ✓ true $8 - 3 \stackrel{?}{=} 5$ ✓ true

The solution set is $\{(2, -3)\}$.

TIP The lines from Example 1 are shown here. The graph shows the point of intersection at $(2, -3)$.

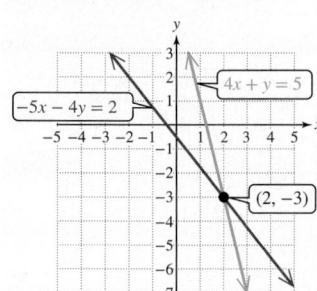

Skill Practice 1   Solve the system by using the substitution method.

$3x + 4y = 5$

$x - 3y = 6$

Now consider the following system of equations.

$$3x - 4y = 5$$
$$5x + 2y = 17$$

None of the variable terms has a coefficient of 1 or -1. Therefore, if we isolate x or y from either equation, the resulting equation will have one or more terms with fractional coefficients. To avoid this scenario, we can use another method called the **addition method** (also called the elimination method). The principle of the addition method is to multiply one or both equations in the system by appropriate constants to create opposite coefficients on one of the variables.

Solving a System of Equations by Using the Addition Method
Step 1 Write both equations in standard form: $Ax + By = C$.
Step 2 Clear fractions or decimals (optional).
Step 3 Multiply one or both equations by nonzero constants to create opposite coefficients for one of the variables.
Step 4 Add the equations from step 3 to eliminate one variable.
Step 5 Solve for the remaining variable.
Step 6 Substitute the known value found in step 5 into one of the original equations to solve for the other variable.
Step 7 Check the ordered pair in each equation and write the solution set.

TIP The addition method is sometimes called the *elimination method*. However, since both the substitution method and addition method eliminate a variable, we use the name addition method to emphasize the technique used.

EXAMPLE 2   **Solving a System of Equations by the Addition Method**

Solve the system by using the addition method.
$$3x - 4y = 5$$
$$5x + 2y = 17$$

Solution:

Step 1: The equations are already written in standard form $Ax + By = C$.

Step 2: There are no decimals or fractions.

Step 3: Notice that the y coefficients in the first and second equations are -4 and 2, respectively. If we multiply the second equation by 2, then the y coefficient in the second equation will be 4, which is the opposite of -4.

$$3x - 4y = 5 \qquad\qquad\qquad 3x - 4y = 5$$
$$5x + 2y = 17 \xrightarrow{\text{Multiply by 2.}} \underline{10x + 4y = 34} \quad \text{Multiply the second equation by 2.}$$
$$13x = 39 \qquad \textbf{Step 4: } \text{Add the equations to eliminate } y.$$

$$\frac{13x}{13} = \frac{39}{13} \qquad \textbf{Step 5: } \text{Divide by 13 to solve for } x.$$

$$x = 3$$

Answer

1. $\{(3, -1)\}$

TIP The lines from Example 2 are shown here. The point of intersection and solution to the system is (3, 1).

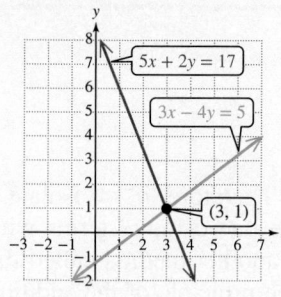

$$3x - 4y = 5$$

$$3(3) - 4y = 5$$

$$9 - 4y = 5$$

$$-4y = -4$$

$$y = 1$$

Step 6: Substitute $x = 3$ into one of the original equations to solve for y.

Step 7: Check the ordered pair (3, 1) in each original equation.

$$3x - 4y = 5 \qquad\qquad 5x + 2y = 17$$

$$3(3) - 4(1) = 5 \qquad\qquad 5(3) + 2(1) = 17$$

$$9 - 4 = 5 \checkmark \qquad\qquad 15 + 2 = 17 \checkmark$$

The solution set is $\{(3, 1)\}$.

Skill Practice 2 Solve the system by using the addition method.

$$-2x + 3y = -14$$

$$6x + 5y = -14$$

TIP In Example 2, the variable y was eliminated. Alternatively, we could have eliminated x by multiplying the first equation by 5 and the second equation by −3. This would create new equations with coefficients of 15 and −15 on the x terms.

$$3x - 4y = 5 \xrightarrow{\text{Multiply by 5.}} 15x - 20y = 25 \qquad 3x - 4y = 5$$

$$5x + 2y = 17 \xrightarrow{\text{Multiply by } -3.} \underline{-15x - 6y = -51} \longrightarrow 3x - 4(1) = 5$$

$$-26y = -26 \qquad 3x - 4 = 5$$

$$y = 1 \longrightarrow 3x = 9$$

$$x = 3$$

The solution set is $\{(3, 1)\}$.

EXAMPLE 3   Solving a System of Equations by the Addition Method

Solve the system by using the addition method.

$$5x = 4y + 6$$

$$-3x + 7y = 1$$

Solution:

$$5x = 4y + 6 \xrightarrow{\text{Subtract } 4y.} 5x - 4y = 6$$

$$-3x + 7y = 1 \qquad\qquad\qquad\quad -3x + 7y = 1$$

Step 1: Write each equation in standard form: $Ax + By = C$.

Step 2: There are no decimals or fractions.

TIP In Example 3, the variable x was eliminated.

Alternatively, we could have eliminated y by multiplying the first equation by 7 and the second equation by 4. This would create new equations with coefficients of −28 and 28 on the y terms.

$$5x - 4y = 6 \xrightarrow{\text{Multiply by 3.}} 15x - 12y = 18$$

$$-3x + 7y = 1 \xrightarrow{\text{Multiply by 5.}} \underline{-15x + 35y = 5}$$

$$23y = 23$$

Step 3: Multiply the first equation by 3. Multiply the second equation by 5.

Step 4: Add the equations to eliminate x.

$$y = 1$$

Step 5: Solve for y.

Answer

2. $\{(1, -4)\}$

$5x = 4y + 6$
$5x = 4(1) + 6$
$5x = 10$
$x = 2$

Step 6: Substitute $y = 1$ into one of the original equations to solve for x.

Step 7: Check the ordered pair (2, 1) in each original equation.

$5x = 4y + 6$	$-3x + 7y = 1$
$5(2) \stackrel{?}{=} 4(1) + 6$	$-3(2) + 7(1) \stackrel{?}{=} 1$
$10 \stackrel{?}{=} 4 + 6 \; ✓ \; \text{true}$	$-6 + 7 \stackrel{?}{=} 1 \; ✓ \; \text{true}$

The solution set is $\{(2, 1)\}$.

Skill Practice 3   Solve the system by using the addition method.

$2x - 9y = 1$
$3x = 17 - 2y$

TECHNOLOGY CONNECTIONS

Solving a System of Linear Equations in Two Variables Using Intersect

The solution to a system of linear equations can be checked on a graphing calculator by first writing the equations in slope-intercept form. From Example 3 we have

$$5x = 4y + 6 \longrightarrow y = \frac{5}{4}x - \frac{3}{2}$$

$$-3x + 7y = 1 \longrightarrow y = \frac{3}{7}x + \frac{1}{7}$$

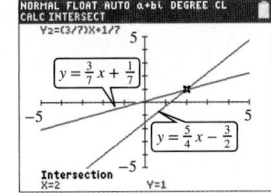

Then graph the equations and use the Intersect feature to approximate the point of intersection. The Intersect feature gives the solution (2, 1).

It is important to write the individual equations in a system of equations in standard form so that the variables line up. Also consider clearing decimals or fractions within an equation to make integer coefficients.

EXAMPLE 4   **Solving a System of Equations by the Addition Method**

Solve the system by using the addition method.

$$\frac{2}{5}x - y = \frac{19}{10}$$

$$5(x + y) = -7y - 41$$

Solution:

$$\frac{2}{5}x - y = \frac{19}{10} \xrightarrow{\text{Multiply by 10.}} 4x - 10y = 19 \qquad \text{Clear the fractions in the first equation. Write the}$$

$$5(x + y) = -7y - 41 \xrightarrow{\text{Simplify.}} 5x + 12y = -41 \qquad \begin{array}{l}\text{second equation in}\\ \text{standard form.}\end{array}$$

Answer
3. $\{(5, 1)\}$

$$4x - 10y = 19 \xrightarrow{\text{Multiply by } -5.} -20x + 50y = -95$$
$$5x + 12y = -41 \xrightarrow{\text{Multiply by 4.}} \underline{20x + 48y = -164}$$
$$98y = -259$$

The LCM of the x coefficients, 4 and 5, is 20. Create opposite coefficients on x of 20 and -20.

$$y = \frac{-259}{98} \qquad \text{Solve for } y.$$

$$y = -\frac{37}{14} \qquad \text{Simplify to lowest terms.}$$

Substituting $y = -\frac{37}{14}$ back into one of the original equations to solve for x would be cumbersome. Alternatively, we can solve for x by repeating the addition method. This time we will eliminate y by creating opposite coefficients on the y terms and then solving for x.

$$4x - 10y = 19 \xrightarrow{\text{Multiply by 6.}} 24x - 60y = 114$$
$$5x + 12y = -41 \xrightarrow{\text{Multiply by 5.}} \underline{25x + 60y = -205}$$
$$49x = -91$$

The LCM of 10 and 12 is 60. Create opposite coefficients of -60 and 60 on the y terms.

$$x = \frac{-91}{49} \qquad \text{Solve for } y.$$

$$x = -\frac{13}{7} \qquad \text{Simplify.}$$

The ordered pair $\left(-\frac{13}{7}, -\frac{37}{14}\right)$ checks in both original equations.

The solution set is $\left\{\left(-\frac{13}{7}, -\frac{37}{14}\right)\right\}$.

Point of Interest

The study of systems of linear equations is a topic in a broader branch of mathematics called linear algebra.

Skill Practice 4   Solve the system by using the addition method.

$$2(x - 2y) = y + 14$$
$$\frac{1}{2}x + \frac{7}{6}y = -\frac{13}{3}$$

The systems in Examples 1–4 each have one unique solution. That is, the lines represented by the two equations intersect in exactly one point. In Examples 5 and 6, we investigate systems with no solution or infinitely many solutions.

Avoiding Mistakes

To verify that the system in Example 5 has no solution, write the equations in slope-intercept form.

$$2x + y = 4 \Rightarrow y = -2x + 4$$
$$6x + 3y = 6 \Rightarrow y = -2x + 2$$

The equations have the same slope, but different y-intercepts. Therefore, the lines are parallel and do not intersect.

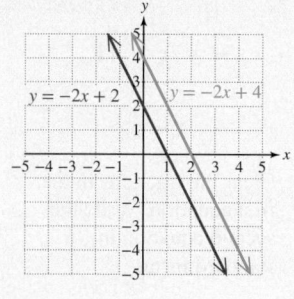

EXAMPLE 5   **Identifying a System of Equations with No Solution**

Solve the system. $2x + y = 4$
$6x + 3y = 6$

Solution:

$$2x + y = 4 \xrightarrow{\text{Multiply by } -3.} -6x - 3y = -12$$
$$6x + 3y = 6 \xrightarrow{\phantom{\text{Multiply by } -3.}} \underline{6x + 3y = 6}$$
$$0 = -6$$

We can eliminate either the x terms or y terms by multiplying the first equation by -3.

Both the x and y terms are eliminated, leading to the contradiction $0 = -6$.

The system of equations reduces to a contradiction. This indicates that there is no solution and the system is inconsistent. The equations represent parallel lines and two parallel lines do not intersect.

The solution set is { }.

Answers

4. $\left\{\left(-\frac{32}{29}, -\frac{94}{29}\right)\right\}$ **5.** { }

Skill Practice 5   Solve the system. $3x - y = 2$
$-9x + 3y = 4$

EXAMPLE 6   Solving a System of Dependent Equations

Solve the system. $y = 2x - 1$
                        $8x - 4y = 4$

Solution:

To verify the solution to Example 6, write the two equations in slope-intercept form.

$y = 2x - 1 \Rightarrow y = 2x - 1$
$8x - 4y = 4 \Rightarrow y = 2x - 1$

The equations have the same slope-intercept form and define the same line.

$y = \overbrace{2x - 1}$
$8x - 4y = 4$ $8x - 4(2x - 1) = 4$

With y already isolated in the first equation, apply the substitution method.

$8x - 8x + 4 = 4$ Solve the resulting equation.

$4 = 4$ Identity.

Notice that both variables were eliminated and the system of equations is reduced to the identity $4 = 4$. Therefore, the two original equations are dependent and represent the same line. The solution set consists of an infinite number of ordered pairs (x, y) that fall on the common line of intersection, $y = 2x - 1$. Therefore, the solutions are ordered pairs of the form $(x, 2x - 1)$.

The solution set is $\{(x, 2x - 1) \mid x$ is any real number$\}$.

Skill Practice 6   Solve the system. $x = 5 - 3y$
                                              $2x + 6y = 10$

TIP Sometimes the solution to a system of dependent equations is written with an arbitrary variable called a **parameter**. For example, letting t represent any real number, then the solution set to Example 6 can be written as $\{(t, 2t - 1) \mid t$ is a real number$\}$.

The solution set $\{(x, 2x - 1) \mid x$ is any real number$)\}$ is called the **general solution** to the system in Example 6. By varying the value of x, we can produce any number of specific solutions to the system of equations. For example:

	$(x, 2x - 1)$	Solution
If $x = 1$	$(1, 2(1) - 1)$	$(1, 1)$
If $x = 2$	$(2, 2(2) - 1)$	$(2, 3)$
If $x = 3$	$(3, 2(3) - 1)$	$(3, 5)$
If $x = -1$	$(-1, 2(-1) - 1)$	$(-1, -3)$

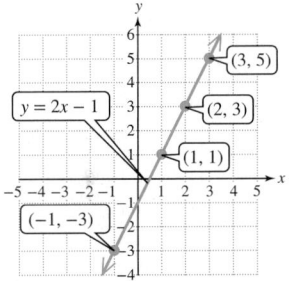

Figure 10-9

TIP The equations $y = 2x - 1$ and $8x - 4y = 4$ represent the same relationship between x and y. Therefore, we have only one unique equation, but two variables. As a result, y depends on the choice of x and vice versa.

Notice that the solutions fall on the common line of intersection as expected (Figure 10-9).

We should also note that the general solution to a system of dependent equations can be written in a variety of forms. For instance, the solutions to the system in Example 6 fall on the line $y = 2x - 1$. By solving the equation for x, we have $x = \dfrac{y + 1}{2}$. Thus, the solution set can also be written as $\left\{ \left(\dfrac{y + 1}{2}, y \right) \,\middle|\, y$ is any real number$\right\}$.

2. Use Systems of Linear Equations in Applications

When solving an application involving two unknowns, sometimes it is convenient to use a system of two independent equations as demonstrated in Examples 7–9.

Answer

6. $\left\{ \left(x, \dfrac{-x + 5}{3} \right) \,\middle|\, x$ is any real number$\right\}$

or $\{(5 - 3y, y) \mid y$ is any real number$\}$

EXAMPLE 7   Solving an Application Involving Mixtures

A hospital uses a 15% bleach solution to disinfect a quarantine area. How much 6% bleach solution must be mixed with an 18% bleach solution to make 50 L of a 15% bleach solution?

Solution:

Let x represent the amount of 6% bleach solution.

Let y represent the amount of 18% bleach solution.

The amount of pure bleach in each mixture is found by multiplying the amount of solution by the concentration rate. This information can be organized in a table.

	6% Solution	18% Solution	15% Solution
Amount of mixture	x	y	50
Amount of pure bleach	$0.06x$	$0.18y$	$0.15(50)$

There are two unknown quantities. We will set up a system of two independent equations relating x and y.

$$\begin{pmatrix} \text{Amount of} \\ \text{6\% mixture} \end{pmatrix} + \begin{pmatrix} \text{Amount of} \\ \text{18\% mixture} \end{pmatrix} = \begin{pmatrix} \text{Amount of} \\ \text{15\% mixture} \end{pmatrix}$$

$$\begin{pmatrix} \text{Amount of} \\ \text{pure bleach in} \\ \text{6\% mixture} \end{pmatrix} + \begin{pmatrix} \text{Amount of} \\ \text{pure bleach in} \\ \text{18\% mixture} \end{pmatrix} = \begin{pmatrix} \text{Amount of} \\ \text{pure bleach in} \\ \text{15\% mixture} \end{pmatrix}$$

$$x + y = 50$$
$$0.06x + 0.18y = 0.15(50)$$

$$\begin{aligned} x + \quad y &= 50 \\ 0.06x + 0.18y &= 7.5 \end{aligned} \xrightarrow{\text{Multiply by 100.}} \begin{aligned} x + \quad y &= 50 \\ 6x + 18y &= 750 \end{aligned} \xrightarrow{\text{Divide by } -6.} \begin{aligned} x + \ y &= \ \ 50 \\ -x - 3y &= -125 \\ \hline -2y &= \ -75 \\ y &= \ \ 37.5 \end{aligned}$$

Substitute $y = 37.5$ back into the equation $x + y = 50$.

$$x + \quad y = 50$$
$$x + 37.5 = 50$$
$$x = 12.5$$

Therefore, 12.5 L of the 6% bleach solution should be mixed with 37.5 L of the 18% bleach solution to make 50 L of 15% bleach solution.

Skill Practice 7   How many ounces of 20% and 35% acid solution should be mixed to produce 15 oz of 30% acid solution?

EXAMPLE 8   Solving an Application Involving Uniform Motion

A riverboat traveling upstream against the current on the Mississippi River takes 3 hr to travel 24 mi. The return trip downstream with the current takes only 2 hr. Find the speed of the boat in still water and the speed of the current.

Answer

7. Mix 5 oz of 20% acid solution with 10 oz of 35% acid solution to make 15 oz of 30% acid solution.

Solution:

Let b represent the speed of the boat in still water.

Let c represent the speed of the current.

The given information can be organized in a table.

	Distance (mi)	Rate (mph)	Time (hr)
Upstream	24	$b - c$	3
Downstream	24	$b + c$	2

Use the relationship $d = rt$; that is, distance = (rate)(time).

$$\left(\begin{array}{c}\text{Distance}\\\text{upstream}\end{array}\right) = \left(\begin{array}{c}\text{Rate}\\\text{upstream}\end{array}\right)\left(\begin{array}{c}\text{Time}\\\text{upstream}\end{array}\right) \longrightarrow 24 = (b - c) \cdot 3$$

$$\left(\begin{array}{c}\text{Distance}\\\text{downstream}\end{array}\right) = \left(\begin{array}{c}\text{Rate}\\\text{downstream}\end{array}\right)\left(\begin{array}{c}\text{Time}\\\text{downstream}\end{array}\right) \longrightarrow 24 = (b + c) \cdot 2$$

$$24 = 3b - 3c \xrightarrow{\text{Divide by 3.}} 8 = b - c$$
$$24 = 2b + 2c \xrightarrow[\text{Divide by 2.}]{} 12 = b + c$$
$$\overline{ \quad 20 = 2b}$$
$$10 = b$$

Substitute $b = 10$ into the equation $12 = b + c$, which gives $c = 2$.

The boat's speed in still water is 10 mph and the speed of the current is 2 mph.

> **Skill Practice 8**   A boat takes 3 hr to go 24 mi upstream against the current. It can go downstream with the current a distance of 48 mi in the same amount of time. Determine the speed of the boat in still water and the speed of the current.

EXAMPLE 9   **Applying a System of Equations in Business**

A lawn service company has fixed monthly costs of $500 and variable costs (labor, gasoline, and depreciation) of $40 per lawn. If the service charges $60 per lawn,

a. Write a cost function representing the cost $C(x)$ to the company to service x lawns per month.

b. Write a revenue function representing the revenue $R(x)$ to service x lawns per month.

c. Determine the number of lawns that must be serviced in a month for the company to break even.

Solution:

Let x represent the number of lawns serviced for a given month.

a. $\left(\begin{array}{c}\text{Monthly}\\\text{cost}\end{array}\right) = \left(\begin{array}{c}\text{Fixed}\\\text{cost}\end{array}\right) + \left(\begin{array}{c}\text{Variable}\\\text{cost}\end{array}\right) \longrightarrow C(x) = 500 + 40x$

b. $\left(\begin{array}{c}\text{Monthly}\\\text{revenue}\end{array}\right) = \left(\begin{array}{c}\text{Revenue}\\\text{per lawn}\end{array}\right)\left(\begin{array}{c}\text{Number}\\\text{of lawns}\end{array}\right) \longrightarrow R(x) = 60x$

c. To break even, the cost must equal revenue, $C(x) = R(x)$.

$$500 + 40x = 60x$$
$$500 = 20x$$
$$x = 25$$

To break even, the company must service 25 lawns.

Skill Practice 9   A storage company rents its units for $120 per month. The company has fixed monthly costs of $2100 and variable costs (air-conditioning and service) of $50 per unit.

a. Write a cost function representing the monthly cost $C(x)$ to the company for x units.

b. Write a revenue function representing the revenue $R(x)$ when x units per month are rented.

c. Determine the number of units that must be rented in a month for the company to break even.

Answers

9. a. $C(x) = 2100 + 50x$
   b. $R(x) = 120x$
   c. 30 units

SECTION 10.2   Practice Exercises

Prerequisite Review

For Exercises R.1–R.4, solve the equation. (See Section 1.2 for review.)

R.1. $4 = 3 + 3(3x + 1)$

R.2. $0.65(x - 6) + 0.35(x + 4) = 0.5$

R.3. $3(7 - 4n) + 2 = -2n - 3 - 10n$

R.4. $-4 + 3x = 3(x - 3) + 5$

For Exercises R.5–R.6,

   a. Solve the equation for x. (See Section 1.2 for review.)

   b. Solve the equation for y.

R.5. $3x - y = 5$

R.6. $x + 4y = 6$

R.7. Suppose that 5% of a solution is fertilizer by volume and 95% is water. How much fertilizer is in a 5-gal bucket?

R.8. Suppose that $4500 is borrowed at 6% simple interest. How much interest is owed after 4 years?

R.9. Suppose that Diana drives 638 mi in 11 hr. What was her average speed?

R.10. A plane's average speed for a trip from Orlando to Albuquerque is 435 mph. If the distance is 1740 mi, how long will the trip take?

Concept Connections

1. Two algebraic methods to solve a system of linear equations in two variables are the _____ method and the _____ method.

2. A system of linear equations in two variables may have no solution. In such a case, the equations represent _____ lines.

3. A system of equations that has no solution is called an _____ system.

4. A system of linear equations in two variables may have infinitely many solutions. In such a case, the equations are said to be _____ .

Objective 1: Solve Systems of Linear Equations in Two Variables Algebraically

For Exercises 5–14, solve the system of equations by using the substitution method. **(See Example 1)**

5. $4x + 12y = 4$
   $y = 5x + 11$

6. $y = -3x - 1$
   $2x - 3y = -8$

7. $x = 2y - 5$
   $3x + 4y = 5$

8. $2x + 5y = 13$
   $x = 4y$

9. $x + 3y = 5$
   $3x - 2y = -18$

10. $2x + y = 2$
   $5x + 3y = 9$

11. $2x + 7y = 1$
   $3y - 7 = 2$

12. $3x = 2y - 11$
   $6 + 5x = 1$

13. $2(x + y) = 2 - y$
   $4x - 1 = 2 - 5y$

14. $5(x + y) = 9 + 2y$
   $6y - 2 = 10 - 7x$

For Exercises 15–28, solve the system of equations by using the addition method. (**See Examples 2–4**)

15. $2x + y = 5$
$-3x - y = -7$

16. $x - 8y = -2$
$-x + 4y = 10$

17. $5x - 3y = 11$
$-10x + 7y = -19$

18. $3x + 2y = -5$
$5x - 8y = -31$

19. $4x - 3y = -3$
$5x - 9y = -9$

20. $2x - 5y = 6$
$6x + 11y = 18$

21. $3x - 7y = 1$
$6x + 5y = -17$

22. $5x - 2y = -2$
$3x + 4y = 30$

23. $11x = -5 - 4y$
$2(x - 2y) = 22 + y$

24. $-3(x - y) = y - 14$
$2x + 2 = 7y$

25. $0.6x + 0.1y = 0.4$
$2x - 0.7y = 0.3$

26. $0.25x - 0.04y = 0.24$
$0.15x - 0.12y = 0.12$

27. $2x + 11y = 4$
$3x - 6y = 5$

28. $3x - 4y = 9$
$2x + 9y = 2$

For Exercises 29–34, solve the system by using any method. If a system does not have one unique solution, state whether the system is inconsistent or whether the equations are dependent. (**See Examples 5–6**)

29. $3x - 4y = 6$
$9x = 12y + 4$

30. $-4x - 8y = 2$
$2x = 8 - 4y$

31. $3x + y = 6$
$x + \dfrac{1}{3}y = 2$

32. $2x - y = 8$
$x - \dfrac{1}{2}y = 4$

33. $2x + 4 = 4 - 5y$
$2 + 4(x + y) = 7y + 2$

34. $3(x - 3y) = 2y$
$2x + 5 = 5 - 7y$

For Exercises 35–36,

   a. Write the general solution.

   b. Find three individual solutions. Answers will vary.

35. $-5x - y = 6$
$10x = -2(y + 6)$

36. $2y = 6 - 4x$
$8x = 12 - 4y$

For Exercises 37–50, solve the system using any method.

37. $3x - 10y = 1900$
$5y + 800 = x$

38. $2x - 7y = 2400$
$-4x + 1800 = y$

39. $5(2x + y) = y - x - 8$
$x - \dfrac{3}{2}y = \dfrac{5}{2}$

40. $3(2x - y) = 2 - x$
$x + \dfrac{5}{4}y = \dfrac{3}{2}$

41. $y = \dfrac{2}{3}x - 1$
$y = \dfrac{1}{6}x + 2$

42. $y = -\dfrac{1}{4}x + 7$
$y = -\dfrac{3}{2}x + 17$

43. $4(x - 2) = 6y + 3$
$\dfrac{1}{4}x - \dfrac{3}{8}y = -\dfrac{1}{2}$

44. $\dfrac{1}{14}x - \dfrac{1}{7}y = \dfrac{1}{2}$
$2(x - 2y) + 3 = 20$

45. $2x = \dfrac{y}{2} + 1$
$0.04x - 0.01y = 0.02$

46. $0.05x + 0.01y = 0.03$
$x + \dfrac{y}{5} = \dfrac{3}{5}$

47. $y = 2.4x - 1.54$
$y = -3.5x + 7.9$

48. $y = -0.18x + 0.129$
$y = -0.15x + 0.1275$

49. $\dfrac{x - 2}{8} + \dfrac{y + 1}{2} = -6$
$\dfrac{x - 2}{2} - \dfrac{y + 1}{4} = 12$

50. $\dfrac{x + 1}{2} - \dfrac{y - 2}{10} = -1$
$\dfrac{x + 1}{6} + \dfrac{y - 2}{2} = 21$

Objective 2: Use Systems of Linear Equations in Applications

51. One antifreeze solution is 36% alcohol and another is 20% alcohol. How much of each mixture should be added to make 40 L of a solution that is 30% alcohol? (**See Example 7**)

52. A pharmacist wants to mix a 30% saline solution with a 10% saline solution to get 200 mL of a 12% saline solution. How much of each solution should she use?

53. A radiator has 16 L of a 36% antifreeze solution. How much must be drained and replaced by pure antifreeze to bring the concentration level up to 50%?

54. Jonas performed an experiment for his science fair project. He learned that rinsing lettuce in vinegar kills more bacteria than rinsing with water or with a popular commercial product. As a follow-up to his project, he wants to determine the percentage of bacteria killed by rinsing with a diluted solution of vinegar.

    a. How much water and how much vinegar should be mixed to produce 10 cups of a mixture that is 40% vinegar?

    b. How much pure vinegar and how much 40% vinegar solution should be mixed to produce 10 cups of a mixture that is 60% vinegar?

55. Michelle borrows a total of $5000 in student loans from two lenders. One charges 4.6% simple interest and the other charges 6.2% simple interest. She is not required to pay off the principal or interest for 3 years. However, at the end of 3 years, she will owe a total of $762 for the interest from both loans. How much did she borrow from each lender?

56. Juan borrows $100,000 to pay for medical school. He borrows part of the money from the school whereby he will pay 4.5% simple interest. He borrows the rest of the money through a government loan that will charge him 6% interest. In both cases, he is not required to pay off the principal or interest during his 4 years of medical school. However, at the end of 4 years, he will owe a total of $19,200 for the interest from both loans. How much did he borrow from each source?

57. Stuart pays back two student loans over a 4-year period. One loan charges the equivalent of 3% simple interest and the other charges the equivalent of 5.5% simple interest. If the total amount borrowed was $24,000 and the total amount of interest paid after 4 years is $3280, find the amount borrowed from each loan.

58. A total of $6000 is invested for 5 years with a total return of $1080. Part of the money is invested in a fund that returns the equivalent of 2% simple interest. The rest of the money is invested at 4% simple interest. Determine the amount invested in each account.

59. Monique and Tara each make an ice cream sundae. Monique gets 2 scoops of Cherry ice cream and 1 scoop of Mint Chocolate Chunk ice cream for a total of 43 g of fat. Tara has 1 scoop of Cherry and 2 scoops of Mint Chocolate Chunk for a total of 47 g of fat. How many grams of fat does 1 scoop of each type of ice cream have?

60. Bryan and Jadyn had barbeque potato chips and soda at a football party. Bryan ate 3 oz of chips and drank 2 cups of soda for a total of 700 mg of sodium. Jadyn ate 1 oz of chips and drank 3 cups of soda for a total of 350 mg of sodium. How much sodium is in 1 oz of chips and how much is in 1 cup of soda?

61. The average weekly salary of two employees is $1350. One makes $300 more than the other. Find their salaries.

62. The average of an electrician's hourly wage and a plumber's hourly wage is $33. One day a contractor hires the electrician for 8 hr of work and the plumber for 5 hr of work and pays a total of $438 in wages. Find the hourly wage for the electrician and for the plumber.

63. A moving sidewalk in an airport moves people between gates. It takes Jason's 9-year-old daughter Josie 40 sec to travel 200 ft walking with the sidewalk. It takes her 30 sec to walk 90 ft against the moving sidewalk (in the opposite direction). Find the speed of the sidewalk and find Josie's speed walking on nonmoving ground. (**See Example 8**)

64. A fishing boat travels along the east coast of the United States and encounters the Gulf Stream current. It travels 44 mi north with the current in 2 hr. It travels 56 mi south against the current in 4 hr. Find the speed of the current and the speed of the boat in still water.

65. Two runners begin at the same point on a 390-m circular track and run at different speeds. If they run in opposite directions, they pass each other in 30 sec. If they run in the same direction, they meet each other in 130 sec. Find the speed of each runner.

66. Two particles begin at the same point and move at different speeds along a circular path of circumference 280 ft. Moving in opposite directions, they pass in 10 sec. Moving in the same direction, they pass in 70 sec. Find the speed of each particle.

67. A cleaning company charges $100 for each office it cleans. The fixed monthly cost of $480 for the company includes telephone service and the depreciation on cleaning equipment and a van. The variable cost is $52 per office and includes labor, gasoline, and cleaning supplies. **(See Example 9)**

  a. Write a linear cost function representing the cost $C(x)$ (in $) to the company to clean x offices per month.

  b. Write a linear revenue function representing the revenue $R(x)$ (in $) for cleaning x offices per month.

  c. Determine the number of offices to be cleaned per month for the company to break even.

  d. If 28 offices are cleaned, will the company make money or lose money?

68. A vendor at a carnival sells cotton candy and caramel apples for $2.00 each. The vendor is charged $100 to set up his booth. Furthermore, the vendor's average cost for each product he produces is approximately $0.75.

  a. Write a linear cost function representing the cost $C(x)$ (in $) to the vendor to produce x products.

  b. Write a linear revenue function representing the revenue $R(x)$ (in $) for selling x products.

  c. Determine the number of products to be produced and sold for the vendor to break even.

  d. If 60 products are sold, will the vendor make money or lose money?

For Exercises 69–70, refer to Figure 10-1 and the narrative on page 714.

69. Suppose that the price p (in $) of theater tickets is influenced by the number of tickets x offered by the theater and demanded by consumers.

  Supply: $p = 0.025x$

  Demand: $p = -0.04x + 104$

  a. Solve the system of equations defined by the supply and demand models.

  b. What is the equilibrium price?

  c. What is the equilibrium quantity?

70. The price p (in $) of a cookbook is determined by the number of cookbooks x demanded by consumers and supplied by the publisher.

  Supply: $p = 0.002x$

  Demand: $p = -0.005x + 70$

  a. Solve the system of equations defined by the supply and demand models.

  b. What is the equilibrium price?

  c. What is the equilibrium quantity?

71. a. Sketch the lines defined by $y = 2x$ and $y = -\frac{1}{2}x + 5$.

  b. Find the area of the triangle bounded by the lines in part (a) and the x-axis.

72. a. Sketch the lines defined by $y = x + 2$ and $y = -\frac{1}{2}x + 2$.

  b. Find the area of the triangle bounded by the lines in part (a) and the x-axis.

73. The **centroid** of a region is the geometric center. For the region shown, the centroid is the point of intersection of the diagonals of the parallelogram.

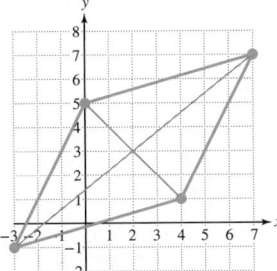

  a. Find an equation of the line through the points $(-3, -1)$ and $(7, 7)$.

  b. Find an equation of the line through the points $(0, 5)$ and $(4, 1)$.

  c. Find the centroid of the region.

74. The centroid of the region shown is the point of intersection of the diagonals of the parallelogram.

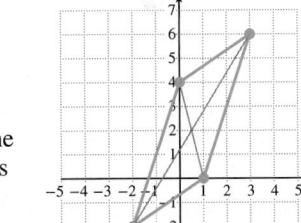

  a. Find an equation of the line through the points $(-2, -2)$ and $(3, 6)$.

  b. Find an equation of the line through the points $(1, 0)$ and $(0, 4)$.

  c. Find the centroid of the region.

75. Two angles are complementary. The measure of one angle is 6° less than twice the measure of the other angle. Find the measure of each angle.

76. Two angles are supplementary. The measure of one angle is 12° more than 5 times the measure of the other angle. Find the measure of each angle.

For Exercises 77–78, find the measure of angles x and y.

77.

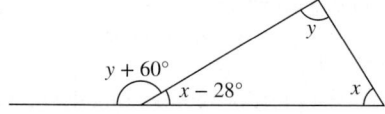

78.

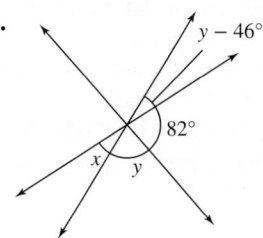

Mixed Exercises

79. Given $f(x) = mx + b$, find m and b if $f(3) = -3$ and $f(-12) = -8$.

80. Given $g(x) = mx + b$, find m and b if $g(2) = 1$ and $g(-4) = 10$.

For Exercises 81–82, use the substitution $u = \frac{1}{x}$ and $v = \frac{1}{y}$ to rewrite the equations in the system in terms of the variables u and v. Solve the system in terms of u and v. Then back substitute to determine the solution set to the original system in terms of x and y.

81. $\dfrac{1}{x} + \dfrac{2}{y} = 1$

$-\dfrac{1}{x} + \dfrac{4}{y} = -7$

82. $-\dfrac{3}{x} + \dfrac{4}{y} = 11$

$\dfrac{1}{x} - \dfrac{2}{y} = -5$

83. During a race, Marta bicycled 12 mi and ran 4 mi in a total of 1 hr 20 min $\left(\frac{4}{3}\,\text{hr}\right)$. In another race, she bicycled 21 mi and ran 3 mi in 1 hr 40 min $\left(\frac{5}{3}\,\text{hr}\right)$. Determine the speed at which she bicycles and the speed at which she runs. Assume that her bicycling speed was the same in each race and that her running speed was the same in each race.

84. Shelia swam 1 mi and ran 6 mi in a total of 1 hr 15 min $\left(\frac{5}{4}\,\text{hr}\right)$. In another training session she swam 2 mi and ran 8 mi in a total of 2 hr. Determine the speed at which she swims and the speed at which she runs. Assume that her swimming speed was the same each day and that her running speed was the same each day.

85. A certain pickup truck gets 16 mpg in the city and 22 mpg on the highway. If a driver drives 254 mi on 14 gal of gas, determine the number of city miles and highway miles that the truck was driven.

86. A sedan gets 12 mpg in the city and 18 mpg on the highway. If a driver drives a total of 420 mi on 26 gal of gas, how many miles in the city and how many miles on the highway did he drive?

Write About It

87. A system of linear equations in x and y can represent two intersecting lines, two parallel lines, or a single line. Describe the solution set to the system in each case.

88. When solving a system of linear equations in two variables using the substitution or addition method, explain how you can detect whether the equations are dependent.

89. When solving a system of linear equations in two variables using the substitution or addition method, explain how you can detect whether the system is inconsistent.

90. Consider a system of linear equations in two variables in which the solution set is $\{(x, x + 2) \mid x \text{ is any real number}\}$. Why do we say that the equations in the system are dependent?

Expanding Your Skills

91. A 50-lb weight is supported from two cables and the system is in equilibrium. The magnitudes of the forces on the cables are denoted by $|F_1|$ and $|F_2|$, respectively. An engineering student knows that the horizontal components of the two forces (shown in red) must be equal in magnitude. Furthermore, the sum of the magnitudes of the vertical components of the forces (shown in blue) must be equal to 50 lb to offset the downward force of the weight. Find the values of $|F_1|$ and $|F_2|$. Write the answers in exact form with no radical in the denominator. Also give approximations to 1 decimal place.

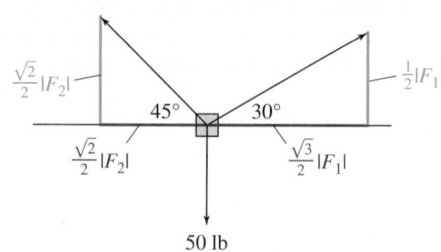

SECTION 10.3	# Systems of Linear Equations in Three Variables and Applications

1. Identify Solutions to a System of Linear Equations in Three Variables

In Section 10.2 we solved systems of linear equations in two variables. In this section, we expand the discussion to solving systems involving three variables. A **linear equation in three variables** is an equation that can be written in the form

$$Ax + By + Cz = D, \text{ where } A, B, \text{ and } C \text{ are not all zero.}$$

For example, $x + 2y + z = 4$ is a linear equation in three variables. A solution to a linear equation in three variables is an ordered triple (x, y, z) that satisfies the equation. For example, several solutions to $x + 2y + z = 4$ are given here.

Solution	Check: $x + 2y + z = 4$
$(1, 1, 1)$	$(1) + 2(1) + (1) \overset{?}{=} 4$ ✓ true
$(4, 0, 0)$	$(4) + 2(0) + (0) \overset{?}{=} 4$ ✓ true
$(0, 2, 0)$	$(0) + 2(2) + (0) \overset{?}{=} 4$ ✓ true
$(0, 0, 4)$	$(0) + 2(0) + (4) \overset{?}{=} 4$ ✓ true

There are infinitely many solutions to the equation $x + 2y + z = 4$. The set of all solutions to a linear equation in three variables can be represented graphically by a plane in space. Figure 10-10 shows a portion of the plane defined by $x + 2y + z = 4$.

In many applications, we are interested in determining the point or points of intersection of two or more planes. This is given by the solutions to a system of linear equations in three variables. For example:

$$
\begin{aligned}
2x + y - 3z &= -2 \\
x - 4y + z &= 24 \\
-3x - y + 4z &= 0
\end{aligned}
$$

A solution to a system of linear equations in three variables is an **ordered triple** (x, y, z) that satisfies each equation in the system. Geometrically, a solution is a point of intersection of the planes represented by the equations in the system (Figure 10-11).

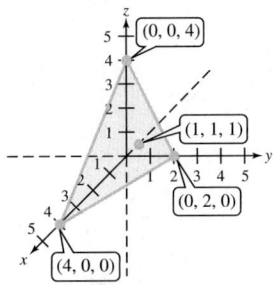

Figure 10-10

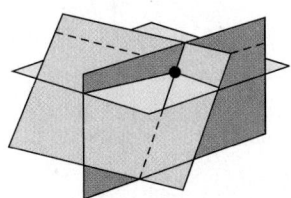

Figure 10-11

EXAMPLE 1	**Determining if an Ordered Triple Is a Solution to a System of Equations**

Determine if the ordered triple is a solution to the system.

$$
\begin{aligned}
2x + y - 3z &= -2 \\
x - 4y + z &= 24 \\
-3x - y + 4z &= 0
\end{aligned}
$$

a. $(3, -5, 1)$ **b.** $(2, -3, 1)$

Solution:

Test the ordered triple in each equation.

a. First equation   Second equation   Third equation

$2x + y - 3z = -2$ $x - 4y + z = 24$ $-3x - y + 4z = 0$

$2(3) + (-5) - 3(1) \overset{?}{=} -2$ $(3) - 4(-5) + (1) \overset{?}{=} 24$ $-3(3) - (-5) + 4(1) \overset{?}{=} 0$

$-2 \overset{?}{=} -2$ ✓ true $24 \overset{?}{=} 24$ ✓ true $0 \overset{?}{=} 0$ ✓ true

The ordered triple $(3, -5, 1)$ is a solution to the system of equations.

b. First equation   Second equation   Third equation

$2x + y - 3z = -2$ $x - 4y + z = 24$ $-3x - y + 4z = 0$

$2(2) + (-3) - 3(1) \overset{?}{=} -2$ $(2) - 4(-3) + (1) \overset{?}{=} 24$ $-3(2) - (-3) + 4(1) \overset{?}{=} 0$

$-2 \overset{?}{=} -2$ ✓ true $15 \overset{?}{=} 24$ false $1 \overset{?}{=} 0$ false

If an ordered triple fails to be a solution to any of the equations in the system, then it is not a solution to the system. The ordered triple $(2, -3, 1)$ is *not* a solution to the second or third equation. Therefore, $(2, -3, 1)$ is *not* a solution to the system of equations.

Skill Practice 1   Determine if the ordered triple is a solution to the system.

$$5x - y + 3z = -7$$
$$3x + 4y - z = 5$$
$$9x + 5y + 7z = 1$$

a. $(-2, -6, 3)$ **b.** $(-1, 2, 0)$

---

2. Solve Systems of Linear Equations in Three Variables

To solve a system of three linear equations in three variables, we first eliminate one variable. The system is then reduced to a two-variable system that can be solved by the techniques learned in Section 10.2.

Solving a System of Three Linear Equations in Three Variables

Step 1 Write each equation in standard form $Ax + By + Cz = D$.

Step 2 Choose a pair of equations and eliminate one of the variables by using the addition method.

Step 3 Choose a different pair of equations and eliminate the *same* variable.

Step 4 Once steps 2 and 3 are complete, the resulting system has two equations in two variables. Solve this system by using the substitution or addition method.

Step 5 Substitute the values of the variables found in step 4 into any of the three original equations that contain the third variable. Solve for the third variable.

Step 6 Check the ordered triple in each original equation. Then write the solution as an ordered triple in set notation.

EXAMPLE 2   **Solving a System of Equations in Three Variables**

Solve the system. $3x - 2y + z = 2$

$5x + y - 2z = 1$

$4x - 3y + 3z = 7$

Answers

1. a. No   **b.** Yes

TIP In Example 2, the y terms can also be eliminated easily because the coefficient on y in equation B is 1. Therefore, y can be eliminated from equations A and B by multiplying equation B by 2. Likewise y can be eliminated from equations B and C by multiplying equation B by 3.

Solution:

A $3x - 2y + z = 2$
B $5x + y - 2z = 1$
C $4x - 3y + 3z = 7$

Step 1: The equations are already in standard form.
• It is helpful to label the equations A, B, and C.
• The z variable can easily be eliminated from equations A and B and from equations A and C. This is accomplished by creating opposite coefficients for the z terms and then adding the equations.

Step 2: Eliminate z from equations A and B.

A $3x - 2y + z = 2$ $\xrightarrow{\text{Multiply by 2.}}$ $6x - 4y + 2z = 4$
B $5x + y - 2z = 1$ $\underline{5x + y - 2z = 1}$
                                          $11x - 3y \quad\; = 5$ D

Step 3: Eliminate z from equations A and C.

A $3x - 2y + z = 2$ $\xrightarrow{\text{Multiply by } -3.}$ $-9x + 6y - 3z = -6$
C $4x - 3y + 3z = 7$ $\underline{4x - 3y + 3z = 7}$
                                           $-5x + 3y \quad\; = 1$ E

FOR REVIEW

Once one of the three variables is eliminated, solve the resulting system of equations in two variables by using either the substitution or addition method. See Section 10.2 for review.

Step 4: D $11x - 3y = 5$
            E $\underline{-5x + 3y = 1}$
               $6x \quad\quad = 6$
               $x = 1$

D $11(1) - 3y = 5$
   $11 - 3y = 5$
   $-3y = -6$
   $y = 2$

Solve the system of equations D and E.

A $3x - 2y + z = 2$
$3(1) - 2(2) + z = 2$
$3 - 4 + z = 2$
$-1 + z = 2$
$z = 3$

Step 5: Substitute the values of the known variables x and y back into one of the original equations. We have chosen equation A.

Step 6: Check the ordered triple $(1, 2, 3)$ in the three original equations.

A $3x - 2y + z = 2$ B $5x + y - 2z = 1$ C $4x - 3y + 3z = 7$
$3(1) - 2(2) + (3) \overset{?}{=} 2$ $5(1) + (2) - 2(3) \overset{?}{=} 1$ $4(1) - 3(2) + 3(3) \overset{?}{=} 7$
$2 \overset{?}{=} 2$ ✓ true $1 \overset{?}{=} 1$ ✓ true $7 \overset{?}{=} 7$ ✓ true

The solution set is $\{(1, 2, 3)\}$.

Skill Practice 2   Solve the system. $2x - y + 5z = -7$
                                          $x + 4y - 2z = 1$
                                          $3x + 2y + z = -7$

In Example 3, we solve a system of linear equations in which one or more equations has a missing term.

EXAMPLE 3   Solving a System of Equations in Three Variables

Solve the system. $2x + y = -2$
                     $3y = 5z - 12$
                     $5(x + z) = 2z + 5$

Solution:

Step 1: Write the equations in standard form.

A $2x + y = -2$ $\longrightarrow$ $2x + y \quad\quad = -2$
B $3y = 5z - 12$ $\longrightarrow$ $3y - 5z = -12$
C $5(x + z) = 2z + 5$ $\longrightarrow$ $5x \quad\; + 3z = 5$

Notice that the equations already have missing variable terms.

Answer
2. $\{(-3, 1, 0)\}$

Steps 2 and 3: This system of equations has several missing terms. For example, equation $\boxed{\text{C}}$ is missing the variable y. If we eliminate y from equations $\boxed{\text{A}}$ and $\boxed{\text{B}}$, then we will have a second equation with variable y missing.

$$\boxed{\text{A}} \quad 2x + y \qquad = -2 \quad \xrightarrow{\text{Multiply by } -3.} \quad -6x - 3y \qquad = 6$$
$$\boxed{\text{B}} \qquad\quad 3y - 5z = -12 \qquad\qquad\qquad\qquad 3y - 5z = -12$$
$$\overline{\qquad -6x \qquad - 5z = -6 \quad \boxed{\text{D}}}$$

Step 4: Pair up equations $\boxed{\text{C}}$ and $\boxed{\text{D}}$. These equations form a system of linear equations in two variables. To solve the system with equations $\boxed{\text{C}}$ and $\boxed{\text{D}}$ we have chosen to eliminate the z variable.

$$\boxed{\text{C}} \quad 5x + 3z = 5 \quad \xrightarrow{\text{Multiply by 5.}} \quad 25x + 15z = 25$$
$$\boxed{\text{D}} \quad -6x - 5z = -6 \quad \xrightarrow[\text{Multiply by 3.}]{} \quad -18x - 15z = -18$$
$$\overline{\qquad\qquad\qquad\qquad\qquad 7x \qquad = 7}$$
$$x = 1$$

$$\boxed{\text{C}} \quad 5x + 3z = 5$$
$$5(1) + 3z = 5$$
$$z = 0$$

$$\boxed{\text{B}} \quad 3y = 5z - 12$$
$$3y = 5(0) - 12$$
$$3y = -12$$
$$y = -4$$

Step 5: Substitute the values of the known variables x and z back into one of the original equations containing y. We have chosen equation $\boxed{\text{B}}$.

Step 6: The ordered triple $(1, -4, 0)$ checks in each original equation.

The solution set is $\{(1, -4, 0)\}$.

Skill Practice 3   Solve the system.
$$a \qquad + 3c = 4$$
$$b + 2c = -1$$
$$2a - 4b \qquad = 14$$

A system of linear equations in three variables may have no solution. This occurs if the equations represent planes that do not all intersect (Figure 10-12). In such a case, we say that the system is **inconsistent**.

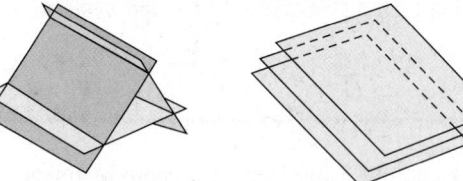 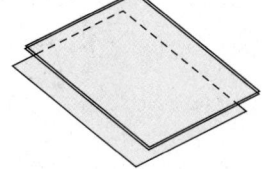

Figure 10-12

A system of linear equations in three variables may also have infinitely many solutions. This occurs if the equations represent planes that intersect in a common line or common plane (Figure 10-13). In such a case, we say that the equations are **dependent**.

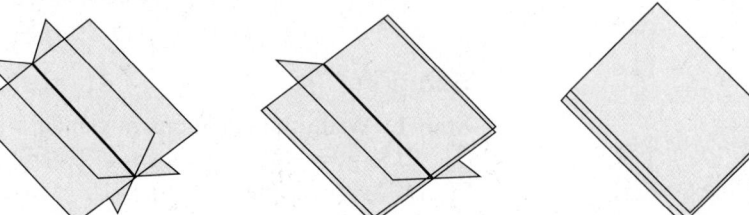

Answer
3. $\{(1, -3, 1)\}$

Figure 10-13

EXAMPLE 4   Determining the Number of Solutions to a System

a. Determine the number of solutions to the system.

b. State whether the system is inconsistent or the equations are dependent.

c. Write the solution set.

$$-x + 6y - 3z = -8$$
$$x - 2y + 2z = 3$$
$$3x + 2y + 4z = -6$$

Solution:

Begin by eliminating a variable from two different pairs of equations. We will eliminate x from $\boxed{A}$ and $\boxed{B}$ and from $\boxed{A}$ and $\boxed{C}$.

$\boxed{A}$ $-x + 6y - 3z = -8$ $\longrightarrow$ $-x + 6y - 3z = -8$

$\boxed{B}$ $x - 2y + 2z = 3$ $\longrightarrow$ $\underline{x - 2y + 2z = 3}$

$\boxed{C}$ $3x + 2y + 4z = -6$ $4y - z = -5$ $\boxed{D}$

Add equations $\boxed{A}$ and $\boxed{B}$ to eliminate x.

$\boxed{A}$ $-x + 6y - 3z = -8$ $\xrightarrow{\text{Multiply by 3.}}$ $-3x + 18y - 9z = -24$

$\boxed{C}$ $3x + 2y + 4z = -6$ $\longrightarrow$ $\underline{3x + 2y + 4z = -6}$

                              $20y - 5z = -30$ $\boxed{E}$

Multiply equation $\boxed{A}$ by 3 and add the result to $\boxed{C}$.

$\boxed{D}$ $4y - z = -5$ $\xrightarrow{\text{Multiply by } -5.}$ $-20y + 5z = 25$

$\boxed{E}$ $20y - 5z = -30$ $\underline{20y - 5z = -30}$

                              $0 = -5$

Solving the system of equations $\boxed{D}$ and $\boxed{E}$ results in a contradiction.

The system of equations reduces to a contradiction.

a. There is no solution.

b. The system is inconsistent.

c. The solution set is { }.

Skill Practice 4   Repeat Example 4 with the given system.

$$x + y + 4z = -1$$
$$3x + y - 4z = 3$$
$$-4x - y + 8z = -2$$

In Example 5, we investigate the case in which a system of equations has infinitely many solutions.

EXAMPLE 5   Determining the Number of Solutions to a System

a. Determine the number of solutions to the system.

b. State whether the system is inconsistent, or the equations are dependent.

c. Write the solution set.

$$2x + y = -3$$
$$2y + 16z = -10$$
$$-7x - 3y + 4z = 8$$

Solution:

Eliminate variable z from equations $\boxed{B}$ and $\boxed{C}$.

$\boxed{A}$ $2x + y = -3$

$\boxed{B}$ $ 2y + 16z = -10$ $\longrightarrow$ $2y + 16z = -10$

$\boxed{C}$ $-7x - 3y + 4z = 8$ $\xrightarrow{\text{Multiply by } -4.}$ $\underline{28x + 12y - 16z = -32}$

                              $28x + 14y = -42$ $\boxed{D}$

Pair up equations $\boxed{A}$ and $\boxed{D}$ to solve for x and y.

$\boxed{A}$ $2x + \quad y = -3$ ⟶ $2x + y = -3$
$\boxed{D}$ $28x + 14y = -42$ ⟶ $-2x - y = 3$
          Divide by -14. $\overline{\quad\quad 0 = 0 \quad}$

The system reduces to the identity $0 = 0$. This implies that

  a. There are infinitely many solutions.

  b. The equations are dependent.

  c. To find the general solution, we need to express the dependency among the variables as an ordered triple. Note that from equation $\boxed{A}$, we can solve for x in terms of y, and from equation $\boxed{B}$, we can solve for z in terms of y. Thus,

Equation $\boxed{A}$: $2x = -y - 3 \Rightarrow x = \dfrac{-y-3}{2} \Rightarrow x = -\dfrac{y+3}{2}$

Equation $\boxed{B}$: $16z = -2y - 10 \Rightarrow z = \dfrac{-2y-10}{16} \Rightarrow z = -\dfrac{y+5}{8}$

Therefore, the solution set is $\left\{ \left(-\dfrac{y+3}{2}, y, -\dfrac{y+5}{8} \right) \middle| y \text{ is any real number} \right\}$.

Skill Practice 5   Repeat Example 5 with the given system.

$$\begin{aligned} 5y + \quad z &= 0 \\ -x \quad\quad + 4z &= 0 \\ -x + 5y + 5z &= 0 \end{aligned}$$

The general solution to the system in Example 5 can be written in a number of forms.

Solve for x and z in terms of y: $\left\{ \left(-\dfrac{y+3}{2}, y, -\dfrac{y+5}{8} \right) \middle| y \text{ is any real number} \right\}$

Solve for y and z in terms of x: $\left\{ \left(x, -2x - 3, \dfrac{x-1}{4} \right) \middle| x \text{ is any real number} \right\}$

Solve for x and y in terms of z: $\left\{ \left(4z + 1, -8z - 5, z \right) \middle| z \text{ is any real number} \right\}$

TIP   Any form of the general solution can be checked by substitution in the original three equations.

Check: $\left\{ \left(-\dfrac{y+3}{2}, y, -\dfrac{y+5}{8} \right) \middle| y \text{ is any real number} \right\}$

$\boxed{A}$ $2x + y = -3$

$2\left(-\dfrac{y+3}{2} \right) + y \overset{?}{=} -3$

$-y - 3 + y = -3$ ✓

$\boxed{B}$ $2y + 16z = -10$

$2y + 16\left(-\dfrac{y+5}{8} \right) \overset{?}{=} -10$

$2y - 2y - 10 = -10$ ✓

$\boxed{C}$ $-7x - 3y + 4z = 8$

$-7\left(-\dfrac{y+3}{2} \right) - 3y + 4\left(-\dfrac{y+5}{8} \right) \overset{?}{=} 8$

$\dfrac{7}{2}y + \dfrac{21}{2} - 3y - \dfrac{1}{2}y - \dfrac{5}{2} \overset{?}{=} 8$

$\dfrac{6}{2}y - 3y + \dfrac{16}{2} = 8$ ✓

Answers

5. a. There are infinitely many solutions.

  b. The equations are dependent.

  c. $\left\{ \left(4z, -\dfrac{1}{5}z, z \right) \middle| z \text{ is any real number} \right\}$ or $\left\{ \left(x, -\dfrac{x}{20}, \dfrac{x}{4} \right) \middle| x \text{ is any real number} \right\}$ or $\{(-20y, y, -5y) \mid y \text{ is any real number}\}$

The topic of dependent equations will be discussed in more detail when we learn matrix methods to solve systems of linear equations. With additional tools available, we can investigate whether the dependent equations represent planes that intersect in a line or whether the equations all represent the same plane (Figure 10-13).

3. Use Systems of Linear Equations in Applications

When solving an application involving three unknowns, sometimes it is convenient to use a system of three independent equations, as demonstrated in Examples 6 and 7.

EXAMPLE 6   Solving an Application Involving Finance

Janette invested a total of $18,000 in three different mutual funds. She invested in a bond fund that returned 4% the first year. An aggressive growth fund lost 8% for the year, and an international fund returned 2%. Janette invested $2000 more in the growth fund than in the other two funds combined. If she had a net loss of −$540 for the year, how much did she invest in each fund?

Solution:

Let x represent the amount invested in the bond fund.   Label the variables.

Let y represent the amount invested in the growth fund.   With three unknowns,
we need three inde-
Let z represent the amount invested in the international fund.   pendent equations.

$x + y + z = 18,000$ ⟶ The total amount invested was $18,000.

$y = (x + z) + 2000$ ⟶ The amount invested in the growth fund was $2000
more than the combined amount in the other two funds.

$0.04x - 0.08y + 0.02z = -540$ ⟶ The sum of the gain and loss from each fund equals
−$540.

|A| $x + y + z = 18,000$ |A| $x + y + z = 18,000$
|B| $y = (x + z) + 2000$ ⟶ Standard form ⟶ |B| $-x + y - z = 2000$
|C| $0.04x - 0.08y + 0.02z = -540$ ⟶ Multiply by 100. ⟶ |C| $4x - 8y + 2z = -54,000$

Eliminate x from equations |A| and |B| and equations |B| and |C|.

|A| $x + y + z = 18,000$ $4 \cdot$ |B| $-4x + 4y - 4z = 8000$
|B| $-x + y - z = 2000$ |C| $\dfrac{4x - 8y + 2z = -54,000}{}$
$\dfrac{}{2y \quad = 20,000}$ $-4y - 2z = -46,000$ |D|
$y = 10,000$

Back substitute.

|D| $-4y - 2z = -46,000$ |A| $x + y + z = 18,000$
$-4(10,000) - 2z = -46,000$ $x + (10,000) + (3000) = 18,000$
$-2z = -6000$ $x + 13,000 = 18,000$
$z = 3000$ $x = 5000$

Janette invested $5000 in the bond fund, $10,000 in the aggressive growth fund, and $3000 in the international fund.

Skill Practice 6   Nicolas mixes three solutions of acid with concentrations of 10%, 15%, and 5%. He wants to make 30 L of a mixture that is 12% acid, and he uses four times as much of the 15% solution as the 5% solution. How much of each of the three solutions must he use?

Answer

6. Nicolas mixes 10 L of the 10% solution, 16 L of the 15% solution, and 4 L of the 5% solution.

4. Modeling with Linear Equations in Three Variables

In Section 7.1, we used regression to find a quadratic model $y = ax^2 + bx + c$ from observed data points. We will now learn how to find a quadratic model by using a system of linear equations in three variables. The premise is that any three noncollinear points (points that do not all fall on the same line) define a unique parabola given by $y = ax^2 + bx + c \, (a \neq 0)$.

EXAMPLE 7   Using a System of Linear Equations to Create a Quadratic Model

Given the noncollinear points $(4, 2)$, $(1, -1)$, and $(-1, 7)$, find an equation of the form $y = ax^2 + bx + c$ that defines the parabola through the points.

Solution:

> **TIP** We can verify that the three points given in Example 7 are *not* collinear by showing that the slopes between two pairs of points are different.
>
> For $(4, 2)$ and $(1, -1)$,
> $$m = \frac{-1 - 2}{1 - 4} = 1$$
> For $(4, 2)$ and $(-1, 7)$,
> $$m = \frac{7 - 2}{-1 - 4} = -1$$

$$y = ax^2 + bx + c$$

Substitute $(4, 2)$: $2 = a(4)^2 + b(4) + c$ $\longrightarrow$ $\boxed{A}$ $16a + 4b + c = 2$

Substitute $(1, -1)$: $-1 = a(1)^2 + b(1) + c$ $\longrightarrow$ $\boxed{B}$ $a + b + c = -1$

Substitute $(-1, 7)$: $7 = a(-1)^2 + b(-1) + c$ $\longrightarrow$ $\boxed{C}$ $a - b + c = 7$

Eliminate variable b from equations $\boxed{A}$ and $\boxed{C}$ and from equations $\boxed{B}$ and $\boxed{C}$.

$$
\begin{array}{ll}
\boxed{A} \quad 16a + 4b + c = 2 \\
4 \cdot \boxed{C} \quad \underline{4a - 4b + 4c = 28} \\
\qquad\quad 20a \quad\;\; + 5c = 30 \quad \boxed{D}
\end{array}
\qquad\qquad
\begin{array}{ll}
\boxed{B} \quad a + b + c = -1 \\
\boxed{C} \quad \underline{a - b + c = 7} \\
\qquad 2a \quad\;\; + 2c = 6 \quad \boxed{E}
\end{array}
$$

$$
\begin{array}{ll}
\boxed{D} \quad 20a + 5c = 30 & \xrightarrow{\text{Divide by 5.}} \quad 4a + c = 6 \\
\boxed{E} \quad 2a + 2c = 6 & \xrightarrow[\text{by } -2.]{\text{Divide}} \quad \underline{-a - c = -3} \\
& \qquad\qquad\qquad\quad 3a \quad\;\; = 3 \\
& \qquad\qquad\qquad\qquad\quad a = 1
\end{array}
$$

Back substitute.

$$
\left\{
\begin{array}{l}
\boxed{E} \quad 2a + 2c = 6 \\
\qquad 2(1) + 2c = 6 \\
\qquad\qquad\quad c = 2 \\[2ex]
\boxed{B} \quad a + b + c = -1 \\
\qquad (1) + b + (2) = -1 \\
\qquad\qquad\qquad b = -4
\end{array}
\right.
$$

Substituting $a = 1$, $b = -4$, and $c = 2$ into the equation $y = ax^2 + bx + c$ gives

$$y = x^2 - 4x + 2$$

The graph of $y = x^2 - 4x + 2$ passes through the points $(4, 2)$, $(1, -1)$, and $(-1, 7)$ as shown in Figure 10-14.

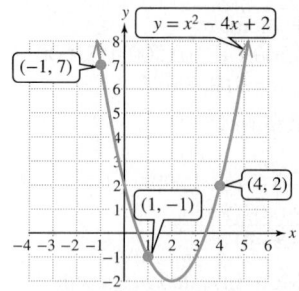

Figure 10-14

The results can also be verified by using the Table feature of a graphing utility. Enter $Y_1 = x^2 - 4x + 2$.

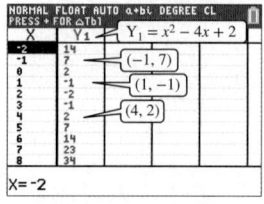

Skill Practice 7   Given the noncollinear points $(-3, 2)$, $(-4, 1)$, and $(-6, -7)$, find an equation of the form $y = ax^2 + bx + c$ that defines the parabola through the points.

Answer

7. $y = -x^2 - 6x - 7$

SECTION 10.3   Practice Exercises

Prerequisite Review

For Exercises R.1–R.2, solve the equation. (See Section 1.2 for review.)

R.1. $\dfrac{3-x}{4} + \dfrac{2x+1}{3} = \dfrac{x}{12} + 1$

R.2. $\dfrac{v-4}{5} + \dfrac{v}{8} = \dfrac{v-2}{5} - 7$

For Exercises R.3–R.8, solve the system. (See Section 10.2 for review.)

R.3. $3x + 4y = 24$
$2x - y = -6$

R.4. $-5x + 3y = -15$
$x - 6y = 3$

R.5. $x = 2y + 4$
$3x - 6y = 12$

R.6. $y = 4x - 7$
$8x - 2y = 14$

R.7. $3x + 5y = 4$
$\dfrac{3}{2}x = 4 - \dfrac{5}{2}y$

R.8. $4x + 6y = 1$
$y = -\dfrac{2}{3}x$

R.9. The measures of two angles in a triangle are 108° and 46°. What is the measure of the third angle?

R.10. The perimeter of a triangle is 58 ft. If the lengths of two sides are 20 ft and 22 ft, find the length of the third side.

Concept Connections

1. The graph of a linear equation in two variables is a line in a two-dimensional coordinate system. The graph of a linear equation in three variables is a _____ in a three-dimensional coordinate system.

2. A solution to a system of linear equations in three variables is an ordered _____ that satisfies each equation in the system. Graphically, this is a point of _____ of three planes.

Objective 1: Identify Solutions to a System of Linear Equations in Three Variables

For Exercises 3–4, find three ordered triples that are solutions to the linear equation in three variables.

3. $2x + 4y - 6z = 12$

4. $3x - 5y + z = 15$

For Exercises 5–8, determine if the ordered triple is a solution to the system of equations. (**See Example 1**)

5. $-x + 3y - 7z = 7$
$2x + 4y + z = 16$
$3x - 5y + 6z = -9$

   a. $(2, 3, 0)$
   b. $(-2, 4, 1)$

6. $2x - 3y + z = -12$
$x + y - 2z = 9$
$-3x + 2y - z = 7$

   a. $(2, 5, -1)$
   b. $(1, 4, -2)$

7. $x + y + z = 2$
$x + 2y - z = 2$
$3x + 5y - z = 6$

   a. $(2, 0, 0)$
   b. $(-1, 2, 1)$

8. $-x - y + z = 3$
$3x + 4y - z = 1$
$5x + 7y - z = -1$

   a. $(1, 2, 6)$
   b. $(3, -1, 5)$

Objective 2: Solve Systems of Linear Equations in Three Variables

For Exercises 9–32, solve the system. If a system has one unique solution, write the solution set. Otherwise, determine the number of solutions to the system, and determine whether the system is inconsistent, or the equations are dependent. (**See Examples 2–5**)

9. $x - 2y + z = -9$
$3x + 4y + 5z = 9$
$-2x + 3y - z = 12$

10. $2x - y + z = 6$
$-x + 5y - z = 10$
$3x + y - 3z = 12$

11. $4x = 3y - 2z - 5$
$2(x + y) = y + z - 6$
$6(x - y) + z = x - 5y - 8$

12. $3x = 5y - z + 13$
   $-(x - y) - z = x - 3$
   $5(x + y) = 3y - 3z - 4$

13. $2x + 5z = 2$
   $3y - 7z = 9$
   $-5x + 9y = 22$

14. $3x - 2y = -8$
   $5y + 6z = 2$
   $7x + 11z = -33$

15. $-4x - 3y \quad\;\; = 0$
   $\qquad\;\; 3y + z = -1$
   $4x \qquad\; - z = 12$

16. $\;\;\; 4x -\;\; y + 2z = 1$
   $3x +\;\; 5y -\;\; z = -2$
   $-9x - 15y + 3z = 0$

17. $2x = 3y - 6z - 1$
   $6y = 12z - 10x + 9$
   $3z = 6y - 3x - 1$

18. $5x = 2y - 3z - 3$
   $4y = -1 - 10x - 5z$
   $2z = 5x - 6y$

19. $x + 2y + 4z = 3$
   $\quad\;\; y + 3z = 5$
   $x \qquad\;\; - 2z = -7$

20. $3x +\;\; 2y + 5z = 6$
   $\qquad\; 3y -\;\; z = 4$
   $3x + 17y \qquad\;\; = 26$

21. $0.2x = 0.1y - 0.6z$
   $0.004x + 0.005y - 0.001z = 0$
   $30x = 50z - 20y$

22. $0.3x = 0.5y - 1.2z$
   $0.05x + 0.1y = 0.04z$
   $100x = 300y - 700z$

23. $\frac{1}{12}x + \frac{1}{4}y + \frac{1}{3}z = \frac{7}{12}$
   $-\frac{1}{10}x + \frac{1}{2}y - \frac{1}{5}z = -\frac{17}{10}$
   $\frac{1}{2}x + \frac{1}{4}y +\;\; z = 3$

24. $\;\; x + \frac{7}{2}y + \frac{1}{2}z = 4$
   $\frac{3}{4}x +\;\; y + \frac{1}{2}z = -1$
   $\frac{1}{10}x - \frac{2}{5}y - \frac{3}{10}z = 1$

25. $3x + 2y + 5z = 12$
   $\qquad 3y + 8z = -8$
   $\qquad\qquad 10z = 20$

26. $-4x + 6y +\;\; z = -4$
   $\qquad\;\; -2y - 4z = -24$
   $\qquad\qquad\; 6z = 48$

27. $\frac{x+2}{3} + \frac{y-4}{2} + \frac{z+1}{6} = 8$
   $-\frac{x+2}{3} \qquad\;\; + \frac{z+1}{2} = 8$
   $\qquad \frac{y-4}{4} - \frac{z+1}{6} = -1$

28. $\frac{x-1}{7} + \frac{y-2}{3} + \frac{z+2}{4} = 13$
   $\qquad\;\; \frac{y-2}{9} - \frac{z+2}{8} = 3$
   $\frac{x-1}{7} \qquad\;\; + \frac{z+2}{2} = 3$

29. $3(x + y) = 6 - 4z + y$
   $4 = 6y + 5z$
   $-3x + 4y + z = 0$

30. $4(x - y) = 8 - z - y$
   $3 = 3x + 4z$
   $-x + 3y + 3z = 1$

31. $-3x +\;\; 4y -\;\; z = -4$
   $\;\; x +\;\; 2y +\;\; z = 4$
   $-12x + 16y - 4z = -16$

32. $\;\;\; x + 2y - 3z = 5$
   $-2x - 5y + 4z = -6$
   $3x + 6y - 9z = 15$

For Exercises 33–36, solve the system from the indicated exercise and write the general solution. **(See Example 5)**

33. Exercise 19   **34.** Exercise 20   **35.** Exercise 31   **36.** Exercise 32

For Exercises 37–38, the general solution is given for a system of linear equations. Find three individual solutions to the system.

37. $-x + 4y + 2z = 4$
   $x - 3y -\;\; z = -2$
   $-x +\;\; y -\;\; z = -2$
   Solution: $\{(-2z + 4, -z + 2, z) \mid z \text{ is any real number}\}$

38. $2x - 3y + z = 1$
   $x + 4y - z = 3$
   $5x - 2y + z = 5$
   Solution: $\{(x, -3x + 4, -11x + 13) \mid x \text{ is any real number}\}$

Objective 3: Use Systems of Linear Equations in Applications

39. Devon invested $8000 in three different mutual funds. A fund containing large cap stocks made 6.2% return in 1 year. A real estate fund lost 13.5% in 1 year, and a bond fund made 4.4% in 1 year. The amount invested in the large cap stock fund was twice the amount invested in the real estate fund. If Devon had a net return of $66 across all investments, how much did he invest in each fund? **(See Example 6)**

40. Pierre inherited $120,000 from his uncle and decided to invest the money. He put part of the money in a money market account that earns 2.2% simple interest. The remaining money was invested in a stock that returned 6% in the first year and a mutual fund that lost 2% in the first year. He invested $10,000 more in the stock than in the mutual fund, and his net gain for 1 year was $2820. Determine the amount invested in each account.

41. A basketball player scored 26 points in one game. In basketball, some baskets are worth 3 points, some are worth 2 points, and free-throws are worth 1 point. He scored four more 2-point baskets than he did 3-point baskets. The number of free-throws equaled the sum of the number of 2-point and 3-point shots made. How many free-throws, 2-point shots, and 3-point shots did he make?

42. A sawmill cuts boards for a lumber supplier. When saws A, B, and C all work for 6 hr, they cut 7200 linear board-ft of lumber. It would take saws A and B working together 9.6 hr to cut 7200 ft of lumber. Saws B and C can cut 7200 ft of lumber in 9 hr. Find the rate (in ft/hr) that each saw can cut lumber.

43. Plant fertilizers are categorized by the percentage of nitrogen (N), phosphorus (P), and potassium (K) they contain, by weight. For example, a fertilizer that has N-P-K numbers of 8-5-5 has 8% nitrogen, 5% phosphorus, and 5% potassium by weight. Suppose that a fertilizer has twice as much potassium by weight as phosphorus. The percentage of nitrogen equals the sum of the percentages of phosphorus and potassium. If nitrogen, phosphorus, and potassium make up 42% of the fertilizer, determine the proper N-P-K label on the fertilizer.

    a. 14-7-14   **b.** 21-7-14   **c.** 14-7-21   **d.** 14-21-21

44. A theater charges $50 per ticket for seats in Section A, $30 per ticket for seats in Section B, and $20 per ticket for seats in Section C. For one play, 4000 tickets were sold for a total of $120,000 in revenue. If 1000 more tickets in Section B were sold than the other two sections combined, how many tickets in each section were sold?

45. The perimeter of a triangle is 55 in. The shortest side is 7 in. less than the longest side. The middle side is 19 in. less than the combined lengths of the shortest and longest sides. Find the lengths of the three sides.

46. A package in the shape of a rectangular solid is to be mailed. The combination of the girth (perimeter of a cross section defined by w and h) and the length of the package is 48 in. The width is 2 in. greater than the height, and the length is 12 in. greater than the width. Find the dimensions of the package.

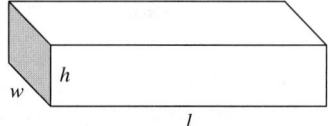

47. The measure of the largest angle in a triangle is 100° larger than the sum of the measures of the other two angles. The measure of the smallest angle is two-thirds the measure of the middle angle. Find the measure of each angle.

48. The measure of the largest angle in a triangle is 18° more than the sum of the measures of the other two angles. The measure of the smallest angle is one-half the measure of the middle angle. Find the measure of each angle.

Objective 4: Modeling with Linear Equations in Three Variables

49. a. Show that the points (1, 0), (3, 10), and (−2, 15) are not collinear by finding the slope between (1, 0) and (3, 10), and the slope between (3, 10) and (−2, 15). **(See Example 7)**

    b. Find an equation of the form $y = ax^2 + bx + c$ that defines the parabola through the points.

    c. Use a graphing utility to verify that the graph of the equation in part (b) passes through the given points.

50. a. Show that the points (2, 9), (−1, −6), and (−4, −3) are not collinear by finding the slope between (2, 9) and (−1, −6), and the slope between (2, 9) and (−4, −3).

    b. Find an equation of the form $y = ax^2 + bx + c$ that defines the parabola through the points.

    c. Use a graphing utility to verify that the graph of the equation in part (b) passes through the given points.

For Exercises 51–52, find an equation of the form $y = ax^2 + bx + c$ that defines the parabola through the three noncollinear points given.

51. (0, 6), (2, −6), (−1, 9)

52. (0, −4), (2, −6), (−3, −31)

The motion of an object traveling along a straight path is given by $s(t) = \frac{1}{2}at^2 + v_0 t + s_0$, where $s(t)$ is the position relative to the origin at time t. For Exercises 53–54, three observed data points are given. Find the values of a, v_0, and s_0.

53. $s(1) = 30$, $s(2) = 54$, $s(3) = 82$

54. $s(1) = -7$, $s(2) = 12$, $s(3) = 37$

Many statistics courses cover a topic called *multiple regression*. This provides a means to predict the value of a dependent variable y based on two or more independent variables x_1, x_2, ..., x_n. The model $y = ax_1 + bx_2 + c$ is a linear model that predicts y based on two independent variables x_1 and x_2. While statistical techniques may be used to find the values of a, b, and c based on a large number of data points, we can form a crude model given three data values (x_1, x_2, y). Use the information given in Exercises 55–56 to form a system of three equations and three variables to solve for a, b, and c.

55. The selling price of a home y (in $1000) is given based on the living area x_1 (in 100 ft^2) and on the lot size x_2 (in acres).

Living Area (100 ft^2) x_1	Lot Size (acres) x_2	Selling Price ($1000) y
28	0.5	225
25	0.8	207
18	0.4	154

  a. Use the data to create a model of the form
  $y = ax_1 + bx_2 + c$.

  b. Use the model from part (a) to predict the selling price of a home that is 2000 ft^2 on a 0.4-acre lot.

56. The gas mileage y (in mpg) for city driving is given based on the weight of the vehicle x_1 (in lb) and on the number of cylinders.

Weight (lb) x_1	Cylinders x_2	Mileage (mpg) y
3500	6	20
3200	4	26
4100	8	18

  a. Use the data to create a model of the form
  $y = ax_1 + bx_2 + c$.

  b. Use the model from part (a) to predict the gas mileage of a vehicle that is 3800 lb and has 6 cylinders.

Write About It

57. Give a geometric description of the solution set to a linear equation in three variables.

58. If a system of linear equations in three variables has no solution, then what can be said about the three planes represented by the equations in the system?

59. Explain the procedure presented in this section to solve a system of linear equations in three variables.

60. Explain how to check a solution to a system of linear equations in three variables.

Expanding Your Skills

For Exercises 61–62, find all solutions of the form (a, b, c, d).

61.
$$2a + b - c + d = 7$$
$$3b + 2c - 2d = -11$$
$$a + 3c + 3d = 14$$
$$4a + 2b - 5c = 6$$

62.
$$3a - 4b + 2c + d = 8$$
$$2a + 3b + 2d = 7$$
$$5b - 3c + 4d = -4$$
$$-a + b - 2c = -7$$

For Exercises 63–64, find all solutions of the form (u, v, w).

63.
$$\frac{u-3}{4} + \frac{v+1}{3} + \frac{w-2}{8} = 1$$
$$\frac{u-3}{2} + \frac{v+1}{2} + \frac{w-2}{4} = 0$$
$$\frac{u-3}{4} - \frac{v+1}{2} + \frac{w-2}{2} = -6$$

64.
$$\frac{u+1}{6} + \frac{v-1}{6} + \frac{w+3}{4} = 11$$
$$\frac{u+1}{3} - \frac{v-1}{2} + \frac{w+3}{4} = 7$$
$$\frac{u+1}{2} - \frac{v-1}{6} + \frac{w+3}{2} = 20$$

Recall that an equation of a circle can be written in the form $(x - h)^2 + (y - k)^2 = r^2$, where (h, k) is the center and r is the radius. Expanding terms, the equation can also be written in the form $x^2 + y^2 + Ax + By + C = 0$. For Exercises 65–66,

  a. Find an equation of the form $x^2 + y^2 + Ax + By + C = 0$ that represents the circle that passes through the given points.

  b. Find the center and radius of the circle.

65. $(2, 2)$, $(6, 0)$, $(7, -3)$

66. $(-1, 12)$, $(5, 10)$, $(9, 2)$

For Exercises 67–68, find the constants A and B so that the two polynomials are equal. (*Hint*: Create a system of linear equations by equating the constant terms and by equating the coefficients on the x terms and x^2 terms.)

67. $11x^2 + 26x - 5 = 2Ax^2 + 5Ax + 3A + Bx^2 - 2Bx - 8B + 2Cx^2 - 7Cx - 4C$

68. $3x^2 + 37x - 82 = Ax^2 + Ax - 12A + 3Bx^2 - 10Bx + 3B + 3Cx^2 + 11Cx - 4C$

1. Set Up a Partial Fraction Decomposition

In Section 4.2 we learned how to add and subtract rational expressions. For example:

$$\frac{5}{x+2} + \frac{3}{x-5} = \frac{5(x-5)}{(x+2)(x-5)} + \frac{3(x+2)}{(x-5)(x+2)}$$

$$= \frac{5(x-5) + 3(x+2)}{(x+2)(x-5)}$$

$$= \frac{8x-19}{(x+2)(x-5)}$$

The fraction $\dfrac{8x-19}{(x+2)(x-5)}$ is the result of adding two simpler fractions, $\dfrac{5}{x+2}$ and $\dfrac{3}{x-5}$. The sum $\dfrac{5}{x+2} + \dfrac{3}{x-5}$ is called the **partial fraction decomposition** of $\dfrac{8x-19}{(x+2)(x-5)}$. In some applications in higher mathematics, it is more convenient to work with the partial fraction decomposition than the more complicated single fraction. Therefore, in this section, we will learn the technique of partial fraction decomposition to write a rational expression as a sum of simpler fractions. That is, we will reverse the process of adding two or more fractions. There are two parts to this process.

I. First we set up the "form" or "structure" for the partial fraction decomposition into simpler fractions. For example, the denominator of $\dfrac{8x-19}{(x+2)(x-5)}$ consists of the distinct linear factors $(x+2)$ and $(x-5)$. From the preceding discussion, the partial fraction decomposition must be of the form:

$$\frac{8x-19}{(x+2)(x-5)} = \frac{A}{x+2} + \frac{B}{x-5}$$

> The expression on the right is the "form" or "structure" for the partial fraction decomposition of $\dfrac{8x-19}{(x+2)(x-5)}$.

II. Next, we solve for the constants A and B. To do so, multiply both sides of the equation by the LCD, and set up a system of linear equations.

$$(x+2)(x-5) \cdot \left[\frac{8x-19}{(x+2)(x-5)}\right] = (x+2)(x-5) \cdot \left[\frac{A}{x+2} + \frac{B}{x-5}\right]$$

Multiply by the LCD to clear fractions.

$$8x - 19 = A(x-5) + B(x+2)$$

$$8x - 19 = Ax - 5A + Bx + 2B \qquad \text{Apply the distributive property.}$$

$$8x - 19 = (A+B)x + (-5A+2B) \qquad \text{Simplify and combine like terms.}$$

x coefficients are equal.

$$8x - 19 = (A+B)x + (-5A+2B)$$

Constants are equal.

Two polynomials are equal if and only if the coefficients on like terms are equal.

$$A + B = 8 \qquad \text{Equate the coefficients on } x.$$
$$-5A + 2B = -19 \qquad \text{Equate the constant terms.}$$

Solve the system of linear equations. Then substitute the values of A and B into the partial fraction decomposition.

$$\boxed{1} \quad A + B = 8 \qquad \xrightarrow{\text{Multiply by 5.}} \quad 5A + 5B = 40$$
$$\boxed{2} \quad -5A + 2B = -19 \qquad\qquad\qquad\quad -5A + 2B = -19$$
$$\overline{\qquad\qquad\qquad\qquad\qquad\qquad 7B = 21}$$
$$\boxed{1} \quad A + B = 8 \qquad\qquad\qquad\qquad\qquad B = 3$$
$$A + 3 = 8$$
$$A = 5$$

$$A = 5 \text{ and } B = 3$$

$$\frac{8x - 19}{(x + 2)(x - 5)} = \frac{A}{x + 2} + \frac{B}{x - 5} = \frac{5}{x + 2} + \frac{3}{x - 5}$$

We begin partial fraction decomposition by factoring the denominator into linear factors $(ax + b)$ and quadratic factors $(ax^2 + bx + c)$ that are irreducible over the integers. A quadratic factor that is irreducible over the integers cannot be factored as a product of binomials with integer coefficients. From the factorization of the denominator, we then determine the proper form for the partial fraction decomposition using the following guidelines.

Decomposition of $\dfrac{f(x)}{g(x)}$ into Partial Fractions

Consider a rational expression $\dfrac{f(x)}{g(x)}$, where $f(x)$ and $g(x)$ are polynomials with real coefficients, $g(x) \neq 0$, and the degree of $f(x)$ is less than the degree of $g(x)$.

PART I:

Step 1 Factor the denominator $g(x)$ completely into linear factors of the form $(ax + b)^m$ and quadratic factors of the form $(ax^2 + bx + c)^n$ that are not further factorible over the integers (irreducible over the integers).

Step 2 Set up the form for the decomposition. That is, write the original rational expression $\dfrac{f(x)}{g(x)}$ as a sum of simpler fractions using these guidelines. Note that $A_1, A_2, \ldots, A_m, B_1, B_2, \ldots, B_n$, and $C_1, C_2, \ldots, C_n$ are constants.

- **Linear factors of $g(x)$:** For each linear factor of $g(x)$, the partial fraction decomposition must include the sum:

$$\frac{A_1}{(ax + b)^1} + \frac{A_2}{(ax + b)^2} + \cdots + \frac{A_m}{(ax + b)^m}$$

- **Quadratic factors of $g(x)$:** For each quadratic factor of $g(x)$, the partial fraction decomposition must include the sum:

$$\frac{B_1x + C_1}{(ax^2 + bx + c)^1} + \frac{B_2x + C_2}{(ax^2 + bx + c)^2} + \cdots + \frac{B_nx + C_n}{(ax^2 + bx + c)^n}$$

PART II:

Step 3 With the form of the partial fraction decomposition set up, multiply both sides of the equation by the LCD to clear fractions.

Step 4 Using the equation from step 3, set up a system of linear equations by equating the constant terms and equating the coefficients of like powers of x.

Step 5 Solve the system of equations from step 4 and substitute the solutions to the system into the partial fraction decomposition.

In Examples 1 and 2, we focus on setting up the proper form for a partial fraction decomposition (Part I). In each example, note that the factors of the denominator will fall into one of the following categories:

$$\text{Linear factors:} \quad ax + b$$

$$\text{Repeated linear factors:} \quad (ax + b)^m \ (m \geq 2, \text{ an integer})$$

$$\text{Quadratic factors (irreducible over the integers):} \quad ax^2 + bx + c$$

$$\text{Repeated quadratic factors (irreducible over the integers):} \quad (ax^2 + bx + c)^n \ (n \geq 2, \text{ an integer})$$

EXAMPLE 1   **Setting Up the Form for a Partial Fraction Decomposition**

Set up the form for the partial fraction decomposition for the given rational expressions.

a. $\dfrac{4x - 15}{(2x + 3)(x - 2)}$ **b.** $\dfrac{4x^2 + 10x + 9}{x^3 + 6x^2 + 9x}$

Solution:

FOR REVIEW

Recall that a perfect square trinomial $a^2 + 2ab + b^2$ factors as $(a + b)^2$. The expression $x^2 + 6x + 9$ can be written as $(x)^2 + 2(x)(3) + (3)^2$ and factors as $(x + 3)^2$. See Section 2.4.

a. $\dfrac{4x - 15}{(2x + 3)^1(x - 2)^1} = \dfrac{A}{(2x + 3)^1} + \dfrac{B}{(x - 2)^1}$

The denominator has distinct linear factors of the form $(2x + 3)^1$ and $(x - 2)^1$. Since each linear factor is raised to the first power, only one fraction is needed for each factor.

b. $\dfrac{4x^2 + 10x + 9}{x^3 + 6x^2 + 9x} = \dfrac{4x^2 + 10x + 9}{x(x^2 + 6x + 9)}$

Factor the denominator completely.

The denominator has a linear factor of x^1 and a *repeated* linear factor $(x + 3)^2$.

$= \dfrac{4x^2 + 10x + 9}{x^1(x + 3)^2} = \dfrac{A}{x^1} + \dfrac{B}{(x + 3)^1} + \dfrac{C}{(x + 3)^2}$

For a repeated factor that occurs m times, one fraction must be given for each power less than or equal to m.

> Include one fraction with $(x + 3)$ raised to each positive integer up to and including 2.

Skill Practice 1 Set up the form for the partial fraction decomposition for the given rational expressions.

a. $\dfrac{-x + 18}{(3x + 1)(x + 4)}$ **b.** $\dfrac{-x^2 + 3x + 8}{x^3 + 4x^2 + 4x}$

Answers

1. a. $\dfrac{A}{3x + 1} + \dfrac{B}{x + 4}$

b. $\dfrac{A}{x} + \dfrac{B}{x + 2} + \dfrac{C}{(x + 2)^2}$

In Example 2, we practice setting up the form for the partial decomposition of a rational expression that contains irreducible quadratic factors in the denominator.

EXAMPLE 2 | **Setting Up the Form for a Partial Fraction Decomposition**

Set up the form for the partial fraction decomposition for the given rational expressions.

a. $\dfrac{2x^2 - 3x + 4}{x^3 + 4x}$ **b.** $\dfrac{3x^2 + 8x + 14}{(x^2 + 2x + 5)^2}$

Solution:

FOR REVIEW

Recall that a sum of squares such as $x^2 + 4$ does not factor over the real numbers. See Section 2.5.

a. $\dfrac{2x^2 - 3x + 4}{x^3 + 4x} = \dfrac{2x^2 - 3x + 4}{x(x^2 + 4)}$

Factor the denominator completely. The denominator has one linear factor x^1 and one irreducible quadratic factor $(x^2 + 4)^1$.

$= \dfrac{2x^2 - 3x + 4}{x^1(x^2 + 4)^1} = \dfrac{A}{x^1} + \dfrac{Bx + C}{(x^2 + 4)^1}$

Since each factor is raised to the first power, only one fraction is needed for each factor.

TIP For a first-degree (linear) denominator, the numerator is constant (degree 0). For a second-degree (quadratic) denominator, the numerator is linear (degree 1).

$= \dfrac{2x^2 - 3x + 4}{x(x^2 + 4)} = \dfrac{A}{x} + \dfrac{Bx + C}{(x^2 + 4)}$

b. $\dfrac{3x^2 + 8x + 14}{(x^2 + 2x + 5)^2}$

The quadratic factor $x^2 + 2x + 5$ does not factor further over the integers.

$= \dfrac{Ax + B}{(x^2 + 2x + 5)^1} + \dfrac{Cx + D}{(x^2 + 2x + 5)^2}$

The factor $(x^2 + 2x + 5)$ appears to the *second* power in the denominator. Therefore, in the partial fraction composition, one fraction must have $(x^2 + 2x + 5)^1$ in the denominator, and one fraction must have $(x^2 + 2x + 5)^2$ in the denominator.

Skill Practice 2 Set up the form for the partial fraction decomposition for the given rational expressions.

a. $\dfrac{7x^2 + 2x + 12}{x^3 + 3x}$ **b.** $\dfrac{-3x^2 - 5x - 19}{(x^2 + 3x + 6)^2}$

---

2. Decompose $\dfrac{f(x)}{g(x)}$, Where $g(x)$ Is a Product of Linear Factors

In Example 3, we find the partial fraction decomposition of a rational expression in which the denominator is a product of distinct linear factors.

EXAMPLE 3 | **Decomposing $\dfrac{f(x)}{g(x)}$, Where $g(x)$ Has Distinct Linear Factors**

Find the partial fraction decomposition. $\dfrac{4x - 15}{(2x + 3)(x - 2)}$

Answers

2. a. $\dfrac{A}{x} + \dfrac{Bx + C}{x^2 + 3}$

b. $\dfrac{Ax + B}{x^2 + 3x + 6} + \dfrac{Cx + D}{(x^2 + 3x + 6)^2}$

Solution:

$$\frac{4x - 15}{(2x + 3)(x - 2)} = \frac{A}{2x + 3} + \frac{B}{x - 2} \qquad \text{From Example 1(a), we have the form for the partial fraction decomposition.}$$

$$(2x + 3)(x - 2)\left[\frac{4x - 15}{(2x + 3)(x - 2)}\right] = (2x + 3)(x - 2)\left[\frac{A}{2x + 3} + \frac{B}{x - 2}\right]$$

To solve for A and B, first multiply both sides by the LCD to clear fractions.

$$4x - 15 = A(x - 2) + B(2x + 3) \qquad \text{Apply the distributive property.}$$

$$4x - 15 = Ax - 2A + 2Bx + 3B$$

$$4x - 15 = (A + 2B)x + (-2A + 3B) \qquad \text{Combine like terms.}$$

$\boxed{1}\quad 4 = \quad A + 2B \qquad$ Equate the x term coefficients.

$\boxed{2}\; -15 = -2A + 3B \qquad$ Equate the constant terms.

Solve the system of linear equations by using the substitution method or addition method.

$$\boxed{1}\quad 4 = \quad A + 2B \xrightarrow{\text{Multiply by 2.}} \quad 2A + 4B = \quad 8$$
$$\boxed{2}\; -15 = -2A + 3B \qquad\qquad\qquad \underline{-2A + 3B = -15}$$
$$7B = -7$$
$$B = -1$$

$$\boxed{1}\quad 4 = A + 2B$$
$$4 = A + 2(-1)$$
$A = 6$ and $B = -1$. $\qquad\qquad A = 6$

$$\frac{4x - 15}{(2x + 3)(x - 2)} = \frac{A}{2x + 3} + \frac{B}{x - 2} \qquad \text{Substitute } A = 6 \text{ and } B = -1 \text{ into the partial fraction decomposition.}$$

$$\frac{4x - 15}{(2x + 3)(x - 2)} = \frac{6}{2x + 3} + \frac{-1}{x - 2} \qquad \text{or equivalently} \qquad \frac{6}{2x + 3} - \frac{1}{x - 2}$$

Skill Practice 3   Find the partial fraction decomposition. $\dfrac{-x + 18}{(3x + 1)(x + 4)}$

FOR REVIEW

Always remember that the result of a partial fraction decomposition can be checked by adding the partial fractions and verifying that the sum equals the original rational expression. See Section 4.2.

To verify the result of Example 3, we can add the rational expressions.

$$\frac{6}{2x + 3} + \frac{-1}{x - 2} = \frac{6(x - 2)}{(2x + 3)(x - 2)} + \frac{-1(2x + 3)}{(x - 2)(2x + 3)}$$

$$= \frac{6(x - 2) - 1(2x + 3)}{(2x + 3)(x - 2)}$$

$$= \frac{4x - 15}{(2x + 3)(x - 2)} \; \checkmark$$

In Example 4, we perform partial fraction decomposition with a rational expression that has repeated linear factors in the denominator.

Answer

3. $\dfrac{5}{3x + 1} + \dfrac{-2}{x + 4}$

EXAMPLE 4   Decomposing $\dfrac{f(x)}{g(x)}$, Where $g(x)$ Has Repeated Linear Factors

Find the partial fraction decomposition. $\dfrac{4x^2 + 10x + 9}{x^3 + 6x^2 + 9x}$

Solution:

$$\frac{4x^2 + 10x + 9}{x(x + 3)^2} = \frac{A}{x} + \frac{B}{(x + 3)^1} + \frac{C}{(x + 3)^2}$$

From Example 1(b), we have the form for the partial fraction decomposition.

$$x(x + 3)^2 \left[\frac{4x^2 + 10x + 9}{x(x + 3)^2}\right] = x(x + 3)^2 \left[\frac{A}{x} + \frac{B}{(x + 3)^1} + \frac{C}{(x + 3)^2}\right]$$

$$4x^2 + 10x + 9 = A(x + 3)^2 + Bx(x + 3) + Cx$$
$$4x^2 + 10x + 9 = A(x^2 + 6x + 9) + Bx^2 + 3Bx + Cx$$
$$4x^2 + 10x + 9 = Ax^2 + 6Ax + 9A + Bx^2 + 3Bx + Cx$$
$$4x^2 + 10x + 9 = (A + B)x^2 + (6A + 3B + C)x + 9A$$

$$
\begin{aligned}
4 &= A + B & &\text{Equate the } x^2 \text{ term coefficients.}\\
10 &= 6A + 3B + C & &\text{Equate the } x \text{ term coefficients.}\\
9 &= 9A & &\text{Equate the constant terms.}\\
A = 1, \; B &= 3, \text{ and } C = -5 & &\text{Solve the system of linear equations.}
\end{aligned}
$$

$$\frac{4x^2 + 10x + 9}{x(x + 3)^2} = \frac{A}{x} + \frac{B}{(x + 3)^1} + \frac{C}{(x + 3)^2}$$

Substitute $A = 1$, $B = 3$, and $C = -5$ into the partial fraction decomposition.

$$\frac{4x^2 + 10x + 9}{x(x + 3)^2} = \frac{1}{x} + \frac{3}{(x + 3)^1} + \frac{-5}{(x + 3)^2} = \frac{1}{x} + \frac{3}{x + 3} - \frac{5}{(x + 3)^2}$$

TIP   Consider the system:

$$4 = A + B$$
$$10 = 6A + 3B + C$$
$$9 = 9A$$

From the third equation $A = 1$.
   Substituting 1 for A in the first equation gives us $B = 3$.
   Substituting 1 for A and 3 for B in the second equation gives us

$$10 = 6(1) + 3(3) + C$$
$$10 = 6 + 9 + C$$
$$-5 = C$$

Skill Practice 4   Find the partial fraction decomposition. $\dfrac{-x^2 + 3x + 8}{x^3 + 4x^2 + 4x}$

3. Decompose $\dfrac{f(x)}{g(x)}$, Where $g(x)$ Has Irreducible Quadratic Factors

We now turn our attention to performing partial fraction decomposition where the denominator of a rational expression contains quadratic factors irreducible over the integers. In Example 5, we also address the situation in which the given rational expression is an **improper rational expression**; that is, the degree of the numerator is greater than or equal to the degree of the denominator. In such a case, we use long division to write the expression in the form:

(polynomial) + (proper rational expression)

where a **proper rational expression** is one in which the degree of the numerator is less than the degree of the denominator.

Answer

4. $\dfrac{2}{x} + \dfrac{-3}{x + 2} + \dfrac{1}{(x + 2)^2}$

EXAMPLE 5   Decomposing $\dfrac{f(x)}{g(x)}$, Where $g(x)$ Has an Irreducible Quadratic Factor

Find the partial fraction decomposition. $\dfrac{x^4 + 3x^3 + 6x^2 + 9x + 4}{x^3 + 4x}$

Solution:

First note that the degree of the numerator is not less than the degree of the denominator. Therefore, perform long division first.

$$\frac{x^4 + 3x^3 + 6x^2 + 9x + 4}{x^3 + 4x} \quad \xrightarrow{\text{Long division}} \quad$$

$$\begin{array}{r} x + 3 \\ x^3 + 4x\,\overline{)\,x^4 + 3x^3 + 6x^2 + 9x + 4} \\ \underline{-(x^4 \qquad\; + 4x^2)} \\ 3x^3 + 2x^2 + 9x \\ \underline{-(3x^3 \qquad + 12x)} \\ 2x^2 - 3x + 4 \end{array}$$

$$= x + 3 + \frac{2x^2 - 3x + 4}{x^3 + 4x} \quad \xleftarrow{\text{Equivalent form}}$$

$$= x + 3 + \overbrace{}^{\text{polynomial}} \overbrace{\frac{2x^2 - 3x + 4}{x(x^2 + 4)}}^{\substack{\text{proper rational} \\ \text{expression}}} \qquad \text{Factor the denominator.}$$

$$\frac{2x^2 - 3x + 4}{x(x^2 + 4)} = \frac{A}{x} + \frac{Bx + C}{(x^2 + 4)}$$

Perform partial fraction decomposition on the proper fraction. From Example 2(a), we have the form for the partial fraction decomposition.

$$x(x^2 + 4)\left[\frac{2x^2 - 3x + 4}{x(x^2 + 4)}\right] = x(x^2 + 4)\left[\frac{A}{x} + \frac{Bx + C}{(x^2 + 4)}\right]$$

To solve for A, B, and C, multiply both sides by the LCD to clear fractions.

$$2x^2 - 3x + 4 = A(x^2 + 4) + (Bx + C)x \qquad \text{Apply the distributive property.}$$
$$2x^2 - 3x + 4 = Ax^2 + 4A + Bx^2 + Cx$$
$$2x^2 - 3x + 4 = (A + B)x^2 + Cx + 4A \qquad \text{Combine like terms.}$$

$$\left.\begin{array}{r} 2 = A + B \\ -3 = C \\ 4 = 4A \end{array}\right\} \begin{array}{l} A = 1,\ B = 1, \\ \text{and } C = -3 \end{array}$$

Equate the x^2 term coefficients.

Equate the x term coefficients.

Equate the constant terms.

Solve the system of linear equations.

$$\frac{2x^2 - 3x + 4}{x(x^2 + 4)} = \frac{A}{x} + \frac{Bx + C}{(x^2 + 4)}$$

Substitute $A = 1$, $B = 1$, and $C = -3$.

$$\frac{2x^2 - 3x + 4}{x(x^2 + 4)} = \frac{1}{x} + \frac{1x + (-3)}{x^2 + 4} \quad \text{or} \quad \frac{1}{x} + \frac{x - 3}{x^2 + 4}$$

Therefore, $\dfrac{x^4 + 3x^3 + 6x^2 + 9x + 4}{x^3 + 4x} = x + 3 + \dfrac{1}{x} + \dfrac{x - 3}{x^2 + 4}$.

TIP   Consider the system:

$$\begin{array}{r} 2 = A + B \\ -3 = C \\ 4 = 4A \end{array}$$

From the second equation, $C = -3$.

From the third equation, $A = 1$.

Substituting 1 for A in the first equation gives us $B = 1$.

Answer

5. $x + 2 + \dfrac{4}{x} + \dfrac{3x + 2}{x^2 + 3}$

Skill Practice 5   Find the partial fraction decomposition.

$$\frac{x^4 + 2x^3 + 10x^2 + 8x + 12}{x^3 + 3x}$$

In Example 6, we demonstrate the case in which a rational expression contains a repeated quadratic factor.

EXAMPLE 6 Decomposing $\dfrac{f(x)}{g(x)}$, Where $g(x)$ Has a Repeated Irreducible Quadratic Factor

Find the partial fraction decomposition. $\dfrac{3x^2 + 8x + 14}{(x^2 + 2x + 5)^2}$

Solution:

$\dfrac{3x^2 + 8x + 14}{(x^2 + 2x + 5)^2} = \dfrac{Ax + B}{(x^2 + 2x + 5)^1} + \dfrac{Cx + D}{(x^2 + 2x + 5)^2}$ From Example 2(b), we have the form for the partial fraction decomposition.

To solve for A, B, C, and D, multiply both sides by the LCD to clear fractions.

$(x^2 + 2x + 5)^2 \left[\dfrac{3x^2 + 8x + 14}{(x^2 + 2x + 5)^2} \right] = (x^2 + 2x + 5)^2 \left[\dfrac{Ax + B}{x^2 + 2x + 5} + \dfrac{Cx + D}{(x^2 + 2x + 5)^2} \right]$

$3x^2 + 8x + 14 = (Ax + B)(x^2 + 2x + 5) + (Cx + D)$

$3x^2 + 8x + 14 = Ax^3 + 2Ax^2 + 5Ax + Bx^2 + 2Bx + 5B + Cx + D$

$3x^2 + 8x + 14 = Ax^3 + (2A + B)x^2 + (5A + 2B + C)x + 5B + D$ Combine like terms.

> **TIP** Consider the system:
>
> $0 = A$
> $3 = 2A + B$
> $8 = 5A + 2B + C$
> $14 = 5B + D$
>
> From the first equation, $A = 0$, and substituting 0 for A in the second equation gives us $B = 3$.
>
> Substituting 0 for A and 3 for B in the third equation gives us $C = 2$.
>
> Substituting 3 for B in the fourth equation gives us $D = -1$.

$0 = A$ Equate the x^3 term coefficients.

$3 = 2A + B$ Equate the x^2 term coefficients.

$8 = 5A + 2B + C$ Equate the x term coefficients.

$14 = 5B + D$ Equate the constant terms.

$A = 0$, $B = 3$, $C = 2$, and $D = -1$ Solve the system of linear equations.

$\dfrac{3x^2 + 8x + 14}{(x^2 + 2x + 5)^2} = \dfrac{Ax + B}{x^2 + 2x + 5} + \dfrac{Cx + D}{(x^2 + 2x + 5)^2}$ Substitute $A = 0$, $B = 3$, $C = 2$, and $D = -1$ into the partial fraction decomposition.

$\dfrac{3x^2 + 8x + 14}{(x^2 + 2x + 5)^2} = \dfrac{(0)x + (3)}{x^2 + 2x + 5} + \dfrac{(2)x + (-1)}{(x^2 + 2x + 5)^2}$ or $\dfrac{3}{x^2 + 2x + 5} + \dfrac{2x - 1}{(x^2 + 2x + 5)^2}$

> **Answer**
>
> **6.** $\dfrac{-3}{x^2 + 3x + 6} + \dfrac{4x - 1}{(x^2 + 3x + 6)^2}$

Skill Practice 6 Find the partial fraction decomposition. $\dfrac{-3x^2 - 5x - 19}{(x^2 + 3x + 6)^2}$

SECTION 10.4   Practice Exercises

Prerequisite Review

For Exercises R.1–R.4, solve the system. (See Sections 10.2 and 10.3 for review.)

R.1. $5A - 4B = -7$
$ 3A = B$

R.2. $2A + 7B = -16$
$ 2A = B$

R.3. $A - 2B = -5$
$ B + 5C = -18$
$ A + C = -5$

R.4. $A = 2B$
$ B + C = 1$
$ 3A + 2B + C = 15$

For Exercises R.5–R.12, factor completely. (See Section 2.4 and Section 2.5 for review.)

R.5. $x^4 + 14x^3 + 45x^2$ **R.6.** $x^5 - 10x^4 + 24x^3$ **R.7.** $2m^2 + m - 3$ **R.8.** $3n^2 + 2n - 1$

R.9. $64u^2 + 80u + 25$ **R.10.** $9t^2 - 60t + 100$ **R.11.** $p^4 - 16$ **R.12.** $q^4 - 625$

For Exercises R.13–R.16, add or subtract as indicated. (See Section 4.2 for review.)

R.13. $\dfrac{2}{x+4} - \dfrac{6}{x}$ **R.14.** $\dfrac{5}{p-3} - \dfrac{4}{p}$

R.15. $\dfrac{6}{y} + \dfrac{3}{y+2} - \dfrac{5}{y^2}$ **R.16.** $\dfrac{2}{t^2} + \dfrac{5}{t-1} - \dfrac{4}{t}$

Concept Connections

1. The process of decomposing a rational expression into two or more simpler fractions is called partial _____.

2. When setting up a partial fraction decomposition, if a fraction has a linear denominator, then the numerator should be (constant/linear). That is, should the numerator be set up as A or $Ax + B$?

3. When setting up a partial fraction decomposition, if the denominator of a fraction is a quadratic polynomial irreducible over the integers, then the numerator should be (constant/linear). That is, should the numerator be set up as A or $Ax + B$?

4. In what situation should long division be used before attempting to decompose a rational expression into partial fractions?

Objective 1: Set Up a Partial Fraction Decomposition

For Exercises 5–20, set up the form for the partial fraction decomposition. Do not solve for A, B, C, and so on.
(**See Examples 1–2**)

5. $\dfrac{-x - 37}{(x+4)(2x-3)}$ **6.** $\dfrac{20x - 4}{(x-5)(3x+1)}$ **7.** $\dfrac{8x - 10}{x^2 - 2x}$

8. $\dfrac{y - 12}{y^2 + 3y}$ **9.** $\dfrac{6w - 7}{w^2 + w - 6}$ **10.** $\dfrac{-10t - 11}{t^2 + 5t - 6}$

11. $\dfrac{x^2 + 26x + 100}{x^3 + 10x^2 + 25x}$ **12.** $\dfrac{-3x^2 + 2x + 8}{x^3 + 4x^2 + 4x}$ **13.** $\dfrac{13x^2 + 2x + 45}{2x^3 + 18x}$

14. $\dfrac{17x^2 - 7x + 18}{7x^3 + 42x}$ **15.** $\dfrac{2x^3 - x^2 + 13x - 5}{x^4 + 10x^2 + 25}$ **16.** $\dfrac{3x^3 - 4x^2 + 11x - 12}{x^4 + 6x^2 + 9}$

17. $\dfrac{5x^2 - 4x + 8}{(x-4)(x^2 + x + 4)}$ **18.** $\dfrac{x^2 + 15x - 6}{(x+6)(x^2 + 2x + 6)}$ **19.** $\dfrac{2x^5 + 3x^3 + 4x^2 + 5}{x(x+2)^3(x^2 + 2x + 7)^2}$

20. $\dfrac{6x^4 - 5x^3 + 2x^2 - 5}{(x-3)(2x+9)^2(x^2 + 1)^2}$

Objectives 2 and 3: Decompose $\dfrac{f(x)}{g(x)}$ into Partial Fractions

For Exercises 21–42, find the partial fraction decomposition. (**See Examples 3–6**)

21. $\dfrac{-x - 37}{(x+4)(2x-3)}$ **22.** $\dfrac{20x - 4}{(x-5)(3x+1)}$ **23.** $\dfrac{8x - 10}{x^2 - 2x}$

24. $\dfrac{y - 12}{y^2 + 3y}$ **25.** $\dfrac{6w - 7}{w^2 + w - 6}$ **26.** $\dfrac{-10t - 11}{t^2 + 5t - 6}$

27. $\dfrac{x^2 + 26x + 100}{x^3 + 10x^2 + 25x}$ **28.** $\dfrac{-3x^2 + 2x + 8}{x^3 + 4x^2 + 4x}$ **29.** $\dfrac{13x^2 + 2x + 45}{2x^3 + 18x}$

30. $\dfrac{17x^2 - 7x + 18}{7x^3 + 42x}$

31. $\dfrac{x^4 - 3x^3 + 13x^2 - 28x + 28}{x^3 + 7x}$

32. $\dfrac{x^4 - 4x^3 + 11x^2 - 13x + 12}{x^3 + 2x}$

33. $\dfrac{2x^3 - x^2 + 13x - 5}{x^4 + 10x^2 + 25}$

34. $\dfrac{3x^3 - 4x^2 + 11x - 12}{x^4 + 6x^2 + 9}$

35. $\dfrac{5x^2 - 4x + 8}{(x - 4)(x^2 + x + 4)}$

36. $\dfrac{x^2 + 15x - 6}{(x + 6)(x^2 + 2x + 6)}$

37. $\dfrac{4x^3 - 4x^2 + 11x - 7}{x^4 + 5x^2 + 6}$

38. $\dfrac{3x^3 - 4x^2 + 6x - 7}{x^4 + 5x^2 + 4}$

39. $\dfrac{2x^3 - 11x^2 - 4x + 24}{x^2 - 3x - 10}$

40. $\dfrac{3x^3 + 11x^2 + x + 10}{x^2 + 3x - 4}$

41. $\dfrac{3x^3 + 2x^2 - x - 5}{x^2 + 2x + 1}$

42. $\dfrac{2x^3 - 17x^2 + 54x - 68}{x^2 - 6x + 9}$

43. a. Factor. $x^3 - x^2 - 21x + 45$
(*Hint*: Use the rational zero theorem.)

    b. Find the partial fraction decomposition for
$$\frac{-3x^2 + 35x - 70}{x^3 - x^2 - 21x + 45}.$$

44. a. Factor. $x^3 + 2x^2 - 7x + 4$

    b. Find the partial fraction decomposition for
$$\frac{10x^2 + 17x - 17}{x^3 + 2x^2 - 7x + 4}.$$

45. a. Factor. $x^3 + 6x^2 + 12x + 8$

    b. Find the partial fraction decomposition for
$$\frac{3x^2 + 8x + 5}{x^3 + 6x^2 + 12x + 8}.$$

46. a. Factor. $x^3 - 9x^2 + 27x - 27$

    b. Find the partial fraction decomposition for
$$\frac{2x^2 - 17x + 37}{x^3 - 9x^2 + 27x - 27}.$$

Write About It

47. Write an informal explanation of partial fraction decomposition.

48. Suppose that a proper rational expression has a single repeated linear factor $(ax + b)^3$ in the denominator. Explain how to set up the partial fraction decomposition.

49. What is meant by a *proper* rational expression?

50. Given an improper rational expression, what must be done first before the technique of partial fraction decomposition may be performed?

Expanding Your Skills

51. a. Determine the partial fraction decomposition for $\dfrac{2}{n(n + 2)}$.

    b. Use the partial fraction decomposition for $\dfrac{2}{n(n + 2)}$ to rewrite the infinite sum
$$\frac{2}{1(3)} + \frac{2}{2(4)} + \frac{2}{3(5)} + \frac{2}{4(6)} + \frac{2}{5(7)} \cdots$$

    c. Determine the value of $\dfrac{1}{n + 2}$ as $n \to \infty$.

    d. Find the value of the sum from part (b).

52. a. Determine the partial fraction decomposition for $\dfrac{3}{n(n + 3)}$.

    b. Use the partial fraction decomposition for $\dfrac{3}{n(n + 3)}$ to rewrite the infinite sum
$$\frac{3}{1(4)} + \frac{3}{2(5)} + \frac{3}{3(6)} + \frac{3}{4(7)} + \frac{3}{5(8)} \cdots$$

    c. Determine the value of $\dfrac{1}{n + 3}$ as $n \to \infty$.

    d. Find the value of the sum from part (b).

For Exercises 53–54, find the partial fraction decomposition. Assume that a and b are nonzero constants.

53. $\dfrac{1}{x(a + bx)}$

54. $\dfrac{1}{a^2 - x^2}$

For Exercises 55–56, find the partial fraction decomposition for the given expression. [*Hint*: Use the substitution $u = e^x$ and recall that $e^{2x} = (e^x)^2$.]

55. $\dfrac{5e^x + 7}{e^{2x} + 3e^x + 2}$

56. $\dfrac{-3e^x - 22}{e^{2x} + 3e^x - 4}$

SECTION 10.5 Systems of Nonlinear Equations in Two Variables

1. Solve Nonlinear Systems of Equations by the Substitution Method

The attending physician in an emergency room treats an unconscious patient suspected of a drug overdose. The physician needs to know the concentration of the drug in the bloodstream at the time the drug was taken to determine the extent of damage to the kidneys. The patient's family does not know the original amount of the drug taken, but believes that he took the drug by injection 3 hr before arriving at the hospital. Blood work at the time of arrival

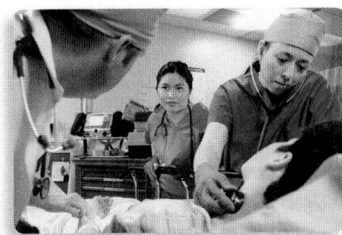

iStock/XiXinXing/Getty Images

($t = 3$ hr after the patient had taken the drug) showed that the drug concentration in the bloodstream was 0.69 μg/dL. One hour later ($t = 4$ hr), the level had dropped to 0.655 μg/dL.

The physician can solve the following system of nonlinear equations to determine the concentration of the drug in the bloodstream at the time of injection. The value A_0 represents the initial concentration of the drug, and the value k is related to the rate at which the kidneys can remove the drug.

$$0.69 = A_0 e^{-3k}$$
$$0.655 = A_0 e^{-4k} \qquad \text{The solution to this problem is discussed in Exercise 59.}$$

A **nonlinear system of equations** is a system in which one or more equations is nonlinear. For example:

$$\begin{array}{ll} -x + 7y = 50 & \text{Second equation} \\ x^2 + y^2 = 100 & \text{nonlinear} \end{array} \qquad \left.\begin{array}{l} 2x^2 + y^2 = 17 \\ x^2 + 2y^2 = 22 \end{array}\right\} \quad \begin{array}{l} \text{Both equations} \\ \text{nonlinear} \end{array}$$

A **solution** to a nonlinear system of equations in two variables is an ordered pair with real-valued coordinates that satisfies each equation in the system. Graphically, these are the points of intersection of the graphs of the equations. A nonlinear system of equations may have no solution or one or more solutions. See Figure 10-15 through Figure 10-17.

TIP A nonlinear system of equations may also have infinitely many solutions. In the graph shown, the "wave" pattern extends infinitely far in both directions.

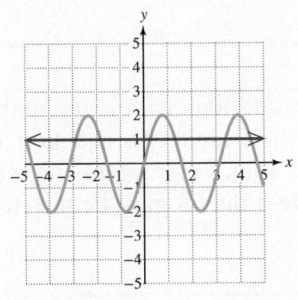

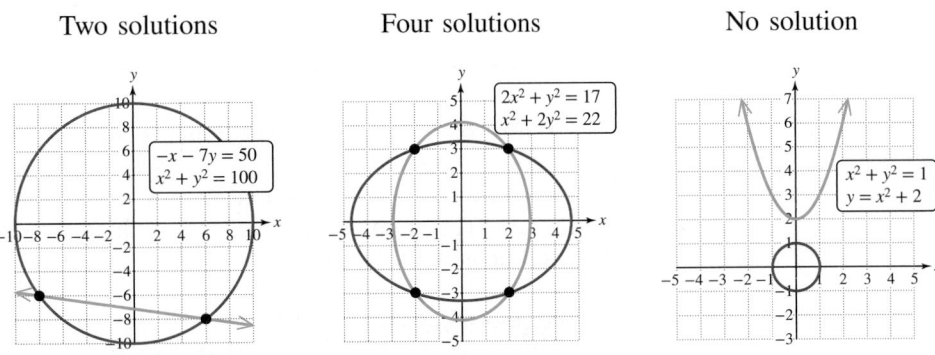

Two solutions	Four solutions	No solution
Figure 10-15	Figure 10-16	Figure 10-17

We will solve nonlinear systems of equations by the substitution method and by the addition method. In Example 1, we begin with the substitution method.

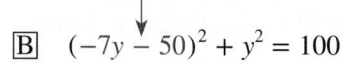

EXAMPLE 1 Solving a System of Nonlinear Equations by Using the Substitution Method

Solve the system by using the substitution method.

$$-x - 7y = 50$$
$$x^2 + y^2 = 100$$

Solution:

A $-x - 7y = 50$
B $x^2 + y^2 = 100$

Equation A is a line and can be written in slope-intercept form as $y = -\frac{1}{7}x - \frac{50}{7}$.

Equation B represents a circle centered at (0, 0) with radius 10.

A sketch of the two equations suggests that the curves intersect at (−8, −6) and (6, −8). See Figure 10-18.

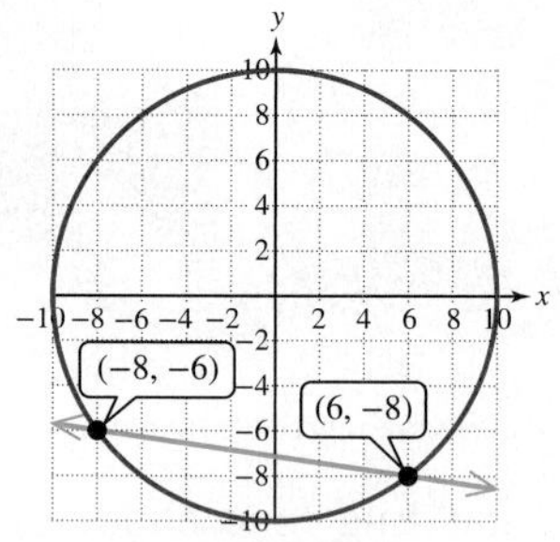

Figure 10-18

A $-x - 7y = 50 \longrightarrow x = -7y - 50$
B $x^2 + y^2 = 100$

To solve the system algebraically by the substitution method, first solve for x or y from either equation.

B $(-7y - 50)^2 + y^2 = 100$

Substitute $x = -7y - 50$ from equation A into equation B.

$$49y^2 + 700y + 2500 + y^2 = 100$$
$$50y^2 + 700y + 2400 = 0 \quad \text{Solve the resulting equation for } y.$$
$$50(y^2 + 14y + 48) = 0 \quad \text{Factor out the GCF of 50.}$$
$$50(y + 6)(y + 8) = 0$$
$$y = -6 \quad \text{or} \quad y = -8$$

FOR REVIEW

The expression $(-7y - 50)^2$ is the square of a binomial. To expand the expression, we have

$(-7y - 50)^2$
$= (-7y)^2 - 2(-7y)(50) + (50)^2$
$= 49y^2 + 700y + 2500$

See Section 2.2 for review.

For each value of y, find the corresponding x value by substituting y into the equation in which x is isolated: $x = -7y - 50$

$y = -6$: $x = -7(-6) - 50 = -8$ The solution is (−8, −6).
$y = -8$: $x = -7(-8) - 50 = 6$ The solution is (6, −8).

Check: (−8, −6)
A $-(-8) - 7(-6) = 50$ ✓
B $(-8)^2 + (-6)^2 = 100$ ✓

Check: (6, −8)
A $-(6) - 7(-8) = 50$ ✓
B $(6)^2 + (-8)^2 = 100$ ✓

The solutions both check in each equation.

The solution set is {(−8, −6), (6, −8)}.

Skill Practice 1 Solve the system by using the substitution method.

$2x + y = 5$
$x^2 + y^2 = 50$

As we solve systems of equations, we will consider only solutions with real coordinates. In Example 2, we have a system of equations in which one equation is $y = 3\sqrt{x - 8}$. The expression $\sqrt{x - 8}$ is a real number for values of x on the interval $[8, \infty)$. Therefore, any ordered pair with an x-coordinate less than 8 must be rejected as a potential solution.

EXAMPLE 2 Solving a System of Nonlinear Equations by Using the Substitution Method

Solve the system by using the substitution method.

$$(x - 5)^2 + y^2 = 25$$
$$y = 3\sqrt{x - 8}$$

Answer

1. {(5, −5), (−1, 7)}

Note that

$(x - 5)^2$
$= (x)^2 - 2(x)(5) + (5)^2$
$= x^2 - 10x + 25$

and

$(3\sqrt{x - 8})^2 = 3^2 \cdot (\sqrt{x - 8})^2$
$\qquad\qquad = 9(x - 8)$

Solution:

$\boxed{A}$ $(x - 5)^2 + y^2 = 25$
$\boxed{B}$ $y = 3\sqrt{x - 8}$

Label the equations. The graphs of the equations are shown in Figure 10-19. The graph suggests that there is only one solution: (9, 3).

$\boxed{A}$ $(x - 5)^2 + (3\sqrt{x - 8})^2 = 25$

$x^2 - 10x + 25 + 9(x - 8) = 25$
$\qquad\qquad x^2 - x - 72 = 0$
$\qquad\quad (x - 9)(x + 8) = 0$
$\qquad\qquad x = 9 \quad \text{or} \quad \cancel{x = -8}$

Substitute $3\sqrt{x - 8}$ for y in equation $\boxed{A}$.

Square each term.

Reject $x = -8$ because $3\sqrt{x - 8}$ is not a real number for $x = -8$.

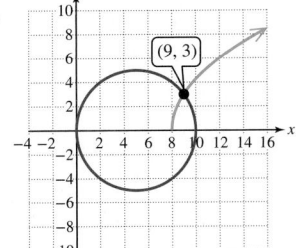

Given $x = 9$, solve for y:

$\boxed{B}$ $y = 3\sqrt{x - 8}$
$\qquad y = 3\sqrt{9 - 8} = 3$

Figure 10-19

The solution is (9, 3) and checks in each original equation.

The solution set is $\{(9, 3)\}$.

Skill Practice 2   Solve the system by using the substitution method.

$x^2 + y^2 = 90$
$\qquad y = \sqrt{x}$

---

2. Solve Nonlinear Systems of Equations by the Addition Method

The substitution method is used most often to solve a system of nonlinear equations. In some situations, however, the addition method is an efficient way to find a solution. Examples 3 and 4 demonstrate that we can eliminate a variable from both equations in a system provided the terms containing the corresponding variables are like terms.

EXAMPLE 3   **Solving a System of Nonlinear Equations by Using the Addition Method**

Solve the system by using the addition method. $2x^2 + y^2 = 17$
$\qquad\qquad\qquad\qquad\qquad\qquad\qquad\qquad\qquad\qquad x^2 + 2y^2 = 22$

Solution:

Using the addition method, the goal is to create opposite coefficients on either the x^2 terms or the y^2 terms. In this case, we have chosen to eliminate the x^2 terms.

$\boxed{A}$ $2x^2 + y^2 = 17$ $\qquad\qquad\qquad\qquad 2x^2 + y^2 = 17$
$\boxed{B}$ $\quad x^2 + 2y^2 = 22$ $\xrightarrow{\text{Multiply by } -2.}$ $-2x^2 - 4y^2 = -44$
$\qquad\qquad\qquad\qquad\qquad\qquad\qquad\qquad\qquad\qquad -3y^2 = -27$
$\qquad\qquad\qquad\qquad\qquad\qquad\qquad\qquad\qquad\qquad\quad y^2 = 9$
$\qquad\qquad\qquad\qquad\qquad\qquad\qquad\qquad\qquad\qquad\quad y = \pm 3$

$y = 3$: $\boxed{B}$ $x^2 + 2(3)^2 = 22$
$\qquad\qquad\qquad\qquad x^2 = 4$
$\qquad\qquad\qquad\qquad x = \pm 2$ The solutions are (2, 3), (−2, 3).

Substitute $y = \pm 3$ into either equation $\boxed{A}$ or $\boxed{B}$ to solve for the corresponding values of x.

Answer
2. $\{(9, 3)\}$

$$y = -3: \quad \boxed{B} \quad x^2 + 2(-3)^2 = 22$$
$$x^2 = 4$$
$$x = \pm 2 \quad \text{The solutions are}$$
$$(2, -3), (-2, -3).$$

The solutions all check in the original equations.

The solution set is $\{(2, 3), (-2, 3), (2, -3), (-2, -3)\}$.

Skill Practice 3 Solve the system by using the addition method.

$$x^2 + y^2 = 17$$
$$x^2 - 2y^2 = -31$$

TECHNOLOGY CONNECTIONS

Solving a System of Nonlinear Equations Using Intersect

The equations in Example 3 each represent a curve called an ellipse. We do not yet know how to graph an ellipse; however, we can graph the curves on a graphing calculator. First solve each equation for y. Enter the resulting functions in the calculator and use the Intersect feature to approximate the points of intersection (Figure 10-20).

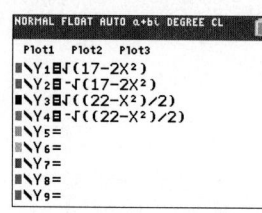

$$2x^2 + y^2 = 17 \xrightarrow{\text{Solve for } y.} y = \pm\sqrt{17 - 2x^2}$$

$$x^2 + 2y^2 = 22 \longrightarrow y = \pm\sqrt{\dfrac{22 - x^2}{2}}$$

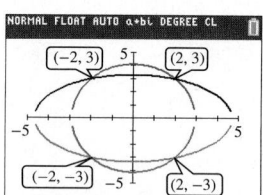

Figure 10-20

TIP The image produced by a graphing calculator in Figure 10-20 does not show the curves touching the x-axis, when indeed they do.

Example 4 illustrates that a nonlinear system of equations may have no solution.

EXAMPLE 4 Solving an Inconsistent System by Using the Addition Method

Solve the system by using the addition method.
$$x^2 + 4y^2 = 4$$
$$x^2 - y^2 = 9$$

TIP Solutions to a system of equations are limited to ordered pairs with real-valued coordinates because we are interested in the points of intersection of the graphs of the equations.

Solution:

$$\begin{array}{l} x^2 + 4y^2 = 4 \\ x^2 - y^2 = 9 \end{array} \xrightarrow[\text{Multiply by } -1.]{} \begin{array}{l} x^2 + 4y^2 = 4 \\ -x^2 + y^2 = -9 \\ \hline 5y^2 = -5 \\ y^2 = -1 \\ y = \pm i \end{array}$$

We use the addition method because like terms are aligned vertically.

The values for y are not real numbers. Therefore, there is no solution to the system of equations over the set of real numbers.

The solution set is $\{\ \}$.

Skill Practice 4 Solve the system by using the addition method.

$$x^2 + y^2 = 16$$
$$4x^2 + 9y^2 = 36$$

Answers

3. $\{(1, 4), (-1, 4), (1, -4), (-1, -4)\}$

4. $\{\ \}$

TECHNOLOGY CONNECTIONS

Solving a Nonlinear System of Equations

The equations in Example 4 represent two curves that have not yet been studied: an ellipse and a hyperbola. However, we can graph the curves on a graphing calculator by solving for y and entering the functions into the calculator.

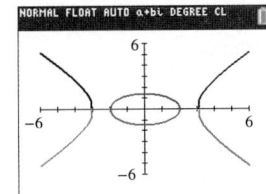

$$x^2 + 4y^2 = 4 \xrightarrow[\text{Solve for } y.]{} y = \pm\sqrt{\frac{4 - x^2}{4}}$$

$$x^2 - y^2 = 9 \xrightarrow{\hspace{2cm}} y = \pm\sqrt{x^2 - 9}$$

From the graph, we see that the curves do not intersect.

3. Use Nonlinear Systems of Equations to Solve Applications

In Example 5, we set up a system of nonlinear equations to model an application involving two independent relationships between two variables.

EXAMPLE 5   Solving an Application of a Nonlinear System

The perimeter of a television screen is 140 in. The area is 1200 in.^{2}.

a. Find the length and width of the screen.

b. Find the length of the diagonal.

Solution:

Let x represent the length of the screen.

Let y represent the width of the screen.

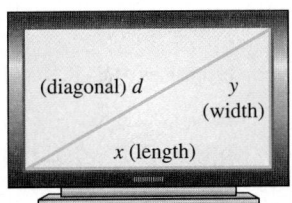

The statement of the problem gives two independent relationships between the length and width of the screen.

The perimeter of a television screen is 140 in. $\longrightarrow$ $2x + 2y = 140$
The area is 1200 in.^{2}. $\longrightarrow$ $xy = 1200$

a. Solve the nonlinear system of equations for x and y.

   $\boxed{A}$ $2x + 2y = 140$

   $\boxed{B}$ $xy = 1200 \xrightarrow[\text{Solve for } y.]{} y = \dfrac{1200}{x}$ Using the substitution method, solve for x or y in either equation.

   $\boxed{A}$ $2x + 2\left(\dfrac{1200}{x}\right) = 140$ Substitute $y = \frac{1200}{x}$ from equation $\boxed{B}$ into equation $\boxed{A}$.

   $2x + \dfrac{2400}{x} = 140$

   $x \cdot \left(2x + \dfrac{2400}{x}\right) = x \cdot (140)$ Multiply both sides by the LCD to clear fractions.

   $2x^2 + 2400 = 140x$ The resulting equation is quadratic.

   $2x^2 - 140x + 2400 = 0$

   $2(x^2 - 70x + 1200) = 0$ Factor the left side.

   $2(x - 40)(x - 30) = 0$

   $x = 40$ or $x = 30$ There are two possible values for the length x.

Substitute $x = 40$ and $x = 30$ into the equation $y = \dfrac{1200}{x}$.

$x = 40$: $\quad y = \dfrac{1200}{40} = 30$ If the length x is 40 in., then the width y is 30 in.

$x = 30$: $\quad y = \dfrac{1200}{30} = 40$ If the length x is 30 in., then the width y is 40 in.

Taking the length to be the longer side, we have that the length is 40 in. and the width is 30 in.

b. $x^2 + y^2 = d^2$ Use the Pythagorean theorem to determine the
$(40)^2 + (30)^2 = d^2$ measure of the diagonal of the screen.
$1600 + 900 = d^2$
$2500 = d^2$
$d = \pm 50$

Excluding the negative solution for d, the diagonal is 50 in.

Answer

5. The length of the rug is 12 ft and the width is 8 ft.

Skill Practice 5 The perimeter of a rectangular rug is 40 ft and the area is 96 ft^2. Find the dimensions of the rug.

SECTION 10.5   Practice Exercises

Prerequisite Review

For Exercises R.1–R.2, perform the indicated operation. (See Section 2.2 for review.)

R.1. $(6x - 5)^2$

R.2. $(7y + 8)^2$

For Exercises R.3–R.8, solve the equation. (See Sections 3.4 and 3.5 for review.)

R.3. $12x^2 + 5x - 2 = 0$

R.4. $10x^2 - 31x - 14 = 0$

R.5. $64t^2 - 16t + 1 = 0$

R.6. $81p^2 + 18p + 1 = 0$

R.7. $y^4 + 5y^2 - 36 = 0$

R.8. $y^4 - 15y^2 - 16 = 0$

For Exercises R.9–R.12, solve the equation. (See Sections 9.5 and 9.6 for review.)

R.9. $\ln x = 17$ (See Section 9.6.)

R.10. $\log x = -5$

R.11. $3^x = 81$ (See Section 9.5.)

R.12. $4^x = \dfrac{1}{256}$

For Exercises R.13–R.18, graph the equation. (See Sections 5.2, 5.4, and 6.1 for review.)

R.13. $2x - y = 4$ (See Section 5.4.)

R.14. $4x + 3y = 9$

R.15. $x^2 + (y + 3)^2 = 9$ (See Section 5.2.)

R.16. $(x + 4)^2 + (y - 1)^2 = 4$

R.17. $y = (x - 2)^2 - 1$ (See Section 6.1.)

R.18. $y = \sqrt{x + 2} + 1$

For Exercises R.19–R.20, let x represent the smaller of two positive numbers and y represent the larger. Translate the English statement into an algebraic equation.

R.19. The difference of the squares of two numbers is 25.

R.20. The ratio of two numbers is 36.

For Exercises R.21–R.22, let x represent the length of a rectangle and y represent the width. Translate the English statement into an algebraic equation.

R.21. The area of the rectangle is 50 m^2.

R.22. The perimeter of the rectangle is 30 m.

Concept Connections

1. A _____ system of equations in two variables is a system in which one or more equations in the system is nonlinear.

2. A solution to a nonlinear system of equations in two variables is an _____ pair with real-valued coordinates that satisfies each equation in the system. Graphically, a solution is a point of _____ of the graphs of the equations.

Objective 1: Solve Nonlinear Systems of Equations by the Substitution Method

For Exercises 3–14,

a. Graph the equations in the system.

b. Solve the system by using the substitution method. (**See Examples 1–2**)

3. $y = x^2 - 2$
$2x - y = 2$

4. $y = -x^2 + 3$
$y - 2x = 0$

5. $x^2 + y^2 = 25$
$x + y = 1$

6. $x^2 + y^2 = 25$
$3y = 4x$

7. $y = \sqrt{x}$
$x^2 + y^2 = 20$

8. $x^2 + y^2 = 10$
$y = \sqrt{x - 2}$

9. $(x + 2)^2 + y^2 = 9$
$y = 2x - 4$

10. $x^2 + (y - 3)^2 = 4$
$y = -x - 4$

11. $y = x^3$
$y = x$

12. $y = \sqrt[3]{x}$
$y = x$

13. $y = -(x - 2)^2 + 5$
$y = 2x + 1$

14. $y = (x + 3)^2 - 1$
$y = 2x + 5$

Objective 2: Solve Nonlinear Systems of Equations by the Addition Method

For Exercises 15–22, solve the system by using the addition method. (**See Examples 3–4**)

15. $2x^2 + 3y^2 = 11$
$x^2 + 4y^2 = 8$

16. $3x^2 + y^2 = 21$
$4x^2 - 2y^2 = -2$

17. $x^2 - xy = 20$
$-2x^2 + 3xy = -44$

18. $4xy + 3y^2 = -9$
$2xy + y^2 = -5$

19. $5x^2 - 2y^2 = 1$
$2x^2 - 3y^2 = -4$

20. $6x^2 + 5y^2 = 38$
$7x^2 - 3y^2 = 9$

21. $x^2 = 1 - y^2$
$9x^2 - 4y^2 = 36$

22. $4x^2 = 4 - y^2$
$16y^2 = 144 + 9x^2$

Mixed Exercises

For Exercises 23–34, solve the system by using any method.

23. $x^2 - 4xy + 4y^2 = 1$
$x + y = 4$

24. $x^2 - 6xy + 9y^2 = 0$
$x - y = 2$

25. $y = x^2 + 4x + 5$
$y = 4x + 5$

26. $y = x^2 - 6x + 9$
$y = -2x + 5$

27. $y = x^2$
$y = \dfrac{1}{x}$

28. $y = \dfrac{1}{x}$
$y = \sqrt{x}$

29. $x^2 + (y - 4)^2 = 25$
$y = -x^2 + 9$

30. $(x - 10)^2 + y^2 = 100$
$x = y^2$

31. $y = -x^2 + 6x - 7$
$y = x^2 - 10x + 23$

32. $y = -x^2 + 6x - 9$
$y = x^2 - 2x - 3$

33. $\dfrac{x^2}{4} + \dfrac{y^2}{16} = 1$
$x^2 + y = 4$

34. $\dfrac{x^2}{4} + y^2 = 1$
$x = -2y^2 + 2$

For Exercises 35–36, use the substitutions $u = \dfrac{1}{x^2}$ and $v = \dfrac{1}{y^2}$ to solve the system of equations.

35. $\dfrac{4}{x^2} - \dfrac{3}{y^2} = -23$

$\dfrac{5}{x^2} + \dfrac{1}{y^2} = 14$

36. $-\dfrac{3}{x^2} + \dfrac{1}{y^2} = 13$

$\dfrac{5}{x^2} - \dfrac{1}{y^2} = -5$

Objective 3: Use Nonlinear Systems of Equations to Solve Applications

37. Find two numbers whose sum is 12 and whose product is 35.

38. Find two numbers whose sum is 9 and whose product is −36.

39. The sum of the squares of two positive numbers is 29 and the difference of the squares of the numbers is 21. Find the numbers.

40. The sum of the squares of two negative numbers is 145 and the difference of the squares of the numbers is 17. Find the numbers.

41. The difference of two positive numbers is 2 and the difference of their squares is 44. Find the numbers.

42. The sum of two numbers is 4 and the difference of their squares is 64. Find the numbers.

43. The ratio of two numbers is 3 to 4 and the sum of their squares is 225. Find the numbers.

44. The ratio of two numbers is 5 to 12 and the sum of their squares is 676. Find the numbers.

45. Find the dimensions of a rectangle whose perimeter is 36 m and whose area is 80 m^2.

46. Find the dimensions of a rectangle whose perimeter is 56 cm and whose area is 192 cm^2.

47. The floor of a rectangular bedroom requires 240 ft^2 of carpeting. Molding is placed around the base of the floor except at two 3-ft doorways. If 58 ft of molding is required around the base of the floor, determine the dimensions of the floor. (**See Example 5**)

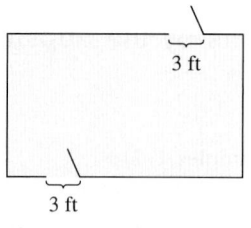

3 ft

3 ft

48. An electronic sign for a grocery store is in the shape of a rectangle. The perimeter of the sign is 72 ft and the area is 320 ft^2. Find the length and width of the sign.

49. A rental truck has a cargo capacity of 288 ft^3. A 10-ft pipe just fits resting diagonally on the floor of the truck. If the cargo space is 6 ft high, find the dimensions of the truck.

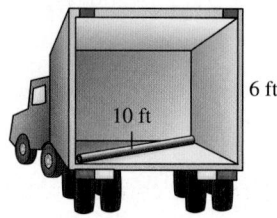

6 ft

10 ft

50. A rectangular window has a 15-yd diagonal and an area of 108 yd^2. Find the dimensions of the window.

51. An aquarium is 16 in. high with volume of 4608 in.^{3} (approximately 20 gal). If the amount of glass used for the bottom and four sides is 1440 in.^{2}, determine the dimensions of the aquarium.

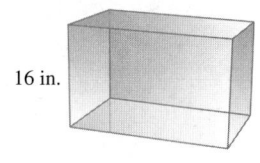

16 in.

52. A closed box is in the shape of a rectangular solid with height 3 m. Its surface area is 268 m^2. If the volume is 240 m^3, find the dimensions of the box.

53. The hypotenuse of a right triangle is $\sqrt{65}$ ft. The sum of the lengths of the legs is 11 ft. Find the lengths of the legs.

54. The hypotenuse of a right triangle is $\sqrt{73}$ in. The sum of the lengths of the legs is 11 in. Find the lengths of the legs.

55. A ball is kicked off the side of a hill at an angle of elevation of 30°. The hill slopes downward 30° from the horizontal. Consider a coordinate system in which the origin is the point on the edge of the hill from which the ball is kicked. The path of the ball and the line of declination of the hill can be approximated by

$$y = -\frac{x^2}{192} + \frac{\sqrt{3}}{3}x \qquad \text{Path of the ball}$$

$$y = -\frac{\sqrt{3}}{3}x \qquad \text{Line of declination of the hill}$$

Solve the system to determine where the ball will hit the ground.

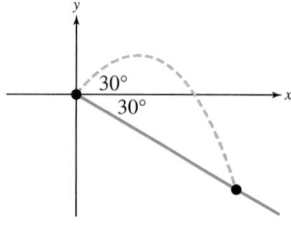

56. A child kicks a rock off the side of a hill at an angle of elevation of 60°. The hill slopes downward 30° from the horizontal. Consider a coordinate system in which the origin is the point on the edge of the hill from which the rock is kicked. The path of the rock and the line of declination of the hill can be approximated by

$$y = -\frac{x^2}{36} + \sqrt{3}x \qquad \text{Path of the rock}$$

$$y = -\frac{\sqrt{3}}{3}x \qquad \text{Line of declination of the hill}$$

Solve the system to determine where the rock will hit the ground.

Write About It

57. What is the difference between a system of linear equations and a system of nonlinear equations?

58. Describe a situation in which the addition method is an efficient technique to solve a system of nonlinear equations.

Expanding Your Skills

59. The attending physician in an emergency room treats an unconscious patient suspected of a drug overdose. The physician does not know the initial concentration A_0 of the drug in the bloodstream at the time of injection. However, the physician knows that after 3 hr, the drug concentration in the blood is 0.69 μg/dL and after 4 hr, the concentration is 0.655 μg/dL. The model $A(t) = A_0e^{-kt}$ represents the drug concentration $A(t)$ (in μg/dL) in the bloodstream t hours after injection. The value of k is a constant related to the rate at which the drug is removed by the body.

  a. Substitute 0.69 for $A(t)$ and 3 for t in the model and write the resulting equation.

  b. Substitute 0.655 for $A(t)$ and 4 for t in the model and write the resulting equation.

  c. Use the system of equations from parts (a) and (b) to solve for k. Round to 3 decimal places.

  d. Use the system of equations from parts (a) and (b) to approximate the initial concentration A_0 (in μg/dL) at the time of injection. Round to 2 decimal places.

  e. Determine the concentration of the drug after 12 hr. Round to 2 decimal places.

60. A patient undergoing a heart scan is given a sample of fluorine-18 (^{18}F). After 4 hr, the radioactivity level in the patient is 44.1 MBq (megabecquerel). After 5 hr, the radioactivity level drops to 30.2 MBq. The radioactivity level $Q(t)$ can be approximated by $Q(t) = Q_0e^{-kt}$, where t is the time in hours after the initial dose Q_0 is administered.

  a. Determine the value of k. Round to 4 decimal places.

  b. Determine the initial dose, Q_0. Round to the nearest whole unit.

  c. Determine the radioactivity level after 12 hr. Round to 1 decimal place.

61. The population $P(t)$ of a culture of bacteria grows exponentially for the first 72 hr according to the model $P(t) = P_0e^{kt}$. The variable t is the time in hours since the culture is started. The population of bacteria is 60,000 after 7 hr. The population grows to 80,000 after 12 hr.

  a. Determine the constant k to 3 decimal places.

  b. Determine the original population P_0. Round to the nearest thousand.

  c. Determine the time required for the population to reach 300,000. Round to the nearest hour.

62. An investment grows exponentially under continuous compounding. After 2 years, the amount in the account is \$7328.70. After 5 years, the amount in the account is \$8774.10. Use the model $A(t) = Pe^{rt}$ to

  a. Find the interest rate r. Round to the nearest percent.

  b. Find the original principal P. Round to the nearest dollar.

  c. Determine the amount of time required for the account to reach a value of \$15,000. Round to the nearest year.

For Exercises 63–64, determine the number of solutions to the system of equations.

63.
$$y = 2^{x+1}$$
$$-1 + \log_2 y = x$$

64. $x^2 - y^2 = 0$
$$|x| = |y|$$

For Exercises 65–70, solve the system.

65. $\log x + 2 \log y = 5$
$2 \log x - \log y = 0$

66. $\log_2 x + 3 \log_2 y = 6$
$\log_2 x - \log_2 y = 2$

67. $2^x + 2^y = 6$
$4^x - 2^y = 14$

68. $3^x - 9^y = 18$
$3^x + 3^y = 30$

69. $(x - 1)^2 + (y + 1)^2 = 5$
$x^2 + (y + 4)^2 = 29$

70. $(x + 3)^2 + (y - 2)^2 = 4$
$(x - 1)^2 + y^2 = 8$

For Exercises 71–72, use substitution to solve the system for the set of ordered triples (x, y, λ) that satisfy the system.

71. $2 = 2\lambda x$
$6 = 2\lambda y$
$x^2 + y^2 = 10$

72. $8 = 4\lambda x$
$2 = 2\lambda y$
$2x^2 + y^2 = 9$

73. Two circles intersect as shown.

  a. Find the points of intersection.
$$x^2 + y^2 = 25$$
$$(x - 4)^2 + (y + 2)^2 = 25$$

  b. Find an equation of the chord common to both circles (shown in black). (*Hint:* A chord is a line segment on the interior of a circle with both endpoints on the circle.)

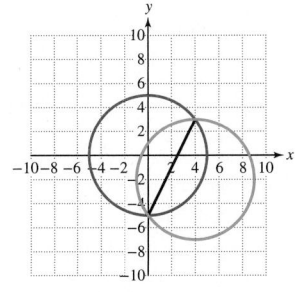

74. The minimum and maximum distances from a point P to a circle are found using the line determined by the given point and the center of the circle. Given the circle defined by $x^2 + y^2 = 9$ and the point $P(4, 5)$,

   a. Find the point on the circle closest to the point $(4, 5)$.

   b. Find the point on the circle farthest from the point $(4, 5)$.

Technology Connections

For Exercises 75–80, use a graphing utility to approximate the solution(s) to the system of equations. Round the coordinates to 3 decimal places.

75. $y = -0.6x + 7$
$y = e^x - 5$

76. $y = -0.7x + 4$
$y = \ln x$

77. $x^2 + y^2 = 40$
$y = -x^2 + 8.5$

78. $x^2 + y^2 = 32$
$y = 0.8x^2 - 9.2$

79. $y = x^2 - 8x + 20$
$y = 4 \log x$

80. $y = 0.2e^x$
$y = -0.6x^2 - 2x - 3$

SECTION 10.6 Inequalities and Systems of Inequalities in Two Variables

OBJECTIVES

1. Solve Linear Inequalities in Two Variables
2. Solve Nonlinear Inequalities in Two Variables
3. Solve Systems of Inequalities in Two Variables

1. Solve Linear Inequalities in Two Variables

Adriana estimates that she has 12 hr of available study time before she takes tests in algebra and biology in back-to-back classes. Suppose that x represents the time she spends studying algebra and y represents the time she spends studying biology. Then the inequality $x + y \leq 12$, where $x \geq 0$ and $y \geq 0$ represents the distribution of time she can allocate studying for each subject.

An inequality of the form $Ax + By < C$, where A and B are not both zero, is called a **linear inequality in two variables**. (Note that the symbols $>$, $\leq$, and $\geq$ can be used in place of $<$ in the definition.) A **solution** to an inequality in two variables is an ordered pair that satisfies the inequality. The set of all such ordered pairs is called the **solution set** to the inequality. Graphically, the solution set is a region in the xy-plane.

iStock/XiXinXing/ Getty Images

To graph the solution set to a linear inequality in two variables, follow these guidelines.

TIP To graph the line in step 1, we can

- Use the slope-intercept form of the equation of the line.
- Graph the x- and y-intercepts.
- Create a table of points.

Graphing a Linear Inequality in Two Variables

Step 1 Graph the related equation. That is, replace the inequality sign with an $=$ sign and graph the line represented by the equation.

- If the inequality is strict (stated with the symbols $<$ or $>$), then draw the line as a dashed line to indicate that the line *is not* part of the solution set.
- If the inequality is stated with the symbols $\leq$ or $\geq$, then draw the line as a solid line to indicate that the line *is* part of the solution set.

Step 2 Choose a test point from either side of the line (not a point on the line itself) and substitute the ordered pair into the inequality.

- If a true statement results, then shade the region (half-plane) from which the test point was taken.
- If a false statement results, then shade the region (half-plane) on the opposite side of the line from which the test point was taken.

EXAMPLE 1   Graphing a Linear Inequality in Two Variables

Graph the solution set. $3x - 2y < 6$

Solution:

$3x - 2y < 6 \xrightarrow[\text{equation}]{\text{related}} 3x - 2y = 6$

Step 1: Graph the related equation $3x - 2y = 6$ using any technique for graphing. In this case, we have chosen to find the x- and y-intercepts.

x-intercept: y-intercept:

$3x - 2(0) = 6$ $3(0) - 2y = 6$

$\quad\quad x = 2$ $\quad\quad y = -3$

The x- and y-intercepts are $(2, 0)$ and $(0, -3)$. Graph the line through the intercepts (Figure 10-21). Because the inequality is strict, draw the line as a dashed line.

Test $(0, 0)$:

$\quad 3x - 2y < 6$

$3(0) - 2(0) \overset{?}{<} 6$

$\quad\quad\quad 0 \overset{?}{<} 6 \checkmark$ true

Step 2: Select a test point either above or below the line and test the ordered pair in the original inequality. In Figure 10-22, we have chosen $(0, 0)$ as a test point.

The test point $(0, 0)$ is a representative point above the line. Since $(0, 0)$ satisfies the original inequality, then it and all other points above the line are solutions.

The solution set is the set of ordered pairs in the region (half-plane) above the line (Figure 10-22).

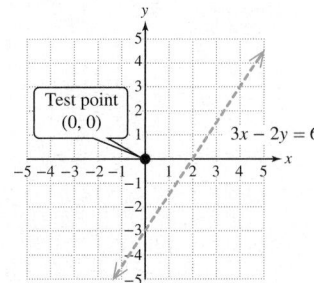

Figure 10-21

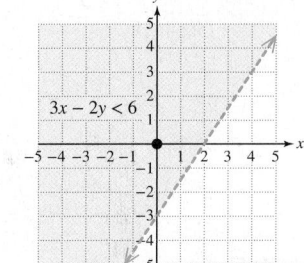

Figure 10-22

TIP   We can also graph the inequality from Example 1 by solving the inequality for y.

$\quad 3x - 2y < 6$

$\quad\quad -2y < -3x + 6$

$\quad\quad\quad y > \tfrac{3}{2}x - 3$

Then shade the half-plane *above* the bounding line because this region contains points with y-coordinates greater than those on the bounding line.

Skill Practice 1   Graph the solution set. $4x - y > 3$

TECHNOLOGY CONNECTIONS

Graphing a System of Inequalities in Two Variables

A graphing utility can be used to graph an inequality in two variables. In most cases, we solve for y first and enter the related equation in the graphing editor. Place the cursor to the left of Y_1 and press **ENTER** two times. This will set the graph style to shade the region above the line ◥. Notice that the calculator image does not differentiate between a solid and dashed bounding line (Figure 10-23).

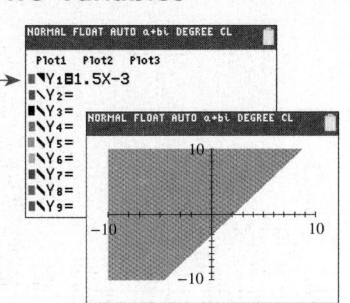

Figure 10-23

Answer

1.

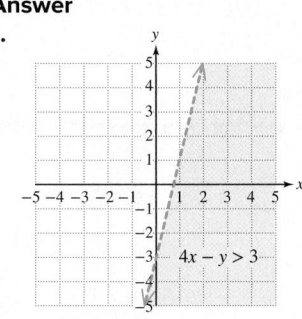

Note: With the cursor placed to the left of Y_1,

- Select the upper right triangle ◥ for inequalities of the form $Y_1 > f(x)$ and $Y_1 \geq f(x)$.
- Select the lower left triangle ◣ for inequalities of the form $Y_1 < f(x)$ and $Y_1 \leq f(x)$.

In Example 2, we graph the solutions to an inequality in which the bounding line passes through the origin.

EXAMPLE 2   Graphing a Linear Inequality

Graph the solution set. $4y \leq 3x$

Solution:

$$4y \leq 3x \longrightarrow 4y = 3x$$
$$\underset{\text{related equation}}{}$$

$$y = \frac{3}{4}x + 0$$

Step 1: Graph the related equation $4y = 3x$. In this case, we have chosen to find the slope-intercept form of the equation and then graph the line using the slope and y-intercept.

Graph the line having y-intercept $(0, 0)$ and slope $\frac{3}{4}$ (Figure 10-24). Because the inequality symbol $\leq$ allows for equality, draw the line as a solid line.

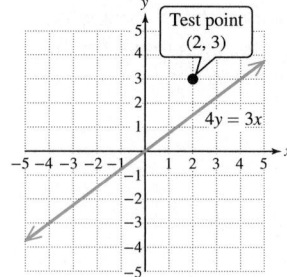

Figure 10-24

TIP As a check, we can select a test point *below* the line, such as (1, −1) and verify that it does indeed satisfy the original inequality.

Test (1, −1):

$4(-1) \overset{?}{\leq} 3(1)$

$-4 \overset{?}{\leq} 3$ ✓ true

Test (2, 3):

$4y \leq 3x$

$4(3) \overset{?}{\leq} 3(2)$

$12 \overset{?}{\leq} 6$ false

Step 2: Select a test point. We have chosen (2, 3).

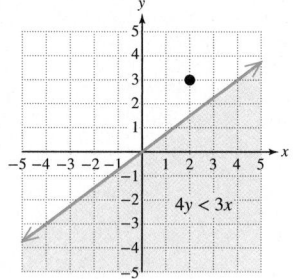

Figure 10-25

The test point (2, 3) is a representative point above the line. Since (2, 3) does *not* satisfy the original inequality, then points on the other side of the line are solutions. Shade below the line.

The solution set is the set of ordered pairs on and below the line (Figure 10-25).

Skill Practice 2   Graph the solution set. $2y \geq 5x$

Answer
2.

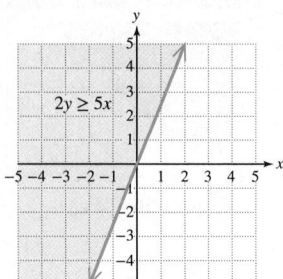

Recall that for a constant k, the equation $x = k$ represents a vertical line in the xy-plane. The inequalities $x < k$ and $x > k$ represent half-planes to the left or right of the vertical line $x = k$ (Figure 10-26). Likewise, $y = k$ represents a horizontal line in the xy-plane. The inequalities $y < k$ and $y > k$ represent half-planes below or above the line $y = k$ (Figure 10-27).

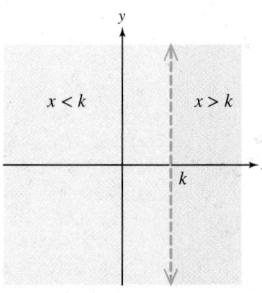

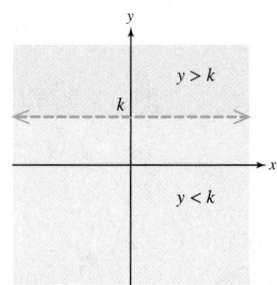

Figure 10-26   **Figure 10-27**

EXAMPLE 3   **Graphing Linear Inequalities with a Horizontal or Vertical Bounding Line**

Graph the solution set.

a. $x \leq -1$ **b.** $3y > 5$

Solution:

a. $x \leq -1$

- The related equation $x = -1$ is a vertical line.

- The inequality $x \leq -1$ represents all points to the *left* of or on the line $x = -1$.

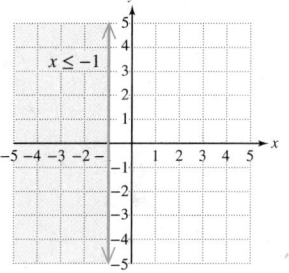

b. $3y > 5$

- The inequality is equivalent to $y > \frac{5}{3}$. The related equation $y = \frac{5}{3}$ is a horizontal line.

- The inequality $y > \frac{5}{3}$ represents all points strictly above the line $y = \frac{5}{3}$.

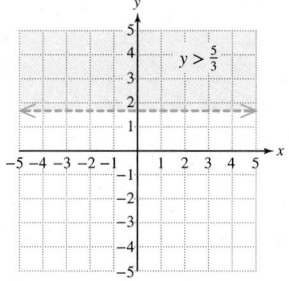

Skill Practice 3   Graph the solution set.

a. $2y < 6$ **b.** $x \geq \frac{9}{4}$

Answers

3. a.

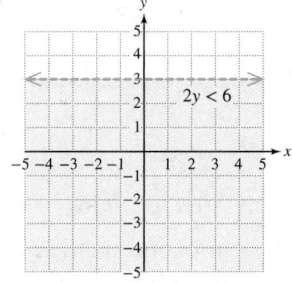

b.

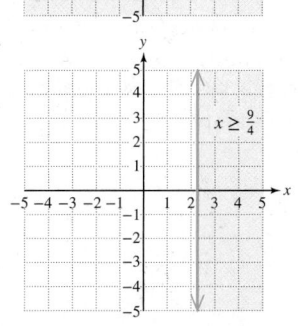

2. Solve Nonlinear Inequalities in Two Variables

The same approach used to graph a linear inequality in two variables is used to graph a *nonlinear* inequality in two variables. This is demonstrated in Example 4.

EXAMPLE 4   Graphing a Nonlinear Inequality

Graph the solution set. $(x - 2)^2 + y^2 > 9$

Solution:

$(x - 2)^2 + y^2 > 9$ $\xrightarrow[\text{equation}]{\text{related}}$ $(x - 2)^2 + y^2 = 9$

Step 1: Graph the related equation. The equation represents a circle centered at (2, 0) with radius 3. Because the inequality is strict, draw the circle as a dashed curve (Figure 10-28).

Center: (2, 0)
Radius: 3

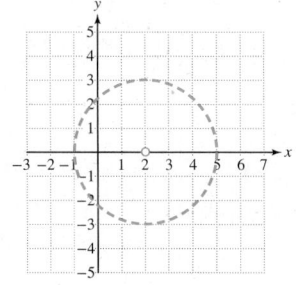

Figure 10-28

Test (2, 0):

$(x - 2)^2 + y^2 > 9$

$(2 - 2)^2 + (0)^2 \overset{?}{>} 9$

$0 \overset{?}{>} 9$ false

Step 2: Select a test point. We have chosen the center (2, 0).

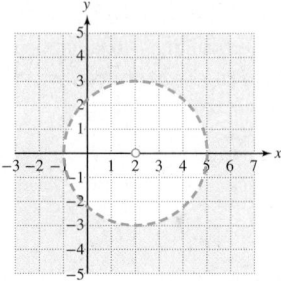

The test point inside the circle does not satisfy the original inequality. Therefore, the solution set consists of the points strictly outside the circle (Figure 10-29).

Figure 10-29

Skill Practice 4   Graph the solution set. $x^2 + (y + 1)^2 < 16$

---

3. Solve Systems of Inequalities in Two Variables

Two or more inequalities in two variables make up a system of inequalities in two variables. The solution set is the set of ordered pairs that satisfy each inequality in the system. To graph the solution set to a system of inequalities, graph the solution sets to the individual inequalities first. The solution to the system of inequalities is the *intersection* of the graphs. This is demonstrated in Example 5.

Answer

4.

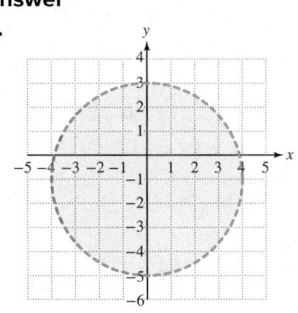

> **EXAMPLE 5**   **Graphing the Solution Set to a System of Linear Inequalities**

Graph the solution set to the system of inequalities. $y \le \frac{1}{2}x + 2$

$3x - y < 3$

Solution:

$y \le \frac{1}{2}x + 2$

$3x - y < 3$

First graph the solutions to the individual inequalities (Figures 10-30 and 10-31). Next, find the intersection (area of overlap) of the solution sets shown in purple in Figure 10-32.

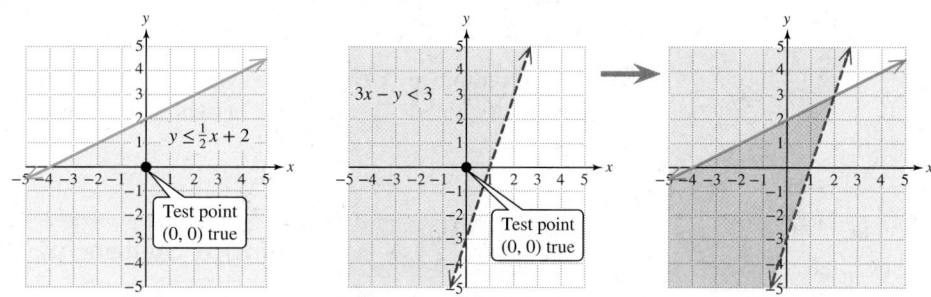

| **Figure 10-30** | **Figure 10-31** | **Figure 10-32** |

Notice that the point of intersection between the two bounding lines is graphed as an open dot (Figure 10-33). This indicates that it is *not* part of the solution set. The reason is that it is not a solution to the strict inequality $3x - y < 3$.

The point of intersection can be found by solving the system of related equations. We have used the substitution method.

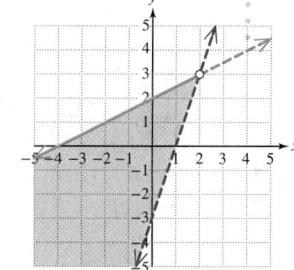

Figure 10-33

$$y = \frac{1}{2}x + 2$$
$$3x - y = 3 \qquad 3x - \left(\frac{1}{2}x + 2\right) = 3$$
$$3x - \frac{1}{2}x - 2 = 3$$
$$\frac{5}{2}x = 5$$
$$x = 2 \longrightarrow y = \frac{1}{2}(2) + 2$$
$$y = 3$$

The point of intersection is (2, 3) and is excluded from the solution set.

> **Skill Practice 5**   Graph the solution set to the system of inequalities.
>
> $y < -\frac{1}{3}x + 1$
>
> $-2x + y \le 1$

Answer

5.

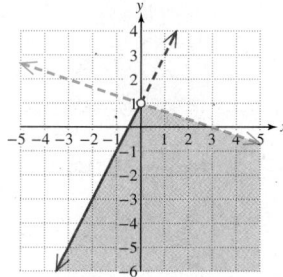

TECHNOLOGY CONNECTIONS

Graphing a System of Inequalities in Two Variables

To graph the system of inequalities from Example 5 on a graphing calculator, solve each inequality for y.

$$y \le \tfrac{1}{2}x + 2$$
$$3x - y < 3 \longrightarrow -y < -3x + 3 \longrightarrow y > 3x - 3$$

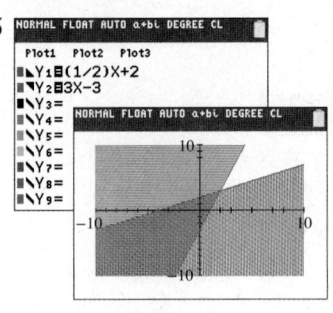

Then enter the related equations in the graphing editor. Choose the appropriate graphing style ◥ or ◣. For $Y_1 \le \tfrac{1}{2}x + 2$, choose ◣. For $Y_2 > 3x - 3$, choose ◥.

In Example 6, we graph a system of nonlinear inequalities in two variables. This is a system of inequalities in which one or more of the individual inequalities is nonlinear.

EXAMPLE 6   Solving a System of Nonlinear Inequalities

Graph the solution set to the system of inequalities.
$$y \le -x^2 + 4$$
$$x - y \ge -2$$
$$y > -5$$

Solution:

To graph the solution set to the given system, first graph each individual inequality.

$$y \le -x^2 + 4 \qquad\qquad x - y \ge -2 \qquad\qquad y > -5$$

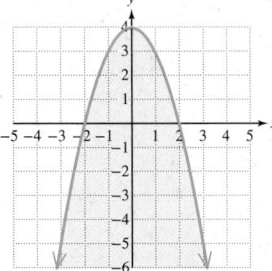   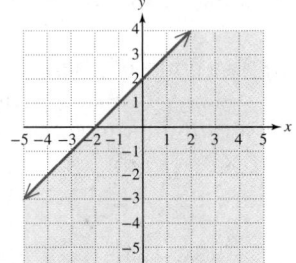   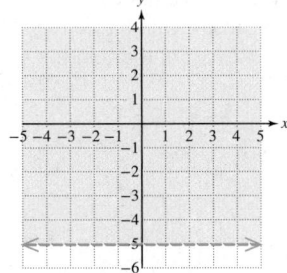

The solution set to the system is the intersection of the three shaded regions (Figure 10-34). The points of intersection are found by pairing up the related equations and solving the system of equations. The inequalities $y \le -x^2 + 4$ and $x - y \ge -2$ include equality. Therefore, the intersection points between the parabola and slanted line are solutions to the system. These points are plotted as closed dots.

On the other hand, because the inequality $y > -5$ is strict, the intersection points between the parabola and horizontal line are *not* solutions to the system. These points are plotted as open dots.

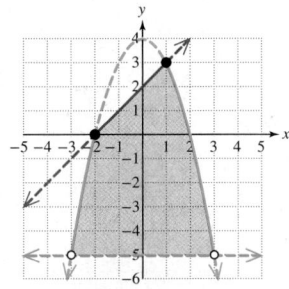

Figure 10-34

To find the points of intersection between the parabola $y = -x^2 + 4$ and the line $x - y = -2$, solve the system:

$$y = -x^2 + 4$$
$$x - y = -2 \qquad \text{The solutions are } (-2, 0) \text{ and } (1, 3).$$

To find the points of intersection between the parabola $y = -x^2 + 4$ and the line $y = -5$, solve the system:

$$y = -x^2 + 4$$
$$y = -5 \qquad \text{The solutions are } (-3, -5) \text{ and } (3, -5).$$

Skill Practice 6   Graph the solution set to the system of inequalities.

$$x^2 + y^2 \le 25$$
$$-x + y < 1$$
$$y \ge -4$$

In Example 7, we refer to the problem addressed at the beginning of this section and set up a system of linear inequalities to model the allocation of time studying algebra and biology.

EXAMPLE 7   **Solving a System of Inequalities in an Application**

Adriana has 12 available study hours for algebra and biology.

- Let x represent the number of hours she spends studying algebra.
- Let y represent the number of hours that she studies biology.

a. Set up an inequality that indicates that the number of hours spent studying algebra cannot be negative.

b. Set up an inequality that indicates that the number of hours spent studying biology cannot be negative.

c. Set up an inequality that indicates that the combined number of hours she spends studying for these two classes is at most 12 hr.

d. Graph the solution set to the system of inequalities from parts (a)–(c).

Solution:

a. $x \ge 0$ The number of hours spent studying algebra is 0 or more.

b. $y \ge 0$ The number of hours spent studying biology is 0 or more.

c. $x + y \le 12$ The sum of time spent studying algebra and the time spent studying biology cannot exceed the maximum number of study hours available. Therefore, the sum is less than or equal to 12 hr.

> **TIP** The points of intersection of the bounding lines are closed dots because they are part of the solution set.

d. $x \ge 0$
$y \ge 0$
$x + y \le 12$

The inequalities $x \ge 0$ and $y \ge 0$ together represent the set of points in the first quadrant, including the bounding axes.

To graph the inequality $x + y \le 12$, we graph the related equation $x + y = 12$. A test point $(0, 0)$ taken below the line results in a true statement. Therefore, the inequality $x + y \le 12$ represents the set of points on and below the line $x + y = 12$.

The solution to the system of inequalities is shown in Figure 10-35.

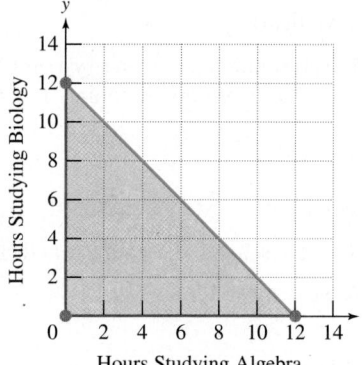

Figure 10-35

Answer

6.

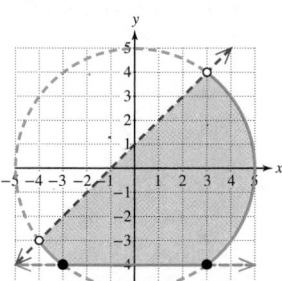

Answers

7. a. $x \geq 0; y \geq 0$
b. $x + y \leq 16$
c.

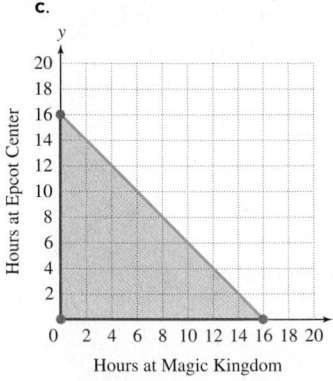

Skill Practice 7 A family plans to spend two 8-hr days at Disney World and will split time between the Magic Kingdom and Epcot Center. Let x represent the number of hours spent at the Magic Kingdom and let y represent the number of hours spent at Epcot Center.

a. Set up two inequalities that indicate that the number of hours spent at the Magic Kingdom and the number of hours spent at Epcot cannot be negative.

b. Set up an inequality that indicates that the combined number of hours spent at the two parks is at most 16 hr.

c. Graph the solution set to the system of inequalities.

SECTION 10.6   Practice Exercises

Prerequisite Review

For Exercises R.1–R.6, graph the equation. (See Sections 5.4 and 5.2 for review.)

R.1. $2x + y = 5$ (See Section 5.4.)

R.2. $-3x + 4y = 6$

R.3. $-4x = y$

R.4. $2x = 3y$

R.5. $x^2 + (y - 4)^2 = 4$ (See Section 5.2.)

R.6. $(x + 3)^2 + y^2 = 16$

For Exercises R.7– R.12, solve the inequality. Write the solution set in interval notation. (See Sections 1.5 and 1.7 for review.)

R.7. $-3(t + 4) < t + 8$ (See Section 1.5.)

R.8. $-4x - 8 \geq -2(x - 6)$

R.9. $2x + 4 \leq -(x + 8)$ and $-\dfrac{1}{4}x < 2$

R.10. $-2.5x < 7.5$ and $4(x - 6) \leq 2x + 6$

R.11. $|x| \geq 3$ (See Section 1.7.)

R.12. $|x| < 2$

Concept Connections

1. An inequality that can be written in the form $Ax + By < C$ (where A and B are not both zero) is called a _____ inequality in two variables.

2. For a constant real number k, the inequality $x < k$ represents the half-plane to the (left/right) of the (horizontal/vertical) line $x = k$.

3. For a constant real number k, the inequality $y > k$ represents the half-plane (above/below) the (horizontal/vertical) line $y = k$.

4. Given the inequality $y \leq 2x + 1$, the bounding line $y = 2x + 1$ is drawn as a (dashed/solid) line.

5. The solution set to the system of inequalities $x < 0$, $y > 0$ represents the points in quadrant (I, II, III, IV).

6. The equation $x^2 + y^2 = 4$ is a circle centered at _____ with radius _____. The solution set to the inequality $x^2 + y^2 < 4$ represents the set of points (inside/outside) the circle $x^2 + y^2 = 4$.

Objective 1: Solve Linear Inequalities in Two Variables

For Exercises 7–10, determine whether the ordered pair is a solution to the inequality.

7. $3x + 4y < 12$
a. $(-1, 3)$
b. $(5, 1)$
c. $(4, 0)$

8. $2x + 3y > 6$
a. $(-3, 3)$
b. $(5, -1)$
c. $(0, 2)$

9. $y \geq (x - 3)^2$
a. $(-3, 30)$
b. $(1, 4)$
c. $(5, 5)$

10. $y \leq x^3 - 1$
a. $(-1, -2)$
b. $(2, 6)$
c. $(-4, -50)$

11. a. Graph the solution set. $4x - 5y \leq 20$
  (See Example 1)

  b. Explain how the graph would differ for the inequality $4x - 5y < 20$.

  c. Explain how the graph would differ for the inequality $4x - 5y > 20$.

12. a. Graph the solution set. $2x + 5y > 10$

  b. Explain how the graph would differ for the inequality $2x + 5y \geq 10$.

  c. Explain how the graph would differ for the inequality $2x + 5y < 10$.

For Exercises 13–24, graph the solution set. **(See Examples 1–3)**

13. $2x + 5y > 5$

14. $-5x + 4y \leq 8$

15. $-30x \geq 20y + 600$

16. $-400x < 100y + 8000$

17. $5x \leq 6y$

18. $3x > 2y$

19. $3 + 2(x + y) > y + 3$

20. $-4 - 3(x - y) < 2y - 4$

21. $x < 6$

22. $y \leq 5$

23. $-\frac{1}{2}y + 4 \leq 5$

24. $-\frac{1}{3}x + 2 < 4$

Objective 2: Solve Nonlinear Inequalities in Two Variables

25. a. Graph the solution set. $x^2 + y^2 < 4$ **(See Example 4)**

  b. Explain how the graph would differ for the inequality $x^2 + y^2 > 4$.

  c. Explain how the graph would differ for the inequality $x^2 + y^2 \geq 4$.

26. a. Graph the solution set. $y \geq x^2 - 1$

  b. Explain how the graph would differ for the inequality $y \leq x^2 - 1$.

  c. Explain how the graph would differ for the inequality $y > x^2 - 1$.

For Exercises 27–36, graph the solution set. **(See Example 4)**

27. $y < -x^2$

28. $x^2 + y^2 \geq 16$

29. $y \leq (x - 2)^2 + 1$

30. $y \geq -(x + 1)^2 - 2$

31. $|x| \leq 3$

32. $|y| \leq 2$

33. $2|y| > 2$

34. $|x| + 1 > 3$

35. $y \geq \sqrt{x}$

36. $y < \sqrt{x - 1}$

Objective 3: Solve Systems of Inequalities in Two Variables

37. a. Is the point $(2, 1)$ a solution to the inequality $y < 2x + 3$?

  b. Is the point $(2, 1)$ a solution to the inequality $x + y \leq 1$?

  c. Is the point $(2, 1)$ a solution to the system of inequalities?

$$y < 2x + 3$$
$$x + y \leq 1$$

38. a. Is the point $(3, 2)$ a solution to the inequality $y < -x + 5$?

  b. Is the point $(3, 2)$ a solution to the inequality $3x + y \geq 11$?

  c. Is the point $(3, 2)$ a solution to the system of inequalities?

$$y < -x + 5$$
$$3x + y \geq 11$$

For Exercises 39–40, determine whether the ordered pair is a solution to the system of inequalities.

39. $x + y < 4$
  $y \leq 2x + 1$
  a. $(0, 1)$ **b.** $(3, 1)$ **c.** $(2, 0)$ **d.** $(1, 4)$

40. $y < -x^2 + 3$
  $x + 2y \leq 2$
  a. $(-2, -1)$ **b.** $(0, -2)$ **c.** $(0, 1)$ **d.** $(3, -6)$

For Exercises 41–58, graph the solution set. If there is no solution, indicate that the solution set is the empty set. **(See Examples 5–6)**

41. $y < \frac{1}{2}x - 4$
  $y > -2x + 1$

42. $y \geq \frac{1}{3}x - 2$
  $y \leq x - 4$

43. $\quad 2x + 5y \leq 5$
  $-3x + 4y \geq 4$

44. $4x - 3y > 3$
  $x + 4y < -4$

45. $x^2 + y^2 \geq 9$
  $x^2 + y^2 \leq 16$

46. $x^2 + y^2 \geq 1$
  $x^2 + y^2 < 25$

47. $y \geq 3x + 3$
  $-3x + y < 1$

48. $y < 2x - 4$
  $-2x + y \geq 2$

49. $|x| < 3$
  $|y| < 3$

50. $|x| \geq 2$
  $|y| \geq 2$

51. $y \geq x^2 - 2$
  $y > x$
  $y \leq 4$

52. $y \leq -x^2 + 7$
  $y \leq -x + 5$
  $y > 1$

53. $x^2 + y^2 \leq 100$

$y < \dfrac{4}{3}x$

$x \leq 8$

54. $x^2 + y^2 < 100$

$y \geq x$

$y \geq 1$

55. $y < e^x$

$y > 1$

$x < 2$

56. $y \leq \dfrac{2}{x}$

$y > 0$

$y < x$

57. $(x + 2)^2 + (y - 3)^2 \leq 9$

$x - y > 2$

58. $(x - 4)^2 + (y + 1)^2 < 25$

$2x - y < -4$

Mixed Exercises

For Exercises 59–64, write an inequality to represent the statement.

59. x is at most 6.

60. y is no more than 7.

61. y is at least -2.

62. x is no less than $\frac{1}{2}$.

63. The sum of x and y does not exceed 18.

64. The difference of x and y is not less than 4.

65. Let x represent the number of hours that Trenton spends studying algebra, and let y represent the number of hours he spends studying history. For parts (a)–(e), write an inequality to represent the given statement. **(See Example 7)**

   a. Trenton has a total of at most 9 hr to study for both algebra and history combined.

   b. Trenton will spend at least 3 hr studying algebra.

   c. Trenton will spend no more than 4 hr studying history.

   d. The number of hours spent studying algebra cannot be negative.

   e. The number of hours spent studying history cannot be negative.

   f. Graph the solution set to the system of inequalities from parts (a)–(e).

66. Let x represent the number of country songs that Sierra puts on a playlist on her portable media player. Let y represent the number of rock songs that she puts on the playlist. For parts (a)–(e), write an inequality to represent the given statement.

   a. Sierra will put at least 6 country songs on the playlist.

   b. Sierra will put no more than 10 rock songs on the playlist.

   c. Sierra wants to limit the length of the playlist to at most 20 songs.

   d. The number of country songs cannot be negative.

   e. The number of rock songs cannot be negative.

   f. Graph the solution set to the system of inequalities from parts (a)–(e).

67. A couple has $60,000 to invest for retirement. They plan to put x dollars in stocks and y dollars in bonds. For parts (a)–(d), write an inequality to represent the given statement.

   a. The total amount invested is at most $60,000.

   b. The couple considers stocks a riskier investment, so they want to invest at least twice as much in bonds as in stocks.

   c. The amount invested in stocks cannot be negative.

   d. The amount invested in bonds cannot be negative.

   e. Graph the solution set to the system of inequalities from parts (a)–(d).

68. A college theater has a seating capacity of 2000. It reserves x tickets for students and y tickets for general admission. For parts (a)–(d), write an inequality to represent the given statement.

   a. The total number of seats available is at most 2000.

   b. The college wants to reserve at least 3 times as many student tickets as general admission tickets.

   c. The number of student tickets cannot be negative.

   d. The number of general admission tickets cannot be negative.

   e. Graph the solution set to the system of inequalities from parts (a)–(d).

69. Write a system of inequalities that represents the points in the first quadrant less than 3 units from the origin.

70. Write a system of inequalities that represents the points in the second quadrant more than 4 units from the origin.

71. Write a system of inequalities that represents the points inside the triangle with vertices $(-3, -4)$, $(3, 2)$, and $(-5, 4)$.

72. Write a system of inequalities that represents the points inside the triangle with vertices $(-4, -4)$, $(1, 1)$, and $(5, -1)$.

73. A weak earthquake occurred roughly 9 km south and 12 km west of the center of Hawthorne, Nevada. The quake could be felt 16 km away. Suppose that the origin of a map is placed at the center of Hawthorne with the positive x-axis pointing east and the positive y-axis pointing north.

   a. Find an inequality that describes the points on the map for which the earthquake could be felt.

   b. Could the earthquake be felt at the center of Hawthorne?

74. A coordinate system is placed at the center of a town with the positive x-axis pointing east, and the positive y-axis pointing north. A cell tower is located 4 mi west and 5 mi north of the origin.

   a. If the tower has a 8-mi range, write an inequality that represents the points on the map serviced by this tower.

   b. Can a resident 5 mi east of the center of town get a signal from this tower?

Write About It

75. Under what circumstances should a dashed line or curve be used when graphing the solution set to an inequality in two variables?

76. Explain how test points are used to determine the region of the plane that represents the solution to an inequality in two variables.

77. Explain how to find the solution set to a system of inequalities in two variables.

78. Describe the solution set to the system of inequalities.
$x \geq 0$, $y \geq 0$, $x \leq 1$, $y \leq 1$

Expanding Your Skills

For Exercises 79–80, graph the solution set.

79. $|x| \geq |y|$

80. $|x| + |y| \leq 1$

Technology Connections

For Exercises 81–82, use a graphing utility to graph the solution set to the system of inequalities.

81. $y \geq 0.4e^x$
$y \leq 0.25x^3 - 4x$

82. $y < \dfrac{4}{x^2 + 1}$
$y > \dfrac{-2}{x^2 + 0.5}$

PROBLEM RECOGNITION EXERCISES

Equations and Inequalities in Two Variables

For Exercises 1–2, for parts (a) and (b), graph the equation. For part (c), solve the system of equations. For parts (d) and (e) graph the solution set to the system of inequalities. If there is no solution, indicate that the solution set is the empty set.

1. a. $y = -3x + 5$

   b. $-2x + y = 0$

   c. $y = -3x + 5$
   $-2x + y = 0$

   d. $y > -3x + 5$
   $-2x + y < 0$

   e. $y < -3x + 5$
   $-2x + y > 0$

2. a. $y = 2x - 3$

   b. $4x - 2y = -2$

   c. $y = 2x - 3$
   $4x - 2y = -2$

   d. $y \geq 2x - 3$
   $4x - 2y \geq -2$

   e. $y \leq 2x - 3$
   $4x - 2y \leq -2$

For Exercises 3–4, for part (a), graph the equations in the system and determine the solution set. For parts (b) and (c), graph the solution set to the inequality.

3. a. $y = x^2$
   $y = \frac{1}{2}x^2$

   b. $y \leq x^2$
   $y \geq \frac{1}{2}x^2$

   c. $y \geq x^2$
   $y \leq \frac{1}{2}x^2$

4. a. $x - y = 1$
   $y = (x - 3)^2$

   b. $x - y \geq 1$
   $y \geq (x - 3)^2$

   c. $x - y \leq 1$
   $y \leq (x - 3)^2$

SECTION 10.7 Linear Programming

OBJECTIVES

1. Write an Objective Function
2. Solve Linear Programming Applications

TIP The notation $z = f(x, y)$ is read as "z is a function of x and y."

1. Write an Objective Function

When a company manufactures a product, the goal is to obtain maximum profit at minimum cost. However, the production process is often limited by certain constraints such as the amount of labor available, the capacity of machinery, and the amount of money available for the company to invest in the process. In this section, we will study a process called **linear programming** that enables us to maximize or minimize a function under specified constraints.

The function to be optimized (maximized or minimized) in a linear programming application is called the **objective function**. The objective function often has two independent variables. For example, given $z = f(x, y)$, the variable z is dependent on the two independent variables x and y.

TIP An objective function represents a quantity that is to be maximized or minimized. In Example 1, the school will want to minimize the cost to transport the students.

EXAMPLE 1 Writing an Objective Function

Suppose that a college wants to rent several buses to transport students to a championship college football game. A large bus costs $1200 to rent and a small bus costs $800 to rent.

Let x represent the number of large buses.

Let y represent the number of small buses.

Write an objective function that represents the total cost z (in $) to rent x large buses and y small buses.

Solution:

$$\begin{pmatrix} \text{Total} \\ \text{cost} \end{pmatrix} = \begin{pmatrix} \text{Cost to rent} \\ x \text{ large buses} \end{pmatrix} + \begin{pmatrix} \text{Cost to rent} \\ y \text{ small buses} \end{pmatrix}$$

$$\text{Cost:} \quad z \quad = \quad 1200x \quad + \quad 800y$$

The objective function is defined by $z = 1200x + 800y$.

Skill Practice 1 An office manager needs to staff the office. She hires full-time employees at $36 per hour and part-time employees at $24 per hour. Write an objective function that represents the total cost (in $) to staff the office with x full-time employees and y part-time employees for 1 hr.

---

2. Solve Linear Programming Applications

In Example 2, we identify constraints imposed on the resources that affect the number of buses that the college can rent.

EXAMPLE 2 Writing a System of Constraints

Refer to the scenario from Example 1. We now set several constraints that affect the number of buses that can be rented.

Let x represent the number of large buses.

Let y represent the number of small buses.

Answer

1. $z = 36x + 24y$

Do not confuse the phrases "at least" and "at most."

The number of students is *at least* 3600 means that the number of students cannot be less than 3600. Equivalently, this is 3600 or more students.

The number of drivers is *at most* 75 means that the number of drivers cannot exceed 75. Equivalently, this is 75 or fewer drivers.

For parts (a)–(d), write an inequality that represents the given statement.

a. The number of large buses cannot be negative.

b. The number of small buses cannot be negative.

c. Large buses can carry 60 people and small buses can carry 45 people. The college must transport at least 3600 students.

d. The number of available bus drivers is at most 75.

Solution:

a. $x \geq 0$ The number of large buses cannot be negative.

b. $y \geq 0$ The number of small buses cannot be negative.

c. $\left(\begin{array}{c} \text{Number of students} \\ \text{carried by large buses} \end{array} \right) + \left(\begin{array}{c} \text{Number of students} \\ \text{carried by small buses} \end{array} \right) \underset{\text{is at least}}{\geq} 3600$ Large buses hold 60 people and small buses hold 45 people.

$60x \qquad + \qquad 45y \qquad \geq 3600$

d. $\left(\begin{array}{c} \text{Number of} \\ \text{large buses} \end{array} \right) + \left(\begin{array}{c} \text{Number of} \\ \text{small buses} \end{array} \right) \underset{\text{is at most}}{\leq} 75$ Because each bus has only one driver, the total number of buses is the same as the total number of drivers.

$x \qquad + \qquad y \qquad \leq 75$

Skill Practice 2   Refer to Skill Practice 1. Suppose that the office manager needs at least 20 employees, but not more than 24 full-time employees. Furthermore, to make the office run smoothly, the manager knows that the number of full-time employees must always be greater than or equal to the number of part-time employees. Write a system of inequalities that represents the constraints on the number of full-time employees x and the number of part-time employees y.

The constraints in Example 2 make up a system of linear inequalities.

The number of large buses is nonnegative.	$x \geq 0$
The number of small buses is nonnegative.	$y \geq 0$
The school must transport at least 3600 students.	$60x + 45y \geq 3600$
The number of drivers (and therefore buses) is at most 75.	$x + y \leq 75$

The region in the plane that represents the solution set to the system of constraints is called the **feasible region** (Figure 10-36). The points of intersection of the bounding lines in the feasible region are called the **vertices** of the feasible region.

The vertices in Figure 10-36 are (60, 0), (75, 0), and (15, 60).

TIP   The point (15, 60) is the point of intersection of the lines $60x + 45y = 3600$ and $x + y = 75$.

The point (60, 0) is the point of intersection of the line $60x + 45y = 3600$ and $y = 0$.

The point (75, 0) is the point of intersection of the line $x + y = 75$ and $y = 0$.

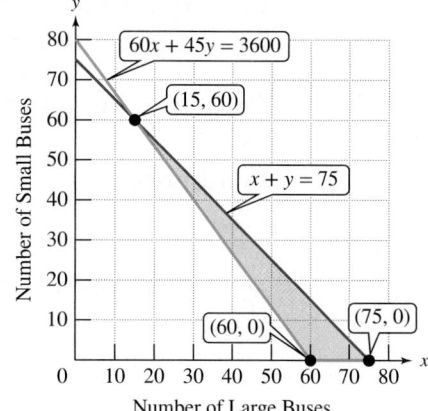

Figure 10-36

Answer

2. $x \geq 0, y \geq 0, x \leq 24,$
$x + y \geq 20, x \geq y$

Points within the shaded region meet all the constraints in the problem. For example, the ordered pair (65, 5) represents 65 large buses and 5 small buses.

65 large buses and 5 small buses ⟶ 70 buses (≤ 75 drivers) ✓
65 large buses and 5 small buses ⟶ transports 4125 students (≥ 3600) ✓

However, points such as (65, 15) and (40, 20) are *outside* the feasible region and do not satisfy all constraints.

65 large buses and 15 small buses ⟶ 80 buses (exceeds the number of drivers)
40 large buses and 20 small buses ⟶ transports only 3300 students

The goal of a linear programming application is to find the maximum or minimum value of the objective function $z = f(x, y)$ when x and y are restricted to the ordered pairs in the feasible region. Fortunately, it has been proven mathematically that if a maximum or minimum value of a function exists, it occurs at one or more of the vertices of the feasible region. This is the basis for the following procedure to solve a linear programming application.

Solving an Application Involving Linear Programming

Step 1 Write an objective function $z = f(x, y)$.
Step 2 Write a system of inequalities defining the constraints on x and y.
Step 3 Graph the feasible region and identify the vertices.
Step 4 Evaluate the objective function at each vertex of the feasible region. Use the results to identify the values of x and y that optimize the objective function. Identify the optimal value of z.

In Example 3, we optimize the objective function found in Example 1 subject to the constraints defined in Example 2.

EXAMPLE 3   Solving a Linear Programming Application

A college wants to rent buses to transport at least 3600 students to a championship college football game. Large buses hold 60 people and small buses hold 45 people. Furthermore, the number of available bus drivers is at most 75. Each large bus costs $1200 to rent and each small bus costs $800 to rent. Find the optimal number of large and small buses that will minimize cost.

Solution:

Let x represent the number of large buses.   Define the relevant variables.
Let y represent the number of small buses.

Cost: $z = 1200x + 800y$ **Step 1:** Write an objective function. Since cost is to be minimized, write a cost function. (See Example 1.)

Constraints: **Step 2:** Write a system of constraints on the relevant variables. (See Example 2.)
$x \geq 0$
$y \geq 0$
$60x + 45y \geq 3600$
$x + y \leq 75$

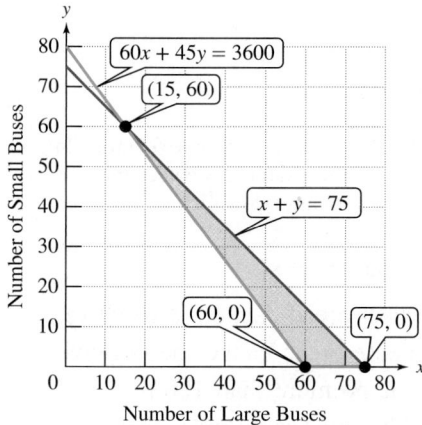

Step 3: Graph the feasible region and identify the vertices.

The vertices are found by identifying the points of intersection between the bounding lines.

Bounding lines:

$60x + 45y = 3600$
$x + y = 75$ Point of intersection $(15, 60)$

$60x + 45y = 3600$
$y = 0$ Point of intersection $(60, 0)$

$x + y = 75$
$y = 0$ Point of intersection $(75, 0)$

Cost function: $z = 1200x + 800y$

at $(15, 60)$ $z = 1200(15) + 800(60) = \boxed{\$66{,}000}$ **Step 4:** Evaluate the objective function $z = 1200x + 800y$ at each vertex.
at $(60, 0)$ $z = 1200(60) + 800(0) = \$72{,}000$
at $(75, 0)$ $z = 1200(75) + 800(0) = \$90{,}000$

The cost would be minimized if the college rents 15 large buses and 60 small buses. The minimum cost is \$66,000.

> **Skill Practice 3**   Refer to Skill Practices 1 and 2. The office manager needs at least 20 employees, but no more than 24 full-time employees. Furthermore, to make the office run smoothly, the manager knows that the number of full-time employees must always be greater than or equal to the number of part-time employees. If she pays full-time employees \$36 per hour and part-time employees \$24 per hour, determine the number of full-time and part-time employees she should hire to minimize total labor cost per hour.

In Example 4, we investigate a situation in which we maximize profit.

EXAMPLE 4   Solving a Linear Programming Application

A baker produces whole wheat bread and cheese bread to sell at the farmer's market. The whole wheat bread is denser and requires more baking time, whereas the cheese bread requires more labor. The baking times and average amount of labor per loaf are given in the table along with the profit for each loaf.

	Time to Bake	Labor	Profit
Wheat bread	1.5 hr	$\frac{1}{3}$ hr	\$1.20
Cheese bread	1 hr	$\frac{1}{2}$ hr	\$1.00

The oven space restricts the baker from baking more than 120 loaves. Furthermore, the amount of oven time for baking is no more than 165 hr and the amount of available labor is at most 55 hr. Determine the number of loaves of each type of bread that the baker should bake to maximize his profit. Assume that all loaves of bread produced are sold.

Point of Interest

Linear programming was first introduced by Russian mathematician Leonid Kantorovich. The technique was used in World War II to minimize costs to the army. Later in 1975, Kantorovich won the Nobel Prize in economics for his contributions to the theory of optimum allocation of resources.

Answer

3. The cost would be minimized at \$600/hr if she hires 10 full-time employees and 10 part-time employees.

Solution:

Let x represent the number of loaves of wheat bread.

Let y represent the number of loaves of cheese bread.

$$\text{Profit} = \begin{pmatrix} \text{Profit from} \\ \text{wheat bread} \end{pmatrix} + \begin{pmatrix} \text{Profit from} \\ \text{cheese bread} \end{pmatrix}$$

Step 1: Write an objective function. In this example, we need to maximize profit.

$$z = 1.20x + 1.00y$$

Step 2: Write a system of constraints on the independent variables.

$x \geq 0$	Number of loaves of wheat bread cannot be negative.
$y \geq 0$	Number of loaves of cheese bread cannot be negative.
$x + y \leq 120$	Total number of loaves is no more than 120.
$1.5x + y \leq 165$	The total amount of baking time is no more than 165 hr.
$\frac{1}{3}x + \frac{1}{2}y \leq 55$	The total amount of available labor is at most 55 hr.

Step 3: Graph the feasible region and identify the vertices. Find the points of intersection between pairs of bounding lines:

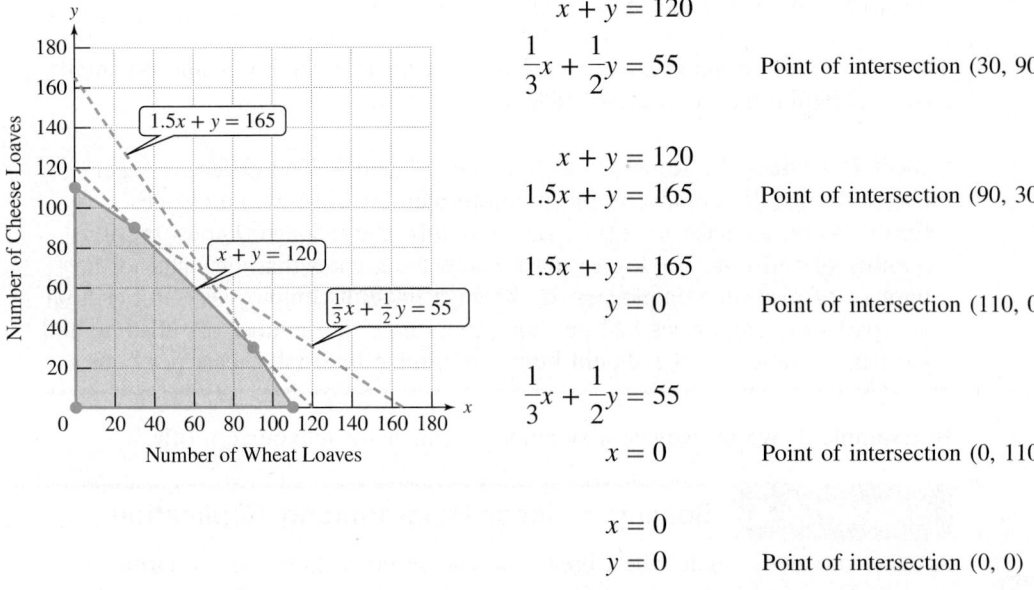

$$x + y = 120$$
$$\frac{1}{3}x + \frac{1}{2}y = 55 \qquad \text{Point of intersection } (30, 90)$$

$$x + y = 120$$
$$1.5x + y = 165 \qquad \text{Point of intersection } (90, 30)$$

$$1.5x + y = 165$$
$$y = 0 \qquad \text{Point of intersection } (110, 0)$$

$$\frac{1}{3}x + \frac{1}{2}y = 55$$
$$x = 0 \qquad \text{Point of intersection } (0, 110)$$

$$x = 0$$
$$y = 0 \qquad \text{Point of intersection } (0, 0)$$

Step 4: Evaluate the objective function at each vertex.

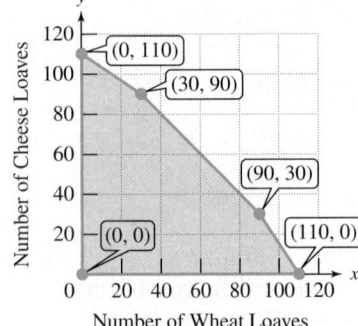

Profit: $\quad z = 1.2x + y$

at $(0, 0)\quad z = 1.2(0) + (0) = \$0$

at $(0, 110)\quad z = 1.2(0) + (110) = \$110$

at $(30, 90)\quad z = 1.2(30) + (90) = \$126$

at $(90, 30)\quad z = 1.2(90) + (30) = \boxed{\$138}$

at $(110, 0)\quad z = 1.2(110) + (0) = \$132$

The profit will be maximized if the baker bakes 90 whole wheat loaves and 30 cheese loaves. The maximum profit is $138.

Skill Practice 4   A manufacturer produces two sizes of leather handbags. It takes longer to cut and dye the leather for the smaller bag, but it takes more time sewing the larger bag. The production constraints and profit for each type of bag are given in the table.

	Cutting and Dying	**Sewing**	**Profit**
Large bag	0.6 hr	2 hr	$30
Small bag	1 hr	1.5 hr	$25

The machinery limits the number of bags produced to at most 1000 per week. If the company has 900 hr per week available for cutting and dying and 1800 hr available per week for sewing, determine the number of each type of bag that should be produced weekly to maximize profit. Assume that all bags produced are also sold.

To find the maximum or minimum value of an objective function, we evaluate the function at the vertices of the feasible region. It seems reasonable that the profit would be maximized at a point on the upper edge of the feasible region. These are the points in the feasible region where the combined values of x and y are the greatest.

The goal of Example 4 was to find the values of x and y that maximized the profit function $z = 1.2x + y$. To see why the profit z was maximized at a vertex of the feasible region, write the equation in slope-intercept form:

$$y = -1.2x + z$$

In this form, the objective function represents a family of parallel lines with slope -1.2 and y-intercept $(0, z)$. To maximize z, we want the line with the greatest y-intercept that still remains in contact with the feasible region. In Figure 10-37, we see that this occurs for the line passing through the point $(90, 30)$ as expected.

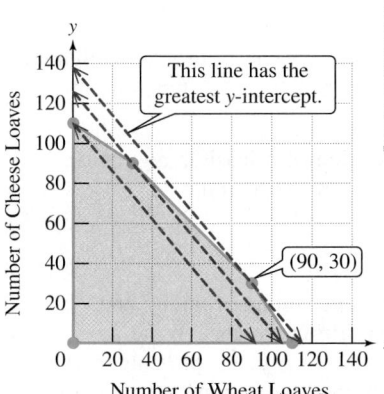

This line has the greatest y-intercept.

(90, 30)

Number of Cheese Loaves

Number of Wheat Loaves

Figure 10-37

Answer

4. The maximum profit of $28,000 is realized when the company produces 600 large bags and 400 small bags.

SECTION 10.7   **Practice Exercises**

Prerequisite Review

For Exercises R.1–R.2, see Section 5.5 for review.

R.1. A salesperson makes a base salary of $500 per week plus 11% commission on sales.

   a. Write a linear function to model the salesperson's weekly salary $S(x)$ for x dollars in sales.

   b. Evaluate $S(8000)$ and interpret the meaning in the context of this problem.

R.2. A luxury car rental company charges $123 per day in addition to a flat fee of $75 for insurance.

   a. Write an equation that represents the cost y (in $) to rent the car for x days.

   b. What is the y-intercept and what does it mean in the context of this problem?

For Exercises R.3–R.6, find the point of intersection of the graphs of the two equations. (See Section 10.2 for review.)

R.3. $4x + y = 6$
     $2y = 4$

R.4. $x - 3y = -3$
     $-4x = -12$

R.5. $x + y = 600$
     $2x - y = 900$

R.6. $3x + 4y = 400$
     $x + y = 120$

For Exercises R.7–R.10, translate the statement to an inequality. (See Section 1.4 for review.)

R.7. In order to vote, the age a of an individual must be at least 18 years.

R.8. To be eligible for a children's discount at the zoo, the age a of the child must be at most 12 years.

R.9. The maximum recommended heart rate r for a 60-year-old adult should be no more than 160 beats per minute.

R.10. During a hot day at the Australian Open tennis tournament, the temperature t on the court was no less than 100°F.

Concept Connections

1. The process that maximizes or minimizes a function subject to linear constraints is called _____ programming.

2. The function to be optimized in a linear programming application is called the _____ function.

3. The region in the plane that represents the solution set to a system of constraints is called the _____ region.

4. The points of intersection of a feasible region are called the _____ of the region.

Objective 1: Write an Objective Function

5. A diner makes a profit of $0.80 for a cup of coffee and $1.10 for a cup of tea. Write an objective function $z = f(x, y)$ that represents the total profit for selling x cups of coffee and y cups of tea. **(See Example 1)**

6. Rita burns 10 calories per minute running and 8 calories per minute lifting weights. Write an objective function $z = f(x, y)$ that represents the total number of calories burned by running for x minutes and lifting weights for y minutes.

7. A courier company makes deliveries with two different trucks. Truck A costs $0.62/mi to operate and truck B costs $0.50/mi to operate. Write an objective function $z = f(x, y)$ that represents the total cost for driving truck A for x miles and driving truck B for y miles.

8. The cost for an animal shelter to spay a female cat is $82 and the cost to neuter a male cat is $55. Write an objective function $z = f(x, y)$ that represents the total cost for spaying x female cats and neutering y male cats.

Julie Miller

Objective 2: Solve Linear Programming Applications

For Exercises 9–12,

a. Determine the values of x and y that produce the maximum or minimum value of the objective function on the given feasible region.

b. Determine the maximum or minimum value of the objective function on the given feasible region.

9. Maximize: $z = 3x + 2y$

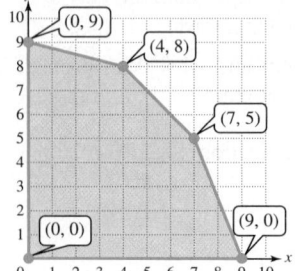

10. Maximize: $z = 1.8x + 2.2y$

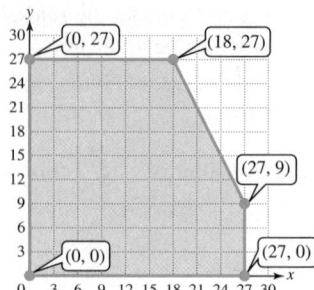

11. Minimize: $z = 1000x + 900y$

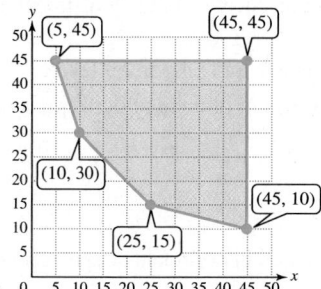

12. Minimize: $z = 6x + 9y$

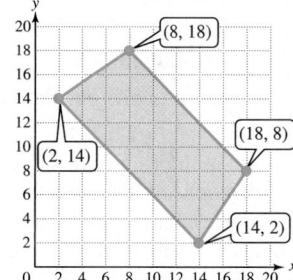

For Exercises 13–18,

   a. For the given constraints, graph the feasible region and identify the vertices.

   b. Determine the values of x and y that produce the maximum or minimum value of the objective function on the feasible region.

   c. Determine the maximum or minimum value of the objective function on the feasible region.

13. $x \geq 0, y \geq 0$
    $x + y \leq 60$
    $y \leq 2x$
    Maximize: $z = 250x + 150y$

14. $x \geq 0, y \geq 0$
    $2x + y \leq 40$
    $x + 2y \leq 50$
    Maximize: $z = 9.2x + 8.1y$

15. $x \geq 0, y \geq 0$
    $3x + y \geq 50$
    $2x + y \geq 40$
    Minimize: $z = 3x + 2y$

16. $x \geq 0, y \geq 0$
    $4x + 3y \geq 60$
    $2x + 3y \geq 36$
    Minimize: $z = 4.5x + 6y$

17. $x \geq 0, y \geq 0$
    $x \leq 36$
    $y \leq 40$
    $x + y \leq 48$
    Maximize: $z = 150x + 90y$

18. $x \geq 0, y \geq 0$
    $x \leq 10$
    $y \leq 8$
    $x + y \leq 12$
    Maximize: $z = 50x + 70y$

For Exercises 19–20, use the given constraints to find the maximum value of the objective function and the ordered pair (x, y) that produces the maximum value.

19. $x \geq 0, y \geq 0$
    $3x + 4y \leq 48$
    $2x + y \leq 22$
    $y \leq 9$
   a. Maximize: $z = 100x + 120y$
   b. Maximize: $z = 100x + 140y$

20. $x \geq 0, y \geq 0$
    $x + y \leq 20$
    $x + 2y \leq 36$
    $x \leq 14$
   a. Maximize: $z = 12x + 15y$
   b. Maximize: $z = 15x + 12y$

21. A furniture manufacturer builds tables. The cost for materials and labor to build a kitchen table is $240 and the profit is $160. The cost to build a dining room table is $320 and the profit is $240. **(See Examples 2–3)**

   Let x represent the number of kitchen tables produced per month. Let y represent the number of dining room tables produced per month.

   a. Write an objective function representing the monthly profit for producing and selling x kitchen tables and y dining room tables.

   b. The manufacturing process is subject to the following constraints. Write a system of inequalities representing the constraints.

     • The number of each type of table cannot be negative.
     • Due to labor and equipment restrictions, the company can build at most 120 kitchen tables.
     • The company can build at most 90 dining room tables.
     • The company does not want to exceed a monthly cost of $48,000.

   c. Graph the system of inequalities represented by the constraints.

   d. Find the vertices of the feasible region.

   e. Test the objective function at each vertex.

   f. How many kitchen tables and how many dining room tables should be produced to maximize profit? (Assume that all tables produced will be sold.)

   g. What is the maximum profit?

22. Guyton makes $24/hr tutoring chemistry and $20/hr tutoring math.

Let x represent the number of hours per week he spends tutoring chemistry. Let y represent the number of hours per week he spends tutoring math.

a. Write an objective function representing his weekly income for tutoring x hours of chemistry and y hours of math.

b. The time that Guyton devotes to tutoring is limited by the following constraints. Write a system of inequalities representing the constraints.

- The number of hours spent tutoring each subject cannot be negative.
- Due to the academic demands of his own classes he tutors at most 18 hr per week.
- The tutoring center requires that he tutors math at least 4 hr per week.
- The demand for math tutors is greater than the demand for chemistry tutors. Therefore, the number of hours he spends tutoring math must be at least twice the number of hours he spends tutoring chemistry.

c. Graph the system of inequalities represented by the constraints.

d. Find the vertices of the feasible region.

e. Test the objective function at each vertex.

f. How many hours tutoring math and how many hours tutoring chemistry should Guyton work to maximize his income?

g. What is the maximum income?

h. Explain why Guyton's maximum income is found at a point on the line $x + y = 18$.

23. A plant nursery sells two sizes of oak trees to landscapers. Large trees cost the nursery $120 from the grower. Small trees cost the nursery $80. The profit for each large tree sold is $35 and the profit for each small tree sold is $30. The monthly demand is at most 400 oak trees. Furthermore, the nursery does not want to allocate more than $43,200 each month on inventory for oak trees.

a. Determine the number of large oak trees and the number of small oak trees that the nursery should have in its inventory each month to maximize profit. (Assume that all trees in inventory are sold.)

b. What is the maximum profit?

c. If the profit on large trees were $50, and the profit on small trees remained the same, then how many of each should the nursery have to maximize profit?

24. A sporting goods store sells two types of exercise bikes. The deluxe model costs the store $400 from the manufacturer and the standard model costs the store $300 from the manufacturer. The profit that the store makes on the deluxe model is $180 and the profit on the standard model is $120. The monthly demand for exercise bikes is at most 30. Furthermore, the store manager does not want to spend more than $9600 on inventory for exercise bikes.

a. Determine the number of deluxe models and the number of standard models that the store should have in its inventory each month to maximize profit. (Assume that all exercise bikes in inventory are sold.)

b. What is the maximum profit?

c. If the profit on the deluxe bikes were $150 and the profit on the standard bikes remained the same, how many of each should the store have to maximize profit?

25. A paving company delivers gravel for a road construction project. The company has a large truck and a small truck. The large truck has a greater capacity, but costs more for fuel to operate. The load capacity and cost to operate each truck per load are given in the table.

	Load Capacity	Cost per Load
Small truck	18 yd^3	$120
Large truck	24 yd^3	$150

The company must deliver at least 288 yd^3 of gravel to stay on schedule. Furthermore, the large truck takes longer to load and cannot make as many trips as the small truck. As a result, the number of trips made by the large truck is at most $\frac{3}{4}$ times the number of trips made by the small truck.

a. Determine the number of trips that should be made by the large truck and the number of trips that should be made by the small truck to minimize cost.

b. What is the minimum cost to deliver gravel under these constraints?

26. A large department store needs at least 3600 labor hours covered per week. It employs full-time staff 40 hr/wk and part-time staff 25 hr/wk. The cost to employ a full-time staff member is more because the company pays benefits such as health care and life insurance.

	Hours per Week	Cost per Hour
Full time	40 hr	$25
Part time	25 hr	$18

The store manager also knows that to make the store run efficiently, the number of full-time employees must be at least 1.25 times the number of part-time employees.

a. Determine the number of full-time employees and the number of part-time employees that should be used to minimize the weekly labor cost.

b. What is the minimum weekly cost to staff the store under these constraints?

27. A manufacturer produces two models of a gas grill. Grill A requires 1 hr for assembly and 0.4 hr for packaging. Grill B requires 1.2 hr for assembly and 0.6 hr for packaging. The production information and profit for each grill are given in the table. **(See Example 4)**

	Assembly	Packaging	Profit
Grill A	1 hr	0.4 hr	$90
Grill B	1.2 hr	0.6 hr	$120

The manufacturer has 1200 hr of labor available for assembly and 540 hr of labor available for packaging.

a. Determine the number of grill A units and the number of grill B units that should be produced to maximize profit assuming that all grills will be sold.

b. What is the maximum profit under these constraints?

c. If the profit on grill A units is $110 and the profit on grill B units is unchanged, how many of each type of grill unit should the manufacturer produce to maximize profit?

29. A farmer has 1200 acres of land and plans to plant corn and soybeans. The input cost (cost of seed, fertilizer, herbicide, and insecticide) for 1 acre for each crop is given in the table along with the cost of machinery and labor. The profit for 1 acre of each crop is given in the last column.

	Input Cost per Acre	Labor/Machinery Cost per Acre	Profit per Acre
Corn	$180	$80	$120
Soybeans	$120	$100	$100

Suppose the farmer has budgeted a maximum of $198,000 for input costs and a maximum of $110,000 for labor and machinery.

a. Determine the number of acres of each crop that the farmer should plant to maximize profit. (Assume that all crops will be sold.)

b. What is the maximum profit?

c. If the profit per acre were reversed between the two crops (that is, $100 per acre for corn and $120 per acre for soybeans), how many acres of each crop should be planted to maximize profit?

28. A manufacturer produces two models of patio furniture. Model A requires 2 hr for assembly and 1.2 hr for painting. Model B requires 3 hr for assembly and 1.5 hr for painting. The production information and profit for selling each model are given in the table.

	Assembly	Painting	Profit
Model A	2 hr	1.2 hr	$150
Model B	3 hr	1.5 hr	$200

The manufacturer has 1200 hr of labor available for assembly and 660 hr of labor available for painting.

a. Determine the number of model A units and the number of model B units that should be produced to maximize profit assuming that all furniture will be sold.

b. What is the maximum profit under these constraints?

c. If the profit on model A units is $180 and the profit on model B units remains the same, how many of each type should the manufacturer produce to maximize profit?

30. To protect soil from erosion, some farmers plant winter cover crops such as winter wheat and rye. In addition to conserving soil, cover crops often increase crop yields in the row crops that follow in spring and summer. Suppose that a farmer has 800 acres of land and plans to plant winter wheat and rye. The input cost for 1 acre for each crop is given in the table along with the cost for machinery and labor. The profit for 1 acre of each crop is given in the last column.

	Input Cost per Acre	Labor/Machinery Cost per Acre	Profit per Acre
Wheat	$90	$50	$42
Rye	$120	$40	$35

Suppose the farmer has budgeted a maximum of $90,000 for input costs and a maximum of $36,000 for labor and machinery.

a. Determine the number of acres of each crop that the farmer should plant to maximize profit. (Assume that all crops will be sold.)

b. What is the maximum profit?

c. If the profit per acre for wheat were $40 and the profit per acre for rye were $45, how many acres of each crop should be planted to maximize profit?

Write About It

31. What is the purpose of linear programming?

33. How is the feasible region determined?

32. What is an objective function?

34. If an optimal value exists for an objective function, it exists at one of the vertices of the feasible region. Explain how to find the vertices.

CHAPTER 10   Detailed Summary

SECTION 10.1 Graphs of Systems of Linear Equations in Two Variables

Key Concepts	Examples	Page
Two or more linear equations taken together form a **system of linear equations**. A **solution** to a system of equations in two variables is an ordered pair that is a solution to each individual equation. Graphically, this is a point of intersection of the graphs of the equations.	**Example 1:** Solve by graphing. $\quad x + y = 3$ $\qquad\qquad\qquad\qquad 2x - y = 0$ Write each equation in slope-intercept form $(y = mx + b)$ to graph the lines. $y = -x + 3$ $y = 2x$	p. 716

• For a system of two linear equations in two variables, there is one solution to the system if the lines intersect in exactly one point.

• If the lines are parallel, then there is no solution to the system, and we say that the system is **inconsistent**.

• If the two equations represent the same line, then there are infinitely many solutions to the system, and we say that the equations are **dependent**.

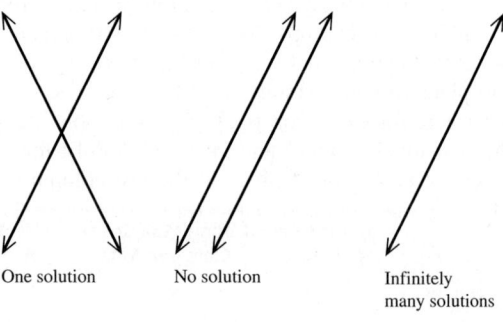

One solution   No solution   Infinitely many solutions

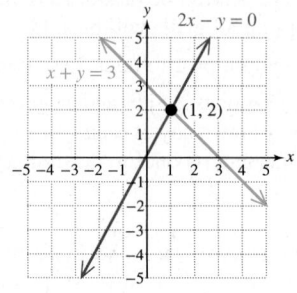

The solution is the point of intersection, (1, 2).
The solution set is {(1, 2)}.

Example 2:   p. 718

Solve by graphing. $\quad y = 2x - 3$
$\qquad\qquad\qquad\qquad -2x + y = 1$

Write each equation in slope-intercept form.

$y = 2x - 3 \longrightarrow y = 2x - 3$
$-2x + y = 1 \longrightarrow y = 2x + 1$

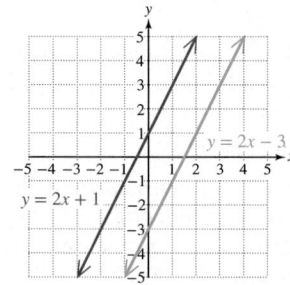

The solution set is { }.

SECTION 10.2 Systems of Linear Equations in Two Variables and Applications

Key Concepts	Examples	Page
The substitution method and the addition method are often used to solve a system of linear equations in two variables. 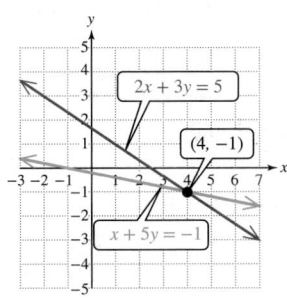	**Example 1:** Solve by using the substitution method. $x + 5y = -1$ $\xrightarrow{\text{Solve for } x.}$ $x = -1 - 5y$ $2x + 3y = 5$ $2(-1 - 5y) + 3y = 5$ Substitute $-1 - 5y$ for x. $-2 - 10y + 3y = 5$ $-7y = 7$ $y = -1$ Back substitution: $x = -1 - 5y$ $= -1 - 5(-1)$ $= 4$ The solution set is $\{(4, -1)\}$.	p. 724
A system of linear equations in two variables will have no solution if the equations in the system represent parallel lines. In such a case, we say that the system is **inconsistent**. 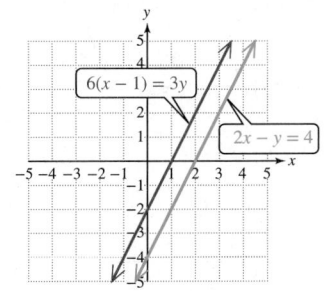	**Example 2:** Solve by using the addition method. $2x - y = 4$ $2x - y = 4$ $6(x - 1) = 3y$ $\xrightarrow{\text{Standard form.}}$ $6x - 3y = 6$ $2x - y = 4$ $\xrightarrow{\text{Multiply by } -3.}$ $-6x + 3y = -12$ $6x - 3y = 6$ $\xrightarrow{}$ $\underline{6x - 3y = 6}$ $0 = -6$ This results in a contradiction. There is no solution to the system, and the system is inconsistent. The solution set is { }.	p. 728
A system of linear equations in two variables will have infinitely many solutions if the equations represent the same line. In such a case, we say that the equations are **dependent**. The solution set to the system is the set of points on the line. 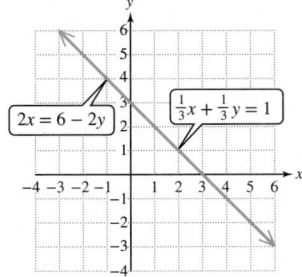	**Example 3:** $\frac{1}{3}x + \frac{1}{3}y = 1$ $\xrightarrow{\text{Multiply by 3.}}$ $x + y = 3$ $2x = 6 - 2y$ $\xrightarrow{}$ $2x + 2y = 6$ Standard form $x + y = 3$ $\xrightarrow{\text{Multiply by } -2.}$ $-2x - 2y = -6$ $2x + 2y = 6$ $\xrightarrow{}$ $\underline{2x + 2y = 6}$ $0 = 0$ The system results in an identity. The equations represent the same line $x + y = 3$. Solving for y in terms of x yields $y = 3 - x$. Thus, the solutions to the system are ordered pairs of the form $(x, 3 - x)$. The solution set can be written as $\{(x, 3 - x) \,|\, x \text{ is any real number}\}$ Likewise, by solving for x in terms of y, the solution set can be written as $\{(3 - y, y) \,|\, y \text{ is any real number}\}$	p. 729

SECTION 10.3 Systems of Linear Equations in Three Variables and Applications

Key Concepts	Examples	Page
A **linear equation in three variables** is an equation that can be written in the form $$Ax + By + Cz = D$$ where A, B, and C are not all zero.	**Example 1:**	p. 737

A **linear equation in three variables** is an equation that can be written in the form

$$Ax + By + Cz = D$$

where A, B, and C are not all zero.

A solution to a system of linear equations in three variables is an **ordered triple** (x, y, z) that satisfies each equation in the system. Geometrically, a solution is a point of intersection of the planes represented by the equations in the system.

Example 1:

$\boxed{A}$ $2x + 4y + 3z = 7$ Eliminate one variable from the
$\boxed{B}$ $3x \quad\ - z = 4$ system to create a system of two
$\boxed{C}$ $\quad y + 5z = -3$ equations and two variables.

$\boxed{A} \quad 2x + 4y + 3z = 7$ We will eliminate
$-4 \cdot \boxed{C} \quad\ \underline{ - 4y - 20z = 12}$ variable y first.
$\phantom{-4\cdot\boxed{C}} \quad 2x \quad\ -17z = 19 \quad \boxed{D}$

$\boxed{B}$ $3x - z = 4$ $\xrightarrow{\text{Multiply by } -2.}$ $-6x + 2z = -8$
$\boxed{D}$ $2x - 17z = 19 \xrightarrow{\text{Multiply by 3.}}$ $\underline{6x - 51z = 57}$
$\phantom{\boxed{D} 2x - 17z = 19} -49z = 49$
$\phantom{\boxed{D} 2x - 17z = 19xxxx} z = -1$

From equation $\boxed{B}$ we have $3x - (-1) = 4$
$ x = 1$

From equation $\boxed{C}$ we have $y + 5(-1) = -3$
$ y = 2$

The ordered triple $(1, 2, -1)$ checks in all three original equations.

The solution set is $\{(1, 2, -1)\}$.

When solving a system of linear equations, if a contradiction arises such as $0 = 4$, the system is **inconsistent** and has no solution.

If an identity arises such as $0 = 0$, the equations are **dependent** and the system has infinitely many solutions.

Example 2:   p. 740

$\boxed{A}$ $3x + y \quad\ = 6$
$\boxed{B}$ $8x + 3y - z = 2$ $8x + 3y - z = 2$
$\boxed{C}$ $-5x - 2y + z = 4$ $\underline{-5x - 2y + z = 4}$
$\phantom{\boxed{C} -5x - 2y + z = 4} 3x + y \quad\ = 6 \quad \boxed{D}$

$\boxed{A} \quad 3x + y = 6$
$-1 \cdot \boxed{D} \quad \underline{-3x - y = -6}$
$\phantom{-1\cdot\boxed{D} -3x - y} 0 = 0$ (Identity)

The equations are dependent. To find the general solution, express the dependency among the variables as an ordered triple. From equation $\boxed{A}$, we can solve for y in terms of x.

$\boxed{A}$ $y = 6 - 3x$

From equation $\boxed{B}$ or $\boxed{C}$ we can substitute $y = 6 - 3x$ and then solve for z in terms of x.

$\boxed{C}$ $-5x - 2(6 - 3x) + z = 4$
$\phantom{\boxed{C} -5x} x - 12 + z = 4$
$\phantom{\boxed{C} -5x - 12} z = 16 - x$

The solution set can be written as

$\{(x, 6 - 3x, 16 - x) | x \text{ is any real number}\}$.

Alternatively, the solution set can be written as

$\left\{ \left(\dfrac{6-y}{3}, y, \dfrac{y+42}{3} \right) \middle| y \text{ is any real number} \right\}$ or

$\{(16 - z, 3z - 42, z) | z \text{ is any real number}\}$.

SECTION 10.4 Partial Fraction Decomposition

Key Concepts	Examples	Page
Partial fraction decomposition is used to write a rational expression as a sum of simpler fractions.	**Example 1:** $$\frac{4x^3 + 4x^2 + 8x + 12}{x^4 + 4x^2} = \frac{4x^3 + 4x^2 + 8x + 12}{x^2(x^2 + 4)}$$ Set up the structure for the decomposition. $$\frac{4x^3 + 4x^2 + 8x + 12}{x^2(x^2 + 4)} = \frac{A}{x} + \frac{B}{x^2} + \frac{Cx + D}{x^2 + 4}$$	p. 750

There are two basic parts to find the partial fraction decomposition of a rational expression.

I. Factor the denominator of the expression into linear factors and quadratic factors that are not further factorable over the integers. Then set up the "form" or "structure" for the partial fraction decomposition into simpler fractions.

II. Next, multiply both sides of the equation by the LCD. Then set up a system of linear equations to find the coefficients of the terms in the numerator of each fraction.

Note: The numerator of the original rational expression must be of lesser degree than the denominator. If this is not the case, first use long division.

Multiply by the LCD to clear fractions.   p. 751

$$x^2(x^2 + 4)\left[\frac{4x^3 + 4x^2 + 8x + 12}{x^2(x^2 + 4)}\right] =$$
$$x^2(x^2 + 4)\left[\frac{A}{x} + \frac{B}{x^2} + \frac{Cx + D}{x^2 + 4}\right]$$

Apply the distributive property and combine like terms.

$$4x^3 + 4x^2 + 8x + 12$$
$$= Ax(x^2 + 4) + B(x^2 + 4) + (Cx + D)x^2$$
$$= Ax^3 + 4Ax + Bx^2 + 4B + Cx^3 + Dx^2$$

Therefore,

$$4x^3 + 4x^2 + 8x + 12$$
$$= (A + C)x^3 + (B + D)x^2 + 4Ax + 4B$$

Equate the coefficients on like terms and solve the resulting system for A, B, C, and D.

$$\left.\begin{array}{l} 4 = A + C \\ 4 = B + D \\ 8 = 4A \\ 12 = 4B \end{array}\right\} \quad A = 2, B = 3, C = 2, D = 1$$

The partial fraction decomposition is:

$$\frac{4x^3 + 4x^2 + 8x + 12}{x^4 + 4x^2} = \frac{2}{x} + \frac{3}{x^2} + \frac{2x + 1}{x^2 + 4}$$

SECTION 10.5 Systems of Nonlinear Equations in Two Variables

Key Concepts	Examples	Page
A **nonlinear system of equations** is a system in which one or more equations is nonlinear.	**Example 1:**	p. 759

The substitution method is often used to solve a nonlinear system of equations.

In some cases, the addition method can be used provided that the terms containing the corresponding variables are like terms.

Use the substitution method to solve the system.

A $y = -x - 1$ A $y = -x - 1$

B $y = x^2 + 6x + 9$ B $y = x^2 + 6x + 9$

B $-x - 1 = x^2 + 6x + 9$

$$0 = x^2 + 7x + 10$$
$$0 = (x + 5)(x + 2)$$
$$x = -5 \quad \text{or} \quad x = -2$$

Example 2:

Use the addition method to solve the system.

$$\boxed{A} \quad x^2 + 2y^2 = 11 \xrightarrow{\text{Multiply by } -1.} -x^2 - 2y^2 = -11$$
$$\boxed{B} \quad x^2 + 3y^2 = 12 \xrightarrow{} \underline{x^2 + 3y^2 = 12}$$
$$y^2 = 1$$
$$y = \pm 1$$

From equation $\boxed{A}$, we have

If $y = 1$, $x^2 + 2(1)^2 = 11$ $\qquad x^2 = 9$
$\qquad\qquad\qquad\qquad\qquad\qquad\qquad x = \pm 3$

If $y = -1$, $x^2 + 2(-1)^2 = 11$ $\qquad x^2 = 9$
$\qquad\qquad\qquad\qquad\qquad\qquad\qquad x = \pm 3$

The solution set is
$\{(3, 1), (3, -1), (-3, 1), (-3, -1)\}$.

From equation $\boxed{A}$, we have

If $x = -5$, $y = -(-5) - 1$ $\qquad y = 4$ $\qquad (-5, 4)$
If $x = -2$, $y = -(-2) - 1$ $\qquad y = 1$ $\qquad (-2, 1)$

The solution set is $\{(-5, 4), (-2, 1)\}$.

The solutions are the points of intersection of the graphs of the two equations.

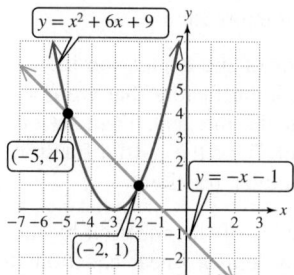

p. 761

SECTION 10.6 Inequalities and Systems of Inequalities in Two Variables

Key Concepts	Examples	Page

Linear inequalities in two variables:

An inequality of the form $Ax + By < C$, where A and B are not both zero, is called a **linear inequality in two variables**. (The symbols $>$, $\leq$, and $\geq$ can be used in place of $<$ in the definition.)

The inequality in Example 1 is a linear inequality.
The inequality in Example 2 is a *nonlinear* inequality.

Example 1:
$x - y > 2$
Graph $x - y = 2$.

Example 2:
$x^2 + y^2 \leq 4$
Graph $x^2 + y^2 = 4$.

p. 768

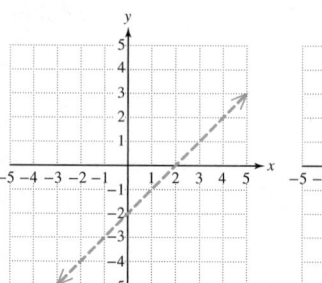

Test a point above or below the line.

Test a point inside or outside the circle.

Solving a linear inequality in two variables:

The basic steps to solve a linear inequality in two variables are as follows.

1. Graph the related equation. The resulting line is drawn as a dashed line if the inequality is strict, and is otherwise drawn as a solid line.
2. Select a test point from either side of the line. If the ordered pair makes the original inequality true, then shade the half-plane from which the point was taken. Otherwise, shade the other half-plane.

A nonlinear inequality in two variables is solved using the same basic procedure.

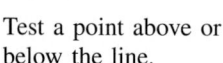

Test $(0, 0)$
$$x - y > 2$$
$$(0) - (0) \overset{?}{>} 2 \text{ false}$$
Shade below the line.

Test $(0, 0)$
$$x^2 + y^2 \leq 4$$
$$(0)^2 + (0)^2 \overset{?}{\leq} 4 \text{ true}$$
Shade inside the circle.

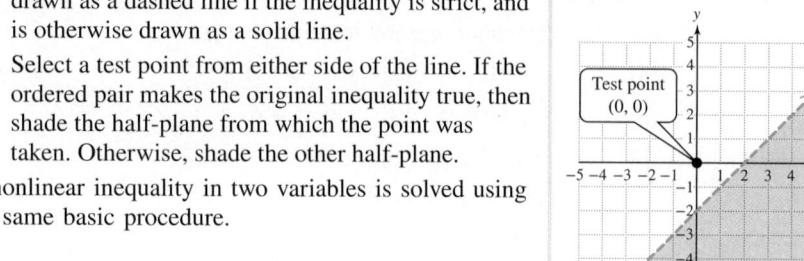

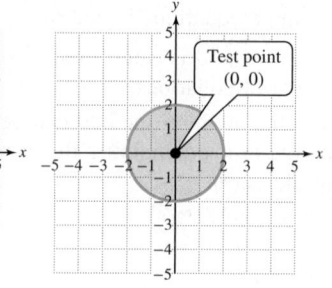

Two or more inequalities in two variables make up a system of inequalities in two variables. The solution set to the system is the region of overlap (intersection) of the solution sets of the individual inequalities.

Example 3:   p. 773

$3x + 2y < 6$

$y \geq 2x + 3$

Graph the related equations and test an ordered pair from a point above or below each line.

$3x + 2y = 6$ $y = 2x + 3$

Test $(-2, 1)$ Test $(-2, 1)$

$3(-2) + 2(1) \overset{?}{<} 6$ Yes $1 \overset{?}{\geq} 2(-2) + 3$ Yes

Shade below the line $3x + 2y = 6$ and above the line $y = 2x + 3$.

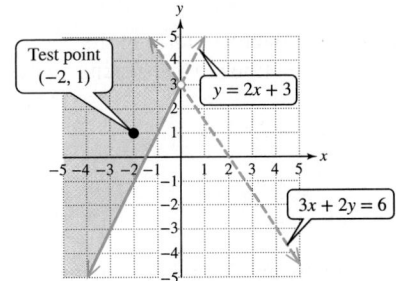

SECTION 10.7 Linear Programming

Key Concepts	Examples	Page

A process called **linear programming** enables us to maximize or minimize a function under specified constraints. The function to be maximized or minimized is called the **objective function**.

The steps to solve a linear programming application are outlined here.

Step 1 Write an objective function, $z = f(x, y)$.

Step 2 Write a system of inequalities defining the constraints on x and y.

Step 3 Graph the feasible region and identify the vertices.

Step 4 Evaluate the objective function at each vertex of the feasible region. Use the results to identify the values of x and y that optimize the objective function, and identify the optimal value of z.

Example 1:   p. 782

An artist makes necklaces and earrings. The cost of materials for a necklace is \$12 and the cost for a pair of earrings is \$6. The artist has time to create at most 40 necklaces per week and at most 60 pairs of earrings per week. Furthermore, the artist does not want to exceed \$540 per week in cost of materials. If the profit on a necklace is \$20 and the profit on a pair of earrings is \$15, how many of each item should the artist make to maximize weekly profit? Assume that all necklaces and earrings made also sell.

Let x represent the number of necklaces.
Let y represent the number of pairs of earrings.

Step 1 Maximize profit: $z = 20x + 15y$

Step 2 $x \geq 0, y \geq 0, x \leq 40, y \leq 60$

$12x + 6y \leq 540$

Step 3

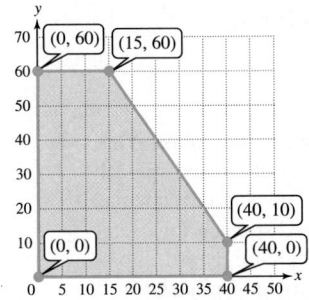

Step 4 Test $z = 20x + 15y$ at each vertex:

at $(0, 0)$: $z = 20(0) + 15(0) = \$0$
at $(0, 60)$: $z = 20(0) + 15(60) = \$900$
at $(15, 60)$: $z = 20(15) + 15(60) = \boxed{\$1200}$
at $(40, 10)$: $z = 20(40) + 15(10) = \$950$
at $(40, 0)$: $z = 20(40) + 15(0) = \$800$

The maximum weekly profit of \$1200 will be obtained if the artist makes 15 necklaces and 60 pairs of earrings.

CHAPTER 10 Review Exercises

SECTION 10.1

1. Determine if the ordered pair is a solution to the system.

$$2x - 3y = 0$$
$$-5x + 6y = -1$$

a. $\left(1, \dfrac{2}{3}\right)$ **b.** $(6, 4)$

For Exercises 2–3, based on the slope-intercept form of the equations, determine the number of solutions to the system.

2. $y = -\dfrac{3}{5}x - 4$

$y = -\dfrac{3}{5}x + 1$

3. $y = 2x + 6$

$y = \dfrac{1}{2}x - 6$

For Exercises 4–7, solve the system by graphing. For systems that do not have one unique solution, also state the number of solutions and whether the system is inconsistent or the equations are dependent.

4. $3x - y = 8$

$x + 2y = -2$

5. $x + 3y = 9$

$2x = 6$

6. $2x = y - 1$

$\dfrac{1}{2}x - \dfrac{1}{4}y = 1$

7. $y = -x + 6$

$3x + 3y = 18$

SECTION 10.2

For Exercises 8–12, solve the system by using any method. If the system does not have one unique solution, state whether the system is inconsistent, or whether the equations are dependent.

8. $4x - y = 7$

$-2x + 5y = 19$

9. $5(x - y) = 19 - 2y$

$0.2x + 0.7y = -1.7$

10. $9x - 2y = 4$

$2x + 4y = 7$

11. $\frac{1}{10}x - \frac{1}{2}y = 1$

$2x = 10y + 6$

12. $y = \frac{3}{4}x$

$4(y - x) = -x$

13. Shenika wants to monitor her daily calcium intake. One day she had 3 cups of milk and 1 cup of cooked spinach for a total of 1140 mg of calcium. The next day, she had 2 cups of milk and $1\frac{1}{2}$ cups of cooked spinach for a total of 960 mg of calcium. How much calcium is in 1 cup of milk and how much is in 1 cup of cooked spinach?

14. How many liters of a 40% acid mixture and how many liters of a 10% acid mixture should be mixed to obtain 20 L of a 22% acid mixture?

15. A plane can travel 960 mi in 2 hr with a tail wind. The return trip against the wind takes 2 hr and 40 min. Find the speed of the plane in still air and the speed of the wind.

16. A fishing boat captain charges $250 for an excursion. His fixed monthly expenses are $1200 for insurance, rent for the dock, and minor office expenses. He also has variable costs of $100 per excursion to cover gasoline, bait, and other equipment.

Paul Yates/Shutterstock

a. Write a linear cost function representing the cost $C(x)$ (in $) for the fishing boat captain to run x excursions per month.

b. Write a linear revenue function representing the revenue $R(x)$ (in $) for x excursions per month.

c. Determine the number of excursions per month for the captain to break even.

d. If 18 excursions are run in a given month, how much money will the fishing boat captain earn or lose?

SECTION 10.3

For Exercises 17–20, solve the system. If a system has one unique solution, write the solution set. Otherwise, determine the number of solutions to the system, and determine whether the system is inconsistent, or the equations are dependent.

17. $3a - 4b + 2c = -17$

$2a + 3b + c = 1$

$4a + b - 3c = 7$

18. $6x = 24 - 5y$

$14 = 7z - 3y$

$4x - 3z = 10$

19. $x + 2y + z = 5$

$x + y - z = 1$

$4x + 7y + 2z = 16$

20. $u + v + 2w = 1$

$2v - 5w = 2$

$3u + 5v + w = 1$

21. Solve the system and write the general solution.

$$5x + 2y + z = 0$$
$$-4x + 3y - z = 0$$
$$6x + 7y + z = 0$$

22. An arena that hosts sporting events and concerts has three sections for three levels of seating. For a basketball game, seats in Section A cost $90, seats in Section B cost $65, and seats in Section C cost $40. The number of seats in Section C equals the number of seats in Sections A and B combined. The arena holds 12,000 seats and the game is sold out. If the total revenue from ticket sales is $655,000, determine the number of seats in each section.

23. Emily receives an inheritance of $20,000 and decides to invest the money. She puts some money in her savings account that earns 1.5% simple interest per year. The remaining money is invested in a bond fund that returns 4.5% and a stock fund that returns 6.2%. She makes a total of $942 at the end of 1 year. If she invested twice as much in the bond fund as the stock fund, determine the amount that she invested in each fund.

For Exercises 24–25, use a system of linear equations in three variables to find an equation of the form $y = ax^2 + bx + c$ that defines the parabola through the points.

24. $(-1, -4), (1, 6), (3, 8)$ **25.** $(1, -2), (2, 1), (3, 10)$

SECTION 10.4

For Exercises 26–31, set up the form for the partial fraction decomposition. Do not solve for A, B, C, and so on.

26. $\dfrac{5x + 22}{x^2 + 8x + 16}$ **27.** $\dfrac{-x - 11}{(x + 2)(x - 1)}$

28. $\dfrac{2x^2 + x - 10}{x^3 + 5x}$ **29.** $\dfrac{7x^2 + 19x + 15}{2x^3 + 3x^2}$

30. $\dfrac{4x^4 - 3x^2 + 2x + 5}{x(2x + 5)^3(x^2 + 2)^2}$ **31.** $\dfrac{2x^3 - x^2 + 8x - 16}{x^4 + 5x^2 + 4}$

For Exercises 32–36, perform the partial fraction decomposition.

32. $\dfrac{-x - 11}{(x + 2)(x - 1)}$ **33.** $\dfrac{5x + 22}{x^2 + 8x + 16}$

34. $\dfrac{2x^4 + 7x^3 + 13x^2 + 19x + 15}{2x^3 + 3x^2}$

35. $\dfrac{2x^2 + x - 10}{x^3 + 5x}$ **36.** $\dfrac{2x^3 - x^2 + 8x - 16}{x^4 + 5x^2 + 4}$

SECTION 10.5

For Exercises 37–38,

a. Graph the equations.

b. Solve the system.

37. $y - x^2 = 1$ **38.** $y = \sqrt{x - 1}$
   $x - y = -3$ $x^2 + y^2 = 5$

For Exercises 39–41, solve the system.

39. $3x^2 - y^2 = -4$ **40.** $2x^2 - xy = 24$
   $x^2 + 2y^2 = 36$ $x^2 + 3xy = -9$

41. $y = \dfrac{8}{x}$
   $y = \sqrt{x}$

42. The sum of the squares of two negative numbers is 97 and the difference of their squares is 65. Find the numbers.

43. The ratio of two numbers is 4 to 3. The sum of the squares of the numbers is 100. Find the numbers.

44. The hypotenuse of a right triangle is $\sqrt{74}$ ft and the sum of the lengths of the legs is 12 ft. Find the lengths of the legs.

45. A rectangular billboard has a perimeter of 72 ft and an area of 288 ft^2. Find the dimensions of the billboard.

SECTION 10.6

46. Graph the solution set to the inequality.

a. $3x + 4y \le 8$

b. $3x + 4y > 8$

47. Graph the solution set to the inequality.

a. $y < (x - 4)^2$

b. $y \ge (x - 4)^2$

For Exercises 48–52, graph the solution set.

48. $5(x + y) \ge 8x + 15$ **49.** $x \le 3.5$

50. $-\dfrac{3}{2}y + 1 < 4$ **51.** $x^2 + (y + 2)^2 < 4$

52. $|y| > 2$

53. Determine if the given ordered pair is a solution to the system of inequalities.

$$x + 2y < 4$$
$$3x - 4y \ge 6$$

a. $(0, 1)$ **b.** $(1, -4)$

For Exercises 54–57, graph the solution set. If there is no solution, indicate that the solution set is the empty set.

54. $y > \dfrac{1}{2}x + 1$ **55.** $x^2 + y^2 \le 9$
   $3x + 2y < 4$ $(x - 1)^2 + y^2 \ge 4$

56. $y \ge x^2 - 3$ **57.** $y > e^x$
   $y > 1$ $y < -x^2 - 1$
   $x + y \le 3$

58. Let x represent the number of hours that Gordon spends tutoring math, and let y represent the number of hours that he spends tutoring English. For parts (a)–(d), write an inequality to represent the given statement.

a. Gordon has at most 12 hr to tutor per week.

b. The amount of time that Gordon spends tutoring English is at least twice the amount of time he spends tutoring math.

c. The number of hours spent tutoring math cannot be negative.

d. The number of hours spent tutoring English cannot be negative.

e. Graph the solution set to the system of inequalities from parts (a)–(d).

SECTION 10.7

59. At a home store, one sheet of $\frac{3}{8}$-in. sanded pine plywood costs \$24. One sheet of $\frac{1}{4}$-in. sanded pine plywood costs \$20. Write an objective function $z = f(x, y)$ that represents the total cost for x $\frac{3}{8}$-in. sheets and y $\frac{1}{4}$-in. sheets.

60. For the feasible region given in the figure and the objective function $z = 36x + 50y$,

   a. Determine the values of x and y that produce the maximum value of the objective function.

   b. Determine the maximum value of the objective function.

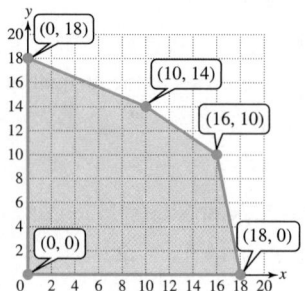

61. For the given constraints and the objective function, $z = 55x + 40y$,

   a. Graph the feasible region and identify the vertices.

$$x \geq 0, y \geq 0$$
$$2x + y \geq 18$$
$$5x + 4y \geq 60$$

   b. Determine the values of x and y that produce the minimum value of the objective function on the feasible region.

c. Determine the minimum value of the objective function on the feasible region.

62. A fitness instructor wants to mix two brands of protein powder to form a blend that limits the amount of fat and carbohydrate but maximizes the amount of fiber.

The nutritional information is given in the table for a single scoop of protein powder.

	Fat	Carbohydrates	Fiber
Brand A	3 g	3 g	10 g
Brand B	2 g	4 g	8 g

Suppose that the fitness instructor wants to make at most 180 scoops of the mixture. She also wants to limit the amount of fat to 480 g and she wants to limit the amount of carbohydrate to 696 g.

   a. Determine the number of scoops of each type of powder that will maximum the amount of fiber.

   b. What is the maximum amount of fiber?

   c. If the fiber content were reversed between the two brands (that is, 8 g for brand A and 10 g for brand B), then how much of each type of protein powder should be used to maximize the amount of fiber?

CHAPTER 10   Test

For Exercises 1–3, determine if the ordered pair or ordered triple is a solution to the system.

1. $x - 5y = -3$
   $y = 2x - 12$

   a. $(7, 2)$

   b. $(-3, 0)$

2. $2x - 3y + z = -5$
   $5x + y - 3z = -18$
   $-x + 2y + 5z = 8$

   a. $(0, 1, -2)$

   b. $(-3, 0, 1)$

3. $2x - 4y < 9$
   $-3x + y \geq 4$

   a. $(-6, 1)$

   b. $(1, 4)$

4. For each pair of systems, write the equations in slope-intercept form and identify the number of solutions to the system

   a. $-3x + y = -4$
      $x + y = 5$

   b. $x + y = 3$
      $4x + 4y = 12$

   c. $2y = 6x$
      $3x - y = 6$

For Exercises 5–6, solve the system by graphing. For systems that do not have one unique solution, also state the number of solutions and whether the system is inconsistent or the equations are dependent.

5. $x + 2y = -6$
  $3x = 2 - y$

6. $2y = x - 5$
  $x = 2y$

For Exercises 7–17, solve the system. If the system does not have one unique solution, also state whether the system is inconsistent or whether the equations are dependent.

7. $x = 5 - 4y$
  $-3x + 7y = 4$

8. $0.2x = 0.35y - 2.5$
  $0.16x + 0.5y = 5.8$

9. $x - \dfrac{2}{5}y = \dfrac{3}{10}$
  $5x = 2y + \dfrac{3}{2}$

10. $7(x - y) = 3 - 5y$
  $4(3x - y) = -2x$

11. $a + 6b + 3c = -14$
  $2a + b - 2c = -8$
  $-3a + 2b + c = -8$

12. $x \quad\quad + 4z = 10$
  $3y - 2z = 9$
  $2x + 5y \quad\quad = 21$

13. $2x - y + z = -3$
  $x - 3y \quad\quad = 2$
  $x + 2y + z = -7$

14. $(x - 4)^2 + y^2 = 25$
  $x - y = 3$

15. $5x^2 + y^2 = 14$
  $x^2 - 2y^2 = -17$

16. $2xy - y^2 = -24$
  $-3xy + 2y^2 = 38$

17. $\dfrac{1}{x + 3} - \dfrac{2}{y - 1} = -7$
  $\dfrac{3}{x + 3} + \dfrac{1}{y - 1} = 7$

18. Solve the system and write the general solution.
  $x \quad\quad -2z = 6$
  $y + 3z = 2$
  $x + y + z = 8$

19. At a candy and nut shop, the manager wants to make a nut mixture that is 56% peanuts. How many pounds of peanuts must be added to an existing mixture of 45% peanuts to make 20 lb of a mixture that is 56% peanuts?

20. Two runners begin at the same point on a 400-m track. If they run in opposite directions, they pass each other in 40 sec. If they run in the same direction, they will meet again in 200 sec. Find the speed of each runner.

21. Dylan invests $15,000 in three different stocks. One stock is very risky and after 1 year loses 8%. The second stock returns 3.2%, and a third stock returns 5.8%. At the end of 1 year, the total return is $274. If he invested $2000 more in the second stock than in the third stock, determine the amount he invested in each stock.

22. The difference of two positive numbers is 3 and the difference of their squares is 33. Find the numbers.

23. A rectangular television screen has a perimeter of 154 in. and an area of 1452 in.^{2}. Find the dimensions of the screen.

24. Use a system of linear equations in three variables to find an equation of the form $y = ax^2 + bx + c$ that defines the parabola through the points $(1, -1)$, $(2, 1)$, and $(-1, 7)$.

For Exercises 25–26, set up the form for the partial fraction decomposition. Do not solve for A, B, C, and so on.

25. $\dfrac{-15x + 15}{3x^2 + x - 2}$

26. $\dfrac{5x^6 + 3x^5 - 4x^3 + x - 3}{x^3(x - 3)(x^2 + 5x + 1)^2}$

For Exercises 27–31, perform the partial fraction decomposition.

27. $\dfrac{-12x - 29}{2x^2 + 11x + 15}$

28. $\dfrac{6x + 8}{x^2 + 4x + 4}$

29. $\dfrac{x^4 - 6x^3 + 4x^2 + 20x - 32}{x^3 - 4x^2}$

30. $\dfrac{x^2 - 2x - 21}{x^3 + 7x}$

31. $\dfrac{7x^3 + 4x^2 + 63x + 15}{x^4 + 11x^2 + 18}$

For Exercises 32–36, graph the solution set.

32. $2(x + y) > 6 - y$

33. $(x + 3)^2 + y^2 \geq 9$

34. $|x| < 4$

35. $x + y \leq 4$
  $2x - y > -2$

36. $y \leq -x^2 + 5$
  $y > 1$
  $x + y \leq 3$

37. A donut shop makes a profit of $2.40 on a dozen donuts and $0.55 per muffin. Write an objective function $z = f(x, y)$ that represents the total profit for selling x dozen donuts and y muffins.

38. For the feasible region given and the objective function $z = 4x + 5y$,

  a. Determine the values of x and y that produce the minimum value of the objective function on the feasible region.

  b. Determine the minimum value of the objective function on the feasible region.

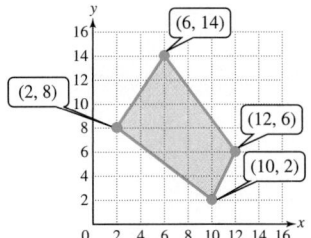

39. For the given constraints and objective function, $z = 600x + 850y$,

    a. Graph the feasible region and identify the vertices.

$$x \geq 0, y \geq 0$$
$$x + y \leq 48$$
$$y \leq 3x$$

    b. Determine the values of x and y that produce the maximum value of the objective function on the feasible region.

    c. Determine the maximum value of the objective function on the feasible region.

40. A weight lifter wants to mix two types of protein powder. One is a whey protein and one is a soy protein. The fat, carbohydrate, and protein content (in grams) for 1 scoop of each powder is given in the table.

	Fat	Carbohydrates	Protein
Whey	3 g	3 g	20 g
Soy	2 g	4 g	18 g

Suppose that the weight lifter wants to make at most 60 scoops of a protein powder mixture. Furthermore, he wants to limit the total fat content to at most 150 g and the total carbohydrate content to at most 216 g.

    a. Determine the number of scoops of each type of powder that will maximize the total protein content under these constraints.

    b. What is the maximum total protein content?

    c. If the protein content were reversed between the two brands (that is, 18 g for the whey protein and 20 g for the soy protein), then how much of each type of protein powder should be used to maximize the amount of protein?

Matrices and Determinants and Applications

Jasper White/Image Source

For matters of security, e-mail messages, websites, and other electronic media often use encryption techniques to encode data into an obscure form. This prevents those who do not know the code from understanding what has been transmitted. A computer program can encrypt words by using a matrix (a two-dimensional array) to encode the message, and then use the inverse of that matrix to decode it.

In this chapter, we will use matrices in a variety of ways. Matrices can be used to model systems of linear equations. Furthermore, a variety of techniques involving the algebra and manipulation of matrices can be used to solve the systems. The first two techniques are called Gaussian elimination and Gauss-Jordan elimination. These methods reduce a system of equations to a simpler system of equations so that the solution can be found by inspection. Next, we use the inverse of a matrix to solve a related matrix equation. Finally, we present Cramer's rule. This method provides convenient formulas to solve a system of linear equations based on the coefficients of the system.

OBJECTIVES

1. Write an Augmented Matrix
2. Use Elementary Row Operations
3. Use Gaussian Elimination and Gauss-Jordan Elimination

1. Write an Augmented Matrix

The data in Table 11-1 represent the body mass index (BMI) for people of selected heights and weights. The National Institutes of Health suggests that a BMI should ideally be between 18.5 and 24.9. A table of data values such as Table 11-1 can be represented as a rectangular array of elements called a **matrix**.

Table 11-1 Body Mass Index

Height \ Weight	140 lb	160 lb	180 lb	200 lb
5'7"	21.9	25.1	28.2	31.3
5'10"	20.1	23.0	25.8	28.7
6'1"	18.5	21.1	23.7	26.4

$$\xrightarrow{\text{matrix}} \begin{bmatrix} 21.9 & 25.1 & 28.2 & 31.3 \\ 20.1 & 23.0 & 25.8 & 28.7 \\ 18.5 & 21.1 & 23.7 & 26.4 \end{bmatrix}$$

(*Source*: National Institutes of Health, www.nih.gov)

A matrix can be used to represent a system of linear equations written in standard form. To do so, we extract the coefficients of each term in the equation to form an **augmented matrix**. For example:

System of Equations	Augmented Matrix
$3x + 2y = 5$	
$x - y + 3z = 1$	
$2x + y + z = 4$	

$$\begin{bmatrix} 3 & 2 & 0 & 5 \\ 1 & -1 & 3 & 1 \\ 2 & 1 & 1 & 4 \end{bmatrix}$$

The vertical bar within the augmented matrix separates the coefficients of the variable terms in the equations from the constant terms.

FOR REVIEW

Recall that the standard form of a linear equation in two variables is $Ax + By = C$. In three variables the standard form is $Ax + By + Cz = D$.

EXAMPLE 1 Writing and Interpreting an Augmented Matrix

a. Write the augmented matrix for the system.
$$2x = 5y + 5$$
$$3(x + y) = 17 + y$$

b. Write a system of linear equations represented by the augmented matrix.
$$\begin{bmatrix} 1 & 0 & 0 & 6 \\ 0 & 1 & 0 & -10 \\ 0 & 0 & 1 & 4 \end{bmatrix}$$

Solution:

a. Write each equation in standard form.
$$2x = 5y + 5 \longrightarrow 2x - 5y = 5$$
$$3(x + y) = 17 + y \longrightarrow 3x + 2y = 17$$
$$\left. \right\} \xrightarrow{\text{augmented matrix}} \begin{bmatrix} 2 & -5 & 5 \\ 3 & 2 & 17 \end{bmatrix}$$

b.
$$\begin{bmatrix} 1 & 0 & 0 & 6 \\ 0 & 1 & 0 & -10 \\ 0 & 0 & 1 & 4 \end{bmatrix}$$
$$\longrightarrow x + 0y + 0z = 6 \qquad\qquad x = 6$$
$$\longrightarrow 0x + y + 0z = -10 \quad \text{or simply} \quad y = -10$$
$$\longrightarrow 0x + 0y + z = 4 \qquad\qquad z = 4$$

Skill Practice 1

a. Write the augmented matrix.
$$7x = 9 + 2y$$
$$2(x - y) = 4$$

b. Write a system of linear equations represented by the augmented matrix.
$$\begin{bmatrix} 1 & 0 & -8 \\ 0 & 1 & 3 \end{bmatrix}$$

Answers

1. **a.** $\begin{bmatrix} 7 & -2 & 9 \\ 2 & -2 & 4 \end{bmatrix}$

   b. $x = -8, y = 3$

2. Use Elementary Row Operations

We can use an augmented matrix to solve a system of linear equations in much the same way as we use the addition method. When using the addition method, notice that the following operations produce an equivalent system of equations. This means that the two systems have the same solution set.

- Interchange two equations.

  Example: $\begin{aligned} -3x - 5y &= -13 \\ x + 2y &= 5 \end{aligned}$ $\xrightarrow[\text{equations.}]{\text{Interchange}}$ $\begin{aligned} x + 2y &= 5 \\ -3x - 5y &= -13 \end{aligned}$

- Multiply an equation by a nonzero constant.

  Example: $\begin{aligned} x + 2y &= 5 \\ -3x - 5y &= -13 \end{aligned}$ $\xrightarrow{\text{Multiply by 3.}}$ $\begin{aligned} 3x + 6y &= 15 \\ -3x - 5y &= -13 \end{aligned}$

- Add a nonzero multiple of one equation to another equation.

  Example: $\begin{aligned} x + 2y &= 5 \\ -3x - 5y &= -13 \end{aligned}$ $\xrightarrow{\text{Multiply by 3.}}$ $\left. \begin{aligned} 3x + 6y &= 15 \\ -3x - 5y &= -13 \end{aligned} \right\}$ Add the equations.
  $$y = 2$$

These same operations performed on a matrix are called **elementary row operations**.

Elementary Row Operations

1. Interchange two rows.

  Example: $\begin{bmatrix} -3 & -5 & | & -13 \\ 1 & 2 & | & 5 \end{bmatrix}$ $\xrightarrow[\substack{\text{Interchange rows} \\ \text{1 and 2.}}]{R_1 \Leftrightarrow R_2}$ $\begin{bmatrix} 1 & 2 & | & 5 \\ -3 & -5 & | & -13 \end{bmatrix}$

2. Multiply a row by a nonzero constant.

  Example: $\begin{bmatrix} 1 & 2 & | & 5 \\ -3 & -5 & | & -13 \end{bmatrix}$ $\xrightarrow[\substack{\text{Multiply} \\ \text{row 1 by 3.}}]{3R_1 \rightarrow R_1}$ $\begin{bmatrix} 3 & 6 & | & 15 \\ -3 & -5 & | & -13 \end{bmatrix}$

3. Add a nonzero multiple of one row to another row.

  Example: $\begin{bmatrix} 1 & 2 & | & 5 \\ -3 & -5 & | & -13 \end{bmatrix}$ $\xrightarrow[\substack{\text{Add 3 times} \\ \text{row 1 to row 2.}}]{3R_1 + R_2 \rightarrow R_2}$ $\begin{bmatrix} 1 & 2 & | & 5 \\ 0 & 1 & | & 2 \end{bmatrix}$

$$\begin{array}{rrr|r} 3R_1 & 3 & 6 & 15 \\ + R_2 & -3 & -5 & -13 \\ \hline \rightarrow R_2 & 0 & 1 & 2 \end{array}$$

> **Avoiding Mistakes**
>
> When we add a constant multiple of one row to another row, we do not change the row that was multiplied by the constant.

Two matrices are said to be **row equivalent** if one matrix can be transformed into the other matrix through a series of elementary row operations.

EXAMPLE 2   Performing Elementary Row Operations

Given $\begin{bmatrix} 4 & 3 & 0 & | & 5 \\ 1 & 4 & -1 & | & 9 \\ 2 & 0 & -3 & | & -8 \end{bmatrix}$, perform the following elementary row operations.

a. $R_1 \Leftrightarrow R_2$ **b.** $\dfrac{1}{4} R_1 \rightarrow R_1$ **c.** $-2R_2 + R_3 \rightarrow R_3$

Solution:

a.
$$\begin{bmatrix} 4 & 3 & 0 & | & 5 \\ 1 & 4 & -1 & | & 9 \\ 2 & 0 & -3 & | & -8 \end{bmatrix} \xrightarrow[\substack{\text{Interchange rows} \\ \text{1 and 2.}}]{R_1 \Leftrightarrow R_2} \begin{bmatrix} 1 & 4 & -1 & | & 9 \\ 4 & 3 & 0 & | & 5 \\ 2 & 0 & -3 & | & -8 \end{bmatrix}$$

> **TIP** The notation $\frac{1}{4}R_1 \to R_1$ means to multiply row 1 by $\frac{1}{4}$ and then *replace* row 1 by the result.

b.
$$\begin{bmatrix} 4 & 3 & 0 & | & 5 \\ 1 & 4 & -1 & | & 9 \\ 2 & 0 & -3 & | & -8 \end{bmatrix} \xrightarrow[\substack{\text{Multiply} \\ \text{row 1 by } \frac{1}{4}.}]{\frac{1}{4}R_1 \to R_1} \begin{bmatrix} 1 & \frac{3}{4} & 0 & | & \frac{5}{4} \\ 1 & 4 & -1 & | & 9 \\ 2 & 0 & -3 & | & -8 \end{bmatrix}$$

c.
$$\begin{bmatrix} 4 & 3 & 0 & | & 5 \\ 1 & 4 & -1 & | & 9 \\ 2 & 0 & -3 & | & -8 \end{bmatrix}$$

$$\begin{array}{rrrrr} -2R_2 & -2 & -8 & 2 & | & -18 \\ +R_3 & 2 & 0 & -3 & | & -8 \\ \hline \to R_3 & 0 & -8 & -1 & | & -26 \end{array}$$

> This now replaces row 3.

$$\begin{bmatrix} 4 & 3 & 0 & | & 5 \\ 1 & 4 & -1 & | & 9 \\ 2 & 0 & -3 & | & -8 \end{bmatrix} \xrightarrow[\substack{\text{Add } -2 \text{ times row 2 to row 3.} \\ \text{The result replaces row 3.}}]{-2R_2 + R_3 \to R_3} \begin{bmatrix} 4 & 3 & 0 & | & 5 \\ 1 & 4 & -1 & | & 9 \\ 0 & -8 & -1 & | & -26 \end{bmatrix}$$

Skill Practice 2 Use the matrix from Example 2 to perform the given row operations.

   a. $R_2 \Leftrightarrow R_3$ **b.** $-\frac{1}{2}R_3$ **c.** $-4R_2 + R_1 \to R_1$

---

3. Use Gaussian Elimination and Gauss-Jordan Elimination

When an elementary row operation is performed on an augmented matrix, a new row-equivalent augmented matrix is obtained that represents an equivalent system of equations. If we perform repeated row operations, we can form an augmented matrix that represents a system of equations that is easier to solve than the original system. In particular, it is easy to solve a system whose augmented matrix is in *row-echelon form* or *reduced row-echelon form*.

> ### Row-Echelon Form and Reduced Row-Echelon Form
>
> A matrix is in **row-echelon form** if it satisfies the following conditions.
>
> 1. Any rows consisting entirely of zeros are at the bottom of the matrix.
> 2. For all other rows, the first nonzero entry is 1. This is called the leading 1.
> 3. The leading 1 in each nonzero row is to the right of the leading 1 in the row immediately above.
>
> *Note*: A matrix is in **reduced row-echelon form** if it is in row-echelon form with the added condition that each row with a leading entry of 1 has zeros above the leading 1.

The following matrices illustrate row-echelon form and reduced row-echelon form.

      Row-Echelon Form

$$\begin{bmatrix} 1 & 5 & | & 3 \\ 0 & 1 & | & 6 \end{bmatrix} \qquad \begin{bmatrix} 1 & -4 & -\frac{1}{2} & | & 6 \\ 0 & 1 & 9 & | & -\frac{1}{3} \\ 0 & 0 & 1 & | & 2 \end{bmatrix} \qquad \begin{bmatrix} 1 & 4.1 & 1.2 & | & 3.1 \\ 0 & 1 & 0.6 & | & 4.7 \\ 0 & 0 & 0 & | & 0 \end{bmatrix}$$

Answers

2. a. $\begin{bmatrix} 4 & 3 & 0 & | & 5 \\ 2 & 0 & -3 & | & -8 \\ 1 & 4 & -1 & | & 9 \end{bmatrix}$

b. $\begin{bmatrix} 4 & 3 & 0 & | & 5 \\ 1 & 4 & -1 & | & 9 \\ -1 & 0 & \frac{3}{2} & | & 4 \end{bmatrix}$

c. $\begin{bmatrix} 0 & -13 & 4 & | & -31 \\ 1 & 4 & -1 & | & 9 \\ 2 & 0 & -3 & | & -8 \end{bmatrix}$

Reduced Row-Echelon Form

$$\begin{bmatrix} 1 & 0 & -27 \\ 0 & 1 & 6 \end{bmatrix} \qquad \begin{bmatrix} 1 & 0 & 0 & 150 \\ 0 & 1 & 0 & -85 \\ 0 & 0 & 1 & 12 \end{bmatrix} \qquad \begin{bmatrix} 1 & 0 & 2 & 25 \\ 0 & 1 & -3 & -4 \\ 0 & 0 & 0 & 0 \end{bmatrix}$$

After writing an augmented matrix in row-echelon form, the corresponding system of linear equations can be solved using back substitution. This method is called **Gaussian elimination**.

Solving a System of Linear Equations Using Gaussian Elimination

1. Write the augmented matrix for the system.
2. Use elementary row operations to write the augmented matrix in row-echelon form.
3. Use back substitution to solve the resulting system of equations.

When writing an augmented matrix in row-echelon form, the goal is to make the elements along the main diagonal 1 and the entries below the main diagonal 0. The **main diagonal** refers to the elements on the diagonal from the upper left to the lower right all to the left of the vertical bar.

The main diagonal stretches from the upper left to the lower right.

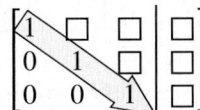

To write an augmented matrix in row-echelon form, work one column at a time from left to right. This is demonstrated in Example 3.

EXAMPLE 3   Solving a System Using Gaussian Elimination

Solve the system by using Gaussian elimination.

$$3x + 7y - 15z = -12$$
$$x + 2y - 4z = -3$$
$$-4x - 6y + 15z = 16$$

Solution:

Need a 1 here

$$\begin{array}{l} 3x + 7y - 15z = -12 \\ x + 2y - 4z = -3 \\ -4x - 6y + 15z = 16 \end{array} \qquad \begin{bmatrix} ③ & 7 & -15 & -12 \\ 1 & 2 & -4 & -3 \\ -4 & -6 & 15 & 16 \end{bmatrix}$$

Set up the augmented matrix.

Use elementary row operations to write the augmented matrix in row-echelon form.

Begin working with column 1.

$$R_1 \Leftrightarrow R_2 \longrightarrow \begin{bmatrix} 1 & 2 & -4 & -3 \\ 3 & 7 & -15 & -12 \\ -4 & -6 & 15 & 16 \end{bmatrix}$$

To obtain a leading 1 in the first row, interchange rows 1 and 2.

TIP To obtain a leading entry of 1 in the first row, we have the option of multiplying the first row by $\frac{1}{3}$. However, this would introduce fractions in the first row. Interchanging rows 1 and 2 is "cleaner."

Now we want zeros below the leading 1 in the first row.

- Multiply row 1 by -3 and add the result to row 2.
- Multiply row 1 by 4 and add the result to row 3.

$$\begin{bmatrix} 1 & 2 & -4 & -3 \\ ③ & 7 & -15 & -12 \\ -4 & -6 & 15 & 16 \end{bmatrix}$$

Need 0's here

$$\begin{array}{llll} -3R_1 & -3 & -6 & 12 & 9 \\ + R_2 & 3 & 7 & -15 & -12 \\ \hline \to R_2 & 0 & 1 & -3 & -3 \end{array} \qquad \begin{array}{llll} 4R_1 & 4 & 8 & -16 & -12 \\ + R_3 & -4 & -6 & 15 & 16 \\ \hline \to R_3 & 0 & 2 & -1 & 4 \end{array}$$

$$\begin{array}{l} -3R_1 + R_2 \to R_2 \\ 4R_1 + R_3 \to R_3 \end{array} \longrightarrow \begin{bmatrix} 1 & 2 & -4 & -3 \\ 0 & 1 & -3 & -3 \\ 0 & 2 & -1 & 4 \end{bmatrix}$$

Next, work with column 2.
We already have a leading 1 in row 2.
We need 0 below the leading 1 in row 2.

$$\begin{bmatrix} 1 & 2 & -4 & -3 \\ 0 & 1 & -3 & -3 \\ 0 & \circled{2} & -1 & 4 \end{bmatrix}$$

Need 0 here

- Multiply row 2 by -2 and add the result to row 3.

$$\begin{array}{r} -2R_2 \\ +R_3 \\ \hline \rightarrow R_3 \end{array} \begin{array}{rrr|r} 0 & -2 & 6 & 6 \\ 0 & 2 & -1 & 4 \\ \hline 0 & 0 & 5 & 10 \end{array}$$

$$-2R_2 + R_3 \rightarrow R_3 \longrightarrow \begin{bmatrix} 1 & 2 & -4 & -3 \\ 0 & 1 & -3 & -3 \\ 0 & 0 & 5 & 10 \end{bmatrix}$$

$$\tfrac{1}{5}R_3 \rightarrow R_3 \longrightarrow \begin{bmatrix} 1 & 2 & -4 & -3 \\ 0 & 1 & -3 & -3 \\ 0 & 0 & 1 & 2 \end{bmatrix}$$

Now work with column 3. We need a leading 1 in row 3.
Multiply row 3 by $\tfrac{1}{5}$.

$$\begin{bmatrix} 1 & 2 & -4 & -3 \\ 0 & 1 & -3 & -3 \\ 0 & 0 & 1 & 2 \end{bmatrix} \begin{array}{l} \longrightarrow x + 2y - 4z = -3 \\ \longrightarrow y - 3z = -3 \\ \longrightarrow z = 2 \end{array}$$

We now have row-echelon form.
The corresponding system of equations has z isolated.

Use back substitution to find x and y.

$$\begin{array}{l} y - 3z = -3 \\ y - 3(2) = -3 \\ y = 3 \end{array}$$

Substitute $z = 2$ in the second equation.

$$\begin{array}{l} x + 2y - 4z = -3 \\ x + 2(3) - 4(2) = -3 \\ x + 6 - 8 = -3 \\ x = -1 \end{array}$$

Substitute $z = 2$ and $y = 3$ into the first equation.

The solution set is $\{(-1, 3, 2)\}$.   The solution $(-1, 3, 2)$ checks in each original equation.

Skill Practice 3   Solve the system by using Gaussian elimination.

$$\begin{array}{rrr} 2x + 7y + & z & = 14 \\ x + 3y - & z & = 2 \\ x + 7y + & 12z & = 45 \end{array}$$

Example 3 illustrates that a system of equations represented by an augmented matrix in row-echelon form is easily solved by using back substitution. If we write an augmented matrix in *reduced* row-echelon form, we can solve the corresponding system of equations by inspection. This is called the Gauss-Jordan elimination method [named after Carl Friedrich Gauss (1777–1855) and German scientist Wilhelm Jordan (1842–1899)].

When writing an augmented matrix in *reduced* row-echelon form, the goal is to make the elements along the main diagonal 1 and the entries above and below the main diagonal 0. This is demonstrated in Examples 4 and 5.

$$\begin{bmatrix} 1 & 0 & \square \\ 0 & 1 & \square \end{bmatrix} \qquad \begin{bmatrix} 1 & 0 & 0 & \square \\ 0 & 1 & 0 & \square \\ 0 & 0 & 1 & \square \end{bmatrix}$$

Answer
3. $\{(2, 1, 3)\}$

Order of Row Operations to Obtain Reduced Row-Echelon Form

To write an augmented matrix with n rows in reduced row-echelon form, transform the entries in the matrix in the following order.

Column 1:   Obtain a leading element of 1 in row 1. Then obtain 0's below this element.

Column 2:   Obtain a leading element of 1 in row 2. Then obtain 0's above and below this element.

Column 3:   Obtain a leading element of 1 in row 3. Then obtain 0's above and below this element.

⋮

Column n: Obtain a leading element of 1 in row n. Then obtain 0's above this element.

EXAMPLE 4   Solving a System Using Gauss-Jordan Elimination

Solve the system by using Gauss-Jordan elimination.
$$x = 17 - 2y$$
$$3(x + 2y) = 47 - y$$

Solution:

$$x = 17 - 2y \longrightarrow x + 2y = 17$$
$$3(x + 2y) = 47 - y \longrightarrow 3x + 7y = 47$$

To use Gaussian elimination or Gauss-Jordan elimination, always begin by writing each equation in standard form.

$$\begin{aligned} x + 2y &= 17 \\ 3x + 7y &= 47 \end{aligned} \qquad \begin{bmatrix} 1 & 2 & | & 17 \\ ③ & 7 & | & 47 \end{bmatrix}$$

Need 0 here

Set up the augmented matrix.
The leading element in row 1 is already 1.
Now we need a zero below the leading 1 in row 1.

Multiply row 1 by -3 and add the result to row 2.

$$\begin{array}{rrr r} -3R_1 & -3 & -6 & | & -51 \\ +R_2 & 3 & 7 & | & 47 \\ \hline \to R_2 & 0 & 1 & | & -4 \end{array} \qquad -3R_1 + R_2 \to R_2 \longrightarrow \begin{bmatrix} 1 & 2 & | & 17 \\ 0 & 1 & | & -4 \end{bmatrix}$$

Need 0 here

$$\begin{bmatrix} 1 & ② & | & 17 \\ 0 & 1 & | & -4 \end{bmatrix}$$

Now work with column 2. The leading element in row 2 is already 1.
Now we need a zero above the leading 1 in row 2.

Multiply row 2 by -2 and add the result to row 1.

$$\begin{array}{rrr r} R_1 & 1 & 2 & | & 17 \\ + -2R_2 & 0 & -2 & | & 8 \\ \hline \to R_1 & 1 & 0 & | & 25 \end{array} \qquad -2R_2 + R_1 \to R_1 \longrightarrow \begin{bmatrix} 1 & 0 & | & 25 \\ 0 & 1 & | & -4 \end{bmatrix}$$

$$\begin{bmatrix} 1 & 0 & | & 25 \\ 0 & 1 & | & -4 \end{bmatrix} \longrightarrow \begin{aligned} x &= 25 \\ y &= -4 \end{aligned}$$

From the corresponding system of equations, we can determine the solution $(25, -4)$ by inspection.

The solution set is $\{(25, -4)\}$.

The solution checks in each original equation.

$$x - 2y = -1$$
$$4x - 7y = 1$$

In Example 5, we use the Gauss-Jordan elimination method to solve a three-variable system of linear equations.

EXAMPLE 5   Solving a System Using Gauss-Jordan Elimination

Solve the system by using Gauss-Jordan elimination.

$$2x - 5y - 21z = 39$$
$$x - 3y - 10z = 22$$
$$x + 3y + 2z = -8$$

Solution:

$$\begin{aligned} 2x - 5y - 21z &= 39 \\ x - 3y - 10z &= 22 \\ x + 3y + 2z &= -8 \end{aligned} \qquad \begin{bmatrix} 2 & -5 & -21 & | & 39 \\ 1 & -3 & -10 & | & 22 \\ 1 & 3 & 2 & | & -8 \end{bmatrix}$$

Set up the augmented matrix.

$$R_1 \Leftrightarrow R_2 \longrightarrow \begin{bmatrix} 1 & -3 & -10 & | & 22 \\ 2 & -5 & -21 & | & 39 \\ 1 & 3 & 2 & | & -8 \end{bmatrix}$$

Use elementary row operations to write the augmented matrix in reduced row-echelon form.

To obtain a leading 1 in the first row, interchange rows 1 and 2.

$$\begin{aligned} -2R_1 + R_2 &\to R_2 \\ -R_1 + R_3 &\to R_3 \end{aligned} \longrightarrow \begin{bmatrix} 1 & -3 & -10 & | & 22 \\ 0 & 1 & -1 & | & -5 \\ 0 & 6 & 12 & | & -30 \end{bmatrix}$$

Multiply row 1 by -2 and add the result to row 2. Multiply row 1 by -1 and add the result to row 3.

This results in zeros below the leading 1 in the first row.

$$\begin{aligned} 3R_2 + R_1 &\to R_1 \\ -6R_2 + R_3 &\to R_3 \end{aligned} \longrightarrow \begin{bmatrix} 1 & 0 & -13 & | & 7 \\ 0 & 1 & -1 & | & -5 \\ 0 & 0 & 18 & | & 0 \end{bmatrix}$$

Row 2 already has a leading 1.

Multiply row 2 by 3 and add the result to row 1. Multiply row 2 by -6 and add the result to row 3.

$$\tfrac{1}{18}R_3 \to R_3 \longrightarrow \begin{bmatrix} 1 & 0 & -13 & | & 7 \\ 0 & 1 & -1 & | & -5 \\ 0 & 0 & 1 & | & 0 \end{bmatrix}$$

Multiply row 3 by $\frac{1}{18}$ to obtain a leading 1 in row 3.

$$\begin{aligned} 13R_3 + R_1 &\to R_1 \\ R_3 + R_2 &\to R_2 \end{aligned} \longrightarrow \begin{bmatrix} 1 & 0 & 0 & | & 7 \\ 0 & 1 & 0 & | & -5 \\ 0 & 0 & 1 & | & 0 \end{bmatrix}$$

Multiply row 3 by 13 and add the result to row 1.

Multiply row 3 by 1 and add the result to row 2.

$$\begin{bmatrix} 1 & 0 & 0 & | & 7 \\ 0 & 1 & 0 & | & -5 \\ 0 & 0 & 1 & | & 0 \end{bmatrix} \longrightarrow \begin{aligned} x &= 7 \\ y &= -5 \\ z &= 0 \end{aligned}$$

The augmented matrix is in reduced row-echelon form.

From the corresponding system of equations, we can determine the solution $(7, -5, 0)$ by inspection.

The solution set is $\{(7, -5, 0)\}$.

The solution checks in each original equation.

$$2x + 7y + 11z = 11$$
$$x + 2y + 8z = 14$$
$$x + 3y + 6z = 8$$

Answers

4. $\{(9, 5)\}$

5. $\{(14, -4, 1)\}$

TECHNOLOGY CONNECTIONS

Finding the Row-Echelon Form and Reduced Row-Echelon Form of a Matrix

Many graphing utilities have the capability to enter and manipulate matrices. To enter the augmented matrix from Example 5, select the MATRIX menu and then the EDIT menu. Select a name for the matrix such as A. Then enter the number of rows and columns of the matrix (in this case 3 × 4) followed by the individual elements.

To access the commands for row-echelon form and reduced row-echelon form, select the **MATH** menu from within the MATRIX menu. Select either *ref* for row-echelon form or *rref* for reduced row-echelon form.

In the home screen, insert the matrix name within the parentheses for the *ref* and *rref* functions.

Alternatively, we can find the row-echelon form or reduced row-echelon form of a matrix if the calculator is in "MATHPRINT" mode. Select **MODE**, and using the arrow keys to navigate through the menu, highlight "MATHPRINT."

In the home screen, select the MATRIX menu, followed by MATH. Then select either *ref* or *rref*. Next, press the **ALPHA** key followed by F3. Enter the dimensions of the matrix (in this case, 3 × 4). Then enter the elements of the matrix using the arrow keys to navigate through the matrix. Hit **ENTER** to complete the calculation.

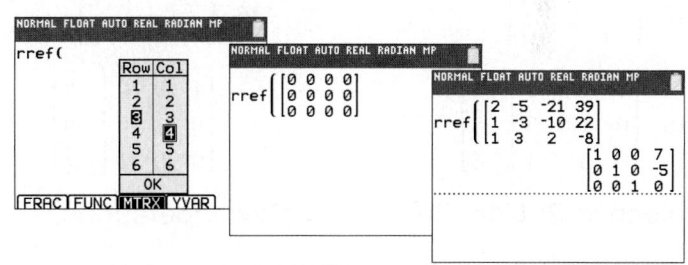

SECTION 11.1   Practice Exercises

Prerequisite Review

For Exercises R.1–R.4, solve the system using any method. (See Sections 10.2 and 10.3 for review.)

R.1. $2(x - 10) = 5y + 3$ (See Section 10.2)
$3x = -4y$

R.2. $2(y + 3) = 28 - 3x$
$3x = 26 - 6y - 2x$

R.3. $3x - 2y + z = -4$ (See Section 10.3)
$2x + 5z = 23$
$x + 4y - 3z = -4$

R.4. $x + 3y - 2z = 8$
$6x + 5y = 24$
$2x + 2y - 3z = 14$

For Exercises R.5–R.6, set up the form for the partial fraction decomposition. Do not solve for A, B, C, and so on. (See Section 10.4 for review.)

R.5. $\dfrac{-12t - 11}{4t^2 - 16t + 15}$

R.6. $\dfrac{-5x^2 - 10x - 13}{7x^3 + 7x}$

Concept Connections

1. A rectangular array of elements is called a _____.

2. Identify the elements on the main diagonal. $\begin{bmatrix} -4 & -1 & 1 & | & 8 \\ 2 & 0 & 5 & | & 11 \\ 0 & 1 & -7 & | & -6 \end{bmatrix}$

3. Explain the meaning of the notation $R_2 \Leftrightarrow R_3$.

4. Explain the meaning of the notation $-R_2 \to R_2$.

5. Explain the meaning of the notation $3R_1 \to R_1$.

6. Explain the meaning of $R_1 \Leftrightarrow R_3$.

7. Explain the meaning of the notation $3R_1 + R_2 \to R_2$.

8. Explain the meaning of the notation $4R_2 + R_3 \to R_3$.

Objective 1: Write an Augmented Matrix

For Exercises 9–14, write the augmented matrix for the given system. **(See Example 1)**

9. $\begin{aligned} -3x + 2y - z &= 4 \\ 8x \quad\ + 4z &= 12 \\ 2y - 5z &= 1 \end{aligned}$

10. $\begin{aligned} -4x - y + z &= 8 \\ 2x \quad\ + 5z &= 11 \\ y - 7z &= -6 \end{aligned}$

11. $\begin{aligned} 4(x - 2y) &= 6y + 2 \\ 3x &= 5y + 7 \end{aligned}$

12. $\begin{aligned} 2(y - x) &= 4 - 8x \\ 5y &= 6 - x \end{aligned}$

13. $\begin{aligned} x &= 2 \\ y &= \tfrac{6}{7} \\ z &= 12 \end{aligned}$

14. $\begin{aligned} x &= 4 \\ y &= 2 \\ z &= -\tfrac{1}{2} \end{aligned}$

For Exercises 15–20, write a system of linear equations represented by the augmented matrix. **(See Example 1)**

15. $\begin{bmatrix} -4 & 6 & | & 11 \\ -3 & 9 & | & 1 \end{bmatrix}$

16. $\begin{bmatrix} -3 & 7 & | & -2 \\ -1 & 1 & | & 4 \end{bmatrix}$

17. $\begin{bmatrix} 1 & 4 & 3 & | & 8 \\ 0 & 1 & 2 & | & 12 \\ 0 & 0 & 1 & | & 6 \end{bmatrix}$

18. $\begin{bmatrix} 1 & 1 & -2 & | & -4 \\ 0 & 1 & 3 & | & 8 \\ 0 & 0 & 1 & | & 2 \end{bmatrix}$

19. $\begin{bmatrix} 1 & 0 & 0 & | & 8 \\ 0 & 1 & 0 & | & -9 \\ 0 & 0 & 1 & | & \tfrac{3}{2} \end{bmatrix}$

20. $\begin{bmatrix} 1 & 0 & 0 & | & 2 \\ 0 & 1 & 0 & | & 6 \\ 0 & 0 & 1 & | & -\tfrac{1}{2} \end{bmatrix}$

Objective 2: Use Elementary Row Operations

For Exercises 21–26, perform the elementary row operations on $\begin{bmatrix} 1 & 4 & | & 2 \\ -3 & 6 & | & 6 \end{bmatrix}$. **(See Example 2)**

21. $R_1 \Leftrightarrow R_2$

22. $-\tfrac{1}{3}R_2 \to R_2$

23. $3R_1 \to R_1$

24. $-3R_1 \to R_1$

25. $\tfrac{1}{3}R_2 + R_1 \to R_1$

26. $3R_1 + R_2 \to R_2$

For Exercises 27–32, perform the elementary row operations on $\begin{bmatrix} 1 & 5 & 6 & | & 2 \\ 2 & 1 & 5 & | & 1 \\ 4 & -2 & -3 & | & 10 \end{bmatrix}$. **(See Example 2)**

27. $R_2 \Leftrightarrow R_3$

28. $R_1 \Leftrightarrow R_2$

29. $\tfrac{1}{4}R_3 \to R_3$

30. $\tfrac{1}{2}R_2 \to R_2$

31. $-2R_1 + R_2 \to R_2$

32. $-4R_1 + R_3 \to R_3$

Objective 3: Use Gaussian Elimination and Gauss-Jordan Elimination

For Exercises 33–36, determine if the matrix is in row-echelon form. If not, explain why.

33. $\begin{bmatrix} 1 & 5 & | & 4 \\ 0 & 2 & | & 6 \end{bmatrix}$

34. $\begin{bmatrix} 1 & 6 & 4 & | & 2 \\ 0 & 1 & 0 & | & -1 \\ 0 & 3 & 1 & | & 3 \end{bmatrix}$

35. $\begin{bmatrix} 1 & 3 & 2 & | & 6 \\ 0 & 1 & 5 & | & 9 \\ 0 & 0 & 0 & | & 0 \end{bmatrix}$

36. $\begin{bmatrix} 1 & 4 & 2 & | & -6 \\ 0 & 1 & -3 & | & 2 \\ 0 & 0 & 1 & | & 0 \end{bmatrix}$

For Exercises 37–40, determine if the matrix is in reduced row-echelon form. If not, explain why.

37. $\begin{bmatrix} 1 & 0 & 0 & | & 3 \\ 1 & 0 & 0 & | & 4 \\ 1 & 0 & 0 & | & 5 \end{bmatrix}$ **38.** $\begin{bmatrix} 1 & 0 & 2 & | & 3 \\ 0 & 1 & 0 & | & 4 \\ 0 & 0 & 1 & | & 5 \end{bmatrix}$ **39.** $\begin{bmatrix} 1 & 0 & 0 & 0 & | & 1 \\ 0 & 1 & 0 & 0 & | & 2 \\ 0 & 0 & 1 & 0 & | & -7 \\ 0 & 0 & 0 & 1 & | & 4 \end{bmatrix}$ **40.** $\begin{bmatrix} 1 & 0 & 0 & -1 & | & 5 \\ 0 & 1 & 0 & 0 & | & 20 \\ 0 & 0 & 1 & 4 & | & -1 \\ 0 & 0 & 0 & 0 & | & 0 \end{bmatrix}$

For Exercises 41–60, solve the system by using Gaussian elimination or Gauss-Jordan elimination. **(See Examples 3–5)**

41. $2x + 3y = -13$
$x + 4y = -14$

42. $-3x + 11y = 58$
$x - 3y = -16$

43. $2x - 7y = -41$
$3x - 9y = -51$

44. $-2x + 15y = 6$
$3x - 12y = -9$

45. $-3(x - 6y) = -167 - y$
$14y = 2x - 122$

46. $2(x - y) = 4x + y - 40$
$9y = 105 - 3x$

47. $3x + 7y + 22z = 83$
$x + 3y + 10z = 37$
$-2x - 5y - 18z = -66$

48. $3x + 5y + 9z = 3$
$x + 3y + 7z = 5$
$-2x - 8y - 15z = -11$

49. $-2x + 4y + z = 7$
$4x - 13y + 10z = 17$
$3x - 9y + 6z = 9$

50. $2x - 8y + 54z = -4$
$x - 2y + 14z = -1$
$x - 3y + 19z = -3$

51. $-3x + 4y - 15z = -44$
$x - y + 4z = 13$
$x - 3y + 14z = 27$

52. $-2x + 5y - 4z = -4$
$x - 2y + z = 3$
$x - 5y + 9z = -5$

53. $2x + 8z = 7y - 46$
$x = 3y - 3z - 18$
$6z = 5y - x - 34$

54. $2x = 7 - y - 3z$
$x + y = z + 5$
$16z = 2y - x - 4$

55. $11y + 65 = 3x + 13z$
$x + 3z = 3y + 15$
$-2x + 4y - 7z = -25$

56. $2(x - 6z) = 3y + x + 17$
$2x - 19 = 3y + 18z$
$-3x + 7y + 36z = -41$

57. $w + 3x - 3z = -5$
$x - 2z = -6$
$-2w - 4x + y + 2z = 1$
$x + y = 5$

58. $w - 2x + 5y = 20$
$-x + 2y = 9$
$x - y = -5$
$2w - x + 7y + z = 25$

59. $x_1 + x_2 + 5x_4 = -4$
$x_2 + 2x_4 = 3$
$-2x_2 + x_3 - 3x_4 = -5$
$3x_1 + 3x_2 + 17x_4 = -10$

60. $x_1 + x_3 - 5x_4 = 1$
$-2x_1 + x_2 - 2x_3 + 16x_4 = -3$
$x_1 + 2x_3 - 10x_4 = 5$
$x_1 - x_3 + 7x_4 = -7$

Mixed Exercises

For Exercises 61–64, set up a system of linear equations to represent the scenario. Solve the system by using Gaussian elimination or Gauss-Jordan elimination.

61. Andre borrowed $20,000 to buy a truck for his business. He borrowed from his parents who charge him 2% simple interest. He borrowed from a credit union that charges 4% simple interest, and he borrowed from a bank that charges 5% simple interest. He borrowed five times as much from his parents as from the bank, and the amount of interest he paid at the end of 1 year was $620. How much did he borrow from each source?

62. Sylvia invested a total of $40,000. She invested part of the money in a certificate of deposit (CD) that earns 2% simple interest per year. She invested in a stock that returns the equivalent of 8% simple interest, and she invested in a bond fund that returns 5%. She invested twice as much in the stock as she did in the CD, and earned a total of $2300 at the end of 1 year. How much principal did she put in each investment?

63. Danielle stayed in three different cities (Washington, D.C., Atlanta, Georgia, and Dallas, Texas) for a total of 14 nights. She spent twice as many nights in Dallas as she did in Washington. The total cost for 14 nights (excluding tax) was $2200. Determine the number of nights that she spent in each city.

City	Cost per Night
Washington	$200
Atlanta	$100
Dallas	$150

64. Three pumps (A, B, and C) work to drain water from a retention pond. Working together, the pumps can pump 1500 gal/hr of water. Pump C works at a rate of 100 gal/hr faster than pump B. In 3 hr, pump C can pump as much water as pumps A and B working together in 2 hr. Find the rate at which each pump works.

For Exercises 65–66, find the partial fraction decomposition for the given rational expression. Use the technique of Gaussian elimination to find A, B, and C.

65. $\dfrac{5x^2 - 6x - 13}{(x + 3)(x - 2)^2} = \dfrac{A}{x + 3} + \dfrac{B}{x - 2} + \dfrac{C}{(x - 2)^2}$

66. $\dfrac{2x^2 + 17x + 3}{(x + 5)(x + 1)^2} = \dfrac{A}{x + 5} + \dfrac{B}{x + 1} + \dfrac{C}{(x + 1)^2}$

Write About It

67. Explain why interchanging two rows of an augmented matrix results in an augmented matrix that represents an equivalent system of equations.

68. Explain why multiplying a row of an augmented matrix by a nonzero constant results in an augmented matrix that represents an equivalent system of equations.

69. Explain the difference between a matrix in row-echelon form and reduced row-echelon form.

70. Consider the matrix $\begin{bmatrix} 5 & -9 & | & -57 \\ 1 & -2 & | & -12 \end{bmatrix}$. Identify two row operations that could be used to obtain a leading entry of 1 in the first row. Also indicate which operation would be less cumbersome as a first step toward writing the matrix in reduced row-echelon form.

Technology Connections

For Exercises 71–72, use a calculator to approximate the reduced row-echelon form of the augmented matrix representing the given system. Give the solution set where x, y, and z are rounded to 2 decimal places.

71. $0.52x - 3.71y - 4.68z = 9.18$

$0.02x + 0.06y + 0.11z = 0.56$

$0.972x + 0.816y + 0.417z = 0.184$

72. $-3.61x + 8.17y - 5.62z = 30.2$

$8.04x - 3.16y + 9.18z = 28.4$

$-0.16x + 0.09y + 0.55z = 4.6$

73. A small grocer finds that the monthly sales y (in \$) can be approximated as a function of the amount spent advertising on the radio x_1 (in \$) and the amount spent advertising in the newspaper x_2 (in \$) according to $y = ax_1 + bx_2 + c$.

The table gives the amounts spent in advertising and the corresponding monthly sales for 3 months.

Radio Advertising, x_1	Newspaper Advertising, x_2	Monthly sales, y
\$2400	\$800	\$36,000
\$2000	\$500	\$30,000
\$3000	\$1000	\$44,000

a. Use the data to write a system of linear equations to solve for a, b, and c.

b. Use a graphing utility to find the reduced row-echelon form of the augmented matrix.

c. Write the model $y = ax_1 + bx_2 + c$.

d. Predict the monthly sales if the grocer spends \$2500 advertising on the radio and \$500 advertising in the newspaper for a given month.

74. The purchase price of a home y (in \$1000) can be approximated based on the annual income of the buyer x_1 (in \$1000) and on the square footage of the home x_2 (in 100 ft^2) according to $y = ax_1 + bx_2 + c$.

The table gives the incomes of three buyers, the square footages of the home purchased, and the corresponding purchase prices of the home.

Income (\$1000) x_1	Square Footage (100 ft^2) x_2	Price (\$1000) y
80	21	180
150	28	250
75	18	160

a. Use the data to write a system of linear equations to solve for a, b, and c.

b. Use a graphing utility to find the reduced row-echelon form of the augmented matrix.

c. Write the model $y = ax_1 + bx_2 + c$.

d. Predict the purchase price for a buyer who makes \$100,000 per year and wants a 2500 ft^2 home.

For Exercises 75–76, the given function values satisfy a function defined by $f(x) = ax^2 + bx + c$.

a. Set up a system of equations to solve for a, b, and c.

b. Use a graphing utility to find the reduced row-echelon form of the augmented matrix.

c. Write a function of the form $f(x) = ax^2 + bx + c$ that fits the data.

75. $f(-3) = -7.28$

$f(-1) = 3.68$

$f(10) = 18.2$

76. $f(3) = 6.95$

$f(-2) = 20.2$

$f(12) = 39.8$

SECTION 11.2   Inconsistent Systems and Dependent Equations

1. Identify Inconsistent Systems

When we studied systems of linear equations in two and three variables in Chapter 10, we learned that a system may have no solution. Such a system is said to be **inconsistent**, and we recognize an inconsistent system if the system reduces to a contradiction.

EXAMPLE 1   Identifying an Inconsistent System

Solve the system.
$$x - 3y - 17z = -59$$
$$x - 2y - 12z = -41$$
$$-2y - 10z = 20$$

Solution:

$$x - 3y - 17z = -59$$
$$x - 2y - 12z = -41$$
$$-2y - 10z = 20$$

$$\begin{bmatrix} 1 & -3 & -17 & -59 \\ 1 & -2 & -12 & -41 \\ 0 & -2 & -10 & 20 \end{bmatrix}$$

Write the augmented matrix.
The leading entry in row 1 is already 1.

$$-R_1 + R_2 \rightarrow R_2 \longrightarrow \begin{bmatrix} 1 & -3 & -17 & -59 \\ 0 & 1 & 5 & 18 \\ 0 & -2 & -10 & 20 \end{bmatrix}$$

Multiply row 1 by -1 and add the result to row 2.

$$\begin{array}{c} 3R_2 + R_1 \rightarrow R_1 \\ 2R_2 + R_3 \rightarrow R_3 \end{array} \longrightarrow \begin{bmatrix} 1 & 0 & -2 & -5 \\ 0 & 1 & 5 & 18 \\ 0 & 0 & 0 & 56 \end{bmatrix}$$

Multiply row 2 by 3 and add the result to row 1.
Multiply row 2 by 2 and add the result to row 3.

The last row of the matrix cannot be written with a leading 1 along the main diagonal. The last row represents a contradiction, $0 = 56$.

$$\begin{bmatrix} 1 & 0 & -2 & -5 \\ 0 & 1 & 5 & 18 \\ 0 & 0 & 0 & 56 \end{bmatrix} \xrightarrow[\text{system}]{\text{equivalent}} \begin{cases} x - 2z = -5 \\ y + 5z = 18 \\ 0 = 56 \end{cases}$$

The contradiction indicates that the system is inconsistent and that there is no solution. The solution set is { }.

Skill Practice 1   Solve the system.
$$5x - 9y - 33z = 3$$
$$x - 2y - 7z = 0$$
$$-2x + y + 8z = -12$$

Answer
1. { }

TECHNOLOGY CONNECTIONS

Recognizing an Inconsistent System on a Calculator

We can verify that the system of equations in Example 1 has no solution by using a calculator to find the reduced row-echelon form of the augmented matrix. Notice that the calculator also displays a contradiction in the third row, $0 = 1$.

In Example 1, once we reached the contradiction $0 = 56$, we stopped manipulating the augmented matrix. However, we have the option of simplifying the augmented matrix to reduced row-echelon form. To match the result given in the calculator, we would multiply row 3 by $\frac{1}{56}$. Then add -18 times row 3 to row 2, and add 5 times row 3 to row 1.

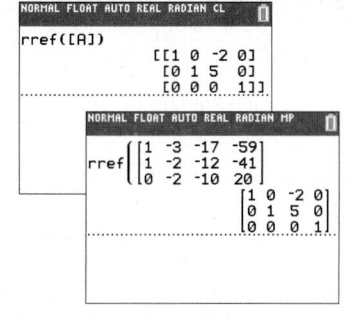

2. Solve Systems with Dependent Equations

Recall that a linear equation in two variables defines a line in a plane. A solution to a system of two linear equations in two variables is a point of intersection of the lines. If the two lines are parallel, the system is inconsistent and has no solution. If the equations in the system represent the same line, we say that the equations are **dependent** and the solution set is the set of all points on the line.

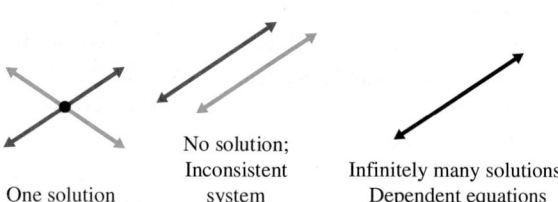

One solution

No solution;
Inconsistent
system

Infinitely many solutions;
Dependent equations

EXAMPLE 2   **Solving a System of Dependent Equations with Two Variables**

Solve the system. $0.25x - 0.75y = 1$
$3y = x - 4$

Solution:

$0.25x - 0.75y = 1$ $\xrightarrow{\text{Multiply by 100.}}$ $25x - 75y = 100$

$3y = x - 4$ $\xrightarrow[\text{Standard form}]{}$ $-x + 3y = -4$

Write the equations in standard form. As an option, consider clearing decimals in the first equation.

$\begin{bmatrix} 25 & -75 & | & 100 \\ -1 & 3 & | & -4 \end{bmatrix}$ $\xrightarrow{\frac{1}{25}R_1 \rightarrow R_1}$ $\begin{bmatrix} 1 & -3 & | & 4 \\ -1 & 3 & | & -4 \end{bmatrix}$

To obtain a 1 in the first row, first column, multiply row 1 by $\frac{1}{25}$.

$\xrightarrow{R_1 + R_2 \rightarrow R_2}$ $\begin{bmatrix} 1 & -3 & | & 4 \\ 0 & 0 & | & 0 \end{bmatrix}$

To obtain a 0 in the second row, first column, add row 1 to row 2.

FOR REVIEW

The topic of dependent equations was first brought up in Sections 10.1 and 10.2. From Example 2, solutions to the system are of the form $(3y + 4, y)$ where y is any real number. To find specific solutions, substitute arbitrary values of y. For example:
    If $y = 0$, then $(3(0) + 4, 0)$ or $(4, 0)$ is a solution.
    If $y = 1$, then $(3(1) + 4, 1)$ or $(7, 1)$ is a solution, and so on.

The second row of the augmented matrix represents the equation $0 = 0$, which is true regardless of the values of x and y. This means that the original system reduces to the single equation $x - 3y = 4$. That is, the two original equations each represent the same line and all points on the line are solutions to the system. Solving the equation $x - 3y = 4$ for x or y illustrates the dependency between x and y.

Solving the equation $x - 3y = 4$ for x gives: $x = 3y + 4$ ◁ *x depends on the choice of y.*

Solving the equation $x - 3y = 4$ for y gives: $y = \dfrac{x - 4}{3}$ ◁ *y depends on the choice of x.*

The solution set can be written as

$$\{(3y + 4, y) \,|\, y \text{ is any real number}\} \quad \text{or} \quad \left\{\left(x, \frac{x - 4}{3}\right) \,\middle|\, x \text{ is any real number}\right\}.$$

Skill Practice 2   Solve the system. $0.3x - 0.1y = -2$
$y - 20 = 3x$

TIP   In Example 2, we wrote the solution set in two ways: one with an arbitrary value of y and one with an arbitrary value of x. However, when using reduced row-echelon form to find the solution set to a system of dependent equations, we generally let the *last* variable in the ordered pair (ordered triple, etc.) be arbitrary. In Example 2, this is $\{(3y + 4, y) \,|\, y \text{ is any real number}\}$.

Answer

2. $\left\{\left(\dfrac{y - 20}{3}, y\right) \,\middle|\, y \text{ is any real number}\right\}$
   or $\{(x, 3x + 20) \,|\, x \text{ is any real number}\}$

A linear equation in three variables represents a plane in space. A solution to a system of equations in three variables is a common point of intersection among all the planes in the system. From Example 1, we have a system with no solution. This means that the planes do not all intersect (Figure 11-1).

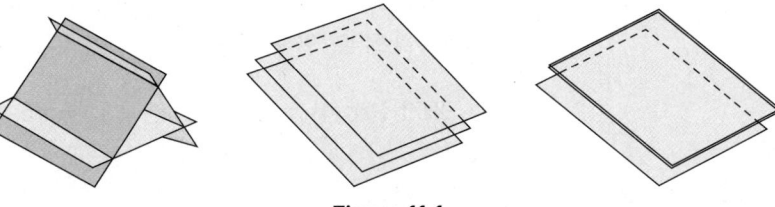

Figure 11-1

A system of linear equations may have infinitely many solutions. In such a case, the equations are dependent. For a system of three equations in three variables, this means that the planes intersect in a common line in space (Figures 11-2 and 11-3), or the three planes all coincide (Figure 11-4).

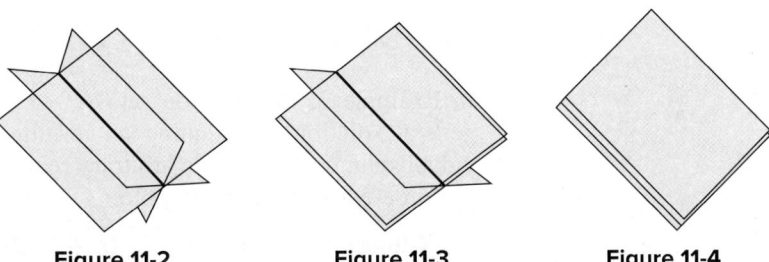

Figure 11-2   **Figure 11-3**   **Figure 11-4**

EXAMPLE 3   **Solving a System of Dependent Equations with Three Variables**

Solve the system.
$$x - 4y + 7z = 14$$
$$-2x + 9y - 16z = -31$$
$$x - 7y + 13z = 23$$

Solution:

$$\begin{aligned} x - 4y + 7z &= 14 \\ -2x + 9y - 16z &= -31 \\ x - 7y + 13z &= 23 \end{aligned} \qquad \begin{bmatrix} 1 & -4 & 7 & | & 14 \\ -2 & 9 & -16 & | & -31 \\ 1 & -7 & 13 & | & 23 \end{bmatrix}$$ Set up the augmented matrix.

$$\xrightarrow[\;-R_1 + R_3 \to R_3\;]{2R_1 + R_2 \to R_2} \begin{bmatrix} 1 & -4 & 7 & | & 14 \\ 0 & 1 & -2 & | & -3 \\ 0 & -3 & 6 & | & 9 \end{bmatrix} \xrightarrow[\;3R_2 + R_3 \to R_3\;]{4R_2 + R_1 \to R_1} \begin{bmatrix} 1 & 0 & -1 & | & 2 \\ 0 & 1 & -2 & | & -3 \\ 0 & 0 & 0 & | & 0 \end{bmatrix}$$

TIP A graphing utility can also be used to find the reduced row-echelon form of a system of dependent equations.

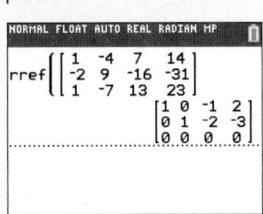

The last row of the matrix cannot be written with a leading 1 along the main diagonal. The last row represents an identity, $0 = 0$.

$$\begin{bmatrix} 1 & 0 & -1 & | & 2 \\ 0 & 1 & -2 & | & -3 \\ 0 & 0 & 0 & | & 0 \end{bmatrix} \xrightarrow[\text{system}]{\text{equivalent}} \begin{cases} x - z = 2 \\ y - 2z = -3 \\ 0 = 0 \end{cases}$$

The system of three equations and three variables reduces to a system of two equations and three variables. The third equation, $0 = 0$, is true regardless of the values of x, y, and z. The top two equations each represent a plane in space. Furthermore, two nonparallel planes intersect in a line. All points on the line are solutions to the system, indicating that there are infinitely many solutions.

TIP In Example 3, we can also write the solution set by using a parameter. By choosing any real number t for z, we have:

$$\{(t + 2, 2t - 3, t)\}$$

Since the top two equations both contain the variable z, we can express x and y in terms of z.

$$x - z = 2 \xrightarrow{\text{Express } x \text{ in terms of } z.} x = z + 2$$
$$y - 2z = -3 \xrightarrow{\qquad\qquad} y = 2z - 3$$
$$\underset{\text{Express } y \text{ in terms of } z.}{}$$

The solution set can be written as $\{(z + 2, 2z - 3, z)\,|\,z \text{ is any real number}\}$.

Skill Practice 3 Solve the system.

$$-4x - 11y + 3z = -24$$
$$x + 3y - z = 7$$
$$3x + 11y - 5z = 29$$

We can verify the solution to Example 3, by substituting $(z + 2, 2z - 3, z)$ back into each original equation.

$$x - 4y + 7z = 14 \longrightarrow (z + 2) - 4(2z - 3) + 7z \overset{?}{=} 14 \longrightarrow 14 = 14 ✓$$
$$-2x + 9y - 16z = -31 \longrightarrow -2(z + 2) + 9(2z - 3) - 16z \overset{?}{=} -31 \longrightarrow -31 = -31 ✓$$
$$x - 7y + 13z = 23 \longrightarrow (z + 2) - 7(2z - 3) + 13z \overset{?}{=} 23 \longrightarrow 23 = 23 ✓$$

In Example 3, the solution set $\{(z + 2, 2z - 3, z)\,|\,z \text{ is any real number}\}$ is the general solution representing an infinite number of ordered triples. To find individual solutions, substitute arbitrary real numbers for z, and construct the corresponding ordered triple. For example:

Choose z Arbitrarily	$(z + 2, 2z - 3, z)$	**Solution**
If $z = 1$ $\longrightarrow$	$(1 + 2, 2(1) - 3, 1)$	$(3, -1, 1)$
If $z = 2$ $\longrightarrow$	$(2 + 2, 2(2) - 3, 2)$	$(4, 1, 2)$
If $z = 3$ $\longrightarrow$	$(3 + 2, 2(3) - 3, 3)$	$(5, 3, 3)$
If $z = -1$ $\longrightarrow$	$(-1 + 2, 2(-1) - 3, -1)$	$(1, -5, -1)$

A system of linear equations that has the same number of equations as variables is called a **square system**. In Example 3, we were presented with three equations and three variables in the original system. However, after writing the augmented matrix in row-echelon form, we see that the system reduces to a system of two equations and three variables. A system of linear equations cannot have a unique solution unless there are at least the same number of equations as variables.

In Example 4, we investigate a nonsquare system that has fewer equations than variables.

EXAMPLE 4   **Solving a System with Fewer Equations than Variables**

Solve the system. $3x - 8y + 18z = 15$
$$x - 3y + 4z = 6$$

Solution:

The two given equations each represent a plane in space. By finding the row-echelon form of the augmented matrix, we can determine the geometrical relationship between the planes.

- If we encounter a contradiction, then the system has no solution and the planes are parallel.
- If we encounter an identity, such as $0 = 0$, then the two equations must represent the same plane.
- Otherwise, the planes must meet in a line.

Answer
3. $\{(-2z - 5, z + 4, z)\,|\,z \text{ is any real number}\}$

$$3x - 8y + 18z = 15 \longrightarrow \begin{bmatrix} 3 & -8 & 18 & | & 15 \\ 1 & -3 & 4 & | & 6 \end{bmatrix} \quad R_1 \Leftrightarrow R_2 \quad \begin{bmatrix} 1 & -3 & 4 & | & 6 \\ 3 & -8 & 18 & | & 15 \end{bmatrix}$$

$$x - 3y + 4z = 6 \longrightarrow$$

$$\xrightarrow{-3R_1 + R_2 \rightarrow R_2} \begin{bmatrix} 1 & -3 & 4 & | & 6 \\ 0 & 1 & 6 & | & -3 \end{bmatrix} \xrightarrow{3R_2 + R_1 \rightarrow R_1} \begin{bmatrix} 1 & 0 & 22 & | & -3 \\ 0 & 1 & 6 & | & -3 \end{bmatrix}$$

The augmented matrix is in reduced row-echelon form. $\begin{bmatrix} 1 & 0 & 22 & | & -3 \\ 0 & 1 & 6 & | & -3 \end{bmatrix} \longrightarrow \begin{cases} x + 22z = -3 \\ y + 6z = -3 \end{cases}$

The system did not reduce to a contradiction or an identity. Therefore, the two planes must intersect in a line and there must be infinitely many solutions.

To write the general solution, write x and y in terms of z.

TIP In parametric form, the solution to Example 4 is given by:

$$\{(-22t - 3, -6t - 3, t)\}$$

$$x + 22z = -3 \xrightarrow{\text{Express } x \text{ in terms of } z.} x = -22z - 3$$

$$y + 6z = -3 \xrightarrow{\text{Express } y \text{ in terms of } z.} y = -6z - 3$$

The solution set is $\{(-22z - 3, -6z - 3, z) \mid z \text{ is any real number}\}$.

> **Skill Practice 4**   Solve the system. $x - 3y - 17z = -17$
> $$-2x + 7y + 38z = 40$$

In Example 5, we investigate a situation in which the solution set to a system of three variables and three equations consists of all points on a common plane in space.

> **EXAMPLE 5**   **Solving a System of Dependent Equations Representing the Same Plane**

Solve the system. $x + 2y + 3z = 6$
$$-x - 2y - 3z = -6$$
$$2x + 4y + 6z = 12$$

Solution:

TIP   The solution to Example 5 tells us that $x = 6 - 2y - 3z$. That is, x depends on both y and z. If we write the solution set in parametric form, we would need *two* parameters.

$$\{(6 - 2s - 3t, s, t)\}$$

A $x + 2y + 3z = 6$
B $-x - 2y - 3z = -6$
C $2x + 4y + 6z = 12$

Upon inspection, you might notice that each equation is a constant multiple of the others. That is, equation B is -1 times equation A. Equation C is 2 times equation A. This means that equations A, B, and C represent the same plane in space.

$$\begin{bmatrix} 1 & 2 & 3 & | & 6 \\ -1 & -2 & -3 & | & -6 \\ 2 & 4 & 6 & | & 12 \end{bmatrix} \xrightarrow[\substack{R_1 + R_2 \rightarrow R_2 \\ -2R_1 + R_3 \rightarrow R_3}]{} \begin{bmatrix} 1 & 2 & 3 & | & 6 \\ 0 & 0 & 0 & | & 0 \\ 0 & 0 & 0 & | & 0 \end{bmatrix}$$

The reduced row-echelon form tells us that the system reduces to an equivalent system of one equation with three variables.

The system is equivalent to the single equation $x + 2y + 3z = 6$ and the solution set is the set of all ordered triples that satisfy this equation. Since we have one unique relationship among three variables, one variable is dependent on the other two variables. Letting two variables be arbitrary real numbers, the common equation defines the value of the third variable. For example, solving the equation $x + 2y + 3z = 6$ for x yields: $x = 6 - 2y - 3z$.

Answer

4. $\{(5z + 1, -4z + 6, z) \mid z \text{ is any real number}\}$

The solution set is $\{(6 - 2y - 3z, y, z) \mid y \text{ and } z \text{ are any real numbers}\}$.

Alternatively, the solution set can be expressed as $\{(x, y, z) \mid x + 2y + 3z = 6\}$.

Skill Practice 5   Solve the system.
$$2x - 3y + 5z = 3$$
$$4x - 6y + 10z = 6$$
$$20x - 30y + 50z = 30$$

3. Solve Applications of Systems of Equations

A city planner can study the flow of traffic through a network of streets by measuring the flow rates (number of vehicles per unit time) at various points. For traffic to flow freely, we use the principle that the flow rate into an intersection is equal to the flow rate out of the intersection. This must be true for all intersections in the network.

EXAMPLE 6   Solving an Application Involving Dependent Equations

Consider the network of four one-way streets shown in Figure 11-5. In the figure, x_1, x_2, x_3, and x_4 indicate flow rates (in vehicles per hour) along the stretches of roads AB, BC, CD, and DA, respectively. The other numbers in the figure indicate other flow rates moving into and out of intersections A, B, C, and D.

a. Set up a system of equations that represents traffic flowing freely.

b. Write the augmented matrix for the system of equations in reduced row-echelon form.

c. If the traffic between intersections D and A is 260 vehicles per hour, determine the flow rates x_1, x_2, and x_3.

d. If the traffic between intersections D and A is between 250 and 300 vehicles per hour, inclusive, determine the flow rates x_1, x_2, and x_3.

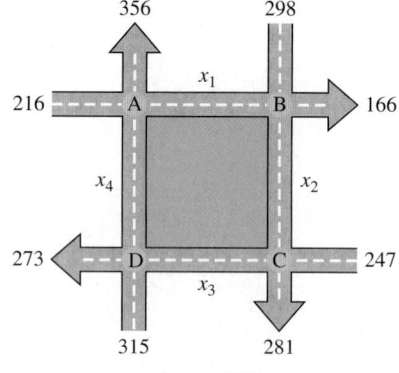

Figure 11-5

> **Point of Interest**
>
> The principle of equal flow rates into and out of a junction applies to many applications of networks, including communications systems, water systems, and even the circulatory system in the human body.

Solution:

a.

Intersection	Flow in	=	Flow out
A	$216 + x_4$	=	$356 + x_1$
B	$298 + x_1$	=	$166 + x_2$
C	$x_2 + 247$	=	$281 + x_3$
D	$x_3 + 315$	=	$273 + x_4$

The number of cars flowing into each intersection must equal the number of cars leaving the intersection for traffic to flow freely.

b.
$$\begin{aligned} -x_1 \qquad\qquad + x_4 &= 140 \\ x_1 - x_2 \qquad\quad &= -132 \\ x_2 - x_3 \quad &= 34 \\ x_3 - x_4 &= -42 \end{aligned}\Bigg\} \rightarrow \left[\begin{array}{cccc|c} -1 & 0 & 0 & 1 & 140 \\ 1 & -1 & 0 & 0 & -132 \\ 0 & 1 & -1 & 0 & 34 \\ 0 & 0 & 1 & -1 & -42 \end{array}\right]$$

Write the equations in standard form. Set up the augmented matrix.

Answer

5. $\left\{\left(\dfrac{3 + 3y - 5z}{2}, y, z\right)\,\middle|\, y \text{ and } z \right.$

$\left. \text{are any real numbers}\right\}$ or

$\{(x, y, z) \mid 2x - 3y + 5z = 3\}$

$$\xrightarrow{-1R_1 \to R_1} \begin{bmatrix} 1 & 0 & 0 & -1 & | & -140 \\ 1 & -1 & 0 & 0 & | & -132 \\ 0 & 1 & -1 & 0 & | & 34 \\ 0 & 0 & 1 & -1 & | & -42 \end{bmatrix} \xrightarrow{-1R_1 + R_2 \to R_2} \begin{bmatrix} 1 & 0 & 0 & -1 & | & -140 \\ 0 & -1 & 0 & 1 & | & 8 \\ 0 & 1 & -1 & 0 & | & 34 \\ 0 & 0 & 1 & -1 & | & -42 \end{bmatrix}$$

$$\xrightarrow{-1R_2 \to R_2} \begin{bmatrix} 1 & 0 & 0 & -1 & | & -140 \\ 0 & 1 & 0 & -1 & | & -8 \\ 0 & 1 & -1 & 0 & | & 34 \\ 0 & 0 & 1 & -1 & | & -42 \end{bmatrix} \xrightarrow{-1R_2 + R_3 \to R_3} \begin{bmatrix} 1 & 0 & 0 & -1 & | & -140 \\ 0 & 1 & 0 & -1 & | & -8 \\ 0 & 0 & -1 & 1 & | & 42 \\ 0 & 0 & 1 & -1 & | & -42 \end{bmatrix}$$

$$\xrightarrow{-1R_3 \to R_3} \begin{bmatrix} 1 & 0 & 0 & -1 & | & -140 \\ 0 & 1 & 0 & -1 & | & -8 \\ 0 & 0 & 1 & -1 & | & -42 \\ 0 & 0 & 1 & -1 & | & -42 \end{bmatrix} \xrightarrow{-1R_3 + R_4 \to R_4} \begin{bmatrix} 1 & 0 & 0 & -1 & | & -140 \\ 0 & 1 & 0 & -1 & | & -8 \\ 0 & 0 & 1 & -1 & | & -42 \\ 0 & 0 & 0 & 0 & | & 0 \end{bmatrix}$$

c. From the reduced row-echelon form, we see that the flow rates x_1, x_2, and x_3 can all be expressed in terms of the flow rate x_4.

$$\begin{bmatrix} 1 & 0 & 0 & -1 & | & -140 \\ 0 & 1 & 0 & -1 & | & -8 \\ 0 & 0 & 1 & -1 & | & -42 \\ 0 & 0 & 0 & 0 & | & 0 \end{bmatrix}$$
$\longrightarrow x_1 - x_4 = -140 \longrightarrow x_1 = x_4 - 140$
$\longrightarrow x_2 - x_4 = -8 \longrightarrow x_2 = x_4 - 8$
$\longrightarrow x_3 - x_4 = -42 \longrightarrow x_3 = x_4 - 42$

If x_4 is 260 vehicles per hour, we have the following values for x_1, x_2, and x_3.

Flow rate between A and B: $x_1 = 260 - 140 \longrightarrow x_1 = 120$ vehicles per hour
Flow rate between B and C: $x_2 = 260 - 8 \longrightarrow x_2 = 252$ vehicles per hour
Flow rate between C and D: $x_3 = 260 - 42 \longrightarrow x_3 = 218$ vehicles per hour

d. If the traffic flow between intersections D and A is given by $250 \le x_4 \le 300$, we can solve the following inequalities to determine the flow rates x_1, x_2, and x_3.

$$250 \le x_1 + 140 \le 300 \longrightarrow 110 \le x_1 \le 160$$
$$250 \le x_2 + 8 \le 300 \longrightarrow 242 \le x_2 \le 292$$
$$250 \le x_3 + 42 \le 300 \longrightarrow 208 \le x_3 \le 258$$

Skill Practice 6   Refer to the figure. Assume that traffic flows freely with flow rates given in vehicles per hour.

a. If the traffic between intersections D and A is 400 vehicles per hour, determine the flow rates x_1, x_2, and x_3.

b. If the traffic between intersections D and A is between 380 and 420 vehicles per hour, determine the flow rates x_1, x_2, and x_3.

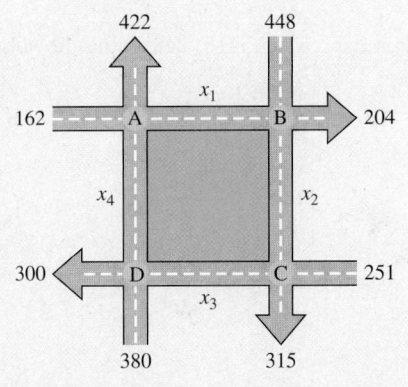

Answers
6. a. $x_1 = 140$, $x_2 = 384$, and
$x_3 = 320$
b. $120 \le x_1 \le 160$,
$364 \le x_2 \le 404$,
$300 \le x_3 \le 340$

SECTION 11.2	Practice Exercises

Prerequisite Review

For Exercises R.1–R.4, solve the one-variable equation. (See Section 1.2 for review.)

R.1. $3(x - 4) - x = 2x - 12$

R.2. $4x - 3 = x - (3 - 3x)$

R.3. $2x + 7 = 2(x + 3) - 1$

R.4. $-x + 5(x - 4) = 4x + 1$

For Exercises R.5–R.8, solve the system of equations. (See Section 10.2 for review.)

R.5. $2x = 12 - 6y$

$y = -\dfrac{1}{3}x + 2$

R.6. $y = \dfrac{1}{2}x - \dfrac{1}{2}$

$x = 2y + 1$

R.7. $\dfrac{1}{3}x - \dfrac{1}{2}y = 1$

$2x = 3y$

R.8. $5x = 15 + 3y$

$6y = 10x + 30$

Concept Connections

1. True or false? A system of linear equations in three variables may have no solution.

2. True or false? A system of linear equations in three variables may have exactly one solution.

3. True or false? A system of linear equations in three variables may have exactly two solutions.

4. True or false? A system of linear equations in three variables may have infinitely many solutions.

5. If a system of linear equations has no solution, then the system is said to be _____.

6. If a system of linear equations has infinitely many solutions, then the equations are said to be

_____.

Objectives 1–2: Identify Inconsistent Systems and Solve Systems with Dependent Equations

For Exercises 7–14, an augmented matrix is given. Determine the number of solutions to the corresponding system of equations.

7. $\left[\begin{array}{cc|c} 1 & 2 & 4 \\ 0 & 0 & 5 \end{array}\right]$

8. $\left[\begin{array}{ccc|c} 1 & 0 & 2 & 5 \\ 0 & 1 & 4 & -2 \\ 0 & 0 & 0 & -1 \end{array}\right]$

9. $\left[\begin{array}{ccc|c} 1 & 0 & 4 & 3 \\ 0 & 1 & -1 & 6 \\ 0 & 0 & 0 & 0 \end{array}\right]$

10. $\left[\begin{array}{cc|c} 1 & 3 & 5 \\ 0 & 0 & 0 \end{array}\right]$

11. $\left[\begin{array}{ccc|c} 1 & 0 & 0 & -3 \\ 0 & 1 & 0 & 4 \\ 0 & 0 & 1 & 0 \end{array}\right]$

12. $\left[\begin{array}{cc|c} 1 & 0 & 3 \\ 0 & 1 & 0 \end{array}\right]$

13. $\left[\begin{array}{ccc|c} 1 & 2 & 5 & -1 \\ 0 & 0 & 0 & 0 \\ 0 & 0 & 0 & 0 \end{array}\right]$

14. $\left[\begin{array}{ccc|c} 1 & 0 & 6 & 7 \\ 0 & 0 & 0 & 0 \\ 0 & 0 & 0 & 0 \end{array}\right]$

For Exercises 15–18, determine the solution set for the system represented by each augmented matrix.

15. **a.** $\left[\begin{array}{cc|c} 1 & 2 & 5 \\ 0 & 1 & 0 \end{array}\right]$

b. $\left[\begin{array}{cc|c} 1 & 2 & 5 \\ 0 & 0 & 0 \end{array}\right]$

c. $\left[\begin{array}{cc|c} 1 & 2 & 5 \\ 0 & 0 & 1 \end{array}\right]$

16. **a.** $\left[\begin{array}{cc|c} 1 & 3 & -4 \\ 0 & 1 & 1 \end{array}\right]$

b. $\left[\begin{array}{cc|c} 1 & 3 & -4 \\ 0 & 0 & 1 \end{array}\right]$

c. $\left[\begin{array}{cc|c} 1 & 3 & -4 \\ 0 & 0 & 0 \end{array}\right]$

17. a. $\begin{bmatrix} 1 & 0 & 6 & | & 3 \\ 0 & 1 & 4 & | & 5 \\ 0 & 0 & 1 & | & 0 \end{bmatrix}$ **b.** $\begin{bmatrix} 1 & 0 & 6 & | & 3 \\ 0 & 1 & 4 & | & 5 \\ 0 & 0 & 0 & | & 1 \end{bmatrix}$ **c.** $\begin{bmatrix} 1 & 0 & 6 & | & 3 \\ 0 & 1 & 4 & | & 5 \\ 0 & 0 & 0 & | & 0 \end{bmatrix}$

18. a. $\begin{bmatrix} 1 & 0 & -2 & | & 3 \\ 0 & 1 & 3 & | & 5 \\ 0 & 0 & 1 & | & 1 \end{bmatrix}$ **b.** $\begin{bmatrix} 1 & 0 & -2 & | & 3 \\ 0 & 1 & 3 & | & 5 \\ 0 & 0 & 0 & | & 0 \end{bmatrix}$ **c.** $\begin{bmatrix} 1 & 0 & -2 & | & 3 \\ 0 & 1 & 3 & | & 5 \\ 0 & 0 & 0 & | & 1 \end{bmatrix}$

For Exercises 19–38, solve the system by using Gaussian elimination or Gauss-Jordan elimination. **(See Exercises 1–5)**

19. $2x + 4y = 5$
$\quad x + 2y = 4$

20. $4x + 16y = 21$
$\quad x + 4y = -1$

21. $2x + 7y = 10$
$\quad \frac{1}{5}x = 1 - \frac{7}{10}y$

22. $4x - 3y = 6$
$\quad y = \frac{4}{3}x - 2$

23. $\quad x - 3y + 14z = -9$
$\quad -2x + 7y - 31z = 21$
$\quad x - 5y + 20z = -14$

24. $\quad x - 3y + 17z = 1$
$\quad x - \ y + \ 7z = 2$
$\quad 2x - 5y + 29z = 5$

25. $5x + 7y - 11z = 45$
$\quad 3x + 5y - \ 9z = 23$
$\quad x + \ y - \ z = 11$

26. $\quad x + \ 3y + \ 9z = 12$
$\quad 2x + \ 7y + 22z = 26$
$\quad -5x - 17y - 53z = -64$

27. $2x = 5y - 16z + 40$
$\quad 2(x + y) = 4z$
$\quad x - 2y + 7z = 18$

28. $x = 2y + 4z + 5$
$\quad y - 4 = -3z$
$\quad 5(x - y) - 9z = 4x - 7$

29. $2x - 5y - 20z = -24$
$\quad x - 3y - 11z = -15$

30. $2x - \ y - 5z = -3$
$\quad x - 2y - 7z = -12$

31. $\quad 2x + \ 3y + 4z = 12$
$\quad -4x - \ 6y - 8z = -24$
$\quad x + 1.5y + 2z = 6$

32. $-x + \ 2y + \ 7z = 14$
$\quad 10x - 20y - 70z = -140$
$\quad -\frac{1}{7}x + \frac{2}{7}y + \ z = 2$

33. $2x - 3y + 9z = -2$
$\quad x = 5y - 8z - 15$
$\quad 3(x - y) + 6z = 2x - 7$

34. $3x + 11y - 3z = -13$
$\quad x + y = z - 15$
$\quad 3(x + y) = z + 2x - 7$

35. $-5x + 12y - 20z = -11$
$\quad x + 4z = 3y + 1$

36. $x = 4y + 20z - 1$
$\quad -2x + 5y + 25z = -1$

37. $x_1 - 3x_2 + 9x_3 - 14x_4 = 32$
$\quad x_2 - 3x_3 + \ 6x_4 = -10$
$\quad x_2 - \ x_3 + \ 2x_4 = -4$
$\quad x_1 - 2x_2 + 8x_3 - 12x_4 = 24$

38. $\quad x_1 - 3x_3 - 12x_4 = -15$
$\quad x_2 + \ x_3 + \ 6x_4 = 8$
$\quad x_2 - 2x_3 - \ 6x_4 = -7$
$\quad -2x_1 + 4x_3 + 16x_4 = 22$

For Exercises 39–40, the solution set to a system of dependent equations is given. Write the specific solutions corresponding to the given values of z.

39. $\{(2z + 1, z - 4, z) \mid z \text{ is any real number}\}$
  a. $z = 1$ **b.** $z = 4$ **c.** $z = -2$

40. $\{(z + 5, 3z - 2, z) \mid z \text{ is any real number}\}$
  a. $z = -3$ **b.** $z = 1$ **c.** $z = 0$

For Exercises 41–44, the solution set to a system of dependent equations is given. Write three ordered triples that are solutions to the system. Answers may vary.

41. $\{(4z, 6 - z, z) \mid z \text{ is any real number}\}$

42. $\{(2z, z - 3, z) \mid z \text{ is any real number}\}$

43. $\left\{ \left(\dfrac{6 - 3y - 6z}{2}, y, z \right) \middle| y \text{ and } z \text{ are any real numbers} \right\}$

44. $\{(4y + 2z - 20, y, z) \mid y \text{ and } z \text{ are any real numbers}\}$

Objective 3: Solve Applications of Systems of Equations

For Exercises 45–48, assume that traffic flows freely through the intersections A, B, C, and D. The values x_1, x_2, x_3, and x_4 and the other numbers in the figures represent flow rates in vehicles per hour. **(See Example 6)**

45.

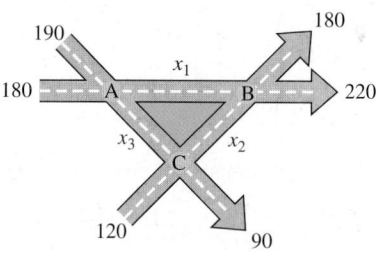

a. Write an equation representing equal flow into and out of intersection A.

b. Write an equation representing equal flow into and out of intersection B.

c. Write an equation representing equal flow into and out of intersection C.

d. Write the system of equations from parts (a)–(c) in standard form.

e. Write the reduced row-echelon form of the augmented matrix representing the system of equations from part (d).

f. If the flow rate between intersections A and C is 120 vehicles per hour, determine the flow rates x_1 and x_2.

g. If the flow rate between intersections A and C is between 100 and 150 vehicles per hour, inclusive, determine the flow rates x_1 and x_2.

46.

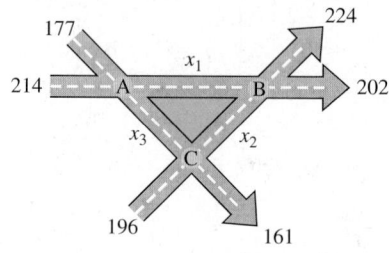

a. Write an equation representing equal flow into and out of intersection A.

b. Write an equation representing equal flow into and out of intersection B.

c. Write an equation representing equal flow into and out of intersection C.

d. Write the system of equations from parts (a)–(c) in standard form.

e. Write the reduced row-echelon form of the augmented matrix representing the system of equations in part (d).

f. If the flow rate between intersections A and C is 156 vehicles per hour, determine the flow rates x_1 and x_2.

g. If the flow rate between intersections A and C is between 100 and 200 vehicles per hour, inclusive, determine the flow rates x_1 and x_2.

47.

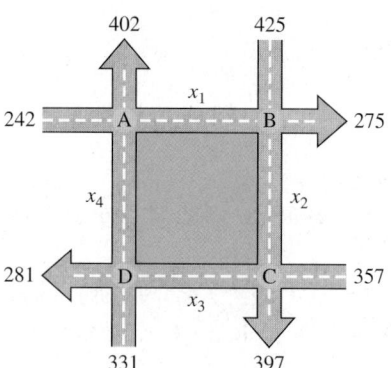

a. Assume that traffic flows at a rate of 220 vehicles per hour on the stretch of road between intersections D and A. Find the flow rates x_1, x_2, and x_3.

b. If traffic flows at a rate of between 200 and 250 vehicles per hour inclusive between intersections D and A, find the flow rates x_1, x_2, and x_3.

48.

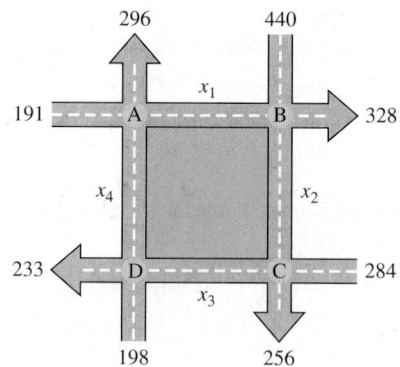

a. Assume that traffic flows at a rate of 180 vehicles per hour on the stretch of road between intersections D and A. Find the flow rates x_1, x_2, and x_3.

b. If traffic flows at a rate of between 150 and 200 vehicles per hour inclusive between intersections D and A, find the flow rates x_1, x_2, and x_3.

49. An accountant checks the reported earnings for a theater for three nightly performances against the number of tickets sold.

Night	Children Tickets	Student Tickets	General Admission	Total Revenue
1	80	400	480	$9280
2	50	350	400	$7800
3	75	525	600	$10,500

a. Let x, y, and z represent the cost for children tickets, student tickets, and general admission tickets, respectively. Set up a system of equations to solve for x, y, and z.

b. Set up the augmented matrix for the system and solve the system. (*Hint*: To make the augmented matrix simpler to work with, consider dividing each linear equation by an appropriate constant.)

c. Explain why the auditor knows that there was an error in the record keeping.

50. A concession stand at a city park sells hamburgers, hot dogs, and drinks. Three patrons buy the following food and drink combinations for the following prices.

Patron	Hamburgers	Hot Dogs	Drinks	Total Revenue
1	1	1	5	$11
2	0	1	2	$5
3	3	1	11	$22

a. Let x, y, and z represent the cost for a hamburger, a hot dog, and a drink, respectively. Set up a system of equations to solve for x, y, and z.

b. Set up the augmented matrix for the system and solve the system.

c. Explain why the concession stand manager knows that there was an error in the record keeping.

Mixed Exercises

51. The solution set to Exercise 25 is $\{(-2z + 16, 3z - 5, z) \mid z$ is any real number$\}$. Verify the solution by substituting the ordered triple into each individual equation.

52. The solution set to Exercise 26 is $\{(3z + 6, -4z + 2, z) \mid z$ is any real number$\}$. Verify the solution by substituting the ordered triple into each individual equation.

The systems in Exercises 53–56 are called **homogeneous** systems because each has $(0, 0, 0)$ as a solution. However, if a system is made up of dependent equations, it will have infinitely many more solutions. For each system, determine whether $(0, 0, 0)$ is the only solution or if the system has infinitely many solutions. If the system has infinitely many solutions, give the solution set.

53.
$$x + 2y - 8z = 0$$
$$-5x - 11y + 43z = 0$$
$$x + 5y - 12z = 0$$

54.
$$x - 2y - 7z = 0$$
$$-3x + 8y + 31z = 0$$
$$-2x + 5y + 22z = 0$$

55.
$$x - 5y + 13z = 0$$
$$-3x + 17y - 45z = 0$$
$$x - 4y + 10z = 0$$

56.
$$-2x + 15y - 83z = 0$$
$$x - 7y + 39z = 0$$
$$x - 5y + 29z = 0$$

Write About It

57. Explain how you can determine from the reduced row-echelon form of a matrix whether the corresponding system of equations is inconsistent.

58. What can you conclude about a system of equations if the corresponding reduced row-echelon form consists of a row entirely of zeros?

59. Consider the following system. By inspection, describe the geometrical relationship among the planes represented by the three equations.
$$x + y + z = 1$$
$$2x + 2y + 2z = 2$$
$$3x + 3y + 3z = 3$$

60. Explain why a system of two equations with three variables cannot have exactly one ordered triple as its solution.

Technology Connections

For Exercises 61–66, use a graphing utility to find the reduced row-echelon form of the augmented matrix for the system in the given exercise. Use the result to verify the answer to the given exercise.

61. Exercise 23

62. Exercise 24

63. Exercise 25

64. Exercise 26

65. Exercise 31

66. Exercise 32

OBJECTIVES

1. Determine the Order of a Matrix
2. Add and Subtract Matrices
3. Multiply a Matrix by a Scalar
4. Multiply Matrices
5. Apply Operations on Matrices

1. Determine the Order of a Matrix

We have seen how the manipulation of an augmented matrix can be used to solve a system of linear equations. Matrices have many other useful mathematical applications, particularly when manipulating tables or databases. In particular, matrices play a useful role in digital photography, film, and computer animation.

Gorodenkoff/Shutterstock

The **order of a matrix** is determined by the number of rows and number of columns. A matrix with m rows and n columns is an $m \times n$ matrix (read as "m by n" matrix).

EXAMPLE 1   Determining the Order of a Matrix

Determine the order of each matrix.

a. $\begin{bmatrix} 3 & \pi & 1.7 \\ -1 & 6 & 10 \end{bmatrix}$ **b.** $\begin{bmatrix} -7 \\ 2 \\ 4 \\ \sqrt{2} \end{bmatrix}$ **c.** $\begin{bmatrix} 1 & 0 & 0 \\ 0 & 1 & 0 \\ 0 & 0 & 1 \end{bmatrix}$ **d.** $[x \quad y \quad z]$

Solution:

a. The matrix has 2 rows and 3 columns. Therefore, it is a 2×3 matrix.

b. The matrix has 4 rows and 1 column. Therefore, it is a 4×1 matrix. A matrix with only 1 column is called a **column matrix**.

c. The matrix has 3 rows and 3 columns. Therefore, it is a 3×3 matrix. A matrix with the same number of rows and columns is called a **square matrix**.

d. The matrix has 1 row and 3 columns. Therefore, it is a 1×3 matrix. A matrix with only 1 row is called a **row matrix**.

Skill Practice 1   Determine the order of each matrix.

a. $\begin{bmatrix} -9 & 2 \\ 4.1 & -3 \\ \sqrt{3} & 4 \end{bmatrix}$ **b.** $\begin{bmatrix} -4 \\ 11 \end{bmatrix}$ **c.** $\begin{bmatrix} -0.1 & 0.4 \\ 0.5 & 0.2 \end{bmatrix}$ **d.** $[105 \quad 311]$

TIP The notation $[a_{ij}]$ is a generic way to denote a matrix with elements of the form a_{ij}, where i and j represent generic row and column numbers.

A matrix can be represented generically as follows:

$$[a_{ij}] = \begin{bmatrix} a_{11} & a_{12} & \cdots & a_{1n} \\ a_{21} & a_{22} & \cdots & a_{2n} \\ \vdots & & & \\ a_{m1} & a_{m2} & \cdots & a_{mn} \end{bmatrix}$$

Using the double subscript notation, a_{43} represents the element in the 4th row, 3rd column. The notation a_{ij} represents the element in the ith row, jth column.

Answers

1. a. 3×2 **b.** 2×1
   c. 2×2 **d.** 1×2

EXAMPLE 2   **Using Matrix Notation**

Consider $A = \begin{bmatrix} -4 & 5 & \sqrt{2} & -\pi \\ 10 & \frac{1}{2} & 0 & 6 \\ -\frac{7}{6} & 8 & 12 & -9 \end{bmatrix}$, where $A = [a_{ij}]$. Determine the value of each element.

  a. a_{32} **b.** a_{23} **c.** a_{14}

TIP

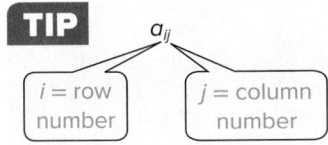

$i = $ row number

$j = $ column number

Solution:

  a. a_{32} represents the element in the 3rd row, 2nd column: 8
  b. a_{23} represents the element in the 2nd row, 3rd column: 0
  c. a_{14} represents the element in the 1st row, 4th column: $-\pi$

Skill Practice 2   Given matrix A from Example 2, determine the value of each element.

  a. a_{13} **b.** a_{31} **c.** a_{34}

2. Add and Subtract Matrices

We now consider the conditions under which two matrices are equal.

Equality of Matrices

Two matrices are equal if and only if they have the same order and if their corresponding elements are equal.

Example:

- The statement $\begin{bmatrix} 3 & -4 \\ x & z \end{bmatrix} = \begin{bmatrix} 3 & y \\ 9 & 7 \end{bmatrix}$ implies that $x = 9$, $y = -4$, and $z = 7$.

- Conversely, if $x = 9$, $y = -4$, and $z = 7$, then $\begin{bmatrix} 3 & -4 \\ x & z \end{bmatrix} = \begin{bmatrix} 3 & y \\ 9 & 7 \end{bmatrix}$.

Next, we define addition and subtraction of matrices.

Addition and Subtraction of Matrices

Let $A = [a_{ij}]$ and $B = [b_{ij}]$ be m by n matrices. Then,

$$A + B = [a_{ij} + b_{ij}] \text{ for } i = 1, 2, \ldots, m \text{ and } j = 1, 2, \ldots, n$$
$$A - B = [a_{ij} - b_{ij}] \text{ for } i = 1, 2, \ldots, m \text{ and } j = 1, 2, \ldots, n$$

That is, to add or subtract two matrices, the matrices must have the same order, and the sum or difference is found by adding or subtracting the corresponding elements.

Answers

2. a. $\sqrt{2}$ **b.** $-\dfrac{7}{6}$ **c.** -9

EXAMPLE 3   Adding and Subtracting Matrices

Given $A = \begin{bmatrix} 3 & -7 & 0 \\ 1 & \frac{4}{3} & -6 \end{bmatrix}$, $B = \begin{bmatrix} 8 & -3 & 4 \\ 0 & 1 & -9 \end{bmatrix}$, and $C = \begin{bmatrix} 3 & -5 \\ 4 & 18 \end{bmatrix}$,

find the sum or difference if possible.

a. $A + B$ **b.** $B - A$ **c.** $A + C$

Solution:

a. $A + B = \begin{bmatrix} 3 & -7 & 0 \\ 1 & \frac{4}{3} & -6 \end{bmatrix} + \begin{bmatrix} 8 & -3 & 4 \\ 0 & 1 & -9 \end{bmatrix}$ Matrix A and matrix B both have the same order (2×3), so the sum is defined.

$= \begin{bmatrix} 3+8 & -7+(-3) & 0+4 \\ 1+0 & \frac{4}{3}+1 & -6+(-9) \end{bmatrix}$ Add the elements in the corresponding positions.

$= \begin{bmatrix} 11 & -10 & 4 \\ 1 & \frac{7}{3} & -15 \end{bmatrix}$ Simplify.

b. $B - A = \begin{bmatrix} 8 & -3 & 4 \\ 0 & 1 & -9 \end{bmatrix} - \begin{bmatrix} 3 & -7 & 0 \\ 1 & \frac{4}{3} & -6 \end{bmatrix}$ Matrix A and matrix B both have the same order (2×3), so the difference is defined.

$= \begin{bmatrix} 8-3 & -3-(-7) & 4-0 \\ 0-1 & 1-\frac{4}{3} & -9-(-6) \end{bmatrix}$ Subtract the elements in the corresponding positions.

$= \begin{bmatrix} 5 & 4 & 4 \\ -1 & -\frac{1}{3} & -3 \end{bmatrix}$ Simplify.

c. Matrix A is a 2×3 matrix, whereas matrix C is a 2×2 matrix. The matrices are of different orders and therefore cannot be added or subtracted.

Skill Practice 3   Given $A = \begin{bmatrix} 3 & 4 & -7 \\ 9 & 2 & 0 \end{bmatrix}$, $B = \begin{bmatrix} -8 & 0 \\ 1 & 5 \\ -4 & \frac{1}{2} \end{bmatrix}$, and $C = \begin{bmatrix} 4 & -5 \\ 12 & 3 \\ -2 & 2 \end{bmatrix}$, find the sum or difference if possible.

a. $A + B$ **b.** $B + C$ **c.** $C - B$

The opposite of a real number a is $-a$. The value $-a$ is also called the additive inverse of a. A matrix A also has an additive inverse, denoted by $-A$.

Additive Inverse of a Matrix

Given $A = [a_{ij}]$, the **additive inverse of A**, denoted by $-A$, is $-A = [-a_{ij}]$.

Verbal Interpretation	Example
The additive inverse of a matrix is found by taking the opposite of each element in the matrix.	Given $A = \begin{bmatrix} -4 & 5 & -1 \\ 0 & -3 & 2 \end{bmatrix}$, $-A = \begin{bmatrix} 4 & -5 & 1 \\ 0 & 3 & -2 \end{bmatrix}$.

Answers

3. a. Not possible

b. $\begin{bmatrix} -4 & -5 \\ 13 & 8 \\ -6 & \frac{5}{2} \end{bmatrix}$

c. $\begin{bmatrix} 12 & -5 \\ 11 & -2 \\ 2 & \frac{3}{2} \end{bmatrix}$

The number 0 is the identity element under the addition of real numbers because $a + (-a) = 0$. A matrix in which all elements are zero is called a **zero matrix** and is denoted by *0*. The sum of a matrix *A* and its additive inverse *−A* is the zero matrix of the same order. For example,

$$\underset{A}{\begin{bmatrix} -4 & 5 & -1 \\ 0 & -3 & 2 \end{bmatrix}} + \underset{(-A)}{\begin{bmatrix} 4 & -5 & 1 \\ 0 & 3 & -2 \end{bmatrix}} = \underset{0}{\begin{bmatrix} 0 & 0 & 0 \\ 0 & 0 & 0 \end{bmatrix}} \quad \text{2 × 3 zero matrix}$$

Properties of Matrix Addition

Let *A*, *B*, and *C* be matrices of order $m \times n$, and let *0* be the zero matrix of order $m \times n$. Then,

1. $A + B = B + A$ Commutative property of matrix addition
2. $A + (B + C) = (A + B) + C$ Associative property of matrix addition
3. $A + (-A) = 0$ Inverse property of matrix addition
4. $A + 0 = 0 + A = A$ Identity property of matrix addition

---

3. Multiply a Matrix by a Scalar

Next we look at the product of a real number *k* and a matrix. This is called **scalar multiplication**. The real number *k* is called a **scalar** to distinguish it from a matrix.

Scalar Multiplication

Let $A = [a_{ij}]$ be an $m \times n$ matrix and let *k* be a real number. Then, $kA = [ka_{ij}]$.

Verbal Interpretation	**Example**
To multiply a matrix *A* by a scalar *k*, multiply each element in the matrix by *k*.	Given $A = \begin{bmatrix} -4 & 5 & -1 \\ 0 & -3 & 2 \end{bmatrix}$, $5A = \begin{bmatrix} -20 & 25 & -5 \\ 0 & -15 & 10 \end{bmatrix}$.

The following properties of scalar multiplication are similar to those of the multiplication of real numbers.

Properties of Scalar Multiplication

Let *A* and *B* be $m \times n$ matrices and let *c* and *d* be real numbers. Then,

1. $c(A + B) = cA + cB$ Distributive property of scalar multiplication
2. $(c + d)A = cA + dA$ Distributive property of scalar multiplication
3. $c(dA) = (cd)A$ Associative property of scalar multiplication

---

EXAMPLE 4   Multiplying a Matrix by a Scalar

Given $A = \begin{bmatrix} 3 & -5 \\ 0 & 1 \\ -4 & 2 \end{bmatrix}$ and $B = \begin{bmatrix} -8 & 9 \\ 4 & -1 \\ 0 & 3 \end{bmatrix}$, find $2A + 4(A + B)$.

Solution:

$$2A + 4(A + B) = 2A + 4A + 4B$$

Apply the distributive property of scalar multiplication.

$$= 6A + 4B$$

$$6A + 4B = 6\begin{bmatrix} 3 & -5 \\ 0 & 1 \\ -4 & 2 \end{bmatrix} + 4\begin{bmatrix} -8 & 9 \\ 4 & -1 \\ 0 & 3 \end{bmatrix} = \begin{bmatrix} 18 & -30 \\ 0 & 6 \\ -24 & 12 \end{bmatrix} + \begin{bmatrix} -32 & 36 \\ 16 & -4 \\ 0 & 12 \end{bmatrix}$$

$$= \begin{bmatrix} -14 & 6 \\ 16 & 2 \\ -24 & 24 \end{bmatrix}$$

Skill Practice 4   Given $A = [4 \quad -3 \quad 9]$ and $B = [-2 \quad 0 \quad 3]$, find $-5A - 2(A + B)$.

TECHNOLOGY CONNECTIONS

Adding Matrices and Multiplying a Scalar by a Matrix

Many graphing utilities have the capability to perform operations on matrices. From Example 4, we can enter matrix A and matrix B into the calculator using the EDIT feature under the MATRIX menu. Then on the home screen enter 6[A] + 4[B] and press [ENTER].

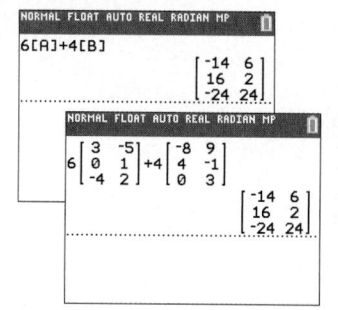

Alternatively, in "MATHPRINT" mode, enter the matrices on the home screen using [ALPHA] F3.

As seen, many of the same properties that are true for addition and multiplication of real numbers are true for the addition of matrices and for scalar multiplication. As a result, we can solve a matrix equation involving these operations in the same way as we solve a linear equation.

EXAMPLE 5   Solving a Matrix Equation

Solve the matrix equation for X, given that $A = \begin{bmatrix} 2 & 5 \\ 1 & -4 \end{bmatrix}$ and $B = \begin{bmatrix} 6 & -1 \\ -3 & -8 \end{bmatrix}$.

$$4X - A = B$$

Solution:

$$4X - A = B$$

$$4X = A + B$$

Add matrix A to both sides.

$$X = \frac{1}{4}(A + B)$$

Perform scalar multiplication by $\frac{1}{4}$ on both sides.

$$X = \frac{1}{4}\left(\begin{bmatrix} 2 & 5 \\ 1 & -4 \end{bmatrix} + \begin{bmatrix} 6 & -1 \\ -3 & -8 \end{bmatrix}\right)$$

Add the matrices within parentheses.

$$X = \frac{1}{4}\begin{bmatrix} 8 & 4 \\ -2 & -12 \end{bmatrix}$$

Multiply the matrix by the scalar $\frac{1}{4}$.

$$X = \begin{bmatrix} 2 & 1 \\ -\frac{1}{2} & -3 \end{bmatrix}$$

Simplify.

TIP   We can substitute matrix X back into the equation $4X - A = B$ to verify that it is a solution to the equation.

Answer

4. $[-24 \quad 21 \quad -69]$

4. Multiply Matrices

Finding the product of two matrices is more complicated than finding the product of a scalar and a matrix. We will demonstrate the process to multiply two matrices and then offer a formal definition.

Consider $A = \begin{bmatrix} 2 & -3 & 1 \\ -4 & 7 & 0 \end{bmatrix}$ and $B = \begin{bmatrix} -1 & -6 \\ 10 & 5 \\ 8 & -2 \end{bmatrix}$.

> **TIP** For matrix multiplication, the number of rows of the first matrix does not need to be equal to the number of columns of the second matrix. For example, the product of a 3×4 and a 4×2 matrix results in a 3×2 matrix.
>
> $3 \times 4 \qquad\qquad 4 \times 2$
>
> equal
>
> product is 3×2

- To multiply AB, we require that the number of columns in A be equal to the number of rows in B.

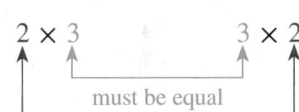

$$
\underset{\substack{\uparrow \\ 2 \times 3}}{\begin{bmatrix} 2 & -3 & 1 \\ -4 & 7 & 0 \end{bmatrix}} \cdot \underset{\substack{\uparrow \\ 3 \times 2}}{\begin{bmatrix} -1 & -6 \\ 10 & 5 \\ 8 & -2 \end{bmatrix}}
$$

- The resulting matrix will have dimensions equal to the number of rows of A by the number of columns of B.

must be equal

The product AB will be a 2×2 matrix.

The product will be a 2×2 matrix.

$$
\begin{bmatrix} 2 & -3 & 1 \\ -4 & 7 & 0 \end{bmatrix} \cdot \begin{bmatrix} -1 & -6 \\ 10 & 5 \\ 8 & -2 \end{bmatrix} = \begin{bmatrix} \square & \square \\ \square & \square \end{bmatrix}
$$

To find the entry in the first row, first column of the product, multiply the corresponding elements in the first row of A by the elements in the first column of B, and take the sum.

$$
\begin{bmatrix} 2 & -3 & 1 \\ -4 & 7 & 0 \end{bmatrix} \cdot \begin{bmatrix} -1 & -6 \\ 10 & 5 \\ 8 & -2 \end{bmatrix} = \begin{bmatrix} 2(-1) + (-3)(10) + 1(8) & \square \\ \square & \square \end{bmatrix}
$$

To find the entry in the first row, second column of the product, take the sum of the products of the corresponding elements in the first row of A and second column of B.

$$
\begin{bmatrix} 2 & -3 & 1 \\ -4 & 7 & 0 \end{bmatrix} \cdot \begin{bmatrix} -1 & -6 \\ 10 & 5 \\ 8 & -2 \end{bmatrix} = \begin{bmatrix} 2(-1) + (-3)(10) + 1(8) & 2(-6) + (-3)(5) + 1(-2) \\ \square & \square \end{bmatrix}
$$

To find the entry in the second row, first column of the product, take the sum of the products of the corresponding elements in the second row of A and first column of B.

Answer

5. $X = \begin{bmatrix} -1 & 1 \\ \frac{5}{2} & -3 \end{bmatrix}$

$$
\begin{bmatrix} 2 & -3 & 1 \\ -4 & 7 & 0 \end{bmatrix} \cdot \begin{bmatrix} -1 & -6 \\ 10 & 5 \\ 8 & -2 \end{bmatrix} = \begin{bmatrix} 2(-1) + (-3)(10) + 1(8) & 2(-6) + (-3)(5) + 1(-2) \\ -4(-1) + 7(10) + 0(8) & \square \end{bmatrix}
$$

To find the entry in the second row, second column of the product, take the sum of the products of the corresponding elements in the second row of A and second column of B.

$$\begin{bmatrix} 2 & -3 & 1 \\ -4 & 7 & 0 \end{bmatrix} \cdot \begin{bmatrix} -1 & -6 \\ 10 & 5 \\ 8 & -2 \end{bmatrix} = \begin{bmatrix} 2(-1) + (-3)(10) + 1(8) & 2(-6) + (-3)(5) + 1(-2) \\ -4(-1) + 7(10) + 0(8) & -4(-6) + 7(5) + 0(-2) \end{bmatrix}$$

Simplifying, we have

$$AB = \begin{bmatrix} 2 & -3 & 1 \\ -4 & 7 & 0 \end{bmatrix} \cdot \begin{bmatrix} -1 & -6 \\ 10 & 5 \\ 8 & -2 \end{bmatrix} = \begin{bmatrix} -24 & -29 \\ 74 & 59 \end{bmatrix}$$

From this example, note that each element in the product AB is of the form $a_1b_1 + a_2b_2 + \cdots + a_nb_n$, where the elements $a_1, a_2, \ldots a_n$ are elements in a row of A and $b_1, b_2, \cdots b_n$ are elements in a column of B. The expression $a_1b_1 + a_2b_2 + \cdots + a_nb_n$ is called an **inner product**.

Matrix Multiplication

Let A be an $m \times p$ matrix and let B be a $p \times n$ matrix, then the product AB is an $m \times n$ matrix. For the matrix AB, the element in the ith row and jth column is the sum of the products of the corresponding elements in the ith row of A and the jth column of B (the inner product of the ith row and jth column).

Formally, if

$$A = \begin{bmatrix} a_{11} & a_{12} & \cdots & a_{1p} \\ a_{21} & a_{22} & \cdots & a_{2p} \\ \vdots & & & \\ a_{m1} & a_{m2} & \cdots & a_{mp} \end{bmatrix} \text{ and } B = \begin{bmatrix} b_{11} & b_{12} & \cdots & b_{1n} \\ b_{21} & b_{22} & \cdots & b_{2n} \\ \vdots & & & \\ b_{p1} & b_{p2} & \cdots & b_{pn} \end{bmatrix}$$

then the elements in the matrix $C = AB$ are given by

$$c_{ij} = a_{i1} \cdot b_{1j} + a_{i2} \cdot b_{2j} + \cdots + a_{ip} \cdot b_{pj}$$

Note: If the number of columns in A does not equal the number of rows in B, then it is not possible to compute the product AB.

EXAMPLE 6   Multiplying Matrices

Given $A = \begin{bmatrix} 2 & 5 & 6 \\ -3 & 0 & 1 \end{bmatrix}$, $B = \begin{bmatrix} 1 \\ 3 \\ -4 \end{bmatrix}$, and $C = \begin{bmatrix} -2 & 10 & 1 \end{bmatrix}$,

find the following products if possible.

a. AB **b.** BC **c.** AC

Solution:

a. A $\cdot$ B

2×3 3×1

equal

The product is a 2×1 matrix.

Matrix A is a 2×3 matrix and matrix B is a 3×1 matrix. The number of columns in A is equal to the number of rows in B. The resulting matrix will be a 2×1 matrix.

$$= \begin{bmatrix} 2 & 5 & 6 \\ -3 & 0 & 1 \end{bmatrix} \cdot \begin{bmatrix} 1 \\ 3 \\ -4 \end{bmatrix} = \begin{bmatrix} 2(1) + 5(3) + 6(-4) \\ -3(1) + 0(3) + 1(-4) \end{bmatrix}$$

Multiply the elements in the first row of A by the elements in the first column of B.

Multiply the elements in the second row of A by the elements in the first column of B.

$$= \begin{bmatrix} -7 \\ -7 \end{bmatrix}$$

b. B $\cdot$ C

3×1 1×3

equal

The product is a 3×3 matrix.

Matrix B is a 3×1 matrix and matrix C is a 1×3 matrix. The number of columns in B is equal to the number of rows in C. The resulting matrix will be a 3×3 matrix.

$$B \cdot C = \begin{bmatrix} 1 \\ 3 \\ -4 \end{bmatrix} \cdot \begin{bmatrix} -2 & 10 & 1 \end{bmatrix} = \begin{bmatrix} 1(-2) & 1(10) & 1(1) \\ 3(-2) & 3(10) & 3(1) \\ -4(-2) & -4(10) & -4(1) \end{bmatrix}$$

$$= \begin{bmatrix} -2 & 10 & 1 \\ -6 & 30 & 3 \\ 8 & -40 & -4 \end{bmatrix}$$

c. A $\cdot$ C

2×3 1×3

not equal

The number of columns in A does not equal the number of rows in C. Therefore, it is not possible to compute the product AC.

Skill Practice 6   Given $A = \begin{bmatrix} 2 & -5 \\ 1 & 0 \end{bmatrix}$, $B = \begin{bmatrix} 3 & 5 & 4 \\ 6 & 0 & -8 \end{bmatrix}$, and $C = \begin{bmatrix} 1 \\ 0 \\ 6 \end{bmatrix}$,

find the products if possible.   **a.** AB **b.** BC **c.** AC

EXAMPLE 7   Multiplying Matrices

Given $A = \begin{bmatrix} 3 & -4 \\ 1 & 5 \end{bmatrix}$ and $B = \begin{bmatrix} 1 & 0 \\ -3 & 6 \end{bmatrix}$, find the following products if possible.

a. AB **b.** BA

Solution:

a. A $\cdot$ B

2×2 2×2

equal

The product is a 2×2 matrix.

$$AB = \begin{bmatrix} 3 & -4 \\ 1 & 5 \end{bmatrix} \cdot \begin{bmatrix} 1 & 0 \\ -3 & 6 \end{bmatrix}$$

$$= \begin{bmatrix} 3(1) + (-4)(-3) & 3(0) + (-4)(6) \\ 1(1) + 5(-3) & 1(0) + 5(6) \end{bmatrix}$$

$$= \begin{bmatrix} 15 & -24 \\ -14 & 30 \end{bmatrix}$$

Avoiding Mistakes

Example 7 shows that for two matrices A and B, the product AB does not necessarily equal BA. That is, matrix multiplication is *not* commutative.

Answers

6. a. $AB = \begin{bmatrix} -24 & 10 & 48 \\ 3 & 5 & 4 \end{bmatrix}$

b. $BC = \begin{bmatrix} 27 \\ -42 \end{bmatrix}$

c. Not possible

b. B $\cdot$ A

    2×2 2×2

    equal

    The product is a
    2×2 matrix.

$$BA = \begin{bmatrix} 1 & 0 \\ -3 & 6 \end{bmatrix} \cdot \begin{bmatrix} 3 & -4 \\ 1 & 5 \end{bmatrix}$$

$$= \begin{bmatrix} 1(3) + 0(1) & 1(-4) + 0(5) \\ -3(3) + 6(1) & -3(-4) + 6(5) \end{bmatrix}$$

$$= \begin{bmatrix} 3 & -4 \\ -3 & 42 \end{bmatrix}$$

Skill Practice 7   Given $A = \begin{bmatrix} 1 & -3 \\ 0 & 5 \end{bmatrix}$ and $B = \begin{bmatrix} 2 & 4 \\ 6 & 1 \end{bmatrix}$, find the following products if possible.   **a.** AB **b.** BA

TECHNOLOGY CONNECTIONS

Multiplying Matrices

To multiply two matrices on a graphing utility, first enter the matrices into the calculator using the EDIT feature under the MATRIX menu. Then on the home screen, enter the product. The products AB and BA from Example 7 are shown here.

Alternatively, in "MATHPRINT" mode, enter the matrices on the home screen using (ALPHA) F3. The products AB and BA are shown here.

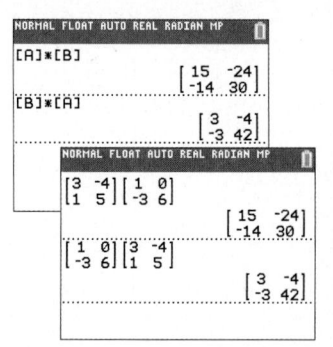

---

5. Apply Operations on Matrices

In applications, a matrix gives mathematicians a systematic format in which to represent data. Multiplication of matrices is a tool to manipulate these data sets. For example, in business, we might multiply the number of items sold by the unit price per item to determine total revenue.

$$\begin{bmatrix} \text{Quantity} \\ \text{matrix} \end{bmatrix} \cdot \begin{bmatrix} \text{Unit price} \\ \text{matrix} \end{bmatrix} = \begin{bmatrix} \text{Total revenue} \\ \text{matrix} \end{bmatrix}$$

    $m \times p$ $p \times n$ $m \times n$

        same

    order of product

> **TIP** The number of columns p of the quantity matrix must equal the number of rows p of the unit price matrix for the product to be defined.

EXAMPLE 8   **Multiplying Matrices in a Business Application**

A company owns two coffee shops. The number of donuts, coffee cakes, hot drinks, and cold drinks sold for each shop is given in matrix Q. The price per item is given in matrix P. Find the product QP and interpret the result.

	Donuts	Coffee cakes	Hot drinks	Cold drinks	
$Q =$	162	34	120	44	Shop 1
	186	50	145	62	Shop 2

$$P = \begin{bmatrix} \$0.40 \\ \$2.50 \\ \$3.50 \\ \$1.50 \end{bmatrix} \begin{matrix} \text{Donuts} \\ \text{Coffee cakes} \\ \text{Hot drinks} \\ \text{Cold drinks} \end{matrix}$$

Answers

7. a. $AB = \begin{bmatrix} -16 & 1 \\ 30 & 5 \end{bmatrix}$

b. $BA = \begin{bmatrix} 2 & 14 \\ 6 & -13 \end{bmatrix}$

Solution:

Matrix Q is a 2×4 matrix and matrix P is a 4×1 matrix. Matrix Q has 4 columns, which match the number of rows of P, so the product QP is defined. The product matrix will be a 2×1 matrix. The product QP is

162 donuts times \$0.40 per donut (Shop 1).

$$\begin{bmatrix} 162 & 34 & 120 & 44 \\ 186 & 50 & 145 & 62 \end{bmatrix} \cdot \begin{bmatrix} 0.40 \\ 2.50 \\ 3.50 \\ 1.50 \end{bmatrix} = \begin{bmatrix} 162(0.40) + 34(2.50) + 120(3.50) + 44(1.50) \\ 186(0.40) + 50(2.50) + 145(3.50) + 62(1.50) \end{bmatrix}$$

$$= \begin{bmatrix} \$635.80 \\ \$799.90 \end{bmatrix} \quad \boxed{\text{Row 1: Revenue for Shop 1}}$$
$$\boxed{\text{Row 2: Revenue for Shop 2}}$$

The product QP is a matrix representing the total revenue from these four items for each shop.

Skill Practice 8   A farmer sells organic zucchini, yellow squash, and corn in two different roadside stands. Matrix Q represents the number of pounds of each type of vegetable sold at each stand. Matrix P gives the price per pound of each item. Find the product QP and interpret the result.

$$Q = \begin{matrix} & \textbf{Zucchini} & \begin{matrix}\textbf{Yellow}\\\textbf{squash}\end{matrix} & \textbf{Corn} \\ & \begin{bmatrix} 42 & 40 & 84 \\ 30 & 36 & 90 \end{bmatrix} & & \end{matrix} \begin{matrix} \textbf{Stand 1} \\ \textbf{Stand 2} \end{matrix}$$

$$P = \begin{bmatrix} \$4.00 \\ \$3.40 \\ \$4.20 \end{bmatrix} \begin{matrix} \textbf{Zucchini} \\ \textbf{Yellow squash} \\ \textbf{Corn} \end{matrix}$$

In Chapter 6 we learned how to apply transformations to the graphs of functions. In Example 9 we see how matrices can also be used to transform images in a rectangular coordinate system. This is particularly helpful to computer programmers who write software for computer gaming.

EXAMPLE 9   Using Matrix Operations to Transform a Graph

Consider the triangle shown in Figure 11-6. The vertices of the triangle can be represented in a matrix where the first row represents the x-coordinates and the second row represents the corresponding y-coordinates.

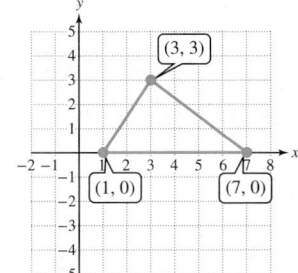

$$A = \begin{bmatrix} 1 & 3 & 7 \\ 0 & 3 & 0 \end{bmatrix} \begin{matrix} \leftarrow x\text{-coordinates} \\ \leftarrow y\text{-coordinates} \end{matrix}$$

Figure 11-6

a. Use addition of matrices to shift the triangle 3 units to the left and 2 units upward.

b. Find the product $\begin{bmatrix} -1 & 0 \\ 0 & 1 \end{bmatrix} \cdot A$ and determine the effect on the graph.

c. Find the product $\begin{bmatrix} 1 & 0 \\ 0 & -1 \end{bmatrix} \cdot A$ and determine the effect on the graph.

Answer

8. $\begin{bmatrix} \$656.80 \\ \$620.40 \end{bmatrix}$; The product QP is a matrix representing the total revenue from these three items for each stand.

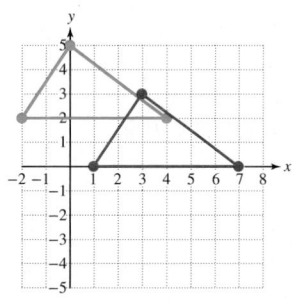

Figure 11-7

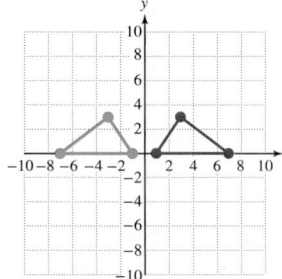

Figure 11-8

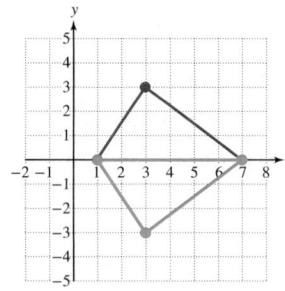

Figure 11-9

Answer

9. $\begin{bmatrix} -1 & -3 & -7 \\ 0 & -3 & 0 \end{bmatrix}$; The triangle is reflected across both the x- and y-axes.

Solution:

a. To shift the graph to the left, subtract 3 from each x-coordinate.

To shift the graph upward, add 2 to each y-coordinate.

$$\begin{bmatrix} 1 & 3 & 7 \\ 0 & 3 & 0 \end{bmatrix} + \begin{bmatrix} -3 & -3 & -3 \\ 2 & 2 & 2 \end{bmatrix} \quad \longleftarrow \text{This row subtracts 3 from each } x\text{-coordinate.}$$
$$\longleftarrow \text{This row adds 2 to each } y\text{-coordinate.}$$

$$= \begin{bmatrix} -2 & 0 & 4 \\ 2 & 5 & 2 \end{bmatrix}$$

The vertices of the new triangle are $(-2, 2)$, $(0, 5)$, and $(4, 2)$. See Figure 11-7.

b. $\begin{bmatrix} -1 & 0 \\ 0 & 1 \end{bmatrix} \cdot A = \begin{bmatrix} -1 & 0 \\ 0 & 1 \end{bmatrix} \begin{bmatrix} 1 & 3 & 7 \\ 0 & 3 & 0 \end{bmatrix}$

$$= \begin{bmatrix} -1(1) + 0(0) & -1(3) + 0(3) & -1(7) + 0(0) \\ 0(1) + 1(0) & 0(3) + 1(3) & 0(7) + 1(0) \end{bmatrix}$$

$$= \begin{bmatrix} -1 & -3 & -7 \\ 0 & 3 & 0 \end{bmatrix} \quad \begin{array}{l} \text{The corresponding points are } (-1, 0), \\ (-3, 3), \text{ and } (-7, 0). \end{array}$$

The triangle has been reflected across the y-axis. See Figure 11-8.

c. $\begin{bmatrix} 1 & 0 \\ 0 & -1 \end{bmatrix} \cdot A = \begin{bmatrix} 1 & 0 \\ 0 & -1 \end{bmatrix} \begin{bmatrix} 1 & 3 & 7 \\ 0 & 3 & 0 \end{bmatrix}$

$$= \begin{bmatrix} 1(1) + 0(0) & 1(3) + 0(3) & 1(7) + 0(0) \\ 0(1) + -1(0) & 0(3) + -1(3) & 0(7) + -1(0) \end{bmatrix}$$

$$= \begin{bmatrix} 1 & 3 & 7 \\ 0 & -3 & 0 \end{bmatrix} \quad \begin{array}{l} \text{The corresponding points are} \\ (1, 0), (3, -3), \text{ and } (7, 0). \end{array}$$

The triangle has been reflected across the x-axis. See Figure 11-9.

Skill Practice 9   Use matrix A from Example 9. Find the product $\begin{bmatrix} -1 & 0 \\ 0 & -1 \end{bmatrix} \cdot A$ and determine the effect on the graph of the triangle in Figure 11-6.

SECTION 11.3   Practice Exercises

Prerequisite Review

For Exercises R.1–R.12, see Section 1.1 for review.

R.1. Apply the commutative property of addition.
$-3 + 4x$

R.2. Apply the commutative property of multiplication.
$t(-2)$

R.3. Apply the associative property of multiplication and simplify. $-5(6m)$

R.4. Apply the associative property of addition and simplify. $6 + (-4 + x)$

R.5. What is the identity element under the addition of real numbers?

R.6. What is the identity element under the multiplication of real numbers?

R.7. Identify the additive inverse of 9.

R.8. Identify the additive inverse of $-4x$.

R.9. Identify the multiplicative inverse of the nonzero real number p.

R.10. Identify the multiplicative inverse of $\dfrac{3}{5}$.

R.11. Apply the distributive property. $\dfrac{1}{2}(4x + 20)$

R.12. Apply the distributive property. $\dfrac{2}{3}(6y + 9)$

For Exercises R.13–R.14, solve for x. (See Section 1.2 for review.)

R.13. $2x - A = B$

R.14. $A + 4x = 2B$

For Exercises R.15–R.16, determine the values of a and b that make the statement true.

R.15. $3x^2 - 4x = ax^2 + bx$

R.16. $-6y^4 + by^2 = ay^4 + 11y^2$

R.17. If a vendor sells T-shirts for \$15 each, what is the revenue for selling 50 T-shirts?

R.18. If a plumber charges \$45/hr, how much will he make for working a 40-hour week?

Concept Connections

1. If the _____ of a matrix is $p \times q$, then p represents the number of _____ and q represents the number of _____.

2. A matrix with the same number of rows and columns is called a _____ matrix.

3. What are the requirements for two matrices to be equal?

4. An $m \times n$ matrix whose elements are all zero is called a _____ matrix.

5. To multiply two matrices A and B, the number of _____ of A must equal the number of _____ of B.

6. If A is a 5×3 matrix and B is a 3×7 matrix, then the product AB will be a matrix of order _____. The product BA (is/is not) defined.

7. True or false: Matrix multiplication is a commutative operation.

8. True or false: If a row matrix A and a column matrix B have the same number of elements, then the product AB is defined.

Objective 1: Determine the Order of a Matrix

9. What is a row matrix?

10. What is a column matrix?

For Exercises 11–16,

a. Give the order of the matrix.

b. Classify the matrix as a square matrix, row matrix, column matrix, or none of these. **(See Example 1)**

11. $\begin{bmatrix} 3 & 5 & -1 \\ \frac{1}{2} & \sqrt{3} & 1.7 \end{bmatrix}$

12. $\begin{bmatrix} 1 & 5 & 6 & 2 \\ -1 & 3 & -\frac{2}{3} & \pi \\ 0 & 1 & -6.1 & 12 \end{bmatrix}$

13. $\begin{bmatrix} 3 \\ 1 \\ 7 \end{bmatrix}$

14. $\begin{bmatrix} 2.4 & 6.9 \end{bmatrix}$

15. $\begin{bmatrix} 4 & 2 \\ 8 & 4 \end{bmatrix}$

16. $\begin{bmatrix} -4 & 10 & 3 & 0 \\ 0 & 1 & 0 & 0 \\ 7 & 0 & 9 & -1 \\ 0 & 2 & 0 & 1 \end{bmatrix}$

For Exercises 17–22, determine the value of the given element of matrix $A = [a_{ij}]$. **(See Example 2)**

$$A = \begin{bmatrix} 3 & -6 & \frac{1}{3} \\ 2 & 4 & 0 \\ \sqrt{5} & 11 & 8.6 \\ \frac{1}{2} & 4 & 2 \end{bmatrix}$$

17. a_{31}

18. a_{32}

19. a_{13}

20. a_{23}

21. a_{43}

22. a_{42}

Objective 2: Add and Subtract Matrices

23. Given $A = \begin{bmatrix} 2 & x \\ z & -5 \end{bmatrix}$ and $B = \begin{bmatrix} y & 4 \\ 10 & -5 \end{bmatrix}$, for what values of x, y, and z will $A = B$?

24. Given $A = \begin{bmatrix} 4 & a \\ b & 7 \end{bmatrix}$ and $B = \begin{bmatrix} c & 12 \\ -1 & 6 \end{bmatrix}$, is it possible that $A = B$? Explain.

25. Given $B = \begin{bmatrix} -4 & 6 & 9 \\ \frac{3}{5} & 1 & 7 \end{bmatrix}$, find the additive inverse of B.

26. Given $C = \begin{bmatrix} -1 & 6 \\ \sqrt{3} & 9 \end{bmatrix}$, find the additive inverse of C.

For Exercises 27–32, add or subtract the given matrices if possible. **(See Example 3)**

$$A = \begin{bmatrix} 6 & -1 \\ 7 & \frac{1}{2} \\ 2 & \sqrt{2} \end{bmatrix} \quad B = \begin{bmatrix} -9 & 2 \\ 6.2 & 2 \\ \frac{1}{3} & \sqrt{8} \end{bmatrix} \quad C = \begin{bmatrix} 11 & 4 \\ 1 & -\frac{1}{3} \\ 1 & 6 \end{bmatrix} \quad D = \begin{bmatrix} 2 & 3 & 8 \\ -1 & 6 & \frac{1}{6} \end{bmatrix}$$

27. $A + B$

28. $A + C$

29. $C - A + B$

30. $B - A - C$

31. $B + D$

32. $C + D$

Objective 3: Multiply a Matrix by a Scalar

33. Explain how to multiply a matrix by a scalar.

34. Given matrix A, explain how to find its additive inverse $-A$.

For Exercises 35–42, use $A = \begin{bmatrix} 2 & 4 & -9 \\ 1 & \sqrt{3} & \frac{1}{2} \end{bmatrix}$ and $B = \begin{bmatrix} -1 & 0 & 4 \\ 2 & 9 & \frac{2}{3} \end{bmatrix}$. **(See Example 4)**

35. $3A$

36. $-6B$

37. $-2A - 7B$

38. $4A - 3B$

39. $-4(A + B)$

40. $2(A - B)$

41. $-3A + 5(A - B)$

42. $-8A - 2(A + B)$

For Exercises 43–48, use $A = \begin{bmatrix} 1 & 6 \\ 4 & -2 \end{bmatrix}$ and $B = \begin{bmatrix} 2 & -4 \\ 6 & 9 \end{bmatrix}$ and solve for X. **(See Example 5)**

43. $2X - B = A$

44. $3X + A = B$

45. $A + 5X = B$

46. $B - 4X = A$

47. $2A - B = 10X$

48. $3B - A = -2X$

Objective 4: Multiply Matrices

49. Given that A is a 4×2 matrix and B is a 2×1 matrix,
   a. Is AB defined? If so, what is the order of AB?
   b. Is BA defined? If so, what is the order of BA?

50. Given that C is a 3×7 matrix and D is a 7×2 matrix,
   a. Is CD defined? If so, what is the order of CD?
   b. Is DC defined? If so, what is the order of DC?

51. Given that E is a 5×1 matrix and F is a 1×5 matrix,
   a. Is EF defined? If so, what is the order of EF?
   b. Is FE defined? If so, what is the order of FE?

52. Given that G is a 1×6 matrix and H is a 6×1 matrix,
   a. Is GH defined? If so, what is the order of GH?
   b. Is HG defined? If so, what is the order of HG?

For Exercises 53–64, **(See Examples 6–7)**
   a. Find AB if possible.
   b. Find BA if possible.
   c. Find A^2 if possible. (*Hint*: $A^2 = A \cdot A$.)

53. $A = \begin{bmatrix} 2 & 3 \\ 5 & 7 \end{bmatrix}$ and $B = \begin{bmatrix} 1 & 4 \\ -1 & 3 \end{bmatrix}$

54. $A = \begin{bmatrix} 1 & -6 \\ 5 & 10 \end{bmatrix}$ and $B = \begin{bmatrix} -2 & 3 \\ 7 & -1 \end{bmatrix}$

55. $A = \begin{bmatrix} 2 & 4 \\ -6 & 3 \\ 1 & 7 \end{bmatrix}$ and $B = \begin{bmatrix} 1 & 4 & -1 \\ -2 & 0 & 10 \end{bmatrix}$

56. $A = \begin{bmatrix} 1 & 3 & 4 \\ -2 & 5 & 6 \end{bmatrix}$ and $B = \begin{bmatrix} 1 & 3 \\ 9 & 2 \\ 0 & 4 \end{bmatrix}$

57. $A = \begin{bmatrix} -9 & 2 & 3 \\ -1 & 5 & 4 \\ 0 & 1 & 7 \end{bmatrix}$ and $B = \begin{bmatrix} -1 \\ 5 \\ 0 \end{bmatrix}$

58. $A = \begin{bmatrix} 2 \\ -1 \\ 3 \end{bmatrix}$ and $B = \begin{bmatrix} 2 & 7 & -1 \\ 0 & 4 & 1 \\ 0 & 3 & -6 \end{bmatrix}$

59. $A = \begin{bmatrix} 1 & \frac{1}{2} \end{bmatrix}$ and $B = \begin{bmatrix} -\frac{1}{3} \\ 2 \end{bmatrix}$

60. $A = \begin{bmatrix} 4 \\ \frac{3}{4} \\ 1 \end{bmatrix}$ and $B = \begin{bmatrix} -5 & \frac{1}{2} & \frac{1}{3} \end{bmatrix}$

61. $A = \begin{bmatrix} 4 \\ -6 \end{bmatrix}$ and $B = \begin{bmatrix} 1 & 2 & 5 & 6 \end{bmatrix}$

62. $A = [4]$ and $B = [-3 \quad 4 \quad 1]$

63. $A = [-5]$ and $B = [5]$

64. $A = [6]$ and $B = [-2]$

For Exercises 65–68, find AB and BA.

65. $A = \begin{bmatrix} 3.1 & -2.3 \\ 1.1 & 6.5 \end{bmatrix}$ and $B = \begin{bmatrix} 1 & 0 \\ 0 & 1 \end{bmatrix}$

66. $A = \begin{bmatrix} 1 & 0 \\ 0 & 1 \end{bmatrix}$ and $B = \begin{bmatrix} 0.05 & -0.07 \\ 0.16 & 0.09 \end{bmatrix}$

67. $A = \begin{bmatrix} 1 & 0 & 0 \\ 0 & 1 & 0 \\ 0 & 0 & 1 \end{bmatrix}$ and $B = \begin{bmatrix} \frac{9}{5} & -3 & \sqrt{6} \\ 5 & \frac{1}{2} & 2 \\ 3 & 0 & 1 \end{bmatrix}$

68. $A = \begin{bmatrix} \frac{2}{3} & 5 & \sqrt{2} \\ 1 & 0 & 3 \\ -\frac{1}{4} & 0 & 8 \end{bmatrix}$ and $B = \begin{bmatrix} 1 & 0 & 0 \\ 0 & 1 & 0 \\ 0 & 0 & 1 \end{bmatrix}$

Objective 5: Apply Operations on Matrices

69. Matrix D gives the dealer invoice prices for sedan and hatchback models of a car with manual transmission or automatic transmission. Matrix M gives the MSRP (manufacturer's suggested retail price) for the cars.

$$D = \begin{array}{c} \\ \\ \end{array} \begin{matrix} \text{Sedan} & \text{Hatchback} \\ \begin{bmatrix} \$29{,}000 & \$27{,}500 \\ \$28{,}500 & \$26{,}900 \end{bmatrix} & \begin{matrix} \textbf{Manual} \\ \textbf{Automatic} \end{matrix} \end{matrix}$$

$$M = \begin{matrix} \text{Sedan} & \text{Hatchback} \\ \begin{bmatrix} \$32{,}600 & \$29{,}900 \\ \$31{,}900 & \$28{,}900 \end{bmatrix} & \begin{matrix} \textbf{Manual} \\ \textbf{Automatic} \end{matrix} \end{matrix}$$

a. Compute $M - D$ and interpret the result.

b. A buyer thinks that a fair price is 6% above dealer invoice. Use scalar multiplication to determine a matrix F that gives the fair price for these cars for each type of transmission.

70. In matrix C, a coffee shop records the cost to produce a cup of standard Columbian coffee and the cost to produce a cup of hot chocolate. Matrix P contains the selling prices to the customer.

$$C = \begin{matrix} \text{Coffee} & \text{Chocolate} \\ \begin{bmatrix} \$0.90 & \$0.84 \\ \$1.26 & \$1.15 \\ \$1.64 & \$1.50 \end{bmatrix} & \begin{matrix} \textbf{Small} \\ \textbf{Medium} \\ \textbf{Large} \end{matrix} \end{matrix}$$

$$P = \begin{matrix} \text{Coffee} & \text{Chocolate} \\ \begin{bmatrix} \$3.05 & \$2.25 \\ \$3.65 & \$3.05 \\ \$4.15 & \$3.65 \end{bmatrix} & \begin{matrix} \textbf{Small} \\ \textbf{Medium} \\ \textbf{Large} \end{matrix} \end{matrix}$$

a. Compute $P - C$ and interpret its meaning.

b. If the tax rate in a certain city is 7%, use scalar multiplication to find a matrix F that gives the final price to the customer (including sales tax) for both beverages for each size. Round each entry to the nearest cent.

71. A street vendor at a parade sells fresh lemonade, soda, bottled water, and iced-tea, and the unit price for each item is given in matrix P. The number of units sold of each item is given in matrix N. Compute NP and interpret the result.

$$N = \begin{matrix} \text{Lemonade} & \text{Soda} & \text{Water} & \text{Tea} \\ [150 & 270 & 440 & 80], \end{matrix}$$

$$P = \begin{bmatrix} \$2.50 \\ \$1.50 \\ \$2.00 \\ \$2.00 \end{bmatrix} \begin{matrix} \textbf{Lemonade} \\ \textbf{Soda} \\ \textbf{Water} \\ \textbf{Tea} \end{matrix}$$

72. A math course has 4 exams weighted 20%, 25%, 25%, and 30%, respectively, toward the final grade. Suppose that a student earns grades of 75, 84, 92, and 86, respectively, on the four exams. The weights are given in matrix W and the test grades are given in matrix G. Compute WG and interpret the result.

$$W = \begin{matrix} \text{Test 1} & \text{Test 2} & \text{Test 3} & \text{Test 4} \\ [0.20 & 0.25 & 0.25 & 0.30], \end{matrix} \quad G = \begin{bmatrix} 75 \\ 84 \\ 92 \\ 86 \end{bmatrix} \begin{matrix} \textbf{Test 1} \\ \textbf{Test 2} \\ \textbf{Test 3} \\ \textbf{Test 4} \end{matrix}$$

73. An electronics store sells three models of tablets. The number of each model sold during "Black Friday" weekend is given in matrix A. The selling price and profit for each model are given in matrix B.
(**See Example 8**)

$$A = \begin{bmatrix} 84 & 70 & 32 \\ 62 & 48 & 16 \\ 70 & 40 & 12 \end{bmatrix} \begin{matrix} \text{Friday} \\ \text{Saturday} \\ \text{Sunday} \end{matrix} \quad B = \begin{bmatrix} \$499 & \$200 \\ \$599 & \$240 \\ \$629 & \$280 \end{bmatrix} \begin{matrix} \text{A} \\ \text{B Model} \\ \text{C} \end{matrix}$$

with column headers Model A B C for matrix A, and Selling Price, Profit for matrix B.

a. Compute AB and interpret the result.

b. Determine the total revenue for Sunday.

c. Determine the total profit for the 3-day period for these three models.

75. The labor costs per hour for an electrician, plumber, and air-conditioning/heating expert are given in matrix L. The time required from each specialist for three new model homes is given in matrix T.

$$L = \begin{bmatrix} \$45 \\ \$38 \\ \$35 \end{bmatrix} \begin{matrix} \text{Electrician} \\ \text{Plumber} \\ \text{AC/heating} \end{matrix}$$

Cost/hr

Time (hr)

	Electrician	Plumber	AC/heating	
$T =$	22	16	14	Model 1
	28	21	18	Model 2
	18	14	9	Model 3

a. Which product LT or TL gives the total cost for these three services for each model?

b. Find a matrix that gives the total cost for these three services for each model.

77. A student researches the cost for three cell phone plans. Matrix C contains the cost per text message and the cost per minute over the maximum number of minutes allowed in each plan. Matrix N_1 contains the number of text messages and the number of minutes over the maximum incurred for 1 month. Matrix N_3 represents the number of text messages and number of minutes over the maximum for 3 months.

	Cost/text	Cost/min	
$C =$	$0.25	$0.40	Plan A
	$0	$0.40	Plan B
	$0.10	$0	Plan C

$$N_1 = \begin{bmatrix} 24 \\ 100 \end{bmatrix} \begin{matrix} \text{Number of texts} \\ \text{Minutes over} \end{matrix}$$

	Month 1	Month 2	Month 3	
$N_3 =$	24	56	30	Number of texts
	100	24	0	Minutes over

a. Find the product CN_1 and interpret its meaning.

b. Find the product CN_3 and interpret its meaning.

74. A gas station manager records the number of gallons of Regular, Plus, and Premium gasoline sold during the week (Monday–Friday) and on the weekends (Saturday–Sunday) in matrix A. The selling price and profit for 1 gal of each type of gasoline are given in matrix B.

	Regular	Plus	Premium	
$A =$	4600	1850	720	Weekdays
	2300	620	480	Weekend

	Selling Price	Profit	
$B =$	$3.59	$0.21	Regular
	$3.79	$0.24	Plus
	$4.19	$0.18	Premium

a. Compute AB and interpret the result.

b. Determine the profit for the weekend.

c. Determine the revenue for the entire week.

76. The number of calories burned per hour for three activities is given in matrix N for a 140-lb woman training for a triathlon. The time spent on each activity for two different training days is given in matrix T.

$$N = \begin{bmatrix} 540 \\ 400 \\ 360 \end{bmatrix} \begin{matrix} \text{Running} \\ \text{Bicycling} \\ \text{Swimming} \end{matrix}$$

Calories/hr

Time

	Running	Bicycling	Swimming	
$T =$	45 min	1 hr	30 min	Day 1
	1 hr	1 hr 30 min	45 min	Day 2

a. Which product NT or TN gives the total number of calories burned from these activities for each day?

b. Find a matrix that gives the total number of calories burned from these activities for each day.

78. Refer to Exercise 77. Suppose that matrix B represents the base cost for each cell phone plan and matrix T represents the tax for each plan.

$$B = \begin{bmatrix} \$39.99 \\ \$49.99 \\ \$59.99 \end{bmatrix} \begin{matrix} \text{Plan A} \\ \text{Plan B} \\ \text{Plan C} \end{matrix}$$

$$T = \begin{bmatrix} \$11.96 \\ \$13.04 \\ \$14.91 \end{bmatrix} \begin{matrix} \text{Plan A} \\ \text{Plan B} \\ \text{Plan C} \end{matrix}$$

a. Compute $B + CN_1 + T$ and interpret its meaning.

b. Which cell phone plan is the least expensive if the student has 60 text messages and talks 20 min more than the maximum?

79. a. Write a matrix A that represents the coordinates of the vertices of the triangle. Place the x-coordinate of each point in the first row of A and the corresponding y-coordinate in the second row of A.
   (See Example 9)

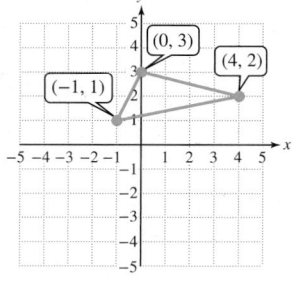

b. Use addition of matrices to shift the triangle 2 units to the right and 4 units downward.

c. Find the product $\begin{bmatrix} -1 & 0 \\ 0 & 1 \end{bmatrix} \cdot A$ and explain the effect on the graph of the triangle.

d. Find the product $\begin{bmatrix} 1 & 0 \\ 0 & -1 \end{bmatrix} \cdot A$ and explain the effect on the graph of the triangle.

e. Find $\begin{bmatrix} 1 & 0 \\ 0 & -1 \end{bmatrix} \cdot A + \begin{bmatrix} -1 & -1 & -1 \\ 2 & 2 & 2 \end{bmatrix}$ and explain the effect on the graph of the triangle.

80. a. Write a matrix A that represents the coordinates of the vertices of the triangle. Place the x-coordinate of each point in the first row of A and the corresponding y-coordinate in the second row of A.

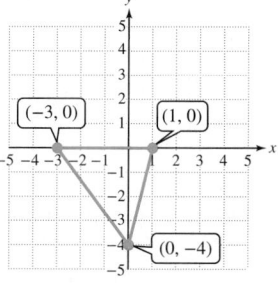

b. Use addition of matrices to shift the triangle 1 unit to the left and 3 units upward.

c. Find the product $\begin{bmatrix} -1 & 0 \\ 0 & 1 \end{bmatrix} \cdot A$ and explain the effect on the graph of the triangle.

d. Find the product $\begin{bmatrix} 1 & 0 \\ 0 & -1 \end{bmatrix} \cdot A$ and explain the effect on the graph of the triangle.

e. Find $\begin{bmatrix} -1 & 0 \\ 0 & 1 \end{bmatrix} \cdot A + \begin{bmatrix} 2 & 2 & 2 \\ -5 & -5 & -5 \end{bmatrix}$ and explain the effect on the graph of the triangle.

81. a. Write a matrix A that represents the coordinates of the vertices of the quadrilateral.

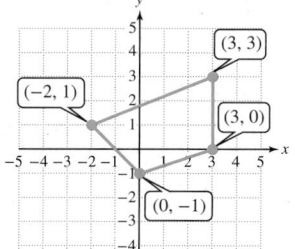

b. What operation on A will shift the graph of the quadrilateral 3 units downward?

c. What operation on A will shift the graph 4 units to the left?

d. Use matrix multiplication to reflect the graph across the x-axis.

e. Use matrix multiplication to reflect the graph across the y-axis.

82. a. Write a matrix A that represents the coordinates of the vertices of the quadrilateral.

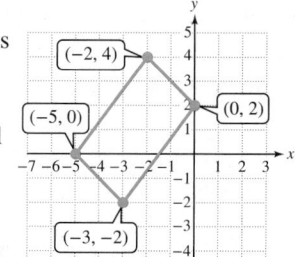

b. What operation on A will shift the graph of the quadrilateral 6 units upward?

c. What operation on A will shift the graph 2 units to the right?

d. Use matrix multiplication to reflect the graph across the x-axis.

e. Use matrix multiplication to reflect the graph across the y-axis.

83. a. Write a matrix A that represents the coordinates of the vertices of the triangle.

b. Multiply $\begin{bmatrix} \frac{\sqrt{3}}{2} & -\frac{1}{2} \\ \frac{1}{2} & \frac{\sqrt{3}}{2} \end{bmatrix} \cdot A$ and round each entry to 1 decimal place.

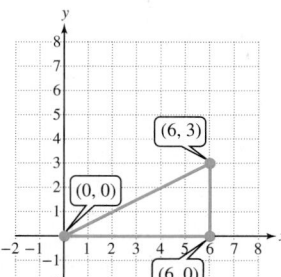

c. Graph the figure represented by the matrix from part (b). What effect does this product have on the graph of the triangle?

84. a. Write a matrix A that represents the coordinates of the vertices of the triangle.

b. Multiply $\begin{bmatrix} \frac{1}{2} & -\frac{\sqrt{3}}{2} \\ \frac{\sqrt{3}}{2} & \frac{1}{2} \end{bmatrix} \cdot A$ and round each entry to 1 decimal place.

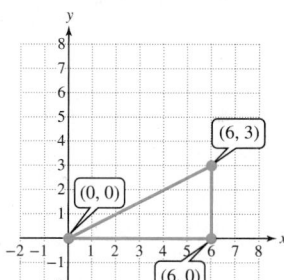

c. Graph the figure represented by the matrix from part (b). What effect does this product have on the graph of the triangle?

For Exercises 85–86, use the following gray scale.

0	1	2	3	4	5	6	7
white	light gray	medium light	gray	medium gray	medium dark	dark gray	black

85. a. Write a 5 × 3 matrix that represents the letter E in dark gray on a white background.

b. Use matrix addition to change the pixels so that the letter E is medium dark on a light gray background.

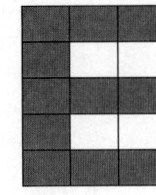

86. a. Write a 5 × 3 matrix that represents the letter T in medium gray on a medium light background.

b. Use matrix addition to change the pixels so that the letter T is dark gray on a light gray background.

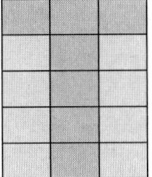

Mixed Exercises

For Exercises 87–92, use matrices A, B, and C to prove the given properties. Assume that the elements within A, B, and C are real numbers.

$$A = \begin{bmatrix} a_1 & a_2 \\ a_3 & a_4 \end{bmatrix} \quad B = \begin{bmatrix} b_1 & b_2 \\ b_3 & b_4 \end{bmatrix} \quad C = \begin{bmatrix} c_1 & c_2 \\ c_3 & c_4 \end{bmatrix}$$

87. Commutative property of matrix addition

$$A + B = B + A$$

88. Associative property of matrix addition

$$A + (B + C) = (A + B) + C$$

89. Inverse property of matrix addition

$$A + (-A) = 0$$

90. Identity property of matrix addition

$$A + 0 = A$$

91. Associative property of scalar multiplication

$$s(tA) = (st)A$$

92. Distributive property of scalar multiplication

$$t(A + B) = tA + tB$$

93. Given $A = \begin{bmatrix} i & 0 \\ 0 & i \end{bmatrix}$, find A^2, A^3, and A^4.

(*Hint*: Recall that $i = \sqrt{-1}$.) Discuss the similarities between A^n and i^n, where n is a positive integer.

94. Given $B = \begin{bmatrix} 0 & -i \\ i & 0 \end{bmatrix}$, find B^2.

95. a. For real numbers a, b, c, and d, find the product.

$$\begin{bmatrix} a & 0 \\ 0 & b \end{bmatrix}\begin{bmatrix} c & 0 \\ 0 & d \end{bmatrix}$$

b. Based on the form of the product of part (a), compute the following product mentally.

$$\begin{bmatrix} 3 & 0 \\ 0 & 7 \end{bmatrix}\begin{bmatrix} 1 & 0 \\ 0 & 2 \end{bmatrix}$$

96. Find the product.

$$\begin{bmatrix} a & a \\ b & b \end{bmatrix}\begin{bmatrix} 1 & -1 \\ -1 & 1 \end{bmatrix}$$

Write About It

97. Explain why the product of a 3 × 2 matrix by a 3 × 2 matrix is undefined.

98. Explain how to add or subtract matrices.

99. Given a matrix, $A = [a_{ij}]$, explain how to find the additive inverse of A.

100. Given two matrices A and B, how can you determine the order of AB, assuming that the product is defined?

Technology Connections

For Exercises 101–104, refer to matrices A, B, and C and perform the indicated operations on a calculator.

$$A = \begin{bmatrix} 1.05 & 3.9 \\ 4.12 & -9.4 \\ -2.4 & 1.5 \end{bmatrix} \quad B = \begin{bmatrix} -10 & 30 \\ 24 & -36 \\ 18 & -8 \end{bmatrix} \quad C = \begin{bmatrix} 6.2 \\ 4.9 \end{bmatrix}$$

101. $2.5A - 3.6B$

102. $-6.4(A + B)$

103. $-3AC$

104. $7.5BC$

SECTION 11.4 **Inverse Matrices and Matrix Equations**

OBJECTIVES

1. Identify Identity and Inverse Matrices
2. Determine the Inverse of a Matrix
3. Solve Systems of Linear Equations Using the Inverse of a Matrix

1. Identify Identity and Inverse Matrices

The identity element under the multiplication of real numbers is 1 because $a \cdot 1 = a$ and $1 \cdot a = a$. We now investigate a similar property for the product of square matrices. The **identity matrix** I_n for matrix multiplication is the $n \times n$ square matrix with 1's along the main diagonal and 0's for all other elements. For example:

The identity matrix of order 2

The identity matrix of order 3

$$I_2 = \begin{bmatrix} 1 & 0 \\ 0 & 1 \end{bmatrix} \quad \text{and} \quad I_3 = \begin{bmatrix} 1 & 0 & 0 \\ 0 & 1 & 0 \\ 0 & 0 & 1 \end{bmatrix}$$

For an $n \times n$ square matrix A, we have that

$$AI_n = A \quad \text{and} \quad I_nA = A \quad \text{(Identity property of matrix multiplication)}$$

In Example 1, we illustrate the identity property of matrix multiplication using a 2×2 matrix.

EXAMPLE 1   **Illustrating the Identity Property of Matrix Multiplication**

Given $A = \begin{bmatrix} a & b \\ c & d \end{bmatrix}$ show that

a. $AI_2 = A$ **b.** $I_2A = A$

Solution:

a. $AI_2 = \begin{bmatrix} a & b \\ c & d \end{bmatrix}\begin{bmatrix} 1 & 0 \\ 0 & 1 \end{bmatrix} = \begin{bmatrix} a(1) + b(0) & a(0) + b(1) \\ c(1) + d(0) & c(0) + d(1) \end{bmatrix} = \begin{bmatrix} a & b \\ c & d \end{bmatrix} \checkmark$

b. $I_2A = \begin{bmatrix} 1 & 0 \\ 0 & 1 \end{bmatrix}\begin{bmatrix} a & b \\ c & d \end{bmatrix} = \begin{bmatrix} 1(a) + 0(c) & 1(b) + 0(d) \\ 0(a) + 1(c) & 0(b) + 1(d) \end{bmatrix} = \begin{bmatrix} a & b \\ c & d \end{bmatrix} \checkmark$

FOR REVIEW

Compare the product of matrix A and the identity matrix from Example 1 to the multiplication of a real number a times the identity element 1.

$$A \cdot I_2 = I_2 \cdot A = A$$
$$a \cdot 1 = 1 \cdot a = a$$

Skill Practice 1   Given $A = \begin{bmatrix} 3 & -4 \\ -2 & 10 \end{bmatrix}$ show that

a. $AI_2 = A$ **b.** $I_2A = A$

For a nonzero real number a, the multiplicative inverse of a is $\frac{1}{a}$ because $a \cdot \frac{1}{a} = 1$ and $\frac{1}{a} \cdot a = 1$. We define the multiplicative inverse of a square matrix in a similar fashion.

Answers

1. a. $\begin{bmatrix} 3 & -4 \\ -2 & 10 \end{bmatrix}\begin{bmatrix} 1 & 0 \\ 0 & 1 \end{bmatrix} = \begin{bmatrix} 3 & -4 \\ -2 & 10 \end{bmatrix}$

   b. $\begin{bmatrix} 1 & 0 \\ 0 & 1 \end{bmatrix}\begin{bmatrix} 3 & -4 \\ -2 & 10 \end{bmatrix} = \begin{bmatrix} 3 & -4 \\ -2 & 10 \end{bmatrix}$

Multiplicative Inverse of a Square Matrix

Let A be an $n \times n$ matrix and let I_n be the identity matrix of order n. If there exists an $n \times n$ matrix A^{-1} such that

$$AA^{-1} = I_n \quad \text{and} \quad A^{-1}A = I_n$$

then A^{-1} (read as "A inverse") is the **multiplicative inverse** of A.

EXAMPLE 2 Determining Whether Two Matrices Are Inverses

Determine whether $A = \begin{bmatrix} 3 & 2 \\ 7 & 5 \end{bmatrix}$ and $B = \begin{bmatrix} 5 & -2 \\ -7 & 3 \end{bmatrix}$ are inverses.

Solution:

We must show that $AB = I_2$ and $BA = I_2$.

$$AB = \begin{bmatrix} 3 & 2 \\ 7 & 5 \end{bmatrix} \begin{bmatrix} 5 & -2 \\ -7 & 3 \end{bmatrix} = \begin{bmatrix} 3(5) + 2(-7) & 3(-2) + 2(3) \\ 7(5) + 5(-7) & 7(-2) + 5(3) \end{bmatrix} = \begin{bmatrix} 1 & 0 \\ 0 & 1 \end{bmatrix} \checkmark$$

$$BA = \begin{bmatrix} 5 & -2 \\ -7 & 3 \end{bmatrix} \begin{bmatrix} 3 & 2 \\ 7 & 5 \end{bmatrix} = \begin{bmatrix} 5(3) + (-2)(7) & 5(2) + (-2)(5) \\ -7(3) + 3(7) & -7(2) + 3(5) \end{bmatrix} = \begin{bmatrix} 1 & 0 \\ 0 & 1 \end{bmatrix} \checkmark$$

Skill Practice 2

> Determine whether $A = \begin{bmatrix} 4 & 3 \\ 13 & 10 \end{bmatrix}$ and $B = \begin{bmatrix} 10 & -3 \\ -13 & 4 \end{bmatrix}$ are inverses.

2. Determine the Inverse of a Matrix

In Example 2, we showed that two given matrices are inverses. Now we look at the task of finding the inverse of a matrix. In Example 3, we will find the inverse of a 2×2 matrix directly from the definition. Then, we develop a general procedure to find the inverse of an $n \times n$ matrix, if the inverse exists.

EXAMPLE 3 Finding the Inverse of a Matrix

Given $A = \begin{bmatrix} 2 & 9 \\ 1 & 5 \end{bmatrix}$, find A^{-1}.

Solution:

We need to find a matrix $A^{-1} = \begin{bmatrix} x_1 & x_2 \\ x_3 & x_4 \end{bmatrix}$ such that $AA^{-1} = I_2$. That is,

$$\begin{bmatrix} 2 & 9 \\ 1 & 5 \end{bmatrix} \begin{bmatrix} x_1 & x_2 \\ x_3 & x_4 \end{bmatrix} = \begin{bmatrix} 1 & 0 \\ 0 & 1 \end{bmatrix} \longrightarrow \begin{bmatrix} 2x_1 + 9x_3 & 2x_2 + 9x_4 \\ x_1 + 5x_3 & x_2 + 5x_4 \end{bmatrix} = \begin{bmatrix} 1 & 0 \\ 0 & 1 \end{bmatrix}$$

Answer

2. Yes; $AB = I_2$ and $BA = I_2$.

For the matrices to be equal, their corresponding elements must be equal. This results in two systems of equations.

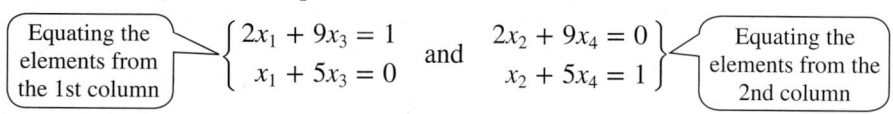

Equating the elements from the 1st column

$$\begin{cases} 2x_1 + 9x_3 = 1 \\ x_1 + 5x_3 = 0 \end{cases}$$

and

$$\begin{cases} 2x_2 + 9x_4 = 0 \\ x_2 + 5x_4 = 1 \end{cases}$$

Equating the elements from the 2nd column

The corresponding augmented matrices are

$$\left[\begin{array}{cc|c} 2 & 9 & 1 \\ 1 & 5 & 0 \end{array}\right] \quad \text{and} \quad \left[\begin{array}{cc|c} 2 & 9 & 0 \\ 1 & 5 & 1 \end{array}\right]$$

We can solve both systems simultaneously by placing the identity matrix to the right of the array of coefficients.

$$\left[\begin{array}{cc|cc} 2 & 9 & 1 & 0 \\ 1 & 5 & 0 & 1 \end{array}\right]$$

Apply Gauss-Jordan elimination.

$$\left[\begin{array}{cc|cc} 2 & 9 & 1 & 0 \\ 1 & 5 & 0 & 1 \end{array}\right] \xrightarrow{R_1 \Leftrightarrow R_2} \left[\begin{array}{cc|cc} 1 & 5 & 0 & 1 \\ 2 & 9 & 1 & 0 \end{array}\right] \xrightarrow{-2R_1 + R_2 \to R_2} \left[\begin{array}{cc|cc} 1 & 5 & 0 & 1 \\ 0 & -1 & 1 & -2 \end{array}\right]$$

$$\xrightarrow{-1R_2 \to R_2} \left[\begin{array}{cc|cc} 1 & 5 & 0 & 1 \\ 0 & 1 & -1 & 2 \end{array}\right] \xrightarrow{-5R_2 + R_1 \to R_1} \left[\begin{array}{cc|cc} 1 & 0 & 5 & -9 \\ 0 & 1 & -1 & 2 \end{array}\right]$$

This result represents the following two matrices and their corresponding systems of equations.

$$\begin{array}{cc} x_1 & x_3 \\ \downarrow & \downarrow \end{array}$$
$$\left[\begin{array}{cc|c} 1 & 0 & 5 \\ 0 & 1 & -1 \end{array}\right] \quad \text{and} \quad \left[\begin{array}{cc|c} 1 & 0 & -9 \\ 0 & 1 & 2 \end{array}\right]$$
$$x_1 = 5, x_3 = -1 \qquad\qquad x_2 = -9, x_4 = 2$$

with $\begin{array}{cc} x_2 & x_4 \\ \downarrow & \downarrow \end{array}$

Therefore, we have: $A^{-1} = \begin{bmatrix} x_1 & x_2 \\ x_3 & x_4 \end{bmatrix} = \begin{bmatrix} 5 & -9 \\ -1 & 2 \end{bmatrix}$

Avoiding Mistakes

The solution to Example 3 can be checked by verifying that $AA^{-1} = I_2$ and $A^{-1}A = I_2$.

Skill Practice 3   Given $B = \begin{bmatrix} 9 & 7 \\ 5 & 4 \end{bmatrix}$, find B^{-1}.

In Example 3, we transformed the left side of the augmented matrix into the identity matrix. That is, we performed row operations to make the array on the left of the vertical bar have 1's along the main diagonal, and 0's elsewhere. In the process, we make the following important observation.

• The array of elements to the right of the vertical bar is the inverse of A.

This observation leads to a general procedure to find the multiplicative inverse of a matrix, provided that the inverse exists.

Finding the Multiplicative Inverse of a Square Matrix

Let A be an $n \times n$ matrix for which A^{-1} exists, and let I_n be the $n \times n$ identity matrix. To find A^{-1},

Step 1 Write a matrix of the form $[A \,|\, I_n]$.
Step 2 Perform row operations to write the matrix in the form $[I_n \,|\, B]$.
Step 3 The matrix B is A^{-1}.

Answer

3. $B^{-1} = \begin{bmatrix} 4 & -7 \\ -5 & 9 \end{bmatrix}$

It is important to note that not all matrices have a multiplicative inverse. If a matrix A is reducible to a row-equivalent matrix with one or more rows of zeros, then the matrix cannot be written in the form $[I_n | B]$. Therefore, the matrix does not have an inverse, and we say that the matrix is **singular**. A matrix that *does* have a multiplicative inverse is said to be **invertible** or **nonsingular**.

In Example 4, we use the general procedure to find the inverse of a 3×3 invertible matrix.

EXAMPLE 4   **Finding the Inverse of a Matrix**

Given $A = \begin{bmatrix} 1 & 3 & 0 \\ 1 & 1 & -2 \\ -3 & -3 & 5 \end{bmatrix}$, find A^{-1} if possible.

Solution:

$$\left[\begin{array}{ccc|ccc} 1 & 3 & 0 & 1 & 0 & 0 \\ 1 & 1 & -2 & 0 & 1 & 0 \\ -3 & -3 & 5 & 0 & 0 & 1 \end{array}\right]$$

Set up the matrix $[A | I_3]$.
Transform the left side of the matrix to the 3×3 identity matrix.

$\xrightarrow[\;3R_1 + R_3 \to R_3\;]{-1R_1 + R_2 \to R_2}$ $\left[\begin{array}{ccc|ccc} 1 & 3 & 0 & 1 & 0 & 0 \\ 0 & -2 & -2 & -1 & 1 & 0 \\ 0 & 6 & 5 & 3 & 0 & 1 \end{array}\right]$ $\xrightarrow{-\frac{1}{2}R_2 \to R_2}$ $\left[\begin{array}{ccc|ccc} 1 & 3 & 0 & 1 & 0 & 0 \\ 0 & 1 & 1 & \frac{1}{2} & -\frac{1}{2} & 0 \\ 0 & 6 & 5 & 3 & 0 & 1 \end{array}\right]$

$\xrightarrow[\;-6R_2 + R_3 \to R_3\;]{-3R_2 + R_1 \to R_1}$ $\left[\begin{array}{ccc|ccc} 1 & 0 & -3 & -\frac{1}{2} & \frac{3}{2} & 0 \\ 0 & 1 & 1 & \frac{1}{2} & -\frac{1}{2} & 0 \\ 0 & 0 & -1 & 0 & 3 & 1 \end{array}\right]$ $\xrightarrow{-1R_3 \to R_3}$ $\left[\begin{array}{ccc|ccc} 1 & 0 & -3 & -\frac{1}{2} & \frac{3}{2} & 0 \\ 0 & 1 & 1 & \frac{1}{2} & -\frac{1}{2} & 0 \\ 0 & 0 & 1 & 0 & -3 & -1 \end{array}\right]$

$\xrightarrow[\;-1R_3 + R_2 \to R_2\;]{3R_3 + R_1 \to R_1}$ $\left[\begin{array}{ccc|ccc} 1 & 0 & 0 & -\frac{1}{2} & -\frac{15}{2} & -3 \\ 0 & 1 & 0 & \frac{1}{2} & \frac{5}{2} & 1 \\ 0 & 0 & 1 & 0 & -3 & -1 \end{array}\right]$

The left side of the matrix is the 3×3 identity matrix. The right side of the matrix represents the inverse of A.

$$A^{-1} = \begin{bmatrix} -\frac{1}{2} & -\frac{15}{2} & -3 \\ \frac{1}{2} & \frac{5}{2} & 1 \\ 0 & -3 & -1 \end{bmatrix}$$

Avoiding Mistakes

As a check, verify that $A \cdot A^{-1} = I_3$ and $A^{-1} \cdot A = I_3$.

Skill Practice 4   Given $A = \begin{bmatrix} -2 & 4 & 1 \\ 4 & -13 & 10 \\ 3 & -9 & 6 \end{bmatrix}$ find A^{-1} if possible.

TECHNOLOGY CONNECTIONS

Finding the Inverse of a Nonsingular Square Matrix

To find the inverse of a nonsingular matrix [A] using a calculator, first enter the matrix into the calculator using the MATRIX menu. Then place [A] on the home screen and use the ⬛ x^{-1} key to take the inverse of [A]. It is very important to realize that in this context, the calculator uses the ⬛ x^{-1} key for the inverse of a matrix, not the reciprocal of a real number.

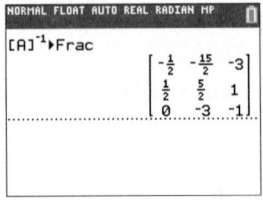

Answer

4. $A^{-1} = \begin{bmatrix} 4 & -11 & \frac{53}{3} \\ 2 & -5 & 8 \\ 1 & -2 & \frac{10}{3} \end{bmatrix}$

In Example 5, we identify a singular matrix—that is, a matrix that has no inverse.

EXAMPLE 5   Identifying a Singular Matrix

Show that matrix A is singular. $A = \begin{bmatrix} 1 & 2 & 0 \\ 4 & 5 & 1 \\ 2 & 1 & 1 \end{bmatrix}$

Solution:

$\begin{bmatrix} 1 & 2 & 0 & | & 1 & 0 & 0 \\ 4 & 5 & 1 & | & 0 & 1 & 0 \\ 2 & 1 & 1 & | & 0 & 0 & 1 \end{bmatrix}$ Set up the matrix $[A \,|\, I_3]$.
Transform the left side of the matrix to the 3×3 identity matrix.

$\begin{matrix} -4R_1 + R_2 \to R_2 \\ \xrightarrow{\hspace{3cm}} \\ -2R_1 + R_3 \to R_3 \end{matrix}$ $\begin{bmatrix} 1 & 2 & 0 & | & 1 & 0 & 0 \\ 0 & -3 & 1 & | & -4 & 1 & 0 \\ 0 & -3 & 1 & | & -2 & 0 & 1 \end{bmatrix}$ On the left of the vertical bar, the elements in rows 2 and 3 are multiples of each other (in fact they are identical). Therefore, $-1R_2 + R_3 \to R_3$ results in zeros in row 3 to the left of the bar.

$\xrightarrow{-1R_2 + R_3 \to R_3}$ $\begin{bmatrix} 1 & 2 & 0 & | & 1 & 0 & 0 \\ 0 & -3 & 1 & | & -4 & 1 & 0 \\ 0 & 0 & 0 & | & 2 & -1 & 1 \end{bmatrix}$ The left side of the matrix cannot be transformed into the 3×3 identity matrix I_3. Therefore, matrix A does not have an inverse.

A is a singular matrix (does not have an inverse).

Skill Practice 5   Show that matrix A is singular. $A = \begin{bmatrix} 1 & -3 & -7 \\ 3 & -7 & -17 \\ 1 & 0 & -1 \end{bmatrix}$

TIP   Example 5 shows us that not all square matrices have a multiplicative inverse. Likewise, the real number 0 does not have a multiplicative inverse because $\dfrac{1}{0}$ is undefined.

TECHNOLOGY CONNECTIONS

Identifying a Singular Matrix

A calculator will return an error message if we attempt to find the inverse of a singular matrix. Consider matrix A from Example 5. The calculator returns the error message shown.

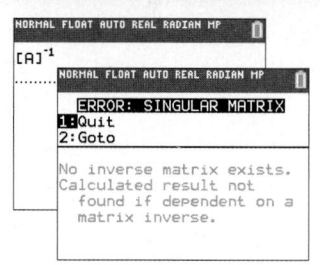

To find the inverse of an invertible 2×2 matrix, we can take a direct approach and apply the following formula.

TIP   If A is an invertible matrix then the restriction $ad - bc \neq 0$ is implied.

Formula for the Inverse of a 2 × 2 Invertible Matrix

Let $A = \begin{bmatrix} a & b \\ c & d \end{bmatrix}$ be an invertible matrix. Then the inverse A^{-1} is given by:

$$A^{-1} = \frac{1}{ad - bc} \cdot \begin{bmatrix} d & -b \\ -c & a \end{bmatrix}$$

Answer
5. The matrix $[A \,|\, I_3]$ cannot be written in the form $[I_3 \,|\, B]$ by using row operations. Therefore, A is a singular matrix.

Avoiding Mistakes

The formula to find the inverse of an invertible 2×2 matrix does not apply to matrices of higher order.

This formula can be verified by applying the general procedure to find the inverse of A. See Exercise 67.

$$\left[\begin{array}{cc|cc} a & b & 1 & 0 \\ c & d & 0 & 1 \end{array}\right] \xrightarrow[\text{matrix}]{\text{row-equivalent}} \left[\begin{array}{cc|cc} 1 & 0 & \frac{d}{ad-bc} & \frac{-b}{ad-bc} \\ 0 & 1 & \frac{-c}{ad-bc} & \frac{a}{ad-bc} \end{array}\right]$$

$$\xrightarrow{\text{inverse of } A} \frac{1}{ad-bc} \cdot \begin{bmatrix} d & -b \\ -c & a \end{bmatrix}$$

EXAMPLE 6 Finding the Inverse of a 2×2 Matrix

Given $A = \begin{bmatrix} 2 & 9 \\ 1 & 5 \end{bmatrix}$, find A^{-1} and compare to the result of Example 3.

Solution:

$$\begin{bmatrix} 2 & 9 \\ 1 & 5 \end{bmatrix} \quad a=2, b=9, c=1, d=5$$

$$A^{-1} = \frac{1}{ad-bc} \cdot \begin{bmatrix} d & -b \\ -c & a \end{bmatrix} = \frac{1}{(2)(5)-(9)(1)} \cdot \begin{bmatrix} 5 & -9 \\ -1 & 2 \end{bmatrix}$$

$$= \frac{1}{1} \cdot \begin{bmatrix} 5 & -9 \\ -1 & 2 \end{bmatrix} = \begin{bmatrix} 5 & -9 \\ -1 & 2 \end{bmatrix} \qquad \text{This is the same result as was found in Example 3.}$$

Skill Practice 6 Given $B = \begin{bmatrix} 9 & 7 \\ 5 & 4 \end{bmatrix}$, find B^{-1}.

---

3. Solve Systems of Linear Equations Using the Inverse of a Matrix

A system of linear equations written in standard form can be represented by using matrix multiplication. For example:

$$\begin{array}{l} 3x - 2y + z = 4 \\ x + 4y - z = 2 \\ 2x + 5y - 2z = 1 \end{array} \xrightarrow[\text{matrix equation}]{\text{corresponding}} \begin{bmatrix} 3 & -2 & 1 \\ 1 & 4 & -1 \\ 2 & 5 & -2 \end{bmatrix}\begin{bmatrix} x \\ y \\ z \end{bmatrix} = \begin{bmatrix} 4 \\ 2 \\ 1 \end{bmatrix}$$

The matrix on the left is called the **coefficient matrix**. It consists of the coefficients of the variable terms.

coefficient matrix   column matrix of variables   column matrix of constants

If we denote the coefficient matrix by A, the column matrix of variables as X, and the column matrix of constant terms as B, we can write the equation as $AX = B$. If A^{-1} exists, then we can solve the equation for X by multiplying both sides by A^{-1}.

$$AX = B \qquad \text{To solve this equation, the goal is to isolate } X.$$
$$A^{-1}AX = A^{-1}B \qquad \text{Multiply both sides by } A^{-1} \text{ (provided that } A^{-1} \text{ exists).}$$
$$(A^{-1}A)X = A^{-1}B$$
$$I_n X = A^{-1}B \qquad \text{The product of a matrix and its inverse is the identity matrix.}$$
$$X = A^{-1}B$$

Answer

6. $B^{-1} = \begin{bmatrix} 4 & -7 \\ -5 & 9 \end{bmatrix}$

This process is summarized as follows.

> ### Solving Systems of Linear Equations Using Inverse Matrices
>
> Suppose that $AX = B$ represents a system of n linear equations in n variables with a unique solution. Then,
>
> $$X = A^{-1}B$$
>
> where A is the coefficient matrix, B is the matrix of constants, and X is the matrix of variables.

EXAMPLE 7   Solving a System of Equations by Using an Inverse Matrix

Use the inverse of the coefficient matrix to solve the system.

$$\begin{aligned} x + 3y \phantom{{}- 2z} &= -10 \\ x + y - 2z &= -4 \\ -3x - 3y + 5z &= 11 \end{aligned}$$

Solution:

$$\begin{aligned} x + 3y \phantom{{}- 2z} &= -10 \\ x + y - 2z &= -4 \\ -3x - 3y + 5z &= 11 \end{aligned} \quad \longrightarrow \quad \begin{matrix} A & \cdot & X & = & B \end{matrix}$$

$$\begin{bmatrix} 1 & 3 & 0 \\ 1 & 1 & -2 \\ -3 & -3 & 5 \end{bmatrix} \begin{bmatrix} x \\ y \\ z \end{bmatrix} = \begin{bmatrix} -10 \\ -4 \\ 11 \end{bmatrix}$$

From Example 4, we have that $A^{-1} = \begin{bmatrix} -\frac{1}{2} & -\frac{15}{2} & -3 \\ \frac{1}{2} & \frac{5}{2} & 1 \\ 0 & -3 & -1 \end{bmatrix}$.

$$X = A^{-1}B$$

$$\begin{bmatrix} x \\ y \\ z \end{bmatrix} = \begin{bmatrix} -\frac{1}{2} & -\frac{15}{2} & -3 \\ \frac{1}{2} & \frac{5}{2} & 1 \\ 0 & -3 & -1 \end{bmatrix} \begin{bmatrix} -10 \\ -4 \\ 11 \end{bmatrix}$$

$$\begin{bmatrix} x \\ y \\ z \end{bmatrix} = \begin{bmatrix} \left(-\frac{1}{2}\right)(-10) + \left(-\frac{15}{2}\right)(-4) + (-3)(11) \\ \left(\frac{1}{2}\right)(-10) + \left(\frac{5}{2}\right)(-4) + (1)(11) \\ (0)(-10) + (-3)(-4) + (-1)(11) \end{bmatrix} = \begin{bmatrix} 2 \\ -4 \\ 1 \end{bmatrix}$$

$$\begin{bmatrix} x \\ y \\ z \end{bmatrix} = \begin{bmatrix} 2 \\ -4 \\ 1 \end{bmatrix} \qquad \begin{array}{l}\text{Therefore, } x = 2,\ y = -4,\text{ and } z = 1. \\ \text{The solution set is } \{(2, -4, 1)\}.\end{array}$$

> **Skill Practice 7**   Use the inverse of the coefficient matrix found in Skill Practice 4 to solve the system.
>
> $$\begin{aligned} -2x + 4y + z &= 0 \\ 4x - 13y + 10z &= 24 \\ 3x - 9y + 6z &= 15 \end{aligned}$$

Answer

7. $\{(1, 0, 2)\}$

SECTION 11.4   Practice Exercises

Prerequisite Review

For Exercises R.1–R.4, match the statement with the property that describes it.

a. Identity property of addition

b. Inverse property of multiplication

c. Identity property of multiplication

d. Inverse property of addition

R.1. $\dfrac{1}{7} \cdot 7 = 1$

R.2. $15 \cdot 1 = 15$

R.3. $-6 + 6 = 0$

R.4. $7 + 0 = 7$

For Exercises R.5–R.8, multiply the matrices. (See Section 11.3 for review.)

R.5. $\begin{bmatrix} -2 & 6 \\ 3 & -1 \end{bmatrix} \cdot \begin{bmatrix} 0 & -3 \\ 4 & -2 \end{bmatrix}$

R.6. $\begin{bmatrix} 5 & -4 \\ -1 & 0 \end{bmatrix} \cdot \begin{bmatrix} -3 & 0 \\ 2 & -5 \end{bmatrix}$

R.7. $\begin{bmatrix} 4 & -5 & 0 \\ 3 & 1 & 6 \\ -3 & -4 & 0 \end{bmatrix} \cdot \begin{bmatrix} 1 & 0 & 0 \\ -1 & 5 & 4 \\ 0 & 3 & 7 \end{bmatrix}$

R.8. $\begin{bmatrix} -9 & -1 & 4 \\ 0 & 0 & 3 \\ 4 & 1 & 0 \end{bmatrix} \cdot \begin{bmatrix} 3 & 8 & 2 \\ -1 & -3 & 4 \\ 5 & 3 & 0 \end{bmatrix}$

For Exercises R.9–R.12, solve the system by using Gauss-Jordan elimination. (See Section 11.1 for review.)

R.9. $\begin{aligned} 2x - 3y &= -10 \\ x + 2y &= 9 \end{aligned}$

R.10. $\begin{aligned} 3x + 3y &= 3 \\ 4x + y &= -5 \end{aligned}$

R.11. $\begin{aligned} 2x + 4y - z &= 0 \\ x + 2y + z &= 0 \\ y + 3z &= -1 \end{aligned}$

R.12. $\begin{aligned} x + 2y + 4z &= 2 \\ 2x - 4y &= -4 \\ 4x + 6z &= -2 \end{aligned}$

Concept Connections

1. The symbol I_n represents the _____ matrix of order n.

2. I_n is an $n \times n$ matrix with _____'s along the main diagonal and _____'s elsewhere.

3. Given an $n \times n$ matrix A, if there exists a matrix A^{-1} such that $A \cdot A^{-1} = I_n$ and $A^{-1} \cdot A = I_n$, then A^{-1} is called the _____ of A.

4. A matrix that does not have an inverse is called a _____ matrix. A matrix that does have an inverse is said to be invertible or _____.

5. Let $A = \begin{bmatrix} a & b \\ c & d \end{bmatrix}$ be an invertible matrix. Then a formula for the inverse A^{-1} is given by _____.

6. Suppose that the matrix equation $AX = B$ represents a system of n linear equations in n variables with a unique solution. Then $X = $ _____.

Objective 1: Identify Identity and Inverse Matrices

7. Write the matrix I_2.

8. Write the matrix I_3.

For Exercises 9–12, verify that

a. $AI_n = A$

b. $I_n A = A$ **(See Example 1)**

9. $A = \begin{bmatrix} -\frac{7}{8} & \sqrt{5} \\ 5.1 & 8 \end{bmatrix}$

10. $A = \begin{bmatrix} \sqrt{3} & 1 \\ \pi & 4 \end{bmatrix}$

11. $A = \begin{bmatrix} 1 & -3 & 4 \\ 9 & 5 & 3 \\ 11 & -6 & -4 \end{bmatrix}$

12. $A = \begin{bmatrix} -3 & 9 & 1 \\ 0 & 4 & -1 \\ 5 & 0 & 3 \end{bmatrix}$

For Exercises 13–18, determine whether A and B are inverses. **(See Example 2)**

13. $A = \begin{bmatrix} 10 & -3 \\ 4 & -2 \end{bmatrix}$ and $B = \begin{bmatrix} \frac{1}{4} & -\frac{3}{8} \\ \frac{1}{2} & -\frac{5}{4} \end{bmatrix}$

14. $A = \begin{bmatrix} 4 & -1 \\ -8 & 6 \end{bmatrix}$ and $B = \begin{bmatrix} \frac{3}{8} & \frac{1}{16} \\ \frac{1}{2} & \frac{1}{4} \end{bmatrix}$

15. $A = \begin{bmatrix} -2 & -3 & 1 \\ -3 & -3 & 1 \\ -2 & -4 & 1 \end{bmatrix}$ and $B = \begin{bmatrix} 1 & -1 & 0 \\ 1 & 0 & -1 \\ 6 & -2 & -3 \end{bmatrix}$

16. $A = \begin{bmatrix} -1 & 2 & 1 \\ -5 & 8 & 2 \\ 7 & -11 & -3 \end{bmatrix}$ and $B = \begin{bmatrix} 2 & 5 & 4 \\ 1 & 4 & 3 \\ 1 & -3 & -2 \end{bmatrix}$

17. $A = \begin{bmatrix} 2 & 1 \\ 3 & 4 \end{bmatrix}$ and $B = \begin{bmatrix} 4 & -3 \\ -7 & 6 \end{bmatrix}$

18. $A = \begin{bmatrix} 3 & 2 \\ 10 & 7 \end{bmatrix}$ and $B = \begin{bmatrix} -7 & 10 \\ 11 & -15 \end{bmatrix}$

Objective 2: Determine the Inverse of a Matrix

For Exercises 19–34, determine the inverse of the given matrix if possible. Otherwise, state that the matrix is singular. **(See Examples 3–6)**

19. $A = \begin{bmatrix} -4 & -3 \\ 6 & 5 \end{bmatrix}$

20. $A = \begin{bmatrix} 5 & -3 \\ 10 & -7 \end{bmatrix}$

21. $A = \begin{bmatrix} -8 & -2 \\ 10 & 5 \end{bmatrix}$

22. $A = \begin{bmatrix} 0 & 1 \\ 6 & -2 \end{bmatrix}$

23. $A = \begin{bmatrix} 3 & 7 \\ 6 & 14 \end{bmatrix}$

24. $A = \begin{bmatrix} 2 & -5 \\ -6 & 15 \end{bmatrix}$

25. $A = \begin{bmatrix} 2 & -7 & 8 \\ 1 & -3 & 3 \\ 1 & -5 & 6 \end{bmatrix}$

26. $A = \begin{bmatrix} 1 & -1 & 1 \\ 1 & 1 & -2 \\ -1 & 0 & 1 \end{bmatrix}$

27. $A = \begin{bmatrix} 1 & -2 & 2 \\ 2 & -3 & 1 \\ 0 & 1 & -1 \end{bmatrix}$

28. $A = \begin{bmatrix} -2 & 5 & -4 \\ 1 & -2 & 1 \\ 1 & -5 & 9 \end{bmatrix}$

29. $A = \begin{bmatrix} 5 & 7 & -11 \\ 3 & 5 & -9 \\ 1 & 1 & -1 \end{bmatrix}$

30. $A = \begin{bmatrix} 1 & 3 & 9 \\ 2 & 7 & 22 \\ -5 & -17 & -53 \end{bmatrix}$

31. $A = \begin{bmatrix} 0 & 1 & 1 \\ \frac{5}{3} & -1 & -3 \\ -1 & 1 & 2 \end{bmatrix}$

32. $A = \begin{bmatrix} 3 & -1 & -\frac{1}{2} \\ 2 & -1 & \frac{1}{2} \\ -5 & 2 & \frac{1}{2} \end{bmatrix}$

33. $A = \begin{bmatrix} 1 & -2 & 5 & 0 \\ 0 & -1 & 2 & 0 \\ 0 & 1 & -1 & 0 \\ 2 & -1 & 7 & 1 \end{bmatrix}$

34. $A = \begin{bmatrix} 1 & 3 & 0 & -3 \\ 0 & 1 & 0 & -2 \\ -2 & -4 & 1 & 2 \\ 0 & 1 & 1 & 0 \end{bmatrix}$

Objective 3: Solve Systems of Linear Equations Using the Inverse of a Matrix

For Exercises 35–38, write the system of equations as a matrix equation of the form $AX = B$, where A is the coefficient matrix, X is the column matrix of variables, and B is the column matrix of constants.

35. $3x - 4y = -1$
$2x + y = 14$

36. $-6x - y = 1$
$2x + 3y = 13$

37. $9x - 6y + 4z = 27$
$4x - z = 1$
$3y + z = 0$

38. $-3x + 8z = 4$
$6y - z = 7$
$2x - y + 6z = -15$

For Exercises 39–50, solve the system by using the inverse of the coefficient matrix. **(See Example 7)**

39. $-4x - 3y = -4$
$6x + 5y = 8$
(See Exercise 19 for A^{-1}.)

40. $5x - 3y = -20$
$10x - 7y = -50$
(See Exercise 20 for A^{-1}.)

41. $3x + 7y = -5$
$4x + 9y = -7$

42. $3x - 8y = -1$
$-4x + 11y = 2$

43. $2x - 7y + 8z = 1$
$x - 3y + 3z = 0$
$x - 5y + 6z = 2$
(See Exercise 25 for A^{-1}.)

44. $x - y + z = -4$
$x + y - 2z = 12$
$-x + z = -5$
(See Exercise 26 for A^{-1}.)

45. $x - 2y + 2z = -12$
$2x - 3y + z = -10$
$y - z = 6$
(See Exercise 27 for A^{-1}.)

46. $-2x + 5y - 4z = -20$
$x - 2y + z = 8$
$x - 5y + 9z = 24$
(See Exercise 28 for A^{-1}.)

47. $r - 2s + t = 2$
$-r + 4s + t = 3$
$2r - 2s - t = -1$

48.
$$a - b + 3c = 2$$
$$2a + b + 2c = 2$$
$$-2a - 2b + c = 3$$

49.
$$w - 2x + 5y = 3$$
$$-x + 2y = 1$$
$$x - y = -1$$
$$2w - x + 7y + z = 5$$
(See Exercise 33 for A^{-1}.)

50.
$$w + 3x - 3z = 8$$
$$x - 2z = 4$$
$$-2w - 4x + y + 2z = -6$$
$$x + y = 0$$
(See Exercise 34 for A^{-1}.)

Mixed Exercises

For Exercises 51–58, determine whether the statement is true or false. If a statement is false, explain why.

51. A 3×2 matrix has a multiplicative inverse.

52. A 2×3 matrix has a multiplicative inverse.

53. Every square matrix has an inverse.

54. Every singular matrix has an inverse.

55. Every nonsingular matrix has an inverse.

56. The matrix $\begin{bmatrix} -a & -b \\ a & b \end{bmatrix}$ is invertible.

57. The matrix $\begin{bmatrix} x & y \\ 2x & 2y \end{bmatrix}$ is invertible.

58. The inverse of the matrix I_4 is itself.

59. Find a 2×2 matrix that is its own inverse. Answers will vary.

60. Given an invertible 2×2 matrix A and the nonzero real number k, find the inverse of kA in terms of A^{-1}.

61. Given $A = \begin{bmatrix} 3 & 2 \\ 5 & 6 \end{bmatrix}$,

  a. Find A^{-1}.   **b.** Find $(A^{-1})^{-1}$.

62. Given $B = \begin{bmatrix} -3 & 2 \\ -5 & 4 \end{bmatrix}$,

  a. Find B^{-1}.   **b.** Find $(B^{-1})^{-1}$.

63. Given $A = \begin{bmatrix} a & 0 \\ 0 & b \end{bmatrix}$, where a and b are nonzero real numbers, find A^{-1}.

64. Given $A = \begin{bmatrix} a & 0 & 0 \\ 0 & b & 0 \\ 0 & 0 & c \end{bmatrix}$, where a, b, and c are nonzero real numbers, find A^{-1}.

Write About It

65. Explain how you can determine if two $n \times n$ matrices, A and B, are inverses.

66. Given a matrix $A = \begin{bmatrix} a & b \\ c & d \end{bmatrix}$, for what conditions on a, b, c, and d will the matrix not have an inverse?

Expanding Your Skills

67. Given $A = \begin{bmatrix} a & b \\ c & d \end{bmatrix}$, perform row operations on the matrix $[A \,|\, I_n]$ to find A^{-1}. This confirms the formula for the inverse of an invertible 2×2 matrix. (See page 846.)

68. Show by counterexample that $(AB)^{-1} \neq A^{-1}B^{-1}$. That is, find two matrices A and B for which $(AB)^{-1} \neq A^{-1}B^{-1}$.

Physicists know that if each edge of a thin conducting plate is kept at a constant temperature, then the temperature at the interior points is the mean (average) of the four surrounding points equidistant from the interior point. Use this principle in Exercise 69 to find the temperature at points x_1, x_2, x_3, and x_4.

69. *Hint*: Set up four linear equations to represent the temperature at points x_1, x_2, x_3, and x_4. Then solve the system. For example, one equation would be:

$$x_1 = \frac{1}{4}(36 + 32 + x_2 + x_3)$$

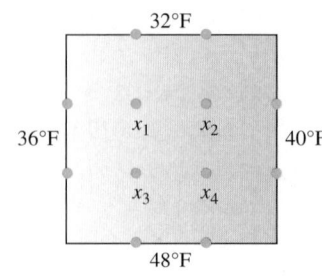

Technology Connections

For Exercises 70–71, use a graphing utility to find the inverse of the given matrix. Round the elements in the inverse to 2 decimal places.

70. $A = \begin{bmatrix} 0.04 & 0.13 & -0.08 & -0.43 \\ 0.19 & 0.33 & 0.06 & -0.84 \\ 0.01 & 0.08 & -0.11 & 0.46 \\ 0.37 & -1.42 & 0.03 & 0.52 \end{bmatrix}$

71. $A = \begin{bmatrix} 3.5 & 2.1 & 1.6 & 2.4 \\ -4.5 & 3.9 & -9.1 & 3.8 \\ 5.7 & 6.8 & -1.4 & -7.7 \\ 3.3 & 2.8 & 4.6 & -1.0 \end{bmatrix}$

For Exercises 72–73, use a graphing calculator and the inverse of the coefficient matrix to find the solution to the given system. Round to 2 decimal places.

72. $(\log 2)x + \left(\sqrt{7}\right)y + z = 4.1$
$(e^2)x - \left(\sqrt{3}\right)y - \pi z = -3.7$
$(\ln 10)\, x + y - 2.2z = 7.2$

73. $\left(\sqrt{11}\right)x + y - (\ln 5)z = 52.3$
$\left(\sqrt{7}\right)x - \pi y + (e^3)z = -27.5$
$-x + (\log 81)\, y - z = 69.8$

SECTION 11.5   Determinants and Cramer's Rule

OBJECTIVES

1. Evaluate the Determinant of a 2 × 2 Matrix
2. Evaluate the Determinant of an $n \times n$ Matrix
3. Apply Cramer's Rule

1. Evaluate the Determinant of a 2 × 2 Matrix

In this section, we present yet another method to solve a linear system with one unique solution. This method, called Cramer's rule, enables us to write a formula to solve a system based on the coefficients of the equations in the system.

Before studying Cramer's rule, we need to become familiar with the determinant of a square matrix. Associated with every square matrix is a real number called the *determinant* of the matrix. A determinant of a square matrix A, denoted $|A|$, is written by enclosing the elements of the matrix within two vertical bars. For example:

Avoiding Mistakes

Notice that a determinant is denoted using vertical bars, whereas a matrix is denoted using brackets.

$$\text{Given } A = \begin{bmatrix} 3 & -1 \\ 4 & 0 \end{bmatrix}, \text{ then } |A| = \begin{vmatrix} 3 & -1 \\ 4 & 0 \end{vmatrix}$$

Determinants have many applications in mathematics, including solving systems of linear equations, finding the area of a triangle, determining whether three points are collinear, and finding an equation of a line through two points (see Exercises 51–58). The determinant of a 2 × 2 matrix is defined as follows.

Determinant of a 2 × 2 Matrix

The **determinant** of the matrix $A = \begin{bmatrix} a & b \\ c & d \end{bmatrix}$ is $ad - bc$. That is,

$$|A| = \begin{vmatrix} a & b \\ c & d \end{vmatrix} = ad - bc.$$

EXAMPLE 1   **Evaluating a 2 × 2 Determinant**

Evaluate the determinant of each matrix.

a. $A = \begin{bmatrix} 6 & -2 \\ 5 & \frac{1}{3} \end{bmatrix}$ **b.** $B = \begin{bmatrix} 2 & -11 \\ 0 & 0 \end{bmatrix}$

Solution:

> **TIP** Example 1(b) illustrates that the determinant of a matrix having a row of all zeros is 0. The same is true for a matrix having a column of all zeros.

a. $|A| = \begin{vmatrix} 6 & -2 \\ 5 & \frac{1}{3} \end{vmatrix} = \overset{a \cdot d}{(6)\left(\frac{1}{3}\right)} - \overset{b \cdot c}{(-2)(5)} = 12$ $a = 6, b = -2, c = 5,$ and $d = \frac{1}{3}$

b. $|B| = \begin{vmatrix} 2 & -11 \\ 0 & 0 \end{vmatrix} = \overset{a \cdot d}{(2)(0)} - \overset{b \cdot c}{(-11)(0)} = 0$ $a = 2, b = -11, c = 0,$ and $d = 0$

Skill Practice 1   Evaluate the determinant.

a. $\begin{vmatrix} 3 & -\frac{1}{2} \\ 10 & 2 \end{vmatrix}$ **b.** $\begin{vmatrix} 9 & 0 \\ -4 & 0 \end{vmatrix}$

---

2. Evaluate the Determinant of an $n \times n$ Matrix

We now develop the tools to compute the determinant of a higher-order square matrix. To find the determinant of an $n \times n$ matrix $A = [a_{ij}]$ we first need to define the minor of an element of the matrix. For any element a_{ij}, the **minor** M_{ij} of that element is the determinant of the resulting matrix obtained by deleting the ith row and jth column. For example, consider the matrix:

$$A = [a_{ij}] = \begin{bmatrix} 5 & -1 & 6 \\ 0 & -7 & 1 \\ 4 & 2 & 6 \end{bmatrix}$$

The element a_{11} is 5. The minor of this element is found by deleting the first row and first column and then evaluating the determinant of the remaining 2×2 matrix:

$$\begin{bmatrix} 5 & -1 & 6 \\ 0 & -7 & 1 \\ 4 & 2 & 6 \end{bmatrix} \qquad M_{11} = \begin{vmatrix} -7 & 1 \\ 2 & 6 \end{vmatrix} = (-7)(6) - (1)(2) = -44$$

The element a_{32} is 2. The minor of this element is found by deleting the third row and second column and then evaluating the determinant of the remaining 2×2 matrix:

$$\begin{bmatrix} 5 & -1 & 6 \\ 0 & -7 & 1 \\ 4 & 2 & 6 \end{bmatrix} \qquad M_{32} = \begin{vmatrix} 5 & 6 \\ 0 & 1 \end{vmatrix} = (5)(1) - (6)(0) = 5$$

EXAMPLE 2   **Determining the Minor for Elements in a Matrix**

Find the minor for each element in the first column of the matrix.

$$\begin{bmatrix} 3 & 4 & -1 \\ 2 & -4 & 5 \\ 0 & 1 & -6 \end{bmatrix}$$

Answers

1. a. 11   **b.** 0

Solution:

For the element a_{11}:

$$\begin{bmatrix} 3 & 4 & -1 \\ 2 & -4 & 5 \\ 0 & 1 & -6 \end{bmatrix} \qquad M_{11} = \begin{vmatrix} -4 & 5 \\ 1 & -6 \end{vmatrix} = (-4)(-6) - (5)(1) = 19$$

For the element a_{21}:

$$\begin{bmatrix} 3 & 4 & -1 \\ 2 & -4 & 5 \\ 0 & 1 & -6 \end{bmatrix} \qquad M_{21} = \begin{vmatrix} 4 & -1 \\ 1 & -6 \end{vmatrix} = (4)(-6) - (-1)(1) = -23$$

For the element a_{31}:

$$\begin{bmatrix} 3 & 4 & -1 \\ 2 & -4 & 5 \\ 0 & 1 & -6 \end{bmatrix} \qquad M_{31} = \begin{vmatrix} 4 & -1 \\ -4 & 5 \end{vmatrix} = (4)(5) - (-1)(-4) = 16$$

Skill Practice 2   Find the minor for each element in the second row of the matrix from Example 2.

Next, we define the determinant for a 3×3 matrix.

> ## Determinant of a 3 × 3 Matrix
>
> $$\begin{vmatrix} a_1 & b_1 & c_1 \\ a_2 & b_2 & c_2 \\ a_3 & b_3 & c_3 \end{vmatrix} = a_1 \begin{vmatrix} b_2 & c_2 \\ b_3 & c_3 \end{vmatrix} - a_2 \begin{vmatrix} b_1 & c_1 \\ b_3 & c_3 \end{vmatrix} + a_3 \begin{vmatrix} b_1 & c_1 \\ b_2 & c_2 \end{vmatrix} \text{ or equivalently,}$$
>
> $$= a_1(b_2c_3 - c_2b_3) - a_2(b_1c_3 - c_1b_3) + a_3(b_1c_2 - c_1b_2) \quad \text{or}$$
> $$= a_1b_2c_3 + b_1c_2a_3 + c_1a_2b_3 - a_3b_2c_1 - b_3c_2a_1 - c_3a_2b_1$$

From this definition, we see that the determinant of the given 3×3 matrix can be written as

$$a_1 \cdot (\text{minor of } a_1) - a_2 \cdot (\text{minor of } a_2) + a_3 \cdot (\text{minor of } a_3)$$

EXAMPLE 3   **Evaluating the Determinant of a 3 × 3 Matrix**

Evaluate the determinant of the matrix. $A = \begin{bmatrix} 2 & 4 & 2 \\ 1 & -3 & 0 \\ -5 & 5 & -1 \end{bmatrix}$

Solution:

$$|A| = \begin{vmatrix} 2 & 4 & 2 \\ 1 & -3 & 0 \\ -5 & 5 & -1 \end{vmatrix} = 2 \cdot \overbrace{\begin{vmatrix} -3 & 0 \\ 5 & -1 \end{vmatrix}}^{\text{minor of 2}} - (1) \overbrace{\begin{vmatrix} 4 & 2 \\ 5 & -1 \end{vmatrix}}^{\text{minor of 1}} + (-5) \cdot \overbrace{\begin{vmatrix} 4 & 2 \\ -3 & 0 \end{vmatrix}}^{\text{minor of } -5}$$

$$= 2[(-3)(-1) - (0)(5)] - 1[(4)(-1) - (2)(5)] + (-5)[(4)(0) - (2)(-3)]$$
$$= 2(3) - 1(-14) - 5(6)$$
$$= -10$$

Answer

2. $M_{21} = -23$; $M_{22} = -18$; $M_{23} = 3$

Skill Practice 3 Evaluate the determinant.

$$\begin{vmatrix} -2 & 4 & 9 \\ 5 & -1 & 2 \\ 1 & 1 & 6 \end{vmatrix}$$

Although we defined the determinant of a 3×3 matrix by expanding the minors of the elements in the first column, *any row or column can be used*. However, we must choose the correct sign to apply to the product of factors of each term. The following array of signs is helpful.

row 1, column 1
$1 + 1 = 2$ (even)

row 1, column 2
$1 + 2 = 3$ (odd)

$$\begin{bmatrix} + & - & + \\ - & + & - \\ + & - & + \end{bmatrix}$$

Notice that the sign is positive if the sum of the row and column numbers is even. The sign is negative if the sum of the row and column numbers is odd.

To evaluate the determinant of an $n \times n$ matrix, first choose any row or column. For each element a_{ij} in the selected row or column, multiply the minor by 1 or -1 depending on whether the sum of the row and column numbers is even or odd. That is, for the element a_{ij}, we have $(-1)^{i+j}M_{ij}$. This product is called the *cofactor* of the element a_{ij}.

Cofactor of an Element of a Matrix

Given a square matrix $A = [a_{ij}]$, the **cofactor** of a_{ij} is $(-1)^{i+j}M_{ij}$, where M_{ij} is the minor of a_{ij}.

Using the definition of the cofactor of an element of an $n \times n$ matrix, we can now present a generic method to find the determinant of the matrix.

Evaluating the Determinant of an $n \times n$ Matrix by Expanding Cofactors

Step 1 Choose any row or column.
Step 2 Multiply each element in the selected row or column by its cofactor.
Step 3 The value of the determinant is the sum of the products from step 2.

It is important to note that this three-step process to evaluate a determinant works for any $n \times n$ matrix, including a 2×2 matrix. For example, given $A = \begin{bmatrix} a & b \\ c & d \end{bmatrix}$, we can evaluate the determinant of A by expanding cofactors about the first row. The cofactor of a is d, and the cofactor of b is $-c$. Therefore,

$$|A| = a(d) + b(-c)$$

$$= ad - bc \text{ as expected.}$$

Answer
3. -42

| EXAMPLE 4 | Evaluating the Determinant of a 3 × 3 Matrix |

Evaluate the determinant of the matrix from Example 3 by expanding cofactors about the elements in the third row.

$$A = \begin{bmatrix} 2 & 4 & 2 \\ 1 & -3 & 0 \\ -5 & 5 & -1 \end{bmatrix}$$

Solution:

$$\begin{vmatrix} 2 & 4 & 2 \\ 1 & -3 & 0 \\ -5 & 5 & -1 \end{vmatrix} = \overbrace{(-5)(-1)^{3+1}\begin{vmatrix} 4 & 2 \\ -3 & 0 \end{vmatrix}}^{\text{cofactor of } -5} + \overbrace{(5)(-1)^{3+2}\begin{vmatrix} 2 & 2 \\ 1 & 0 \end{vmatrix}}^{\text{cofactor of } 5} + \overbrace{(-1)(-1)^{3+3}\begin{vmatrix} 2 & 4 \\ 1 & -3 \end{vmatrix}}^{\text{cofactor of } -1}$$

$$= (-5)(1)[(4)(0) - (2)(-3)] + (5)(-1)[(2)(0) - (2)(1)] + (-1)(1)[(2)(-3) - (4)(1)]$$

$$= -5(6) - 5(-2) - 1(-10)$$

$$= -30 + 10 + 10$$

$$= -10$$

This is the same result as in Example 3.

> **Skill Practice 4**   Evaluate the determinant by expanding cofactors about the elements in the third column.
>
> $$\begin{vmatrix} -2 & 4 & 9 \\ 5 & -1 & 2 \\ 1 & 1 & 6 \end{vmatrix}$$

TECHNOLOGY CONNECTIONS

Evaluating a Determinant

To find the determinant of a matrix on a graphing calculator, first enter the matrix in the calculator by using the MATRIX menu and EDIT menu. Select a name for the matrix such as [A]. Then access the determinant function det(under the MATRIX and **MATH** menus. Enter det([A]) in the home screen and press **ENTER**.

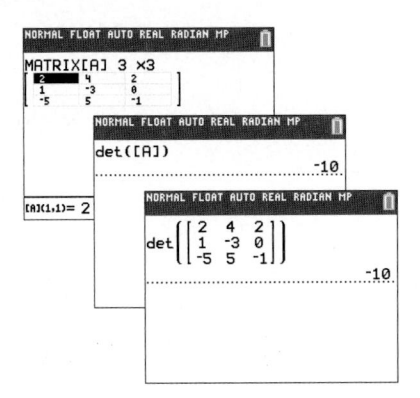

Alternatively, if the calculator is in "MATHPRINT" mode, enter the matrix by selecting **ALPHA** F3.

Answer

4. −42

TIP   The determinant of a 3×3 matrix can also be evaluated by using the "method of diagonals."

Step 1: Recopy columns 1 and 2 to the right of the matrix.

Step 2: Multiply the elements on the diagonals labeled d_1 through d_6 (each diagonal has three elements).

Step 3: The value of the determinant is $(d_1 + d_2 + d_3) - (d_4 + d_5 + d_6)$.

The determinant from Example 4 is evaluated as follows:

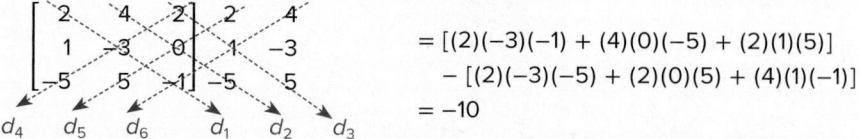

$$= [(2)(-3)(-1) + (4)(0)(-5) + (2)(1)(5)]$$
$$- [(2)(-3)(-5) + (2)(0)(5) + (4)(1)(-1)]$$
$$= -10$$

Some students find the method of diagonals to be a faster technique to find the determinant of a 3×3 matrix. However, it is critical to note that the method of diagonals only works for the determinant of a 3×3 matrix.

Perhaps one of the most important applications of the determinant of a matrix is to determine whether the matrix has an inverse. In Section 11.4, we found the inverse of a 2×2 matrix $A = \begin{bmatrix} a & b \\ c & d \end{bmatrix}$ by applying the formula $A^{-1} = \dfrac{1}{ad - bc}\begin{bmatrix} d & -b \\ -c & a \end{bmatrix}$. This is equivalent to $A^{-1} = \dfrac{1}{|A|}\begin{bmatrix} d & -b \\ -c & a \end{bmatrix}$. Furthermore, the inverse exists if $ad - bc \neq 0$ or equivalently, if $|A| \neq 0$. This is true in general for any $n \times n$ matrix.

Using Determinants to Determine if a Matrix Is Invertible

Let A be an $n \times n$ matrix. Then A is invertible if and only if $|A| \neq 0$.

In Example 5, we will use the determinant of A to determine if A has an inverse.

EXAMPLE 5   **Using a Determinant to Determine if a Matrix Is Invertible**

Use $|A|$ to determine if A is invertible. $A = \begin{vmatrix} -2 & 0 & 1 & 2 \\ 6 & 2 & -2 & 5 \\ 5 & 3 & -1 & 1 \\ 0 & 4 & 2 & 1 \end{vmatrix}$

Solution:

To evaluate a determinant, we can simplify the arithmetic by expanding around the row or column that contains the greatest number of 0 elements. In this case, we have chosen row 1.

$$\begin{vmatrix} -2 & 0 & 1 & 2 \\ 6 & 2 & -2 & 5 \\ 5 & 3 & -1 & 1 \\ 0 & 4 & 2 & 1 \end{vmatrix} = -2(-1)^{1+1}\begin{vmatrix} 2 & -2 & 5 \\ 3 & -1 & 1 \\ 4 & 2 & 1 \end{vmatrix} + 0(-1)^{1+2}\begin{vmatrix} 6 & -2 & 5 \\ 5 & -1 & 1 \\ 0 & 2 & 1 \end{vmatrix}$$

$$+ 1(-1)^{1+3}\begin{vmatrix} 6 & 2 & 5 \\ 5 & 3 & 1 \\ 0 & 4 & 1 \end{vmatrix} + 2(-1)^{1+4}\begin{vmatrix} 6 & 2 & -2 \\ 5 & 3 & -1 \\ 0 & 4 & 2 \end{vmatrix}$$

The second term is zero. Evaluating the determinants in the first, third, and fourth terms (shown in red), we have:

$$\begin{vmatrix} -2 & 0 & 1 & 2 \\ 6 & 2 & -2 & 5 \\ 5 & 3 & -1 & 1 \\ 0 & 4 & 2 & 1 \end{vmatrix} = -2(1)(42) + 0 + 1(1)(84) + 2(-1)(0) = 0$$

The determinant of A is zero. Therefore, A is not invertible.

Skill Practice 5   Use $|A|$ to determine if A is invertible. $A = \begin{vmatrix} 3 & 2 & 1 & 0 \\ 4 & 0 & 3 & 1 \\ 2 & 0 & 0 & 5 \\ 3 & -1 & 0 & 9 \end{vmatrix}$

3. Apply Cramer's Rule

We have learned several methods to solve a system of linear equations: the substitution method, the addition method, Gaussian elimination, Gauss-Jordan elimination, and the application of matrix inverses. We now present another method called Cramer's rule.

Cramer's rule involves finding the ratio of several determinants derived from the coefficients of the equations within the system. For example, consider the following system of equations.

$$a_1x + b_1y = c_1$$
$$a_2x + b_2y = c_2$$

Using the addition method to solve for x, we have

$$a_1x + b_1y = c_1 \quad \underrightarrow{\text{Multiply by } b_2.} \qquad a_1b_2x + b_1b_2y = c_1b_2$$
$$a_2x + b_2y = c_2 \quad \underrightarrow{\text{Multiply by } -b_1.} \qquad \underline{-a_2b_1x - b_1b_2y = -c_2b_1}$$
$$(a_1b_2 - a_2b_1)x \qquad = c_1b_2 - c_2b_1$$

$$x = \frac{c_1b_2 - c_2b_1}{a_1b_2 - a_2b_1} = \frac{\begin{vmatrix} c_1 & b_1 \\ c_2 & b_2 \end{vmatrix}}{\begin{vmatrix} a_1 & b_1 \\ a_2 & b_2 \end{vmatrix}}$$

Using similar logic, we can show that

$$y = \frac{a_1c_2 - a_2c_1}{a_1b_2 - a_2b_1} = \frac{\begin{vmatrix} a_1 & c_1 \\ a_2 & c_2 \end{vmatrix}}{\begin{vmatrix} a_1 & b_1 \\ a_2 & b_2 \end{vmatrix}}$$

These results are summarized as Cramer's rule for a system of linear equations in two variables.

Cramer's Rule for a System of Two Linear Equations in Two Variables

Given the system $\begin{aligned} a_1x + b_1y &= c_1 \\ a_2x + b_2y &= c_2 \end{aligned}$

let $D = \begin{vmatrix} a_1 & b_1 \\ a_2 & b_2 \end{vmatrix}$, $D_x = \begin{vmatrix} c_1 & b_1 \\ c_2 & b_2 \end{vmatrix}$, and $D_y = \begin{vmatrix} a_1 & c_1 \\ a_2 & c_2 \end{vmatrix}$.

Then if $D \neq 0$, the system has the unique solution: $x = \dfrac{D_x}{D}$ and $y = \dfrac{D_y}{D}$

Answer

5. $|A| = -9$; Since $|A| \neq 0$, A is invertible.

TIP   Here are some memory tips to remember the patterns presented in Cramer's rule.

$$a_1 x + b_1 y = c_1$$
$$a_2 x + b_2 y = c_2$$

Coefficients of
x terms y terms

1. The determinant D is the determinant of the coefficients of x and y.

$$D = \begin{vmatrix} a_1 & b_1 \\ a_2 & b_2 \end{vmatrix}$$

x-coefficients
replaced by c_1 and c_2

2. The determinant D_x has the column of x term coefficients replaced by c_1 and c_2.

$$D_x = \begin{vmatrix} c_1 & b_1 \\ c_2 & b_2 \end{vmatrix}$$

y-coefficients
replaced by c_1 and c_2

3. The determinant D_y has the column of y term coefficients replaced by c_1 and c_2.

$$D_y = \begin{vmatrix} a_1 & c_1 \\ a_2 & c_2 \end{vmatrix}$$

EXAMPLE 6   Solving a 2 × 2 System by Using Cramer's Rule

Solve the system by using Cramer's rule. $-11x + 6y = 4$
$$2x - 5y = -3$$

Solution:

For this system, $a_1 = -11$, $b_1 = 6$, $c_1 = 4$, $a_2 = 2$, $b_2 = -5$, and $c_2 = -3$.

$$D = \begin{vmatrix} a_1 & b_1 \\ a_2 & b_2 \end{vmatrix} \longrightarrow D = \begin{vmatrix} -11 & 6 \\ 2 & -5 \end{vmatrix} = (-11)(-5) - (6)(2) = 43$$

$$D_x = \begin{vmatrix} c_1 & b_1 \\ c_2 & b_2 \end{vmatrix} \longrightarrow D_x = \begin{vmatrix} 4 & 6 \\ -3 & -5 \end{vmatrix} = (4)(-5) - (6)(-3) = -2$$

$$D_y = \begin{vmatrix} a_1 & c_1 \\ a_2 & c_2 \end{vmatrix} \longrightarrow D_y = \begin{vmatrix} -11 & 4 \\ 2 & -3 \end{vmatrix} = (-11)(-3) - (4)(2) = 25$$

Therefore, $x = \dfrac{D_x}{D} = \dfrac{-2}{43}$ and $y = \dfrac{D_y}{D} = \dfrac{25}{43}$.

The solution set is $\left\{ \left(-\dfrac{2}{43}, \dfrac{25}{43} \right) \right\}$.

Skill Practice 6   Solve the system by using Cramer's rule. $3x - 4y = 9$
$$-5x + 6y = 2$$

The patterns associated with Cramer's rule can be generalized to solve a system of n linear equations in n variables.

Answer

6. $\left\{ \left(-31, -\dfrac{51}{2} \right) \right\}$

Cramer's Rule for a System of n Linear Equations in n Variables

Consider the following system of n linear equations in n variables.

$$a_{11}x_1 + a_{12}x_2 + \cdots + a_{1n}x_n = b_1$$
$$a_{21}x_1 + a_{22}x_2 + \cdots + a_{2n}x_n = b_2$$
$$\vdots$$
$$a_{n1}x_1 + a_{n2}x_2 + \cdots + a_{nn}x_n = b_n$$

If the system has a unique solution, then the solution is $(x_1, x_2, \ldots, x_i, \ldots, x_n)$, where

$$x_1 = \frac{D_1}{D}, x_2 = \frac{D_2}{D}, \ldots, x_i = \frac{D_i}{D}, \ldots, x_n = \frac{D_n}{D}.$$

$D \neq 0$ is the determinant of the coefficient matrix, and D_i is the determinant formed by replacing the ith column of the coefficient matrix by the column of constants $b_1, b_2, \ldots, b_n$.

In Example 7, we apply Cramer's rule to a system of equations with three variables.

EXAMPLE 7   Solving a 3 × 3 System by Using Cramer's Rule

Solve the system by using Cramer's rule.

$$2x - 3y + 5z = 11$$
$$-5x + 7y - 2z = -6$$
$$9x - 2y + 3z = 4$$

Solution:

Evaluate the determinants D, D_x, D_y, and D_z.

$$D = \begin{vmatrix} 2 & -3 & 5 \\ -5 & 7 & -2 \\ 9 & -2 & 3 \end{vmatrix} = -222 \qquad D_x = \begin{vmatrix} 11 & -3 & 5 \\ -6 & 7 & -2 \\ 4 & -2 & 3 \end{vmatrix} = 77$$

$$D_y = \begin{vmatrix} 2 & 11 & 5 \\ -5 & -6 & -2 \\ 9 & 4 & 3 \end{vmatrix} = 117 \qquad D_z = \begin{vmatrix} 2 & -3 & 11 \\ -5 & 7 & -6 \\ 9 & -2 & 4 \end{vmatrix} = -449$$

Avoiding Mistakes

It is important to note that each equation in the system of equations should be written in standard form before applying Cramer's rule.

Therefore, $x = \dfrac{D_x}{D} = -\dfrac{77}{222}$, $y = \dfrac{D_y}{D} = -\dfrac{117}{222}$ or $-\dfrac{39}{74}$, and $z = \dfrac{D_z}{D} = \dfrac{449}{222}$.

The solution set is $\left\{ \left(-\dfrac{77}{222}, -\dfrac{39}{74}, \dfrac{449}{222} \right) \right\}$.

Skill Practice 7   Solve the system by using Cramer's rule.

$$5x + 3y - 3z = -14$$
$$3x - 4y + z = 2$$
$$x + 7y + z = 6$$

Although Cramer's rule may seem cumbersome for solving a 3 × 3 system of linear equations, it provides convenient formulas that can be programmed into a computer or calculator to solve the system. However, it is important to remember that Cramer's rule does not apply if $D = 0$. In such a case, the system of equations is either inconsistent (has no solution) or the equations are dependent (the system has infinitely many solutions).

Answer

7. $\left\{ \left(-\dfrac{13}{34}, \dfrac{5}{17}, \dfrac{147}{34} \right) \right\}$

| EXAMPLE 8 | Identifying Whether Cramer's Rule Applies |

Solve the system by using Cramer's rule if possible. Otherwise, use a different method.

$$x + 3y = 6$$
$$-2x - 6y = -12$$

Solution:

Evaluate D. $D = \begin{vmatrix} 1 & 3 \\ -2 & -6 \end{vmatrix} = 1(-6) - 3(-2) = -6 + 6 = 0$

Since $D = 0$, Cramer's rule does not apply. Using Gauss-Jordan elimination, we have

$$\begin{bmatrix} 1 & 3 & | & 6 \\ -2 & -6 & | & -12 \end{bmatrix} \xrightarrow{2R_1 + R_2 \to R_2} \begin{bmatrix} 1 & 3 & | & 6 \\ 0 & 0 & | & 0 \end{bmatrix}$$

The last row of the augmented matrix represents the equation $0 = 0$, which is true for all values of x and y. The system reduces to the single equation $x + 3y = 6$ and there are infinitely many solutions.

The solution set is $\{(6 - 3y, y) \mid y \text{ is any real number}\}$.

Skill Practice 8   Solve the system by using Cramer's rule if possible. Otherwise, use a different method.

$$x + 4y = 2$$
$$3x + 12y = 4$$

Answer

8. $\{\,\}$

| SECTION 11.5 | Practice Exercises |

Prerequisite Review

For Exercises R.1–R.4, determine the value of the given element of matrix $A = [a_{ij}]$. (See Section 11.3 for review.)

$$A = \begin{bmatrix} -\sqrt{2} & 4 & -3 \\ 2.4 & \frac{1}{2} & 8 \\ -1 & 0 & -6 \end{bmatrix}$$

R.1. a_{23} **R.2** a_{31} **R.3.** a_{32} **R.4** a_{13}

For Exercises R.5–R.6, simplify the exponential expression. (See Section 2.1 for review.)

R.5. $(-1)^4$ **R.6.** $(-1)^7$

For Exercises R.7–R.10, solve the system. (See Sections 10.2 and 10.3 for review.)

R.7. $3(x - 1) - 7y = 3$ **R.8.** $5(y - 1) = 5 - 2x$
    $14y = 6x - 12$ $2x = 10 - 5y$

R.9. $x - y = 4$ **R.10.** $-x + 2y - 5z = 3$
    $x + 2y - 2z = 4$ $5x - 13y + 13z = 8$
    $x - 4y + 2z = -2$ $x - 3y + z = 4$

Concept Connections

1. Associated with every square matrix A is a real number denoted by $|A|$ called the _____ of A.

2. For a 2×2 matrix $A = \begin{bmatrix} a & b \\ c & d \end{bmatrix}$, $|A| = \begin{vmatrix} a & b \\ c & d \end{vmatrix} = $ _____.

3. Given $A = [a_{ij}]$, the _____ of the element a_{ij} is the determinant obtained by deleting the ith row and jth column.

4. Given $A = [a_{ij}]$, then the value $(-1)^{i+j} M_{ij}$ is called the _____ of the element a_{ij}.

5. The determinant of a 3×3 matrix

$$A = \begin{bmatrix} a_1 & b_1 & c_1 \\ a_2 & b_2 & c_2 \\ a_3 & b_3 & c_3 \end{bmatrix} \text{ is given by}$$

$$|A| = a_1 \begin{vmatrix} \square & \square \\ \square & \square \end{vmatrix} - a_2 \begin{vmatrix} \square & \square \\ \square & \square \end{vmatrix} + a_3 \begin{vmatrix} \square & \square \\ \square & \square \end{vmatrix}$$

6. Suppose that the given system has one solution.

$$a_1 x + b_1 y = c_1$$
$$a_2 x + b_2 y = c_2$$

Cramer's rule gives the solution as $x = \dfrac{\square}{\square}$ and $y = \dfrac{\square}{\square}$,

where $D = \begin{vmatrix} \square & \square \\ \square & \square \end{vmatrix}$, $D_x = \begin{vmatrix} \square & \square \\ \square & \square \end{vmatrix}$, and

$$D_y = \begin{vmatrix} \square & \square \\ \square & \square \end{vmatrix}.$$

Objective 1: Evaluate the Determinant of a 2 × 2 Matrix

For Exercises 7–16, evaluate the determinant of the matrix. **(See Example 1)**

7. $A = \begin{bmatrix} 3 & -2 \\ 6 & 5 \end{bmatrix}$

8. $B = \begin{bmatrix} 7 & 12 \\ -1 & 4 \end{bmatrix}$

9. $C = \begin{bmatrix} \frac{2}{3} & \frac{1}{5} \\ 10 & 12 \end{bmatrix}$

10. $D = \begin{bmatrix} \frac{8}{9} & 4 \\ \frac{5}{2} & 18 \end{bmatrix}$

11. $E = \begin{bmatrix} -3 & 0 \\ 4 & 0 \end{bmatrix}$

12. $F = \begin{bmatrix} 0 & 9 \\ 0 & 4 \end{bmatrix}$

13. $G = \begin{bmatrix} x & 4 \\ 9 & x \end{bmatrix}$

14. $H = \begin{bmatrix} y & 16 \\ 4 & y \end{bmatrix}$

15. $T = \begin{bmatrix} e^x & e^{2x} \\ 4 & -e^x \end{bmatrix}$

16. $V = \begin{bmatrix} \log x & \log x \\ 2 & 5 \end{bmatrix}$

Objective 2: Evaluate the Determinant of an n × n Matrix

For Exercises 17–22, refer to the matrix $A = [a_{ij}] = \begin{bmatrix} -6 & 11 & 8 \\ 4 & -2 & -5 \\ -3 & 7 & 10 \end{bmatrix}$.

   a. Find the minor of the given element. **(See Example 2)**

   b. Find the cofactor of the given element.

17. a_{12}

18. a_{23}

19. a_{31}

20. a_{13}

21. a_{22}

22. a_{33}

For Exercises 23–32, evaluate the determinant of the matrix and state whether the matrix is invertible. **(See Examples 3–5)**

23. $A = \begin{bmatrix} 4 & 1 & 3 \\ 0 & -1 & 2 \\ 5 & 8 & 0 \end{bmatrix}$

24. $B = \begin{bmatrix} 9 & 5 & -1 \\ 2 & 0 & 4 \\ 7 & -2 & 0 \end{bmatrix}$

25. $C = \begin{bmatrix} 5 & 1 & 6 \\ 2 & 3 & 4 \\ 8 & -1 & 7 \end{bmatrix}$

26. $D = \begin{bmatrix} -3 & 1 & -2 \\ 10 & 5 & 8 \\ 6 & 7 & -4 \end{bmatrix}$

27. $E = \begin{bmatrix} 2 & 0 & 1 \\ 1 & -1 & 2 \\ 3 & 1 & 0 \end{bmatrix}$

28. $F = \begin{bmatrix} 1 & -3 & 17 \\ 1 & -1 & 7 \\ 2 & -5 & 29 \end{bmatrix}$

29. $G = \begin{bmatrix} 5 & 6 & 4 & 1 \\ 2 & 0 & 3 & 0 \\ 0 & 1 & 4 & 0 \\ -1 & 2 & 0 & 0 \end{bmatrix}$

30. $H = \begin{bmatrix} 8 & 0 & 5 & 1 \\ 0 & 3 & 4 & -2 \\ 2 & 6 & 3 & 0 \\ -1 & 0 & 0 & 0 \end{bmatrix}$

31. $T = \begin{bmatrix} 3 & 8 & 1 & 4 \\ -2 & 4 & 0 & 5 \\ -1 & 1 & 0 & -1 \\ 0 & 5 & 2 & 3 \end{bmatrix}$

32. $W = \begin{bmatrix} 2 & 5 & 2 & 4 \\ 0 & 0 & -3 & 1 \\ 4 & 8 & 0 & 1 \\ -1 & 2 & 0 & 5 \end{bmatrix}$

Objective 3: Apply Cramer's Rule

For Exercises 33–48, solve the system if possible by using Cramer's rule. If Cramer's rule does not apply, solve the system by using another method. **(See Examples 6–8)**

33. $2x + 10y = 11$
$3x - 5y = 6$

34. $-5x - 8y = 3$
$4x + 7y = 13$

35. $-10x + 4y = 7$
$6x = 7y + 2$

36. $11x + 6y = 8$
$2x = 9y + 5$

37. $3(x - y) = y + 8$
$y = \frac{3}{4}x - 2$

38. $5(x + y) = 7x + 4$
$4x = 10y - 8$

39. $y = -3x + 7$
$\frac{1}{2}x + \frac{1}{6}y = 1$

40. $x = 4y + 5$
$3(x - 4) = 12y$

41. $11x \qquad - 3z = 1$
$\qquad 2y + 9z = 6$
$4x + 5y \qquad = -9$

42. $-2x + 6y \qquad = 9$
$\qquad 5y + 7z = 1$
$4x \qquad - 3z = -8$

43. $2x - 5y + z = 11$
$3x + 7y - 4z = 8$
$x - 9y + 2z = 4$

44. $-5x - 6y + 8z = 1$
$2x + y - 4z = 5$
$3x - 4y - z = -2$

45. $2x - 3y + z = 6$
$-4x + 6y - 2z = -12$
$6x - 9y + 3z = 18$

46. $-x + y - 3z = 4$
$3x - 3y + 9z = -12$
$-2x + 2y - 6z = 8$

47. $x - 2y + 3z = -1$
$5x - 7y + 3z = 1$
$x \qquad - 5z = 2$

48. $x + 3y - 5z = 10$
$-2x - 4y + 8z = -14$
$x + y - 3z = 5$

For Exercises 49–50, use Cramer's rule to solve for the indicated variable.

49. $x_1 + 2x_2 + 3x_3 - 4x_4 = 3$
$\qquad 5x_2 \qquad + x_4 = 9$
$x_1 \qquad + 4x_3 \qquad = -1$
$\qquad 5x_3 - 2x_4 = 8$ Solve for x_2.

50. $-2x_1 - x_2 + x_3 + 3x_4 = 10$
$x_1 \qquad + 5x_3 \qquad = 4$
$\qquad 2x_2 \qquad + x_4 = -1$
$\qquad 4x_3 + 2x_4 = 7$ Solve for x_3.

Mixed Exercises

Determinants can be used to determine whether three points are collinear (lie on the same line). Given the ordered pairs (x_1, y_1), (x_2, y_2), and (x_3, y_3), the points are collinear if the determinant to the right equals zero. For Exercises 51–54, determine if the points are collinear.

$$\begin{vmatrix} x_1 & y_1 & 1 \\ x_2 & y_2 & 1 \\ x_3 & y_3 & 1 \end{vmatrix}$$

51. $(3, 6), (6, 10), (-3, -2)$

52. $(-2, 1), (-4, -4), (4, 16)$

53. $(4, -3), (5, -7), (8, -14)$

54. $(0, 6), (1, 4), (4, -6)$

The equation at the right represents an equation of the line passing through the distinct points (x_1, y_1) and (x_2, y_2). For Exercises 55–56,

$$\begin{vmatrix} x & y & 1 \\ x_1 & y_1 & 1 \\ x_2 & y_2 & 1 \end{vmatrix} = 0$$

   a. Use the determinant equation to write an equation of the line passing through the given points.

   b. Write the equation of the line in slope-intercept form.

55. $(-3, 2)$ and $(-4, 6)$

56. $(-4, 1)$ and $(-5, 4)$

For Exercises 57–58, use the formula at the right to find the area of a triangle with vertices (x_1, y_1), (x_2, y_2), and (x_3, y_3). Choose the + or − sign so that the value of the area is positive.

$$\text{Area} = \pm\frac{1}{2}\begin{vmatrix} x_1 & y_1 & 1 \\ x_2 & y_2 & 1 \\ x_3 & y_3 & 1 \end{vmatrix}$$

57. $(1, 0), (7, -2), (4, -5)$

58. $(-2, 1), (-1, -6), (-8, -5)$

Given a square matrix A, elementary row operations (or column operations) performed on A affect the value of $|A|$ in the following ways:

- Interchanging any two rows (or columns) of A will change the sign of $|A|$.
- Multiplying a row (or column) of A by a constant real number k multiplies $|A|$ by k.
- Adding a multiple of a row (or column) of A to another row (or column) of A does not change the value of $|A|$.

For Exercises 59–64, demonstrate these three properties.

59. Given $A = \begin{bmatrix} 5 & 2 \\ -3 & 6 \end{bmatrix}$ and $B = \begin{bmatrix} -3 & 6 \\ 5 & 2 \end{bmatrix}$,

    a. Evaluate $|A|$.

    b. Evaluate $|B|$.

    c. How are A and B related and how are $|A|$ and $|B|$ related?

61. Given $A = \begin{bmatrix} 1 & -3 \\ 4 & 1 \end{bmatrix}$ and $B = \begin{bmatrix} 2 & -6 \\ 4 & 1 \end{bmatrix}$,

    a. Evaluate $|A|$.

    b. Evaluate $|B|$.

    c. How are A and B related and how are $|A|$ and $|B|$ related?

63. Given $A = \begin{bmatrix} 1 & 2 \\ 3 & 4 \end{bmatrix}$ and $B = \begin{bmatrix} 1 & 2 \\ 6 & 10 \end{bmatrix}$,

    a. Evaluate $|A|$.

    b. Evaluate $|B|$.

    c. How are A and B related and how are $|A|$ and $|B|$ related?

60. Given $A = \begin{bmatrix} 2 & -7 \\ 4 & 10 \end{bmatrix}$ and $B = \begin{bmatrix} 4 & 10 \\ 2 & -7 \end{bmatrix}$,

    a. Evaluate $|A|$.

    b. Evaluate $|B|$.

    c. How are A and B related and how are $|A|$ and $|B|$ related?

62. Given $A = \begin{bmatrix} 2 & 1 \\ 5 & 7 \end{bmatrix}$ and $B = \begin{bmatrix} -6 & -3 \\ 5 & 7 \end{bmatrix}$,

    a. Evaluate $|A|$.

    b. Evaluate $|B|$.

    c. How are A and B related and how are $|A|$ and $|B|$ related?

64. Given $A = \begin{bmatrix} 1 & 1 \\ 5 & 6 \end{bmatrix}$ and $B = \begin{bmatrix} 1 & 1 \\ 3 & 4 \end{bmatrix}$,

    a. Evaluate $|A|$.

    b. Evaluate $|B|$.

    c. How are A and B related and how are $|A|$ and $|B|$ related?

Given a square matrix A, if either of the following conditions are true, then $|A| = 0$.

- A row (or column) of A consists entirely of zeros.
- One row (or column) is a constant multiple of another row (or column).

For Exercises 65–68, demonstrate these two properties.

65. Given $A = \begin{bmatrix} 3 & 5 \\ 6 & 10 \end{bmatrix}$, find $|A|$.

67. Given $A = \begin{bmatrix} 4 & -5 & 0 \\ 3 & -1 & 0 \\ 0 & 1 & 0 \end{bmatrix}$, find $|A|$.

69. Evaluate $|I_2|$.

71. Evaluate $\begin{vmatrix} a & 0 & 0 \\ 0 & b & 0 \\ 0 & 0 & c \end{vmatrix}$.

66. Given $A = \begin{bmatrix} -2 & 7 \\ -6 & 21 \end{bmatrix}$, find $|A|$.

68. Given $A = \begin{bmatrix} 5 & 3 & 1 \\ -1 & 0 & 3 \\ 0 & 0 & 0 \end{bmatrix}$, find $|A|$.

70. Evaluate $|I_3|$.

72. Evaluate $\begin{vmatrix} x & 0 \\ 0 & x \end{vmatrix}$.

73. If A and B are square matrices, then the product property of determinants indicates that $|AB| = |A| \cdot |B|$. Use matrix A and matrix B to demonstrate this property.

$A = \begin{bmatrix} 4 & -2 \\ 3 & 1 \end{bmatrix}$ and $B = \begin{bmatrix} -5 & 1 \\ 3 & 2 \end{bmatrix}$

74. The **transpose** of a square matrix A, denoted as A^T, is a square matrix that results by writing the rows of A as the columns of A^T.

    a. Given $A = \begin{bmatrix} 1 & 2 & 5 \\ 0 & 8 & 4 \\ 3 & 7 & 6 \end{bmatrix}$, find A^T.

    b. Show that $|A| = |A^T|$.

Write About It

75. What is the difference between the minor of an element a_{ij} and the cofactor of the element?

76. Explain the difference between the notation $\begin{bmatrix} a & b \\ c & d \end{bmatrix}$ and $\begin{vmatrix} a & b \\ c & d \end{vmatrix}$.

77. The determinant of a square matrix can be computed by expanding the cofactors of the elements in any row or column. How would you choose which row or column?

78. Consider the system shown here. Describe the pattern associated with constructing the determinants used with Cramer's rule.

$$a_1x + b_1y = c_1$$
$$a_2x + b_2y = c_2$$

Technology Connections

For Exercises 79–82, use a graphing utility to evaluate the determinant of the matrix. Round to the nearest whole unit.

79. $\begin{bmatrix} \sqrt{3} & e & 1.6 \\ \log 5 & -2\pi & \ln 3 \\ -4 & 8.4 & -\sqrt{6} \end{bmatrix}$

80. $\begin{bmatrix} 8.9 & -2.3 & 3.8 \\ -1.7 & 0.9 & 4.6 \\ 2.7 & 10.1 & 14.9 \end{bmatrix}$

81. $A = \begin{bmatrix} -0.4 & 1.5 & 9 & 11.3 \\ -3.5 & 0.2 & -1.1 & 3 \\ 8 & 9.4 & -5.4 & 2 \\ -1 & 4.6 & 10.8 & -9.7 \end{bmatrix}$

82. $B = \begin{bmatrix} -2\pi & e^2 & 9.1 & \log 2 \\ \log 50 & -\sqrt{11} & 4.3 & \pi \\ -4.9 & 0 & e^2 & 8.1 \\ \sqrt{7} & \ln 7 & -9.7 & 0 \end{bmatrix}$

PROBLEM RECOGNITION EXERCISES

Using Multiple Methods to Solve Systems of Linear Equations

For Exercises 1–4, solve the system of equations using

a. The substitution method or the addition method (see Sections 10.2 and 10.3).

b. Gaussian elimination (see Section 11.1).

c. Gauss-Jordan elimination (see Section 11.1).

d. The inverse of the coefficient matrix (see Section 11.4).

e. Cramer's rule (see Section 11.5).

1. $x = -3y - 10$
$-3x - 7y = 22$

2. $2x = 2 - 8y$
$3x + 10y = 5$

3. $x + 2y - z = 0$
$2x + z = 4$
$2x - y + 2z = 5$

4. $x + 4y + 2z = 10$
$2y + z = 4$
$x + y = 2$

For Exercises 5–8,

a. Evaluate the determinant of the coefficient matrix.

b. Based on the value of the determinant from part (a), can an inverse matrix or Cramer's rule be used to solve the system?

c. Solve the system using an appropriate method.

5. $1.5x - 2y = 3$
$-3x + 4y = 12$

6. $5x - 2y = 1$
$x - 0.4y = 4$

7. $x - 3y + 7z = 1$
$-2x + 5y - 11z = -3$
$x - 5y + 13z = -1$

8. $x - 2y + 3z = -7$
$-2x + y = -1$
$x - z = 3$

CHAPTER 11   Detailed Summary

SECTION 11.1 Solving Systems of Linear Equations Using Matrices

Key Concepts	Examples	Page
A system of linear equations can be represented by an **augmented matrix**. **Elementary row operations** performed on an augmented matrix result in a **row-equivalent** matrix that represents an equivalent system of equations. Equivalent systems have the same solution set.	**Example 1:** $$\begin{aligned} x - y + 5z &= 6 \\ y - 2z &= 10 \\ 2x - 5y + 18z &= -8 \end{aligned} \xrightarrow[\text{matrix}]{\text{augmented}} \begin{bmatrix} 1 & -1 & 5 & 6 \\ 0 & 1 & -2 & 10 \\ 2 & -5 & 18 & -8 \end{bmatrix}$$	p. 802

Elementary row operations:
  1. Interchange two rows.
  2. Multiply a row by a nonzero constant.
  3. Add a nonzero multiple of one row to another row.
 	
Elementary row operations can be used to write a matrix in row-echelon form or reduced row-echelon form.
 	
To solve a system of linear equations:
  • The method of Gaussian elimination (Example 2) uses elementary row operations to write an augmented matrix in row-echelon form so that the system can be solved by back substitution.
	Example 2:
	Solve the system from Example 1 by Gaussian elimination.
	
	$$\begin{bmatrix} 1 & -1 & 5 & 6 \\ 0 & 1 & -2 & 10 \\ 2 & -5 & 18 & -8 \end{bmatrix} \xrightarrow{-2R_1 + R_3 \to R_3} \begin{bmatrix} 1 & -1 & 5 & 6 \\ 0 & 1 & -2 & 10 \\ 0 & -3 & 8 & -20 \end{bmatrix}$$
	$$\xrightarrow{3R_2 + R_3 \to R_3} \begin{bmatrix} 1 & -1 & 5 & 6 \\ 0 & 1 & -2 & 10 \\ 0 & 0 & 2 & 10 \end{bmatrix} \xrightarrow{\frac{1}{2}R_3 \to R_3} \begin{bmatrix} 1 & -1 & 5 & 6 \\ 0 & 1 & -2 & 10 \\ 0 & 0 & 1 & 5 \end{bmatrix}$$
	 row-echelon form
	$$\begin{aligned} x - y + 5z &= 6 \\ y - 2z &= 10 \\ z &= 5 \end{aligned}$$ Using back substitution, the solution set is $\{(1, 20, 5)\}$.
		p. 803

• With the method of Gauss-Jordan elimination (Example 3) the augmented matrix is written in *reduced* row-echelon form. Then the solution to the system can be found by inspection.	**Example 3:** Use Gauss-Jordan elimination to solve the system from Examples 1 and 2. $$\begin{bmatrix} 1 & -1 & 5 & 6 \\ 0 & 1 & -2 & 10 \\ 0 & 0 & 1 & 5 \end{bmatrix} \xrightarrow{R_2 + R_1 \to R_1} \begin{bmatrix} 1 & 0 & 3 & 16 \\ 0 & 1 & -2 & 10 \\ 0 & 0 & 1 & 5 \end{bmatrix}$$ $$\xrightarrow[\substack{2R_3 + R_2 \to R_2}]{-3R_3 + R_1 \to R_1} \begin{bmatrix} 1 & 0 & 0 & 1 \\ 0 & 1 & 0 & 20 \\ 0 & 0 & 1 & 5 \end{bmatrix} \begin{aligned} x &= 1 \\ y &= 20 \\ z &= 5 \end{aligned}$$ The solution set is $\{(1, 20, 5)\}$. <div align="center">*reduced* row-echelon form</div>	p. 804

SECTION 11.2 Inconsistent Systems and Dependent Equations

Key Concepts	Examples	Page
A system of equations that has no solution is called an **inconsistent system**. An inconsistent system is detected algebraically if a contradiction is reached when solving the system.	**Example 1:** $$2x + y = 4$$ $$-6x - 3y = 6$$ $\longrightarrow$ $\begin{bmatrix} 2 & 1 & \vert & 4 \\ -6 & -3 & \vert & 6 \end{bmatrix}$ $\frac{1}{2}R_1 \to R_1$ $\qquad$ $6R_1 + R_2 \to R_2$ $\longrightarrow \begin{bmatrix} 1 & \frac{1}{2} & \vert & 2 \\ -6 & -3 & \vert & 6 \end{bmatrix} \longrightarrow \begin{bmatrix} 1 & \frac{1}{2} & \vert & 2 \\ 0 & 0 & \vert & 18 \end{bmatrix}$ The last row of the augmented matrix represents the contradiction $0 = 18$. The system is inconsistent, and the solution set is { }.	p. 812
A system of linear equations may have infinitely many solutions. In such a case, the equations are said to be **dependent**. Dependent equations are detected algebraically if an identity is reached when solving the system.	**Example 2:** $$x + 3y + 3z = 7$$ $$-2x - 5y - 4z = -13$$ $$x + y - z = 5$$ $\longrightarrow \begin{bmatrix} 1 & 3 & 3 & \vert & 7 \\ -2 & -5 & -4 & \vert & -13 \\ 1 & 1 & -1 & \vert & 5 \end{bmatrix}$ $2R_1 + R_2 \to R_2$ $\qquad$ $-3R_2 + R_1 \to R_1$ $-1R_1 + R_3 \to R_3$ $\qquad$ $2R_2 + R_3 \to R_3$ $\longrightarrow \begin{bmatrix} 1 & 3 & 3 & \vert & 7 \\ 0 & 1 & 2 & \vert & 1 \\ 0 & -2 & -4 & \vert & -2 \end{bmatrix} \longrightarrow \begin{bmatrix} 1 & 0 & -3 & \vert & 4 \\ 0 & 1 & 2 & \vert & 1 \\ 0 & 0 & 0 & \vert & 0 \end{bmatrix}$ The corresponding system of equations is $$x - 3z = 4$$ $$y + 2z = 1$$ $$0 = 0 \quad \text{(identity)}$$ The first two equations can both be represented in terms of z. The solution set is $\{(3z + 4, -2z + 1, z) \vert z \text{ is any real number}\}$.	p. 814

SECTION 11.3 Operations on Matrices

Key Concepts	Examples	Page
An $m \times n$ matrix has m rows and n columns.	**Example 1:** $$A = \begin{bmatrix} -3 & 6.1 & \pi & 0 \\ \sqrt{3} & 4.1 & e & 7 \end{bmatrix}$$ is a 2 × 4 matrix.	p. 824
The notation a_{ij} represents the element of a matrix in the ith row and jth column.	The notation a_{21} represents the element in the second row and first column. That is, $a_{21} = \sqrt{3}$.	
A matrix with only one row is called a **row matrix**. A matrix with only one column is called a **column matrix**. A matrix with the same number of rows and columns is called a **square matrix**.	**Example 2:** $[3 \quad -5 \quad 1]$ is a row matrix. $\begin{bmatrix} 4 \\ -5 \end{bmatrix}$ is a column matrix. $\begin{bmatrix} 0.4 & 1.2 \\ 9.3 & 6.1 \end{bmatrix}$ is a square matrix.	p. 824

Adding and subtracting matrices: If A and B represent two matrices of the same order, then $A + B$ and $A - B$ are found by adding or subtracting the corresponding elements.	**Example 3:** $$\begin{bmatrix} 3 & 5 \\ -2 & 1.4 \\ \sqrt{5} & -4 \end{bmatrix} + \begin{bmatrix} -1 & 0 \\ 8 & 3.2 \\ 6\sqrt{5} & 3 \end{bmatrix} = \begin{bmatrix} 2 & 5 \\ 6 & 4.6 \\ 7\sqrt{5} & -1 \end{bmatrix}$$	p. 825
Scalar multiplication: Let $A = [a_{ij}]$ be an $m \times n$ matrix and let k be a real number. Then, $kA = [ka_{ij}]$.	**Example 4:** Given $A = \begin{bmatrix} 3 & 9 & -4 \\ 0 & -1 & 2 \end{bmatrix}$, $$4A = 4\begin{bmatrix} 3 & 9 & -4 \\ 0 & -1 & 2 \end{bmatrix} = \begin{bmatrix} 12 & 36 & -16 \\ 0 & -4 & 8 \end{bmatrix}.$$	p. 827
Solving a matrix equation: The same principles apply to solve a matrix equation as with a linear equation.	**Example 5:** Given $A = \begin{bmatrix} 9 & 1 \\ 5 & -2 \end{bmatrix}$ and $B = \begin{bmatrix} 6 & -5 \\ 2 & 1 \end{bmatrix}$, solve for X. $$3X + A = B$$ $$3X = B - A$$ $$X = \frac{1}{3}(B - A)$$ $$X = \frac{1}{3}\left(\begin{bmatrix} 6 & -5 \\ 2 & 1 \end{bmatrix} - \begin{bmatrix} 9 & 1 \\ 5 & -2 \end{bmatrix}\right)$$ $$X = \frac{1}{3}\begin{bmatrix} -3 & -6 \\ -3 & 3 \end{bmatrix}$$ $$X = \begin{bmatrix} -1 & -2 \\ -1 & 1 \end{bmatrix}$$	p. 828
Matrix multiplication: Let A be an $m \times p$ matrix and let B be a $p \times n$ matrix, then the product AB is an $m \times n$ matrix. For the matrix AB, the element in the ith row and jth column is the sum of the products of the corresponding elements in the ith row of A and the jth column of B (the inner product of the ith row and jth column). *Note*: If the number of columns in A does not equal the number of rows in B, then it is not possible to compute the product AB.	**Example 6:** Let $A = \begin{bmatrix} 1 & 0 & -3 \\ 4 & -2 & 4 \end{bmatrix}$ and $B = \begin{bmatrix} 6 & 8 \\ -3 & 1 \\ -1 & 4 \end{bmatrix}$. Then $AB = \begin{bmatrix} 1 & 0 & -3 \\ 4 & -2 & 4 \end{bmatrix} \cdot \begin{bmatrix} 6 & 8 \\ -3 & 1 \\ -1 & 4 \end{bmatrix}$ $\qquad \underset{2 \times 3}{} \qquad \underset{3 \times 2}{}$ equal The product is a 2×2 matrix. $$= \begin{bmatrix} (1)(6) + (0)(-3) + (-3)(-1) & (1)(8) + (0)(1) + (-3)(4) \\ (4)(6) + (-2)(-3) + (4)(-1) & (4)(8) + (-2)(1) + (4)(4) \end{bmatrix}$$ $$= \begin{bmatrix} 9 & -4 \\ 26 & 46 \end{bmatrix}$$	p. 829

SECTION 11.4 Inverse Matrices and Matrix Equations

Key Concepts	Examples	Page
The **identity matrix** I_n for the multiplication of matrices is the $n \times n$ square matrix with 1's along the main diagonal and 0's for all other elements.	**Example 1:** $$I_2 = \begin{bmatrix} 1 & 0 \\ 0 & 1 \end{bmatrix}$$	p. 841
Inverse of a square matrix: Let A be an $n \times n$ matrix. If there exists an $n \times n$ matrix A^{-1} such that $$AA^{-1} = I_n \quad \text{and} \quad A^{-1}A = I_n$$ then A^{-1} is the **multiplicative inverse** of A. A matrix that does not have an inverse is called a **singular matrix**. A matrix that does have an inverse is said to be **invertible** or **nonsingular**.	**Example 2:** $$A = \begin{bmatrix} 5 & 3 \\ 3 & 2 \end{bmatrix} \text{ and } B = \begin{bmatrix} 2 & -3 \\ -3 & 5 \end{bmatrix} \text{ are inverses because}$$ $AB = I_2$ and $BA = I_2$.	p. 842
Finding the inverse of a nonsingular matrix: Let A be an $n \times n$ matrix for which A^{-1} exists. To find A^{-1}: **Step 1** Write a matrix of the form $[A \,|\, I_n]$. **Step 2** Perform row operations to write the matrix in the form $[I_n \,|\, B]$. **Step 3** The matrix B is A^{-1}. The inverse of a nonsingular 2×2 matrix A can also be found as follows. If $A = \begin{bmatrix} a & b \\ c & d \end{bmatrix}$, then $A^{-1} = \dfrac{1}{ad - bc} \cdot \begin{bmatrix} d & -b \\ -c & a \end{bmatrix}$, or equivalently $A^{-1} = \dfrac{1}{|A|} \begin{bmatrix} d & -b \\ -c & a \end{bmatrix}$.	**Example 3:** To find the inverse of $A = \begin{bmatrix} 4 & 8 \\ 2 & 5 \end{bmatrix}$, we have $$\begin{bmatrix} 4 & 8 & | & 1 & 0 \\ 2 & 5 & | & 0 & 1 \end{bmatrix} \xrightarrow{\frac{1}{4}R_1 \to R_1} \begin{bmatrix} 1 & 2 & | & \frac{1}{4} & 0 \\ 2 & 5 & | & 0 & 1 \end{bmatrix}$$ $$\xrightarrow{-2R_1 + R_2 \to R_2} \begin{bmatrix} 1 & 2 & | & \frac{1}{4} & 0 \\ 0 & 1 & | & -\frac{1}{2} & 1 \end{bmatrix} \xrightarrow{-2R_2 + R_1 \to R_1} \begin{bmatrix} 1 & 0 & | & \frac{5}{4} & -2 \\ 0 & 1 & | & -\frac{1}{2} & 1 \end{bmatrix}$$ $$A^{-1} = \begin{bmatrix} \frac{5}{4} & -2 \\ -\frac{1}{2} & 1 \end{bmatrix}$$	p. 843
Solving a system using inverse matrices: Suppose that $AX = B$ represents a system of n linear equations in n variables with a unique solution. Then, $$X = A^{-1}B$$ where A is the coefficient matrix, B is the matrix of constants, and X is the matrix of variables.	**Example 4:** $$A \cdot X = B$$ $\begin{aligned} 4x + 8y &= 28 \\ 2x + 5y &= 16 \end{aligned} \qquad \begin{bmatrix} 4 & 8 \\ 2 & 5 \end{bmatrix}\begin{bmatrix} x \\ y \end{bmatrix} = \begin{bmatrix} 28 \\ 16 \end{bmatrix}$ $$X = A^{-1}B = \begin{bmatrix} \frac{5}{4} & -2 \\ -\frac{1}{2} & 1 \end{bmatrix}\begin{bmatrix} 28 \\ 16 \end{bmatrix}$$ $$X = \begin{bmatrix} \left(\frac{5}{4}\right)(28) + (-2)(16) \\ \left(-\frac{1}{2}\right)(28) + (1)(16) \end{bmatrix} = \begin{bmatrix} 3 \\ 2 \end{bmatrix}$$ The solution set is $\{(3, 2)\}$.	p. 847

SECTION 11.5 Determinants and Cramer's Rule

Key Concepts	Examples	Page
Associated with every square matrix is a real number called the determinant of the matrix. **Determinant of a 2 × 2 matrix:** Given $A = \begin{bmatrix} a & b \\ c & d \end{bmatrix}$, the determinant of A is denoted by $\lvert A \rvert$ or $\begin{vmatrix} a & b \\ c & d \end{vmatrix}$ and the value is $ad - bc$.	**Example 1:** The determinant of $A = \begin{bmatrix} 3 & -4 \\ 2 & 10 \end{bmatrix}$ is $\lvert A \rvert = \begin{vmatrix} 3 & -4 \\ 2 & 10 \end{vmatrix} = (3)(10) - (-4)(2)$ $= 38$	p. 850
The minor and cofactor of an element of a matrix: Given an $n \times n$ matrix $A = [a_{ij}]$, the minor M_{ij} of an element a_{ij} is the determinant of the resulting matrix obtained by deleting the ith row and jth column. The **cofactor** of a_{ij} is $(-1)^{i+j} M_{ij}$ where M_{ij} is the minor of a_{ij}.	**Example 2:** Given $A = \begin{bmatrix} 3 & 4 & -2 \\ 9 & -8 & 5 \\ -6 & 1 & 0 \end{bmatrix}$, **a.** The minor of a_{21} is $\begin{vmatrix} 4 & -2 \\ 1 & 0 \end{vmatrix} = 2$. **b.** The cofactor of a_{21} is $(-1)^{2+1} \begin{vmatrix} 4 & -2 \\ 1 & 0 \end{vmatrix} = -2$.	p. 852
Determinant of a 3 × 3 matrix: $\begin{vmatrix} a_1 & b_1 & c_1 \\ a_2 & b_2 & c_2 \\ a_3 & b_3 & c_3 \end{vmatrix} = a_1 \begin{vmatrix} b_2 & c_2 \\ b_3 & c_3 \end{vmatrix} - a_2 \begin{vmatrix} b_1 & c_1 \\ b_3 & c_3 \end{vmatrix} + a_3 \begin{vmatrix} b_1 & c_1 \\ b_2 & c_2 \end{vmatrix}$ **Find the determinant of an $n \times n$ matrix:** **Step 1** Choose any row or column. **Step 2** Multiply each element in the selected row or column by its cofactor. **Step 3** The value of the determinant is the sum of the products from step 2.	**Example 3:** $\begin{vmatrix} 3 & 1 & -4 \\ 6 & 0 & -1 \\ 0 & 5 & 8 \end{vmatrix} = 3 \begin{vmatrix} 0 & -1 \\ 5 & 8 \end{vmatrix} - 6 \begin{vmatrix} 1 & -4 \\ 5 & 8 \end{vmatrix} + 0 \begin{vmatrix} 1 & -4 \\ 0 & -1 \end{vmatrix}$ $= 3(5) - 6(28) + 0(-1)$ $= -153$	p. 853
Using a determinant to determine if a matrix is invertible: An $n \times n$ matrix A is invertible if and only if $\lvert A \rvert \neq 0$.	**Example 4:** The matrix $A = \begin{bmatrix} 3 & 1 & -4 \\ 6 & 0 & -1 \\ 0 & 5 & 8 \end{bmatrix}$ is invertible because $\lvert A \rvert = -153 \neq 0$ (See Example 3).	p. 856
Cramer's rule is a method of solving a system of linear equations with a unique solution by finding ratios of determinants derived from the coefficients of the equations within the system. **Cramer's rule for a system of two equations and two variables:** Given $\begin{aligned} a_1 x + b_1 y &= c_1 \\ a_2 x + b_2 y &= c_2 \end{aligned}$ if $D \neq 0$, then $x = \dfrac{D_x}{D}$ and $y = \dfrac{D_y}{D}$, where $D = \begin{vmatrix} a_1 & b_1 \\ a_2 & b_2 \end{vmatrix}$, $D_x = \begin{vmatrix} c_1 & b_1 \\ c_2 & b_2 \end{vmatrix}$, and $D_y = \begin{vmatrix} a_1 & c_1 \\ a_2 & c_2 \end{vmatrix}$. The patterns associated with Cramer's rule can be generalized to solve a system of n linear equations in n variables. When applying Cramer's rule, if $D = 0$, then the system has either no solution or infinitely many solutions.	**Example 5:** Solve the system by using Cramer's rule. $3x + 7y = 12$ $4x - 5y = -7$ $D = \begin{vmatrix} 3 & 7 \\ 4 & -5 \end{vmatrix} = -43$ $D_x = \begin{vmatrix} 12 & 7 \\ -7 & -5 \end{vmatrix} = -11$; $D_y = \begin{vmatrix} 3 & 12 \\ 4 & -7 \end{vmatrix} = -69$ $x = \dfrac{D_x}{D} = \dfrac{-11}{-43}$ and $y = \dfrac{D_y}{D} = \dfrac{-69}{-43}$ $x = \dfrac{11}{43}$ and $y = \dfrac{69}{43}$ The solution set is $\left\{ \left(\dfrac{11}{43}, \dfrac{69}{43} \right) \right\}$.	p. 857

CHAPTER 11   Review Exercises

SECTION 11.1

1. Write a system of linear equations represented by the augmented matrix. Then write the solution set.

$$\begin{bmatrix} 1 & -2 & 3 & | & -1 \\ 0 & 1 & 4 & | & -11 \\ 0 & 0 & 1 & | & -2 \end{bmatrix}$$

For Exercises 2–4, perform the elementary row operations on the matrix. $\begin{bmatrix} 2 & -3 & | & 1 \\ 5 & 6 & | & -4 \end{bmatrix}$

2. $R_1 \Leftrightarrow R_2$

3. $\frac{1}{2} R_1 \rightarrow R_1$

4. $-2R_1 + R_2 \rightarrow R_2$

For Exercises 5–8, solve the system by using Gaussian elimination or Gauss-Jordan elimination.

5. $-2x + y = -16$
   $x - 2y = 17$

6. $2(x - 6y) = 36$
   $47y = 7x - 141$

7. $2x - 5y + 18z = 44$
   $x - 3y + 11z = 27$
   $x - 2y + 11z = 29$

8. $w + x \quad - 2z = 3$
   $2x \quad - 3z = 3$
   $2w \quad + y + z = 3$
   $4y - z = 9$

9. Lily borrowed a total of $10,000. She borrowed part of the money from her friend Sly who did not charge her interest. She borrowed part of the money from a credit union at 5% simple interest, and she borrowed the rest of the money from a bank at 7.5% interest. At the end of 1 year, she owed $500 in interest. If she borrowed $1000 less from her friend than she did from the bank, determine how much she borrowed from each source.

SECTION 11.2

For Exercises 10–13, determine the solution set for the system represented by each augmented matrix.

10. $\begin{bmatrix} 1 & 0 & | & 4 \\ 0 & 1 & | & 0 \end{bmatrix}$

11. $\begin{bmatrix} 1 & -2 & | & 6 \\ 0 & 0 & | & 1 \end{bmatrix}$

12. $\begin{bmatrix} 1 & 4 & | & 0 \\ 0 & 0 & | & 0 \end{bmatrix}$

13. $\begin{bmatrix} 1 & 0 & -3 & | & 0 \\ 0 & 1 & 2 & | & 1 \\ 0 & 0 & 0 & | & 0 \end{bmatrix}$

For Exercises 14–19, solve the system by using Gaussian elimination or Gauss-Jordan elimination.

14. $3x + 6y = -9$
    $x + 2y = -3$

15. $-(2x - y) = 8 - y$
    $y = x - 6$

16. $x - 2y = 3z - 10$
    $x - y = z - 7$
    $3x - 7y - 11z = -320$

17. $x \quad - 3z = 5$
    $-2x + y + 10z = -7$
    $x + y + z = 8$

18. $2x = 3y - z - 4$
    $x - 2y = z + 2y - 2$
    $x + y = 2z - 2$

19. $5y = x + 2z + 1$
    $2(x - 5y) + 4z = -2$
    $3(x + 2z) = 15y - 3$

20. The solution set to a system of dependent equations is given. Write three ordered triples that are solutions to the system. Answers may vary.

    $\{(2z - 3, z + 2, z) \,|\, z \text{ is any real number}\}$

21. **a.** Assume that traffic flows freely around the traffic circle. The flow rates given are measured in vehicles per hour. If the flow rate x_3 is 130 vehicles per hour, determine the flow rates x_1 and x_2.

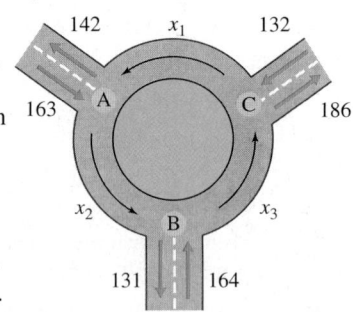

b. If traffic between intersections B and C flows at a rate of between 100 and 150 vehicles per hour, inclusive, find the range of values for x_1 and x_2.

22. **a.** Assume that traffic flows freely through intersections A, B, C, and D and that all flow rates are measured in vehicles per hour. If the flow rate x_4 is 220 vehicles per hour, find the flow rates x_1, x_2, and x_3.

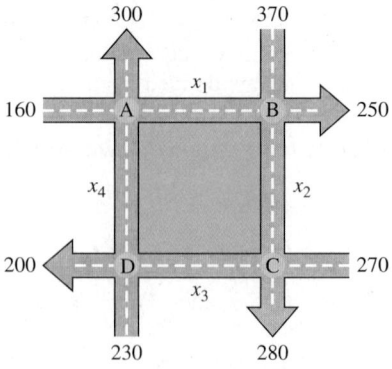

b. If the flow rate x_4 is between 200 and 250 vehicles per hour, inclusive, find the range of values x_1, x_2, and x_3.

SECTION 11.3

For Exercises 23–26,

  a. Give the order of the matrix.

  b. Classify the matrix as a square matrix, row matrix, column matrix, or none of these.

23. $\begin{bmatrix} 1 & 2 \\ -3 & \pi \\ 4.1 & \sqrt{2} \end{bmatrix}$ **24.** $[8 \quad 4 \quad 1 \quad -6]$

25. $\begin{bmatrix} -3.1 \\ 8.7 \end{bmatrix}$ **26.** $\begin{bmatrix} -3 & 8 \\ 0 & 0 \end{bmatrix}$

For Exercises 27–28, determine the value of the given element of the matrix $A = [a_{ij}]$.

$$A = \begin{bmatrix} -1 & 8 & -3 \\ 4 & 6 & 9 \end{bmatrix}$$

27. a_{21} **28.** a_{12}

29. For what value of x, y, and z will $A = B$?

$$A = \begin{bmatrix} 3 & -4 \\ x & z \end{bmatrix} \quad \text{and} \quad B = \begin{bmatrix} y & -4 \\ 6 & 8 \end{bmatrix}$$

30. Solve the equation $-3X + A = B$ for X, given that

$$A = \begin{bmatrix} 2 & -7 \\ 2 & -5 \end{bmatrix} \text{ and } B = \begin{bmatrix} 5 & 2 \\ -1 & 7 \end{bmatrix}.$$

For Exercises 31–40, perform the indicated operations if possible.

$$A = \begin{bmatrix} -4 & 1 \\ 6 & -2 \\ 1 & 3 \end{bmatrix} \quad B = \begin{bmatrix} 2 & 3 & -7 \\ 1 & 5 & -6 \end{bmatrix} \quad C = \begin{bmatrix} \pi & 4 \\ -3 & 1 \\ 0 & 5 \end{bmatrix}$$

31. $3A$ **32.** $-2B$ **33.** $A + B$

34. $B + C$ **35.** $2A - C$ **36.** $4A + 3B$

37. AB **38.** BC **39.** AC

40. CA

For Exercises 41–44, perform the indicated operations if possible.

$$A = \begin{bmatrix} 2 & 6 \\ -1 & 4 \end{bmatrix} \quad B = \begin{bmatrix} 1 \\ -3 \end{bmatrix} \quad C = [2 \quad 7]$$

41. A^2 **42.** AB **43.** BC **44.** CB

45. A company owns two movie theaters in town. The number of popcorns and drinks sold for each theater is given in matrix Q. The price per item is given in matrix P. Find the product QP and interpret the result.

	Popcorn (small)	Popcorn (large)	Drinks (small)	Drinks (large)	
$Q =$	386	244	418	216	Theater 1
	450	382	476	262	Theater 2

$$P = \begin{bmatrix} \$8.50 \\ \$6.50 \\ \$5.50 \\ \$3.50 \end{bmatrix} \begin{matrix} \textbf{Popcorn (small)} \\ \textbf{Popcorn (large)} \\ \textbf{Drinks (small)} \\ \textbf{Drinks (large)} \end{matrix}$$

46. Matrix M gives the manufacturer price for four models of dining room tables. Matrix P gives the retail price to the customer.

	Wood	Metal	
$M =$	\$1050	\$940	Large
	\$890	\$800	Small

	Wood	Metal	
$P =$	\$1365	\$1222	Large
	\$1157	\$1040	Small

  a. Compute $P - M$ and interpret its meaning.

  b. If the tax rate in a certain city is 6%, use scalar multiplication to find a matrix F that gives the final price (including sales tax) to the customer for each model.

47. a. Write a matrix A that represents the coordinates of the vertices of the triangle. Place the x-coordinate of each point in the first row of A and the corresponding y-coordinate in the second row of A.

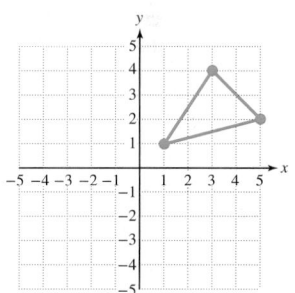

  b. Use addition of matrices to shift the triangle 3 units to the left and 1 unit downward.

  c. Find the product $\begin{bmatrix} -1 & 0 \\ 0 & 1 \end{bmatrix} \cdot A$ and explain the effect on the graph of the triangle.

  d. Find the product $\begin{bmatrix} 1 & 0 \\ 0 & -1 \end{bmatrix} \cdot A$ and explain the effect on the graph of the triangle.

SECTION 11.4

48. Given two $n \times n$ matrices A and B, what are the criteria for the matrices to be inverses?

49. Determine whether A and B are inverses.

$$A = \begin{bmatrix} 4 & 1 \\ 3 & 2 \end{bmatrix} \quad B = \begin{bmatrix} 2 & -3 \\ -1 & 4 \end{bmatrix}$$

50. Determine whether A and B are inverses.

$$A = \begin{bmatrix} 1 & 3 \\ 2 & 1 \end{bmatrix} \quad B = \begin{bmatrix} -\frac{1}{5} & \frac{3}{5} \\ \frac{2}{5} & -\frac{1}{5} \end{bmatrix}$$

For Exercises 51–56, determine the inverse of the given matrix if possible. Otherwise state that the matrix is singular.

51. $A = \begin{bmatrix} 5 & -2 \\ 1 & 2 \end{bmatrix}$ **52.** $A = \begin{bmatrix} \frac{1}{4} & \frac{3}{8} \\ \frac{1}{8} & -\frac{1}{16} \end{bmatrix}$

53. $A = \begin{bmatrix} 2 & 3 \\ 16 & 24 \end{bmatrix}$ **54.** $A = \begin{bmatrix} 1 & 0 & 1 \\ -1 & 5 & -3 \\ 1 & -3 & 2 \end{bmatrix}$

55. $A = \begin{bmatrix} -5 & 4 & 1 \\ 15 & -12 & -4 \\ 4 & -3 & -1 \end{bmatrix}$ **56.** $A = \begin{bmatrix} 1 & -3 & -17 \\ 1 & -2 & -12 \\ 0 & -2 & -10 \end{bmatrix}$

57. Write the system of equations as a matrix equation of the form $AX = B$, where A is the coefficient matrix, X is the column matrix of variables, and B is the column matrix of constants.

$$\begin{aligned} -3x + 7y \quad\;\; &= 6 \\ 4x \quad\;\; + 2z &= -3 \\ 2x - y + 5z &= -13 \end{aligned}$$

For Exercises 58–61, solve the system using the inverse of the coefficient matrix.

58. $\dfrac{1}{4}x + \dfrac{3}{8}y = 4$ (See Exercise 52 for A^{-1}.)

$\dfrac{1}{8}x - \dfrac{1}{16}y = -2$

59. $5x - 2y = 26$ (See Exercise 51 for A^{-1}.)

$x + 2y = -2$

60. $x \quad\;\; + z = 2$ (See Exercise 54 for A^{-1}.)

$-x + 5y - 3z = -6$

$x - 3y + 2z = 3$

61. $-5x + 4y + z = 6$ (See Exercise 55 for A^{-1}.)

$15x - 12y - 4z = -21$

$4x - 3y - z = -5$

SECTION 11.5

For Exercises 62–65, refer to the matrix

$$A = [a_{ij}] = \begin{bmatrix} -5 & 4 & -2 \\ 1 & 0 & 6 \\ 8 & -9 & 0 \end{bmatrix}.$$

a. Find the minor of the given element.

b. Find the cofactor of the given element.

62. a_{13} **63.** a_{31}

64. a_{32} **65.** a_{23}

For Exercises 66–71, evaluate the determinant of the given matrix.

66. $A = \begin{bmatrix} 9 & -4 \\ 2 & -3 \end{bmatrix}$ **67.** $B = \begin{bmatrix} 3 & x \\ x & 27 \end{bmatrix}$

68. $C = \begin{bmatrix} 9 & -15 \\ -3 & 5 \end{bmatrix}$ **69.** $D = \begin{bmatrix} 4 & -1 & 0 \\ 6 & 8 & -2 \\ 1 & 5 & 3 \end{bmatrix}$

70. $E = \begin{bmatrix} 4 & -9 & 0 \\ -3 & 8 & 0 \\ 6 & 1 & 0 \end{bmatrix}$ **71.** $F = \begin{bmatrix} -2 & 0 & 3 & 1 \\ 1 & 1 & 0 & 5 \\ 4 & 0 & 0 & -2 \\ 0 & -3 & 0 & 6 \end{bmatrix}$

For Exercises 72–76, solve the system by using Cramer's rule if possible. If Cramer's rule does not apply, use another method to solve the system.

72. $3x - 7y = 11$ **73.** $9x = 3y + 5$

$4x + 2y = 3$ $-2(x + 3y) = 4$

74. $2x + 5y = 10$ **75.** $3x - 2y + z = 4$

$10y = -4(x - 5)$ $5x + 3y + 6z = 1$

$-2x \quad\;\; + 5z = 7$

76. $2x + y - z = 5$

$7x + 7y - 6z = 5$

$3x + 5y - 4z = -1$

For Exercises 77–78, use Cramer's rule to solve for the indicated variable.

77. $-6x + 7y \quad\;\; = 8$ Solve for y.

$2x + 5y + z = -3$

$3x \quad\;\; + 2z = 11$

78. $3x_1 \quad\;\; + 4x_3 \quad\;\; = 6$ Solve for x_4.

$4x_1 \quad\;\; + 2x_3 + x_4 = -7$

$x_2 \quad\;\; - 3x_4 = 2$

$5x_3 + x_4 = 1$

CHAPTER 11   Test

For Exercises 1–3, perform the elementary row operations

on the matrix $A = \begin{bmatrix} 3 & 1 & 4 & | & -2 \\ 1 & 5 & -3 & | & 1 \\ 0 & 4 & 2 & | & 6 \end{bmatrix}$.

1. $R_1 \Leftrightarrow R_2$ **2.** $-3R_2 + R_1 \to R_1$ **3.** $\frac{1}{4}R_3 \to R_3$

4. Explain why the matrix is not in reduced row-echelon form.

$$\begin{bmatrix} 1 & 0 & 6 & | & 4 \\ 0 & 1 & 0 & | & 2 \\ 0 & 0 & 1 & | & -3 \end{bmatrix}$$

For Exercises 5–7, write the solution set for the system represented by each augmented matrix.

5. $\begin{bmatrix} 1 & 4 & | & 2 \\ 0 & 1 & | & 3 \end{bmatrix}$

6. $\begin{bmatrix} 1 & 0 & 0 & | & 4 \\ 0 & 1 & 0 & | & 2 \\ 0 & 0 & 0 & | & 1 \end{bmatrix}$

7. $\begin{bmatrix} 1 & 0 & -3 & | & 0 \\ 0 & 1 & 2 & | & 5 \\ 0 & 0 & 0 & | & 0 \end{bmatrix}$

For Exercises 8–9, refer to the matrix $A = \begin{bmatrix} -1 & 3 & 0 \\ 2 & 5 & -4 \\ 6 & 9 & -8 \end{bmatrix}$.

a. Find the minor of the given element.

b. Find the cofactor of the given element.

8. a_{12}

9. a_{31}

For Exercises 10–11, evaluate the determinant of the matrix.

10. $A = \begin{bmatrix} 3 & 7 \\ 4 & -1 \end{bmatrix}$

11. $B = \begin{bmatrix} -3 & 4 & 7 \\ 1 & -2 & 3 \\ 6 & 5 & 0 \end{bmatrix}$

12. Use the determinant of A to determine whether A has an inverse.

$$A = \begin{bmatrix} 1 & 2 & 3 \\ 1 & 4 & 11 \\ 2 & 5 & 10 \end{bmatrix}$$

For Exercises 13–16, solve the system by using Gaussian elimination or Gauss-Jordan elimination.

13. $-3(x + y) = 3y - 12$
    $-3x = 4y - 6$

14. $-3x = 11 - 18y$
    $x - 6y = 2$

15. $x - 2y = 5z + 4$
    $6y + 18z = 2x - 8$
    $-3x + 8y + 20z = -18$

16. $2x + 6y + 30z = 2$
    $x + 2y + 11z = 0$
    $-3x - 6y - 33z = 0$

17. Solve the system by using Cramer's rule.

$$3x - 5y = 7$$
$$11x + 2y = 8$$

18. Use Cramer's rule to solve for x.

$$2x = 3y - 4z + 11$$
$$9x + y = z - 1$$
$$3x - 4y = 7$$

19. Solve the equation $A - 4X = B$ for X, given that

$$A = \begin{bmatrix} 2 & 5 \\ -2 & -3 \end{bmatrix} \text{ and } B = \begin{bmatrix} 6 & -3 \\ 14 & 5 \end{bmatrix}.$$

For Exercises 20–23, perform the indicated operations if possible.

$$A = \begin{bmatrix} 4 & 1 & -3 \\ 2 & 4 & 6 \end{bmatrix} \quad B = \begin{bmatrix} 1 & 9 \\ 0 & -1 \\ 3 & 5 \end{bmatrix} \quad C = \begin{bmatrix} 0 & 1 & -4 \\ 2 & -1 & 8 \end{bmatrix}$$

20. $2A - 3C$

21. $A + B$

22. AB

23. BA

24. Determine whether A and B are inverses.

$$A = \begin{bmatrix} -1 & \frac{2}{3} & -\frac{2}{3} \\ 1 & -\frac{1}{3} & \frac{1}{3} \\ 1 & -\frac{2}{3} & \frac{5}{3} \end{bmatrix} \text{ and } B = \begin{bmatrix} 1 & 2 & 0 \\ 4 & 3 & 1 \\ 1 & 0 & 1 \end{bmatrix}$$

For Exercises 25–27, determine the inverse of the matrix if possible. Otherwise, state that the matrix is singular.

25. $A = \begin{bmatrix} 3 & 2 \\ 5 & 4 \end{bmatrix}$

26. $A = \begin{bmatrix} 2 & 5 & 10 \\ 1 & 3 & 7 \\ 1 & 4 & 11 \end{bmatrix}$

27. $A = \begin{bmatrix} 3 & -1 & -1 \\ 2 & -1 & 1 \\ -5 & 2 & 1 \end{bmatrix}$

For Exercises 28–29, solve the system by using the inverse of the coefficient matrix.

28. $3x + 2y = 13$ (See Exercise 25 for A^{-1}.)
    $5x + 4y = 25$

29. $3x - y - z = 8$ (See Exercise 27 for A^{-1}.)

$2x + z = y$

$-5x + 2y + z = -11$

30. a. Assume that traffic flows freely around the traffic circle. The flow rates given are measured in vehicles per hour. If the flow rate x_3 is 210 vehicles per hour, determine the flow rates x_1 and x_2.

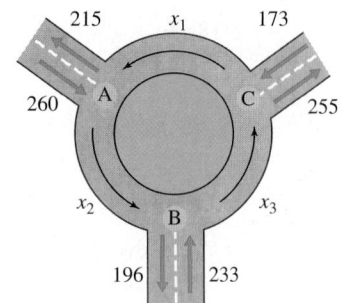

b. If traffic between intersections B and C flows at a rate of between 200 and 250 vehicles per hour, inclusive, find the range of values for x_1 and x_2.

31. Matrix C represents the number of calories burned per hour of exercise riding a bicycle, running, and walking for individuals of two different weights. Matrix N represents the number of hours spent working out with each type of activity for a given week. Find the product CN and interpret its meaning.

$$C = \begin{matrix} \textbf{Bike} & \textbf{Run} & \textbf{Walk} & \\ \begin{bmatrix} 400 & 500 & 320 \\ 550 & 780 & 480 \end{bmatrix} & & & \begin{matrix} \textbf{120-lb person} \\ \textbf{180-lb person} \end{matrix} \end{matrix}$$

$$N = \begin{bmatrix} 6 \\ 3 \\ 5 \end{bmatrix} \begin{matrix} \textbf{Bike} \\ \textbf{Run} \\ \textbf{Walk} \end{matrix}$$

32. Given three points (x_1, y_1), (x_2, y_2), and (x_3, y_3), the points are collinear if the following determinant is equal to zero.

$$\begin{vmatrix} x_1 & y_1 & 1 \\ x_2 & y_2 & 1 \\ x_3 & y_3 & 1 \end{vmatrix}$$

Determine if the points $(4, -11)$, $(-1, -1)$, and $(-5, 7)$ are collinear.

Prerequisite Review for Calculus

SECTION A.1 Algebra for Calculus

For Exercises 1–2, write an inequality using an absolute value that represents the given condition.

1. The distance between y and L is less than ε (epsilon).

2. The distance between x and c is less than δ (delta).

3. Simplify $\dfrac{x + 8}{|x + 8|}$

    a. for $x > -8$.   **b.** for $x < -8$.

4. Simplify $\dfrac{14 - x}{|x - 14|}$

    a. for $x > 14$.   **b.** for $x < 14$.

For Exercises 5–10,
    a. Simplify the expression.
    b. Substitute 0 for h in the simplified expression.

5. $\dfrac{2(x + h)^2 + 3(x + h) - (2x^2 + 3x)}{h}$

6. $\dfrac{3(x + h)^2 - 4(x + h) - (3x^2 - 4x)}{h}$

7. $\dfrac{\dfrac{1}{(x + h) - 2} - \dfrac{1}{x - 2}}{h}$

8. $\dfrac{\dfrac{1}{2(x + h) + 5} - \dfrac{1}{2x + 5}}{h}$

9. $\dfrac{(x + h)^3 - x^3}{h}$

10. $\dfrac{(x + h)^4 - x^4}{h}$

For Exercises 11–12,
    a. Rationalize the numerator of the expression and simplify.
    b. Substitute 0 for h in the simplified expression.

11. $\dfrac{\sqrt{x + h} + 1 - (\sqrt{x} + 1)}{h}$

12. $\dfrac{\sqrt{2(x + h)} - \sqrt{2x}}{h}$

For Exercises 13–22, factor completely and write the answer with no negative exponents. Do not rationalize the denominator.

13. $\dfrac{3}{2}x^{1/2} + \dfrac{5}{2}x^{3/2}$

14. $\dfrac{7}{6}x^{1/6} - \dfrac{1}{6}x^{-5/6}$

15. $4(3x + 1)^3(3)(x^2 + 2)^3 + (3x + 1)^4(3)(x^2 + 2)^2(2x)$

16. $3(-2x + 3)^2(-2)(4x^2 - 5)^2 + (-2x + 3)^3(2)(4x^2 - 5)(8x)$

17. $\dfrac{6(t - 1)^5(2t + 5)^6 - 6(2t + 5)^5(2)(t - 1)^6}{[(2t + 5)^6]^2}$

18. $\dfrac{6x^5(x^2 + 4)^3 - 3(x^2 + 4)^2(2x)x^6}{[(x^2 + 4)^3]^2}$

19. $(x^2 + 4)^{1/2} + x \cdot \dfrac{1}{2}(x^2 + 4)^{-1/2}(2x)$

20. $(2 - x^2)^{1/2} + x \cdot \dfrac{1}{2}(2 - x^2)^{-1/2}(-2x)$

21. $(x^2 + 1)^{-1/2} + x\left(-\dfrac{1}{2}\right)(x^2 + 1)^{-3/2}(2x)$

22. $6(3x - 1)^{1/3} + 6x\left(-\dfrac{1}{3}\right)(3x - 1)^{-2/3}(3)$

For Exercises 23–28, write the answer as a single term and simplify. It is not necessary to rationalize the denominator.

23. $\dfrac{2\sqrt{x + 4} - \dfrac{x}{\sqrt{x + 4}}}{(\sqrt{x + 4})^2}$

24. $\dfrac{2x\sqrt{16 - x^2} + \dfrac{x^3}{\sqrt{16 - x^2}}}{(\sqrt{16 - x^2})^2}$

25. $(x - 4)^{1/3} + \dfrac{x}{3(x - 4)^{2/3}}$

26. $(x + 5)^{1/4} + \dfrac{x}{4(x + 5)^{3/4}}$

27. $\dfrac{1}{2}\left(\dfrac{2x}{x + 1}\right)^{-1/2}\left[\dfrac{(x + 1)(2) - 2x(1)}{(x + 1)^2}\right]$

28. $\dfrac{1}{3}\left(\dfrac{3x}{x^2 + 1}\right)^{-2/3}\left[\dfrac{(x^2 + 1) \cdot 3 - 3x(2x)}{(x^2 + 1)^2}\right]$

SECTION A.2   Equations and Inequalities for Calculus

In calculus you will see the symbol y'. For Exercises 1–4, treat y' as a variable and solve the equation for y'.

1. $\dfrac{2x}{25} + \dfrac{2y}{9}y' = 0$

2. $2xy^3 + 3x^2y^2y' - y' = 1$

3. $3y^2y' + 6xy + 3x^2y' = 2y^2 + 4xyy'$

4. $3(x + y)^2 + 3(x + y)^2 y' - 3y^2y' = 3x^2$

For Exercises 5–10,

    a. Simplify the expression. Do not rationalize the denominator.
    b. Find the values of x for which the expression equals zero.
    c. Find the values of x for which the denominator is zero.

5. $2x\sqrt{2x - 3} + x^2\left(\dfrac{1}{2}\right)\dfrac{1}{\sqrt{2x - 3}}(2)$

6. $\dfrac{2x(2x - 7)^{1/2} - x^2\left(\dfrac{1}{2}\right)(2x - 7)^{-1/2}(2)}{\left[(2x - 7)^{1/2}\right]^2}$

7. $\dfrac{(1)(x^2 - 9)^{1/2} - x\left(\dfrac{1}{2}\right)(x^2 - 9)^{-1/2}(2x)}{\left[(x^2 - 9)^{1/2}\right]^2}$

8. $\dfrac{4x(4x - 5) - 2x^2(4)}{(4x - 5)^2}$

9. $\dfrac{-6x(6x + 1) - (-3x^2)(6)}{(6x + 1)^2}$

10. $\sqrt{4 - x^2} - x\left(\dfrac{1}{2}\right)\dfrac{1}{\sqrt{4 - x^2}}(2x)$

Some applications of calculus use a mathematical structure called a power series. To find the interval of convergence of a power series, it is often necessary to solve an absolute value inequality. For Exercises 11–12, solve the absolute value inequality to find the interval of convergence.

11. $\left|\dfrac{x + 1}{2}\right| < 1$

12. $\left|-\dfrac{x}{2}\right| < 1$

13. A 6-ft man walks away from a lamppost. At the instant the man is 14 ft away from the lamppost, his shadow is 10 ft long. Find the height of the lamppost.

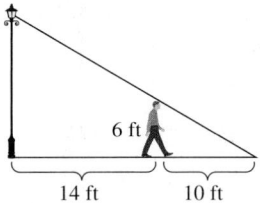

14. A water trough has a cross section in the shape of an equilateral triangle with sides of length 1 m. The length is 3 m. Determine the volume of water when the water level is $\frac{1}{2}$ m.

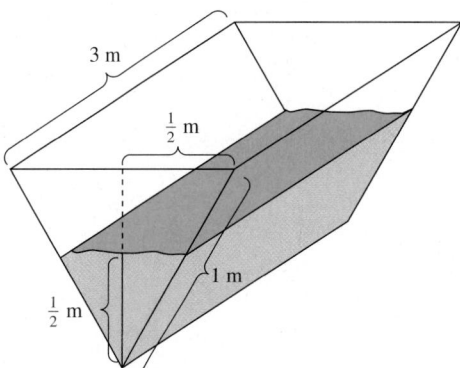

15. A contractor builds a swimming pool with cross section in the shape of a trapezoid. The deep end is 8 ft deep and the shallow end is 3 ft deep. The length of the pool is 50 ft and the width is 20 ft. As the pool is being filled, find the volume of water when the depth is 4 ft.

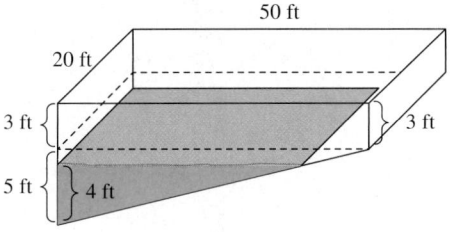

Student Answer Appendix

Section R.1 Practice Exercises, pp. 9–11

R.1. Prime **R.3.** Composite **R.5.** Composite **R.7.** Prime
R.9. $2 \cdot 2 \cdot 3 \cdot 3$ **R.11.** $2 \cdot 3 \cdot 7$ **R.13.** $2 \cdot 5 \cdot 11$ **R.15.** $3 \cdot 3 \cdot 3 \cdot 5$

1. numerator; b **3.** $1; 4$ **5.** multiple **7.** Numerator: 7;
denominator: 8; proper **9.** Numerator: 9; denominator: 5;
improper **11.** Numerator: 6; denominator: 6; improper
13. Numerator: 12; denominator: 1; improper

15. $\dfrac{1}{5}$ **17.** $\dfrac{8}{3}$ or $2\dfrac{2}{3}$ **19.** $\dfrac{7}{8}$ **21.** $\dfrac{3}{4}$ **23.** $\dfrac{5}{8}$ **25.** $\dfrac{4}{3}$ or $1\dfrac{1}{3}$

27. $\dfrac{4}{3}$ or $1\dfrac{1}{3}$ **29.** $\dfrac{2}{3}$ **31.** $\dfrac{9}{2}$ or $4\dfrac{1}{2}$ **33.** $\dfrac{3}{5}$ **35.** $\dfrac{5}{3}$ or $1\dfrac{2}{3}$

37. $\dfrac{90}{13}$ or $6\dfrac{12}{13}$ **39.** 1 **41.** 1 **43.** 10 **45.** $\dfrac{1}{6}$ **47.** $704

49. 35 students **51.** 8 pieces **53.** 8 jars

55. $\dfrac{3}{7}$ **57.** $\dfrac{1}{2}$ **59.** 3 **61.** $\dfrac{16}{y}$ **63.** $\dfrac{7}{8}$ **65.** $\dfrac{6}{5}$ or $1\dfrac{1}{5}$

67. $\dfrac{43}{40}$ or $1\dfrac{3}{40}$ **69.** $\dfrac{3}{26}$ **71.** $\dfrac{29}{36}$ **73.** $\dfrac{35}{48}$ **75.** $\dfrac{7}{24}$

77. $\dfrac{51}{28}$ or $1\dfrac{23}{28}$ **79.** $\dfrac{11}{12}$ cup sugar **81.** $\dfrac{7}{50}$ in.

83. $\dfrac{46}{5}$ or $9\dfrac{1}{5}$ **85.** $\dfrac{1}{6}$ **87.** $\dfrac{11}{54}$ **89.** $\dfrac{7}{2}$ or $3\dfrac{1}{2}$ **91.** $\dfrac{13}{8}$ or $1\dfrac{5}{8}$

93. $\dfrac{59}{12}$ or $4\dfrac{11}{12}$ **95.** $\dfrac{1}{8}$ **97.** $8\frac{19}{24}$ in. **99.** $1\frac{7}{12}$ hr **101.** $2\frac{1}{4}$ lb

103. 25 in.

Section R.2 Practice Exercises, pp. 19–22

R.1. terminating **R.3.** nonterminating **R.5.** terminating
R.7. nonterminating **R.9.** 2.45 **R.11.** 2.455

1. set **3.** real; numbers **5.** opposites **7.** $\{5, 6, 7, 8, 9, 10\}$
9. Yes **11.** 0, 1 **13.** $\sqrt{11}, \sqrt{7}$ **15.** a; rational
17. b; rational **19.** a; rational **21.** c; irrational
23. a; rational **25.** a; rational **27.** b; rational **29.** c; irrational
31. a. 3 is an element of the set of natural numbers.
b. -3.1 is not an element of the set of whole numbers.
33. a. False **b.** False **c.** True **d.** True
35. a. 6 **b.** 6 **c.** $-12, 6$ **d.** $0.\overline{3}, 0.33, -0.9, -12, \frac{11}{4}, 6$
e. $\sqrt{5}, \frac{\pi}{6}$ **f.** $\sqrt{5}, 0.\overline{3}, 0.33, -0.9, -12, \frac{11}{4}, 6, \frac{\pi}{6}$
37. -18 **39.** 6.1 **41.** $\dfrac{5}{8}$ **43.** $-\dfrac{7}{3}$ **45.** 3 **47.** $-\dfrac{7}{3}$

49. 8 **51.** -72.1 **53.** 2 **55.** 1.5 **57.** -1.5 **59.** $\dfrac{3}{2}$

61. -10 **63.** $-\dfrac{1}{2}$ **65.** -4 **67.** -19 **69.** -7 **71.** 14

73. -22.1 **75.** -8.1 **77.** $-\dfrac{5}{3}$ or $-1\dfrac{2}{3}$ **79.** $-\dfrac{67}{45}$ or $-1\dfrac{22}{45}$

81. a. 6 **b.** -6 **c.** -22 **d.** 22 **e.** -22
83. a. 0 **b.** 0 **c.** 50 **d.** 0 **e.** -50
85. a. -3.6 **b.** 10.6 **c.** 3.6 **d.** 3.6 **e.** -10.6
87. -32 **89.** $\dfrac{8}{21}$ **91.** $\dfrac{3}{5}$ **93.** $-\dfrac{18}{5}$ or $-3\dfrac{3}{5}$

95. Undefined **97.** 0 **99.** 3.72 **101.** $\dfrac{5}{11}$

103. -27 **105.** $-\dfrac{1}{25}$ **107.** For example: $\pi, -\sqrt{2}, \sqrt{3}$
109. For example: $-5, -2, -1$ **111.** False, $|n|$ is never negative.
113. For all negative values of a **115. a.** 2.65 **b.** 4.36 **c.** 6.28

Section R Problem Recognition Exercises, p. 22

1. a. -4 **b.** 32 **c.** -12 **d.** 2
2. a. 10 **b.** 14 **c.** -24 **d.** -6
3. a. -27 **b.** -324 **c.** -4 **d.** -45
4. a. 30 **b.** 24 **c.** -81 **d.** -9
5. a. 50 **b.** -15 **c.** $\dfrac{1}{2}$ **d.** 5
6. a. -5 **b.** -24 **c.** -16 **d.** -80
7. a. 64 **b.** 12 **c.** $\dfrac{1}{4}$ **d.** -20
8. a. -7 **b.** -24 **c.** -63 **d.** -18
9. a. -400 **b.** 85 **c.** -16 **d.** 75
10. a. 7 **b.** 294 **c.** $\dfrac{2}{3}$ **d.** -35
11. a. 8 **b.** 4 **c.** 4 **d.** 8
12. a. -16 **b.** 2 **c.** -2 **d.** 16
13. a. $\dfrac{23}{12}$ or $1\dfrac{11}{12}$ **b.** $-\dfrac{41}{12}$ or $-3\dfrac{5}{12}$ **c.** -2 **d.** $-\dfrac{9}{32}$
14. a. $-\dfrac{7}{12}$ **b.** $-\dfrac{2}{15}$ **c.** $\dfrac{23}{15}$ or $1\dfrac{8}{15}$ **d.** $-\dfrac{21}{25}$
15. a. $\dfrac{44}{17}$ or $2\dfrac{10}{17}$ **b.** $\dfrac{27}{10}$ or $2\dfrac{7}{10}$ **c.** $\dfrac{61}{10}$ or $6\dfrac{1}{10}$ **d.** $\dfrac{187}{25}$ or $7\dfrac{12}{25}$
16. a. $-\dfrac{61}{8}$ or $-7\dfrac{5}{8}$ **b.** $-\dfrac{23}{38}$ **c.** $\dfrac{15}{8}$ or $1\dfrac{7}{8}$
d. $-\dfrac{437}{32}$ or $-13\dfrac{21}{32}$

Section R.3 Practice Exercises, pp. 28–31

R.1. $\dfrac{1}{40}$ **R.3.** $\dfrac{3}{5}$ **R.5.** 14

1. base; exponent; power **3.** radicand; n; nth

5. $\left(\dfrac{1}{6}\right)^4$ **7.** $x \cdot x \cdot x$ **9.** $2b \cdot 2b \cdot 2b$

11. $10 \cdot y \cdot y \cdot y \cdot y \cdot y$ **13.** 36 **15.** $\dfrac{1}{49}$ **17.** 0.008 **19.** 64

21. -49 **23.** 49 **25.** $\dfrac{125}{27}$ **27.** 6 **29.** 3

31. Not a real number **33.** $\dfrac{1}{2}$ **35.** -7 **37.** 0.2

39. -3 **41.** 10 **43.** Not a real number **45.** $\dfrac{3}{4}$ **47.** 32

49. 40 **51.** 25 **53.** 13 **55.** 8 **57.** 78 **59.** 0

61. $\dfrac{7}{6}$ or $1\dfrac{1}{6}$ **63.** -603 **65.** $\dfrac{109}{150}$ **67.** 5.4375 **69.** $\dfrac{2}{3}$

71. -1 **73.** 21 **75.** Undefined **77.** $\dfrac{9}{10}$ **79.** 9 in.^{2}

81. 8.06 cm^2 **83.** 14.1 ft^3 **85.** 26.8 ft^3 **87.** 141.4 in.^3

89. a. $25°C$ **b.** $100°C$ **c.** $0°C$ **d.** $-40°C$

91. $4\frac{1}{6} \text{ gal}$ **93. a.** 16 **b.** 16 **c.** -16 **d.** 2

    e. -2 **f.** Not a real number

95. a. 2 **b.** -2 **c.** -2 **d.** 10 **e.** Not a real number

    f. -10

97. $-10.1°C$ **99.** -3 **101.** $\sqrt{34}$ **103.** $12/(6-2)$

105. $\frac{2}{3}$ or 0.6666666667 **107.** 6744.25 **109.** 0.58

Chapter R Review Exercises, pp. 33–34

1. Improper   **3.** Improper   **5.** $\frac{5}{3}$ or $1\frac{2}{3}$ **7.** $\frac{35}{36}$ **9.** $\frac{2}{7}$

11. $\frac{6}{5}$ or $1\frac{1}{5}$ **13.** $\frac{17}{10}$ or $1\frac{7}{10}$ **15.** 357 million km^2

17. a. $7, 1$ **b.** $7, -4, 0, 1$ **c.** $7, 0, 1$ **d.** $7, \frac{1}{3}, -4, 0, -0.\overline{2}, 1$

    e. $-\sqrt{3}, \pi$ **f.** $7, \frac{1}{3}, -4, 0, -\sqrt{3}, -0.\overline{2}, \pi, 1$

19. 2 **21.** $\frac{1}{2}$ **23.** -4 **25.** 2 **27.** 15 **29.** $\frac{11}{63}$

31. $-\frac{14}{15}$ **33.** -2.15 **35.** -12 **37.** -1

39. $-\frac{29}{18}$ or $-1\frac{11}{18}$ **41.** -1.2 **43.** -10.2 **45.** -170

47. -2 **49.** $-\frac{1}{6}$ **51.** 0 **53.** 0 **55.** $-\frac{3}{2}$

57. 216 **59.** 6 **61.** $\frac{1}{16}$ **63.** Not a real number

65. -10 **67.** -625 **69.** 625 **71.** $-\frac{2}{5}$ **73.** 17

75. $-\frac{7}{120}$ **77.** $-\frac{1}{3}$ **79.** 3 **81.** 6 **83.** 11 **85.** 75

87. 256 **89.** 70.6

Section R Test, pp. 34–35

1. $\frac{15}{4}$ or $3\frac{3}{4}$ **2.** $\frac{3}{2}$ or $1\frac{1}{2}$ **3.** $\frac{49}{16}$ or $3\frac{1}{16}$ **4.** $\frac{1}{8}$

5. $\frac{19}{8}$ or $2\frac{3}{8}$ **6. a.** 8 **b.** $-6, 8, 0$ **c.** $8, 0$

    d. $-6, \frac{3}{5}, 8, 0, 0.\overline{5}$ **e.** $4\pi, \sqrt{7}$ **f.** $-6, \frac{3}{5}, 8, 0, 4\pi, 0.\overline{5}, \sqrt{7}$

7. a. 13 **b.** 13 **c.** $-\frac{1}{13}$ **8. a.** 9 **b.** -9

9. a. 1 **b.** 1 **10. a.** -49 **b.** 49

11. a. 2 **b.** 16 **12. a.** 8 **b.** 4

13. a. -5 **b.** Not a real number   **14.** 40 cm^2

15. $-\frac{7}{8}$ **16.** -12 **17.** 28 **18.** -12

19. -32 **20.** 96 **21.** 0 **22.** Undefined

23. 6 **24.** 4.66 **25.** -28 **26.** $\frac{2}{3}$ **27.** $\frac{1}{3}$ **28.** 9

29. -8 **30.** 6 **31.** -17 **32.** $\frac{1}{4}$

CHAPTER 1

Section 1.1 Practice Exercises, pp. 43–46

R.1. 1 **R.3.** 1 **R.5.** 0 **R.7. a.** 6 **b.** 6 **R.9. a.** -18
b. -18 **R.11. a.** -8 **b.** -8 **R.13. a.** -40 **b.** -40
R.15. a. 30 **b.** 30

1. constant   **3.** $1; 1$ **5. a.** 3 terms   **b.** 6 **c.** $2, -5, 6$
    7. a. 5 terms   **b.** -7 **c.** $1, -7, 1, -4, 1$ **9.** a   **11.** f   **13.** e

15. i   **17.** b   **19.** g   **21.** d   **23.** h   **25.** a

27. $2x - 6y + 16$ **29.** $-40s + 90t + 30$ **31.** $7w - 5z$

33. $\frac{1}{2}a - 2b + \frac{8}{5}$ **35.** $7.8x - 12.3$ **37.** $14c - 16 - 30d + 5f$

39. $14y - 2x$ **41.** $6p^2 + p - 6$ **43.** $-p^2 - 3p$ **45.** $n^3 + m - 6$

47. $7ab + 8a$ **49.** $16xy^2 - 5y^2$ **51.** $-4c - 6$ **53.** $-9w + 10$

55. $4z - 16$ **57.** $7s - 26$ **59.** $-12w + 13$ **61.** 0 **63.** $4c + 2$

65. $1.4x + 10.2$ **67.** $-2a^2 + 3a + 38$ **69.** $2y^2 - 3y - 5$

71. $-62.7x + 220$ **73.** $-4m + 15n + 2$ **75.** $88v + 59w - 20$

77. $6y^2$ **79.** $t - 5$ **81.** $x - 25$ **83.** $x + 2t$ **85.** $2(3t)$

87. $8(y - 3)$ **89.** $\frac{5}{t+2}$ **91.** $0.12t$ **93.** $\frac{1}{3}(m - 10)$

95. a. $J = C - 1$ **b.** $C = J + 1$ **97. a.** $T = 4N$ **b.** $N = \frac{1}{4}T$

99. $D = P - 0.25P$ or $D = 0.75P$ **101.** $v = \sqrt{2gh}$ **103.** $\frac{5\pi}{6}$

105. $\frac{\pi}{2}$ **107.** 0, for example: $3 + 0 = 3$ **109.** Reciprocal

111. No, for example: $6 - 5 \neq 5 - 6; 1 \neq -1$

113. a. $C = 0.12k + 14.89$ **b.** $\$158.89$

115. a. $C = 640m + 200n + 500$ **b.** $\$8780$

117. a. $C = 159n + 0.11(159n)$ or $C = 176.49n$ **b.** $\$705.96$

Section 1.2 Practice Exercises, pp. 55–57

R.1. x **R.3.** t **R.5.** $19z - 43$ **R.7.** $20x - 100$
R.9. $13x - 10$ **R.11.** 45

1. equation   **3.** linear   **5.** solution; set   **7.** conditional
9. empty set; $\{ \}$ or $\varnothing$ **11.** Linear   **13.** Nonlinear   **15.** Linear

17. b   **19.** $\{12\}$ **21.** $\{-2\}$ **23.** $\left\{\frac{21}{20}\right\}$ **25.** $\{-40\}$

27. $\{-10.9\}$ **29.** $\{11\}$ **31.** $\{13\}$ **33.** $\{13\}$ **35.** $\left\{\frac{3}{2}\right\}$

37. $\{0\}$ **39.** $\{3\}$ **41.** $\left\{\frac{7}{4}\right\}$ **43.** $\{0\}$ **45.** $\left\{\frac{4}{3}\right\}$

47. $\{-4\}$ **49.** $\{3\}$ **51.** $\{-6\}$ **53.** $\{1\}$ **55.** $\{2\}$

57. It is an equation that is true for some values of the variable but false for other values.

59. Identity; $\{x \mid x \text{ is a real number}\}$ **61.** Conditional equation; $\{0\}$

63. Contradiction; $\{ \}$ **65.** $P = \frac{I}{rt}$ **67.** $l = \frac{A}{w}$

69. $c = P - a - b$ **71.** $s_1 = s_2 - \Delta s$ **73.** $y = -\frac{7}{2}x + 4$

75. $y = \frac{5}{4}x - \frac{1}{2}$ **77.** $y = -\frac{3}{2}x + 3$

79. $d = \frac{2S}{n} - a$ or $d = \frac{2S - an}{n}$

81. $h = \frac{3V}{\pi r^2}$ **83.** $2x + (-6) = -24$; the number is -9.

85. $\frac{x}{12} = \frac{1}{3}$; the number is 4.   **87.** $5(x + 3) = x + 7$; the number is -2.

89. 7.5% **91.** 33 mi   **93. a.** $C = 7x$ **b.** The motorist will save money beginning on the 16th working day.

95. a. $S_1 = 45{,}000 + 2250x$ **b.** $S_2 = 48{,}000 + 2000x$ **c.** 12 years

97. The equation is an identity. The solution set is $\mathbb{R}$.

99. $a = 6$ **101.** $a = 3$

Problem Recognition Exercises, p. 58

1. Expression; $-4x + 4$ **2.** Expression; $-7y + 5$

3. Equation; $\{1\}$ **4.** Equation; $\{0\}$ **5.** Expression; $-10a - 39$

6. Expression; $28x - 10$ **7.** Equation; $\left\{\frac{1}{2}\right\}$ **8.** Equation; $\left\{\frac{3}{4}\right\}$

9. Equation; $\left\{-\dfrac{19}{6}\right\}$ **10.** Equation; $\left\{\dfrac{25}{22}\right\}$

11. Expression; $-\dfrac{1}{6}v - \dfrac{1}{10}$ **12.** Expression; $-\dfrac{17}{8}t + \dfrac{1}{2}u$

13. Equation; { }   **14.** Equation; { }   **15.** Equation; $\left\{\dfrac{39}{8}\right\}$

16. Equation; $\left\{\dfrac{2}{25}\right\}$ **17.** Expression; $0.17c + 4.495$

18. Expression; $-1.006k - 0.78$ **19.** Equation; $\{p \mid p \text{ is a real number}\}$
20. Equation; $\{u \mid u \text{ is a real number}\}$

Section 1.3 Practice Exercises, pp. 65–69

R.1. $[-7 + (-2)] + 5$; -4 **R.3.** $-1 - (-13)$; 12
R.5. $5(-9 - 2)$; -55 **R.7. a.** 19.2 **b.** $0.32x$ **R.9.** $2(x + 10)$
R.11. $t - 13$

1. $437.50   **3.** $10,400

5. 1.8 gal of bleach and 10.2 gal of water   **7.** $0.33x$ **9.** $\dfrac{d}{t}; \dfrac{d}{r}$

11. The computer costs $489 and the printer costs $249.
13. 13 ft, 15 ft, and 22 ft   **15.** Jenna prepared 280 and Sam prepared 230.
17. The width is 9 ft and the length is 11 ft.   **19.** The sides are 4.5 ft.
21. The angles are 30°, 30°, and 120°.
23. The angles are 15° and 75°.   **25.** She would pay $5100 for 4 years at 8.5% and $5812.50 for 5 years at 7.75%; the 8.5% option for 4 years requires less interest.
27. She must sell $60,000.   **29.** The total for merchandise was $1197.02 and the sales tax was $96.36.
31. The price before markup was $35.90.
33. He invested $8500 in the 2% account and $4000 in the 5% account.
35. $12,000 was borrowed at 6% and $6000 was borrowed at 11%.
37. She invested $12,000 in the 4% account and $8000 in the 3% account.
39. Fernando invested $4500 in the 3-year CD and $2500 in the 18-month CD.
41. 8 oz should be used.   **43.** 2 L should be used.
45. 12.5 L of 18% solution should be added to 7.5 L of 10% solution.
47. 1250 gal of E5   **49.** 96 ft³ of sand
51. The plane flies 300 mph from Atlanta to Fort Lauderdale and 240 mph on the return trip.
53. The speeds are 46 mph and 50 mph.
55. The plane to Los Angeles travels 400 mph and the plane to New York City travels 460 mph.
57. The distance is 24 mi.   **59.** 300 km   **61.** $336
63. a. $C = 110 + 60x$ **b.** 4 hr   **65.** 10 L should be drained and replaced by water.   **67.** The easement is 8 ft.
69. a. The kitchen is 14 ft by 10 ft.   **b.** 154 ft²   **c.** $1958.88
71. Aliyah invested $2760 in the stock returning 11% and $3000 in the stock returning 5%.
73. The original number is 68.   **75.** $x_2 = 1.8$ m   **77.** 4 kg

Section 1.4 Practice Exercises, pp. 77–81

R.1. 18, 1   **R.3.** $-2, \frac{3}{5}, 18, 1, 3.\overline{5}, 9.6$ **R.5.** True   **R.7.** False
R.9. False   **R.11.** False   **R.13.** True

1. c is greater than or equal to d **3.** $\{x \mid x > 5\}$; interval
5. parenthesis   **7.** intersection; $A \cap B$ **9.** $(-\infty, \infty)$
11. $a \geq 5$ **13.** $3c \leq 9$ **15.** $m + 4 > 70$ **17.** $a \geq 18$
19. $c \leq 25$ **21.** $s \geq 261$ **23.** $r \leq 4.5$ **25.** $18 \leq a \leq 25$
27. $<$ **29.** $>$ **31.** $>$ **33.** $<$ **35.** True   **37.** True
39. $(-7, \infty)$; $\{x \mid x > -7\}$ **41.** $(-\infty, 4.1]$; $\{x \mid x \leq 4.1\}$
43. $[-6, 0)$; $\{x \mid -6 \leq x < 0\}$ **45.** ⟵———————┤———⟶; $(-\infty, 6]$ 6

47. ⟵—(———————┤—⟶; $\left(-\dfrac{7}{6}, \dfrac{1}{3}\right]$ $-\frac{7}{6}$ $\frac{1}{3}$

49. ———(———————⟶; $(4, \infty)$ 4

51. ———(———————┤—⟶; $\{x \mid -3 < x \leq 7\}$ -3 7

53. ⟵———————┤—⟶; $\{x \mid x \leq 6.7\}$ 6.7

55. ———[———————⟶; $\{x \mid x \geq -\frac{3}{5}\}$ $-\frac{3}{5}$

57. ⟵———————⟶; $(-\infty, -3)$ -3

59. ———————(———⟶; $\left(\frac{5}{2}, \infty\right)$ $\frac{5}{2}$

61. ———[———————⟶; $[2, \infty)$ 2

63. ———(———————)———⟶; $(-4, 4)$ -4 4

65. ———[———————]———⟶; $[-3, 0]$ -3 0

67. a. $\{0, 3, 4, 6, 8, 9, 12\}$ **b.** $\{0, 12\}$ **c.** $\{-2, 0, 4, 8, 12\}$
   d. $\{4, 8\}$ **e.** $\{-2, 0, 3, 4, 6, 8, 9, 12\}$ **f.** { }
69. a. $\{0, 1, 2, 3, 4, 5, 8, 12\}$ **b.** $\{8\}$ **c.** { }
   d. $\{3\}$ **e.** $\{-2, 4, 6, 7, 8\}$ **f.** $\{0, 4, 6, 7, 8, 12\}$
71. a. $\mathbb{R}$ **b.** $\{x \mid -1 \leq x < 9\}$ **c.** $\{x \mid x < 9\}$
   d. $\{x \mid x < -8\}$ **e.** $\{x \mid x < -8 \text{ or } x \geq -1\}$ **f.** { }
73. $[-7, -4)$ **75.** $(-\infty, -4) \cup (2, \infty)$ **77.** { }   **79.** $[-7, \infty)$
81. $[0, 5)$ **83.** $[-7, \infty)$ **85. a.** $(-\infty, 4)$ **b.** $(-2, 1]$
87. a. $(-\infty, \infty)$ **b.** $[3, 5)$ **89. a.** $[-1, 5)$ **b.** $(-2, \infty)$
91. a. $(-1, 3)$ **b.** $\left(-\dfrac{5}{2}, \dfrac{9}{2}\right)$ **93. a.** $(0, 2]$ **b.** $(-4, 5]$
95. All real numbers less than -4
97. All real numbers greater than -2 and less than or equal to 7
99. All real numbers between -180 and 90, inclusive
101. All real numbers   **103.** $p < 130$ **105.** $130 \leq p \leq 139$
107. $2.2 \leq \text{pH} \leq 2.4$ acidic   **109.** $3.0 \leq \text{pH} \leq 3.5$ acidic
111. A parenthesis is used with ∞ and $-\infty$ and when an endpoint to an interval is not included in the set.
113. $[1, 2)$ **115.** $[-5, -2)$
117. Anatomy only: 23; Biology only: 15

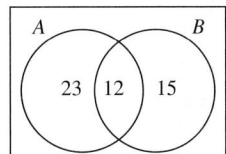

Section 1.5 Practice Exercises, pp. 88–92

R.1. $<$ **R.3.** $>$ **R.5.** $<$ **R.7.** $>$ **R.9.** $(-\infty, 5)$
R.11. $[-1, \infty)$ **R.13.** $\left[-1, \dfrac{5}{6}\right)$ **R.15.** $C \cup D = \{x \mid x < 5\}$
R.17. $D \cap F = \{x \mid 0 \leq x < 3\}$ **R.19.** $C \cup F = \mathbb{R}$

1. inequality   **3.** $a < x < b$
5. a. $\{-3\}$; n/a; ——————•——————⟶ -3
   b. $\{x \mid x > -3\}$; $(-3, \infty)$; ———(———————⟶ -3
   c. $\{x \mid x < -3\}$; $(-\infty, -3)$; ⟵———)———————⟶ -3

7. $\{y \mid y \le -1\}$; $(-\infty, -1]$;
-1

9. $\{x \mid x < -11\}$; $(-\infty, -11)$;
-11

11. $\{w \mid w \le 3\}$; $(-\infty, 3]$;
3

13. $\{a \mid a \le 8.5\}$; $(-\infty, 8.5]$;
8.5

15. $\{c \mid c < 2\}$; $(-\infty, 2)$;
2

17. $\left\{x \mid x < -\frac{13}{2}\right\}$; $\left(-\infty, -\frac{13}{2}\right)$;
$-\frac{13}{2}$

19. $\left\{x \mid x \le \frac{17}{6}\right\}$; $\left(-\infty, \frac{17}{6}\right]$;   **21.** { }
$\frac{17}{6}$

23. $\left\{x \mid x \ge -\frac{5}{6}\right\}$; $\left[-\frac{5}{6}, \infty\right)$;
$-\frac{5}{6}$

25. $\mathbb{R}$; $(-\infty, \infty)$;

27. a. $[-2, 4)$
-2 $\qquad$ 4

b. $(-\infty, \infty)$

29. a. $(-\infty, 5]$
5

b. $(-\infty, -6)$
-6

31. a. $(-\infty, 3.2]$
3.2

b. $(-\infty, 18)$
18

33. a. $(-\infty, -2] \cup \left(-\frac{1}{3}, \infty\right)$ **b.** { }
-2 $\;$ $-\frac{1}{3}$

35. $-2.8 < y$ and $y \le 15$

37. $\left[\frac{5}{2}, 7\right)$
$\frac{5}{2}$ $\qquad$ 7

39. $(-6, 6]$
-6 $\qquad$ 6

41. $[-4, 2)$
-4 $\qquad$ 2

43. $\left[\frac{6}{5}, 2\right)$
$\frac{6}{5}$ $\qquad$ 2

45. $\left[-\frac{5}{2}, \frac{13}{2}\right]$
$-\frac{5}{2}$ $\qquad$ $\frac{13}{2}$

47. $(-\infty, -0.5]$ **49.** $\left(-\infty, -\frac{5}{11}\right]$ **51.** $[8, \infty)$ **53.** { }

55. $[-6, \infty)$ **57.** $(-\infty, \infty)$ **59.** $\left(-\frac{11}{2}, \frac{7}{2}\right]$

61. $(-\infty, 1) \cup (5, \infty)$

63. a. $4800 \le x \le 10,800$ **b.** $x < 4800$ or $x > 10,800$
65. a. $44\% < x < 48\%$ **b.** $x \le 44\%$ or $x \ge 48\%$
67. $12.0 \le x \le 15.2$ g/dL   **69.** $90 \le d \le 110$ yd
71. Marilee needs to score at least 96 on the final exam.
73. Rita needs to score between 77 and 100, inclusive.
75. Boys 8 years old or older will be on average at least 51 in. tall.
77. Boys 6 years old or younger will be on average no more than 46 in. tall.
79. a. She needs to sell in excess of $375,000.
b. She needs to sell in excess of $1,375,000.
c. The base salary is still the same. The increase comes solely from commission.
81. It will take more than 1.6 hr or 1 hr 36 min.
83. The steps are the same with the following exception. If both sides of an inequality are multiplied or divided by a negative real number, then the direction of the inequality sign must be reversed.

85. The statement $-5 > y > -2$ is equivalent to $-5 > y$ and $y > -2$. However, no real number is less than -5 and also greater than -2.
87. More than 73 jackets must be sold.
89. The length must be 300 ft or less.
91. An average score in league play between 140 and 220, inclusive, would produce a handicap of 72 or less.
93. a. Donovan would need to sell more than $250,000 in merchandise.
b. Job A

Section 1.6 Practice Exercises, pp. 97–99

R.1. 8   **R.3.** $-\frac{3}{4}$ **R.5.** -2.7 **R.7.** 6

R.9. $\left\{\frac{11}{3}\right\}$ **R.11.** $\mathbb{R}$ **R.13.** { }

1. absolute; $\{k, -k\}$ **3.** $w; -w$ **5.** negative; positive
7. a. $t - 4$ **b.** $4 - t$ **9. a.** $x - 4.6$ **b.** $4.6 - x$
11. a. $x + 2$ **b.** $-x - 2$
13. a. 1   **b.** -1 **15. a.** $\pi - 3$ **b.** $\pi - 3$
17. $|1 - 6|$ or $|6 - 1|$; 5   **19.** $|3 - (-4)|$ or $|-4 - 3|$; 7
21. $|6 - 2\pi|$ or $|2\pi - 6|$; $2\pi - 6$ **23.** $|\sqrt{21} - 4|$ or $|4 - \sqrt{21}|$; $\sqrt{21} - 4$
25. a. $\{6, -6\}$ **b.** $\{0\}$ **c.** { }
27. a. $\{7, -1\}$ **b.** $\{3\}$ **c.** { }
29. $\{6, -6\}$ **31.** { }   **33.** $\{-2, 2\}$ **35.** $(\sqrt{2}, -\sqrt{2})$

37. { }   **39.** $\{0\}$ **41.** $\left\{4, -\frac{4}{3}\right\}$ **43.** $\left\{\frac{9}{2}, -\frac{1}{2}\right\}$

45. $\left\{-\frac{28}{3}, 4\right\}$ **47.** $\left\{\frac{5}{3}, 1\right\}$ **49.** $\{11, 3\}$ **51.** { }

53. $\left\{\frac{19}{3}, \frac{29}{3}\right\}$ **55.** $\left\{\frac{7}{3}\right\}$ **57.** { }   **59.** $\left\{\frac{3}{2}\right\}$

61. $\left\{-2, -\frac{3}{2}\right\}$ **63.** $\{-5, 1\}$ **65.** $\{0\}$ **67.** $\left\{\frac{3}{2}\right\}$ **69.** $\mathbb{R}$

71. $\{-1.44, -0.4\}$ **73.** { }   **75.** $\{w \mid w$ is a real number$\}$
77. { }   **79. a.** $|x - 4| = 6$ or equivalently $|4 - x| = 6$ **b.** $\{-2, 10\}$
81. a. $|y - (-8)| = 3$ or $|-8 - y| = 3$ **b.** $\{-11, -5\}$

83. $x \le \frac{1}{2}$ **85.** $t \le \frac{6}{5}$ **87.** $n \le -\frac{5}{3}$

89. After isolating the absolute value, the equation becomes $|3x + 7| = -4$. An absolute value cannot be negative.
91. 0   **93.** $2n$ **95.** $-n$

Section 1.7 Practice Exercises, pp. 105–107

R.1.
-7 $\qquad$ 7

R.3.
-7 $\qquad$ 7

R.5. $(-\infty, 1) \cup (2, \infty)$ **R.7.** $[1, 2]$ **R.9.** $\left(\frac{35}{8}, \frac{37}{8}\right)$

1. $-k; k$ **3.** $\mathbb{R}$ **5.** { }; $(-\infty, \infty)$ **7.** 2; 5; 9
9. a. $\{-5, 5\}$ **b.** $(-\infty, -5) \cup (5, \infty)$
-5 $\qquad$ 5

c. $(-5, 5)$
-5 $\qquad$ 5

11. a. $\{10, -4\}$ **b.** $(-\infty, -4) \cup (10, \infty)$
-4 $\qquad$ 10

c. $(-4, 10)$
-4 $\qquad$ 10

13. a. { }   **b.** $(-\infty, \infty)$
c. { }

15. a. { }   **b.** $(-\infty, \infty)$
c. { }

17. a. $\{-3\}$ **b.** $(-\infty, -3) \cup (-3, \infty)$
-3
c. { }

19. a. $\{-13, -5\}$

b. $(-13, -5)$

-13 -5

c. $(-\infty, -13) \cup (-5, \infty)$

-13 -5

21. $\left(-2, -\dfrac{1}{2}\right)$ **23.** $(-\infty, -8] \cup [18, \infty)$

25. $(-\infty, -2] \cup [3, \infty)$ **27.** $(-\infty, \infty)$ **29.** $\{\,\}$

31. $\left(-\infty, -\dfrac{5}{4}\right] \cup \left[\dfrac{23}{4}, \infty\right)$ **33.** $(-2, 10)$ **35.** $(-\infty, -8] \cup [2, \infty)$

37. $\{\,\}$ **39.** $\mathbb{R}; (-\infty, \infty)$ **41.** $[-10, 6]$

43. $\left(-\infty, -\dfrac{12}{5}\right) \cup \left(\dfrac{4}{5}, \infty\right)$ **45.** $(-15, 9)$ **47.** $(-\infty, -15] \cup [9, \infty)$

49. $(-10.5, 4.5)$ **51.** $|x| > 7$ **53.** $|x + 3| < 4$ **55.** $|x - c| < \delta$

57. a. $|v - 16| < 0.01$ **b.** $(15.99, 16.01)$

59. a. $|x - 4| > 1$ **b.** $(-\infty, 3) \cup (5, \infty)$

61. a. $|t - 36.5| \le 1.5$

b. $[35, 38]$; If the refrigerator is set to 36.5°F, the actual temperature would be between 35°F and 38°F, inclusive.

63. a. $|x - 0.51| \le 0.03$

b. $[0.48, 0.54]$; The candidate is expected to receive between 48% of the vote and 54% of the vote, inclusive.

65. a. $|x - 32| \le 0.05$

b. $[31.95, 32.05]$; the actual volume of juice may be between 31.95 oz and 32.05 oz, inclusive.

67. a. $\{\,\}$ **b.** $\{\,\}$ **c.** $\mathbb{R}; (-\infty, \infty)$ **69. a.** $\{\,\}$ **b.** $\{\,\}$
c. $\mathbb{R}; (-\infty, \infty)$

71. a. $\{0\}$ **b.** $\{\,\}$ **c.** $\{0\}$ **d.** $(-\infty, 0) \cup (0, \infty)$
e. $\mathbb{R}; (-\infty, \infty)$

73. a. $\{-4\}$ **b.** $\{\,\}$ **c.** $\{-4\}$ **d.** $(-\infty, -4) \cup (-4, \infty)$
e. $\mathbb{R}; (-\infty, \infty)$

75. $|x - 2| \le 5$ **77.** $|x - 7| > 3$ **79.** $\left(-\infty, \dfrac{11}{2}\right)$

81. $(-9, -1) \cup (1, 9)$ **83.** $[-4, -3] \cup [2, 3]$

85. $\hat{p} - z\sqrt{\dfrac{\hat{p}\hat{q}}{n}} < p < \hat{p} + z\sqrt{\dfrac{\hat{p}\hat{q}}{n}}$

Problem Recognition Exercises, p. 108

1. a. $\{9\}$ **b.** $\{9, -3\}$ **c.** $(-3, 9)$ **d.** $(-\infty, -3] \cup [9, \infty)$

2. a. $\left\{-\dfrac{22}{5}\right\}$ **b.** $\{\,\}$ **c.** $\{\,\}$ **d.** $(-\infty, \infty)$

3. a. $\{-7\}$ **b.** $(-\infty, -7)$ **c.** $[-7, \infty)$

4. a. $\{29\}$ **b.** $[29, \infty)$ **c.** $(-\infty, 29)$

5. a. $\left\{\dfrac{1}{2}, -\dfrac{1}{6}\right\}$ **b.** $\left\{\dfrac{1}{2}\right\}$ **6. a.** $(-7, 6]$ **b.** $(-7, 6]$

7. a. $(-5, \infty)$ **b.** $[1, \infty)$ **8. a.** $(-\infty, -7) \cup (2, \infty)$ **b.** $\{\,\}$

9. a. Linear equation   **b.** $\{-6\}$ **10. a.** Linear equation   **b.** $\{9\}$

11. a. Absolute value inequality   **b.** $[-6, -2]$

12. a. Absolute value inequality   **b.** $\{\,\}$

13. a. Compound inequality   **b.** $(-6, 9)$

14. a. Compound inequality   **b.** $(-\infty, 2) \cup [7, \infty)$

15. a. Absolute value equation   **b.** $\{4, -16\}$

16. a. Absolute value equation   **b.** $\{1, -6\}$

17. a. Linear inequality   **b.** $(-\infty, 12]$

18. a. Linear inequality   **b.** $[-1, \infty)$

19. a. Absolute value inequality   **b.** $(-\infty, -3] \cup [12, \infty)$

20. a. Absolute value inequality   **b.** $(-5, 25)$

21. a. Absolute value equation   **b.** $\{\,\}$

22. a. Absolute value equation   **b.** $\left\{-\dfrac{1}{5}\right\}$

23. a. Linear equation   **b.** $\{-16\}$

24. a. Linear equation   **b.** $\{-11\}$

25. a. Compound inequality   **b.** $(8, 12]$

26. a. Compound inequality   **b.** $[-6, -3)$

27. a. Linear equation   **b.** $(-\infty, \infty)$

28. a. Linear equation   **b.** $\{\,\}$

Chapter 1 Review Exercises, pp. 114–117

1. d   **3.** c   **5.** e   **7.** a   **9.** b   **11.** $\dfrac{1}{2}x + 4y - \dfrac{5}{2}$

13. $-13a + b + 5c$ **15.** $9p + 3$ **17.** $9x - 1$

19. a. $J = E + 150$ **b.** $E = J - 150$

21. a. $C = 3.6s + 50p + 250n$ **b.** \$8260

23. $\left\{-\dfrac{40}{3}\right\}$ **25.** $\{\,\}$ **27.** $y = \dfrac{4}{3}x - 2$

29. $x = \dfrac{c - 6y}{4}$ **31. a.** \$23,856   **b.** \$61,344

33. There were 16,600 deaths due to alcohol-related accidents.

35. Shawna invested \$8500 in the international fund and \$3500 in the real estate fund.

37. 700 cc   **39.** 15 mi   **41.** $6y \le 8$

43. a. $\{10, 11, 12, 13, 14, 16\}$ **b.** $\{10, 12\}$
c. $\{7, 8, 9, 10, 11, 12, 13\}$ **d.** $\{10, 11\}$
e. $\{7, 8, 9, 10, 11, 12, 14, 16\}$ **f.** $\{10\}$

45. $[-10, 1)$ **47.** $(-\infty, \infty)$ **49.** $(-\infty, -3] \cup (-1, \infty)$

51. $-\dfrac{4}{3}$ **a.** $\left\{x | x < -\dfrac{4}{3}\right\}$ **b.** $\left(-\infty, -\dfrac{4}{3}\right)$

53. $-\dfrac{8}{13}$ **a.** $\left\{x | x < -\dfrac{8}{13}\right\}$ **b.** $\left(-\infty, -\dfrac{8}{13}\right)$

55. $-\dfrac{67}{4}$ **a.** $\left\{x | x \le -\dfrac{67}{4}\right\}$ **b.** $\left(-\infty, -\dfrac{67}{4}\right]$

57. Dave must earn at least 95 on his fifth test.   **59.** $[-5, 2)$

61. $\{\,\}$ **63.** $(-\infty, -2) \cup [2, \infty)$ **65.** $(-5, \infty)$ **67.** $(3, 11)$

69. $[-7, -2)$ **71. a.** $125 \le x \le 200$ **b.** $x < 125$ or $x > 200$

73. More than 8.36 in. is needed.   **75. a.** $4 - w$ **b.** $w - 4$

77. a. $5x + 8$ **b.** $-5x - 8$

79. $|-12 - (-8)|$ or $|-8 - (-12)|$; 4

81. $\left|\sqrt{19} - 5\right|$ or $\left|5 - \sqrt{19}\right|$; $5 - \sqrt{19}$

83. $\{17, -17\}$ **85.** $\{-0.44, 2.54\}$ **87.** $\{3, 1\}$ **89.** $\{\,\}$

91. $\left\{-\dfrac{5}{4}\right\}$ **93.** $\left\{-\dfrac{1}{2}\right\}$ **95.** $\{\,\}$

97. a. $\{\,\}$ **b.** $(-\infty, \infty)$ **c.** $\{\,\}$

99. a. $\{-3, -1\}$ **b.** $\{-2\}$ **c.** $(-\infty, \infty)$

101. $\left(-\infty, -\dfrac{1}{5}\right) \cup \left(-\dfrac{1}{5}, \infty\right)$ **103.** $\left[0, \dfrac{6}{5}\right]$

105. $(-12, 0)$ **107.** $(-\infty, \infty)$ **109.** $(-\infty, \infty)$ **111.** $\{\,\}$

113. $3\dfrac{1}{8} \le L \le 3\dfrac{5}{8}$ or, equivalently in interval notation, $\left[3\dfrac{1}{8}, 3\dfrac{5}{8}\right]$. This means that the actual length of the screw may be between $3\dfrac{1}{8}$ in. and $3\dfrac{5}{8}$ in., inclusive.

Chapter 1 Test, pp. 117–118

1. $-2b - 6$ **2.** $2x - 1$ **3.** $-2x + 1$ **4.** $\{133\}$

5. $\left\{\dfrac{4}{5}\right\}$ **6.** $\{142,500\}$ **7.** $\left\{\dfrac{14}{9}\right\}$ **8.** $\{10, -22\}$

9. $\{-8, 2\}$ **10.** $\{3, 0\}$ **11.** $\{\,\}$ **12.** $\{-1\}$

13. a. Identity   **b.** Contradiction   **c.** Conditional equation

14. a. It took 0.4 hr (24 min).   **b.** The distance is 1.8 mi.

15. She invested \$1000 in the 5% account.

16. 24 gal of 20% solution must be used.

17. $y = -2x + 3$ **18.** $z = \dfrac{x - \mu}{\sigma}$ **19. a.** $M = 2J$ **b.** $J = \dfrac{1}{2}M$

20. a. $\mathbb{R}$ **b.** $\{x | 0 \le x < 2\}$ **c.** $\{x | x < 2\}$ **d.** $\{x | x < -1\}$
e. $\{x | x < -1$ or $x \ge 0\}$ **f.** $\{\,\}$

21. $t - 5$　　**22. a.** $|3\pi - 9|$ or $|9 - 3\pi|$　　**b.** $3\pi - 9$

23. $(34, \infty)$

24. $\left(-\infty, \frac{18}{5}\right]$

25. $(8, 24]$

26. $\left[-\frac{1}{3}, 2\right]$　　**27.** $(-2, 13]$　　**28.** $(-\infty, -24] \cup [-15, \infty)$

29. $[-3, 0)$　　**30.** $(-\infty, \infty)$　　**31.** $\{\ \}$　　**32.** $\{\ \}$

33. $\left(-\infty, -\frac{1}{3}\right] \cup \left[\frac{17}{3}, \infty\right)$　　**34.** $(-18.75, 17.25)$

35. $(-\infty, \infty)$　　**36.** $[-3, 8]$　　**37. a.** $\{\ \}$　　**b.** $\{\ \}$　　**c.** $\mathbb{R}$

38. a. $\{13\}$　　**b.** $\{\ \}$　　**c.** $\{13\}$　　**d.** $(-\infty, 13) \cup (13, \infty)$

e. $(-\infty, \infty)$

39. It can carry at most seven more passengers.

40. a. $9 \le x \le 33$　　**b.** $x < 9$　or　$x > 33$

41. a. $|x - 15.4| \le 0.1$　　**b.** $[15.3, 15.5]$; the actual mass is somewhere between 15.3 g and 15.5 g, inclusive.

CHAPTER 2

Section 2.1 Practice Exercises, pp. 127–130

R.1. $2x^3y^2$　　**R.3.** $\dfrac{6m^5}{n}$　　**R.5.** -64　　**R.7.** $\dfrac{16}{81}$

1. 1　　**3.** scientific　　**5.** $m - n$　　**7. a.** 49　　**b.** -49

c. $-\dfrac{1}{49}$　　**d.** $\dfrac{1}{49}$　　**e.** -343　　**f.** -343

9. a. 1　　**b.** $-\dfrac{1}{3}$　　**c.** $-\dfrac{2}{3}$　　**d.** 1

11. a. $\dfrac{1}{64}$　　**b.** $\dfrac{8}{x^2}$　　**c.** $\dfrac{1}{64x^2}$　　**d.** $-\dfrac{1}{64}$

13. a. q^2　　**b.** $\dfrac{1}{q^2}$　　**c.** $\dfrac{5p^3}{q^2}$　　**d.** $\dfrac{5q^2}{p^3}$　　**15.** y^8

17. 13^2 or 169　　**19.** y^8　　**21.** 3^4x^8 or $81x^8$　　**23.** $\dfrac{1}{p^3}$

25. $\dfrac{1}{7^3}$ or $\dfrac{1}{343}$　　**27.** $\dfrac{1}{w^2}$　　**29.** $\dfrac{1}{a^7}$　　**31.** r^2　　**33.** $\dfrac{1}{z^4}$

35. a^3b^2　　**37.** 1　　**39.** $\dfrac{65}{4}$　　**41.** $\dfrac{26}{25}$　　**43.** 3　　**45.** $\dfrac{5}{2}$

47. $\dfrac{q^2}{p^3}$　　**49.** $-\dfrac{3b^7}{2a^3}$　　**51.** $\dfrac{x^{16}}{81y^{20}z^8}$　　**53.** $-\dfrac{4m^4}{n^2}$　　**55.** $\dfrac{4q^{11}}{p^4}$

57. $\dfrac{5x^8}{y^2}$　　**59.** $\dfrac{16a^2}{b^6}$　　**61.** $-\dfrac{27y^{27}}{8x^{24}}$　　**63.** $\dfrac{4x^{18}}{y^{10}}$　　**65.** $27x^3y^9$

67. a. 3.5×10^5　　**b.** 3.5×10^{-5}　　**c.** 3.5×10^0

69. a. 8.6×10^{-1}　　**b.** 8.6×10^0　　**c.** 8.6×10^1

71. 2.998×10^{10} cm/sec　　**73.** 1.0×10^{-5} cm　　**75.** 4.2×10^0 L

77. a. 0.00000261　　**b.** 2,610,000　　**c.** 2.61

79. a. 0.6718　　**b.** 6.718　　**c.** 67.18

81. 1,670,000,000,000,000,000,000 molecules　　**83.** 0.000007 m

85. 3.5×10^5　　**87.** Proper　　**89.** Proper　　**91.** 8×10^5

93. 4×10^{-4}　　**95.** 1.86×10^{16}　　**97.** 7.2×10^{-20}

99. 1.24×10^{11}　　**101. a.** 3.1536×10^7 sec　　**b.** 3.1536×10^{11} gal

103. 2×10^4 songs　　**105.** 2.5×10^{13} red blood cells

107. There are 2×10^4 or 20,000 people per square mile.

109. a. $<$　　**b.** $>$　　**111. a.** $>$　　**b.** $=$　　**113.** $(3x + 5)^{12}$

115. $(6v - 7)^{90}$　　**117.** $\dfrac{31}{4}$　　**119.** x^{m+4}　　**121.** x^{2m+7}

123. x^{m-8}　　**125.** x^{m+2}　　**127.** x^{12mn}　　**129.** $x^{3m+4}y^{2n+5}$

131. In the expression $6x^0$, the exponent 0 applies to x only. In the expression $(6x)^0$, the exponent 0 applies to a base of $(6x)$. The first expression simplifies to 6, and the second expression simplifies to 1.

133. Yes; $(-4)^2 = 16 > 0$　　**135.** No　　**137.** 1.55×10^{18} N

139. a. 540 months　　**b.** \$10,800　　**c.** \$55,395.45

Section 2.2 Practice Exercises, pp. 139–143

R.1. $-10y^2 + 6x - 2$　　**R.3.** $3m - 4n + 5r$　　**R.5.** $5x^2 + x$

R.7. $-3.8m^4 + 0.9m^2$　　**R.9.** $19x + 22y - 2$　　**R.11.** $4x - \dfrac{5}{4}$

1. polynomial　　**3.** 1; 1　　**5.** binomial

7. leading; leading; coefficient　　**9.** zero

11. distributive　　**13.** squares; $a^2 - b^2$

15. a. Yes　　**b.** Yes　　**c.** No　　**d.** No

17. $-18x^7 + 7.2x^3 - 4.1$; Leading coefficient -18; Degree 7

19. $-y^2 + \dfrac{1}{3}y$; Leading coefficient -1; Degree 2　　**21.** 11

23. For example: $3x^5$　　**25.** For example: $x^2 + 2x + 1$

27. For example: $6x^4 - x^2$　　**29.** $m^2 + 10m$

31. $3x^4 + 2x^3 - 8x^2 + 2x$　　**33.** $2w^3 + \dfrac{1}{9}w^2 + 0.9w$

35. $9x^3 - 5x^2 + 2x - 8$　　**37.** $30y^3$　　**39.** $-4p^3 - 2p + 12$

41. $11ab^2 - a^2b$　　**43.** $-2x^3 + 4x^2 + 5$　　**45.** $-2xy^3 + 3x^2y + xy + 5$

47. $t^3 - 13t^2 - 9t - 13$　　**49.** $\dfrac{1}{2}a^2 - \dfrac{9}{10}ab + \dfrac{3}{5}b^2 + 8$

51. $-x^2 + 6x - 16$　　**53.** $-42x^5y^6$　　**55.** $2m^5n^5 - 6m^4n^4 + 8m^3n^3$

57. $3x^2y^2 - 4x^2y^3$　　**59.** $12x^2 + 28x - 5$　　**61.** $2y^4 - 21y^2 - 36$

63. $25s^2 + 5st - 6t^2$　　**65.** $5n^3 + 3n^2 + 50n + 30$　　**67.** $x^3 - 343$

69. $6x^3 + 7x^2y + 4xy^2 + y^3$　　**71.** $4a^4 - 17a^3b + 8a^2b^2 - 5ab^3 + b^4$

73. $\dfrac{1}{2}a^2 + ab + \dfrac{1}{2}ac - 12b^2 + 8bc - c^2$　　**75.** $a^2 - 64$

77. $9p^2 - 1$　　**79.** $9h^2 - k^2$　　**81.** $t^2 - 14t + 49$

83. $9h^2 - 6hk + k^2$　　**85.** $4z^4 - w^6$　　**87.** $25x^4 - 30x^2y + 9y^2$

89. $\dfrac{1}{25}c^2 - \dfrac{4}{9}d^6$　　**91.** $16t^4 + 24t^2p^3 + 9p^6$

93. a. $A^2 - B^2$　　**b.** $x^2 + 2xy + y^2 - B^2$; Both are examples of multiplying conjugates to get a difference of squares.

95. $w^2 + 2wv + v^2 - 4$　　**97.** $4 - x^2 - 2xy - y^2$

99. $9a^2 - 24a + 16 - b^2$

101 Write $(x + y)^3$ as $(x + y)^2(x + y)$. Square the binomial and then use the distributive property to multiply the resulting trinomial by the remaining factor of $(x + y)$.

103. $8x^3 + 12x^2y + 6xy^2 + y^3$　　**105.** $64a^3 - 48a^2b + 12ab^2 - b^3$

107. Multiply and simplify the first two binomials. Then multiply the resulting trinomial by the third binomial, using the distributive property.

109. $6a^4 + 32a^3 + 10a^2$　　**111.** $x^3 + 5x^2 - 9x - 45$

113. $-3y^2 - 10y - 8$　　**115.** $x^{2n} - 4x^n - 21$　　**117.** $z^{2n} + 2w^mz^n + w^{2m}$

119. $a^{2n} - 25$　　**121.** $-60x - 50$　　**123.** $10x + 22$

125. $25y^2 - 4x^2 - 12x - 9$　　**127.** $x^3 + 14x^2 + 64x + 96$

129. $ac + bc$　　**131.** $x^2 + 6x - 27$

133. a. $x + 1$　　**b.** $x + (x + 1)$; $2x + 1$　　**c.** $x(x + 1)$; $x^2 + x$

d. $x^2 + (x + 1)^2$; $2x^2 + 2x + 1$

135. $x^4 + 8x^3 + 24x^2 + 32x + 16$　　**137.** $(5x - 6)$　　**139.** $(2y - 1)$

141. False　　**143.** True

Problem Recognition Exercises, p. 143

1. a. $9x^2 + 6x + 1$　　**b.** $9x^2 - 1$　　**c.** 2

2. a. -10　　**b.** $81m^2 - 25$　　**c.** $81m^2 - 90m + 25$

3. a. $16x^4y^2$　　**b.** $x^4 + 2x^2y + 8x^2 + y^2 + 8y + 16$

4. a. a^4b^6　　**b.** $a^4 - 2a^2b^3 + b^6$

5. a. $-12c^5d^5$　　**b.** $c^5 - c^3d^5 + 4c^3 - 3c^2 + 3d^5 - 12$

6. a. x^{12}　　**b.** x^{48}　　**7. a.** -36　　**b.** 36　　**c.** $\dfrac{1}{36}$

8. a. 10　　**b.** 1　　**c.** -10　　**9. a.** -30　　**b.** $-10p - 50$　　**c.** 0

10. a. $-8x - 32$ **b.** -20 **c.** 0 **11.** $2t^2 + t - 1$
12. $-15x^4 - 5x^3 + 10x^2$ **13.** $36z^2 - 25$
14. $9y^3 + 2y^2 - 3y + 1$ **15.** $6b^2 - 11b + 4$
16. $10a^3 + 19a^2 + 11a + 2$ **17.** $t^3 - 6t^2 + 8t + 3$
18. $k^2 + 4k + 25$ **19.** $7t^3 - 12t^2 + 2t$ **20.** $\dfrac{11}{12}p^3 - \dfrac{1}{2}p^2 + \dfrac{1}{5}p + 5$
21. $-30.6w^6 + 15.6w^5 - 7.2w^4$ **22.** $36a^4 - 48a^2b + 16b^2$
23. $\dfrac{1}{4}z^4 - \dfrac{1}{9}$ **24.** $m^2 - 8m - 7$ **25.** $x^2 + 3x - 14$
26. $2m^4 - 8m^3 - 13m^2 + 46m - 21$
27. $25 - 10a - 10b + a^2 + 2ab + b^2$ **28.** $a^2 - x^2 + 2xy - y^2$
29. $4xy$ **30.** $a^3 - 12a^2 + 48a - 64$ **31.** $-\dfrac{1}{8}x^2 + \dfrac{1}{3}x - \dfrac{1}{6}$
32. $-\dfrac{1}{2}x^6y^4z^5w^3$

Section 2.3 Practice Exercises, pp. 149–151

R.1. 5 **R.3.** 4 **R.5.** x^2 **R.7.** $5c^2d$ **R.9.** $3(2m + n)$

1. product   **3.** greatest; common; factor   **5.** $3(x + 4)$
7. $2z(3z + 2)$ **9.** $4p(p^5 - 1)$ **11.** $12x^2(x^2 - 3)$ **13.** $9t(st + 3)$
15. $9a^2b^3(a^2 + 3ab - 2b^2)$ **17.** $5xy(2x + 3y - 1)$
19. $b(13b - 11a^2 - 12a)$ **21.** $(3z - 2b)(2a - 5)$
23. $(2x - 3)(2x^2 + 1)$ **25.** $(2x + 1)^2(y - 3)$
27. $3(x - 2)^2(y + 2)$ **29.** $-1(x^2 + 10x - 7)$
31. $-3xy(4x^2 + 2x + 1)$ **33.** $-t(2t^2 - 11t + 3)$
35. a. $(2x - y)(a + 3b)$ **b.** $(2w - 1)(5w - 3b)$
   c. In part (b), $-3b$ was factored out so that the signs in the last two
   terms would be changed. The resulting binomial factor matches the
   binomial factor from the first two terms.
37. $(y + 4)(y^2 + 3)$ **39.** $(p - 7)(6 + q)$ **41.** $(m + n)(2x + 3y)$
43. $(2x - 3y)(5a - 4b)$ **45.** $(x^2 - 3)(x - 1)$ **47.** $6p(p + 3)(q - 5)$
49. $100(x - 3)(x^2 + 2)$ **51.** $(3a + b)(2x - y)$ **53.** $(4 - b)(a + 3)$
55. Cannot be factored   **57.** It is not possible to get a common
binomial factor regardless of the order of terms.
59. $x = \dfrac{6}{4 + t}$ **61.** $x = \dfrac{5 - ay}{6 - b}$ or $x = \dfrac{ay - 5}{b - 6}$ **63.** $P = \dfrac{A}{1 + rt}$
65. $A = \dfrac{U}{v + cw}$ **67.** $y = \dfrac{bx}{c - a}$ or $y = -\dfrac{bx}{a - c}$
69. Length $= 2w + 1$ **71.** $\dfrac{x^2 - 7x + 2}{x^4}$ **73.** $\dfrac{5x^2 - 11x + 1}{x^5}$
75. $-\dfrac{3x^2}{(5x - 1)^4}$ **77.** $\dfrac{6m}{(m + 3)^3}$ **79.** $12x^4 + 3x^3 + 6x^2$
81. $5(3t - 4)^2 - 3(3t - 4)$
83. The GCF in both cases is a squared factor. In the first expression, the
GCF is x^2 and in the second expression, the GCF is $(t + 5)^2$.
85. $x^n(x^{2n} + 2x^n - 5)$ **87.** $\dfrac{2(7a^{2n} - 3a^n + 1)}{a^{5n}}$

Section 2.4 Practice Exercises, pp. 162–164

R.1. $35x^2 + 9x - 18$ **R.3.** $t^2 - 22t + 121$ **R.5.** $9a^2 + 30ab^2 + 25b^4$
R.7. $5ab^2(5a^2b^2 - 3ab - 4)$ **R.9.** $3(2a + 1)(a - 2)$
R.11. $(3x + 7)(2x + 3)$ **R.13.** $10m(m + 4)(m - 9)$

1. positive   **3.** Both are correct.   **5.** $(a + b)^2$; $(a - b)^2$
7. prime   **9.** $(b - 8)(b - 4)$ **11.** $(y + 12)(y - 2)$
13. $(x + 10)(x + 3)$ **15.** $(c - 8)(c + 2)$ **17.** $(2x + 3)(x - 5)$
19. $(6a - 5)(a + 1)$ **21.** $(s + 3t)(s - 2t)$ **23.** $3(x - 18)(x - 2)$
25. $2(c - 4)(c + 3)$ **27.** $2(x - y)(x + 5y)$ **29.** Prime
31. $(3x + 5y)(x + 3y)$ **33.** $5uv(u - 3v)^2$ **35.** $x(x - 7)(x + 2)$
37. $(2z - 5)(5z + 1)$ **39.** Prime   **41.** $-2(t + 4)(t - 10)$
43. $(7a - 4)(2a + 3)$ **45.** $2b(3a + 2)(a + 3)$
47. a. $x^2 + 10x + 25$ **b.** $(x + 5)^2$

49. a. $9x^2 - 12xy + 4y^2$ **b.** $(3x - 2y)^2$ **51.** $30x$ **53.** $16z^2t$
55. $(y - 4)^2$ **57.** $(8m + 5)^2$ **59.** Not a perfect square trinomial
61. $(3a - 5b)^2$ **63.** $4(4t^2 - 20tv + 5v^2)$; Not a perfect square trinomial
65. $5(b^2 - 2)^2$ **67. a.** $(u - 5)^2$ **b.** $(x^2 - 5)^2$ **c.** $(a - 4)^2$
69. a. $(u + 13)(u - 2)$ **b.** $(w^3 + 13)(w^3 - 2)$ **c.** $(y + 9)(y - 6)$
71. $(3x - 4)(3x + 1)$ **73.** $(2x - 9)(x - 1)$ **75.** $(3y + 11)(y + 6)$
77. $(3y^3 + 2)(y^3 + 3)$ **79.** $(4p^2 + 1)(p^2 + 1)$ **81.** $(x^2 + 12)(x^2 + 3)$
83. The factorization $(2y - 1)(2y - 4)$ is not factored completely because
the factor $2y - 4$ has a GCF of 2.
85. $(w^2 + 6)^2$ **87.** $(9w + 5)^2$ **89.** $3(a + b)(x - 2)$
91. $2abc^2(6a + 2b - 3c)$ **93.** $-2x(5x - 6)(2x - 5)$ **95.** Prime
97. $(2w^2 - 15)(w^2 - 2)$ **99.** $(1 - d)(1 - 3d)$ or $(3d - 1)(d - 1)$
101. $(a + 2b)(x - 5a)$ **103.** $8(z - 4w)(z + 7w)$
105. $\dfrac{2(x - 4)(x + 3)}{x^5}$ **107.** $\dfrac{(2x - 5)(x - 4)}{x^7}$ **109.** $\dfrac{(2x + 5)^2}{(2x + 3)^3}$
111. No   **113.** Yes   **115.** No

Section 2.5 Practice Exercises, pp. 171–173

R.1. $16x^2 - 121$ **R.3.** $m^6 - 36n^4$ **R.5.** $t^3 + 125$ **R.7.** $8x^6 - y^9$

1. difference; $(a + b)(a - b)$ **3.** is not   **5.** sum; cubes
7. $(a - b)(a^2 + ab + b^2)$ **9.** $4, 16, 25, 64, x^2, x^4$
11. $(x - 3)(x + 3)$ **13.** $(4 - 7w)(4 + 7w)$ **15.** $2(2a - 9b)(2a + 9b)$
17. Prime   **19.** $2(a^2 + 4)(a - 2)(a + 2)$ **21.** $(7 - k^3)(7 + k^3)$
23. $(x - 4)(x + 4)(x - 1)$ **25.** $(2x + 1)(2x - 1)(x + 3)$
27. $(9y + 7)(y - 2)(y + 2)$ **29.** $(7x + 2 - y)(7x + 2 + y)$
31. $(w - 3n + 1)(w + 3n - 1)$ **33.** $(p^2 - 5 - t^2)(p^2 - 5 + t^2)$
35. $(3u^2 - 2v^2 + 5)(3u^2 + 2v^2 - 5)$
37. Look for a binomial of the form $a^3 + b^3$. This factors as
$(a + b)(a^2 - ab + b^2)$.
39. $(2x - 1)(4x^2 + 2x + 1)$ **41.** $(5c + 3)(25c^2 - 15c + 9)$
43. $(x - 10)(x^2 + 10x + 100)$ **45.** $(4t^2 + 1)(16t^4 - 4t^2 + 1)$
47. $2(10y^2 + x)(100y^4 - 10y^2x + x^2)$ **49.** $2z(2z - 3)(4z^2 + 6z + 9)$
51. $(p^4 - 5)(p^8 + 5p^4 + 25)$ **53.** $\left(6y - \dfrac{1}{5}\right)\left(6y + \dfrac{1}{5}\right)$
55. $2(3d^6 - 4)(3d^6 + 4)$ **57.** $2(121v^2 + 16)$ **59.** $4(x - 2)(x + 2)$
61. $(5 - 7q)(5 + 7q)$ **63.** $(t + 2s - 6)(t + 2s + 6)$
65. $(3 - t)(9 + 3t + t^2)$ **67.** $\left(3a + \dfrac{1}{2}\right)\left(9a^2 - \dfrac{3}{2}a + \dfrac{1}{4}\right)$
69. $2(m + 2)(m^2 - 2m + 4)$
71. $(x - y)(x + y)(x^2 + y^2)$ **73.** $(a + b)(a^2 - ab + b^2)(a^6 - a^3b^3 + b^6)$
75. $\left(\dfrac{1}{2}p - \dfrac{1}{5}\right)\left(\dfrac{1}{4}p^2 + \dfrac{1}{10}p + \dfrac{1}{25}\right)$ **77.** Prime
79. $\left(\dfrac{1}{5}x - \dfrac{1}{2}y\right)\left(\dfrac{1}{5}x + \dfrac{1}{2}y\right)$
81. $(a + b)(a^2 - ab + b^2)(a - b)(a^2 + ab + b^2)$
83. $(2 + y)(4 - 2y + y^2)(2 - y)(4 + 2y + y^2)$
85. $(h^2 + k^2)(h^4 - h^2k^2 + k^4)$ **87.** $(2x^2 + 5)(4x^4 - 10x^2 + 25)$
89. $4x^2 - 9$ **91.** $8a^3 - 27$ **93.** $64x^6 + y^3$
95. a. $x^2 - y^2$ **b.** $(x + y)(x - y)$ **c.** 20 in.^{2}
97. $\dfrac{4}{3}\pi(R - r)(R^2 + Rr + r^2)$ **99.** $(21 + 19)(21 - 19)$; 80
101. $(x + y)(x - y + 1)$ **103.** $(x + y)(x^2 - xy + y^2 + 1)$
105. $(3a - c)(3a + c)(4a - 1)(16a^2 + 4a + 1)$
107. a. $(x - 1)(x^4 + x^3 + x^2 + x + 1)$ **b.** $(x - 1)(x^{n-1} + x^{n-2} + \cdots + 1)$

Problem Recognition Exercises, pp. 174–178

1. A polynomial whose only factors are 1 and itself
2. Factor out the GCF.
3. Difference of squares $a^2 - b^2$, difference of cubes $a^3 - b^3$, or sum of
cubes $a^3 + b^3$

4. Look for a perfect square trinomial, $a^2 + 2ab + b^2$ or $a^2 - 2ab + b^2$.

5. Try factoring by grouping 2 terms by 2 terms or by grouping 3 terms by 1 term.

6. Let $u = (4x^2 + 1)$. The polynomial becomes $3u^2 + 20u + 12$. Factor this simpler expression and then back substitute.

7. a. Trinomial   **b.** $3(2x + 3)(x - 5)$

8. a. Trinomial   **b.** $m(4m + 1)(2m - 3)$

9. a. Difference of squares   **b.** $2(2a - 5)(2a + 5)$

10. a. Grouping   **b.** $(b + y)(a - b)$

11. a. Trinomial   **b.** $(2u - v)(7u - 2v)$

12. a. Perfect square trinomial   **b.** $(3p - 2q)^2$

13. a. Difference of cubes   **b.** $2(2x - 1)(4x^2 + 2x + 1)$

14. a. Sum of squares   **b.** Prime

15. a. Sum of cubes   **b.** $(3y + 5)(9y^2 - 15y + 25)$

16. a. None of these   **b.** Prime

17. a. Sum of cubes   **b.** $2(4p^2 + 3q)(16p^4 - 12p^2q + 9q^2)$

18. a. Perfect square trinomial   **b.** $5(b - 3)^2$

19. a. Difference of squares   **b.** $(2a - 1)(2a + 1)(4a^2 + 1)$

20. a. Perfect square trinomial   **b.** $(9u - 5v)^2$

21. a. None of these   **b.** $-10x(10x + 1)$

22. a. Sum of squares   **b.** $4(x^2 + 4)$

23. a. Grouping   **b.** $2(2x - y)(3a + b)$

24. a. Difference of cubes   **b.** $(5y - 2)(25y^2 + 10y + 4)$

25. a. Trinomial   **b.** $(5y - 1)(y + 3)$

26. a. Difference of squares   **b.** $2(m^2 - 8)(m^2 + 8)$

27. a. Difference of squares   **b.** $(t - 10)(t + 10)$

28. a. Difference of squares   **b.** $(2m - 7n)(2m + 7n)$

29. a. Sum of cubes   **b.** $(y + 3)(y^2 - 3y + 9)$

30. a. Sum of cubes   **b.** $(x + 1)(x^2 - x + 1)$

31. a. Trinomial   **b.** $(d - 4)(d + 7)$

32. a. Trinomial   **b.** $(c + 8)(c - 3)$

33. a. Perfect square trinomial   **b.** $(x - 6)^2$

34. a. Perfect square trinomial   **b.** $(p + 8)^2$

35. a. Grouping   **b.** $(ax + b)(2x - 5)$

36. a. Grouping   **b.** $(4x + a)(2x - b)$

37. a. Trinomial   **b.** $(2y - 1)(5y + 4)$

38. a. Trinomial   **b.** $(3z + 2)(4z + 1)$

39. a. Difference of squares   **b.** $10(p - 8)(p + 8)$

40. a. Difference of squares   **b.** $2(5a - 6)(5a + 6)$

41. a. Difference of cubes   **b.** $z(z - 4)(z^2 + 4z + 16)$

42. a. Difference of cubes   **b.** $t(t - 2)(t^2 + 2t + 4)$

43. a. Trinomial   **b.** $b(b + 5)(b - 9)$

44. a. Trinomial   **b.** $y(y - 4)(y - 10)$

45. a. Perfect square trinomial   **b.** $(3w + 4x)^2$

46. a. Perfect square trinomial   **b.** $(2k - 5p)^2$

47. a. Grouping   **b.** $10(2x + a)(3x - 1)$

48. a. Grouping   **b.** $10(5x + c)(x - 4)$

49. a. Difference of squares   **b.** $(w^2 + 4)(w - 2)(w + 2)$

50. a. Difference of squares   **b.** $(k^2 + 9)(k - 3)(k + 3)$

51. a. Difference of cubes   **b.** $(t^2 - 2)(t^4 + 2t^2 + 4)$

52. a. Sum of cubes   **b.** $(p^2 + 3)(p^4 - 3p^2 + 9)$

53. a. Trinomial   **b.** $(4p - 1)(2p - 5)$

54. a. Trinomial   **b.** $(3m + 4)(3m - 5)$

55. a. Perfect square trinomial   **b.** $(6y - 1)^2$

56. a. Perfect square trinomial   **b.** $(3a + 7)^2$

57. a. Sum of squares   **b.** $2(x^2 + 25)$

58. a. Sum of squares   **b.** $4(y^2 + 16)$

59. a. Trinomial   **b.** $s^2(3r - 2)(4r + 5)$

60. a. Trinomial   **b.** $w^2(7z + 4)(z - 2)$

61. a. Trinomial   **b.** $(x - 3y)(x + 11y)$

62. a. Trinomial   **b.** $(s + 3t)(s - 12t)$

63. a. Sum of cubes   **b.** $(m^2 + n)(m^4 - m^2n + n^2)$

64. a. Difference of cubes   **b.** $(a - b^2)(a^2 + ab^2 + b^4)$

65. a. None of these   **b.** $x(x - 4)$

66. a. None of these   **b.** $y(y - 9)$

67. $(x - y)(x + y)^2$ **68.** $(u - v)^2(u + v)$ **69.** $(a + 3)^4(6a + 19)$

70. $(4 - b)^3(2 - b)$ **71.** $18(3x + 5)^2(4x + 5)$

72. $5(2y + 3)^2(6y + 11)$ **73.** $\left(\frac{1}{10}x + \frac{1}{7}\right)^2$ **74.** $\left(\frac{1}{5}a + \frac{1}{6}\right)^2$

75. $5x^2(5x^2 - 6)$ **76.** $x^3(x^3 - 2)$ **77.** $(4p^2 + q^2)(2p - q)(2p + q)$

78. $(s^2t^2 + 9)(st - 3)(st + 3)$ **79.** $\left(y + \frac{1}{4}\right)\left(y^2 - \frac{1}{4}y + \frac{1}{16}\right)$

80. $\left(z + \frac{1}{5}\right)\left(z^2 - \frac{1}{5}z + \frac{1}{25}\right)$ **81.** $(a + b)(a - b)(6a + b)$

82. $(2p - q)(2p + q)(p + 3q)$ **83.** $\left(\frac{1}{3}t + \frac{1}{4}\right)^2$ **84.** $\left(\frac{1}{5}y + \frac{1}{2}\right)^2$

85. $(x + 6 - a)(x + 6 + a)$ **86.** $(a + 5 + b)(a + 5 - b)$

87. $(p + q - 9)(p + q + 9)$ **88.** $(m - n - 3)(m - n + 3)$

89. $(b - x - 2)(b + x + 2)$ **90.** $(p - y + 3)(p + y - 3)$

91. $(2 + u - v)(2 - u + v)$ **92.** $(5 - a - b)(5 + a + b)$

93. $(3a + b)(2x - y)$ **94.** $(q + 3)(5p - 4)$

95. $(u - 2)(u + 2)(u^2 + 2u + 4)(u^2 - 2u + 4)$

96. $(1 - v)(1 + v)(1 + v + v^2)(1 - v + v^2)$

97. $(x^4 + 1)(x^2 + 1)(x + 1)(x - 1)$ **98.** $(y^4 + 16)(y^2 + 4)(y + 2)(y - 2)$

99. $(a + b)(a - b + 1)$ **100.** $(5c - 3d)(5c + 3d + 1)$

101. $(x + y)(x^2 - xy + y^2)(5w - 2z)$ **102.** $(x^2 + 2)(x + 3)(x - 3)$

103. $(y^2 + 10)(y + 1)(y - 1)$ **104.** $(x^3 + 16)(x + 2)(x^2 - 2x + 4)$

105. $(y^3 + 41)(y + 3)(y^2 - 3y + 9)$

106. $(x + y + z)(x^2 + 2xy + y^2 - xz - yz + z^2)$

107. $(a + 5 - b)(a^2 + 10a + 25 + ab + 5b + b^2)$

108. $(3m + 21n + 7)^2$ **109.** $(2x + 63y - 9)^2$

110. $(3c - 8)(-c + 2)$ or $-(3c - 8)(c - 2)$

111. $3(5d + 3)(-d + 3)$ or $-3(5d + 3)(d - 3)$

112. $(p - 4)(p^2 + 4p + 16)(p^4 + 1)(p^2 + 1)(p + 1)(p - 1)$

113. $(t + 3)(t^2 - 3t + 9)(t^2 + 1)(t - 1)(t + 1)$

114. $(m + 3)(m^2 - 3m + 9)(m - 1)(m^2 + m + 1)$

115. $(n - 2)(n^2 + 2n + 4)(n + 1)(n^2 - n + 1)$

116. $2z(2x + 3)(4x^2 - 6x + 9)(x - 1)(x^2 + x + 1)$

117. $3y(2y - 1)(4y^2 + 2y + 1)(y + 1)(y^2 - y + 1)$

118. $(x - y)(x + y - 1)$ **119.** $(a + b)(a - b - 1)$

120. $(a - c)(a + 2c + 1)$ **121.** $(x - y)(x + 3y + 1)$

Chapter 2 Review Exercises, pp. 182–183

1. a. 1   **b.** -1 **c.** 9   **d.** 1   **3.** p^3 **5.** $\dfrac{144b^8}{a^6}$

7. $\dfrac{2u^{12}}{v^4}$ **9. a.** 0.98   **b.** 9.8   **c.** 98

11. 1.763×10^{12} **13.** 0.4 people per acre

15. $2.5y^{11} + 7.61y^9 - y^5$; Leading coefficient: 2.5; Degree: 11

17. $-2x^2 - 3x - xy - 1$ **19.** $3a^3 + 5a^2 + 6a$ **21.** $x^4 - x^2 - 1$

23. $6x - 6y$ **25.** $-11x + 1$ **27.** $-3x - 11$ **29.** $2x^3 - 14x^2 - 8x$

31. $x^2 - x - 42$ **33.** $\dfrac{1}{4}x^2 - 2x - 5$ **35.** $27x^3 + 125$

37. $4x^2 - 20x + 25$ **39.** $9y^2 - 121$ **41.** $\dfrac{4}{9}t^2 - 16$

43. $x^2 + 4x + 4 - b^2$ **45.** $8x^3 + 12x^2 + 6x + 1$

47. a. $4x^2 + 12x + 9$ **b.** $16x^2 + 24x + 9$ **c.** $12x^2 + 12x$

49. $7(3w^3 - w + 2)$ **51.** $(t + 4)(3t + 5)$

53. $12(2x - 3)(x^2 + 3)$ **55.** $(y - 6)(y^2 + 1)$

57. $y = \dfrac{a + d}{c - b}$ or $y = -\dfrac{a + d}{b - c}$ **59.** $\dfrac{7 + 5w + w^2}{w^8}$

61. $(3m - 5t)(m + 2t)$ **63.** $k^2(3k + 2)(2k + 1)$

65. $2(5z + 4)^2$ **67.** $(4x - 3)^2$ **69.** $(w^2 + 1)(3w^2 - 5)$

71. $11p(2p + 1)(p - 2)$ **73.** $(4t - 3)(3t + 2)$

75. $(5 - y)(5 + y)$ **77.** Prime   **79.** $(a + 4)(a^2 - 4a + 16)$

81. $y(3y - 2)(3y + 2)$ **83.** $(a + 6 - b)(a + 6 + b)$
85. $(y - 3 - 4x)(y - 3 + 4x)$ **87.** $3k(k - 3)(k^2 + 3k + 9)$
89. $(5n + m + 6)(5n - m - 6)$ **91.** $-4(3p - 2)(p + 3)$
93. $(x^2 + 3y)(x^2 + 3y - 1)$

Chapter 2 Test, pp. 183–184

1. $5a^{13}$ **2.** x^{11} **3.** $\dfrac{9x^{12}}{25y^{14}}$ **4.** $\dfrac{8y^{12}}{x^7}$

5. 4.5×10^{10}; 1.66×10^6 **6.** 0.0000008 **7.** 1.2×10^{14}

8. $8x^2 - 8x + 8$ **9.** $2a^3 - 13a^2 + 2a + 45$ **10.** $2x^2 - \dfrac{23}{3}x - 6$

11. $25x^2 - 16y^4$ **12.** $9a^2 + 6ab + b^2 - c^2$ **13.** $49x^2 - 56x + 16$

14. $x^2 + 11x - 34$ **15.** $y = \dfrac{C}{a - b}$ or $y = -\dfrac{C}{b - a}$

16. $x = \dfrac{ct + 7}{5 - d}$ or $x = -\dfrac{ct + 7}{d - 5}$ **17.** $(3y - 4)(y + 7)$ **18.** Prime

19. $3(a + 6b)(a + 3b)$ **20.** $(c - 1)(c + 1)(c^2 + 1)$
21. $(y - 7)(x + 3)$ **22.** Prime   **23.** $-10(u - 2)(u - 1)$
24. $3(2t - 5)(2t + 5)$ **25.** $5(y - 5)^2$ **26.** $7q(3q + 2)$
27. $(2x + 1)(x - 2)(x + 2)$ **28.** $(y - 5)(y^2 + 5y + 25)$
29. $(x + 4 - y)(x + 4 + y)$ **30.** $r^2(r^2 + 16)(r - 4)(r + 4)$
31. $(x^2 + 3)(x^2 + 2)$ **32.** $(2 - c)(6a + b)$
33. $(x^2 + 9)(x - 3)(x + 3)(x + 2)$ **34.** $(c - 2a - 11)(c + 2a + 11)$
35. $(3u - v^2)(9u^2 + 3uv^2 + v^4)$ **36.** $\dfrac{7w^2 + 2w + 4}{w^6}$

37. $\dfrac{(2x + 1)(7x + 3)}{x^3}$

CHAPTER 3

Section 3.1 Practice Exercises, pp. 193–195

R.1. a. 10 and -10 **b.** 10 **R.3.** 15 **R.5.** $\dfrac{1}{5}$ **R.7.** -1

R.9. a. 4 **b.** 4 **R.11. a.** x **b.** $-x$ **R.13.** x^9 **R.15.** y^4

R.17. d^{10}

1. odd; even   **3.** $\sqrt[n]{\dfrac{a}{b}}$ **5.** t^{12} **7.** $2\sqrt{3x}$ **9.** -11

11. 11 **13.** $|y|$ **15.** y **17.** $|2x - 5|$ **19.** w^6 **21.** $\dfrac{|a|}{b^2}$

23. cd^2 **25. a.** $t \ge 0$ **b.** All real numbers
27. a. $c^3\sqrt{c}$ **b.** $c^2\sqrt[3]{c}$ **c.** $c\sqrt[4]{c^3}$ **d.** $\sqrt[9]{c^7}$
29. a. $2\sqrt{6}$ **b.** $2\sqrt[3]{3}$ **31.** $x^2\sqrt{x}$ **33.** $a^2b^2\sqrt{a}$
35. $-x^2y^3\sqrt[4]{y}$ **37.** $2\sqrt{7}$ **39.** $15\sqrt{2}$ **41.** $3\sqrt[3]{2}$
43. $5b\sqrt{ab}$ **45.** $2x^2\sqrt[3]{5x}$ **47.** $-2x^2z\sqrt[3]{2y}$ **49.** $2wz\sqrt[4]{5z^3}$

51. x **53.** p^2 **55.** 5 **57.** $\dfrac{1}{2}$ **59.** $\dfrac{5\sqrt[3]{2}}{3}$ **61.** $\dfrac{5\sqrt[3]{9}}{6}$

63. a. Not like radicals   **b.** Like radicals   **c.** Not like radicals
65. a. Both expressions can be simplified by using the distributive property.
b. Neither expression can be simplified because the terms do not
contain like radicals or like terms.
67. $9\sqrt{5}$ **69.** $2\sqrt[3]{tw}$ **71.** $5\sqrt{10}$ **73.** $8\sqrt[4]{3} - \sqrt[4]{14}$
75. $2\sqrt{x} + 2\sqrt{y}$ **77.** Cannot be simplified further

79. Cannot be simplified further   **81.** $\dfrac{29}{18}z\sqrt[3]{6}$ **83.** $0.7x\sqrt{y}$

85. Simplify each radical: $3\sqrt{2} + 35\sqrt{2}$. Then add like radicals: $38\sqrt{2}$
87. $8\sqrt{3}$ **89.** $-5\sqrt{2} + 3\sqrt{3}$ **91.** $-5\sqrt{2a}$ **93.** $6x\sqrt[3]{x}$
95. $14p^2\sqrt{5}$ **97.** $-\sqrt[3]{a^2b}$ **99.** $(5x + 6)\sqrt{x}$
101. $(5x - 6)\sqrt{2}$ **103.** $33d\sqrt[3]{2c}$
105. False; $\sqrt{9} + \sqrt{16} \ne \sqrt{9 + 16}$; $7 \ne 5$
107. True   **109.** False; $\sqrt{y} + \sqrt{y} = 2\sqrt{y} \ne \sqrt{2y}$

111. False; $2w\sqrt{5} + 4w\sqrt{5} = 6w\sqrt{5} \ne 6w^2\sqrt{5}$
113. $9\sqrt{6}$ cm ≈ 22.0 cm   **115.** $x = 2\sqrt{2}$ ft
117. 2.0×10^4 **119.** $2\sqrt{2}$

Section 3.2 Practice Exercises, pp. 201–203

R.1. $2t^2 + \dfrac{5}{3}t$ **R.3.** $18p^2 - 59p + 35$ **R.5.** $-6k^3 + 19k^2 - 28k + 45$

R.7. $9d^2 - 48d + 64$ **R.9.** $25m^2 - 49n^2$ **R.11.** x^2

R.13. $6y^3$ **R.15.** $\dfrac{x + 2}{2}$ or $\dfrac{1}{2}x + 1$ **R.17.** $3 - 10\sqrt{7}$

1. x **3.** $m - n$ **5.** rationalizing   **7.** $\sqrt{x} - 2$ **9.** $\sqrt{15}$

11. $2\sqrt{5}$ **13.** $-24\sqrt{2}$ **15.** $-15a$ **17.** $5a\sqrt[3]{b^2}$
19. $-24m\sqrt[4]{3}$ **21.** $12 - 6\sqrt{3}$ **23.** $16\sqrt{5} + 32\sqrt{15}$
25. $-42a + 28\sqrt{ab}$ **27.** $-8 + 7\sqrt{30}$ **29.** $x - 5\sqrt{x} - 36$
31. $9a - 28\sqrt{ab} + 3b$ **33.** $8\sqrt{p} + 3p + 5\sqrt{pq} + 16\sqrt{q} - 2q$
35. a. $9 - x^2$ **b.** $3 - x^2$
37. a. $p^2 - 14p + 49$ **b.** $p - 2\sqrt{7p} + 7$
39. a. $25c^2 + 20cd + 4d^2$ **b.** $25c + 20\sqrt{cd} + 4d$
41. a. $49x^2 - 16y^2$ **b.** 115 **43.** $5a - 2\sqrt{15ab} + 3b$

45. $x - 24$ **47.** $x - 8$ **49.** $\dfrac{\sqrt{3}}{3}$ **51.** $\dfrac{3\sqrt{2y}}{y}$ **53.** $\dfrac{a\sqrt{2a}}{2}$

55. $\dfrac{3\sqrt{2}}{2}$ **57.** $\dfrac{3\sqrt[3]{4}}{2}$ **59.** $\dfrac{-6\sqrt[4]{x^3}}{x}$ **61.** $\dfrac{7\sqrt[3]{2}}{2}$

63. $\dfrac{\sqrt[3]{4w}}{w}$ **65.** $\dfrac{\sqrt[3]{2x}}{x}$ **67.** $\sqrt{2} + \sqrt{6}$ **69.** $\sqrt{x} - 23$

71. $\dfrac{4\sqrt{2} - 12}{-7}$ or $\dfrac{-4\sqrt{2} + 12}{7}$ **73.** $4\sqrt{6} + 8$ **75.** $-\sqrt{21} + 2\sqrt{7}$

77. $\dfrac{-\sqrt{p} + \sqrt{q}}{p - q}$ **79.** $\dfrac{w + 11\sqrt{w} + 18}{81 - w}$ **81.** True

83. False; the expression is of the form $(a - b)^2 = a^2 - 2ab + b^2$
where $a = x$ and $b = \sqrt{5}$. Therefore,
$(x - \sqrt{5})^2 = x^2 - 2x\sqrt{5} + (\sqrt{5})^2 = x^2 - 2\sqrt{5}x + 5$.
85. False; 5 is multiplied by 3 only.   **87.** True   **89.** $6x$
91. $3x + 1$ **93.** $x + 19 - 8\sqrt{x + 3}$ **95.** $2t + 10\sqrt{2t - 3} + 22$

97. $12\sqrt{5}$ ft^2 **99.** $18\sqrt{15}$ in.^{2} **101. a.** $\dfrac{\sqrt{2}}{2}$ **b.** $\dfrac{\sqrt[3]{4}}{2}$

103. a. $\dfrac{\sqrt{5a}}{5a}$ **b.** $\dfrac{\sqrt{5} - a}{5 - a^2}$ **105.** $\dfrac{-33}{2\sqrt{3} - 12}$

107. $\dfrac{a - b}{a + 2\sqrt{ab} + b}$ **109.** $\dfrac{5}{\sqrt{4 + 5h} + 2}$

111. a. $\dfrac{14}{3} \cdot \dfrac{30}{7} = \dfrac{420}{21} = 20$

b. $(5 - \sqrt{5})(5 + \sqrt{5})$
$= (5)^2 - (\sqrt{5})^2$
$= 25 - 5$
$= 20$

Problem Recognition Exercises, pp. 203–204

1. a. $3\sqrt{2}$ **b.** Cannot be simplified further.   **c.** $\sqrt{2}$
2. a. $\sqrt{7}$ **b.** $2\sqrt{7}$ **c.** Cannot be simplified further.
3. a. $9z$ **b.** $9 + 6\sqrt{z} + z$ **c.** $9 - z$
4. a. $16 - 8\sqrt{x} + x$ **b.** $16 - x$ **c.** $16x$

5. a. $\dfrac{6\sqrt{2x}}{x}$ **b.** $\dfrac{\sqrt{6x}}{x}$ **c.** $\dfrac{12\sqrt{2} - 12x}{2 - x^2}$

6. a. $\dfrac{45 + 15\sqrt{y}}{9 - y}$ **b.** $\dfrac{5\sqrt{3y}}{y}$ **c.** $\dfrac{\sqrt{5y}}{y}$

7. a. $3\sqrt{5} - 1$ **b.** $8 - 3\sqrt{5}$ **c.** $10 - 4\sqrt{5}$
8. a. $-8 + 11\sqrt{3}$ **b.** $12 + 16\sqrt{3}$ **c.** $3\sqrt{3} - 9$
9. a. $4a^7\sqrt{a}$ **b.** $2a^5\sqrt[3]{2}$ **10. a.** $3y^3$ **b.** $3y^4\sqrt[3]{3y}$

11. a. $2\sqrt{6}$ **b.** $2\sqrt[3]{3}$ **12. a.** $3\sqrt{6}$ **b.** $3\sqrt[3]{2}$
13. a. $10y^3\sqrt{2}$ **b.** $2y^2\sqrt[3]{25}$ **14. a.** $4z^7\sqrt{2z}$ **b.** $2z^5\sqrt[3]{4}$
15. a. $4\sqrt{5}$ **b.** $2\sqrt[3]{10}$ **c.** $2\sqrt[4]{5}$
16. a. $4\sqrt{3}$ **b.** $2\sqrt[3]{6}$ **c.** $2\sqrt[4]{3}$
17. a. $x^2y^3\sqrt{x}$ **b.** $xy^2\sqrt[3]{x^2}$ **c.** $xy\sqrt[4]{xy^2}$
18. a. $a^5b^4\sqrt{b}$ **b.** $a^3b^3\sqrt[3]{a}$ **c.** $a^2b^2\sqrt[4]{a^2b}$
19. a. $2st^2\sqrt[3]{4s^2}$ **b.** $2st\sqrt[4]{2st^2}$ **c.** $2st\sqrt[5]{t}$
20. a. $2v^2w^6\sqrt[3]{12vw^2}$ **b.** $2vw^5\sqrt[4]{6v^3}$ **c.** $2vw^4\sqrt[5]{3v^2}$
21. a. $2\sqrt{5}$ **b.** 5   **22. a.** $2\sqrt{10}$ **b.** 10
23. a. $-3\sqrt{6}$ **b.** 60   **24. a.** $-7\sqrt{7}$ **b.** 210
25. a. $3\sqrt{2}$ **b.** 4   **26. a.** $3\sqrt{3}$ **b.** 6
27. a. $7\sqrt{2}$ **b.** 240   **28. a.** $-\sqrt{2}$ **b.** 60
29. a. $14\sqrt[3]{3}$ **b.** $48\sqrt[3]{9}$ **30. a.** $\sqrt[3]{2}$ **b.** $30\sqrt[3]{4}$

Section 3.3 Practice Exercises, pp. 211–214

R.1. y^{15} **R.3.** y^{19} **R.5.** $\dfrac{1}{p^7}$ **R.7.** $\dfrac{3}{28}$

R.9. $\dfrac{1-3x}{2}$ or $\dfrac{1}{2}-\dfrac{3}{2}x$ **R.11.** $-3x-11$ **R.13.** $-2x^2-6x$

R.15. $-20p^2-2p+6$ **R.17.** $6.25-0.64y^2$ **R.19.** $9c^2-60c+100$

1. -1 **3.** real; imaginary   **5.** $11i$ **7.** $7i\sqrt{2}$ **9.** $i\sqrt{19}$
11. $-4i$ **13.** -6 **15.** $-5\sqrt{2}$ **17.** $-2\sqrt{21}$
19. 7   **21.** $3i$ **23.** Real part: 3; Imaginary part: -7
25. Real part: 0; Imaginary part: 19
27. Real part: $-\frac{1}{4}$; Imaginary part: 0
29. $0+8i$ **31.** $2+2\sqrt{3}i$ or $2+2i\sqrt{3}$ **33.** $\dfrac{4}{7}+\dfrac{3}{14}i$
35. $2+3i$ **37.** $-\dfrac{9}{2}+\sqrt{3}i$ or $-\dfrac{9}{2}+i\sqrt{3}$
39. $-2+i\sqrt{2}$ **41. a.** 1   **b.** i **c.** -1 **d.** $-i$
43. a. i **b.** $-i$ **c.** -1 **d.** -1
45. a. $10-10x$ **b.** $10-10i$ **47. a.** $15n-24n^2$ **b.** $24+15i$
49. a. $-2x^2-x+36$ **b.** $38-i$ **51.** $-3+61i$
53. $-\dfrac{1}{3}+\dfrac{7}{12}i$ **55.** $-2-3i$ **57.** $-2+10i$ **59.** $\sqrt{21}+i\sqrt{33}$
61. $36-57i$ **63. a.** $3+6i$ **b.** 45   **65. a.** $0-8i$ **b.** 64
67. a. $9-16p^2$ **b.** 25   **69. a.** $25z^2-60z+36$ **b.** $11-60i$
71. 116   **73.** 49   **75.** 5   **77.** $-40-42i$ **79.** $-3+4\sqrt{7}i$
81. $6+0i$ **83.** $\dfrac{8}{5}+\dfrac{6}{5}i$ **85.** $\dfrac{94}{173}-\dfrac{81}{173}i$ **87.** $\dfrac{6}{41}-\dfrac{\sqrt{5}}{41}i$
89. $0-\dfrac{5}{13}i$ **91.** $4i\sqrt{2}$ **93.** $2i$
95. a. $(5i)^2+25=0$ ✓   **b.** $(-5i)^2+25=0$ ✓
97. a. $(2+i\sqrt{3})^2-4(2+i\sqrt{3})+7=0$ ✓
  b. $(2-i\sqrt{3})^2-4(2-i\sqrt{3})+7=0$ ✓
99. $(a+bi)(c+di)$
  $= ac+adi+bci+bdi^2$
  $= ac+(ad+bc)i+bd(-1)$
  $= (ac-bd)+(ad+bc)i$
101. Any real number. For example: 5.   **103.** a^2+b^2
105. a. $(x+3)(x-3)$ **b.** $(x+3i)(x-3i)$
107. a. $(x+8)(x-8)$ **b.** $(x+8i)(x-8i)$
109. a. $(x+\sqrt{3})(x-\sqrt{3})$ **b.** $(x+i\sqrt{3})(x-i\sqrt{3})$

Section 3.4 Practice Exercises, pp. 221–224

R.1. 12 cm^2 **R.3.** 36 in.^2 **R.5.** $x+2$ **R.7.** $5(x+4)$
R.9. $(b-5)(b-9)$ **R.11.** $3(z-3)(4z-1)$
R.13. $(5-9w)(5+9w)$ **R.15.** $(x-8)(x+8)(x-1)$
R.17. $x(x-8)(x+2)$ **R.19.** $-2(t+5)(t-8)$

1. quadratic; second   **3.** 0   **5.** Linear   **7.** Quadratic
9. Neither   **11.** Correct form
13. Incorrect form. Polynomial is not factored.
15. Incorrect form. Equation is not set equal to 0.
17. a. $(w+9)(w-9)$ **b.** $\{-9,9\}$
19. a. $(3x-1)(x+5)$ **b.** $\left\{\dfrac{1}{3},-5\right\}$ **21.** $\{-3,-5\}$
23. $\left\{-\dfrac{9}{2},\dfrac{1}{5}\right\}$ **25.** $\left\{0,-4,\dfrac{3}{10}\right\}$ **27.** $\{0.4,-2.1\}$
29. $\{-9,3\}$ **31.** $\left\{-3,\dfrac{1}{2}\right\}$ **33.** $\left\{0,\dfrac{3}{2}\right\}$ **35.** $\left\{\dfrac{23}{3}\right\}$
37. $\{-3\}$ **39.** $\left\{-\dfrac{1}{3},2\right\}$ **41.** $\{-5,4\}$ **43.** $\left\{\dfrac{5}{2},-1\right\}$
45. $\{-12,5\}$ **47.** $\left\{-\dfrac{1}{11}\right\}$ **49.** $\left\{-1,\dfrac{1}{2},3\right\}$
51. $\{0,6,-2\}$ **53.** $\{0,4,-4\}$ **55.** $\left\{3,-3,-\dfrac{5}{2}\right\}$
57. $\left\{0,\dfrac{1}{2},-6\right\}$ **59.** $\left\{-\dfrac{1}{5},\dfrac{1}{5},-\dfrac{4}{3}\right\}$ **61.** $\{-7,-2,2\}$
63. a. $x(2x+3)=629$ **b.** The width is 17 yd and the length is 37 yd.
65. a. $\dfrac{1}{2}x(x-2)=40$ **b.** The base is 10 ft and the height is 8 ft.
67. a. $\dfrac{8}{5}x^2=640$
  b. The length is 20 in., the width is 8 in., and the height is 4 in.
69. a. $x^2+(x+2)^2=(2x-2)^2$
  b. The legs are 6 ft and 8 ft, and the hypotenuse is 10 ft.
71. The length is 7 ft, and the width is 5 ft.
73. The length is 20 yd, and the width is 15 yd.
75. a. The base is 5 in., and the height is 6 in.   **b.** The area is 15 in.²
77. The base is 10 ft, and the height is 5 ft.
79. a. yes   **b.** no   **c.** yes   **81.** 9 cm   **83.** 20 ft
85. They were 5 mi apart.
87. a. 14 mi   **b.** The alternative route using superhighways
89. The lengths are 6 m, 8 m, and 10 m.
91. a. $x(x+2)=120$ **b.** The integers are 10 and 12 or -10 and -12.
93. a. $x^2+(x+1)^2=113$ **b.** The integers are 7 and 8 or -7 and -8.
95. The dimensions of the cargo space are 6 ft by 7 ft by 12 ft.
97. The base is 9 ft and the height is 12 ft.
99. a. The lengths of the sides of the lower triangle are 6 ft, 8 ft, and 10 ft.
  b. The total area is 44 ft².

Section 3.5 Practice Exercises, pp. 231–233

R.1. $\dfrac{\sqrt{17}}{2}$ **R.3.** $2\sqrt{7}$ **R.5. a.** $\dfrac{7}{6}$ **b.** $\dfrac{17}{6}$

R.7. a. $\dfrac{20}{21}$ **b.** 6   **R.9.** $a^2+1.2a+0.36$

R.11. $h^2+\dfrac{1}{2}hk+\dfrac{1}{16}k^2$ **R.13.** $(x+10)^2$ **R.15.** $\left(\dfrac{1}{2}t+\dfrac{3}{5}\right)^2$

1. $\sqrt{k}$; $-\sqrt{k}$ **3.** completing   **5.** 4; 1   **7.** $\{\pm12\}$
9. $\{\pm2\}$ **11.** $\{\pm\sqrt{7}\}$ **13.** $\left\{\pm\dfrac{11}{6}\right\}$ **15.** $\{\pm5i\}$
17. $\{-1,-5\}$ **19.** $\left\{\dfrac{-3\pm\sqrt{7}}{2}\right\}$ **21.** $\{-5\pm3i\sqrt{2}\}$
23. $\left\{-\dfrac{4}{5}\pm\dfrac{\sqrt{3}}{5}i\right\}$ **25.** $\{\pm4i\}$ **27.** $n=9; (x-3)^2$
29. $n=16; (t+4)^2$ **31.** $n=\dfrac{1}{4}; \left(c-\dfrac{1}{2}\right)^2$ **33.** $n=\dfrac{25}{4}; \left(y+\dfrac{5}{2}\right)^2$
35. $n=\dfrac{1}{25}; \left(b+\dfrac{1}{5}\right)^2$ **37.** $n=\dfrac{1}{9}; \left(p-\dfrac{1}{3}\right)^2$

39. 1. Divide both sides by a to make the leading coefficient 1.
2. Isolate the variable terms on one side of the equation.
3. Complete the square.   4. Apply the square root property and solve for x.

41. $\{-3, -5\}$ **43.** $\{-3 \pm i\sqrt{7}\}$ **45.** $\{-2 \pm i\sqrt{2}\}$

47. $\{5, -2\}$ **49.** $\left\{-1 \pm \dfrac{\sqrt{6}}{2}i\right\}$ **51.** $\left\{2 \pm \dfrac{2}{3}i\right\}$

53. $\left\{\dfrac{1}{5} \pm \dfrac{\sqrt{3}}{5}\right\}$ **55.** $\left\{-\dfrac{3}{4} \pm \dfrac{\sqrt{65}}{4}\right\}$ **57.** $\{2 \pm \sqrt{11}\}$

59. $\{1, -7\}$ **61.** 8 in.   **63.** The shelf extends 4.2 ft.
65. They are 25 mi apart.   **67.** The radius is approximately 25 yd.
69. The distance is $90\sqrt{2}$ ft or approximately 127.3 ft.
71. **a.** The length is approximately 2.91 in. and the width is approximately 1.94 in.

   b. Using the rounded values from part (a), the screen is approximately 949 pixels by 632 pixels.

73. **a.** 4.5 thousand textbooks or 35.5 thousand textbooks

   b. From the graph, profit increases to a point as more books are produced. Beyond that point, the market is "flooded," and profit decreases. There are two points at which the profit is $20,000. Producing 4.5 thousand books makes the same profit using fewer resources than producing 35.5 thousand books.

Section 3.6 Practice Exercises, pp. 242–246

R.1. $i\sqrt{3}$ **R.3.** $-2i\sqrt{5}$ **R.5.** $\dfrac{1}{4} + \dfrac{1}{2}i$ **R.7.** $-1 + \sqrt{2}i$

R.9. $\left\{\dfrac{16}{7}\right\}$ **R.11.** $\{3.6\}$ **R.13.** $r = \dfrac{I}{Pt}$ **R.15.** $a = \dfrac{v - v_0}{t}$

1. quadratic; $\dfrac{-b \pm \sqrt{b^2 - 4ac}}{2a}$ **3.** 8; -42; -27

5. $b^2 - 4ac$; discriminant   **7.** real

9. $\{-12, 1\}$ **11.** $\left\{\dfrac{3 \pm \sqrt{37}}{2}\right\}$ **13.** $\{-2 \pm \sqrt{2}i\}$

15. $\{3 \pm i\}$ **17.** $\left\{\dfrac{7}{10} \pm \dfrac{\sqrt{11}}{10}i\right\}$ **19.** $\left\{\dfrac{1}{2}, -\dfrac{3}{10}\right\}$

21. $\left\{\pm\dfrac{7}{3}i\right\}$ **23.** $\left\{\dfrac{5 \pm \sqrt{137}}{14}\right\}$ **25.** $\left\{\dfrac{5}{2}\right\}$ **27.** $\{-5, 7\}$
29. a. -56 **b.** 2 nonreal solutions   **31. a.** 40   **b.** 2 real solutions
33. a. 121   **b.** 2 real solutions   **35. a.** 0   **b.** 1 real solution

37. $r = \sqrt{\dfrac{A}{\pi}}$ or $r = \dfrac{\sqrt{A\pi}}{\pi}$ **39.** $t = \sqrt{\dfrac{2s}{g}}$ or $t = \dfrac{\sqrt{2sg}}{g}$

41. $a = \sqrt{c^2 - b^2}$ **43.** $I = \dfrac{1}{c}\sqrt{\dfrac{L}{Rt}}$ or $I = \dfrac{\sqrt{LRt}}{cRt}$

45. $w = \dfrac{c \pm \sqrt{c^2 + 4kr}}{2k}$ **47.** $t = \dfrac{-v_0 \pm \sqrt{v_0^2 + 2as}}{a}$

49. $I = \dfrac{-CR \pm \sqrt{C^2R^2 - 4CL}}{2CL}$

51. a. $s = -16t^2 + 16t$ **b.** It would take Michael Jordan 0.5 sec to reach his maximum height of 4 ft.
53. a. $s = -16t^2 + 75t + 4$ **b.** The ball will be at an 80-ft height 1.5 sec and 3.2 sec after being kicked.
55. There were 8 players.
57. There were 600,000 organisms approximately 9 hr and 39 hr after the culture was started.
59. a. 235 ft   **b.** 62 mph
61. a. Approximately 4.5 fatalities per 100 million miles driven
   b. Approximately 1 fatality per 100 million miles driven
   c. Approximately 4.2 fatalities per 100 million miles driven
   d. For drivers 26 years old and 71 years old

63. Linear; $\{-2\}$ **65.** Quadratic; $\{0, -2\}$ **67.** Linear; $\{1\}$
69. Neither

71. $\left\{\dfrac{4}{3}\right\}$ **73.** $\{-2 \pm \sqrt{2}\}$ **75.** $\{7 \pm \sqrt{55}\}$

77. $\mathbb{R}$ **79.** $\left\{1, -\dfrac{5}{6}\right\}$ **81.** $\{\ \}$ **83.** $\{-2\}$

85. $\left\{\pm\dfrac{\sqrt{35}}{7}i\right\}$ **87.** $x = 2y$ or $x = -y$

89. $x^2 - 2x - 8 = 0$ **91.** $12x^2 - 11x + 2 = 0$
93. $x^2 - 5 = 0$ **95.** $x^2 + 4 = 0$ **97.** $x^2 - 2x + 5 = 0$

99. $x_1 + x_2 = \dfrac{-b + \sqrt{b^2 - 4ac}}{2a} + \dfrac{-b - \sqrt{b^2 - 4ac}}{2a}$

$$= \dfrac{-b + \sqrt{b^2 - 4ac} + (-b) - \sqrt{b^2 - 4ac}}{2a}$$

$$= \dfrac{-2b}{2a} = -\dfrac{b}{a}$$

101. a. $L = \dfrac{1 + \sqrt{5}}{2} \approx 1.62$ **b.** 14.6 ft

103. a. $y = \dfrac{160 - 4x}{6}$ or $y = \dfrac{80 - 2x}{3}$ **b.** $A = x\left(\dfrac{80 - 2x}{3}\right)$

   c. Each pen can be 25 yd by 10 yd, or it can be 15 yd by $\dfrac{50}{3}$ yd.

Chapter 3 Review Exercises, pp. 251–253

1. a. $|x|$ **b.** x **c.** $|x|$ **d.** $x + 1$ **3.** $6\sqrt{3}$

5. $-10ab^3\sqrt[3]{2b}$ **7.** $3y^4z^2\sqrt[3]{2xz^2}$ **9.** $\dfrac{p^6\sqrt{p}}{3}$

11. False; 5 and $3\sqrt{x}$ are not like radicals.   **13.** $5\sqrt{7}$ **15.** $10\sqrt{2}$
17. $(5 + 6b)\sqrt{3a}$ **19.** 6   **21.** $-2\sqrt{21} + 6\sqrt{33}$ **23.** 54

25. $7y - 2\sqrt{21xy} + 3x$ **27.** $-2z - 9\sqrt{6z} - 42$ **29.** $\dfrac{\sqrt{14y}}{2y}$

31. $\dfrac{4\sqrt[3]{3p}}{3p}$ **33.** $\sqrt{10} - \sqrt{15}$ **35.** $4i$ **37.** -15 **39.** -1

41. $-i$ **43.** $-5 + 5i$ **45.** $25 + 0i$

47. $-\dfrac{17}{4} + i$; real part: $-\dfrac{17}{4}$; imaginary part: 1   **49.** $\dfrac{4}{13} - \dfrac{7}{13}i$

51. $-\dfrac{3}{2} + \dfrac{5}{2}i$ **53.** $-\dfrac{2}{3} + \dfrac{\sqrt{10}}{6}i$ **55.** Quadratic   **57.** Linear

59. a. $(5x - 4)(x + 2)$ **b.** $\left\{\dfrac{4}{5}, -2\right\}$ **61.** $\{-3, 5\}$

63. $\left\{-\dfrac{4}{3}, -1\right\}$ **65.** $\{-4, -1, 4\}$ **67.** Length 15 ft; width 8 ft; height 10 ft   **69.** $\{\pm\sqrt{5}\}$ **71.** $\{\pm 9\}$ **73.** $\{2 \pm 6\sqrt{2}\}$

75. $\left\{\dfrac{1 \pm \sqrt{3}}{3}\right\}$ **77.** 9 in.   **79.** $n = 64$; $(x + 8)^2$

81. $n = \dfrac{1}{16}$; $\left(y + \dfrac{1}{4}\right)^2$ **83.** $\{-2 \pm 3i\}$ **85.** $\{3 \pm 2\sqrt{3}\}$

87. $\left\{\dfrac{1}{3}, -1\right\}$ **89.** $\{2 \pm \sqrt{3}\}$ **91.** $\left\{\dfrac{5}{2} \pm \dfrac{5\sqrt{3}}{2}i\right\}$

93. $\left\{2, -\dfrac{5}{6}\right\}$ **95.** $\left\{\dfrac{4}{5}, -\dfrac{1}{5}\right\}$ **97. a.** 0   **b.** 1 real solution

99. a. -219 **b.** 2 nonreal solutions

101. $r = \sqrt{\dfrac{V}{\pi h}}$ or $r = \dfrac{\sqrt{V\pi h}}{\pi h}$ **103.** $y = k \pm \sqrt{r^2 - (x - h)^2}$

105. a. $s = -16t^2 + 200t + 2$ **b.** 0.4 sec   **107.** $\left\{\dfrac{2 \pm \sqrt{46}}{6}\right\}$

109. $\left\{\pm\dfrac{5}{3}\right\}$ **111.** $\left\{-3, \dfrac{5}{2}\right\}$ **113.** $\{5 \pm 2i\}$

Chapter 3 Test, pp. 253–254

1. a. y **b.** $|y|$ **2.** 3   **3.** $\dfrac{4}{3}$ **4.** $2\sqrt[3]{4}$ **5.** $a^2bc^2\sqrt{bc}$

6. $3x^2yz^2\sqrt{2xy}$ **7.** $4w^2\sqrt{w}$ **8.** $\dfrac{x^2}{5y}$ **9.** $\dfrac{3\sqrt{2}}{2}$ **10.** $3\sqrt{5}$

11. $3\sqrt{2x} - 3\sqrt{5x}$ **12.** $40 - 10\sqrt{5x} - 3x$ **13.** 83

14. $7y - 8\sqrt{7yz} + 16z$ **15.** $\dfrac{-2\sqrt[3]{x^2}}{x}$ **16.** $\dfrac{x + 6 + 5\sqrt{x}}{9 - x}$

17. $2i\sqrt{2}$ **b.** $8i$ **c.** $\dfrac{1}{2} + \dfrac{\sqrt{2}}{2}i$ **18.** $1 - 11i$

19. $30 + 16i$ **20.** -28 **21.** $-33 - 56i$ **22.** 104

23. $\dfrac{17}{25} + \dfrac{6}{25}i$ **24.** $n = \dfrac{121}{4}; \left(d + \dfrac{11}{2}\right)^2$ **25.** $\{-3 \pm 3\sqrt{3}\}$

26. $\left\{\dfrac{3}{4} \pm \dfrac{\sqrt{47}}{4}i\right\}$ **27.** $\left\{1, \dfrac{1}{3}\right\}$ **28.** $\left\{\dfrac{-7 \pm \sqrt{5}}{2}\right\}$

29. a. -40 **b.** 2 nonreal solutions   **30. a.** 0
b. 1 real solution   **31. a.** 76   **b.** 2 real solutions

32. $t = \dfrac{-v_0 \pm \sqrt{v_0^2 + 128}}{-32}$ or $t = \dfrac{v_0 \pm \sqrt{v_0^2 + 128}}{32}$

33. $c = \sqrt{49 - a^2 - b^2}$ **34.** $\{-5, 0, 4\}$ **35.** $\{6 \pm 3i\}$

36. $\left\{\dfrac{-3 \pm \sqrt{65}}{2}\right\}$ **37.** $\{\pm\sqrt{6}\}$ **38.** $\left\{-\dfrac{3}{2}, -1, \dfrac{3}{2}\right\}$

39. a. $s = -16t^2 + 60t + 2$ **b.** The ball will be at a height of 52 ft at times 1.25 sec and 2.5 sec after being kicked.   **40.** The base of the triangular portions is 5 ft and the height is 12 ft.

CHAPTER 4

Section 4.1 Practice Exercises, pp. 265–268

R.1. $-\dfrac{20}{3}$ **R.3.** x^{16} **R.5.** $\dfrac{1}{m^5}$ **R.7.** -1

R.9. $-\dfrac{8}{5}$ **R.11.** $\dfrac{8}{7}$ **R.13.** $-\dfrac{6}{7}$

1. rational   **3.** $\dfrac{p}{q}$ **5.** $\dfrac{pr}{qs}$ **7.** $\dfrac{5}{2}$ **9.** $-\dfrac{1}{8}$

11. 0   **13.** Undefined
15. a. At 1 hr: 4.8 ng/mL; At 12 hr: 3.9 ng/mL; At 24 hr: 1.0 ng/mL; At 48 hr: 0.3 ng/mL   **b.** 0 ng/mL

17. $k = -2$ **19.** $x = \dfrac{5}{2}, x = -8$ **21.** $m = -2, m = -3$

23. There are no restricted values.   **25.** There are no restricted values.

27. $t = 0, t = 5$ **29.** For example: $\dfrac{1}{x - 2}$

31. For example: $\dfrac{1}{(x + 3)(x - 7)}$ **33. a.** $\dfrac{2}{5}$ **b.** $\dfrac{2}{5}$

35. a and b   **37. a.** $\dfrac{2x}{y}$ **b.** Cannot be simplified

39. a. $\dfrac{(x + 4)(x + 2)}{(x + 4)(x - 1)}$ **b.** $x \neq -4, x \neq 1$ **c.** $\dfrac{x + 2}{x - 1}$

41. a. $\dfrac{(x - 9)^2}{(x - 9)(x + 9)}$ **b.** $x \neq 9, x \neq -9$ **c.** $\dfrac{x - 9}{x + 9}$

43. $-\dfrac{1}{4m^2n^3}$ **45.** $\dfrac{2}{3}$ **47.** $\dfrac{1}{x + 5}$ **49.** $-\dfrac{1}{3c - 5}$

51. $-\dfrac{t + 4}{t + 3}$ **53.** $\dfrac{(2p - 1)^2}{p + 1}$ **55.** $\dfrac{3 - z}{2z - 5}$ **57.** $\dfrac{2(z^2 - 4z + 16)}{z + 4}$

59. $-\dfrac{2x - 5}{2}$ **61.** 1   **63.** -1 **65.** -1 **67.** -2

69. $\dfrac{c + 4}{c - 4}$; cannot be simplified   **71.** $-\dfrac{1}{12(x + y)}$ **73.** $\dfrac{4}{3w^2}$

75. $\dfrac{p^3}{8q}$ **77.** $\dfrac{2}{3}$ **79.** $\dfrac{1}{2}$ **81.** $-\dfrac{y - x}{2x}$ or $\dfrac{x - y}{2x}$

83. $\dfrac{2x(x + 5)}{x - 5}$ **85.** $\dfrac{3}{2x}$ **87.** $2x(x - 2)$ **89.** $\dfrac{t}{t + 1}$

91. $\dfrac{1}{a - 4}$ **93.** $\dfrac{x + y}{x - y}$ **95.** $\dfrac{x(x - 1)}{x + 1}$ **97.** -1

99. $\dfrac{24b}{a^3}$ **101.** $\dfrac{1}{4}$ **103.** $\dfrac{5x + y}{x^2 + y^2}$ **105.** $3(m - 4)(m - 2)$

107. $(a + b)(a - b)$ **109.** $\dfrac{2y}{x + 2y}$ **111.** $\dfrac{1}{2}$

113. $\dfrac{(m - n)(m + 5)}{(m + 4)(m - 2)}$ **115.** $\dfrac{3(x - 3)}{2}$ **117.** $\dfrac{3(y + 1)}{(3y + 4)(y - 5)}$

119. $\dfrac{2ab}{5}$ in.^{2} **121.** $\dfrac{2x}{x - 2}$ m^2

123. If $x = y$, then the denominator $x - y$ will equal zero. Division by zero is undefined.

125. x^n **127.** $\dfrac{w^{2n}}{w + z}$

Section 4.2 Practice Exercises, pp. 278–281

R.1. $\dfrac{1}{6}$ **R.3.** $-\dfrac{91}{90}$ **R.5.** $-\dfrac{3}{2}$ **R.7.** $\dfrac{3y^4}{x^2}$

1. $\dfrac{p + r}{q}; \dfrac{p - r}{q}$ **3.** $\dfrac{2}{x}$ **5.** $\dfrac{1}{x + 1}$ **7.** $\dfrac{2x + 5}{(2x + 9)(x - 6)}$

9. 2   **11.** $40x$ **13.** $30m^4n^7$ **15.** $(x - 4)(x + 2)(x - 6)$

17. $x^2(x - 1)(x + 7)^2$ **19.** $(x - 6)(x - 2)$ **21.** $a - 4$ or $4 - a$

23. $\dfrac{8p - 15}{6p^2}$ **25.** $\dfrac{-t - s}{st}$ **27.** $\dfrac{1}{3}$ **29.** $\dfrac{2b + 20}{b(b + 5)}$

31. $\dfrac{2x}{x - 6}$ or $\dfrac{-2x}{6 - x}$ **33.** $\dfrac{6b^2 + 5b + 4}{(b - 4)(b + 1)}$

35. $\dfrac{10x}{(2x + 1)(x - 2)}$ **37.** $\dfrac{3y - 1}{y + 4}$ **39.** $\dfrac{11x + 6}{(x - 6)(x + 6)(x + 3)}$

41. $-\dfrac{3}{w}$ **43.** 1   **45.** $\dfrac{10 - x}{3(x - 5)(x + 5)}$ or $\dfrac{x - 10}{3(5 - x)(5 + x)}$

47. $\dfrac{m - 5}{(m + 5)(m + 3)}$ **49.** $\dfrac{5x^2}{27}$ **51.** $\dfrac{1}{x}$ **53.** $\dfrac{10}{3}$

55. -4 **57.** $28y$ **59.** -8 **61.** $\dfrac{3 - p}{p - 1}$ **63.** $\dfrac{2a + 3}{4 - 9a}$

65. $-\dfrac{4(w - 1)}{w + 2}$ **67.** $\dfrac{1}{y + 4}$ **69.** $\dfrac{x + 2}{x - 1}$ **71.** $\dfrac{t^2}{(t + 1)^2}$

73. $\dfrac{-a + 2}{-a - 3}$ **75.** $-\dfrac{y + 1}{y - 5}$ **77.** $\dfrac{w^2 - 3}{w - 2}$ **79.** $-\dfrac{3}{t + 3}$

81. $-\dfrac{t}{t + 1}$ **83.** $-\dfrac{2}{x(x + h)}$ **85.** $\dfrac{x^2 + x + 6}{2(x + 1)}$

87. $\dfrac{-2y}{(x - y)(x + y)}$ or $\dfrac{2y}{(y - x)(y + x)}$ **89.** $\dfrac{2p^2 + 2p - 3}{3(p + 4)}$

91. $\dfrac{x^2 + 13x + 38}{(x + 5)^3}$ **93.** $\dfrac{1}{x(x^2 + 3)}$ **95.** $\dfrac{2b^2 + 3a}{b(b - a)}$

97. $\dfrac{-1}{4(4 + h)}$ **99.** $\dfrac{-6}{x(x + h)}$ **101.** $\dfrac{5x^2 + 12x + 24}{(x - 2)(x + 2)(x^2 + 2x + 4)}$

103. $\dfrac{3x^2 + 5x + 18}{3x^2}$ cm   **105.** $\dfrac{4x^2 - 2x + 50}{(x - 3)(x + 5)}$ m

107. $x(x + 1)$ **109.** $\dfrac{x^3 - x - 2}{x^3}$ **111.** $\dfrac{3(2x + 1)}{2\sqrt{3x}}$

113. $\dfrac{2x^2 + 1}{(x^2 + 1)\sqrt{x^2 + 1}}$ **115.** $\dfrac{2(8x^2 + 9)}{\sqrt{4x^2 + 9}}$

Problem Recognition Exercises, pp. 281–282

1. $\dfrac{4y^2 - 8y + 9}{2y(2y - 3)}$ **2.** $\dfrac{x^2 + x - 13}{x - 4}$ **3.** $\dfrac{4x - 1}{4x - 3}$ **4.** $\dfrac{a - 5}{3a}$

5. $\dfrac{2y - 5}{(y - 1)(y + 1)}$ **6.** $\dfrac{4}{w + 4}$ **7.** $\dfrac{a + 4}{2}$ **8.** $\dfrac{(t - 3)(t + 2)}{t}$

9. $\dfrac{a}{2a - 1}$ **10.** $\dfrac{2xy^2}{5}$ **11.** $\dfrac{-x^2 + 4xy - y^2}{(x - y)(x + y)}$ or $\dfrac{x^2 - 4xy + y^2}{(y - x)(y + x)}$

12. $3(x - 4)$ **13.** $\dfrac{-x^2 + 4x + 14}{6(x - 2)}$ **14.** $\dfrac{x^2 + 14x + 99}{10(x + 7)}$

15. $-\dfrac{1}{3}$ **16.** $\dfrac{1}{9}$ **17.** $y - 1$ **18.** $\dfrac{5t^2 - 6t - 17}{(t + 2)(t - 3)}$

19. $\dfrac{3(x + 4)}{2x - 3}$ **20.** $\dfrac{3z^3}{2x^2y^2}$ **21.** $\dfrac{9x^2 - 6x + 15}{4(3x - 1)}$ **22.** $\dfrac{2x + 3}{1 - 5x}$

23. $\dfrac{y^2 + 11y - 1}{(y + 3)(y - 2)}$ **24.** $a - 10$

Section 4.3 Practice Exercises, pp. 287–289

R.1. $\left\{\dfrac{5}{2}\right\}$ **R.3.** $\{3\}$ **R.5.** $\{995\}$

R.7. $x = \dfrac{t}{a - b}$ or $x = -\dfrac{t}{b - a}$ **R.9.** $(n - 3)(n + 3)$

R.11. $6x(x - 3)$ **R.13.** $(2z + 5)(z - 8)$

R.15. a. $-\dfrac{1}{2}$ **b.** Undefined

1. rational
3. No; 4 is a restricted value in the second and third expressions in the equation.

5. $x \neq 5, x \neq -4$ **7.** $x \neq \dfrac{3}{2}, x \neq 0, x \neq 3$ **9.** $\{-14\}$

11. $\{-24\}$ **13.** $\left\{-\dfrac{15}{22}\right\}$ **15.** $\{6\}$ **17.** $\{5\}$ **19.** $\{6\}$

21. $\{-25\}$ **23.** $\{3, 1\}$ **25.** $\{4, -4\}$ **27.** $\{8, -2\}$

29. $\{-2\}$ (The value 2 does not check.)   **31.** $\{60\}$

33. { } (The value 5 does not check.)   **35.** $\left\{\dfrac{31}{5}\right\}$

37. $\{5\}$ (The value -2 does not check.)

39. { } (The value 4 does not check.)   **41.** $\left\{-\dfrac{11}{4}\right\}$

43. $m = \dfrac{FK}{a}$ **45.** $E = \dfrac{IR}{K}$ **47.** $R = \dfrac{E - Ir}{I}$ or $R = \dfrac{E}{I} - r$

49. $B = \dfrac{2A - hb}{h}$ or $B = \dfrac{2A}{h} - b$ **51.** $t = \dfrac{b}{x - a}$ or $t = \dfrac{-b}{a - x}$

53. $x = \dfrac{y}{1 - yz}$ or $x = \dfrac{-y}{yz - 1}$ **55.** $h = \dfrac{2A}{a + b}$ **57.** $R = \dfrac{R_1 R_2}{R_2 + R_1}$

59. $t_2 = \dfrac{s_2 - s_1 + vt_1}{v}$ or $t_2 = \dfrac{s_2 - s_1}{v} + t_1$ **61.** $\dfrac{1}{a + 1}$

63. $\{1\}$ **65.** $\dfrac{x^2 - 12}{x(x - 1)}$ **67.** $\{-1\}$ **69.** $\dfrac{8p - 11}{4(2p - 3)}$

71. $\dfrac{x + 3}{6x^2}$ **73.** $\left\{-\dfrac{4}{15}\right\}$ **75.** $\dfrac{c + 7}{(c + 3)^2(c + 1)}$

77. { } (The value 4 does not check.)   **79.** $\left\{-\dfrac{5}{2}\right\}$

81. 16 years

Section 4.4 Practice Exercises, pp. 296–299

R.1. $\left\{-\dfrac{3}{4}\right\}$ **R.3.** $\left\{-2, -\dfrac{5}{6}\right\}$ **R.5.** $\{-4, 4\}$

1. proportion   **3.** $\{21\}$ **5.** $\left\{\dfrac{1}{3}\right\}$ **7.** $\{5, -1\}$

9. $\left\{\dfrac{3}{7}, -\dfrac{3}{7}\right\}$ **11.** A 14-oz box contains 84 g of fat.

13. Pam needs 11.5 gal of gas.
15. 62.5 lb of cement and 225 lb of gravel must be used.
17. LDL is 144 mg/dL and the total cholesterol is 204 mg/dL.
19. There are approximately 4000 bison in the park.
21. There are 31 men.
23. $a = 8$ ft; $z = 8.4$ ft
25. $x = 12$ in., $y = 13$ in., $a = 4.2$ in.

27. a. $x + 7$ **b.** $\dfrac{48}{x}$ **c.** $\dfrac{83}{x + 7}$

29. The motorist drives 40 mph in the rain and 60 mph in sunny weather.
31. The Broadmoor truck travels 70.4 mph and the Wescott truck travels 76.8 mph.
33. The cyclist rode 10 mph against the wind.
35. Celeste walks 5 ft/sec on the ground and 7 ft/sec on the moving walkway.
37. Joe runs 6 mph and Beatrice runs 8 mph.
39. It will take them $\dfrac{24}{7}$ hr.
41. It will take them $\dfrac{20}{3}$ hr.
43. a. The new pump will take 20 hr.
   b. The technician should return at noon on Friday.
45. Gus would take 6 hr; Sid would take 12 hr.
47. The Washington Monument is approximately 555 ft tall.
49. It will take $\dfrac{220}{7}$ sec or approximately 31.4 sec.
51. The plane flies 300 mph in still air.
53. There are 480 deer.
55. It will take 15 hr.

Section 4.5 Practice Exercises, pp. 305–308

R.1. -5 **R.3.** Not a real number   **R.5.** 32   **R.7.** x^{20}

R.9. 5^3 or 125   **R.11.** $\dfrac{27x^6y^{15}}{z^3}$ **R.13.** $\dfrac{7}{a^4}$ **R.15.** $\dfrac{9}{a^7b^2}$

1. $\left(\sqrt[n]{a}\right)^m$ or $\sqrt[n]{a^m}$ **3.** $\dfrac{1}{\sqrt[3]{x}}$ **5. a.** 5   **b.** Undefined   **c.** -5

7. a. 3   **b.** -3 **c.** -3 **9. a.** $\dfrac{11}{13}$ **b.** $\dfrac{13}{11}$

11. a. 8   **b.** $\dfrac{1}{8}$ **c.** -8 **d.** $-\dfrac{1}{8}$ **e.** Undefined   **f.** Undefined

13. a. 16   **b.** $\dfrac{1}{16}$ **c.** -16 **d.** $-\dfrac{1}{16}$ **e.** 16   **f.** $\dfrac{1}{16}$

15. a. $\sqrt[11]{y^4}$ or $\left(\sqrt[11]{y}\right)^4$ **b.** $6\sqrt[11]{y^4}$ or $6\left(\sqrt[11]{y}\right)^4$ **c.** $\sqrt[11]{(6y)^4}$ or $\left(\sqrt[11]{6y}\right)^4$

17. $\sqrt[3]{q^2}$ **19.** $6\sqrt[4]{y^3}$ **21.** $\sqrt[3]{x^2y}$ **23.** $\dfrac{6}{\sqrt[5]{r^2}}$ **25.** $a^{3/5}$

27. $(6x)^{1/2}$ **29.** $6x^{1/2}$ **31.** $(a^5 + b^5)^{1/5}$ **33.** $\dfrac{1}{x}$ **35.** p

37. y^2 **39.** $6^{2/5}$ **41.** $\dfrac{4}{t^{5/3}}$ **43.** a^7 **45.** $\dfrac{25a^4d}{c}$ **47.** $\dfrac{y^9}{x^8}$

49. $\dfrac{2z^3}{w}$ **51.** $5xy^2z^3$ **53.** $x^{13}z^{4/3}$ **55.** $\dfrac{x^3y^2}{z^5}$ **57.** $\dfrac{(m + n)^{1/2}}{m}$

59. a. 10.9%   **b.** 8.8%   **c.** The account in part (a).

61. 2.7 in.   **63.** 15.9%   **65.** $2c^{3/4}(c + 2)$

67. $m^{1/2}(m + 4)(m - 9)$ **69.** $\dfrac{8x + 5}{3x^{2/3}}$ **71.** $(3x + 1)^{2/3}(8x + 1)$

73. $\dfrac{2(2x + 1)}{(3x + 2)^{2/3}}$ **75.** $\dfrac{x(13x - 80)}{(x - 8)^{2/5}}$ **77.** $\dfrac{1}{512}$ **79.** 27

81. $-\dfrac{1}{81}$ **83.** $\dfrac{27}{1000}$ **85.** Not a real number   **87.** -2

89. -2 **91.** 6   **93.** 10   **95.** $\dfrac{3}{4}$ **97.** 1   **99.** $\dfrac{9}{2}$

101. $\left(\sqrt[3]{8}\right)^4$ works within the parentheses first. Taking the root first results in a smaller number that must be raised to the fourth power. $\sqrt[3]{8^4}$ takes the fourth power of 8 first. This results in a larger number whose cube root may be difficult to identify.

103. $\sqrt[20]{x^{17}y^8}$ **105.** $\sqrt[18]{m^5}$ **107.** $\sqrt[8]{x^7}$

109. Yes; The model gives a mean surface temperature of approximately 29.1°C.

111. 316.2278   **113.** 34.2990   **115.** 0.1170

Section 4.6 Practice Exercises, pp. 314–316

R.1. $x+4$ **R.3.** 5   **R.5.** $a^2+10a+25$
R.7. $5a-6\sqrt{5a}+9$ **R.9.** $r+22+10\sqrt{r-3}$

R.11. $\{3\}$ **R.13.** $\{4, 11\}$ **R.15.** $\left\{-2, \dfrac{7}{3}\right\}$

1. radical   **3.** extraneous   **5.** $\{100\}$ **7.** $\{4\}$ **9.** $\{3\}$
11. $\{42\}$ **13.** $\{29\}$ **15.** $\{140\}$
17. $\{\ \}$ (The value 25 does not check.)
19. $\{7\}$ (The value -1 does not check.)
21. $\{\ \}$ (The value 3 does not check.)
23. $\{-32, 22\}$ **25.** $\{5\}$ **27.** $\{5\sqrt[3]{5}\}$ or $\{5^{4/3}\}$
29. $\left\{\pm\dfrac{1}{32}\right\}$ **31.** $\{2\}$ **33.** $\{-3\}$
35. $\{\ \}$ (The value $\frac{1}{2}$ does not check.)
37. $\{2\}$ (The value 18 does not check.)   **39.** $\{-1, 4\}$
41. $\{7\}$; The value -2 does not check.   **43.** $\{9\}$ **45.** $\{-4\}$
47. $\left\{\dfrac{9}{5}\right\}$ **49.** $\{2\}$ **51.** $\{\ \}$ (The value 9 does not check.)
53. $\left\{-\dfrac{11}{4}\right\}$ **55.** $\{\ \}$ (The value $\frac{8}{3}$ does not check.)
57. $\{-1\}$ (The value 3 does not check.)
59. $\left\{\dfrac{1}{3}, -1\right\}$ **61.** $\{\ \}$ (The values 3 and 23 do not check.)
63. $\{6\}$ **65.** $\{2\}$ (The value 18 does not check.)
67. $\{-10\}$ **69. a.** 30.25 ft   **b.** 34.5 m
71. a. \$2 million   **b.** \$1.2 million   **c.** 50,000 passengers
73. 288π in.^{3} **75. a.** 55%   **b.** 9.3 hr
77. a. 12.6% per year   **b.** \$24,800
79. $x = \pm\sqrt{(z-16)^2 + y^2}$ **81.** $g = \dfrac{4\pi^2 L}{T^2}$
83. $b = \sqrt{y^2 - 4}$ **85.** $a = \sqrt{64 - x^2}$
87. $\left\{\dfrac{19 - 3\sqrt{5}}{2}\right\}$; The value $\dfrac{19 + 3\sqrt{5}}{2}$ does not check.
89. Pam can row to a point $166\frac{2}{3}$ ft down the beach or to a point 300 ft down the beach to be home in 5 min.

Section 4.7 Practice Exercises, pp. 320–321

R.1. $(x-10)(x+7)$ **R.3.** $(2z+7)(4z-3)$

R.5. $(2t-11)(2t+11)$ **R.7.** $\{6\}$ **R.9.** $\{\pm 12\}$

R.11. $\{\pm 6i\}$ **R.13.** $\{\pm 2\sqrt{2}\}$ **R.15.** $\left\{\dfrac{1}{2}\right\}$

R.17. $\{13\}$ **R.19.** $\left\{\dfrac{5}{2}\right\}$ **R.21.** $\{64\}$

R.23. $\left\{\dfrac{5}{4}, -\dfrac{1}{2}\right\}$ **R.25.** $\left\{\dfrac{1}{4} \pm \dfrac{\sqrt{39}}{4}i\right\}$

1. quadratic; $m^{1/3}$ **3.** $3x-1$
5. a. $\{-4, -6\}$ **b.** $\{-4, -1, -2, -3\}$ **7.** $\{-2, 1, -3, 2\}$
9. $\{\pm 3i, \pm 2\}$ **11.** $\{27, -8\}$ **13.** $\left\{-\dfrac{1}{32}, -243\right\}$
15. $\{4\}$ (The value 64 does not check.)
17. $\left\{\dfrac{1}{4}\right\}$ (The value 4 does not check.)   **19.** $\{-7\}$
21. $\{4\}$; The value 64 does not check.   **23.** $\{\pm\sqrt{3}, \pm 2i\}$
25. $\left\{\pm\dfrac{\sqrt{2}}{3}, \pm i\right\}$ **27.** $\left\{-4, \dfrac{5}{2}\right\}$ **29.** $\{-5, -3, 1, 3\}$
31. $\{4, 6\}$ **33.** $\left\{-3, -\dfrac{8}{5}\right\}$ **35.** $\{-5, -1, 2, 10\}$
37. $\left\{\dfrac{3}{2}, -\dfrac{3}{5}\right\}$ **39.** $\left\{\dfrac{1}{3125}, 32\right\}$ **41.** $\{81\}$
43. $\{-3, -1, 1, 3\}$ **45.** $\left\{0, \dfrac{625}{16}\right\}$ **47.** $\left\{\pm\sqrt{\dfrac{-5 \pm \sqrt{53}}{2}}\right\}$
49. $\left\{\dfrac{3}{2}, 10\right\}$ **51.** $\{-2, 1, 4\}$

Problem Recognition Exercises, pp. 321–322

1. a. Quadratic   **b.** $\{2, -7\}$ **2. a.** Quadratic   **b.** $\{4, 5\}$
3. a. Quadratic form and higher degree polynomial   **b.** $\{-3, -1, 1, 3\}$
4. a. Quadratic form and higher degree polynomial   **b.** $\{2, -2, i, -i\}$
5. a. Quadratic form and radical   **b.** $\{16\}$ (The value 1 does not check.)
6. a. Quadratic form and radical   **b.** $\{81\}$ (The value 1 does not check.)
7. a. Linear   **b.** $\{4\}$ **8. a.** Linear   **b.** $\{-1\}$
9. a. Quadratic   **b.** $\{\pm i\sqrt{2}\}$ **10. a.** Quadratic   **b.** $\{\pm i\sqrt{6}\}$
11. a. Rational   **b.** $\{2, -1\}$ **12. a.** Rational   **b.** $\{-2, 4\}$
13. a. Quadratic   **b.** $\left\{\dfrac{1 \pm \sqrt{17}}{2}\right\}$
14. a. Quadratic   **b.** $\left\{\dfrac{1 \pm \sqrt{13}}{2}\right\}$
15. a. Radical   **b.** $\{3\}$ (The value -1 does not check.)
16. a. Radical   **b.** $\{6\}$ (The value -1 does not check.)
17. a. Quadratic form and radical   **b.** $\{-125, 27\}$
18. a. Quadratic form and radical   **b.** $\{-64, -1\}$
19. a. Absolute value   **b.** $\{-1, 9\}$
20. a. Absolute value   **b.** $\{-2, 6\}$
21. a. Quadratic in form and higher degree polynomial   **b.** $\{\pm 2, \pm 2i\}$
22. a. Quadratic in form and higher degree polynomial   **b.** $\{\pm 5, \pm 5i\}$
23. a. Quadratic form and quadratic   **b.** $\left\{-\dfrac{7}{4}, -\dfrac{3}{2}\right\}$
24. a. Quadratic form and quadratic   **b.** $\left\{-\dfrac{13}{10}, \dfrac{1}{5}\right\}$
25. a. Quadratic form and higher degree polynomial
    b. $\left\{\dfrac{1}{2}, -\dfrac{1}{2}, \sqrt{2}, -\sqrt{2}\right\}$
26. a. Quadratic form and higher degree polynomial
    b. $\{2, -2, \sqrt{3}, -\sqrt{3}\}$

27. a. Quadratic form and higher degree polynomial

$\quad$ **b.** $\left\{ 2, 1, -1 \pm i\sqrt{3}, -\dfrac{1}{2} \pm \dfrac{\sqrt{3}}{2} i \right\}$

28. a. Quadratic form and higher degree polynomial

$\quad$ **b.** $\left\{ 3, -1, -\dfrac{3}{2} \pm \dfrac{3\sqrt{3}}{2} i, \dfrac{1}{2} \pm \dfrac{\sqrt{3}}{2} i \right\}$

29. a. Radical $\quad$ **b.** $\{5, 4\}$ $\quad$ **30. a.** Radical $\quad$ **b.** $\{6, 10\}$

31. a. Radical $\quad$ **b.** $\{0, 2\}$

32. a. Radical $\quad$ **b.** $\{-1\}$ (The value -6 does not check.)

33. a. Absolute value $\quad$ **b.** $\{\ \}$ $\quad$ **34. a.** Absolute value $\quad$ **b.** $\{\ \}$

35. a. Quadratic, quadratic form, and rational $\quad$ **b.** $\left\{ 1, \dfrac{17}{2} \right\}$

36. a. Quadratic, quadratic form, and rational $\quad$ **b.** $\{24, -11\}$

37. a. Quadratic form and radical $\quad$ **b.** $\{64, -125\}$

38. a. Quadratic form and radical $\quad$ **b.** $\{32, 1\}$

39. a. Quadratic form and higher degree polynomial $\quad$ **b.** $\left\{ \pm\sqrt{2}, \pm 2i \right\}$

40. a. Quadratic form and higher degree polynomial $\quad$ **b.** $\left\{ \pm\dfrac{\sqrt{2}}{2}, \pm i \right\}$

41. a. Higher degree polynomial $\quad$ **b.** $\{\pm 4i, 1\}$

42. a. Higher degree polynomial $\quad$ **b.** $\{\pm 3i, 1\}$

43. a. Higher degree polynomial $\quad$ **b.** $\left\{ 4, \pm i\sqrt{5} \right\}$

44. a. Higher degree polynomial $\quad$ **b.** $\left\{ 3, \pm 2i\sqrt{2} \right\}$

45. a. Quadratic form and rational $\quad$ **b.** $\left\{ \dfrac{8}{3}, 2 \right\}$

46. a. Quadratic form and rational $\quad$ **b.** $\left\{ -\dfrac{7}{2}, -\dfrac{3}{8} \right\}$

Chapter 4 Review Exercises, pp. 328–329

1. a. $w = 2, w = -2$ $\quad$ **b.** There are no restricted values.

3. $2a$ $\quad$ **5.** $x - 1$ $\quad$ **7.** $-\dfrac{x^2 + 3x + 9}{3 + x}$ $\quad$ **9.** $-\dfrac{2t + 5}{t + 7}$

11. $\dfrac{a}{2}$ $\quad$ **13.** $-\dfrac{x - y}{5x}$ or $\dfrac{y - x}{5x}$ $\quad$ **15.** $(x + 8)^2$ $\quad$ **17.** $-\dfrac{1}{b}$

19. $\dfrac{5(y^2 + 1)}{7(y^2 - 2y + 4)}$ $\quad$ **21.** $\dfrac{x^2 + x - 1}{x^3}$ $\quad$ **23.** $\dfrac{y - 3}{2y - 1}$ or $\dfrac{3 - y}{1 - 2y}$

25. $\dfrac{4k^2 - k + 3}{(k + 1)^2(k - 1)}$ $\quad$ **27.** $\dfrac{2(a^2 - 5)}{(a - 5)(a + 3)}$ $\quad$ **29.** $\dfrac{9a^2 + a + 4}{(3a - 1)(a - 2)}$

31. $\dfrac{x(2y + 1)}{4y}$ $\quad$ **33.** $\dfrac{1 + a}{1 - a}$ or $-\dfrac{a + 1}{a - 1}$ $\quad$ **35.** $\dfrac{y}{x - y}$

37. $\{3\}$ (The value 1 does not check.) $\quad$ **39.** $\{0, 17\}$ $\quad$ **41.** $\{5, 1\}$

43. $x = \dfrac{b}{c - a}$ or $x = \dfrac{-b}{a - c}$ $\quad$ **45.** $\left\{ \dfrac{15}{2} \right\}$ $\quad$ **47.** $\left\{ -\dfrac{7}{11} \right\}$

49. The quarterback would gain 231 yd.

51. Tony averaged 20 mph the first day and 15 mph the second day.

53. It would take $\frac{40}{9}$ hr.

55. a. 1000 $\quad$ **b.** $\dfrac{1}{1000}$ $\quad$ **c.** -1000 $\quad$ **d.** $-\dfrac{1}{1000}$

$\quad$ **e.** Undefined $\quad$ **f.** Undefined

57. $\dfrac{1}{2}$ $\quad$ **59.** p

61. a. $\sqrt[7]{x^2}$ or $\left(\sqrt[7]{x}\right)^2$ $\quad$ **b.** $9\sqrt[7]{x^2}$ or $9\left(\sqrt[7]{x}\right)^2$

$\quad$ **c.** $\sqrt[7]{(9x)^2}$ or $\left(\sqrt[7]{9x}\right)^2$

63. 2.1544 $\quad$ **65.** $x^{3/2}(x - 9)(x - 10)$ $\quad$ **67.** $\dfrac{3x + 5}{(2x + 5)^{3/4}}$

69. $\left\{ \dfrac{49}{2} \right\}$ $\quad$ **71.** $\{-12\}$ $\quad$ **73.** $\{9\}$ $\quad$ **75.** $\left\{ \dfrac{1}{2}, 4 \right\}$

77. a. 25.3 ft/sec $\quad$ **b.** 8 ft $\quad$ **79.** $\{4, 16\}$ $\quad$ **81.** $\left\{ \pm\dfrac{\sqrt{6}}{2}, \pm i \right\}$

83. $\{32, 1\}$ $\quad$ **85.** $\left\{ \dfrac{1 \pm \sqrt{181}}{6} \right\}$ $\quad$ **87.** $\{\pm\sqrt{7}, \pm 2\}$

Chapter 4 Test, p. 330

1. $x = \dfrac{1}{5}, x = -5$ $\quad$ **2.** $\dfrac{2m^2}{3n}$ $\quad$ **3.** $\dfrac{9(x + 1)}{3x + 5}$

4. $-(x - 3)$ or $-x + 3$ $\quad$ **5.** $x - 4$ $\quad$ **6.** $\dfrac{x^2 + 5x + 2}{x + 1}$

7. $\dfrac{3}{4}$ $\quad$ **8.** $\dfrac{u^2 v^2}{2(v^2 - uv + u^2)}$ $\quad$ **9.** $-a$ $\quad$ **10.** $\dfrac{1}{(x + 5)(x + 3)}$

11. $\dfrac{3x}{2(x + 7)}$ $\quad$ **12.** $\dfrac{y - 3}{y^2}$ $\quad$ **13.** $x + 3$ $\quad$ **14.** $\{3\}$

15. $\{0, 4\}$ $\quad$ **16.** $\{\ \}$ (The value 4 does not check.)

17. $T = \dfrac{1}{p - v}$ or $T = \dfrac{-1}{v - p}$ $\quad$ **18.** $m_1 = \dfrac{Fr^2}{Gm_2}$

19. $\dfrac{1}{6}$ or 2 $\quad$ **20.** $a = 14$ m, $y = 15$ m

21. The cities are 1960 mi apart.

22. Lance rides 16 mph against the wind and 20 mph with the wind.

23. It would take $\frac{20}{7}$ hr working together.

24. a. 10 $\quad$ **b.** $\dfrac{1}{1000}$ $\quad$ **c.** 10,000 $\quad$ **d.** Undefined because $\sqrt{-1,000,000}$ is not a real number. $\quad$ **e.** 10,000 $\quad$ **f.** $-\dfrac{1}{100}$

25. $\dfrac{1}{t^{12}}$ $\quad$ **26.** $(2x)^{1/3}$ $\quad$ **27.** $5\sqrt[3]{y}$ $\quad$ **28.** $x^{5/2}(2x + 3)^2$

29. $\dfrac{12(x + 2)}{(x + 4)^{1/2}}$ $\quad$ **30.** $\{-16\}$ $\quad$ **31.** $\left\{ \dfrac{17}{5} \right\}$

32. $\{2\}$ (The value 42 does not check.) $\quad$ **33.** 2 in.

34. $\{9\}$ (The value 4 does not check.) $\quad$ **35.** $\{8, -64\}$

36. $\left\{ \dfrac{11}{3}, 6 \right\}$ $\quad$ **37.** $\left\{ \pm\sqrt{6}, \pm 3 \right\}$

CHAPTER 5

Section 5.1 Practice Exercises, pp. 340–344

R.1. $4\sqrt{3}$ $\quad$ **R.3. a.** No $\quad$ **b.** Yes $\quad$ **R.5. a.** Yes $\quad$ **b.** No

R.7. $y = \dfrac{c - ax}{b}$ or $y = \dfrac{c}{b} - \dfrac{ax}{b}$ $\quad$ **R.9.** 10 $\quad$ **R.11.** 2 $\quad$ **R.13.** $\{-11, 9\}$

R.15. $\{\pm 3\}$ $\quad$ **R.17.** $\{x \mid x \geq 7\}$ $\quad$ **R.19.** $\{x \mid x < -9\}$

1. origin $\quad$ **3.** $d = \sqrt{(x_2 - x_1)^2 + (y_2 - y_1)^2}$ $\quad$ **5.** solution $\quad$ **7.** 0

9.

11. a. Quadrant II $\quad$ **b.** Quadrant III $\quad$ **c.** x-axis $\quad$ **d.** y-axis

13. a. $2\sqrt{5}$ $\quad$ **b.** $(-3, 9)$ $\quad$ **15. a.** $9\sqrt{2}$ $\quad$ **b.** $\left(-\dfrac{5}{2}, \dfrac{1}{2} \right)$

17. a. 5 $\quad$ **b.** $(3.7, -4.4)$ $\quad$ **19.** Yes $\quad$ **21.** No

23. a. Yes   **b.** No   **c.** Yes   **25.** $\{x \mid x \neq 3\}$ **27.** $\{x \mid x \geq 10\}$
29. $\{x \mid x \leq 1.5\}$

31. 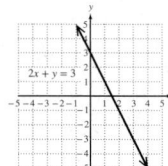   **33.**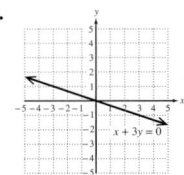

35. 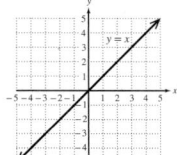   **37.**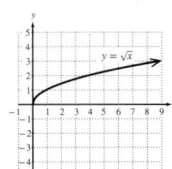

39. 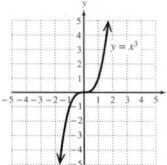   **41.**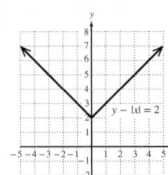

43. 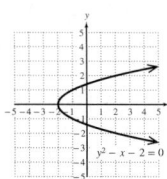   **45.**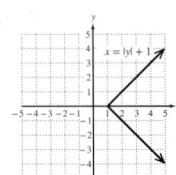

47.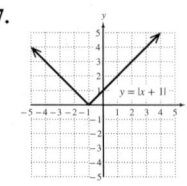

49. x-intercepts: $(-1, 0)$, $(9, 0)$; y-intercepts: $(0, -3)$, $(0, 3)$
51. x-intercept: $(-2, 0)$; y-intercept: None
53. x-intercept: $(0, 0)$; y-intercept: $(0, 0)$
55. x-intercept: $(-6, 0)$; y-intercept: $(0, 3)$
57. x-intercepts: $(-3, 0)$, $(3, 0)$; y-intercept: $(0, 9)$
59. x-intercepts: $(3, 0)$, $(7, 0)$; y-intercept: $(0, 3)$
61. x-intercept: $(-1, 0)$; y-intercepts: $(0, -1)$, $(0, 1)$
63. x-intercept: $(0, 0)$; y-intercept: $(0, 0)$
65. x-intercept: None; y-intercept: None
67. $d(A, C) = 2\sqrt{26}$ mi and $d(B, C) = 2\sqrt{17}$ mi
69. a. 457 pixels   **b.** $(223, 184)$ **c.** $(317, 119)$
71. From the distance formula,
$d(A, B) = |x|$, $d(A, C) = |x|$, and $d(B, C) = |x|$.
73. a. Length: $3\sqrt{2}$ ft; Width: $2\sqrt{2}$ ft
    b. Perimeter: $10\sqrt{2}$ ft; Area: 12 ft²
75. Center: $(1, 2)$; Radius: $\sqrt{10}$ **77.** Area: 25 m²   **79.** Collinear
81. Not collinear
83. The points (x_1, y_1) and (x_2, y_2) define the endpoints of the hypotenuse
d of a right triangle. The lengths of the legs of the triangle are
$|x_2 - x_1|$ and $|y_2 - y_1|$. Applying the Pythagorean theorem produces
$d^2 = |x_2 - x_1|^2 + |y_2 - y_1|^2$, or equivalently
$d = \sqrt{(x_2 - x_1)^2 + (y_2 - y_1)^2}$ for $d \geq 0$.
85. To find the x-intercept(s), substitute 0 for y and solve for x. To find the
y-intercept(s), substitute 0 for x and solve for y.   **87.** $\sqrt{91}$ **89.** $9\sqrt{2}$
91. The viewing window is part of the Cartesian plane shown in the dis-
play screen of a calculator. The boundaries of the window are often
denoted by [Xmin, Xmax, Xscl] by [Ymin, Ymax, Yscl].

93. 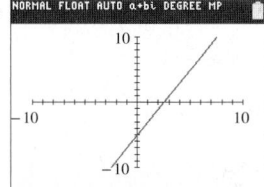   **95.**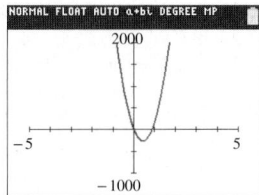

97.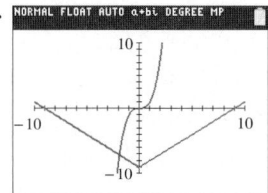

Section 5.2 Practice Exercises, pp. 349–351

R.1. $n = 196$; $(x + 14)^2$ **R.3.** $n = \dfrac{121}{4}$; $\left(a - \dfrac{11}{2}\right)^2$

R.5. $n = \dfrac{1}{81}$; $\left(m - \dfrac{1}{9}\right)^2$ **R.7.** $6\sqrt{5}$ **R.9.** $x^2 - 18x + 81$

R.11. $x^2 + \dfrac{4}{3}x + \dfrac{4}{9}$ **R.13.** $(x - 13)^2$ **R.15.** $\left(\dfrac{1}{10}x + \dfrac{1}{7}\right)^2$

R.17. $y = \pm\sqrt{9 - x^2}$

1. circle; center   **3.** $(x - h)^2 + (y - k)^2 = r^2$ **5.** No   **7.** Yes
9. Center: $(4, -2)$; Radius: 9   **11.** Center: $(0, 2.5)$; Radius: 2.5
13. Center: $(0, 0)$; Radius: $2\sqrt{5}$ **15.** Center: $\left(\dfrac{3}{2}, -\dfrac{3}{4}\right)$; Radius: $\dfrac{9}{7}$
17. a. $(x + 2)^2 + (y - 5)^2 = 1$ **19. a.** $(x + 4)^2 + (y - 1)^2 = 9$
    b. 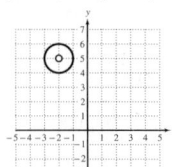   **b.**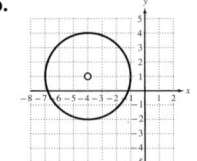

21. a. $(x + 4)^2 + (y + 3)^2 = 11$ **23. a.** $x^2 + y^2 = 6.76$
    b. 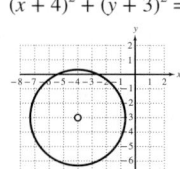   **b.**

25. a. $(x - 2)^2 + (y - 1)^2 = 25$ **27. a.** $(x + 2)^2 + (y + 1)^2 = 100$
    b. 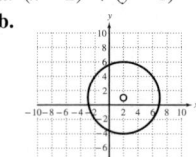   **b.**

29. a. $(x - 4)^2 + (y - 6)^2 = 16$ **31. a.** $(x - 5)^2 + (y + 5)^2 = 25$
    b. 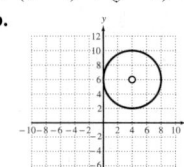   **b.**

33. $(x - 8)^2 + (y + 11)^2 = 25$ **35.** $\left(x - \sqrt{7}\right)^2 + \left(y - \sqrt{7}\right) = 7$
37. $\{(-1, 5)\}$ **39.** $\{\ \}$
41. $(x + 3)^2 + (y - 1)^2 = 4$; Center: $(-3, 1)$; Radius: 2
43. $(x - 11)^2 + (y + 3)^2 = 1$; Center: $(11, -3)$; Radius: 1

45. $x^2 + (y - 10)^2 = 104$; Center: $(0, 10)$; Radius: $2\sqrt{26}$

47. $(x - 4)^2 + (y + 10)^2 = 24$; Center: $(4, -10)$; Radius: $2\sqrt{6}$

49. $(x - 2)^2 + (y - 9)^2 = -4$; Degenerate case: { }

51. $x^2 + \left(y - \dfrac{5}{2}\right)^2 = 0$; Degenerate case; (single point): $\left\{ \left(0, \dfrac{5}{2}\right) \right\}$

53. $\left(x - \dfrac{1}{2}\right)^2 + \left(y - \dfrac{3}{4}\right)^2 = \dfrac{25}{16}$; Center: $\left(\dfrac{1}{2}, \dfrac{3}{4}\right)$; Radius: $\dfrac{5}{4}$

55. $(x - 4)^2 + (y - 6)^2 = 2.25$

57.

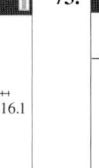

59. A circle is the set of all points in a plane that are equidistant from a fixed point called the center.

61. $y = -2$ and $y = 14$

63. $\left(3 + \sqrt{17}, 3 + \sqrt{17}\right)$ and $\left(3 - \sqrt{17}, 3 - \sqrt{17}\right)$

65. a. 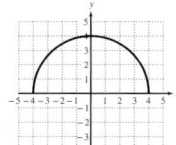   **b.**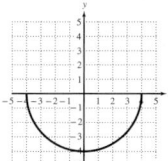

c. 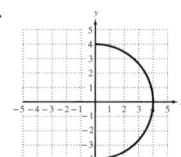   **d.**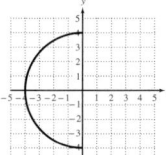

67. a. 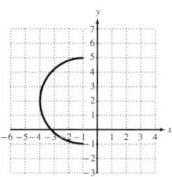   **b.**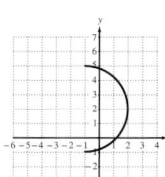

c. 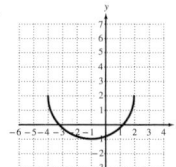   **d.**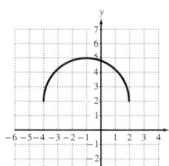

69. $\sqrt{49 - 12\sqrt{5}}$ or $3\sqrt{5} - 2$

71. 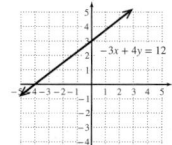   **73.**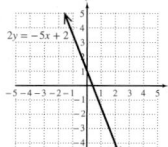

Section 5.3 Practice Exercises, pp. 360–366

R.1. $\{\pm\sqrt{5}\}$ **R.3.** $\{-4, -1\}$ **R.5.** $\{-10, 6\}$ **R.7.** $[-8, \infty)$

R.9. $(-\infty, 3)$ **R.11. a.** $(5, 0)$ **b.** $(0, -4)$

R.13. a. $(4, 0), (-4, 0)$ **b.** $(0, -4)$ **R.15.** $t^2 + 2th + h^2$

1. relation; domain; y **3.** y **5.** -5

7. Yes. For a given time after the tree is planted, there cannot be two or more different heights. That is, the height is unique at any given time.

9. a. {(Tom Hanks, 5), (Jack Nicholson, 12), (Sean Penn, 5), (Dustin Hoffman, 7)}   **b.** {Tom Hanks, Jack Nicholson, Sean Penn, Dustin Hoffman}   **c.** {5, 12, 7}   **d.** Yes

11. a. {(-4, 3), (-2, -3), (1, 4), (3, -2), (3, 1)}   **b.** {-4, -2, 1, 3}   **c.** {3, -3, 4, -2, 1}   **d.** No   **13.** False   **15.** Yes   **17.** No

19. Yes   **21.** Yes   **23.** No   **25.** Yes   **27.** No   **29.** Yes

31. a. Yes   **b.** No   **33.** (4, 1)

35. a. -2 **b.** -2 **c.** 0   **d.** 4   **e.** 10

37. a. 5   **b.** 5   **c.** 5   **d.** 5   **e.** 5   **39.** $\dfrac{1}{3}$ **41.** 3   **43.** Undefined

45. 3   **47.** $\dfrac{1}{t}$ **49.** $\sqrt{x + h + 1}$ **51.** $a^2 + 11a + 28$ **53.** Undefined

55. $x^2 + 2xh + h^2 + 3x + 3h$ **57.** $-4x^2 - 8xh - 4h^2 - 5x - 5h + 2$

59. $-3x^2 - 6xh - 3h^2 + 7$ **61.** $x^3 + 3x^2h + 3xh^2 + h^3 + 2x + 2h - 5$

63. 7   **65.** 4   **67.** -1 **69.** 2

71. a. $d(2) = 36$; Joe rides 36 mi in 2 hr.   **b.** 12 mi

73. $C(225) = 279$; If the cost of the food is \$225, then the total bill including tax and tip is \$279.

75. x-intercept: $(2, 0)$; y-intercept: $(0, -4)$

77. x-intercepts: $(8, 0), (-8, 0)$; y-intercept: $(0, -8)$

79. x-intercepts: $\left(2\sqrt{3}, 0\right), \left(-2\sqrt{3}, 0\right)$; y-intercept: $(0, 12)$

81. x-intercept: $(8, 0)$; y-intercept: $(0, 8)$

83. x-intercept: $(4, 0)$; y-intercept: $(0, -2)$

85. The y-intercept is $(0, 14{,}820)$ and means that the amount owed after the initial down payment is \$14,820. The t-intercept is $(60, 0)$ and means that after 60 months, the amount owed is \$0.

87. Domain: $\{-3, -2, -1, 2, 3\}$; Range: $\{-4, -3, 3, 4, 5\}$

89. Domain: $(-3, \infty)$; Range: $[1, \infty)$

91. Domain: $(-\infty, \infty)$; Range: $(-\infty, \infty)$

93. Domain: $(-\infty, \infty)$; Range: $[-3, \infty)$

95. Domain: $(-5, 1]$; Range: $\{-1, 1, 3\}$

97. a. $(-\infty, 4) \cup (4, \infty)$ **b.** $(-\infty, -2) \cup (-2, 2) \cup (2, \infty)$
   c. $(-\infty, \infty)$ **99. a.** $[-9, \infty)$ **b.** $(-\infty, 9]$ **c.** $(-9, \infty)$

101. a. $(-\infty, \infty)$ **b.** $(-\infty, \infty)$ **c.** $(-\infty, 5) \cup (5, \infty)$

103. a. $(-\infty, \infty)$ **b.** $(-\infty, -4) \cup (-4, 7) \cup (7, \infty)$
   c. $(-\infty, -2) \cup (-2, \infty)$

105. a. $(-\infty, \infty)$ **b.** $(-\infty, \infty)$ **c.** $(-\infty, -5) \cup (-5, 3) \cup (3, \infty)$

107. a. $(-\infty, \infty)$ **b.** $\left[\dfrac{5}{7}, \infty\right)$ **c.** $\left(-\infty, \dfrac{5}{7}\right) \cup \left(\dfrac{5}{7}, \infty\right)$

109. a. $[-15, \infty)$ **b.** $[-15, \infty)$ **c.** $[-15, -11) \cup (-11, \infty)$

111. a. $(-\infty, \infty)$ **b.** $[0, \infty)$ **c.** $[0, 7)$

113. a. -4 **b.** 2   **c.** $x = -3, x = -1, x = 1$ **d.** $x = -2, x = 2$
   e. $(0, 0)$ and $\left(-\dfrac{10}{3}, 0\right)$ **f.** $(0, 0)$ **g.** $(-\infty, \infty)$ **h.** $[-4, \infty)$

115. a. 0   **b.** 5   **c.** $x = -3, x = -1, x = 1$ **d.** $x = -4, x = 0$
   e. $(-2, 0)$ and $\left(\dfrac{4}{3}, 0\right)$ **f.** $(0, -4)$ **g.** $[-4, \infty)$ **h.** $[-4, 5]$

117. $r(x) = 400 - x$ **119.** $P(x) = 3x$ **121.** $C(x) = 90 - x$

123. $f(x) = 3x^2 - 2$

125. a. $P(s) = 4s$ **b.** $A(s) = s^2$ **c.** $A(P) = \left(\dfrac{P}{4}\right)^2$ or $A(P) = \dfrac{P^2}{16}$
   d. $P(A) = 4\sqrt{A}$ **e.** $d(s) = \sqrt{2}s$ **f.** $s(d) = \dfrac{d}{\sqrt{2}}$ or $s(d) = \dfrac{d\sqrt{2}}{2}$
   g. $P(d) = 2\sqrt{2}d$ **h.** $A(d) = \dfrac{d^2}{2}$

Section 5.4 Practice Exercises, pp. 376–381

R.1. $y = 3x - 5$ **R.3. a.** -4 **b.** $x = \dfrac{4}{5}$

R.5. x-intercepts: $(-3, 0), (3, 0)$; y-intercept: $(0, 9)$ **R.7.** $(-2, \infty)$

R.9. $(-\infty, -10]$ **R.11.** -4

1. linear   **3.** horizontal   **5.** zero; undefined   **7.** x

9. x-intercept: $(-4, 0)$; y-intercept: $(0, 3)$

11. x-intercept: $\left(\dfrac{2}{5}, 0\right)$; y-intercept: $(0, 1)$

13. *x*-intercept: $(-6, 0)$;
y-intercept: None

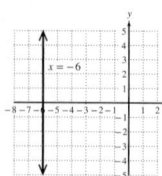

15. *x*-intercept: None;
y-intercept: $(0, 2)$

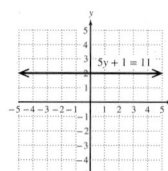

17. *x*-intercept: $(5, 0)$;
y-intercept: $(0, 2)$

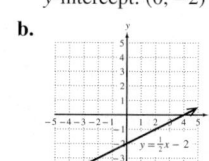

19. *x*-intercept: $(0, 0)$;
y-intercept: $(0, 0)$

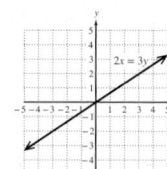

21. $m = \dfrac{3}{10}$ **23.** $m = \dfrac{1}{40}$ **25.** $m = -3$ **27.** $m = -\dfrac{3}{5}$ **29.** $m = -\dfrac{1}{4}$

31. $m = -\dfrac{26}{23}$ **33.** $m = -\dfrac{20}{7}$ **35.** Undefined   **37.** $m = 3$

39. $m = -\dfrac{1}{3}$ **41.** $m = 0$ **43.** Undefined   **45.** 0   **47.** 41.6 ft

49. Change in population over change in time

51. a. $y = \frac{1}{2}x - 2; m = \frac{1}{2}$;
y-intercept: $(0, -2)$
b.

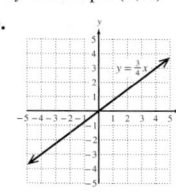

53. a. $y = \frac{3}{2}x + 2; m = \frac{3}{2}$;
y-intercept: $(0, 2)$
b.

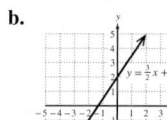

55. a. $y = \frac{3}{4}x; m = \frac{3}{4}$;
y-intercept: $(0, 0)$
b.

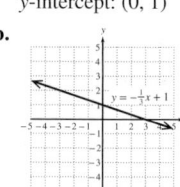

57. a. $y = 7; m = 0$;
y-intercept: $(0, 7)$
b.

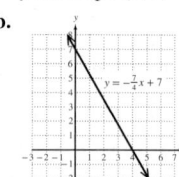

59. a. $y = -\frac{1}{3}x + 1; m = -\frac{1}{3}$;
y-intercept: $(0, 1)$
b.

61. a. $y = -\frac{7}{4}x + 7; m = -\frac{7}{4}$;
y-intercept: $(0, 7)$
b.

63. a. Linear   **b.** Linear   **c.** Neither   **d.** Constant

65. a. $y = \dfrac{1}{2}x + 9$ **b.** $f(x) = \dfrac{1}{2}x + 9$

67. a. $y = -3x - 3$ **b.** $f(x) = -3x - 3$

69. a. $y = \dfrac{2}{3}x + \dfrac{1}{3}$ **b.** $f(x) = \dfrac{2}{3}x + \dfrac{1}{3}$

71. a. $y = 5$ **b.** $f(x) = 5$

73. a. $y = 1.2x + 0.78$ **b.** $f(x) = 1.2x + 0.78$

75. a. $y = 2x - 6$ **b.** $f(x) = 2x - 6$

77. a. $y = -\dfrac{4}{3}x + \dfrac{19}{3}$ **b.** $f(x) = -\dfrac{4}{3}x + \dfrac{19}{3}$

79. *x*-intercept: $\left(-\dfrac{9}{5}, 0\right)$; y-intercept: $(0, -9)$; $m = -5$

81. *x*-intercept: $(0, 0)$; y-intercept: $(0, 0)$; $m = \dfrac{2}{11}$

83. *x*-intercept: None; y-intercept: $(0, 6)$; $m = 0$

85. a. $x = 3$ **b.** $x = 2$ **c.** 4   **d.** $x = 6$ **87. a.** $(-\infty, 2]$ **b.** $(2, \infty)$

89. a. $(1, \infty)$ **b.** $(-\infty, 1]$ **91. a.** $\{-1\}$ **b.** $(-\infty, -1)$ **c.** $[-1, \infty)$

93. a. $\{2\}$ **b.** $(-\infty, 2)$ **c.** $[2, \infty)$

95. a. $\{-5\}$ **b.** $[-5, \infty)$ **c.** $(-\infty, -5]$

97. a. $\{14\}$ **b.** $(14, \infty)$ **c.** $(-\infty, 14)$

99. The line will be slanted if both A and B are nonzero. If A is zero and B is not zero, then the equation can be written in the form $y = k$ and the graph is a horizontal line. If B is zero and A is not zero, then the equation can be written in the form $x = k$, and the graph is a vertical line.

101. 4 units^2 **103.** 10 units^2

105. a. $y = -\dfrac{A}{B}x + \dfrac{C}{B}$ **b.** $m = -\dfrac{A}{B}$ **c.** $\left(0, \dfrac{C}{B}\right)$

107. a. $\{-1.5\}$ **b.** $(-\infty, -1.5)$ **c.** $(-1.5, \infty)$

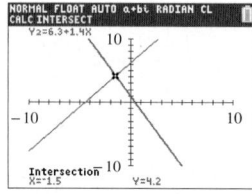

109. a. $\{-4, 7.8\}$ **b.** $(-\infty, -4] \cup [7.8, \infty)$ **c.** $[-4, 7.8]$

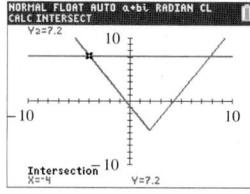   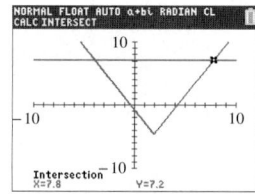

111. The lines are not exactly the same. The slopes are different.

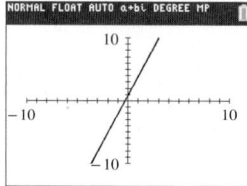

Section 5.5 Practice Exercises, pp. 389–396

R.1. Opposite: 4; reciprocal: $-\dfrac{1}{4}$ **R.3.** Opposite: $-\dfrac{2}{3}$; reciprocal: $\dfrac{3}{2}$

R.5. $y = -5x + 8$ **R.7.** $y = \dfrac{3}{5}x + 3$; slope $= \dfrac{3}{5}$; y-intercept is $(0, 3)$

R.9. $m = -3$ **R.11.** Slope is undefined   **R.13.** $C = 0.12k + 14.89$

1. $y - y_1 = m(x - x_1)$ **3.** -1 **5.** $y = -2x - 1$ **7.** $y = \dfrac{2}{3}x + \dfrac{2}{3}$

9. $y = 1.2x - 1.48$ **11.** $y = \dfrac{1}{9}x + \dfrac{4}{3}$ **13.** $y = -\dfrac{8}{5}x + 8$

15. $y = 3.5x - 2.95$ **17.** $y = -4$ **19.** $x = \dfrac{2}{3}$ **21.** Undefined

23. a. $m = \dfrac{3}{11}$ **b.** $m = -\dfrac{11}{3}$ **25. a.** $m = -6$ **b.** $m = \dfrac{1}{6}$

27. a. $m = 1$ **b.** $m = -1$ **29.** Perpendicular   **31.** Parallel

33. Perpendicular   **35.** Neither   **37.** $y = -2x + 9; 2x + y = 9$

39. $y = -5x + 26; 5x + y = 26$ **41.** $y = \dfrac{3}{7}x + \dfrac{38}{7}; 3x - 7y = -38$

43. $y = 0.5x + 5.3; 5x - 10y = -53$ **45.** $y = 6$

47. $y = -\dfrac{3}{4}$ **49.** $x = -61.5$

51. a. $S(x) = 0.12x + 400$ for $x \geq 0$ **b.** $S(8000) = 1360$ means that the sales person will make \$1360 if \$8000 in merchandise is sold for the week.

53. a. $T(x) = 0.019x + 172$ for $x > 0$
b. $T(80,000) = 1692$ means that the property tax is \$1692 for a home with a taxable value of \$80,000.

55. a. $C(x) = 34.5x + 2275$ **b.** $R(x) = 80x$ **c.** $P(x) = 45.5x - 2275$
d. 50 items   **57. a.** {730}   **b.** [0, 730)   **c.** (730, ∞)

59. a. $C(x) = 2.88x + 790$ **b.** $R(x) = 6x$ **c.** $P(x) = 3.12x - 790$
d. The business will make a profit if it produces and sells 254 dozen or more cookies.   **e.** The business will lose \$322.

61. a. $y = -1.25x + 1312.5$
b. $m = -1.25$ mph/mb means that for an increase of 1 mb in pressure, the wind speed decreases by 1.25 mph.
c. 125 mph
d. No. There is no guarantee that the linear trend continues outside the interval of the observed data points.

63. a. $y = 2.75x + 29.5$
b. $m = 2.75$ means that the average height of girls increased by 2.75 in. per year during this time period.
c. 59.75 in.   **d.** 66.4 in.

65. a.

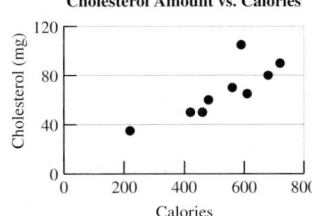

b. $c(x) = 0.125x$
c. $m = 0.125$ means that the amount of cholesterol increases at an average rate of 0.125 mg per calorie of hamburger.
d. 81.25 mg

67. Yes   **69.** No

71. a. $y = -1.4x + 1456$
b.

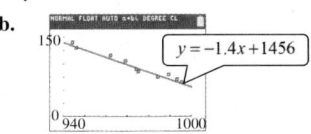

c. 126 mph   **d.** The results differ by 1 mph.

73. a. $y = 2.48x + 31.0$
b.

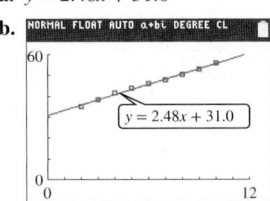

c. 58.28 in.   **d.** 64.8 in.   **e.** 1.6 in.

75. a. $y = 0.118x + 4.97$
b.

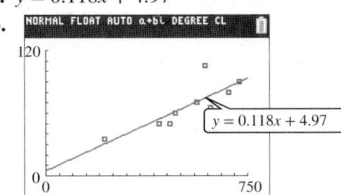

c. Approximately 82 mg

77. $\left(\frac{8}{5}, 0\right)$ **79.** $f(x) = \frac{7}{3}x + 4$ **81.** $h(x) = x + 5$

83. If the slopes of the two lines are the same and the y-intercepts are different, then the lines are parallel. If the slope of one line is the opposite of the reciprocal of the slope of the other line, then the lines are perpendicular.

85. Profit is equal to revenue minus cost.   **87.** $y = -x + 4$

89. $y = 12x + 17$ **91.** $y = -\frac{5}{2}x + \frac{21}{2}$ for $1 \leq x \leq 5$

Problem Recognition Exercises, p. 396

1. b, f   **2.** a, c, d, h   **3.** a   **4.** b, g   **5.** c, e

6. a   **7.** c, h   **8.** b   **9.** e   **10.** g   **11.** c, h

12. a   **13.** g   **14.** e   **15.** h   **16.** f   **17.** e

18. g   **19.** d, h   **20.** b, f

Problem Recognition Exercises, p. 397

1. 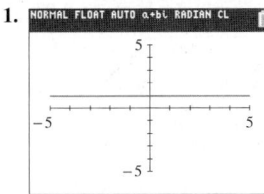   **2.**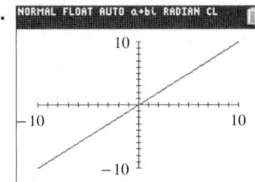

3. 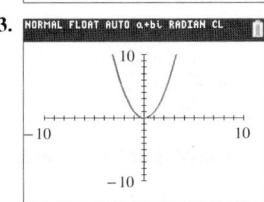   **4.**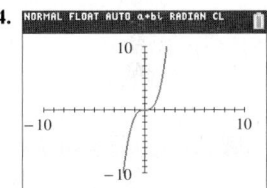

5. 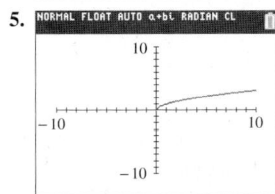   **6.**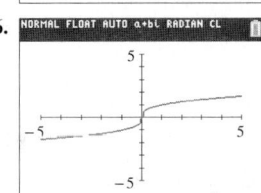

7. 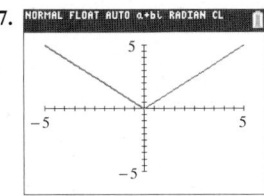   **8.**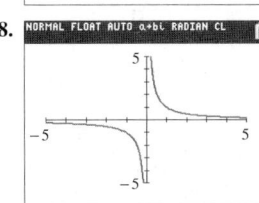

9. The graphs have the shape of $y = x^2$ with a vertical shift.

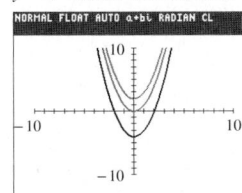

10. The graphs have the shape of $y = |x|$ with a vertical shift.

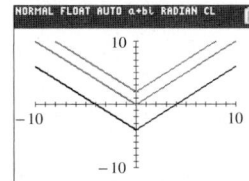

11. The graphs have the shape of $y = \sqrt{x}$ with a horizontal shift.

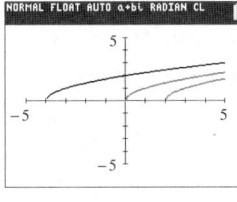

12. The graphs have the shape of $y = x^2$ with a horizontal shift.

13. The graph of $g(x) = -|x|$ has the shape of the graph of $y = |x|$ but is reflected across the x-axis.

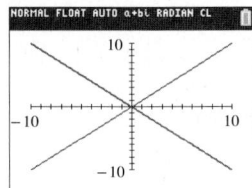

15. The graphs have the shape of $y = x^2$ but show a vertical shrink or stretch.

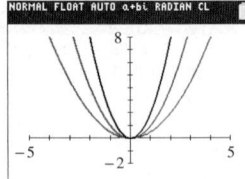

17. The graph of $g(x) = \sqrt{-x}$ has the shape of the graph of $y = \sqrt{x}$ but is reflected across the y-axis.

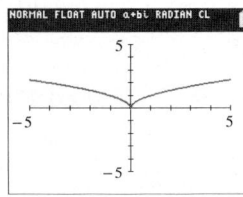

14. The graph of $g(x) = -\sqrt{x}$ has the shape of the graph of $y = \sqrt{x}$ but is reflected across the x-axis.

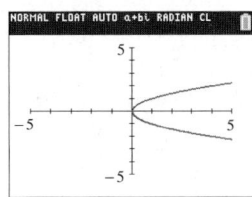

16. The graphs have the shape of $y = |x|$ but show a vertical shrink or stretch.

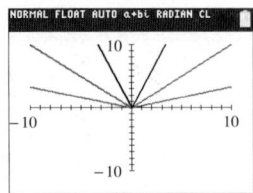

18. The graph of $g(x) = \sqrt[3]{-x}$ has the shape of the graph of $y = \sqrt[3]{x}$ but is reflected across the y-axis.

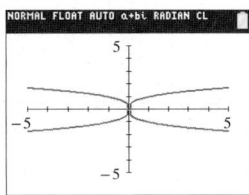

21. a. {(Dara Torres, 12), (Carl Lewis, 10), (Bonnie Blair, 6), (Michael Phelps, 16)}
   b. {Dara Torres, Carl Lewis, Bonnie Blair, Michael Phelps}
   c. {12, 10, 6, 16}   **d.** Yes

23. No   **25.** Yes   **27. a.** 5   **b.** 4   **c.** $x = 3$

29. x-intercepts: (4, 0), (2, 0); y-intercept: (0, 2)

31. Domain: $\{-4, -2, 0, 2, 3, 5\}$; Range: $\{-3, 0, 2, 1\}$

33. $\left(-\infty, -\dfrac{9}{2}\right) \cup \left(-\dfrac{9}{2}, \infty\right)$ **35.** $[2, \infty)$

37. a. -2 **b.** -1 **c.** $x = -1, x = 3$
   d. $x = -4$ **e.** (0, 0), (2, 0)
   f. (0, 0)   **g.** $(-\infty, \infty)$ **h.** $(-\infty, 1]$

39. x-intercept: $(-4, 0)$;   **41.** x-intercept: None;
   y-intercept: (0, 2) y-intercept: (0, 2)

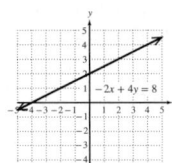   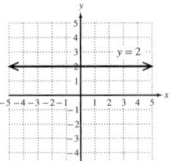

43. $m = \dfrac{1}{8}$ **45.** 0   **47.** Undefined

49. $y = -\dfrac{2}{3}x - \dfrac{13}{3}$; $f(x) = -\dfrac{2}{3}x - \dfrac{13}{3}$

51. a. {1}   **b.** $(-\infty, 1)$ **c.** $[1, \infty)$ **53. a.** $\dfrac{2}{3}$ **b.** $-\dfrac{3}{2}$

55. $y = 3x - 1$ **57.** $y = -0.9x + 6.29$ **59.** $y = 2x - 10$ **61.** $y = 7$

63. a. $C(x) = 1500 + 35x$ **b.** $R(x) = 60x$ **c.** $P(x) = 25x - 1500$
   d. The studio needs more than 60 private lessons per month to make a profit.   **e.** The studio will make $550.

65. a. $y = 11.9x + 169$
   b. 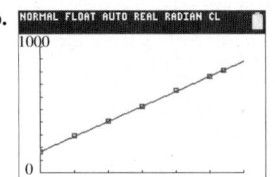   **c.** 1002 m

Chapter 5 Review Exercises, pp. 402–405

1. a. $5\sqrt{5}$ **b.** $\left(\dfrac{3}{2}, 3\right)$ **3. a.** Yes   **b.** No

5. x-intercept: (4, 0); y-intercepts: (0, −4), (0, −10)

7.

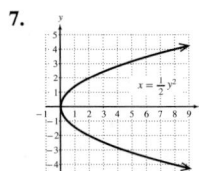

9. $4\sqrt{5}$ units   **11.** Center: $\left(0, \dfrac{3}{2}\right)$; Radius: $\sqrt{17}$

13. a. $(x + 3)^2 + (y - 1)^2 = 11$ **15. a.** $(x - 4)^2 + (y - 1)^2 = 25$
   b.   **b.**

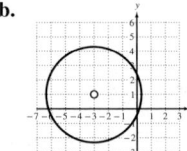   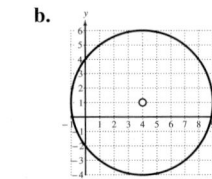

17. a. $(x + 5)^2 + (y - 1)^2 = 9$ **b.** Center: $(-5, 1)$; Radius: 3

19. $\{(-3, 5)\}$

Chapter 5 Test, pp. 405–406

1. a. $2\sqrt{13}$ **b.** $(2, -4)$

2. a. $(3, -1)$ **b.** $\sqrt{41}$ **c.** $(x - 3)^2 + (y + 1)^2 = 41$

3. a. x-intercept: $(-4, 0)$; y-intercepts: (0, 4), (0, −4)   **b.** No

4. a. $(x + 7)^2 + (y - 5)^2 = 4$ **b.** Center: $(-7, 5)$; Radius: 2

5. Yes   **6.** No

7. a. -12 **b.** $-2x^2 - 4xh - 2h^2 + 7x + 7h - 3$
   c. $\left(\dfrac{1}{2}, 0\right)$ and (3, 0)   **d.** (0, −3)

8. a. -2 **b.** 0   **c.** $x = -2$ and $x = 2$
   d. $(-\infty, \infty)$ **e.** $(-\infty, 2]$

9. $\left(-\infty, -\dfrac{7}{3}\right) \cup \left(-\dfrac{7}{3}, \infty\right)$ **10** $(-\infty, 4]$

11. a. $m = -\dfrac{3}{4}$ **b.** (0, 2)

   c. 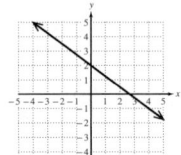   **d.** $\dfrac{4}{3}$ **e.** $-\dfrac{3}{4}$

12. $y = 3x + 12$ **13.** $y = 4x - 18$

14. 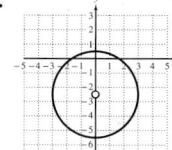   **15.**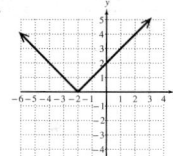

16. a. $\{-2\}$ **b.** $(-\infty, -2)$ **c.** $[-2, \infty)$

17. a.

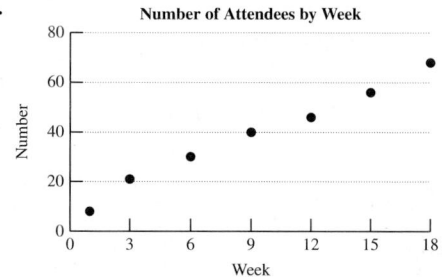

b. $y = 4x + 4$

c. The slope is 4 and means that the number of attendees has increased by approximately 4 people per week.

d. 100 people

18. a. $y = 3.3x + 8.5$

b. 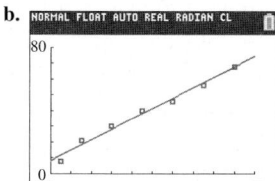   **c.** 88 people

CHAPTER 6

Section 6.1 Practice Exercises, pp. 419–423

R.1. 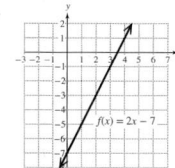   **R.3.**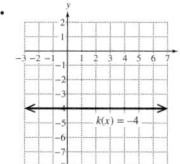

R.5. Each function represents a line with a slope of 1. However, the y-intercept of f is $(0, 0)$ whereas the y-intercept of g is $(0, -1)$. This means that the two lines have an identical "slant," but the graph of g is shifted down one unit from the graph of f.

1. left   **3.** down   **5.** horizontal shrink

7. vertical shrink   **9.** e   **11.** b   **13.** a

15. 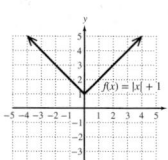   **17.**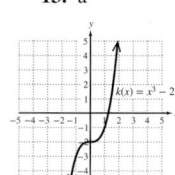

19. 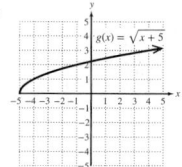   **21.**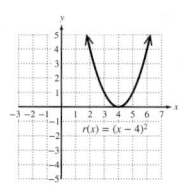

23. 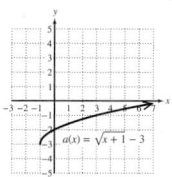   **25.**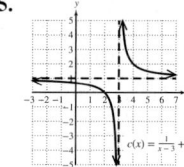

27. 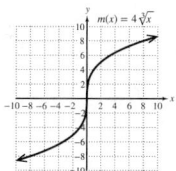   **29.**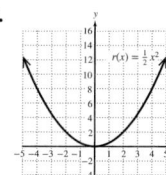

31. 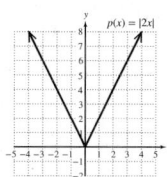   **33.**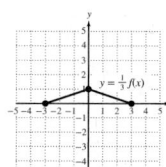

35. 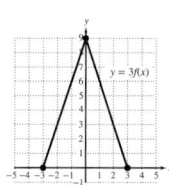   **37.**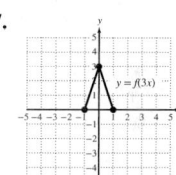

39. 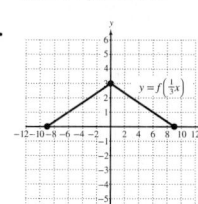   **41.**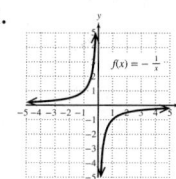

43. 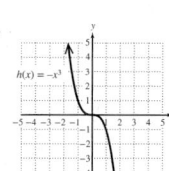   **45.**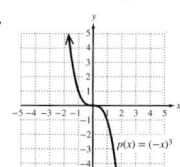

47. 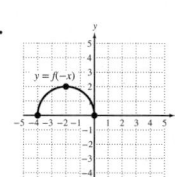   **49.**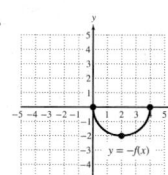

51. 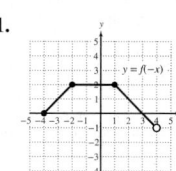   **53.**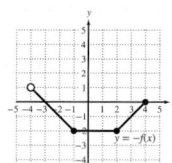

55. Parent function: $f(x) = \dfrac{1}{x}$. Shift the graph of f to the left 1 unit, stretch the graph vertically by a factor of 3, and shift the graph downward by 2 units.

57. Parent function: $f(x) = x^2$. Shift the graph of f to the right 2.1 units, shrink the graph vertically by a factor of $\frac{1}{3}$, and shift the graph upward 7.9 units.

59. Parent function: $f(x) = \sqrt{x}$. Shift the graph of f to the left 5 units, shrink the graph horizontally by a factor of $\frac{1}{2}$, stretch the graph vertically by a factor of 2, and reflect the graph across the y-axis.

61. Parent function: $f(x) = \sqrt{x}$. Stretch the graph of f horizontally by a factor of 3, reflect across the x-axis, and shift the graph downward 6 units.

63.
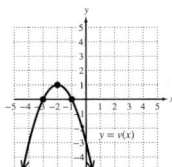

65.
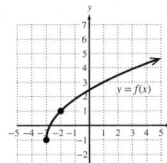

67.
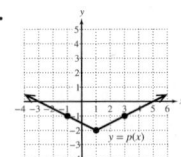

69.
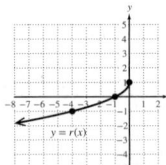

71.
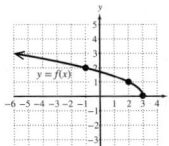

73.
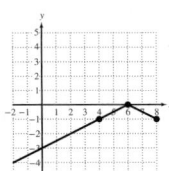

75.
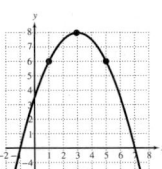

77.
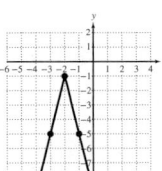

79.
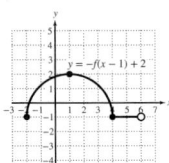

81.
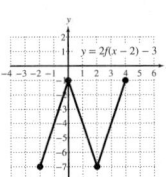

83.
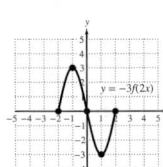

85.
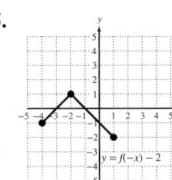

87. $y = (-x + 4.5)^3 + 2.1$

89. $y = -\dfrac{2}{x} - 3$

91. As written, $g(x) = |2x|$ is in the form $g(x) = f(ax)$ with $a > 1$. This indicates a horizontal shrink. However, $g(x)$ can also be written as $g(x) = |2| \cdot |x| = 2|x|$. This is written in the form $g(x) = af(x)$ with $a > 1$. This represents a vertical stretch.

93. The graph of f is the same as the graph of $y = |x|$ with a horizontal shift to the right 2 units and a vertical shift downward 3 units. By contrast, the graph of g is the graph of $y = |x|$ with a horizontal shift to the right 3 units and a vertical shift downward 2 units.

95. $f(x) = (x - 2)^2 - 3$

97. $f(x) = \dfrac{1}{x + 3}$

99. $f(x) = -x^3 + 1$

101. $y = 2t^2 + 1$

103. a.
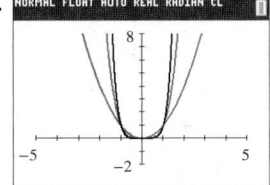
b.
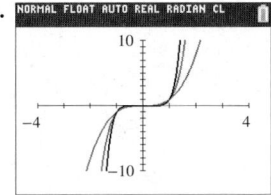

c. The general shape of $y = x^n$ is similar to the graph of $y = x^2$ for even values of n greater than 1.

d. The general shape of $y = x^n$ is similar to the graph of $y = x^3$ for odd values of n greater than 1.

Section 6.2 Practice Exercises, pp. 432–437

R.1. The points are vertically aligned and equidistant from the x-axis. The graph of the points is symmetric with respect to the x-axis.

R.3. The points are equidistant from the origin along a common line through the origin. The graph of the points is symmetric with respect to the origin.

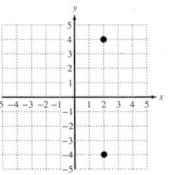
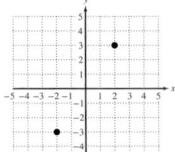

R.5. a. 4   **b.** 4   **c.** 28   **d.** 28

R.7. a. -99 **b.** 99   **c.** -1 **d.** 1

R.9. Interval notation: $(-\infty, 8)$

R.11. Interval notation: $[-2.4, 5.8)$

1. y **3.** origin   **5.** origin   **7.** y-axis   **9.** x-axis

11. x-axis, y-axis, and origin   **13.** None of these

15. x-axis, y-axis, and origin   **17.** None of these

19. y-axis symmetry   **21.** Odd   **23.** Even

25. Neither even nor odd

27. a. $f(-x) = 4x^2 - 3|x|$ **b.** Yes   **c.** Even

29. a. $h(-x) = -4x^3 + 2x$ **b.** $-h(x) = -4x^3 + 2x$
 c. Yes   **d.** Odd

31. a. $m(-x) = 4x^2 - 2x - 3$ **b.** $-m(x) = -4x^2 - 2x + 3$
 c. No   **d.** No   **e.** Neither

33. Even   **35.** Odd   **37.** Neither   **39.** Even

41. Odd   **43.** Neither   **45.** Odd

47. a. 12   **b.** 13   **c.** 4   **d.** 5   **e.** 5

49. a. 1   **b.** 2   **c.** -1 **d.** 1   **e.** -1

51. 15 ft/sec; 30 ft/sec; 2.7 ft/sec; and 1.4 ft/sec

53. c   **55.** d

57. a.   **b.**   **c.**
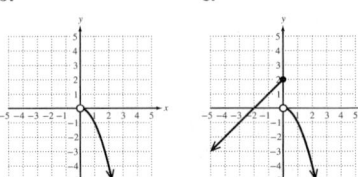

59. a.   **b.**   **c.**

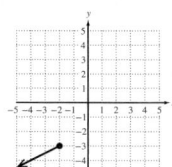   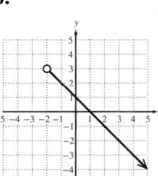

61.   **63.**

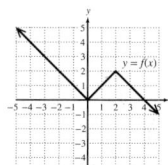   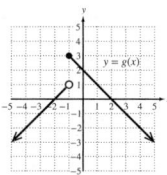

65.   **67.**

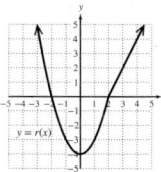   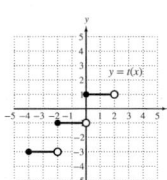

69.   **71. a.**   **b.** $y = |x|$

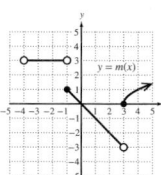   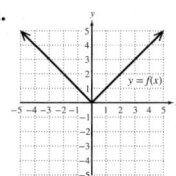

73. -5 **75.** -1

77. 0 **79.** -9

81.   **83.**

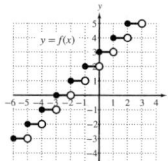   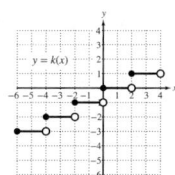

85. $C(x) = \begin{cases} 0.44 & \text{for } 0 < x \le 1 \\ 0.61 & \text{for } 1 < x \le 2 \\ 0.78 & \text{for } 2 < x \le 3 \\ 0.95 & \text{for } 3 < x \le 3.5 \end{cases}$

87. $S(x) = \begin{cases} 2000 & \text{for } 0 \le x < 40{,}000 \\ 2000 + 0.05(x - 40{,}000) & \text{for } x \ge 40{,}000 \end{cases}$

89. $f(x) = \begin{cases} -2 & \text{for } x < 1 \\ 3 & \text{for } x \ge 1 \end{cases}$ **91.** $f(x) = \begin{cases} -|x| & \text{for } x < 2 \\ -2 & \text{for } x \ge 2 \end{cases}$

93. $f(x) = \begin{cases} \dfrac{1}{x} & \text{for } x < 0 \\ x & \text{for } x > 0 \end{cases}$

95. a.   **b.** $(-\infty, \infty)$ **c.** $(-\infty, 1] \cup (2, \infty)$

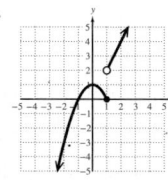   **d.** $f(-1) = 0, f(1) = 0,$ and $f(2) = 4$
e. $x = 3$ **f.** $x = -2$

97. a. 2 **b.** -4 **c.** 4 **d.** 3 **e.** -3 **f.** 4

99. If replacing y by $-y$ in the equation results in an equivalent equation, then the graph is symmetric to the x-axis. If replacing x by $-x$ in the equation results in an equivalent equation, then the graph is symmetric to the y-axis. If replacing both x by $-x$ and y by $-y$ results in an equivalent equation, then the graph is symmetric to the origin.

101. At $x = 1$, there are two different y values. The relation contains the ordered pairs $(1, 2)$ and $(1, 3)$.

103. $f(x) = \begin{cases} 0.10x & \text{if } 0 < x \le 9525 \\ 952.50 + 0.12(x - 9525) & \text{if } 9525 < x \le 38{,}700 \\ 4453.50 + 0.22(x - 38{,}700) & \text{if } 38{,}700 < x \le 82{,}500 \end{cases}$

or

$f(x) = \begin{cases} 0.10x & \text{if } 0 < x \le 9525 \\ 0.12x - 190.50 & \text{if } 9525 < x \le 38{,}700 \\ 0.22x - 4060.50 & \text{if } 38{,}700 < x \le 82{,}500 \end{cases}$

105. 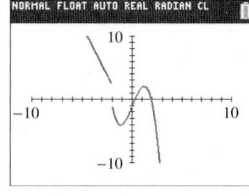   **107.**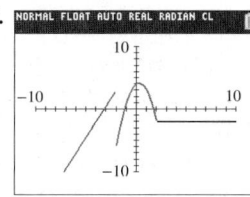

Section 6.3 Practice Exercises, pp. 445–450

R.1. c   **R.3.** b   **R.5.** down   **R.7.** 3 and 4

R.9. 5 and 6   **R.11.** $4x + 4h - 5$ **R.13.** $2x^2 + 4xh + 2h^2$

1. $m = \dfrac{f(x_2) - f(x_1)}{x_2 - x_1}$ **3.** decreasing   **5.** $m = \dfrac{3}{2}$

7. a. \$364.80/year   **b.** \$772.20/year   **c.** Increasing

9. a. 1   **b.** 4   **c.** -2 **11. a.** 1   **b.** 1   **c.** 7

13. a. 1   **b.** $\dfrac{1}{3}$ **c.** $\dfrac{1}{5}$

15. a. $(2, \infty)$ **b.** $(-3, -2)$ **c.** $(-2, 2)$

17. a. $(-\infty, \infty)$ **b.** Never decreasing   **c.** Never constant

19. a. $(1, \infty)$ **b.** $(-\infty, 1)$ **c.** Never constant

21. a. $(-\infty, -2)$ and $(2, \infty)$ **b.** Never decreasing   **c.** $(-2, 2)$

23. At $x = 1$, the function has a relative minimum of -3.

25. At $x = -2$, the function has a relative minimum of 0. At $x = 0$, the function has a relative maximum of 2. At $x = 2$, the function has a relative minimum of 0.

27. At $x = -2$, the function has a relative minimum of -4. At $x = 0$, the function has a relative maximum of 0. At $x = 2$, the function has a relative minimum of -4.

29. a. $(8, 12)$ and $(18, 20)$ **b.** $(0, 8)$ and $(12, 18)$

c. The function has relative minima of 3 ft and 3.5 ft at approximately 8 days and 18 days after recording began. The function has a relative maximum of 4.5 ft at a time 12 days after recording began.

d. The weather was dry on the intervals of decreasing depth, and water from the pond evaporated. The weather was rainy during intervals of increasing depth.

31. a. $2x + 2h - 3$ **b.** 2   **33. a.** $5x + 5h + 9$ **b.** 5

35. a. $x^2 + 2xh + h^2 + 4x + 4h$ **b.** $2x + h + 4$

37. -2 **39.** $-10x - 5h - 4$

41. $3x^2 + 3xh + h^2$ **43.** $-\dfrac{1}{x(x + h)}$

45. a. $\dfrac{4\sqrt{x + h} - 4\sqrt{x}}{h}$ **b.** 1.6569; 1.9524; 1.9950; 1.9995   **c.** 2

47. a. −262; The number of new flu cases dropped by 262 per month during this time interval.

b. Between months 4 and 6: −683 cases/month; Between months 10 and 12: −110/month

c.

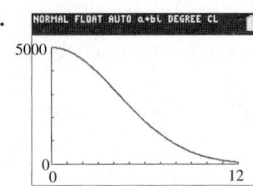

The number of new flu cases dropped slowly during the first two months. Then the rate of new cases dropped more rapidly between months 4 and 6 (perhaps as health department officials managed the outbreak). Finally, the rate of new cases dropped more slowly toward the end of the outbreak.

49. Relative minimum

51. a. Concave down   **b.** Decreasing

53. a. Concave up   **b.** Decreasing

55. The average rate of change of f on the interval $[x_1, x_2]$ is the slope of the secant line passing through the points $(x_1, f(x_1))$ and $(x_2, f(x_2))$.

57. A function is increasing on an interval I if for all $x_1 < x_2$ on I, $f(x_1) < f(x_2)$. In other words, the function exclusively "rises" from left to right over the interval.

59. a. $\dfrac{\sqrt{x + h + 3} - \sqrt{x + 3}}{h}$ **b.** $\dfrac{1}{\sqrt{x + h + 3} + \sqrt{x + 3}}$

c. $\dfrac{1}{2\sqrt{x + 3}}$

61. a. $-9.68t - 4.84h + 88$ **b.** 78.32 ft/sec   **c.** 58.96 ft/sec

d. 39.6 ft/sec   **e.** 20.24 ft/sec

63. a. Relative maximum of 4.667 at $x = 1.667$

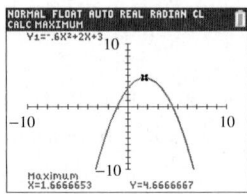

b. Increasing on $(-\infty, 1.667)$; Decreasing on $(1.667, \infty)$

65. a. Relative maximum of 7.824 at $x = -3.390$; Relative minimum of −7.936 at $x = 0.590$

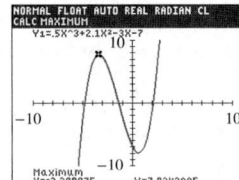 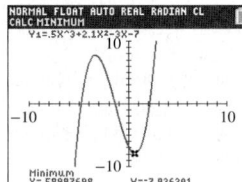

b. Increasing on $(-\infty, -3.390)$ and $(0.590, \infty)$ Decreasing on $(-3.390, 0.590)$

Problem Recognition Exercises, p. 450

1. a. $(-\infty, \infty)$

b. $[-3, \infty)$

c. −3

d. $x = -3$ and $x = 5$

e. $(-2, 0)$ and $(4, 0)$

f. $(0, -2)$

g. Increasing on $(1, \infty)$

h. Decreasing on $(-\infty, 1)$

i. Never constant

j. Relative minimum of −3 at $x = 1$

2. a. $(-\infty, \infty)$

b. $(-\infty, 5]$

c. 1

d. $x = -3$ and $x = 1$

e. $\left(-1 - \sqrt{5}, 0\right)$ and $\left(-1 + \sqrt{5}, 0\right)$

f. $(0, 4)$

g. Increasing on $(-\infty, -1)$

h. Decreasing on $(-1, \infty)$

i. Never constant

j. Relative maximum of 5 at $x = -1$

3. a. $(-\infty, \infty)$

b. $(-\infty, \infty)$

c. $-\dfrac{3}{2}$

d. $x = 6$

e. $(4, 0)$

f. $(0, -2)$

g. Increasing on $(-\infty, \infty)$

h. Never decreasing

i. Never constant

j. No relative minima or maxima

4. a. $(-\infty, \infty)$

b. $(-\infty, \infty)$

c. −1

d. $x = 0$

e. $\left(\dfrac{1}{2}, 0\right)$

f. $(0, 1)$

g. Never increasing

h. Decreasing on $(-\infty, \infty)$

i. Never constant

j. No relative minima or maxima

5. a. $(-\infty, \infty)$

b. $[0, \infty)$

c. 1

d. $x = -1$ and $x = 1$

e. $(0, 0)$

f. $(0, 0)$

g. Increasing on $(0, 2)$

h. Decreasing on $(-\infty, 0)$

i. Constant on $(2, \infty)$

j. Relative minimum of 0 at $x = 0$

6. a. $(-\infty, 1) \cup (1, \infty)$

b. $(-\infty, 0) \cup (0, \infty)$

c. Undefined

d. $x = 2$

e. None

f. $(0, -1)$

g. Never increasing

h. Decreasing on $(-\infty, 1)$ and $(1, \infty)$

i. Never constant

j. No relative maximum or minimum

Section 6.4 Practice Exercises, pp. 458–462

R.1. $(-\infty, -2) \cup (-2, \infty)$ **R.3.** $(-\infty, 2]$ **R.5.** $(-\infty, 3)$

R.7. $(-\infty, \infty)$ **R.9. a.** −2   **b.** 1   **R.11. a.** 5   **b.** 2

1. $f(x); g(x)$ **3.** $f(g(x))$ **5.** $(f + g)(x) = |x| + 3$; Graph d

7. $(f + g)(x) = x^2 - 4$; Graph a

9. −13   **11.** 6   **13.** $\dfrac{11}{3}$ **15.** $-\dfrac{4}{3}$ **17.** Undefined

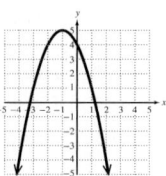

19. $(r - p)(x) = -x^2 - 6x; (-\infty, \infty)$

21. $(p \cdot q)(x) = (x^2 + 3x)\sqrt{1 - x}; (-\infty, 1]$

23. $\left(\dfrac{q}{p}\right)(x) = \dfrac{\sqrt{1 - x}}{x^2 + 3x}; (-\infty, -3) \cup (-3, 0) \cup (0, 1]$

25. $\left(\dfrac{p}{q}\right)(x) = \dfrac{x^2 + 3x}{\sqrt{1 - x}}; (-\infty, 1)$

27. $(s \cdot t)(x) = \dfrac{-1}{x + 3}$; Domain: $(-\infty, -3) \cup (-3, 2) \cup (2, 3) \cup (3, \infty)$

29. $(s + t)(x) = -\dfrac{x^3 - 4x^2 - 5x + 23}{(x + 3)(x - 3)(x - 2)}$

Domain: $(-\infty, -3) \cup (-3, 2) \cup (2, 3) \cup (3, \infty)$

31. $(s \cdot v)(x) = \dfrac{(x - 2)\sqrt{x + 3}}{(x - 3)(x + 3)}$; Domain: $(-3, 3) \cup (3, \infty)$

33. 48   **35.** -3 **37.** 192   **39.** $\sqrt{210}$ **41.** -27

43. Undefined (not a real number)   **45.** 3315   **47.** 315

49. a. $(f \circ g)(x) = 2x^2 + 4$ **b.** $(g \circ f)(x) = 4x^2 + 16x + 16$

c. No

51. $(n \circ p)(x) = x^2 - 9x - 5; (-\infty, \infty)$

53. $(m \circ n)(x) = \sqrt{x + 3}; [-3, \infty)$

55. $(q \circ n)(x) = \dfrac{1}{x - 15}; (-\infty, 15) \cup (15, \infty)$

57. $(q \circ r)(x) = \dfrac{1}{|2x + 3| - 10}; \left(-\infty, -\dfrac{13}{2}\right) \cup \left(-\dfrac{13}{2}, \dfrac{7}{2}\right) \cup \left(\dfrac{7}{2}, \infty\right)$

59. $(n \circ r)(x) = |2x + 3| - 5; (-\infty, \infty)$

61. $(n \circ n)(x) = x - 10; (-\infty, \infty)$

63. $(f \circ g)(x) = \dfrac{9x^2}{9 - x^2}; (-\infty, -3) \cup (-3, 0) \cup (0, 3) \cup (3, \infty)$

65. $(f \circ g)(x) = \dfrac{2}{\sqrt{x - 1} - 4}; [1, 17) \cup (17, \infty)$

67. $(f \circ f)(x) = \dfrac{x - 2}{-2x + 5}; (-\infty, 2) \cup \left(2, \dfrac{5}{2}\right) \cup \left(\dfrac{5}{2}, \infty\right)$

69. $(f \circ g \circ h)(x) = 2\left(\sqrt[3]{x}\right)^2 + 1$

71. $(h \circ g \circ f)(x) = \sqrt[3]{(2x + 1)^2}$

73. a. $C(x) = 21.95x$ **b.** $T(a) = 1.06a + 10.99$

c. $(T \circ C)(x) = 23.267x + 10.99$

d. $(T \circ C)(4) = 104.058$; The total cost to purchase 4 boxes of stationery is $104.06.

75. a. $r(t) = 80t$ **b.** $d(r) = 7.2r$

c. $(d \circ r)(t) = 576t$ represents the distance traveled (in ft) in t minutes.

d. $(d \circ r)(30) = 17,280$ means that the bicycle will travel 17,280 ft (approximately 3.27 mi) in 30 min.

77. $f(x) = x^2$ and $g(x) = x + 7$ **79.** $f(x) = \sqrt[3]{x}$ and $g(x) = 2x + 1$

81. $f(x) = |x|$ and $g(x) = 2x^2 - 3$ **83.** $f(x) = \dfrac{5}{x}$ and $g(x) = x + 4$

85. a. 1   **b.** -1 **c.** -6 **d.** $-\dfrac{1}{2}$ **e.** 1   **f.** 1   **g.** -2

87. a. -1 **b.** 0   **c.** Undefined   **d.** -3

e. -3 **f.** 0   **g.** 2

89. 1   **91.** 3   **93.** 6   **95.** Undefined

97. a. $d(r) = 2r$ **b.** $r(d) = \dfrac{d}{2}$

c. $(V \circ r)(d) = \dfrac{1}{6}\pi d^3$ is the volume of the sphere as a function of its diameter.

99. $(A \circ A)(x) = (1.045)^2 x$ represents the amount of money in the account after 2 years compounded annually.

101. a. $A_1(x) = \pi(x + 5)^2$ represents the area of the outer circle in terms of the radius of the inner circle, x.

b. $A_2(x) = \pi x^2$ represents the area of the inner circle based on its radius, x.

c. $(A_1 - A_2)(x) = \pi(x + 5)^2 - \pi x^2$ or $10\pi x + 25\pi$ represents the area of the region outside the inner circle and inside the outer circle.

103. $c(b) = (b + 8)^2$ **105.** $z(y) = 2y - 4$

107. $m(x) = \sqrt[3]{x}, n(x) = x + 1, h(x) = 4x, k(x) = x^2$

109. $(g \circ g)(x) = \sqrt{\sqrt{x - 1} - 1}; [2, \infty)$

111. $(f \circ g)(x) = -\dfrac{3}{x + 14}; (-\infty, -14) \cup (-14, 2]$

113. $(f \circ g)(x) = \dfrac{9}{25 - x^2}; (-\infty, -5) \cup (-5, -4) \cup (-4, 4) \cup (4, 5) \cup (5, \infty)$

Chapter 6 Review Exercises, pp. 467–469

1. $y = (-x + 5)^2 - 2$

3. 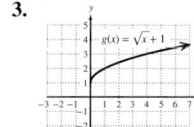   **5.**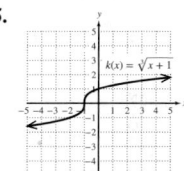

7. 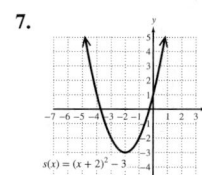   **9.**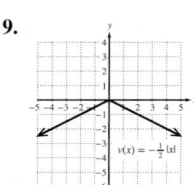

11. 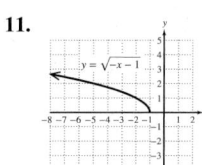   **13.**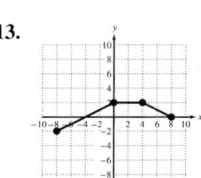

15. 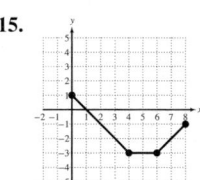   **17.**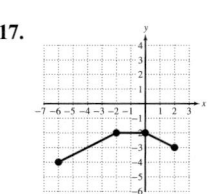

19. x-axis   **21.** y-axis, x-axis, and origin   **23.** y-axis

25. Odd   **27.** Even   **29.** Neither

31. a. 18   **b.** 1 **c.** 5   **d.** 4

33. 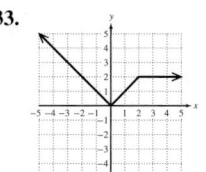   **35. a.** -3 **b.** -3 **c.** -1 **d.** 5

37. $m = \dfrac{3}{7}$ **39. a.** 0   **b.** 2

41. a. \$755.60 per year   **b.** \$1858.60 per year   **c.** Increasing

43. a. $(-\infty, -3)$ and $(-2, 0)$ **b.** $(-3, -2)$ and $(0, 3)$ **c.** $(3, \infty)$

45. At $x = -2$, the function has a relative minimum of -2. At $x = 4$, the function has a relative minimum of -1. At $x = 2$, the function has a relative maximum of 1.

47. -6 **49.** $-4x - 2h + 6$ **51.** $\dfrac{1}{4}$ **53.** -9 **55.** -1

57. a. 1   **b.** -2 **c.** Undefined   **d.** 3   **e.** 1   **f.** Undefined

59. $(n - m)(x) = x^2$; Domain: $(-\infty, \infty)$

61. $\left(\dfrac{n}{p}\right)(x) = \dfrac{x^2 - 4x}{\sqrt{x - 2}}$; Domain: $(2, \infty)$

63. $(q \circ n)(x) = \dfrac{1}{x^2 - 4x - 5}$; Domain: $(-\infty, -1) \cup (-1, 5) \cup (5, \infty)$

65. $f(x) = x^2$ and $g(x) = x - 4$

67. a. $d(t) = 60t$ **b.** $n(d) = \dfrac{d}{28}$

   c. $(n \circ d)(t) = \dfrac{15t}{7}$ represents the number of gallons of gasoline used in t hours.

   d. $(n \circ d)(7) = 15$ means that 15 gal of gasoline is used in 7 hr.

Chapter 6 Test, pp. 469–470

1. a. 2   **b.** -1 **c.** $x = -3, x = 1$ **d.** $(-\infty, \infty)$
   e. $(-\infty, \infty)$ **f.** $(-1, 1)$ **g.** $(-\infty, -1)$ and $(1, \infty)$
   h. Never constant   **i.** Relative maximum of 1 at $x = 1$
   j. Relative minimum of -1 at $x = -1$ **k.** Odd function
   l. $(-2, 0), (0, 0),$ and $(2, 0)$ **m.** $(0, 0)$

2. 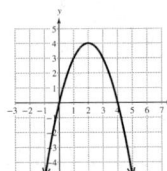   **3.**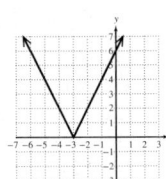

4. 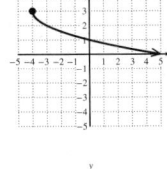   **5.**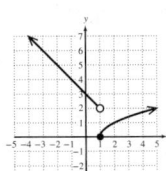

6. 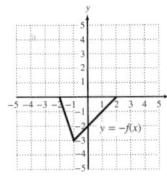   **7.**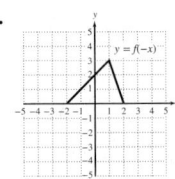

8. 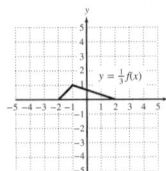   **9.**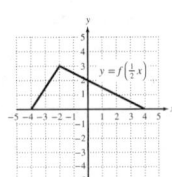

10. a. 3   **b.** $-2x - h + 5$ **11.** x-axis   **12.** x-axis, y-axis, origin
13. Odd   **14.** Neither   **15. a.** 4   **b.** -5
16. a. -4 **b.** -4 **c.** 0   **d.** 4   **e.** 6   **17.** 1   **18.** 0
19. Undefined (not a real number)
20. $(f \cdot g)(x) = \dfrac{x - 4}{x - 3}$; Domain: $(-\infty, 3) \cup (3, \infty)$

21. $\left(\dfrac{g}{f}\right)(x) = \dfrac{1}{(x - 3)(x - 4)}$; Domain: $(-\infty, 3) \cup (3, 4) \cup (4, \infty)$

22. $(g \circ h)(x) = \dfrac{1}{\sqrt{x - 5} - 3}$; Domain: $[5, 14) \cup (14, \infty)$

23. $f(x) = \sqrt[3]{x}$ and $g(x) = x - 7$

24. a. -1 **b.** -1 **c.** -2 **d.** Undefined
   e. $(-3, 0)$ **f.** $(-\infty, 2)$

CHAPTER 7

Section 7.1 Practice Exercises, pp. 481–487

R.1. 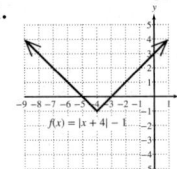   **R.3.**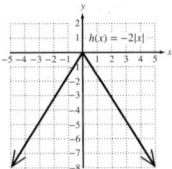

R.5. $n = 81; (x + 9)^2$ **R.7.** $n = \dfrac{121}{4}; \left(x - \dfrac{11}{2}\right)^2$ **R.9.** $\left\{-4, \dfrac{1}{5}\right\}$

R.11. $\left\{\dfrac{1 \pm \sqrt{13}}{3}\right\}$ **R.13.** $\dfrac{5}{6}$ **R.15.** x-intercept: $\left(-\dfrac{7}{2}, 0\right)$;
y-intercept: $(0, 7)$ **R.17.** $x = 2$ **R.19.** Domain: $(-\infty, \infty)$;
Range: $(-\infty, 3]$

1. quadratic   **3.** (h, k) **5.** upward; minimum; k
7. a. Downward   **b.** $(4, 1)$ **c.** $(3, 0)$ and $(5, 0)$ **d.** $(0, -15)$
   e. 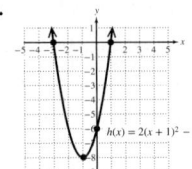   **f.** $x = 4$ **g.** Maximum: 1
   h. Domain: $(-\infty, \infty)$; Range: $(-\infty, 1]$

9. a. Upward   **b.** $(-1, -8)$ **c.** $(-3, 0)$ and $(1, 0)$ **d.** $(0, -6)$
   e. 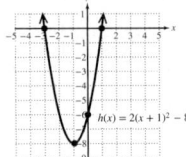   **f.** $x = -1$ **g.** Minimum: -8
   h. Domain: $(-\infty, \infty)$; Range: $[-8, \infty)$

11. a. Upward   **b.** $(1, 0)$ **c.** $(1, 0)$ **d.** $(0, 3)$
   e. 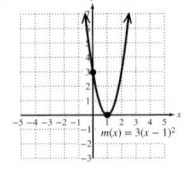   **f.** $x = 1$ **g.** Minimum: 0
   h. Domain: $(-\infty, \infty)$; Range: $[0, \infty)$

13. a. Downward   **b.** $(-4, 1)$ **c.** $\left(-4 + \sqrt{5}, 0\right)$ and $\left(-4 - \sqrt{5}, 0\right)$
   d. $\left(0, -\dfrac{11}{5}\right)$
   e. 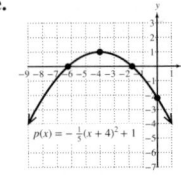   **f.** $x = -4$ **g.** Maximum: 1
   h. Domain: $(-\infty, \infty)$; Range: $(-\infty, 1]$

15. a. $f(x) = (x + 3)^2 - 4$ **b.** $(-3, -4)$ **c.** $(-1, 0)$ and $(-5, 0)$
d. $(0, 5)$
e.

$f(x) = x^2 + 6x + 5$

f. $x = -3$ **g.** Minimum: -4
h. Domain: $(-\infty, \infty)$; Range: $[-4, \infty)$

17. a. $p(x) = 3(x - 2)^2 - 19$ **b.** $(2, -19)$
c. $\left(\dfrac{6 + \sqrt{57}}{3}, 0\right)$ and $\left(\dfrac{6 - \sqrt{57}}{3}, 0\right)$ **d.** $(0, -7)$
e.

$p(x) = 3x^2 - 12x - 7$

f. $x = 2$ **g.** Minimum: -19
h. Domain: $(-\infty, \infty)$; Range: $[-19, \infty)$

19. a. $c(x) = -2\left(x + \dfrac{5}{2}\right)^2 + \dfrac{33}{2}$ **b.** $\left(-\dfrac{5}{2}, \dfrac{33}{2}\right)$
c. $\left(\dfrac{-5 + \sqrt{33}}{2}, 0\right)$ and $\left(\dfrac{-5 - \sqrt{33}}{2}, 0\right)$ **d.** $(0, 4)$
e.

$c(x) = -2x^2 - 10x + 4$

f. $x = -\dfrac{5}{2}$ **g.** Maximum: $\dfrac{33}{2}$
h. Domain: $(-\infty, \infty)$;
Range: $\left(-\infty, \dfrac{33}{2}\right]$

21. a. $h(x) = -2\left(x - \dfrac{7}{4}\right)^2 + \dfrac{49}{8}$ **b.** $\left(\dfrac{7}{4}, \dfrac{49}{8}\right)$
c. $(0, 0)$ and $\left(\dfrac{7}{2}, 0\right)$ **d.** $(0, 0)$
e.

f. $x = \dfrac{7}{4}$ **g.** Maximum: $\dfrac{49}{8}$
h. Domain: $(-\infty, \infty)$;
Range: $\left(-\infty, \dfrac{49}{8}\right]$

23. a. $p(x) = \left(x + \dfrac{9}{2}\right)^2 - \dfrac{13}{4}$ **b.** $\left(-\dfrac{9}{2}, -\dfrac{13}{4}\right)$
c. $\left(\dfrac{-9 + \sqrt{13}}{2}, 0\right)$ and $\left(\dfrac{-9 - \sqrt{13}}{2}, 0\right)$ **d.** $(0, 17)$
e.

f. $x = -\dfrac{9}{2}$ **g.** Minimum: $-\dfrac{13}{4}$
h. Domain: $(-\infty, \infty)$;
Range: $\left[-\dfrac{13}{4}, \infty\right)$

25. $(7, -238)$ **27.** $(9, 28)$ **29.** $(0, -5)$ **31.** $(-0.75, -4.275)$

33. a. Downward   **b.** $(1, -3)$ **c.** None   **d.** $(0, -4)$
e.

$g(x) = -x^2 + 2x - 4$

f. $x = 1$ **g.** Maximum: -3
h. Domain: $(-\infty, \infty)$;
Range: $(-\infty, -3]$

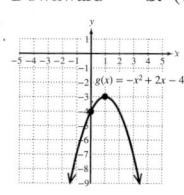

35. a. Upward   **b.** $\left(\dfrac{3}{2}, -\dfrac{33}{4}\right)$
c. $\left(\dfrac{15 + \sqrt{165}}{10}, 0\right)$ and $\left(\dfrac{15 - \sqrt{165}}{10}, 0\right)$ **d.** $(0, 3)$
e.

$f(x) = 5x^2 - 15x + 3$

f. $x = \dfrac{3}{2}$ **g.** Minimum: $-\dfrac{33}{4}$
h. Domain: $(-\infty, \infty)$;
Range: $\left[-\dfrac{33}{4}, \infty\right)$

37. a. Upward   **b.** $(0, 3)$ **c.** None   **d.** $(0, 3)$
e.

f. $x = 0$ **g.** Minimum: 3
h. Domain: $(-\infty, \infty)$; Range: $[3, \infty)$

39. a. Downward   **b.** $(-5, 0)$ **c.** $(-5, 0)$ **d.** $(0, -50)$
e.

f. $x = -5$ **g.** Maximum: 0
h. Domain: $(-\infty, \infty)$; Range: $(-\infty, 0]$

41. a. Upward   **b.** $\left(\dfrac{1}{2}, \dfrac{11}{4}\right)$ **c.** None   **d.** $(0, 3)$
e.

f. $x = \dfrac{1}{2}$ **g.** Minimum: $\dfrac{11}{4}$
h. Domain: $(-\infty, \infty)$;
Range: $\left[\dfrac{11}{4}, \infty\right)$

43. a. \$17   **b.** \$2450   **c.** \$10 and \$24   **45. a.** 3.96 m
b. 0.72 m   **c.** 7.9 m   **47. a.** 24 hr   **b.** 988,000
49. The numbers are 12 and 12.   **51.** The numbers are 5 and -5.
53. a. 40 ft by 80 ft   **b.** 3200 ft^2 **55. a.** $V(x) = -40x^2 + 240x$
b. $x = 3$; The sheet of aluminum should be folded 3 in. from each end.
c. 360 in.^{3} **57. a.** $y = -0.000838t^2 + 0.0812t + 0.040$
b. 48 hr   **c.** 2 g   **59. a.** 15.3 sec   **b.** 16.7 sec
c. $v(x) = -0.1436x^2 - 0.4413x + 195.7$ **d.** 188.9 cm/sec
61. False. If a relation has two y-intercepts, then it would fail the vertical line test and the relation would not define y as a function of x.
63. True   **65.** Discriminant is 0; one x-intercept   **67.** Discriminant is 57; two x-intercepts   **69.** Discriminant is -96; no x-intercepts
71. Graph g   **73.** Graph h   **75.** Graph f   **77.** Graph c
79. For a parabola opening upward, such as the graph of $f(x) = x^2$, the minimum value is the y-coordinate of the vertex. There is no maximum value because the y values of the function become arbitrarily large for large values of $|x|$.
81. No function defined by $y = f(x)$ can have two y-intercepts because the graph would fail the vertical line test.
83. Because a parabola is symmetric with respect to the vertical line through the vertex, the x-coordinate of the vertex must be equidistant from the x-intercepts. Therefore, given $y = f(x)$, the x-coordinate of the vertex is 4 because 4 is midway between 2 and 6. The y-coordinate of the vertex is $f(4)$.
85. $f(x) = 2(x - 2)^2 - 3$ **87.** $f(x) = -\dfrac{1}{3}(x - 4)^2 + 6$
89. $c = 9$ **91.** $b = 4$ or $b = -4$

Section 7.2 Practice Exercises, pp. 499–504

R.1. $-2x^5 - 4x^3 + 9x^2 + 7$, leading term: $-2x^5$; leading coefficient: -2

R.3. $\{-9, 0, 2\}$ **R.5.** $\{\pm 4, \pm\sqrt{2}\}$ **R.7.** x-intercepts: $\left(\dfrac{5}{3}, 0\right)$ and $(-7, 0)$; y-intercept: $(0, -35)$ **R.9. a.** -3 **b.** 11

R.11. even   **R.13.** odd   **R.15.** neither   **R.17.** y-axis

1. polynomial   **3.** is not   **5.** 1   **7.** 5   **9.** cross

11. f has at least one zero on the interval $[a, b]$.   **13.** Down left and down right. As $x \to -\infty, f(x) \to -\infty$, and as $x \to \infty, f(x) \to -\infty$.

15. Down left and up right. As $x \to -\infty, f(x) \to -\infty$, and as $x \to \infty$, $f(x) \to \infty$.   **17.** Up left and down right. As $x \to -\infty, f(x) \to \infty$, and as $x \to \infty, f(x) \to -\infty$.   **19.** Up left and up right. As $x \to -\infty$, $f(x) \to \infty$, and as $x \to \infty, f(x) \to \infty$.   **21.** 3; 4   **23.** $-2, 5, -5$; each of multiplicity 1   **25.** $0, -4, \dfrac{5}{2}$; each of multiplicity 1

27. 0 (multiplicity 3), 5 (multiplicity 2)   **29.** 0 (multiplicity 1), -2 (multiplicity 3), -4 (multiplicity 1)   **31.** $0, \dfrac{5}{3}, -\dfrac{9}{2}, \pm\sqrt{3}$; each of multiplicity 1   **33.** $3 \pm \sqrt{5}$; each of multiplicity 1

35. $-3, -\dfrac{1}{2}, \dfrac{1}{2}, 3$; each of multiplicity 1   **37.** $-\sqrt{7}$ (multiplicity 1), 0 (multiplicity 4), $\sqrt{7}$ (multiplicity 1)

39. a. Yes   **b.** No   **c.** No   **d.** Yes

41. a. Yes   **b.** Yes   **c.** No   **d.** No

43. a. Yes   **b.** $-\dfrac{5}{2}$ **45.** Not a polynomial function. The graph is not smooth.

47. Polynomial function   **a.** Minimum degree 3   **b.** Leading coefficient positive; degree odd   **c.** -4 (odd multiplicity), 1 (odd multiplicity), 3 (odd multiplicity)

49. Polynomial function   **a.** Minimum degree 6   **b.** Leading coefficient negative; degree even   **c.** -4 (odd multiplicity), -3 (odd multiplicity), -1 (even multiplicity), 2 (odd multiplicity), $\dfrac{7}{2}$ (odd multiplicity)

51. Not a polynomial function. The graph is not continuous.

53. a. $y = x^6$ **b.** Shrink $y = x^6$ vertically by a factor of $\dfrac{1}{3}$. Reflect across the x-axis. Shift downward 2 units.   **c.** Graph iii   **55. a.** $y = x^3$ **b.** Shift $y = x^3$ to the left 2 units. Reflect across the x-axis. Shift upward 3 units.   **c.** Graph i   **57. a.** $y = x^5$ **b.** Shift $y = x^5$ to the right 3 units. Reflect across the y-axis. Shift upward 1 unit.   **c.** Graph iv

59.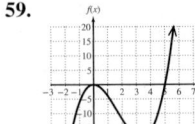

61.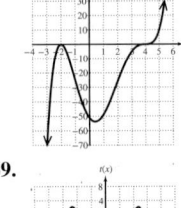

63.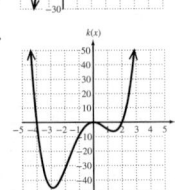

65.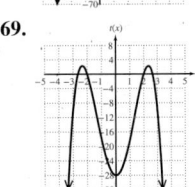

67.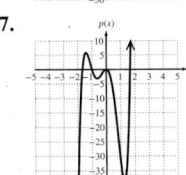

69.

71.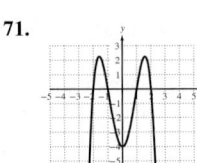

73.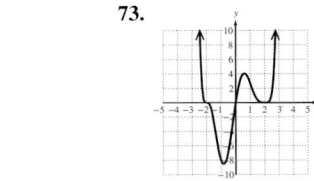

75.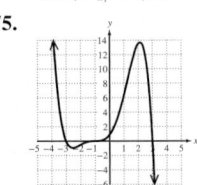

77. False. The value 5 is a zero with even multiplicity. Therefore, the graph touches but does not cross the x-axis at 5.

79. False. An nth-degree polynomial has at most $n - 1$ turning points. Therefore, a third-degree polynomial has at most 2 turning points.

81. True   **83.** False. If the leading coefficient is negative, the graph will be down to the far left and down to the far right.

85. False. The only real solution to the equation $x^3 - 27 = 0$ is $x = 3$. Therefore, the graph of $f(x) = x^3 - 27$ has only one x-intercept.

87. True   **89. a.** $(0, 12)$ and $(68, 184)$ **b.** $(12, 68)$ and $(184, 200)$ **c.** 3   **d.** Degree 4; leading coefficient negative   **e.** 184 sec after launch   **f.** 2.85 G-forces

91. The x-intercepts are the real solutions to the equation $f(x) = 0$.

93. A function is continuous if its graph can be drawn without lifting the pencil from the paper.

95. a. $f(3) = 2; f(4) = 6$ **b.** By the intermediate value theorem, because $f(3) = 2$ and $f(4) = 6$, then f must take on every value between 2 and 6 on the interval $[3, 4]$.   **c.** $x = \dfrac{3 + \sqrt{17}}{2} \approx 3.56$

97. $V(t) = -0.0406t^3 + 0.154t^2 + 0.173t - 0.0024$

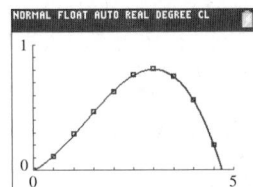

99. a. $V = lwh$
$= (30 - 2x)(24 - 2x)(x)$
$= 4x^3 - 108x^2 + 720x$
The domain is restricted to $0 < x < 12$ because the width of the rectangular sheet is 24 in. The maximum amount that can be removed from each end would be half of 24 in.

b. 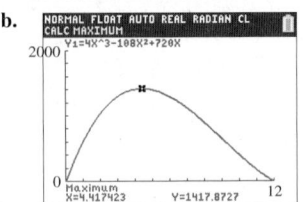   **c.** 1418 in.^{3}

4.4 in.

101. Window b is better.

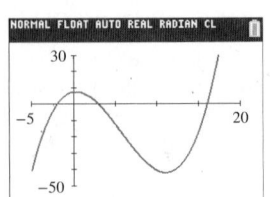

103.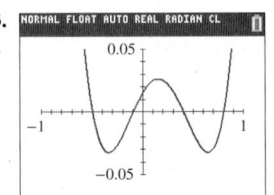

Section 7.3 Practice Exercises, pp. 515–519

R.1. $322\frac{3}{8}$ **R.3.** $4x^4 + 20x^3 + x^2 - 13x - 3$ **R.5. a.** 0   **b.** 0

c. 30   **R.7.** $\{\pm 2i\}$ **R.9.** $\{1 \pm i\sqrt{2}\}$ **R.11.** x-intercepts:

$(3, 0), (-4, 0),$ and $\left(\frac{1}{2}, 0\right)$; y-intercept: $(0, 24)$ **R.13.** $5 + 3i$

R.15. $6 + 8i$ **R.17.** $-14 + 2i$ **R.19.** $-7 + 24i$

R.21. a. yes   **b.** yes

1. Dividend: $f(x)$; Divisor: $d(x)$; Quotient: $q(x)$; Remainder: $r(x)$

3. $f(c)$ **5.** True   **7. a.** $3x + 12 + \dfrac{65}{2x - 5}$ **b.** Dividend:

$6x^2 + 9x + 5$; Divisor: $2x - 5$; Quotient: $3x + 12$; Remainder: 65

c. $(2x - 5)(3x + 12) + 65 = 6x^2 + 9x + 5$ ✓

9. $3x^2 + x + 4 + \dfrac{6}{x - 4}$ **11.** $4x^3 - 20x^2 + 13x + 4$

13. $3x^3 - 6x^2 + 2x - 4$ **15.** $x^3 + 4x^2 - 5x - 2 + \dfrac{5x}{x^2 + 5}$

17. $3x^2 + 1 + \dfrac{5x - 2}{2x^2 + x - 3}$ **19.** $x^2 + 3x + 9$

21. $\dfrac{5}{2}x^2 + \dfrac{1}{4}x + \dfrac{1}{8} + \dfrac{\frac{25}{8}}{2x - 1}$ **23. a.** $2x^4 - 5x^3 - 5x^2 - 4x + 29$

b. $x - 3$ **c.** $2x^3 + x^2 - 2x - 10$ **d.** -1

25. a. $x^3 - 2x^2 - 25x - 4$ **b.** $x + 4$ **c.** $x^2 - 6x - 1$

d. 0   **27.** $4x - 9 + \dfrac{55}{x + 6}$ **29.** $5x + 3$

31. $-5x^3 + 10x^2 - 23x + 38 + \dfrac{-72}{x + 2}$

33. $4x^4 - 13x^3 - 97x^2 - 59x + 21$ **35.** $x^4 - 2x^3 + 4x^2 - 8x + 16$

37. $2x^3 - 4x^2 - 62x - 56$ **39.** The remainder is 39.

41. a. 201   **b.** 201   **43. a.** -112 **b.** 0   **c.** 123   **d.** 0

45. a. -2 **b.** 0   **c.** 0   **d.** 18   **47. a.** No   **b.** Yes

49. a. No   **b.** Yes   **51. a.** Yes   **b.** Yes   **53. a.** Yes

b. Yes   **55. a.** Yes   **b.** No   **57. a.** No   **b.** Yes

59. a. No   **b.** Yes   **61. a.** 0   **b.** 0   **c.** $\{-3, -\sqrt{5}, \sqrt{5}, 3\}$

63. a. Yes   **b.** Yes   **c.** $\{2 \pm 5i\}$ **d.** $2 \pm 5i$

65. a. $f(x) = (x + 1)(2x - 9)(x + 4)$ **b.** $\left\{-1, \frac{9}{2}, -4\right\}$

67. a. $f(x) = 4\left(x - \frac{1}{4}\right)(5x + 1)(x + 2)$ or $f(x) = (4x - 1)(5x + 1)(x + 2)$

b. $\left\{\frac{1}{4}, -\frac{1}{5}, -2\right\}$ **69. a.** $f(x) = (x - 3)(3x - 1)^2$ **b.** $\left\{3, \frac{1}{3}\right\}$

71. $f(x) = x^3 - x^2 - 14x + 24$

73. $f(x) = x^4 - \frac{5}{2}x^3 + \frac{3}{2}x^2$ or $f(x) = 2x^4 - 5x^3 + 3x^2$

75. $f(x) = x^2 - 44$ **77.** $f(x) = x^3 + 2x^2 + 9x + 18$

79. $f(x) = 6x^3 - 23x^2 - 6x + 8$ **81.** $f(x) = x^2 - 14x + 113$

83. Direct substitution; -2 **85. a.** Yes   **b.** Yes   **c.** Yes

d. No   **e.** Yes   **f.** No

87. False   **89.** $m = 28$ **91.** $m = -5$ **93.** $r = 0$

95. a. $V(x) = 2x^3 + 3x^2 - x$ **b.** 534 cm^3 **97.** The divisor must be

of the form $(x - c)$, where c is a constant.   **99.** Compute $f(c)$ either

by direct substitution or by using the remainder theorem. The remainder

theorem states that $f(c)$ is equal to the remainder obtained after dividing

$f(x)$ by $(x - c)$.   **101. a.** $f(x) = (x - 5)(x - i)(x + i)$

b. $\{5, i, -i\}$ **103. a.** $f(x) = (x + 1)^2(x - \sqrt{3})(x + \sqrt{3})$

b. $\{-1, \sqrt{3}, -\sqrt{3}\}$ **105. a.** 3   **b.** 3 is a solution.

c. $\left\{3, -4, -\frac{2}{5}\right\}$

Section 7.4 Practice Exercises, pp. 532–536

R.1. $81 + 0i$ **R.3.** $49 + 0i$ **R.5.** $-22 + 0i$

R.7. $2x^3 + x^2 + 18x + 9$ **R.9.** $\{\pm 2\sqrt{7}\}$ **R.11.** $\{\pm\sqrt{10}i\}$

R.13. $-\dfrac{2}{9}, 7, 0, -0.\overline{9}, 6.25$ **R.15.** $f(-x) = 2x^4 - 3x^2 - 5x + 1$

1. zeros   **3.** $a - bi$ **5.** greater than   **7.** $\pm 1, \pm 2, \pm 4$

9. $\pm 1, \pm 2, \pm 3, \pm 6, \pm\dfrac{1}{2}, \pm\dfrac{3}{2}, \pm\dfrac{1}{4}, \pm\dfrac{3}{4}$

11. $\pm 1, \pm 2, \pm 4, \pm 8, \pm\dfrac{1}{2}, \pm\dfrac{1}{3}, \pm\dfrac{2}{3}, \pm\dfrac{4}{3}, \pm\dfrac{8}{3}, \pm\dfrac{1}{4}, \pm\dfrac{1}{6}, \pm\dfrac{1}{12}$

13. 7 and $\dfrac{5}{3}$ **15.** $-\dfrac{1}{2}, 1$ **17.** $-3, \dfrac{1}{2}, 2, 4$ **19.** $5, 1 \pm \sqrt{5}$

21. $\dfrac{1}{5}, \pm\sqrt{7}$ **23.** 2 (multiplicity 2), $\dfrac{1}{3}, -4$ **25.** $-2, 3 \pm i$

27. $\pm\sqrt{10}, \pm 3i$ **29.** one   **31.** 7   **33. a.** $2 \pm 5i, \pm\sqrt{7}$

b. $[x - (2 + 5i)][x - (2 - 5i)](x - \sqrt{7})(x + \sqrt{7})$

c. $\{2 \pm 5i, \pm\sqrt{7}\}$ **35. a.** $4 \pm i, \dfrac{4}{3}$

b. $[x - (4 + i)][x - (4 - i)](3x - 4)$ **c.** $\left\{4 \pm i, \dfrac{4}{3}\right\}$

37. a. $-3 \pm 2i, -\dfrac{1}{4}, 1, -4$ **b.** $[x - (-3 + 2i)][x - (-3 - 2i)](4x + 1)$

$(x - 1)(x + 4)$ **c.** $\left\{-3 \pm 2i, -\dfrac{1}{4}, 1, -4\right\}$

39. $f(x) = 5x^3 - 4x^2 + 180x - 144$

41. $f(x) = -5x^4 + 10x^3 + 60x^2 - 200x + 160$

43. $f(x) = 18x^3 + 39x^2 + 8x - 16$ **45.** $f(x) = x^6 - 14x^5 + 65x^4$

47. $f(x) = x^4 - 12x^3 + 62x^2 - 300x + 925$

49. Positive: 3 or 1; Negative: 3 or 1   **51.** Positive: 6, 4, 2, or 0;

Negative: 1   **53.** Positive: 0; Negative: 4, 2, or 0

55. Positive: 0; Negative: 0   **57.** 4 real zeros; $f(x)$ has 1 positive

real zero, no negative real zeros, and the number 0 is a zero of

multiplicity 3.   **59. a.** Yes   **b.** No   **61. a.** Yes   **b.** Yes

63. a. Yes   **b.** Yes   **65.** True   **67.** False. Only numbers less

than -3 are also guaranteed to be lower bounds.

69. $\dfrac{7}{4}, -\dfrac{1}{2},$ and 4 (each with multiplicity 1)   **71.** $\pm\sqrt{5}, \dfrac{1}{2}, -2,$ and -4

(each with multiplicity 1)   **73.** -3 (multiplicity 2) and $\dfrac{1}{2}$

(multiplicity 2)   **75.** $\dfrac{1}{2}, 3, 1 \pm 2i$ (each with multiplicity 1)

77. $\dfrac{5}{2}$ (multiplicity 2), $1 \pm i$ (each multiplicity 1)

79. 0 (multiplicity 2), -1 (multiplicity 2), and $\pm i\sqrt{10}$ (each multiplicity 1)

81. 0 (multiplicity 3) and $5 \pm 3i$ (each multiplicity 1)

83. -1 and $2 \pm 3i$ (each multiplicity 1)   **85.** False. For example, the

graph of $f(x) = x^4 + 1$ has no x-intercepts. Thus, $x^4 + 1$ has no real

zeros.   **87.** False. For example, the graph of $f(x) = x^{10} + 1$ has no

x-intercepts.   **89.** True   **91.** All statements are true.

93. a. $f(2) = -2$ and $f(3) = 1$. Because $f(2)$ and $f(3)$ have opposite signs,

the intermediate value theorem guarantees that f has at least one real

zero between 2 and 3.

b. $\dfrac{7 \pm \sqrt{17}}{4}$; Furthermore, $\dfrac{7 + \sqrt{17}}{4} \approx 2.78$ is on the interval $[2, 3]$.

95. If a polynomial has real coefficients, then all nonreal zeros must come

in conjugate pairs. This means that if the polynomial has nonreal zeros,

there would be an even number of them. A third-degree polynomial

has 3 zeros (including multiplicities). Therefore, it would have either

2 or 0 nonreal zeros, leaving room for either 1 or 3 real zeros.

97. $f(x)$ has no variation in sign, nor does $f(-x)$. By Descartes' rule of

signs, there are no positive or negative real zeros. Furthermore,

0 itself is not a zero of $f(x)$ because x is not a factor of $f(x)$. Therefore,

there are no real zeros of $f(x)$.   **99.** $n - 2$ possible nonreal zeros

101. The triangular front has a base of 6 ft and a height of 4 ft. The length

is 9 ft.   **103.** Each dimension was decreased by 1 in.

105. The dimensions are either 2 cm by 3 cm or $-1 + \sqrt{13}$ cm by

$\dfrac{1 + \sqrt{13}}{2}$ cm.

107. a. $(x + 3)(x - 1)(x^2 + 4)$ **b.** $(x + 3)(x - 1)(x + 2i)(x - 2i)$

109. a. $(x - \sqrt{5})(x + \sqrt{5})(x^2 + 7)$

b. $(x - \sqrt{5})(x + \sqrt{5})(x + \sqrt{7}i)(x - \sqrt{7}i)$

111. The fourth roots of 1 are $1, -1, i,$ and $-i$.

113. The number $\sqrt{5}$ is a real solution to the equation $x^2 - 5 = 0$ and a zero of the polynomial $f(x) = x^2 - 5$. However, by the rational zeros theorem, the only possible rational zeros of $f(x)$ are ± 1 and ± 5. This means that $\sqrt{5}$ is irrational.   **115.** -2

Section 7.5 Practice Exercises, pp. 544–548

R.1. a. $A \cup B = \{-4, -2, -1, 0, 2, 3, 4, 6, 9\}$ **b.** $A \cap B = \{-2, 0, 6\}$

R.3. $(-\infty, -4) \cup [5, \infty)$ **R.5.** $\left[-\dfrac{9}{5}, 2\right)$ **R.7.** $\{-5\}$

R.9. a. negative   **b.** positive   **R.11. a.** positive   **b.** positive

R.13. $\left\{-2, 2, \dfrac{5}{2}\right\}$ **R.15.** $\{-2, -1, 4\}$ **R.17.** $\left\{-\dfrac{9}{2}\right\}$

R.19. $\left\{\dfrac{3 \pm \sqrt{93}}{6}\right\}$

1. polynomial; 2   **3.** $(-\infty, \infty); \{\ \}$ **5. a.** $(-4, -1)$

b. $[-4, -1]$ **c.** $(-\infty, -4) \cup (-1, \infty)$ **d.** $(-\infty, -4] \cup [-1, \infty)$

7. a. $(-\infty, -3) \cup (-3, \infty)$ **b.** $(-\infty, \infty)$ **c.** $\{\ \}$ **d.** $\{-3\}$

9. a. $(-\infty, -2) \cup (0, 3)$ **b.** $(-\infty, -2] \cup [0, 3]$

c. $(-2, 0) \cup (3, \infty)$ **d.** $[-2, 0] \cup [3, \infty)$

11. a. $(0, 3) \cup (3, \infty)$ **b.** $[0, \infty)$ **c.** $(-\infty, 0)$

d. $(-\infty, 0] \cup \{3\}$ **13. a.** $(-\infty, \infty)$ **b.** $(-\infty, \infty)$

c. $\{\ \}$ **d.** $\{\ \}$

15. a. $\left\{\dfrac{3}{5}, 5\right\}$ **b.** $\left(\dfrac{3}{5}, 5\right)$ **c.** $\left[\dfrac{3}{5}, 5\right]$

d. $\left(-\infty, \dfrac{3}{5}\right) \cup (5, \infty)$ **e.** $\left(-\infty, \dfrac{3}{5}\right] \cup [5, \infty)$

17. a. $\{-3, 4\}$ **b.** $(-\infty, -3) \cup (4, \infty)$ **c.** $(-\infty, -3] \cup [4, \infty)$

d. $(-3, 4)$ **e.** $[-3, 4]$ **19. a.** $\{-6\}$ **b.** $\{\ \}$ **c.** $\{-6\}$

d. $(-\infty, -6) \cup (-6, \infty)$ **e.** $(-\infty, \infty)$

21. $\left(-\infty, -\dfrac{3}{2}\right] \cup \left[\dfrac{3}{2}, \infty\right)$ **23.** $\left(-1, \dfrac{4}{3}\right)$ **25.** $(-\infty, 0] \cup [3, \infty)$

27. $\left(\dfrac{-3 - \sqrt{59}}{5}, \dfrac{-3 + \sqrt{59}}{5}\right)$ **29.** $(-7, 7)$

31. $\left(-\infty, -\dfrac{\sqrt{2}}{4}\right] \cup \left[\dfrac{\sqrt{2}}{4}, \infty\right)$ **33.** $[-4, 1] \cup [3, \infty)$

35. $(-\infty, -2) \cup (-2, 0) \cup (4, \infty)$ **37.** $[-3, -1] \cup [1, 3]$

39. $\left(-\infty, -\dfrac{5}{2}\right) \cup (-2, 2)$ **41.** $[-1, 0] \cup \{3\}$

43. $\left(-\infty, \dfrac{3}{5}\right] \cup [5, \infty)$ **45.** $\left(-\infty, -\dfrac{1}{3}\right) \cup \left(0, \dfrac{5}{2}\right) \cup \left(\dfrac{5}{2}, 4\right)$

47. $(-\infty, \infty)$ **49.** $\{\ \}$ **51.** $\left(-\infty, \dfrac{3}{4}\right) \cup \left(\dfrac{3}{4}, \infty\right)$

53. $\{-2\}$ **55. a.** $s(t) = -16t^2 + 216t$ **b.** The shell will explode 6.75 sec after launch.   **c.** The spectators can see the shell between 1 sec and 6.75 sec after launch.   **57.** The car will stop within 250 ft if the car is traveling less than 50 mph.   **59.** The width should be between $2\sqrt{15}$ ft and $4\sqrt{5}$ ft. This is between approximately 7.7 ft and 8.9 ft.

61. a. $(-2, 0), (2, 0)$ **b.** $(0, -4)$

c.   **d.** $(-\infty, -2) \cup (2, \infty)$

e. $(-2, 2)$

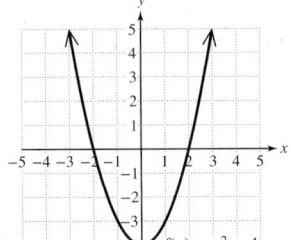

$f(x) = x^2 - 4$

63. a. $(-2, 0), (1, 0)$ **b.** $(0, -4)$

c.   **d.** $(1, \infty)$

e. $(-\infty, -2) \cup (-2, 1)$

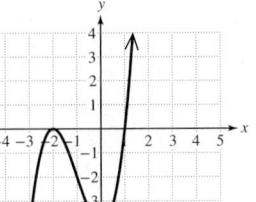

$f(x) = x^3 + 3x^2 - 4$

65. a. $(0, 0), (-\sqrt{3}, 0), (-\sqrt{2}, 0), (\sqrt{2}, 0), (\sqrt{3}, 0)$

b. $(0, 0)$

c. $(-\sqrt{3}, -\sqrt{2}) \cup (\sqrt{2}, \sqrt{3})$

d. $(-\infty, -\sqrt{3}) \cup (-\sqrt{2}, 0) \cup (0, \sqrt{2}) \cup (\sqrt{3}, \infty)$

67. $[-3, 3]$ **69.** $(-\infty, -\sqrt{5}] \cup [\sqrt{5}, \infty)$

71. $(-\infty, -6] \cup \left[\dfrac{3}{2}, \infty\right)$

73. a.

	a		b		c
Sign of $(x - a)^2$:	$+$		$+$	$+$	$+$
Sign of $(b - x)$:	$+$		$+$	$-$	$-$
Sign of $(x - c)^3$:	$-$		$-$	$-$	$+$
Sign of $(x - a)^2(b - x)(x - c)^3$:	$-$		$-$	$+$	$-$

b. (b, c) **c.** $(-\infty, a) \cup (a, b) \cup (c, \infty)$

75. The solution set to the inequality $f(x) < 0$ corresponds to the values of x for which the graph of $y = f(x)$ is below the x-axis.

77. a. $0.552x^3 + 4.13x^2 - 1.84x - 10.2 < 0$

b.

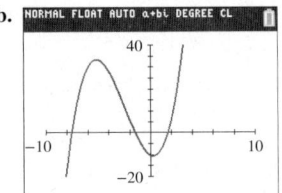

c. The real zeros are approximately $-7.6, -1.5,$ and 1.6.

d. $(-\infty, -7.6) \cup (-1.5, 1.6)$

79. a.

b. $(1.9, 90)$

c. The radius should be no more than 1.9 in. to keep the amount of aluminum to at most 90 in.^{2}.

Problem Recognition Exercises, p. 548

1. a. $\{2, 4\}$ **b.** $(-\infty, 2] \cup [4, \infty)$ **c.** $(2, 4)$

2. a. $\{-4, 0, 4\}$ **b.** $(-\infty, -4) \cup (4, \infty)$ **c.** $[-4, 4]$

3. a. $\left\{-\dfrac{8}{7}\right\}$ **b.** $\left(-\dfrac{8}{7}, \infty\right)$ **c.** $\left(-\infty, -\dfrac{8}{7}\right)$

4. a. $\left\{-\dfrac{5}{2}, 3, 7\right\}$ **b.** $(3, 7)$ **c.** $(-\infty, 3] \cup [7, \infty)$

5. a. $\left\{\dfrac{2}{5}\right\}$ **b.** $\left\{\dfrac{2}{5}\right\}$ **c.** $\left(-\infty, \dfrac{2}{5}\right) \cup \left(\dfrac{2}{5}, \infty\right)$

6. a. $\left\{-5, \dfrac{7}{3}, 5\right\}$ **b.** $\left[-5, \dfrac{7}{3}\right] \cup [5, \infty)$ **c.** $(-\infty, -5] \cup \left[\dfrac{7}{3}, 5\right]$

7. a. $(x + 2)^2(3x - 1)(x - 1)$ **b.** $\left\{-2, \dfrac{1}{3}, 1\right\}$

c. $(-2, 0), \left(\dfrac{1}{3}, 0\right), (1, 0)$

8. a. $x(x + 3)^3(x - i)(x + i)$ **b.** $\{0, -3, \pm i\}$ **c.** $(0, 0), (-3, 0)$

Chapter 7 Review Exercises, pp. 553–555

1. $(-5, 2)$
3. a. $f(x) = -2(x - 1)^2 + 8$ **f.**
   b. Downward
   c. $(1, 8)$
   d. $(-1, 0)$ and $(3, 0)$
   e. $(0, 6)$ **g.** $x = 1$
   h. Maximum value: 8
   i. Domain: $(-\infty, \infty)$; Range: $(-\infty, 8]$

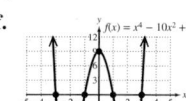

5. a. 45 yd by 90 yd   **b.** 4050 yd^2
7. a. $E(a) = -0.476a^2 + 37.0a - 44.6$ **b.** 39 years   **c.** $674
9. a. Up to the left, up to the right;
   As $x \to -\infty, f(x) \to \infty$, and as $x \to \infty, f(x) \to \infty$.
   b. Zeros: 3, −3, 1, −1 (each with   **f.**
   multiplicity 1)
   c. $(3, 0), (-3, 0), (1, 0), (-1, 0)$
   d. $(0, 9)$ **e.** Even function

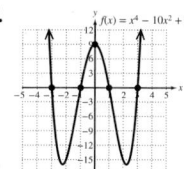

11. a. Down to the left, up to the right;
   As $x \to -\infty, f(x) \to -\infty$, and as $x \to \infty, f(x) \to \infty$.
   b. Zeros: 0 (with multiplicity 3) and   **f.**
   $4 \pm \sqrt{3}$ (each with multiplicity 1)
   c. $(0, 0), (4 + \sqrt{3}, 0), (4 - \sqrt{3}, 0)$
   d. $(0, 0)$
   e. Neither even nor odd

13. False. It may have three or fewer turning points.
15. False. There are infinitely many such polynomials. For example, any polynomial of the form $f(x) = a(x - 2)(x - 3)(x - 4)$ has the required zeros.
17. a. $-2x^2 + 3x - 9 + \dfrac{22x - 28}{x^2 + x - 3}$
   b. Dividend: $-2x^4 + x^3 + 4x - 1$; Divisor: $x^2 + x - 3$;
   Quotient: $-2x^2 + 3x - 9$; Remainder: $22x - 28$
19. $2x^4 - 4x^3 + 8x^2 - 15x + 25 + \dfrac{-49}{x + 2}$
21. 65   **23. a.** No   **b.** Yes   **25. a.** Yes   **b.** Yes
27. $f(x) = (3x - 2)(5x + 1)(x - 4)$ **29.** $f(x) = 8x^3 - 22x^2 - 7x + 3$
31. a. 4   **b.** $\pm 1, \pm 2, \pm 4, \pm 8$ **c.** -2 (multiplicity 2)
   d. -2 (multiplicity 2), $\pm \sqrt{2}$ **33. a.** $11 \pm i, \pm \sqrt{3}$
   b. $[x - (11 - i)][x - (11 + i)](x - \sqrt{3})(x + \sqrt{3})$
   c. $\{11 \pm i, \pm \sqrt{3}\}$ **35.** $f(x) = 3x^3 - 5x^2 + 12x - 20$
37. Positive: 0; Negative: 2 or 0   **39. a.** Yes   **b.** Yes
41. a. $(-2, 0)$ **b.** $[-2, 0] \cup \{3\}$ **c.** $(-\infty, -2) \cup (0, 3) \cup (3, \infty)$
   d. $(-\infty, -2] \cup [0, \infty)$ **43.** $(-\infty, -3] \cup [6, \infty)$ **45.** $\{1\}$
47. $(-3, 2) \cup (4, \infty)$ **49. a.** $s(t) = -16t^2 + 40t + 2$
   b. $0.5 < t < 2$ sec

Chapter 7 Test, pp. 555–556

1. a. $f(x) = 2(x - 3)^2 - 2$ **b.** Upward   **c.** $(3, -2)$
   d. $(2, 0), (4, 0)$ **e.** $(0, 16)$ **f.**
   g. $x = 3$ **h.** Minimum value: −2
   i. Domain: $(-\infty, \infty)$;
   Range: $[-2, \infty)$

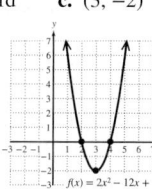

2. a. Up to the left and up to the right; As $x \to -\infty, f(x) \to \infty$, and
   as $x \to \infty, f(x) \to \infty$.   **b.** $\pm 1, \pm 3, \pm 7, \pm 21, \pm\dfrac{1}{2}, \pm\dfrac{3}{2}, \pm\dfrac{7}{2}, \pm\dfrac{21}{2}$
   c. $\dfrac{7}{2}, -3$ (each multiplicity 1), and 1 (multiplicity 2)   **d.** $\left(\dfrac{7}{2}, 0\right),$
   $(-3, 0), (1, 0)$ **e.** $(0, -21)$ **f.** Neither even nor odd
   g.

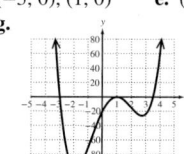

3. a. $-0.25x^9$ **b.** Up to the left and down to the right; As $x \to -\infty$,
   $f(x) \to \infty$, and as $x \to \infty, f(x) \to -\infty$.   **c.** 0 (multiplicity 3), 2
   (multiplicity 2), −1 (multiplicity 4)
4. a. 4   **b.** $2, -2, 3i, -3i$ **c.** $(2, 0)$ and $(-2, 0)$ **d.** Even
5. a. No   **b.** Yes   **c.** No   **d.** Yes
6. a. $2x^2 + 2x + 4 + \dfrac{11x - 9}{x^2 - 3x + 1}$ **b.** Dividend: $2x^4 - 4x^3 + x - 5$;
   Divisor: $x^2 - 3x + 1$; Quotient: $2x^2 + 2x + 4$; Remainder: $11x - 9$
7. a. Yes   **b.** No   **c.** No   **d.** Yes   **e.** 117
8. a. $\pm 2i, 4 \pm i$ **b.** $(x - 2i)(x + 2i)[x - (4 + i)][x - (4 - i)]$
   c. $\{\pm 2i, 4 \pm i\}$
9. a. 4   **b.** $\pm 1, \pm 2, \pm 3, \pm 4, \pm 6, \pm 12, \pm\dfrac{1}{3}, \pm\dfrac{2}{3}, \pm\dfrac{4}{3}$ **c.** Yes
   d. Yes   **e.** $\pm 1, \pm\dfrac{1}{3}, \pm\dfrac{2}{3}, \pm\dfrac{4}{3}, -2, -3$; From part (c), the value 2
   itself is not a zero of $f(x)$. Likewise, from part (d), the value −4 itself
   is not a zero. Therefore, 2 and −4 are also eliminated from the list of
   possible rational zeros.   **f.** $\dfrac{2}{3}$ and -3 **g.** $\dfrac{2}{3}, -3, \sqrt{2}, -\sqrt{2}$
   h.

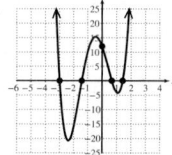

10. $f(x) = 15x^3 - 53x^2 - 30x + 8$ **11.** Positive: 3 or 1; Negative: 2 or 0
12. $(-4, 5)$ **13.** $(-4, 1) \cup (3, \infty)$ **14.** $(-\infty, -1] \cup [0, \infty)$
15. $\left(-\infty, -\dfrac{7}{3}\right) \cup \left(-\dfrac{7}{3}, \infty\right)$
16. a. $y(20) = 140.3$ means that with 20,000 plants per acre, the yield
   will be 140.3 bushels per acre; $y(30) = 172$ means that with 30,000
   plants per acre, the yield will be 172 bushels per acre; $y(60) = 143.5$
   means that with 60,000 plants per acre, the yield will be 143.5
   bushels per acre.   **b.** 40,400   **c.** 183 bushels per acre
17. a. $s(t) = -4.9t^2 + 98t$ **b.** 10 sec after launch   **c.** 490 m
   d. $2.3 < t < 17.7$ sec   **18. a.** $n(a) = 0.0011a^2 - 0.027a + 2.46$
   b. 12 years   **c.** 2.3 visits per year

CHAPTER 8

Section 8.1 Practice Exercises, pp. 568–572

R.1. $\dfrac{q+6}{q+2}; q \neq 6, q \neq -2$ **R.3.** $-\dfrac{v+4}{5v}; v \neq 0, v \neq 4$

R.5. $3x^2 + 7x + 21 + \dfrac{64x - 4}{x^2 - 3x}$ **R.7.** $6x^2 - 21x + 64 + \dfrac{-194}{x + 3}$

R.9. $-\dfrac{1}{2}x$ **R.11.** $\dfrac{2}{7}$ **R.13.** $\dfrac{2}{x^3}$ **R.15.** $\left\{\dfrac{3}{2}\right\}$ **R.17.** $\{-3\}$

1. $q(x)$ **3.** x approaches 5 from the left   **5.** nonzero

7. $(-\infty, 5) \cup (5, \infty)$ **9.** $(-\infty, -1) \cup \left(-1, \dfrac{1}{4}\right) \cup \left(\dfrac{1}{4}, \infty\right)$

11. $(-\infty, \infty)$ **13. a.** 2   **b.** $-\infty$ **c.** ∞ **d.** 2
  e. Never increasing   **f.** $(-\infty, 4)$ and $(4, \infty)$
  g $(-\infty, 4) \cup (4, \infty)$ **h.** $(-\infty, 2) \cup (2, \infty)$ **i.** $x = 4$
  j. $y = 2$
15. a. -1 **b.** ∞ **c.** ∞ **d.** -1 **e.** $(-\infty, -3)$
  f. $(-3, \infty)$ **g.** $(-\infty, -3) \cup (-3, \infty)$ **h.** $(-1, \infty)$
  i. $x = -3$ **j.** $y = -1$

17. $x = 4$ **19.** $x = 5$ and $x = -\dfrac{1}{2}$ **21.** None

23. $t = \dfrac{-2 + \sqrt{10}}{2}$ and $t = \dfrac{-2 - \sqrt{10}}{2}$ **25.** a   **27.** d

29. a. $y = 0$ **b.** Graph does not cross $y = 0$.

31. a. $y = 3$ **b.** $\left(\dfrac{7}{4}, 3\right)$

33. a. No horizontal asymptote   **b.** Not applicable
35. a. $y = 0$ **b.** $(-2, 0)$

37. a. $\dfrac{1 + \dfrac{3}{x} + \dfrac{1}{x^2}}{2 + \dfrac{5}{x^2}}$ **b.** 0   **c.** $y = \dfrac{1}{2}$ **39.** $y = x + 1$

41. $y = 2x^2 - 9$ **43.** $y = 3x - 1$ **45.** $y = -x^2 - 4$
47. Vertical asymptote: $x = 0$; Slant asymptote: $y = 2x$
49. Vertical asymptote: $x = -6$; Slant asymptote: $y = -3x + 22$
51. Vertical asymptotes: $x = \sqrt{5}$ and $x = -\sqrt{5}$; Slant asymptote: $y = x + 5$
53. Vertical asymptotes: $x = 2$, $x = -2$, and $x = -1$; Horizontal
  asymptote: $y = 0$ **55.** Slant asymptote: $y = 2x - 5$

57. a. 4 hr   **b.** 3 hr   **c.** 2 hr   **d.** $D(x) = \dfrac{120}{x}$
  e. No, because the speed, x, would have to approach infinity.
59. a. \$200,000   **b.** \$600,000; \$1,800,000; \$5,400,000   **c.** 70%

61. a. $R(x) = \dfrac{6x}{x + 6}$
  b.

x	6	12	18	30
$R(x)$	3	4	4.5	5

  c. 6 Ω; Even for large values of x, the total resistance will always be
  less than 6 Ω. This is consistent with the statement that the total
  resistance is always less than the resistance in any individual
  branch of the circuit.
63. The numerator and denominator share a common factor of $x + 2$.
  The value -2 is not in the domain of f. The graph will have a "hole" at
  $x = -2$ rather than a vertical asymptote.
65. a. $w(t) = 20 + 2t$ **b.** $s(t) = 240 + 40t$ **c.** $C(t) = \dfrac{240 + 40t}{20 + 2t}$
  d. 12 g/L   **e.** $C(10) = 16$ g/L; $C(20) = 17.3$ g/L; $C(60) = 18.9$ g/L
  f. The horizontal asymptote is $y = 20$, and it represents the limiting
  concentration rate of 20 g/L.

67. 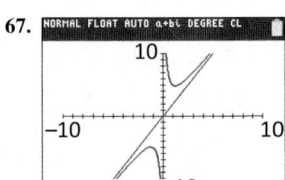   **69.**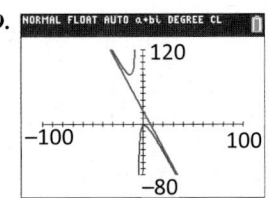

71. 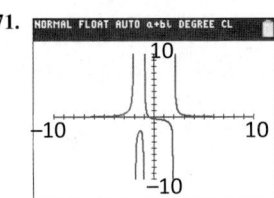   **73.**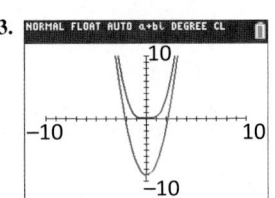

Section 8.2 Practice Exercises, pp. 581–585

R.1. x-intercepts: $(-1, 0)$ and $\left(\dfrac{7}{2}, 0\right)$; y-intercept: $(0, 7)$

R.3. $2x - 3 + \dfrac{-10}{x - 1}$ **R.5. a.** odd   **b.** origin

R.7. a. neither   **b.** neither   **R.9. a.** even   **b.** y-axis

R.11. $(2, 5)$ **R.13. a.** $\{\pm 2\}$ **b.** $\{5\}$ **R.15.** $\left\{\dfrac{5}{2}\right\}$

R.17. $x = 0, x \pm \sqrt{2}$ **R.19.** $x = -\dfrac{7}{2}$

1. No   **3.** The graph of g is the same as the graph of f with a shift to
  the left 4 units and a shift downward 5 units.   **5.** numerator

7. 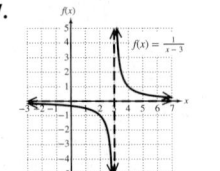   **9.**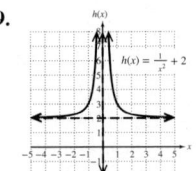

11. 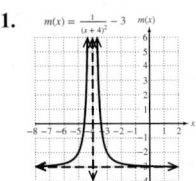   **13.**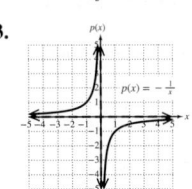

15. a. $(-3, 0)$ and $\left(\dfrac{7}{2}, 0\right)$ **b.** $x = -2$ and $x = -\dfrac{1}{4}$
  c. Horizontal asymptote: $y = \dfrac{1}{2}$ **d.** $\left(0, -\dfrac{21}{2}\right)$

17. a. $\left(\dfrac{9}{4}, 0\right)$ **b.** $x = 3$ and $x = -3$
  c. Horizontal asymptote: $y = 0$ **d.** $(0, 1)$

19. a. $\left(\dfrac{1}{5}, 0\right)$ and $(-3, 0)$ **b.** $x = -2$
  c. Slant asymptote: $y = 5x + 4$ **d.** $\left(0, -\dfrac{3}{2}\right)$

21. 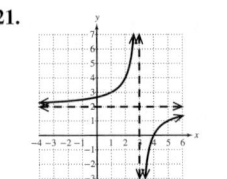   **23.**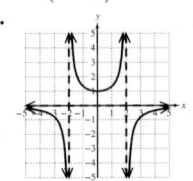

25.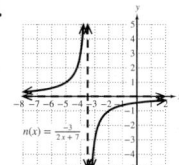

27.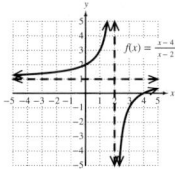

29.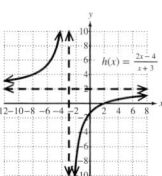

31.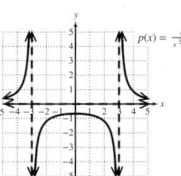

33.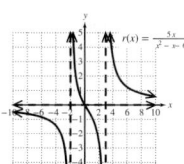

35.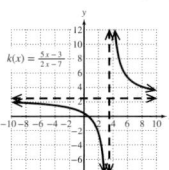

37.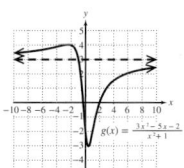

39.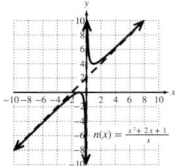

41.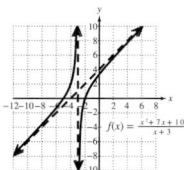

43.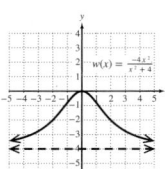

45.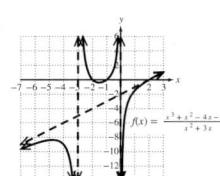

47.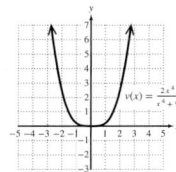

49. a. $C(x) = 109.94 + 20x$ **b.** $\overline{C}(x) = \dfrac{109.94 + 20x}{x}$

c. $\overline{C}(5) = 41.99$; $\overline{C}(30) = 23.67$; $\overline{C}(120) = 20.92$

d. The average cost would approach \$20 per session. This is the same as the fee paid to the gym in the absence of fixed costs.

51. a. $C(10) = 10.0$ mcg/mL; $C(60) = 6.2$ mcg/mL; $C(120) = 3.3$ mcg/mL

b.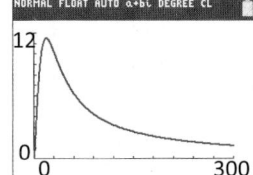

c. The maximum concentration is approximately 11.5 mcg/mL, and this occurs approximately 17.3 min after ingestion.

d. $C(240) = 1.7$ mcg/mL

53. a. 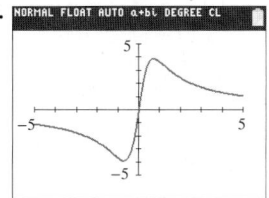   **b.** $t \geq 0$

c. 4 mg/L

d. 0 mg/L

55. a. $F(v) = 560\left(\dfrac{772.4}{772.4 - v}\right)$ **b.**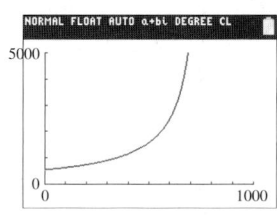

c. The frequency increases, making the pitch of the siren higher to the observer.

57. a. $f(x) = \dfrac{1}{(x + 1)^2} + 3$ **b.** Domain: $(-\infty, -1) \cup (-1, \infty)$; Range: $(3, \infty)$

59. a. $f(x) = 2 + \dfrac{1}{x + 3}$ **b.**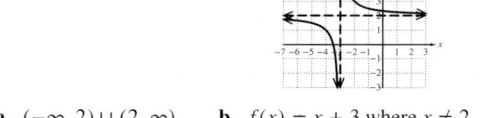

61. a. $(-\infty, 2) \cup (2, \infty)$ **b.** $f(x) = x + 3$ where $x \neq 2$ **c.** None
d. $x = 2$ **e.** Graph iii
63. a. $(-\infty, -5) \cup (-5, -4) \cup (-4, \infty)$ **b.** $f(x) = \dfrac{2}{x + 4}$ where $x \neq -5$
c. $x = -4$ **d.** $x = -5$ **e.** Graph iv

65. $f(x) = \dfrac{x^2 + 4x + 3}{x^2 - 4x + 4}$ **67.** $f(x) = \dfrac{20x - 30}{x^2 - 3x - 10}$

Problem Recognition Exercises, p. 586

1. 2, -1, and -4 **2.** -2, 1, and 3   **3.** $(-2, 0)$, $(1, 0)$, and $(3, 0)$
4. $(2, 0)$, $(-1, 0)$, and $(-4, 0)$ **5.** $(2, 0)$, $(-1, 0)$, and $(-4, 0)$
6. $x = -2$, $x = 1$, and $x = 3$ **7.** Horizontal asymptote: $y = 1$

8. $\left(\dfrac{1 + \sqrt{281}}{10}, 1\right) \approx (1.78, 1)$ and $\left(\dfrac{1 - \sqrt{281}}{10}, 1\right) \approx (-1.58, 1)$

9. 4, $\sqrt{2}$, and $-\sqrt{2}$ **10.** 1, -2 (multiplicity 2)
11. $(1, 0)$ and $(-2, 0)$ **12.** $(4, 0)$, $\left(\sqrt{2}, 0\right)$, and $\left(-\sqrt{2}, 0\right)$
13. $(4, 0)$, $\left(\sqrt{2}, 0\right)$, and $\left(-\sqrt{2}, 0\right)$ **14.** $x = 1$ and $x = -2$
15. Horizontal asymptote: $y = 1$

16. $\left(\dfrac{-1 + \sqrt{85}}{7}, 1\right) \approx (1.17, 1)$ and $\left(\dfrac{-1 - \sqrt{85}}{7}, 1\right) \approx (-1.46, 1)$

17. Graph b   **18.** Graph a   **19. a.** $q(x) = 2x - 4$
b. $r(x) = 12x - 32$ **20.** $y = 2x - 4$ **21.** $\left(\dfrac{8}{3}, \dfrac{4}{3}\right)$
22. $\left\{\dfrac{8}{3}\right\}$; The solution to $r(x) = 0$ gives the x-coordinate of the point where the graph of f crosses its slant asymptote.

Section 8.3 Practice Exercises, pp. 592–596

R.1. $\left(-\infty, -\dfrac{8}{9}\right]$ **R.3.** $(-\infty, -1) \cup [6, \infty)$
R.5. a. $(-7, 3)$ **b.** $[-7, -3]$ **c.** $(-\infty, -7) \cup (3, \infty)$
d. $(-\infty, -7] \cup [3, \infty)$
R.7. a. $\{\ \}$ **b.** $\{-4\}$ **c.** $(-\infty, -4) \cup (-4, \infty)$ **d.** $(-\infty, \infty)$
R.9. $\{5\}$ **R.11.** $\left\{-\dfrac{5}{2}\right\}$ **R.13. a.** $x = 36$ **b.** $x = -6, x = 6$
c. None
1. positive; positive; negative   **3.** negative; positive; positive; negative
5. a. $(2, 3)$ **b.** $[2, 3)$ **c.** $(-\infty, 2) \cup (3, \infty)$ **d.** $(-\infty, 2] \cup (3, \infty)$
7. a. $(-\infty, 2) \cup (2, \infty)$ **b.** $(-\infty, 2) \cup (2, \infty)$ **c.** $\{\ \}$ **d.** $\{\ \}$
9. a. $(-\infty, -5) \cup (0, 2)$ **b.** $(-\infty, -5] \cup [0, 2)$
c. $(-5, 0) \cup (2, \infty)$ **d.** $[-5, 0] \cup (2, \infty)$
11. a. $(-\infty, -2) \cup (2, 3.5)$ **b.** $(-\infty, -2) \cup (2, 3.5]$
c. $(-2, 2) \cup (3.5, \infty)$ **d.** $(-2, 2) \cup [3.5, \infty)$

13. a. $[-2, 3)$ **b.** $(-2, 3)$ **c.** $(-\infty, -2] \cup (3, \infty)$
     d. $(-\infty, -2) \cup (3, \infty)$

15. a. $\{0\}$ **b.** $\{\ \}$ **c.** $(-\infty, \infty)$ **d.** $(-\infty, 0) \cup (0, \infty)$

17. $(-1, 5]$ **19.** $[2, \infty)$ **21.** $(-3, -1] \cup [2, \infty)$ **23.** $\left(\dfrac{7}{2}, 6\right)$

25. $(-\infty, 2)$ **27.** $(-5, -2]$ **29.** $(-\infty, 2]$ **31.** $(-2, \infty)$

33. $(-3, -1) \cup (0, \infty)$ **35.** $(-\infty, 1) \cup (4, 7]$

37. $\left\{-\dfrac{1}{2}\right\} \cup [2, 4) \cup (4, \infty)$

39. a. The horizontal asymptote is $y = 0$ and means that the temperature will approach 0°C as time increases without bound.
   b. More than 6 hr is required for the temperature to fall below 5°C.

41. The smoker would need another cigarette 51 min after the previous cigarette.

43. $(-\infty, -6) \cup \left(\dfrac{3}{2}, \infty\right)$ **45.** $(-\infty, -2) \cup [0, \infty)$

47. a.

Sign of $(a - x)$:	+	−	−	−
Sign of $(x - b)^2$:	+	+	+	+
Sign of $(c - x)^5$:	+	+	+	−
Sign of $\dfrac{(a - x)(x - b)^2}{(c - x)^5}$:	+	−	−	+

           a b c

   b. $(-\infty, a) \cup (c, \infty)$ **c.** $(a, b) \cup (b, c)$

49. Both the numerator and denominator of the rational expression are positive for all real numbers x. Therefore, the expression cannot be negative for any real number.

51. $[3, 5)$ **53.** $(-\infty, -32]$ **55.** $[2, 18]$ **57.** $[0, 2) \cup (4, 6]$

59. $(-3, 3)$ **61.** $\left(-\infty, -2\sqrt{5}\right) \cup (-4, 4) \cup \left(2\sqrt{5}, \infty\right)$

63. a.

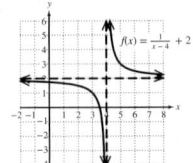

   b. $(0.8, 4)$ and $(10.8, 4)$
   c. It is safe to give a second dose approximately 10.8 hr after the first dose.

Section 8.4 Practice Exercises, pp. 601–605

R.1. $\{99\}$ **R.3.** $\{18\}$ **R.5.** $\left\{\dfrac{200}{11}\right\}$

R.7. $r = \dfrac{A}{Pt}$ **R.9.** $E = \dfrac{IR}{K}$

1. directly   **3.** constant; variation

5. a. 2; 4; 6; 8; 10   **b.** y is also doubled.   **c.** y is also tripled.
   d. increases   **e.** decreases

7. inversely   **9.** jointly   **11.** $C = kr$ **13.** $\overline{C} = \dfrac{k}{n}$

15. $V = khr^2$ **17.** $E = \dfrac{ks}{\sqrt{n}}$ **19.** $c = \dfrac{kmn}{t^3}$ **21.** $k = \dfrac{5}{2}$

23. $k = 972$ **25.** $k = 5$ **27. a.** 2   **b.** 8

29. a. 225 mg   **b.** 270 mg   **c.** 315 mg   **d.** 30 lb

31. a. \$0.40 per mile   **b.** \$0.27 per mile
   c. \$0.20 per mile   **d.** 500 mi

33. 638 ft   **35. a.** 333.2 ft   **b.** 60 mph

37. a. 6.4 days   **b.** 12 people   **39.** 32 A   **41.** \$1440

43. 27.37   **45.** 7.75 mph   **47.** \$11,145.60   **49.** $y = 3.2x$

51. $y = \dfrac{12}{x}$ **53.** **a** and **c**

55. The variable P varies directly as the square of v and inversely as t.

57. a. $1600\pi \text{ m}^2$
   b. The surface area is 4 times as great. Doubling the radius results in $(2)^2$ times the surface area of the sphere.
   c. The intensity at 20 m should be $\frac{1}{4}$ the intensity at 10 m. This is because the energy from the light is distributed across an area 4 times as great.
   d. 50 lux

59. The intensity is $\frac{1}{100}$ as great.   **61.** y will be $\frac{1}{4}$ its original value.

63. y will be 9 times its original value.

Problem Recognition Exercises, p. 605

1. a. Equation quadratic in form and polynomial equation
   b. $\{\pm 2\sqrt{2}, \pm 1\}$
2. a. Absolute value inequality   **b.** $\left(-\frac{1}{2}, 1\right)$
3. a. Polynomial inequality   **b.** $\left[-\frac{1}{2}, 5\right]$
4. a. Radical equation   **b.** $\{1\}$
5. a. Absolute value equation   **b.** $\{9, -1\}$
6. a. Rational equation   **b.** $\{4 \pm 2\sqrt{6}\}$
7. a. Polynomial inequality   **b.** $[-5, -2] \cup [2, \infty)$
8. a. Compound inequality   **b.** $(-\infty, 1)$
9. a. Linear inequality   **b.** $[-23, \infty)$
10. a. Absolute value equation   **b.** $\{9, 1\}$
11. a. Rational inequality   **b.** $(-\infty, 2) \cup [5, \infty)$
12. a. Absolute value inequality   **b.** $(-\infty, -13) \cup (5, \infty)$
13. a. Radical equation   **b.** $\{-4\}$ (The value -7 does not check.)
14. a. Quadratic equation   **b.** $\left\{\dfrac{3}{4} \pm \dfrac{\sqrt{10}}{4}i\right\}$
15. a. Compound inequality   **b.** $(-\infty, -6) \cup (4, \infty)$
16. a. Linear equation and rational equation   **b.** $\left\{-\dfrac{11}{2}\right\}$
17. a. Polynomial inequality   **b.** $\{5\}$
18. a. Rational inequality   **b.** $\{\ \}$
19. a. Radical equation and equation quadratic in form   **b.** $\{16, 81\}$
20. a. Polynomial equation and equation quadratic in form
   b. $\{\pm 2, \pm 3\}$

Chapter 8 Review Exercises, pp. 609–610

1. a. -3 **b.** ∞ **c.** $-\infty$ **d.** -3
   e. $(-\infty, -2)$ and $(-2, \infty)$ **f.** Never decreasing
   g. $(-\infty, -2) \cup (-2, \infty)$ **h.** $(-\infty, -3) \cup (-3, \infty)$
   i. $x = -2$ **j.** $y = -3$

3. $x = \dfrac{5}{2}, x = -3$ **5. a.** $y = 0$ **b.** Graph does not cross $y = 0$.

7. a. $y = -2$ **b.** $(2, -2)$

9. Vertical asymptotes: $x = \sqrt{3}, x = -\sqrt{3}$; Slant asymptote: $y = 2x - 1$

11. Vertical asymptote: $x = 0$; Nonlinear asymptote: $y = -x^2 + 1$

13.

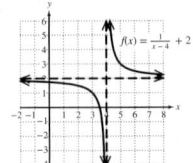

15.

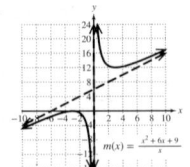

17. a. $P(1) = 78\%$ $P(4) = 57\%$ $P(6) = 49\%$
   b. $P(t)$ will approach 6.25%.

19. a. $(0, 2)$ **b.** $[0, 2)$ **c.** $(-\infty, 0) \cup (2, \infty)$
   d. $(-\infty, 0] \cup (2, \infty)$

21. a. $\{7\}$ **b.** $(-\infty, 5) \cup (7, \infty)$ **c.** $(5, 7)$

23. a. $\{4\}$ **b.** $[4, 6)$ **c.** $(-\infty, 4] \cup (6, \infty)$

25. $(-\infty, 0) \cup (0, 3]$ **27.** $(-\infty, 0) \cup \left(\frac{4}{5}, 2\right)$

29. a. $\overline{C}(x) = \dfrac{120 + 15x}{x}$

b. The trainer must have more than 120 sessions with his clients for his average cost to drop below $16 per session.

31. $m = kw$ **33.** $y = \dfrac{kx\sqrt{z}}{t^3}$ **35.** $k = 2.4$ **37.** 5 lb

39. The force will be 16 times as great.

Chapter 8 Test, p. 611

1. Vertical asymptote: $x = 7$; Slant asymptote: $y = 2x + 11$

2. Vertical asymptotes: $x = \dfrac{1}{2}$, $x = -\dfrac{1}{2}$; Horizontal asymptote: $y = 0$

3. Horizontal asymptote: $y = \dfrac{5}{3}$

4. Nonlinear asymptote: $y = x^2 - 6$

5.   **6.**   **7.**

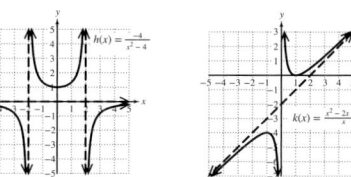

8. a. $(1, 2)$ **b.** $[1, 2)$ **c.** $(-\infty, 1) \cup (2, \infty)$
d. $(-\infty, 1] \cup (2, \infty)$
9. a. $[-1, 5)$ **b.** $(-1, 5)$ **c.** $(-\infty, -1] \cup (5, \infty)$
d. $(-\infty, -1) \cup (5, \infty)$
10. $(-\infty, -3] \cup (2, \infty)$ **11.** $(-3, 3)$
12. $(-\infty, 0) \cup \left(\dfrac{3}{7}, 1\right)$ **13.** $E = kv^2$ **14.** $k = 4.2$
15. 294 ft²   **16.** 178 lb   **17.** The pressure is 9 times as great.
18. a. 1000 rabbits after 1 year, 1667 rabbits after 5 years, and 1818 after 10 years.
b. The rabbit population will approach 2000 as t increases.

CHAPTER 9

Section 9.1 Practice Exercises, pp. 622–627

R.1. $(-\infty, -1) \cup (-1, \infty)$ **R.3.** $\left(-\infty, \dfrac{1}{4}\right)$

R.5. a. $f(-2) = -14$ **b.** $f(t) = 5t - 4$ **c.** $f\left(\dfrac{x + 4}{5}\right) = x$

R.7. $(n \circ p)(x) = x^2 - 3x + 1$ **R.9.** $(f \circ g)(x) = x$

R.11. $(p \circ p)(x) = x^4 - 6x^3 + 6x^2 + 9x$ **R.13.** $y = \dfrac{x - 2}{7}$

R.15. $y = \dfrac{5 - 5x}{x + 1}$ **R.17.** No

R.19.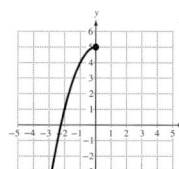

1. $\{(2, 1), (3, 2), (4, 3)\}$
3. one; to; one   **5.** x; x **7.** Yes
9. No   **11.** No   **13.** No
15. Yes   **17.** No   **19.** No
21. No
23. Yes; If $f(a) = f(b)$, then $4a - 7 = 4b - 7$, which implies that $a = b$.

25. Yes; If $g(a) = g(b)$, then $a^3 + 8 = b^3 + 8$, which implies that $a = b$.

27. No; For example the points $(1, -3)$ and $(-1, -3)$ have the same y values but different x values. That is, $m(a) = m(b) = -3$, but $a \neq b$.

29. No; For example, the points $(2, 3)$ and $(-4, 3)$ have the same y values but different x values. That is, $p(a) = p(b) = 3$, but $a \neq b$.
31. Yes   **33.** No   **35.** Yes

37. a. Yes   **b.** The value $g(x)$ represents the number of years since the year 2010 based on the number of applicants to the freshman class, x.

39. a. If $f(a) = f(b)$, then $2a - 3 = 2b - 3$, which implies that $a = b$. The function is one-to-one.

b. $f^{-1}(x) = \dfrac{x + 3}{2}$ **c.**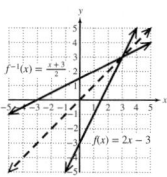

41. $f^{-1}(x) = 4 - 9x$ **43.** $h^{-1}(x) = x^3 + 5$

45. $m^{-1}(x) = \sqrt[3]{\dfrac{x - 2}{4}}$ **47.** $c^{-1}(x) = \dfrac{5 - 2x}{x}$

49. $t^{-1}(x) = -\dfrac{2x + 4}{x - 1}$ **51.** $f^{-1}(x) = \sqrt[3]{b(x + c)} + a$

53. a.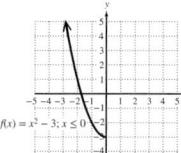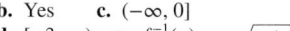
b. Yes   **c.** $(-\infty, 0]$
d. $[-3, \infty)$ **e.** $f^{-1}(x) = -\sqrt{x + 3}$

f.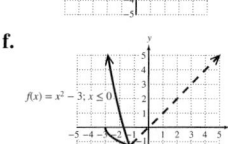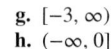
g. $[-3, \infty)$
h. $(-\infty, 0]$

55. a.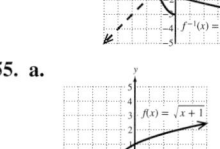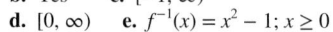
b. Yes   **c.** $[-1, \infty)$
d. $[0, \infty)$ **e.** $f^{-1}(x) = x^2 - 1$; $x \geq 0$
f. The range of f is $[0, \infty)$. Therefore, the domain of f^{-1} must be $[0, \infty)$.

g. 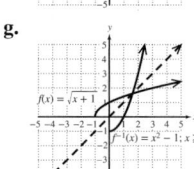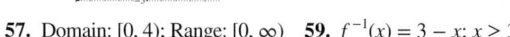   **h.** $[0, \infty)$ **i.** $[-1, \infty)$

57. Domain: $[0, 4)$; Range: $[0, \infty)$ **59.** $f^{-1}(x) = 3 - x$; $x \geq 3$

61. subtracts; $x - 6$ **63.** $f^{-1}(x) = \dfrac{x + 4}{7}$

65. $f^{-1}(x) = \sqrt[3]{x - 20}$ **67.** $f^{-1}(x) = \dfrac{x - 1}{8}$

69. $q^{-1}(x) = (x - 1)^5 + 4$

71. 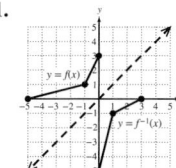   **73.**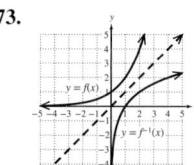

75. a. 12   **b.** 0.5   **c.** 10   **77.** True

79. False. The range of a one-to-one function is the same as the domain of its inverse.

81. a. 50 mph   **b.** $w^{-1}(x) = \dfrac{x - 1220}{-1.17}$; The inverse gives the barometric pressure $w^{-1}(x)$ for a given wind speed x.   **c.** 957 mb

83. a. $T(x) = 6.33x$ **b.** $T^{-1}(x) = \dfrac{x}{6.33}$

    c. $T^{-1}(x)$ represents the mass of a mammal based on the amount of air inhaled per breath, x.

    d. $T^{-1}(170) = 27$ means that a mammal that inhales 170 mL of air per breath during normal respiration is approximately 27 kg (this is approximately 60 lb—the size of a Labrador retriever).

85. a. $T(x) = 24x + 108$ **b.** $T^{-1}(x) = \dfrac{x - 108}{24}$

    c. $T^{-1}(x)$ represents the taxable value of a home (in $1000) based on x dollars of property tax paid on the home.

    d. $T^{-1}(2988) = 120$ means that if a homeowner is charged $2998 in property taxes, then the taxable value of the home is $120,000.

87. The domain and range of a function and its inverse are reversed.

89. If a horizontal line intersects the graph of a function in more than one point, then the function has at least two ordered pairs with the same y-coordinate but different x-coordinates. This conflicts with the definition of a one-to-one function.

91. a. $f(8) = 3$ **b.** $f(32) = 5$ **c.** $f(2) = 1$ **d.** $f\left(\frac{1}{8}\right) = -3$

93. Let f be an increasing function. Then for every value a and b in the domain of f such that $a < b$ we have $f(a) < f(b)$. Now if $u \neq v$, then either $u < v$ or $v < u$. Then either $f(u) < f(v)$ or $f(v) < f(u)$. In either case, $f(u) \neq f(v)$, and f is one-to-one.

Section 9.2 Practice Exercises, pp. 636–641

R.1. a. 25   **b.** 1   **c.** $\dfrac{1}{25}$ **R.3. a.** $\dfrac{1}{25}$ **b.** 1   **c.** 25

R.5. a. -2 **b.** $-\infty$ **c.** -2 **d.** ∞ **e.** Never decreasing

    f. $(-\infty, -1)$ and $(-1, \infty)$ **g.** $(-\infty, -1) \cup (-1, \infty)$

    h. $(-\infty, -2) \cup (-2, \infty)$

R.7. The graph of g is the same as the graph of f but with a vertical stretch by a factor of 2, a shift to the right 1 unit, and a shift upward 4 units.

R.9. 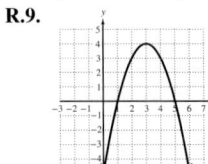   **R.11.** $2640

1. is not; is   **3.** decreasing   **5.** $(0, \infty)$ **7.** $y = 0$

9. a. 0.2   **b.** 2264.9364   **c.** 9.7385   **d.** 156.9925

11. a. 64   **b.** 0.1436   **c.** 0.0906   **d.** 0.1520

13. a, d

15. Domain: $(-\infty, \infty)$;   **17.** Domain: $(-\infty, \infty)$;
Range: $(0, \infty)$ Range: $(0, \infty)$

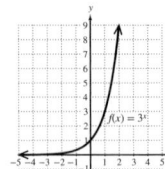   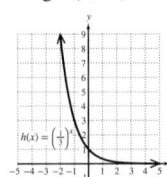

19. Domain: $(-\infty, \infty)$;   **21.** Domain: $(-\infty, \infty)$;
Range: $(0, \infty)$ Range: $(0, \infty)$

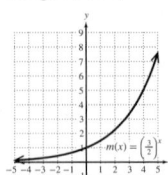   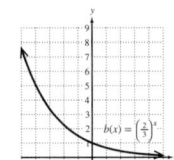

23. a. 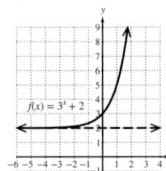   **25. a.**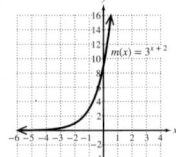

    b. Domain: $(-\infty, \infty)$;   **b.** Domain: $(-\infty, \infty)$;
    Range: $(2, \infty)$ Range: $(0, \infty)$
    c. $y = 2$ **c.** $y = 0$

27. a. 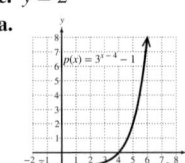   **29. a.**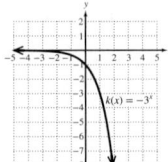

    b. Domain: $(-\infty, \infty)$;   **b.** Domain: $(-\infty, \infty)$;
    Range: $(-1, \infty)$ Range: $(-\infty, 0)$
    c. $y = -1$ **c.** $y = 0$

31. a. 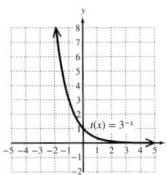   **33. a.**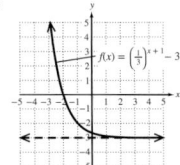

    b. Domain: $(-\infty, \infty)$;   **b.** Domain: $(-\infty, \infty)$;
    Range: $(0, \infty)$ Range: $(-3, \infty)$
    c. $y = 0$ **c.** $y = -3$

35. a. 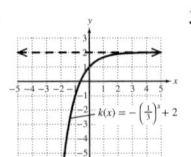   **37. a.** 54.5982   **39. a.**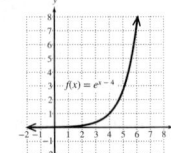
                       b. 0.0408
                       c. 36.8020
                       d. 23.1407

    b. Domain: $(-\infty, \infty)$;   **b.** Domain: $(-\infty, \infty)$;
    Range: $(-\infty, 2)$ Range: $(0, \infty)$
    c. $y = 2$ **c.** $y = 0$

41. a. 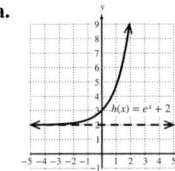   **43. a.**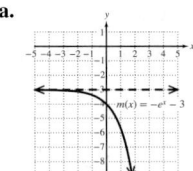

    b. Domain: $(-\infty, \infty)$;   **b.** Domain: $(-\infty, \infty)$;
    Range: $(2, \infty)$ Range: $(-\infty, -3)$
    c. $y = 2$ **c.** $y = -3$

45.

	Compounding Option	n Value	Result
a.	Annually	1	$12,166.53
b.	Quarterly	4	$12,201.90
c.	Monthly	12	$12,209.97
d.	Daily	365	$12,213.89
e.	Continuously	n/a	$12,214.03

47. a. $26,997.18   **b.** $29,836.49   **c.** $34,665.06

49. a. $2200   **b.** $2214.03

    c. 5.5% simple interest results in less interest.

51. 3.8% compounded continuously for 30 years results in more interest.

53. a. $A(28.9) = 5$ means that after 28.9 years, the amount of ^{90}Sr remaining is 5 μg. After one half-life, the amount of substance has been halved. **b.** $A(57.8) = 2.5$ means that after 57.8 years, the amount of ^{90}Sr remaining is 2.5 μg. After two half-lives, the amount of substance has been halved, twice. **c.** $A(100) = 0.909$ means that after 100 years, the amount of ^{90}Sr remaining is approximately 0.909 μg.

55. a. Increasing

b. $P(0) = 310$ means that in the year 2010, the U.S. population was approximately 310 million. This is the initial population in 2010.

c. $P(10) = 341$ means that in the year 2020, the U.S. population was approximately 341 million.

d. $P(20) = 376$; $P(30) = 414$

e. $P(200) = 2137$; In the year 2210 the U.S population will be approximately 2.137 billion. The model cannot continue indefinitely because the population will become too large to be sustained from the available resources.

57. a. 760 mmHg **b.** 241 mmHg

59. a. $T(t) = 78 + 272e^{-0.046t}$ **b.** 250°F

c. Yes; after 60 min, the cake will be approximately 95.2°F.

61. a. $39,000 **b.** It costs the farmer $84,800 to run the tractor for 800 hr during the first year.

63. a. $\{2\}$ **b.** $\{3\}$ **c.** $\{4\}$ **d.** x is between 2 and 3.

e. x is between 3 and 4.

65. a. and **d.**

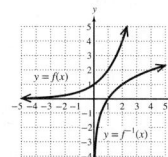

b. Yes
c. Domain: $(-\infty, \infty)$; Range: $(0, \infty)$
e. Domain: $(0, \infty)$; Range: $(-\infty, \infty)$
f. $f^{-1}(1) = 0$; $f^{-1}(2) = 1$; $f^{-1}(4) = 2$

67. a. ∞ **b.** 0 **c.** ∞ **d.** $-\infty$

69. The range of an exponential function is the set of positive real numbers; that is, 2^x is nonnegative for all values of x in the domain.

71. $\{0, 2\}$ **73. a.** e^{x+h} **b.** e^{2x} **c.** e^{x-h} **d.** 1 **e.** $\dfrac{1}{e^{2x}}$

75. $e^{2x} + 2 + e^{-2x}$ or $\dfrac{e^{4x} + 2e^{2x} + 1}{e^{2x}}$

77. $\left(\dfrac{e^x + e^{-x}}{2}\right)^2 - \left(\dfrac{e^x - e^{-x}}{2}\right)^2$

$= \dfrac{1}{4}[(e^{2x} + 2 + e^{-2x}) - (e^{2x} - 2 + e^{-2x})]$

$= \dfrac{1}{4}(4) = 1$

79. $\dfrac{e^x(e^h - 1)}{h}$

81. The graphs of Y_2 and Y_3 are close approximations of $Y_1 = e^x$ near $x = 0$.

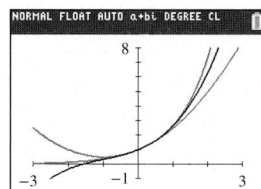

Section 9.3 Practice Exercises, pp. 652–656

R.1. 4 **R.3.** $\dfrac{1}{27}$ **R.5.** 3 **R.7.** -6 **R.9.** $(-\infty, -4)$

R.11. $(-\infty, -15] \cup [-5, \infty)$ **R.13.** $(-\infty, 6]$

R.15. $\left(-\infty, -\dfrac{7}{2}\right) \cup \left(-\dfrac{7}{2}, \infty\right)$ **R.17.** $m^{-1}(x) = \sqrt[3]{x} + 4$

R.19. $(f \circ g)(x) = x$; $(g \circ f)(x) = x$

1. logarithmic **3.** common; natural **5.** 0; 0

7. x; x **9.** $8^2 = 64$ **11.** $10^{-4} = \dfrac{1}{10,000}$ **13.** $e^0 = 1$

15. $a^c = b$ **17.** $\log_5 125 = 3$

19. $\log_{1/5} 125 = -3$ **21.** $\log 1,000,000,000 = 9$

23. $\log_a b = 7$ **25.** 2 **27.** 1 **29.** 8 **31.** -4

33. -1 **35.** 6 **37.** -3 **39.** -2 **41.** 5

43. -5 **45.** -2 **47.** $\dfrac{1}{5}$ **49.** $-\dfrac{1}{2}$

51. a. Between 4 and 5; 4.6705 **b.** Between 6 and 7; 6.0960
c. Between -1 and 0; -0.6198 **d.** Between -6 and -5; -5.4949
e. Between 5 and 6; 5.7482 **f.** Between -3 and -2; -2.2924

53. a. 4.5433 **b.** -1.7037 **c.** 2.5217 **d.** 2.5310
e. 22.0842 **f.** -7.2502

55. 11 **57.** 1 **59.** $x + y$ **61.** $a + b$ **63.** 0

65. 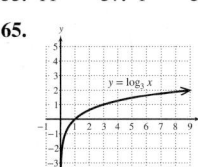 **67.**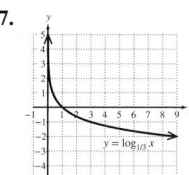

69. 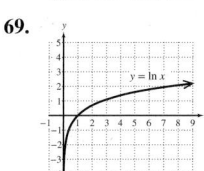 **71. a.**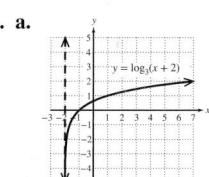

b. Domain: $(-2, \infty)$;
Range: $(-\infty, \infty)$
c. $x = -2$

73. a. 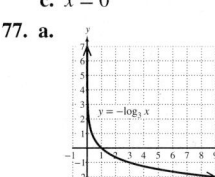 **75. a.**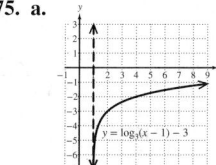

b. Domain: $(0, \infty)$; **b.** Domain: $(1, \infty)$;
Range: $(-\infty, \infty)$ Range: $(-\infty, \infty)$
c. $x = 0$ **c.** $x = 1$

77. a. **b.** Domain: $(0, \infty)$;
Range: $(-\infty, \infty)$
c. $x = 0$

79. $(-\infty, 8)$ **81.** $\left(-\dfrac{7}{6}, \infty\right)$ **83.** $(-\infty, \infty)$

85. $(-\infty, -4) \cup (4, \infty)$ **87.** $(-\infty, 11)$

89. $(-\infty, -3) \cup (4, \infty)$ **91.** $(-\infty, 4) \cup (4, \infty)$

93. a. 6.9 **b.** 3.2 **c.** Approximately 5012 times more intense

95. a. 150 dB **b.** 90 dB **c.** 1,000,000 times more intense

97. a. 2.3 **b.** 2 **c.** Lemon juice is more acidic.

99. a. $3^4 = x + 1$ **b.** $\{80\}$ **c.** $\log_3 (80 + 1) = \log_3 81 = 4$ ✓

101. a. $4^3 = 7x - 6$ **b.** $\{10\}$ **c.** $\log_4 (7 \cdot 10 - 6) = \log_4 64 = 3$ ✓

103. 1 **105.** $-\dfrac{1}{2}$ **107. a.** 3 **b.** 3 **c.** They are the same.

109. a. 2 **b.** 2 **c.** They are the same.

111. a. 5 **b.** 5 **c.** They are the same.

113. a. 19.8 years **b.** $t(0.04) = 17.3$; $t(0.06) = 11.6$; $t(0.08) = 8.7$

115. a. 23.7797 **b.** -33.1787 **c.** Given a number $a \times 10^n$, $\log(a \times 10^n)$ is between n and $n + 1$, inclusive.

117. $(-\infty, 1) \cup (3, \infty)$ **119.** $(-4, \infty)$ **121.** $(6, \infty)$

123. a. The graphs match closely   **b.** $\ln 1.5 \approx 0.4010$
on the interval $(0, 2)$.

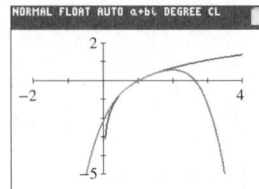

125. The graphs are the same.

Problem Recognition Exercises, p. 657

1. a. $(-\infty, \infty)$ **b.** $\{3\}$ **c.** No x-intercept
    d. $(0, 3)$ **e.** No asymptotes   **f.** Never increasing
    g. Never decreasing   **h.** Graph E

2. a. $(-\infty, \infty)$ **b.** $(-\infty, \infty)$ **c.** $\left(\dfrac{3}{2}, 0\right)$
    d. $(0, -3)$ **e.** No asymptotes   **f.** $(-\infty, \infty)$
    g. Never decreasing   **h.** Graph G

3. a. $(-\infty, \infty)$ **b.** $[-4, \infty)$ **c.** $(1, 0)$ and $(5, 0)$ **d.** $(0, 5)$
    e. No asymptotes   **f.** $(3, \infty)$ **g.** $(-\infty, 3)$ **h.** Graph N

4. a. $(-\infty, \infty)$ **b.** $(-\infty, \infty)$ **c.** $(2, 0)$ **d.** $\left(0, -\sqrt[3]{2}\right)$
    e. No asymptotes   **f.** $(-\infty, \infty)$ **g.** Never decreasing
    h. Graph B

5. a. $(-\infty, 1) \cup (1, \infty)$ **b.** $(-\infty, 0) \cup (0, \infty)$ **c.** None
    d. $(0, -2)$ **e.** Vertical asymptote: $x = 1$; Horizontal asymptote:
    $y = 0$ **f.** Never increasing   **g.** $(-\infty, 1) \cup (1, \infty)$
    h. Graph L

6. a. $(-\infty, -2) \cup (-2, \infty)$ **b.** $(-\infty, 3) \cup (3, \infty)$ **c.** $(0, 0)$
    d. $(0, 0)$ **e.** Vertical asymptote: $x = -2$; Horizontal asymptote:
    $y = 3$ **f.** $(-\infty, -2)$ and $(-2, \infty)$ **g.** Never decreasing
    h. Graph A

7. a. $(-\infty, \infty)$ **b.** $(0, \infty)$ **c.** No x-intercept   **d.** $(0, 1)$
    e. Horizontal asymptote: $y = 0$ **f.** $(-\infty, \infty)$
    g. Never decreasing   **h.** Graph M

8. a. $(-\infty, \infty)$ **b.** $(-\infty, 0]$ **c.** $(-3, 0)$ **d.** $(0, -9)$
    e. No asymptotes   **f.** $(-\infty, -3)$ **g.** $(-3, \infty)$ **h.** Graph C

9. a. $(-\infty, \infty)$ **b.** $[-1, \infty)$ **c.** $(3, 0)$ and $(5, 0)$ **d.** $(0, 3)$
    e. No asymptotes   **f.** $(4, \infty)$ **g.** $(-\infty, 4)$ **h.** Graph I

10. a. $(-\infty, \infty)$ **b.** $(-\infty, 3]$ **c.** $(-3, 0)$ and $(3, 0)$ **d.** $(0, 3)$
    e. No asymptotes   **f.** $(-\infty, 0)$ **g.** $(0, \infty)$ **h.** Graph D

11. a. $(-\infty, 3]$ **b.** $[0, \infty)$ **c.** $(3, 0)$ **d.** $(0, \sqrt{3})$
    e. No asymptotes   **f.** Never increasing   **g.** $(-\infty, 3)$ **h.** Graph F

12. a. $[3, \infty)$ **b.** $[0, \infty)$ **c.** $(3, 0)$ **d.** No y-intercept
    e. No asymptotes   **f.** $(3, \infty)$ **g.** Never decreasing   **h.** Graph K

13. a. $(-\infty, \infty)$ **b.** $(2, \infty)$ **c.** No x-intercept   **d.** $(0, 3)$
    e. Horizontal asymptote: $y = 2$ **f.** $(-\infty, \infty)$ **g.** Never decreasing
    h. Graph H

14. a. $(-2, \infty)$ **b.** $(-\infty, \infty)$ **c.** $(-1, 0)$ **d.** $(0, \ln 2)$
    e. Vertical asymptote: $x = -2$ **f.** $(-2, \infty)$ **g.** Never decreasing
    h. Graph J

Section 9.4 Practice Exercises, pp. 664–667

R.1. x^9 **R.3.** $\dfrac{16z^8}{w^6}$ **R.5.** 4 **R.7.** -3 **R.9.** $\dfrac{1}{2}$

R.11. 4 **R.13.** $\dfrac{8y^7}{x^2}$

1. $\log_b x + \log_b y$ **3.** $p \log_b x$ **5.** $10; e$ **7.** $3 + \log_5 z$
9. $\log 8 + \log c + \log d$ **11.** $\log_2(x + y) + \log_2 z$
13. $\log_{12} p - \log_{12} q$ **15.** $1 - \ln 5$ **17.** $\log(m^2 + n) - 2$
19. $4 \log(2x - 3)$ **21.** $\dfrac{3}{7}\log_6 x$ **23.** $kt \ln 2$

25. $\log_4 7 + \log_4 y + \log_4 z$ **27.** $-1 + \log_7 m + 2 \log_7 n$
29. $10 \log_2 x - \log_2 y - \log_2 z$ **31.** $5 \log_6 p - \log_6 q - 3 \log_6 t$
33. $1 - \dfrac{1}{2}\log(a^2 + b^2)$ **35.** $\dfrac{1}{3}\ln x + \dfrac{1}{3}\ln y - \ln w - 2 \ln z$
37. $\dfrac{1}{4}\ln(a^2 + 4) - \dfrac{3}{4}$
39. $\log 2 + \log x + 8 \log(x^2 + 3) - \dfrac{1}{2}\log(4 - 3x)$ **41.** $\dfrac{1}{3}\log_5 x + \dfrac{1}{6}$
43. $2 + 2 \log_2 a + \dfrac{1}{2}\log_2 (3 - b) - \log_2 c - 2 \log_2 (b + 4)$
45. $\ln 4y$ **47.** 1 **49.** 2 **51.** 1 **53.** $\log_2(x^2 t)$
55. $\log_8\left(\dfrac{m^4}{n^3 p^2}\right)$ **57.** $\ln\left(\dfrac{x}{x^2 - 9}\right)^3$ **59.** $\ln\sqrt{\dfrac{x + 1}{x - 1}}$
61. $\log\left(\dfrac{x^6}{\sqrt[3]{yz^2}}\right)$ **63.** $\log_4[\sqrt[3]{p}(q + 4)]$ **65.** $\ln\left[\dfrac{(x + 2)^3}{\sqrt{x}}\right]$
67. $\log(8y - 7)$ **69.** 1.392 **71.** 2.26 **73.** 2.01
75. 1.036 **77.** 2.366
79. a. Between 3 and 4   **b.** 3.9069 **c.** $2^{3.9069} \approx 15$
81. a. Between 0 and 1   **b.** 0.6826 **c.** $5^{0.6826} \approx 3$
83. a. Between -2 and -1 **b.** -1.7370 **c.** $2^{-1.7370} \approx 0.3$
85. $25.4800; 2^{25.4800} \approx 4.68 \times 10^7$
87. $-8.7128; 4^{-8.7128} \approx 5.68 \times 10^{-6}$ **89.** True
91. False; $\log_5\left(\dfrac{1}{125}\right) \neq \dfrac{1}{\log_5 125}$ (The left side is -3 and the right
side is $\frac{1}{3}$.)
93. True   **95.** False; $\log(10 \cdot 10) \neq (\log 10)(\log 10)$ (The left side is
2 and the right side is 1.)
97. True   **99.** The given statement $\log_5(-5) + \log_5(-25)$ is not
defined because the arguments to the logarithmic expressions are not
positive real numbers.
101. a. $\dfrac{\ln(x + h) - \ln x}{h}$
    b. $\dfrac{1}{h}[\ln(x + h) - \ln x] = \dfrac{1}{h}\ln\left(\dfrac{x + h}{x}\right) = \ln\left(\dfrac{x + h}{x}\right)^{1/h}$
103. $\log\left(\dfrac{-b + \sqrt{b^2 - 4ac}}{2a}\right) + \log\left(\dfrac{-b - \sqrt{b^2 - 4ac}}{2a}\right)$
$= \log\left(\dfrac{-b + \sqrt{b^2 - 4ac}}{2a} \cdot \dfrac{-b - \sqrt{b^2 - 4ac}}{2a}\right)$
$= \log\left[\dfrac{b^2 - (b^2 - 4ac)}{4a^2}\right]$
$= \log\left(\dfrac{4ac}{4a^2}\right)$
$= \log\left(\dfrac{c}{a}\right) = \log c - \log a$
105. $\log_2 9$
107. Let $M = \log_b x$ and $N = \log_b y$, which implies that $b^M = x$ and $b^N = y$.
Then $\dfrac{x}{y} = \dfrac{b^M}{b^N} = b^{M-N}$. Writing the expression $\dfrac{x}{y} = b^{M-N}$ in
logarithmic form, we have $\log_b\left(\dfrac{x}{y}\right) = M - N$, or equivalently,
$\log_b\left(\dfrac{x}{y}\right) = \log_b x - \log_b y$ as desired.

109. 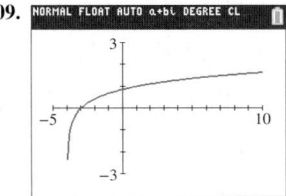   **111.**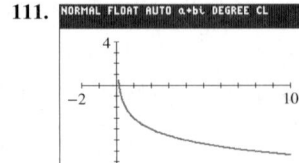

113. a. The graphs are the same.

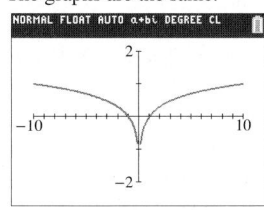

b. $\frac{1}{2}\log x^2 = \log(x^2)^{1/2} = \log\sqrt{x^2} = \log|x|$

Section 9.5 Practice Exercises, pp. 673–676

R.1. 5^{2x} **R.3.** 3^{5x-20} **R.5.** between 2 and 3
R.7. between 6 and 7   **R.9.** $7^{1/3}$ **R.11.** $3x$ **R.13.** 5

R.15. $(x-4)\log 3$ **R.17.** $g^{-1}(x) = \frac{x^3+1}{2}$ **R.19.** $\{-2, 1, 4\}$

1. exponential   **3.** Subtract 3 from both sides and divide by 2.

5. $\{4\}$ **7.** $\left\{\frac{1}{3}\right\}$ **9.** $\{-1\}$ **11.** $\{1\}$ **13.** $\{-15\}$

15. $\left\{\frac{19}{9}\right\}$ **17.** $\left\{\frac{\ln 87}{\ln 6}\right\}$; $t \approx 2.4925$

19. $\left\{\frac{\ln 1020}{\ln 19}\right\}$; $x \approx 2.3528$

21. $\left\{\frac{\log 128{,}100 - 3}{4}\right\}$; $x \approx 0.5269$

23. $\left\{\frac{\ln 3}{0.2}\right\}$ or $\{5 \ln 3\}$; $t \approx 5.4931$

25. $\left\{\frac{5 + \ln 2}{2}\right\}$; $n \approx 2.8466$ **27.** $\left\{\frac{5 \ln 3}{2 \ln 5 - 6 \ln 3}\right\}$; $x \approx -1.6286$

29. $\left\{\frac{\ln 2 - 4 \ln 7}{3 \ln 7 + 6 \ln 2}\right\}$; $x \approx -0.7093$

31. $\{\ln 11\}$; $x \approx 2.3979$ **33.** $\{\ \}$ **35.** 20 years   **37.** 4%
39. 5 years   **41.** 5%   **43.** 10 years, 2 months
45. a. 72 mCi   **b.** 5.3 days
47. Ocean: 14.1 m; Tahoe: 8.7 m; Erie: 3.5 m
49. Ocean: 93.8 m   **51.** 69 min (1 hr 9 min)

53. $\{4\}$ **55.** $\left\{\frac{1}{2}\right\}$ **57.** $\left\{\frac{\ln 21}{\ln 8}\right\}$ **59.** $\{\log 0.0138\}$

61. $\left\{\frac{\ln 15}{0.07}\right\}$ **63.** $\left\{\frac{\log 25}{0.03}\right\}$ **65.** $\left\{\frac{11}{12}\right\}$ **67.** $\{-2\}$

69. $\left\{\frac{\ln 3}{\ln 5 - \ln 3}\right\}$ **71.** $\left\{\frac{2 \ln 2}{\ln 6 - \ln 2}\right\}$ **73.** $\{\ln 10\}$

75. $\{\log 9\}$ **77.** $\{-\ln 2, \ln 3\}$ **79.** $\left\{\log \frac{3}{2}, \log 4\right\}$

81. $f^{-1}(x) = \log_2(x+7)$ **83.** $f^{-1}(x) = \log(x-1) + 3$
85. $\{\ln(4 \pm \sqrt{10})\}$ **87.** $\{\ln 2, 0\}$ **89.** $\{\log 13\}$
91. $\{\ln(4 \pm \sqrt{15})\}$ **93.** $\{-1, 0\}$ **95.** $\{0, 3\}$
97. $\{-1.4408, 2.8584\}$

Section 9.6 Practice Exercises, pp. 681–684

R.1. $(-\infty, 3)$ **R.3.** $\log_b[(x-1)(x+2)]$

R.5. $\log_b\left(\frac{x}{1-x}\right)$ **R.7.** $\log\left[\frac{x^2(x-4)^3}{(x+3)^4}\right]$

R.9. $x^2 - 2x = 2^3$ **R.11.** $x^2 = e^4$

1. x; y
3. Apply the equivalence property of logarithms and equate the arguments. Then solve the equation $5x - 3 = 4x + 2$. Check the potential solutions in the original equation.
5. a. No   **b.** Yes   **c.** No   **7.** $\{6\}$ **9.** $\{\ \}$ **11.** $\{-2\}$
13. $\{-9, 2\}$ **15.** $\{9\}$ **17.** $\{10^{42}\}$ **19.** $\{e^{0.08}\}$; $x \approx 1.0833$

21. $\{35\}$ **23.** $\{32\}$ **25.** $\{23\}$
27. $\{10^{4.1} - 17\}$; $p \approx 12{,}572.2541$

29. $\left\{\frac{4 - e^3}{3}\right\}$; $t \approx -5.3618$

31. $\{2\}$; The value -4 does not check.   **33.** $\{32\}$
35. $\{25\}$; The value -5 does not check.
37. $\{5\}$; The value 2 does not check.
39. $\{\ \}$; The values 5 and 3 do not check.
41. $\{-2\}$; The value 7 does not check.

43. $\{5\}$ **45.** $\{10\}$ **47.** $\{25\}$ **49.** $\left\{-\frac{37}{9}\right\}$

51. $\{3\}$ (The value -3 does not check.)   **53.** $\{4\}$

55. $\left\{\frac{5}{4}\right\}$ (The value -1 does not check.)

57. $\{\ \}$ (The value -3 does not check.)
59. a. 3.4×10^{-8} W/m^2 corresponds to 45.3 dB, which indicates a moderate hearing impairment.   **b.** 10^{-9} W/m^2
61. a. 3.16×10^{-9} mol/L   **b.** 5.01×10^{-3} mol/L
63. 4 months   **65. a.** 11,000   **b.** 32
67. $f^{-1}(x) = e^x - 5$ **69.** $f^{-1}(x) = 10^{x+9} - 7$ **71.** $\{-3, 3\}$
73. $\{900\}$ **75.** $\{2, -2\}$ **77.** $\{-10, 2\}$ **79.** $\{3, -3\}$

81. $\{3\}$ **83.** $\left\{e^4, \frac{1}{e^4}\right\}$; $x \approx 54.5982$, $x \approx 0.0183$

85. $\{\ \}$; The value -4 does not check.

87. $\left\{\ln\frac{1}{2}, \ln 4\right\}$; $x \approx -0.6931$, $x \approx 1.3863$
89. a. $\{1\}$ **b.** $\{1, 10\}$
91. a. $\{1\}$ **b.** $\{1, e\}$; $x = 1, x \approx 2.7183$
93. $\{e, e^4\}$; $x \approx 2.7183$, $x \approx 54.5982$ **95.** $\{100, 1\}$

97. $\{10{,}000\}$ **99.** $\left\{\frac{5}{3}\right\}$; The value $-\frac{5}{2}$ does not check.

101. $\{2.0960\}$

Section 9.7 Practice Exercises, pp. 695–702

R.1. 4.1918   **R.3.** $t = \frac{5c - b}{a}$ **R.5.** $p = \frac{1}{n}\sqrt{\frac{t}{cs}}$ or $p = \frac{\sqrt{tcs}}{ncs}$

R.7. $\left\{\frac{\ln 2}{0.12}\right\}$; $t \approx 5.7762$ **R.9.** $\left\{-\frac{1}{2}\right\}$

R.11. a. decreasing   **b.** $P(0) = 10$; $P(20) \approx 7.7724$

1. growth; decay   **3.** logistic

5. $k = -\frac{\ln\left(\frac{Q}{Q_0}\right)}{t}$ or $\frac{\ln Q_0 - \ln Q}{t}$

7. $D = 10^{(M-8.8)/5.1}$ **9.** $[H^+] = 10^{-pH}$

11. $t = \frac{\ln\left(\frac{A}{P}\right)}{\ln(1+r)}$ or $\frac{\ln A - \ln P}{\ln(1+r)}$ **13.** $k = Ae^{-E/(RT)}$

15. a. 4.4%   **b.** 11.6 years   **17. a.** \$8000   **b.** 7 years

19. a.–b.

Country	Population (in millions) when $t = 0$	Population (in millions) when $t = 10$	$P(t) = P_0 e^{kt}$
Asian	22.9	23.7	$P(t) = 22.9e^{0.00343t}$
South Pacific	19.0	22.6	$P(t) = 19e^{0.01735t}$

c. Asian: 24.5 million; South Pacific: 26.9 million
d. The population growth rate for the South Pacific country is greater.
e. Asian country: approximately 79 years; South Pacific country: approximately 26 years

21. a.

Country	$P(t) = ab^t$	$P(t) = P_0 e^{kt}$	Baseline Population
Central American	$P(t) = 4.3(1.0135)^t$	$P(t) = 4.3e^{0.01341t}$	4.3 million
Northern European	$P(t) = 4.6(1.0062)^t$	$P(t) = 4.6e^{0.00618t}$	4.6 million

  b. Central American country: approximately 11 years;
    Northern European country: approximately 13 years
  c. The population growth rate for the Central American country
    is greater.

23. 2053 years   **25. a.** $Q(t) = 2e^{-0.0079t}$ **b.** 28 years
27. a. $Q(t) = 300e^{-0.0063t}$ **b.** 110 min
29. a. $P(t) = 2,000,000e^{-0.1155t}$
  b. $P(0) = 2,000,000$; $P(6) = 1,000,000$;
    $P(12) = 500,000$; $P(60) = 1953$
31. a. $P(0) = 78$ means that on January 1, 1900, the U.S. population was
    approximately 78 million.
  b. 338 million   **c.** 427 million   **d.** 2076
  e. 0   **f.** 725 million
33. a. 150,000   **b.** 2,000,000   **c.** 3.3 months   **d.** 2,400,000
35. exponential   **37.** logarithmic
39. a. exponential   **41. a.** linear

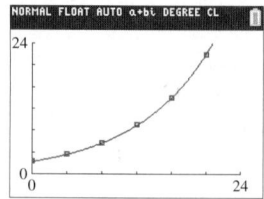
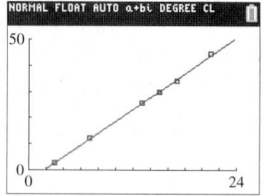

  b. $y = 2.3(1.12)^x$ **b.** $y = 2.28x - 4.08$

43. a. logarithmic   **45. a.** logistic

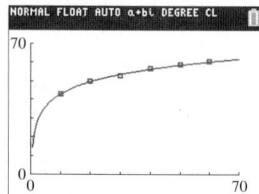
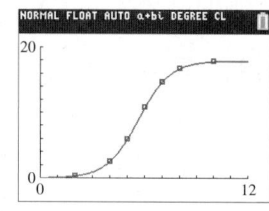

  b. $y = 20.7 + 9.72 \ln x$ **b.** $y = \dfrac{18}{1 + 496e^{-1.1x}}$

47. a. $y = 34.9(2.134)^t$ **b.** $y = 34.9e^{0.758t}$ **c.** 15,000 cases
  d. No, eventually the number of cases would exceed the human
    population.
  e. $y = \dfrac{11,731}{1 + 205e^{-0.67t}}$ **f.** 6000 cases
49. a. $H(t) = 4.86 + 6.35 \ln t$ **b.** 24 years
  c. No, the tree will eventually die.
51. a. $y = -2920t + 29,200$ **b.** $y = 29,200(0.8)^t$
  c. $14,600 and $0   **d.** $9568 and $3135
53. A visual representation of the data can be helpful in determining the
  type of equation or function that best models the data.
55. An exponential growth model has unbounded growth, whereas a
  logistic growth model imposes a limiting value on the dependent
  variable. That is, a logistic growth model has an upper bound
  restricting the amount of growth.
57. a. $t = -\dfrac{\ln\left(1 - \dfrac{Ar}{12P}\right)}{12 \ln\left(1 + \dfrac{r}{12}\right)}$ **b.** This represents the amount of time
    (in years) required to completely pay
    off a loan of A dollars at interest rate
    r, by paying P dollars per month.

Chapter 9 Review Exercises, pp. 708–711

1. No
3. Yes; If $f(a) = f(b)$, then $a^3 - 1 = b^3 - 1$, which implies that $a = b$.
5. Yes, because $(f \circ g)(x) = (g \circ f)(x) = x$

7. $f^{-1}(x) = \sqrt[3]{\dfrac{x + 5}{2}}$

9. a. 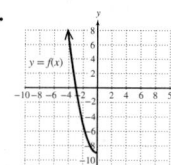   **b.** Yes   **c.** $(-\infty, 0]$
                d. $[-9, \infty)$
                e. $f^{-1}(x) = -\sqrt{x + 9}$

  f. 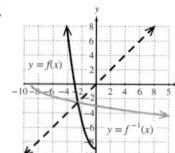   **g.** $[-9, \infty)$
                h. $(-\infty, 0]$

11. a. $f^{-1}(x) = \dfrac{x}{5280}$
  b. f^{-1} represents the conversion from x feet to $f^{-1}(x)$ miles.
  c. 4.2 mi
13. a. 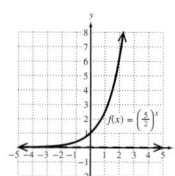   **15. a.**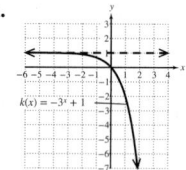

  b. $(-\infty, \infty)$ **c.** $(0, \infty)$ **b.** $(-\infty, \infty)$ **c.** $(-\infty, 1)$
  d. $y = 0$ **d.** $y = 1$
17. Increasing   **19. a.** $3456   **b.** $3563.16   **c.** 7.2% simple
  interest results in less interest.
21. $b^4 = x^2 + y^2$ **23.** $\log 1,000,000 = 6$ **25.** 4   **27.** -6
29. 0   **31.** 7   **33.** $(4, \infty)$ **35.** $(-\infty, \infty)$ **37.** $(-\infty, 4) \cup (4, \infty)$
39. a. 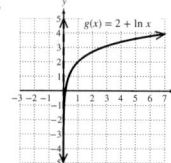   **b.** $(0, \infty)$ **c.** $(-\infty, \infty)$
                  d. $x = 0$

41. pH ≈ 4.5; acidic   **43.** 1   **45.** x **47.** $\log_b x + \log_b y$
49. $2 - \dfrac{1}{2}\log(c^2 + 10)$ **51.** $\dfrac{1}{3}\ln a + \dfrac{2}{3}\ln b - \ln c - 5 \ln d$
53. $\log_5\left(\dfrac{y^4\sqrt{z}}{x^3}\right)$ **55.** $\ln\sqrt[4]{x + 3}$ **57.** 1.587
59. 3.2839; $7^{3.2839} \approx 596$ **61.** $\{3\}$ **63.** $\{2\}$
65. $\left\{\dfrac{\ln 51}{\ln 7}\right\}$; $x \approx 2.0206$ **67.** $\left\{\dfrac{\ln 537}{\ln 11}\right\}$; $w \approx 2.6215$
69. $\left\{\dfrac{\ln 3}{3 \ln 4 - 2 \ln 3}\right\}$; $x \approx 0.5600$
71. $\left\{\dfrac{\ln\left(\frac{2.989}{400}\right)}{-2}\right\}$ or $\left\{\dfrac{\ln 400 - \ln 2.989}{2}\right\}$; $t \approx 2.4483$
73. $\{\ln 8\}$; $x \approx 2.0794$ **75.** $\{-1\}$ **77.** $\{-11, 1\}$
79. $\{-4\}$ **81.** $\{e^{2.1} + 8\}$; $n \approx 16.1662$
83. $\{2\}$; The value -8 does not check.
85. $\{2\}$; The value -3 does not check.   **87.** $\{32\}$
89. $\left\{10,000,000, \dfrac{1}{100,000}\right\}$

91. $f^{-1}(x) = \log_4 x$ **93. a.** 1.39 m; murky   **b.** 9.2 m

95. $t = -\dfrac{1}{k}\ln\left(\dfrac{T-T_f}{T_0}\right)$ or $\dfrac{1}{k}[\ln T_0 - \ln(T-T_f)]$

97. a. 85.5 million   **b.** decreasing   **c.** Approximately 32 years
99. 2800 years
101. a. $Y_1 = 2.38(1.5)^x$ **b.**

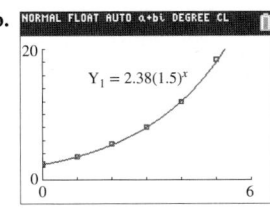

Chapter 9 Test, pp. 711–712

1. a. $f^{-1}(x) = \sqrt[3]{\dfrac{x+1}{4}}$

   b. $(f \circ f^{-1})(x) = 4\left(\sqrt[3]{\dfrac{x+1}{4}}\right)^3 - 1 = 4\left(\dfrac{x+1}{4}\right) - 1 = x+1-1 = x$

   $(f^{-1} \circ f)(x) = \sqrt[3]{\dfrac{4x^3 - 1 + 1}{4}} = \sqrt[3]{\dfrac{4x^3}{4}} = \sqrt[3]{x^3} = x$

2. a. Yes   **b.**   **3.** $f^{-1}(x) = \dfrac{4x+3}{x-1}$

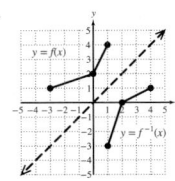

4. a. Domain: $(-\infty, 0]$; Range: $(-\infty, 1]$ **b.** $f^{-1}(x) = -\sqrt{1-x}$
   c. Domain: $(-\infty, 1]$; Range: $(-\infty, 0]$
5. a. Domain: $(0, \infty)$; Range: $(-\infty, \infty)$
   b. $f^{-1}(x) = 10^x$ **c.** Domain: $(-\infty, \infty)$; Range: $(0, \infty)$
6. a. Domain: $(-\infty, \infty)$; Range: $(1, \infty)$ **b.** $f^{-1}(x) = \log_3(x-1)$
   c. Domain: $(1, \infty)$; Range: $(-\infty, \infty)$
7. a. Domain: $[-5, \infty)$; Range $[0, \infty)$ **b.** $f^{-1}(x) = x^2 - 5; x \geq 0$
   c. Domain: $[0, \infty)$; Range: $[-5, \infty)$

8. a.   **9. a.**

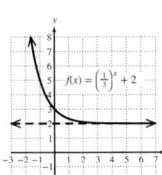   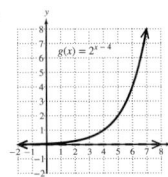

   b. $(-\infty, \infty)$ **c.** $(2, \infty)$ **b.** $(-\infty, \infty)$ **c.** $(0, \infty)$
   d. $y = 2$ **d.** $y = 0$

10. a.   **11. a.**

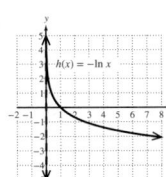   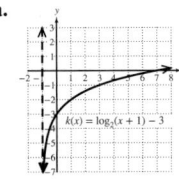

   b. $(0, \infty)$ **c.** $(-\infty, \infty)$ **b.** $(-1, \infty)$ **c.** $(-\infty, \infty)$
   d. $x = 0$ **d.** $x = -1$
12. $e^a = x + y$ **13.** -2 **14.** 3   **15.** 8   **16.** -4
17. $a^2 + b^2$ **18.** 0   **19.** $\left(-\infty, \dfrac{7}{2}\right)$ **20.** $(-\infty, -5) \cup (5, \infty)$
21. $5\ln x + 2\ln y - \ln w - \dfrac{1}{3}\ln z$ **22.** $\dfrac{1}{2}\log(a^2 + b^2) - 4$
23. $\log_2\left(\dfrac{a^6\sqrt[3]{c^2}}{b^4}\right)$ **24.** $\ln\sqrt{x+3}$ **25.** 1.783   **26.** -2.013

27. $\{6\}$ **28.** $\left\{-\dfrac{7}{3}\right\}$ **29.** $\left\{\dfrac{\ln 53}{\ln 5} - 3\right\}; x \approx -0.5331$

30. $\left\{\dfrac{7\ln 2 - 3\ln 3}{2\ln 3 - \ln 2}\right\}; c \approx 1.0346$ **31.** $\left\{\dfrac{\ln 2}{4}\right\}; x \approx 0.1733$

32. $\{2\log 3\}; x \approx 0.9542$ **33.** $\{0\}$ **34.** $\{1\}$
35. $\{e^3 - 2\}; x \approx 18.0855$ **36.** $\{4\}$; The value -3 does not check.
37. $\{2\}$; The value -32 does not check.   **38.** $\{4\}$
39. $t = e^{(92-S)/k} - 1$ **40.** $t = \dfrac{\ln\left(\frac{A}{P}\right)}{n\ln\left(1 + \frac{r}{n}\right)}$ or $\dfrac{\ln A - \ln P}{n\ln\left(1 - \frac{r}{n}\right)}$

41. a. 6.1%   **b.** 26.4 years
42. a. $P(t) = 10,000e^{0.1386t}$ **b.** Approximately 45 hr
43. a. 400 deer were present when the park service began tracking the herd.
   b. 536 deer   **c.** 680 deer   **d.** 15 years   **e.** 0
   f. 1200 deer
44. a. $N = 23.1(2.283)^t$
   b.   **c.** 89,000

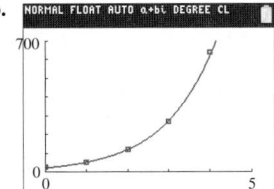

CHAPTER 10

Section 10.1 Practice Exercises, pp. 720–723

R.1. x-intercept: $\left(\dfrac{15}{2}, 0\right)$; y-intercept: $(0, 3)$

R.3. $y = -\dfrac{7}{2}x - 7$; slope: $-\dfrac{7}{2}$; y-intercept: $(0, -7)$

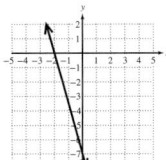

R.5. Vertical   **R.7.** Horizontal   **R.9.** Slanted

R.11.   **R.13.**

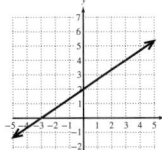   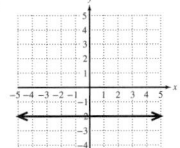

R.15. Perpendicular   **R.17.** Neither   **R.19.** Perpendicular
R.21. Parallel

1. system   **3.** intersect   **5.** the empty set, $\{\ \}$
7. independent   **9.** $(2, 11)$ is a solution.
11. $(-1, 4)$ is a solution.   **13.** None
15. a. Consistent   **b.** Independent   **c.** One solution
17. a. Inconsistent   **b.** Independent   **c.** Zero solutions
19. a. Consistent   **b.** Dependent   **c.** Infinitely many solutions
21. One solution   **23.** No solution; The system is inconsistent.
25. Infinitely many solutions; The equations are dependent.

27. $\{(-2, 1)\}$

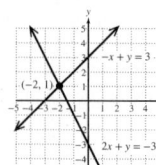

29. $\{(1, 1)\}$

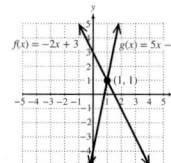

31. $\{(3, -4)\}$

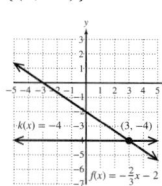

33. $\{(2, -2)\}$

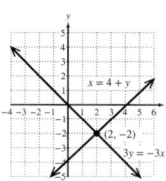

35. No solution; $\{\ \}$; inconsistent system

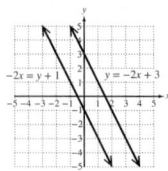

37. Infinitely many solutions;

$$\left\{\left(x, \frac{2}{3}x - 1\right)\middle| x \text{ is any real number}\right\} \text{ or}$$

$$\left\{\left(\frac{3}{2}y + \frac{3}{2}, y\right)\middle| y \text{ is any real number}\right\};$$

dependent equations

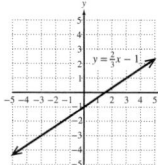

39. $\{(2, -2)\}$

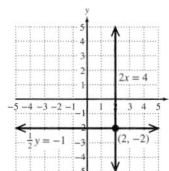

41. Infinitely many solutions;

$$\left\{\left(x, \frac{1}{3}x + 2\right)\middle| x \text{ is any real number}\right\} \text{ or}$$

$$\{(3y - 6, y)|y \text{ is any real number}\};$$

dependent equations

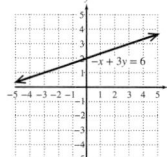

43. No solution; $\{\ \}$; inconsistent system

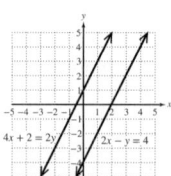

45. For example: $x + y = 2$; $2x + y = -1$ **47.** $C = 2$ and $D = 3$

49. $\{(2.470, -2.253)\}$ **51.** $\{(-0.529, -13.806)\}$

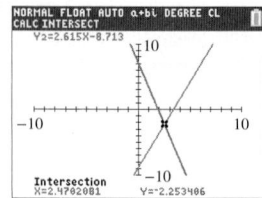
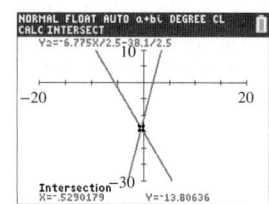

Section 10.2 Practice Exercises, pp. 732–736

R.1. $\left\{-\frac{2}{9}\right\}$ **R.3.** $\{\ \}$

R.5. a. $x = \dfrac{y + 5}{3}$ **b.** $y = 3x - 5$

R.7. 0.25 gallon   **R.9.** 58 mph

1. substitution; addition   **3.** inconsistent   **5.** $\{(-2, 1)\}$
7. $\{(-1, 2)\}$ **9.** $\{(-4, 3)\}$ **11.** $\{(-10, 3)\}$ **13.** $\left\{\left(-\frac{1}{2}, 1\right)\right\}$
15. $\{(2, 1)\}$ **17.** $\{(4, 3)\}$ **19.** $\{(0, 1)\}$ **21.** $\{(-2, -1)\}$
23. $\{(1, -4)\}$ **25.** $\left\{\left(\frac{1}{2}, 1\right)\right\}$ **27.** $\left\{\left(\frac{79}{45}, \frac{2}{45}\right)\right\}$
29. $\{\ \}$; The system is inconsistent.

31. $\{(x, -3x + 6)|x \text{ is any real number}\}$

or $\left\{\left(\dfrac{6 - y}{3}, y\right)\middle| y \text{ is any real number}\right\}$;

The equations are dependent.

33. $\{(0, 0)\}$

35. a. $(x, -5x - 6)|x \text{ is any real number}\}$

or $\left\{\left(-\dfrac{y + 6}{5}, y\right)\middle| y \text{ is any real number}\right\}$

b. For example: $(0, -6)$, $(1, -11)$, $(-2, 4)$

37. $\{(300, -100)\}$ **39.** $\left\{\left(-\dfrac{4}{41}, -\dfrac{71}{41}\right)\right\}$ **41.** $\{(6, 3)\}$ **43.** $\{\ \}$

45. $\{(x, 4x - 2)|x \text{ is any real number}\}$

or $\left\{\left(\dfrac{y + 2}{4}, y\right)\middle| y \text{ is any real number}\right\}$

47. $\{(1.6, 2.3)\}$ **49.** $\{(18, -17)\}$
51. 25 L of 36% solution and 15 L of 20% solution should be mixed.
53. 3.5 L should be replaced.   **55.** She borrowed $3500 at 4.6% and $1500 at 6.2%.   **57.** He borrowed $20,000 at 3% and $4000 at 5.5%.
59. Cherry has 13 g of fat and Mint Chocolate Chunk has 17 g of fat.
61. One makes $1200 and the other makes $1500.
63. The sidewalk moves at 1 ft/sec and Josie walks 4 ft/sec on nonmoving ground.   **65.** The speeds are 8 m/sec and 5 m/sec.
67. a. $C(x) = 52x + 480$ **b.** $R(x) = 100x$ **c.** 10 offices
d. The company will make money.
69. a. $\{(1600, 40)\}$ **b.** $40   **c.** 1600 tickets

71. a.

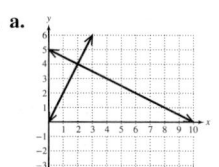

b. 20 square units

73. a. $y = \dfrac{4}{5}x + \dfrac{7}{5}$ **b.** $y = -x + 5$ **c.** $(2, 3)$

75. The angles are 32° and 58°.

77. $x = 60°$, $y = 88°$ **79.** $m = \dfrac{1}{3}$ and $b = -4$ **81.** $\left\{\left(\frac{1}{3}, -1\right)\right\}$

83. Marta bicycles 18 mph and runs 6 mph.
85. The truck was driven 144 mi in the city and 110 mi on the highway.
87. If the system represents two intersecting lines, then the lines intersect in exactly one point. The solution set consists of the ordered pair representing that point. If the lines in the system are parallel, then the lines do not intersect and the system has no solution. If the equations in a system of linear equations represent the same line, then the solution set is the set of points on the line.
89. If the system of equations reduces to a contradiction such as $0 = 1$, then the system has no solution and is said to be inconsistent.
91. $|F_1| = 50(\sqrt{3} - 1)$ lb ≈ 36.6 lb and
$|F_2| = 25\sqrt{2}(3 - \sqrt{3})$ lb ≈ 44.8 lb

Section 10.3 Practice Exercises, pp. 745–748

R.1. $\left\{-\dfrac{1}{4}\right\}$ **R.3.** $\{(0, 6)\}$

R.5. $\left\{\left(x, \dfrac{x-4}{2}\right) \middle| x \text{ is any real number}\right\}$ or

$\{(2y + 4, y) | y \text{ is any real number}\}$

R.7. $\{\ \}$ **R.9.** $26°$

1. plane   **3.** For example: $(0, 0, -2)$, $(0, 3, 0)$, and $(6, 0, 0)$
5. a. Yes   **b.** No
7. a. Yes   **b.** Yes
9. $\{(1, 4, -2)\}$ **11.** $\{(-2, 1, 3)\}$ **13.** $\{(1, 3, 0)\}$
15. No solution; The system is inconsistent.   **17.** $\left\{\left(\frac{1}{2}, \frac{1}{3}, -\frac{1}{6}\right)\right\}$
19. Infinitely many solutions; The equations are dependent.
21. $\{(0, 0, 0)\}$ **23.** $\{(1, -2, 3)\}$ **25.** $\{(6, -8, 2)\}$
27. $\{(1, 12, 17)\}$ **29.** No solution; The system is inconsistent.
31. Infinitely many solutions; The equations are dependent.

33. $\left\{\left(x, \dfrac{-3x - 11}{2}, \dfrac{x + 7}{2}\right) \middle| x \text{ is any real number}\right\}$ or

$\left\{\left(\dfrac{-2y - 11}{3}, y, \dfrac{5 - y}{3}\right) \middle| y \text{ is any real number}\right\}$ or

$\{(2z - 7, -3z + 5, z) | z \text{ is any real number}\}$

35. $\left\{\left(x, \dfrac{1}{3}x, \dfrac{12 - 5x}{3}\right) \middle| x \text{ is any real number}\right\}$ or

$\{(3y, y, 4 - 5y) | y \text{ is any real number}\}$ or

$\left\{\left(\dfrac{12 - 3z}{5}, \dfrac{4 - z}{5}, z\right) \middle| z \text{ is any real number}\right\}$

37. For example: $(2, 1, 1)$, $(0, 0, 2)$, $(4, 2, 0)$ **39.** He invested $4000 in the large cap fund, $2000 in the real estate fund, and $2000 in the bond fund.   **41.** He made eight free-throws, six 2-point shots, and two 3-point shots.   **43.** b   **45.** The sides are 15 in., 18 in., and 22 in.   **47.** The angles are $16°$, $24°$, and $140°$.
49. a. The slopes are 5 and -1.   **b.** $y = 2x^2 - 3x + 1$

c.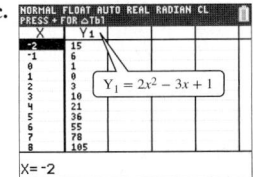

51. $y = -x^2 - 4x + 6$ **53.** $a = 4$, $v_0 = 18$, and $s_0 = 10$
55. a. $y = 7x_1 + 10x_2 + 24$ **b.** $168,000
57. The set of all ordered pairs that are solutions to a linear equation in three variables forms a plane in space.
59. Pair up two equations in the system and eliminate a variable. Choose a different pair of two equations from the system and eliminate the same variable. The result should be a system of two linear equations in two variables. Solve this system using either the substitution or addition method. Then back substitute to find the third variable.
61. $\{(2, -1, 0, 4)\}$ **63.** $\{(-13, 11, 10)\}$
65. a. $x^2 + y^2 - 4x + 6y - 12 = 0$ **b.** Center: $(2, -3)$; Radius: 5
67. $A = 5$, $B = 3$, $C = -1$

Section 10.4 Practice Exercises, pp. 756–758

R.1. $\{(1, 3)\}$ **R.3.** $\{(-1, 2, -4)\}$ **R.5.** $x^2(x + 5)(x + 9)$
R.7. $(2m + 3)(m - 1)$ **R.9.** $(8u + 5)^2$ **R.11.** $(p^2 + 4)(p - 2)(p + 2)$

R.13. $\dfrac{-4x - 24}{x(x + 4)}$ **R.15.** $\dfrac{9y^2 + 7y - 10}{y^2(y + 2)}$

1. fraction decomposition   **3.** linear; $Ax + B$

5. $\dfrac{A}{x + 4} + \dfrac{B}{2x - 3}$ **7.** $\dfrac{A}{x} + \dfrac{B}{x - 2}$ **9.** $\dfrac{A}{w - 2} + \dfrac{B}{w + 3}$

11. $\dfrac{A}{x} + \dfrac{B}{x + 5} + \dfrac{C}{(x + 5)^2}$ **13.** $\dfrac{A}{2x} + \dfrac{Bx + C}{x^2 + 9}$

15. $\dfrac{Ax + B}{x^2 + 5} + \dfrac{Cx + D}{(x^2 + 5)^2}$ **17.** $\dfrac{A}{x - 4} + \dfrac{Bx + C}{x^2 + x + 4}$

19. $\dfrac{A}{x} + \dfrac{B}{x + 2} + \dfrac{C}{(x + 2)^2} + \dfrac{D}{(x + 2)^3} + \dfrac{Ex + F}{x^2 + 2x + 7} + \dfrac{Gx + H}{(x^2 + 2x + 7)^2}$

21. $\dfrac{3}{x + 4} + \dfrac{-7}{2x - 3}$ **23.** $\dfrac{5}{x} + \dfrac{3}{x - 2}$ **25.** $\dfrac{1}{w - 2} + \dfrac{5}{w + 3}$

27. $\dfrac{4}{x} + \dfrac{-3}{x + 5} + \dfrac{1}{(x + 5)^2}$ **29.** $\dfrac{5}{2x} + \dfrac{4x + 1}{x^2 + 9}$

31. $x - 3 + \dfrac{4}{x} + \dfrac{2x - 7}{x^2 + 7}$ **33.** $\dfrac{2x - 1}{x^2 + 5} + \dfrac{3x}{(x^2 + 5)^2}$

35. $\dfrac{3}{x - 4} + \dfrac{2x + 1}{x^2 + x + 4}$ **37.** $\dfrac{3x + 1}{x^2 + 2} + \dfrac{x - 5}{x^2 + 3}$

39. $2x - 5 + \dfrac{4}{x + 2} + \dfrac{-3}{x - 5}$ **41.** $3x - 4 + \dfrac{4}{x + 1} + \dfrac{-5}{(x + 1)^2}$

43. a. $(x - 3)^2(x + 5)$ **b.** $\dfrac{2}{x - 3} + \dfrac{1}{(x - 3)^2} + \dfrac{-5}{x + 5}$

45. a. $(x + 2)^3$ **b.** $\dfrac{3}{x + 2} + \dfrac{-4}{(x + 2)^2} + \dfrac{1}{(x + 2)^3}$

47. Partial fraction decomposition is a procedure in which a rational expression is written as a sum of two or more simpler rational expressions.   **49.** A proper rational expression is a rational expression in which the degree of the numerator is less than the degree of the denominator.

51. a. $\dfrac{1}{n} - \dfrac{1}{n + 2}$

b. $\left(\dfrac{1}{1} - \dfrac{1}{3}\right) + \left(\dfrac{1}{2} - \dfrac{1}{4}\right) + \left(\dfrac{1}{3} - \dfrac{1}{5}\right) + \left(\dfrac{1}{4} - \dfrac{1}{6}\right) + \left(\dfrac{1}{5} - \dfrac{1}{7}\right) + \cdots$

c. 0 **d.** $\dfrac{3}{2}$

53. $\dfrac{1}{ax} - \dfrac{b}{a(a + bx)}$ **55.** $\dfrac{2}{e^x + 1} + \dfrac{3}{e^x + 2}$

Section 10.5 Practice Exercises, pp. 764–768

R.1. $36x^2 - 60x + 25$ **R.3.** $\left\{-\dfrac{2}{3}, \dfrac{1}{4}\right\}$ **R.5.** $\left\{\dfrac{1}{8}\right\}$

R.7. $\{\pm 2, \pm 3i\}$ **R.9.** $\{e^{17}\}$ **R.11.** $\{4\}$

R.13.   **R.15.**

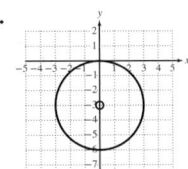

R.17.

R.19. $y^2 - x^2 = 25$ **R.21.** $xy = 50$

1. nonlinear

3. a. 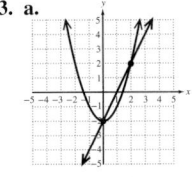   **5. a.**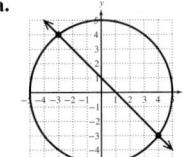

b. $\{(2, 2), (0, -2)\}$ **b.** $\{(-3, 4), (4, -3)\}$

7. a.

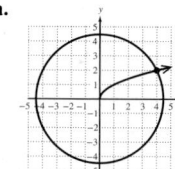

9. a.

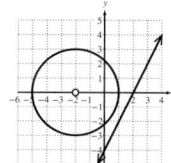

b. $\{(4, 2)\}$ **b.** $\{\ \}$

11. a.

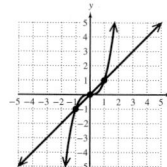

13. a.

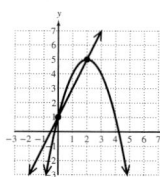

b. $\{(-1, -1), (0, 0), (1, 1)\}$ **b.** $\{(0, 1), (2, 5)\}$

15. $\{(2, 1), (2, -1), (-2, 1), (-2, -1)\}$ **17.** $\{(4, -1), (-4, 1)\}$

19. $\left\{\left(1, \sqrt{2}\right), \left(1, -\sqrt{2}\right), \left(-1, \sqrt{2}\right), \left(-1, -\sqrt{2}\right)\right\}$ **21.** $\{\ \}$

23. $\left\{(3, 1), \left(\frac{7}{3}, \frac{5}{3}\right)\right\}$ **25.** $\{(0, 5)\}$ **27.** $\{(1, 1)\}$

29. $\{(0, 9), (-3, 0), (3, 0)\}$ **31.** $\{(3, 2), (5, -2)\}$

33. $\{(-2, 0), (0, 4), (2, 0)\}$ **35.** $\left\{\left(1, \frac{1}{3}\right), \left(-1, \frac{1}{3}\right), \left(1, -\frac{1}{3}\right), \left(-1, -\frac{1}{3}\right)\right\}$

37. The numbers are 5 and 7.   **39.** The numbers are 5 and 2.

41. The numbers are 12 and 10.   **43.** The numbers are 9 and 12 or -9 and -12.   **45.** The rectangle is 10 m by 8 m.   **47.** The floor is 20 ft by 12 ft.   **49.** The truck is 6 ft by 6 ft by 8 ft.   **51.** The aquarium is 24 in. by 12 in. by 16 in.   **53.** The legs are 4 ft and 7 ft.

55. The ball will hit the ground at the point $\left(128\sqrt{3}, -128\right)$ or approximately $(221.7, -128)$.   **57.** A system of linear equations contains only linear equations, whereas a nonlinear system has one or more equations that are nonlinear.

59. a. $0.69 = A_0 e^{-3k}$ **b.** $0.655 = A_0 e^{-4k}$ **c.** $k \approx 0.052$
    d. $A_0 \approx 0.81$ μg/dL   **e.** 0.43 μg/dL

61. a. $k \approx 0.058$ **b.** The original population is 40,000.
    c. The population will reach 300,000 approximately 35 hr after the culture is started.

63. Infinitely many solutions   **65.** $\{(10, 100)\}$ **67.** $\{(2, 1)\}$

69. $\left\{(2, 1), \left(\frac{7}{5}, \frac{6}{5}\right)\right\}$ **71.** $\{(1, 3, 1), (-1, -3, -1)\}$

73. a. $(0, -5)$ and $(4, 3)$ **b.** $y = 2x - 5$ for $0 \leq x \leq 4$

75. $\{(2.359, 5.584)\}$ **77.** $\{(1.538, 6.135), (-1.538, 6.135), (3.693, -5.135), (-3.693, -5.135)\}$ **79.** $\{\ \}$

Section 10.6 Practice Exercises, pp. 776–779

R.1.

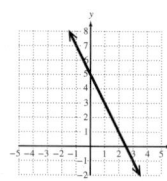

R.3.

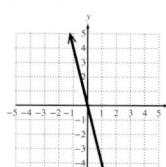

R.5.

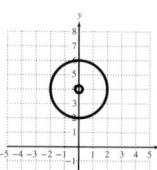

R.7. $(-5, \infty)$ **R.9.** $(-8, -4]$ **R.11.** $(-\infty, -3] \cup [3, \infty)$

1. linear   **3.** above; horizontal   **5.** II
7. a. Yes   **b.** No   **c.** No
9. a. No   **b.** Yes   **c.** Yes

11. a.

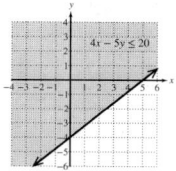

b. The bounding line would be drawn as a dashed line.
c. The bounding line would be dashed and the graph would be shaded strictly below the line.

13.

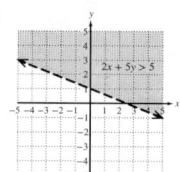

15.

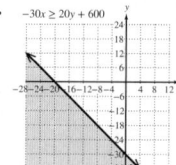

17.

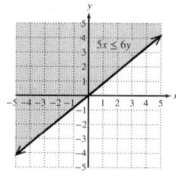

19.

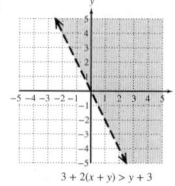

21.

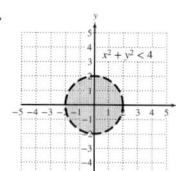

23.

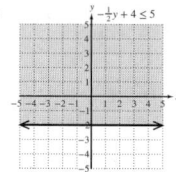

25. a.

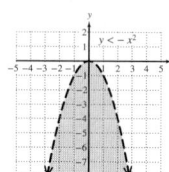

b. The region outside the circle would be shaded.
c. The shaded region would contain points on the circle (solid curve) and points outside the circle.

27.

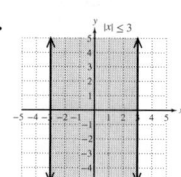

29.

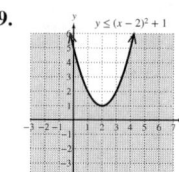

31.

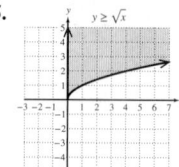

33.

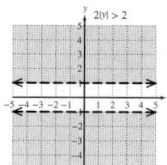

35.

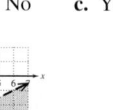

37. a. Yes   **b.** No   **c.** No
39. a. Yes   **b.** No   **c.** Yes   **d.** No

41.

43.

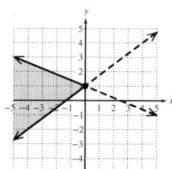

45.

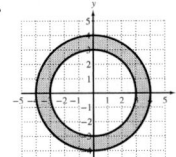

47. The solution set is { }.

49.

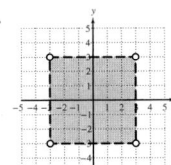

51.

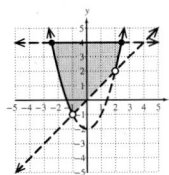

53.

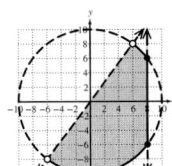

55.

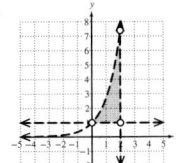

57. The solution set is { }.　　**59.** $x \le 6$

61. $y \ge -2$　　**63.** $x + y \le 18$

65. a. $x + y \le 9$　　**b.** $x \ge 3$
c. $y \le 4$　　**d.** $x \ge 0$　　**e.** $y \ge 0$
f.

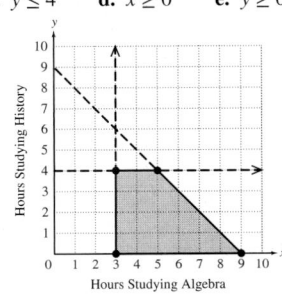

67. a. $x + y \le 60{,}000$　　**b.** $y \ge 2x$　　**c.** $x \ge 0$　　**d.** $y \ge 0$
e.

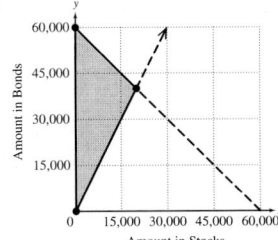

69. $x^2 + y^2 < 9$,　　$x > 0$,　　$y > 0$

71. $y > x - 1$
$y > -4x - 16$
$y < -\frac{1}{4}x + \frac{11}{4}$

73. a. $(x + 12)^2 + (y + 9)^2 \le 256$
b. Yes; The center of Hawthorne is 15 km from the earthquake.
75. If the inequality is strict—that is, posed with < or >—then the bounding line or curve should be dashed.
77. Find the solution set to each individual inequality in the system. Then to find the solution set for the system of inequalities, take the intersection of the solution sets to the individual inequalities.

79.

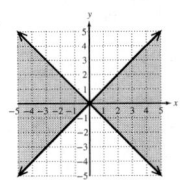

81.

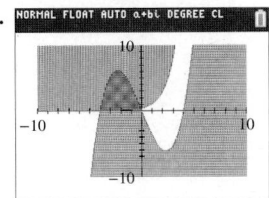

Problem Recognition Exercises, p. 779

1. a.

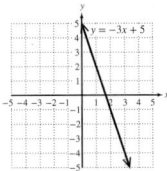

b.

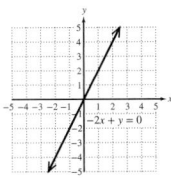

c. $\{(1, 2)\}$

d.

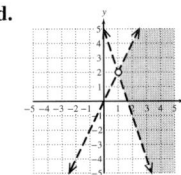

e.

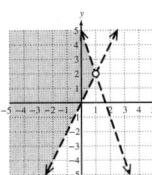

2. a.

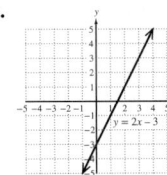

b.

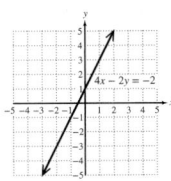

c. { }

d.

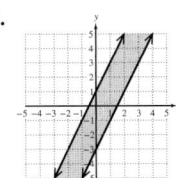

e. { }

3. a. $\{(0, 0)\}$

b.

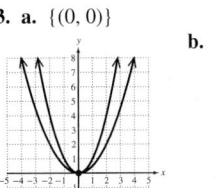

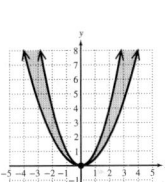

c.

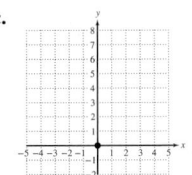

4. a. $\{(2, 1), (5, 4)\}$

b.

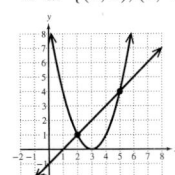

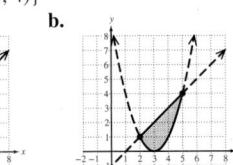

c.

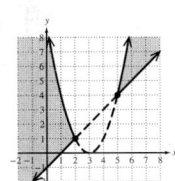

Section 10.7 Practice Exercises, pp. 785–789

R.1. a. $S(x) = 0.11x + 500$, for $x \ge 0$
b. $S(8000) = 1380$; The salesperson will make \$1380 if \$8000 in merchandise is sold for the week.

R.3. $(1, 2)$　　**R.5.** $(500, 100)$　　**R.7.** $a \ge 18$　　**R.9.** $r \le 160$

1. linear　　**3.** feasible　　**5.** $z = 0.80x + 1.10y$
7. $z = 0.62x + 0.50y$
9. a. $x = 7, y = 5$　　**b.** Maximum value: 31
11. a. $x = 10, y = 30$　　**b.** Minimum value: 37,000

13. a. Vertices:

$(0, 0), (20, 40), (60, 0)$

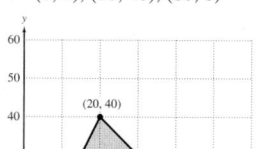

b. $x = 60, y = 0$
c. Maximum: 15,000

15. a. Vertices:

$(0, 50), (10, 20), (20, 0)$

b. $x = 20, y = 0$
c. Minimum: 60

17. a. Vertices:

$(0, 0), (0, 40), (8, 40),$
$(36, 12), (36, 0)$

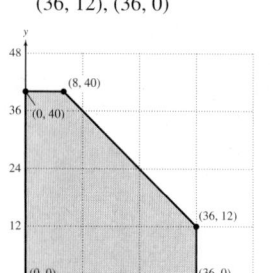

b. $x = 36, y = 12$
c. Maximum: 6480

19. a. 1520 at $(8, 6)$
b. 1660 at $(4, 9)$

21. a. Profit: $z = 160x + 240y$
b. $x \geq 0; y \geq 0; x \leq 120; y \leq 90; 240x + 320y \leq 48{,}000$
c.

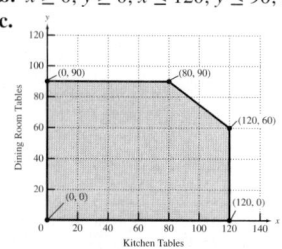

d. $(0, 0), (0, 90), (80, 90), (120, 60), (120, 0)$
e. Profit at $(0, 0)$: $z = 0$
Profit at $(0, 90)$: $z = 21{,}600$
Profit at $(80, 90)$: $z = 34{,}400$
Profit at $(120, 60)$: $z = 33{,}600$
Profit at $(120, 0)$: $z = 19{,}200$
f. The greatest profit is realized when 80 kitchen tables and 90 dining room tables are produced.
g. The maximum profit is $34,400.

23. a. 280 large trees and 120 small trees would maximize profit.
b. The maximum profit is $13,400.
c. In this case, the nursery should have 360 large trees and no small trees.

25. a. The company should make 8 trips with the small truck and 6 trips with the large truck.
b. The minimum cost is $1860.

27. a. The manufacturer should produce 600 grill A units and 500 grill B units to maximize profit.
b. The maximum profit is $114,000.
c. In this case, the manufacturer should produce 1200 grill A units and 0 grill B units.

29. a. The farmer should plant 900 acres of corn and 300 acres of soybeans.
b. The maximum profit is $138,000.
c. In this case, 500 acres of corn and 700 acres of soybeans should be planted.

31. Linear programming is a technique that enables us to maximize or minimize a function under specific constraints.
33. The feasible region for a linear programming application is found by first identifying the constraints on the relevant variables. Then the regions defined by the individual constraints are graphed. The intersection of the constraints defines the feasible region.

Chapter 10 Review Exercises, pp. 796–798

1. a. Yes **b.** No **3.** One solution
5. $\{(3, 2)\}$

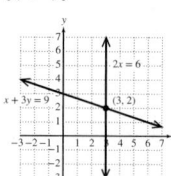

7. $\{(x, -x + 6) \mid x \text{ is any real number}\}$ or $\{(-y + 6, y) \mid y \text{ is any real number}\}$; infinitely many solutions; The equations are dependent.

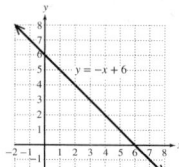

9. $\{(2, -3)\}$ **11.** $\{ \ \}$; The system is inconsistent.
13. Milk has 300 mg per cup and spinach has 240 mg per cup.
15. The speed of the plane in still air is 420 mph and the speed of the wind is 60 mph. **17.** $\{(-1, 2, -3)\}$
19. Infinitely many solutions; The equations are dependent.

21. $\left\{ \left(x, -\dfrac{1}{5}x, -\dfrac{23}{5}x \right) \Big| x \text{ is any real number} \right\}$ or

$\{(-5y, y, 23y) \mid y \text{ is any real number}\}$ or

$\left\{ \left(-\dfrac{5}{23}z, \dfrac{1}{23}z, z \right) \Big| z \text{ is any real number} \right\}$

23. She put $2000 in savings, and invested $12,000 in the bond fund and $6000 in the stock fund. **25.** $y = 3x^2 - 6x + 1$

27. $\dfrac{A}{x + 2} + \dfrac{B}{x - 1}$ **29.** $\dfrac{A}{x} + \dfrac{B}{x^2} + \dfrac{C}{2x + 3}$

31. $\dfrac{Ax + B}{x^2 + 1} + \dfrac{Cx + D}{x^2 + 4}$ **33.** $\dfrac{5}{x + 4} + \dfrac{2}{(x + 4)^2}$ **35.** $\dfrac{-2}{x} + \dfrac{4x + 1}{x^2 + 5}$

37. a.

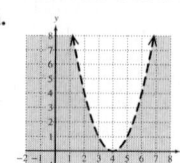

39. $\{(2, 4), (2, -4), (-2, 4), (-2, -4)\}$

b. $\{(-1, 2), (2, 5)\}$

41. $\{(4, 2)\}$ **43.** The numbers are 8 and 6 or -8 and -6.
45. The billboard is 12 ft by 24 ft.
47. a.

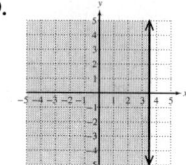

b.

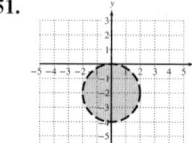

49. **51.**

53. a. No   **b.** Yes

55.

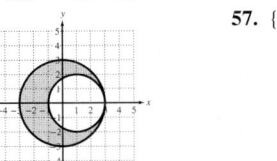

57. { }

59. $z = 24x + 20y$

61. a.

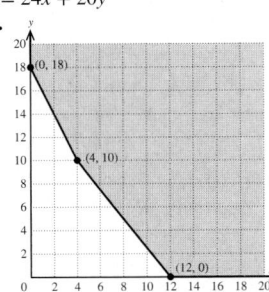

b. $x = 4, y = 10$ **c.** 620

Chapter 10 Test, pp. 798–800

1. a. Yes   **b.** No
2. a. No   **b.** Yes
3. a. Yes   **b.** No
4. a. $y = 3x - 4; y = -x + 5$; one solution
   b. $y = -x + 3; y = -x + 3$; infinitely many solutions
   c. $y = 3x; y = 3x - 6$; no solution
5. $\{(2, -4)\}$

6. { }; No solution; The system is inconsistent

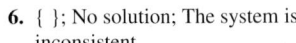

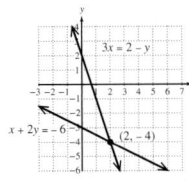

7. $\{(1, 1)\}$ **8.** $\{(5, 10)\}$

9. $\left\{ \left(x, \dfrac{10x - 3}{4} \right) \middle| x \text{ is any real number} \right\}$
   or $\left\{ \left(\dfrac{4y + 3}{10}, y \right) \middle| y \text{ is any real number} \right\}$;
   The equations are dependent.
10. { }; The system is inconsistent.   **11.** $\{(1, -4, 3)\}$
12. $\{(-2, 5, 3)\}$ **13.** { }; The system is inconsistent.
14. $\{(0, -3), (7, 4)\}$ **15.** $\{(1, 3), (1, -3), (-1, 3), (-1, -3)\}$
16. $\{(5, -2), (-5, 2)\}$ **17.** $\left\{ \left(-2, \dfrac{5}{4} \right) \right\}$
18. $\left\{ \left(x, \dfrac{22 - 3x}{2}, \dfrac{x - 6}{2} \right) \middle| x \text{ is any real number} \right\}$ or
   $\left\{ \left(\dfrac{22 - 2y}{3}, y, \dfrac{2 - y}{3} \right) \middle| y \text{ is any real number} \right\}$ or
   $\{(2z + 6, 2 - 3z, z) \mid z \text{ is any real number}\}$
19. The manager should mix 4 lb of peanuts with 16 lb of the 45% mixture.   **20.** One runner runs 6 m/sec and the other runs 4 m/sec.
21. Dylan invested $3000 in the risky stock, $7000 in the second stock, and $5000 in the third stock.
22. The numbers are 7 and 4.   **23.** The screen is 44 in. by 33 in.
24. $y = 2x^2 - 4x + 1$ **25.** $\dfrac{A}{x + 1} + \dfrac{B}{3x - 2}$
26. $\dfrac{A}{x} + \dfrac{B}{x^2} + \dfrac{C}{x^3} + \dfrac{D}{x - 3} + \dfrac{Ex + F}{x^2 + 5x + 1} + \dfrac{Gx + H}{(x^2 + 5x + 1)^2}$

27. $\dfrac{-7}{x + 3} + \dfrac{2}{2x + 5}$ **28.** $\dfrac{6}{x + 2} + \dfrac{-4}{(x + 2)^2}$

29. $x - 2 - \dfrac{3}{x} + \dfrac{8}{x^2} - \dfrac{1}{x - 4}$

30. $\dfrac{-3}{x} + \dfrac{4x - 2}{x^2 + 7}$ **31.** $\dfrac{7x + 1}{x^2 + 2} + \dfrac{3}{x^2 + 9}$

32.

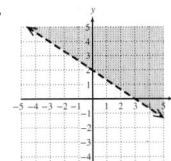

33.

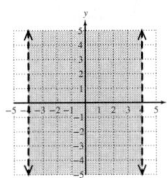

34.

35.

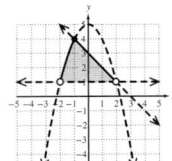

36.

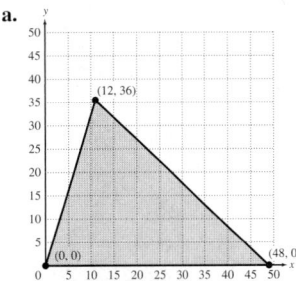

37. $z = 2.4x + 0.55y$

38. a. $x = 2, y = 8$ **b.** Minimum value: 48

39. a.

b. $x = 12, y = 36$
c. Maximum value: 37,800
40. a. 30 scoops of each type of protein powder should be mixed to maximize protein content.
   b. The maximum protein content is 1140 g.
   c. In this case, 24 scoops of whey protein should be mixed with 36 scoops of soy protein.

CHAPTER 11

Section 11.1 Practice Exercises, pp. 809–812

R.1. $\{(4, -3)\}$ **R.3.** $\{(-1, 3, 5)\}$ **R.5.** $\dfrac{A}{2t - 5} + \dfrac{B}{2t - 3}$

1. matrix   **3.** Interchange rows 2 and 3.
5. Multiply row 1 by 3 and replace the original row 1 with the result.
7. Add 3 times row 1 to row 2 and replace the original row 2 with the result.

9. $\begin{bmatrix} -3 & 2 & -1 & 4 \\ 8 & 0 & 4 & 12 \\ 0 & 2 & -5 & 1 \end{bmatrix}$ **11.** $\begin{bmatrix} 4 & -14 & 2 \\ 3 & -5 & 7 \end{bmatrix}$

13. $\begin{bmatrix} 1 & 0 & 0 & 2 \\ 0 & 1 & 0 & \frac{6}{7} \\ 0 & 0 & 1 & 12 \end{bmatrix}$ **15.** $-4x + 6y = 11$
   $-3x + 9y = 1$

17. $\begin{aligned} x + 4y + 3z &= 8 \\ y + 2z &= 12 \\ z &= 6 \end{aligned}$ 19. $\begin{aligned} x &= 8 \\ y &= -9 \\ z &= \tfrac{3}{2} \end{aligned}$ 21. $\left[\begin{array}{cc|c} -3 & 6 & 6 \\ 1 & 4 & 2 \end{array}\right]$

23. $\left[\begin{array}{cc|c} 3 & 12 & 6 \\ -3 & 6 & 6 \end{array}\right]$ 25. $\left[\begin{array}{cc|c} 0 & 6 & 4 \\ -3 & 6 & 6 \end{array}\right]$ 27. $\left[\begin{array}{ccc|c} 1 & 5 & 6 & 2 \\ 4 & -2 & -3 & 10 \\ 2 & 1 & 5 & 1 \end{array}\right]$

29. $\left[\begin{array}{ccc|c} 1 & 5 & 6 & 2 \\ 2 & 1 & 5 & 1 \\ 1 & -\tfrac{1}{2} & -\tfrac{3}{4} & \tfrac{5}{2} \end{array}\right]$ 31. $\left[\begin{array}{ccc|c} 1 & 5 & 6 & 2 \\ 0 & -9 & -7 & -3 \\ 4 & -2 & -3 & 10 \end{array}\right]$

33. No; The element on the main diagonal in the second row is not 1.
35. Yes
37. No; The elements on the main diagonal are not 1 with zeros above and below.
39. Yes   41. $\{(-2, -3)\}$ 43. $\{(4, 7)\}$ 45. $\{(5, -8)\}$
47. $\{(1, 2, 3)\}$ 49. $\{(0, 1, 3)\}$ 51. $\{(7, -2, 1)\}$
53. $\{(0, 2, -4)\}$ 55. $\{(-6, -4, 3)\}$ 57. $\{(1, 2, 3, 4)\}$
59. $\{(-10, 1, 0, 1)\}$
61. He borrowed $10,000 from his parents, $8000 from the credit union, and $2000 from the bank.
63. She spent 4 nights in Washington, 2 nights in Atlanta, and 8 nights in Dallas.
65. $\dfrac{2}{x + 3} + \dfrac{3}{x - 2} + \dfrac{-1}{(x - 2)^2}$
67. Interchanging two rows in an augmented matrix represents interchanging two equations in a system of equations. This operation does not affect the solution set of the system.
69. Reduced row-echelon form is the same format as row-echelon form with the added condition that all elements above the leading 1's must be 0's.
71. $\{(9.32, -17.48, 12.93)\}$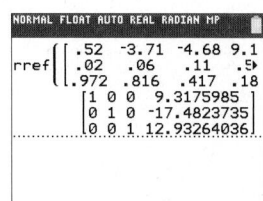

73. **a.** $\begin{aligned} 2400a + 800b + c &= 36,000 \\ 2000a + 500b + c &= 30,000 \\ 3000a + 1000b + c &= 44,000 \end{aligned}$ **b.**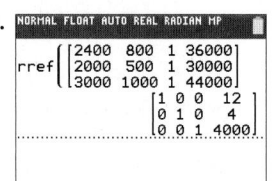

c. $y = 12x_1 + 4x_2 + 4000$ **d.** $36,000

75. **a.** $\begin{aligned} 9a - 3b + c &= -7.28 \\ a - b + c &= 3.68 \\ 100a + 10b + c &= 18.2 \end{aligned}$ **b.**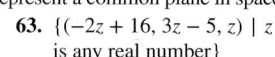
c. $f(x) = -0.32x^2 + 4.2x + 8.2$

Section 11.2 Practice Exercises, pp. 820–823

R.1. $\mathbb{R}$ **R.3.** $\{\ \}$ **R.5.** $\left\{\left(x, -\tfrac{1}{3}x + 2\right) \mid x \text{ is any real number}\right\}$ or $\{(6 - 3y, y) \mid y \text{ is any real number}\}$
R.7. $\{\ \}$

1. True   3. False   5. inconsistent   7. No solution
9. Infinitely many solutions   11. One solution

13. Infinitely many solutions   15. **a.** $\{(5, 0)\}$
    b. $\{(5 - 2y, y) \mid y \text{ is any real number}\}$ **c.** $\{\ \}$
17. **a.** $\{(3, 5, 0)\}$ **b.** $\{\ \}$
    c. $\{(-6z + 3, -4z + 5, z) \mid z \text{ is any real number}\}$
19. $\{\ \}$ 21. $\left\{\left(\dfrac{10 - 7y}{2}, y\right) \Big| y \text{ is any real number}\right\}$
23. $\{\ \}$ 25. $\{(-2z + 16, 3z - 5, z) \mid z \text{ is any real number}\}$
27. $\{(4, 0, 2)\}$ 29. $\{(5z + 3, -2z + 6, z) \mid z \text{ is any real number}\}$
31. $\left\{\left(\dfrac{12 - 3y - 4z}{2}, y, z\right) \Big| y \text{ and } z \text{ are any real numbers}\right\}$ or
    $\{(x, y, z) \mid 2x + 3y + 4z = 12\}$
33. $\{(-3z + 5, z + 4, z) \mid z \text{ is any real number}\}$
35. $\{(-4z + 7, 2, z) \mid z \text{ is any real number}\}$ 37. $\{\ \}$
39. **a.** $(3, -3, 1)$ **b.** $(9, 0, 4)$ **c.** $(-3, -6, -2)$
41. For example: $(0, 6, 0), (4, 5, 1), (8, 4, 2)$
43. For example: $(0, 0, 1), (0, 2, 0), (3, 0, 0)$
45. **a.** $180 + 190 = x_1 + x_3$ **b.** $x_1 + x_2 = 180 + 220$
    c. $x_3 + 120 = x_2 + 90$
    d. $\begin{aligned} x_1 \quad\ + x_3 &= 370 \\ x_1 + x_2 \quad\ &= 400 \\ x_2 - x_3 &= 30 \end{aligned}$ **e.** $\left[\begin{array}{ccc|c} 1 & 0 & 1 & 370 \\ 0 & 1 & -1 & 30 \\ 0 & 0 & 0 & 0 \end{array}\right]$
    f. $x_1 = 250$ vehicles per hour;   **g.** $220 \le x_1 \le 270$ vehicles per hour;
    $x_2 = 150$ vehicles per hour $130 \le x_2 \le 180$ vehicles per hour
47. **a.** $x_1 = 60$ vehicles per hour;   **b.** $40 \le x_1 \le 90$ vehicles per hour;
    $x_2 = 210$ vehicles per hour; $190 \le x_2 \le 240$ vehicles per hour;
    $x_3 = 170$ vehicles per hour $150 \le x_3 \le 200$ vehicles per hour
49. **a.** $\begin{aligned} 80x + 400y + 480z &= 9280 \\ 50x + 350y + 400z &= 7800 \\ 75x + 525y + 600z &= 10,500 \end{aligned}$
    b. $\{\ \}$ **c.** The system of equations reduces to a contradiction. There are no values for x, y, and z that can simultaneously meet the conditions of this problem.
51. $5(-2z + 16) + 7(3z - 5) - 11z = 45$ ✓
    $3(-2z + 16) + 5(3z - 5) - 9z = 23$ ✓
    $(-2z + 16) + (3z - 5) - z = 11$
53. $(0, 0, 0)$ is the only solution.
55. Infinitely many solutions; $\{(2z, 3z, z) \mid z \text{ is any real number}\}$
57. If a row of the reduced row-echelon form results in a contradiction (that is, zeros to the left of the vertical bar and a nonzero element to the right), then the system is inconsistent.
59. The equations are equivalent, meaning that they all have the same solution set. The points in the solution set represent a common plane in space.
61. $\{\ \}$ 63. $\{(-2z + 16, 3z - 5, z) \mid z \text{ is any real number}\}$

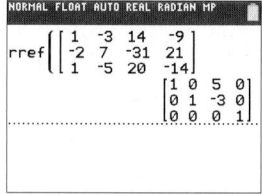   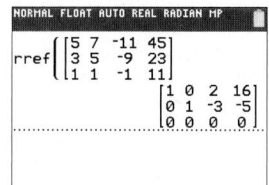

65. $\left\{\left(\dfrac{12 - 3y - 4z}{2}, y, z\right) \Big| y \text{ and } z \text{ are any real numbers}\right\}$ or
    $\{(x, y, z) \mid 2x + 3y + 4z = 12\}$

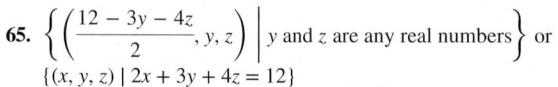

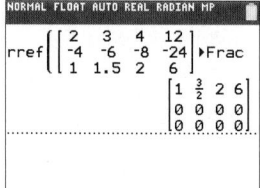

Section 11.3 Practice Exercises, pp. 834–840

R.1. $4x + (-3)$ or $4x - 3$ **R.3.** $(-5 \cdot 6)m; -30m$ **R.5.** 0

R.7. -9 **R.9.** $\dfrac{1}{p}$ **R.11.** $2x + 10$ **R.13.** $x = \dfrac{A + B}{2}$

R.15. $a = 3, b = -4$ **R.17.** $750

1. order; rows; columns

3. The order of the matrices must be the same, and the corresponding elements must be equal.

5. columns; rows   **7.** False

9. A row matrix is a matrix with only one row.

11. a. 2×3 **b.** None of these   **13. a.** 3×1 **b.** Column matrix

15. a. 2×2 **b.** Square matrix   **17.** $\sqrt{5}$ **19.** $\frac{1}{3}$ **21.** 2

23. $x = 4, y = 2, z = 10$ **25.** $\begin{bmatrix} 4 & -6 & -9 \\ -\frac{3}{5} & -1 & -7 \end{bmatrix}$

27. $\begin{bmatrix} -3 & 1 \\ 13.2 & \frac{5}{2} \\ \frac{7}{3} & 3\sqrt{2} \end{bmatrix}$ **29.** $\begin{bmatrix} -4 & 7 \\ 0.2 & \frac{7}{6} \\ -\frac{2}{3} & 6 + \sqrt{2} \end{bmatrix}$ **31.** Not possible

33. Multiply each element in the matrix by the scalar.

35. $\begin{bmatrix} 6 & 12 & -27 \\ 3 & 3\sqrt{3} & \frac{3}{2} \end{bmatrix}$ **37.** $\begin{bmatrix} 3 & -8 & -10 \\ -16 & -2\sqrt{3} - 63 & -\frac{17}{3} \end{bmatrix}$

39. $\begin{bmatrix} -4 & -16 & 20 \\ -12 & -4\sqrt{3} - 36 & -\frac{14}{3} \end{bmatrix}$ **41.** $\begin{bmatrix} 9 & 8 & -38 \\ -8 & 2\sqrt{3} - 45 & -\frac{7}{3} \end{bmatrix}$

43. $X = \begin{bmatrix} \frac{3}{2} & 1 \\ 5 & \frac{7}{2} \end{bmatrix}$ **45.** $X = \begin{bmatrix} \frac{1}{5} & -2 \\ \frac{2}{5} & \frac{11}{5} \end{bmatrix}$ **47.** $X = \begin{bmatrix} 0 & \frac{8}{5} \\ \frac{1}{5} & -\frac{13}{10} \end{bmatrix}$

49. a. Yes; 4×1 **b.** No   **51. a.** Yes; 5×5 **b.** Yes; 1×1

53. a. $\begin{bmatrix} -1 & 17 \\ -2 & 41 \end{bmatrix}$ **b.** $\begin{bmatrix} 22 & 31 \\ 13 & 18 \end{bmatrix}$ **c.** $\begin{bmatrix} 19 & 27 \\ 45 & 64 \end{bmatrix}$

55. a. $\begin{bmatrix} -6 & 8 & 38 \\ -12 & -24 & 36 \\ -13 & 4 & 69 \end{bmatrix}$ **b.** $\begin{bmatrix} -23 & 9 \\ 6 & 62 \end{bmatrix}$ **c.** Not possible

57. a. $\begin{bmatrix} 19 \\ 26 \\ 5 \end{bmatrix}$ **b.** Not possible   **c.** $\begin{bmatrix} 79 & -5 & 2 \\ 4 & 27 & 45 \\ -1 & 12 & 53 \end{bmatrix}$

59. a. $\begin{bmatrix} \frac{2}{3} \end{bmatrix}$ **b.** $\begin{bmatrix} -\frac{1}{3} & -\frac{1}{6} \\ 2 & 1 \end{bmatrix}$ **c.** Not possible

61. a. $\begin{bmatrix} 4 & 8 & 20 & 24 \\ -6 & -12 & -30 & -36 \end{bmatrix}$ **b.** Not possible

c. Not possible

63. a. $[-25]$ **b.** $[-25]$ **c.** $[25]$

65. $\begin{bmatrix} 3.1 & -2.3 \\ 1.1 & 6.5 \end{bmatrix}$ **67.** $\begin{bmatrix} \frac{9}{5} & -3 & \sqrt{6} \\ 5 & \frac{1}{2} & 2 \\ 3 & 0 & 1 \end{bmatrix}$

69. a. $M - D = \begin{bmatrix} \$3600 & \$2400 \\ \$3400 & \$2000 \end{bmatrix}$;

This represents the profit that the dealer clears for each model.

b. $F = 1.06D = \begin{bmatrix} \$30,740 & \$29,150 \\ \$30,210 & \$28,514 \end{bmatrix}$

71. [$1820]; The value $1820 represents the total revenue from the sale of these four items.

73. a. $\begin{bmatrix} \$103,974 & \$42,560 \\ \$69,754 & \$28,400 \\ \$66,438 & \$26,960 \end{bmatrix}$; The first column gives the total

revenue for Friday, Saturday, and Sunday, respectively. The second column gives the profit for Friday, Saturday, and Sunday, respectively.

b. $66,438   **c.** $97,920

75. a. TL (The product LT is not possible.)   **b.** $TL = \begin{bmatrix} \$2088 \\ \$2688 \\ \$1657 \end{bmatrix}$

77. a. $CN_1 = \begin{bmatrix} \$46 \\ \$40 \\ \$2.40 \end{bmatrix}$; The matrix CN_1 represents the additional cost for

24 text messages and 100 extra minutes for each of the cell phone plans.

b. $CN_3 = \begin{bmatrix} \$46 & \$23.60 & \$7.50 \\ \$40 & \$9.60 & \$0 \\ \$2.40 & \$5.60 & \$3 \end{bmatrix}$; The matrix CN_3 represents the

additional cost per month for each plan. For example, row 1 represents the cost for plan A for months 1, 2, and 3, respectively.

79. a. $A = \begin{bmatrix} -1 & 0 & 4 \\ 1 & 3 & 2 \end{bmatrix}$ **b.** $\begin{bmatrix} -1 & 0 & 4 \\ 1 & 3 & 2 \end{bmatrix} + \begin{bmatrix} 2 & 2 & 2 \\ -4 & -4 & -4 \end{bmatrix}$

$= \begin{bmatrix} 1 & 2 & 6 \\ -3 & -1 & -2 \end{bmatrix}$

c. $\begin{bmatrix} 1 & 0 & -4 \\ 1 & 3 & 2 \end{bmatrix}$; This matrix represents the reflection of the

triangle across the y-axis.   **d.** $\begin{bmatrix} -1 & 0 & 4 \\ -1 & -3 & -2 \end{bmatrix}$; This matrix

represents the reflection of the triangle across the x-axis.

e. $\begin{bmatrix} -2 & -1 & 3 \\ 1 & -1 & 0 \end{bmatrix}$; This matrix represents the reflection of the

triangle across the x-axis, followed by a shift to the left 1 unit and a shift upward 2 units.

81. a. $A = \begin{bmatrix} -2 & 3 & 3 & 0 \\ 1 & 3 & 0 & -1 \end{bmatrix}$ **b.** $A + \begin{bmatrix} 0 & 0 & 0 & 0 \\ -3 & -3 & -3 & -3 \end{bmatrix}$

c. $A + \begin{bmatrix} -4 & -4 & -4 & -4 \\ 0 & 0 & 0 & 0 \end{bmatrix}$

d. $\begin{bmatrix} 1 & 0 \\ 0 & -1 \end{bmatrix} \cdot A = \begin{bmatrix} -2 & 3 & 3 & 0 \\ -1 & -3 & 0 & 1 \end{bmatrix}$

e. $\begin{bmatrix} -1 & 0 \\ 0 & 1 \end{bmatrix} \cdot A = \begin{bmatrix} 2 & -3 & -3 & 0 \\ 1 & 3 & 0 & -1 \end{bmatrix}$

83. a. $\begin{bmatrix} 0 & 6 & 6 \\ 0 & 3 & 0 \end{bmatrix}$ **b.** $\begin{bmatrix} 0 & 3.7 & 5.2 \\ 0 & 5.6 & 3 \end{bmatrix}$

c. It appears that the triangle was rotated approximately $30°$ counterclockwise.

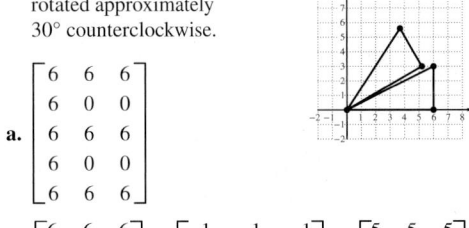

85. a. $\begin{bmatrix} 6 & 6 & 6 \\ 6 & 0 & 0 \\ 6 & 6 & 6 \\ 6 & 0 & 0 \\ 6 & 6 & 6 \end{bmatrix}$

b. $\begin{bmatrix} 6 & 6 & 6 \\ 6 & 0 & 0 \\ 6 & 6 & 6 \\ 6 & 0 & 0 \\ 6 & 6 & 6 \end{bmatrix} + \begin{bmatrix} -1 & -1 & -1 \\ -1 & 1 & 1 \\ -1 & -1 & -1 \\ -1 & 1 & 1 \\ -1 & -1 & -1 \end{bmatrix} = \begin{bmatrix} 5 & 5 & 5 \\ 5 & 1 & 1 \\ 5 & 5 & 5 \\ 5 & 1 & 1 \\ 5 & 5 & 5 \end{bmatrix}$

87. $A + B = \begin{bmatrix} a_1 & a_2 \\ a_3 & a_4 \end{bmatrix} + \begin{bmatrix} b_1 & b_2 \\ b_3 & b_4 \end{bmatrix}$

$= \begin{bmatrix} a_1 + b_1 & a_2 + b_2 \\ a_3 + b_3 & a_4 + b_4 \end{bmatrix} = \begin{bmatrix} b_1 + a_1 & b_2 + a_2 \\ b_3 + a_3 & b_4 + a_4 \end{bmatrix}$

$= B + A$

89. $A + (-A) = \begin{bmatrix} a_1 & a_2 \\ a_3 & a_4 \end{bmatrix} + \begin{bmatrix} -a_1 & -a_2 \\ -a_3 & -a_4 \end{bmatrix}$

$= \begin{bmatrix} a_1 + (-a_1) & a_2 + (-a_2) \\ a_3 + (-a_3) & a_4 + (-a_4) \end{bmatrix} = \begin{bmatrix} 0 & 0 \\ 0 & 0 \end{bmatrix} = \mathbf{0}$

91. $s(tA) = s \cdot \left(t \begin{bmatrix} a_1 & a_2 \\ a_3 & a_4 \end{bmatrix} \right) = s \cdot \begin{bmatrix} ta_1 & ta_2 \\ ta_3 & ta_4 \end{bmatrix}$

$= \begin{bmatrix} sta_1 & sta_2 \\ sta_3 & sta_4 \end{bmatrix} = (st) \begin{bmatrix} a_1 & a_2 \\ a_3 & a_4 \end{bmatrix} = (st)A$

93. $A^2 = \begin{bmatrix} -1 & 0 \\ 0 & -1 \end{bmatrix}$, $A^3 = \begin{bmatrix} -i & 0 \\ 0 & -i \end{bmatrix}$, $A^4 = \begin{bmatrix} 1 & 0 \\ 0 & 1 \end{bmatrix}$; The entries along the main diagonal in matrix A^n are the same as the value of i^n.

95. a. $\begin{bmatrix} ac & 0 \\ 0 & bd \end{bmatrix}$ **b.** $\begin{bmatrix} 3 & 0 \\ 0 & 14 \end{bmatrix}$

97. The number of columns in the first matrix is not equal to the number of rows in the second matrix.

99. To find $-A$, take the additive inverse of each individual element of A. That is, $-A = [-a_{ij}]$.

101.
```
NORMAL FLOAT AUTO REAL RADIAN MP
2.5[A]-3.6[B]
         [38.625  -98.25]
          -76.1   106.1
          -70.8   32.55
```

103.
```
NORMAL FLOAT AUTO REAL RADIAN MP
-3[A][C]
                [-76.86]
                 61.548
                 22.59
```

Section 11.4 Practice Exercises, pp. 848–851

R.1. b   **R.3.** d   **R.5.** $\begin{bmatrix} 24 & -6 \\ -4 & -7 \end{bmatrix}$ **R.7.** $\begin{bmatrix} 9 & -25 & -20 \\ 2 & 23 & 46 \\ 1 & -20 & -16 \end{bmatrix}$

R.9. $\{(1, 4)\}$ **R.11.** $\{(2, -1, 0)\}$

1. identity   **3.** inverse   **5.** $\dfrac{1}{ad - bc} \begin{bmatrix} d & -b \\ -c & a \end{bmatrix}$

7. $I_2 = \begin{bmatrix} 1 & 0 \\ 0 & 1 \end{bmatrix}$

9. a. $AI_2 = \begin{bmatrix} -\frac{7}{8} & \sqrt{5} \\ 5.1 & 8 \end{bmatrix} \begin{bmatrix} 1 & 0 \\ 0 & 1 \end{bmatrix} = \begin{bmatrix} -\frac{7}{8}(1) + \sqrt{5}(0) & -\frac{7}{8}(0) + \sqrt{5}(1) \\ 5.1(1) + 8(0) & 5.1(0) + 8(1) \end{bmatrix}$

$= \begin{bmatrix} -\frac{7}{8} & \sqrt{5} \\ 5.1 & 8 \end{bmatrix}$ ✓

b. $I_2A = \begin{bmatrix} 1 & 0 \\ 0 & 1 \end{bmatrix} \begin{bmatrix} -\frac{7}{8} & \sqrt{5} \\ 5.1 & 8 \end{bmatrix} = \begin{bmatrix} 1\left(-\frac{7}{8}\right) + 0(5.1) & 1(\sqrt{5}) + 0(8) \\ 0\left(-\frac{7}{8}\right) + 1(5.1) & 0(\sqrt{5}) + 1(8) \end{bmatrix}$

$= \begin{bmatrix} -\frac{7}{8} & \sqrt{5} \\ 5.1 & 8 \end{bmatrix}$ ✓

11. a. $AI_3 = \begin{bmatrix} 1 & -3 & 4 \\ 9 & 5 & 3 \\ 11 & -6 & -4 \end{bmatrix} \begin{bmatrix} 1 & 0 & 0 \\ 0 & 1 & 0 \\ 0 & 0 & 1 \end{bmatrix}$

$= \begin{bmatrix} 1(1) + -3(0) + 4(0) & 1(0) + -3(1) + 4(0) & 1(0) - 3(0) + 4(1) \\ 9(1) + 5(0) + 3(0) & 9(0) + 5(1) + 3(0) & 9(0) + 5(0) + 3(1) \\ 11(1) - 6(0) - 4(0) & 11(0) - 6(1) - 4(0) & 11(0) - 6(0) - 4(1) \end{bmatrix}$

$= \begin{bmatrix} 1 & -3 & 4 \\ 9 & 5 & 3 \\ 11 & -6 & -4 \end{bmatrix}$ ✓

b. $I_3A = \begin{bmatrix} 1 & 0 & 0 \\ 0 & 1 & 0 \\ 0 & 0 & 1 \end{bmatrix} \begin{bmatrix} 1 & -3 & 4 \\ 9 & 5 & 3 \\ 11 & -6 & -4 \end{bmatrix}$

$= \begin{bmatrix} 1(1) + 0(9) + 0(11) & 1(-3) + 0(5) + 0(-6) & 1(4) + 0(3) + 0(-4) \\ 0(1) + 1(9) + 0(11) & 0(-3) + 1(5) + 0(-6) & 0(4) + 1(3) + 0(-4) \\ 0(1) + 0(9) + 1(11) & 0(-3) + 0(5) + 1(-6) & 0(4) + 0(3) + 1(-4) \end{bmatrix}$

$= \begin{bmatrix} 1 & -3 & 4 \\ 9 & 5 & 3 \\ 11 & -6 & -4 \end{bmatrix}$ ✓

13. Yes   **15.** Yes   **17.** No   **19.** $A^{-1} = \begin{bmatrix} -\frac{5}{2} & -\frac{3}{2} \\ 3 & 2 \end{bmatrix}$

21. $A^{-1} = \begin{bmatrix} -\frac{1}{4} & -\frac{1}{10} \\ \frac{1}{2} & \frac{2}{5} \end{bmatrix}$ **23.** Singular matrix

25. $A^{-1} = \begin{bmatrix} 3 & -2 & -3 \\ 3 & -4 & -2 \\ 2 & -3 & -1 \end{bmatrix}$ **27.** $A^{-1} = \begin{bmatrix} 1 & 0 & 2 \\ 1 & -\frac{1}{2} & \frac{3}{2} \\ 1 & -\frac{1}{2} & \frac{1}{2} \end{bmatrix}$

29. Singular matrix   **31.** $A^{-1} = \begin{bmatrix} 3 & -3 & -6 \\ -1 & 3 & 5 \\ 2 & -3 & -5 \end{bmatrix}$

33. $A^{-1} = \begin{bmatrix} 1 & -3 & -1 & 0 \\ 0 & 1 & 2 & 0 \\ 0 & 1 & 1 & 0 \\ -2 & 0 & -3 & 1 \end{bmatrix}$ **35.** $\begin{bmatrix} 3 & -4 \\ 2 & 1 \end{bmatrix} \begin{bmatrix} x \\ y \end{bmatrix} = \begin{bmatrix} -1 \\ 14 \end{bmatrix}$

37. $\begin{bmatrix} 9 & -6 & 4 \\ 4 & 0 & -1 \\ 0 & 3 & 1 \end{bmatrix} \begin{bmatrix} x \\ y \\ z \end{bmatrix} = \begin{bmatrix} 27 \\ 1 \\ 0 \end{bmatrix}$ **39.** $\{(-2, 4)\}$ **41.** $\{(-4, 1)\}$

43. $\{(-3, -1, 0)\}$ **45.** $\{(0, 2, -4)\}$

47. $\left\{ \left(1, \dfrac{1}{2}, 2\right) \right\}$ **49.** $\{(1, -1, 0, 2)\}$

51. False. If a matrix has an inverse, it must be a square matrix.

53. False. For example, a square matrix with a row or column of all zeros does not have an inverse.

55. True

57. False. If one row (or column) of a matrix is a multiple of another row (or column), then the matrix does not have an inverse.

59. For example: $\begin{bmatrix} 0 & 1 \\ 1 & 0 \end{bmatrix}$

61. a. $A^{-1} = \begin{bmatrix} \frac{3}{4} & -\frac{1}{4} \\ -\frac{5}{8} & \frac{3}{8} \end{bmatrix}$ **b.** $(A^{-1})^{-1} = A = \begin{bmatrix} 3 & 2 \\ 5 & 6 \end{bmatrix}$

63. $A^{-1} = \begin{bmatrix} \frac{1}{a} & 0 \\ 0 & \frac{1}{b} \end{bmatrix}$

65. If $AB = I_n$ and $BA = I_n$, then the matrices are inverses.

67. $\begin{bmatrix} a & b & | & 1 & 0 \\ c & d & | & 0 & 1 \end{bmatrix} \overset{\frac{1}{a}R_1 \to R_1}{=} \begin{bmatrix} 1 & \frac{b}{a} & | & \frac{1}{a} & 0 \\ c & d & | & 0 & 1 \end{bmatrix}$

$\overset{-c \cdot R_1 + R_2 \to R_2}{=} \begin{bmatrix} 1 & \frac{b}{a} & | & \frac{1}{a} & 0 \\ 0 & d - \frac{cb}{a} & | & -\frac{c}{a} & 1 \end{bmatrix}$

$= \begin{bmatrix} 1 & \frac{b}{a} & | & \frac{1}{a} & 0 \\ 0 & \frac{ad - bc}{a} & | & -\frac{c}{a} & 1 \end{bmatrix}$

$\overset{\frac{a}{ad-bc} R_2 \to R_2}{=} \begin{bmatrix} 1 & \frac{b}{a} & | & \frac{1}{a} & 0 \\ 0 & 1 & | & -\frac{c}{ad - bc} & \frac{a}{ad - bc} \end{bmatrix}$

$\overset{-\frac{b}{a} R_2 + R_1 \to R_1}{=} \begin{bmatrix} 1 & 0 & | & \frac{d}{ad - bc} & -\frac{b}{ad - bc} \\ 0 & 1 & | & -\frac{c}{ad - bc} & \frac{a}{ad - bc} \end{bmatrix}$

Therefore, $A^{-1} = \dfrac{1}{ad - bc} \begin{bmatrix} d & -b \\ -c & a \end{bmatrix}$, provided $ad - bc \neq 0$.

69. $x_1 = 36.5°F$, $x_2 = 37.5°F$, $x_3 = 40.5°F$, $x_4 = 41.5°F$

71. $A^{-1} = \begin{bmatrix} 0.29 & -0.10 & 0.08 & -0.27 \\ -0.09 & 0.12 & 0.00 & 0.27 \\ -0.12 & 0.00 & -0.07 & 0.25 \\ 0.15 & 0.03 & -0.06 & -0.01 \end{bmatrix}$

73. $\{(5.35, 41.71, 4.45)\}$

Section 11.5 Practice Exercises, pp. 860–864

R.1. 8   **R.3.** 0   **R.5.** 1

R.7. $\left\{\left(x, \frac{3}{7}x - \frac{6}{7}\right) \mid x \text{ is any real number}\right\}$ or

$\left\{\left(\frac{7}{3}y + 2, y\right) \mid y \text{ is any real number}\right\}$ **R.9.** $\{\ \}$

1. determinant   **3.** minor   **5.** $\begin{vmatrix} b_2 & c_2 \\ b_3 & c_3 \end{vmatrix}; \begin{vmatrix} b_1 & c_1 \\ b_3 & c_3 \end{vmatrix}; \begin{vmatrix} b_1 & c_1 \\ b_2 & c_2 \end{vmatrix}$

7. 27   **9.** 6   **11.** 0   **13.** $x^2 - 36$ **15.** $-5e^{2x}$

17. a. 25   **b.** -25 **19. a.** -39 **b.** -39

21. a. -36 **b.** -36 **23.** -39; Yes   **25.** -13; Yes

27. 0; No   **29.** 13; Yes   **31.** 121; Yes

33. $\left\{\left(\frac{23}{8}, \frac{21}{40}\right)\right\}$ **35.** $\left\{\left(-\frac{57}{46}, -\frac{31}{23}\right)\right\}$

37. $\left\{\left(\frac{4y + 8}{3}, y\right) \mid y \text{ is any real number}\right\}$

39. $\{\ \}$ **41.** $\left\{\left(\frac{63}{157}, -\frac{333}{157}, \frac{536}{471}\right)\right\}$ **43.** $\left\{\left(\frac{13}{2}, \frac{3}{2}, \frac{11}{2}\right)\right\}$

45. $\left\{\left(\frac{3y - z + 6}{2}, y, z\right) \mid y \text{ and } z \text{ are any real numbers}\right\}$ or

$\{(x, y, z) \mid 2x - 3y + z = 6\}$

47. $\{\ \}$ **49.** $x_2 = \frac{113}{60}$ **51.** Yes   **53.** No

55. a. $\begin{vmatrix} x & y & 1 \\ -3 & 2 & 1 \\ -4 & 6 & 1 \end{vmatrix} = 0$ **b.** $y = -4x - 10$

57. 12 square units

59. a. 36   **b.** -36
c. Rows 1 and 2 are interchanged between matrix A and matrix B. The determinants are opposite in sign.

61. a. 13   **b.** 26
c. Row 1 of matrix B is 2 times row 1 of matrix A. The value $|B| = 2|A|$.

63. a. -2 **b.** -2
c. Row 2 of matrix B is the same as the sum of 3 times row 1 of A and row 2 of A. The value $|A|$ equals $|B|$.

65. 0   **67.** 0   **69.** 1   **71.** abc

73. $|AB| = -130$; $|A| = 10$ and $|B| = -13$.
So, $|A| \cdot |B| = (10)(-13) = -130$ and, therefore, $|A| \cdot |B| = |AB|$.

75. The minor is the determinant of the matrix obtained by deleting the ith row and jth column of the original matrix. The cofactor is the product of the minor and the factor $(-1)^{i+j}$.

77. Choose the row or column with the greatest number of zero elements.

79. -27 **81.** 10,112

Problem Recognition Exercises, p. 864

1. $\{(2, -4)\}$ **2.** $\{(5, -1)\}$ **3.** $\{(2, -1, 0)\}$ **4.** $\{(2, 0, 4)\}$

5. a. $\begin{vmatrix} 1.5 & -2 \\ -3 & 4 \end{vmatrix} = 0$ **b.** No   **c.** $\{\ \}$

6. a. $\begin{vmatrix} 5 & -2 \\ 1 & -0.4 \end{vmatrix} = 0$ **b.** No   **c.** $\{\ \}$

7. a. $\begin{vmatrix} 1 & -3 & 7 \\ -2 & 5 & -11 \\ 1 & -5 & 13 \end{vmatrix} = 0$ **b.** No

c. $\{(2z + 4, 3z + 1, z) \mid z \text{ is any real number}\}$

8. a. $\begin{vmatrix} 1 & -2 & 3 \\ -2 & 1 & 0 \\ 1 & 0 & -1 \end{vmatrix} = 0$ **b.** No

c. $\{(z + 3, 2z + 5, z) \mid z \text{ is any real number}\}$

Chapter 11 Review Exercises, pp. 870–872

1. $\begin{aligned} x - 2y + 3z &= -1 \\ y + 4z &= -11 \\ z &= -2; \end{aligned}$ Solution set: $\{(-1, -3, -2)\}$

3. $\begin{bmatrix} 1 & -\frac{3}{2} & \mid & \frac{1}{2} \\ 5 & 6 & \mid & -4 \end{bmatrix}$ **5.** $\{(5, -6)\}$ **7.** $\{(0, 2, 3)\}$

9. Lily borrowed \$1000 from her friend, \$7000 from the credit union, and \$2000 from the bank.

11. $\{\ \}$ **13.** $\{(3z, -2z + 1, z) \mid z \text{ is any real number}\}$ **15.** $\{\ \}$

17. $\{(3z + 5, -4z + 3, z) \mid z \text{ is any real number}\}$

19. $\{(5y - 2z - 1, y, z) \mid y \text{ and } z \text{ are any real numbers}\}$ or
$\{(x, y, z) \mid x - 5y + 2z = -1\}$

21. a. $x_1 = 76$ vehicles per hour; $x_2 = 97$ vehicles per hour
b. $46 \le x_1 \le 96$ vehicles per hour; $67 \le x_2 \le 117$ vehicles per hour

23. a. 3×2 **b.** None of these

25. a. 2×1 **b.** Column matrix   **27.** 4

29. $x = 6, y = 3, z = 8$ **31.** $3A = \begin{bmatrix} -12 & 3 \\ 18 & -6 \\ 3 & 9 \end{bmatrix}$ **33.** Not possible

35. $2A - C = \begin{bmatrix} -8 - \pi & -2 \\ 15 & -5 \\ 2 & 1 \end{bmatrix}$ **37.** $AB = \begin{bmatrix} -7 & -7 & 22 \\ 10 & 8 & -30 \\ 5 & 18 & -25 \end{bmatrix}$

39. Not possible   **41.** $A^2 = \begin{bmatrix} -2 & 36 \\ -6 & 10 \end{bmatrix}$ **43.** $BC = \begin{bmatrix} 2 & 7 \\ -6 & -21 \end{bmatrix}$

45. $QP = \begin{bmatrix} \$7922 \\ \$9843 \end{bmatrix}$; QP is the matrix representing the total revenue from these four items for each theater.

47. a. $A = \begin{bmatrix} 1 & 3 & 5 \\ 1 & 4 & 2 \end{bmatrix}$ **b.** $\begin{bmatrix} 1 & 3 & 5 \\ 1 & 4 & 2 \end{bmatrix} + \begin{bmatrix} -3 & -3 & -3 \\ -1 & -1 & -1 \end{bmatrix}$
$= \begin{bmatrix} -2 & 0 & 2 \\ 0 & 3 & 1 \end{bmatrix}$

c. $\begin{bmatrix} -1 & -3 & -5 \\ 1 & 4 & 2 \end{bmatrix}$; This matrix represents the reflection of the triangle across the y-axis.   **d.** $\begin{bmatrix} 1 & 3 & 5 \\ -1 & -4 & -2 \end{bmatrix}$; This matrix represents the reflection of the triangle across the x-axis.

49. No   **51.** $A^{-1} = \begin{bmatrix} \frac{1}{6} & \frac{1}{6} \\ -\frac{1}{12} & \frac{5}{12} \end{bmatrix}$ **53.** Singular matrix

55. $A^{-1} = \begin{bmatrix} 0 & -1 & 4 \\ 1 & -1 & 5 \\ -3 & -1 & 0 \end{bmatrix}$

57. $\begin{bmatrix} -3 & 7 & 0 \\ 4 & 0 & 2 \\ 2 & -1 & 5 \end{bmatrix} \begin{bmatrix} x \\ y \\ z \end{bmatrix} = \begin{bmatrix} 6 \\ -3 \\ -13 \end{bmatrix}$ **59.** $\{(4, -3)\}$

61. $\{(1, 2, 3)\}$ **63. a.** 24   **b.** 24   **65. a.** 13   **b.** -13

67. $81 - x^2$ **69.** 156   **71.** -270

73. $\left\{\left(\frac{3}{10}, -\frac{23}{30}\right)\right\}$ **75.** $\left\{\left(-\frac{7}{25}, -\frac{222}{125}, \frac{161}{125}\right)\right\}$ **77.** $y = -\frac{94}{67}$

Chapter 11 Test, pp. 873–874

1. $\begin{bmatrix} 1 & 5 & -3 & | & 1 \\ 3 & 1 & 4 & | & -2 \\ 0 & 4 & 2 & | & 6 \end{bmatrix}$ **2.** $\begin{bmatrix} 0 & -14 & 13 & | & -5 \\ 1 & 5 & -3 & | & 1 \\ 0 & 4 & 2 & | & 6 \end{bmatrix}$

3. $\begin{bmatrix} 3 & 1 & 4 & | & -2 \\ 1 & 5 & -3 & | & 1 \\ 0 & 1 & \frac{1}{2} & | & \frac{3}{2} \end{bmatrix}$

4. The elements above the leading 1 in the third column are not all zero. Specifically, the element in row 1, column 3 should be 0.

5. $\{(-10, 3)\}$ **6.** $\{\ \}$ **7.** $\{(3z, -2z + 5, z)\,|\,z \text{ is any real number}\}$

8. a. 8   **b.** -8 **9. a.** -12 **b.** -12 **10.** -31 **11.** 236

12. $|A| = 0$. Therefore, A is singular (does not have an inverse).

13. $\{(-2, 3)\}$ **14.** $\{\ \}$ **15.** $\{(-2, -8, 2)\}$

16. $\{(-3z - 2, -4z + 1, z)\,|\,z \text{ is any real number}\}$

17. $\left\{\left(\dfrac{54}{61}, -\dfrac{53}{61}\right)\right\}$ **18.** $x = \dfrac{7}{31}$ **19.** $X = \begin{bmatrix} -1 & 2 \\ -4 & -2 \end{bmatrix}$

20. $2A - 3C = \begin{bmatrix} 8 & -1 & 6 \\ -2 & 11 & -12 \end{bmatrix}$ **21.** Not possible

22. $AB = \begin{bmatrix} -5 & 20 \\ 20 & 44 \end{bmatrix}$ **23.** $BA = \begin{bmatrix} 22 & 37 & 51 \\ -2 & -4 & -6 \\ 22 & 23 & 21 \end{bmatrix}$ **24.** Yes

25. $A^{-1} = \begin{bmatrix} 2 & -1 \\ -\frac{5}{2} & \frac{3}{2} \end{bmatrix}$ **26.** Singular matrix

27. $A^{-1} = \begin{bmatrix} 3 & 1 & 2 \\ 7 & 2 & 5 \\ 1 & 1 & 1 \end{bmatrix}$ **28.** $\{(1, 5)\}$ **29.** $\{(2, 1, -3)\}$

30. a. $x_1 = 128$ vehicles per hour; $x_2 = 173$ vehicles per hour
   b. $118 \leq x_1 \leq 168$ vehicles per hour; $163 \leq x_2 \leq 213$ vehicles per hour

31. $CN = \begin{bmatrix} 5500 \\ 8040 \end{bmatrix}$; This represents the total number of calories burned by two individuals with different weights after biking 6 hr, running 3 hr, and walking 5 hr. For example, the element 5500 in the first row tells us that 5500 cal would be burned by a 120-lb individual who biked 6 hr, ran 3 hr, and walked 5 hr in a given week.

32. Yes

APPENDIX A

Section A.1 Algebra for Calculus, pp. A-1–A-2

1. $|y - L| < \varepsilon$ or $|L - y| < \varepsilon$ **2.** $|x - c| < \delta$ or $|c - x| < \delta$

3. a. 1   **b.** -1 **4. a.** -1 **b.** 1

5. a. $4x + 2h + 3$ **b.** $4x + 3$ **6. a.** $6x + 3h - 4$ **b.** $6x - 4$

7. a. $\dfrac{-1}{(x - 2)(x + h - 2)}$ **b.** $\dfrac{-1}{(x - 2)^2}$

8. a. $\dfrac{-2}{(2x + 5)(2x + 2h + 5)}$ **b.** $\dfrac{-2}{(2x + 5)^2}$

9. a. $3x^2 + 3xh + h^2$ **b.** $3x^2$

10. a. $4x^3 + 6x^2h + 4xh^2 + h^3$ **b.** $4x^3$

11. a. $\dfrac{1}{\sqrt{x} + \sqrt{x + h}}$ **b.** $\dfrac{1}{2\sqrt{x}}$

12. a. $\dfrac{2}{\sqrt{2x + 2h} + \sqrt{2x}}$ **b.** $\dfrac{1}{\sqrt{2x}}$

13. $\dfrac{x^{1/2}(5x + 3)}{2}$ **14.** $\dfrac{7x - 1}{6x^{5/6}}$

15. $6(3x + 1)^3(x^2 + 2)^2(5x^2 + x + 4)$

16. $-2(-2x + 3)^2(4x^2 - 5)(28x^2 - 24x - 15)$

17. $\dfrac{42(t - 1)^5}{(2t + 5)^7}$ **18.** $\dfrac{24x^5}{(x^2 + 4)^4}$ **19.** $\dfrac{2(x^2 + 2)}{(x^2 + 4)^{1/2}}$

20. $-\dfrac{2(x - 1)(x + 1)}{(2 - x^2)^{1/2}}$ **21.** $\dfrac{1}{(x^2 + 1)^{3/2}}$ **22.** $\dfrac{6(2x - 1)}{(3x - 1)^{2/3}}$

23. $\dfrac{x + 8}{(x + 4)\sqrt{x + 4}}$ **24.** $-\dfrac{x(x^2 - 32)}{(16 - x^2)\sqrt{16 - x^2}}$

25. $\dfrac{4(x - 3)}{3(x - 4)^{2/3}}$ **26.** $\dfrac{5(x + 4)}{4(x + 5)^{3/4}}$

27. $\dfrac{1}{(2x)^{1/2}(x + 1)^{3/2}}$ **28.** $-\dfrac{(x - 1)(x + 1)}{(3x)^{2/3}(x^2 + 1)^{4/3}}$

Section A.2 Equations and Inequalities for Calculus, pp. A-2–A-3

1. $y' = -\dfrac{9x}{25y}$ **2.** $y' = \dfrac{1 - 2xy^3}{3x^2y^2 - 1}$

3. $y' = \dfrac{2y(y - 3x)}{3x^2 - 4xy + 3y^2}$ **4.** $y' = -\dfrac{y(2x + y)}{x(2y + x)}$

5. a. $\dfrac{x(5x - 6)}{\sqrt{2x - 3}}$ **b.** $x = 0, x = \dfrac{6}{5}$ **c.** $x = \dfrac{3}{2}$

6. a. $\dfrac{x(3x - 14)}{(2x - 7)^{3/2}}$ **b.** $x = 0, x = \dfrac{14}{3}$ **c.** $x = \dfrac{7}{2}$

7. a. $-\dfrac{9}{(x^2 - 9)^{3/2}}$ **b.** None   **c.** $x = -3, x = 3$

8. a. $\dfrac{4x(2x - 5)}{(4x - 5)^2}$ **b.** $x = 0, x = \dfrac{5}{2}$ **c.** $x = \dfrac{5}{4}$

9. a. $-\dfrac{6x(3x + 1)}{(6x + 1)^2}$ **b.** $x = 0, x = -\dfrac{1}{3}$ **c.** $x = -\dfrac{1}{6}$

10. a. $\dfrac{2(2 - x^2)}{\sqrt{4 - x^2}}$ **b.** $x = \pm\sqrt{2}$ **c.** $x = 2$ and $x = -2$

11. $(-3, 1)$ **12.** $(-2, 2)$ **13.** 14.4 ft

14. $\dfrac{\sqrt{3}}{4}$ m^3 **15.** 1600 ft^3

Subject Index

Factoring and Special Case Products

$a^2 + 2ab + b^2 = (a + b)^2$

$a^2 - 2ab + b^2 = (a - b)^2$

$a^2 - b^2 = (a + b)(a - b)$

$a^3 + b^3 = (a + b)(a^2 - ab + b^2)$

$a^3 - b^3 = (a - b)(a^2 + ab + b^2)$

Complex Numbers

$i = \sqrt{-1}$ and $i^2 = -1$

For a real number $b > 0$, $\sqrt{-b} = i\sqrt{b}$

The complex numbers $a + bi$ and $a - bi$ are conjugates, and $(a + bi)(a - bi) = a^2 + b^2$.

Quadratic Formula

Given $ax^2 + bx + c = 0$, $a \neq 0$, the solutions are

$$x = \frac{-b \pm \sqrt{b^2 - 4ac}}{2a}$$

Absolute Value Equations/Inequalities

If $k \geq 0$, then

$|u| = k$ is equivalent to $u = k$ or $u = -k$.

$|u| = |w|$ is equivalent to $u = w$ or $u = -w$.

$|u| < k$ is equivalent to $-k < u < k$.

$|u| > k$ is equivalent to $u < -k$ or $u > k$.

Distance Formulas

The distance between two points a and b on a number line is given by $|a - b|$ or $|b - a|$.

The distance between (x_1, y_1) and (x_2, y_2) is

$$d = \sqrt{(x_2 - x_1)^2 + (y_2 - y_1)^2}$$

Midpoint Formula

The midpoint of the line segment between (x_1, y_1) and (x_2, y_2) is

$$M = \left(\frac{x_1 + x_2}{2}, \frac{y_1 + y_2}{2} \right)$$

Linear and Quadratic Functions

$f(x) = b$ constant function: slope 0 and y-intercept $(0, b)$

$f(x) = mx + b$ linear function: slope m and y-intercept $(0, b)$

$f(x) = ax^2 + bx + c$ $(a \neq 0)$ quadratic function: vertex $\left(\frac{-b}{2a}, f\left(\frac{-b}{2a}\right) \right)$

Slope and Average Rate of Change

Slope of a line through (x_1, y_1) and (x_2, y_2):

$$m = \frac{\Delta y}{\Delta x} = \frac{y_2 - y_1}{x_2 - x_1}$$

Average rate of change of $f(x)$ between (x_1, y_1) and (x_2, y_2): $\dfrac{f(x_2) - f(x_1)}{x_2 - x_1}$

Difference quotient: $\dfrac{f(x + h) - f(x)}{h}$

Circle

$(x - h)^2 + (y - k)^2 = r^2$ where $r > 0$

Parabola

Vertical Axis of Symmetry
$(x - h)^2 = 4p(y - k)$

Horizontal Axis of Symmetry
$(y - k)^2 = 4p(x - h)$

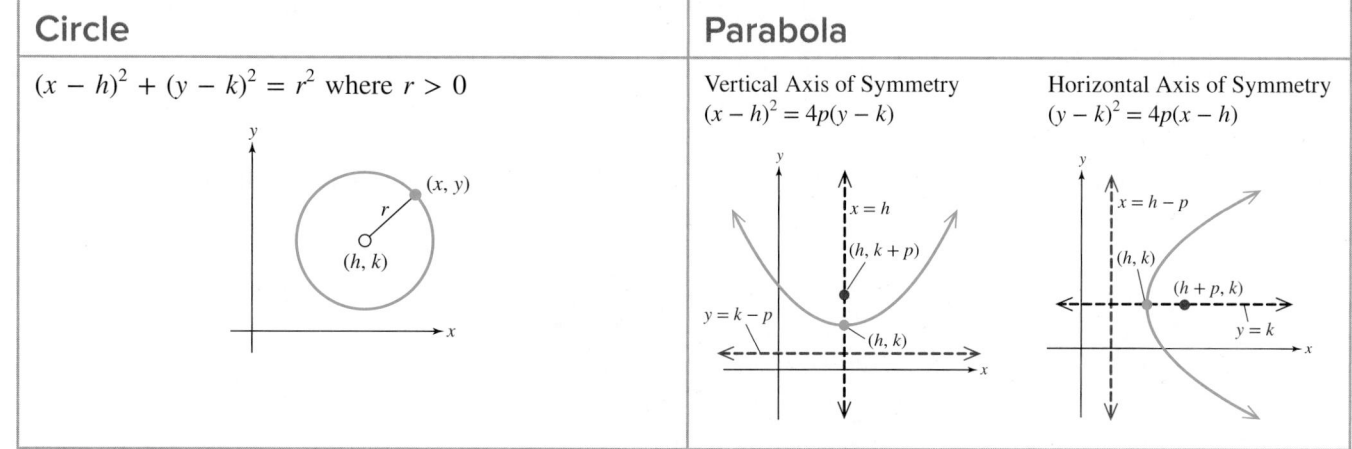

Graphs of Basic Functions

Constant Function
$f(x) = b$

Linear Function
$f(x) = mx + b$
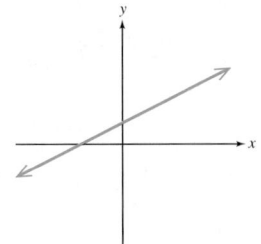

Identity Function
$f(x) = x$
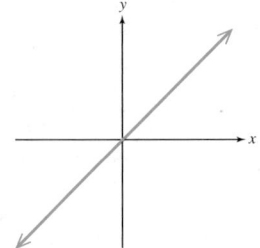

Quadratic Function
$f(x) = x^2$
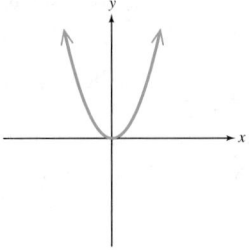

Cubic Function
$f(x) = x^3$
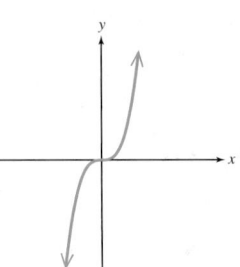

Absolute Value Function
$f(x) = |x|$
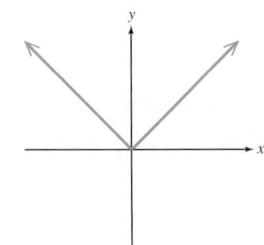

Square Root Function
$f(x) = \sqrt{x}$
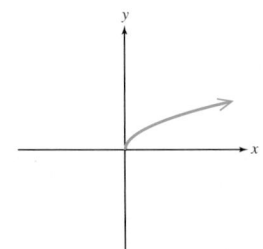

Cube Root Function
$f(x) = \sqrt[3]{x}$
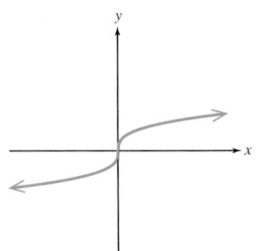

Reciprocal Function
$f(x) = \dfrac{1}{x}$
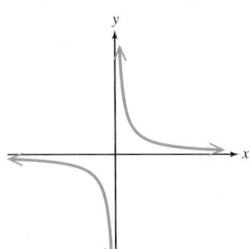

Greatest Integer Function
$f(x) = [\![x]\!]$
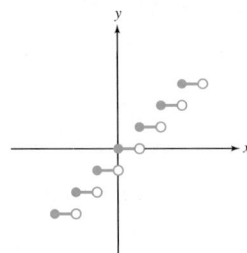

Exponential Function
$f(x) = b^x$, where $b > 0$ and $b \neq 1$
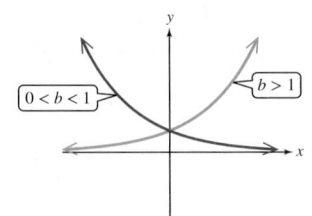

Logarithmic Function
$f(x) = \log_b x$, where $b > 0$ and $b \neq 1$
$y = \log_b x \Leftrightarrow b^y = x$
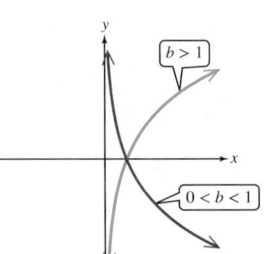

Ellipse

Major Axis Horizontal
$$\frac{(x-h)^2}{a^2} + \frac{(y-k)^2}{b^2} = 1$$

Major Axis Vertical
$$\frac{(x-h)^2}{b^2} + \frac{(y-k)^2}{a^2} = 1$$

$a > b$, and $c^2 = a^2 - b^2$, where $c > 0$.

Foci are c units from the center on the major axis.

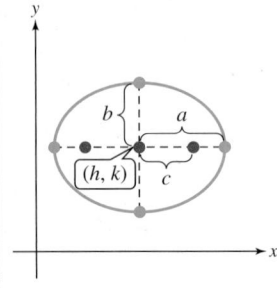
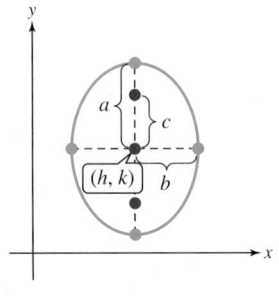

Hyperbola

Transverse Axis Horizontal
$$\frac{(x-h)^2}{a^2} - \frac{(y-k)^2}{b^2} = 1$$

Transverse Axis Vertical
$$\frac{(y-k)^2}{a^2} - \frac{(x-h)^2}{b^2} = 1$$

$c^2 = a^2 + b^2$, where $c > 0$.

Foci are c units from the center on the transverse axis.

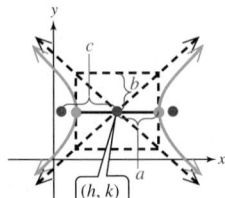
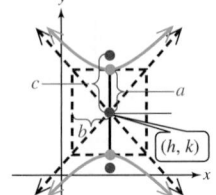

Tests for Symmetry

Consider the graph of an equation in x and y. The graph of the equation is

- Symmetric to the **y-axis** if substituting $-x$ for x results in an equivalent equation.

- Symmetric to the **x-axis** if substituting $-y$ for y results in an equivalent equation.

- Symmetric to the **origin** if substituting $-x$ for x and $-y$ for y results in an equivalent equation.

Even and Odd Functions

- f is an **even function** if $f(-x) = f(x)$ for all x in the domain of f.

- f is an **odd function** if $f(-x) = -f(x)$ for all x in the domain of f.

Properties of Logarithms

$\log_b 1 = 0$ $\qquad \log_b (xy) = \log_b x + \log_b y$

$\log_b b = 1$ $\qquad \log_b\left(\dfrac{x}{y}\right) = \log_b x - \log_b y$

$\log_b b^x = x$ $\qquad \log_b x^p = p \log_b x$

$b^{\log_b x} = x$

Change-of-base formula: $\qquad \log_b x = \dfrac{\log_a x}{\log_a b}$

$b^x = b^y$ implies that $x = y$.

$\log_b x = \log_b y$ implies that $x = y$.

Variation

$\left.\begin{array}{l} y \text{ varies } \textbf{directly} \text{ as } x. \\ y \text{ is } \textbf{directly} \text{ proportional to } x. \end{array}\right\} y = kx$

$\left.\begin{array}{l} y \text{ varies } \textbf{inversely} \text{ as } x. \\ y \text{ is } \textbf{inversely} \text{ proportional to } x. \end{array}\right\} y = \dfrac{k}{x}$

$\left.\begin{array}{l} y \text{ varies } \textbf{jointly} \text{ as } w \text{ and } x. \\ y \text{ is } \textbf{jointly} \text{ proportional to } \\ w \text{ and } x. \end{array}\right\} y = kwx$

Arithmetic Sequences and Series

nth term: $a_n = a_1 + (n - 1)d$

Sum of a finite arithmetic series:

$$S_n = \frac{n}{2}(a_1 + a_n)$$

Geometric Sequences and Series

nth term: $a_n = a_1 r^{n-1}$

Finite geometric series: $\displaystyle\sum_{i=1}^{n} a_1 r^{i-1} = \dfrac{a_1(1 - r^n)}{1 - r}$

Infinite geometric series:

$$\sum_{i=1}^{\infty} a_1 r^{i-1} = \frac{a_1}{1 - r} \text{ provided that } |r| < 1$$

Binomial Theorem

Factorial notation: $n! = n(n - 1)(n - 2)...(2)(1)$ and $0! = 1$

Binomial coefficients: $\dbinom{n}{r} = {}_nC_r = \dfrac{n!}{r! \cdot (n - r)!}$

Binomial theorem: $(a + b)^n = \dbinom{n}{0}a^n + \dbinom{n}{1}a^{n-1}b + \dbinom{n}{2}a^{n-2}b^2 + \cdots + \dbinom{n}{n-1}ab^{n-1} + \dbinom{n}{n}b^n = \displaystyle\sum_{r=0}^{n}\dbinom{n}{r}a^{n-r}b^r$

Perimeter and Circumference

Rectangle
$P = 2l + 2w$

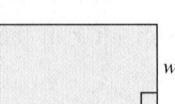

Square
$P = 4s$

Triangle
$P = a + b + c$

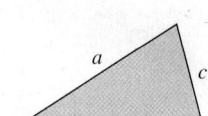

Circle
Circumference: $C = 2\pi r$

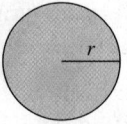

Area

Rectangle
$A = lw$

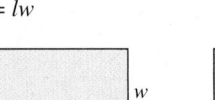

Square
$A = s^2$

Parallelogram
$A = bh$

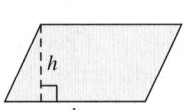

Triangle
$A = \frac{1}{2}bh$

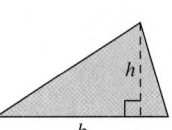

Trapezoid
$A = \frac{1}{2}(b_1 + b_2)h$

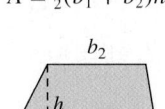

Circle
$A = \pi r^2$

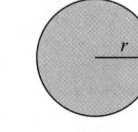

Volume

Rectangular Solid
$V = lwh$

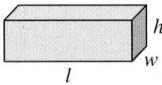

Right Circular Cylinder
$V = \pi r^2 h$

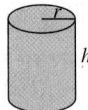

Right Circular Cone
$V = \frac{1}{3}\pi r^2 h$

Sphere
$V = \frac{4}{3}\pi r^3$

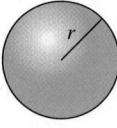

Angles

- Two angles are complementary if the sum of their measures is 90°.

$x + y = 90°$

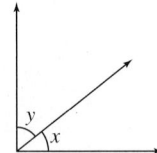

- Two angles are supplementary if the sum of their measures is 180°.

$x + y = 180°$

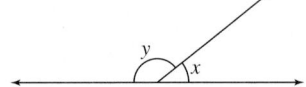

Triangles

- The sum of the measures of the angles of a triangle is 180°.

$x + y + z = 180°$

- Given a right triangle with legs of length a and b, and hypotenuse of length c, the Pythagorean theorem indicates that

$a^2 + b^2 = c^2$

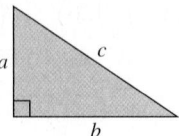